The Tumor Immunoenvironment

Michael R. Shurin · Viktor Umansky
Anatoli Malyguine
Editors

The Tumor Immunoenvironment

Springer

Editors
Michael R. Shurin
Department of Pathology
University of Pittsburgh Medical Center
Pittsburgh, PA
USA

Anatoli Malyguine
SAIC-Frederick, Inc.
National Cancer Institute at Frederick
Frederick, MD
USA

Viktor Umansky
Clinical Cooperation Unit
Dermato-Oncology
German Cancer Research Center
Heidelberg
Germany

ISBN 978-94-007-6216-9 ISBN 978-94-007-6217-6 (eBook)
DOI 10.1007/978-94-007-6217-6
Springer Dordrecht Heidelberg New York London

Library of Congress Control Number: 2013932859

Printed on acid-free paper

Springer is part of Springer Science+Business Media (www.springer.com)

Preface

The recent decade brought a tectonic shift in our understanding of the mechanisms regulating tumor development, progression, and metastases. During the majority of the last century, it was widely believed that these processes are governed mainly by genetic alterations in tumor cells. An incredible effort was expended to uncover the molecular mechanisms responsible for genomic instability, tumor cell survival, invasion, metastases, etc. Many transcription factors and signal transduction pathways were implicated in these processes. Not surprisingly, all six of the hallmark capabilities of cancer, suggested by Hanahan and Weinberg in their seminal review in 2000, included traits associated only with tumor cells. However, at the end of the last century, it became increasingly clear that the molecular abnormalities associated with tumor cells could not explain the complexity of the events involved in the regulation of tumor progression. It is now evident that the tumor microenvironment plays a major role in these processes. Epidemiological and experimental data have directly implicated inflammation as one of the major factors responsible for tumor development. The host immune system was shown to play a major role in control of tumor progression. Myeloid cells were demonstrated to be a critical factor in promoting tumor cell invasion and metastases. Tumor development and progression represent intricately connected circuits of intrinsic (associated with tumor cells) and extrinsic (associated with tumor microenvironment) factors. The understanding of tumor biology is impossible without a clear understanding of the role of tumor microenvironment. In 2011, Hanahan and Weinberg revisited those hallmarks of cancer and added the evasion of immune destruction as an emerging new hallmark, and tumor-promoting inflammation as one of the enabling characteristics of cancer. It is evident that, in the near future, tumor microenvironment will occupy an even more prominent role in our understanding of tumor biology.

The cells of the immune system represent, arguably, the most critical element of tumor microenvironment. They are not only responsible for the immune control of tumor progression, but are also involved in tumor cell invasion, conditioning of the metastatic niche, angiogenesis, etc. This book is focused on the analysis of the different components of the immune system, in the regulation of tumor progression.

It presents a unique opportunity for readers to put together the complex and often convoluted relationship between different immune cells and tumors. The editors and contributors effectively presented a logical and comprehensive overview of this complex issue. Readers will find information about the role of inflammation in promoting tumors and the regulation of antitumor immune responses; the analysis of the different immune suppressive mechanisms responsible for tumor escape; the evaluation of abnormalities in different immune cells in cancer including dendritic cells, natural killer cells and T cells, as well as the contribution of regulatory T cells, myeloid-derived suppressor cells, granulocytes, mast cells, and macrophages into tumor progression.

However, this book goes far beyond just a description of the immunological abnormalities in cancer. It presents an overview of therapeutic strategies in targeting both tumor cells and tumor microenvironment. The unique value of this volume is that cancer immune therapy is discussed in the context of the regulation of tumor microenvironment. Finally, this book offers the analysis of the biomarkers of immune responses in cancer, the field that is extremely important for the design and evaluation of numerous immune therapeutic strategies.

I believe this book provides a rare example of the synthetic approach to complex biological problems and is a must read for people interested in the role of the immune system in tumor–stroma interaction.

Dmitry I. Gabrilovich MD, Ph.D.
Robert Rothman Endowed Chair in Cancer Research
Head, Section of Dendritic Cell Biology
H. Lee Moffitt Cancer Center and Research Institute,
Professor of Oncologic Sciences and Molecular Medicine
University of South Florida
Tampa, FL, USA

References

Hanahan D, Weinberg RA (2000) The hallmarks of cancer. Cell 100(1):57–70
Hanahan D, Weinberg RA (2011) Hallmarks of cancer: the next generation. Cell 144(5):646–674

Contents

Contributors

Mads Hald Andersen Department of Hematology, Center for Cancer Immune Therapy, University Hospital Herlev, Herlev, Denmark

Ron N. Apte Faculty of Health Sciences, The Shraga Segal Department of Microbiology, Immunology and Genetics, Ben-Gurion University of the Negev, Beer-sheva, Israel

Natalia Aptsiauri Department of Cliniclas Analysis and Immunology, Virgen de las Nieves University Hospital, University of Granada Medical School, Granada, Spain

Maria Libera Ascierto Infectious Disease and Immunogenetics Section, Department of Transfusion Medicine, Clinical Center and Trans-NIH Center for Human Immunology, National Institutes of Health, Bethesda, MD, USA; Department of Internal Medicine, University of Genoa, Genoa, Italy

Pamela L. Beatty Department of Immunology, University of Pittsburgh School of Medicine, Pittsburgh, PA, USA

Adit Ben-Baruch Head, Ela Kodesz Institute for Research on Cancer Development and Prevention, Department of Cell Research and Immunology, George S. Wise Faculty of Life Sciences, Tel Aviv University, Tel Aviv, Israel

Philipp Beckhove German Cancer Research Center, Heidelberg, Germany

Jürgen C. Becker Department of Dermatology, Medical University of Graz, Graz, Austria

Arya Biragyn Immunotherapeutics Section, Laboratory of Molecular Biology and Immunology, National Institute on Aging, Baltimore, MD, USA

Monica Bodogai Immunoregulations Section, Laboratory of Molecular Biology and Immunology, National Institute on Aging, Baltimore, MD, USA

Lisa H. Butterfield UPCI Immunologic Monitoring and Cellular Products Laboratory, Hillman Cancer Center, University of Pittsburgh, Pittsburgh, PA, USA

Teresa Cabrera Department of Cliniclas Analysis and Immunology, Virgen de las Nieves University Hospital, University of Granada Medical School, Granada, Spain

Shu-Hsia Chen Department of Oncological Sciences, Mount Sinai School of Medicine, New York, NY, USA

Kimberly Dunham Laboratory of Cell Mediated Immunity, SAIC-Frederick, Inc., Frederick, MD, USA

Olivera J. Finn Department of Immunology, University of Pittsburgh School of Medicine, Pittsburgh, PA, USA

Zvi G. Fridlender Laboratory of Lung Cancer Research, Center for COPD and Smoking Damages, Institute of Pulmonary Medicine, Hadassah-Hebrew University Medical Center, Jerusalem, Israel

Angel Garcia-Lora Department of Clinical Analysis and Immunology, Virgen de las Nieves University Hospital, Granada, Spain

Federico Garrido Department of Cliniclas Analysis and Immunology, Virgen de las Nieves University Hospital, University of Granada Medical School, Granada, Spain

Dmitriy W. Gutkin Department of Pathology, Pittsburgh VA Hospital, University of Pittsburgh Medical Center, Pittsburgh, PA, USA

James L. Gulley Medical Oncology Branch and Laboratory of Tumor Immunology and Biology, Center for Cancer Research, National Cancer Institute, National Institutes of Health, Bethesda, MD, USA

James Hodge Laboratory of Tumor Immunology and Biology, Center for Cancer Research, National Cancer Institute, National Institutes of Health, Bethesda, MD, USA

Arthur A. Hurwitz Tumor Immunity and Tolerance Section, Laboratory of Molecular Immunoregulation, Cancer and Inflammation Program, Frederick National Laboratory for Cancer Research, Frederick, MD, USA

Tomoko Iwata-Kajihara Division of Cellular Signaling, Institute for Advanced Medical Research, School of Medicine, Keio University, Tokyo, Japan

Anahid Jewett Tumor Immunology Laboratory, Division of Oral Biology and Medicine, The Jane and Jerry Weintraub Center for Reconstructive Biotechnology, Jonsson Comprehensive Cancer Center, UCLA School of Dentistry, Los Angeles, CA, USA

Hajime Kamijuku Division of Cellular Signaling, Institute for Advanced Medical Research, School of Medicine, Keio University, Tokyo, Japan

Yutaka Kawakami Division of Cellular Signaling, Institute for Advanced Medical Research, School of Medicine, Keio University, Tokyo, Japan

Naoshi Kawamura Division of Cellular Signaling, Institute for Advanced Medical Research, School of Medicine, Keio University, Tokyo, Japan

Asuka Kobayashi Division of Cellular Signaling, Institute for Advanced Medical Research, School of Medicine, Keio University, Tokyo, Japan

Chie Kudo-Saito Division of Cellular Signaling, Institute for Advanced Medical Research, School of Medicine, Keio University, Tokyo, Japan

Yaeli Lebel-Haziv Department of Cell Research and Immunology, George S. Wise Faculty of Life Sciences, Tel Aviv University, Tel Aviv, Israel

Catalina Lee-Chang Immunoregulations Section, Laboratory of Molecular Biology and Immunology, National Institute on Aging, Baltimore, MD, USA

Paul V. Lehmann Cellular Technology Limited, Shaker Heights, OH, USA

Tal Leibovich-Rivkin Department of Cell Research and Immunology, George S. Wise Faculty of Life Sciences, Tel Aviv University, Tel Aviv, Israel

Shalom Lerrer Department of Cell Research and Immunology, George S. Wise Faculty of Life Sciences, Tel Aviv University, Tel Aviv, Israel

Yong Lu Department of Lymphoma and Myeloma, Center for Cancer Immunology Research, University of Texas MD Anderson Cancer Center, Houston, TX, USA

Ravi Madan Medical Oncology Branch and Laboratory of Tumor Immunology and Biology, Center for Cancer Research, National Cancer Institute, National Institutes of Health, Bethesda, MD, USA

Anatoli Malyguine Laboratory of Cell Mediated Immunity, SAIC-Frederick, Inc., National Cancer Institute at Frederick, Frederick, MD, USA

Shelly Maman Department of Cell Research and Immunology, George S. Wise Faculty of Life Sciences, Tel Aviv University, Tel Aviv, Israel; Institute of Human Virology, University of Maryland School of Medicine, Baltimore, MD, USA

Francesco M. Marincola Infectious Disease and Immunogenetics Section, Department of Transfusion Medicine, Clinical Center and Trans-NIH Center for Human Immunology, National Institutes of Health, Bethesda, MD, USA

Jean S. Marshall Department of Microbiology and Immunology, Faculty of Medicine, Dalhousie University, Halifax, NS, Canada

Annie Mirsoian Department of Dermatology, Davis School of Medicine, University of California, Sacramento, CA, USA

William J. Murphy Department of Dermatology, Davis School of Medicine, University of California, Sacramento, CA, USA

Shoko Nakamura Division of Cellular Signaling, Institute for Advanced Medical Research, School of Medicine, Keio University, Tokyo, Japan

Hiroshi Nishio Division of Cellular Signaling, Institute for Advanced Medical Research, School of Medicine, Keio University, Tokyo, Japan

Sharon A. Oldford Department of Microbiology and Immunology, Faculty of Medicine, Dalhousie University, Halifax, NS, Canada

Suzanne Ostrand-Rosenberg Department of Biological Sciences, University of Maryland Baltimore County, Baltimore, MD, USA

Angela Pardee University of Pittsburgh Cancer Institute, University of Pittsburgh, Pittsburgh, PA, USA

Jeong Hoon Park Division of Cellular Signaling, Institute for Advanced Medical Research, School of Medicine, Keio University, Tokyo, Japan

Francisco Ruiz-Cabello Department of Cliniclas Analysis and Immunology, Virgen de las Nieves University Hospital, University of Granada Medical School, Granada, Spain

Ryosuke Satomi Division of Cellular Signaling, Institute for Advanced Medical Research, School of Medicine, Keio University, Tokyo, Japan

Thomas Sayers Basic Sciences Program, Cancer and Inflammation Program, SAIC-Frederick, Inc., Frederick National Laboratory, Frederick, MD, USA

Gail D. Sckisel Department of Dermatology, Davis School of Medicine, University of California, Sacramento, CA, USA

Yevgeniya V. Segal Department of Pathology, University of Pittsburgh Medical Center, Pittsburgh, PA, USA

Andy Sikora Departments of Otolaryngology and Dermatology, Mount Sinai School of Medicine, Tisch Cancer Institute and Immunology Institute, New York, NY, USA

Nishith Singh Medical Oncology Branch, Center for Cancer Research, National Cancer Institute, National Institutes of Health, Bethesda, MD, USA

Pratima Sinha Department of Biological Sciences, University of Maryland Baltimore County, Baltimore, MD, USA

Galina V. Shurin Division of Experimental Pathology, Department of Pathology, University of Pittsburgh Medical Center, Pittsburgh, PA, USA

Michael R. Shurin Division of Clinical Immunopathology, Department of Pathology, University of Pittsburgh Medical Center, University of Pittsburgh Cancer Institute, Pittsburgh, PA, USA

Per thor Straten Department of Hematology, Center for Cancer Immune Therapy, University Hospital Herlev, Herlev, Denmark

Srividya Sundararaman Cellular Technology Limited, Shaker Heights, OH, USA

Hidetoshi Sumimoto Division of Cellular Signaling, Institute for Advanced Medical Research, School of Medicine, Keio University, Tokyo, Japan

Peter Svider Department of Otolaryngology, Mount Sinai School of Medicine, New York, NY, USA

Mayuri Tanaka Division of Cellular Signaling, Institute for Advanced Medical Research, School of Medicine, Keio University, Tokyo, Japan

Han-Ching Tseng Division of Oral Biology and Medicine, UCLA School of Dentistry, The Jane and Jerry Weintraub Center for Reconstructive Biotechnology, University of California, Los Angeles, CA, USA

Nobuo Tsukamoto Division of Cellular Signaling, Institute for Advanced Medical Research, School of Medicine, Keio University, Tokyo, Japan

Takahiro Tsujikawa Division of Cellular Signaling, Institute for Advanced Medical Research, School of Medicine, Keio University, Tokyo, Japan

Viktor Umansky German Cancer Research Center, Heidelberg, Germany

Jo Van Ginderachter Myeloid Cell Immunology Laboratory, Cellular and Molecular Immunology, Vrije Universiteit Brussel, Brussels, Belgium

S. H. van der Burg Department of Clinical Oncology, Leiden University Medical Center, Leiden, The Netherlands

Elena Voronov Faculty of Health Sciences, The Shraga Segal Department of Microbiology, Immunology and Genetics, Ben-Gurion University of the Negev, Beer-sheva, Israel

Lazar Vujanovic University of Pittsburgh Cancer Institute, University of Pittsburgh, Pittsburgh, PA, USA

Ena Wang Infectious Disease and Immunogenetics Section, Department of Transfusion Medicine, Clinical Center and Trans-NIH Center for Human Immunology, National Institutes of Health, Bethesda, MD, USA

Stephanie K. Watkins Tumor Immunity and Tolerance Section, Laboratory of Molecular Immunoregulation, Cancer and Inflammation Program, Frederick National Laboratory for Cancer Research, Frederick, MD, USA; Department of

Surgery, Loyola University Medical Center, Cardinal Bernardin Cancer Center, Maywood, IL, USA

Jonathan Weiss Cancer and Inflammation Program, Frederick National Laboratory for Cancer Research, Frederick, MD, USA

Polina Weitzenfeld Department of Cell Research and Immunology, George S. Wise Faculty of Life Sciences, Tel Aviv University, Tel Aviv, Israel

M. J. P. Welters Department of Clinical Oncology, Leiden University Medical Center, Leiden, The Netherlands

Isaac P. Witz Department of Cell Research and Immunology, George S. Wise Faculty of Life Sciences, Tel Aviv University, Tel Aviv, Israel; Institute of Human Virology, University of Maryland School of Medicine, Baltimore, MD, USA

Theresa L. Whiteside Departments of Pathology, Immunology and Otolaryngology, University of Pittsburgh School of Medicine, University of Pittsburgh Cancer Institute, Pittsburgh, PA, USA

Tomonori Yaguchi Division of Cellular Signaling, Institute for Advanced Medical Research, School of Medicine, Keio University, Tokyo, Japan

Jing Yang Department of Lymphoma and Myeloma, Center for Cancer Immunology Research, University of Texas MD Anderson Cancer Center, Houston, TX, USA

Eitan Yefenof Department of Cancer Immunology, The Lautenberg Research Center, Hebrew University-Hadassah Medical School, Jerusalem, Israel

Qing Yi Department of Lymphoma and Myeloma, MD Anderson Cancer Center, Houston, TX, USA

Anthony E. Zamora Department of Dermatology, Davis School of Medicine, University of California, Sacramento, CA, USA

Peiyuan Zhu Department of Blood Transfusion, School of Medicine, Jinling Hospital, Nanjing University, Nanjing, Jiangsu, China

Chapter 1
Role of the Immunological Environment in Cancer Initiation, Development and Progression

Anatoli Malyguine, Viktor Umansky and Michael R. Shurin

The last two decades have been characterized by a substantial progress in our understanding of the role of the immune system in tumor progression. We have learned how to manipulate the immune system to generate measurable tumor-specific immune responses. On the other hand, cancer cells induce malfunctions in immunity, as they manage to escape recognition and elimination by immune cells and factors. Chronic inflammation associated with a strong immunosuppression was also found to contribute to tumor initiation, progression and metastatic process. The tumor immunoenvironment represents specific conditions and elements that support cancer cell survival, proliferation and spreading. Understanding the role of the immune system in controlling and supporting tumor initiation, formation, growth and progression has crucial implications for cancer therapy.

Cancer represents more than 200 different diseases and is a major public health problem in the United States and other parts of the world. Some of the earliest evidence of cancer is found among fossilized bone tumors, human mummies in ancient Egypt, and ancient manuscripts. The oldest description of cancer called the Edwin Smith Papyrus was discovered in Egypt and dates back to about 3000 BC. It describes 8 cases of tumors or ulcers of the breast that were treated by

A. Malyguine
Laboratory of Cell Mediated Immunity, SAIC-Frederick, Inc, Frederick National Laboratory, Frederick, MD, USA

V. Umansky
Skin Cancer Unit, German Cancer Research Center (DKFZ), Heidelberg and Department of Dermatology, Venereology and Allergology, University Medical Center Mannheim, Ruprecht-Karl University of Heidelberg, Mannheim, Heidelberg, Germany

M. R. Shurin (✉)
Department of Pathology, Division of Clinical Immunopathology, University of Pittsburgh Medical Center, University of Pittsburgh Cancer Institute, Pittsburgh, PA, USA
e-mail: shurinmr@upmc.edu

M. R. Shurin et al. (eds.), *The Tumor Immunoenvironment*,
DOI: 10.1007/978-94-007-6217-6_1,

cauterization (www.Cancer.gov). Hippocrates named the disease "karkinos" (the Greek name for crab) to describe tumors.

At the present time, one in eight deaths worldwide is due to cancer. Cancer causes more deaths than AIDS, tuberculosis, and malaria combined and is the leading cause of death in developed countries and the second leading cause of death in developing countries (following heart diseases) (ACS 2012). According to estimates from the International Agency for Research on Cancer (IARC), there were 12.7 million new cancer cases in 2008 (the most recent year of available data) in the world (Ferlay et al.2008). The total cancer deaths estimate in 2008 was 7.6 million (about 21,000 cancer deaths a day). By 2030, worldwide 21.4 million new cancer cases and 13.2 million cancer deaths expected due to the growth and aging of the population, as well as reductions in childhood mortality and deaths from infectious diseases in developing countries (ACS 2012).

Although the ability of the immune system to effectively respond to tumor growth is now recognized, its role in controlling tumor initiation, expansion, and progression is a matter of long-term controversy. Understanding how the immune system affects cancer development and progression has been one of the most challenging questions in immunology.

Cancer biotherapy began around 1768 when Dr. G. White reported "the wonderful method of curing cancers by means of toads" (Goldsmith 1774; Hoption Cann et al. 2002). He described a woman from Hungerford, England, who treated patients with breast cancer (*In many cultures, animals such as guinea pigs or pigeons are applied to diseased parts of the body*). The method required that a toad be applied to the breast lesion until its death. One patient treated by this unorthodox method had a regression of her metastatic lesions following the "*toad cure*". It is possible that the skin of the toad contains some poisonous substances that might adversely affect cancer cells. Since the dead toad was affixed to the breast lesion for several weeks, it also provided an excellent breeding ground for local infections. Although, surgeon to the Duke of Kent injected himself with malignant tissue as a prophylaxis against development of cancer in 1777 and doctor to Louis XVII inoculated himself with breast cancer in hope of reversing a soft-tissue sarcoma in 1808, the principle that the immune system can recognize and respond to neoplastic cells was first proposed in the 19th century.

In 1890s, William Coley, a surgeon from Memorial Sloan Kettering Cancer Institute in New York, reported that using heat-killed endotoxin-containing bacteria (a combination of *Streptococcus pyogenes* and *Serratia marcescens*) resulted in a cure rate of 10 % in soft-tissue sarcoma patients (Coley 1891). A key aspect that Coley found to be necessary for tumor regression was the induction of a mild to moderate fever. At present, the only conventional treatment analogous to Coley's technique is bacillus Calmette-Guerin (BCG) treatment of bladder cancer. Yet unlike Coley's approach, BCG therapy uses a live bacterium (Rakoff-Nahoum and Medzhitov 2009).

The concept that the immune system surveys the body and prevents the outgrowth of carcinomas that would otherwise occur with high frequency was first suggested by Ehrlich (1909). With the better understanding of the mechanisms of immune response, Frank Macfarlane Burnett in 1957 proposed his cancer immunosurveillance theory which underpins the current belief that tumors can be recognized and eliminated by the immune system and proposed that tumor-specific neo-antigens were capable of eliciting a protective immunity (Burnet 1957a, b). Lewis Thomas speculated that complex and long-lived organisms should possess mechanisms capable of protecting against tumors (Thomas 1959).

These initial observations and hypotheses were confirmed in numerous experimental models demonstrating that the immune system can identify and destroy cancerous cells in a process termed cancer immunosurveillance, which functions as an important defense mechanism against cancer. Numerous reports of increased incidence and aggressiveness of a variety of cancers in immunodeficient patients or in patients receiving immunosuppressive therapy have further supported the hypothesis that the immune system plays a critical role in controlling the generation of malignant tumors. For instance, a systematic review of studies evaluating the incidence of cancer in both organ recipients and people with HIV/AIDS compared with the general population suggests that the weakening of the immune system may result in the increase of new cases of cancer in immunocompromised populations (Cobucci et al. 2012). The ability of immune cells to recognize and destroy cancerous cells has been directly documented both in vitro and in vivo, suggesting the role of cellular mechanisms in tumor immunosurveillance. Cytokines such as interleukin-2 are now established agents for the treatment of tumors. The description of a wide variety of human cancer antigens that are expressed on multiple cancer types, including many common epithelial cancers, presents new opportunities for the development of cancer immunotherapies (Vanneman and Dranoff 2012). Thus, data obtained from various studies in animal tumor models and in cancer patients offer ample evidence that several innate and adaptive immune cell types, specific effector molecules and definite pathways can collectively function as tumor-suppressor mechanisms (Vesely et al. 2011).

There are a large number of examples of how the immune system is able to recognize tumor antigens and eliminate or control tumor cell growth and spreading. As a result, we have learned how to manipulate the immune system to generate measurable tumor-specific immune responses (Rosenberg 2012). Unfortunately, the results of the numerous cancer vaccine clinical trials were mostly disappointing, and although immunotherapy of cancer is still being considered as an attractive therapeutic approach, its impact on clinical practice, with the exception of several antibodies, cytokines and dendritic cell (DC) vaccines, is very limited (Prestwich et al. 2008). Moreover, clinical studies demonstrated that the therapy-induced tumor-specific immune responses do not always correlate with clinical responses regardless of the generation of tumor-specific cytotoxic lymphocytes recognizing and efficiently killing tumor cells ex vivo, showing that somehow the anti-tumor immunity is often ineffective (Shurin et al. 2010). It is also obvious that though theoretically the immune reaction is responsible for controlling nascent cancer through immunosurveillance, tumors are able to escape this control and develop into clinical cancer.

Immune responses against cancer, including those induced by vaccination, depend on a balance between functional activity of various subsets of effector and suppressor T cells. While suppressor cells represent an important mechanism by which the immune system regulates specific immune responses, expansion of these cells in cancer patients interferes with the antitumor immunity and responses to therapy. In an immunocompetent cancer patient, the immune system may suppress effector cell attack against tumor antigens, especially in the tumor microenvironment. The suppressive compartment of the immune system includes several heterogeneous subsets of immune cells, including regulatory T cells (Tregs), myeloid-derived suppressor cells (MDSCs), alternatively activated (M2) or regulatory subsets of tumor-associated macrophages (TAMs), protumorigenic neutrophils (N2), tolerogenic or regulatory tumor-associated DCs (regDCs), regulatory B cells and possibly specific subsets of natural killer T (NKT) cells (Byrne et al. 2011; Montero et al. 2012; Shurin et al. 2011; Allavena and Mantovani 2012; Gregory and Houghton 2011).

Immune escape is the result of tumor-induced changes in cancer cells themselves, as well as the surrounding stromal tissues and the immune system. Cancers have been found to utilize diverse mechanisms to avoid, suppress and polarize both innate and adaptive anti-tumor immune responses. There is a significant number of identified mechanisms leading to immune unresponsiveness associated with the immunosuppressive tumor microenvironment.

Down-regulation of antigen processing and presentation by malignant cells, altered expression of certain chemokines and cytokines, induction of apoptosis in immune cells and suppression of immune cell function have been implicated in tumor escape from immune recognition and elimination (Coley 1891; Condamine and Gabrilovich 2011; Goldsmith 1774; Gregory and Houghton 2011). Importantly, both adaptive and innate responses might be dysfunctional in the tumor microenvironment. For instance, several identified tumor-derived factors have been reported to block the generation of DCs and their ability to uptake, process and present tumor antigens to T cells (Shurin et al. 2006). Furthermore, up-regulation of the immunosuppressive cell surface glycoprotein CD200 on acute myeloid leukemia (AML) cells specifically compromises NK cell anti-tumor responses. Patients with high CD200 expression on their AML cells exhibited a reduced frequency of activated NK cells and a lowered lytic activity and IFN-γ response against autologous CD200-expressing leukemic cells (Coles et al. 2011; Lion et al. 2012).

Tumor-redirected differentiation and functional polarization of immune cells results in accumulation of specific immune cell subsets with pro-tumorigenic potential, which support tumor development, growth and progression through different mechanisms. Thus, the immune system plays a dual role in cancer. It can not only suppress tumor growth by destroying cancer cells or inhibiting their outgrowth but also promote tumor progression either by selecting tumor cells that can survive in an immunocompetent host or by establishing conditions within the tumor microenvironment that facilitate tumor outgrowth (Gregory and Houghton 2011). For instance, antigen-specific Tregs primarily target DCs and inhibit DC functions including the expression of costimulatory molecules and the presentation of antigen early during the generation of the immune response. The end result is a complete inhibition of both the expansion and differentiation of T effector cells. Polyclonal Tregs also act on DCs, but at a later phase, and fail to inhibit expansion of T effector cells, but appear to modulate cell differentiation and trafficking (Shevach 2011). MDSCs represent a heterogeneous cell population composed mainly of myeloid progenitor cells that do not differentiate into mature macrophages, DCs or granulocytes. The tumor microenvironment effects the composition of cancer-induced MDSCs through the release of various tumor-derived factors, including cyclooxygenase 2, prostaglandins, granulocyte-macrophage colony stimulating factor (GM-CSF), macrophage CSF (M-CSF), IL-6, IL-10, vascular endothelial growth factor (VEGF), stem-cell factor, IL-3, FMS-related tyrosine kinase 3 (FLT3), and cell-expressed molecules (such as Notch). MDSCs are characterized by combinations of different surface markers and can be divided into two major subsets: granulocytic PMN- and monocytic MO-MDSCs (Hussain and Harris 2007; Ismail and Shurin (2012) Jain 2005).

MDSCs also exert their direct immunosuppressive function on antigen-specific T cell responses but also on mitogen-activated T lymphocytes, therefore bypassing the antigen dependency (Solito et al. 2011). In addition to being potent suppressors of T cell function, recent studies have demonstrated the ability of MDSCs to modulate activity of NK and myeloid cells and have implicated MDSCs in the induction of Tregs (Condamine and Gabrilovich 2011). Regulatory DCs in cancer may directly and indirectly maintain antigen-specific and non-specific T cell unresponsiveness by controlling T cell polarization, MDSC and Treg differentiation and activity, and affecting specific microenvironmental conditions in premalignant niches (Ma et al. 2012). Tumor-associated macrophages (TAMs) are also significant for fostering tumor progression. Up to 50 % of a malignant tumor mass can be composed of TAMs. While classical macrophages (M1) uptake antigens and play an important role in control of infections, TAMs can be reprogrammed in the tumor microenvironment in M2 cells as a result of tumor-driven 'alternative' activation (Daurkin et al. 2011). M2 are able to inhibit functions of immune cells and promote tumor survival, progression, angiogenesis and metastasis by releasing IL-10, PGE2, NO, high amounts of TGF-β or reactive oxygen species (ROS) (Whiteside 2010; Talmadge 2011). TAMs also contribute to immune evasion via induction of tolerogenic forkhead box P3 ($FOXP3^+$) and IL-10—secreting T cells as well as via upregulation of inhibitory receptor cytotoxic T lymphocyte antigen 4 (CTLA-4) expression in effector T cells (Daurkin et al. 2011).

Although neutrophils are traditionally considered in the context of their antibacterial functions, it is becoming increasingly clear that tumor-associated neutrophiles (TANs) play an important role in cancer biology (Fridlender and Albelda 2012). Many cancers are capable of recruiting neutrophiles to sites of tumorigenesis where they enhance tumor growth (Houghton 2010). N2 neutrophiles can inhibit effector T cell functions by the secretion of stored arginase 1 (ARG1) that degrades extracellular arginine, a factor needed for the proper activity of T cells (Fridlender and Albelda 2012). Additionally, products secreted from TANs, such as ROS and proteinases, have defined and specific roles in regulating tumor cell proliferation, angiogenesis, and metastasis (Gregory and Houghton 2011). Neutrophiles can also have a significant impact on the tumor microenvironment via produced cytokines and chemokines, which influence inflammatory cell recruitment and activation (Sansone and Bromberg 2011).

A pathophysiological association between inflammation and cancer has already been proposed in the 19th century, when in 1863 Rudolf Virchow noted leuçocytes in neoplastic tissues and made a connection between inflammation and cancer (Virchow 1863). He suggested that the "lymphoreticular infiltrate" reflected the origin of cancer at sites of chronic inflammation. Later, numerous laboratory and population-based studies suggested that certain malignancies arise at tissues severely damaged by chronic inflammation (Jochems and Schlom 2011). For example, cancers of stomach, liver, gallbladder, prostate, and pancreas are causally linked to gastric inflammation, chronic hepatitis, cholecystitis, inflammatory atrophy of the prostate, and chronic pancreatitis, respectively (Aggarwal et al. 2009). Colitis, a condition characterized by persistent colonic mucosal inflammation, often progresses to colorectal cancer; inflammatory bowel disease increases the risk of colorectal cancer by 10-fold and the management of colitis with anti-inflammatory therapy reduces this risk (Kundu and Surh 2012). Although approximately 25 % of all cancers have a proven etiologic background of chronic inflammation and/or infection (Mantovani et al. 2008; Montero et al. 2012), 90–95 % of neoplasia are linked to obesity, tobacco smoke, environmental pollutants, radiation and chronic infections, which all have in common a chronic inflammatory state (Grivennikov et al. 2010).

The role of inflammation in tumorigenesis is now accepted, and it is likely that an inflammatory microenvironment is an important cofactor for the development of all tumors, including those in which a direct causal relationship with inflammation is not yet confirmed (Chow et al. 2012). In the case of infection, host cells synthesize and release a number of antimicrobial factors, which include reactive oxygen species (ROS) and nitrogen intermediates (RNI), cytokines and chemokines, which recruit and activate protective effector cells such as macrophages, neutrophils, mast cells and DCs. If infection still persists, negative condition develops as a result of the continuous attack of infected tissues by immune cells and may promote cancer growth (Ismail and Shurin 2012).

Some of the mechanisms of tumor promotion by an inflammatory microenvironment are an increase of mutation rates and proliferation of mutated cells. Activated inflammatory cells provide ROS and RNI which induce DNA damage and genomic instability (Grivennikov et al. 2010; Lowe and Storkus 2011). Also inflammatory cells may promote ROS accumulation in neighboring epithelial cells as a result of production of cytokines as TNF-α. Furthermore, DNA damage can lead to inflammation and in turn promote tumorigenesis (Grivennikov et al. 2010). The production of pro-inflammatory cytokines and chemokines (IL-6, IL-8, IL-1β, CCL2, CCL20) may be activated through signal pathways of several oncoproteins such as Ras, Myc and RET (Mantovani et al. 2008). Production of tumor promoting cytokines that activate transcription factors, such as NF-κB, STAT3 and

AP-1, in pre-malignant cells, induce genes that stimulate cell proliferation and survival (Grivennikov et al. 2010).

Since intensive tumor growth requires additional blood supply, at some point the tumor becomes oxygen and nutrition deficient. As a result of tumor hypoxia and necrosis, the pro-inflammatory mediators are released enabling neoangiogenesis in tumor microenvironment (Vakkila and Lotze 2004). Important role in this process is played by *RAS, MYC* and *RET* oncogene family members. They activate a transcriptional program resulting in transformation of the tumor microenvironment through the recruitment of inflammatory cells and production of inflammation- and tumor-promoting chemokines and cytokines, metalloproteinases or adhesion molecules (Soucek et al. 2007; Sparmann and Bar-Sagi 2004). In addition, mutations in Von Hippel-Lindau tumor suppressor (VHL), transforming growth factor-β (TGF-β), and phosphatase and tensin homologue (PTEN), may activate transcription factors involved in inflammation and vascularization, particularly NF-κB, hypoxia inducible factor 1α (HIF-1α), and STAT3 (Mantovani et al. 2008).

Current studies show that NF-κB plays a fundamental role in the formation and development of malignant tissue caused by inflammation. As an ubiquitous central transcription factor, NF-κB plays a role both in the transformation of tissue cells to cancer cells, and in the regulation of the immune cell activity (Pikarsky et al. 2004; Karin 2006). The stimulation of immune cells by inflammatory cytokines such as interferon, TNF-α or IL-1β also leads ultimately to the activation of NF-κB and thereby to nonspecific inflammatory reactions. In tumor cells, the continued activation of NF-κB leads to the increased expression of genes which encode inflammation-promoting cytokines, adhesion molecules, angiogenic factors, etc. (Karin 2006). Furthermore, through the increased expression of anti-apoptotic genes such as *BCL2*, NF-κB activation promotes the survival of cancer cells (Van Waes 2007). There are emerging indications of an interaction between the NF-κB and HIF-1α systems (Rius et al. 2008).

New blood vessels growing in tumor site are often functionally impaired, leading to an increased interstitial fluid pressure, hypoxia and low pH within the tumor microenvironment (TME) that negatively influence lymphocyte homing, extravasation and function (Pardoll and Drake 2012; Schafer and Werner 2008). As Virchow already described over a 100 years ago, in terms of tissue morphology, tumor tissue resembles a chronically infected non-healing wound (Schafer and Werner 2008). Cancer cell hypoxic signals induce the expression and release of VEGF and PDGF promoting a local chronic inflammation, which supports tumor growth and progression (Rini 2009). Tumor cell hypoxia can also enable the migration of inflammatory cells, such as TAMs into tumor, which boost angiogenesis further by secreting such factors as VEGF (Allen and Louise Jones 2011; Finger and Giaccia 2010).

Therefore, all these events limit immune reactions (i.e., a immunosurveillance) mediated by immune effector cells like CD8 and NK cells that protect the host against premalignant and cancer cells. It is reasonable to assume that chronic inflammation helps the creation of an early primary tumor lesion that is less sensitive to type 1 immune response, allowing the tumor progression and metastatic spread.

Over 90 % of cancer patients die not from a primary lesion but from metastases to organs such as the brain, liver, lung and bones (Shurin et al. 2011). Metastatic process requires close interaction of cancer cells, stromal elements, and immune and inflammatory cells. The process of metastasis starts from epithelial-mesenchymal transition that permits cancer cells to enter blood and lymphatic vessels. Structural alterations in the extracellular matrix (EM) of the tumor microenvironment, which allow invasion and metastasis, are carried out mostly by stromal-derived matrix metalloproteinases (MMP), which degrade EM substrates like collagen. Moreover, TAMs and neutrophils are also important producers of matrix MMP within the TME (Lowe and Storkus 2011; Solinas et al. 2010; Kalluri and Weinberg 2009). IL-1, TNF-α and IL-6 promote MMP expression, invasiveness, and metastasis via NF-κB and STAT3 (Yu et al. 2007). EM expression of integrins and other cell surface receptors also increase tumor cells migratory capacity. In addition, inflammatory infiltrates such as TAMs, MDSCs, and cancer-associated fibroblasts could provide significant levels of TGF-β, an important regulator of the epithelial-mesenchymal transition and metastasis (Yang and Weinberg 2008).

Once metastatic cells enter the circulation, they need to survive in suspension. The survival of these cells is affected by inflammatory mediators released by immune cells activated by cancer- or pathogen-derived stimuli (Luo et al. 2004; Kim et al. 2009) and depends on activation of NF-κB. A variety of cytokines, including TNF-α and IL-6, can also promote circulating cancer cell survival (Nguyen et al. 2009) and some of these cytokines can physically link cancer cells to TAMs, allowing them to travel together (Condeelis and Pollard 2006). Circulating cancer cells may overcome immunosurveillance by interaction with platelets or macrophages which results in protection of cancer cells from NK mediated killing (Palumbo et al. 2007). Interestingly, tumor cells co-cultivated with macrophages develop an increasingly metastatic phenotype, comparable with that induced by the activation of the NF-κB pathway or TNF-α activation (Wyckoff et al. 2007). The migration of metastasis initiating cells is directed by chemokine gradients via CXCR4, CCR4, CCR7, CCR9 and CCR10 (Bonecchi et al. 2009). To colonize distant sites/organs, cancer cells becoming trapped in capillary beds resulting in integrin-dependent attachment to endothelium (Chaffer and Weinberg 2011). Several proinflammatory cytokines that are elevated in the circulation of cancer patients up regulate expression of adhesion molecules on the endothelium or in target organs and facilitate metastatic cell attachment (Mantovani et al. 2008). The homing is followed by extravasation into the tissue, and quick adaptation of malignant cells to a foreign environment by interaction with immune, inflammatory, and stromal cells (Polyak and Weinberg 2009). A state of chronic inflammation may provide a hospitable environment to incoming cancer cells by preventing apoptosis and inducing epigenetic and mutational effects that would favor cancer progression within the distal tissue location. In addition, the various factors secreted by locally recruited inflammatory cells, such as TAMs, could provide the protumorigenic effect (Sansone and Bromberg 2011; Lowe and Storkus 2011).

In summary, we now appreciate that the immune system, in addition to tumor-suppressive function by eliminating nascent transformed tumor cells, can also facilitate tumor initiation and progression by providing a complimentary TME through the maintenance of chronic inflammatory state in the tumor mass and by inducing polarized immunosuppressive regulatory cells. However, the distinctions between tumor-promoting inflammation and tumor-suppressive immunity are still not clear due to the dual role of some cytokines and other molecules in the immune system. Recently it was shown that interaction between tumor cells and DCs, but not monocytes, leads to rapid induction of the genomic mutator activation-induced cytidine deaminase (AID) and AID-dependent DNA double-strand breaks (DSBs) in tumor cell lines and primary tumor cells (Koduru et al. 2012). AID-mediated genomic damage led to altered tumorigenicity and indolent behavior of tumor cells in vivo. These data show a novel pathway for the capacity of immune cells to regulate genomic integrity (Koduru et al. 2012).

Understanding the role of the immune system in controlling and supporting tumor initiation, formation, growth and progression has crucial implications for cancer therapy since immunomodulatory interventions aimed at early pathogenic events may no longer be efficient when these pathways have altered due to a different effects of the immune response (Schreiber et al. 2011). Therefore, it is critical to recognize why and how the cancer-associated immune activities evolve over time, so that time-dependent therapies may be rationally implemented for an improved clinical outcome. These new insights in evolving interactions of different cell subsets in the tumor immunoenvironment are constantly improving the design and efficacy of modern cancer immunotherapy protocols, as reviewed elsewhere (Whiteside 2010; Wyckoff et al. 2007; Yang and Weinberg 2008; Yu et al. 2007). Deciphering the interaction between immune cells, malignant cells, stromal elements and treatment modalities will therefore guide the future combination of immunotherapy with conventional therapies to achieve optimal anti-tumor effects in cancer patients.

References

Aggarwal BB, Vijayalekshmi RV, Sung B (2009) Targeting inflammatory pathways for prevention and therapy of cancer: short-term friend, long-term foe. Clin Cancer Res 15(2):425–430

Allavena P, Mantovani A (2012) Immunology in the clinic review series; focus on cancer: tumour-associated macrophages: undisputed stars of the inflammatory tumour microenvironment. Clin Exp Immunol 167(2):195–205

Allen M, Louise Jones J (2011) Jekyll and Hyde: the role of the microenvironment on the progression of cancer. J Pathol 223(2):162–176

Bonecchi R, Galliera E, Borroni EM, Corsi MM, Locati M, Mantovani A (2009) Chemokines and chemokine receptors: an overview. Front Biosci 14:540–551

Burnet M (1957a) Cancer; a biological approach. I. The processes of control. Br Med J 1(5022):779–786

Burnet M (1957b) Biology and medicine. Eugenics Rev 49(3):127–135

Byrne WL, Mills KH, Lederer JA, O'Sullivan GC (2011) Targeting regulatory T cells in cancer. Cancer Res 71(22):6915–6920

Chaffer CL, Weinberg RA (2011) A perspective on cancer cell metastasis. Science 331(6024):1559–1564

Chow MT, Moller A, Smyth MJ (2012) Inflammation and immune surveillance in cancer. Semin Cancer Biol 22(1):23–32

Cobucci RN, Saconato H, Lima PH, Rodrigues HM, Prudencio TL, Junior JE, Giraldo PC, Goncalves AK (2012) Comparative incidence of cancer in HIV-AIDS patients and transplant recipients. Cancer Epidemiol 36(2):69–73

Coles SJ, Wang EC, Man S, Hills RK, Burnett AK, Tonks A, Darley RL (2011) CD200 expression suppresses natural killer cell function and directly inhibits patient anti-tumor response in acute myeloid leukemia. Leukemia 25(5):792–799

Coley WB (1891) II. Contribution to the knowledge of Sarcoma. Ann Surg 14(3):199–220

Condamine T, Gabrilovich DI (2011) Molecular mechanisms regulating myeloid-derived suppressor cell differentiation and function. Trends Immunol 32(1):19–25

Condeelis J, Pollard JW (2006) Macrophages: obligate partners for tumor cell migration, invasion, and metastasis. Cell 124(2):263–266

Daurkin I, Eruslanov E, Stoffs T, Perrin GQ, Algood C, Gilbert SM, Rosser CJ, Su LM, Vieweg J, Kusmartsev S (2011) Tumor-associated macrophages mediate immunosuppression in the renal cancer microenvironment by activating the 15-lipoxygenase-2 pathway. Cancer Res 71(20):6400–6409

Ehrlich P (1909) Über den jetzigen Stand der Karzinomforschung. Beiträge zur experimentellen Pathologie und Chemotherapie, pp 117–164

Ferlay J, Shin H, Bray F, Forman D, Mathers C, Parkin D (2010) GLOBOCAN 2008, Cancer incidence and mortality Worldwide http://globocaniarc.fr

Finger EC, Giaccia AJ (2010) Hypoxia, inflammation, and the tumor microenvironment in metastatic disease. Cancer Metastasis Rev 29(2):285–293

Fridlender ZG, Albelda SM (2012) Tumor-associated neutrophils: friend or foe? Carcinogenesis

Goldsmith O (1774) An history of the frog kind: a history of the earth and animated nature. J Nourse 7:102–107

Gregory AD, Houghton AM (2011) Tumor-associated neutrophils: new targets for cancer therapy. Cancer Res 71(7):2411–2416

Grivennikov SI, Greten FR, Karin M (2010) Immunity, inflammation, and cancer. Cell 140(6):883–899

Hoption Cann SA, van Netten JP, van Netten C, Glover DW (2002) Spontaneous regression: a hidden treasure buried in time. Med Hypotheses 58(2):115–119

Houghton AM (2010) The paradox of tumor-associated neutrophils: fueling tumor growth with cytotoxic substances. Cell Cycle 9(9):1732–1737

Hussain SP, Harris CC (2007) Inflammation and cancer: an ancient link with novel potentials. Int J Cancer 121(11):2373–2380

Ismail N, Shurin MR (2012) Cancer and infection: friends or foes? Future Oncol 8(9):1061–1064

Jain RK (2005) Normalization of tumor vasculature: an emerging concept in antiangiogenic therapy. Science 307(5706):58–62

Jochems C, Schlom J (2011) Tumor-infiltrating immune cells and prognosis: the potential link between conventional cancer therapy and immunity. Exp Biol Med 236(5):567–579

Kalluri R, Weinberg RA (2009) The basics of epithelial-mesenchymal transition. J Clin Investig 119(6):1420–1428

Karin M (2006) Nuclear factor-kappaB in cancer development and progression. Nature 441(7092):431–436

Kim S, Takahashi H, Lin WW, Descargues P, Grivennikov S, Kim Y, Luo JL, Karin M (2009) Carcinoma-produced factors activate myeloid cells through TLR2 to stimulate metastasis. Nature 457(7225):102–106

Koduru S, Wong E, Strowig T, Sundaram R, Zhang L, Strout MP, Flavell RA, Schatz DG, Dhodapkar KM, Dhodapkar MV (2012) Dendritic cell-mediated activation-induced cytidine

deaminase (AID)-dependent induction of genomic instability in human myeloma. Blood 119(10):2302–2309
Kundu JK, Surh YJ (2012) Emerging avenues linking inflammation and cancer. Free Radic Biol Med 52(9):2013–2037
Lion E, Willemen Y, Berneman ZN, Van Tendeloo VF, Smits EL (2012) Natural killer cell immune escape in acute myeloid leukemia. Leukemia 26(9):2019–2026
Lowe DB, Storkus WJ (2011) Chronic inflammation and immunologic-based constraints in malignant disease. Immunotherapy 3(10):1265–1274
Luo JL, Maeda S, Hsu LC, Yagita H, Karin M (2004) Inhibition of NF-kappaB in cancer cells converts inflammation: induced tumor growth mediated by TNFalpha to TRAIL-mediated tumor regression. Cancer Cell 6(3):297–305
Ma Y, Shurin GV, Gutkin DW, Shurin MR (2012) Tumor associated regulatory dendritic cells. Semin Cancer Biol 22(4):298–306
Mantovani A, Allavena P, Sica A, Balkwill F (2008) Cancer-related inflammation. Nature 454(7203):436–444
Montero AJ, Diaz-Montero CM, Kyriakopoulos CE, Bronte V, Mandruzzato S (2012) Myeloid-derived suppressor cells in cancer patients: a clinical perspective. J Immunother 35(2):107–115
Nguyen DX, Bos PD, Massague J (2009) Metastasis: from dissemination to organ-specific colonization. Nat Rev 9(4):274–284
Palumbo JS, Talmage KE, Massari JV, La Jeunesse CM, Flick MJ, Kombrinck KW, Hu Z, Barney KA, Degen JL (2007) Tumor cell-associated tissue factor and circulating hemostatic factors cooperate to increase metastatic potential through natural killer cell-dependent and-independent mechanisms. Blood 110(1):133–141
Pardoll D, Drake C (2012) Immunotherapy earns its spot in the ranks of cancer therapy. J Exp Med 209(2):201–209
Pikarsky E, Porat RM, Stein I, Abramovitch R, Amit S, Kasem S, Gutkovich-Pyest E, Urieli-Shoval S, Galun E, Ben-Neriah Y (2004) NF-kappaB functions as a tumour promoter in inflammation-associated cancer. Nature 431(7007):461–466
Polyak K, Weinberg RA (2009) Transitions between epithelial and mesenchymal states: acquisition of malignant and stem cell traits. Nat Rev 9(4):265–273
Prestwich RJ, Errington F, Hatfield P, Merrick AE, Ilett EJ, Selby PJ, Melcher AA (2008) The immune system: is it relevant to cancer development, progression and treatment? Clin Oncol (R Coll Radiol) 20(2):101–112
Rakoff-Nahoum S, Medzhitov R (2009) Toll-like receptors and cancer. Nat Rev 9(1):57–63
Rini BI (2009) Metastatic renal cell carcinoma: many treatment options, one patient. J Clin Oncol 27(19):3225–3234
Rius J, Guma M, Schachtrup C, Akassoglou K, Zinkernagel AS, Nizet V, Johnson RS, Haddad GG, Karin M (2008) NF-kappaB links innate immunity to the hypoxic response through transcriptional regulation of HIF-1alpha. Nature 453(7196):807–811
Rosenberg SA (2012) Raising the bar: the curative potential of human cancer immunotherapy. Sci Transl Med 4(127):127ps8
Sansone P, Bromberg J (2011) Environment, inflammation, and cancer. Curr Opin Genet Dev 21(1):80–85
Schafer M, Werner S (2008) Cancer as an overhealing wound: an old hypothesis revisited. Nat Rev Mol Cell Biol 9(8):628–638
Schreiber RD, Old LJ, Smyth MJ (2011) Cancer immunoediting: integrating immunity's roles in cancer suppression and promotion. Science 331(6024):1565–1570
Shevach EM (2011) Biological functions of regulatory T cells. Adv Immunol 112:137–176
Shurin MR, Shurin GV, Lokshin A, Yurkovetsky ZR, Gutkin DW, Chatta G, Zhong H, Han B, Ferris RL (2006) Intratumoral cytokines/chemokines/growth factors and tumor infiltrating dendritic cells: friends or enemies? Cancer Metastasis Rev 25(3):333–356
Shurin MR, Gregory M, Morris JC, Malyguine AM (2010) Genetically modified dendritic cells in cancer immunotherapy: a better tomorrow? Expert Opin Biol Ther 10(11):1539–1553

Shurin GV, Ouellette CE, Shurin MR (2011) Regulatory dendritic cells in the tumor immuno environment. Cancer Immunol Immunother 61(2):223–230
ACS, American Cancer Society (2012) Global Cancer facts and Figures 2012. Available from: http://wwwcancerorg/research/cancerfactsfigures/index
Solinas G, Marchesi F, Garlanda C, Mantovani A, Allavena P (2010) Inflammation-mediated promotion of invasion and metastasis. Cancer Metastasis Rev 29(2):243–248
Solito S, Bronte V, Mandruzzato S (2011) Antigen specificity of immune suppression by myeloid-derived suppressor cells. J Leukoc Biol 90(1):31–36
Soucek L, Lawlor ER, Soto D, Shchors K, Swigart LB, Evan GI (2007) Mast cells are required for angiogenesis and macroscopic expansion of Myc-induced pancreatic islet tumors. Nat Med 13(10):1211–1218
Sparmann A, Bar-Sagi D (2004) Ras-induced interleukin-8 expression plays a critical role in tumor growth and angiogenesis. Cancer Cell 6(5):447–458
Talmadge JE (2011) Immune cell infiltration of primary and metastatic lesions: mechanisms and clinical impact. Semin Cancer Biol 21(2):131–138
Thomas L (1959) Discussion. In: Lawrence HS (ed) Cellular and humoral aspects of the hypersensitive states. Hoeber-Harper, New York, pp 529–532
Vakkila J, Lotze MT (2004) Inflammation and necrosis promote tumour growth. Nat Rev Immunol 4(8):641–648
Van Waes C (2007) Nuclear factor-kappaB in development, prevention, and therapy of cancer. Clin Cancer Res 13(4):1076–1082
Vanneman M, Dranoff G (2012) Combining immunotherapy and targeted therapies in cancer treatment. Nat Rev 12(4):237–251
Vesely MD, Kershaw MH, Schreiber RD, Smyth MJ (2011) Natural innate and adaptive immunity to cancer. Annu Rev Immunol 29:235–271
Virchow R (1863) Die Krankhaften Geschwülste. August Hirschwald, Berlin
Whiteside TL (2010) Inhibiting the inhibitors: evaluating agents targeting cancer immunosuppression. Expert Opinion Biol Ther 10(7):1019–1035
Wyckoff JB, Wang Y, Lin EY, Li JF, Goswami S, Stanley ER, Segall JE, Pollard JW, Condeelis J (2007) Direct visualization of macrophage-assisted tumor cell intravasation in mammary tumors. Cancer Res 67(6):2649–2656
Yang J, Weinberg RA (2008) Epithelial-mesenchymal transition: at the crossroads of development and tumor metastasis. Dev Cell 14(6):818–829
Yu H, Kortylewski M, Pardoll D (2007) Crosstalk between cancer and immune cells: role of STAT3 in the tumour microenvironment. Nat Rev Immunol 7(1):41–51

Part I
Tumor Microenvironment and Immunoenvironment

Chapter 2
The Metastatic Microenvironment

Shelly Maman and Isaac P. Witz

Abstract Metastasis is the major killer of cancer patients. Although increased understanding of the metastatic process was achieved in recent years, the mechanisms underlying the progression of cancer cells to form site-specific metastasis are still awaiting complete elucidation. The current consensus is that circulating tumor cells disseminate into future metastatic sites and that these disseminated tumor cells form micrometastasis in these sites. The micrometastases remain in a state of dormancy in these sites until "awakened" to progress towards overt metastases. Whereas the evidence implicating chemokine–chemokine receptor interactions as the mechanism responsible for the targeted migration of tumor cells to future metastatic sites is quite strong, the mechanisms that maintain dormancy of disseminated tumor cells and the mechanisms that awaken these dormant micrometastases, driving their progression towards frank metastasis, are still obscure. It is clear, however, that the metastatic microenvironment plays a major role in these events. Three topics are discussed in this review: Mechanisms that are involved in the targeted migration of tumor cells to future metastatic sites; Specific molecular signatures expressed by metastases and micrometastases and interactions between metastatic and micrometastatic cells with the metastatic microenvironment. In reviewing these topics we focused on studies performed in our lab with neuroblatoma lung and melanoma brain metastasis.

S. Maman · I. P. Witz
Department of Cell Research and Immunology George S. Wise Faculty of Life Sciences,
Tel Aviv University, Tel Aviv, Israel

S. Maman · I. P. Witz
Institute of Human Virology, University of Maryland School of Medicine,
Baltimore, MD, USA

I. P. Witz (✉)
Department of Cell Research and Immunology George S, Wise Faculty of Life Sciences,
Tel Aviv University, Tel Aviv 69978, Israel
e-mail: isaacw@tauex.tau.ac.il

M. R. Shurin et al. (eds.), *The Tumor Immunoenvironment*,
DOI: 10.1007/978-94-007-6217-6_2,

Keywords Tumor · Metastatic microenvironment · Metastasis · Site-specific metastasis · Micrometastasis · Dormancy · Chemokines · Neuroblastoma · Melanoma · Molecular signature

Abbreviations

TME	Tumor microenvironment
CAF	cancer-associated fibroblast
CXCL	chemokine (C-X-C motif) ligand
SDF	Stromal cell-derived actor
CCL	Chemokine (C–C motif) ligand
TNF	Tumor necrosis factor
MCP	Monocyte chemotactic protein
CXCR	Chemokine (C-X-C motif) receptor
IFN	Interferon
TEM	Transendothelial migration
CCR	Chemokine (C–C) motif receptor
CTC	Circulating tumor cells
DTC	Disseminated tumor cells
PHOX	Paired-like homeobox
MMP	Matrix metalloproteinase
ERK	Extracellular Signal-Regulated Kinase
PCR	Polymerase chain reaction

2.1 Introduction

Stephen Paget, about 120 years ago, conceptualized in his "seed and soil" theory the idea that the tumor microenvironment (TME) plays an important role in site-specific metastasis (Paget 1889). This idea was revitalized in the early seventies of the last century, marking the onset of the post Paget TME research era (Fidler 2001; Onuigbo 1975; Hart and Fidler 1980; Hart et al. 1981; Hart 1982; Weiss et al. 1984; 1988; Nicolson 1988; Pauli and Lee 1988; Pauli and Augustin-Voss 1990; Togo et al. 1995; Kuo et al. 1995). Histological, molecular and cellular studies indicated that the interface between tumor cells and the stroma of the TME is a multi-component interactive arena (Fidler 2001; Jung et al. 2002; Mueller and Fusenig 2002; Lynch and Matrisian 2002; Ben-Baruch 2003; Cunha et al. 2003; Mantovani et al. 2004; Park et al. 2000; McCawley and Matrisian 2001; Rubin 2001; Tlsty 2001; Liotta and Kohn 2001; Fidler 2002; Cunha et al. 2002; van Kempen et al. 2002; Unger and Weaver 2003; Hendrix et al. 2003; Eshel et al. 2002). These studies also indicated that the TME does not operate as a binary neutral growth medium that either supports or does not support metastasis as conceived by Paget, but rather as an active regulator of the malignancy phenotype of cancer cells (Witz and Levy-Nissenbaum 2006).

There is a general consensus in the TME field that the progression of cancer towards metastasis is regulated to a large extent by interactions of the cancer cells with non-tumor cells in their vicinity and with soluble factors released or secreted from the cancer cells and from the non-tumor cells in the microenvironment. These tumor-microenvironment interactions are bidirectional and each interaction partner regulates and shapes the phenotype of the other (Witz and Levy-Nissenbaum 2006; Witz 2008b; Kopfstein and Christofori 2006; Weinberg 2008; Gupta and Massague 2006).

Although both the non-tumor cells in the TME as well as the tumor cells themselves are accessories in tumor progression towards metastasis, the tumor is undoubtedly the original perpetuator. On the one hand it evolves into an increasing malignant entity and at the same time recruits non-tumor cells to the TME and programs these cells as well as resident non-tumor cells to promote tumor progression (Chaput et al. 2008; Klebanoff et al. 2011; Goetz 2012). These corrupted non-tumor cells fulfill inductive, adaptive and selective functions. Signals delivered by such cells may direct the tumor towards one or several possible molecular evolution pathways. Many of these pathways may lead to metastasis (Witz 2008b).

Metastasis is the major cause of death in cancer patients (Mehlen and Puisieux 2006). However, only recently did the scientific community demonstrate an increased interest and research efforts in this important aspect of oncology. This point is illustrated in Fig. 2.1, which shows PubMed data on the number of published "metastasis"-related papers in each of the years between 1995 and 2011. Among these publications are numerous informative reviews (Fidler 2001, 2002; Witz 2008b; Kopfstein and Christofori 2006; Gupta and Massague 2006; Mehlen and Puisieux 2006; Weinberg 1995; Ruiz and Gunthert 1996; Lawrence and Steeg 1996; Meyer and Hart 1998; Yokota 2000; Pass 2002; Hunter 2004; Pantel and Brakenhoff 2004; Nguyen 2004; Braun and Naume 2005; Zigrino et al. 2005; Steeg 2005; DiMeo and Kuperwasser 2006; Dai et al. 2006; Palmieri et al. 2007; Langley and Fidler 2007; Nguyen and Massague 2007; Albini et al. 2008; Hu and Polyak 2008; Kumar and Weaver 2009; Joyce and Pollard 2009; Egeblad et al. 2010; Valastyan and Weinberg 2011).

As already recognized by Paget, there is a predilection of tumors to metastasize to specific organ sites. However, the metastatic capacity of a certain tumor is not restricted to a single organ site. For example breast cancer metastasizes to bone, lungs, regional lymph nodes, liver and brain while prostate cancer metastasizes to bones and lymph nodes. Melanoma spreads mainly to lymph nodes, liver and brain. Each tumor type has therefore several different metastatic microenvironments. Since the tumor and its microenvironment regulate and shape each other's phenotype (Witz 2008b), it is to be expected that the metastases arising in one organ site be different from metastases derived from the very same tumor developing in a different organ site. It is also to be expected that different reciprocal signaling cascades take place between metastases and non-tumor microenvironmental cells in different metastatic microenvironments.

Arriving at a secondary organ site, metastatic cells have several possible fates: proliferation, entrance to a dormant state and initiation of apoptosis. It is the

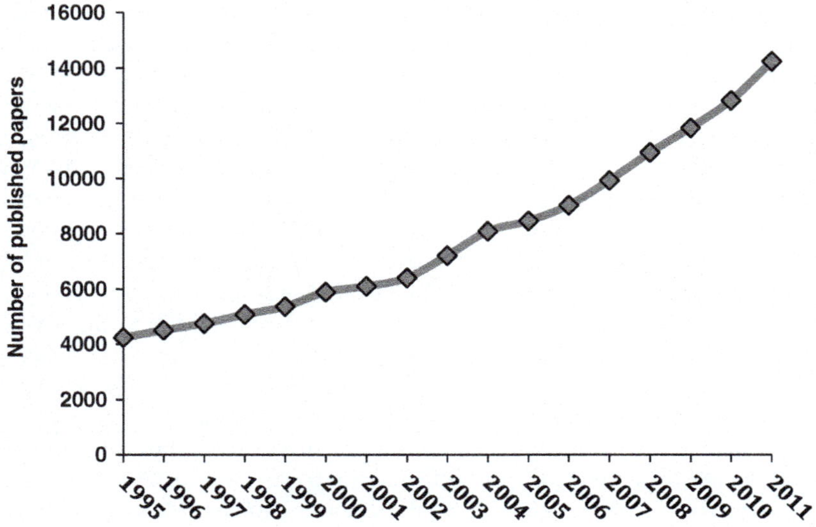

Fig. 2.1 Number of "metastasis"-related published papers in each of the years between 1995 and 2011 according to Pubmed database

interactions with the microenvironment that will determine whether cancer cells will progress towards metastasis or whether they will stay dormant or disappear altogether. Thus, tumor-microenvironment interactions regulate either anti- or pro-malignancy functions. These and related issues are still in the infancy stage of research and remain to be elucidated. For example: What attracts tumor cells to specific metastatic microenvironment? What sustains the survival of disseminated tumor cells in a particular organ site? What induces them to proliferate? Are the survival and growth factors for metastases in a particular metastatic microenvironment similar or different from survival and growth factors for metastases from the same tumor in a different metastatic microenvironment?

This chapter addresses briefly some of these issues with a particular emphasis to our work on neuroblastoma lung metastasis and melanoma brain metastasis.

2.2 Attraction of Tumor Cells to Metastatic Sites: The Role of Chemokine–Chemokine Receptor Axes

Chemokines are involved in site-specific metastasis (Ben-Baruch 2008; Takeuchi et al. 2007; Fulton 2009; Zlotnik et al. 2011). This involvement occurs at different levels:

1. Secretion of chemokines from tumor cells and from non-tumor cells in the TME

2. Expression of chemokine receptors by tumor cells
3. Expression of chemokine receptors by non-tumor cells in the TME.

Homeostatic and inflammatory chemokines are secreted from a large variety of tumor cells and from non-tumor cells in the TME. Such chemokines can mobilize chemokine-receptor-expressing cells such as myeloid or lymphatic cells to the TME with wide ranging biological consequences manifested *inter alia* by tumor destruction, angiogenesis and metastasis enhancement (Mantovani et al. 1992; Bar-Eli 1999; Wang et al. 2006; Burger and Kipps 2006; Soria et al. 2008; Navarini-Meury and Conrad 2009; Schmid and Varner 2010; Sapoznik et al. 2012; Soria et al. 2012; Umansky and Sevko 2012). Below are some examples of biological activities mediated by TME-derived chemokines.

CAF-derived CXCL12 (SDF-1) enhanced tumor growth through the CXCR4 receptor expressed by breast carcinoma cells. The CXCL12–CXCR4 axis also supported angiogenesis by recruiting endothelial progenitor cells into the carcinomas. Interestingly, the myofibroblastic phenotype and the ability to enhance tumor growth in vivo were stably maintained in the CAFs even in the absence of contact between them and the tumor cells (Orimo et al. 2005).

Chemokine-driven vicious cycles that enhance tumor progression operate in the TME of mammary carcinomas in mice and in breast cancer in humans. We showed for example that mouse mammary carcinoma cells secreted high levels of CCL2 (MCP-1) known for its capacity to attract monocytes to the TME. Monocyte-derived TNF-α up-regulated CCL2 secretion from the tumor cells, and CCL2 in turn promoted the secretion of TNF-α from monocytes. In this vicious cycle, the tumor cells and the monocytes in the TME promoted each other's ability to express and secrete pro-malignancy factors (Neumark et al. 2003). A similar situation exists in breast cancer in humans (Ben-Baruch 2003). Monocyte chemoattractants CCL5 and CCL2 secreted by breast tumor cells may induce monocyte infiltration to the microenvironment of breast tumors. The resulting tumor-associated macrophages may secrete TNF-α, which induces or up-regulates the secretion of several pro-malignancy factors from the tumor cells such as matrix metalloproteinases. TNF-α also further up-regulates the secretion of CCL5 and CCL2, which drive the merry-go-round for another cycle (Ben-Baruch 2003). It is not unlikely that similar cycles operate also in other types of cancer.

In some tumor types, the CXCR3-CXCL10 axis is considered to antagonize tumor growth and progression (Chakraborty et al. 2008; Agostini et al 2001). This axis may, however, also engage in pro-malignancy activities (Maru et al. 2008). In a study performed in our lab it was shown that the interaction of the CXCL10 chemokine with its CXCR3 receptor expressed by colorectal carcinoma cells promotes, rather than antagonizes, tumor progression (Zipin-Roitman et al. 2007). It was also indicated that a vicious cycle involving the CXCR3-CXCL10 axis and IFN-γ operates in colorectal carcinoma progression (Zipin-Roitman et al. 2007). CXCL10 secreted from CXCR3-expressing colorectal carcinoma cells promotes, by an autocrine mechanism, progression-promoting functions in these tumor cells. CXCL10, at the same time, attracts CXCR3-expressing Th1 cells to the tumor site. The infiltrating

Th1 cells secrete IFN-γ, which, in addition to its immune functions, promotes the release of CXCL10 from IFN-γ receptor–expressing colorectal carcinoma cells while up-regulating CXCR3 expression. This further promotes the capacity of the colorectal carcinoma cells to respond to CXCL10-mediated pro-malignancy functions. The expression of chemokine receptors by tumors cells enables their targeted migration to specific organ sites expressing the corresponding chemokine ligands. This targeted migration strategy generating site-specific metastasis was "hijacked" from normal migratory mechanisms operating in organogenesis, leukocyte migration and lymphoid tissue neogenesis (extensively reviewed by Zlotnik et al. 2011).

The chemokine receptor CXCR4 is expressed by many cancer types of humans and animals. One of the first studies showing that this receptor is involved in site-specific metastasis was performed by Müller et al. (2001). These authors demonstrated that the expression in the lung of CXCL12 (SDF1), the chemokine ligand of CXCR4, attracts breast cancer cells to this metastatic site.

We evaluated the possibility that neuroblastoma cells, similar to hemopoietic stem cells, use chemokine–chemokine receptor interactions to home to the bone-marrow, a primary metastatic site for such cancer cells. The results of this study demonstrated that CXCR4 expression might be a general characteristic of neuroblastoma cells (Geminder et al. 2001). Such cells express not only CXCR4, but also its ligand, CCL12. CXCR4 expression by neuroblastoma cells is tightly regulated by tumor cell-derived autocrine CCL12, as demonstrated by the ability of neutralizing antibodies against human CCL12 to up-regulate CXCR4 expression on the tumor cells. Conversely CXCR4 expression by neuroblastoma cells was reduced following short-term exposure to recombinant human CCL12. These and additional results strongly suggested that the ability of neuroblastoma tumors to preferentially form metastases in the bone-marrow might be facilitated by a set of complex CXCR4-CCL12 interactions.

Clinical studies supported the above conclusion. It was reported that the clinical outcome in patients with tumors highly expressing CXCR4 was significantly worse than in those patients with a low-expression of CXCR4. It was concluded that CXCR4 expression in neuroblastoma primary tumors is significantly correlated with the pattern of metastatic spread (Russell et al. 2004).

Apart from chemotaxis, chemokine–chemokine receptor interactions have additional functions. They activate various signaling pathways and alter gene expression profiles resulting, for example, in promotion of growth factors of tumor cells (Eshel et al. 2002; Zhang et al. 2010; Richmond et al. 2009). Overexpressing CXCR4 in neuroblastoma cells, we found that gene expression patterns in these cells differed considerably from those in control cells. We hypothesized that these differences were due to an autocrine CCL12-CXCR4 interaction (Nevo et al. 2004).

Fractalkine (CX3CL1) is a chemokine that is expressed either as a soluble molecule or as a membrane-bound molecule, which functions also as an adhesion molecule. Soluble CX3CL1 is capable of attracting fractalkine receptor (CX3CR1)-expressing cells. There is evidence that CX3CL1 and its CX3CR1 receptor are involved in cancer, especially in that of neural origin as well as in prostate, pancreas and breast carcinoma. Such cancer cells express high levels of CX3CR1, which is involved with migration and site-specific dissemination (Marchesi et al. 2010).

Transendothelial migration (TEM) of tumor cells is a crucial step in metastasis formation, involving adhesion molecules and chemokines. Since CX3CL1 takes part in both adhesion and chemotaxis and since bone-marrow is the first metastatic site of neuroblastoma, we asked if the CX3CR1-CX3CL1 axis is involved in the transmigration of neuroblastoma cells across bone-marrow endothelium (Nevo et al. 2009). We first demonstrated that functional CX3CR1and its membrane CX3CL1 ligand are expressed by several neuroblastoma cell lines. It was then demonstrated that CX3CR1-expressing neuroblastoma cells were stimulated by CX3CL1 to transmigrate across human bone-marrow endothelial cells. These results led us to hypothesize that the CX3CR1-CX3CL1 axis participates in bone-marrow metastasis of neuroblastoma.

With a few exceptions, the information about the expression and function of chemokine receptors on melanoma cells and their role in melanoma metastasis is rather fragmented (Richmond et al. 2009; Somasundaram and Herlyn 2009). CCR7 was implicated in lymph node metastasis, CCR9 was shown to be involved in metastasis to the small intestine and CCR10 in metastasis to the skin (Kakinuma and Hwang 2006).

The frequency of brain metastasis in melanoma is increasing and such metastases represent a significant cause of death in melanoma patients. Of all human solid tumors, melanoma has one of the highest risks to develop brain metastasis. More than 40 % of advance stage melanoma patients are treated for complications due to brain metastasis (Denkins et al. 2004).

The mechanisms underlying the targeted migration of melanoma cells to the brain are yet to be discovered. Hypothesizing that melanoma cells employ chemokine receptor-ligand axes to migrate to the brain, we established a chemokine receptor profile of cultured melanoma cells (3 cell lines of cutaneous melanoma and 5 cell lines of melanoma brain metastasis) (Izraely et al. 2010). This profile indicated that cultured melanoma cells express CCR3, CCR4, CXCR3, CXCR7, and CX3CR1. Utilizing cells from newly created variants of human melanoma xenografts, we found that the expression of CCR4 was significantly higher in a brain metastatic variant compared to its expression in the corresponding local variant. AKT phosphorylation patterns in melanoma cells were influenced by exposure of such cells to the CCR4 ligand, CCL22, which is expressed in brain. We hypothesize that CCR4 may be involved in melanoma brain metastasis and that this chemokine receptor may be a novel molecular biomarker for the identification of melanoma cells likely to metastasize to the brain (Izraely et al. 2010).

Concluding this section it is important to note that given the multiple steps in the metastatic cascade, the mechanism for the involvement of chemokine–chemokine receptor axes in site-specific metastasis is undoubtedly considerably more complex than receptor-ligand interactions.

2.3 Molecular Determinants of Metastasis

Discovering molecules that could serve as novel biomarkers and therapy targets for metastatic diseases is an important goal. For example, prevention strategies for metastasis could be developed if cells expressing metastatic biomarkers would be identified in the primary tumor. Currently, the availability of *bona fide* metastatic biomarkers is rather limited. Many more molecules associated with tumor progression should be identified and characterized.

Metastasis, a multistep process that requires the coordinated action of many genes, is the primary cause of mortality of cancer patients and in spite of the recent augmented interest in and understanding of this process (Valastyan and Weinberg 2011; Zlotnik et al. 2011; Fidler 2011; Langley and Fidler 2011; Coghlin and Murray 2010; Chaffer and Weinberg 2011; Shibue and Weinberg 2011; Gupta et al. 2005), it is still incompletely understood. The identification of genes that promote or suppress tumor metastasis is an essential requisite for the understanding of this process. The development of microarray technologies had a huge impact on many disciplines of biomedicine including cancer research. Cancer researchers used these technologies to determine the metastatic potential of tumors (Budhu et al. 2005; Adler and Chang 2006; Glinsky 2006; Fingleton 2007; Sarasin and Kauffmann 2008; Sabbah et al. 2008; Woo et al. 2011). However, several investigators expressed concern about results of microarray assays. For example, attempts to link expression profiles and molecular markers to liver metastases in colorectal cancer were not successfully validated as a diagnostic or prognostic tool applicable to routine clinical practice (Nadal et al. 2007). These authors advocated improving reproducibility, increasing consistency, and validating results. In another study, concern was expressed related to the lack of progress in defining markers or gene signatures in metastasis of malignant melanoma (Timar et al. 2010). These authors suggested "that only efficient inter-disciplinary collaboration throughout genomic analysis of human skin melanoma could lead to major advances in defining relevant gene-sets appropriate for clinical prognostication or revealing basic molecular pathways of melanoma progression".

Some of the above studies can be also criticized for not addressing the issue of organ specificity. After all, metastasis is an organ-specific event. The identification of genes associated with site-specific metastasis was addressed by the group of Massague, who identified groups of genes linked to breast cancer metastasis to various organs (Kang et al. 2003; Minn et al. 2005a, b; Bos et al. 2009). Whereas some of these genes were specifically linked to metastasis in specific organ sites, others were also associated with metastasis to other sites.

Much like other malignancies, neuroblastoma and melanoma metastasis are complex, multistep processes. We elected to study the various gene products involved in metastasis of these tumors by employing xenograft models, which recapitulate the phenotypes seen in the clinic. In order to eliminate "background noise" due to genetic differences between metastatic and non-metastatic cells or between metastases of one organ to those of another organ, our xenograft models consisted of human metastatic and non-metastatic cell variants of the same genetic background. Such models exist for several types of cancer but none for neuroblastoma or melanoma metastasis. We generated such variants for neuroblastoma lung metastasis (Nevo and Sagi-Assif 2008) and for melanoma brain metastasis (Izraely et al. 2012).

Chronologically the neuroblastoma metastasis model was developed first. An orthotopic implantation of human neuroblastoma cell lines into the adrenal gland of athymic nude mice yielded local adrenal tumors, as well as lung metastases. After repeated cycles of in vivo passages, local adrenal and lung metastatic variants were generated. The human origin and the metastatic phenotype of these variants were confirmed (Nevo and Sagi-Assif 2008). The melanoma metastasis model was developed pretty much along the same scheme (Izraely et al. 2012) except that the inoculation of the melanoma cells to nude mice was via the intra cardiac route, which is used by other investigators studying brain metastasis (Weil et al. 2005; Palmieri et al. 2006).

The various human tumor and metastasis variants generated in our lab, comprising tumor cells propagating in the local, orthotopic site (adrenal gland for neuroblastoma and skin for melanoma) and 2 corresponding metastatic sites (lung for neuroblastoma and brain for melanoma), share a common genetic background. Genetic, proteomic and transcriptomic differences between the variants may thus be ascribed to their differential malignant phenotype and to the different (local versus metastatic) microenvironments they reside in. These reproducible models can also serve as an unlimited source of biological material to be used in various types of investigations facilitating, for example, the identification of novel metastasis biomarkers and targets for therapy.

Analyzing gene expression of cultured cells will obviously reveal only genes that preserved their expression during and after the transition from the in vivo metastatic microenvironment to culture conditions. The possibility cannot be excluded that the expression of certain genes requires constant signaling from the particular in vivo microenvironment and that the expression of these genes will fade away following explantation. However, several studies indicated that the downstream effects of exogenous signals could endure for extended periods of time or even be permanent (Hardy et al. 2010; Matsumiya and Stafforini 2010; Khoo et al. 2011).

Neuroblastoma lung metastasis. Neuroblastoma is the most commonly occurring extracranial tumor in children. It is initiated most frequently in the adrenal gland and accounts for approximately 8 % of all malignancies in patients younger than 15 years (Brodeur and Castleberry 1997). More than half of these patients have a metastatic disease at diagnosis. Children older than 1 year with a widespread metastatic disease or with a large, aggressive, localized tumor, have an extremely poor prognosis (Modak and Cheung 2010; Mullassery et al. 2009). The lung-metastasizing human neuroblastoma variants described above exhibited an aggressive and metastatic phenotype in vivo and a malignant phenotype in vitro (Nevo and Sagi-Assif 2008).

A robust gene-expression based classifier, which reliably predicts neuroblastoma tumor behavior and can aid physicians in choosing the most appropriate form of first-line treatment, was developed several years ago (Oberthuer et al. 2006).

This neuroblastoma-specific oligonucleotide-array utilizing several platforms of gene-expression data comprises of 10,163 (11 K) probes for the 8,155 Unigene Cluster considered to be important in the development and progression of neuroblastoma. Aiming to identify molecular correlates of neuroblastoma metastasis and to determine the clinical relevance of these molecules, the genetically identical local and lung metastasizing human neuroblastoma variants described above were subjected to genome-wide expression profiling using the neuroblastoma-specific array (Nevo et al. 2010).

Our filtering and statistical comparison criteria revealed 112 genes that were differentially expressed in local and lung metastatic variant. These differentially expressed genes were intersected with genes differentially expressed in stage 1 and stage 4 primary tumors of neuroblastoma patients. By using the same gene-expression platform, molecular correlates associated with metastatic progression in primary neuroblastoma tumors were identified. The resulting smaller gene set was clinically relevant as it discriminated between high- and low-risk neuroblsatoma patients, suggesting that these genes could be used as therapy targets or prognostic markers in neuroblsatoma (Nevo et al. 2010).

Melanoma brain metastasis. Patients with malignant melanoma have a very high risk to develop brain metastasis. Greater than 40 % of advance stage melanoma patients have such metastasis (Denkins et al. 2004; Soffietti et al. 2002). Treatment options for melanoma patients with brain metastasis are limited (Bafaloukos and Gogas 2004). Tumor cells with the potential to metastasize to and colonize the brain may express distinctive metastasis-promoting molecular determinants. The results of gene expression profiling experiments performed in our lab (Izraely et al. 2012) demonstrated that about 40 genes were differentially expressed in brain-metastasizing human melanoma variants and in the corresponding local, sub-dermal variants. The functional significance of the genes differentially expressed in the brain-metastasizing and the sub-dermal melanoma cells, to brain metastasis is investigated at present. For example, Claudin-1, a tight junction protein, whose expression was significantly higher in the sub-dermal melanoma cells compared to the brain metastasizing cells, turned out to be a melanoma metastasis-suppressor gene.

A recent report identified a group of about 20 genes linked to breast cancer brain metastasis (Bos et al. 2009). We found that several of these genes were more highly expressed in brain metastasizing melanoma cells than in the corresponding cutaneous variants. The existence of a molecular signature of brain metastasis common to several types of cancer may thus be postulated.

2.4 Micrometastasis and Dormancy

Circulating tumor cells (CTC) were described for the first time in the middle of last century (Romsdahl et al. 1960). CTC are capable of disseminating primarily to regional lymph nodes and bone-marrow and persist in these organs in a state of dormancy for long periods. It is postulated that "awakening" of the dormant disseminated tumor cells (DTC) or micrometastases would lead to full blown, overt metastasis (Balic et al. 2010; Riethdorf et al. 2008; Alix-Panabieres et al. 2008). The research on micrometastasis became intense in the eighties and nineties of last century, when epithelial cells were detected in the bone-marrow of patients with epithelial cancers such as colorectal, breast and lung cancer (Dearnaley et al. 1981; Schlimok et al. 1986; Schlimok et al. 1991; Schlimok and Riethmuller 1990; Schlimok et al. 1990; Riethmuller and Johnson 1992; Lindemann et al. 1992; Pantel et al. 1993a, b; Cote et al. 1991). Nowadays micrometastasis has become an integral phase of the metastatic cascade (Riethdorf et al. 2008; Alix-Panabieres et al. 2008; Goss and Chambers 2010; Hedley and Chambers 2009). Micrometastatic cells remain as solitary cells or as small, steady state cell clusters, either due to a balance between proliferation and apoptosis or due to cell cycle arrest (Chaffer and Weinberg 2011; Chambers et al. 2002). For example, micrometastatic cells of breast cancer are in a state of dynamic dormancy, i.e., cell division and cell death are balanced (Meng et al. 2004). In view of the strong possibility that such cells are precursors for metastasis, it has been proposed that these cells could serve as targets for therapy (Goss and Chambers 2010). It is therefore logical to search for specific molecular targets on micrometastatic cells (Hedley and Chambers 2009; Ringel 2011; Vera-Ramirez et al. 2010).

The existence of micrometastatic cells could also be used to evaluate cancer outcome. For example, in a recent study the authors identified a dormancy-associated gene signature in breast cancer determining that tumors that exhibited a high dormancy score showed a significant correlation with low metastasis, since these tumors were more likely to undergo prolonged dormancy before resuming metastatic growth (Kim et al. 2012).

As mentioned above, regional lymph nodes and bone-marrow are major target sites for DTC (Balic et al. 2010). If micrometastasis indeed progress towards frank metastasis in a given organ site, it is logical to assume that micrometastases are present in this particular organ site. However, with some exceptions e.g. Yokoyama et al. (2012) the experimental evidence to support this assumption is rather limited. A possible reason for that is that detection of micrometastsis represents a great technical challenge (Riethdorf et al. 2008). Employing the xenograft models of human neuroblastoma lung and melanoma brain metastasis described above (Nevo and Sagi-Assif 2008; Izraely et al. 2012) which consisted of local and metastatic variants with an identical genetic background, we detected the presence of dormant micrometastases that formed spontaneously in lungs and brain following an orthotopic inoculation of neuroblastoma (Edry Botzer et al. 2011) and melanoma (Izraely et al. 2012) cells respectively. These systems allowed for a comparison of characteristics between metastatic cells in a specific organ and micrometastatic cells appearing in the same organ. Both metastatic and micrometastatic cells of the two tumor systems generated local tumors when implanted in the orthotopic sites, demonstrating that the intrinsic autonomous proliferative capacity of these cells remained intact except in the corresponding metastatic microenvironment.

A comparative in vitro characterization of metastatic and micrometastatic neuroblsatoma cells revealed similarities and differences. Micrometastatic, but not metastatic, neuroblsatoma cells expressed the minimal residual disease markers PHOX2B and tyrosine hydroxylase. The metastatic neuroblsatoma cells demonstrated a higher migratory capacity, an elevated MMP secretion, and a higher constitutive ERK phosphorylation than micrometastatic cells (Edry Botzer et al. 2011). A preliminary comparative in vitro characterization of metastatic and micrometastatic melanoma cells demonstrated that the gene expression pattern of both cells was in general similar (Izraely et al. 2012). However, in view of the biological differences between these 2 types of brain-localizing melanoma cells, a thorough comparative analysis between these cells is warranted.

Concluding this part of the review, it is our opinion that studying metastases and micrometastases developing in the same organ site may lead to a better understanding of the role of the metastatic microenvironment in tumor dormancy, to solving possible mechanisms underlying the transition of micro- to macrometastases and to finding ways to induce or prolong tumor dormancy.

2.5 Cross-Talk Between Tumor Cells and the Metastatic Microenvironment

Different subsets of cells in the primary tumor are genetically pre-destined to metastasize to specific organs (Dai et al. 2006; Kang et al. 2003; Ring and Ross 2005). Genes coding for: growth and angiogenesis factors; adhesion molecules and receptors for such molecules and for capacities to migrate to and invade specific organ target sites are responsible for site-specific metastasis. In addition to the genetic makeup of the cancer cells, the microenvironment of the organ to which cancer cells metastasize (referred hereafter as the metastatic microenvironment) plays a crucial role in the establishment, maintenance and further progression of metastasis (Pratap et al. 2011; Croci 2007; Kaplan et al. 2006; Harlozinska 2005; Cairns et al. 2003; Radinsky 1995; Radinsky and Fidler 1992). However, and although recent studies shed some light on the contribution of the metastatic microenvironment to site-specific metastasis (Lorusso and Ruegg 2012; Spano et al. 2012; Koh and Kang 2012; Lukanidin and Sleeman 2012; Sleeman et al. 2012; Spano and Zollo 2012; Taylor et al. 2011; Mathot and Stenninger 2012; Friedl and Alexander 2011; Cirri and Chiarugi 2012), there is still quite a lot to discover.

In attempts to comprehend the role of the metastatic microenvironment on site-specific metastasis, one should consider the following possible scenarios. Since the microenvironments of different organs differ in their cellular and molecular composition, it is to be expected that different interactions will take place between tumor cells metastasizing to organ A with the corresponding microenvironment, and the tumor-microenvironment interactions of cells from the same tumor that metastasize to organ B. The consequence of these differences would be the generation of a different phenotype of metastases from the same tumor in different organ sites. Using this scenario, one could imagine that cells from different tumor types metastasizing to the same organ site would share certain genetic and/or phenotypic traits. Contributing to the complexity of the interactions taking place in the metastatic microenvironment is the fact, that the microenvironment is an ever changing milieu; At times, it will enhance the malignancy of metastasizing cancer cells, and at other times, it will inhibit tumor progression (Witz 2008; Klein et al. 2007; Lin et al. 2009).

Above we summarized the role of chemokine–chemokine receptor axes in the targeted migration of tumor cells to selective organ sites. The mechanisms that sustain tumor cells in these sites and those promoting their progression and further dissemination to additional sites are still incompletely deciphered.

Focusing on the earliest steps in site-specific metastasis, it was demonstrated (Peinado et al. 2011) that cancer cells from the primary tumor communicate with bone-marrow-derived hematopoietic progenitor cells. This cross-talk is mediated by cytokines and chemokines secreted from the tumor cells and/or by tumor-derived exosomes. These tumor-derived factors are involved in the recruitment of bone-marrow-derived hematopoietic progenitor cells to future metastatic micro-environments by up-regulating the expression of fibronectin, matrix metalloproteinases or S100A8 and S100A9 proteins in these microenvironments. These and other molecules are those which directly mediate this recruitment. The recruited hematopoietic cells generate a pro-angiogenic, pre-metastatic niche, which supports the sustainability of cancer cells in this niche (Peinado et al. 2011). It should be noted that bone-marrow-derived cells could be detected in the pre-metastatic niche prior to the arrival of tumor cells at that niche (Psaila et al. 2006).

How do factors derived from the primary tumor select a particular organ to serve as a future organ-specific metastatic site? This is still an open question. A possible answer to this question is that cancer cells released into the circulation from non-metastatic primary tumors disseminate to future metastatic sites but are unable to progress further towards metastasis. Such disseminated cells, which may be dormant and therefore hard to detect, could be those that release the factors which will subsequently recruit the bone-marrow-derived cells, forming the pre-metastatic niche hospitable for colonization by subsequent waves of released tumor cells and for the propagation of already present tumor cells (Bidard et al. 2008).

Results obtained in our lab may support this possibility. Working with the xenograft models of human neuroblastoma lung and melanoma brain metatastasis, we asked if non-metastatic neuroblastoma or melanoma cells inoculated orthotopically (neuroblastoma into the adrenal gland and melanoma subcutaneously) to nude mice would disseminate to the corresponding metastatic sites. No overt metastasis was formed and standard detection methods failed to detect disseminated tumor cells in these organ sites. However, if the lungs of neuroblastoma-inoculated mice or the brains of melanoma-inoculated mice were cultured in vitro for a few weeks, human neuroblastoma and melanoma cells could be observed in the corresponding organ culture. Real-time PCR using human-specific probes confirmed the organ culture results (Izraely et al. 2012; Edry Botzer et al. 2011). These results indicate that tumor cells may disseminate to future metastatic sites and persist in these sites as undetectable "sleepers" without progressing to frank metastasis. Whether these micrometastatic "sleepers" are able to attract hematopoietic progenitor cells to the corresponding organ sites is unanswered as yet.

Interactions between cancer cells metastasizing to a specific organ site and the microenvironment of that site is subject to active research efforts. The working hypothesis of all these studies is that interactions between the microenvironment and tumor cells determine metastasis formation at this organ site (Langley and Fidler 2011; Sleeman et al. 2012; Rowley 2012; Cuiffo and Karnoub 2012; Krishnan et al. 2012; Nishimori et al. 2012; Reddy et al. 2012; Korkaya et al. 2011; Achyut and Yang 2011; St Hill 2011). Indeed these studies and others demonstrate that the microenvironment of the metastatic organ functions at several levels to facilitate metastatic growth of tumor cells that disseminated to that site. These functions include the creation of a pre-metastatic niche (Psaila et al. 2006), the delivery of site-specific chemo-attractants for tumor cells (Zlotnik et al. 2011) and the formation of a favorable milieu to sustain metastatic cells and promote their propagation by providing survival and proliferation signals (Langley and Fidler 2011; Rowley 2012; Cuiffo and Karnoub 2012; Krishnan et al. 2012; Nishimori et al. 2012; Reddy et al. 2012; Korkaya et al. 2011; Achyut and Yang 2011; St Hill 2011).

Based on the assumption that the microenvironment of future metastatic sites exerts far reaching influences on the ability of tumor cells to metastasize to that site, employing the human melanoma brain xenograft model developed in our lab (Izraely et al. 2012), we assessed the influence of brain-derived soluble factors on several malignancy traits of melanoma cells (Klein et al. 2012). It was found that brain-derived soluble factors enhanced the migration of melanoma cells metastasizing to the brain, but did not affect the migration of melanoma cells growing locally under the skin. This differential influence on brain-metastasizing cells could enhance the generation of new metastases from existing ones (metastasis-derived metastasis) (Langley and Fidler 2007).

Brain-derived factors also up-regulated the expression of the chemokine receptor CCR4 on melanoma cells. This finding is interesting in view of our previous findings that the CCR4-CCR4 ligand axis may be involved in the targeted migration of melanoma cells to the brain (Izraely et al. 2010). It is not unlikely that CCR4 ligands secreted from the brain interact with the CCR4-expressing melanoma cells, thereby directing them to the brain. Brain-derived soluble factors also enhanced the transmigration of melanoma cells growing locally under the skin across human brain endothelium. This activity could promote the capacity of these cells to metastasize to the brain.

An interesting finding was that brain-derived soluble factors, while enhancing the viability of melanoma cells growing locally, caused an S phase arrest followed by apoptosis of brain-metastasizing melanoma cells (Klein 2012). This represents another example of the fact that the TME may exert yin-yang activities i.e. opposing functions on interacting tumor cells (Witz 2008).

Asking what keeps micrometastatic neuroblastoma cells residing in the lungs from progressing to overt metastasis, we hypothesize that the lung microenvironment

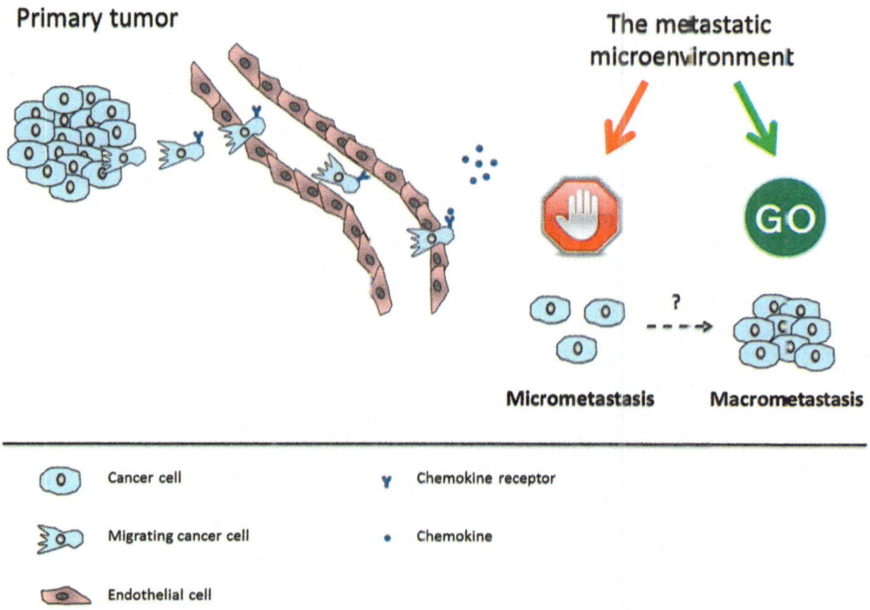

Fig. 2.2 A suggested model for the influence of the metastatic microenvironment on cancer progression and dormancy. Cancer cells expressing chemokine receptors detach from the primary tumor and metastasize to distant organs expressing the appropriate chemokine ligands. Once lodged in a secondary organ, the interactions with the metastatic microenvironments will determine whether these metastasizing cancer cells remain dormant or progress towards frank metastasis. It would be interesting to find out if the breakdown of this microenvironmental control mechanism is responsible for the awakening of dormant micrometastasis

contains factors that restrain the propagation of such cells. It was indeed found that a lung-derived soluble factor (or factors) caused a G0-G1 arrest followed by a decrease in cell viability of neuroblastoma lung metastases. This cytotoxic effect was significantly greater on micrometastatic lung-residing neuroblastoma cells. The fact that the lung contains a factor that restrains the proliferation of neurolastoma cells, may explain the fact that lung metastasis in neuroblatoma is a late event in the progression of this disease (Cowie et al. 1997; Kammen et al. 2001). The fact that normal organs express metastasis-restraining factors may constitute a hitherto un-described manifestation of intercellular surveillance, or microenvironmental control (Klein et al. 2007; Flaberg et al. 2011; Flaberg et al. 2012; Allen 2011; Bissell and Hines 2011). We suggest that such a mechanism (Fig. 2.2) would explain micrometastasis dormancy.

2.6 Conclusion

Metastatic tumor cells are endowed with characteristics conferring upon them the general capacity to migrate and disseminate to distant organs. Different types of cancer have multiple favorite metastatic organ sites. For example breast cancer metastasizes to bone, lungs, liver and brain, while colorectal cancer metastasizes mainly to liver and also to lungs. Each tumor type may therefore encounter several different metastatic microenvironments. It is therefore to be expected that the cross-talk between tumor cells infiltrating a certain secondary organ site be different from the cross talk between cells originating in the same tumor but infiltrating a different secondary organ. Since the microenvironment regulates and shapes the phenotype of tumor cells (see above), the result of this difference may be the emergence of multiple metastatic variants each expressing a different phenotype. The impact of this variability on cancer therapy is still largely unknown and should be pursued further.

Acknowledgments The studies performed in Dr. Witz's laboratory are generously supported by the following foundations and individuals: The Dr. Miriam and Sheldon G. Adelson Medical Research Foundation (Needham, MA, USA); Bonnie and Steven Stern (New York, NY, USA); The Deutsche Forschungsgemeinschaft (DFG); The Fred August and Adele Wolpers Charitable Fund (Clifton, NJ, USA); the late Natan Blutinger (West Orange, NJ, USA); The Pikovsky Fund (Jerusalem, Israel); and James J. Leibman and Rita S. Leibman Endowment Fund for Cancer Research (New York, NY, USA).

References

Achyut BR, Yang L (2011) Transforming growth factor-beta in the gastrointestinal and hepatic tumor microenvironment. Gastroenterology 141:1167–1178

Adler AS, Chang HY (2006) From description to causality: mechanisms of gene expression signatures in cancer. Cell Cycle 5:1148–1151

Agostini C, Calabrese F, Rea F, Facco M, Tosoni A, Loy M et al (2001) Cxcr3 and its ligand CXCL10 are expressed by inflammatory cells infiltrating lung allografts and mediate chemotaxis of T cells at sites of rejection. Am J Pathol 158:1703–1711

Albini A, Mirisola V, Pfeffer U (2008) Metastasis signatures: genes regulating tumor-microenvironment interactions predict metastatic behavior. Cancer Metastasis Rev 27:75–83

Alix-Panabieres C, Riethdorf S, Pantel K (2008) Circulating tumor cells and bone marrow micrometastasis. Clin Cancer Res 14:5013–5021

Allen M, Louise Jones J (2011) Jekyll and Hyde: the role of the microenvironment on the progression of cancer. J Pathol 223:162–176

Bafaloukos D, Gogas H (2004) The treatment of brain metastases in melanoma patients. Cancer Treat Rev 30:515–520

Balic M, Williams A, Dandachi N, Cote RJ (2010) Micrometastasis: detection methods and clinical importance. Cancer Biomark 9:397–419

Bar-Eli M (1999) Role of interleukin-8 in tumor growth and metastasis of human melanoma. Pathobiology 67:12–18

Ben-Baruch A (2003a) Host microenvironment in breast cancer development: inflammatory cells, cytokines and chemokines in breast cancer progression: reciprocal tumor-microenvironment interactions. Breast Cancer Res: BCR 5:31–36

Ben-Baruch A (2003b) Host microenvironment in breast cancer development: inflammatory cells, cytokines and chemokines in breast cancer progression: reciprocal tumor-microenvironment interactions. Breast Cancer Res 5:31–36

Ben-Baruch A (2008) Organ selectivity in metastasis: regulation by chemokines and their receptors. Clin Exp Metastasis 25:345–356

Bidard FC, Pierga JY, Vincent-Salomon A, Poupon MF (2008) A "class action" against the microenvironment: do cancer cells cooperate in metastasis? Cancer Metastasis Rev 27:5–10

Bissell MJ, Hines WC (2011) Why don't we get more cancer? A proposed role of the microenvironment in restraining cancer progression. Nat Med 17:320–329

Bos PD, Zhang XH, Nadal C, Shu W, Gomis RR, Nguyen DX et al (2009) Genes that mediate breast cancer metastasis to the brain. Nature 459:1005–1009

Braun S, Naume B (2005) Circulating and disseminated tumor cells. J Clin Oncol 23:1623–1626

Brodeur GM, Castleberry RP (1997) Neuroblastoma. In: Pizzo PA, Poplack DG (eds) Principles and practice of pediatric oncology, 3rd edn. Lippincott-Raven Publishers, Philadelphia, pp 761–783

Budhu AS, Zipser B, Forgues M, Ye QH, Sun Z, Wang XW (2005) The molecular signature of metastases of human hepatocellular carcinoma. Oncology 69(Suppl 1):23–27

Burger JA, Kipps TJ (2006) CXCR4: a key receptor in the crosstalk between tumor cells and their microenvironment. Blood 107:1761–1767

Cairns RA, Khokha R, Hill RP (2003) Molecular mechanisms of tumor invasion and metastasis: an integrated view. Curr Mol Med 3:659–671

Chaffer CL, Weinberg RA (2011) A perspective on cancer cell metastasis. Science 331:1559–1564

Chakraborty K, Bose A, Pal S, Sarkar K, Goswami S, Ghosh D et al (2008) Neem leaf glycoprotein restores the impaired chemotactic activity of peripheral blood mononuclear cells from head and neck squamous cell carcinoma patients by maintaining CXCR3/CXCL10 balance. Int Immunopharmacol 8:330–340

Chambers AF, Groom AC, MacDonald IC (2002) Dissemination and growth of cancer cells in metastatic sites. Nat Rev Cancer 2:563–572

Chaput N, Conforti R, Viaud S, Spatz A, Zitvogel L (2008) The Janus face of dendritic cells in cancer. Oncogene 27:5920–5931

Cirri P, Chiarugi P (2012) Cancer-associated-fibroblasts and tumour cells: a diabolic liaison driving cancer progression. Cancer Metastasis Rev 31:195–208

Coghlin C, Murray GI (2010) Current and emerging concepts in tumour metastasis. J Pathol 222:1–15

Cote RJ, Rosen PP, Lesser ML, Old LJ, Osborne MP (1991) Prediction of early relapse in patients with operable breast cancer by detection of occult bone marrow micrometastases. J Clin Oncol 9:1749–1756

Cowie F, Corbett R, Pinkerton CR (1997) Lung involvement in neuroblastoma: incidence and characteristics. Med Pediatr Oncol 28:429–432

Croci DO, Zacarias Fluck MF, Croci DO, Rico MJ, Matar P, Rabinovich GA, Scharovsky OG (2007) Dynamic cross-talk between tumor and immune cells in orchestrating the immunosuppressive network at the tumor microenvironment. Cancer Immunol Immunother 56:1687–1700

Cuiffo BG, Karnoub AE (2012) Mesenchymal stem cells in tumor development: emerging roles and concepts. Cell Adhes Migr 6:220–230

Cunha GR, Hayward SW, Wang YZ (2002) Role of stroma in carcinogenesis of the prostate. Differentiation; Res Biol Divers 70:473–485

Cunha GR, Hayward SW, Wang YZ, Ricke WA (2003) Role of the stromal microenvironment in carcinogenesis of the prostate. Int j Cancer J Int du Cancer 107:1–10

Dai CY, Haqq CM, Puzas JE (2006) Molecular correlates of site-specific metastasis. Semin Radiat Oncol 16:102–110

Dearnaley DP, Sloane JP, Ormerod MG, Steele K, Coombes RC, Clink HM et al (1981) Increased detection of mammary carcinoma cells in marrow smears using antisera to epithelial membrane antigen. Br J Cancer 44:85–90

Denkins Y, Reiland J, Roy M, Sinnappah-Kang ND, Galjour J, Murry BP et al (2004) Brain metastases in melanoma: roles of neurotrophins. Neuro Oncol 6:154–165

DiMeo TA, Kuperwasser C (2006) The evolving paradigm of tissue-specific metastasis. Breast Cancer Res 8:301

Edry Botzer L, Maman S, Sagi-Assif O, Meshel T, Nevo I, Bauerle T et al (2011) Lung-residing metastatic and dormant neuroblastoma cells. Am J Pathol 179:524–536
Egeblad M, Nakasone ES, Werb Z (2010) Tumors as organs: complex tissues that interface with the entire organism. Dev Cell 18:884–901
Eshel R, Neumark E, Sagi-Assif O, Witz IP (2002) Receptors involved in microenvironment-driven molecular evolution of cancer cells. Semin Cancer Biol 12:139–147
Fidler IJ (2001) Seed and soil revisited: contribution of the organ microenvironment to cancer metastasis. Surgical oncology clinics of North America 10:257–69, vii-viiii
Fidler IJ (2002) Critical determinants of metastasis. Semin Cancer Biol 12:89–96
Fidler IJ (2011) The role of the organ microenvironment in brain metastasis. Semin Cancer Biol 21:107–112
Fingleton B (2007) Molecular targets in metastasis: lessons from genomic approaches. Cancer Genomics Proteomics 4:211–221
Flaberg E, Markasz L, Petranyi G, Stuber G, Dicso F, Alchihabi N et al (2011) High-throughput live-cell imaging reveals differential inhibition of tumor cell proliferation by human fibroblasts. Int J Cancer J Int du Cancer 128:2793–2802
Flaberg E, Guven H, Savchenko A, Pavlova T, Kashuba V, Szekely L (2012) et al. The architecture of fibroblast monolayers of different origin differentially influences tumor cell growth. Int J Cancer J Int du Cancer 131:2274–2283
Friedl P, Alexander S (2011) Cancer invasion and the microenvironment: plasticity and reciprocity. Cell 147:992–1009
Fulton AM (2009) The chemokine receptors CXCR4 and CXCR3 in cancer. Curr Oncol Rep 11:125–131
Geminder H, Sagi-Assif O, Goldberg L, Meshel T, Rechavi G, Witz IP et al (2001) A possible role for CXCR4 and its ligand, the CXC chemokine stromal cell-derived factor-1, in the development of bone marrow metastases in neuroblastoma. J Immunol 167:4747–4757
Glinsky GV (2006) Genomic models of metastatic cancer: functional analysis of death-from-cancer signature genes reveals aneuploid, anoikis-resistant, metastasis-enabling phenotype with altered cell cycle control and activated Polycomb Group (PcG) protein chromatin silencing pathway. Cell Cycle 5:1208–1216
Goetz JG (2012) Tumor microenvironment indoctrination: An emerging hallmark of cancer. Cell Adhes Migr 6:190–192
Goss PE, Chambers AF (2010) Does tumour dormancy offer a therapeutic target? Nat Rev Cancer 10:871–877
Gupta GP, Massague J (2006) Cancer metastasis: building a framework. Cell 127:679–695
Gupta PB, Mani S, Yang J, Hartwell K, Weinberg RA (2005) The evolving portrait of cancer metastasis. Cold Spring Harb Symp Quant Biol 70:291–297
Hardy KM, Booth BW, Hendrix MJ, Salomon DS, Strizzi L (2010) ErbB/EGF signaling and EMT in mammary development and breast cancer. J Mammary Gland Biol Neoplasia 15:191–199
Harlozinska A (2005) Progress in molecular mechanisms of tumor metastasis and angiogenesis. Anticancer Res 25:3327–3333
Hart IR (1982) 'Seed and soil' revisited: mechanisms of site-specific metastasis. Cancer Metastasis Rev 1:5–16
Hart IR, Fidler IJ (1980) Role of organ selectivity in the determination of metastatic patterns of B16 melanoma. Cancer Res 40:2281–2287
Hart IR, Talmadge JE, Fidler IJ (1981) Metastatic behavior of a murine reticulum cell sarcoma exhibiting organ-specific growth. Cancer Res 41:1281–1287
Hedley BD, Chambers AF (2009) Tumor dormancy and metastasis. Adv Cancer Res 102:67–101
Hendrix MJ, Seftor EA, Kirschmann DA, Quaranta V, Seftor RE (2003) Remodeling of the microenvironment by aggressive melanoma tumor cells. Ann NY Acad Sci 995:151–161
Hu M, Polyak K (2008) Microenvironmental regulation of cancer development. Curr Opin Genet Dev 18:27–34
Hunter KW (2004) Host genetics and tumour metastasis. Br J Cancer 90:752–755

Izraely S, Klein A, Sagi-Assif O, Meshel T, Tsarfaty G, Hoon DS et al (2010) Chemokine-chemokine receptor axes in melanoma brain metastasis. Immunol Lett 130:107–114

Izraely S, Sagi-Assif O, Klein A, Meshel T, Tsarfaty G, Pasmanik-Chor M et al (2012) The metastatic microenvironment: brain-residing melanoma metastasis and dormant micrometastasis. Int J Cancer J Int du Cancer 131:1071–1082

Joyce JA, Pollard JW (2009) Microenvironmental regulation of metastasis. Nat Rev Cancer 9:239–252

Jung YD, Ahmad SA, Liu W, Reinmuth N, Parikh A, Stoeltzing O et al (2002) The role of the microenvironment and intercellular cross-talk in tumor angiogenesis. Semin Cancer Biol 12:105–112

Kakinuma T, Hwang ST (2006) Chemokines, chemokine receptors, and cancer metastasis. J Leukoc Biol 79:639–651

Kammen BF, Matthay KK, Pacharn P, Gerbing R, Brasch RC, Gooding CA (2001) Pulmonary metastases at diagnosis of neuroblastoma in pediatric patients: CT findings and prognosis. AJR Am J Roentgenol 176:755–759

Kang Y, Siegel PM, Shu W, Drobnjak M, Kakonen SM, Cordon-Cardo C et al (2003) A multigenic program mediating breast cancer metastasis to bone. Cancer Cell 3:537–549

Kaplan RN, Rafii S, Lyden D (2006) Preparing the "soil": the premetastatic niche. Cancer Res 66:11089–11093

Khoo JJ, Forster S, Mansell A (2011) Toll-like receptors as interferon-regulated genes and their role in disease. J Interferon Cytokine Res: The Official J Int Soc Interferon Cytokine Res 31:13–25

Kim RS, Avivar-Valderas A, Estrada Y, Bragado P, Sosa MS, Aguirre-Ghiso JA et al (2012) Dormancy signatures and metastasis in estrogen receptor positive and negative breast cancer. PLoS ONE 7:e35569

Klebanoff CA, Acquavella N, Yu Z, Restifo NP (2011) Therapeutic cancer vaccines: are we there yet? Immunol Rev 239:27–44

Klein G, Imreh S, Zabarovsky ER (2007) Why do we not all die of cancer at an early age? Adv Cancer Res 98:1–16

Klein A, Sagi-Assif O, Izraely S, Meshel T, Pasmanik-Chor M, Nahmias C, et al. (2012) The metastatic microenvironment: brain-derived soluble factors alter the malignant phenotype of cutaneous and brain-metastasizing melanoma cells. Int J Cancer 131:2509–2518

Koh BI, Kang Y (2012) The pro-metastatic role of bone marrow-derived cells: a focus on MSCs and regulatory T cells. EMBO Rep 13:412–422

Kopfstein L, Christofori G (2006) Metastasis: cell-autonomous mechanisms versus contributions by the tumor microenvironment. Cell Mol Life Sci: CMLS 63:449–468

Korkaya H, Liu S, Wicha MS (2011) Breast cancer stem cells, cytokine networks, and the tumor microenvironment. J Clin Investig 121:3804–3809

Krishnan V, Stadick N, Clark R, Bainer R, Veneris JT, Khan S, et al. (2012) Using MKK4's metastasis suppressor function to identify and dissect cancer cell-microenvironment interactions during metastatic colonization Cancer Metastasis Rev 31:605–613

Kumar S, Weaver VM (2009) Mechanics, malignancy, and metastasis: the force journey of a tumor cell. Cancer Metastasis Rev 28:113–127

Kuo TH, Kubota T, Watanabe M, Furukawa T, Teramoto T, Ishibiki K et al (1995) Liver colonization competence governs colon cancer metastasis. Proc Natl Acad Sci USA 92:12085–12089

Langley RR, Fidler IJ (2007) Tumor cell-organ microenvironment interactions in the pathogenesis of cancer metastasis. Endocr Rev 28:297–321

Langley RR, Fidler IJ (2011) The seed and soil hypothesis revisited–the role of tumor-stroma interactions in metastasis to different organs. Int J Cancer J Int du Cancer 128:2527–2535

Lawrence JA, Steeg PS (1996) Mechanisms of tumor invasion and metastasis. World J Urol 14:124–130

Lin HJ, Zuo T, Chao JR, Peng Z, Asamoto LK, Yamashita SS et al (2009) Seed in soil, with an epigenetic view. Biochim Biophys Acta 1790:920–924

Lindemann F, Schlimok G, Dirschedl P, Witte J, Riethmuller G (1992) Prognostic significance of micrometastatic tumour cells in bone marrow of colorectal cancer patients. Lancet 340:685–689
Liotta LA, Kohn EC (2001) The microenvironment of the tumour-host interface. Nature 411:375–379
Lorusso G, Ruegg C (2012) New insights into the mechanisms of organ-specific breast cancer metastasis. Semin Cancer Biol 22:226–233
Lukanidin E, Sleeman JP (2012) Building the niche: the role of the S100 proteins in metastatic growth. Semin Cancer Biol 22:216–225
Lynch CC, Matrisian LM (2002) Matrix metalloproteinases in tumor-host cell communication. Differ; Res Biol Divers 70:561–573
Mantovani A, Bottazzi B, Colotta F, Sozzani S, Ruco L (1992) The origin and function of tumor-associated macrophages. Immunol Today 13:265–270
Mantovani A, Allavena P, Sica A (2004) Tumour-associated macrophages as a prototypic type II polarised phagocyte population: role in tumour progression. Eur J Cancer 40:1660–1667
Marchesi F, Locatelli M, Solinas G, Erreni M, Allavena P, Mantovani A (2010) Role of CX3CR1/CX3CL1 axis in primary and secondary involvement of the nervous system by cancer. J Neuroimmunol 224:39–44
Maru SV, Holloway KA, Flynn G, Lancashire CL, Loughlin AJ, Male DK et al (2008) Chemokine production and chemokine receptor expression by human glioma cells: role of CXCL10 in tumour cell proliferation. J Neuroimmunol 199:35–45
Mathot L, Stenninger J (2012) Behavior of seeds and soil in the mechanism of metastasis: a deeper understanding. Cancer Sci 103:626–631
Matsumiya T, Stafforini DM (2010) Function and regulation of retinoic acid-inducible gene-I. Crit Rev Immunol 30:489–513
McCawley LJ, Matrisian LM (2001) Tumor progression: defining the soil round the tumor seed. Curr Biol: CB 11:R25–R27
Mehlen P, Puisieux A (2006) Metastasis: a question of life or death. Nat Rev Cancer 6:449–458
Meng S, Tripathy D, Frenkel EP, Shete S, Naftalis EZ, Huth JF et al (2004) Circulating tumor cells in patients with breast cancer dormancy. Clin Cancer Res 10:8152–8162
Meyer T, Hart IR (1998) Mechanisms of tumour metastasis. Eur J Cancer 34:214–221
Minn AJ, Gupta GP, Siegel PM, Bos PD, Shu W, Giri DD et al (2005a) Genes that mediate breast cancer metastasis to lung. Nature 436:518–524
Minn AJ, Kang Y, Serganova I, Gupta GP, Giri DD, Doubrovin M et al (2005b) Distinct organ-specific metastatic potential of individual breast cancer cells and primary tumors. J Clin Invest 115:44–55
Modak S, Cheung NK (2010) Neuroblastoma: therapeutic strategies for a clinical enigma. Cancer Treat Rev 36:307–317
Mueller MM, Fusenig NE (2002) Tumor-stroma interactions directing phenotype and progression of epithelial skin tumor cells. Differe; Res Biolo Divers 70:486–497
Mullassery D, Dominici C, Jesudason EC, McDowell HP, Losty PD (2009) Neuroblastoma: contemporary management. Archives Dis Childhood Educ Pract Ed 94:177–185
Muller A, Homey B, Soto H, Ge N, Catron D, Buchanan ME et al (2001) Involvement of chemokine receptors in breast cancer metastasis. Nature 410:50–56
Nadal C, Maurel J, Gascon P (2007) Is there a genetic signature for liver metastasis in colorectal cancer? World J Gastroenterol: WJG 13:5832–5844
Navarini-Meury AA, Conrad C (2009) Melanoma and innate immunity–aActive inflammation or just erroneous attraction? Melanoma as the source of leukocyte-attracting chemokines. Semin Cancer Biol 19:84–91
Neumark E, Sagi-Assif O, Shalmon B, Ben-Baruch A, Witz IP (2003) Progression of mouse mammary tumors: MCP-1-TNFalpha cross-regulatory pathway and clonal expression of promalignancy and antimalignancy factors. Int J Cancer 106:879–886

Nevo I, Sagi-Assif O, Meshel T, Geminder H, Goldberg-Bittman L, Ben-Menachem S et al (2004) The tumor microenvironment: CXCR4 is associated with distinct protein expression patterns in neuroblastoma cells. Immunol Lett 92:163–169
Nevo I, Sagi-Assif O, Edry Botzer L, Amar D, Maman S, Kariv N et al (2008) Generation and characterization of novel local and metastatic human neuroblastoma variants. Neoplasia 10:816–827
Nevo I, Sagi-Assif O, Meshel T, Ben-Baruch A, Johrer K, Greil R et al (2009) The involvement of the fractalkine receptor in the transmigration of neuroblastoma cells through bone-marrow endothelial cells. Cancer Lett 273:127–139
Nevo I, Oberthuer A, Botzer E, Sagi-Assif O, Maman S, Pasmanik-Chor M et al (2010) Gene-expression-based analysis of local and metastatic neuroblastoma variants reveals a set of genes associated with tumor progression in neuroblastoma patients. Int J Cancer 126:1570–1581
Nguyen TH (2004) Mechanisms of metastasis Clin Dermatol 22:209–216
Nguyen DX, Massague J (2007) Genetic determinants of cancer metastasis. Nat Rev Genet 8:341–352
Nicolson GL (1988) Organ specificity of tumor metastasis: role of preferential adhesion, invasion and growth of malignant cells at specific secondary sites. Cancer Metastasis Rev 7:143–188
Nishimori H, Ehata S, Suzuki HI, Katsuno Y, Miyazono K (2012) Prostate cancer cells and bone stromal cells mutually interact with each other through bone morphogenetic protein-mediated signals. J Biol Chem 287:20037–20046
Oberthuer A, Berthold F, Warnat P, Hero B, Kahlert Y, Spitz R et al (2006) Customized oligonucleotide microarray gene expression-based classification of neuroblastoma patients outperforms current clinical risk stratification. J Clin Oncol 24:5070–5078
Onuigbo WI (1975) Human model for studying seed-soil factors in blood-borne metastasis. Arch Pathol 99:342–343
Orimo A, Gupta PB, Sgroi DC, Arenzana-Seisdedos F, Delaunay T, Naeem R et al (2005) Stromal fibroblasts present in invasive human breast carcinomas promote tumor growth and angiogenesis through elevated SDF-1/CXCL12 secretion. Cell 121:335–348
Paget S (1889) The distribution of secondary growths in cancer of the breast. The Lancet 133:571–573
Palmieri D, Smith QR, Lockman PR, Bronder J, Gril B, Chambers AF et al (2006) Brain metastases of breast cancer. Breast Dis 26 139–147
Palmieri D, Chambers AF, Felding-Habermann B, Huang S, Steeg PS (2007) The biology of metastasis to a sanctuary site. Clin Cancer Res 13:1656–1662
Pantel K, Brakenhoff RH (2004) Dissecting the metastatic cascade. Nat Rev Cancer 4:448–456
Pantel K, Izbicki JR, Angstwurm M, Braun S, Passlick B, Karg O et al (1993a) Immunocytological detection of bone marrow micrometastasis in operable non-small cell lung cancer. Cancer Res 53:1027–1031
Pantel K, Braun S, Schlimok G, Riethmuller G (1993b) Micrometastatic tumour cells in bone marrow in colorectal cancer. Lancet 341:501
Park CC, Bissell MJ, Barcellos-Hoff MH (2000) The influence of the microenvironment on the malignant phenotype. Mol Med Today 6:324–329
Pass HI (2002) Biology of metastatic disease. Semin Thorac Cardiovasc Surg 14:10–17
Pauli BU, Lee CL (1988) Organ preference of metastasis. The role of organ-specifically modulated endothelial cells. Lab Invest; J Tech Methods Pathol 58:379–387
Pauli BU, Augustin-Voss HG, Augustin-Voss HG, el-Sabban ME, Johnson RC, Hammer DA (1990) Organ-preference of metastasis. The role of endothelial cell adhesion molecules. Cancer Metastasis Rev 9:175–189
Peinado H, Lavotshkin S, Lyden D (2011) The secreted factors responsible for pre-metastatic niche formation: old sayings and new thoughts. Semin Cancer Biol 21:139–146
Pratap J, Lian JB, Stein GS (2011) Metastatic bone disease: role of transcription factors and future targets. Bone 48:30–36

Psaila B, Kaplan RN, Port ER, Lyden D (2006) Priming the 'soil' for breast cancer metastasis: the pre-metastatic niche. Breast Dis 26:65–74
Radinsky R (1995) Modulation of tumor cell gene expression and phenotype by the organ-specific metastatic environment. Cancer Metastasis Rev 14:323–338
Radinsky R, Fidler IJ (1992) Regulation of tumor cell growth at organ-specific metastases. In Vivo 6:325–331
Reddy BY, Lim PK, Silverio K, Patel SA, Won BW, Rameshwar P (2012) The microenvironmental effect in the progression, metastasis, and dormancy of breast cancer: a model system within bone marrow. Int J Breast Cancer 2012:721659
Richmond A, Yang J, Su Y (2009) The good and the bad of chemokines/chemokine receptors in melanoma. Pigment Cell Melanoma Res 22:175–186
Riethdorf S, Wikman H, Pantel K (2008) Review: Biological relevance of disseminated tumor cells in cancer patients. Int J Cancer 123:1991–2006
Riethmuller G, Johnson JP (1992) Monoclonal antibodies in the detection and therapy of micrometastatic epithelial cancers. Curr Opin Immunol 4:647–655
Ring BZ, Ross DT (2005) Predicting the sites of metastases. Genome Biol 6:241
Ringel MD (2011) Metastatic dormancy and progression in thyroid cancer: targeting cells in the metastatic frontier. Thyroid 21:487–492
Romsdahl MM, Potter JF, Malmgren RA, Chu EW, Brindley CO, Smith RR (1960) A clinical study of circulating tumor cells in malignant melanoma. Surg Gynecol Obstet 111:675–681
Rowley D, Barron DA (2012) The reactive stroma microenvironment and prostate cancer progression. Endocrine-related cancer 19:R187–204
Rubin H (2001) Selected cell and selective microenvironment in neoplastic development. Cancer Res 61:799–807
Ruiz P, Gunthert U (1996) The cellular basis of metastasis. World J Urol 14:141–150
Russell HV, Hicks J, Okcu MF, Nuchtern JG (2004) CXCR4 expression in neuroblastoma primary tumors is associated with clinical presentation of bone and bone marrow metastases. J Pediatr Surg 39:1506–1511
Sabbah M, Emami S, Redeuilh G, Julien S, Prevost G, Zimber A et al (2008) Molecular signature and therapeutic perspective of the epithelial-to-mesenchymal transitions in epithelial cancers. Drug resistance updates : reviews and commentaries in antimicrobial and anticancer chemotherapy 11:123–151
Sapoznik S, Ortenberg R, Galore-Haskel G, Kozlovski S, Levy D, Avivi C, et al. (2012) CXCR1 as a novel target for directing reactive T cells toward melanoma: implications for adoptive cell transfer immunotherapy. Cancer Immunol Immunother 61:1833–1847
Sarasin A, Kauffmann A (2008) Overexpression of DNA repair genes is associated with metastasis: a new hypothesis. Mutat Res 659:49–55
Schlimok G, Riethmuller G (1990) Detection, characterization and tumorigenicity of disseminated tumor cells in human bone marrow. Semin Cancer Biol 1:207–215
Schlimok G, Gottlinger H, Funke I, Swierkot S, Hauser H, Riethmuller G (1986) In vivo and in vitro labelling of epithelial tumor cells with anti 17–1A monoclonal antibodies in bone marrow of cancer patients. Hybridoma 5(Suppl 1):S163–S170
Schlimok G, Funke I, Bock B, Schweiberer B, Witte J, Riethmuller G (1990) Epithelial tumor cells in bone marrow of patients with colorectal cancer: immunocytochemical detection, phenotypic characterization, and prognostic significance. J Clin Oncol 8:831–837
Schlimok G, Funke I, Pantel K, Strobel F, Lindemann F, Witte J et al (1991) Micrometastatic tumour cells in bone marrow of patients with gastric cancer: methodological aspects of detection and prognostic significance. Eur J Cancer 27:1461–1465
Schmid MC, Varner JA (2010) Myeloid cells in the tumor microenvironment: modulation of tumor angiogenesis and tumor inflammation. J Oncol 2010:201026
Shibue T, Weinberg RA (2011) Metastatic colonization: settlement, adaptation and propagation of tumor cells in a foreign tissue environment. Semin Cancer Biol 21:99–106
Sleeman JP, Christofori G, Fodde R, Collard JG, Berx G, Decraene C et al (2012) Concepts of metastasis in flux: the stromal progression model. Semin Cancer Biol 22:174–186

Soffietti R, Ruda R, Mutani R (2002) Management of brain metastases. J Neurol 249:1357–1369
Somasundaram R, Herlyn D (2009) Chemokines and the microenvironment in neuroectodermal tumor-host interaction. Semin Cancer Biol 19:92–96
Soria G, Yaal-Hahoshen N, Azenshtein E, Shina S, Leider-Trejo L, Ryvo L et al (2008) Concomitant expression of the chemokines RANTES and MCP-1 in human breast cancer: a basis for tumor-promoting interactions. Cytokine 44:191–200
Soria G, Lebel-Haziv Y, Ehrlich M, Meshel T, Suez A, Avezov E et al (2012) Mechanisms regulating the secretion of the promalignancy chemokine CCL5 by breast tumor cells: CCL5's 40 s loop and intracellular glycosaminoglycans. Neoplasia 14:1–19
Spano D, Zollo M (2012) Tumor microenvironment: a main actor in the metastasis process. Clin Exp Metastasis 29:381–395
Spano D, Heck C, De Antonellis P, Christofori G, Zollo M (2012) Molecular networks that regulate cancer metastasis. Semin Cancer Biol 22:234–249
St Hill CA (2011) Interactions between endothelial selectins and cancer cells regulate metastasis. Front Biosci: J Virtual Libr 16:3233–3251
Steeg PS (2005) New insights into the tumor metastatic process revealed by gene expression profiling. Am J Pathol 166:1291–1294
Takeuchi H, Kitago M, Hoon DS (2007) Effects of chemokines on tumor metastasis. Cancer Treat Res 135:177–184
Taylor MA, Lee YH, Schiemann WP (2011) Role of TGF-beta and the tumor microenvironment during mammary tumorigenesis. Gene Expr 15:117–132
Timar J, Gyorffy B, Raso E (2010) Gene signature of the metastatic potential of cutaneous melanoma: too much for too little? Clin Exp Metastasis 27:371–387
Tlsty TD (2001) Stromal cells can contribute oncogenic signals. Semin Cancer Biol 11:97–104
Togo S, Shimada H, Kubota T, Moossa AR, Hoffman RM (1995) "Seed" to "soil" is a return trip in metastasis. Anticancer Res 15:791–794
Umansky V, Sevko A (2012) Melanoma-induced immunosuppression and its neutralization. Semin Cancer Biol 22:319–326
Unger M, Weaver VM (2003) The tissue microenvironment as an epigenetic tumor modifier. Methods Mol Biol 223:315–347
Valastyan S, Weinberg RA (2011) Tumor metastasis: molecular insights and evolving paradigms. Cell 147:275–292
van Kempen LC, Rhee JS, Dehne K, Lee J, Edwards DR, Coussens LM (2002) Epithelial carcinogenesis: dynamic interplay between neoplastic cells and their microenvironment. Differ; Res Biol Divers 70:610–623
Vera-Ramirez L, Sanchez-Rovira P, Ramirez-Tortosa CL, Quiles JL, Ramirez-Tortosa MC, Alvarez JC et al (2010) Gene-expression profiles, tumor microenvironment, and cancer stem cells in breast cancer: latest advances towards an integrated approach. Cancer Treat Rev 36:477–484
Wang J, Loberg R, Taichman RS (2006) The pivotal role of CXCL12 (SDF-1)/CXCR4 axis in bone metastasis. Cancer Metastasis Rev 25:573–587
Weil RJ, Palmieri DC, Bronder JL, Stark AM, Steeg PS (2005) Breast cancer metastasis to the central nervous system. American J Pathol 167:913–920
Weinberg RA (1995) Prospects for cancer genetics. Cancer Surv 25:3–12
Weinberg RA (2008) Coevolution in the tumor microenvironment. Nat Genet 40:494–495
Weiss L, Voit A, Lane WW (1984) Metastatic patterns in patients with carcinomas of the lower esophagus and upper rectum. Invasion Metastasis 4:47–60
Weiss L, Harlos JP, Torhorst J, Gunthard B, Hartveit F, Svendsen E et al (1988) Metastatic patterns of renal carcinoma: an analysis of 687 necropsies. J Cancer Res Clin Oncol 114:605–612
Witz IP (2008a) Yin-Yang activities and vicious cycles in the tumor microenvironment. Cancer Res 68:9–13
Witz IP (2008b) Tumor-microenvironment interactions: dangerous liaisons. Adv Cancer Res 100:203–229

Witz IP, Levy-Nissenbaum O (2006) The tumor microenvironment in the post-PAGET era. Cancer Lett 242:1–10
Woo HG, Park ES, Thorgeirsson SS, Kim YJ (2011) Exploring genomic profiles of hepatocellular carcinoma. Mol Carcinog 50:235–243
Yokota J (2000) Tumor progression and metastasis. Carcinogenesis 21:497–503
Yokoyama N, Otani T, Hashidate H, Maeda C, Katada T, Sudo N et al (2012) Real-time detection of hepatic micrometastases from pancreatic cancer by intraoperative fluorescence imaging: preliminary results of a prospective study. Cancer 118:2813–2819
Zhang J, Lu Y, Pienta KJ (2010) Multiple roles of chemokine (C–C motif) ligand 2 in promoting prostate cancer growth. J Natl Cancer Inst 102:522–528
Zigrino P, Loffek S, Mauch C (2005) Tumor-stroma interactions: their role in the control of tumor cell invasion. Biochimie 87:321–328
Zipin-Roitman A, Meshel T, Sagi-Assif O, Shalmon B, Avivi C, Pfeffer RM et al (2007) CXCL10 promotes invasion-related properties in human colorectal carcinoma cells. Cancer Res 67:3396–3405
Zlotnik A, Burkhardt AM, Homey B (2011) Homeostatic chemokine receptors and organ-specific metastasis. Nat Rev Immunol 11:597–606

Chapter 3
Tumor Infiltration by Immune Cells: Pathologic Evaluation and a Clinical Significance

Dmitriy W. Gutkin

Abstract Tumor infiltrating leukocytes comprise a significant population of cells in the tumor microenvironment. Pathologic evaluation of tumors by methods of light microscopy and immunohistochemistry allow identification of all types of leukocytes, including the cells of myeloid, monocytoid and lymphoid origin. The density of these cells, their spatial distribution in tumor islets and stroma, their level of maturation and functional status vary considerably in the tumors of different nature, grade and stage. By the analysis of clinical data it has been shown that specific types of tumor infiltrating leukocytes have a major impact on the clinical course and the outcome of malignant diseases. This chapter describes an approach used by pathologists to gain insight into morphologic and functional properties of tumor infiltrating leukocytes, as well as their prognostic significance.

Keywords Tumor infiltrating leukocytes · Lymphocytes · Dendritic cells · Macrophages · Neutrophils · Mast cells · NK cells · Eosinophils · Tumor stroma · Immunohistochemistry

3.1 Introduction

While evaluating malignant neoplasms, pathologists primarily focus on architectural patterns and cytological characteristics of malignant cells which determine the type of tumor, its grade and stage. These are the main factors determining

D. W. Gutkin (✉)
Department of Pathology, University of Pittsburgh Medical Center, Pittsburgh VA Hospital, Pittsburgh, PA, USA
e-mail: dwgutkin@hotmail.com

M. R. Shurin et al. (eds.), *The Tumor Immunoenvironment*,
DOI: 10.1007/978-94-007-6217-6_3,

biologic behavior of the tumor, as well as cancer therapy and prognosis. That is not to say that other types of cells in the tumor mass are completely ignored. In fact, the first mentioning of the presence of inflammatory cells in malignant neoplasm was made by the "father of pathology", Rudolf Virchow, in 1863 (Balkwill and Mantovani 2001). However, a systematic evaluation of tumor-infiltrating leukocytes started only four decades ago (Adelstein et al. 1978).

With the emergence of the concept of immune surveillance, it became clear that immune competent cells play a major role in tumor biology. According to this concept, the immune system is able to detect and eliminate emerging malignant cells to prevent their uncontrolled proliferation. On the basis of this concept, an established malignant tumor can be viewed as a failure of the immune system to launch a successful anti-tumor attack, and at the same time, as a success of an emerging tumor to escape such an attack. In addition, accumulating evidence suggests that the local inflammatory process, previously believed to be the host response against cancer, might actually contribute to the development of malignancy (Balkwill and Mantovani 2001; Coussens and Werb 2001; Almholt and Johnsen 2003; de Visser et al. 2006; Murdoch et al. 2008).

On average only about half of the cells within the tumor mass are actually malignant cells. In addition to the cancerous cells themselves, tumors are comprised of stroma cells, fibroblasts, endothelial cells, pericytes, adipocytes, and numerous types of tumor infiltrating leukocytes. The latter commonly include myeloid cells, such as neutrophils, eosinophils, mast cells, macrophages, dendritic cells, as well as B lymphocytes, T lymphocytes and natural killer cells (Coussens and Werb 2002). An approach that is commonly used by pathologists to gain insight in the in vivo interaction between tumors and the immune system is to quantify the numbers of tumor infiltrating leukocytes and to correlate these numbers with tumor characteristics and prognostic outcome. These studies have been carried out across many types of cancer.

Evaluation of tumor infiltrating leukocytes is performed by microscopic examination of frozen or fixed tumor tissue. Some of the cells, particularly neutrophils, eosinophils, and mast cells, can be easily recognized by routine histochemical stains (e.g., H&E, Giemsa, toluidine blue, etc.). However, these histochemical stains do not allow recognizing tumor infiltrating leukocytes with certainty. In addition, they cannot discriminate between different subpopulations of cells or determine their state of maturation. Thus, evaluation of tumor infiltrating leukocytes frequently requires utilization of methods that can determine not only morphology of the cells, but also their molecular phenotype. In pathology practice it is usually done by immunohistochemistry that detects specific protein expression in the cells of interest.

Immunohistochemistry can be performed on fresh or fixed tissue. Tumor tissue is cut into slices 5–10 μm thick and specially treated to make cellular proteins more accesible. Specific, usually monoclonal, antibodies against proteins of interest are applied to the tissue and the resulting antigen–antibody complexes are visualized with fluorescent or color stains (Taylor et al. 2006). Stained cells are counted under high magnification (usually, 400x or 1000x), and the results are presented in a quantitative manner (e.g., number of cells per high power field) or in a semi-quantitative manner (e.g., absent, weak, moderate, brisk infiltration).

Upon microscopic analysis, the first thing that pathologists notice is the presence or absence of a specific type of tumor infiltrating leukocytes in a tumor tissue. As discussed below, there are enormous variations in the density of immune cells in different types of tumor. Quite often there is a predilection of a specific type of leukocytes to a specific type of cancer (e.g., eosinophils to squamous cell carcinoma). Since numerous chemokines regulating immune cell recruitment and retention in tissue are known, it is possible to correlate the density of a specific tumor infiltrating leukocyte type with chemokine expression profile in the tumor. The latter can be studied by immunohistochemical or molecular methods.

Second, microscopic analysis can determine a spatial distribution of immune cells in the tumor mass. Tumor infiltrating leukocytes can be located within cancer cell nests (intratumoral distribution), in the central cancer stroma (stromal distribution) and along the invasive tumor margins (peritumoral distribution). Since immune cells can have a dissimilar effect on malignant cells and stromal cells, the exact location of leukocytes is very important in the evaluation of their role. At the same time, the effect of leukocytes on tumor depends on their functional status and the level of maturation. These parameters can also be tested for some of the cell types by immunohistochemistry.

Next, density of tumor infiltrating leukocytes can be correlated with the stage of a tumor. The process of tumor development, especially for epithelial tumors (carcinomas), includes such steps as cellular dysplasia, carcinoma in situ (non-invasive), locally invasive neoplasm, and metastatic dissemination. Several studies, discussed below, describe the correlation between tumor infiltrating leukocyte density and the tumor stage, which helps to elucidate the involvement of immune cells in the tumor progression.

Another correlation that is frequently performed during pathologic examination is a correlation of leukocytes density with the tumor grade. Depending on the level of morphologic atypia, malignant tumors are classified as well-differentiated, moderately differentiated and poorly differentiated. Consistent correlations are found between tumor infiltrating leukocytes density and tumor grade for many neoplasms.

Finally, tumor infiltrating leukocyte density can be correlated with the disease progression, clinical course, outcome and response to treatment From a practical standpoint these are the most interesting findings that can be made by pathologists working in this field. Not surprisingly, the number of publications focused on the correlation between specific tumor infiltrating leukocytes and patients' prognosis is rapidly growing and currently exceeds 5,000.

In summary, when studying tumor infiltrating leukocytes pathologists aim to answer the following questions:

1. Are these cells present in the tumor tissue and in what numbers?
2. What are their functional status and the level of maturation?

3. What is the spatial distribution of leukocytes in correlation with tumor cell features (i.e.,. proliferation, apoptosis, necrosis) and stromal features (i.e., angiogenesis)?
4. Is there a correlation of density, state of maturation, or spatial distribution of tumor infiltrating leukocytes with tumor grade, stage and prognosis?

As it will be shown below, today there are no solid answers to these questions, and in many instances the results of various studies are controversial. Additionally, there are no uniformly acceptable methods of studying tumor infiltrating leukocytes, therefore careful attention to the techniques is warranted to compare results of different studies.

3.2 Neutrophils

Neutrophils, like all myeloid lineage cells, originate from hematopoietic stem cells in the bone marrow. They represent the main population of leukocytes in blood and are considered to be the first line of immune response to tissue injury. Neutrophils make up a significant portion of the inflammatory cell infiltrate found in a wide variety of human cancers, including pulmonary (Bellocq et al. 1998), gastric (Eck et al. 2003; Griffiths et al. 1998; Rice et al. 2000), renal carcinoma (Jensen et al. 2009), etc. (Wislez et al. 2003). Neutrophils can be easily recognized in tissue using routine histochemical staining (e.g., H&E, Fig. 3.1a). In addition, more advanced stains targeting neutrophil-specific proteins, such as CD15 or CD66b, can be utilized (Table 3.1, Fig. 3.1b) (Jensen et al. 2009).

It seems reasonable to believe that neutrophils, representing the first line of immune response, are recruited into the tumor mass in order to destroy emerging malignant cells. However, it appears that the recruitment of neutrophils can be a tumor-driven process, and doesn't necessarily represent means of host defense (Houghton 2010). Recruitment of neutrophils in tissue is regulated by chemokines, and many cell types within the tumor microenvironment, are capable of secreting

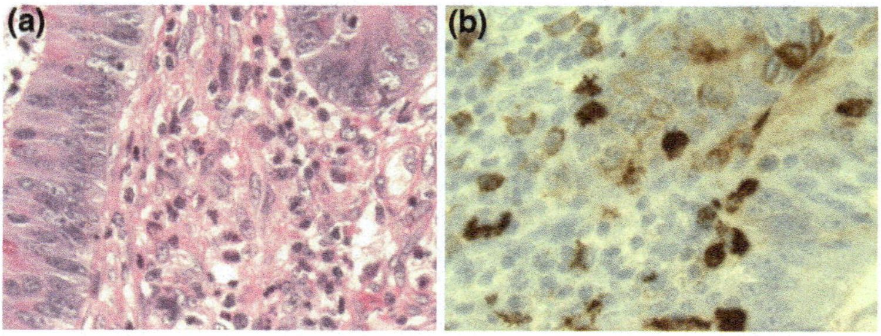

Fig. 3.1 Tumor-infiltrating neutrophils in colonic adenocarcinoma. **a** H&E staining, 600x; **b** CD15 immunostain, 600x. Neutrophils are present within the tumor nests and in the tumor stroma

Table 3.1 Commonly used immunohistochemical markers of tumor-associated myeloid cells

Target	Staining	Property
Neutrophils		
	Myeloperoxidase	Enzyme in the granules of neutrophils and to a lesser extent the granules of monocytes
	CD15	Cell surface membrane protein, expressed on neutrophils, a subset of tissue macrophages, and activated T lymphocytes
	CD66b	A member of the immunoglobulin superfamily, expressed on neutrophils
Mast cells	Tryptase	Enzyme in mast cell granules
	CD117	Tyrosine kinase receptor (c-Kit)
	Luna	Cytoplasmic granules of mast cells
	CD68	Glycoprotein of cytoplasmic granules
Macrophages	HLA-DR	Major histocompatibility complex, class II, cell surface receptor, marker of M1 activation
	CD163	Transmembrane protein, marker of M2 activation
	CD204	Macrophage scavenger receptor 1, marker of M2 activation

neutrophil chemotactic substances. It was shown that, at least in some cases, the tumor cells themselves mediate neutrophil recruitment to sites of tumorogenesis by secreting various chemokines (Gregory and Houghton 2011). Chemokines that are known to recruit neutrophils include CXCL1 and CXCL5 (Walz et al. 1997; Bechara et al. 2007). Tumor cells (i.e., pulmonary carcinoma) can also produce interleukin-8 (IL-8), a known chemoattractant for neutrophils (Sparmann and Bar-Sagi 2004; Ji et al. 2006). Another chemokine that has been implicated in this process is interleukin-10 (IL-10) (Di Carlo et al. 1998; Vora et al. 1996), which contributes to the adhesion of neutrophils to intratumoral microvessels and their intratumoral accumulation (Ma et al. 1993; Vinten-Johansen et al. 1999).

An interesting observation was recently made by Reid and colleagues (Reid et al. 2011), who found preferential chemotaxis of neutrophils to certain types of pancreatic neoplasms. They showed that among invasive pancreatic carcinomas, tumor-infiltrating neutrophils occur almost exclusively in micropapillary or undifferentiated carcinomas. All of these neutrophil-rich pancreatic tumors expressed MUC1, a powerful chemoattractant and marker of aggressive behavior. Density of tumor-infiltrating neutrophils may also depend on the tumor grade and stage. Specifically, an increase in neutrophil infiltration has been demonstrated in colorectal cancer progression from premalignant aberrant crypt foci to adenomas to carcinomas (Roncucci et al. 2008).

Thus, the presence of neutrophils in various malignant neoplasms is well recognized and can be attributed to the intrinsic tumor properties. But what is the role of these cells in tumor? Do they have the anti-tumor or pro-tumor effects? Today there is no simple answer to this question since interactions of neutrophils with different elements of the tumor microenvironment are very complex.

On one hand, neutrophils are cytolytic and can eliminate tumor cell populations (Di Carlo et al. 2001). Neutrophils produce several cytotoxic mediators, including reactive oxygen species (ROS), proteases, membrane-perforating agents and soluble mediators of cell killing, such as TNF-α, IL-1β and interferons. High quantities of neutrophil-derived ROS are tumoricidal. The most injurious of the ROS is hypochlorous acid (HOCl) which acts through cell lysis or induction of apoptosis (Dallegri et al. 1991; Simon et al. 2000). Morphologic examination of tumor tissue shows that the damage produced by neutrophils takes two forms: predominantly liquefactive necrosis when their cytotoxicity against tumor cells prevails (Musiani et al. 1996) and predominantly ischemic and/or hemorrhagic necrosis when their main target is the vascular endothelium (Di Carlo et al. 1998; Westlin and Gimbrone 1993; Bratt and Palmblad 1997).

On the other hand, neutrophils can have an opposite effect and directly induce tumor cell proliferation through the expression of growth-promoting bioactive molecules. Specifically, neutrophil-derived hepatocyte growth factor has been correlated with increased tumor growth in lung cancers (Wislez et al. 2003).

These opposite effects on tumor cells require direct interaction of neutrophils with tumor cells and therefore are produced by neutrophils infiltrating tumor islands. Even more important for tumor development are the effects of neutrophils infiltrating central tumor stroma and peritumoral invasive margin. These cells definitely promote tumor progression through remodeling of the extracellular matrix, enhancing tumor cell migration and invasion and modulating angiogenesis (Gregory and Houghton 2011; Shojaei et al. 2008; Tazzyman et al. 2009; Coffelt et al. 2010). Among these processes, tumor angiogenesis is considered critical not only for providing nutrients to developing tumors, but also for tumor cell dissemination via the hematogenous route. Neutrophils are well equipped for extracellular matrix degradation, being the main source of matrix metalloproteinases, especially MMP-9. This enzyme causes proteolysis of extracellular matrix and subsequent release and activation of major proangiogenic factors, such as VEGF and fibroblast growth factor-2 (FGF-2) (Murdoch et al. 2008; Belotti et al. 2003; Ardi et al. 2009; Ebrahem et al. 2010; Kessenbrock et al. 2010; Jodele et al. 2005; Bergers et al. 2000; Coussens et al. 2000; Ardi et al. 2007).

Another potent enzyme that may be involved in neutrophil-induced matrix degradation is neutrophil elastase. It exhibits broad substrate specificity, which includes nearly all components of extracellular matrix. It was recently shown that neutrophil elastase promotes tumor growth (Houghton et al. 2010). Accordingly, simple neutrophil depletion experiments using Gr-1 antibodies have been shown to inhibit tumor growth (Pekarek et al. 1995), limit metastasis number (Tazawa et al. 2003) and reduce endothelial cell recruitment to tumors (Sparmann and Bar-Sagi 2004).

Taken together, these data demonstrate a dual potential of tumor-infiltrating neutrophils, indicating that neutrophils present in a close proximity to malignant cells may predominantly exert tumoricidic effect, whereas neutrophils in the tumor stroma or at the invasive margin may be pro-tumorigenic. Not surprisingly, this controversy reflects in the numerous studies dealing with the prognostic significance of tumor-infiltrating neutrophils. The majority of clinical studies regarding tumor-infiltrating neutrophils have demonstrated that their presence and high density are associated with poor clinical outcomes, including decreased patients' survival. This correlation has been shown for pulmonary adenocarcinoma (Bellocq et al. 1998), gastric adenocarcinoma (Zhao et al. 2012), colorectal carcinoma (Rao et al. 2012), renal cell carcinoma (Jensen et al. 2009), etc. For example, in the study of Jensen et al. (2009) intratumoral neutrophils decreased the five-year recurrence free survival rate from 87 to just 53 %. However, in some studies tumor-infiltrating neutrophils were not found to be associated with prognosis in colorectal cancer (Nielsen et al. 1999), gastric carcinoma (Rice et al. 2000 Nielsen et al. 1999) and non–small cell lung cancer (Lee et al. 1989), or even associated with reduced mortality risk (Caruso et al. 2002).

In summary, numerous reports of tumor-infiltrating neutrophils demonstrate their complex and controversial role. It is likely that pro-tumorigenic and anti-tumorigenic properties of these cells depend on their localization in the tumor mass (intratumoral vs. peritumoral), their functional status and the stage of the tumor development (emerging neoplasm vs. invasive cancer). This biological controversy reflects ithe prognostic significance of tumor-infiltrating neutrophils. At the same time, it opens an opportunity to influence the development of tumor through modulating neutrophil recruitment and function.

3.3 Mast Cells

Mast cells originate from hematopoietic stem cells in the bone marrow and exit the bone marrow as committed, but undifferentiated, precursors before trafficking through the circulation to their target tissues, where they ultimately undergo terminal differentiation. Mature mast cells populate most tissues, but are found in highest numbers in the skin, airways and digestive tract (Grimbaldeston et al. 2006). Mast cells secretory granules store a complex mixture of preformed effector molecules—such as biogenic amines, proteoglycans, cytokines and neutral serine proteases, which are released upon appropriate cell stimulation (Crivellato et al. 2009).

Peripheral mast cells can be detected by variety of histochemical stains such as Giemsa, toluidine blue and chloracetate esterase, and by using immunostaining for tryptase, VEGF and CD117 (Fig. 3.2) and other markers (Table 3.1). Histochemical stains predominantly identify intact mast cells, whereas immunostains indicates the presence of both intact and degranulated mast cells in tissues (Ishibashi et al. 2006a; Al-Shibli et al. 2010; Amini et al. 2007). Remarkably, mast cells were recognized to infiltrate the interface between developing tumors and healthy tissues as early as 1891 (Westphal 1891). An increased number of mast cells has been found in a variety of malignant neoplasms including pulmonary non-small cell carcinoma (Al-Shibli et al. 2010), pancreatic adenocarcinoma (Esposito et al. 2004), breast carcinoma (Amini et al. 2007), colorectal carcinoma (Gulubova and Vlaykova 2009), malignant melanoma (Ch'ng et al. 2006), etc.

Several factors released by tumor cells, such as stem cell factor (SCF) and adrenomedullin, CCL2, CCL5 are thought to be responsible for the recruitment of

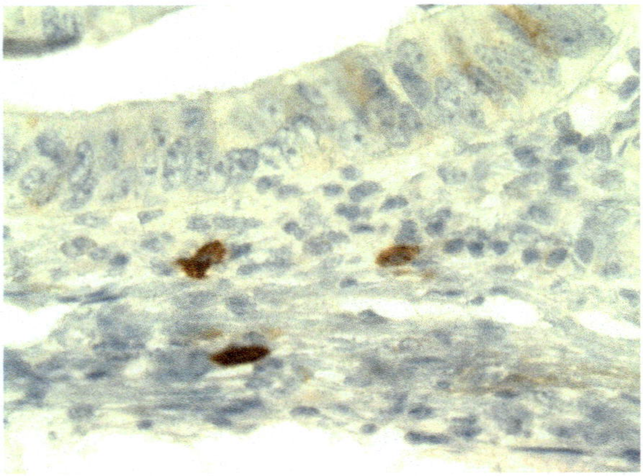

Fig. 3.2 Tumor-infiltrating mast cells in colonic adenocarcinoma. CD117 immunostain, 600x. Mast cells are present in the tumoral stroma

mast cells into tumors (Meininger et al. 1992; Zudaire et al. 2006; Zhang et al. 2000; Huang et al. 2008; Soucek et al. 2007). Specifically, it was shown that overexpression of SCF (which receptor CD117 is a characteristic protein of mast cells) increases mast cell accumulation in developing mammary tumors, whereas inhibition of SCF expression prevents this effect (Zhang et al. 2000). Mast cells also express a number of chemokine receptors, including CXCR4, CCR3 and CCR5, the ligands of which are up-regulated by various cell types in most forms of tumors (Juremalm and Nilsson 2005) It was shown, for instance, that CCL5 production in developing tumors triggers mast cell recruitment to pancreatic tumors (Soucek et al. 2007).

In regard of spatial distribution of tumor-infiltrating mast cells, the majority of reports show their presence predominantly in the tumor stroma adjacent to the neoplastic cells (Ch'ng et al. 2006; Soucek et al. 2007; Sawatsubashi et al. 2000; Toth et al. 2000; Samoszuk and Corwin 2003; Coussens et al. 1999; Humphreys et al. 2000; Aoki et al. 2003), but in some of them mast cells were detected within the islands of tumor cells (Amini et al. 2007; Welsh et al. 2005).

Mast cells represent an early infiltrating cell type in many tumors, appearing in the initial stages of tumor development. For instance, they have been shown to accumulate in and around adenomatous polyps (precursors to invasive colon cancer), where their density is significantly higher than in neighboring healthy tissue (Gounaris et al. 2007). Similarly, mast cell density was increasing in progression from dysplasia to invasive squamous cancer of the uterine cervix (Benitez-Bribiesca et al. 2001; Cabanillas-Saez et al. 2002). In oral squamous cell carcinoma, the density of mast cells also increases with disease progression and was significantly higher in oral squamous cell carcinoma than in hyperkeratosis and normal oral mucosa (Iamaroon et al. 2003; Sharma et al. 2010). In hepatocellular carcinoma mast cell density was higher in poorly differentiated than in well differentiated tumors (Peng et al. 2005).

As it was discussed above for neutrophils, tumor-infiltrating mast cells have direct effects on tumor cells, such as mast cell-mediated cytotoxicity, and indirect effects such as mast cell-directed angiogenesis, tissue remodeling and immune cell recruitment (Theoharides and Conti 2004). Mast cells could be detrimental to tumor growth by secreting several cytokines and proteolytic enzymes, inducing apoptosis of the malignant cells, such as IL-4 (Gooch et al. 1998). In addition mast cells can recruit other immune cells (like eosinophils and dendritic cells) into developing tumors and modulate their ability to kill tumor cells (Maltby et al. 2009; Suto et al. 2006).

On the other hand, mast cells can promote tumor development through the secretion of numerous pro-angiogenic molecules, including VEGF (Grutzkau et al. 1998), bFGF (Qu et al. 1998), TNF-α (Okayama et al. 1998) and angiopoetin-1 (Nakayama et al. 2004; Norrby 2002). For example, in malignant melanoma mast cell accumulation was correlated with increased expression of VEGF (Toth-Jakatics et al. 2000), FGF-2 (Ribatti et al. 2003a) and neovascularization. The same effect was reported in lung carcinoma (Tataroglu et al. 2004), gastric carcinoma (Yano et al. 1999; Kondo et al. 2006), colorectal carcinoma (Gulubova and Vlaykova 2009; Acikalin et al. 2005), endometrial cancer (Ribatti et al. 2005), etc. Mast cells can also facilitate tumor angiogenesis through the release of heparin and heparin-like molecules (Samoszuk et al. 2005). In addition, mast cells are the major sources of proteases, which act on the surrounding extracellular matrix (Coussens and Werb 1996; Meininger 1995) as was discussed above.

Similarly to the situation with tumor infiltrating neutrophils, there are conflicting reports on the prognostic value of mast cell density within tumors on patient survival. Generally, mast cell density correlates with angiogenesis and poor tumor outcome. This correlation was demonstrated in gastric carcinoma (Yano et al. 1999; Kondo et al. 2006), colorectal carcinoma (Gulubova and Vlaykova 2009; Acikalin et al. 2005), pulmonary adenocarcinoma (Ishibashi et al. 2006a; Takanami et al. 2000), pancreatic ductal adenocarcinoma (Esposito et al. 2004), esophageal squamous cell carcinoma (Elpek et al. 2001) and melanoma (Ribatti et al. 2003b). In some studies, however, increased mast cell numbers have been found either to correlate with improved clinical survival or not to correlate with survival at all. For instance, high mast cell density in tumor stroma correlated with a favorable prognosis in breast carcinoma (Dabiri et al. 2004; Rajput et al. 2008) and colon carcinoma (Tan et al. 2005).

At the same time, several studies of pulmonary non-small cell carcinoma, renal cell carcinoma, breast carcinoma, esophageal squamous cell carcinoma showed no significant impact of mast cell density on patient' prognosis (Al-Shibli et al. 2010; Amini et al. 2007; Tuna et al. 2006). Confirming an importance of a spatial distribution of mast cells, a study of pulmonary non-small-cell lung carcinoma demonstrated a correlation between favorable prognosis and the density of mast cells in tumor islets, but not in surrounding stroma (Welsh et al. 2005).

In summary, mast cells are recruited early in the developing tumor and preferentially accumulate in the tumor stroma. They may play an important role in anti-tumor immunity by either direct tumoricidal effect or by facilitating other immune cell homing. However, they may have a pro-tumorogenic effect, mainly through their role in angiogenesis. Accordingly, the ultimate result of these effects on tumor growth will vary depending on specific circumstances, which explains conflicting data on the prognostic significance.

3.4 Eosinophils

Other members of the granulocyte family of cells, eosinophils, emerge in a bone marrow from a common precursor and reach peripheral tissue through blood circulation. Similarly to neutrophils and mast cells, the presence of eosinophils in tumors was recognized a long time ago (Lowe et al. 1981a). Specific factors, facilitating recruitment of eosinophils into the tumor masses are not well known. One of the chemokines most likely involved in this process is interleukin 5 (IL-5), which was shown to be expressed in several malignant neoplasms (Pandit et al. 2007).

Due to their characteristic morphologic appearance, eosinophils can be reliably identified in tissue section by simple histochemical stains (H&E, Luna, Fig. 3.3). (Ishibashi et al. 2006a; Amini et al. 2007). They are present both within and around tumor islets, most often at the invasive tumor margin (Sato et al. 1981; Lowe et al. 1981b). Remarkably, tumor infiltrating eosinophils are most often found in squamous cell carcinomas of various organs. Significant density of eosinophils was reported in squamous cell carcinoma of head and neck (Sato et al. 1981; Goldsmith et al. 1992; Fujii et al. 2002; Thompson et al. 1994), oral cavity (Dorta et al. 2002), esophagus (Ishibashi et al. 2006a), uterine cervix (Bostrom and Hart 1981; Lowe 1988; Bethwaite et al. 1993), skin, vagina, etc., (Lowe et al. 1981a; Ono et al. 2002).

However, even in carcinomas with similar cell morphology variations of eosinophil density are quite remarkable. For example, in a study of cervical carcinoma, Lowe and colleagues found a very marked tissue eosinophilia with over

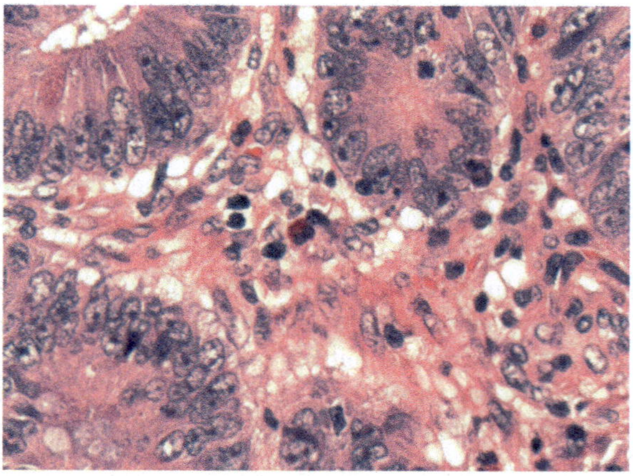

Fig. 3.3 Tumor-infiltrating eosinophils in colonic adenocarcinoma. H&E stain, 600x; Eosinophils are present in the tumor stroma

100 eosinophils per high power field in 3 % of cases, minor degree of tissue eosinophilia in 37 % of cases, while in 60 % there were no eosinophils in the tumor (Lowe et al. 1981b). The authors suggested that the tumors with high eosinophil density, usually of the large cell non-keratinizing type may be regarded as a distinct histopathological entity.

Tumor infiltrating eosinophils have also been found in pulmonary non-small cell carcinoma, colon adenocarcinoma, gastric adenocarcinoma, uterine carcinoma, and in transitional cell carcinoma of the bladder (Lowe et al. 1981a; Pandit et al. 2007; Moezzi et al. 2000; Pretlow et al. 1983). At the same time, eosinophils are not present in pulmonary small cell carcinoma or invasive breast carcinoma (Amini et al. 2007; Lowe et al. 1981a)

Although the exact mechanisms of eosinophils interaction with the tumor microenvironment are not known, majority of studies indicate that tumor-infiltrating eosinophils are associated with favorable clinical outcome. This correlation was shown in patients with gastric and colorectal adenocarcinoma (Nielsen et al. 1999; Moezzi et al. 2000; Pretlow et al. 1983), esophageal squamous cell carcinoma (Ishibashi et al. 2006a), cervical squamous cell carcinoma (Bostrom and Hart 1981; Lowe 1988; Bethwaite et al. 1993; van Driel et al. 1996), and other cancers (Lowe et al. 1981a; Goldsmith et al. 1992; Fujii et al. 2002; Thompson et al. 1994; Dorta et al. 2002; Ono et al. 2002). For instance, Ishibashi et al. reported that in esophageal squamous cell carcinoma the number of tumor-associated eosinophils was significantly higher in the cases without venous invasion, lymph node metastasis and clinical recurrence. These findings suggest the possible correlation between tumor-associated tissue eosinophils and a less aggressive biological behavior of the tumor (Ishibashi et al. 2006a).

In summary, eosinophils show a marked predilection to particular types of tumor, present predominantly in tumor stroma, and a high density of tumor-infiltrating eosinophils is associated with a good clinical prognosis.

3.5 Macrophages

Macrophages emerge in the tissue from hematopoetic blood-born precursors of the monocytic lineage. They are specialized phagocytic cells that attack foreign substances, infectious agents and cancer cells through destruction and ingestion (Nagorsen et al. 2007; Baay et al. 2011).

Macrophages are polarized into two functionally distinct forms M1 and M2. The M1 macrophages produce high levels of interleukin (IL)-12, IL-23, TNF-α, IL-1, IL-6, CXC ligand 10 (CXCL10), inducible nitric oxide synthase, human leukocyte antigen (HLA)-DR and reactive oxygen and nitrogen intermediates. M2 macrophages express high levels of IL-10, IL-1 receptor antagonist, CC ligand 22 (CCL22), scavenger mannose receptor, galactose receptor, arginase I and CD163 antigen (Mantovani et al. 2007; Mantovani et al. 2005; Baj-Krzyworzeka et al. 2007; Mantovani et al. 2002; Gordon 2003).

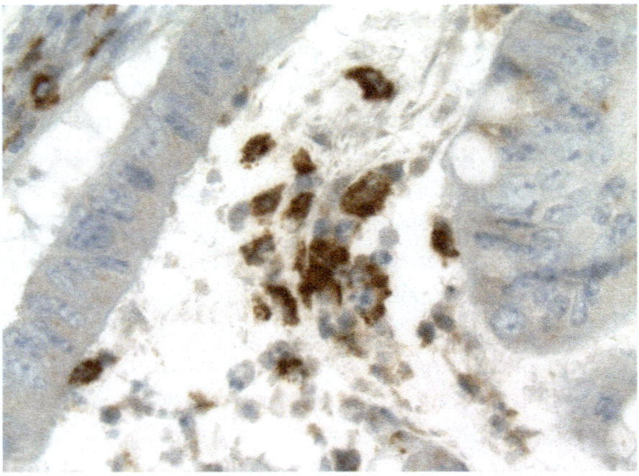

Fig. 3.4 Tumor-infiltrating macrophages in colonic adenocarcinoma. CD68 immunostain, 600x. Macrophages are present in the tumor stroma

Macrophages cannot be reliably identified in tissue by routine histochemical stains, but instead requires immunostaining for macrophages-associated antigens (e.g., CD68: Fig. 3.4). To distinguish M1 and M2 macrophages even more sophisticated techniques are necessary, such as double immunohistochemical staining for CD68/HLA-DR (markers for M1 macrophages) and CD163 or CD204 (markers for M2 macrophages) (Kurahara et al. 2009; Ma et al. 2010) (Table 3.1).

The number of macrophages is often increased in tumors compared to healthy tissue (Allen and Hogg 1987; Sickert et al. 2005). Tumor-infiltrating macrophages constitute a major component of the leukocyte infiltrate in malignant tumors (Talmadge et al. 2007), but their density varies widely even in the tumors of the same origin. For example, in colonic adenocarcinoma their density ranged from a very low rate of 1.7 cell/HPF up to 46/HPF (Nagorsen et al. 2007).

Several tumor-associated chemokines can be responsible for macrophage homing at the tumor site. Chemotactic cytokine ligand 2 (CCL2, also known as monocyte chemoattractant protein 1 (MCP1)), is produced in a wide range of tumors and its expression correlates with the density of tumor-infiltrating macrophages (Negus et al. 1995; Mazzucchelli et al. 1996; Marcus et al. 2004). VEGF and related molecules are also potent macrophage attractants contributing to their recruitment (Sica and Bronte 2007; Fischer et al. 2007; Kaplan et al. 2005). Another chemokine, macrophage colony-stimulating factor (M-CSF) is expressed in tumor cells and promotes differentiation and survival of macrophages (Chambers et al. 1997; Toy et al. 2001; Kluger et al. 2004).

The spatial distribution of tumor infiltrating macrophages can vary, but usually they are more frequent in tumor stroma than in tumor islets (Nagorsen et al. 2007). Studies that specifically identified M1 and M2 phenotypes in tumors found a predominance of cells with M2 profile. For instance, in pulmonary non-small cell carcinoma 70 % of tumor infiltrating macrophages were M2 macrophages and the remaining 30 % were M1 macrophages (Ma et al. 2010). The distinction of M1 and M2 in tumors is very important since these cells have an opposite effect on tumors. M1 macrophages produce effector molecules such as reactive oxygen intermediates, reactive nitrogen intermediates and TNF-αa, and thus have the anti-tumor effect. In contrast, M2 macrophages can promote tumor growth and metastasis by secretion of matrix-degrading enzymes (including MMP-9), angiogenic factors and immunosuppressive cytokines. The balance of these macrophage forms determines the anti- or pro-tumor effects of the macrophage population (Mantovani et al. 2005; Gordon 2003; Solinas et al. 2009; Giraudo et al. 2004; Pollard 2004; Condeelis and Pollard 2006; Lewis and Pollard 2006; De Palma et al. 2005; Mantovani et al. 2004; Mosser and Edwards 2008).

Heterogeneity of tumor-infiltrating macrophages can explain conflicting results of the studies correlating their density with the clinical course and prognosis. For example, assessing the total counts of tumor-infiltrating macrophages (CD68+) several groups have reported a favorable association between the density of tumor islet macrophages and survival of patients with pulmonary non-small cell carcinoma (Welsh et al. 2005; Kawai et al. 2008), colorectal carcinoma (Dadabayev et al. 2004; Funada et al. 2003; Khorana et al. 2003) and hepatocellular carcinoma (Li et al. 2009a). In contrast, Al-Shibli et al. did not find CD68+ macrophages in neither tumor nor stromal compartments of pulmonary carcinoma specimens to correlate with patients' survival (Al-Shibli et al. 2009). Similar results were also reported by Toomey et al. (2003). Yet another groups found the presence of macrophages to be linked to poor prognosis in numerous types of malignancy, including gastric carcinoma (Kawahara et al. 2010), pancreatic carcinoma (Kurahara et al. 2009), thyroid carcinoma (Ryder et al. 2008), leiomyosarcoma (Lee et al. 2008) and other cancers (Kang et al. 2010; Bingle et al. 2002).

One of the reasons for this discrepancy is a spatial distribution of macrophages in tumor tissue. In several studies an increased density of macrophages within the tumor islets conferred a marked survival advantage, whereas increased number of macrophages in the tumor stroma was associated with poor prognosis (Welsh et al. 2005; Dai et al. 2010). Therefore, the positive and negative associations can antagonize each other and reduce the net predictive value when the two macrophage pools are counted together.

Since the majority of macrophages located in tumor islets are of M1 phenotype and the majority of macrophages in tumor stroma are of M2 phenotype, it is reasonable to expect that the difference in microlocalization can reflect the difference in macrophage activation pattern. In fact, when differentiating between M1 and M2 phenotypes in pulmonary carcinoma, high density of M1 macrophages was associated with a significantly better prognosis (Ma et al. 2010; Ohri et al. 2009). Confirming these findings, Ohtaki et al. (Ohtaki et al. 2010) reported that whereas the density of CD68+ macrophages was of marginal prognostic significance in lung adenocarcinoma, M2 macrophages (identified with the use of CD204) showed a strong association with poor outcome. Unfavorable prognostic role of tumor infiltrating M2 macrophages was also demonstrated in pancreatic carcinoma (Kurahara et al. 2009), renal cell carcinoma (Komohara et al. 2011) and endometrioid carcinoma (Espinosa et al. 2010).

In summary, macrophages represent a major population of tumor-infiltrating leukocytes. They can be present either within the tumor islets or in peri-tumoral stroma and their spatial distribution has important prognostic implications. The discovery of functional heterogeneity of tumor-infiltrating macrophages is a great example of our scientific advances in this field of oncologic research. It is now accepted that M1 macrophages, predominantly situated within tumor islets, have the anti-tumor effect, whereas M2 cells, located in the stroma, promote tumor growth. Since the specific factors, regulating M1 vs. M2 differentiation are being actively studied, it opens a door to a new way of targeted therapeutic interactions.

3.6 Conventional (Myeloid) Dendritic Cells

Conventional dendritic cells (DC) emerge in the tissue from hematopoetic blood-born precursors. In their immature state they express proteins that allow them to uptake and process antigens. Immature DC (iDC) are characterized by high endocytic activity and low T cell activation potential. They express chemokine receptors (e.g., CCR6) that allow migration into tissue. After uptaking an antigen DC undergo maturation steps, up-regulate proteins of antigen processing machinery (APM) and travel to the regional lymph nodes. Simultaneously, they up-regulate cell surface receptors, such as CD80, CD86 and CD40 that act as co-receptors in T cell activation (Hashimoto et al. 2000). These processes are regulated by cytokines, some of which, like IL-4 and GM-CSF, direct differentiation of DC, while other, like TNF-α, induce their maturation (Caux et al. 2000; Shurin et al. 2006). Finally, mature DC (mDC) present processed antigens to CD8 and CD4 T cells in the context of MHC I and MHC II, respectively.

Tumor-infiltrating DC, similarly to macrophages, cannot be identified in tissue by routine histochemical techniques, thus requiring the use of immunohistochemistry. For evaluation of the total number of DC the most often used marker is S-100 protein, although it is rather non-specific. Immunostains usually utilized to identify iDC include CD1a, CD209/DC-SIGN and CD207/Langerin. For mDC this list consists of CD83, CD86, CD208/LAMP and HLA-DR (Fig. 3.5; Table 3.2).

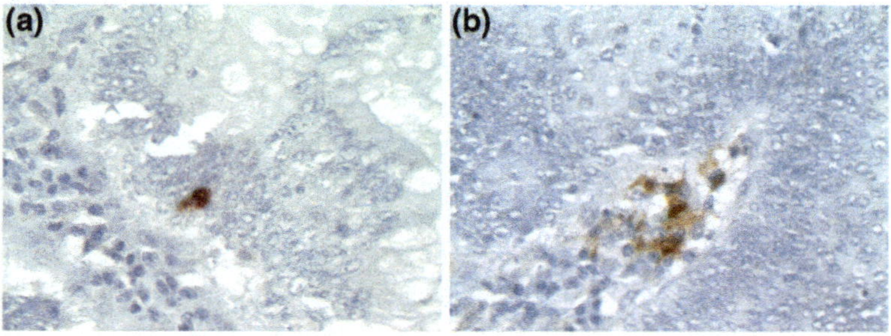

Fig. 3.5 Tumor-infiltrating myeloid dendritic cells in colonic adenocarcinoma. **a** CD1a immunostain, 600x; **b** CD83 immunostain, 600x. iDC (CD1a+) are present within epithelial nests, mDC (CD83+) are present in tumor stroma

Table 3.2 Commonly used immunohistochemical markers of tumor-associated dendritic cells

Antibody	Properties	Target
S-100 protein	Low molecular weight protein characterized by two calcium binding sites of the helix-loop-helix conformation	All types of DC (mature and immature DCs). Cells derived from neural crest: (Schwann cells, melanocytes, glial cells), chondrocytes, adipocytes, myoepithelial cells, macrophages
CD1a	49 kDa cell surface glycoprotein expressed in association with beta-2-microglobulin. Expressed predominantly in early steps of DC maturation	Myeloid immature DCs
CD 209/DC-SIGN (DC-specific ICAM-3-grabbing nonintegrin)	DC-specific adhesion receptor that mediates DC binding to ICAM-3. Presumably mediates the recognition of non-self and the presentation of foreign antigens. Can regulate important adhesion processes	Myeloid immature DCs
CD 207/Langerin	C-type lectin responsible for the formation of Birbeck granules, a typical hallmark for DC of Langerhans type	Myeloid immature DCs
CD 83	40–45 kDa glycoprotein expressed predominantly in the late steps of DC maturation. CD 83+ DC co-express the highest levels of HLA II	Myeloid mature DCs
CD86	Membrane protein of the immunoglobulin superfamily, which provides a co-stimulatory signal necessary for T cell activation and survival	Myeloid mature DCs
CD208/DC-LAMP (Dendritic cell-lysosomal associated membrane protein),	Member of the lysosomal associated membrane protein (LAMP) family. Plays an important role in antigen processing and MHC-II restricted antigen presentation	Myeloid mature DCs
CD 123	IL-3 receptor α-chain involved in cell signaling for cell growth and differentiation	Plasmacytoid DC

As it was described earlier for other tumor-infiltrating leukocytes, density of DC in the tumor mass varies widely depending on the type of malignancy. For instance, in breast carcinoma tumor-infiltrating DC are detected in 30–50 % of tumors (Bell et al. 1999). In two main types of pulmonary non-small cell carcinoma (adenocarcinoma and squamous cell carcinoma) DC are found in 60–80 % of tumors (Nakajima et al. 1985; Colasante et al. 1993; Colasante et al. 1995; Inoshima et al. 2002). At the same time, DC density in two types of pulmonary neuroendocrine tumors (small cell carcinoma and carcinoid tumor) is usually very low (Nakajima et al. 1985; Coli et al. 1990; Zeid and Muller 1993; Katsenelson et al. 2001). Katsenelson and co-workers found different populations of DC, including CD1a+ iDC and CD83+ mDC, in small cell carcinoma, but samples of carcinoid tumor were devoid of DC. Thus it is possible that pulmonary neuroendocrine carcinoma produced factors that inhibit DC generation, maturation or induce DC apoptosis (Katsenelson et al. 2001). In transitional cell carcinoma of the urinary bladder a dense infiltrate of S-100+ DC is detected in 50 % of cases (Inoue et al. 1993). In oral squamous cell carcinoma density of DC infiltrates was low in 20 % of specimens, intermediate in 42 % of specimens and high in 37 % of specimens (Reichert et al. 2001). Pancreatic carcinoma is characterized by a paucity of tumor-infiltrating DC: significant numbers of S100+ DC and CD1a+ iDC were found in only 4 % of tumors (Dallal et al. 2002).

Unlike other tumor-infiltrating leukocytes, the density of tumor-infiltrating DC is lower in tumor than in the corresponding normal tissue. For example, Troy and co-workers (Troy et al. 1998a) compared the number of DC in prostate carcinoma and adjacent normal prostatic tissue and found that there were significantly less CD1a+ iDC in prostate cancer compared with normal prostatic tissue and only a small subset of DC expressed markers of activation, such as CD83, CD86. The density of CD 83 + mDC density is also significantly lower in gastric cancer tissue than in normal gastric tissue (Tsukayama et al. 2005). As discussed later, the low density of tumor infiltrating DC may present a survival advantage to malignant tumors and thus be a mechanism of immune escape.

Very interesting study was performed by Vakkila and co-workers to compare DC density in pediatric and adult tumors (Vakkila et al. 2006). While DC were present in adult tumors (colon carcinoma, breast carcinoma, esophageal carcinoma), tumor-infiltrating DC were virtually absent in pediatric malignancies (Ewing's sarcoma, rhabdomyosarcoma, hepatoblastoma, neuroblastoma, Wilms' tumor). Inflammatory infiltrate in pediatric tumors was composed mainly of macrophages, whereas in adult tumors, DC formed 37 % of leukocytes within the tumor islands and 25 % around the tumors. The reason for this striking difference merits further investigation.

Recruitment of DC into the peripheral tissue is dependent on the action of chemokines, "inviting" them to survey the tissue microenvironment (Vicari et al. 2004). Human myeloid DC express CCR2, CCR6 and CXCR2, the receptors for CCL2, CCL20, CXCL1 and CXCL5, that are likely to be responsible for their homing. Upon maturation DC also up-regulate expression of CCR7, a chemotactic receptor that directs their trafficking through the lymphatic system, in response to chemokines CCL19 and CCL21 (Sozzani et al. 1997; Feijoo et al. 2005). Another chemokine that can chemoattract DC is CX3CL1. In breast carcinoma the number of intratumoral CD1a+ DC was significantly increased in the high CX3CL1 expression group compared with those in the low CX3CL1 expression group (Park et al. 2012).

While numerous chemokines may be involved in DC attraction, one of them, called BRAK (CXC14), has recently become a focus of a particular interest. CXC14 is steadily expressed in normal tissue of different types (Hromas et al. 1999; Frederick et al. 2000) and has been shown to be a potent iDC chemoattractant and activator (Shellenberger et al. 2004; Schaerli et al. 2005; Shurin et al. 2005). Notably, its expression is markedly reduced in several malignant neoplasms (Frederick et al. 2000; Shurin et al. 2005), and this phenomenon may explain a paucity of DC in many advanced tumors.

When present in the tumor mass, DC can be seen within cancer nests, in tumor stroma and in peritumoral areas. Their spatial distribution seems to depend on the type of the tumor. In colorectal carcinoma, an infiltration of tumor stroma by DC was significantly higher than in tumor islets (Nagorsen et al. 2007). In contrast, in pulmonary non-small cell carcinoma, DC were located predominantly in cancer nests and their number correlates with the extent of cancer cell apoptosis. In the areas of scattered DC distribution, only a few apoptotic tumor cells can be detected, while in the areas of DC aggregations, apoptotic tumor cells were significantly more abundant (Kurabayashi et al. 2004).

Spatial distribution of tumor-infiltrating DC seems to depend on their level of maturation. It was demonstrated that in majority of tumors iDC are located within the tumor nests, while mDC are present in the stroma. For example, in breast carcinoma CD1a+ iDC, were retained predominantly within the tumor epithelium, whereas CD83+ and LAMP+ mDC were confined to peritumoral areas (Hillenbrand et al. 1999; Lespagnard et al. 1999; Coventry et al. 2002). Similar data were reported for colonic adenocarcinoma (Dadabayev et al. 2004; Suzuki et al. 2002; Miyagawa et al. 2004), oral squamous cell carcinoma (O'Donnell et al. 2007), biliary carcinoma (Takagi et al. 2004; Furihata et al. 2005), transitional cell carcinoma of the urinary bladder (Ioachim-Velogianni et al. 1995; Troy et al. 1999), renal cell carcinoma (Troy et al. 1998b; Schwaab et al. 1999; Aso et al. 2004) and melanoma (Vermi et al. 2003).

Considering the current paradigm of the role of myeloid conventional DC in antigen uptake and presentation to T cells, it appears reasonable that iDC are located in a close proximity to the surveyed tumor cells, whereas mDC are situated in the tumor stroma, probably on their way to the regional lymph nodes, or presenting the processed tumor-associated antigens to the effector cells locally. At the same time, it is clear that in order to perform immune survaillance tumor-infiltrating DC have to be completely functional, yet numerous studies show that they are functionally impaired and incompletely activated, resulting in inefficient T cell priming (Troy et al. 1998b; Schwaab et al. 1999; Aso et al. 2004; Vermi et al. 2003).

Recently it was shown that they are also morphologically abnormal. Jia and colleagues (Jia et al. 2012) studied the ultrastructure of DC in human endometrioid adenocarcinoma by electron microscopy and found that compared to DC in normal tissue, tumor infiltrating DC were small in volume, have abnormal nuclear chromatin, and decreased cytoplasmic processes. Interestingly, these morphologic changes correlated with low expression of maturation markers (CD80, CD86 and CD40).

Several tumor-associated factors can be responsible for this morphologic and functional impairment of tumor infiltrating DC, including VEGF and IL-10 (Gabrilovich 2004; Della Porta et al. 2005; Osada et al. 2008; Mimura et al. 2007; Gabrilovich et al. 1996; Michielsen et al. 2011).

In regard to correlation of DC density with tumor grade, majority of studies showed higher DC density in well differentiated than in poorly differentiated neoplasms. This correlation was reported in pulmonary non-small cell carcinoma (Zeid and Muller 1993), head and neck carcinoma (Chen et al. 2005), prostate carcinoma (Bigotti et al. 1991), endometrial carcinoma (Coppola et al. 1998) and other tumors. However, in breast carcinoma the number of tumor infiltrating DC was higher in high-grade tumors (Lespagnard et al. 1999).

Correlation of DC density with tumor stage was performed by Kikuchi and co-workers (2002). They found that in head and neck cancer the numbers of iDC were greater in patients with lower stage of the disease and decreased with tumor progression. Interestingly, mDC density showed the reverse correlation (Kikuchi et al. 2002). Significant decrease of iDC with simultaneous increase of mDC was also demonstrated in the progression steps of cervical squamous cell carcinoma (Hubert et al. 2005; Hayati and Zulkarnaen 2007).

Correlations of the density of tumor infiltrating DC with clinical outcome were extensively studied for numerous types of tumors. In the majority of them high density of tumor infiltrating DC (especially mDC) was a favorable prognostic feature. In fact, some of the studies found the density of tumor-infiltrating mDC to be a better predictor of clinical outcome than other well-established parameters (Reichert et al. 2001; Dieu-Nosjean et al. 2008). In a large cohort of patients with pulmonary non small cell carcinoma increasing density of stromal DCs were associated with increased disease-specific survival (DSS) (Al-Shibli et al. 2009; Dai et al. 2010; Inoshima et al. 2002). In breast carcinoma high mDC density was also a favorable prognostic marker. At the same time, no such correlation was found for total DC and iDC density (Iwamoto et al. 2003; Treilleux et al. 2004; Coventry and Morton 2003).

The same correlations were found in colonic carcinoma (Nagorsen et al. 2007; Ambe et al. 1989; Nakayama et al. 2003), gastric carcinoma (Tsukayama et al. 2005; Ishigami et al. 2000a; Takahashi et al. 2002), hepatocellular carcinoma, (Yin et al. 2003; Cai et al. 2006), biliary carcinoma (Furihata et al. 2005; Nakakubo et al. 2003), oral squamous cell carcinoma (Reichert et al. 2001), melanoma (Ladanyi et al. 2007; Simonetti et al. 2007) and other tumors.

In summary, DC comprise a small, but clinically significant population of tumor-infiltrating immune cells. Unlike other types of immune cells they are decreased in numbers in established malignant neoplasms. Their spatial distribution pattern is consistent with their presumptive functional role: iDC are predominantly situated in a close proximity to tumor cells, and mDC are present in the stroma. Although functionally impaired, DC are generally associated with favorable prognostic outcome.

3.7 Plasmacytoid DC

Plasmacytoid dendritic cells (pDC) originate in the bone marrow from DC progenitors common to pDC and mDC, and can be identified by their unique immunoprofile (Liu et al. 2009; Naik et al. 2007). They are recognized as the main source of type I IFN after challenge with pathogens (Liu 2005; Asselin-Paturel et al. 2001; Ma et al. 2012). Similarly to conventional DC, tumor-infiltrating pDC cannot be identified in tissue by routine histochemical techniques, thus requiring the use of immunohistochemistry. The most often used marker for this purpose is CD123 (Table 3.2). While present in tissues at low numbers in the healthy steady state, pDC accumulate in lymphoid and non-lymphoid tissues under different pathological conditions (Matta et al. 2010). They commonly represent a minor fraction (10–15 %) of the infiltrating immune cells (Vermi et al. 2011; Labidi-Galy et al. 2011), but at least in some tumors they were found to be the most abundant DC subset (Labidi-Galy et al. 2011).

Circulating pDCs express multiple chemotactic receptors, including CXCR4 and ChemR23 (CMKLR1) that can trigger they recruitment into peripheral tissues (Vermi et al. 2011). Malignant tumors have been shown to express high levels of CXCR4 ligand, stromal-derived factor-1 (CXCL12), which likely represents the main axis for pDC accumulation in tumors (Vermi et al. 2003; Zou et al. 2001; Wilke et al. 2011). Another chemokine that can direct pDC homing into the tumor mass is CCL20 acting through CCR6 receptor on pDCs (Charles et al. 2010).

Accumulation of pDC in tumors has been directly demonstrated in primary carcinomas of different organs (breast, ovary, head and neck, lung, skin, cervix, prostate and liver), as well as cutaneous melanoma (Vermi et al. 2003; Treilleux et al. 2004; Labidi-Galy et al. 2011; Wilke et al. 2011; Thiel et al. 2009; Watkins et al. 2011; Hartmann et al. 2003). While the importance of pDCs in innate and adaptive immune responses against pathogens is well established, their role in anti-tumor immunity is not clear. Theoretically, pDC have the ability to orchestrate the local immune response to cancer cells by producing IFN, recruiting other immune cells via amplification of the proinflammatory chemokine network and cross-presenting antigens to CD8+ T cells (Penna et al. 2002; Mouries et al. 2008; Hoeffel et al. 2007; Dunn et al. 2006). They can also induce apoptosis of tumor cell lines either directly by secreting TRAIL or indirectly via the effect of IFN-α on other cytotoxic cells (Chaperot et al. 2006).

However, numerous experimental and clinical evidences show that pDC possess immunosuppressive and tolerogenic property and, thus, promote tumor growth and progression (see below). This apparent discrepancy might be explained, at least in part, by the properties of tumor-infiltrating pDC.

Since well-established methods of identification of pDC are now available, several studies were focused on a functional characterization of these cells. Practically all of the currently existing results point to the same conclusion: tumor-infiltrating pDC are defective in IFN production and instead secrete immunosuppressive soluble factors responsible for tumor progression (Labidi-Galy et al. 2011; Watkins et al. 2011; Hartmann et al. 2003).

Although the exact mechanisms of this phenomenon are not known, several possible scenarios can explain its origin. First, it is known that pDC express several receptors that negatively regulate the amplitude of the IFN response. One of them, ILT7, recognizes BST2, the protein that is strongly expressed on tumor cell lines and carcinomas (Cao and Bover 2010; Cai et al. 2009). This interaction may reduce pDC ability to produce the required amount of IFN-α to sustain elimination of cancer cells (Vermi et al. 2011). Other possible tolerogenic mechanism of tumor infiltrating pDC includes expression of IDO and secretion of Granzyme B, which leads to inhibition of T cell activation and immunosuppression (Muller and Prendergast 2007; Norian et al. 2009; Jahrsdorfer et al. 2010). In addition, pDC can drive CD4+ T cells to the generation of CD4+CD25+Foxp3+ Treg cells (see below), which leads to anergy and immune suppression, favoring the immune escape of tumor cells (Sharma et al. 2007a; Ito et al. 2007).

These findings have strong clinical correlations: the results of several studies indicate that prognosis of different types of tumors is inversely related to the density of tumor-infiltrating pDC. Negative prognostic influence of pDC has been demonstrated in ovarian cancer (Labidi-Galy et al. 2011), breast carcinoma (Treilleux et al. 2004), oral squamous cell carcinoma (O'Donnell et al. 2007) and other tumors. For instance, CD123+ pDC infiltration was found in 13 % of the breast carcinoma. Their presence was strongly associated with shorter overall patients' survival and relapse-free survival and was found to be an independent adverse prognostic factor (Treilleux et al. 2004).

3.8 Tumor Infiltrating Lymphocytes

Lymphocytes usually represent the main population of infiltrating immune cells in malignant tumors. Their role in tumor development is considered so significant, that modern cancer staging includes lymphocytic infiltration as one of the required parameters of pathologic evaluation (Edge 2010).

On routine histochemical stain (H&E) lymphocytes are easily identified as small discohesive cells with distinct cellular morphology, however, specific types of lymphocytes (B cells, T cells, NK cells) cannot be recognized. Since different types of lymphocytes have different roles in tumor immunity, evaluation of total lymphocytic density in tumors is of limited significance. Instead, modern morphologic techniques, specifically, immunohistochemistry, provides valuable information regarding the role of each type of lymphocytes in tumor development. Detailed description of immunologic role of lymphocytes, their development and

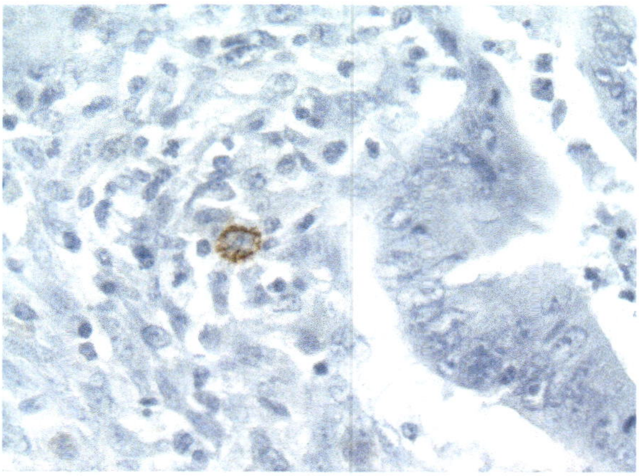

Fig. 3.6 Tumor-infiltrating B lymphocytes in colonic adenocarcinoma. CD20 immunostain, 600x. Both intraepithelial and stromal B cells are present

homing is discussed in other chapters below. Here we focus on morphologic identification of lymphocytes in tumor tissue and the results of clinical-pathologic correlations related to the density of different types of tumor-infiltrating lymphocytes.

3.9 B Lymphocytes

Tumor-infiltrating B lymphocytes are usually identified by immohistochemic stain for CD20, a phosphoprotein expressed on the surface of all mature B cells (Fig. 3.6; Table 3.3). B lymphocytes can be present within tumor islets, in tumor stroma and at the tumor invasive margin. They are typically less frequent than T

Table 3.3 Commonly used immunohistochemical markers of tumor-associated lymphocytes

Target	Staining	Properties
B lymphocytes	CD20	33 kD protein, pan-B lymphocyte marker
T lymphocytes	CD3	Part of T-cell receptor complex
T helpers	CD4	Cell-surface glycoprotein, co-receptor for the T Cell receptor complex
CTL (cytotoxic T cells)	CD8	Cell-surface glycoprotein, co-receptor for the T Cell receptor complex
Tregs (regulatory T cells)	CD25	Interleukin-2 receptor alpha chain (IL-2Ralpha), expressed by early progenitors of the T and B lineage as well as by activated mature T and B lymphocytes
	FoxP3	A member of the forkhead/winged-helix family of transcriptional regulators, in CD25+CD4+ positive regulatory T cells
Natural killer cells	CD56	Glycosylated transmembrane protein, expressed by NK cells, a subset of T cells, and neuroectodermal-derived cells
	CD 57	Human natural killer-1, expressed on NK cells

lymphocytes, comprising up to 40 % of tumor infiltrating lymphocyte population (Coronella-Wood and Hersh 2003; Chin et al. 1992; Marsigliante et al. 1999; Nelson 2010).

It is assumed that tumor-infiltrating B cells mediate their effects through autoantibodies, which could directly modulate the function of target proteins or promote tumor immunity through the opsonization of tumor antigens, complement-mediated destruction of tumor cells or antibody-dependent cytotoxicity (Punt et al. 1994; Yasuda et al. 2006). These autoantibodies can be produced by B cells situated outside the tumor mass, but by virtue of location tumor-infiltrating B cells can raise the local concentration of autoantibodies to physiologically significant levels (Nelson 2010). In addition, B cells can directly kill tumor cells through antibody-independent mechanisms (Lundy and Killer 2009) and serve as antigen-presenting cells for T lymphocytes (Yanaba et al. 2008; Rodriguez-Pinto 2005). It is worth mentioning that tumor-infiltrating B cells usually are found in close association with T lymphocytes and DC. (Dieu-Nosjean et al. 2008; Al-Shibli et al. 2008).

Correlation of tumor-infiltrating B lymphocyte density with clinical outcomes was performed in numerous types of cancer. Consistent with their important immune-mediating functions, tumor-infiltrating B cells are associated with a better survival in patients with lung cancer (Al-Shibli et al. 2008; Pelletier et al. 2001), cervical cancer (Ancuta et al. 2009; Nedergaard et al. 2008), prostate cancer (Karja et al. 2005), ovarian cancer (Milne et al. 2009), soft tissue sarcoma (Sorbye et al. 2012a), melanoma (Oble et al. 2009) and other tumors (Lim et al. 2010).

3.10 T Lymphocytes

T lymphocytes represent the main population of tumor-infiltrating immune cells and are considered necessary for efficient tumor eradication (de Visser et al. 2006; Pages et al. 2010). Their detection in tumor mass requires a use of immunostains directed towards specific cellular antigens, expressed by particular subtypes of T cells (Table 3.3). Since CD3 is a part of the T cell receptor complex and is expressed on all mature T lymphocytes, the immunostain for CD3 is used most often to detect a density of the entire T cell population. Two major T lymphocyte subsets, T helper cells and cytotoxic T cells, can be identified by their expression of CD4 and CD8, respectively (Fig. 3.7). Detection of more specific subtypes of T cells often require double immunostaining to identify the concomitant expression of two specific antigens on the cells of interest. For instance, this approach helps to identify natural killer T cells by double staining for CD3 and CD56 (Vivier and Anfossi 2004). The same approach makes it possible to distinguish regulatory T cells (Treg), by double staining for CD4/CD25 or CD4/FoxP3 antigens. These cells are extensively studied today, since they have been implicated in suppressing anti-tumor immune response (Hori et al. 2003). CD4+ T helper lymphocyte population is a heterogeneous class of cells, including T helper type 1 (Th1) and type 2 lymphocytes (Th2). Th1 cells have a principal role in activating cytotoxic T

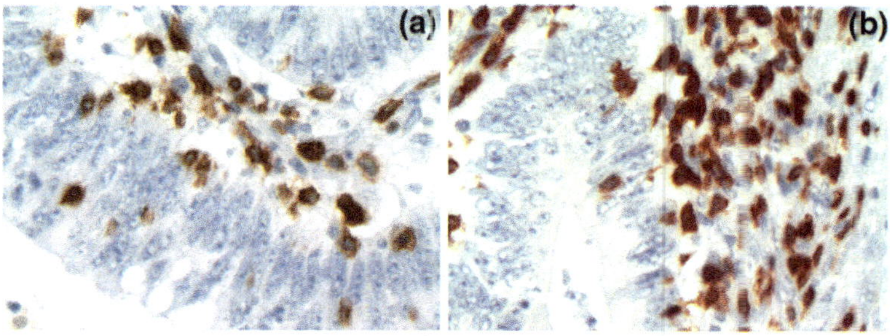

Fig. 3.7 Tumor-infiltrating T lymphocytes in colonic adenocarcinoma. **a** CD4 immunostain, 600x; **b** CD8 immunostain, 600x. Both intraepithelial and stromal T helper cells (CD4+) and cytotoxic T cells (CD8+) are present

cells, while T helper type 2 lymphocytes (Th2) stimulate humoral immunity and activate eosinophils. These two types of cells in tumor tissue can be recognized by immunostains for cellular antigens T-bet expressed on Th1 and GATA-3 expressed on Th2 (De Monte et al. 2011). Thus, the current panel of available immunostains can distinguish different subtypes of tumor-infiltrating T lymphocytes. In addition, attempts at a more functional evaluation of these cells have been made by staining for activation markers on lymphocytes such as Granzyme B, CD25, OX40 (CD134), CD69 and other molecules (Hillen et al. 2008; Gao et al. 2007; Ladanyi et al. 2004).

As expected based on their role in anti-tumor immunity, tumor-infiltrating T cells generally are associated with favorable prognosis. However, it is important to remember that different types of T lymphocytes have different functions in the tumor microenvironment. Thus, the pathologic and clinical correlation has to be evaluated with caution and take into consideration the methods that were used for the detection of tumor-infiltrating T cells. Therefore, studies that utilized one immunostain (e.g., CD3, detecting the entire population of T cells) would be difficult to compare with the studies evaluating specific subtypes of T cells (e.g., T helpers, Tregs, etc.). With that in mind we first review the data published for the total T cell population (CD3+), and then for individual subtypes of cells.

High density of tumor-infiltrating CD3+ T cells has been associated with favorable prognosis in various types of cancers. For instance, it was reported for pulmonary non-small cell carcinoma (Al-Shibli et al. 2010; Johnson et al. 2000), colorectal carcinoma (Ropponen et al. 1997; Galon et al. 2006), gastric carcinoma (Lee et al. 2008), ovarian carcinoma (Nelson 2008; Tomsova et al. 2008) and cervical cancers (Ancuta et al. 2009). Recently, Gooden and colleagues (2011) performed a meta-analysis of 33 large clinical studies and found a strong positive effect of CD3+ cells on cancer patients' survival (Gooden et al. 2011).

Among specific subtypes of tumor infiltrating lymphocytes, cytotoxic T cells (CD8+) have also been associated with a better patients' survival in many types of cancer, including pulmonary non-small cell carcinoma (Kawai et al. 2008; Dai et al.

2010; Al-Shibli et al. 2008; Zhuang et al. 2010), colorectal carcinoma (Funada et al. 2003; Naito et al. 1998; Diederichsen et al. 2003; Chiba et al. 2004), esophageal carcinoma (Schumacher et al. 2001), urothelial carcinoma (Sharma et al. 2007b), cholangiocellular carcinoma (Oshikiri et al. 2003), endometrial carcinoma (Kondratiev et al. 2004) and ovarian cancer (Sato et al. 2005). However, in other studies, CD8+ T cell density was not found to correlate with prognosis in pulmonary non-small cell carcinoma (Mori et al. 2000; Wakabayashi et al. 2003), esophageal squamous cell carcinoma (Ishibashi et al. 2006a), and in soft tissue sarcoma (Sorbye et al. 2011). In the meta-analysis of Gooden and colleagues (2011) CD8+ T cells had a positive effect of on patients' survival (Gooden et al. 2011).

Tumor-infiltrating T helper cells (CD4+) are not studied as extensively as cytotoxic CD8+ T cells, however, several reports indicate their favorable prognostic significance. Remarkably, this effect depends on a spatial distribution of cells. Thus, studying the prognostic role of epithelial and stromal CD4+ T cells in patients with resected non-small cell carcinoma, Al-Shibli and colleagues found that increasing numbers of CD4+ in tumor stroma, but not in cancer islets, correlated significantly with improved disease-specific survival (Al-Shibli et al. 2008). Other research groups reported similar results (Pelletier et al. 2001; Wakabayashi et al. 2003). High density of CD4+ T cells also correlated significantly with an improved survival in patients with soft tissue sarcoma (Sorbye et al. 2012b).

A few studies addressed clinical significance of T helper subtypes (Th1 and Th2). It is known than in terms of anti-tumor immunity, Th2 activation is less effective than Th1 activation (Gooden et al. 2011; Yu and Fu 2006). Recently, De Monte et al. analyzed these subtypes of T helper cells in pancreatic carcinoma. They found that Th2 cells (GATA-3+) were significantly more frequent than Th1 cells (T-bet+) and that Th2/Th1 ration was an independent predictor of a poor survival (De Monte et al. 2011).

In addition to the reports on the individual role of T cell types, several studies found a favorable prognostic effect of concurrent infiltration by CD8+ cell and CD4+ cells. Specifically, this effect was shown in non-small cell carcinoma (Hiraoka et al. 2006) and esophageal squamous cell carcinoma (Cho et al. 2003).

Another subset of CD4+ T cells variably present within the tumor environment are the CD4+ CD25+ regulatory T cells (Ichihara et al. 2003; Liyanage et al. 2002). They usually represent a small fraction of tumor-infiltrating lymphocytes (5–10 % of CD4+ cells), but may have a significant influence on tumor development (Gao et al. 2007; Sakaguchi et al. 2001; Woo et al. 2002; Bates et al. 2006). It has been shown that the amount of Treg cells are higher in tumors than in normal tissues due to an active recruitment of theses cells by factors present in the tumor microenvironment (Ishibashi et al. 2006b; Curiel et al. 2004). Accumulation of Tregs may be associated with disease progression (De Panfilis et al. 2008; Hussein et al. 2006; Miracco et al. 2007). High percentage of Treg cells in various neoplasms creates an immune-suppressive microenvironment that curbs anti-tumor immunity and promotes tumor growth (Gao et al. 2007; Woo et al. 2002; Bates et al. 2006).

The prognostic significance of Treg cells at the tumor site has been studied extensively showing, however, conflicting results. It has been associated with poor prognosis is some malignancies (Bates et al. 2006; Wolf et al. 2005; Perez et al. 2007; Shimizu et al. 2010; Petersen et al. 2006; Jordanova et al. 2008;Li et al. 2009b; Shen et al. 2010), favorable prognosis in other (Salama et al. 2009) or no clinical significance (Nosho et al. 2010; Ladanyi et al. 2010). For instance, Gooden and colleagues (Gooden et al. 2011) performed a meta-analysis of 22 studies analyzing the role of tumor-infiltrating Tregs, and 18 of these showed no statistically significant impact on overall, disease-specific or progression-free survival. They determined that the CD8/FoxP3 ratio, which was used in relatively few studies, correlates with a favorable outcome. These results underline the need to examine FoxP3 and CD8 simultaneously and analyze the cell density ratios in different types of tumor.

In summary, T lymphocytes represent a major population of tumor-infiltrating immune cells. They are present in all types of malignant neoplasms, infiltrating all of their spatial domains. Their diversity makes it difficult to evaluate the significance of these cells in tumor development, since individual subtypes of T cells have opposite effects on tumor progression. In general, high density of tumor-infiltrating T cells is associated with favorable prognosis, which can be probably attributed to Th1 helper cells and cytotoxic CD8+ T cells. The role of tumor-infiltrating Treg cells is controversial and needs to be further clarified.

3.11 Natural Killer Cells

Natural killer (NK) cells play a major role in the elimination of tumor cells that have lost self-MHC expression during a transformation process (Papamichail et al. 2004; Walzer et al. 2007; Orange and Ballas 2006; Karre 2002). Altered HLA-class I expression has been frequently observed in malignant tumors (Passlick et al. 1996; Restifo et al. 1993; Purdy and Campbell 2009). It is a tumor escape mechanism from the cytotoxic T cell response and therefore, NK cells may play an important role in the control and eradication of cancer cells. NK cell recognition of target cells is guided by the balance of activating and inhibitory signals given by different groups of surface receptors (Moretta et al. 2003; Moretta et al. 2002). Although originally named for their capacity to elicit cytotoxicity, NK cells are also a potent source of cytokines and chemokines, especially IFN-γ, TNF-α and GM-CSF. In addition to direct effects on tumor cells, these cytokines can promote the differentiation, activation and/or recruitment of other immune cells to the tumor site (Perussia 1996; Degli-Esposti and Smyth 2005; Di Santo 2008; French and Yokoyama 2003).

Morphologic identification of tumor-infiltrating NK cells requires the use of immunochemical stains. NK cells do not express B cell markers (CD20) and T cell markers (CD3), but can be detected by monoclonal antibodies targeting their specific antigens (e.g., CD56 or CD57) (Table 3.3) (Al-Shibli et al. 2009; Ishigami et al. 2000b). These markers are not unique for NK cells as the expression of CD56 was shown on cells with neuroendocrine differentiation and CD57 was found on some T-lineage lymphocytes (Focosi et al. 2010).

In general, NK cells density is low in human neoplasms, with the exception of some renal cell carcinomas (Ishigami et al. 2000b; Villegas et al. 2002; Coca et al. 1997; Schleypen et al. 2003; Schleypen et al. 2006; Ralfkiaer et al. 1987). Several studies have demonstrated that NK cells are primarily confined to the tumor stroma and to the interfaces between stromal cells and surrounding tumor cells. Remarkably, NK cells detected within tumor tissues did not appear to be in direct contact with cancerous cells (Al-Shibli et al. 2009; Carrega et al. 2008; Esendagli et al. 2008). This pattern of spatial distribution seems to conflict with an idea of direct NK cell interaction with malignant cells. In addition, experimental data reveal functional deficiency of tumor-infiltrating NK, likely caused by tumor-associated cytokines (Carrega et al. 2008; Doubrovina et al. 2003). Based on these findings, Carrega et al. (2008) suggested that NK cell cytokine production might be as relevant as cytotoxic functions in controlling cancer cell growth (Carrega et al. 2008).

In regard to correlation of NK cell density with the clinical course and prognosis, the majority of studies showed their favorable prognostic value. Specifically, it was demonstrated for gastric carcinoma (Ishigami et al. 2000b), colorectal carcinoma(Coca et al. 1997), pulmonary adeno- and squamous cell carcinoma (Al-Shibli et al. 2009; Villegas et al. 2002; Takanami et al. 2001). Confirming the importance of a spatial distribution of NK cells in tumor tissue, Al-Shibli et al. (2009) showed that high density of stromal NK cells was an independent, positive prognostic factor for disease-specific survival in patients with pulmonary carcinoma, whereas high density of NK cells within tumor islets was not (Al-Shibli et al. 2009). On the other hand, in soft tissue sarcomas there was no correlation of NK cell density in tumor or peritumoral tissue with prognosis (Sorbye et al. 2012c).

In summary, NK cells represent a small population of tumor-infiltrating lymphocytes, have predominantly stromal distribution in tumor tissue and in general are associated with a favorable prognosis, especially in tumors of epithelial origin (carcinomas).

3.12 Conclusions

Although extensive work is being done in the field of the tumor microenvironment and, specifically, immunoenvironment, it is clear that there is a significant gap between the profound knowledge on cancer cell biology and more limited knowledge on the roles of other cells in the tumor milieu. Morphologic analysis of cellular populations, performed by pathologists, adds significant value and agreeably complements available data received by immunologic, molecular and other methods. Such an analysis allows correlating the type and density of tumor-infiltrating leukocytes with pathologic characteristics of malignant neoplasm, its type, grade and stage. This correlation helps understand the involvement of individual types of immune cells in tumor development and progression. The most valuable part of morphologic evaluation of tumor-infiltrating leukocytes is determining their spatial distribution. As it was shown in numerous studies, the same cells in tumor islets and tumoral or peritumoral stroma may have different and often even opposite effects on tumor biology. In addition, it is now possible to assess the maturation stage and functional capacity of tumor-infiltrating leukocytes in conjunction with their spatial distribution.

Future directions in this field most likely will be addressing the following issues:

1. Functional alterations of tumor-infiltrating leukocytes that prevent them from eliminating malignant cells. With the development of new monoclonal antibodies for immunohistochemical evaluation of infiltrating immune cells, it becomes possible to ascertain more precisely the functional status of these cells and their level of maturation.
2. Interactions between different types and subtypes of tumor-infiltrating leukocytes. Increasing number of studies evaluate several types of immune cells simultaneously by applying different immunostains. This approach allows to estimate the relative densities of cells in the tumor (i.e., T cell/B cell or Teffector/Treg cell ratios), which provides clinically valuable information.
3. Correlations of tumor-infiltrating leukocytes density, spatial distribution and functional status with cancer therapy. The majority of clinico-pathologic correlations discussed above were performed on the surgically resected specimens of tumors, without consideration of chemo- or radio-therapy used before the resection (neo-adjuvant therapy). However, such considerations have a significant importance, since neo-adjuvant therapy may influence the parameters of the tumor microenvironment, including types, densities and functional characteristics of tumor-associated leukocytes.

All of this information is essential for determining both the biological and clinical significance of immune cells within the tumor microenvironment. Abundant clinical studies discussed above helped to clarify which types and subtypes of immune cells are beneficial and detrimental for a clinical course and outcome of particular malignant neoplasms. It opens a possibility to develop new methods of therapeutic interventions targeting "benign" cells of the tumor immunoenvironment in addition to malignant cells.

References

Acikalin MF, Oner U, Topcu I, Yasar B, Kiper H, Colak E (2005) Tumour angiogenesis and mast cell density in the prognostic assessment of colorectal carcinomas. Dig Liver Dis 37(3):162–169

Adelstein EH, Davis WE, Oxenhandler RW, Templer JW, Barrett BA (1978) Lymphocyte-tumor cell interaction in patients with head and neck cancer. Laryngoscope 88(4):575–581

Allen C, Hogg N (1987) Elevation of infiltrating mononuclear phagocytes in human colorectal tumors. J Natl Cancer Inst 78(3):465–470

Almholt K, Johnsen M (2003) Stromal cell involvement in cancer. Recent results in cancer research Fortschritte der Krebsforschung 162:31–42

Al-Shibli KI, Donnem T, Al-Saad S, Persson M, Bremnes RM, Busund LT (2008) Prognostic effect of epithelial and stromal lymphocyte infiltration in non-small cell lung cancer. Clin Cancer Res 14(16):5220–5227

Al-Shibli K, Al-Saad S, Donnem T, Persson M, Bremnes RM, Busund LT (2009) The prognostic value of intraepithelial and stromal innate immune system cells in non-small cell lung carcinoma. Histopathology 55(3):301–312

Al-Shibli K, Al-Saad S, Andersen S, Donnem T, Bremnes RM, Busund LT (2010) The prognostic value of intraepithelial and stromal CD3-, CD117- and CD138-positive cells in non-small cell lung carcinoma. Apmis 118(5):371–382

Ambe K, Mori M, Enjoji M (1989) S-100 protein-positive dendritic cells in colorectal adenocarcinomas. Distribution and relation to the clinical prognosis. Cancer 63(3):496–503

Amini RM, Aaltonen K, Nevanlinna H et al (2007) Mast cells and eosinophils in invasive breast carcinoma. BMC cancer 7:165

Ancuta E, Ancuta C, Zugun-Eloae F, Iordache C, Chirieac R, Carasevici E (2009) Predictive value of cellular immune response in cervical cancer. Rom J Morphol Embryol = Revue roumaine de morphologie et embryologie 50(4):651–655

Aoki M, Pawankar R, Niimi Y, Kawana S (2003) Mast cells in basal cell carcinoma express VEGF, IL-8 and RANTES. Int Arch Allergy Immunol 130(3):216–223

Ardi VC, Kupriyanova TA, Deryugina EI, Quigley JP (2007) Human neutrophils uniquely release TIMP-free MMP-9 to provide a potent catalytic stimulator of angiogenesis. Proc Natl Acad Sci USA 104(51):20262–20267

Ardi VC, Van den Steen PE, Opdenakker G, Schweighofer B, Deryugina EI, Quigley JP (2009) Neutrophil MMP-9 proenzyme, unencumbered by TIMP-1, undergoes efficient activation in vivo and catalytically induces angiogenesis via a basic fibroblast growth factor (FGF-2)/FGFR-2 pathway. J Biol Chem 284(38):25854–25866

Aso T, Ogawa Y, Naoe M, Fukagai T, Yoshida H, Kushima M (2004) Immunohistochemical analysis of CD83, CD8 and CD4 positive cells in renal cell carcinoma. Nihon Hinyokika Gakkai zasshi 95(4):645–650

Asselin-Paturel C, Boonstra A, Dalod M et al (2001) Mouse type I IFN-producing cells are immature APCs with plasmacytoid morphology. Nat Immunol 2(12):1144–1150

Baay M, Brouwer A, Pauwels P, Peeters M, Lardon F (2011) Tumor cells and tumor-associated macrophages: secreted proteins as potential targets for therapy. Clin Dev Immunol 2011:565187

Baj-Krzyworzeka M, Szatanek R, Weglarczyk K, Baran J, Zembala M (2007) Tumour-derived microvesicles modulate biological activity of human monocytes. Immunol Lett 113(2):76–82

Balkwill F, Mantovani A (2001) Inflammation and cancer: back to Virchow? Lancet 357(9255):539–545

Bates GJ, Fox SB, Han C et al (2006) Quantification of regulatory T cells enables the identification of high-risk breast cancer patients and those at risk of late relapse. J Clin Oncol 24(34):5373–5380

Bechara C, Chai H, Lin PH, Yao Q, Chen C (2007) Growth related oncogene-alpha (GRO-alpha): roles in atherosclerosis, angiogenesis and other inflammatory conditions. Med Sci Monit 13(6):RA87–RA90

Bell D, Chomarat P, Broyles D et al (1999) In breast carcinoma tissue, immature dendritic cells reside within the tumor, whereas mature dendritic cells are located in peritumoral areas. J Exp Med 190(10):1417–1426

Bellocq A, Antoine M, Flahault A et al (1998) Neutrophil alveolitis in bronchioloalveolar carcinoma: induction by tumor-derived interleukin-8 and relation to clinical outcome. Am J Pathol 152(1):83–92

Belotti D, Paganoni P, Manenti L et al (2003) Matrix metalloproteinases (MMP9 and MMP2) induce the release of vascular endothelial growth factor (VEGF) by ovarian carcinoma cells: implications for ascites formation. Cancer Res 63(17):5224–5229
Benitez-Bribiesca L, Wong A, Utrera D, Castellanos E (2001) The role of mast cell tryptase in neoangiogenesis of premalignant and malignant lesions of the uterine cervix. J Histochem Cytochem 49(8):1061–1062
Bergers G, Brekken R, McMahon G et al (2000) Matrix metalloproteinase-9 triggers the angiogenic switch during carcinogenesis. Nat Cell Biol 2(10):737–744
Bethwaite PB, Holloway LJ, Yeong ML, Thornton A (1993) Effect of tumour associated tissue eosinophilia on survival of women with stage IB carcinoma of the uterine cervix. J Clin Pathol 46(11):1016–1020
Bigotti G, Coli A, Castagnola D (1991) Distribution of Langerhans cells and HLA class II molecules in prostatic carcinomas of different histopathological grade. Prostate 19(1):73–87
Bingle L, Brown NJ, Lewis CE (2002) The role of tumour-associated macrophages in tumour progression: implications for new anticancer therapies. J Pathol 196(3):254–265
Bostrom SG, Hart WR (1981) Carcinomas of the cervix with intense stromal eosinophilia. Cancer 47(12):2887–2893
Bratt J, Palmblad J (1997) Cytokine-induced neutrophil-mediated injury of human endothelial cells. J Immunol 159(2):912–918
Cabanillas-Saez A, Schalper JA, Nicovani SM, Rudolph MI (2002) Characterization of mast cells according to their content of tryptase and chymase in normal and neoplastic human uterine cervix. Int J Gynecol Cancer 12(1):92–98
Cai XY, Gao Q, Qiu SJ et al (2006) Dendritic cell infiltration and prognosis of human hepatocellular carcinoma. J Cancer Res Clin Oncol 132(5):293–301
Cai D, Cao J, Li Z et al (2009) Up-regulation of bone marrow stromal protein 2 (BST2) in breast cancer with bone metastasis. BMC cancer 9:102
Cao W, Bover L (2010) Signaling and ligand interaction of ILT7: receptor-mediated regulatory mechanisms for plasmacytoid dendritic cells. Immunol Rev 234(1):163–176
Carrega P, Morandi B, Costa R et al (2008) Natural killer cells infiltrating human nonsmall-cell lung cancer are enriched in CD56 bright CD16(-) cells and display an impaired capability to kill tumor cells. Cancer 112(4):863–875
Caruso RA, Bellocco R, Pagano M, Bertoli G, Rigoli L, Inferrera C (2002) Prognostic value of intratumoral neutrophils in advanced gastric carcinoma in a high-risk area in northern Italy. Mod Pathol 15(8):831–837
Caux C, Ait-Yahia S, Chemin K et al (2000) Dendritic cell biology and regulation of dendritic cell trafficking by chemokines. Springer Semin Immunopathol 22(4):345–369
Chambers SK, Kacinski BM, Ivins CM, Carcangiu ML (1997) Overexpression of epithelial macrophage colony-stimulating factor (CSF-1) and CSF-1 receptor: a poor prognostic factor in epithelial ovarian cancer, contrasted with a protective effect of stromal CSF-1. Clin Cancer Res 3(6):999–1007
Chaperot L, Blum A, Manches O et al (2006) Virus or TLR agonists induce TRAIL-mediated cytotoxic activity of plasmacytoid dendritic cells. J Immunol 176(1):248–255
Charles J, Di Domizio J, Salameire D et al (2010) Characterization of circulating dendritic cells in melanoma: role of CCR6 in plasmacytoid dendritic cell recruitment to the tumor. J Invest Dermatol 130(6):1646–1656
Chen WK, Chen FJ, Zeng ZY et al (2005) [Expression and clinical significance of CD1a+ dendritic cells in glottic squamous cell carcinoma]. Ai zheng = Aizheng = Chin J Cancer 24(9):1124–1126
Chiba T, Ohtani H, Mizoi T et al (2004) Intraepithelial CD8+ T-cell-count becomes a prognostic factor after a longer follow-up period in human colorectal carcinoma: possible association with suppression of micrometastasis. Br J Cancer 91(9):1711–1717
Chin Y, Janseens J, Vandepitte J, Vandenbrande J, Opdebeek L, Raus J (1992) Phenotypic analysis of tumor-infiltrating lymphocytes from human breast cancer. Anticancer Res 12(5):1463–1466

Ch'ng S, Wallis RA, Yuan L, Davis PF, Tan ST (2006) Mast cells and cutaneous malignancies. Mod Pathol 19(1):149–159

Cho Y, Miyamoto M, Kato K et al (2003) CD4+ and CD8+ T cells cooperate to improve prognosis of patients with esophageal squamous cell carcinoma. Cancer Res 63(7):1555–1559

Coca S, Perez-Piqueras J, Martinez D et al (1997) The prognostic significance of intratumoral natural killer cells in patients with colorectal carcinoma. Cancer 79(12):2320–2328

Coffelt SB, Lewis CE, Naldini L, Brown JM, Ferrara N, De Palma M (2010) Elusive identities and overlapping phenotypes of proangiogenic myeloid cells in tumors. Am J Pathol 176(4):1564–1576

Colasante A, Poletti V, Rosini S, Ferracini R, Musiani P (1993) Langerhans cells in Langerhans cell histiocytosis and peripheral adenocarcinomas of the lung. Am Rev Respir Dis 148(3):752–759

Colasante A, Castrilli G, Aiello FB, Brunetti M, Musiani P (1995) Role of cytokines in distribution and differentiation of dendritic cell/Langerhans' cell lineage in human primary carcinomas of the lung. Hum Pathol 26(8):866–872

Coli A, Bigotti G, Cione T (1990) Langerhans cell and HLA class II molecules distribution in lung carcinoma. J Nucl Med Allied Sci 34(4 Suppl):185–189

Condeelis J, Pollard JW (2006) Macrophages: obligate partners for tumor cell migration, invasion, and metastasis. Cell 124(2):263–266

Coppola D, Fu L, Nicosia SV, Kounelis S, Jones M (1998) Prognostic significance of p53, bcl-2, vimentin, and S100 protein-positive Langerhans cells in endometrial carcinoma. Hum Pathol 29(5):455–462

Coronella-Wood JA, Hersh EM (2003) Naturally occurring B-cell responses to breast cancer. Cancer Immunol Immunother 52(12):715–738

Coussens LM, Werb Z (1996) Matrix metalloproteinases and the development of cancer. Chem Biol 3(11):895–904

Coussens LM, Werb Z (2001) Inflammatory cells and cancer: think different! J Exp Med 193(6):F23–F26

Coussens LM, Werb Z (2002) Inflammation and cancer. Nature 420(6917):860–867

Coussens LM, Raymond WW, Bergers G et al (1999) Inflammatory mast cells up-regulate angiogenesis during squamous epithelial carcinogenesis. Genes Dev 13(11):1382–1397

Coussens LM, Tinkle CL, Hanahan D, Werb Z (2000) MMP-9 supplied by bone marrow-derived cells contributes to skin carcinogenesis. Cell 103(3):481–490

Coventry BJ, Morton J (2003) CD1a-positive infiltrating-dendritic cell density and 5-year survival from human breast cancer. Br J Cancer 89(3):533–538

Coventry BJ, Lee PL, Gibbs D, Hart DN (2002) Dendritic cell density and activation status in human breast cancer—CD1a, CMRF-44, CMRF-56 and CD-83 expression. Br J Cancer 86(4):546–551

Crivellato E, Nico B, Ribatti D (2009) Mast cell contribution to tumor angiogenesis: a clinical approach. Eur Cytokine Netw 20(4):197–206

Curiel TJ, Coukos G, Zou L et al (2004) Specific recruitment of regulatory T cells in ovarian carcinoma fosters immune privilege and predicts reduced survival. Nat Med 10(9):942–949

Dabiri S, Huntsman D, Makretsov N et al (2004) The presence of stromal mast cells identifies a subset of invasive breast cancers with a favorable prognosis. Mod Pathol 17(6):690–695

Dadabayev AR, Sandel MH, Menon AG et al (2004) Dendritic cells in colorectal cancer correlate with other tumor-infiltrating immune cells. Cancer Immunol Immunother 53(11):978–986

Dai F, Liu L, Che G et al (2010) The number and microlocalization of tumor-associated immune cells are associated with patient's survival time in non-small cell lung cancer. BMC Cancer 10:220

Dallal RM, Christakos P, Lee K, Egawa S, Son YI, Lotze MT (2002) Paucity of dendritic cells in pancreatic cancer. Surgery 131(2):135–138

Dallegri F, Ottonello L, Ballestrero A et al (1991) Tumor cell lysis by activated human neutrophils: analysis of neutrophil-delivered oxidative attack and role of leukocyte function-associated antigen 1. Inflammation 15(1):15–30

De Monte L, Reni M, Tassi E et al (2011) Intratumor T helper type 2 cell infiltrate correlates with cancer-associated fibroblast thymic stromal lymphopoietin production and reduced survival in pancreatic cancer. J Exp Med 208(3):469–478

De Palma M, Venneri MA, Galli R et al (2005) Tie2 identifies a hematopoietic lineage of proangiogenic monocytes required for tumor vessel formation and a mesenchymal population of pericyte progenitors. Cancer Cell 8(3):211–226

De Panfilis G, Campanini N, Santini M et al (2008) Phase- and stage-related proportions of T cells bearing the transcription factor FOXP3 infiltrate primary melanoma. J Invest Dermatol 128(3):676–684

de Visser KE, Eichten A, Coussens LM (2006) Paradoxical roles of the immune system during cancer development. Nat Rev 6(1):24–37

Degli-Esposti MA, Smyth MJ (2005) Close encounters of different kinds: dendritic cells and NK cells take centre stage. Nat Rev Immunol 5(2):112–124

Della Porta M, Danova M, Rigolin GM et al (2005) Dendritic cells and vascular endothelial growth factor in colorectal cancer: correlations with clinicobiological findings. Oncology 68(2–3):276–284

Di Carlo E, Coletti A, Modesti A, Giovarelli M, Forni G, Musiani P (1998) Local release of interleukin-10 by transfected mouse adenocarcinoma cells exhibits pro- and anti-inflammatory activity and results in a delayed tumor rejection. Eur Cytokine Netw 9(1):61–68

Di Carlo E, Forni G, Lollini P, Colombo MP, Modesti A, Musiani P (2001) The intriguing role of polymorphonuclear neutrophils in antitumor reactions. Blood 97(2):339–345

Di Santo JP (2008) Natural killer cells: diversity in search of a niche. Nat Immunol 9(5):473–475

Diederichsen AC, Hjelmborg JB, Christensen PB, Zeuthen J, Fenger C (2003) Prognostic value of the CD4+/CD8+ ratio of tumour infiltrating lymphocytes in colorectal cancer and HLA-DR expression on tumour cells. Cancer Immunol Immunother 52(7):423–428

Dieu-Nosjean MC, Antoine M, Danel C et al (2008) Long-term survival for patients with non-small-cell lung cancer with intratumoral lymphoid structures. J Clin Oncol 26(27):4410–4417

Dorta RG, Landman G, Kowalski LP, Lauris JR, Latorre MR, Oliveira DT (2002) Tumour-associated tissue eosinophilia as a prognostic factor in oral squamous cell carcinomas. Histopathology 41(2):152–157

Doubrovina ES, Doubrovin MM, Vider E et al (2003) Evasion from NK cell immunity by MHC class I chain-related molecules expressing colon adenocarcinoma. J Immunol 171(12):6891–6899

Dunn GP, Koebel CM, Schreiber RD (2006) Interferons, immunity and cancer immunoediting. Nat Rev Immunol 6(11):836–848

Ebrahem Q, Chaurasia SS, Vasanji A et al (2010) Cross-talk between vascular endothelial growth factor and matrix metalloproteinases in the induction of neovascularization in vivo. Am J Pathol 176(1):496–503

Eck M, Schmausser B, Scheller K, Brandlein S, Muller-Hermelink HK (2003) Pleiotropic effects of CXC chemokines in gastric carcinoma: differences in CXCL8 and CXCL1 expression between diffuse and intestinal types of gastric carcinoma. Clin Exp Immunol 134(3):508–515

Edge SB (ed) (2010) AJCC cancer staging manual/American Joint Committee on Cancer, 7th edn. Springer, New York

Elpek GO, Gelen T, Aksoy NH et al (2001) The prognostic relevance of angiogenesis and mast cells in squamous cell carcinoma of the oesophagus. J Clin Pathol 54(12):940–944

Esendagli G, Bruderek K, Goldmann T et al (2008) Malignant and non-malignant lung tissue areas are differentially populated by natural killer cells and regulatory T cells in non-small cell lung cancer. Lung cancer (Amsterdam, Netherlands) 59(1): 32–40

Espinosa I, Jose Carnicer M, Catasus L et al (2010) Myometrial invasion and lymph node metastasis in endometrioid carcinomas: tumor-associated macrophages, microvessel density, and HIF1A have a crucial role. Am J Surg Pathol 34(11):1708–1714

Esposito I, Menicagli M, Funel N et al (2004) Inflammatory cells contribute to the generation of an angiogenic phenotype in pancreatic ductal adenocarcinoma. J Clin Pathol 57(6):630–636

Feijoo E, Alfaro C, Mazzolini G et al (2005) Dendritic cells delivered inside human carcinomas are sequestered by interleukin-8. Int J Cancer 116(2):275–281
Fischer C, Jonckx B, Mazzone M et al (2007) Anti-PlGF inhibits growth of VEGF(R)-inhibitor-resistant tumors without affecting healthy vessels. Cell 131(3):463–475
Focosi D, Bestagno M, Burrone O, Petrini M (2010) CD57+ T lymphocytes and functional immune deficiency. J Leukoc Biol 87(1):107–116
Frederick MJ, Henderson Y, Xu X et al (2000) In vivo expression of the novel CXC chemokine BRAK in normal and cancerous human tissue. Am J Pathol 156(6):1937–1950
French AR, Yokoyama WM (2003) Natural killer cells and viral infections. Curr Opin Immunol 15(1):45–51
Fujii M, Yamashita T, Ishiguro R, Tashiro M, Kameyama K (2002) Significance of epidermal growth factor receptor and tumor associated tissue eosinophilia in the prognosis of patients with nasopharyngeal carcinoma. Auris Nasus Larynx 29(2):175–181
Funada Y, Noguchi T, Kikuchi R, Takeno S, Uchida Y, Gabbert HE (2003) Prognostic significance of CD8 + T cell and macrophage peritumoral infiltration in colorectal cancer. Oncol Rep 10(2):309–313
Furihata M, Ono Y, Ichikawa K, Tomita S, Fujimori T, Kubota K (2005) Prognostic significance of CD83 positive, mature dendritic cells in the gallbladder carcinoma. Oncol Rep 14(2):353–356
Gabrilovich D (2004) Mechanisms and functional significance of tumour-induced dendritic-cell defects. Nat Rev Immunol 4(12):941–952
Gabrilovich DI, Chen HL, Girgis KR et al (1996) Production of vascular endothelial growth factor by human tumors inhibits the functional maturation of dendritic cells. Nat Med 2(10):1096–1103
Galon J, Costes A, Sanchez-Cabo F et al (2006) Type, density, and location of immune cells within human colorectal tumors predict clinical outcome. Science (New York) 313(5795): 1960–1964
Gao Q, Qiu SJ, Fan J et al (2007) Intratumoral balance of regulatory and cytotoxic T cells is associated with prognosis of hepatocellular carcinoma after resection. J Clin Oncol 25(18):2586–2593
Giraudo E, Inoue M, Hanahan D (2004) An amino-bisphosphonate targets MMP-9-expressing macrophages and angiogenesis to impair cervical carcinogenesis. J Clin Investig 114(5):623–633
Goldsmith MM, Belchis DA, Cresson DH, Merritt WD 3rd, Askin FB (1992) The importance of the eosinophil in head and neck cancer. Otolaryngol Head Neck Surg 106(1):27–33
Gooch JL, Lee AV, Yee D (1998) Interleukin 4 inhibits growth and induces apoptosis in human breast cancer cells. Cancer Res 58(18):4199–4205
Gooden MJ, de Bock GH, Leffers N, Daemen T, Nijman HW (2011) The prognostic influence of tumour-infiltrating lymphocytes in cancer: a systematic review with meta-analysis. Br J Cancer 105(1):93–103
Gordon S (2003) Alternative activation of macrophages. Nat Rev Immunol 3(1):23–35
Gounaris E, Erdman SE, Restaino C et al (2007) Mast cells are an essential hematopoietic component for polyp development. Proc Natl Acad Sci USA 104(50):19977–19982
Gregory AD, Houghton AM (2011) Tumor-associated neutrophils: new targets for cancer therapy. Cancer Res 71(7):2411–2416
Griffiths AP, Rice A, Dixon MF (1998) Anaplastic gastric adenocarcinoma with extensive neutrophilic infiltration. Histopathology 33(4):392–393
Grimbaldeston MA, Metz M, Yu M, Tsai M, Galli SJ (2006) Effector and potential immunoregulatory roles of mast cells in IgE-associated acquired immune responses. Curr Opin Immunol 18(6):751–760
Grutzkau A, Kruger-Krasagakes S, Baumeister H et al (1998) Synthesis, storage, and release of vascular endothelial growth factor/vascular permeability factor (VEGF/VPF) by human mast cells: implications for the biological significance of VEGF206. Mol Biol Cell 9(4):875–884

Gulubova M, Vlaykova T (2009) Prognostic significance of mast cell number and microvascular density for the survival of patients with primary colorectal cancer. J Gastroenterol Hepatol 24(7):1265–1275
Hartmann E, Wollenberg B, Rothenfusser S et al (2003) Identification and functional analysis of tumor-infiltrating plasmacytoid dendritic cells in head and neck cancer. Cancer Res 63(19):6478–6487
Hashimoto SI, Suzuki T, Nagai S, Yamashita T, Toyoda N, Matsushima K (2000) Identification of genes specifically expressed in human activated and mature dendritic cells through serial analysis of gene expression. Blood 96(6):2206–2214
Hayati AR, Zulkarnaen M (2007) An immunohistochemical study of CD1a and CD83-positive infiltrating dendritic cell density in cervical neoplasia. Int J Gynecol Pathol 26(1):83–88
Hillen F, Baeten CI, van de Winkel A et al (2008) Leukocyte infiltration and tumor cell plasticity are parameters of aggressiveness in primary cutaneous melanoma. Cancer Immunol Immunother 57(1):97–106
Hillenbrand EE, Neville AM, Coventry BJ (1999) Immunohistochemical localization of CD1a-positive putative dendritic cells in human breast tumours. Br J Cancer 79(5–6):940–944
Hiraoka K, Miyamoto M, Cho Y et al (2006) Concurrent infiltration by CD8+ T cells and CD4+ T cells is a favourable prognostic factor in non-small-cell lung carcinoma. Br J Cancer 94(2):275–280
Hoeffel G, Ripoche AC, Matheoud D et al (2007) Antigen crosspresentation by human plasmacytoid dendritic cells. Immunity 27(3):481–492
Hori S, Nomura T, Sakaguchi S (2003) Control of regulatory T cell development by the transcription factor Foxp3. Science (New York) 299(5609): 1057–1061
Houghton AM (2010) The paradox of tumor-associated neutrophils: fueling tumor growth with cytotoxic substances. Cell cycle (Georgetown, Tex) 9(9):1732–1737
Houghton AM, Rzymkiewicz DM, Ji H et al (2010) Neutrophil elastase-mediated degradation of IRS-1 accelerates lung tumor growth. Nat Med 16(2):219–223
Hromas R, Broxmeyer HE, Kim C et al (1999) Cloning of BRAK, a novel divergent CXC chemokine preferentially expressed in normal versus malignant cells. Biochem Biophys Res Commun 255(3):703–706
Huang B, Lei Z, Zhang GM et al (2008) SCF-mediated mast cell infiltration and activation exacerbate the inflammation and immunosuppression in tumor microenvironment. Blood 112(4):1269–1279
Hubert P, Caberg JH, Gilles C et al (2005) E-cadherin-dependent adhesion of dendritic and Langerhans cells to keratinocytes is defective in cervical human papillomavirus-associated (pre)neoplastic lesions. J Pathol 206(3):346–355
Humphreys TR, Monteiro MR, Murphy GF (2000) Mast cells and dendritic cells in basal cell carcinoma stroma. Dermatol Surg 26(3):200–203; discussion 3–4
Hussein MR, Elsers DA, Fadel SA, Omar AE (2006) Immunohistological characterisation of tumour infiltrating lymphocytes in melanocytic skin lesions. J Clin Pathol 59(3):316–324
Iamaroon A, Pongsiriwet S, Jittidecharaks S, Pattanaporn K, Prapayasatok S, Wanachantararak S (2003) Increase of mast cells and tumor angiogenesis in oral squamous cell carcinoma. J Oral Pathol Med 32(4):195–199
Ichihara F, Kono K, Takahashi A, Kawaida H, Sugai H, Fujii H (2003) Increased populations of regulatory T cells in peripheral blood and tumor-infiltrating lymphocytes in patients with gastric and esophageal cancers. Clin Cancer Res 9(12):4404–4408
Inoshima N, Nakanishi Y, Minami T et al (2002) The influence of dendritic cell infiltration and vascular endothelial growth factor expression on the prognosis of non-small cell lung cancer. Clin Cancer Res 8(11):3480–3486
Inoue K, Furihata M, Ohtsuki Y, Fujita Y (1993) Distribution of S-100 protein-positive dendritic cells and expression of HLA-DR antigen in transitional cell carcinoma of the urinary bladder in relation to tumour progression and prognosis. Virchows Arch 422(5):351–355

Ioachim-Velogianni E, Stavropoulos NE, Kitsiou E, Stefanaki S, Agnantis NJ (1995) Distribution of CD1a-positive Langerhans cells and lymphocyte subsets in transitional cell carcinoma of the urinary bladder. An immunohistological study on frozen sections. J Pathol 177(4):401–406
Ishibashi S, Ohashi Y, Suzuki T et al (2006a) Tumor-associated tissue eosinophilia in human esophageal squamous cell carcinoma. Anticancer Res 26(2B):1419–1424
Ishibashi Y, Tanaka S, Tajima K, Yoshida T, Kuwano H (2006b) Expression of Foxp3 in non-small cell lung cancer patients is significantly higher in tumor tissues than in normal tissues, especially in tumors smaller than 30 mm. Oncol Rep 15(5):1315–1319
Ishigami S, Natsugoe S, Tokuda K et al (2000a) Clinical impact of intratumoral natural killer cell and dendritic cell infiltration in gastric cancer. Cancer Lett 159(1):103–108
Ishigami S, Natsugoe S, Tokuda K et al (2000b) Prognostic value of intratumoral natural killer cells in gastric carcinoma. Cancer 88(3):577–583
Ito T, Yang M, Wang YH et al (2007) Plasmacytoid dendritic cells prime IL-10-producing T regulatory cells by inducible costimulator ligand. J Exp Med 204(1):105–115
Iwamoto M, Shinohara H, Miyamoto A et al (2003) Prognostic value of tumor-infiltrating dendritic cells expressing CD83 in human breast carcinomas. Int J Cancer 104(1):92–97
Jahrsdorfer B, Vollmer A, Blackwell SE et al (2010) Granzyme B produced by human plasmacytoid dendritic cells suppresses T-cell expansion. Blood 115(6):1156–1165
Jensen HK, Donskov F, Marcussen N, Nordsmark M, Lundbeck F, von der Maase H (2009) Presence of intratumoral neutrophils is an independent prognostic factor in localized renal cell carcinoma. J Clin Oncol 27(28):4709–4717
Ji H, Houghton AM, Mariani TJ et al (2006) K-ras activation generates an inflammatory response in lung tumors. Oncogene 25(14):2105–2112
Jia J, Wang Z, Li X, Wang Z, Wang X (2012) Morphological characteristics and co-stimulatory molecule (CD80, CD86, CD40) expression in tumor infiltrating dendritic cells in human endometrioid adenocarcinoma. Eur J Obstet Gynecol Reprod Biol 160(2):223–227
Jodele S, Chantrain CF, Blavier L et al (2005) The contribution of bone marrow-derived cells to the tumor vasculature in neuroblastoma is matrix metalloproteinase-9 dependent. Cancer Res 65(8):3200–3208
Johnson SK, Kerr KM, Chapman AD et al (2000) Immune cell infiltrates and prognosis in primary carcinoma of the lung. Lung Cancer (Amsterdam, Netherlands) 27(1): 27–35
Jordanova ES, Gorter A, Ayachi O et al (2008) Human leukocyte antigen class I, MHC class I chain-related molecule A, and CD8+/regulatory T-cell ratio: which variable determines survival of cervical cancer patients? Clin Cancer Res 14(7):2028–2035
Juremalm M, Nilsson G (2005) Chemokine receptor expression by mast cells. Chem Immunol Allergy 87:130–144
Kang JC, Chen JS, Lee CH, Chang JJ, Shieh YS (2010) Intratumoral macrophage counts correlate with tumor progression in colorectal cancer. J Surg Oncol 102(3):242–248
Kaplan RN, Riba RD, Zacharoulis S et al (2005) VEGFR1-positive haematopoietic bone marrow progenitors initiate the pre-metastatic niche. Nature 438(7069):820–827
Karja V, Aaltomaa S, Lipponen P, Isotalo T, Talja M, Mokka R (2005) Tumour-infiltrating lymphocytes: a prognostic factor of PSA-free survival in patients with local prostate carcinoma treated by radical prostatectomy. Anticancer Res 25(6C):4435–4438
Karre K (2002) NK cells, MHC class I molecules and the missing self. Scand J Immunol 55(3):221–228
Katsenelson NS, Shurin GV, Bykovskaia SN, Shogan J, Shurin MR (2001) Human small cell lung carcinoma and carcinoid tumor regulate dendritic cell maturation and function. Mod Pathol 14(1):40–45
Kawahara A, Hattori S, Akiba J et al (2010) Infiltration of thymidine phosphorylase-positive macrophages is closely associated with tumor angiogenesis and survival in intestinal type gastric cancer. Oncol Rep 24(2):405–415
Kawai O, Ishii G, Kubota K et al (2008) Predominant infiltration of macrophages and CD8(+) T Cells in cancer nests is a significant predictor of survival in stage IV nonsmall cell lung cancer. Cancer 113(6):1387–1395

Kessenbrock K, Plaks V, Werb Z (2010) Matrix metalloproteinases: regulators of the tumor microenvironment. Cell 141(1):52–67
Khorana AA, Ryan CK, Cox C, Eberly S, Sahasrabudhe DM (2003) Vascular endothelial growth factor, CD68, and epidermal growth factor receptor expression and survival in patients with stage II and Stage III colon carcinoma: a role for the host response in prognosis. Cancer 97(4):960–968
Kikuchi K, Kusama K, Taguchi K et al (2002) Dendritic cells in human squamous cell carcinoma of the oral cavity. Anticancer Res 22(2A):545–557
Kluger HM, Dolled-Filhart M, Rodov S, Kacinski BM, Camp RL, Rimm DL (2004) Macrophage colony-stimulating factor-1 receptor expression is associated with poor outcome in breast cancer by large cohort tissue microarray analysis. Clin Cancer Res 10(1 Pt 1):173–177
Komohara Y, Hasita H, Ohnishi K et al (2011) Macrophage infiltration and its prognostic relevance in clear cell renal cell carcinoma. Cancer Sci 102(7):1424–1431
Kondo K, Muramatsu M, Okamoto Y et al (2006) Expression of chymase-positive cells in gastric cancer and its correlation with the angiogenesis. J Surg Oncol 93(1):36–42; discussion-3
Kondratiev S, Sabo E, Yakirevich E, Lavie O, Resnick MB (2004) Intratumoral CD8+ T lymphocytes as a prognostic factor of survival in endometrial carcinoma. Clin Cancer Res 10(13):4450–4456
Kurabayashi A, Furihata M, Matsumoto M, Hayashi H, Ohtsuki Y (2004) Distribution of tumor-infiltrating dendritic cells in human non-small cell lung carcinoma in relation to apoptosis. Pathol Int 54(5):302–310
Kurahara H, Shinchi H, Mataki Y et al (2009) Significance of M2-polarized tumor-associated macrophage in pancreatic cancer. J Surg Res 167(2):e211–e219
Labidi-Galy SI, Sisirak V, Meeus P et al (2011) Quantitative and functional alterations of plasmacytoid dendritic cells contribute to immune tolerance in ovarian cancer. Cancer Res 71(16):5423–5434
Ladanyi A, Somlai B, Gilde K, Fejos Z, Gaudi I, Timar J (2004) T-cell activation marker expression on tumor-infiltrating lymphocytes as prognostic factor in cutaneous malignant melanoma. Clin Cancer Res 10(2):521–530
Ladanyi A, Kiss J, Somlai B et al (2007) Density of DC-LAMP(+) mature dendritic cells in combination with activated T lymphocytes infiltrating primary cutaneous melanoma is a strong independent prognostic factor. Cancer Immunol Immunother 56(9):1459–1469
Ladanyi A, Mohos A, Somlai B et al (2010) FOXP3+ cell density in primary tumor has no prognostic impact in patients with cutaneous malignant melanoma. Pathol Oncol Res 16(3):303–309
Lee TK, Horner RD, Silverman JF, Chen YH, Jenny C, Scarantino CW (1989) Morphometric and morphologic evaluations in stage III non-small cell lung cancers. Prognostic significance of quantitative assessment of infiltrating lymphoid cells. Cancer 63(2):309–316
Lee CH, Espinosa I, Vrijaldenhoven S et al (2008) Prognostic significance of macrophage infiltration in leiomyosarcomas. Clin Cancer Res 14(5):1423–1430
Lespagnard L, Gancberg D, Rouas G et al (1999) Tumor-infiltrating dendritic cells in adenocarcinomas of the breast: a study of 143 neoplasms with a correlation to usual prognostic factors and to clinical outcome. Int J Cancer 84(3):309–314
Lewis CE, Pollard JW (2006) Distinct role of macrophages in different tumor microenvironments. Cancer Res 66(2):605–612
Li YW, Qiu SJ, Fan J et al (2009a) Tumor-infiltrating macrophages can predict favorable prognosis in hepatocellular carcinoma after resection. J Cancer Res Clin Oncol 135(3):439–449
Li JF, Chu YW, Wang GM et al (2009b) The prognostic value of peritumoral regulatory T cells and its correlation with intratumoral cyclooxygenase-2 expression in clear cell renal cell carcinoma. BJU Int 103(3):399–405
Lim KH, Telisinghe PU, Abdullah MS, Ramasamy R (2010) Possible significance of differences in proportions of cytotoxic T cells and B-lineage cells in the tumour-infiltrating lymphocytes of typical and atypical medullary carcinomas of the breast. Cancer Immun 10:3
Liu YJ (2005) IPC: professional type 1 interferon-producing cells and plasmacytoid dendritic cell precursors. Annu Rev Immunol 23:275–306

Liu K, Victora GD, Schwickert TA et al (2009) In vivo analysis of dendritic cell development and homeostasis. Science (New York, NY) 324(5925):392–397
Liyanage UK, Moore TT, Joo HG et al (2002) Prevalence of regulatory T cells is increased in peripheral blood and tumor microenvironment of patients with pancreas or breast adenocarcinoma. J Immunol 169(5):2756–2761
Lowe DG (1988) Carcinoma of the cervix with massive eosinophilia. Br J Obstet Gynaecol 95(4):393–401
Lowe D, Jorizzo J, Hutt MS (1981a) Tumour-associated eosinophilia: a review. J Clin Pathol 34(12):1343–1348
Lowe D, Jorizzo J, Chiphangwi J, Hutt MS (1981b) Cervical carcinoma in Malawi: a histopathologic study of 260 cases. Cancer 47(10):2493–2495
Lundy SK, Killer B (2009) Lymphocytes: the evidence and the potential. Inflamm Res 58(7):345–357
Ma XL, Weyrich AS, Lefer DJ, Lefer AM (1993) Diminished basal nitric oxide release after myocardial ischemia and reperfusion promotes neutrophil adherence to coronary endothelium. Circ Res 72(2):403–412
Ma J, Liu L, Che G, Yu N, Dai F, You Z (2010) The M1 form of tumor-associated macrophages in non-small cell lung cancer is positively associated with survival time. BMC cancer 10:112
Ma Y, Shurin GV, Gutkin DW, Shurin MR (2012) Tumor associated regulatory dendritic cells. Semin Cancer Biol 22(4):298–306
Maltby S, Khazaie K, McNagny KM (2009) Mast cells in tumor growth: angiogenesis, tissue remodelling and immune-modulation. Biochim Biophys Acta 1796(1):19–26
Mantovani A, Sozzani S, Locati M, Allavena P, Sica A (2002) Macrophage polarization: tumor-associated macrophages as a paradigm for polarized M2 mononuclear phagocytes. Trends Immunol 23(11):549–555
Mantovani A, Sica A, Sozzani S, Allavena P, Vecchi A, Locati M (2004) The chemokine system in diverse forms of macrophage activation and polarization. Trends Immunol 25(12):677–686
Mantovani A, Sica A, Locati M (2005) Macrophage polarization comes of age. Immunity 23(4):344–346
Mantovani A, Sica A, Locati M (2007) New vistas on macrophage differentiation and activation. Eur J Immunol 37(1):14–16
Marcus B, Arenberg D, Lee J et al (2004) Prognostic factors in oral cavity and oropharyngeal squamous cell carcinoma. Cancer 101(12):2779–2787
Marsigliante S, Biscozzo L, Marra A et al (1999) Computerised counting of tumour infiltrating lymphocytes in 90 breast cancer specimens. Cancer Lett 139(1):33–41
Matta BM, Castellaneta A, Thomson AW (2010) Tolerogenic plasmacytoid DC. Eur J Immunol 40(10):2667–2676
Mazzucchelli L, Loetscher P, Kappeler A et al (1996) Monocyte chemoattractant protein-1 gene expression in prostatic hyperplasia and prostate adenocarcinoma. Am J Pathology 149(2):501–509
Meininger CJ (1995) Mast cells and tumor-associated angiogenesis. Chem Immunol 62:239–257
Meininger CJ, Yano H, Rottapel R, Bernstein A, Zsebo KM, Zetter BR (1992) The c-kit receptor ligand functions as a mast cell chemoattractant. Blood 79(4):958–963
Michielsen AJ, Hogan AE, Marry J et al (2011) Tumour tissue microenvironment can inhibit dendritic cell maturation in colorectal cancer. PLoS ONE 6(11):e27944
Milne K, Kobel M, Kalloger SE et al (2009) Systematic analysis of immune infiltrates in high-grade serous ovarian cancer reveals CD20, FoxP3 and TIA-1 as positive prognostic factors. PLoS ONE 4(7):e6412
Mimura K, Kono K, Takahashi A, Kawaguchi Y, Fujii H (2007) Vascular endothelial growth factor inhibits the function of human mature dendritic cells mediated by VEGF receptor-2. Cancer Immunol Immunother 56(6):761–770
Miracco C, Mourmouras V, Biagioli M et al (2007) Utility of tumour-infiltrating CD25+ FOXP3+ regulatory T cell evaluation in predicting local recurrence in vertical growth phase cutaneous melanoma. Oncol Rep 18(5):1115–1122

Miyagawa S, Soeda J, Takagi S, Miwa S, Ichikawa E, Noike T (2004) Prognostic significance of mature dendritic cells and factors associated with their accumulation in metastatic liver tumors from colorectal cancer. Hum Pathol 35(11):1392–1396

Moezzi J, Gopalswamy N, Haas RJ Jr, Markert RJ, Suryaprasad S, Bhutani MS (2000) Stromal eosinophilia in colonic epithelial neoplasms. Am J Gastroenterol 95(2):520–523

Moretta L, Bottino C, Pende D, Mingari MC, Biassoni R, Moretta A (2002) Human natural killer cells: their origin, receptors and function. Eur J Immunol 32(5):1205–1211

Moretta L, Ferlazzo G, Mingari MC, Melioli G, Moretta A (2003) Human natural killer cell function and their interactions with dendritic cells. Vaccine 21(Suppl 2):S38–S42

Mori M, Ohtani H, Naito Y et al (2000) Infiltration of CD8+ T cells in non-small cell lung cancer is associated with dedifferentiation of cancer cells, but not with prognosis. Tohoku J Exp Med 191(2):113–118

Mosser DM, Edwards JP (2008) Exploring the full spectrum of macrophage activation. Nat Rev Immunol 8(12):958–969

Mouries J, Moron G, Schlecht G, Escriou N, Dadaglio G, Leclerc C (2008) Plasmacytoid dendritic cells efficiently cross-prime naive T cells in vivo after TLR activation. Blood 112(9):3713–3722

Muller AJ, Prendergast GC (2007) Indoleamine 2, 3-dioxygenase in immune suppression and cancer. Curr Cancer Drug Targets 7(1):31–40

Murdoch C, Muthana M, Coffelt SB, Lewis CE (2008) The role of myeloid cells in the promotion of tumour angiogenesis. Nat Rev 8(8):618–631

Musiani P, Allione A, Modica A et al (1996) Role of neutrophils and lymphocytes in inhibition of a mouse mammary adenocarcinoma engineered to release IL-2, IL-4, IL-7, IL-10, IFN-alpha, IFN-gamma, and TNF-alpha. Lab Invest; J Tech Methods Pathol 74(1):146–157

Nagorsen D, Voigt S, Berg E, Stein H, Thiel E, Loddenkemper C (2007) Tumor-infiltrating macrophages and dendritic cells in human colorectal cancer: relation to local regulatory T cells, systemic T-cell response against tumor-associated antigens and survival. J Transl Med 5:62

Naik SH, Sathe P, Park HY et al (2007) Development of plasmacytoid and conventional dendritic cell subtypes from single precursor cells derived in vitro and in vivo. Nat Immunol 8(11):1217–1226

Naito Y, Saito K, Shiiba K et al (1998) CD8+ T cells infiltrated within cancer cell nests as a prognostic factor in human colorectal cancer. Cancer Res 58(16):3491–3494

Nakajima T, Kodama T, Tsumuraya M, Shimosato Y, Kameya T (1985) S-100 protein-positive Langerhans cells in various human lung cancers, especially in peripheral adenocarcinomas. Virchows Arch 407(2):177–189

Nakakubo Y, Miyamoto M, Cho Y et al (2003) Clinical significance of immune cell infiltration within gallbladder cancer. Br J Cancer 89(9):1736–1742

Nakayama Y, Inoue Y, Minagawa N et al (2003) Relationships between S-100 protein-positive cells and clinicopathological factors in patients with colorectal cancer. Anticancer Res 23(6a):4423–4426

Nakayama T, Yao L, Tosato G (2004) Mast cell-derived angiopoietin-1 plays a critical role in the growth of plasma cell tumors. J Clin Investig 114(9):1317–1325

Nedergaard BS, Ladekarl M, Nyengaard JR, Nielsen K (2008) A comparative study of the cellular immune response in patients with stage IB cervical squamous cell carcinoma. Low numbers of several immune cell subtypes are strongly associated with relapse of disease within 5 years. Gynecol Oncol 108(1):106–111

Negus RP, Stamp GW, Relf MG et al (1995) The detection and localization of monocyte chemoattractant protein-1 (MCP-1) in human ovarian cancer. J Clin Investig 95(5):2391–2396

Nelson BH (2008) The impact of T-cell immunity on ovarian cancer outcomes. Immunol Rev 222:101–116

Nelson BH (2010) CD20+ B cells: the other tumor-infiltrating lymphocytes. J Immunol 185(9):4977–4982

Nielsen HJ, Hansen U, Christensen IJ, Reimert CM, Brunner N, Moesgaard F (1999) Independent prognostic value of eosinophil and mast cell infiltration in colorectal cancer tissue. J Pathol 189(4):487–495

Norian LA, Rodriguez PC, O'Mara LA et al (2009) Tumor-infiltrating regulatory dendritic cells inhibit CD8+ T cell function via L-arginine metabolism. Cancer Res 69(7):3086–3094

Norrby K (2002) Mast cells and angiogenesis. Apmis 110(5):355–371

Nosho K, Baba Y, Tanaka N et al (2010) Tumour-infiltrating T-cell subsets, molecular changes in colorectal cancer, and prognosis: cohort study and literature review. J Pathol 222(4):350–366

Oble DA, Loewe R, Yu P, Mihm MC Jr (2009) Focus on TILs: prognostic significance of tumor infiltrating lymphocytes in human melanoma. Cancer Immun 9:3

O'Donnell RK, Mick R, Feldman M et al (2007) Distribution of dendritic cell subtypes in primary oral squamous cell carcinoma is inconsistent with a functional response. Cancer Lett 255(1):145–152

Ohri CM, Shikotra A, Green RH, Waller DA, Bradding P (2009) Macrophages within NSCLC tumour islets are predominantly of a cytotoxic M1 phenotype associated with extended survival. Eur Respir J 33(1):118–126

Ohtaki Y, Ishii G, Nagai K et al (2010) Stromal macrophage expressing CD204 is associated with tumor aggressiveness in lung adenocarcinoma. J Thorac Oncol 5(10):1507–1515

Okayama Y, Ono Y, Nakazawa T, Church MK, Mori M (1998) Human skin mast cells produce TNF-alpha by substance P. Int Arch Allergy Immunol 117(Suppl 1):48–51

Ono Y, Ozawa M, Tamura Y et al (2002) Tumor-associated tissue eosinophilia of penile cancer. Int J Urol 9(2):82–87

Orange JS, Ballas ZK (2006) Natural killer cells in human health and disease. Clin Immunol (Orlando, Fla) 118(1): 1–10

Osada T, Chong G, Tansik R et al (2008) The effect of anti-VEGF therapy on immature myeloid cell and dendritic cells in cancer patients. Cancer Immunol Immunother 57(8):1115–1124

Oshikiri T, Miyamoto M, Shichinohe T et al (2003) Prognostic value of intratumoral CD8+ T lymphocyte in extrahepatic bile duct carcinoma as essential immune response. J Surg Oncol 84(4):224–228

Pages F, Galon J, Dieu-Nosjean MC, Tartour E, Sautes-Fridman C, Fridman WH (2010) Immune infiltration in human tumors: a prognostic factor that should not be ignored. Oncogene 29(8):1093–1102

Pandit R, Scholnik A, Wulfekuhler L, Dimitrov N (2007) Non-small-cell lung cancer associated with excessive eosinophilia and secretion of interleukin-5 as a paraneoplastic syndrome. Am J Hematol 82(3):234–237

Papamichail M, Perez SA, Gritzapis AD, Baxevanis CN (2004) Natural killer lymphocytes: biology, development, and function. Cancer Immunol Immunother 53(3):176–186

Park MH, Lee JS, Yoon JH (2012) High expression of CX3CL1 by tumor cells correlates with a good prognosis and increased tumor-infiltrating CD8+ T cells, natural killer cells, and dendritic cells in breast carcinoma. J Surg Oncol 106(4):386–392

Passlick B, Pantel K, Kubuschok B et al (1996) Expression of MHC molecules and ICAM-1 on non-small cell lung carcinomas: association with early lymphatic spread of tumour cells. Eur J Cancer 32A(1):141–145

Pekarek LA, Starr BA, Toledano AY, Schreiber H (1995) Inhibition of tumor growth by elimination of granulocytes. J Exp Med 181(1):435–440

Pelletier MP, Edwardes MD, Michel RP, Halwani F, Morin JE (2001) Prognostic markers in resectable non-small cell lung cancer: a multivariate analysis. Can J Surg 44(3):180–188

Peng SH, Deng H, Yang JF et al (2005) Significance and relationship between infiltrating inflammatory cell and tumor angiogenesis in hepatocellular carcinoma tissues. World J Gastroenterol 11(41):6521–6524

Penna G, Vulcano M, Roncari A, Facchetti F, Sozzani S, Adorini L (2002) Cutting edge: differential chemokine production by myeloid and plasmacytoid dendritic cells. J Immunol 169(12):6673–6676

Perez SA, Karamouzis MV, Skarlos DV et al (2007) CD4+CD25+ regulatory T-cell frequency in HER-2/neu (HER)-positive and HER-negative advanced-stage breast cancer patients. Clin Cancer Res 13(9):2714–2721

Perussia B (1996) The cytokine profile of resting and activated NK cells. Methods (San Diego, Calif) 9(2): 370–378

Petersen RP, Campa MJ, Sperlazza J et al (2006) Tumor infiltrating Foxp3+ regulatory T-cells are associated with recurrence in pathologic stage I NSCLC patients. Cancer 107(12):2866–2872

Pollard JW (2004) Tumour-educated macrophages promote tumour progression and metastasis. Nat Rev 4(1):71–78

Pretlow TP, Keith EF, Cryar AK et al (1983) Eosinophil infiltration of human colonic carcinomas as a prognostic indicator. Cancer Res 43(6):2997–3000

Punt CJ, Barbuto JA, Zhang H, Grimes WJ, Hatch KD, Hersh EM (1994) Anti-tumor antibody produced by human tumor-infiltrating and peripheral blood B lymphocytes. Cancer Immunol Immunother 38(4):225–232

Purdy AK, Campbell KS (2009) Natural killer cells and cancer: regulation by the killer cell Ig-like receptors (KIR). Cancer Biol Ther 8(23):2211–2220

Qu Z, Huang X, Ahmadi P et al (1998) Synthesis of basic fibroblast growth factor by murine mast cells. Regulation by transforming growth factor beta, tumor necrosis factor alpha and stem cell factor. Int Arch Allergy Immunol 115(1):47–54

Rajput AB, Turbin DA, Cheang MC et al (2008) Stromal mast cells in invasive breast cancer are a marker of favourable prognosis: a study of 4,444 cases. Breast Cancer Res Treat 107(2):249–257

Ralfkiaer E, Hou-Jensen K, Gatter KC, Drzewiecki KT, Mason DY (1987) Immunohistological analysis of the lymphoid infiltrate in cutaneous malignant melanomas. Virchows Arch 410(4):355–361

Rao HL, Chen JW, Li M et al (2012) Increased intratumoral neutrophil in colorectal carcinomas correlates closely with malignant phenotype and predicts patients' adverse prognosis. PLoS ONE 7(1):e30806

Reichert TE, Scheuer C, Day R, Wagner W, Whiteside TL (2001) The number of intratumoral dendritic cells and zeta-chain expression in T cells as prognostic and survival biomarkers in patients with oral carcinoma. Cancer 91(11):2136–2147

Reid MD, Basturk O, Thirabanjasak D et al (2011) Tumor-infiltrating neutrophils in pancreatic neoplasia. Mod Pathol 24(12):1612–1619

Restifo NP, Kawakami Y, Marincola F et al (1993) Molecular mechanisms used by tumors to escape immune recognition: immunogenetherapy and the cell biology of major histocompatibility complex class I. J Immunother Emphasis Tumor Immunol 14(3):182–190

Ribatti D, Vacca A, Ria R et al (2003a) Neovascularisation, expression of fibroblast growth factor-2, and mast cells with tryptase activity increase simultaneously with pathological progression in human malignant melanoma. Eur J Cancer 39(5):666–674

Ribatti D, Ennas MG, Vacca A et al (2003b) Tumor vascularity and tryptase-positive mast cells correlate with a poor prognosis in melanoma. Eur J Clin Invest 33(5):420–425

Ribatti D, Finato N, Crivellato E et al (2005) Neovascularization and mast cells with tryptase activity increase simultaneously with pathologic progression in human endometrial cancer. Am J Obstet Gynecol 193(6):1961–1965

Rice AJ, Griffiths AP, Martin IG, Dixon MF (2000) Gastric carcinoma with prominent neutrophil infiltration. Histopathology 37(3):289–290

Rodriguez-Pinto D (2005) B cells as antigen presenting cells. Cell Immunol 238(2):67–75

Roncucci L, Mora E, Mariani F et al (2008) Myeloperoxidase-positive cell infiltration in colorectal carcinogenesis as indicator of colorectal cancer risk. Cancer Epidemiol Biomarkers Prev 17(9):2291–2297

Ropponen KM, Eskelinen MJ, Lipponen PK, Alhava E, Kosma VM (1997) Prognostic value of tumour-infiltrating lymphocytes (TILs) in colorectal cancer. J Pathol 182(3):318–324

Ryder M, Ghossein RA, Ricarte-Filho JC, Knauf JA, Fagin JA (2008) Increased density of tumor-associated macrophages is associated with decreased survival in advanced thyroid cancer. Endocr Relat Cancer 15(4):1069–1074
Sakaguchi S, Sakaguchi N, Shimizu J et al (2001) Immunologic tolerance maintained by CD25+ CD4+ regulatory T cells: their common role in controlling autoimmunity, tumor immunity, and transplantation tolerance. Immunol Rev 182:18–32
Salama P, Phillips M, Grieu F et al (2009) Tumor-infiltrating FOXP3+ T regulatory cells show strong prognostic significance in colorectal cancer. J Clin Oncol 27(2):186–192
Samoszuk M, Corwin MA (2003) Acceleration of tumor growth and peri-tumoral blood clotting by imatinib mesylate (Gleevec). Int J Cancer 106(5):647–652
Samoszuk M, Kanakubo E, Chan JK (2005) Degranulating mast cells in fibrotic regions of human tumors and evidence that mast cell heparin interferes with the growth of tumor cells through a mechanism involving fibroblasts. BMC cancer 5:121
Sato M, Yoshida H, Yanagawa T et al (1981) Carcinoma of the maxillary sinus with eosinophilia. Report of a case. Int J Oral Sur0067 10(1):62–67
Sato E, Olson SH, Ahn J et al (2005) Intraepithelial CD8+ tumor-infiltrating lymphocytes and a high CD8+/regulatory T cell ratio are associated with favorable prognosis in ovarian cancer. Proc Natl Acad Sci USA 102(51):18538–18543
Sawatsubashi M, Yamada T, Fukushima N, Mizokami H, Tokunaga O, Shin T (2000) Association of vascular endothelial growth factor and mast cells with angiogenesis in laryngeal squamous cell carcinoma. Virchows Arch 436(3):243–248
Schaerli P, Willimann K, Ebert LM, Walz A, Moser B (2005) Cutaneous CXCL14 targets blood precursors to epidermal niches for Langerhans cell differentiation. Immunity 23(3):331–342
Schleypen JS, Von Geldern M, Weiss EH et al (2003) Renal cell carcinoma-infiltrating natural killer cells express differential repertoires of activating and inhibitory receptors and are inhibited by specific HLA class I allotypes. Int J Cancer 106(6):905–912
Schleypen JS, Baur N, Kammerer R et al (2006) Cytotoxic markers and frequency predict functional capacity of natural killer cells infiltrating renal cell carcinoma. Clin Cancer Res 12(3 Pt 1):718–725
Schumacher K, Haensch W, Roefzaad C, Schlag PM (2001) Prognostic significance of activated CD8(+) T cell infiltrations within esophageal carcinomas. Cancer Res 61(10):3932–3936
Schwaab T, Schned AR, Heaney JA et al (1999) In vivo description of dendritic cells in human renal cell carcinoma. J Urol 162(2):567–573
Sharma MD, Baban B, Chandler P et al (2007a) Plasmacytoid dendritic cells from mouse tumor-draining lymph nodes directly activate mature Tregs via indoleamine 2, 3-dioxygenase. J Clin Investig 117(9):2570–2582
Sharma P, Shen Y, Wen S et al (2007b) CD8 tumor-infiltrating lymphocytes are predictive of survival in muscle-invasive urothelial carcinoma. Proc Natl Acad Sci USA 104(10): 3967–3972
Sharma B, Sriram G, Saraswathi TR, Sivapathasundharam B (2010) Immunohistochemical evaluation of mast cells and angiogenesis in oral squamous cell carcinoma. Indian J Dent Res 21(2):260–265
Shellenberger TD, Wang M, Gujrati M et al (2004) BRAK/CXCL14 is a potent inhibitor of angiogenesis and a chemotactic factor for immature dendritic cells. Cancer Res 64(22): 8262–8270
Shen Z, Zhou S, Wang Y et al (2010) Higher intratumoral infiltrated Foxp3+ Treg numbers and Foxp3+/CD8+ ratio are associated with adverse prognosis in resectable gastric cancer. J Cancer Res Clin Oncol 136(10):1585–1595
Shimizu K, Nakata M, Hirami Y, Yukawa T, Maeda A, Tanemoto K (2010) Tumor-infiltrating Foxp3+ regulatory T cells are correlated with cyclooxygenase-2 expression and are associated with recurrence in resected non-small cell lung cancer. J Thorac Oncol 5(5):585–590
Shojaei F, Zhong C, Wu X, Yu L, Ferrara N (2008) Role of myeloid cells in tumor angiogenesis and growth. Trends Cell Biol 18(8):372–378

Shurin GV, Ferris RL, Tourkova IL et al (2005) Loss of new chemokine CXCL14 in tumor tissue is associated with low infiltration by dendritic cells (DC), while restoration of human CXCL14 expression in tumor cells causes attraction of DC both in vitro and in vivo. J Immunol 174(9):5490–5498

Shurin MR, Shurin GV, Lokshin A et al (2006) Intratumoral cytokines/chemokines/growth factors and tumor infiltrating dendritic cells: friends or enemies? Cancer Metastasis Rev 25(3):333–356

Sica A, Bronte V (2007) Altered macrophage differentiation and immune dysfunction in tumor development. J Clin Investig 117(5):1155–1166

Sickert D, Aust DE, Langer S, Haupt I, Baretton GB, Dieter P (2005) Characterization of macrophage subpopulations in colon cancer using tissue microarrays. Histopathology 46(5):515–521

Simon HU, Haj-Yehia A, Levi-Schaffer F (2000) Role of reactive oxygen species (ROS) in apoptosis induction. Apoptosis 5(5):415–418

Simonetti O, Goteri G, Lucarini G et al (2007) In melanoma changes of immature and mature dendritic cell expression correlate with tumor thickness: an immunohistochemical study. Int J Immunopathol Pharmacol 20(2):325–333

Solinas G, Germano G, Mantovani A, Allavena P (2009) Tumor-associated macrophages (TAM) as major players of the cancer-related inflammation. J Leukoc Biol 86(5):1065–1073

Sorbye SW, Kilvaer T, Valkov A et al (2011) Prognostic impact of lymphocytes in soft tissue sarcomas. PLoS ONE 6(1):e14611

Sorbye SW, Kilvaer TK, Valkov A et al (2012a) Prognostic impact of peritumoral lymphocyte infiltration in soft tissue sarcomas. BMC Clin Pathol 12:5

Sorbye SW, Kilvaer T, Valkov A et al (2012b) High expression of CD20+ lymphocytes in soft tissue sarcomas is a positive prognostic indicator. Oncoimmunology 1(1):75–77

Sorbye SW, Kilvaer TK, Valkov A et al (2012c) Prognostic impact of CD57, CD68, M-CSF, CSF-1R, Ki67 and TGF-beta in soft tissue sarcomas. BMC Clin Pathol 12:7

Soucek L, Lawlor ER, Soto D, Shchors K, Swigart LB, Evan GI (2007) Mast cells are required for angiogenesis and macroscopic expansion of Myc-induced pancreatic islet tumors. Nat Med 13(10):1211–1218

Sozzani S, Luini W, Borsatti A et al (1997) Receptor expression and responsiveness of human dendritic cells to a defined set of CC and CXC chemokines. J Immunol 159(4):1993–2000

Sparmann A, Bar-Sagi D (2004) Ras-induced interleukin-8 expression plays a critical role in tumor growth and angiogenesis. Cancer Cell 6(5):447–458

Suto H, Nakae S, Kakurai M, Sedgwick JD, Tsai M, Galli SJ (2006) Mast cell-associated TNF promotes dendritic cell migration. J Immunol 176(7):4102–4112

Suzuki A, Masuda A, Nagata H et al (2002) Mature dendritic cells make clusters with T cells in the invasive margin of colorectal carcinoma. J Pathol 196(1):37–43

Takagi S, Miyagawa S, Ichikawa E et al (2004) Dendritic cells, T-cell infiltration, and Grp94 expression in cholangiocellular carcinoma. Hum Pathol 35(7):881–886

Takahashi A, Kono K, Itakura J et al (2002) Correlation of vascular endothelial growth factor-C expression with tumor-infiltrating dendritic cells in gastric cancer. Oncology 62(2):121–127

Takanami I, Takeuchi K, Naruke M (2000) Mast cell density is associated with angiogenesis and poor prognosis in pulmonary adenocarcinoma. Cancer 88(12):2686–2692

Takanami I, Takeuchi K, Giga M (2001) The prognostic value of natural killer cell infiltration in resected pulmonary adenocarcinoma. J Thoracic Cardiovasc Surg 121(6):1058–1063

Talmadge JE, Donkor M, Scholar E (2007) Inflammatory cell infiltration of tumors: Jekyll or Hyde. Cancer Metastasis Rev 26(3–4):373–400

Tan SY, Fan Y, Luo HS, Shen ZX, Guo Y, Zhao LJ (2005) Prognostic significance of cell infiltrations of immunosurveillance in colorectal cancer. World J Gastroenterol 11(8):1210–1214

Tataroglu C, Kargi A, Ozkal S, Esrefoglu N, Akkoclu A (2004) Association of macrophages, mast cells and eosinophil leukocytes with angiogenesis and tumor stage in non-small cell lung carcinomas (NSCLC). Lung cancer (Amsterdam, Netherlands) 43(1):47–54

Taylor CH, Shi SR, Barr NJ, Wu N (2006) Techniques of immunohistochemistry: principles, pitfalls and standardization. In: DD J (ed) Diagnostic immunohistochemistry. Churchill Livingstone Elsevier, Philadelphia

Tazawa H, Okada F, Kobayashi T et al (2003) Infiltration of neutrophils is required for acquisition of metastatic phenotype of benign murine fibrosarcoma cells: implication of inflammation-associated carcinogenesis and tumor progression. Am J Pathol 163(6):2221–2232

Tazzyman S, Lewis CE, Murdoch C (2009) Neutrophils: key mediators of tumour angiogenesis. Int J Exp Pathol 90(3):222–231

Theoharides TC, Conti P (2004) Mast cells: the Jekyll and Hyde of tumor growth. Trends Immunol 25(5):235–241

Thiel A, Pries R, Jeske S, Trenkle T, Wollenberg B (2009) Effect of head and neck cancer supernatant and CpG-oligonucleotides on migration and IFN-alpha production of plasmacytoid dendritic cells. Anticancer Res 29(8):3019–3025

Thompson AC, Bradley PJ, Griffin NR (1994) Tumor-associated tissue eosinophilia and long-term prognosis for carcinoma of the larynx. Am J Surg 168(5):469–471

Tomsova M, Melichar B, Sedlakova I, Steiner I (2008) Prognostic significance of CD3+ tumor-infiltrating lymphocytes in ovarian carcinoma. Gynecol Oncol 108(2):415–420

Toomey D, Smyth G, Condron C et al (2003) Infiltrating immune cells, but not tumour cells, express FasL in non-small cell lung cancer: No association with prognosis identified in 3-year follow-up. Int J Cancer 103(3):408–412

Toth T, Toth-Jakatics R, Jimi S, Takebayashi S (2000) Increased density of interstitial mast cells in amyloid A renal amyloidosis. Mod Pathol 13(9):1020–1028

Toth-Jakatics R, Jimi S, Takebayashi S, Kawamoto N (2000) Cutaneous malignant melanoma: correlation between neovascularization and peritumor accumulation of mast cells over-expressing vascular endothelial growth factor. Hum Pathol 31(8):955–960

Toy EP, Chambers JT, Kacinski BM, Flick MB, Chambers SK (2001) The activated macrophage colony-stimulating factor (CSF-1) receptor as a predictor of poor outcome in advanced epithelial ovarian carcinoma. Gynecol Oncol 80(2):194–200

Treilleux I, Blay JY, Bendriss-Vermare N et al (2004) Dendritic cell infiltration and prognosis of early stage breast cancer. Clin Cancer Res 10(22):7466–7474

Troy A, Davidson P, Atkinson C, Hart D (1998a) Phenotypic characterisation of the dendritic cell infiltrate in prostate cancer. J Urol 160(1):214–219

Troy AJ, Summers KL, Davidson PJ, Atkinson CH, Hart DN (1998b) Minimal recruitment and activation of dendritic cells within renal cell carcinoma. Clin Cancer Res 4(3):585–593

Troy AJ, Davidson PJ, Atkinson CH, Hart DN (1999) CD1a dendritic cells predominate in transitional cell carcinoma of bladder and kidney but are minimally activated. J Urol 161(6):1962–1967

Tsukayama S, Omura K, Yoshida K, Tanaka Y, Watanabe G (2005) Prognostic value of CD83-positive mature dendritic cells and their relation to vascular endothelial growth factor in advanced human gastric cancer. Oncol Rep 14(2):369–375

Tuna B, Yorukoglu K, Unlu M, Mungan MU, Kirkali Z (2006) Association of mast cells with microvessel density in renal cell carcinomas. Eur Urol 50(3):530–534

Vakkila J, Jaffe R, Michelow M, Lotze MT (2006) Pediatric cancers are infiltrated predominantly by macrophages and contain a paucity of dendritic cells: a major nosologic difference with adult tumors. Clin Cancer Res 12(7 Pt 1):2049–2054

van Driel WJ, Hogendoorn PC, Jansen FW, Zwinderman AH, Trimbos JB, Fleuren GJ (1996) Tumor-associated eosinophilic infiltrate of cervical cancer is indicative for a less effective immune response. Hum Pathol 27(9):904–911

Vermi W, Bonecchi R, Facchetti F et al (2003) Recruitment of immature plasmacytoid dendritic cells (plasmacytoid monocytes) and myeloid dendritic cells in primary cutaneous melanomas. J Pathol 200(2):255–268

Vermi W, Soncini M, Melocchi L, Sozzani S, Facchetti F (2011) Plasmacytoid dendritic cells and cancer. J Leukoc Biol 90(4):681–690

Vicari AP, Treilleux I, Lebecque S (2004) Regulation of the trafficking of tumour-infiltrating dendritic cells by chemokines. Semin Cancer Biol 14(3):161–169

Villegas FR, Coca S, Villarrubia VG et al (2002) Prognostic significance of tumor infiltrating natural killer cells subset CD57 in patients with squamous cell lung cancer. Lung cancer (Amsterdam, Netherlands) 35(1): 23–28

Vinten-Johansen J, Zhao ZQ, Nakamura M et al (1999) Nitric oxide and the vascular endothelium in myocardial ischemia-reperfusion injury. Ann N Y Acad Sci 874:354–370

Vivier E, Anfossi N (2004) Inhibitory NK-cell receptors on T cells: witness of the past, actors of the future. Nat Rev Immunol 4(3):190–198

Vora M, Romero LI, Karasek MA (1996) Interleukin-10 induces E-selectin on small and large blood vessel endothelial cells. J Exp Med 184(3):821–829

Wakabayashi O, Yamazaki K, Oizumi S et al (2003) CD4+ T cells in cancer stroma, not CD8+ T cells in cancer cell nests, are associated with favorable prognosis in human non-small cell lung cancers. Cancer Sci 94(11):1003–1009

Walz A, Schmutz P, Mueller C, Schnyder-Candrian S (1997) Regulation and function of the CXC chemokine ENA-78 in monocytes and its role in disease. J Leukoc Biol 62(5):604–611

Walzer T, Jaeger S, Chaix J, Vivier E (2007) Natural killer cells: from CD3(-)NKp46(+) to post-genomics meta-analyses. Curr Opin Immunol 19(3):365–372

Watkins SK, Zhu Z, Riboldi E et al (2011) FOXO3 programs tumor-associated DCs to become tolerogenic in human and murine prostate cancer. J Clin Investig 121(4):1361–1372

Welsh TJ, Green RH, Richardson D, Waller DA, O'Byrne KJ, Bradding P (2005) Macrophage and mast-cell invasion of tumor cell islets confers a marked survival advantage in non-small-cell lung cancer. J Clin Oncol 23(35):8959–8967

Westlin WF, Gimbrone MA Jr (1993) Neutrophil-mediated damage to human vascular endothelium. Role of cytokine activation. Am J Pathol 142(1):117–128

Westphal E (1891) Ferbenanlytische Untersuchungen Zur Histologie Und Klinik Des Blutes. In: Ehrlich P (ed) Uber mastzellen. Hirschwald, Berlin, p 17

Wilke CM, Kryczek I, Zou W (2011) Antigen-presenting cell (APC) subsets in ovarian cancer. Int Rev Immunol 30(2–3):120–126

Wislez M, Rabbe N, Marchal J et al (2003) Hepatocyte growth factor production by neutrophils infiltrating bronchioloalveolar subtype pulmonary adenocarcinoma: role in tumor progression and death. Cancer Res 63(6):1405–1412

Wolf D, Wolf AM, Rumpold H et al (2005) The expression of the regulatory T cell-specific forkhead box transcription factor FoxP3 is associated with poor prognosis in ovarian cancer. Clin Cancer Res 11(23):8326–8331

Woo EY, Yeh H, Chu CS et al (2002) Cutting edge: regulatory T cells from lung cancer patients directly inhibit autologous T cell proliferation. J Immunol 168(9):4272–4276

Yanaba K, Bouaziz JD, Matsushita T, Magro CM (2008) St Clair EW, Tedder TF. B-lymphocyte contributions to human autoimmune disease. Immunol Rev 223:284–299

Yano H, Kinuta M, Tateishi H et al (1999) Mast cell infiltration around gastric cancer cells correlates with tumor angiogenesis and metastasis. Gastric Cancer 2(1):26–32

Yasuda M, Mizukami M, Hanagiri T et al (2006) Antigens recognized by IgG derived from tumor-infiltrating B lymphocytes in human lung cancer. Anticancer Res 26(5A):3607–3611

Yin XY, Lu MD, Lai YR, Liang LJ, Huang JF (2003) Prognostic significances of tumor-infiltrating S-100 positive dendritic cells and lymphocytes in patients with hepatocellular carcinoma. Hepatogastroenterology 50(53):1281–1284

Yu P, Fu YX (2006) lymphocytes: friends or foes? Laboratory Invest; J Tech Methods Pathol 86(3):231–245

Zeid NA, Muller HK (1993) S100 positive dendritic cells in human lung tumors associated with cell differentiation and enhanced survival. Pathology 25(4):338–343

Zhang W, Stoica G, Tasca SI, Kelly KA, Meininger CJ (2000) Modulation of tumor angiogenesis by stem cell factor. Cancer Res 60(23):6757–6762

Zhao JJ, Pan K, Wang W et al (2012) The prognostic value of tumor-infiltrating neutrophils in gastric adenocarcinoma after resection. PLOS ONE 7(3):e33655

Zhuang X, Xia X, Wang C et al (2010) A high number of CD8+ T cells infiltrated in NSCLC tissues is associated with a favorable prognosis. Appl Immunohistochem Mol Morphol 18(1):24–28

Zou W, Machelon V, Coulomb-L'Hermin A et al (2001) Stromal-derived factor-1 in human tumors recruits and alters the function of plasmacytoid precursor dendritic cells. Nat Med 7(12):1339–1346

Zudaire E, Martinez A, Garayoa M et al (2006) Adrenomedullin is a cross-talk molecule that regulates tumor and mast cell function during human carcinogenesis. Am J Pathol 168(1):280–291

Chapter 4
Immunologic Interpretation of Cancer Biology: Impact on Clinical Outcome

Maria Libera Ascierto, Francesco M. Marincola and Ena Wang

Abstract The biology of tumors cannot be understood by simply studying the traits of the neoplastic cells in isolation. Instead, the contributions of the "tumor microenvironment" to tumor genesis must be considered. The complexity of the interaction between not transformed host cells with cancer cells and the molecular pathways involved in such interaction could be recently appreciated by high-throughput tools capable of providing a global view of biological processes. Based on this approach numerous studies have defined a new portrait of cancer, describing a linkage between the immune infiltrate, prognosis and response to therapy. Here, we provide an overview of the current status of the field describing the immune cells involved in this phenomenon. Molecular events and biomarkers associated with the favorable clinic outcome will be described, analyzing also commonalities and discrepancies among studies.

Keywords Tumor microenvironment · Tumor immune infiltrate · Prognostic value of immune infiltrate · Genetic alterations of host · Genetic alterations of tumor

M. L. Ascierto (✉) · E. Wang (✉)
National Institutes of Health, Building 10, Room 1N224, 10 Center Drive, Bethesda, MD 20892, USA
e-mail: asciertoml@cc.nih.gov; marialiberaasc@yahoo.it

E. Wang
e-mail: EWang@cc.nih.gov

M. L. Ascierto · F. M. Marincola · E. Wang
Infectious Disease and Immunogenetics Section (IDIS), Department of Transfusion Medicine, Clinical Center and Trans-NIH Center for Human Immunology (CHI), National Institutes of Health, Bethesda, MD 20892, USA
e-mail: FMarincola@cc.nih.gov

M. L. Ascierto
Department of Internal Medicine, University of Genoa, Genoa, Italy

M. R. Shurin et al. (eds.), *The Tumor Immunoenvironment*,
DOI: 10.1007/978-94-007-6217-6_4,

4.1 Introduction

Traditionally, anti-cancer therapies have targeted exclusively transformed cancer cells. However, more recently tumors are not only considered as masses of proliferating cells, but it is widely accepted that non-transformed host cells, which include both innate and acquired immunity, such as endothelial, fibroblasts, mesenchymal cells, MDSCs, macrophages, DCs, mast cells, eosinophils, neutrophils, NK cells, and lymphocytes, interact with malignant tumor cells to form a dynamic environment in which the non-transformed cells exert immune surveillance or promote spread of cancer cells. In turn, also cancer cells affect the phenotype of the non transformed host cells by express suppressive molecules or provide negative feedback to compromise their function.

Viewed from this prospective, the biology of tumors can no longer be restricted to the study of the cancer cells. Instead, the contributions of the "tumor microenvironment" to tumor genesis must be considered. Due to the high complexity of the tumor microenvironment, understanding the networking between not transformed host's and cancer cells and the molecular pathways involved in such interaction is a "must". Such complexity could be recently appreciated in its extent by high-throughput tools, such as gene profiling or DNA sequencing capable of providing a global view of biological processes defining possible biomarkers of tumor regression or therapy's response. Based on this approach oncologists and immunologists are combining their efforts to reshape the biology of cancer to provide better diagnostic, prognostic and predictive power.

4.2 The Role of Immunity in Cancer

The first studies able to re-classify cancer biology according to global transcriptional analysis were published about a decade ago (Alizadeh et al. 2000; Lossos et al. 2000; Perou et al. 2000). Since then, numerous studies described a cancer phenotype where the infiltration of immune cells of both the innate and adaptive arms of the immune system as well as the combined with the expression of interferon–stimulated genes (ISGs) and immune effectors functions (IEFs), sustain a self-perpetuating inflammatory process which influences tumor growth and/or responsiveness to anti-cancer immunotherapy (Ascierto et al. 2011; Disis 2010).

These observations emphasized the host immune cells both within and surrounding tumors as critical determinants of cancer biology and key factors for the success or failure of human cancer therapy. Those statements changed the field of tumor immunology as pointed out by a recent review of (Hanahan and Weinberg 2011) in which "avoiding immune destruction" is being considered an emerging hallmark of cancer. However, it is also quite confusing when we contemplate that just a decade ago the same authors claimed that the presence of immune cells and tumor-associated inflammatory responses promote tumor growth (Hanahan and Weinberg 2000).

To better understand this apparent paradox the intensity of inflammatory processes must be considered. Usually the inflammation associated with cancer is similar to a chronic inflammatory process, where the production of growth and angiogenic factors stimulates tissue repair and growth. Occasionally, however, it is observed a cancer inflammatory process similar to acute inflammation characterized by the presence of innate and adaptive T cells responses which favor the activation of immune effector mechanisms able to generate spontaneous or treatment induced cancer regression.

This phenomenon strongly indicates that even established cancer hold the characteristic of plasticity of cancer-related inflammation and should be considered as a dynamic event (Mantovani et al. 2008). Before tumor establishment, immune surveillance can control or eliminate some premalignant lesions. However, tumor cells can become resistant to the first line of defense and develop a phenotype able of manipulating immune cells, inducing a process referred to as "immune editing". Trough the immune editing mechanism, the tumor cells are able to escape immune recognition, by down regulate the expression of major histocompatibility complex, by decrease expression of co-stimulatory molecules important for T cell activation, and by enhance surface expression of molecules, which suppress the activation of T cells, (PD-L1/B7-H1 and B7-H4). Cancer cells can also limit the function of the immune system through the secretion of soluble factors able to inhibit the activation, proliferation, and differentiation of the various components of the immune response (Mantovani et al. 2008). However, at one point a threshold is reached: the immune cells and the secreted chemokines and cytokines turn an indolent process that favors tumor growth into an acute process that promotes tumor destruction.

In theory lesions can regress and all cancer cells are eliminated, disappear. Immunotherapy aimed at manipulate and optimized immune cells by blockade immune checking points, induce production of pro inflammatory cytokines and chemokines which in turn helps this process through the recruiting and activation of T cells and natural killer cells and gear toward reaching the invisible threshed to tip the balance in favor of tumor elimination. In some cases tumor cells are still able to escape these "acute" immune—related destructive phenomena, thus leading to tumor progression or recurrence.

The observation of a similar behavior in other tissues undergoing different types of immune-mediated tissue-specific destruction (TSD) such as flares of autoimmunity, clearance of pathogen infected cells during acute infection, allograft rejection and graft versus host disease, leaded us to propose that distinct immune processes ultimately converge into a "immunological constant of rejection" mechanism (Wang et al. 2008) (Fig. 4.1). Thus, although distinct immune phenomena are prompted by different mechanisms, the requirements for endogenous or exogenous inflammatory stimuli reach at one point a threshold which turns a chronic inflammatory process into an acute one, converging into in a conserved tissue destruction mechanism.

In the following discussion, we provide a description of the immune cells which are mostly involved in these phenomena; our attention will be particularly focused

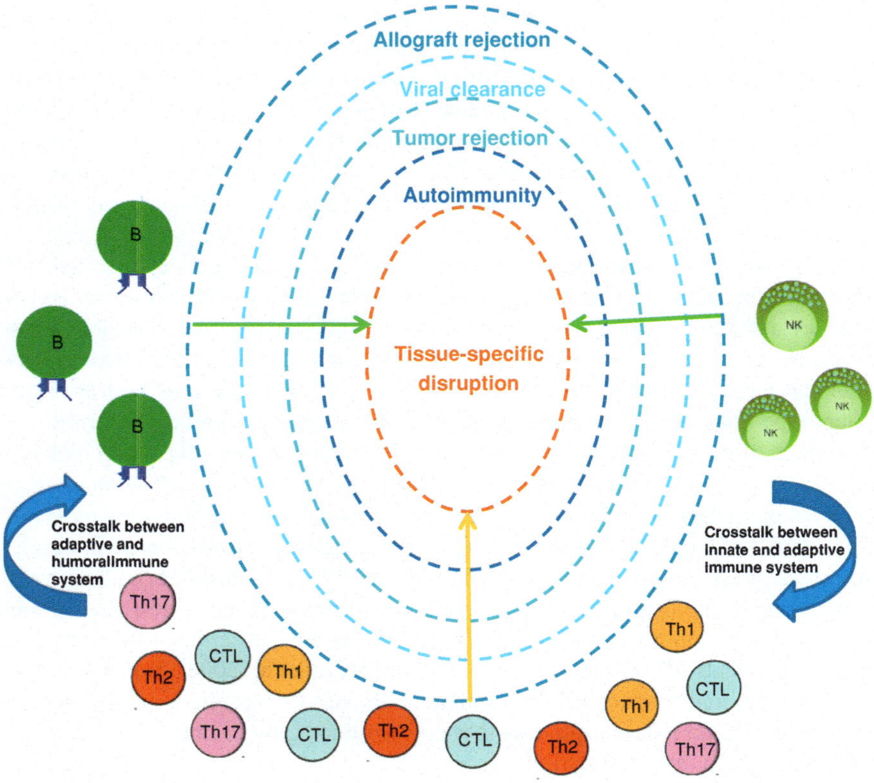

Fig. 4.1 Immunological constant of rejection. Immune molecular pathways observed during tissue rejection and inflammatory responses to cancer are also observed in curse of the inflammatory response against viral infection, allograft rejection as well as autoimmunity disease. This suggests that although the tissue-specific destruction varies in distinct pathologic states, the effector immune response converges into a single mechanism that includes the activation of adaptive and innate cytotoxic mechanisms and the crosstalk between them. The keys genes involved in those processes we can observed that they are mostly *centered* around IFN-γ signaling, TNF signaling, Interferon Regulatory Factors (*IRFs*) as well as Immune Effector Function (*IEF*) genes such as granzyme B, perforin. Those inflammatory key genes, in several condition are associate with better prognosis and in other conditions are instead associated with negative prognosis. To better understand this apparent paradox the intensity of inflammatory processes must be considered

on the infiltration of T cells, NK cells and B cell in tumor side. Molecular events and biomarkers associated with the favorable clinic outcome will be described, analyzing also commonalities and discrepancies among studies.

Moreover, studies supporting interplay between immune contexture of tumor microenvironment, genetic background of the host as well as genetic instability of neoplastic cells, and their association with cancer patients' outcome will be discussed. It is important to point out that while many of the results reported below are quite impressive in terms of similar conclusions, there are also some conflicting observations. These can potentially be explained by differences in methodologies used. Older studies used light microscopy, and in some cases no distinction was made between different cell types. The advancement of technologies such as Polymerase Chain Reaction (PCR), and gene expression profiling has addressed the study of tumor-infiltrating immune cell populations in more details reducing the number of conflicting observations (Bedognetti et al. 2010).

However, we need to consider that sometimes it is possible to observe discrepancies between these similar techniques mostly due to the usage of different standards for expression normalization. Moreover, if we consider that sometimes immune cell types found in tumors can vary in degree of maturation and/or activation, and many diverse cell types can share similar markers, the description of an objective and real picture becomes even more challenging.

4.3 Prognostic Significance of T Cells

Increasing evidence indicates that some patients with cancer can generate an adaptive immune response specifically directed against antigenic proteins expressed by tumors (Marincola 2005). In particular, an adaptive T cell response, which is composed of both cytotoxic CD8+ T cells (CTLs) and CD4+ T cells, can promote the secretion of cytokines such as interferon gamma (IFN-γ) and tumor necrosis factor alpha (TNF-α) generating an acute inflammation which results in expansion of cytotoxic CD8+ T cells, tissue destruction and control of cancer growth. Although in animal models the in vivo eradication of tumors is mostly dependent by CTLs, there has been increased appreciation over the last years for the importance of the CD4+ T cells, which through the secretion of cytokines involved in the regulation of acute inflammatory responses, are able to enhance or limit CTLs function or proliferation (Disis 2010). Thus, CD4+ cells are referred to as T-helper (Th) cells which manifest with different phenotypes.

Th1 cells are stimulated by type I dendritic cells (DCs). Together these cells, generate a tumor environment rich of cytokines such as IFN-γ, TNF-α, Interleukin (IL)-12 (IL-12), IL-2 and support CTL and tissue destruction; the results is a potential control or elimination of tumor growth. Th2 cells are stimulated by tumor associated macrophages (TMAs) and myeloid–derived suppressor cells (MDSCs). Together these cells types generate a tumor environment rich of cytokines such as IL-10, IL-4, IL-5, IL-6, which enhance B cell responses and limit CTL proliferation. Th17 cells secrete IL-17 and are operative in pathologic autoimmune disease. Regulatory T cells (Tregs) secrete IL-10 and Tumor Growth Factor (TGF-β), which limit CTL response resulting in potential proliferation of tumor cells. We could assume that the overall immune response elicited within each individual tumor results from the balance of the type of cells and cytokines secreted.

4.3.1 CTLs, Th1 Cells and T Memory Cells

Gene expression studies performed in distinct solid tumors types indicated that an immune response characterized by the expression of gene related to adaptive and innate effector immunity such as antigen presentation, IFN-γ production, activation of T cell receptor signaling, is associated with improved prognosis in melanoma, head and neck, breast, bladder, urothelial, ovarian, colorectal, renal, prostatic, and lung cancer (Ascierto et al. 2011). One important investigation demonstrating this association was performed in colon cancer patients through the identification of a cluster of genes inversely correlated with tumor recurrence. The identified predictive signatures, encoding Interferon (IFN) regulatory factor 1 (IRF1), IFN-γ, CD3, CD8, granulysin, and grazyme B is mostly associated with Th1 immunity and CTLs generation (Galon et al. 2006).

The same observation was also previously described by Guidoboni et al. (2001) in colorectal cancer where high numbers of activated CTLs correlated with improved overall and disease-free survival, particularly in patients with stage III tumors ($n = 109$, $P < 0.001$). Moreover, a high frequency of microsatellite instability was correlated to survival. Multivariate analysis revealed that patients with both features had a risk of death and relapse markedly lower than that associated with microsatellite status or intra-tumoral cytotoxic lymphocytes separately. Similar observations were then reported in other cancer types (Ascierto et al. 2011).

In melanoma, a pilot study of 19 patients vaccinated with a combination of four tumor antigens plus IL-12, (Gajewski 2011) observed by global transcriptional analysis that tumors of patients likely to respond to therapy have a pre-existing over expression of interferon stimulated genes (ISGs) and Th1 cell attracting chemokines.

These findings were associated with the histopathological demonstration of the presence of CTLs in the same tumors. Interestingly, together with the effector component of the immune response, the tumors derived from the responding patients also displayed the presence of immune inhibitory mechanisms such as expression of Programmed Death Ligand 1 (PD-L1) which upon binding with PD-1 mediate the de-phosphorylation of signaling molecules downstream of the T cell receptor, thus dampening T cell sensitivity to antigenic stimulation (Ahmadzadeh et al. 2009). The co-expression of indoleamine-2,3-dioexygenase (IDO), which causes immune suppression through breakdown of tryptophan in the tumor microenvironment producing toxic tryptophan catabolites which consequently cause growth arrest and functional suppression of effector T cells, was also observed in the same group of patients (Masferrer et al. 2000; Sinha et al. 2007). These findings can be interpreted as demonstration that tumors phenotype do not discriminate among various components of the immune response as they can sustain immune effector and immune regulatory functions simultaneously; it is probably the overall balance between the immunosuppressive and immune-active factors that determines the ultimate fate of individual cancers.

In addition to the Th1 signature, the second key feature of a potentially effective immune response is the capacity of T cells to travel into the site of the tumor and infiltrate deeply into the tumor parenchyma. Although a connotation of lymphocytic infiltrates as a favorable prognostic biomarker in cancer was originally reported by Cochran (1969), much progress has been made in the last decade. Clemente et al. (1996), reported that the presence of tumor infiltrating lymphocytes in the vertical growth phase of primary cutaneous melanoma was an independently favorable prognostic factor.

Similar observations were subsequently showed in a study conducted on 186 patients with ovarian cancer were T cells infiltrates turned out to be a significant prognostic factor (Zhang et al. 2003). The authors observed significant differences in the distributions of progression-free survival and overall survival according to the presence or absence of intratumoral T cells ($P < 0.001$). The presence of intratumoral T cells was independently correlated with delayed recurrence or delayed death in multivariate analysis and was associated with increased expression of IFN-γ, IL2, and lymphocyte-attracting chemokines within the tumor. On the other end the absence of intratumoral T cells was associated with increased levels of vascular endothelial growth factor.

However, the study that best characterized the phenotype of the T cells able to penetrate into tumor parenchyma came from gene expression analysis performed on colon cancer patients. Investigators showed that high levels of intra-tumoral CD45RO+ (memory) T cells correlated with better survival. Also the localization was important for the prognostic significance which was very favorable when the T cells were accumulating in the center and in the invasive margins of the tumors. Gene expression analysis performed on the same cohort of patients demonstrated that the Th1 signature, described above, facilitate the infiltration of intra-tumoral CD45+RO T cells (Pages et al. 2005).

All together, these investigations suggested that prognosis in patients with cancer is positively affected by the presence of type I adaptive immune response in tumor environment and by the ability of T cells to penetrate through tumor stroma and infiltrate deeply into parenchyma.

4.3.2 T Regulatory Cells

In contrast to CTLs, Th1 CD4+ cells and memory T cells, the relevance of other CD4+ T cell populations on clinical outcome has been more controversial (Fridman et al. 2012). The case of regulatory T (Treg) cells serves as a good example of conflicting data that lead to difficult interpretation. Although, there are different subpopulations of Treg, most studies define them as a population of CD4+ T cells expressing high levels of CD25 and the transcription factor forkhead box protein P3 (FOXP3). However, none of these markers is fully restricted to Treg cells. Indeed, CD25 and FOXP3 are also expressed by activated effector T cells, and there are also FOXP3^{-} suppressor cells. Thus, the initial report by Curiel et al. (2004) which demonstrated a correlation of intra-tumoural Treg cells and poor survival in ovarian cancer and which have been followed by similar observations in other cancer type such as hepatocellular carcinoma (Gao et al. 2007) has to be carefully reconsidered. In fact, several analyses of other cancer types have found no impact of Treg infiltration on survival (Ascierto et al. 2011; Heimberger et al. 2008; Hillen et al. 2008; Mahmoud et al. 2011). By contrast, other reports have demonstrated a positive clinical correlation between the density of intra-tumoral Treg cells and the local immune control of tumors (Badoual et al. 2006; Carreras et al. 2006; Frey et al. 2010; Salama et al. 2009). In colorectal cancer, Tosolini et al. (2011) recently described two clusters of genes associated with regulatory functions. Although the first cluster (IL-10/TGF-β) was not associated with a favorable outcome, FOXP3 (second cluster) mRNA expression and the presence of high density FOXP3 positive cells were associated with better survival. We recently described similar observation in patients with metastatic melanoma receiving high-dose interleukin-2 plus the gp100:209–217(210 M) peptide vaccine melanoma patients. The results showed that the vaccine–interleukin-2 group, as compared with the interleukin-2–only group, experienced a significant improvement in overall clinical response and longer progression-free survival. In study, an increase in T regulatory cells (CD4+FOXP3+) was observed in patients responding to treatment (Schwartzentruber et al. 2011).

The reasons for Treg discrepancies are not evident; they may be due the imperfect markers used to phenotype suppressive cells or due to technical differences. Regulatory T cells can lose FOXP3 (Hoffmann et al. 2009), effector T cells can transiently express FOXP3 without acquisition of suppressive functions and FOXP3 acts as a tumor suppressor gene when expressed by tumor cells. These observations complicate the interpretation of the aforementioned studies in absence of functional and cell-specific analyses (Walker et al. 2003; Zuo et al. 2007; Wang et al. 2009).

4.3.3 Th17 Cells

The analysis of other CD4+ T cell populations has also yielded apparently contradictory results (Wilke et al. 2011). Th17 cells have been reported associated with poor prognosis in colorectal, lung and hepatocellular carcinoma (HCC) (Jochems and Schlom 2011). Zhang et al. (2009) have examined IL-17+ cells in patients with HCC and suggested a potential pro-tumor role for IL-17. In this case increased IL-17− producing cell density within the tumors of HCC patients was correlated with both micro vessel density and poor prognosis. Notably, HCC is frequently associated with chronic viral hepatitis which can strongly affect the generation and function of Th17 cells in cancer patients.

In non-small-cell lung cancer patients, higher levels of IL-17 within the tumor correlated with higher blood vessel density and shorter survival (Chen et al. 2010). On the contrary in other studies Th17 cells have been reported to predict better survival (Tassi et al. 2008). In ovarian cancer, tumor-associated Th17 levels correlate positively with microenvironment Th1 cells, cytotoxic CD8+ T cells and natural killer cells (Kryczek et al. 2006). The same group also observed a combined infiltration of IL-17 expressing CD4+ T cells and CD8+ effector T cells; through synergistic action between IL-17 and IFN-γ expression Th17 cells were observed to stimulate the expression of CXCL-9 and -10 by ovarian cancer cells as well as tumor infiltrating macrophages in order to recruit more effector T and NK cells to the tumor microenvironment. This combination was observed to positively predict patient outcome in the context of ovarian, colon, melanoma and pancreatic carcinoma. Altogether these studies demonstrated that Th17 cell infiltration in several tumor types was quantitatively and positively correlated with NK cell-mediated innate and adaptive immune responses (Kryczek et al. 2009). Still very few studies have focused on primary Th17 cells in the human tumor microenvironment, so it is difficult to deduce the exact role(s) they may have in cancer patients. Moreover, Th17 cell biology has been partially examined in patients with well-established cancer. It may be important to investigate the roles of Th17 cells and IL-17 in the very early phases of human tumor growth to better understand how these roles may change during disease progression (Wilke et al. 2011).

4.3.4 Th2 Cells

Th2 cells, through the activation of B cells or through the production of the immunosuppressive cytokine IL-10, seem to be associated with aggressive tumours. Recently (De Monte et al. 2011) showed that in pancreatic cancer the Th2 immune deviation has an active role in tumor progression. Moreover, the quantity of Th2 with respect to Th1 cells present in the tumor stroma has a direct correlation with prognosis in surgically resected patients. However, this is not a general phenomenon, as Th2 cells are also associated with favorable outcome in Hodgkin's lymphoma and breast cancer, which suggests a protective effect of antibodies in these diseases (Schreck et al. 2009; Yoon et al. 2010).

4.4 Prognostic Significance of Innate and Humoral Immune Response: Focus on NK Cells and B Cells

Numerous lines of evidence suggest that presence of T cells identifies cancer patients with an improved prognosis while the role of B cells and that of innate immune effector mechanisms and their cross talk with adaptive immune responses as played a Cinderella role (Shanker and Marincola 2011). This may also be due to the fact that cells of the innate immune system are involved in tissue repair and remodeling and the factors secreted by these cells are usually believed to enhance rather than inhibit tumor growth. Indeed, several types of innate immune cells have been shown to be independent predictors of poor prognosis and cancer progression (Martuza et al. 1991; Kelley et al. 2007; Zhu et al. 2008). These include mast cells, eosinophils, neutrophils, type 2 macrophages, and MDSCs (Ostrand-Rosenberg and Sinha 2009; Allavena et al. 2008). Even DCs, as professional antigen-presenting cells (APCs), may participate in tumor immune evasion by failing to stimulate effective adaptive responses (Jensen et al. 2011; Chaput et al. 2008).

However, experimental as well as clinical observations suggest that immune-mediated tumor destruction is dependent upon coordinate activation of immune effector genes expressed by cells of the innate and also humoral immune systems (Wang et al. 2008; Zuo et al. 2007; Shanker et al. 2007). Thus, a description of the effect of these parallel arms of the immune system is necessary and an association with patient's clinical outcome needs to be evaluated.

4.4.1 NK Cells

Among cells of the innate immune system, Natural Killer (NK) cells play a major role against tumors and they participate in the shaping of the adaptive immune response thought the secretion of cytokines such as IFN-γ (Vivier et al. 2011, 2012; Moretta et al. 1994). An important feature of NK cells is their capacity to distinguish stressed (such as tumor cells, infected cells and cells have undergone physical or chemical injuries) from healthy cells. NK cells were initially identified through their ability to kill tumor cell, hence their name (Oldham and Herberman 1973; Herberman et al. 1975; Kiessling et al. 1975). Recognition of tumor cells by NK cells is mediated by the interaction of activating receptors with ligands expressed on tumor target cells which at the same time do not express ligands specific for the NK inhibitory receptors. NK cells also express adhesion molecules, thereby interacting with tumor cells mediating their disruption (Moretta and Moretta 2004).

However, the role of NK cells in controlling the growth of human tumors has been less extensively explored. In mice, tumor rejection is positively correlated with precursor frequency of both tumor-specific CD8+ T cells and NK effector cells are able to interact with tumor cells via the activating receptor killer cell lectin-like receptor subfamily K, member 1 (KLRK1 also known as NKG2D) (Shanker et al. 2010). Additionally, our group observed that nude mice treated with oncolytic viruses can reject tumor xenografts (Worschech et al. 2009). This rejection was associated with the activation of ISGs (both IFN-γ and IFN-α stimulated genes), upregulation of CXCR3 and CCR5 ligands, and activation of IEF genes (granzyme B, caspase 8). Since these mice lack T cells and secondarily lack B cell responses, this immune-mediated tissue destruction is thought to be induced by innate immune effectors such as NK cells and activated macrophages. This study suggested that, at least in this model, innate immunity can be an independent effector of tissue-specific destruction not requiring adaptive immunity.

In human, it was a decade ago when it has been showed for the first time a correlation between a partial regression of primary tumor growth in melanoma patients and NK cell activity, suggesting that NK cells may represent a potential prognostic marker (Jochems and Schlom 2011). Subsequently, Coca et al. (1997) showed that intratumoral natural killer cells analyzed in a cohort of 157 patients with colorectal cancer was positively correlated with favorable prognosis; patients with little and moderate NK infiltration showed significantly shorter survival rates (overall and disease free survival) than those with extensive infiltration ($P < 0.01$). Three significant factors affecting survival were selected in a stepwise fashion in increasing order as follows: TNM stage, NK infiltration, and lymphocytic infiltration. Patients with TNM Stage III disease and extensive NK infiltration showed significantly longer survival rates than those with little or moderate infiltration ($P < 0.001$). The same observations were described in other cancer types such as in renal, lung cancer and hepatocellular carcinoma (Ishigami et al. 2000; Villegas et al. 2002; Donskov and von der Maase 2006). However, in the majority of these studies NK cells have been detected according to the expression of CD57 or CD56 markers which has been shown to inaccurately identify NK cells since they are also expressed by other tumor-infiltrating lymphoid cells. The advent of high throughput technology, mostly based on gene expression profiling has shown the relevance of the prognostic infiltration of NK cell markers at the tumor site.

In breast cancer, a meta analysis conducted on a large publicly available set of microarray data from primary tumors suggested that all the major histological subtypes of breast cancer displayed variable expression of ligands for NK cell receptors. In particular, NKG2D-ligands and DNAX Accessory Molecule-1 (DNAM1) ligands, known for their NK cell activating function, were found to be widely expressed across all breast cancer subtypes (Mamessier et al. 2011). The same group reported that a decrease in the expression of the activating receptor NKG2D, DNAM1 and NKp46 (considered a gene uniquely expressed by NK cells) was associated with tumor progression.

Recently, our group also observed that NK cell molecular signatures are predictive of relapse free survival of favorable prognosis of breast cancer patients (manuscript in preparation). Tumors were obtained from patients experiencing either 5–9 years relapse-free survival or tumor relapse within 1–6 years following initial treatment. Based on differential expression of CD56 and CD16, NK cells did not vary between relapse free and progressing groups. However, tumors from patients with no recurrence were characterized by up-regulation of activating receptors NKp46, NKp30, NKG2D and DNAM1 as well as molecules involved in the interaction of NK cells with tumor cells. On the contrary, expression of NK cells inhibitory receptors transcripts was not significantly different in patients with widely diverging.

An involvement of activating receptors in cancer prognosis has also been recently observed in gastrointestinal stromal tumor, where an increased expression of the NK activating receptor NKp30 (also known as NCR3) was associated with prolonged survival of patients as well as in renal cancer, where high expression of the activating receptor NKp46 was associated with higher survival probability (Delahaye et al. 2011; Eckl et al. 2012).

Taken together, the following observations suggest that although NK cells represent only a minor component of tumor microenvironments and of lymphocytes infiltrates, they play an important role in cancer immune surveillance as they are able to interact with tumor cells mediating their description or with other target cells, such as DCs, favoring their activation which positively affect the adaptive immune response.

4.4.2 B Cells

The prognostic significance of intra-tumoral B cells remains unclear. Mouse models support a negative role of B cells in cancer immune surveillance. Initially, this activity was mostly associated with an increased production of IL-10 by B cells, which is considered an immunosuppressive cytokine (Wong et al. 2010). However, intravenous administration of recombinant IL-10 to humans produces pro-inflammatory effects by enhancing release of IFN-γ, IFN-inducible protein 10, TNF-α, and IL-1 and appears to induce activation of CTL and NK cells, as reflected by increased plasma levels of granzyme-B (Mocellin et al. 2004). These observations lead to the hypothesis that IL-10 might contribute to the immune-mediated rejection of cancer, at least under some circumstances. Thus, a deleterious role of B cells based on the production of IL-10 is not fully supported by experimental observations (Mocellin et al. 2003; Rossi et al. 2004).

Another negative effect of intra-tumoral B cells could be due to the production of IgG, forming antigen-IgG antibody complexes which may activate an M2 protoumor phenotype in macrophages promoting the early stage of carcinogenesis (Andreu et al. 2010). However, recent evidence, suggest a favorable association between B cell infiltration and prognosis in epithelial ovarian cancer. In particular, tissue microarray (TMA) performed in a cohort of 198 patients showed that intraepithelial CD20+ cells occurred in 41.9 % (83/198) of evaluable tumors. Moreover, presence of CD20+ infiltrates was strongly associated with that of T cell subsets (CD3, CD4, and CD8), with the activation markers CD45RO and CD25 and with Granzyme B and FoxP3. Finally, CD20+ infiltrates were associated with increased disease specific survival (DSS). This effect was mostly ascribed to the promotion of the opsonization of tumor antigens, complement-mediated destruction of tumor cells, or antibody-dependent cellular cytotoxicity and to the fact that B cells can also present antigen to both CD4+ and CD8+ T cells (Milne et al. 2009).

A similar portrait was recently described by our group for breast cancer patients where the expression of the B cells marker immunoglobulin K C (IGKC), predicted relapse-free survival with higher than 85 % accuracy (Ascierto et al. 2012). Interestingly, genes involved in primary immunodeficiency signaling, T cell apoptosis,

CTLA-4 signaling and production of NO and reactive oxygen species were also up-regulated in the tumor specimens of relapse-free patients. Such paradoxical findings as to simultaneous up-regulation of immune effector genes and immune suppressor genes may suggest that tumor-derived factors were responsible for the expression of immune suppressor genes thereby facilitating cancer progression even in the presence of the increased immune effector genes.

A subsequent comprehensive analysis of human gene expression profiles confirmed the stromal IGKC as a prognostic marker in breast cancer (Schmidt et al. 2012) and other solid tumor types including lung adenocarcinoma and colorectal carcinoma. Interestingly, any significant association instead was found for ovarian carcinoma (Schmidt et al. 2012).

The observations that IGKC in tumor is dependent upon plasma cells contradict the assumption that tumor cells are capable of producing immunoglobulins to promote tumor growth and survival (Qiu et al. 2003). Rather it supports a previous report that cancer specimens typically have tumor infiltration of IgG-positive plasma cells (Ito et al. 1986). Although the biological role of the IGK have to be further addressed, the prognostic impact shared by breast, lung and colorectal carcinoma represents a comprehensive biomarker predicting of the immune system in a variety of cancer types. Moreover, the evidences that the presence of both CD8+ and CD20+ Tumor-Infiltrating Lymphocytes (TILs) was associated with increased survival compared with CD8+ TILs alone provides evidence that CD8+ and CD20+ TIL act cooperatively to promote immunity (Nielsen et al. 2012). The authors proposed three roles for CD20+ TILs in promoting antitumor immunity. First, B cell can bind tumor antigens via surface Ig molecules, process them and then present peptides to CD8+ and CD4+ T cells via MHC Class I and Class II, respectively. Second, B cells trough the secretion of lymphotoxin, can induce stromal cells to express adhesion molecules, cytokines and chemokines which in turn can recruit and retain other lymphocytes. Third, Type -1 and Type 2 B effector cells (Be1 and Be2) can secrete cytokines such as IFN-γ, and IL-4, which can activate T-cell responses toward Th1, Th2 or other functional states.

In summary cells of the innate, adaptive and humoral immunity and the cytokines that they produce are associated with good clinical outcome for all cancer types. In Fig. 4.2 it is possible to note that although this concept is strongly evident and supported for Th1 cells, CTLs and CD8 memory cells, is less appreciated for NK cells and remain still controversial for other T helper cell populations and for B cells. This may be due to the different status of their maturation, on the balance between immune cells in the tumor microenvironment and on the tumor type in consideration.

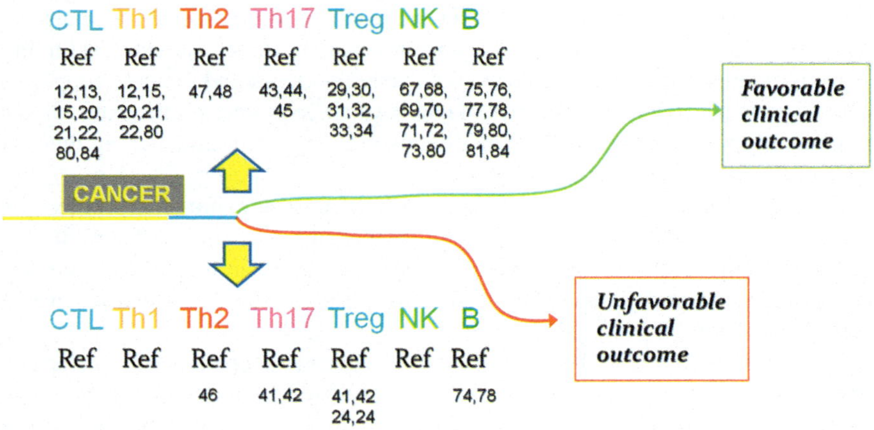

Fig. 4.2 Association of immune cell infiltrates with prognosis in cancer

4.5 The Immune Score Approach: A Novel Approach for Cancer Classification

Usually, outcome prediction in cancer is based on evaluation of tissue samples obtained during surgical removal of the primary tumor focusing on their histo-pathological characteristics. In addition, histological as well as radiological analysis of tumor draining, regional lymph nodes and distant organs are performed to identity evidence of metastases. Until now tumor staging (AJCC/UICC-TNM classification) summarizes data on tumor burden (T), presence of cancer cells in draining and regional lymph nodes (N), and evidence for metastases (M). However, this classification provides inaccurate information for prognosis since cancer outcomes can vary significantly among patients within the same stage and does not predict response to therapy.

Based on numerous reports suggesting that cancer development is controlled by the host's immune system the importance of including immunological biomarkers for the prediction of prognosis and response to therapy appeared to be necessary. By analyzing 400 patients with colorectal cancer, Mlecnik et al. (2011) suggest that immune cell infiltration by cytotoxic CD8-positive and memory CD45RO-positive T cells has prognostic discriminatory power that is superior to standard staging. The results demonstrate two key findings: patients with high immune scores have increased disease-free and overall survival as compared with patients whose tumors demonstrate low immune scores, and the immune score was superior in predicting disease outcome as compared with a host of important prognostic clinical parameters, including TNM staging. Thus, although there are still few issues that need to be addressed (Broussard and Disis 2011), it seems important to consider immune scoring as a prognostic factor and to introduce this parameter as a marker to classify cancers, as part of the routine diagnostic and prognostic assessment of tumors. At the same time, the inherent complexity of quantitative immunohistochemistry, in conjunction with variable assay protocols across laboratories, the different immune cell types analyzed different region selection criteria, and variable ways to quantify immune infiltration underscore the urgent need to reach assay harmonization (Emens et al. 2012).

In an effort to promote the immunoscore in routine clinical settings worldwide, the Society for Immunotherapy of Cancer (SITC), the European Academy of Tumor Immunology, the Cancer and Inflammation Program, the National Cancer Institute, National Institutes of Health, USA and "La Fondazione Melanoma" initiated on February 2012 in Naples a task force on Immunoscoring as a New Possible Approach for the Classification of Cancer. An immune-classification of tumors has been proposed based on an immune score, performed by the quantification of two lymphocyte populations (CD3/CD8, or CD3/CD45RO, or CD8/CD45RO), both in the core of the tumor and the invasive margin of the tumor, to establish prognosis of clinical outcome in patients. Initially, the Immune score screening, which involves the participation of 21 international centers, will be performed on colon cancer patients where high densities of T cells (CD3+), of cytotoxic T cells (CD8+), and of memory T cells (CD45RO+) are clearly associated with a longer disease-free (after surgical resection of the primary tumor) and/or overall survival (Pages et al. 2009). However, the validation of Immune score on other cancer types is under discussion.

A "Workshop on Immune Scoring" organized in Naples in December of 2012 will lead to the preparation of a summary document providing recommendations for the harmonization and implementation of the Immune Score as a new component for the classification of cancer (Galon 2012).

4.6 The Immune Phenotype Associated with Immune Responsiveness in Partly Dependent Upon Genetic Background of the Host and Somatic Alteration of the Tumor

All together the results derived from the studies mentioned above strongly underline the relative role played by the innate, adaptive a humoral immune systems in the tumor milieu and their coordinated action in tumor rejection which in our idea is limited not only by immunosuppressant induced by tumor growth but also by a natural immune response of our system that try to prevent a perceived autoimmune reaction.

There have been several demonstrations of a linkage between the induction of autoimmunity with immune based cancer therapies and antitumor response (Disis 2010; Wang et al. 2008). Autoimmunity and tumor regression have occurred after the use of a variety of immune based treatments. The development of *vitiligo* and uveitis has been reported after tumor regression was induced with the adoptive transfer of T cells expanded ex vivo from tumor-infiltrating lymphocytes in patients with metastatic melanoma (Dudley et al. 2002).

In a study performed on 200 patients with melanoma undergoing high dose IFN-α-2β adjuvant treatment, it was observed that 26 % developed autoantibodies and clinical symptoms consist in autoimmunity. Moreover, autoimmunity in this study was showed to be predictive of response free survival (RFS) and overall survival (OR) ($P < 0.001$) (Gogas et al. 2006). If we assume that if autoimmunity is associated with cancer immune responsiveness, polymorphisms of genes related to autoimmunity itself might be associated with cancer regression (Wang et al. 2012). A recent observation supporting this hypothesis came from a group of patients with melanoma treated with the adoptive transfer of ex vivo activated tumor infiltrating lymphocytes (Uccellini et al. 2012).

Consistent with the concept that autoimmunity and cancer rejection might represent different facets of the same phenomenon, we observed that polymorphisms protecting against the susceptibility to develop systemic lupus erythematosus (SLE) such as the *IRF5* rs10954213 GG genotype, were significantly more prevalent among patients who did not respond to adoptive TIL therapy. It is interesting to observe that the *IRF5* genotype appeared to segregate two different patterns. When genes differentiating melanoma cell lines in vitro according to genotype were applied for class prediction, a segregation of responding and non responding cases was observed and it was only partially predictive of the *IRF5* genotype in vivo. The resulting segregation of cases according to genotype was associated with likelihood of responsiveness suggesting that germline variants can affect directly the intrinsic biology of cancer cells besides affecting the behavior of host's cells (Uccellini et al. 2012).

Thus, we can conclude that the immune phenotype associated with immune responsiveness in partly dependent genetic background of the host (Wang et al. 2012).

Another demonstration, of the impact of genetic alteration of the host on tumor regression and response to cancer therapy was also showed by our group in the evaluation of polymorphisms of CXCR3 and CCR3,-ligand chemokines, (which are considered critical in tumor microenvironment for the recruitment of T lymphocytes and subsequent immune mediated rejection), as predictive biomarkers of clinical response to adoptive therapy in melanoma patients (manuscript in preparation). A common single nucleotide polymorphism of CXCR3 (rs2280964) has been associated with the variation in chemotactic activity. CCR5, polymorphism Δ32, (a deletion of 32 bases encoding a protein not expressed on cell surface), has been correlated with poor prognosis in metastatic melanoma patients. In our study it was postulated that polymorphisms of CXCR3 and CCR5 genes may influence the migration of TILs on tumor side and eventually tumor regression. The results showed that no significant correlation was detected between CXCR3 polymorphisms and response. Surprisingly CCR5-Δ32 carriers had a better overall survival compared to wild type patients. These results allowed us to generate new hypotheses on the role of this molecule in the modulation of stimulatory conditions.

Together with genetic alteration of the host, we need to consider that the immune responsiveness and in general tumor regression is also related to acquired alterations of cancer cells genetics. This explains the phenomenon of mixed responses. In fact, a proportion of cancer patients experience a mix response to therapy characterized by simultaneous regression of the some metastases, while other progress. These cases are very rare and unique because they only consider the tumor's aspects of immune responsiveness excluding the genetic background of the patients.

We recently performed a comparative gene expression analysis of 15 metastases (10 regressing and 5 progressing) obtained from 2 melanoma patients experiencing a mixed response following immunotherapy. The obtained results indicated that the regression of melanoma metastases is associated with acute inflammation mediated by up regulation of genes involved in antigen presentation. This unique study provided a clear demonstration that within the same genetic background tumor can behave differently supporting the hypothesis that tumor rejection is in part dependent upon tumor biology.

Based on these last mentioned observations, it looks like one side the genetic background of the host's have an important impact on immune responsiveness, on the other side it looks like tumor can behave differently within the same genetic background.

This apparent paradox can be explained by a multifactor model of cancer immune responsiveness.

It should be emphasized that host and cancer genetics are mostly overlapping since cancer cells carry the majority of the host genetics. Thus, inherited genetic factors may affect the biology of the cancer besides that of normal cells. Then, it could be hypothesized that some patients carry a genetic background that make them resistant to immunotherapy by affecting either the biology of immune response, the biology of cancer cells, or both.

On the other hand, an "immune –responsive genotype" may still be limited by the genetics of the tumors: although the patients may be predisposed to cancer rejection the tumor lacks additional properties necessary for its recognition by the immune system. In the model we propose a favorable genetic background of the host is necessary but not sufficient for tumor rejection. A good example is provided by the analysis of patients with IRF5 polymorphisms where the immune resistant phenotypes appears to exclusively preclude cancer rejection during adoptive therapy with tumor infiltrating lymphocytes; however the "immune responsive phenotype" can be segregated into two categories. One enriched in patients responding to therapy and the other of non-responding.

4.7 Conclusions

It is becoming clear that tumors can be segregated into at least two categories independently of their histology. Of them, one is characterized by a molecular signature consistent with a Th1 type of immune activation which is also associated with, lymphocytic infiltrate, better prognosis and enhanced likelihood to respond to therapy. This dichotomy may be determined by a continuum interaction of a multitude of factors including the host's genetic background, somatic mutations and external factors such as intensity and effectiveness of treatment or general condition of the patient. This multistep inference may also explain why it is generally easier to predict accurately lack of responsiveness than responsiveness.

Although it is still unclear whether future progresses will cause drastic changes in understanding the biology of cancer or will just add new details in order to better elaborate regulatory circuits that have been already mapped out, looking ahead we prospect significant advances during the coming decade in our understanding of biology of cancer.

Acknowledgments Researcher in the Infectious Disease and Immunogenetics Section (FM Marincola's laboratory) are supported by NIH Intramural Research Program. ML Ascierto thanks Dr. Ermelinda Ascierto and Dr. Domenico Ascierto for insightful suggestions.

References

Ahmadzadeh M, Johnson LA, Heemskerk B, Wunderlich JR, Dudley ME, White DE et al (2009) Tumor antigen-specific CD8 T cells infiltrating the tumor express high levels of PD-1 and are functionally impaired. Blood 114:44–1537

Alizadeh AA, Eisen MB, Davis RE, Ma C, Lossos IS, Rosenwald A et al (2000) Distinct types of diffuse large B-cell lymphoma identified by gene expression profiling. Nature 403:11–503

Allavena P, Sica A, Solinas G, Porta C, Mantovani A (2008) The inflammatory microenvironment in tumor progression: the role of tumor-associated macrophages. Crit Rev Oncol Hematol 66:1–9

Andreu P, Johansson M, Affara NI, Pucci F, Tan T, Junankar S et al (2010) FcRgamma activation regulates inflammation-associated squamous carcinogenesis. Cancer Cell 17:121–134

Ascierto ML, De Giorgi V, Liu Q, Bedognetti D, Spivey TL, Murtas D et al (2011) An immunologic portrait of cancer. J Transl Med 9:146

Ascierto ML, Kmieciak M, Idowu MO, Manjili R, Zhao YD, Grimes M et al (2012) A signature of immune function genes associated with recurrence-free survival in breast cancer patients. Breast Cancer Res Tr 131:871–880

Badoual C, Hans S, Rodriguez J, Peyrard S, Klein C, Agueznay Nel H et al (2006) Prognostic value of tumor-infiltrating CD4 + T-cell subpopulations in head and neck cancers. Clin Cancer Res 12:72–465

Bedognetti D, Wang E, Sertoli MR, Marincola FM (2010) Gene-expression profiling in vaccine therapy and immunotherapy for cancer. Expert Rev Vaccines 9:65–555

Broussard EK, Disis ML (2011) TNM staging in colorectal cancer: T is for T cell and M is for memory. J Clin Oncol 29:601–603

Carreras J, Lopez-Guillermo A, Fox BC, Colomo L, Martinez A, Roncador G et al (2006) High numbers of tumor-infiltrating FOXP3-positive regulatory T cells are associated with improved overall survival in follicular lymphoma. Blood 108:64–2957

Chaput N, Conforti R, Viaud S, Spatz A, Zitvogel L (2008) The Janus face of dendritic cells in cancer. Oncogene 27:5920–5931

Chen X, Wan J, Liu J, Xie W, Diao X, Xu J et al (2010) Increased IL-17-producing cells correlate with poor survival and lymphangiogenesis in NSCLC patients. Lung Cancer 69:348–354

Clemente CG, Mihm MC Jr, Bufalino R, Zurrida S, Collini P, Cascinelli N (1996) Prognostic value of tumor infiltrating lymphocytes in the vertical growth phase of primary cutaneous melanoma. Cancer 77:10–1303

Coca S, Perez-Piqueras J, Martinez D, Colmenarejo A, Saez MA, Vallejo C et al (1997) The prognostic significance of intratumoral natural killer cells in patients with colorectal carcinoma. Cancer 79:2320–2328

Cochran AJ (1969) Histology and prognosis in malignant melanoma. J Pathol 97:68–459

Curiel TJ, Coukos G, Zou L, Alvarez X, Cheng P, Mottram P et al (2004) Specific recruitment of regulatory T cells in ovarian carcinoma fosters immune privilege and predicts reduced survival. Nat Med 10:9–942

De Monte L, Reni M, Tassi E, Clavenna D, Papa I, Recalde H et al (2011) Intratumor T helper type 2 cell infiltrate correlates with cancer-associated fibroblast thymic stromal lymphopoietin production and reduced survival in pancreatic cancer. J Exp Med 208:469–478

Delahaye NF, Rusakiewicz S, Martins I, Menard C, Roux S, Lyonnet L et al (2011) Alternatively spliced NKp30 isoforms affect the prognosis of gastrointestinal stromal tumors. Nat Med 17:700–707

Disis ML (2010) Immune regulation of cancer. J Clin Oncol 28:8–4531

Donskov F, von der Maase H (2006) Impact of immune parameters on long-term survival in metastatic renal cell carcinoma. J Clin Oncol 24:1997–2005

Dudley ME, Wunderlich JR, Robbins PF, Yang JC, Hwu P, Schwartzentruber DJ et al (2002) Cancer regression and autoimmunity in patients after clonal repopulation with antitumor lymphocytes. Science 298:850–854

Eckl J, Buchner A, Prinz PU, Riesenberg R, Siegert SI, Kammerer R et al (2012) Transcript signature predicts tissue NK cell content and defines renal cell carcinoma subgroups independent of TNM staging. J Mol Med 90:55–66 (Berl)

Emens LA, Silverstein SC, Khleif S, Marincola FM, Galon J (2012) Toward integrative cancer immunotherapy: targeting the tumor microenvironment. J Transl Med 10:70

Frey DM, Droeser RA, Viehl CT, Zlobec I, Lugli A, Zingg U et al (2010) High frequency of tumor-infiltrating FOXP3(+) regulatory T cells predicts improved survival in mismatch repair-proficient colorectal cancer patients. Int J Cancer 126:2635–2643

Fridman WH, Pages F, Sautes-Fridman C, Galon J (2012) The immune contexture in human tumours: impact on clinical outcome. Nat Rev Cancer 12:298–306

Gajewski TF (2011) Molecular profiling of melanoma and the evolution of patient-specific therapy. Semin Oncol 38:42–236

Galon J, Costes A, Sanchez-Cabo F, Kirilovsky A, Mlecnik B, Lagorce-Pages C et al (2006) Type, density, and location of immune cells within human colorectal tumors predict clinical outcome. Science 313:4–1960

Galon J, Pages F, Marincola FM, Thurin M, Trinchieri G, Fox BA et al (2012) The immune score as a new possible approach for the classification of cancer. J Transl Med 3:10–11

Gao Q, Qiu SJ, Fan J, Zhou J, Wang XY, Xiao YS et al (2007) Intratumoral balance of regulatory and cytotoxic T cells is associated with prognosis of hepatocellular carcinoma after resection. J Clin Oncol 25:93–2586

Gogas H, Ioannovich J, Dafni U, Stavropoulou-Giokas C, Frangia K, Tsoutsos D et al (2006) Prognostic significance of autoimmunity during treatment of melanoma with interferon. N Engl J Med 354:709–718

Guidoboni M, Gafa R, Viel A, Doglioni C, Russo A, Santini A et al (2001) Microsatellite instability and high content of activated cytotoxic lymphocytes identify colon cancer patients with a favorable prognosis. Am J Pathol 159:297–304

Hanahan D, Weinberg RA (2000) The hallmarks of cancer. Cell 100:57–70

Hanahan D, Weinberg RA (2011) Hallmarks of cancer: the next generation. Cell 144:74–646

Heimberger AB, Abou-Ghazal M, Reina-Ortiz C, Yang DS, Sun W, Qiao W et al (2008) Incidence and prognostic impact of FoxP3 + regulatory T cells in human gliomas. Clin Cancer Res 14:72–5166

Herberman RB, Nunn ME, Holden HT, Lavrin DH (1975) Natural cytotoxic reactivity of mouse lymphoid cells against syngeneic and allogeneic tumors. II. Characterization of effector cells. Int J Cancer 16:230–239

Hillen F, Baeten CI, van de Winkel A, Creytens D, van der Schaft DW, Winnepenninckx V et al (2008) Leukocyte infiltration and tumor cell plasticity are parameters of aggressiveness in primary cutaneous melanoma. Cancer Immunol Immunother 57:97–106

Hoffmann P, Boeld TJ, Eder R, Huehn J, Floess S, Wieczorek G et al (2009) Loss of FOXP3 expression in natural human CD4 + CD25 + regulatory T cells upon repetitive in vitro stimulation. Eur J Immunol 39:1088–1097

Ishigami S, Natsugoe S, Tokuda K, Nakajo A, Che X, Iwashige H et al (2000) Prognostic value of intratumoral natural killer cells in gastric carcinoma. Cancer 88:577–583

Ito T, Saga S, Nagayoshi S, Imai M, Aoyama A, Yokoi T et al (1986) Class distribution of immunoglobulin-containing plasma cells in the stroma of medullary carcinoma of breast. Breast Cancer Res Treat 7:97–103

Jensen TO, Schmidt H, Moller HJ, Donskov F, Hoyer M, Sjoegren P et al (2012) Intratumoral neutrophils and plasmacytoid dendritic cells indicate poor prognosis and are associated with pSTAT3 expression in AJCC stage I/II melanoma. Cancer 118(9):2476–2485

Jochems C, Schlom J (2011) Tumor-infiltrating immune cells and prognosis: the potential link between conventional cancer therapy and immunity. Exp Biol Med 236:567–579 (Maywood)

Kelley T, Beck R, Absi A, Jin T, Pohlman B, Hsi E (2007) Biologic predictors in follicular lymphoma: importance of markers of immune response. Leuk Lymphoma 48:2403–2411

Kiessling R, Klein E, Pross H, Wigzell H (1975) "Natural" killer cells in the mouse. II. Cytotoxic cells with specificity for mouse Moloney leukemia cells. Characteristics of the killer cell. Eur J Immunol 5:117–121

Kryczek I, Wei S, Zou L, Zhu G, Mottram P, Xu H et al (2006) Cutting edge: induction of B7–H4 on APCs through IL-10: novel suppressive mode for regulatory T cells. J Immunol 177:40–44

Kryczek I, Banerjee M, Cheng P, Vatan L, Szeliga W, Wei S et al (2009) Phenotype, distribution, generation, and functional and clinical relevance of Th17 cells in the human tumor environments. Blood 114:1141–1149

Lossos IS, Alizadeh AA, Eisen MB, Chan WC, Brown PO, Botstein D et al (2000) Ongoing immunoglobulin somatic mutation in germinal center B cell-like but not in activated B cell-like diffuse large cell lymphomas. Proc Natl Acad Sci U S A 97:13–10209

Mahmoud SM, Paish EC, Powe DG, Macmillan RD, Lee AH, Ellis IO et al (2011) An evaluation of the clinical significance of FOXP3 + infiltrating cells in human breast cancer. Breast Cancer Res Treat 127:99–108

Mamessier E, Sylvain A, Thibult ML, Houvenaeghel G, Jacquemier J, Castellano R et al (2011) Human breast cancer cells enhance self tolerance by promoting evasion from NK cell antitumor immunity. J Clin Invest 121:3609–3622

Mantovani A, Romero P, Palucka AK, Marincola FM (2008) Tumour immunity: effector response to tumour and role of the microenvironment. Lancet 371:83–771

Marincola FM (2005) A balanced review of the status T cell-based therapy against cancer. J Transl Med 3:16

Martuza RL, Malick A, Markert JM, Ruffner KL, Coen DM (1991) Experimental therapy of human glioma by means of a genetically engineered virus mutant. Science 252:854–856

Masferrer JL, Leahy KM, Koki AT, Zweifel BS, Settle SL, Woerner BM et al (2000) Antiangiogenic and antitumor activities of cyclooxygenase-2 inhibitors. Cancer Res 60:11–1306

Milne K, Kobel M, Kalloger SE, Barnes RO, Gao D, Gilks CB et al (2009) Systematic analysis of immune infiltrates in high-grade serous ovarian cancer reveals CD20, FoxP3 and TIA-1 as positive prognostic factors. PLoS ONE 4:e6412

Mlecnik B, Tosolini M, Kirilovsky A, Berger A, Bindea G, Meatchi T et al (2011) Histopathologic-based prognostic factors of colorectal cancers are associated with the state of the local immune reaction. J Clin Oncol 29:610–618

Mocellin S, Panelli MC, Wang E, Nagorsen D, Marincola FM (2003) The dual role of IL-10. Trends Immunol 24:36–43

Mocellin S, Panelli M, Wang E, Rossi CR, Pilati P, Nitti D et al (2004) IL-10 stimulatory effects on human NK cells explored by gene profile analysis. Genes Immun 5:621–630

Moretta L, Moretta A (2004) Unravelling natural killer cell function: triggering and inhibitory human NK receptors. EMBO J 23:255–259

Moretta L, Ciccone E, Mingari MC, Biassoni R, Moretta A (1994) Human natural killer cells: origin, clonality, specificity, and receptors. Adv Immunol 55:341–380

Nielsen JS, Sahota RA, Milne K, Kost SE, Nesslinger NJ, Watson PH et al (2012) CD20 + tumor-infiltrating lymphocytes have an atypical CD27- memory phenotype and together with CD8 + T cells promote favorable prognosis in ovarian cancer. Clin Cancer Res 18:3281–3292

Oldham RK, Herberman RB (1973) Evaluation of cell-mediated cytotoxic reactivity against tumor associated antigens with 125I-iododeoxyuridine labeled target cells. J Immunol 111:862–871

Ostrand-Rosenberg S, Sinha P (2009) Myeloid-derived suppressor cells: linking inflammation and cancer. J Immunol 182:4499–4506

Pages F, Berger A, Camus M, Sanchez-Cabo F, Costes A, Molidor R et al (2005) Effector memory T cells, early metastasis, and survival in colorectal cancer. N Engl J Med 353:66–2654

Pages F, Kirilovsky A, Mlecnik B, Asslaber M, Tosolini M, Bindea G et al (2009) In situ cytotoxic and memory T cells predict outcome in patients with early-stage colorectal cancer. J Clin Oncol 27:5944–5951

Perou CM, Sorlie T, Eisen MB, van de Rijn M, Jeffrey SS, Rees CA et al (2000) Molecular portraits of human breast tumours. Nature 406:52–747

Qiu X, Zhu X, Zhang L, Mao Y, Zhang J, Hao P et al (2003) Human epithelial cancers secrete immunoglobulin g with unidentified specificity to promote growth and survival of tumor cells. Cancer Res 63:6488–6495

Rossi CR, Testori A, Mocellin S, Campana L, Lejeune F (2004) Melanoma—what is new in sentinel node biopsy and locoregional treatments in 2003? Report of a workshop at the third research meeting on Melanoma, Milan, Italy, May 2003. Melanoma Res 14:329–332

Salama P, Phillips M, Grieu F, Morris M, Zeps N, Joseph D et al (2009) Tumor-infiltrating FOXP3 + T regulatory cells show strong prognostic significance in colorectal cancer. J Clin Oncol 27:186–192

Schmidt M, Hellwig B, Hammad S, Othman A, Lohr M, Chen Z et al (2012) A comprehensive analysis of human gene expression profiles identifies stromal immunoglobulin kappa C as a compatible prognostic marker in human solid tumors. Clin Cancer Res 18:2695–2703

Schreck S, Friebel D, Buettner M, Distel L, Grabenbauer G, Young LS et al (2009) Prognostic impact of tumour-infiltrating Th2 and regulatory T cells in classical Hodgkin lymphoma. Hematol Oncol 27:31–39

Schwartzentruber DJ, Lawson DH, Richards JM, Conry RM, Miller DM, Treisman J et al (2011) Gp100 peptide vaccine and interleukin-2 in patients with advanced melanoma. N Engl J Med 364:2119–2127

Shanker A, Marincola FM (2011) Cooperativity of adaptive and innate immunity: implications for cancer therapy. Cancer Immunol Immunother 60:1061–1074

Shanker A, Verdeil G, Buferne M, Inderberg-Suso EM, Puthier D, Joly F et al (2007) CD8 T cell help for innate antitumor immunity. J Immunol 179:6651–6662

Shanker A, Buferne M, Schmitt-Verhulst AM (2010) Cooperative action of CD8 T lymphocytes and natural killer cells controls tumour growth under conditions of restricted T-cell receptor diversity. Immunology 129:41–54

Sinha P, Clements VK, Fulton AM, Ostrand-Rosenberg S (2007) Prostaglandin E2 promotes tumor progression by inducing myeloid-derived suppressor cells. Cancer Res 67:13–4507

Tassi E, Gavazzi F, Albarello L, Senyukov V, Longhi R, Dellabona P et al (2008) Carcinoembryonic antigen-specific but not antiviral CD4 + T cell immunity is impaired in pancreatic carcinoma patients. J Immunol 181:6595–6603

Tosolini M, Kirilovsky A, Mlecnik B, Fredriksen T, Mauger S, Bindea G et al (2011) Clinical impact of different classes of infiltrating T cytotoxic and helper cells (Th1, th2, treg, th17) in patients with colorectal cancer. Cancer Res 71:1263–1271

Uccellini L, De Giorgi V, Zhao Y, Tumaini B, Erdenebileg N, Dudley ME et al (2012) IRF5 gene polymorphisms in melanoma. J Transl Med 10:170

Villegas FR, Coca S, Villarrubia VG, Jimenez R, Chillon MJ, Jareno J et al (2002) Prognostic significance of tumor infiltrating natural killer cells subset CD57 in patients with squamous cell lung cancer. Lung Cancer 35:23–28

Vivier E, Raulet DH, Moretta A, Caligiuri MA, Zitvogel L, Lanier LL et al (2011) Innate or adaptive immunity? The example of natural killer cells. Science 331:44–49

Vivier E, Ugolini S, Blaise D, Chabannon C, Brossay L (2012) Targeting natural killer cells and natural killer T cells in cancer. Nat Rev Immunol 12:239–252

Walker MR, Kasprowicz DJ, Gersuk VH, Benard A, Van Landeghen M, Buckner JH et al (2003) Induction of FoxP3 and acquisition of T regulatory activity by stimulated human CD4 + CD25- T cells. J Clin Invest 112:1437–1443

Wang E, Worschech A, Marincola FM (2008) The immunologic constant of rejection. Trends Immunol 29:62–256

Wang XY, Han M, Xiao ZP, Chen L, Kong LH, Fu K et al (2009) The comparison of CD4(+);CD25(high); Foxp3(+); regulatory T cells between neonatal cord blood and adult peripheral blood. Bao X, Yu Fen Zi Mian Yi Xue Za Zhi. 25:1152–1154

Wang E, Uccellini L, Marincola FM (2012) A genetic inference on cancer immune responsiveness. Oncoimmunology 1:520–525

Wilke CM, Kryczek I, Wei S, Zhao E, Wu K, Wang G et al (2011) Th17 cells in cancer: help or hindrance? Carcinogenesis 32:643–649

Wong SC, Puaux AL, Chittezhath M, Shalova I, Kajiji TS, Wang X et al (2010) Macrophage polarization to a unique phenotype driven by B cells. Eur J Immunol 40:2296–2307

Worschech A, Chen N, Yu YA, Zhang Q, Pos Z, Weibel S et al (2009) Systemic treatment of xenografts with vaccinia virus GLV-1h68 reveals the immunologic facet of oncolytic therapy. BMC Genomics 10:301

Yoon NK, Maresh EL, Shen D, Elshimali Y, Apple S, Horvath S et al (2010) Higher levels of GATA3 predict better survival in women with breast cancer. Hum Pathol 41:1794–1801

Zhang L, Conejo-Garcia JR, Katsaros D, Gimotty PA, Massobrio M, Regnani G et al (2003) Intratumoral T cells, recurrence, and survival in epithelial ovarian cancer. N Engl J Med 348:13–203

Zhang JP, Yan J, Xu J, Pang XH, Chen MS, Li L et al (2009) Increased intratumoral IL-17-producing cells correlate with poor survival in hepatocellular carcinoma patients. J Hepatol 50:980–989

Zhu XD, Zhang JB, Zhuang PY, Zhu HG, Zhang W, Xiong YQ et al (2008) High expression of macrophage colony-stimulating factor in peritumoral liver tissue is associated with poor survival after curative resection of hepatocellular carcinoma. J Clin Oncol 26:2707–2716

Zuo T, Wang L, Morrison C, Chang X, Zhang H, Li W et al (2007) FOXP3 is an X-linked breast cancer suppressor gene and an important repressor of the HER-2/ErbB2 oncogene. Cell 129:1275–1286

Part II
Developmental Characteristics of the Tumor Immunoenvironment

Chapter 5
Development of Antitumor Cellular Immunity

M. J. P. Welters and S. H. van der Burg

Abstract A dazzling picture of many different types of innate and adaptive immune cells that have infiltrated a patient's tumor emerges when a tumor section is studied under the microscope. There is good evidence that patient survival depends on the numbers, type, character and localization of particular tumor-infiltrating immune cells, in particular T cells and macrophages. Here we discuss the events governing the arousal of a spontaneous tumor-specific T cell response and how the tumor-rejecting efficacy of this T cell response is regulated by the intratumoral cytokine milieu, the expression of inhibitory molecules and co-infiltrating immune cells. Finally, we describe approaches to change the local micromilieu so that the net outcome is a strong induction of an anti-tumor immune response coupled to a better infiltration of tumors under conditions that allows these immune cells to exert their function and to control tumor outgrowth.

Keywords Tumor microenvironment · Immune infiltration · Tumor escape mechanism · Combination therapy · Tumorigenesis · Antitumor immunity · T cells

---

M. J. P. Welters · S. H. van der Burg
Department of Clinical Oncology, Leiden University Medical Center,
Albinusdreef 2, 2333 ZA, Leiden, The Netherlands

S. H. van der Burg (✉)
Department of Clinical Oncology, Leiden University Medical Center,
Building 1, K1-P, PO box 9600, 2300 RC Leiden, The Netherlands
e-mail: shvdburg@lumc.nl

M. R. Shurin et al. (eds.), *The Tumor Immunoenvironment*,
DOI: 10.1007/978-94-007-6217-6_5,

5.1 Immune Infiltration of Tumors is Associated with Clinical Outcome

The transformation of cells and the outgrowth of carcinomas take place in the face of the immune system. Nevertheless, the immune system is able to eradicate tumors, to stop them in their growth and/or to prevent its progression to metastasis, a concept which is called immunosurveillance (Swann and Smyth 2007; Vesely et al. 2011; Zitvogel et al. 2006; Smyth et al. 2006). The effector arm of the immune system is an effective suppressor mechanism of tumor outgrowth but it is opposed by other parts of the immune system displaying tumor promoting functions. The combination of these two acting sides of the immune system may determine the way a tumor develops and is called cancer immunoediting (Dunn et al. 2002; Schreiber et al. 2011). The role of the immune system in cancer is demonstrated by a multitude of immunohistochemical studies on different types of tumors showing that the magnitude and type of immune cell infiltrating a patient's tumor is correlated with the final clinical outcome of the cancer patients. To be more precise, patients with tumors evidently infiltrated with CD8+ T lymphocytes [in cervical cancer: (Piersma et al. 2007), in breast cancer: (Marrogi et al. 1997; Menegaz et al. 2008; Mahmoud et al. 2011), in ovarian cancer: (Zhang et al. 2003), in non small cell lung cancer (NSCLC): (Al-Shibli et al. 2008; Dieu-Nosjean et al. 2008; Hiraoka et al. 2006; Nelson 2008), in melanoma: (Clemente et al. 1996; Haanen et al. 2006) and memory T cells (i.e. CD45RO+) as described for colorectal cancer (Pages et al. 2005; Galon et al. 2006; Nosho et al. 2010)], but also other immune cells such as dendritic cells (DC) (Eisenthal et al. 2001; Zeid and Muller 1993), type 1 macrophages (M1) (Algars et al. 2012; Kinouchi et al. 2011; Erreni et al. 2011; Ohri et al. 2009; Forssell et al. 2007; Ohno et al. 2003) and B cells (Nielsen et al. 2012; Martinet et al. 2011; Ladanyi et al. 2011; Schmidt et al. 2008; Erdag et al. 2012) form groups of patients that have a favorable prognosis in terms of disease free period and survival (Zhang et al. 2003; Pages et al. 2005; Galon et al. 2006; Erdag et al. 2012; Oble et al. 2009; Sato et al. 2005; Denkert et al. 2010; Wang et al. 2012; Kashimura et al. 2012; Nelson 2010).

In contrast, a dense infiltration of tumors CD4+ forkhead box P3 (FoxP3)-positive regulatory T cells (Tregs) is related to a worse prognosis of the cancer patient (Curiel et al. 2004; Shen et al. 2010; Jacobs et al. 2010; Yamagami et al. 2011; Elkord et al. 2010; Raghavan and Quiding-Jarbrink 2011; Mathai et al. 2012; Kim et al. 2012). The exception to the rule might be the infiltration of Tregs in colorectal cancer, where these cells are believed to suppress tumor promoting interleukin 17 (IL-17)-producing T cells (Correale et al. 2010; Ladoire et al. 2011; Whiteside 2012). IL-17-producing T cells have been discovered only 6 years ago and their function was heavily debated as investigators showed tumor promoting and tumor suppressive functions of T cells producing IL-17. However, it is imperative to understand that the production of IL17 is not synonymous to a T helper 17 (Th17) cell, and that other cells of the immune system can produce IL-17 as well. The few reports on Th17 cells in the human microenvironment suggest that Th17 are correlated with improved patient survival (Wilke et al. 2011a, b; Zamarron and Chen 2011). Most recently, Th17 were reported to have stem cell-like features that allow them to promote long-term anti tumor immunity (Wei et al. 2012). These Th17 are polyclonal functional [produce not only IL-17 but also interferon alpha (IFN-γ), tumor necrosis factor alpha (TNF-α) and granulocyte macrophage colony-stimulating factor (GM-CSF)], are highly resistant to apoptosis, have a highly proliferative renewal capacity and are able to persist in time (Kryczek et al. 2011). Importantly, in many tumor types (among which cervical cancer, breast cancer, colon carcinoma, tonsillar carcinoma and gastric cancer) the ratio between the effector T cells (i.e. CD8+ T cells) and the inhibitory T cells, such as Tregs, has been found to form an (independent) prognostic factor for the patients' outcome (Shen et al. 2010; Jordanova et al. 2008; Liu et al. 2011; Shah et al. 2011; Yoon et al. 2012; Suzuki et al. 2010; Nasman et al. 2012). Notably, the capacity of T cell to infiltrate the tumor can also be hampered as a result of an elevated lymph drainage from tumors to the lymph node (Harrell et al. 2007; Thomas et al. 2012) through mechanical stress acting on stromal cell functions and on the extracellular matrix (Swartz and Lund 2012) and may lead to an overall poor immune infiltration of tumors. In addition to Tregs, the infiltration of tumors with type 2 macrophages (M2) (van Dongen et al. 2010; Bronkhorst et al. 2011; Kurahara et al. 2011; Heusinkveld and van der Burg 2011; Allavena and Mantovani 2012), immature myeloid cells and myeloid-derived suppressor cells (MDCSs) (Montero et al. 2012; Poschke and Kiessling 2012; Gabrilovich et al. 2012) also has a negative impact on the clinical outcome of the patient.

5.2 The Influence of the Intratumoral Cytokine Milieu

An important aspect of the tumor is the cytokine profile within the tumor microenvironment. Some of these cytokines may help to reject tumor cells whereas others promote tumor growth. The critical role of the inflammatory cytokine IFN-γ and the cytotoxic components such as granzymes and granulysin in controlling the tumor growth has been demonstrated in mice studies (Koebel et al. 2007) as well as in patient cohorts (Galon et al. 2006; Tosolini et al. 2011). IFN-γ is produced by tumor infiltrating natural killer (NK), T and NKT cells and the levels are even further increased through IL-12, produced by well activated antigen presenting cells (APCs) (Yoshimoto et al. 1998; Okamura et al. 1995). It upregulates major histocompatibility complex (MHC) on tumor cells, induces the immunoproteasome in APCs, recruits lymphocytes and skews naïve T cells into Th1 cells via IL-12 producing APCs. Furthermore, IFN-γ produced by Th1 cells at the tumor site is found to be very important for the recruitment of CD8+ cytotoxic T cells (CTLs) into the tumor as well as for the sustainment of CD8+ T cell effector function (Bos and Sherman 2010; Wong et al. 2008). In addition, the cytokines IL-2 and type 1 interferons (i.e. IFN-α) are proinflammatory cytokines that via activated T cells (producing IL-2) and/or proper APC activation (by IFN-α

and IL-2) endow CTLs with the ability to kill tumor cells (Mocellin et al. 2002; Gajewski 2012). The chemokines CXCL9/10/11 and/or CCL5 (RANTES) are held responsible for the attraction of effector lymphocytes (expressing CXCR3 and/or CCR5, respectively) from the circulation into the tumor (Rahir and Moser 2012; Abastado 2012; Koizumi et al. 2007; Verbeke et al. 2011).

Good examples of tumor promoting cytokines, that work either by inducing an immune suppressive environment for the infiltrated effector cells and/or by alteration of the tumor vasculature, are transforming growth factor β (TGF-β), IL-10, vascular endothelial growth factor (VEGF), IL-6, prostaglandin E2 (PGE2) and the enzyme indoleamine 2,3-dioxygenase (IDO). TGF-β, released upon tissue damage in order to remodel/repair the tumor tissue, has multiple roles inside the tumor. It regulates the suppression of infiltrated effector T and B cells (Flavell et al. 2010), activates Tregs via tolerogenic DC (Yang et al. 2010; Bierie and Moses 2010) and recruits NK cells, neutrophils and/or macrophages to the tumor but inhibits their function (Bierie and Moses 2010). TGF-β enhances tumor cell migration via chemotaxis of fibroblasts that facilitate the invasion of tumor cells into the normal tissue (Shieh et al. 2011; Padua and Massague 2009), and it also alters the extracellular matrix in the tumor (Swartz and Lund 2012). Another tumor promoting cytokine, IL-10, is produced by intratumoral DCs, macrophages, Tregs and/or Th2 cells and is able to inhibit the cytokine production and function of tumor-specific effector T cells (Moore et al. 2001; Joss et al. 2000). IL-10 downregulates cell surface MHC expression on APCs and on tumor cells resulting in non-optimal stimulation of CTLs and a less effective attack of tumor cells (Kim et al. 1995; Steinbrink et al. 1999). In addition, IL-10 also decreases the expression of B7 costimulatory molecules on APCs and inhibits the production of proinflammatory cytokines and chemokines by these APCs (de Waal Malefyt et al. 1991; Ding et al. 1993). Furthermore, IL-10 is involved in the development of Tregs and their suppressive action (Wei et al. 2005; Zou 2006).

PGE2 is produced by several inflamed tumors and promotes tumor angiogenesis, metastasis and converts the differentiation of immune stimulatory APCs (monocytes, M1 macrophages, DCs) towards tumor promoting APCs (Herfs et al. 2009; Obermajer et al. 2011; Heusinkveld et al. 2011). VEGF, produced by both tumor cells and APCs, stimulates angiogenesis within the tumor tissue but also stimulates myeloid cell differentiation and function towards so called type 2 DCs (DC2; CD11c-CD123+), which are involved in skewing the T cell response towards Th2 (Osada et al. 2008; Sheng et al. 2011; Schmid and Varner 2010). The cytokine IL-6 can be produced by tumor cells, tumor resident macrophages as well as by T cells upon trauma. Similar to PGE2, IL-6 skews monocyte differentiation towards M2 macrophages (Sheng et al. 2011; Dijkgraaf et al. 2012a; Waetzig and Rose-John 2012). IDO produced by tumor infiltrating APCs and certain tumor cells can inhibit the effector function of T cells by starving them from tryptophan (Muller and Prendergast 2007; Singer et al. 2011; Soliman et al. 2010). IDO expression in APCs is induced as a result of Forkhead box O3 (FOXO3) activation (Watkins et al. 2011). Of note, Tregs can be induced by the kynurenine system activated by IDO expressing APCs (Watkins et al. 2011; Mandi and Vecsei 2012).

The macrophages, MDSCs and Tregs are attracted to the tumor by the chemokines CCL1, CCL2 and CXCL8, which also promote angiogenesis and by CCL21, CCL22 and/or hypoxia-induced CCL28 (Facciabene et al. 2011; Shields et al. 2010; Hoelzinger et al. 2010). In cervical cancer patients it has been observed that the expression of CCL2 within the tumor correlates with decreased survival (Zijlmans et al. 2006; Kleine-Lowinski et al. 1999), however, in primary ovarian tumors the overexpression of CCL2 is related to a higher susceptibility of tumor cells to chemotherapy regimens and associated with increased survival (Fader et al. 2010), indicating that not simply the expression of CCL2 but also co-conditioning factors are likely to determine the outcome of CCL2 expression. Furthermore, the chemokines CXCR4 (Mishra et al. 2011) and CXCR16/CXCL16 (Deng et al. 2010a) can promote tumor growth and metastasis by cross talk with the cancer-associated fibroblasts (Lazennec and Richmond 2010). These fibroblasts are also involved in regulating tumor growth by the secretion of soluble factors that are either pro-tumorigenic (e.g.IL-10 and TGF-β), enhancing tumor growth, vascularization and invasion, or by the secretion of factors that indirectly can suppress tumor growth by activating immune cells (Rasanen and Vaheri 2010; Joyce and Pollard 2009).

So in summary, the balance between the tumor promoting and anti-tumor facets of the immune system is dictating whether the tumor is progressively growing or controlled and perhaps eradicated. The mechanism behind the infiltration of the tumor by immune cells is believed to be an orchestration of several factors, each of which may influence the attraction and entrance of immune cells to the tumor. Several factors have been described to be responsible for this infiltration [excellently reviewed by Rahir and Moser (2012)]: (1) the vasculature of the tumor tissue, (2) the guidance of blood cells in their extravasation (leaving the blood stream into the tissue) by adhesion molecules, (3) the attraction of the cells by chemokines produced within the tumor microenvironment, (4) tumor antigens expressed on tumor cells or APCs within the tumor or (5) at the lymphoid site and (6) the presence of CD4+ Th1 and/or Th17 cells not only helping the optimal priming of CD8+ T lymphocytes but also having an important role within the tumor, namely as helper cells favoring the entry into and the accumulation of CD8+ T cells in the tumor tissue by inducing a strong inflammatory environment. Overall, there is clear evidence that tumor infiltrating immune cells can dictate the final outcome of a developing tumor. The mechanisms underlying tumor-induced initiation of immune cell infiltration and the spontaneous priming of tumor-specific T cells, however, are still not very well understood.

5.3 Development of Tumor-Specific T Cell Immunity

The primary site for the induction of an adaptive immune response against pathogens is the lymph node (LN) (Martin-Fontecha et al. 2009; Breart and Bousso 2006). Consequently, it has always been assumed that priming of T cells either following vaccination or by exposure to tumor-derived antigens also occurs in the LN. Indeed, tumor-derived antigens were taken up within the tumor by APCs, carried to the tumor-draining lymph node (TDLN), processed and presented in MHC class II to CD4+ T cells and in MHC class I—via a process called cross-presentation—to CD8+ T cells (van Mierlo et al. 2002; Melief 2003; Melief 2008). However, as most tumor types first metastasize to the TDLN one can envisage alternative routes such as the cross-presentation of tumor antigens by APCs that have ingested antigens from dying/dead tumor cells directly within this TDLN. Alternatively, these LN resident tumor cells might also directly present tumor antigens to activate T cells as was shown in mouse models (Zinkernagel 2002; Ochsenbein et al. 2001) and in vitro for melanoma (Verdegaal et al. 2011). In agreement with the common believe that T cells are primed in the LN is the fact that especially TDLN harbor tumor-specific T cells. In early stage melanoma, tumor-reactive T cells are detectable at higher frequencies in the TDLN than in blood (Molenkamp et al. 2006; Vuylsteke et al. 2006). Similarly, in cervical cancer we readily detected polyclonal populations of human papillomavirus type 16 (HPV16) oncoproteins E6- and E7-specific CD4+ and CD8+ T cells in almost all TDLN (de Vos van Steenwijk et al. 2010; van Poelgeest et al. 2012).

Interestingly, accumulating evidence suggests that immune responses may also be directly induced within the tumor environment. Immunohistochemistry studies have revealed the presence of LN-like structures, called ectopic LNs or tertiary lymphoid structures, present within tumors of the lung, colon and oropharynx (i.e. those originating from the tonsils in the head and neck region) (Dieu-Nosjean et al. 2008; Coppola et al. 2011; Randall 2010). In human NSCLC these ectopic LNs were found to harbor high density of Lamp-positive DCs (i.e. matured DCs), which have the capacity to appropriately prime a tumor-specific T cell response. Indeed, the presence of such tertiary lymphoid structures was associated with better clinical outcome (Dieu-Nosjean et al. 2008). Studies of these LN-like structures in colorectal cancer showed that they comprised follicles with CD3+ T cells in the cortex like zone and CD20+ B cells within the follicular structure as well as CD21+ DCs in the follicular germinal centers. Moreover, the T and B cell areas displayed proliferative cells as determined by Ki-67 suggesting that these structures indeed resemble secondary and/or tertiary lymph nodes (Coppola et al. 2011). These phenomena were recapitulated in a tumor mouse model where the injection of DCs—genetically modified to produce the chemokine (C–C motif) ligand 21 (CCL-21) that induces the homing and localization of lymphocytes to the lymphoid organs by binding to CCR7—resulted in lymphoid structure formation within the tumor mass, priming of naïve T cells and subsequent tumor regression (Kirk et al. 2001). These findings suggest that the site of T cell priming is not exclusive to the normal LNs and confirm that ectopic LNs may play an important role in the development of a clinically effective antitumor response. Indeed just recently, a 12-chemokine gene signature in metastatic melanoma was correlated to the presence of these ectopic LN as well as a better clinical outcome for these patients (Messina et al. 2012). Interestingly, the development of tertiary LN-like structures is not exclusively found in solid tumors (Dieu-Nosjean et al. 2008; Coppola et al. 2011; Randall 2010;

Coppola and Mule 2008), but can also be found in autoimmune diseases (such as rheumatoid arthritis) (Timmer et al. 2007; Takemura et al. 2001), chronic inflammation (Olszewski 2002; Winter et al. 2010), transplantation (Sato et al. 2011) and even develops at the site of vaccination (Harris et al. 2012).

Irrespective of the site of activation, an important aspect of T cell priming is the context in which this occurs. One can envisage that the activation and functional polarization of T cells directly by tumors cells in LNs is determined by the immunogenicity of the antigen, co-stimulatory and inhibitory molecules on tumor cells and by tumor-produced immunosuppressants (IL-10, IDO, Galectins) (Joss et al. 2000; de Waal Malefyt et al. 1991; Singer et al. 2011; Soliman et al. 2010; Wilke et al. 2011; Yang et al. 2008; Rabinovich and Croci 2012). In the cross-presentation route the outcome is dependent on the activation status of the DCs that carry, process and present the tumor antigens to the T cells (Melief 2003; Steinman et al. 2003). In addition, the local micromilieu may have an impact. The presence of metastatic tumor cells has been reported to favor a tolerogenic milieu for example in early cervical cancer (Battaglia et al. 2009). However, one should bear in mind that also the lymph drainage from tumors to the LN is elevated when compared to drainage from normal tissues (Harrell et al. 2007; Thomas et al. 2012). This suggests that the TDLN microenvironment can be shaped by tumor-derived cytokines, chemokines and other compounds that may support or suppress an antitumor response.

For instance, suppression may occur through the specific accumulation of Tregs in TDLN as was found in patients with an unfavorable course of colorectal cancer (Deng et al. 2010b) but it may also directly influence the capacity to elicit tumor immunity in TDLN as was demonstrated in a mouse model where tumor cells directly injected in the LN of naïve mice were readily eradicated whereas tumor cells injected in the TDLN of tumor-bearing mice continued to grow in the face of the immune system (Preynat-Seauve et al. 2007). In line with this notion are reports showing that TDLN can harbor functionally impaired T cells (Baitsch et al. 2011; Contassot et al. 2009; Mantovani et al. 2008) and even antigen-specific Tregs (van der Burg et al. 2007).

Clearly, if tumors can affect T cell priming at a distance they for sure can accomplish this at the short range. Therefore, it is highly likely that the quality of the T cell response elicited in the ectopic LNs may also be determined by the local milieu. The previously mentioned positive correlation between the cellular content of these ectopic LNs and the clinical outcome of patients was associated with the presence of CD4+ Th1-specific T box transcription factor (Tbet) positive (IFN-γ-associated) T cells in the tumor (Dieu-Nosjean et al. 2008; Coppola et al. 2011; Randall 2010). In contrast, these lymphoid tissues have also been detected in breast cancers but here these structures comprised DCs that induced IL-13 producing CD4+ Th2 cells and this was associated with a non-beneficial clinical response. These Th2 cells promoted tumor growth potentially via the enhanced differentiation of macrophages to a M2 phenotype (Asprod et al. 2007; DeNardo and Coussens 2007). Thus, whereas ectopical LNs can elicit tumor-immunity they are likely under the same control of the tumor with respect to the micromilieu that determines final outcome. The questions on why, how and at what site LN-like structures are generated, still have to be answered but it seems that they develop at pathological sites where an adaptive immune response is needed (e.g. the tumor, the inflammation site, the transplanted organ, the vaccine injection site). B cells have also been detected in ectopic LN structures found in tumor of colorectal carcinoma patients (Coppola et al. 2011) and the presence of tumor resident B cells is related to better prognosis for patients with breast, ovarian, colorectal, cervical and non-small cell lung cancer (Nelson 2010). Activated B cells can facilitate T cell responses but resting B cells are likely to inhibit the development of effective T cell responses (Nelson 2010; Qin et al. 1998; DiLillo et al. 2010), however their exact role is still unclear.

Thus, tumor-specific T cell responses are not necessarily induced in the TDLN but may also happen directly within LN structures in the tumor. In both cases it is likely to occur under conditions that are controlled by all kinds of tumor-derived and immune cell produced factors together determining the type and efficacy of these activated T cells.

5.4 Initiation of the Immune Response

The next question is what factors are responsible for the priming of a tumor-specific immune response during tumor development? Several mechanisms have been described.

Upon exposure to carcinogens or genotoxic events cellular DNA may get damaged resulting in a cell cycle arrest (senescence). This DNA damage response (DDR) pathway dependents on the activation of the DNA double-strand break checkpoint kinase ataxia telangiectasia (ATM) mutated kinase and checkpoint kinase 2 (CHK2) (Di Leonardo et al. 1994; Lombard et al. 2005; Zhan et al. 2010; Gordon and Nelson 2012) as well as the accumulation of the tumor suppressor gene p53 and the upregulation of the tumor suppressors $p16^{ink4a}$ and $P19^{Arf}$ (Braig and Schmitt 2006; Hornsby 2007; Mallette et al. 2007). The DDR pathway is activated to allow the cell to repair the damage or to force it into programmed cellular death (i.e. apoptosis) when the damage is beyond repair [reviewed in (Gordon and Nelson 2012; Zhou and Elledge 2000)]. Importantly, activation of the DDR pathway can result in NF-κB activation leading to the production of inflammatory cytokines (e.g. IL-1β, IL-6, IL-15), chemokines (e.g. MCP-1, CXCL-1), and adhesion molecules (e.g. ICAM-1) which may attract immune cells such as monocytes (by MCP-1) and neutrophils (by CXCL-1) and help T cells (by ICAM-1 and IL-15) (Stagg et al. 2007; Biton and Ashkenazi 2011; Fumagalli and d'Adda di Fagagna 2009; Kloster et al. 2011; Rodier et al. 2009).

Whether the transition to senescence upon acute oncogenic stress is mediated through the autophagy pathway remains under debate (Young et al. 2009) but this pathway is a cellular survival mechanism that limits cellular damage upon the cellular stress and is generally known to be the first defense mechanism of cells to internal stress. Autophagy may also result in cell death either via apoptosis or necrosis. The necrotic cell death is mediated by cell death ligands such as TNF-α and Fas ligand (FasL, CD95L) (Shen and Codogno 2012). Nowadays, it is believed that the tumor cells need to undergo apoptosis and not senescence as in the latter case they remain in an unresponsive state but are still able to secrete all kind of tumor promoting factors (Kahlem et al. 2004). One of the endpoints of apoptosis is an efficient engulfment of the intact cell corpse by professional phagocytes. DCs and macrophages attracted by the release of lysophosphatidylcholine from apoptotic cells (Lauber et al. 2003) are able to do this and to present the ingested tumor antigens to B and/or T cells (Albert et al. 1998).

In addition, exposure to carcinogens or genotoxic events may result in the expression of NK cell- and CD8+ T cell-expressed NKG2D ligands, such as MHC class I chain-related chain A (MICA) and MICB, ringing the alarm bells of the immune system and facilitating the killing of tumor cells by these effector cells (Gasser et al. 2005; Hayakawa and Smyth 2006).

During necrotic cell death a number of damage (or danger)-associated molecular pattern molecules (DAMPs) can be released. DAMPs can initiate and perpetuate immune responses in the absence of infections with pathogens. The most common known DAMPs are RNA, DNA, adenosine-5′-triphosphate (ATP), uric acid, high mobility group box 1 (HMBG1), heat shock proteins (hsp) and hyaluronic acid (Tang et al. 2012). For instance, HMGB-1 is released by damaged cells and necrotic cells but not by apoptotic cells (Scaffidi et al. 2002) and can bind to TLR2, TLR4 and the receptor for advanced glycation endproducts (RAGE), which are all implicated in inflammatory reactions (Sims et al. 2010). Notably, HMGB-1 mediated inflammation is repressed by the co-expression of CD24 (Chen et al. 2009). The soluble HMBG1 was shown to activate DCs via its binding to TLR4. This interaction results in the inhibition of lysosomal degradation of tumor antigens leaving the antigen intact for the cross-presentation route and subsequent presentation in MHC at the cell surface as well as upregulates the production of pro-IL-1β. Furthermore, dying tumor cells release ATP which can act on a series of purinergic receptors, among which P2X7 has the highest affinity for ATP. When present on APCs, the receptor ligation causes the K+ efflux-dependent assembly of the inflammasome, which in turn activates caspase-1 required for the proteolytic processing of pro-IL-1β and the secretion of mature IL-1β and subsequently to induce adaptive immunity (Franchi et al. 2009; Kepp et al. 2011; Martins et al. 2009; Schroder and Tschopp 2010). Importantly, the production of IL-1β was found to be essential for the efficient priming of T cells (Ghiringhelli et al. 2009). Similarly, uric acid can also activate the NALP3 inflammasome. It activates the inflammasome pathway in DCs resulting in the production of active IL-1β and IL-18 (Martinon et al. 2006). Furthermore, tumor-derived DNA can be ingested by professional APCs where it binds to the intracellular DNA sensor (i.e. IFI/p204 or DDX41) starting a signaling cascade via the endoplasmic reticulum situated stimulator of interferon genes (STING), the activation of IRF3 and the type I IFN transcription pathway to induce IFN-β production (Fuertes et al. 2011;

Gajewski et al. 2012; Romano et al. 2012; Zhang et al. 2011). Consequently, cross presentation of tumor antigens to CD8+ T cells by CD8α+ DC occurs after the initial production of IFN-β by plasmacytoid DC (Gajewski et al. 2012; Di Domizio et al. 2012).

More recently it has been demonstrated that not only dying/dead tumor cells can elicit an anti-tumor response. Tumor-derived exosomes, which are micro-vessels of the tumor cells armed with tumor antigens presented by MHC, integrins and cytokines, can activate T cells once taken up by APCs, processed and optimally presented to T cells (Wieckowski et al. 2009). However, it should be noted that when these exosomes (also) harbor FasL and cytokines like IL-10 and TGF-β, they can stimulate cancer associated fibroblasts and Tregs (Webber et al. 2010; Szajnik et al. 2010).

A more indirect way of tumors to attract the immune system is by its growth. An expanding tumor requires increasing amounts of nutrients and oxygen and thus starts to support the formation of blood vessels. This vascularization as well as the invasion of the tumor cells into the surrounding tissue provokes pro-inflammatory signals which may lead to the activation of DCs and the induction of an immune response (Fuchs and Matzinger 1996). Tumor growth and associated tissue remodeling utilizes proteases to cleave components of the extracellular matrix. One of these is biglycan. This protease is able to trigger TLR-2 and TLR-4 on macrophages and DCs thereby inducing pro-inflammatory cytokine production (Schaefer et al. 2005; Edwards 2012). In line with this is the observation that the group of patients of whom their HPV-induced cervical tumor deeply invaded the surrounding tissue not only displayed an HPV-specific T cell response but also was the group of patients who benefitted most from concurrent radiotherapy (Heusinkveld et al. 2012).

In conclusion, during the development of the tumor, its growth and the invasion of the surrounding tissue the immune system is alarmed via several mechanisms. The result can be a response of the adaptive immune system which not necessarily may be effective as tumors raise several hurdles to suppress immunity. These hurdles need to be overcome before an effective control of the tumor ensues.

5.5 Improving the Effect of Spontaneous Tumor-Specific Immune Responses

The chronic inflammatory nature of the tumor microenvironment, with high numbers of tumor associated M2 macrophages, tolerogenic DCs, MDSCs, and Tregs, is not likely to sustain the capacity of effector cells to exert their anti-tumor function (Wang et al. 2008). In addition, the beneficial clinical effect of T cells can be impaired via downregulation of cell surface MHC class I on tumor cells (Jordanova et al. 2008; Garrido et al. 2010; Maleno et al. 2011; del Campo et al. 2012), the upregulation of the non-classical MHC class I molecule HLA-E

(Gooden et al. 2011) as well as by T cell expressed inhibitory molecules of which programmed cell death protein 1 (PD-1) (Weber 2010; Topalian et al. 2012; Karim et al. 2009), cytotoxic T lymphocyte antigen 4 (CTLA-4) (Weber 2010; Hodi et al. 2010), T cell immunoglobulin and mucin domain 3 (TIM-3) (Ngiow et al. 2011), lymphocyte activation gene 3 (LAG3), CD94/NKG2A (Gooden et al. 2011), V-domain Ig suppressor of T cell activation (VISTA) (Wang et al. 2011), CD200 and BTLA (Pardoll 2012; Haymaker et al. 2012; Fourcade et al. 2012) are known to impair T cell stimulation and effector function (Pardoll 2012; Pentcheva-Hoang et al. 2009; Gajewski et al. 2006; Sakuishi et al. 2010; Wherry 2011; Woo et al. 2012), thereby allowing the tumor cells to escape from the immune attack. The impairment of T cell expansion and their function via these inhibitory molecules can readily be relieved by blocking the inhibitory molecules expressed by T cells or their ligands on APC or tumor cells with monoclonal antibodies. The anti-CTLA-4 antibody ipilimumab has been approved for the treatment of melanoma (Hodi et al. 2010) and there is also strong evidence that antibodies blocking the inhibitory molecule PD-1 or its ligand PD-L1 enhance the anti-tumor immune response (Topalian et al. 2012; Brahmer et al. 2012).

Initial and ongoing efforts in the development of active immunotherapeutic approaches concern the enhancement of effector cell function and frequencies, for instance by vaccination (Melief and van der Burg 2008). Strategies to boost the spontaneous anti-tumor response are the use of chemotherapeutic agents. The group of Zitvogel et al. incisively studied the response of tumors beyond the stereotypical apoptotic pathway and found that a number of chemotherapeutic agents rendered tumor-cell death immunogenic resulting in the uptake of tumor antigen by local APC, activation of these APC via concomitant release of danger signals and subsequently the activation of an anti-tumor response (Kepp et al. 2011; Green et al. 2009; Hannani et al. 2011; Locher et al. 2010). Also radiation has been found to induce this immunogenic cell death and to elicit a vigorous response of the immune system (Kepp et al. 2011; Hannani et al. 2011 Golden et al. 2012). Radiation is likely to facilitate the tumor-specific immune response as cervical cancer patients with pre-existing tumor-immunity displayed clinical better responses upon radiotherapy (Heusinkveld et al. 2012). Recent studies show that indeed high numbers of IFN-γ-producing effector cells are required (Kenter et al. 2008; Welters et al. 2010; Porter et al. 2011; Powell et al. 2006; Restifo et al. 2012) but generally are not sufficient as these effector cells need to travel to the tumor and, within this microenvironment, should not encounter too much suppression. New efforts, therefore, aim at simultaneously changing the tumor microenvironment by shifting the balance towards the tumor-rejecting immune cells. Several common approaches are being explored by many research groups. Interestingly, some types of chemotherapy may have side effects that stimulate the immune system, for instance by the depletion of Tregs (Ghiringhelli et al. 2007; Vermeij et al. 2012) and MDSCs (Suzuki et al. 2005), thereby alleviating a number of immunosuppressive mechanisms and/or through direct and indirect stimulatory effects of immune effectors (Zitvogel et al. 2008). In addition, tumor-promoting M2 macrophages were found to be more susceptible to chemotherapy than tumor-rejecting M1 macrophages or DC (Dijkgraaf et al. 2012b). Notably, the effects of chemotherapy may differ per patient as we recently found that chemotherapy applied to tumor cells with an activated NF-κB signaling pathway in fact enhanced their capacity to drive monocyte to M2 macrophage differentiation in an IL-6 and/or PGE2 dependent manner (Dijkgraaf et al. 2012b). There are numerous other pharmacological approaches to overcome the immunosuppressive mechanisms of myeloid cells which aim at the inhibition of immunosuppressive function, blocking their recruitment, and forcing their maturation (Gabrilovich et al. 2012). Preferably one would like to re-polarize the suppressive myeloid cells towards activated M1 macrophages as can be achieved by certain chemotherapeutic compounds (Kodumudi et al. 2010), the combined treatments with immune potentiating compounds (IL-12, CpG, IL10-blocking antibodies), agonistic CD40 antibodies or by inhibition of NF-κB signaling (Gabrilovich et al. 2012). Activation of M1 macrophages by anti-CD40 has resulted in tumor control both in mice and humans (Beatty et al. 2011; Lum et al. 2006). Alternatively, one may re-polarize these suppressive myeloid cells via the cognate interaction with Th1 cells (Heusinkveld et al. 2011), but this requires enough tumor-specific Th1 cells to be aroused and homed to the tumor. Such a clinically active Th1 response can be achieved through vaccination (Kenter et al. 2009; Gao et al. 2012). The inhibitory effect of Tregs may be counteracted by low doses of cyclophosphamide, denileukin diftitox (Ontak), dacluzimab (anti-CD25 antibody) or other drugs (Vermeij et al. 2012; Jacobs et al. 2012; Powell et al. 2008; Rasku et al. 2008; Rech and Vonderheide 2009). Improved T cell responses have been seen when these treatments were combined with vaccination (Vermeij et al. 2012; Rech and Vonderheide 2009; Dannull et al. 2005; Morse et al. 2008; Walter et al. 2012).

Apart from removing or inhibiting suppressive mechanisms the polarization of the spontaneously induced immune responses and their capacity to exert their cancer-rejecting function will benefit from a change of the local micromilieu towards what is seen during acute inflammation or tissue rejection (Wang et al. 2008). Currently, topical application of the TLR-7 ligand imiquimod can result in the elimination of pre-malignant and malignant lesions, including high grade vulvar intraepithelial neoplasia (van Seters et al. 2008; Winters et al. 2008) and basal cell carcinoma (Ghafouri-Fard 2012; Roozeboom et al. 2012). Imiquimod has also been applied to melanoma in situ when presented in the head and neck region (Ellis et al. 2012). Topical treatment with imiquimod results in the increased infiltration of lesions by CD4+ and CD8+ T cells as well as DCs (Daayana et al. 2010; Hermanns-Le et al. 2003; Ooi et al. 2006). More experimental approaches are the intratumoral injections with pro-inflammatory agents such as CpG, Poly I:C, CD40L plasmid DNA, which alone or in combinations were shown to improve T cell homing, activation of local APCs and the induction of a local cytokine milieu that favors the induction and antitumor effects of a tumor-specific Th1/CTL response (Amos et al. 2011; Fan et al. 2012; Grauer et al. 2008; Stone et al. 2009), known to be important for tumor control (Fridman et al. 2011). Similar effects can be observed when cytokines cocktails are injected near TDLNs (Berinstein et al. 2012).

A powerful option to create such a tumor rejecting milieu that is already applied in clinical trials is the use of the immune modulator IFN-α. This cytokine is used to treat chronic viral hepatitis infection and malignancies (Pasquali and Mocellin 2010). It enhances the differentiation of antigen-specific Th1 cells, promotes the generation of CTL and sustains the survival of T cells (Belardelli et al. 2002; Huber and Farrar 2011). Moreover, type I IFNs promote the differentiation of monocytes into DCs and enhance DC activity (Mohty et al. 2003; Parlato et al. 2001; Santini et al. 2002; Santini et al. 2000; Santodonato et al. 2003; Tosi et al. 2004). In addition, IFN-α enhances the expression of HLA class I and II on tumor cells (Beniers et al. 1991; Cangemi et al. 2003), and high doses of IFN-α intravenously given to melanoma patients results in decreased levels of Tregs (Mozzillo and Ascierto 2012) and reduces the numbers of neutrophils (Verdegaal et al. 2011), which are known independent prognostic factors for short survival (Schmidt et al. 2007). Moreover, daily injections of IFN-α complemented with the adoptive transfer of tumor-reactive T cells in metastatic melanoma patients can lead to clinical success (Verdegaal et al. 2011). In addition, the co-injection of IFN-α with vaccines resulted in a consistent enhancement of vaccine-specific CD4+ and CD8+ T cells and increased the percentage of blood circulating DC precursors in mice and men (Sikora et al. 2009; Di Pucchio et al. 2006; Zeestraten et al. 2012). Another cytokine which is considered to be used as an immunomodulatory therapy is the use of IFN-γ. This cytokine is known to induce M1 macrophages and the expression of cytotoxic ligands on tumor cells, to enhance tumor antigen processing and presentation as well as to steer and sustain the adoptive immune responses against tumors (Ikeda et al. 2002). In addition, it can directly affect tumor growth and tumor angiogenesis (Ikeda et al. 2002). Recombinant IFN-γ has been used in the past to treat cutaneous T cell lymphoma (Kaplan et al. 1990) and more recently for both cutaneous T- and B-cell lymphomas by the intratumoral injection of adenovirus-IFN-γ. This resulted in systemic immune activation that polarized the immune response to Th1 responses and increased the antibody response to tumor antigens (Dummer et al. 2010). However, IFN-γ may also have some unwanted immunological effects (Wilke et al. 2011) warranting careful immunomonitoring of local events.

In summary, there are a couple of non-exclusive treatment options that potentially can boost the expansion and efficacy of spontaneously aroused, vaccine-induced, and ex vivo expanded infused tumor-specific T cell by modulation of the systemic and local immune environment. It is highly likely that the best control of tumors is only achieved when a number of modalities are used together.

References

Abastado JP (2012) The next challenge in cancer immunotherapy: controlling T-cell traffic to the tumor. Cancer Res 72:2159–2161

Albert ML, Sauter B, Bhardwaj N (1998) Dendritic cells acquire antigen from apoptotic cells and induce class I-restricted CTLs. Nature 392:86–89

Algars A, Irjala H, Vaittinen S, Huhtinen H, Sundstrom J, Salmi M et al (2012) Type and location of tumor-infiltrating macrophages and lymphatic vessels predict survival of colorectal cancer patients. Int J Cancer 131:864–873

Allavena P, Mantovani A (2012) Immunology in the clinic review series; focus on cancer: tumour-associated macrophages: undisputed stars of the inflammatory tumour microenvironment. Clin Exp Immunol 167:195–205

Al-Shibli KI, Donnem T, Al-Saad S, Persson M, Bremnes RM, Busund LT (2008) Prognostic effect of epithelial and stromal lymphocyte infiltration in non-small cell lung cancer. Clin Cancer Res 14:5220–5227

Amos SM, Pegram HJ, Westwood JA, John LB, Devaud C, Clarke CJ et al (2011) Adoptive immunotherapy combined with intratumoral TLR agonist delivery eradicates established melanoma in mice. Cancer Immunol Immunother 60:671–683

Aspord C, Pedroza-Gonzalez A, Gallegos M, Tindle S, Burton EC, Su D et al (2007) Breast cancer instructs dendritic cells to prime interleukin 13-secreting CD4+ T cells that facilitate tumor development. J Exp Med 204:1037–1047

Baitsch L, Baumgaertner P, Devevre E, Raghav SK, Legat A, Barba L et al (2011) Exhaustion of tumor-specific CD8(+) T cells in metastases from melanoma patients. J Clin Invest 121:2350–2360

Battaglia A, Buzzonetti A, Baranello C, Ferrandina G, Martinelli E, Fanfani F et al (2009) Metastatic tumour cells favour the generation of a tolerogenic milieu in tumour draining lymph node in patients with early cervical cancer. Cancer Immunol Immunother 58:1363–1373

Beatty GL, Chiorean EG, Fishman MP, Saboury B, Teitelbaum UR, Sun W et al (2011) CD40 agonists alter tumor stroma and show efficacy against pancreatic carcinoma in mice and humans. Science 331:1612–1616

Belardelli F, Ferrantini M, Proietti E, Kirkwood JM (2002) Interferon-alpha in tumor immunity and immunotherapy. Cytokine Growth Factor Rev 13:119–134

Beniers AJ, Peelen WP, Debruyne FM, Schalken JA (1991) HLA-class-I and -class-II expression on renal tumor xenografts and the relation to sensitivity for alpha-IFN, gamma-IFN and TNF. Int J Cancer 48:709–716

Berinstein NL, Wolf GT, Naylor PH, Baltzer L, Egan JE, Brandwein HJ et al (2012) Increased lymphocyte infiltration in patients with head and neck cancer treated with the IRX-2 immunotherapy regimen. Cancer Immunol Immunother 61:771–782

Bierie B, Moses HL (2010) Transforming growth factor beta (TGF-beta) and inflammation in cancer. Cytokine Growth Factor Rev 21:49–59

Biton S, Ashkenazi A (2011) NEMO and RIP1 control cell fate in response to extensive DNA damage via TNF-alpha feedforward signaling. Cell 145:92–103

Bos R, Sherman LA (2010) CD4+ T-cell help in the tumor milieu is required for recruitment and cytolytic function of CD8+ T lymphocytes. Cancer Res 70:8368–8377

Brahmer JR, Tykodi SS, Chow LQ, Hwu WJ, Topalian SL, Hwu P et al (2012) Safety and activity of anti-PD-L1 antibody in patients with advanced cancer. N Engl J Med 366:2455–2465

Braig M, Schmitt CA (2006) Oncogene-induced senescence: putting the brakes on tumor development. Cancer Res 66:2881–2884

Breart B, Bousso P (2006) Cellular orchestration of T cell priming in lymph nodes. Curr Opin Immunol 18:483–490

Bronkhorst IH, Ly LV, Jordanova ES, Vrolijk J, Versluis M, Luyten GP et al (2011) Detection of M2-macrophages in uveal melanoma and relation with survival. Invest Ophthalmol Vis Sci 52:643–650

Cangemi G, Morandi B, D'Agostino A, Peri C, Conte R, Damonte G et al (2003) IFN-alpha mediates the up-regulation of HLA class I on melanoma cells without switching proteasome to immunoproteasome. Int Immunol 15:1415–1421

Chen GY, Tang J, Zheng P, Liu Y (2009) CD24 and Siglec-10 selectively repress tissue damage-induced immune responses. Science 323:1722–1725

Clemente CG, Mihm MC Jr, Bufalino R, Zurrida S, Collini P, Cascinelli N (1996) Prognostic value of tumor infiltrating lymphocytes in the vertical growth phase of primary cutaneous melanoma. Cancer 77:1303–1310

Contassot E, Preynat-Seauve O, French L, Huard B (2009) Lymph node tumor metastases: more susceptible than primary tumors to CD8(+) T-cell immune destruction. Trends Immunol 30:569–573

Coppola D, Mule JJ (2008) Ectopic lymph nodes within human solid tumors. J Clin Oncol 26:4369–4370

Coppola D, Nebozhyn M, Khalil F, Dai H, Yeatman T, Loboda A et al (2011) Unique ectopic lymph node-like structures present in human primary colorectal carcinoma are identified by immune gene array profiling. Am J Pathol 179:37–45

Correale P, Rotundo MS, Del Vecchio MT, Remondo C, Migali C, Ginanneschi C et al (2010) Regulatory (FoxP3+) T-cell tumor infiltration is a favorable prognostic factor in advanced colon cancer patients undergoing chemo or chemoimmunotherapy. J Immunother 33:435–441

Curiel TJ, Coukos G, Zou L, Alvarez X, Cheng P, Mottram P et al (2004) Specific recruitment of regulatory T cells in ovarian carcinoma fosters immune privilege and predicts reduced survival. Nat Med 10:942–949

Daayana S, Elkord E, Winters U, Pawlita M, Roden R, Stern PL et al (2010) Phase II trial of imiquimod and HPV therapeutic vaccination in patients with vulval intraepithelial neoplasia. Br J Cancer 102:1129–1136

Dannull J, Su Z, Rizzieri D, Yang BK, Coleman D, Yancey D et al (2005) Enhancement of vaccine-mediated antitumor immunity in cancer patients after depletion of regulatory T cells. J Clin Invest 115:3623–3633

de Vos van Steenwijk PJ, Heusinkveld M, Ramwadhdoebe TH, Lowik MJ, van der Hulst JM, Goedemans R et al (2010) An unexpectedly large polyclonal repertoire of HPV-specific T cells is poised for action in patients with cervical cancer. Cancer Res 70:2707–2717

de Waal Malefyt R, Haanen J, Spits H, Roncarolo MG, te Velde A, Figdor C (1991) Interleukin 10 (IL-10) and viral IL-10 strongly reduce antigen-specific human T cell proliferation by diminishing the antigen-presenting capacity of monocytes via downregulation of class II major histocompatibility complex expression. J Exp Med 174:915–924

del Campo AB, Carretero J, Aptsiauri N, Garrido F (2012) Targeting HLA class I expression to increase tumor immunogenicity. Tissue Antigens 79:147–154

DeNardo DG, Coussens LM (2007) Inflammation and breast cancer. Balancing immune response: crosstalk between adaptive and innate immune cells during breast cancer progression. Breast Cancer Res 9:212

Deng L, Chen N, Li Y, Zheng H, Lei Q (2010a) CXCR6/CXCL16 functions as a regulator in metastasis and progression of cancer. Biochim Biophys Acta 1806:42–49

Deng L, Zhang H, Luan Y, Zhang J, Xing Q, Dong S et al (2010b) Accumulation of foxp3+ T regulatory cells in draining lymph nodes correlates with disease progression and immune suppression in colorectal cancer patients. Clin Cancer Res 16:4105–4112

Denkert C, Loibl S, Noske A, Roller M, Muller BM, Komor M et al (2010) Tumor-associated lymphocytes as an independent predictor of response to neoadjuvant chemotherapy in breast cancer. J Clin Oncol 28:105–113

Di Domizio J, Dorta-Estremera S, Gagea M, Ganguly D, Meller S, Li P et al (2012) Nucleic acid-containing amyloid fibrils potently induce type I interferon and stimulate systemic autoimmunity. Proc Natl Acad Sci U S A 109:14550–14555

Di Leonardo A, Linke SP, Clarkin K, Wahl GM (1994) DNA damage triggers a prolonged p53-dependent G1 arrest and long-term induction of Cip1 in normal human fibroblasts. Genes Dev 8:2540–2551

Di Pucchio T, Pilla L, Capone I, Ferrantini M, Montefiore E, Urbani F et al (2006) Immunization of stage IV melanoma patients with Melan-A/MART-1 and gp100 peptides plus IFN-alpha results in the activation of specific CD8(+) T cells and monocyte/dendritic cell precursors. Cancer Res 66:4943–4951

Dieu-Nosjean MC, Antoine M, Danel C, Heudes D, Wislez M, Poulot V et al (2008) Long-term survival for patients with non-small-cell lung cancer with intratumoral lymphoid structures. J Clin Oncol 26:4410–4417

Dijkgraaf EM, Welters MJ, Nortier JW, van der Burg SH, Kroep JR (2012a) Interleukin-6/interleukin-6 receptor pathway as a new therapy target in epithelial ovarian cancer. Curr Pharm Des 18:3816–3827

Dijkgraaf EM, Heusinkveld M, Tummers B, Vogelpoel LTC, Goedemans R, Nortier JWR et al (2012b) Chemotherapy alters monocyte differentiation to favor generation of cancer-supporting M2 macrophages in the tumor microenvironment. Cancer Res (accepted for publication)

DiLillo DJ, Yanaba K, Tedder TF (2010) B cells are required for optimal CD4+ and CD8+ T cell tumor immunity: therapeutic B cell depletion enhances B16 melanoma growth in mice. J Immunol 184:4006–4016

Ding L, Linsley PS, Huang LY, Germain RN, Shevach EM (1993) IL-10 inhibits macrophage costimulatory activity by selectively inhibiting the up-regulation of B7 expression. J Immunol 151:1224–1234

Dummer R, Eichmuller S, Gellrich S, Assaf C, Dreno B, Schiller M et al (2010) Phase II clinical trial of intratumoral application of TG1042 (adenovirus-interferon-gamma) in patients with advanced cutaneous T-cell lymphomas and multilesional cutaneous B-cell lymphomas. Mol Ther 18:1244–1247

Dunn GP, Bruce AT, Ikeda H, Old LJ, Schreiber RD (2002) Cancer immunoediting: from immunosurveillance to tumor escape. Nat Immunol 3:991–998

Edwards IJ (2012) Proteoglycans in prostate cancer. Nat Rev Urol 9:196–206

Eisenthal A, Polyvkin N, Bramante-Schreiber L, Misonznik F, Hassner A, Lifschitz-Mercer B (2001) Expression of dendritic cells in ovarian tumors correlates with clinical outcome in patients with ovarian cancer. Hum Pathol 32:803–807

Elkord E, Alcantar-Orozco EM, Dovedi SJ, Tran DQ, Hawkins RE, Gilham DE (2010) T regulatory cells in cancer: recent advances and therapeutic potential. Expert Opin Biol Ther 10:1573–1586

Ellis LZ, Cohen JL, High W, Stewart L (2012) Melanoma in situ treated successfully using imiquimod after nonclearance with surgery: review of the literature. Dermatol Surg 38:937–946

Erdag G, Schaefer JT, Smolkin ME, Deacon DH, Shea SM, Dengel LT et al (2012) Immunotype and immunohistologic characteristics of tumor-infiltrating immune cells are associated with clinical outcome in metastatic melanoma. Cancer Res 72:1070–1080

Erreni M, Mantovani A, Allavena P (2011) Tumor-associated macrophages (TAM) and inflammation in colorectal cancer. Cancer Microenviron 4:141–154

Facciabene A, Peng X, Hagemann IS, Balint K, Barchetti A, Wang LP et al (2011) Tumour hypoxia promotes tolerance and angiogenesis via CCL28 and T(reg) cells. Nature 475:226–230

Fader AN, Rasool N, Vaziri SA, Kozuki T, Faber PW, Elson P et al (2010) CCL2 expression in primary ovarian carcinoma is correlated with chemotherapy response and survival outcomes. Anticancer Res 30:4791–4798

Fan H, Zhang I, Chen X, Zhang L, Wang H, Da Fonseca A et al (2012) Intracerebral CpG Immunotherapy with carbon nanotubes abrogates growth of subcutaneous melanomas in mice. Clin Cancer Res 18:5628–5638

Flavell RA, Sanjabi S, Wrzesinski SH, Licona-Limon P (2010) The polarization of immune cells in the tumour environment by TGFbeta. Nat Rev Immunol 10:554–567

Forssell J, Oberg A, Henriksson ML, Stenling R, Jung A, Palmqvist R (2007) High macrophage infiltration along the tumor front correlates with improved survival in colon cancer. Clin Cancer Res 13:1472–1479

Fourcade J, Sun Z, Pagliano O, Guillaume P, Luescher IF, Sander C et al (2012) CD8(+) T cells specific for tumor antigens can be rendered dysfunctional by the tumor microenvironment through upregulation of the inhibitory receptors BTLA and PD-1. Cancer Res 72:887–896

Franchi L, Eigenbrod T, Munoz-Planillo R, Nunez G (2009) The inflammasome: a caspase-1-activation platform that regulates immune responses and disease pathogenesis. Nat Immunol 10:241–247

Fridman WH, Galon J, Pages F, Tartour E, Sautes-Fridman C, Kroemer G (2011) Prognostic and predictive impact of intra- and peritumoral immune infiltrates. Cancer Res 71:5601–5605

Fuchs EJ, Matzinger P (1996) Is cancer dangerous to the immune system? Semin Immunol 8:271–280

Fuertes MB, Kacha AK, Kline J, Woo SR, Kranz DM, Murphy KM et al (2011) Host type I IFN signals are required for antitumor CD8+ T cell responses through CD8{alpha}+ dendritic cells. J Exp Med 208:2005–2016

Fumagalli M, d'Adda di Fagagna F (2009) SASPense and DDRama in cancer and ageing. Nat Cell Biol 11:921–923

Gabrilovich DI, Ostrand-Rosenberg S, Bronte V (2012) Coordinated regulation of myeloid cells by tumours. Nat Rev Immunol 12:253–268

Gajewski TF (2012) Cancer immunotherapy. Mol Oncol 6:242–250

Gajewski TF, Meng Y, Harlin H (2006) Immune suppression in the tumor microenvironment. J Immunother 29:233–240

Gajewski TF, Fuertes MB, Woo SR (2012) Innate immune sensing of cancer: clues from an identified role for type I IFNs. Cancer Immunol Immunother 61:1343–1347

Galon J, Costes A, Sanchez-Cabo F, Kirilovsky A, Mlecnik B, Lagorce-Pages C et al (2006) Type, density, and location of immune cells within human colorectal tumors predict clinical outcome. Science 313:1960–1964

Gao J, Bernatchez C, Sharma P, Radvanyi LG, Hwu P (2012) Advances in the development of cancer immunotherapies. Trends Immunol pii: S1471-4906(12)00142-1. doi:10.1016/j.it.2012.08.004. [Epub ahead of print]. PMID: 23031830

Garrido F, Algarra I, Garcia-Lora AM (2010) The escape of cancer from T lymphocytes: immunoselection of MHC class I loss variants harboring structural-irreversible "hard" lesions. Cancer Immunol Immunother 59:1601–1606

Gasser S, Orsulic S, Brown EJ, Raulet DH (2005) The DNA damage pathway regulates innate immune system ligands of the NKG2D receptor. Nature 436:1186–1190

Ghafouri-Fard S (2012) Immunotherapy in nonmelanoma skin cancer. Immunotherapy 4:499–510

Ghiringhelli F, Menard C, Puig PE, Ladoire S, Roux S, Martin F et al (2007) Metronomic cyclophosphamide regimen selectively depletes CD4+ CD25+ regulatory T cells and restores T and NK effector functions in end stage cancer patients. Cancer Immunol Immunother 56:641–648

Ghiringhelli F, Apetoh L, Tesniere A, Aymeric L, Ma Y, Ortiz C et al (2009) Activation of the NLRP3 inflammasome in dendritic cells induces IL-1beta-dependent adaptive immunity against tumors. Nat Med 15:1170–1178

Golden EB, Pellicciotta I, Demaria S, Barcellos-Hoff MH, Formenti SC (2012) The convergence of radiation and immunogenic cell death signaling pathways. Front Oncol 2:88

Gooden M, Lampen M, Jordanova ES, Leffers N, Trimbos JB, van der Burg SH et al (2011) HLA-E expression by gynecological cancers restrains tumor-infiltrating CD8(+) T lymphocytes. Proc Natl Acad Sci U S A 108:10656–10661

Gordon RR, Nelson PS (2012) Cellular senescence and cancer chemotherapy resistance. Drug Resist Updat 15:123–131

Grauer OM, Molling JW, Bennink E, Toonen LW, Sutmuller RP, Nierkens S et al (2008) TLR ligands in the local treatment of established intracerebral murine gliomas. J Immunol 181:6720–6729

Green DR, Ferguson T, Zitvogel L, Kroemer G (2009) Immunogenic and tolerogenic cell death. Nat Rev Immunol 9:353–363

Haanen JB, Baars A, Gomez R, Weder P, Smits M, de Gruijl TD et al (2006) Melanoma-specific tumor-infiltrating lymphocytes but not circulating melanoma-specific T cells may predict survival in resected advanced-stage melanoma patients. Cancer Immunol Immunother 55:451–458

Hannani D, Sistigu A, Kepp O, Galluzzi L, Kroemer G, Zitvogel L (2011) Prerequisites for the antitumor vaccine-like effect of chemotherapy and radiotherapy. Cancer J 17:351–358

Harrell MI, Iritani BM, Ruddell A (2007) Tumor-induced sentinel lymph node lymphangiogenesis and increased lymph flow precede melanoma metastasis. Am J Pathol 170:774–786

Harris RC, Chianese-Bullock KA, Petroni GR, Schaefer JT, Brill LB 2nd, Molhoek KR et al (2012) The vaccine-site microenvironment induced by injection of incomplete Freund's adjuvant, with or without melanoma peptides. J Immunother 35:78–88

Hayakawa Y, Smyth MJ (2006) NKG2D and cytotoxic effector function in tumor immune surveillance. Semin Immunol 18:176–185

Haymaker C, Wu R, Bernatchez C, Radvanyi L (2012) PD-1 and BTLA and CD8(+) T-cell "exhaustion" in cancer: "exercising" an alternative viewpoint. Oncoimmunology 1:735–738

Herfs M, Herman L, Hubert P, Minner F, Arafa M, Roncarati P et al (2009) High expression of PGE2 enzymatic pathways in cervical (pre)neoplastic lesions and functional consequences for antigen-presenting cells. Cancer Immunol Immunother 58:603–614

Hermanns-Le T, Paquet P, Nikkels AF, Pierard-Franchimont C, Pierard GE (2003) Prolonged imiquimod treatment and graft-versus-host reaction: histological mimicry in the skin infiltration pattern of the monocyte-macrophage-dendrocyte lineage. Dermatology 206:361–365

Heusinkveld M, van der Burg SH (2011) Identification and manipulation of tumor associated macrophages in human cancers. J Transl Med 9:216

Heusinkveld M, de Vos van Steenwijk PJ, Goedemans R, Ramwadhdoebe TH, Gorter A, Welters MJ et al (2011) M2 macrophages induced by prostaglandin E2 and IL-6 from cervical carcinoma are switched to activated M1 macrophages by CD4+ Th1 cells. J Immunol 187:1157–1165

Heusinkveld M, Goedemans R, Briet RJ, Gelderblom H, Nortier JW, Gorter A et al (2012) Systemic and local human papillomavirus 16-specific T-cell immunity in patients with head and neck cancer. Int J Cancer 131:E74–E85

Hiraoka K, Miyamoto M, Cho Y, Suzuoki M, Oshikiri T, Nakakubo Y et al (2006) Concurrent infiltration by CD8+ T cells and CD4+ T cells is a favourable prognostic factor in non-small-cell lung carcinoma. Br J Cancer 94:275–280

Hodi FS, O'Day SJ, McDermott DF, Weber RW, Sosman JA, Haanen JB et al (2010) Improved survival with ipilimumab in patients with metastatic melanoma. N Engl J Med 363:711–723

Hoelzinger DB, Smith SE, Mirza N, Dominguez AL, Manrique SZ, Lustgarten J (2010) Blockade of CCL1 inhibits T regulatory cell suppressive function enhancing tumor immunity without affecting T effector responses. J Immunol 184:6833–6842

Hornsby PJ (2007) Senescence as an anticancer mechanism. J Clin Oncol 25:1852–1857

Huber JP, Farrar JD (2011) Regulation of effector and memory T-cell functions by type I interferon. Immunology 132:466–474

Ikeda H, Old LJ, Schreiber RD (2002) The roles of IFN gamma in protection against tumor development and cancer immunoediting. Cytokine Growth Factor Rev 13:95–109

Jacobs JF, Idema AJ, Bol KF, Grotenhuis JA, de Vries IJ, Wesseling P et al (2010) Prognostic significance and mechanism of Treg infiltration in human brain tumors. J Neuroimmunol 225:195–199

Jacobs JF, Nierkens S, Figdor CG, de Vries IJ, Adema GJ (2012) Regulatory T cells in melanoma: the final hurdle towards effective immunotherapy? Lancet Oncol 13:e32–e42

Jordanova ES, Gorter A, Ayachi O, Prins F, Durrant LG, Kenter GG et al (2008) Human leukocyte antigen class I, MHC class I chain-related molecule A, and CD8+/regulatory T-cell ratio: which variable determines survival of cervical cancer patients? Clin Cancer Res 14:2028–2035

Joss A, Akdis M, Faith A, Blaser K, Akdis CA (2000) IL-10 directly acts on T cells by specifically altering the CD28 co-stimulation pathway. Eur J Immunol 30:1683–1690

Joyce JA, Pollard JW (2009) Microenvironmental regulation of metastasis. Nat Rev Cancer 9:239–252

Kahlem P, Dorken B, Schmitt CA (2004) Cellular senescence in cancer treatment: friend or foe? J Clin Invest 113:169–174

Kaplan EH, Rosen ST, Norris DB, Roenigk HH Jr, Saks SR, Bunn PA Jr (1990) Phase II study of recombinant human interferon gamma for treatment of cutaneous T-cell lymphoma. J Natl Cancer Inst 82:208–212

Karim R, Jordanova ES, Piersma SJ, Kenter GG, Chen L, Boer JM et al (2009) Tumor-expressed B7-H1 and B7-DC in relation to PD-1+ T-cell infiltration and survival of patients with cervical carcinoma. Clin Cancer Res 15:6341–6347

Kashimura S, Saze Z, Terashima M, Soeta N, Ohtani S, Osuka F et al (2012) CD83(+) dendritic cells and Foxp3(+) regulatory T cells in primary lesions and regional lymph nodes are inversely correlated with prognosis of gastric cancer. Gastric Cancer 15:144–153

Kenter GG, Welters MJ, Valentijn AR, Lowik MJ, Berends-van der Meer DM, Vloon AP et al (2008) Phase I immunotherapeutic trial with long peptides spanning the E6 and E7 sequences of high-risk human papillomavirus 16 in end-stage cervical cancer patients shows low toxicity and robust immunogenicity. Clin Cancer Res 14:169–717

Kenter GG, Welters MJ, Valentijn AR, Lowik MJ, Berends-van der Meer DM, Vloon AP et al (2009) Vaccination against HPV-16 oncoproteins for vulvar intraepithelial neoplasia. N Engl J Med 361:1838–1847

Kenter GG, Welters MJ, Valentijn AR, Lowik MJ, Berends-van der Meer DM, Vloon AP (2009) Vaccination against HPV-16 oncoproteins for vulvar intraepithelial neoplasia. N Engl J Med 361:1838–1847

Kepp O, Galluzzi L, Martins I, Schlemmer F, Adjemian S, Michaud M et al (2011) Molecular determinants of immunogenic cell death elicited by anticancer chemotherapy. Cancer Metastasis Rev 30:61–69

Kim J, Modlin RL, Moy RL, Dubinett SM, McHugh T, Nickoloff BJ et al (1995) IL-10 production in cutaneous basal and squamous cell carcinomas. A mechanism for evading the local T cell immune response. J Immunol 155:2240–2247

Kim ST, Jeong H, Woo OH, Seo JH, Kim A, Lee ES (2012) Tumor-infiltrating lymphocytes, tumor characteristics, and recurrence in patients with early breast cancer. Am J Clin Oncol [Epub ahead of print] PMID: 22495453

Kinouchi M, Miura K, Mizoi T, Ishida K, Fujibuchi W, Ando T et al (2011) Infiltration of CD14-positive macrophages at the invasive front indicates a favorable prognosis in colorectal cancer patients with lymph node metastasis. Hepatogastroenterology 58:352–358

Kirk CJ, Hartigan-O'Connor D, Mule JJ (2001) The dynamics of the T-cell antitumor response: chemokine-secreting dendritic cells can prime tumor-reactive T cells extranodally. Cancer Res 61:8794–8802

Kleine-Lowinski K, Gillitzer R, Kuhne-Heid R, Rosl F (1999) Monocyte-chemo-attractant-protein-1 (MCP-1)-gene expression in cervical intra-epithelial neoplasias and cervical carcinomas. Int J Cancer 82:6–11

Kloster MM, Naderi EH, Carlsen H, Blomhoff HK, Naderi S (2011) Hyperactivation of NF-kappaB via the MEK signaling is indispensable for the inhibitory effect of cAMP on DNA damage-induced cell death. Mol Cancer 10:45

Kodumudi KN, Woan K, Gilvary DL, Sahakian E, Wei S, Djeu JY (2010) A novel chemoimmunomodulating property of docetaxel: suppression of myeloid-derived suppressor cells in tumor bearers. Clin Cancer Res 16:4583–4594

Koebel CM, Vermi W, Swann JB, Zerafa N, Rodig SJ, Old LJ et al (2007) Adaptive immunity maintains occult cancer in an equilibrium state. Nature 450:903–907

Koizumi K, Hojo S, Akashi T, Yasumoto K, Saiki I (2007) Chemokine receptors in cancer metastasis and cancer cell-derived chemokines in host immune response. Cancer Sci 98:1652–1658

Kryczek I, Zhao E, Liu Y, Wang Y, Vatan L, Szeliga W (2011) Human TH17 cells are long-lived effector memory cells. Sci Transl Med 3:104ra0

Kurahara H, Shinchi H, Mataki Y, Maemura K, Noma H, Kubo F et al (2011) Significance of M2-polarized tumor-associated macrophage in pancreatic cancer. J Surg Res 167:e211–e219
Ladanyi A, Kiss J, Mohos A, Somlai B, Liszkay G, Gilde K et al (2011) Prognostic impact of B-cell density in cutaneous melanoma. Cancer Immunol Immunother 60:1729–1738
Ladoire S, Martin F, Ghiringhelli F (2011) Prognostic role of FOXP3+ regulatory T cells infiltrating human carcinomas: the paradox of colorectal cancer. Cancer Immunol Immunother 60:909–918
Lauber K, Bohn E, Krober SM, Xiao YJ, Blumenthal SG, Lindemann RK et al (2003) Apoptotic cells induce migration of phagocytes via caspase-3-mediated release of a lipid attraction signal. Cell 113:717–730
Lazennec G, Richmond A (2010) Chemokines and chemokine receptors: new insights into cancer-related inflammation. Trends Mol Med 16:133–144
Liu F, Lang R, Zhao J, Zhang X, Pringle GA, Fan Y et al (2011) CD8(+) cytotoxic T cell and FOXP3(+) regulatory T cell infiltration in relation to breast cancer survival and molecular subtypes. Breast Cancer Res Treat 130:645–655
Locher C, Conforti R, Aymeric L, Ma Y, Yamazaki T, Rusakiewicz S et al (2010) Desirable cell death during anticancer chemotherapy. Ann N Y Acad Sci 1209:99–108
Lombard DB, Chua KF, Mostoslavsky R, Franco S, Gostissa M, Alt FW (2005) DNA repair, genome stability, and aging. Cell 120:497–512
Lum HD, Buhtoiarov IN, Schmidt BE, Berke G, Paulnock DM, Sondel PM et al (2006) In vivo CD40 ligation can induce T-cell-independent antitumor effects that involve macrophages. J Leukoc Biol 79:1181–1192
Mahmoud SM, Paish EC, Powe DG, Macmillan RD, Grainge MJ, Lee AH et al (2011) Tumor-infiltrating CD8+ lymphocytes predict clinical outcome in breast cancer. J Clin Oncol 29:1949–1955
Maleno I, Aptsiauri N, Cabrera T, Gallego A, Paschen A, Lopez-Nevot MA et al (2011) Frequent loss of heterozygosity in the beta2-microglobulin region of chromosome 15 in primary human tumors. Immunogenetics 63:65–71
Mallette FA, Gaumont-Leclerc MF, Ferbeyre G (2007) The DNA damage signaling pathway is a critical mediator of oncogene-induced senescence. Genes Dev 21:43–48
Mandi Y, Vecsei L (2012) The kynurenine system and immunoregulation. J Neural Transm 119:197–209
Mantovani A, Romero P, Palucka AK, Marincola FM (2008) Tumour immunity: effector response to tumour and role of the microenvironment. Lancet 371:771–783
Marrogi AJ, Munshi A, Merogi AJ, Ohadike Y, El-Habashi A, Marrogi OL et al (1997) Study of tumor infiltrating lymphocytes and transforming growth factor-beta as prognostic factors in breast carcinoma. Int J Cancer 74:492–501
Martinet L, Garrido I, Filleron T, Le Guellec S, Bellard E, Fournie JJ et al (2011) Human solid tumors contain high endothelial venules: association with T- and B-lymphocyte infiltration and favorable prognosis in breast cancer. Cancer Res 71:5678–5687
Martin-Fontecha A, Lanzavecchia A, Sallusto F (2009) Dendritic cell migration to peripheral lymph nodes. Handb Exp Pharmacol 31–49
Martinon F, Petrilli V, Mayor A, Tardivel A, Tschopp J (2006) Gout-associated uric acid crystals activate the NALP3 inflammasome. Nature 440:237–241
Martins I, Tesniere A, Kepp O, Michaud M, Schlemmer F, Senovilla L et al (2009) Chemotherapy induces ATP release from tumor cells. Cell Cycle 8:3723–3728
Mathai AM, Kapadia MJ, Alexander J, Kernochan LE, Swanson PE, Yeh MM (2012) Role of Foxp3-positive tumor-infiltrating lymphocytes in the histologic features and clinical outcomes of hepatocellular carcinoma. Am J Surg Pathol 36:980–986
Melief CJ (2003) Mini-review: Regulation of cytotoxic T lymphocyte responses by dendritic cells: peaceful coexistence of cross-priming and direct priming? Eur J Immunol 33:2645–2654
Melief CJ (2008) Cancer immunotherapy by dendritic cells. Immunity 29:372–383

Melief CJ, van der Burg SH (2008) Immunotherapy of established (pre)malignant disease by synthetic long peptide vaccines. Nat Rev Cancer 8:351–360

Menegaz RA, Michelin MA, Etchebehere RM, Fernandes PC, Murta EF (2008) Peri- and intratumoral T and B lymphocytic infiltration in breast cancer. Eur J Gynaecol Oncol 29:321–326

Messina JL, Fenstermacher DA, Eschrich S, Qu X, Berglund AE, Lloyd MC et al (2012) 12-chemokine gene signature identifies lymph node-like structures in melanoma: potential for patient selection for immunotherapy? Sci Rep 2:765

Mishra P, Banerjee D, Ben-Baruch A (2011) Chemokines at the crossroads of tumor-fibroblast interactions that promote malignancy. J Leukoc Biol 89:31–39

Mocellin S, Panelli M, Wang E, Rossi CR, Marincola FM (2002) Tumor microenvironment: what have we learned studying the immune response in this puzzling battlefield? Tumori 88:437–444

Mohty M, Vialle-Castellano A, Nunes JA, Isnardon D, Olive D, Gaugler B (2003) IFN-alpha skews monocyte differentiation into toll-like receptor 7-expressing dendritic cells with potent functional activities. J Immunol 171:3385–3393

Molenkamp BG, van Leeuwen PA, van den Eertwegh AJ, Sluijter BJ, Scheper RJ, Meijer S et al (2006) Immunomodulation of the melanoma sentinel lymph node a novel adjuvant therapeutic option. Immunobiology 211:651–661

Montero AJ, Diaz-Montero CM, Kyriakopoulos CE, Bronte V, Mandruzzato S (2012) Myeloid-derived suppressor cells in cancer patients: a clinical perspective. J Immunother 35:107–115

Moore KW, de Waal Malefyt R, Coffman RL, O'Garra A (2001) Interleukin-10 and the interleukin-10 receptor. Annu Rev Immunol 19:683–765

Morse MA, Hobeika AC, Osada T, Serra D, Niedzwiecki D, Lyerly HK et al (2008) Depletion of human regulatory T cells specifically enhances antigen-specific immune responses to cancer vaccines. Blood 112:610–618

Mozzillo N, Ascierto P (2012) Reduction of circulating regulatory T cells by intravenous high-dose interferon alfa-2b treatment in melanoma patients. Clin Exp Metastasis 29:801–805

Muller AJ, Prendergast GC (2007) Indoleamine 2,3-dioxygenase in immune suppression and cancer. Curr Cancer Drug Targets 7:31–40

Nasman A, Romanitan M, Nordfors C, Grun N, Johansson H, Hammarstedt L et al (2012) Tumor infiltrating CD8+ nand Foxp3+ lymphocytes correlate to clinical outcome and human papillomavirus (HPV) status in tonsillar cancer. PLoS ONE 7:e38711

Nelson BH (2008) The impact of T-cell immunity on ovarian cancer outcomes. Immunol Rev 222:101–116

Nelson BH (2010) CD20+ B cells: the other tumor-infiltrating lymphocytes. J Immunol 185:4977–4982

Ngiow SF, Teng MW, Smyth MJ (2011) Prospects for TIM3-targeted antitumor immunotherapy. Cancer Res 71:6567–6571

Nielsen JS, Sahota RA, Milne K, Kost SE, Nesslinger NJ, Watson PH et al (2012) CD20+ tumor-infiltrating lymphocytes have an atypical CD27- memory phenotype and together with CD8+ T cells promote favorable prognosis in ovarian cancer. Clin Cancer Res 18:3281–3292

Nosho K, Baba Y, Tanaka N, Shima K, Hayashi M, Meyerhardt JA et al (2010) Tumour-infiltrating T-cell subsets, molecular changes in colorectal cancer, and prognosis: cohort study and literature review. J Pathol 222:350–366

Obermajer N, Muthuswamy R, Lesnock J, Edwards RP, Kalinski P (2011) Positive feedback between PGE2 and COX2 redirects the differentiation of human dendritic cells toward stable myeloid-derived suppressor cells. Blood 118:5498–5505

Oble DA, Loewe R, Yu P, Mihm MC Jr (2009) Focus on TILs: prognostic significance of tumor infiltrating lymphocytes in human melanoma. Cancer Immun 9:3

Ochsenbein AF, Sierro S, Odermatt B, Pericin M, Karrer U, Hermans J et al (2001) Roles of tumour localization, second signals and cross priming in cytotoxic T-cell induction. Nature 411:1058–1064

Ohno S, Inagawa H, Dhar DK, Fujii T, Ueda S, Tachibana M et al (2003) The degree of macrophage infiltration into the cancer cell nest is a significant predictor of survival in gastric cancer patients. Anticancer Res 23:5015–5022

Ohri CM, Shikotra A, Green RH, Waller DA, Bradding P (2009) Macrophages within NSCLC tumour islets are predominantly of a cytotoxic M1 phenotype associated with extended survival. Eur Respir J 33:118–126

Okamura H, Tsutsi H, Komatsu T, Yutsudo M, Hakura A, Tanimoto T et al (1995) Cloning of a new cytokine that induces IFN-gamma production by T cells. Nature 378:88–91

Olszewski WL (2002) De novo lymph node formation in chronic inflammation of the human leg. Ann N Y Acad Sci 979:166–177, discussion 88–96

Ooi T, Barnetson RS, Zhuang L, McKane S, Lee JH, Slade HB et al (2006) Imiquimod-induced regression of actinic keratosis is associated with infiltration by T lymphocytes and dendritic cells: a randomized controlled trial. Br J Dermatol 154:72–78

Osada T, Chong G, Tansik R, Hong T, Spector N, Kumar R et al (2008) The effect of anti-VEGF therapy on immature myeloid cell and dendritic cells in cancer patients. Cancer Immunol Immunother 57:1115–1124

Padua D, Massague J (2009) Roles of TGFbeta in metastasis. Cell Res 19:89–102

Pages F, Berger A, Camus M, Sanchez-Cabo F, Costes A, Molidor R et al (2005) Effector memory T cells, early metastasis, and survival in colorectal cancer. N Engl J Med 353:2654–2666

Pardoll DM (2012) The blockade of immune checkpoints in cancer immunotherapy. Nat Rev Cancer 12:252–264

Parlato S, Santini SM, Lapenta C, Di Pucchio T, Logozzi M, Spada M et al (2001) Expression of CCR-7, MIP-3beta, and Th-1 chemokines in type I IFN-induced monocyte-derived dendritic cells: importance for the rapid acquisition of potent migratory and functional activities. Blood 98:3022–3029

Pasquali S, Mocellin S (2010) The anticancer face of interferon alpha (IFN-alpha): from biology to clinical results, with a focus on melanoma. Curr Med Chem 17:3327–3336

Pentcheva-Hoang T, Corse E, Allison JP (2009) Negative regulators of T-cell activation: potential targets for therapeutic intervention in cancer, autoimmune disease, and persistent infections. Immunol Rev 229:67–87

Piersma SJ, Jordanova ES, van Poelgeest MI, Kwappenberg KM, van der Hulst JM, Drijfhout JW et al (2007) High number of intraepithelial CD8+ tumor-infiltrating lymphocytes is associated with the absence of lymph node metastases in patients with large early-stage cervical cancer. Cancer Res 67:354–361

Porter DL, Levine BL, Kalos M, Bagg A, June CH (2011) Chimeric antigen receptor-modified T cells in chronic lymphoid leukemia. N Engl J Med 365:725–733

Poschke I, Kiessling R (2012) On the armament and appearances of human myeloid-derived suppressor cells. Clin Immunol 144:250–268

Powell DJ Jr, Dudley ME, Hogan KA, Wunderlich JR, Rosenberg SA (2006) Adoptive transfer of vaccine-induced peripheral blood mononuclear cells to patients with metastatic melanoma following lymphodepletion. J Immunol 177:6527–6539

Powell DJ Jr, Attia P, Ghetie V, Schindler J, Vitetta ES, Rosenberg SA (2008) Partial reduction of human FOXP3+ CD4 T cells in vivo after CD25-directed recombinant immunotoxin administration. J Immunother 31:189–198

Preynat-Seauve O, Contassot E, Schuler P, Piguet V, French LE, Huard B (2007) Extralymphatic tumors prepare draining lymph nodes to invasion via a T-cell cross-tolerance process. Cancer Res 67:5009–5016

Qin Z, Richter G, Schuler T, Ibe S, Cao X, Blankenstein T (1998) B cells inhibit induction of T cell-dependent tumor immunity. Nat Med 4:627–630

Rabinovich GA, Croci DO (2012) Regulatory circuits mediated by lectin-glycan interactions in autoimmunity and cancer. Immunity 36:322–335

Raghavan S, Quiding-Jarbrink M (2011) Regulatory T cells in gastrointestinal tumors. Expert Rev Gastroenterol Hepatol 5:489–501

Rahir G, Moser M (2012) Tumor microenvironment and lymphocyte infiltration. Cancer Immunol Immunother 61:751–759

Randall TD (2010) Bronchus-associated lymphoid tissue (BALT) structure and function. Adv Immunol 107:187–241

Rasanen K, Vaheri A (2010) Activation of fibroblasts in cancer stroma. Exp Cell Res 316:2713–2722

Rasku MA, Clem AL, Telang S, Taft B, Gettings K, Gragg H et al (2008) Transient T cell depletion causes regression of melanoma metastases. J Transl Med 6:12

Rech AJ, Vonderheide RH (2009) Clinical use of anti-CD25 antibody daclizumab to enhance immune responses to tumor antigen vaccination by targeting regulatory T cells. Ann N Y Acad Sci 1174:99–106

Restifo NP, Dudley ME, Rosenberg SA (2012) Adoptive immunotherapy for cancer: harnessing the T cell response. Nat Rev Immunol 12:269–281

Rodier F, Coppe JP, Patil CK, Hoeijmakers WA, Munoz DP, Raza SR et al (2009) Persistent DNA damage signalling triggers senescence-associated inflammatory cytokine secretion. Nat Cell Biol 11:973–979

Romano E, Nardelli-Haefliger D, Donda A, Corgnac S, Romero P (2012) SITC 26th annual meeting–summary. J Transl Med 10:105

Roozeboom MH, Arits AH, Nelemans PJ, Kelleners-Smeets NW (2012) Overall treatment success after treatment of primary superficial basal cell carcinoma: a systematic review and meta-analysis of randomized and nonrandomized trials. Br J Dermatol 167:733–756

Sakuishi K, Apetoh L, Sullivan JM, Blazar BR, Kuchroo VK, Anderson AC (2010) Targeting Tim-3 and PD-1 pathways to reverse T cell exhaustion and restore anti-tumor immunity. J Exp Med 207:2187–2194

Santini SM, Lapenta C, Logozzi M, Parlato S, Spada M, Di Pucchio T et al (2000) Type I interferon as a powerful adjuvant for monocyte-derived dendritic cell development and activity in vitro and in Hu-PBL-SCID mice. J Exp Med 191:1777–1788

Santini SM, Di Pucchio T, Lapenta C, Parlato S, Logozzi M, Belardelli F (2002) The natural alliance between type I interferon and dendritic cells and its role in linking innate and adaptive immunity. J Interferon Cytokine Res 22:1071–1080

Santodonato L, D'Agostino G, Nisini R, Mariotti S, Monque DM, Spada M et al (2003) Monocyte-derived dendritic cells generated after a short-term culture with IFN-alpha and granulocyte-macrophage colony-stimulating factor stimulate a potent Epstein-Barr virus-specific CD8+ T cell response. J Immunol 170:5195–5202

Sato E, Olson SH, Ahn J, Bundy B, Nishikawa H, Qian F et al (2005) Intraepithelial CD8+ tumor-infiltrating lymphocytes and a high CD8+/regulatory T cell ratio are associated with favorable prognosis in ovarian cancer. Proc Natl Acad Sci U S A 102:18538–18543

Sato M, Hirayama S, Matsuda Y, Wagnetz D, Hwang DM, Guan Z et al (2011) Stromal activation and formation of lymphoid-like stroma in chronic lung allograft dysfunction. Transplantation 91:1398–1405

Scaffidi P, Misteli T, Bianchi ME (2002) Release of chromatin protein HMGB1 by necrotic cells triggers inflammation. Nature 418:191–195

Schaefer L, Babelova A, Kiss E, Hausser HJ, Baliova M, Krzyzankova M et al (2005) The matrix component biglycan is proinflammatory and signals through Toll-like receptors 4 and 2 in macrophages. J Clin Invest 115:2223–2233

Schmid MC, Varner JA (2010) Myeloid cells in the tumor microenvironment: modulation of tumor angiogenesis and tumor inflammation. J Oncol 2010:201026

Schmidt H, Suciu S, Punt CJ, Gore M, Kruit W, Patel P et al (2007) Pretreatment levels of peripheral neutrophils and leukocytes as independent predictors of overall survival in patients with American joint committee on cancer stage IV melanoma: results of the EORTC 18951 biochemotherapy trial. J Clin Oncol 25:1562–1569

Schmidt M, Bohm D, von Torne C, Steiner E, Puhl A, Pilch H et al (2008) The humoral immune system has a key prognostic impact in node-negative breast cancer. Cancer Res 68:5405–5413

Schreiber RD, Old LJ, Smyth MJ (2011) Cancer immunoediting: integrating immunity's roles in cancer suppression and promotion. Science 331:1565–1570

Schroder K, Tschopp J (2010) The inflammasomes. Cell 140:821–832

Shah W, Yan X, Jing L, Zhou Y, Chen H, Wang Y (2011) A reversed CD4/CD8 ratio of tumor-infiltrating lymphocytes and a high percentage of CD4(+)FOXP3(+) regulatory T cells are significantly associated with clinical outcome in squamous cell carcinoma of the cervix. Cell Mol Immunol 8:59–66

Shen HM, Codogno P (2012) Autophagy is a survival force via suppression of necrotic cell death. Exp Cell Res 318:1304–1308

Shen Z, Zhou S, Wang Y, Li RL, Zhong C, Liang C et al (2010) Higher intratumoral infiltrated Foxp3+ Treg numbers and Foxp3+/CD8+ ratio are associated with adverse prognosis in resectable gastric cancer. J Cancer Res Clin Oncol 136:1585–1595

Sheng KC, Wright MD, Apostolopoulos V (2011) Inflammatory mediators hold the key to dendritic cell suppression and tumor progression. Curr Med Chem 18:5507–5518

Shieh AC, Rozansky HA, Hinz B, Swartz MA (2011) Tumor cell invasion is promoted by interstitial flow-induced matrix priming by stromal fibroblasts. Cancer Res 71:790–800

Shields JD, Kourtis IC, Tomei AA, Roberts JM, Swartz MA (2010) Induction of lymphoidlike stroma and immune escape by tumors that express the chemokine CCL21. Science 328:749–752

Sikora AG, Jaffarzad N, Hailemichael Y, Gelbard A, Stonier SW, Schluns KS et al (2009) IFN-alpha enhances peptide vaccine-induced CD8+ T cell numbers, effector function, and antitumor activity. J Immunol 182:7398–7407

Sims GP, Rowe DC, Rietdijk ST, Herbst R, Coyle AJ (2010) HMGB1 and RAGE in inflammation and cancer. Annu Rev Immunol 28:367–388

Singer K, Gottfried E, Kreutz M, Mackensen A (2011) Suppression of T-cell responses by tumor metabolites. Cancer Immunol Immunother 60:425–431

Smyth MJ, Dunn GP, Schreiber RD (2006) Cancer immunosurveillance and immunoediting: the roles of immunity in suppressing tumor development and shaping tumor immunogenicity. Adv Immunol 90:1–50

Soliman H, Mediavilla-Varela M, Antonia S (2010) Indoleamine 2,3-dioxygenase: is it an immune suppressor? Cancer J 16:354–359

Stagg J, Johnstone RW, Smyth MJ (2007) From cancer immunosurveillance to cancer immunotherapy. Immunol Rev 220:82–101

Steinbrink K, Jonuleit H, Muller G, Schuler G, Knop J, Enk AH (1999) Interleukin-10-treated human dendritic cells induce a melanoma-antigen-specific anergy in CD8(+) T cells resulting in a failure to lyse tumor cells. Blood 93:1634–1642

Steinman RM, Hawiger D, Nussenzweig MC (2003) Tolerogenic dendritic cells. Annu Rev Immunol 21:685–711

Stone GW, Barzee S, Snarsky V, Santucci C, Tran B, Langer R et al (2009) Nanoparticle-delivered multimeric soluble CD40L DNA combined with toll-like receptor agonists as a treatment for melanoma. PLoS ONE 4:e7334

Suzuki E, Kapoor V, Jassar AS, Kaiser LR, Albelda SM (2005) Gemcitabine selectively eliminates splenic Gr-1+/CD11b+ myeloid suppressor cells in tumor-bearing animals and enhances antitumor immune activity. Clin Cancer Res 11:6713–6721

Suzuki H, Chikazawa N, Tasaka T, Wada J, Yamasaki A, Kitaura Y et al (2010) Intratumoral CD8(+) T/FOXP3 (+) cell ratio is a predictive marker for survival in patients with colorectal cancer. Cancer Immunol Immunother 59:653–661

Swann JB, Smyth MJ (2007) Immune surveillance of tumors. J Clin Invest 117:1137–1146

Swartz MA, Lund AW (2012) Lymphatic and interstitial flow in the tumour microenvironment: linking mechanobiology with immunity. Nat Rev Cancer 12:210–219

Szajnik M, Czystowska M, Szczepanski MJ, Mandapathil M, Whiteside TL (2010) Tumor-derived microvesicles induce, expand and up-regulate biological activities of human regulatory T cells (Treg). PLoS ONE 5:e11469

Takemura S, Braun A, Crowson C, Kurtin PJ, Cofield RH, O'Fallon WM et al (2001) Lymphoid neogenesis in rheumatoid synovitis. J Immunol 167:1072–1080

Tang D, Kang R, Coyne CB, Zeh HJ, Lotze MT (2012) PAMPs and DAMPs: signal 0s that spur autophagy and immunity. Immunol Rev 249:158–175

Thomas SN, Rutkowski JM, Pasquier M, Kuan EL, Alitalo K, Randolph GJ et al (2012) Impaired humoral immunity and tolerance in K14-VEGFR-3-Ig mice that lack dermal lymphatic drainage. J Immunol 189:2181–2190

Timmer TC, Baltus B, Vondenhoff M, Huizinga TW, Tak PP, Verweij CL et al (2007) Inflammation and ectopic lymphoid structures in rheumatoid arthritis synovial tissues dissected by genomics technology: identification of the interleukin-7 signaling pathway in tissues with lymphoid neogenesis. Arthritis Rheum 56:2492–2502

Topalian SL, Hodi FS, Brahmer JR, Gettinger SN, Smith DC, McDermott DF et al (2012) Safety, activity, and immune correlates of anti-PD-1 antibody in cancer. N Engl J Med 366:2443–2454

Tosi D, Valenti R, Cova A, Sovena G, Huber V, Pilla L et al (2004) Role of cross-talk between IFN-alpha-induced monocyte-derived dendritic cells and NK cells in priming CD8+ T cell responses against human tumor antigens. J Immunol 172:5363–5370

Tosolini M, Kirilovsky A, Mlecnik B, Fredriksen T, Mauger S, Bindea G et al (2011) Clinical impact of different classes of infiltrating T cytotoxic and helper cells (Th1, th2, treg, th17) in patients with colorectal cancer. Cancer Res 71:1263–1271

van der Burg SH, Piersma SJ, de Jong A, van der Hulst JM, Kwappenberg KM, van den Hende M et al (2007) Association of cervical cancer with the presence of CD4+ regulatory T cells specific for human papillomavirus antigens. Proc Natl Acad Sci U S A 104:12087–12092

van Dongen M, Savage ND, Jordanova ES, Briaire-de Bruijn IH, Walburg KV, Ottenhoff TH et al (2010) Anti-inflammatory M2 type macrophages characterize metastasized and tyrosine kinase inhibitor-treated gastrointestinal stromal tumors. Int J Cancer 127:899–909

van Mierlo GJ, den Boer AT, Medema JP, van der Voort EI, Fransen MF, Offringa R et al (2002) CD40 stimulation leads to effective therapy of CD40(−) tumors through induction of strong systemic cytotoxic T lymphocyte immunity. Proc Natl Acad Sci U S A 99:5561–5566

van Poelgeest MIE, Welters MJP, Heusinkveld M, Aghai Z, van Ham JJ, Oostendorp J et al (2012) Lymph node-derived HPV-specific T-cells for the use of adoptive cell therapy in cervical cancer (submitted)

van Seters M, van Beurden M, ten Kate FJ, Beckmann I, Ewing PC, Eijkemans MJ et al (2008) Treatment of vulvar intraepithelial neoplasia with topical imiquimod. N Engl J Med 358:1465–1473

Verbeke H, Struyf S, Laureys G, Van Damme J (2011) The expression and role of CXC chemokines in colorectal cancer. Cytokine Growth Factor Rev 22:345–358

Verdegaal EM, Visser M, Ramwadhdoebe TH, van der Minne CE, van Steijn JA, Kapiteijn E et al (2011) Successful treatment of metastatic melanoma by adoptive transfer of blood-derived polyclonal tumor-specific CD4+ and CD8+ T cells in combination with low-dose interferon-alpha. Cancer Immunol Immunother 60:953–963

Vermeij R, Leffers N, Hoogeboom BN, Hamming IL, Wolf R, Reyners AK et al (2012) Potentiation of a p53-SLP vaccine by cyclophosphamide in ovarian cancer: a single-arm phase II study. Int J Cancer 131:E670–E680

Vesely MD, Kershaw MH, Schreiber RD, Smyth MJ (2011) Natural innate and adaptive immunity to cancer. Annu Rev Immunol 29:235–271

Vuylsteke RJ, Molenkamp BG, van Leeuwen PA, Meijer S, Wijnands PG, Haanen JB et al (2006) Tumor-specific CD8+ T cell reactivity in the sentinel lymph node of GM-CSF-treated stage I melanoma patients is associated with high myeloid dendritic cell content. Clin Cancer Res 12:2826–2833

Waetzig GH, Rose-John S (2012) Hitting a complex target: an update on interleukin-6 trans-signalling. Expert Opin Ther Targets 16:225–236

Walter S, Weinschenk T, Stenzl A, Zdrojowy R, Pluzanska A, Szczylik C et al (2012) Multipeptide immune response to cancer vaccine IMA901 after single-dose cyclophosphamide associates with longer patient survival. Nat Med. doi:10.1038/nm.2883. [Epub ahead of print]. PMID: 22842478

Wang E, Worschech A, Marincola FM (2008) The immunologic constant of rejection. Trends Immunol 29:256–262
Wang L, Rubinstein R, Lines JL, Wasiuk A, Ahonen C, Guo Y et al (2011) VISTA, a novel mouse Ig superfamily ligand that negatively regulates T cell responses. J Exp Med 208:577–592
Wang G, Liu G, Liu Y, Li X, Su Z (2012) FOXP3 expression in esophageal cancer cells is associated with poor prognosis in esophageal cancer. Hepatogastroenterology 59(119). doi:10.5754/hge11961. [Epub ahead of print]. PMID: 22389297
Watkins SK, Zhu Z, Riboldi E, Shafer-Weaver KA, Stagliano KE, Sklavos MM et al (2011) FOXO3 programs tumor-associated DCs to become tolerogenic in human and murine prostate cancer. J Clin Invest 121:1361–1372
Webber J, Steadman R, Mason MD, Tabi Z, Clayton A (2010) Cancer exosomes trigger fibroblast to myofibroblast differentiation. Cancer Res 70:9621–9630
Weber J (2010) Immune checkpoint proteins: a new therapeutic paradigm for cancer–preclinical background: CTLA-4 and PD-1 blockade. Semin Oncol 37:430–439
Wei S, Kryczek I, Zou L, Daniel B, Cheng P, Mottram P et al (2005) Plasmacytoid dendritic cells induce CD8+ regulatory T cells in human ovarian carcinoma. Cancer Res 65:5020–5026
Wei S, Zhao E, Kryczek I, Zou W (2012) Th17 cells have stem cell-like features and promote long-term immunity. Oncoimmunology 1:516–519
Welters MJ, Kenter GG, de Vos van Steenwijk PJ, Lowik MJ, Berends-van der Meer DM, Essahsah F et al (2010) Success or failure of vaccination for HPV16-positive vulvar lesions correlates with kinetics and phenotype of induced T-cell responses. Proc Natl Acad Sci U S A 107:11895–11899
Wherry EJ (2011) T cell exhaustion. Nat Immunol 12:492–499
Whiteside TL (2012) What are regulatory T cells (Treg) regulating in cancer and why? Semin Cancer Biol 22:327–334
Wieckowski EU, Visus C, Szajnik M, Szczepanski MJ, Storkus WJ, Whiteside TL (2009) Tumor-derived microvesicles promote regulatory T cell expansion and induce apoptosis in tumor-reactive activated CD8+ T lymphocytes. J Immunol 183:3720–3730
Wilke CM, Bishop K, Fox D, Zou W (2011a) Deciphering the role of Th17 cells in human disease. Trends Immunol 32:603–611
Wilke CM, Kryczek I, Wei S, Zhao E, Wu K, Wang G et al (2011b) Th17 cells in cancer: help or hindrance? Carcinogenesis 32:643–649
Wilke CM, Wei S, Wang L, Kryczek I, Kao J, Zou W (2011c) Dual biological effects of the cytokines interleukin-10 and interferon-gamma. Cancer Immunol Immunother 60:1529–1541
Winter S, Loddenkemper C, Aebischer A, Rabel K, Hoffmann K, Meyer TF et al (2010) The chemokine receptor CXCR5 is pivotal for ectopic mucosa-associated lymphoid tissue neogenesis in chronic Helicobacter pylori-induced inflammation. J Mol Med (Berlin) 88:1169–1180
Winters U, Daayana S, Lear JT, Tomlinson AE, Elkord E, Stern PL et al (2008) Clinical and immunologic results of a phase II trial of sequential imiquimod and photodynamic therapy for vulval intraepithelial neoplasia. Clin Cancer Res 14:5292–5299
Wong SB, Bos R, Sherman LA (2008) Tumor-specific CD4+ T cells render the tumor environment permissive for infiltration by low-avidity CD8+ T cells. J Immunol 180:3122–3131
Woo SR, Turnis ME, Goldberg MV, Bankoti J, Selby M, Nirschl CJ et al (2012) Immune inhibitory molecules LAG-3 and PD-1 synergistically regulate T-cell function to promote tumoral immune escape. Cancer Res 72:917–927
Yamagami W, Susumu N, Tanaka H, Hirasawa A, Banno K, Suzuki N et al (2011) Immunofluorescence-detected infiltration of CD4+ FOXP3+ regulatory T cells is relevant to the prognosis of patients with endometrial cancer. Int J Gynecol Cancer 21:1628–1634
Yang RY, Rabinovich GA, Liu FT (2008) Galectins: structure, function and therapeutic potential. Expert Rev Mol Med 10:e17
Yang L, Pang Y, Moses HL (2010) TGF-beta and immune cells: an important regulatory axis in the tumor microenvironment and progression. Trends Immunol 31:220–227

Yoon HH, Orrock JM, Foster NR, Sargent DJ, Smyrk TC, Sinicrope FA (2012) Prognostic impact of FoxP3+ regulatory T cells in relation to CD8+ T lymphocyte density in human colon carcinomas. PLoS ONE 7:e42274

Yoshimoto T, Takeda K, Tanaka T, Ohkusu K, Kashiwamura S, Okamura H et al (1998) IL-12 up-regulates IL-18 receptor expression on T cells, Th1 cells, and B cells: synergism with IL-18 for IFN-gamma production. J Immunol 161:3400–3407

Young AR, Narita M, Ferreira M, Kirschner K, Sadaie M, Darot JF et al (2009) Autophagy mediates the mitotic senescence transition. Genes Dev 23:798–803

Zamarron BF, Chen W (2011) Dual roles of immune cells and their factors in cancer development and progression. Int J Biol Sci 7:651–658

Zeestraten EC, Speetjens FM, Welters MJ, Saadatmand S, Stynenbosch LF, Jongen R et al (2012) Addition of interferon-alpha to the p53-SLP((R)) vaccine results in increased production of interferon-gamma in vaccinated colorectal cancer patients: a phase I/II clinical trial. Int J Cancer. doi:10.1002/ijc.27819. [Epub ahead of print]. PMID: 22948952

Zeid NA, Muller HK (1993) S100 positive dendritic cells in human lung tumors associated with cell differentiation and enhanced survival. Pathology 25:338–343

Zhan H, Suzuki T, Aizawa K, Miyagawa K, Nagai R (2010) Ataxia telangiectasia mutated (ATM)-mediated DNA damage response in oxidative stress-induced vascular endothelial cell senescence. J Biol Chem 285:29662–29670

Zhang L, Conejo-Garcia JR, Katsaros D, Gimotty PA, Massobrio M, Regnani G et al (2003) Intratumoral T cells, recurrence, and survival in epithelial ovarian cancer. N Engl J Med 348:203–213

Zhang Z, Yuan B, Bao M, Lu N, Kim T, Liu YJ (2011) The helicase DDX41 senses intracellular DNA mediated by the adaptor STING in dendritic cells. Nat Immunol 12:959–965

Zhou BB, Elledge SJ (2000) The DNA damage response: putting checkpoints in perspective. Nature 408:433–439

Zijlmans HJ, Fleuren GJ, Baelde HJ, Eilers PH, Kenter GG, Gorter A (2006) The absence of CCL2 expression in cervical carcinoma is associated with increased survival and loss of heterozygosity at 17q11.2. J Pathol 208:507–517

Zinkernagel RM (2002) On cross-priming of MHC class I-specific CTL: rule or exception? Eur J Immunol 32:2385–2392

Zitvogel L, Tesniere A, Kroemer G (2006) Cancer despite immunosurveillance: immunoselection and immunosubversion. Nat Rev Immunol 6:715–727

Zitvogel L, Apetoh L, Ghiringhelli F, Kroemer G (2008) Immunological aspects of cancer chemotherapy. Nat Rev Immunol 8:59–73

Zou W (2006) Regulatory T cells, tumour immunity and immunotherapy. Nat Rev Immunol 6:295–307

Chapter 6
The Versatile World of Inflammatory Chemokines in Cancer

Tal Leibovich-Rivkin, Yaeli Lebel-Haziv, Shalom Lerrer, Polina Weitzenfeld and Adit Ben-Baruch

Abstract Until recently, inflammatory chemokines were viewed mainly as indispensable "gate keepers" of immunity and inflammation. However, updated research indicates that members of this chemokine sub-family are important constituents of the tumor microenvironment, having multifaceted tumor-promoting roles in cancer. A very large number of studies indicate that many of the inflammatory chemokines are exploited by the tumor cells for their own benefit, and are actually skewed to the pro-malignancy phenotype. The different chemokines may be simultaneously expressed at the tumor site, having overlapping but also distinct tumor-promoting impacts. In general (except for the axis of CXCR3 and its ligands, that acts as a "double-edged sword"), the inflammatory chemokines induce immune imbalance at primary tumors and metastatic sites, doing so by promoting the presence and activation of tumor-associated macrophages (TAM), myeloid-derived suppressor cells (MDSC) and/or T regulatory cells (Treg). In parallel, immune suppression is ensued due to inhibition of Th1 cells and cytotoxic T lymphocytes (CTL). The chemokines also elevate metastasis-related processes, such as angiogenesis and osteoclastogenesis in the bone. Furthermore, they act directly on the tumor cells, promoting their proliferation, migration and invasion properties. Obviously, not all chemokines have the same pro-malignancy roles; however, chemokines that share the same receptor tend to have much in common in terms of their pro-cancerous activities. Accordingly, we will describe the roles of inflammatory chemokines in malignancy by using a receptor-based categorization: (1) CXCR1 and CXCR2 with their ELR$^+$ CXC

Yaeli Lebel-Haziv, Shalom Lerrer, Polina Weitzenfeld are equal contributors to this chapter.

T. Leibovich-Rivkin · Y. Lebel-Haziv · S. Lerrer · P. Weitzenfeld · A. Ben-Baruch (✉)
Department of Cell Research and Immunology, George S. Wise Faculty of Life Sciences, Tel Aviv University, 69978 Tel Aviv, Israel
e-mail: aditbb@tauex.tau.ac.il

M. R. Shurin et al. (eds.), *The Tumor Immunoenvironment*,
DOI: 10.1007/978-94-007-6217-6_6,

chemokine ligands, primarily CXCL8 (IL-8) but also CXCL1 (MGSA, GROα) and CXCL5 (ENA-78); (2) CXCR3 with its non-ELR CXC chemokine ligands: CXCL9 (Mig), CXCL10 (IP-10) and CXCL11 (I-Tac); (3) CCR2 and its ligands, mainly CCL2 (MCP-1); (4) CCR5 and its corresponding chemokines, with major emphasis on CCL5 (RANTES) and CCL3 (MIP-1α). Based on the findings obtained so far, we propose that inflammatory chemokines and their receptors are attractive therapeutic targets in malignancy, and discuss the expected difficulties in translating such approaches in the clinic.

Keywords Inflammatory chemokines · CXCR1/CXCR2 ligands · CXCR3 ligands · CCR2 ligands · CCR5 ligands, TAM, MDSC, T cells

6.1 Introduction

The tumor microenvironment is a complex entity that brings together several sub-environments, interacting and affecting each other. One of the most influential tumor sub-environments is the immune system, containing diverse cell populations and soluble factors. These components may exert immune surveillance functions and thus have the ability to protect the host against the developing tumor; however, very often they are skewed by the cancer cells to a pro-tumor phenotype. Under such circumstances, cells and soluble mediators of the immune system are exploited by the cancer cells, leading to formation of a tumor immune-environment that is favorable for the malignant cells, promoting their propagation and spreading capabilities.

Under physiological conditions, leukocytes, cytokines and chemokines cooperate, eventually mounting inflammatory reactions that protect the host against invading pathogens. These same elements join forces also in many pathological conditions, including cancer. Extensive research of the last two decades suggests that tumors are inflammatory organs, in which the tumor immune-environment has been diverted to a cancer-supporting entity (Balkwill and Mantovani 2011; Hanahan and Weinberg 2011; Hagemann et al. 2007; Joyce and Pollard 2009).

In this context, major roles have been attributed to inflammatory chemokines in cancer. The term "Inflammatory chemokines" denotes a functional categorization, describing chemokines that attract leukocytes to infected and damaged sites, playing key roles in the fight against foreign entities and enabling tissue repair. At such inflammatory sites, these chemokines are inducibly expressed in response to exposure of the tissue to inflammatory insults. In parallel, other chemokines known by the name "homeostatic chemokines" attract leukocytes to primary and secondary lymphoid organs, and are therefore involved in normal hematopoietic processes (Allen et al. 2007; Mantovani et al. 2006; Zlotnik et al. 2006).

Both these two functional sub-groups of chemokines include many members, divided by structural criteria to the CXC, CC, C and CX_3C sub-groups. The structural categorization is based on the number and the location of conserved cysteine residues in the N' terminus of the chemokine molecules. The CXC structural sub-group is further divided to chemokines expressing an ELR motif prior to the CXC sequence and therefore are termed "ELR^+ CXC chemokines", and to those that do not express such a motif, and are known by the name "non-ELR CXC chemokines" (Allen et al. 2007; Mantovani et al. 2006; Zlotnik et al. 2006).

The chemokine family includes around 50 different proteins in human, sharing the fundamental activity of chemoattracting leukocytes. In immune related-activities, chemokines that are presented by endothelial cell-expressed glycosaminoglycans (GAG) to leukocytes, activate high affinity heterotrimeric G protein-coupled receptors (GPCR) that are expressed by target immune and inflammatory cells. Many of the chemokines bind with high affinity several receptors, and similarly most of the chemokine receptors are activated by several chemokines. Very often, those chemokines that bind the same receptor have much in common, in terms of target cell preference and impact on immune and inflammatory activities (Allen et al. 2007; Mantovani et al. 2006; Zlotnik et al. 2006).

To follow on the above, in this chapter we will discuss the roles of inflammatory chemokines in cancer, using a receptor-based classification: (1) The receptors CXCR1 and CXCR2 and their ELR^+ CXC chemokines; (2) The CXCR3 receptor and its non-ELR CXC chemokines; (3) CCR2 and its chemokine ligands; (4) CCR5 and its corresponding chemokines. In each sub-section, the chapter will provide a general overview of the activities of the relevant receptors and their chemokines in cancer, their impact on events taking place at the tumor microenvironment, and their direct effects on tumor cells (summarized in Fig. 6.1).

Prior to addressing the four chemokine receptor sub-groups indicated above, it is important to put things in a broader perspective. Generally speaking, different inflammatory chemokines may be simultaneously expressed by the tumor cells and by stroma cells in their vicinity. The receptors corresponding to these chemokines are obviously expressed by specific leukocyte sub-types, but in addition functional chemokine receptors are expressed also by many different types of tumor cells. As expected from their roles in immune-related activities, inflammatory chemokines induce the migration of inflammatory cells such as monocytes and neutrophils to tumor sites, and also of cells exerting acquired immune activities, of which the most relevant ones to cancer are T cells. Therefore, there are specific circumstances in which these inflammatory chemokines mount protective immune activities against the tumor cells, and are thus beneficial for the host, as has been well documented for CXCR3 and its ligands. However, in the majority of cases, under the influence of the tumor cells and their products, most of the inflammatory chemokines are diverted towards the tumor-promoting phenotype, including the same CXCR3 ligands mentioned above. In general, the inflammatory chemokines lead to positioning of leukocytes with detrimental effects in the tumors, and this activity is fundamental to the way the chemokines affect cancer development and

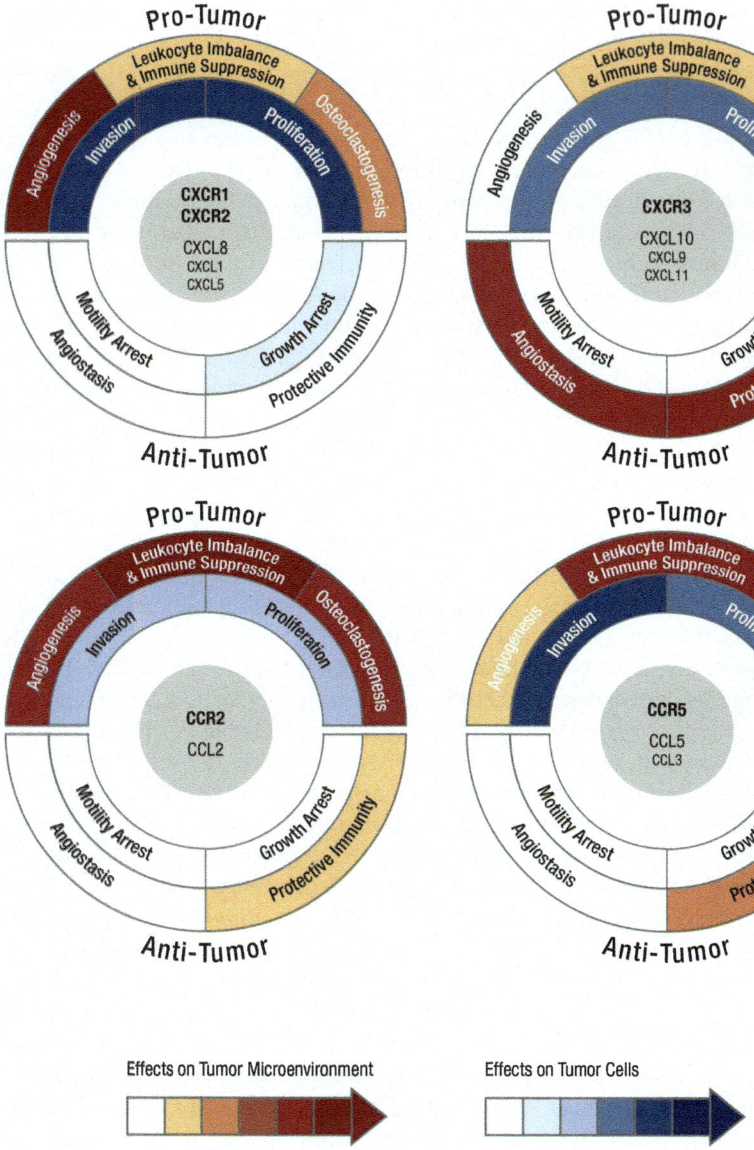
Pro-Tumor
Leukocyte Imbalance & Immune Suppression
Angiogenesis
Invasion
Proliferation
Osteoclastogenesis
CXCR1
CXCR2
CXCL8
CXCL1
CXCL5
Motility Arrest
Angiostasis
Growth Arrest
Protective Immunity
Anti-Tumor
Pro-Tumor
Leukocyte Imbalance & Immune Suppression
Angiogenesis
Invasion
Proliferation
Osteoclastogenesis
CXCR3
CXCL10
CXCL9
CXCL11
Motility Arrest
Angiostasis
Growth Arrest
Protective Immunity
Anti-Tumor
Pro-Tumor
Leukocyte Imbalance & Immune Suppression
Angiogenesis
Invasion
Proliferation
Osteoclastogenesis
CCR2
CCL2
Motility Arrest
Angiostasis
Growth Arrest
Protective Immunity
Anti-Tumor
Pro-Tumor
Leukocyte Imbalance & Immune Suppression
Angiogenesis
Invasion
Proliferation
Osteoclastogenesis
CCR5
CCL5
CCL3
Motility Arrest
Angiostasis
Growth Arrest
Protective Immunity
Anti-Tumor
Effects on Tumor Microenvironment
Effects on Tumor Cells

◄ **Fig. 6.1** The versatile world of inflammatory chemokines in cancer. The tumor microenvironment contains simultaneously a large variety of inflammatory chemokines, which may have overlapping but also distinct activities, as illustrated in the figure. Here, we demonstrate the roles of inflammatory chemokines in cancer, using a receptor-based categorization according to the following receptors: CXCR1/CXCR2, CXCR3, CCR2 and CCR5. An individual chart has been ascribed to each of the receptors, describing the impact of the receptor/s and their most relevant chemokines on cancer. As has been indicated in the chapter, inflammatory chemokines may exert anti-tumor activities (demonstrated in the lower half of each chart, termed *Anti-Tumor*), primarily mediated by recruitment of leukocytes with protective activities to the tumor site. However, generally-speaking, the effects of the inflammatory chemokines in malignancy are dominated by their pro-tumoral functions (upper half of the charts, termed *Pro-Tumor*). The activities of the chemokines are exerted on cells of the tumor microenvironment (external circle of each receptor's chart, carrying red color code; each activity has been given a specific intensity value, according to its relative contribution to the activities of the chemokines and their receptors) and on the tumor cells themselves (Internal circle, carrying blue color code, intensified according to the same guidelines as in the red color code). The shift in equilibrium towards the pro-tumoral direction is manifested in the figure by (1) Strong colors dominating the upper, pro-malignancy part of the receptor charts; (2) Minimal coloring in the lower, anti-tumor part of the charts, indicating that such activities are actually hardly exerted by the chemokines (except for the specific case of CXCR3 ligands). In the receptor-based charts, the activities of the chemokines were summed up by using brief terms of functional categorization. Each categorization includes several functions, of which all or only some are exerted by the chemokines/receptors that are included in each chart. The categorizations include the following functions: (1) At the upper part of the receptor charts, describing *Pro-Tumor* activities: *Leukocyte Imbalance and Immune Suppression* = Induction of high TAM levels at the tumor site, elevated MDSC and Treg presence and activities, reduced localization of Th1 cells and lower tumor cell killing by CTL and/or shift in neutrophil balance (due to yet undefined roles of neutrophils in cancer, aspects related to neutrophils were taken into account only when the chemokines had direct impacts on neutrophil levels or functions); *Angiogenesis* = Increased proliferation and migration of endothelial cells, tube formation and neovascularization; *Osteoclastogenesis* = Elevated processes of bone osteolysis and resorption; *Proliferation* = Increased tumor cell survival, proliferation and/or cancer stem cell-related functions; *Invasion* = Induction of tumor cell adhesion, migration and/or invasion, that may lead to metastatic spread in remote organs. (2) At the lower part of the receptor charts, describing *Anti-Tumor* activities: *Protective Immunity* = Promotion of immune surveillance, exerted by CTL and NK cells (for neutrophils the considerations were as above, in the *Pro-Tumor* part). *Angiostasis* = Inhibition of angiogenic processes; *Growth Arrest* = Induction of apoptosis or senescence in response to stress; *Motility Arrest* = Prevention of tumor cell adhesion, migration and invasion, that may limit metastasis formation. For more details on the specific functions that adhere to each of the receptor-chemokine axes, the readers are referred to the relevant sections of the chapter

progression. In addition, these chemokines can act on stroma cells and promote angiogenic processes and osteoclastogenesis, while in parallel they can also induce processes of tumor cell proliferation and invasion.

On the whole, the activities of inflammatory chemokines and their receptors are mostly diverted towards tumor promotion and progression. Here, each of the receptors—CXCR1/CXCR2, CXCR3, CCR2 and CCR5—and the corresponding chemokine ligands, has its own "flavor" and preferable mode of action, by which it impacts malignant processes (Fig. 6.1). In this chapter we will describe those features that are the most characteristics for each receptor and for its most effective ligands. However, due to space and length limitations, not all the relevant studies would be mentioned; rather, in each sub-section we will refer the readers to recent review papers, and will provide selected specific references, mainly of publications of the last several years.

6.2 Inflammatory Chemokines in Cancer

6.2.1 CXCR1 and CXCR2 Ligands

In human, CXCR1 and CXCR2 are the prototype receptors for ELR^+ CXC chemokines. In immune-related inflammatory conditions they are known as powerful attractants of neutrophils to sites of acute inflammation. Accordingly, the highest levels of constitutive expression of CXCR1 and CXCR2 are denoted on neutrophils, but both receptors are also expressed by monocytes and by other specific leukocyte subtypes (Allen et al. 2007; Mantovani et al. 2006; Zlotnik et al. 2006).

CXCR1 and CXCR2 show 77 % identity at the amino acid level, they have many shared characteristics and both bind CXCL8 with high affinity. However, these two receptors diverge in the spectrum of other chemokines they bind, and thus show some differences in their in vivo activities. CXCR2 is a promiscuous receptor also for the ELR^+ CXC chemokines CXCL1, CXCL2, CXCL3, CXCL5, CXCL6 and CXCL7, whereas CXCR1 binds with high affinity also CXCL6 and possibly CXCL7 (Allen et al. 2007; Mantovani et al. 2006; Zlotnik et al. 2006). The different ELR^+ CXC chemokines and their two receptors exemplify a complex net of interactions that is fundamental to the immune integrity of the host. Extending beyond the immune context, the activation of CXCR1 and CXCR2 has been shown to regulate additional physiological and also pathological conditions, of which malignancy is a major one.

Based on extensive research, ELR^+ CXC chemokines and their CXCR1 and CXCR2 receptors, are considered powerful pro-tumorigenic components in many cancer diseases [reviewed in (Dhawan and Richmond 2002; Ijichi 2012; Singh et al. 2010a; Vandercappellen et al. 2008; Waugh and Wilson 2008)]. The chemokines act on leukocytes and endothelial cells (EC) in manners that promote tumor development and metastasis, and in addition they induce tumor cell proliferation and invasion. The number of reports opposing this view is rather small, and there is a general consensus that the axis of CXCR1/CXCR2-ELR^+ CXC chemokines has detrimental pro-tumoral impacts in a very large number of malignancies. The effects of this axis in cancer will be described below, but there is a need to indicate that in one specific aspect, which is the process of senescence that takes place in response to stresses, this axis protects the cells against progression to a malignant state. It has been recently found by several investigators that CXCR2 and its ligands reinforce senescence early in cancer (Acosta et al. 2008; Ruan et al. 2012; Acosta and Gil 2009). These findings indicate that the CXCR1/CXCR2-ELR^+ CXC chemokine axis exerts pro-malignancy effects on one hand but pro-senescent functions on the other, emphasizing the need to identify the mechanisms involved in such opposing activities of these factors.

Going back to the tumor-promoting roles of ELR$^+$ CXC chemokines and their receptors, most of the research on human cancers has focused on the powerful chemokine CXCL8, and its two receptors, CXCR1 and CXCR2. The studies on human cancer biopsies and tumor extracts have been relatively limited; although the different studies have exemplified a general trend for associations between elevated expression of the receptors and of ELR$^+$ CXC chemokines with disease [for example (Grepin et al. 2011; Xu et al. 2012; Dimberg et al. 2012; Sunaga et al. 2012; Huang et al. 2010; Yang et al. 2010)], there are human malignancies in which the results were not fully conclusive, such as breast cancer (Green et al. 1997; Rody et al. 2011; Snoussi et al. 2010; Zuccari et al. 2012). However, determination of such chemokines and primarily of CXCL8 in patient serum yielded more definite findings, showing significant associations between high CXCL8 serum levels and increased tumor load, metastasis, disease progression and reduced survival. This pattern has been demonstrated in the past for example in melanoma, colon cancer and breast cancer (Benoy et al. 2004; Yokoe et al. 1997; Singh et al. 2010a; Ning et al. 2011).

Furthermore, the roles of ELR$^+$ CXC chemokines in malignancy have been extensively addressed in murine model systems, with two important limitations: the first is that the existence of a functional murine CXCR1 has been controversial and non-conclusive, and the second is that a mouse counterpart to human CXCL8 was not identified. Therefore, studies of leukocyte/stroma receptors in the mouse have actually addressed only CXCR2, and this was done in parallel to targeted reduction or over-expression of both CXCR1 and CXCR2 in human tumor cells. Also, CXCL8 was the subject of research when human tumor cells were analyzed in immune-deficient mice, but when murine host chemokines were studied, mainly CXCL1 and CXCL5 were addressed.

Investigations in animal model systems have provided a strong support for the causative tumor-promoting roles of ELR$^+$ CXC chemokines and their receptors in malignancy. Modalities such as neutralizing antibodies, siRNA, pharmacological inhibitors of CXCR1 and CXCR2, and approaches of over-expression, have shown direct roles for these chemokines and for CXCR1 and CXCR2 in promoting tumor growth and/or metastasis in many malignant diseases [several examples out of many: (Matsuo et al. 2009; Merritt et al. 2008; Rolny et al. 2008; Shamaladevi et al. 2009; Yamamoto et al. 2008; Grepin et al. 2011; Agarwal et al. 2010; Bandapalli et al. 2012; Huh et al. 2010; Chen et al. 2011; Varney et al. 2011; Vegran et al. 2011)]. Particular emphasis was put on melanoma, where extensive research has identified roles for tumor cell-expressed CXCL1, CXCL8, CXCR1 and CXCR2, as well as for host CXCR2 in advancing disease course (Dhawan and Richmond 2002; Haghnegahdar et al. 2000; Luan et al. 1997; Singh et al. 2009a; Singh et al. 2009b; Singh et al. 2009c; Gabellini et al. 2009; Huh et al. 2010; Singh et al. 2010b). Furthermore, direct activities of the chemokines were identified in the bone, where CXCL1 has elevated tumor cell adhesion in the bone matrix (Li et al. 2012), and both CXCL1 and CXCL8 stimulated osteoclastogenesis and bone resorption

(Bendre et al. 2003; Li et al. 2012; Pathi et al. 2010). Together, such chemokine-induced mechanisms may lead to increased bone metastasis and osteolysis.

The pro-tumoral activities of ELR^+ CXC chemokines are exerted on many different leukocyte, stroma and tumor cell properties. In view of the powerful chemotactic roles of ELR^+ CXC chemokines on neutrophils in physiological inflammatory conditions, studies of the last several years have questioned the effects of these chemokines on neutrophil recruitment and activation in cancer. In general, the roles of neutrophils in malignancy are debatable, with findings showing that these cells can inhibit or promote disease course, depending on the circumstances (Fridlender et al. 2012; Gregory et al. 2011; Piccard et al. 2012). For example, a recent study by Granot et al. 2011 has shown that neutrophils inhibited the metastatic seeding of breast tumor cells in the lungs, and that they did so by generating H_2O_2 (Granot et al. 2011). Alongside with that study, Fridlender and his colleagues have demonstrated that the activities of neutrophils in cancer are regulated by transforming growth factor beta (TGFβ) (Fridlender et al. 2009), and that after TGFβ blockade tumor-associated neutrophils acquired an anti-tumor phenotype. This latter study has also shown that ELR^+ CXC chemokines were expressed at the tumor site, and that recruitment of neutrophils to the site was followed by oxygen-mediated cytotoxicity mechanisms (Fridlender et al. 2009). However, the same study has found that a process largely driven by TGFβ has led to pro-cancer activities of neutrophils, illustrated by experiments in control tumors, where neutrophil depletion decreased tumor growth and resulted in more activated $CD8^+$ T cells in the tumors (Fridlender et al. 2009).

In parallel to these latter observations, many past studies demonstrated pro-malignancy roles for neutrophil-mediated processes that take place in the context of ELR^+ CXC chemokines (De Larco et al. 2004; Strell et al. 2010; Huh et al. 2010), including promotion of tumor cell motility, angiogenesis and mutagenesis through reactive oxygen species (De Larco et al. 2004). Along the same lines, it was demonstrated that following CXCL1/CXCL8-induced recruitment of neutrophils to tumors, the neutrophils interacted with the tumor cells, leading to ICAM-1-mediated motility of the cancer cells (Strell et al. 2010). In another study, following their CXCL8-induced migration to tumors, neutrophils interacted with tumor cells, eventually potentiating the anchoring of the cancer cells to vascular endothelium, again in an ICAM-1-dependent process (Huh et al. 2010).

The activities described above for ELR^+ CXC chemokine-affected neutrophils may account for elevated malignancy and metastasis, and they are complemented by the very powerful angiogenic effects of these chemokines. Chemokines and their receptors are important regulators of neovascularization in inflammatory processes and wound healing, as well as in diverse pathological conditions including cancer (Keeley et al. 2008; Keeley et al. 2010). Actually, this field of research has been overloaded with studies demonstrating that ELR^+ CXC chemokines, mainly CXCL8, are correlatively and causatively linked with high vascularization in cancers, and this is a major mode by which they contribute to tumor growth and metastasis [e.g. (Merritt et al. 2008; Keeley et al. 2008; Yoo et al. 2008; Agarwal et al. 2010; Vegran et al. 2011; Keeley et al. 2010; Wang et al. 2012c)].

The angiogenic effects of ELR^+ CXC chemokines are exerted by direct activity on EC, mainly through CXCR2. Although it was shown that both CXCR1 and CXCR2 can mediate the vascularizing activities of the chemokines on EC [migration and/or survival; (Gabellini et al. 2009; Li et al. 2005)], similar effects could be induced on EC via CXCR2 only (Keeley et al. 2008; Keeley et al. 2010). Moreover, in addition to CXCL8, CXCL1-3 and CXCL5 were found to have vascularizing activities [e.g. (Matsuo et al. 2009; Haghnegahdar et al. 2000; Luan et al. 1997; Xu et al. 2012)]; because these latter chemokines are not acting through CXCR1, it is assumed that CXCR2 is the major receptor mediating the angiogenic activities of ELR^+ CXC chemokines. Supporting this view are many studies showing that inhibition of CXCR2 activities has given rise to reduced vascularization in tumors (Matsuo et al. 2009; Singh et al. 2009c; Addison et al. 2000; Ning et al. 2012).

CXCL8 and the other ELR^+ CXC chemokines directly induce a large array of angiogenic functions in EC, including migration, proliferation, tube formation and the release of vascular endothelial growth factor (VEGF) [for example: (Matsuo et al. 2009; Yoo et al. 2008; Martin et al. 2009; Xu et al. 2012; Agarwal et al. 2010)]. In parallel, CXCL8 was shown to promote neovascularization indirectly, for instance by inducing VEGF production by the tumor cells (Li et al. 2008; Yang et al. 2010; Wang et al. 2012c). Thus, by acting on EC and also on the tumor cells, CXCL8 and other members of the ELR^+ CXC sub-family of chemokines contribute to the very important aspect of angiogenesis, which is fundamental to increased growth of tumors, and to their ability to spread to remote organs.

To follow on the observations showing that CXCL8 up-regulated VEGF production by tumor cells, it is important to indicate that ELR^+ CXC chemokines are very potent inducers of tumor-promoting functions in the cancer cells themselves, including invasion-related properties and proliferation. In this context, mainly CXCL8 but also other ELR^+ CXC chemokines (e.g. CXCL1, CXCL5) were shown to induce adhesion of tumor cells to EC, extracellular matrix and bone matrix (Warner et al. 2008; Huh et al. 2010; Li et al. 2012; Ju et al. 2012). The chemokines also potently promoted tumor cell migration and invasion [for example: (Araki et al. 2009; Neiva et al. 2009; Singh and Lokeshwar 2009; Bandapalli et al. 2012; Wang et al. 2012c; Ju et al. 2012; Yeudall et al. 2012; Lee et al. 2011; Halpern et al. 2011; Kuai et al. 2012; Nieman et al. 2011; Welte et al. 2012)], and amplified the expression of ICAM-1 in the tumor cells, as well as of other molecules involved in adhesive/migratory processes, such as VCAM-1, CD44 and the integrin $\alpha V\beta 3$ (Lee et al. 2012a; Kuai et al. 2012; Ju et al. 2012). In parallel, CXCL8 was shown to induce epithelial-to-mesenchymal transition in tumor cells (Bates et al. 2004; Fernando et al. 2011; Li et al. 2012) and elevations were found in the expression levels of matrix metalloproteinases (MMP) (Merritt et al. 2008; Wang et al. 2012c; Ju et al. 2012).

Very often, the migratory and invasive properties of tumor cells that were induced by the chemokines have been causatively linked to increased tumor load and metastasis formation. Depending on the cancer cell type and the experimental system, these functional events have been mediated by tumor cell expressed

CXCR1, CXCR2 or both receptors together (Singh et al. 2009a; Singh et al. 2009b; Gabellini et al. 2009; Warner et al. 2008; Bates et al. 2004; Varney et al. 2011; Singh et al. 2010b; Lee et al. 2011; Nieman et al. 2011; Fernando et al. 2011). Often, interactions between the tumor cells and stroma cells were stimulating tumor cell migration. For example, adipocytes were shown to release CXCL8, whose activities have led to adhesion of ovarian tumor cells to human omentum, and to migration towards human omental adipocytes and towards mouse omentum in vivo, in a process mediated by CXCR1 (Nieman et al. 2011). Cross-talks were also observed between tumor cells and mesenchymal stem cells [MSC, known as precursors to deleterious cancer-associated fibroblasts, CAF (Kalluri and Zeisberg 2006; Mishra et al. 2011; Shimoda et al. 2010)] or adipose tissue derived stromal cells. These interactions involved processes mediated through ELR^+ CXC chemokines and CXCR2, leading to tumor cell migration (Halpern et al. 2011; Welte et al. 2012), and such events could take place after the tumor cells attracted the stroma cells in their direction (Welte et al. 2012). Additional interactions were denoted between EC and the tumor cells, in which co-culturing of these two cell populations has induced CXCL8, and furthermore, Bcl-2 has induced in EC the release of CXCL8, leading to increased migration of the tumor cells (Neiva et al. 2009).

As already mentioned above, one additional very important activity of CXCL8, and in specific malignancies such as melanoma also of CXCL1, is induction of proliferation, anchorage-independent growth, increased survival and lower levels of apoptosis in the tumor cells. Several studies have shown that ELR^+ CXC chemokines, of which CXCL8 is the most prominent, promote proliferation and expression of cyclins such as cyclin D and E1, modifies cell cycle control, and shifts the balance between pro-apoptotic and anti-apoptotic proteins, eventually leading to inhibition of apoptotic processes. Moreover, these activities were detected not only by in vitro tests, but also in vivo, in tumors that have developed in mice (Dhawan and Richmond 2002; Shamaladevi et al. 2009; Singh and Lokeshwar 2009; Yang et al. 2010; Bandapalli et al. 2012; Wang et al. 2012c; Welte et al. 2012). As with other effects mediated by these chemokines, depending on the tumor context and experimental design, the direct activities of the chemokines were mediated by tumor cell-expressed CXCR1 or CXCR2, and in some cases by both receptors together (Shamaladevi et al. 2009; Singh et al. 2009a; Singh et al. 2009b; Gabellini et al. 2009; Zhong et al. 2008; Yang et al. 2010; Varney et al. 2011; Singh et al. 2010b; Ning et al. 2012; Singh et al. 2010b).

To follow on the above findings, it is interesting to note that CXCL8 and its proliferation-inducing functions have been shown to be of major relevance to cancer stem cells (CSC) [otherwise termed tumor-initiating cells (TIC)]. These cells were shown to express CXCL8, and the chemokine was up-regulated by elements such as neurotensin and the EMT-related protein snail (Tang et al. 2012b; Hwang et al. 2011), or after exposure to chemotherapy (Levina et al. 2008). Several studies reported the expression of CXCR1 and/or CXCR2 by these cells (Levina et al. 2008; Ginestier et al. 2010); CXCR1 blockade in breast CSC has led to massive apoptosis in the bulk tumor population via FasL/Fas signaling, has reduced CSC in the tumors and has inhibited metastasis formation (Ginestier et al. 2010). Evidently, the production of CXCL8 by CSC was required for self renewal of these cells, and has contributed to tumor growth and angiogenesis (Tang et al. 2012b; Hwang et al. 2011).

The major tumor-promoting roles that were identified for ELR^+ CXC chemokines and their receptors in malignancy have been the basis for studies on the regulation of this axis. In the limits of the present chapter, we would illustrate the complexity of this issue by providing several representative examples only. First, an important aspect is the control of CXCL8 expression and activities by cells of the tumor microenvironment, to which several examples have been already given above. In addition, fibroblasts are major regulators of CXCL8 expression and activities. For example, co-culturing of tumor cells with fibroblasts (as is the case also for co-culturing with macrophages and EC), has given rise to substantial elevations in the release of CXCL1 and CXCL8 [depending on the cell system; (Zhong et al. 2008; Knowles et al. 2009; Tjomsland et al. 2011)]. Increased CXCL8 expression could be induced also through the activation of tumor-expressed c-Met by fibroblast-derived HGF (Knowles et al. 2009). The tumor-promoting activities of fibroblasts were mediated through CXCR2-dependent mechanisms (Ijichi et al. 2011).

Being inflammatory mediators that are regulated in the immune system by inflammatory cytokines, the expression of ELR^+ CXC chemokines by malignant cells was induced by cytokines such as tumor necrosis factor α (TNFα) and interleukin 1β (IL-1β). It has been shown in the past that these two cytokines promoted the release of CXCL8 by tumor cells, and that TNFα acted on the tumor cells through CXCR1- and CXCR2-mediated autocrine loops (Kreeger et al. 2009; De Larco et al. 2001; Pantschenko et al. 2003). Furthermore, it was recently described by Massague's group that paracrine interactions, mediated via CXCR1, promoted resistance to chemotherapy and metastasis in breast tumors. In that study, genotoxic agents have limited the survival of cancer cells but also increased the production of TNFα by EC, which then has enhanced CXCL1 and CXCL2 expression in cancer cells. The chemokines recruited $CD11b^+$ $Gr1^+$ myeloid cells that expressed CXCR2, which in turn enhanced the viability of the cancer cells through S100A8/9 (calcium binding proteins that are associated with chronic inflammation and cancer) (Acharyya et al. 2012). In addition, another recent study suggested that induction of HIF-1α by IL-1β has led to increased tumor cell migration in a process possibly mediated by the CXCL8-CXCR1 pathway (Naldini et al. 2010).

Another interesting regulatory mode of CXCL8 which was characterized in breast cancer, is the associations between high CXCL8 expression and reduced expression of estrogen receptor α [(ERα); (Lin et al. 2004; Freund et al. 2003)], an indicator for poor prognosis in this disease. In contrast to ERα, estrogen is a powerful tumor-promoting factor in breast cancer, inducing tumor cell proliferation. Estrogen was shown to promote CXCL8 release by breast cancer cells (Yang et al. 2009; Bendrik and Dabrosin 2009; Haim et al. 2011), doing so in cooperativity with epidermal growth factor (EGF) through combined activation of ERα and AP-1 (Haim et al. 2011). Furthermore, the EGF-signaling receptor ErbB2 induced the expression of ERβ in breast tumor cells, which has then led to elevations in CXCL8 (Chen et al. 2011).

In addition, it was found that constitutive activation of EGF receptor (EGFR, ErbB1) and the activation of Ras induced the expression of CXCL8 in tumor cells, as indicated in several research systems [e.g. (Cataisson et al. 2009; O'Hayer et al. 2009; Bonavia et al. 2012; Kim et al. 2011)]. Accordingly, inhibition of EGFR has potentiated the anti-tumor activities of antibodies against CXCL8 in a murine model of breast cancer (Salcedo et al. 2002). However, it was revealed that in non-transformed cells, constitutively activated Ras could induce the release of CXCL8 only when the protective activities of p53 were diminished (Leibovich-Rivkin et al. 2012). In addition, while in the non-transformed cells, down-regulation of p53 alone did not induce the release of CXCL8 in the absence of Ras hyper-activation (Leibovich-Rivkin et al. 2012), p53 mutant having gain-of-function characteristics has obtained powerful abilities to promote the expression of the chemokine in tumor cells (Fontemaggi et al. 2009; Yeudall et al. 2012).

To conclude, the information that has been obtained on the CXCR1/CXCR2-ELR$^+$ CXC chemokine axis in cancer strongly points to very prominent tumor-elevating roles of these components (Fig. 6.1), mediated mostly by increased angiogenesis and induction of tumor cell proliferation and invasion. Thus, this axis is an attractive target for inhibition, possibly with very many inhibitors that are now being developed. Here, in view of its being shared by all ELR$^+$ CXC chemokines and its roles in angiogenesis, CXCR2 may be a good candidate for inhibition. However, it is essential to further elucidate the mechanisms regulating the activities of this axis, and to find out whether its pro-senescence activities could be used in favor of the host, for the prevention of cancer progression.

6.2.2 CXCR3 Ligands

In response to chemokines of the non-ELR CXC subfamily, CXCR3-expressing leukocytes are recruited to infected and inflamed sites, including primarily Th1 and natural killer (NK) cells (Vandercappellen et al. 2008; Lacotte et al. 2009; Groom et al. 2010). In addition to the CXCR3 receptor that was originally identified, now termed CXCR3-A, two additional variants have been characterized in human cells but not in mice: CXCR3-B that has a longer N'-terminal extracellular domain due to alternative splicing (Lasagni et al. 2003), and CXCR3-alt that is truncated at the carboxyl terminus and is predicted to have four or five trans-membrane domains (Ehlert et al. 2004) instead of the seven domains characterizing other GPCR.

CXCR3 binds the IFNγ-induced non-ELR CXC chemokines CXCL9, CXCL10 and CXCL11, which are typical inflammatory chemokines (Vandercappellen et al. 2008; Lacotte et al. 2009). In line with the high promiscuity of the chemokine world, other interactions also exist: CXCL11 is a functional ligand of CXCR7 (Burns et al. 2006), and CXCL4 was proposed to signal through CXCR3-B (Lasagni et al. 2003). However, within the limits of this chapter, emphasis will be given to CXCR3 and its three predominant ligands—CXCL9, CXCL10, CXCL11 (to be called herein CXCL9-11) - which have been well studied in many malignancies [reviewed in (Vandercappellen et al. 2008; Keeley et al. 2008; Lacotte et al. 2009; Fulton 2009; Ben-Baruch 2007; Keeley et al. 2010; Groom et al. 2011)].

When one considers the roles of the CXCR3-CXCL9-11 axis in cancer, a very complex picture is obtained, demonstrating a typical "double-edged sword" mode of action. To provide a general view of the activities of CXCL9-11 in malignancy, we will first describe their potent anti-tumor properties, and then will discuss their opposite effects through which they support growth and progression of tumors.

Early studies on this axis in cancer have given rise to the "Immunoangiostasis" theory, based on observations showing that the three CXCR3 ligands have powerful anti-tumorigenic functions due to two complementing activities: The first is promotion of immune responses which are mediated mainly by CXCR3-expressing Th1 cells and NK cells, and the second is inhibition of angiogenesis (Keeley et al. 2008; Keeley et al. 2010). These two distinct activities of the CXCR3-binding non-ELR CXC ligands have been documented in a large number of malignant diseases, and were found to be pivotal in protection against tumor growth and metastasis, as has been previously reviewed (Vandercappellen et al. 2008; Keeley et al. 2008; Lacotte et al. 2009; Fulton 2009; Ben-Baruch 2007; Keeley et al. 2010; Groom et al. 2011). The prominent roles of CXCR3 and its ligands in anti-tumor activities were proven in many in vivo studies, using a number of inhibiting modalities or neutralizing antibodies. In parallel, studies of the last several years have revealed additional facets of the anti-tumor CXCR3-CXCL9-11 axis in cancer, showing that it is regulated by other factors of the tumor microenvironment, and by a complex net of interactions between many cells of the immune system, to be illustrated below.

Fundamentally, tumor cells are an important source for the chemokines, expressing them in response to IFNγ but also independently of this cytokine [e.g. (Wendel et al. 2008; Przewoznik et al. 2012; Zhu et al. 2010; Andersson et al. 2011; Bronger et al. 2012)]. Recent studies indicate that the ability of CXCR3 ligands to induce protective anti-tumor responses requires, in addition to infiltration of Th1 and NK cells to the tumors (Wendel et al. 2008; Fujihara et al. 2008; Przewoznik et al. 2012), the recruitment of CD8^{+} T cells and CTL activities (Andersson et al. 2009; Zhu et al. 2010; Andersson et al. 2012; Hong et al. 2011; Wang et al. 2011). In addition, an important role was found to dendritic cells (DC) and antigen-presenting cells (APC) (Fujita et al. 2009; Andersson et al. 2012; Tanese et al. 2012). For example, in a murine glioma model, antigen-loaded Th1-polarizing DC were shown to induce CTL responses in a CXCL10-depenent mechanism, resulting in strong anti-tumor effects (Fujita et al. 2009). Moreover, several reports have suggested a role for CXCL9 and CXCL10 in reducing Treg (Andersson et al. 2009; Fujita et al. 2009; Muthuswamy et al. 2012), thus possibly shifting the balance in favor of protective mechanisms. The importance of T cell infiltration was reinforced by recent studies showing that in melanoma patients, CXCL9 and CXCL10 were up-regulated in chemotherapy-sensitive lesions, and they correlated with T cell infiltration, improved tumor control, and improved patient survival (Hong et al. 2011).

The findings described above illustrate the impact of CXCR3 and its ligands on the balance between different leukocyte populations, and on their consequent ability to mount an effective immune response against tumors. However, recent studies suggest that in contrast to the inflammatory nature of CXCR3 ligands and their participation in physiologically-related inflammatory processes (Lacotte et al. 2009; Groom et al. 2011), these chemokines are suppressed at the setting of cancer-related inflammation. Thus, the inflammatory conditions that prevail in many tumors and are usually leading to enhancement of disease course, may lead to inhibition of anti-tumor immune activities. For example, the cyclooxygenase (COX) system that synthesizes prostaglandins (PGE), contributed to the inflammatory immune suppressive nature of the tumor microenvironment. Recent studies indicate that PGE2 reduced the release of CXCL9 and CXCL10 from tumor cells, and that inhibition of COX has led to up-regulation of these two chemokines, giving rise to increased attraction of T effector cells and reduced migration of Treg to the tumors (Bronger et al. 2012; Muthuswamy et al. 2012). The clinical relevance of these observations was exemplified by findings showing inverse correlation between COX-2 over-expression and CXCL9 levels in biopsies of breast cancer patients (Bronger et al. 2012). Also, the production of nitric oxide (NO) by iNOS—which are key molecules in the inflammatory nature of many malignancies—was found to inhibit the expression of CXCL10 in melanoma, and thus was suggested to lead to a pro-cancerous tumor milieu and to poor prognostic outcome (Tanese et al. 2012).

Being the second fundamental element of the immunoangiostatic activity of CXCR3 ligands, inhibition of angiogenesis has been shown in many tumor types [reviewed in (Vandercappellen et al. 2008; Keeley et al. 2008; Lacotte et al. 2009; Fulton 2009; Ben-Baruch 2007; Keeley et al. 2010; Groom et al. 2011)]. Using human EC, it was found that these chemokines have induced angiostatic effects by acting through CXCR3-B (Lasagni et al. 2003). A study of CXCR3-B transfected cells has revealed that the p38 MAPK pathway was a downstream effector of CXCR3-B, mediating the angiostatic action of this chemokine receptor (Petrai et al. 2008). Moreover, studies with CXCL10 mutants in human cells suggested that binding to CXCR3 and not GAG, was essential for the tumor angiostatic activity of this chemokine (Yang and Richmond 2004). However, the involvement of GAG in the angiostatic activities of these chemokines is currently under debate, because another study has shown that the angiostatic functions of CXCL10 could take place independently of CXCR3, in a mechanism requiring GAG (Campanella et al. 2010). Such a mechanism may very well explain the anti-angiogenic activities of the non-ELR CXC chemokines on mouse EC, which do not express CXCR3-B. Here, it is important to note that hetero-dimerization of the chemokines (specifically CXCL4) with angiogenic factors (e.g. basic fibroblast growth factor (bFGF)) have been also shown to take place, and were suggested to contribute to inhibition of the angiogenic properties of factors such as bFGF, CXCL8 and VEGF (Keeley et al. 2008). The connections of the non-ELR CXC chemokines and the VEGF pathway were also illustrated in a renal cancer model, where intratumoral injection of CXCL9 combined with anti VEGFR2 therapy resulted in delayed resistance to the anti-angiogenic therapy, and had a beneficial impact of restoring angiostasis (Bhatt et al. 2010).

The observations described so far have illustrated the immunoangiostatic anti-tumorigenic functions of non-ELR CXC chemokines; however, the activities of the CXCR3-CXCL9-11 axis have been lately revealed to be more complex than originally expected, because the members of this axis can be used by the tumor cells for their own needs, and may therefore be deleterious. The members of this axis can promote malignancy by acting on cells of the tumor microenvironment, but mainly through direct activities on the tumor cells. In terms of the tumor milieu, recent studies suggest that cancer cells deviate immune cells from the protective phenotype, towards the pro-malignancy type. This has been shown in epidermal carcinogenesis, where it was suggested that recruitment of CXCR3-expressing $CD4^+$ and $CD8^+$ cells to the skin promoted keratinocyte proliferation (Winkler et al. 2011). In addition, in melanoma CXCL9 and CXCL10 induced disruption of endothelial cell barrier, possibly paving the way towards more efficient transendothelial migration of the tumor cells (Amatschek et al. 2010), that might result in enhanced invasiveness and metastasis formation

In parallel, the non-ELR CXC chemokines act very potently on the tumor cells, and accordingly CXCR3 is expressed by many types of cancer cells as well (Fulton 2009; Ben-Baruch 2007; Ma et al. 2009; Cambien et al. 2009; Murakami et al. 2013). This has been shown not only in malignant cell lines, but also in clinical samples of cancer patients, where CXCR3 expression by the tumor cells was correlated with poor survival, and CXCR3 was higher in metastatic foci within lymph nodes and liver compared to primary tumors [e.g. (Ma et al. 2009; Murakami et al. 2013)].

The ectopic expression of CXCR3 on tumor cells may endow them selective advantages, as they may passively sequester chemokines that are anti-tumorigenic in nature. But probably this is not the whole story, because a growing number of studies indicate that the non-ELR CXC chemokines actively stimulate pro-malignancy functions in CXCR3-expressing tumor cells, leading primarily to increased tumor cell proliferation and migration (Fulton 2009; Ben-Baruch 2007; Cambien et al. 2009; Pradelli et al. 2009; Murakami et al. 2013; Liu et al. 2011; Lee et al. 2012b; Shin et al. 2011). In line with such cancer-enhancing activities, studies in animal models have revealed a direct and causative role for the CXCR3-CXCL9-11 axis in promoting metastasis formation in many cancer types (Fulton 2009; Ben-Baruch 2007; Cambien et al. 2009; Pradelli et al. 2009; Walser et al. 2006; Murakami et al. 2013). Moreover, following the CXCL10-induced recruitment of cancer cells to the bone, chemokines of this subfamily were found to support osteoclast differentiation and to promote the formation of osteolytic bone metastases (Lee et al. 2012a). Also, it was found that the migration of the tumor cells may be enhanced by cells of their intimate milieu. For example, monocytes "conditioned" by co-culturing with human B cell precursor acute lymphoblastic leukemia cells released CXCL10, which in turn acted back on the tumor cells, and promoted their migration and invasion (Lee et al. 2012b).

The direct activities of the chemokines on the tumor cells bring about an important issue, related to the types of CXCR3 receptors expressed by the tumor cells, and their potential implications. In EC, CXCR3-B is the receptor inhibiting growth, while CXCR3-A increases survival (Lasagni et al. 2003). Along the same lines, recent data suggest that CXCR3-B is an "inhibitory" receptor, while CXCR3-A has an "inducing" phenotype not only in EC but also in tumor cells. Several reports indicate that CXCR3-B has anti-tumorigenic effects, and that inhibition of its activities lead to increased proliferation and migration of the tumor cells (Datta et al. 2006; Datta et al. 2008; Gacci et al. 2009; Datta et al. 2010).

To conclude, in this part of the chapter we have enlightened the complex nature of the CXCR3-CXCL9-11 axis in malignancy, having opposing activities that are difficult to expect in advance (Fig. 6.1). In many cancer types, contradicting results have been obtained, showing tumor-inhibiting as well as tumor-supporting activities for this axis. Eventually, the impact of CXCR3 and its ligands on disease course reflects equilibrium between effects that are exerted on many different cell types, not only of the tumor microenvironment but also directly on the tumor cells themselves. This equilibrium may be dictated by the differential response of specific target cells to the chemokines, by the ability of the chemokines to amplify the immune-potentiating modalities, and by the type of receptor expressed by different cell types.

The emerging literature on this topic indicates that the type of receptor—be it CXCR3-A, CXCR3-B and/or CXCR3-alt—is a crucial determinant of the overall impact of the CXCR3-CXCL9-11 axis in cancer. This fact has major implications, because it raises the possibility that it would be difficult to implement the findings obtained in mouse models—where CXCR3-B and CXCR3-alt are not expressed—to human patients. Together with findings suggesting that CXCR3-A is the receptor mediating the recruitment of anti-tumor Th1, CTL and NK cells, it is possible that the equilibrium in mouse models is biased to the immune-potentiating arm of CXCR3-CXCL9-11 activities. If so, this would suggest that the cancer-promoting activities of CXCR3 ligands are more potent than currently assumed based on murine models.

The above information leaves us with uncertainty regarding the therapeutic implications of the CXCR3-CXCL9-11 axis in cancer. To date, studies in animal model systems have used a large diversity of approaches whose aim was to increase the activities of CXCR3 ligands in malignancy, believing that such modalities would strengthen immunoangiostasis. However, the pro-malignancy activities of this axis suggest that implementing such approaches in human clinical settings may pose tumor-promoting threats. Thus, the picture in this case is far from being resolved, emphasizing the need for improved research and understanding of the complex implications of this axis in cancer.

6.2.3 CCR2 Ligands

CCR2 is best known for its major roles in mediating the migration of monocytes in response to chemokines of the CC group, specifically those termed monocyte chemotactic proteins (MCP). These chemokines share structural and genetic properties and thus have much in common in terms of target cell specificity; however, their in vivo activities do not fully overlap, possibly due to their different expression patterns in the organism and because some of them use other receptors besides CCR2 (Allen et al. 2007; Mantovani et al. 2006; Zlotnik et al. 2006; Conti and Rollins 2004; Yadav et al. 2010).

Of the different members of this group—that includes the chemokines CCL2, CCL7, CCL8, CCL13 (and murine CCL12)—CCL2 is the most potent activator of CCR2. Accordingly, the CCR2-CCL2 pair is the one dictating many of the biological responses in which monocytes are involved. In parallel to its strong impact on monocyte migration, CCL2 induces chemotactic responses also of DC, NK cells and T lymphocytes (Conti and Rollins 2004; Deshmane et al. 2009; Yadav et al. 2010).

In view of their fundamental roles in regulating monocyte recruitment, CCL2 and other MCP chemokines are essential for mounting effective physiological inflammatory responses, but they are also strongly involved in diverse pathological conditions (Deshmane et al. 2009; Yadav et al. 2010). Specifically in cancer, elevated expression of MCP chemokines and mainly of CCL2 was denoted in different malignancies; the chemokines were associated with advanced disease course and metastasis, as was shown to be the case for example in breast, colorectal and gastric cancers, while chemokine expression was hardly detected in normal epithelial cells (Soria et al. 2008; Ben-Baruch 2012a; Soria and Ben-Baruch 2008; Fujimoto et al. 2009; Hu et al. 2009; Ueno et al. 2000; Soria et al. 2011; Hwang et al. 2012).

These observations have led many investigators to question the roles of the MCP chemokines in malignancy, and whether they exert anti-tumor responses as would have been expected from the leukocyte target cells they act on. In the malignancy context, most studies have focused on the CCR2-CCL2 pair because of its outmost effects on monocyte migration. Several studies revealed anti-malignancy activities for this pair, increasing the potency of tumor vaccines or chemotherapy (Huang et al. 1994; Manome et al. 1995; Rollins and Sunday 1991; Tsuchiyama et al. 2008; Berencsi et al. 2011); however, these investigations are outnumbered by numerous findings providing evidence to a strong pro-tumorigenic impact of CCR2 and CCL2 in cancer [reviewed in (Keeley et al. 2008; Yadav et al. 2010; Conti and Rollins 2004; Deshmane et al. 2009; Ben-Baruch 2012a; Soria and Ben-Baruch 2008; Craig and Loberg 2006; Yadav et al. 2010; Ben-Baruch 2012b; Verma et al. 2012)].

Improved understanding of the roles played by CCR2-CCL2 in malignancy was provided by studies in tumor model systems in mice. Here, the different investigations used knockout (KO) of CCR2 in the animals, CCR2 antagonists and to other modalities that reduced CCR2 or CCL2, such as siRNA/shRNA or neutralizing antibodies [examples of the last several years: (Hart et al. 2009; Lu and Kang 2009; Pahler et al. 2008; Popivanova et al. 2009; Koga et al. 2008; Mizutani et al. 2009; Baba et al. 2012; Izhak et al. 2012; Leuschner et al. 2011; Nakasone et al. 2012; Wolf et al. 2012; Chiu et al. 2012; Cortez-Retamozo et al. 2012; Fridlender et al. 2011; Jin et al. 2010; Lesokhin et al. 2012; Park et al. 2012; Qian et al. 2011; Tsuyada et al. 2012; Wang et al. 2012a)]. Joined by over-expression approaches (Lu and Kang 2009; Mizutani et al. 2009; Stathopoulos et al. 2008), these studies pointed out to direct and causative roles for CCR2 and CCL2 in elevating malignancy and metastasis formation, and have revealed many of their mechanisms of action.

Based on murine studies and clinical investigations, it is now clear that the CCR2-CCL2 pair promotes malignancy primarily by shifting the immune balance towards leukocyte sub-populations that support tumor growth and metastasis. Acting primarily on myeloid cells, CCL2 has the strongest impact on disease course by recruiting and activating two myeloid sub-populations: the first is of monocytes that turn at the tumor site to tumor-associated M2 macrophages (TAM), which have been long ago characterized as cells capable of releasing a large variety of tumor-supporting factors (Allavena et al. 2008; Biswas and Mantovani 2010); the second sub-population is of MDSC, recently identified for their ability to down-regulate potential anti-tumor T cell activities (Murdoch et al. 2008; Greten et al. 2011; Youn and Gabrilovich 2010). The shift in immunological balance imposed by CCR2-CCL2 is achieved by elevated levels of these two cell populations, in parallel to reduced activities of CTL and additional modifications in other leukocyte sub-populations, as would be described below (for specific references—see below).

As indicated above, many lines of evidence indicate that the CCR2-CCL2 axis plays major roles in positioning of monocytes at tumor sites, and leads to their skewing to the M2 phenotype, eventually giving rise to high presence of TAM in the tumors and potentiating tumor growth (Fujimoto et al. 2009; Ueno et al. 2000; Pahler et al. 2008; Popivanova et al. 2009; Koga et al. 2008; Mizutani et al. 2009; Stathopoulos et al. 2008; Allavena et al. 2008; Leuschner et al. 2011; Nakasone et al. 2012; Wolf et al. 2012; Cortez-Retamozo et al. 2012; Fridlender et al. 2011; Biswas and Mantovani 2010; Diaz-Valdes et al. 2011). In addition, driven by the CCR2-CCL2 axis, monocyte recruitment is also fundamental to metastasis formation (Lu and Kang 2009; Qian et al. 2011). For example, the study by Pollard and his colleagues has shown that CCR2-expressing inflammatory monocytes infiltrated breast metastasis, in response to CCL2 synthesized by the tumor cells and by stroma cells (Qian et al. 2011). That study also indicated that the CCL2-recurited monocytes promoted the metastatic seeding of the tumor cells, doing so in a VEGF-mediated manner (Qian et al. 2011). Also, Part et al. have shown that following priming of the murine host with the chemotherapeutic agent cyclophosphamide, an abrupt expansion of myeloid cells has taken place in the bone marrow and the circulation, and has promoted metastasis formation. This process was mediated by host-derived CCL2, whose inhibition significantly reduced the pro-metastatic effects of cyclophosphamide (Park et al. 2012).

Alongside with the above findings, much evidence has been recently provided on the infiltration of tumors by CCR2-expressing MDSC, whose activities have led to suppression of CTL responses against tumor cells. The recruitment and/or activation of MDSC were shown to be induced by CCL2, and also by tumor cell-derived β-defensin 3 (Hart et al. 2009; Huang et al. 2007; Umemura et al. 2008; Lesokhin et al. 2012; Gehad et al. 2012). A recent study of ovarian tumor progression has shown that two monocyte subsets, differing in their markers, were present at the peritoneum at different tumor stages. These two monocyte subpopulations had immune suppressive activities towards naïve $CD8^+$ and $CD4^+$ T cells. CCR2 was a critical factor in recruiting these suppressive cells to the ovarian tumor microenvironment, as indicated by genetic ablation of CCR2 in the mice, leading to lower tumor burden (Hart et al. 2009). Moreover, CCR2-expressing MDSC limited the efficacy of immune-therapy by down-regulating the migration of $CD8^+$ T cells to the tumor site (Lesokhin et al. 2012). Also, in parallel to inducing monocyte polarization to the M2 phenotype, CCL2 was connected to reduced levels of active $CD8^+$ CTL (Fridlender et al. 2011).

These recent studies have also shown that the inhibitory activities of MDSC were mediated through the cytokine TGFβ, and also via the activation of arginase-1 and production of NO (Hart et al. 2009; Umemura et al. 2008; Lesokhin et al. 2012; Gehad et al. 2012), which are important mediators of T cell inhibition (Greten et al. 2011; Youn and Gabrilovich 2010; Schaer et al. 2011). To follow on the above, the tumor-related activities of CCL2 were regulated by its nitration (Molon et al. 2011). The recent study by Viola's group has shown that reactive nitrogen species (RNS) induced nitration of CCL2, and that this form of the chemokine was ineffective in inducing T cell infiltration to tumor sites. Accordingly, prevention of RNS production has improved intratumoral T cell migration, and has enhanced tumor eradication through CTL-mediated responses (Molon et al. 2011).

Adding to the suppressive influence of MDSC on anti-tumor cell responses, the CCR2-CCL2 axis contributed to the generation of tolerized DC: Tumor-bearing mice deposited CCL2 in interlobular vascular-rich regions of the thymus, where Sirpα+ conventional DC have accumulated. The CCR2-CCL2 pair was involved in enhanced capacity of the DC to take up antigens, resulting in a shift to negative selection (Baba et al. 2012).

As expected from its inflammatory nature, CCL2 is also associated with the inflammatory phenotype of the tumor microenvironment. CCL2 was found to induce the recruitment to tumors of Th17 cells (Su et al. 2010), considered as cells that may contribute to the pro-malignancy and inflammatory nature of the tumor milieu. Interactions between CCL2 and inflammatory factors residing at the tumor microenvironment have also been revealed: TNFα and IL-1β were shown to promote the release of CCL2 by tumor cells [e.g. of the breast (Neumark et al. 2002; Neumark et al. 2003; Seeger et al. 2006; Seeger et al. 2008; Soria et al. 2011)]; furthermore, the ability of the CCR2 antagonist 7ND to inhibit tumor growth and monocyte infiltration to tumors was accompanied by reduced expression of the inflammatory cytokines TNFα and IL-1α, presumably from monocytes (Koga et al. 2008). Further adding to the network that may exist between CCL2, TNFα and IL-1β are findings showing that in breast cancer patient biopsies, TNFα and IL-1β expression in the tumor cells was coordinated with high abundance of CCL2 throughout different stages of disease (Soria et al. 2011).

Relating further to the CCR2-CCL2-induced immune imbalance is the impact of this axis on neutrophil responses (Pahler et al. 2008; Granot et al. 2011; Cortez-Retamozo et al. 2012). As previously mentioned, the implications of neutrophils on malignancy are not yet fully resolved; however, improved insights to this issue were recently provided by the study of Granot et al. 2011 described above, indicating that CCL2 secretion has enhanced tumor growth at the primary site, but at the same time the chemokine took role in neutrophil entrainment, so that neutrophils exerted an anti-metastatic response that inhibited tumor cell seeding at distant sites (Granot et al. 2011).

Overall, modification of immunological balance is a major pathway through which the CCR2-CCL2 axis diverts the immune-environment towards the pro-malignancy phenotype. However, CCL2 has additional activities on the tumor microenvironment, mediated by inducing angiogenesis and bone osteolysis, thus contributing to metastasis formation. Several recent reviews and updated publications documented the strong angiogenic activities of CCL2 in many tumor systems [e.g. (Keeley et al. 2008; Conti and Rollins 2004; Stathopoulos et al. 2008; Salcedo et al. 2000; Goede et al. 1999; Niu et al. 2008; Verma et al. 2012; Izhak et al. 2010)]. Here, the chemokine was shown to act indirectly by promoting the abundance of TAM at tumor sites. These cells were correlated with increased angiogenesis due to their ability to release a large variety of angiogenic factors, such as VEGF (Conti and Rollins 2004; Stathopoulos et al. 2008; Allavena et al. 2008; Goede et al. 1999; Verma et al. 2012; Qian et al. 2011; Biswas and Mantovani 2010; Izhak et al. 2010). Also, it has been well established that EC express CCR2, and that CCL2 directly promotes angiogenesis by inducing the proliferation and migration of these cells (Keeley et al. 2008; Niu et al. 2008; Weber et al. 1999; Wang et al. 2012a; Roy and Kolattukudy 2012). These direct activities were mediated by the transcription factor MCP-1-induced protein (MCPIP) (Niu et al. 2008; Roy and Kolattukudy 2008). A recent study by Roy & Kolattukudy suggested that MCPIP induced EC differentiation via induction of oxidative stress that has led to endoplasmic reticulum stress, and thereafter to autophagy which was involved in tube formation (Roy and Kolattukudy 2008).

An additional important tumor-promoting mechanism induced by CCL2 is osteoclastogenesis and bone loss, demonstrated for example in breast cancer [reviewed in (Ben-Baruch 2012a; Soria and Ben-Baruch 2008; Craig and Loberg 2006)]. CCL2 was shown to regulate pathological conditions in the bone, to exhibit chemotactic activities towards osteoclasts and to be expressed by osteoblasts (Craig and Loberg 2006; Fritz et al. 2002; Kinder et al. 2008; Wright and Friedland 2004; Bussard et al. 2010a; Bussard et al. 2010b). These observations have led researchers to speculate that CCL2 activities may be connected to the fact that the bone is a preferred metastatic site in breast cancer, which is highly colonized by the tumor cells. Indeed, studies performed in this direction have led to the conclusion that CCL2 is responsible for "conditioning" of the bone microenvironment (Kinder et al. 2008; Chen et al. 2009; Molloy et al. 2009; Zhu et al. 2007; Bussard et al. 2012a). The release of CCL2 by osteoblasts, in response to cancer cells, has led to osteoclast activation and to bone loss (Kinder et al. 2008; Zhu et al. 2007), thus creating a unique niche that favored breast tumor growth. This was indicated by the fact that metastasis in the bone was correlated with osteoclast formation and recruitment of osteoclasts to the bone, and that CCL2-expressing tumor cells engaged $CCR2^+$-expressing monocytic cells, including preosteoclasts and macrophages, leading to elevated tumor cell localization in the bone (Lu and Kang 2009; Mizutani et al. 2009). In addition, it was found that MSC that have matured to osteoblasts released CCL2 that has induced migration of breast tumor cells (Molloy et al. 2009), and it was proposed that the angiogenic activities of CCL2 potentiated vascularization at the bone (Wilson et al. 2010).

Overall, the findings described above illustrate the pro-cancerous effects of the CCR2-CCL2 axis on the tumor microenvironment, in primary tumors and metastatic sites. However, CCR2 is also expressed by the tumor cells. Accordingly, CCL2-mediated signals promoted tumor cell migration, invasion and MMP production [examples: (Kawai et al. 2009; Mestdagt et al. 2006; Nam et al. 2006; Youngs et al. 1997; Chiu et al. 2012; Tang and Tsai 2012a)]. Although such activities of CCL2 are important and may have substantial roles in elevating metastasis, the major effects of the CCR2-CCL2 pair are dictated by its ability to act on immune cells and stroma cells, and less so by its direct activities on the tumor cells (Fig. 6.1).

The above-mentioned findings set CCR2 and its MCP ligands—of which CCL2 is the most prominent—as attractive therapeutic targets in malignancy. To date, a large variety of approaches are available for inhibition of this axis, including CCR2 antagonists, antibodies and siRNA modalities that were used in animal model systems. The efficacy of these measures in some of the experimental systems was strong, but in others less pronounced, emphasizing the challenge that we are about to face when trying to introduce these applications to the clinic. In addition, in view of the major roles of CCR2 in regulating monocyte recruitment to inflammatory sites, it may be expected that inhibition of this receptor would yield undesired side effects on immune activities following exposure to pathogens.

6.2.4 CCR5 Ligands

CCR5 is a promiscuous receptor that binds several chemokines of the CC family, including CCL3, CCL4 and CCL5. CCR5 and these three ligands are only part of a more complex net of interactions existing between these chemokines, several additional chemokines and CCR5, CCR1 and CCR3 (Levy 2009; Oppermann 2004; Mueller and Strange 2004). However, most of the studies related to these chemokines and their receptors in solid malignancies have focused on the CCR5-CCL5 pair. In parallel, in hematological cancers and mainly in multiple myeloma (MM), most findings were obtained with CCL3 (see below). Therefore, this chapter will discuss mainly the CCR5-CCL5 axis in cancer, and will also describe CCL3 roles in promoting MM.

CCR5 is expressed by a large number of leukocytes, primarily effector T cells, monocytes and macrophages. Accordingly, CCR5 and its ligands play important regulatory roles in infection and inflammation. CCR5 has been extensively studied for its impact on immune activities, and has been at the center of AIDS research because of its being an HIV co-receptor. Particular interest has been put on the mutated Δ32 CCR5 receptor which is not expressed at the cell surface, providing protection against R5 HIV viruses (Levy 2009; Oppermann 2004; Mueller and Strange 2004); by analyzing this mutation of CCR5 in cancer, efforts were made to decipher the roles of this receptor in malignancy, as will be discussed below.

The key roles played by CCL5-induced leukocyte chemotaxis in protective immunity have given rise to new research directions in malignancy. Based on the premises that the chemokine may be mediating anti-tumor effects but that such activities are actually suppressed in cancer, hope was raised that potentiation of the CCR5-CCL5 axis may inhibit tumor formation and metastasis. As a result, the chemokine, and its three receptors were studied in many cancer diseases, where their expression patterns, roles and modes of action have been partly identified thus far [reviewed in (Ben-Baruch 2012a; Soria and Ben-Baruch 2008; Soria 2009; Ben-Baruch 2012b; Lapteva and Huang 2010; Suffee et al. 2011)].

As expected, CCL5 and its receptors have been found to lead to protective anti-cancer immunity when appropriate stimulatory conditions were provided. Specifically, when the activities of DC and lymphocytes were boosted by a variety of manipulations, CCL5 acted as an adjuvant that has potentiated anti-tumor activities, primarily those mediated by T lymphocytes [(Lapteva and Huang 2010); specific examples: (Nesbeth et al. 2009; Song et al. 2009; Inoue et al. 2008; Nesbeth et al. 2010; Gonzalez-Martin et al. 2011)]. These findings suggest that conditioning of the tumor microenvironment can serve as a good platform in which CCL5 can strengthen anti-cancer activities; however, they also testify to "failure" of acquired immunity to act against arising tumors, possibly because of tumor-induced suppression of protective immune mechanisms. Along these lines, associations between high expression levels of CCL5 in tumors, protective immune infiltrates and improved disease course are not many, suggesting that in most cases the potential beneficial effects of the CCR5-CCL5 axis are not appropriately activated in cancer.

Rather, it has been revealed that very often the tumor cells take advantage of CCR5 and of its corresponding ligands. Dominating the field are studies on CCR5-CCL5-mediated responses, showing that this axis is skewed in cancer to the pro-malignancy phenotype, and that these two components actively promote tumor growth and spread (Ben-Baruch 2012a; Soria and Ben-Baruch 2008; Soria 2009; Ben-Baruch 2012b; Lapteva and Huang 2010; Suffee et al. 2011). Many studies indicate that cancer cells constitute a major source for the chemokine and that its elevated levels are associated with poor prognosis and advanced disease [Reviewed in (Ben-Baruch 2012a; Soria and Ben-Baruch 2008; Soria 2009; Ben-Baruch 2012b); specific publications as example: (Soria et al. 2008; Sugasawa et al. 2008a; Sugasawa et al. 2008b; Tan et al. 2009; Wu et al. 2008a; Wigler et al. 2002; Azenshtein et al. 2002; Luboshits et al. 1999; Yaal-Hahoshen et al. 2006; Borczuk et al. 2008; Zhang et al. 2009; Soria et al. 2012; Su et al. 2010; Velasco-Velazquez et al. 2012; Chang et al. 2012a; Soria et al. 2011)]. In addition, many cells of the tumor microenvironment can contribute to high CCL5 load at the tumor site, including leukocytes, EC, fibroblasts and MSC [e.g. (Sugasawa et al. 2008a; Laubli et al. 2009; Karnoub et al. 2007; Su et al. 2010; Mi et al. 2011; Gallo et al. 2012)].

The direct and causative pro-malignancy functions of the CCR5-CCL5 pair were proven in a large number of studies using animal model systems. Here, when different approaches were taken to reduce the expression or activities of each of these two components, tumor load and metastasis formation were significantly inhibited, and survival was often increased. The CCR5/CCL5 inhibitory modalities that were used included siRNA/shRNA to CCR5 or CCL5, neutralizing antibodies against these components, the CCR1/CCR5 antagonist met-CCL5, CCR5 KO mice and different pharmacological inhibitors of CCR5 such as maraviroc and TAK-779 [representative studies, of the last several years: (Sugasawa et al. 2008a; Wu et al. 2008a; Borczuk et al. 2008; Laubli et al. 2009; Karnoub et al. 2007; Adler et al. 2003; Mi et al. 2011; Robinson et al. 2003; Velasco-Velazquez et al. 2012; Chang et al. 2012a; Chang et al. 2012b; Song et al. 2012; Cambien et al. 2011; Wang et al. 2012b)].

Together with extensive in vitro analyses and studies of patient samples, two non-mutually exclusive tumor-promoting pathways have been identified for the CCR5-CCL5 axis, one acting on cells of the tumor microenvironment and the other directly affecting the malignant cells. When the intimate milieu of the tumor cells is concerned, a major pro-tumoral activity of CCL5 is mediated by its chemotactic properties towards leukocytes. Evidently, CCL5 changes the balance between different types of leukocyte infiltrates, leading to predominance of cells whose activities support malignancy, rather than exerting anti-tumor immune activities.

Supporting such tumor-promoting roles of CCL5 are studies showing that the chemokine has led to increased accumulation of Treg in the tumors. These studies have also shown that Treg recruitment was dependent on CCR5 expression by these cells, and that inhibition of this process has given rise to reduced tumor growth (Tan et al. 2009; Chang et al. 2012a; Chang et al. 2012b). In parallel, apoptosis of $CD8^+$ T cells and suppression of their activities were denoted (Sugasawa et al. 2008a; Chang et al. 2012a; Chang et al. 2012b). In the specific case of colorectal cancer, CCR5-CCL5 signaling increased the synthesis of TGFβ in Treg cells, in turn enhancing the cytolysis by $CD8^+$ T cells (Chang et al. 2012a). Emphasizing the roles of the CCR5-CCL5 axis in down-regulating the potential anti-tumor activities of $CD8^+$ T cells are studies showing that inhibition of this pair leads to increased presence of cytotoxic T cells at tumor sites, and that reduced suppression of $CD8^+$ cells was associated with lower tumor volume (Chang et al. 2012a; Song et al. 2012; Chang et al. 2012b).

Studies on tumor microenvironments indicate that immune suppression and inflammatory conditions are often connected in malignancy (Ben-Baruch 2006). Accordingly, the CCR5-CCL5 axis is involved not only in mediating immune suppression, but also is associated with tumor inflammation. By virtue of its strong chemotactic activities towards monocytes, the CCR5/CCR1-CCL5 axis actively induced high presence of TAM in the tumors, and elevated macrophage content was correlated with high malignancy [e.g. (Ben-Baruch 2012a; Soria and Ben-Baruch 2008; Wu et al. 2008a; Laubli et al. 2009; Adler et al. 2003; Robinson et al. 2003; Wu et al. 2008b)]. Moreover, CCL5 released from tumor cells and from tumor-derived fibroblasts has been shown to induce the migration of Th17 cells, and thus may contribute to the inflammatory nature of the tumor milieu by yet another aspect (Su et al. 2010). In addition, the expression of CCL5 by tumor cells was up-regulated by inflammatory cytokines such as TNFα and IL-1β [e.g. (Soria et al. 2008; Azenshtein et al. 2002; Ali et al. 2000)]. By doing so, the inflammatory cytokines promoted the abundance of CCL5 at the tumor site, and accordingly recent findings demonstrated associations between CCL5, TNFα and IL-1β along malignancy course in breast cancer (Soria et al. 2011).

The above findings testify for the ability of the CCR5-CCL5 pair to skew the immune balance towards immune/inflammatory activities that support malignancy. Recent studies show that CCL5, CCR5 and other chemokines/receptors associated with them, control additional cells and events at the tumor site. Recent findings point to novel and yet minimally addressed angiogenic functions of CCL5 (Wu et al. 2008a; Azenshtein et al. 2002; Suffee et al. 2011; Suffee et al. 2012). Specifically, CCL5 was shown to have angiogenic effects that were mediated by VEGF, in mechanisms depending on CCR5, CCR1 and GAG (Suffee et al. 2012).

Here, it is important to note that the studies of hematological malignancies have provided evidence to additional cancer cell-microenvironment interactions—specifically osteoclastogenesis—that were induced by CCL5, and also very strongly by CCL3. Particularly in MM, tumor cell-derived CCL3 was shown to act on osteoclasts and promote osteoclastogenesis (Reviewed in Soria 2009). Moreover, it was recently demonstrated that CCL3 repressed mineralization and osteocalcin production by primary human bone marrow stromal cells. A CCL5/CCL3-CCR5/CCR1-depenedent process promoted the migration of MM cells to the bone marrow (Reviewed in Soria 2009); more recent references (Dairaghi et al. 2012; Vallet et al. 2011). By acting in these two complementary pathways, CCL3, CCL5 and their corresponding receptors play major roles in potentiating disease course in MM.

The diverse tumor-promoting functions of CCL5 on the tumor microenvironment are complemented by direct stimulation of cancer cells, leading to increased tumor cell proliferation and of migratory/invasive functions that are required for metastasis. Although, in general, the extent to which CCL5 promotes tumor cell proliferation is not very high, such effects were reported in quite a large number of tumor cell types [e.g. (Sugasawa et al. 2008a; Aldinucci et al. 2008; Murooka et al. 2009;

Cambien et al. 2011; Zhang et al. 2010)]. CCL5 was also shown to increase the proportion of CD44^{+}/CD24^{-} breast CSC (Zhang et al. 2009), thus revealing yet another important mode of activity that may support malignancy. Of the three CCL5 receptors, it was mainly tumor-cell expressed CCR5 that mediated the proliferating activities of the chemokine (Aldinucci et al. 2008; Murooka et al. 2009; Cambien et al. 2011; Zhang et al. 2010). In line with such roles of CCR5, melanoma tumors which have been developed in CCR5 KO mice were enriched with apoptotic proteins, whereas the expression of survival proteins was reduced (Song et al. 2012). Also, CCL5 activities have been shown to induce the mTOR pathway of increased translation, leading to elevated protein expression for cyclin D1, c-Myc and Dad-1, without affecting their mRNA levels (Murooka et al. 2009).

As already indicated, another key determinant of the pro-tumoral activities of CCR5-CCL5 is induction of tumor cell migration and invasion. This important mode of action has been established with respect to CCL5 already at the very initial stages of research on its roles in malignancy (Ben-Baruch 2012a; Soria and Ben-Baruch 2008; Soria 2009; Ben-Baruch 2012b). Recent studies provide further evidence to this CCL5 function, in many different model systems [e.g. (Zhang et al. 2009; Karnoub et al. 2007; Pinilla et al. 2009; Makinoshima and Dezawa 2009; Velasco-Velazquez et al. 2012; Mi et al. 2011; Gallo et al. 2012; Cambien et al. 2011; Wang et al. 2012b)]. In line with the above, CCR5- and CCL5-induced tumor cell migration and invasion have been shown to be causally linked with elevated metastasis in mice (Karnoub et al. 2007; Velasco-Velazquez et al. 2012; Mi et al. 2011; Cambien et al. 2011). From the mechanistic point of view, it was demonstrated that the roles of CCL5 in invasion were regulated by c-Myc, and roles for $\alpha V\beta 3$ integrins and MMP were found in such processes. Actually, MMP have been induced by CCL5 not only in tumor cells, but also in EC and leukocytes (Wu et al. 2008a; Azenshtein et al. 2002; Chuang et al. 2009; Cappellen et al. 2007; Wang et al. 2012b; Suffee et al. 2012). Here again, induction of migratory and invasive properties in the tumor cells was mostly mediated by tumor cell-expressed CCR5 (Borczuk et al. 2008; Velasco-Velazquez et al. 2012; Wang et al. 2012b). Of interest in this respect are recent findings showing that CCR5 expression and CCL5-induced migration were more prominent in CD44^{+}/CD24^{-} breast CSC (Zhang et al. 2009).

Another important aspect related to CCL5-induced tumor cell invasion is that the process can be mediated not only by tumor cell-derived CCL5, but also by CCL5 produced by stroma cells, such as MSC. Following co-culturing with tumor cells, MSC have produced elevated levels of CCL5, that in turn has up-regulated the migratory properties of the tumor cells, in a process leading to increased metastasis formation (Karnoub et al. 2007; Pinilla et al. 2009; Mi et al. 2011; Gallo et al. 2012). In breast cancer, this enhanced metastatic ability was reversible and depended on CCL5 signaling through CCR5 (Karnoub et al. 2007). Another study of breast cancer has shown that osteopontin-CCL5 interactions between MSC and tumor cells contributed to metastasis formation (Mi et al. 2011).

Overall, CCL5 (and other CCR5 ligands) act on cells of the tumor microenvironment and on the tumor cells in many different mechanisms, leading to elevation in malignancy-related properties. Specifically concerning CCR5, hopes were raised that the Δ32 mutation in this receptor will provide improved understanding of the roles played by CCR5 in malignancy. Many studies in this respect were performed in breast cancer; however, non-conclusive findings were obtained [reviewed in (Ben-Baruch 2012a; Soria and Ben-Baruch 2008)]. Nevertheless, evidence for dominance of CCR5 over CCR3 and CCR1 in mediating CCL5 activities in cancer has emerged from a large number of studies (some cited above), suggesting that CCR5 is indeed the receptor mediating most of the tumor-promoting activities of CCL5.

To conclude, the above summary suggests that the CCR5-CCL5 pair is diverted by tumor cells to entities with ability to improve tumor growth and metastasis. The potential anti-tumor activities of immune cells which may have been recruited to the tumor site are inhibited, and the activities of the CCR5-CCL5 pair skew the tumor microenvironment towards immune suppression and inflammatory nature (Fig. 6.1). These mechanisms are complemented to some extent by the ability of CCR5-CCL5 to promote angiogenesis and tumor cell proliferation, and are strengthened mainly by the pivotal roles of this axis in promoting tumor cell invasion (Fig. 6.1). Accordingly, the activities of CCL5 and CCR5 have been found not only to actively promote tumor growth, but also to advance the most important and devastating step of metastasis (Wu et al. 2008a; Borczuk et al. 2008; Karnoub et al. 2007; Velasco-Velazquez et al. 2012; Mi et al. 2011; Song et al. 2012; Cambien et al. 2011; Wang et al. 2012b; Zhang et al. 2010).

As a whole, these findings pose the CCR5-CCL5 axis, and also CCL3 as preferable therapeutic targets in cancer. This research direction is attractive because of the extensive efforts taken by pharmaceutical companies to produce CCR5 antagonists for use in AIDS, and because of the availability of maraviroc for clinical use (Wasmuth 2012). However, when the complex roles of CCL5 in malignancy are taken into account, it is expected that CCR5-CCL5/CCL3 shut-off would lead to inhibition of potential protective immune mechanisms that could have been possibly effective against the tumor, primarily following appropriate boosting of the immune system.

6.3 Concluding Remarks

This review has demonstrated the powerful cancer-promoting activities of inflammatory chemokines, and their multiple impacts on the tumor microenvironment and on the cancer cells. These factors, that in principle could have protected the individual against the arising and developing tumor, are being used by the tumor cells for their own propagation, motility and spread (Fig. 6.1). To translate these findings in the clinic, we need to develop improved manners for targeting the chemokine-related pathways. Here, it is important to remember that many chemokines are expressed simultaneously at the tumor site, thus we need to shift the equilibrium between different chemokines and their receptors, so that several pathways will be affected concomitantly.

In view of the pro-tumoral impacts of the inflammatory chemokines in cancer, we face two major challenges. The first is to identify modes for skewing the immune balance back towards the anti-malignancy phenotype. To this end, it is essential to identify the tumor-chemokine networks operating in each tumor type. Based on this information, we need to develop means for potentiating the migration of Th1, CTL and NK cells to the tumors, while inhibiting inflammatory infiltrates of monocytes, MDSC, neutrophils and Treg that exert tumor-promoting functions. However, as is the case with many other components of the immune system, these goals are difficult to achieve. To give an example, potentiation of Th1 and NK activities might be achieved by elevating the expression of CXCR3 or CCR5 ligands; however, these same chemokines could act on the tumor cells, to induce their proliferation and migration.

The second challenge is to develop means that will block chemokine receptors on deleterious immune cells and on tumor cells. Because of the functional redundancies of chemokines belonging to the same sub-group, their receptors may be superior therapeutic targets. Here, we need to understand which of the chemokine receptors are mediating most potently the abilities of the chemokines to induce tumor cell proliferation and invasion, and what are the regulatory pathways controlling these events. But again, difficulties are expected in implementing such measures. For example, modalities that will block CXCR1/CXCR2-, CCR2- or CCR5-mediated responses may inhibit the pro-tumor activities of their ligands on the cancer cells and possibly inhibit the recruitment of detrimental myeloid cells to the tumor site; however, simultaneously they may prevent the functions of monocytes, neutrophils and T cells that may have potential protective activities.

Thus, changing the balance between different chemokine types or targeting specific chemokine receptors may prove beneficial in terms of tumor eradication and inhibition of metastasis, but may prove counter-productive if the same measures would inhibit anti-tumor processes. In addition, such interventions in immune activities may decrease the ability of the host to mount protective immune and inflammatory responses against pathogens. Here, it is possible that deleterious effects would be partly prevented due to the promiscuity and redundancy of the "chemokine world", because specific chemokine-receptor axes can be backed up by others. Nevertheless, the numerous difficulties expected in implementing chemokine-based therapeutics in cancer would require new approaches, which will enable direct targeting of chemokines or chemokine receptors in specific cells, thus increasing specificity and reducing undesired side effects.

To conclude, inflammatory chemokines are highly relevant factors in malignancy, they have most important impacts on disease course, and are potential targets for therapeutic implications. While our understanding of these factors has been dramatically improved over the last several years, much is still to be learnt about the tumor microenvironment-chemokine network. Moreover, in view of the complex and redundant nature of the "chemokine world" and of inflammatory chemokines in particular, there is a need for novel research directions that will be based on a more comprehensive view of the interactions between the different chemokines, between them and other inflammatory factors at the tumor microenvironment, and of the impact that different therapeutic modalities may have not only on cancer cells, but also on the immune integrity of the host.

Acknowledgments The authors thank the following organizations for supporting the studies related to this review, performed in Dr. Ben-Baruch's laboratory: Israel Science Foundation, The Cooperation Program in Cancer Research of the Deutsches Krebsforschungszentrum (DKFZ) and Israel's Ministry of Science and Technology (MOST), Israel Ministry of Health, Israeli Cancer Association and Federico Foundation.

References

Acharyya S, Oskarsson T, Vanharanta S et al (2012) A CXCL1 Paracrine Network Links Cancer Chemoresistance and Metastasis. Cell 150:165–178

Acosta JC, Gil J (2009) A role for CXCR2 in senescence, but what about in cancer? Cancer Res 69:2167–2170

Acosta JC, O'Loghlen A, Banito A et al (2008) Chemokine signaling via the CXCR2 receptor reinforces senescence. Cell 133:1006–1018

Addison CL, Daniel TO, Burdick MD et al (2000) The CXC chemokine receptor 2, CXCR2, is the putative receptor for ELR^+ CXC chemokine-induced angiogenic activity. J Immunol 165:5269–5277

Adler EP, Lemken CA, Katchen NS, Kurt RA (2003) A dual role for tumor-derived chemokine RANTES (CCL5). Immunol Lett 90:187–194

Agarwal A, Tressel SL, Kaimal R et al (2010) Identification of a metalloprotease-chemokine signaling system in the ovarian cancer microenvironment: implications for antiangiogenic therapy. Cancer Res 70:5880–5890

Aldinucci D, Lorenzon D, Cattaruzza L et al (2008) Expression of CCR5 receptors on reed-sternberg cells and Hodgkin lymphoma cell lines: involvement of CCL5/Rantes in tumor cell growth and microenvironmental interactions. Int J Cancer 122:769–776

Ali S, Kaur J, Patel KD (2000) Intercellular cell adhesion molecule-1, vascular cell adhesion molecule-1, and regulated on activation normal T cell expressed and secreted are expressed by human breast carcinoma cells and support eosinophil adhesion and activation. Am J Pathol 157:313–321

Allavena P, Sica A, Solinas G, Porta C, Mantovani A (2008) The inflammatory microenvironment in tumor progression: the role of tumor-associated macrophages. Crit Rev Oncol Hematol 66:1–9

Allen SJ, Crown SE, Handel TM (2007) Chemokine: receptor structure, interactions, and antagonism. Annu Rev Immunol 25:787–820

Andersson A, Srivastava MK, Harris-White M et al (2011) Role of CXCR3 ligands in IL-7/IL-7R alpha-Fc-mediated antitumor activity in lung cancer. Clin Cancer Res 17:3660–3672

Amatschek S, Lucas R, Eger A et al (2010) CXCL9 induces chemotaxis, chemorepulsion and endothelial barrier disruption through CXCR3-mediated activation of melanoma cells. Br J Cancer 104:469–479

Andersson A, Yang SC, Huang M et al (2009) IL-7 promotes CXCR3 ligand-dependent T cell antitumor reactivity in lung cancer. J Immunol 182:6951–6958

Araki K, Shimura T, Yajima T et al (2009) Phosphoglucose isomerase/autocrine motility factor promotes melanoma cell migration through ERK activation dependent on autocrine production of interleukin-8. J Biol Chem 284:32305–32311

Azenshtein E, Luboshits G, Shina S et al (2002) The CC chemokine RANTES in breast carcinoma progression: regulation of expression and potential mechanisms of promalignant activity. Cancer Res 62:1093–1102

Baba T, Badr Mel S, Tomaru U, Ishizu A, Mukaida N (2012) Novel process of Intrathymic tumor-immune tolerance through CCR2-mediated recruitment of Sirpalpha(+) dendritic cells: a murine model. PLoS One 7:e41154

Balkwill FR, Mantovani A (2011) Cancer-related inflammation: common themes and therapeutic opportunities. Semin Cancer Biol 22:33–40

Bandapalli OR, Ehrmann F, Ehemann V et al (2012) Down-regulation of CXCL1 inhibits tumor growth in colorectal liver metastasis. Cytokine 57:46–53

Bates RC, DeLeo MJ 3rd, Mercurio AM (2004) The epithelial-mesenchymal transition of colon carcinoma involves expression of IL-8 and CXCR-1-mediated chemotaxis. Exp Cell Res 299:315–324

Ben-Baruch A (2006) Inflammation-associated immune suppression in cancer: the roles played by cytokines, chemokines and additional mediators. Semin Cancer Biol 16:38–52

Ben-Baruch A (2007) Expert commentary: the chemokine receptor CXCR3 and its ligands in malignancy: do they act as double edged swords? In: Grinwald LR (ed) Chemokine research trends. Nova Science Publishers, Inc, New York, pp 1–6

Ben-Baruch A (2012a) The inflammatory CC chemokines CCL2 and CCL5 in malignancy: leukocyte migration and beyond. In: Ben-Baruch A (ed) The inflammatory milieu of tumors: cytokines and chemokines that affect tumor growth and metastasis, Bentham Science Publishers, pp 99–114

Ben-Baruch A (2012b) The tumor-promoting flow of cells into, within and out of the tumor site: regulation by the inflammatory axis of TNF alpha and chemokines. Cancer Microenviron 5:151–164

Bendre MS, Montague DC, Peery T, Akel NS, Gaddy D, Suva LJ (2003) Interleukin-8 stimulation of osteoclastogenesis and bone resorption is a mechanism for the increased osteolysis of metastatic bone disease. Bone 33:28–37

Bendrik C, Dabrosin C (2009) Estradiol increases IL-8 secretion of normal human breast tissue and breast cancer in vivo. J Immunol 182:371–378

Benoy IH, Salgado R, Van Dam P et al (2004) Increased serum interleukin-8 in patients with early and metastatic breast cancer correlates with early dissemination and survival. Clin Cancer Res 10:7157–7162

Berencsi K, Rani P, Zhang T et al (2011) In vitro migration of cytotoxic T lymphocyte derived from a colon carcinoma patient is dependent on CCL2 and CCR2. J Transl Med 9:33

Bhatt RS, Wang X, Zhang L et al (2010) Renal cancer resistance to antiangiogenic therapy is delayed by restoration of angiostatic signaling. Mol Cancer Ther 9:2793–2802

Biswas SK, Mantovani A (2010) Macrophage plasticity and interaction with lymphocyte subsets: cancer as a paradigm. Nat Immunol 11:889–896

Bonavia R, Inda MM, Vandenberg S et al (2012) EGFRvIII promotes glioma angiogenesis and growth through the NF-kappaB, interleukin-8 pathway. Oncogene 4054–4066

Borczuk AC, Papanikolaou N, Toonkel RL et al (2008) Lung adenocarcinoma invasion in TGFbetaRII-deficient cells is mediated by CCL5/RANTES. Oncogene 27:557–564

Bronger H, Kraeft S, Schwarz-Boeger U et al (2012) Modulation of CXCR3 ligand secretion by prostaglandin E2 and cyclooxygenase inhibitors in human breast cancer. Breast Cancer Res 14:R30

Burns JM, Summers BC, Wang Y et al (2006) A novel chemokine receptor for SDF-1 and I-TAC involved in cell survival, cell adhesion, and tumor development. J Exp Med 203:2201–2213

Bussard KM, Okita N, Sharkey N, Neuberger T, Webb A, Mastro AM (2010a) Localization of osteoblast inflammatory cytokines MCP-1 and VEGF to the matrix of the trabecula of the femur, a target area for metastatic breast cancer cell colonization. Clin Exp Metastasis 27:331–340

Bussard KM, Venzon DJ, Mastro AM (2010b) Osteoblasts are a major source of inflammatory cytokines in the tumor microenvironment of bone metastatic breast cancer. J Cell Biochem 111:1138–1148

Cambien B, Richard-Fiardo P, Karimdjee BF et al (2011) CCL5 neutralization restricts cancer growth and potentiates the targeting of PDGFR beta in colorectal carcinoma. PLoS One 6:e28842

Cambien B, Karimdjee BF, Richard-Fiardo P et al (2009) Organ-specific inhibition of metastatic colon carcinoma by CXCR3 antagonism. Br J Cancer 100:1755–1764

Campanella GS, Colvin RA, Luster AD (2010) CXCL10 can inhibit endothelial cell proliferation independently of CXCR3. PLoS One 5:e12700

Cappellen D, Schlange T, Bauer M, Maurer F, Hynes NE (2007) Novel c-MYC target genes mediate differential effects on cell proliferation and migration. EMBO Rep 8:70–76

Cataisson C, Ohman R, Patel G et al (2009) Inducible cutaneous inflammation reveals a protumorigenic role for keratinocyte CXCR2 in skin carcinogenesis. Cancer Res 69:319–328

Chang LY, Lin YC, Mahalingam J et al (2012a) Tumor-derived chemokine CCL5 enhances TGF-beta-mediated killing of CD8(+) T cells in colon cancer by T-regulatory cells. Cancer Res 72:1092–1102

Chang LY, Lin YC, Kang CW et al (2012b) The indispensable role of CCR5 for in vivo suppressor function of tumor-derived CD103 + effector/memory regulatory T cells. J Immunol 189:567–574

Chen YC, Sosnoski DM, Gandhi UH, Novinger LJ, Prabhu KS, Mastro AM (2009) Selenium modifies the osteoblast inflammatory stress response to bone metastatic breast cancer. Carcinogenesis 30:1941–1948

Chen Y, Chen L, Li JY et al (2011) ERbeta and PEA3 co-activate IL-8 expression and promote the invasion of breast cancer cells. Cancer Biol Ther 11:497–511

Chiu HY, Sun KH, Chen SY et al (2012) Autocrine CCL2 promotes cell migration and invasion via PKC activation and tyrosine phosphorylation of paxillin in bladder cancer cells. Cytokine 59:423–432

Chuang JY, Yang WH, Chen HT et al (2009) CCL5/CCR5 axis promotes the motility of human oral cancer cells. J Cell Physiol 220:418–426

Conti I, Rollins BJ (2004) CCL2 (monocyte chemoattractant protein-1) and cancer. Semin Cancer Biol 14:149–154

Cortez-Retamozo V, Etzrodt M, Newton A et al (2012) Origins of tumor-associated macrophages and neutrophils. Proc Natl Acad Sci U S A 109:2491–2496

Craig MJ, Loberg RD (2006) CCL2 (monocyte chemoattractant protein-1) in cancer bone metastases. Cancer Metastasis Rev 25:611–619

Dairaghi DJ, Oyajobi BO, Gupta A et al (2012) CCR1 blockade reduces tumor burden and osteolysis in vivo in a mouse model of myeloma bone disease. Blood 1449–1457

Datta D, Banerjee P, Gasser M, Waaga-Gasser AM, Pal S (2010) CXCR3-B can mediate growth-inhibitory signals in human renal cancer cells by down-regulating the expression of heme oxygenase-1. J Biol Chem 285:36842–36848

Datta D, Flaxenburg JA, Laxmanan S et al (2006) Ras-induced modulation of CXCL10 and its receptor splice variant CXCR3-B in MDA-MB-435 and MCF-7 cells: relevance for the development of human breast cancer. Cancer Res 66:9509–9518

Datta D, Contreras AG, Grimm M, Waaga-Gasser AM, Briscoe DM, Pal S (2008) Calcineurin inhibitors modulate CXCR3 splice variant expression and mediate renal cancer progression. J Am Soc Nephrol 19:2437–2446

De Larco JE, Wuertz BR, Rosner KA et al (2001) A potential role for interleukin-8 in the metastatic phenotype of breast carcinoma cells. Am J Pathol 158:639–646

De Larco JE, Wuertz BR, Furcht LT (2004) The potential role of neutrophils in promoting the metastatic phenotype of tumors releasing interleukin-8. Clin Cancer Res 10:4895–4900

Deshmane SL, Kremlev S, Amini S, Sawaya BE (2009) Monocyte chemoattractant protein-1 (MCP-1): an overview. J Interferon Cytokine Res 29:313–326

Dhawan P, Richmond A (2002) Role of CXCL1 in tumorigenesis of melanoma. J Leukoc Biol 72:9–18

Diaz-Valdes N, Basagoiti M, Dotor J et al (2011) Induction of monocyte chemoattractant protein-1 and interleukin-10 by TGFbeta1 in melanoma enhances tumor infiltration and immunosuppression. Cancer Res 71:812–821

Dimberg J, Strom K, Lofgren S, Zar N, Lindh M, Matussek A (2012) DNA promoter methylation status and protein expression of interleukin-8 in human colorectal adenocarcinomas. Int J Colorectal Dis 27:709–714

Ehlert JE, Addison CA, Burdick MD, Kunkel SL, Strieter RM (2004) Identification and partial characterization of a variant of human CXCR3 generated by posttranscriptional exon skipping. J Immunol 173:6234–6240

Fernando RI, Castillo MD, Litzinger M, Hamilton DH, Palena C (2011) IL-8 signaling plays a critical role in the epithelial-mesenchymal transition of human carcinoma cells. Cancer Res 71:5296–5306

Fontemaggi G, Dell'Orso S, Trisciuoglio D et al (2009) The execution of the transcriptional axis mutant p53, E2F1 and ID4 promotes tumor neo-angiogenesis. Nat Struct Mol Biol 16:1086–1093

Freund A, Chauveau C, Brouillet JP et al (2003) IL-8 expression and its possible relationship with estrogen-receptor-negative status of breast cancer cells. Oncogene 22:256–265

Fridlender ZG, Sun J, Kim S et al (2009) Polarization of tumor-associated neutrophil phenotype by TGF-beta: "N1" versus "N2" TAN. Cancer Cell 16:183–194

Fridlender ZG, Albelda SM (2012) Tumor-associated neutrophils: friend or foe? Carcinogenesis 33:949–955

Fridlender ZG, Kapoor V, Buchlis G et al (2011) Monocyte chemoattractant protein-1 blockade inhibits lung cancer tumor growth by altering macrophage phenotype and activating CD8^{+} cells. Am J Respir Cell Mol Biol 44:230–237

Fritz EA, Glant TT, Vermes C, Jacobs JJ, Roebuck KA (2002) Titanium particles induce the immediate early stress responsive chemokines IL-8 and MCP-1 in osteoblasts. J Orthop Res 20:490–498

Fujihara A, Kurooka M, Miki T, Kaneda Y (2008) Intratumoral injection of inactivated sendai virus particles elicits strong antitumor activity by enhancing local CXCL10 expression and systemic NK cell activation. Cancer Immunol Immunother 57:73–84

Fujimoto H, Sangai T, Ishii G et al (2009) Stromal MCP-1 in mammary tumors induces tumor-associated macrophage infiltration and contributes to tumor progression. Int J Cancer 125:1276–1284

Fujita M, Zhu X, Ueda R et al (2009) Effective immunotherapy against murine gliomas using type 1 polarizing dendritic cells–significant roles of CXCL10. Cancer Res 69:1587–1595

Fulton AM (2009) The chemokine receptors CXCR4 and CXCR3 in cancer. Curr Oncol Rep 11:125–131

Gabellini C, Trisciuoglio D, Desideri M et al (2009) Functional activity of CXCL8 receptors, CXCR1 and CXCR2, on human malignant melanoma progression. Eur J Cancer 45:2618–2627

Gacci M, Serni S, Lapini A et al (2009) CXCR3-B expression correlates with tumor necrosis extension in renal cell carcinoma. J Urol 181:843–848

Gallo M, De Luca A, Lamura L, Normanno N (2012) Zoledronic acid blocks the interaction between mesenchymal stem cells and breast cancer cells: implications for adjuvant therapy of breast cancer. Ann Oncol 23:597–604

Gehad AE, Lichtman MK, Schmults CD et al (2012) Nitric oxide-producing myeloid-derived suppressor cells inhibit vascular E-selectin expression in human squamous cell carcinomas. J Invest Dermatol

Ginestier C, Liu S, Diebel ME et al (2010) CXCR1 blockade selectively targets human breast cancer stem cells in vitro and in xenografts. J Clin Invest 120:485–497

Goede V, Brogelli L, Ziche M, Augustin HG (1999) Induction of inflammatory angiogenesis by monocyte chemoattractant protein-1. Int J Cancer 82:765–770

Gonzalez-Martin A, Gomez L, Lustgarten J, Mira E, Manes S (2011) Maximal T cell-mediated antitumor responses rely upon CCR5 expression in both CD4(+) and CD8(+) T cells. Cancer Res 71:5455–5466

Granot Z, Henke E, Comen EA, King TA, Norton L, Benezra R (2011) Tumor entrained neutrophils inhibit seeding in the premetastatic lung. Cancer Cell 20:300–314

Green AR, Green VL, White MC, Speirs V (1997) Expression of cytokine messenger RNA in normal and neoplastic human breast tissue: identification of interleukin-8 as a potential regulatory factor in breast tumours. Int J Cancer 72:937–941

Gregory AD, Houghton AM (2011) Tumor-associated neutrophils: new targets for cancer therapy. Cancer Res 71:2411–2416

Grepin R, Guyot M, Jacquin M et al (2011) Acceleration of clear cell renal cell carcinoma growth in mice following bevacizumab/Avastin treatment: the role of CXCL cytokines. Oncogene 31:1683–1694

Greten TF, Manns MP, Korangy F (2011) Myeloid derived suppressor cells in human diseases. Int Immunopharmacol 11:802–807

Groom JR, Luster AD (2011) CXCR3 ligands: redundant, collaborative and antagonistic functions. Immunol Cell Biol 89:207–215

Hanahan D, Weinberg RA (2011) Hallmarks of cancer: the next generation. Cell 144:646–74

Hagemann T, Balkwill F, Lawrence T (2007) Inflammation and cancer: a double-edged sword. Cancer Cell 12:300–301

Haghnegahdar H, Du J, Wang D et al (2000) The tumorigenic and angiogenic effects of MGSA/GRO proteins in melanoma. J Leukoc Biol 67:53–62

Haim K, Weitzenfeld P, Meshel T, Ben-Baruch A (2011) Epidermal growth factor and estrogen act by independent pathways to additively promote the release of the angiogenic chemokine CXCL8 by breast tumor cells. Neoplasia 13:230–243

Halpern JL, Kilbarger A, Lynch CC (2011) Mesenchymal stem cells promote mammary cancer cell migration in vitro via the CXCR2 receptor. Cancer Lett 308:91–99

Hart KM, Bak SP, Alonso A, Berwin B (2009) Phenotypic and functional delineation of murine CX(3)CR1 monocyte-derived cells in ovarian cancer. Neoplasia 11:564–573, 1 p following 73

Hong M, Puaux AL, Huang C et al (2011) Chemotherapy induces intratumoral expression of chemokines in cutaneous melanoma, favoring T-cell infiltration and tumor control. Cancer Res 71:6997–7009

Hu H, Sun L, Guo C et al (2009) Tumor cell-microenvironment interaction models coupled with clinical validation reveal CCL2 and SNCG as two predictors of colorectal cancer hepatic metastasis. Clin Cancer Res 15:5485–5493

Huang D, Ding Y, Zhou M et al (2010) Interleukin-8 mediates resistance to antiangiogenic agent sunitinib in renal cell carcinoma. Cancer Res 70:1063–1071

Huang S, Singh RK, Xie K et al (1994) Expression of the JE/MCP-1 gene suppresses metastatic potential in murine colon carcinoma cells. Cancer Immunol Immunother 39:231–238

Huang B, Lei Z, Zhao J et al (2007) CCL2/CCR2 pathway mediates recruitment of myeloid suppressor cells to cancers. Cancer Lett 252:86–92

Huh SJ, Liang S, Sharma A, Dong C, Robertson GP (2010) Transiently entrapped circulating tumor cells interact with neutrophils to facilitate lung metastasis development. Cancer Res 70:6071–6082

Hwang TL, Lee LY, Wang CC, Liang Y, Huang SF, Wu CM (2012) CCL7 and CCL21 overexpression in gastric cancer is associated with lymph node metastasis and poor prognosis. World J Gastroenterol 18:1249–1256

Hwang WL, Yang MH, Tsai ML et al (2011) SNAIL regulates interleukin-8 expression, stem cell-like activity, and tumorigenicity of human colorectal carcinoma cells. Gastroenterology 141:279–291, 91 e1–5

Ijichi H, Chytil A, Gorska AE et al (2011) Inhibiting Cxcr2 disrupts tumor-stromal interactions and improves survival in a mouse model of pancreatic ductal adenocarcinoma. J Clin Invest 121:4106–4117

Ijichi H (2012) Inhibition of CXCLs/CXCR2 axis in the tumor microenvironment might be a potent therapeutics for pancreatic cancer. Oncoimmunology 1:569–71

Inoue H, Iga M, Xin M et al (2008) TARC and RANTES enhance antitumor immunity induced by the GM-CSF-transduced tumor vaccine in a mouse tumor model. Cancer Immunol Immunother 57:1399–1411

Izhak L, Wildbaum G, Jung S, Stein A, Shaked Y, Karin N (2012) Dissecting the autocrine and paracrine roles of the CCR2-CCL2 axis in tumor survival and angiogenesis. PLoS One 7:e28305

Izhak L, Wildbaum G, Weinberg U et al (2010) Predominant expression of CCL2 at the tumor site of prostate cancer patients directs a selective loss of immunological tolerance to CCL2 that could be amplified in a beneficial manner. J Immunol 184:1092–1101

Jin G, Kawsar HI, Hirsch SA et al (2010) An antimicrobial peptide regulates tumor-associated macrophage trafficking via the chemokine receptor CCR2, a model for tumorigenesis. PLoS One 5:e10993

Joyce JA, Pollard JW (2009) Microenvironmental regulation of metastasis. Nat Rev Cancer 9:239–252

Ju D, Sun D, Xiu L, Meng X, Zhang C, Wei P (2012) Interleukin-8 is associated with adhesion, migration and invasion in human gastric cancer SCG-7901 cells. Med Oncol 29:91–99

Kalluri R, Zeisberg M (2006) Fibroblasts in cancer. Nat Rev Cancer 6:392–401

Karnoub AE, Dash AB, Vo AP et al (2007) Mesenchymal stem cells within tumour stroma promote breast cancer metastasis. Nature 449:557–563

Kawai Y, Kaidoh M, Yokoyama Y, Sano K, Ohhashi T (2009) Chemokine CCL2 facilitates ICAM-1-mediated interactions of cancer cells and lymphatic endothelial cells in sentinel lymph nodes. Cancer Sci 100:419–428

Keeley EC, Mehrad B, Strieter RM (2008) Chemokines as mediators of neovascularization. Arterioscler Thromb Vasc Biol 28:1928–1936

Keeley EC, Mehrad B, Strieter RM (2010) CXC chemokines in cancer angiogenesis and metastases. Adv Cancer Res 106:91–111

Keklikoglou I, Koerner C, Schmidt C et al (2012) MicroRNA-520/373 family functions as a tumor suppressor in estrogen receptor negative breast cancer by targeting NF-kappaB and TGF-beta signaling pathways. Oncogene 31:4150–4163

Kim S, Lewis C, Nadel JA (2011) Epidermal growth factor receptor reactivation induced by E-prostanoid-3 receptor- and tumor necrosis factor-alpha-converting enzyme-dependent feedback exaggerates interleukin-8 production in airway cancer (NCI-H292) cells. Exp Cell Res 317:2650–2660

Kinder M, Chislock E, Bussard KM, Shuman L, Mastro AM (2008) Metastatic breast cancer induces an osteoblast inflammatory response. Exp Cell Res 314:173–183

Knowles LM, Stabile LP, Egloff AM et al (2009) HGF and c-met participate in paracrine tumorigenic pathways in head and neck squamous cell cancer. Clin Cancer Res 15:3740–3750

Koga M, Kai H, Egami K et al (2008) Mutant MCP-1 therapy inhibits tumor angiogenesis and growth of malignant melanoma in mice. Biochem Biophys Res Commun 365:279–284

Kreeger PK, Mandhana R, Alford SK, Haigis KM, Lauffenburger DA (2009) RAS mutations affect tumor necrosis factor-induced apoptosis in colon carcinoma cells via ERK-modulatory negative and positive feedback circuits along with non-ERK pathway effects. Cancer Res 69:8191–8199

Kuai WX, Wang Q, Yang XZ, Zhao Y, Yu R, Tang XJ (2012) Interleukin-8 associates with adhesion, migration, invasion and chemosensitivity of human gastric cancer cells. World J Gastroenterol 18:979–985

Lacotte S, Brun S, Muller S, Dumortier H (2009) CXCR3, inflammation, and autoimmune diseases. Ann N Y Acad Sci 1173:310–317

Lapteva N, Huang XF (2010) CCL5 as an adjuvant for cancer immunotherapy. Expert Opin Biol Ther 10:725–733

Lasagni L, Francalanci M, Annunziato F et al (2003) An alternatively spliced variant of CXCR3 mediates the inhibition of endothelial cell growth induced by IP-10, Mig, and I-TAC, and acts as functional receptor for platelet factor 4. J Exp Med 197:1537–1549

Laubli H, Spanaus KS, Borsig L (2009) Selectin-mediated activation of endothelial cells induces expression of CCL5 and promotes metastasis through recruitment of monocytes. Blood 114:4583–4591

Lee CY, Huang CY, Chen MY, Lin CY, Hsu HC, Tang CH (2011) IL-8 increases integrin expression and cell motility in human chondrosarcoma cells. J Cell Biochem 112:2549–2557

Lee JH, Kim HN, Kim KO et al (2012a) CXCL10 promotes osteolytic Bone metastasis by enhancing cancer outgrowth and osteoclastogenesis. Cancer Res 72:3175–3186

Lee Y, Chittezhath M, Andre V et al (2012b) Protumoral role of monocytes in human B-cell precursor acute lymphoblastic leukemia: involvement of the chemokine CXCL10. Blood 119:227–237

Leibovich-Rivkin T, Buganim Y, Solomon H, Meshel T, Rotter V, Ben-Baruch A (2012) Tumor-promoting circuits that regulate a cancer-related chemokine cluster: Dominanc eof inflammatory mediators over oncogenic alerations. Cancers 4:55–76

Lesokhin AM, Hohl TM, Kitano S et al (2012) Monocytic CCR2(+) myeloid-derived suppressor cells promote immune escape by limiting activated CD8 T-cell infiltration into the tumor microenvironment. Cancer Res 72:876–886

Leuschner F, Dutta P, Gorbatov R et al (2011) Therapeutic siRNA silencing in inflammatory monocytes in mice. Nat Biotechnol 29:1005–1010

Levina V, Marrangoni AM, DeMarco R, Gorelik E, Lokshin AE (2008) Drug-selected human lung cancer stem cells: cytokine network, tumorigenic and metastatic properties. PLoS ONE 3:e3077

Levy JA (2009) The unexpected pleiotropic activities of RANTES. J Immunol 182:3945–3946

Li A, Varney ML, Valasek J, Godfrey M, Dave BJ, Singh RK (2005) Autocrine role of interleukin-8 in induction of endothelial cell proliferation, survival, migration and MMP-2 production and angiogenesis. Angiogenesis 8:63–71

Li M, Zhang Y, Feurino LW et al (2008) Interleukin-8 increases vascular endothelial growth factor and neuropilin expression and stimulates ERK activation in human pancreatic cancer. Cancer Sci 99:733–737

Li X, Sterling JA, Fan KH et al (2012) Loss of TGF-beta responsiveness in prostate stromal cells alters chemokine levels and facilitates the development of mixed osteoblastic/osteolytic bone lesions. Mol Cancer Res 10:494–503

Lin Y, Huang R, Chen L et al (2004) Identification of interleukin-8 as estrogen receptor-regulated factor involved in breast cancer invasion and angiogenesis by protein arrays. Int J Cancer 109:507–515

Liu C, Luo D, Reynolds BA et al (2011) Chemokine receptor CXCR3 promotes growth of glioma. Carcinogenesis 32:129–137

Lu X, Kang Y (2009) Chemokine (C-C motif) ligand 2 engages CCR2 + stromal cells of monocytic origin to promote breast cancer metastasis to lung and bone. J Biol Chem 284:29087–29096

Luan J, Shattuck-Brandt R, Haghnegahdar H et al (1997) Mechanism and biological significance of constitutive expression of MGSA/GRO chemokines in malignant melanoma tumor progression. J Leukoc Biol 62:588–597

Luboshits G, Shina S, Kaplan O et al (1999) Elevated expression of the CC chemokine regulated on activation, normal T cell expressed and secreted (RANTES) in advanced breast carcinoma. Cancer Res 59:4681–4687

Ma X, Norsworthy K, Kundu N et al (2009) CXCR3 expression is associated with poor survival in breast cancer and promotes metastasis in a murine model. Mol Cancer Ther 8:490–498

Makinoshima H, Dezawa M (2009) Pancreatic cancer cells activate CCL5 expression in mesenchymal stromal cells through the insulin-like growth factor-I pathway. FEBS Lett 583:3697–3703

Manome Y, Wen PY, Hershowitz A et al (1995) Monocyte chemoattractant protein-1 (MCP-1) gene transduction: an effective tumor vaccine strategy for non-intracranial tumors. Cancer Immunol Immunother 41:227–235
Mantovani A, Bonecchi R, Locati M (2006) Tuning inflammation and immunity by chemokine sequestration: decoys and more. Nat Rev Immunol 6:907–918
Martin D, Galisteo R, Gutkind JS (2009) CXCL8/IL8 stimulates vascular endothelial growth factor (VEGF) expression and the autocrine activation of VEGFR2 in endothelial cells by activating NFkappaB through the CBM (Carma3/Bcl10/Malt1) complex. J Biol Chem 284:6038–6042
Matsuo Y, Raimondo M, Woodward TA et al (2009) CXC-chemokine/CXCR2 biological axis promotes angiogenesis in vitro and in vivo in pancreatic cancer. Int J Cancer 125:1027–1037
Merritt WM, Lin YG, Spannuth WA et al (2008) Effect of interleukin-8 gene silencing with liposome-encapsulated small interfering RNA on ovarian cancer cell growth. J Natl Cancer Inst 100:359–372
Mestdagt M, Polette M, Buttice G et al (2006) Transactivation of MCP-1/CCL2 by beta-catenin/TCF-4 in human breast cancer cells. Int J Cancer 118:35–42
Mi Z, Bhattacharya SD, Kim VM, Guo H, Talbot LJ, Kuo PC (2011) Osteopontin promotes CCL5-mesenchymal stromal cell-mediated breast cancer metastasis. Carcinogenesis 32:477–487
Mishra P, Banerjee D, Ben-Baruch A (2011) Chemokines at the crossroads of tumor-fibroblast interactions that promote malignancy. J Leukoc Biol 89:31–39
Mizutani K, Sud S, McGregor NA et al (2009) The chemokine CCL2 increases prostate tumor growth and bone metastasis through macrophage and osteoclast recruitment. Neoplasia 11:1235–1242
Molloy AP, Martin FT, Dwyer RM et al (2009) Mesenchymal stem cell secretion of chemokines during differentiation into osteoblasts, and their potential role in mediating interactions with breast cancer cells. Int J Cancer 124:326–332
Molon B, Ugel S, Del Pozzo F et al (2011) Chemokine nitration prevents intratumoral infiltration of antigen-specific T cells. J Exp Med 208:1949–1962
Mueller A, Strange PG (2004) The chemokine receptor, CCR5. Int J Biochem Cell Biol 36:35–38
Murakami T, Kawada K, Iwamoto M et al (2013) The role of CXCR3 and CXCR4 in colorectal cancer metastasis. Int J Cancer 276–287
Murdoch C, Muthana M, Coffelt SB, Lewis CE (2008) The role of myeloid cells in the promotion of tumour angiogenesis. Nat Rev Cancer 8:618–631
Murooka TT, Rahbar R, Fish EN (2009) CCL5 promotes proliferation of MCF-7 cells through mTOR-dependent mRNA translation Biochem Biophys Res Commun 387:381–386
Muthuswamy R, Berk E, Junecko BF et al (2012) NF-kappaB hyperactivation in tumor tissues allows tumor-selective reprogramming of the chemokine microenvironment to enhance the recruitment of cytolytic T effector cells. Cancer Res 3735–3743
Nakasone ES, Askautrud HA, Kees T et al (2012) Imaging tumor-stroma interactions during chemotherapy reveals contributions of the microenvironment to resistance. Cancer Cell 21:488–503
Nam JS, Kang MJ, Suchar AM et al (2006) Chemokine (C-C motif) ligand 2 mediates the prometastatic effect of dysadherin in human breast cancer cells. Cancer Res 66:7176–7184
Naldini A, Filippi I, Miglietta D, Moschetta M, Giavazzi R, Carraro F (2010) Interleukin-1beta regulates the migratory potential of MDAMB231 breast cancer cells through the hypoxia-inducible factor-1alpha. Eur J Cancer 46:3400–3408
Neiva KG, Zhang Z, Miyazawa M, Warner KA, Karl E, Nor JE (2009) Cross talk initiated by endothelial cells enhances migration and inhibits anoikis of squamous cell carcinoma cells through STAT3/Akt/ERK signaling. Neoplasia 11:583–593
Nesbeth Y, Scarlett U, Cubillos-Ruiz J et al (2009) CCL5-mediated endogenous antitumor immunity elicited by adoptively transferred lymphocytes and dendritic cell depletion. Cancer Res 69:6331–6338

Nesbeth YC, Martinez DG, Toraya S et al (2010) $CD4^+$ T cells elicit host immune responses to MHC class II-negative ovarian cancer through CCL5 secretion and CD40-mediated licensing of dendritic cells. J Immunol 184:5654–5662

Neumark E, Cohn MA, Lukanidin E, Witz IP, Ben-Baruch A (2002) Possible co-regulation of genes associated with enhanced progression of mammary adenocarcinomas. Immunol Lett 82:111–121

Neumark E, Sagi-Assif O, Shalmon B, Ben-Baruch A, Witz IP (2003) Progression of mouse mammary tumors: MCP-1-TNFalpha cross-regulatory pathway and clonal expression of promalignancy and ant malignancy factors. Int J Cancer 106:879–886

Nieman KM, Kenny HA, Penicka CV et al (2011) Adipocytes promote ovarian cancer metastasis and provide energy for rapid tumor growth. Nat Med 17:1498–1503

Ning Y, Labonte MJ, Zhang W et al (2012) The CXCR2 antagonist, SCH-527123, shows antitumor activity and sensitizes cells to oxaliplatin in preclinical colon cancer models. Mol Cancer Ther 11:1353–1364

Ning Y, Manegold PC, Hong YK, et al (2011) Interleukin-8 is associated with proliferation, migration, angiogenesis and chemosensitivity in vitro and in vivo in colon cancer cell line models. Int J Cancer 128:2038–2049

Niu J, Azfer A, Zhelyabovska O, Fatma S, Kolattukudy PE (2008) Monocyte chemotactic protein (MCP)-1 promotes angiogenesis via a novel transcription factor, MCP-1-induced protein (MCPIP). J Biol Chem 283:14542–14551

O'Hayer KM, Brady DC, Counter CM (2009) ELR^+ CXC chemokines and oncogenic ras-mediated tumorigenesis. Carcinogenesis 30:1841–1847

Oppermann M (2004) Chemokine receptor CCR5: insights into structure, function, and regulation. Cell Signal 16:1201–1210

Pahler JC, Tazzyman S, Erez N et al (2008) Plasticity in tumor-promoting inflammation: impairment of macrophage recruitment evokes a compensatory neutrophil response. Neoplasia 10:329–340

Pantschenko AG, Pushkar I, Miller LJ et al (2003) In vitro demonstration of breast cancer tumor cell sub-populations based on interleukin-1/tumor necrosis factor induction of interleukin-8 expression. Oncol Rep 10:1011–1017

Park SI, Liao J, Berry JE et al (2012) Cyclophosphamide creates a receptive microenvironment for prostate cancer skeletal metastasis. Cancer Res 72:2522–2532

Pathi SP, Kowalczewski C, Tadipatri R, Fischbach C (2010) A novel 3-D mineralized tumor model to study breast cancer bone metastasis. PLoS One 5:e8849

Petrai I, Rombouts K, Lasagni L et al (2008) Activation of p38 (MAPK) mediates the angiostatic effect of the chemokine receptor CXCR3-B. Int J Biochem Cell Biol 40:1764–1774

Piccard H, Muschel RJ, Opdenakker G (2012) On the dual roles and polarized phenotypes of neutrophils in tumor development and progression. Crit Rev Oncol Hematol 82:296–309

Pinilla S, Alt E, Abdul Khalek FJ et al (2009) Tissue resident stem cells produce CCL5 under the influence of cancer cells and thereby promote breast cancer cell invasion. Cancer Lett 284:80–85

Popivanova BK, Kostadinova FI, Furuichi K et al (2009) Blockade of a chemokine, CCL2, reduces chronic colitis-associated carcinogenesis in mice. Cancer Res 69:7884–7892

Pradelli E, Karimdjee-Soilihi B, Michiels JF et al (2009) Antagonism of chemokine receptor CXCR3 inhibits osteosarcoma metastasis to lungs. Int J Cancer 125:2586–2594

Przewoznik M, Homberg N, Naujoks M et al (2012) Recruitment of natural killer cells in advanced stages of endogenously arising B-cell lymphoma: implications for therapeutic cell transfer. J Immunother 35:217–222

Qian BZ, Li J, Zhang H et al (2011) CCL2 recruits inflammatory monocytes to facilitate breast-tumour metastasis. Nature 475:222–225

Robinson SC, Scott KA, Wilson JL, Thompson RG, Proudfoot AE, Balkwill FR (2003) A chemokine receptor antagonist inhibits experimental breast tumor growth. Cancer Res 63:8360–8365

Rody A, Karn T, Liedtke C, et al (2011) A clinically relevant gene signature in triple negative and basal-like breast cancer. Breast Cancer Res 13:R97

Rollins BJ, Sunday ME (1991) Suppression of tumor formation in vivo by expression of the JE gene in malignant cells. Mol Cell Biol 11:3125–3131

Rolny C, Capparuccia L, Casazza A et al (2008) The tumor suppressor semaphorin 3B triggers a prometastatic program mediated by interleukin 8 and the tumor microenvironment. J Exp Med 205:1155–1171

Roy A, Kolattukudy PE (2012) Monocyte chemotactic protein-induced protein (MCPIP) promotes inflammatory angiogenesis via sequential induction of oxidative stress, endoplasmic reticulum stress and autophagy. Cell Signal 24:2123–2131

Ruan JW, Liao YC, Lua I, Li MH, Hsu CY, Chen JH (2012) Human pituitary tumor-transforming gene 1 overexpression reinforces oncogene-induced senescence through CXCR2/p21 signaling in breast cancer cells. Breast Cancer Res 14:R106

Salcedo R, Ponce ML, Young HA et al (2000) Human endothelial cells express CCR2 and respond to MCP-1: direct role of MCP-1 in angiogenesis and tumor progression. Blood 96:34–40

Salcedo R, Martins-Green M, Gertz B, Oppenheim JJ, Murphy WJ (2002) Combined administration of antibodies to human interleukin 8 and epidermal growth factor receptor results in increased antimetastatic effects on human breast carcinoma xenografts. Clin Cancer Res 8:2655–2665

Schaer DA, Lesokhin AM, Wolchok JD (2011) Hiding the road signs that lead to tumor immunity. J Exp Med 208:1937–1940

Seeger H, Wallwiener D, Mueck AO (2006) Different effects of estradiol and various antiestrogens on TNF-alpha-induced changes of biochemical markers for growth and invasion of human breast cancer cells. Life Sci 78:1464–1468

Seeger H, Wallwiener D, Mueck AO (2008) Effects of estradiol and progestogens on tumor-necrosis factor-alpha-induced changes of biochemical markers for breast cancer growth and metastasis. Gynecol Endocrinol 24:576–579

Shamaladevi N, Lyn DA, Escudero DO, Lokeshwar BL (2009) CXC receptor-1 silencing inhibits androgen-independent prostate cancer. Cancer Res 69:8265–8274

Shimoda M, Mellody KT, Orimo A (2010) Carcinoma-associated fibroblasts are a rate-limiting determinant for tumour progression. Semin Cell Dev Biol 21:19–25

Shin SY, Hyun J, Lim Y, Lee YH (2011) 3'-Chloro-5, 7-dimethoxyisoflavone inhibits TNFalpha-induced CXCL10 gene transcription by suppressing the NF-kappaB pathway in HCT116 human colon cancer cells. Int Immunopharmacol 11:2104–2111

Singh RK, Lokeshwar BL (2009) Depletion of intrinsic expression of Interleukin-8 in prostate cancer cells causes cell cycle arrest, spontaneous apoptosis and increases the efficacy of chemotherapeutic drugs. Mol Cancer 8:57

Singh S, Nannuru KC, Sadanandam A, Varney ML, Singh RK (2009a) CXCR1 and CXCR2 enhances human melanoma tumourigenesis, growth and invasion. Br J Cancer 100:1638–1646

Singh S, Sadanandam A, Nannuru KC et al (2009b) Small-molecule antagonists for CXCR2 and CXCR1 inhibit human melanoma growth by decreasing tumor cell proliferation, survival, and angiogenesis. Clin Cancer Res 15:2380–2386

Singh S, Varney M, Singh RK (2009c) Host CXCR2-dependent regulation of melanoma growth, angiogenesis, and experimental lung metastasis. Cancer Res 69:411–415

Singh S, Singh AP, Sharma B, Owen LB, Singh RK (2010a) CXCL8 and its cognate receptors in melanoma progression and metastasis. Future Oncol 6:111–6

Singh S, Sadanandam A, Varney ML, Nannuru KC, Singh RK (2010b) Small interfering RNA-mediated CXCR1 or CXCR2 knock-down inhibits melanoma tumor growth and invasion. Int J Cancer 126:328–336

Snoussi K, Mahfoudh W, Bouaouina N et al (2010) Combined effects of IL-8 and CXCR2 gene polymorphisms on breast cancer susceptibility and aggressiveness. BMC Cancer 10:283

Song JK, Park MH, Choi DY et al (2012) Deficiency of C-C chemokine receptor 5 suppresses tumor development via inactivation of NF-kappaB and upregulation of IL-1Ra in melanoma model. PLoS One 7:e33747

Song S, Liu C, Wang J et al (2009) Vaccination with combination of Fit3L and RANTES in a DNA prime-protein boost regimen elicits strong cell-mediated immunity and antitumor effect. Vaccine 27:1111–1118

Soria G (2009) AB-B the CCL5/CCR5 axis in cancer. In: Fulton A (ed) Chemokine receptors in cancer, Humana Press, New York

Soria G, Ben-Baruch A (2008) The inflammatory chemokines CCL2 and CCL5 in breast cancer. Cancer Lett 267:271–285

Soria G, Yaal-Hahoshen N, Azenshtein E et al (2008) Concomitant expression of the chemokines RANTES and MCP-1 in human breast cancer: a basis for tumor-promoting interactions. Cytokine 44:191–200

Soria G, Lebel-Haziv Y, Ehrlich M et al (2012) Mechanisms regulating the secretion of the promalignancy chemokine CCL5 by breast tumor cells: CCL5's 40s loop and intracellular glycosaminoglycans. Neoplasia 14:1–19

Soria G, Ofri-Shahak M, Haas I et al (2011) Inflammatory mediators in breast cancer: coordinated expression of TNFalpha and IL-1beta with CCL2 and CCL5 and effects on epithelial-to-mesenchymal transition. BMC Cancer 11:130

Strell C, Lang K, Niggemann B, Zaenker KS, Entschladen F (2010) Neutrophil granulocytes promote the migratory activity of MDA-MB-468 human breast carcinoma cells via ICAM-1. Exp Cell Res 316:138–148

Stathopoulos GT, Psallidas I, Moustaki A et al (2008) A central role for tumor-derived monocyte chemoattractant protein-1 in malignant pleural effusion. J Natl Cancer Inst 100:1464–1476

Su X, Ye J, Hsueh EC, Zhang Y, Hoft DF, Peng G (2010) Tumor microenvironments direct the recruitment and expansion of human Th17 cells. J Immunol 184:1630–1641

Suffee N, Hlawaty H, Meddahi-Pelle A et al (2012) RANTES/CCL5-induced pro-angiogenic effects depend on CCR1, CCR5 and glycosaminoglycans. Angiogenesis 727–744

Suffee N, Richard B, Hlawaty H, Oudar O, Charnaux N, Sutton A (2011) Angiogenic properties of the chemokine RANTES/CCL5. Biochem Soc Trans 39:1649–1653

Sugasawa H, Ichikura T, Kinoshita M et al (2008a) Gastric cancer cells exploit $CD4^+$ cell-derived CCL5 for their growth and prevention of $CD8^+$ cell-involved tumor elimination. Int J Cancer 122:2535–2541

Sugasawa H, Ichikura T, Tsujimoto H et al (2008b) Prognostic significance of expression of CCL5/RANTES receptors in patients with gastric cancer. J Surg Oncol 97:445–450

Sunaga N, Imai H, Shimizu K et al (2012) Oncogenic KRAS-induced interleukin-8 overexpression promotes cell growth and migration and contributes to aggressive phenotypes of non-small cell lung cancer. Int J Cancer 130:1733–1744

Tan MC, Goedegebuure PS, Belt BA et al (2009) Disruption of CCR5-dependent homing of regulatory T cells inhibits tumor growth in a murine model of pancreatic cancer. J Immunol 182:1746–1755

Tanese K, Grimm EA, Ekmekcioglu S (2012) The role of melanoma tumor-derived nitric oxide in the tumor inflammatory microenvironment: Its impact on the chemokine expression profile, including suppression of CXCL10. Int J Cancer 131:891–901

Tang CH, Tsai CC (2012b) CCL2 increases MMP-9 expression and cell motility in human chondrosarcoma cells via the Ras/Raf/MEK/ERK/NF-kappaB signaling pathway. Biochem Pharmacol 83:335–344

Tang KH, Ma S, Lee TK et al (2012b) CD133(+) liver tumor-initiating cells promote tumor angiogenesis, growth, and self-renewal through neurotensin/interleukin-8/CXCL1 signaling. Hepatology 55:807–820

Tjomsland V, Spangeus A, Valila J et al (2011) Interleukin 1alpha sustains the expression of inflammatory factors in human pancreatic cancer microenvironment by targeting cancer-associated fibroblasts. Neoplasia 13:664–675

Tsuchiyama T, Nakamoto Y, Sakai Y, Mukaida N, Kaneko S (2008) Optimal amount of monocyte chemoattractant protein-1 enhances antitumor effects of suicide gene therapy against hepatocellular carcinoma by M1 macrophage activation. Cancer Sci 99:2075–2082
Tsuyada A, Chow A, Wu J et al (2012) CCL2 mediates cross-talk between cancer cells and stromal fibroblasts that regulates breast cancer stem cells. Cancer Res 72:2768–2779
Ueno T, Toi M, Saji H et al (2000) Significance of macrophage chemoattractant protein-1 in macrophage recruitment, angiogenesis, and survival in human breast cancer. Clin Cancer Res 6:3282–3289
Umemura N, Saio M, Suwa T et al (2008) Tumor-infiltrating myeloid-derived suppressor cells are pleiotropic-inflamed monocytes/macrophages that bear M1- and M2-type characteristics. J Leukoc Biol 83:1136–1144
Vallet S, Pozzi S, Patel K et al (2011) A novel role for CCL3 (MIP-1alpha) in myeloma-induced bone disease via osteocalcin downregulation and inhibition of osteoblast function. Leukemia 25:1174–1181
Vandercappellen J, Van Damme J, Struyf S (2008) The role of CXC chemokines and their receptors in cancer. Cancer Lett 267:226–244
Varney ML, Singh S, Li A, Mayer-Ezell R, Bond R, Singh RK (2011) Small molecule antagonists for CXCR2 and CXCR1 inhibit human colon cancer liver metastases. Cancer Lett; 300: 180–188.
Vegran F, Boidot R, Michiels C, Sonveaux P, Feron O (2011) Lactate influx through the endothelial cell monocarboxylate transporter MCT1 supports an NF-kappaB/IL-8 pathway that drives tumor angiogenesis. Cancer Res 71:2550–2560
Velasco-Velazquez M, Jiao X, De La Fuente M et al (2012) CCR5 antagonist blocks metastasis of basal breast cancer cells. Cancer Res 3839–3850
Verma R, Bowen RL, Slater SE, Mihaimeed F, Jones JL (2012) Pathological and epidemiological factors associated with advanced stage at diagnosis of breast cancer. Br Med Bull 129–145
Walser TC, Rifat S, Ma X et al (2006) Antagonism of CXCR3 inhibits lung metastasis in a murine model of metastatic breast cancer. Cancer Res 66:7701–7707
Wang P, Yang X, Xu W, Li K, Chu Y, Xiong S (2011) Integrating individual functional moieties of CXCL10 and CXCL11 into a novel chimeric chemokine leads to synergistic antitumor effects: a strategy for chemokine-based multi-target-directed cancer therapy. Cancer Immunol Immunother 59:1715–1726
Wang S, Xu M, Li F et al (2012a) Ethanol promotes mammary tumor growth and angiogenesis: the involvement of chemoattractant factor MCP-1. Breast Cancer Res Treat 133:1037–1048
Wang SW, Wu HH, Liu SC et al (2012b) CCL5 and CCR5 interaction promotes cell motility in human osteosarcoma. PLoS One 7:e35101
Wang Y, Xu RC, Zhang XL et al (2012c) Interleukin-8 secretion by ovarian cancer cells increases anchorage-independent growth, proliferation, angiogenic potential, adhesion and invasion. Cytokine 59:145–155
Warner KA, Miyazawa M, Cordeiro MM et al (2008) Endothelial cells enhance tumor cell invasion through a crosstalk mediated by CXC chemokine signaling. Neoplasia 10:131–139
Wasmuth JC, Rockstroh JK, Hardy WD (2012) Drug safety evaluation of maraviroc for the treatment of HIV infection. Expert Opin Drug Saf 11:161–174
Waugh DJ, Wilson C (2008) The interleukin-8 pathway in cancer. Clin Cancer Res 14:6735–6741
Weber KS, Nelson PJ, Grone HJ, Weber C (1999) Expression of CCR2 by endothelial cells: implications for MCP-1 mediated wound injury repair and In vivo inflammatory activation of endothelium. Arterioscler Thromb Vasc Biol 19:2085–2093
Welte G, Alt E, Devarajan E, Krishnappa S, Jotzu C, Song YH (2012) Interleukin-8 derived from local tissue-resident stromal cells promotes tumor cell invasion. Mol Carcinog
Wendel M, Galani IE, Suri-Payer E, Cerwenka A (2008) Natural killer cell accumulation in tumors is dependent on IFN-gamma and CXCR3 ligands. Cancer Res 68:8437–8445
Wigler N, Shina S, Kaplan O et al (2002) Breast carcinoma: a report on the potential usage of the CC chemokine RANTES as a marker for a progressive disease. Isr Med Assoc J 4:940–943

Wilson TJ, Nannuru KC, Futakuchi M, Singh RK (2010) Cathepsin G-mediated enhanced TGF-beta signaling promotes angiogenesis via upregulation of VEGF and MCP-1. Cancer Lett 288:162–169

Winkler AE, Brotman JJ, Pittman ME et al (2011) CXCR3 enhances a T-cell-dependent epidermal proliferative response and promotes skin tumorigenesis. Cancer Res 71:5707–5716

Wolf MJ, Hoos A, Bauer J et al (2012) Endothelial CCR2 signaling induced by colon carcinoma cells enables extravasation via the JAK2-Stat5 and p38MAPK pathway. Cancer Cell 22:91–105

Wright KM, Friedland JS (2004) Regulation of chemokine gene expression and secretion in staphylococcus aureus-infected osteoblasts. Microbes Infect 6:844–852

Wu Y, Li YY, Matsushima K, Baba T, Mukaida N (2008a) CCL3-CCR5 axis regulates intratumoral accumulation of leukocytes and fibroblasts and promotes angiogenesis in murine lung metastasis process. J Immunol 181:6384–6393

Wu FY, Ou ZL, Feng LY et al (2008b) Chemokine decoy receptor d6 plays a negative role in human breast cancer. Mol Cancer Res 6:1276–1288

Xu J, Zhang C, He Y et al (2012) Lymphatic endothelial cell-secreted CXCL1 stimulates lymphangiogenesis and metastasis of gastric cancer. Int J Cancer 130:787–797

Yang J, Richmond A (2004) The angiostatic activity of interferon-inducible protein-10/CXCL10 in human melanoma depends on binding to CXCR3 but not to glycosaminoglycan. Mol Ther 9:846–855

Yang J, Wang Y, Gao Y, Shao J, Zhang XJ, Yao Z (2009) Reciprocal regulation of 17beta-estradiol, interleukin-6 and interleukin-8 during growth and progression of epithelial ovarian cancer. Cytokine 46:382–391

Yang G, Rosen DG, Liu G et al (2010) CXCR2 promotes ovarian cancer growth through dysregulated cell cycle, diminished apoptosis, and enhanced angiogenesis. Clin Cancer Res 16:3875–3886

Yaal-Hahoshen N, Shina S, Leider-Trejo L et al (2006) The chemokine CCL5 as a potential prognostic factor predicting disease progression in stage II breast cancer patients. Clin Cancer Res 12:4474–4480

Yadav A, Saini V, Arora S (2010) MCP-1: chemoattractant with a role beyond immunity: a review. Clin Chim Acta 411:1570–1579

Yamamoto M, Kikuchi H, Ohta M et al (2008) TSU68 prevents liver metastasis of colon cancer xenografts by modulating the premetastatic niche. Cancer Res 68:9754–9762

Yeudall WA, Vaughan CA, Miyazaki H et al (2012) Gain-of-function mutant p53 upregulates CXC chemokines and enhances cell migration. Carcinogenesis 33:442–451

Yokoe T, Iino Y, Takei H et al (1997) Changes of cytokines and thyroid function in patients with recurrent breast cancer. Anticancer Res 17:695–699

Yoo JY, Kim JH, Kim J et al (2008) Short hairpin RNA-expressing oncolytic adenovirus-mediated inhibition of IL-8: effects on antiangiogenesis and tumor growth inhibition. Gene Ther 15:635–651

Youn JI, Gabrilovich DI (2010) The biology of myeloid-derived suppressor cells: the blessing and the curse of morphological and functional heterogeneity. Eur J Immunol 40:2969–2975

Youngs SJ, Ali SA, Taub DD, Rees RC (1997) Chemokines induce migrational responses in human breast carcinoma cell lines. Int J Cancer 71:257–266

Zhang X, Haney KM, Richardson AC et al (2010) Anibamine, a natural product CCR5 antagonist, as a novel lead for the development of anti-prostate cancer agents. Bioorg Med Chem Lett 20:4627–4630

Zhang Y, Yao F, Yao X et al (2009) Role of CCL5 in invasion, proliferation and proportion of $CD44^{+}$/CD24- phenotype of MCF-7 cells and correlation of CCL5 and CCR5 expression with breast cancer progression. Oncol Rep 21:1113–1121

Zhong L, Roybal J, Chaerkady R et al (2008) Identification of secreted proteins that mediate cell-cell interactions in an in vitro model of the lung cancer microenvironment. Cancer Res 68:7237–7245

Zhu X, Fallert-Junecko BA, Fujita M et al (2010) Poly-ICLC promotes the infiltration of effector T cells into intracranial gliomas via induction of CXCL10 in IFN-alpha and IFN-gamma dependent manners. Cancer Immunol Immunother 59:1401–1409

Zhu J, Jia X, Xiao G, Kang Y, Partridge NC, Qin L (2007) EGF-like ligands stimulate osteoclastogenesis by regulating expression of osteoclast regulatory factors by osteoblasts: implications for osteolytic bone metastases. J Biol Chem 282:26656–26664

Zlotnik A, Yoshie O, Nomiyama H (2006) The chemokine and chemokine receptor superfamilies and their molecular evolution. Genome Biol 7:243

Zuccari DA, Leonel C, Castro R, et al (2012) An immunohistochemical study of interleukin-8 (IL-8) in breast cancer. Acta Histochem 114:571–576

Chapter 7
Inflammation, Tumor Progression, and Immune Suppression

Suzanne Ostrand-Rosenberg and Pratima Sinha

Abstract Solid tumors consist of tumor cells and a heterogeneous mixture of host cells of both hematopoietic and non-hematopoietic origin. Although some of the host cells have anti-tumor activity, many have been co-opted by tumor-secreted factors and are immune suppressive. Two populations of cells of myeloid origin, myeloid-derived suppressor cells (MDSC) and M2-type macrophages, are potent immune suppressive cells that are particularly prevalent in solid tumors. MDSC and M2 macrophages use multiple mechanisms to individually promote immune suppression and amplify their effects through cross-talk. This chapter will summarize the characteristics of these two myeloid cell populations and then focus on the role of tumor-associated inflammation in inducing MDSC and M2 macrophage development and function.

Keywords MDSC · M2 macrophage · Inflammation · Immunosuppression · Tumor-derived factors · Tumor microenvironment · Hypoxia

Abbreviations

ADAM17	Disintegrin and metalloproteinase 17
DC	Dendritic cells
MDSC	Myeloid-derived suppressor cells
MO-MDSC	Monocytic MDSC
NK	Natural killer cells
PGE2	Prostaglandin E2
PMN-MDSC	Granulocytic/neutrophilic MDSC
TLR	Toll-like receptor

S. Ostrand-Rosenberg (✉) · P. Sinha
Department of Biological Sciences, University of Maryland Baltimore County, Baltimore, MD 21250, USA
e-mail: srosenbe@umbc.edu

M. R. Shurin et al. (eds.), *The Tumor Immunoenvironment*,
DOI: 10.1007/978-94-007-6217-6_7,

7.1 Introduction

Chronic inflammation and immune suppression are now recognized as "hallmarks" of cancer that not only accompany tumor progression but also contribute to the malignant phenotype and tumor growth (Hanahan and Weinberg 2011). Chronic inflammation has long been recognized as a facilitator of malignant transformation (Balkwill and Mantovani 2001); however, the immune system has only more recently been appreciated as also being a major contributor (Ostrand-Rosenberg and Sinha 2009). Although the immune system is a systemic organ, many immune suppressive functions are mediated locally by host immune cells within the tumor microenvironment. These immune cells are activated by a variety of tumor-produced factors, as well as by factors produced by host cells. The immune system consists of a diversity of cell populations that mediate both adaptive immunity (B cells and T cells) and innate immunity [macrophages, natural killer (NK), NK T cells, dendritic cells (DC), and myeloid-derived suppressor cells (MDSC)]. Individual cell populations facilitate carcinogenesis and malignancy through redundant and unique mechanisms, including: (1) Cytotoxic $CD8^{+}$T effector cells (Matsushita et al. 2012) and innate effector cells such as NK cells and macrophages (O'Sullivan et al. 2012) promote the expansion of malignant cells that are resistant to immune-mediated cytotoxicity through a process known as immunoediting; (2) $CD4^{+}$ T cells drive cancer cell invasion and metastasis by inducing macrophages that activate growth factor receptors on tumor cells (DeNardo et al. 2009); (3) $CD4^{+}$ T regulatory cells inhibit the activation of $CD8^{+}$ T cytotoxic cells (Josefowicz et al. 2012) and block cytotoxicity by NK cells (Ralainirina et al. 2007); (4) B lymphocytes support a chronic inflammatory state that facilitates carcinogenesis by activating myeloid cells via their Fc receptors(Andreu et al. 2010; de Visser et al. 2005); (5) Type II NKT cells (NKTII) induce production of TGFβ, which in turn prevents activation of cytotoxic $CD8^{+}$ T cells (Terabe and Berzofsky 2008; Terabe et al. 2005); (6) Tumor-associated macrphages (TAMs; also known as M2 macrophages) produce type 2 cytokines which promote tumor progression by driving angiogenesis and matrix remodeling, as well as facilitating metastasis (Mantovani and Sica 2010; Wyckoff et al. 2007); and (7) Myeloid-derived suppressor cells impair both adaptive and innate immunity through multiple mechanisms. Not all of these suppressive cell populations are present in all patients or experimental animals with cancer; however, MDSC and macrophages have been found in most cancer patients and mice with transplanted or experimental tumors. Because of their widespread presence, their potent immune suppressive function against both innate and adaptive anti-tumor immunity, and their induction by inflammation, this chapter will review the role of inflammation in driving MDSC and macrophage accumulation and function, and the interactions between these cell populations that foster tumor progression.

7.2 Early Studies Identified MDSC as Immune Suppressive Cells, but their Widespread Presence and Central Role in Inhibiting Anti-Tumor Immunity has Only Recently been Appreciated

Unlike cells of the adaptive immune system, MDSC are neither MHC-restricted nor antigen-specific in their suppressive activity. Suppressive cells with similar characteristics were first identified in the early 1980s in newborn mice and were called "natural" suppressor cells (Okada and Strober 1982a, b; Strober 1984). More than a decade later, studies identified cells with similar suppressive activity in patients with squamous cell carcinoma of the head and neck (Garrity et al. 1997; Young et al. 1996, 2001). These cells were characterized by their expression of CD34 and myeloid cell markers, the absence of markers characteristic of lymphoid or NK cells, their immature phenotype, their ability to differentiate under certain conditions to DC, and their chemoattraction by vascular endothelial growth factor (VEGF) (Garrity et al. 1997; Young et al. 1997, 1999). The role of these cells as potent immune suppressive agents in cancer patients began to gain acceptance when it was realized that their levels were significantly increased with tumor progression and that they impacted the immune cell infiltrate of solid tumors (Young et al. 1996; Almand et al. 2001).

Appreciation of the significance of these suppressive cells in patients led to a re-examination of suppressor cell populations in mice and the findings that cells with similar suppressive function accumulated in mice with transplanted and spontaneous (transgenic) tumors (Bronte et al. 1998; Gabrilovich et al. 2001; Melani et al. 2003). Murine MDSC were characterized by their expression of the granulocytic marker Gr1 and the monocytic/macrophage marker CD11b (also known as Mac-1). Origin of the cells was attributed to dysregulation of myelopoiesis (Bronte et al. 1999). It became apparent in the mid 2000s that the immune suppressor cell populations identified in multiple studies were most likely related. This realization led to adaptation of a unifying nomenclature and the introduction of the term "myeloid-derived suppressor cells (Gabrilovich et al. 2007)."

The critical role of MDSC as regulatory cells that impede anti-tumor immunity and promote the progression of primary and metastatic tumors was initially established in mouse tumor systems. These studies unequivocally demonstrated that in vivo reduction of MDSC in tumor-bearing mice delayed tumor progression, restored immune competence, enabled vaccine-induced activation of tumor-reactive T cells, and extended survival time (De Santo et al. 2005; Kusmartsev and Gabrilovich 2003; Li et al. 2004; Sinha et al. 2005a, b; Wiers et al. 2000; Young et al. 1995). Subsequent studies with cancer patients using inhibitors that impair MDSC function have confirmed that MDSC similarly play a key regulatory role in humans (Kusmartsev et al. 2008; Mirza et al. 2006; Solito et al 2011; Srivastava et al. 2008). Further studies have shown that MDSC levels in the blood of cancer patients positively correlate with clinical cancer stage and extent of metastasis, and may be useful diagnostic indicators of tumor progression (Diaz-Montero et al. 2008; Dumitru et al. 2012; Montero et al. 2012).

7.3 MDSC are a Heterogeneous Population of Immature Myeloid Cells

MDSC are immature myeloid cells and their accumulation is the result of impaired myelopoiesis due to tumor-secreted factors. There are two dominant subpopulations of MDSC: monocytic (MO-MDSC) and granulocytic/neutrophilic (PMN-MDSC) MDSC (Movahedi et al. 2008). MDSC in mice were originally characterized by their expression of the plasma membrane markers Gr1 (marker of granulocytes) and CD11b (marker of monocytes and macrophages). Gr1 consists of two molecules: Ly6G and Ly6C. MO-MDSC are mononuclear and express Ly6C, while PMN-MDSC are polymorphonuclear and express Ly6G. Since these characteristics are shared with neutrophils (Youn et al. 2012), definitive identification of MDSC requires demonstrating suppressive activity in addition to documenting phenotype (Ostrand-Rosenberg and Sinha 2009).

Additional plasma membrane markers, including CD80, IL-4Rα, F4/80, PD-L1, and CD115 have been attributed to mouse MDSC. However, these markers are not expressed by all mouse MDSC. Cell-specific markers (e.g. macrophage F4/80 and CD115) are typically expressed by terminally differentiated cells and cells in the end stages of differentiation. Therefore, expression of these markers by some MDSC reflects that MDSC are a continuum of immature myeloid cells that are in various stages of differentiation into mature dendritic cells, macrophages, and/or granulocytes.

There is no equivalent molecule to Gr1 in humans, complicating the identification of MDSC in patients and making it essential that phenotyping be accompanied by functional studies demonstrating suppressive activity. Similar to the mouse situation, a limited number of markers are universally expressed by human MDSC, and additional markers are sporadically expressed. Currently, human MO-MDSC are phenotypically identified by their expression of CD11b, CD14 (monocyte marker), IL-4Rα, and CD33 (myeloid lineage marker), and the absence of HLA-DR and lineage markers of lymphocytes and NK cells. In contrast, human PMN-MDSC are phenotypically identified by their expression of CD11b, CD15, and the absence of CD14 and other lineage markers. As for mouse MDSC, the definitive identification of human MDSC requires demonstration of suppressive activity (Dumitru et al. 2012; Montero et al. 2012). Table 7.1 lists the characteristics of murine and human MDSC.

Table 7.1 Characteristics of immune suppressive M2 macrophages and monocytic and granulocytic MDSC[a]

Species	Cells	Characteristics
Mouse	Monocytic MDSC	$Gr1^{+}CD11b^{+}Ly6C^{+}Ly6G^{low/-}CD115^{+}F4/80^{+}IL\text{-}4R\alpha^{+}$(CD124)
		Lin^{-} for T, NK, B
		$Arg1^{+}iNOS^{-}$(NOS2)$IL\text{-}10^{+}$
		Mononuclear
	Granulocytic MDSC	$Gr1^{+}CD11b^{+}Ly6C^{-}Ly6G^{+}CD115^{+/-}F4/80^{low}IL\text{-}4R\alpha^{+/-}$(CD124)
		Lin^{-} for T, NK, B
		$Arg1^{+}iNOS^{-}IL\text{-}10^{+}$
		ROS^{high}
		Polymorphonuclear
	M2 macrophages/ TAMs	$F4/80^{-}CD11b^{+}CD115^{+}IL\text{-}4R\alpha^{+}$ $Gr1^{-}$
		$Arg1^{+}IL\text{-}10^{hi}IL\text{-}12^{low}$
		$CD206^{+}$ (mannose receptor) $CD36^{+}$ (scavenger receptor)
Human	Monocytic MDSC	$CD11b^{+}CD33^{+}CD14^{+}HLA\text{-}DR^{low/-}CD15^{low/-}$
		Lin^{-} T, NK, B
		Mononuclear
	Granulocytic MDSC	$CD11b^{+}CD33^{+}CD14^{-}HLA\text{-}DR^{low/-}CD15^{+}$
		Lin^{-} T, NK, B
		Polymorphonuclear
	M2 macrophages/ TAMs	$CD68^{-}CD14^{+}CD36^{+}CD11b^{+}IL\text{-}4R\alpha^{+}$ $HLA\text{-}DR^{low}$
		$IL\text{-}10^{high}IL\text{-}12^{low}$
		$CD206^{+}CD36^{+}$

[a] adapted from (Gabrilovich et al. 2012)

7.4 MDSC Inhibit both Adaptive and Innate Immunity Through a Variety of Disparate Mechanisms

MDSC display extreme diversity in the mechanisms they use to inhibit anti-tumor immunity. MDSC were originally identified by their ability to block the activation of $CD8^{+}$ T lymphocytes (Gabrilovich et al. 2001). They mediate this suppressive activity through multiple distinct mechanisms ranging from the release of inhibitor molecules and deprivation of amino acids to modification of receptors on target cells.

MDSC release of arginase I suppresses T cell activation by depriving T lymphocytes of L-arginine. This deprivation limits expression of the T cell receptor-associated ζ chain which is an essential component for inducing T cell proliferation (Ezernitchi et al. 2006; Rodriguez et al. 2005, 2007). Nitric oxide synthase degrades L-arginine to yield nitric oxide (NO). NO subsequently inhibits T cell activation by preventing the phosphorylation of multiple transcription factors and destabilizing IL-2 mRNA (Bronte and Zanovello 2005; Rodriguez and Ochoa 2008).

MDSC also produce and release reactive oxygen and nitrogen species (ROS, RNS) which inhibit T cell activation (Corzo et al. 2009; Kusmartsev et al. 2004). The potent oxidant peroxynitrite, nitrates the T cell receptor of potentially reactive T cells, and the MHC class I molecules of target cells, thereby inhibiting T cell recognition of tumor cells and masking tumor cell recognition by activated T cells (Nagaraj et al. 2007; Lu et al. 2011).

In addition to depriving T cells of arginine, MDSC also inhibit T cell activation by depriving T cells of cystine/cysteine (Srivastava et al. 2010). Cysteine is an essential amino acid for all cells. Most cells generate their cysteine by importing cystine from their extracellular oxidizing environment and reducing it intracellularly to cysteine. Alternatively, cells convert intracellular methionine to cysteine using the enzyme cystathionase. T cells, however, lack cystathionase and do not have the plasma membrane transporter to import cystine. Therefore, T cells must generate cysteine by direct import of cysteine. MDSC prevent this importation by scavenging extracellular cysteine and thereby starving T cells of cysteine.

As well as direct effects of MDSC products, MDSC also indirectly prevent T cell activation. L-selectin (also known as CD62L) is expressed by naïve T cells and is required for T cell homing to lymph nodes, where T cells are usually activated. Following activation, T cells translocate intracellular disintegrin and metalloproteinase 17 (ADAM 17) from their cytosol to their cell surface, and ADAM 17 then cleaves cell surface CD62L so activated T cells can exit the lymph node. Many cells contain intracellular ADAM17, however, MDSC constitutively express cell surface ADAM17. As a result, T cells in tumor-bearing mice with high levels of MDSC are deficient for CD62L and are unable to enter lymph nodes and become activated (Hanson et al. 2009).

Mouse and human MDSC also indirectly impair T cell function by inducing $CD4^+$ T regulatory cells (T regs) that inhibit the activation of $CD8^+$ T cells (Huang et al. 2006; Hoechst et al. 2008). MDSC-mediated induction of T regs requires arginase 1, interferon-γ (IFNγ), and IL-10. As discussed in the following section, interactions between macrophages and MDSC increase MDSC production of IL-10, indicating that close cell encounters within the tumor microenvironment play an important role in exacerbating the immune suppressive effects of MDSC and probably other suppressive cells as well. Table 7.2 summarizes the suppressive mechanisms used by MDSC.

7.5 Tumor-Promoting Macrophages

The tumor microenvironment is a complex mileau of tumor cells and host cells. It is frequently a highly pro-inflammatory environment due to inflammatory mediators produced by tumor cells and infiltrating host cells. In addition to MDSC, other myeloid cells, including macrophages, dendritic cells, and cancer promoting fibroblasts are also present. Macrophages are dynamic cells with extreme plasticity. They adapt their phenotype and function in response to inflammatory

Table 7.2 Major effector mechanisms used by MDSC to suppress anti-tumor immunity

Suppressive effect	Mechanism
Depletion of arginine needed for T cell activation and function	Arginase
T cell toxicity	Reactive oxygen species
Nitration of MHC class I molecules of tumor cells; nitration of T cell receptor	Reactive nitrogen species
Sequestration of cysteine/cystine, an essential amino acid for T cell activation and function	Absence of cysteine exporter
Promotion of T regulatory cells	IL-10
Inhibition of T cell trafficking to lymph nodes	ADAM17-mediated cleavage of CD62L
Polarization of macrophages towards an M2 phentoype	IL-10
Inhibition of NK cell development and function	Down-regulation of macrophage and DC production of IL-12
Promotion of angiogenesis	VEGF
Reduction of DC	??

environmental cues. At the M1 end of the spectrum under conditions of acute inflammation, macrophages express high levels of IL-12 and nitric oxide, low levels of IL-10 and arginase I, and are bacteriocidal and tumoricidal. At the M2 end of the spectrum under conditions of chronic inflammation, macrophages have the reciprocal phenotype of IL-10^{high}, IL-12^{low}, ArgI^{high}, and are pro-tumorigenic (Balkwill et al. 2005; Mantovani et al. 2002; Biswas and Mantovani 2010). A major function of M2 macrophages is their production of high levels of VEGF that drives tumor angiogenesis (Lin et al. 2006). Multiple subpopulations of macrophages between the M1 and M2 extremes have been identified (Movahedi et al. 2010; Sonda et al. 2011; Torroella-Kouri et al. 2009). These subpopulations localize to distinct areas within solid tumors where their specific phenotype and function are driven by the local microenvironment.

The pockets of hypoxia within solid tumors exacerbate the M2 phenotype of macrophages and thereby increase the immune suppression and tumor-promoting functions of macrophages. Hypoxia up-regulates macrophage expression of Tie-2, the receptor for angiopoietin-2. Angiopoietin-2 increases synthesis of thymidine phosphorylase and cathepsin B, two pro-angiogenic enzymes. Increased expression of Tie-2 also leads to higher macrophage production of IL-10 and matrix metalloproteinase-9, which facilitates tumor invasion and metastasis (Coffelt et al. 2010).

Multiple cell populations within the inflammatory tumor microenvironment polarize macrophages. For example, B lymphocytes increase macrophage production of IL-10 (de Visser et al. 2005), and autoantibodies have been shown to bridge interactions between macrophages and tumor-promoting leukocytes by binding to macrophages and Fc receptors on the leukocytes (Andreu et al. 2010). Type 2 CD4^{+} T cells also enhance the pro-tumorigenic activity of macrophages by increasing macrophage production of epidermal growth factor which drives tumor invasion and metastasis (DeNardo et al. 2009). $CD4^+$ T reg production of IL-10, IL-4, and IL-13 drives macrophages towards a M2 phenotype by rendering them less susceptible to lipopolysaccharide, a bacterial product that induces M1 characteristics (Tiemessen et al. 2007). Tumor-secreted factors also directly increase macrophage production of the tumor-promoting factors IL-10, IL-1β, CCL5, CCL22, and MMP's 7 and 9 (Hagemann et al. 2006).

Many studies have confirmed that macrophages are potent drivers of tumor progression (Mantovani and Sica 2010; Qian and Pollard 2010). Recent genomic studies have capitalized on this knowledge and have demonstrated that a specific gene signature of TAMs is associated with the failure of patients with Hodgkin's lymphoma to respond to therapy (Steidl et al. 2010).

Overall, the inflammatory tumor microenvironment drives and maintains the M2 phenotype of macrophages consistent with these cells dominating the tissue remodeling and angiogenesis that continually occur within the rapidly evolving tumor microenvironment. Table 7.1 lists the predominant characteristics of immune suppressive M2 macrophages/TAMs.

7.6 Inflammation Drives Tumor Progression and Immune Suppressive Cells

As MDSC were gaining recognition as significant immune suppressive cells, cancer biologists were proving the hypothesis that inflammation facilitates the onset and progression of cancer. The concept that inflammation predisposes towards cancer was first articulated by the German pathologist Rudolf Virchow in the late 1800s (Balkwill and Mantovani 2001). This concept is supported by extensive epidemiological data demonstrating strong correlations between numerous inflammatory diseases and the subsequent onset of cancer. Chronic inflammation that predisposes towards cancer is of two types: chronic inflammation associated with bacterial or viral infection, and "sterile inflammation" which results from chronic physical or chemical stress. An example of the former is inflammatory bowel disease caused by commensal microbes and the development of colorectal cancer. An example of the latter is lung cancer and mesothelioma caused by asbestos irritation (Ameille et al. 2011). Sterile inflammation as caused by obesity also increases cancer risk (reviewed by Trinchieri 2012).

The first experimental evidence suggesting a causative linkage between inflammation and myeloid cells came from experiments with MDSC derived from mice with lung cancer (Rodriguez et al. 2005). In this report, blockade of cyclooxygenase 2 (COX2) decreased arginase I content of myeloid cells from tumor-bearing patients. Since arginase I is a dominant effector molecule used by granulocytic MDSC and M2 macrophages to suppress T cell activation, this finding indicated that COX2 increases myeloid cell-mediated immune suppression. Subsequent studies in mouse models demonstrated that prostaglandin E2 (PGE2) directly mediates the differentiation of

MDSC from bone marrow progenitor cells (Sinha et al. 2007), and that in vivo inhibition of PGE2 by specific antagonists or antibodies decreased MDSC accumulation and delayed tumor progression (Sinha et al. 2007; Fujita et al. 2011; Talmadge et al. 2007; Veltman et al. 2010). These findings identified COX2 and PGE2 as potential therapeutic targets for reducing MDSC levels and potency. However, since MDSC were neither eliminated nor fully inactivated, these studies also demonstrated that pro-inflammatory mediators other than PGE2 and COX2 also contribute to the development and maintenance of MDSC.

IL-6 (Neurath and Finotto 2011) and IL-1β (Dinarello 2011a, b) are increasingly being recognized as dominant pro-inflammatory cytokines. Many tumors, as well as macrophages and MDSC produce IL-6 and/or IL-1β. In addition, elevated IL-6 and/or IL-1β levels are characteristic of some inflammatory diseases that precede malignancy. These observations led to studies examining the role of these cytokines in the induction of MDSC. Studies with IL-6 and IL-1β-transfected tumor cells demonstrated that the rate of accumulation and quantity of tumor-induced MDSC was proportional to cytokine production by the tumor cells, identifying IL-6 and IL-1β as drivers of MDSC (Bunt et al. 2006, 2007; Song et al. 2005). Likewise, MDSC suppressive potency against T cells is regulated by IL-6 and IL-1β (Bunt et al. 2007), and suppressive potency against NK cells is regulated by IL-1β (Elkabets et al. 2010). Tumor-bearing mice deficient for the IL-1 receptor (Bunt et al. 2007), or treated with the natural inhibitor of IL-1, the IL-1 receptor antagonist (IL-1Ra) (Song et al. 2005), developed less suppressive MDSC. In contrast, mice deficient for IL-1Ra developed elevated levels of more suppressive MDSC (Bunt et al. 2007). IL-1β also increased the in vivo half life of MDSC by increasing MDSC resistance to Fas-FasL-mediated apoptosis (Chornoguz et al. 2010; Sinha et al. 2011).

IL-6 similarly increases MDSC levels in tumor-bearing mice (Bunt et al. 2007), and in combination with GM-CSF drives the differentiation of bone marrow progenitor cells to MDSC (Marigo et al. 2010). IL-6 also regulates MDSC suppressive potency via indoleamine 2,3 dioxygenase (IDO) (Smith et al. 2012). MDSC generated in IDO-deficient mice have decreased suppressive potency. Supplementation with IL-6 reconstitutes MDSC function, indicating that IDO drives MDSC activity through an IL-6-regulated mechanism. Since IL-6 is downstream of IL-1β, IL-6 and IL-1β most likely activate MDSC through a common pathway.

The heterodimeric S100A8/A9 complex is another pro-inflammatory mediator that is a potent driver of MDSC. S100A8/A9 molecules are calcium-binding proteins that are liberally expressed in inflammatory environments and are released by many tumor cells and by leukocytes. They are inherently inflammatory and amplify inflammation by chemoattracting leukocytes that produce other inflammatory mediators. S100A8/A9 complexes activate MDSC by signaling via STAT3 (Cheng et al. 2008) and NF-κB (Sinha et al. 2008). This action causes the accumulation of MDSC by blocking differentiation of progenitor cells to DC and macrophages (Cheng et al. 2008). S100A8/A9 complexes are chemoattractants for MDSC and recruit MDSC to tumor sites by binding to N-glycan receptors including receptor for advanced glycation endproducts (RAGE) on MDSC (Sinha et al. 2008). The combined release of S100A8/A9 by both tumor cells and MDSC facilitates the recruitment and retention of MDSC within the inflammatory tumor microenvironment. Figure 7.1a illustrates the inflammatory mediators that increase MDSC accumulation and suppressive potency.

Unlike MDSC which only have an immune suppressive, pro-tumor function, macrophages and DC can be either pro-tumor or anti-tumor depending on environmental cues. As discussed in Sect. 7.3, pro-tumor M2 macrophages/TAMs are a product of the pro-inflammatory tumor microenvironment. DC are similarly polarized by the inflammatory tumor mileau towards a pro-tumor phenotype.

DC are the most effective antigen presenting cells (APC) and as such are critically important for generating immune effector cells. In healthy individuals, DC produce IL-12. IL-12 drives the activation of cytotoxic $CD8^+$ T cells and type

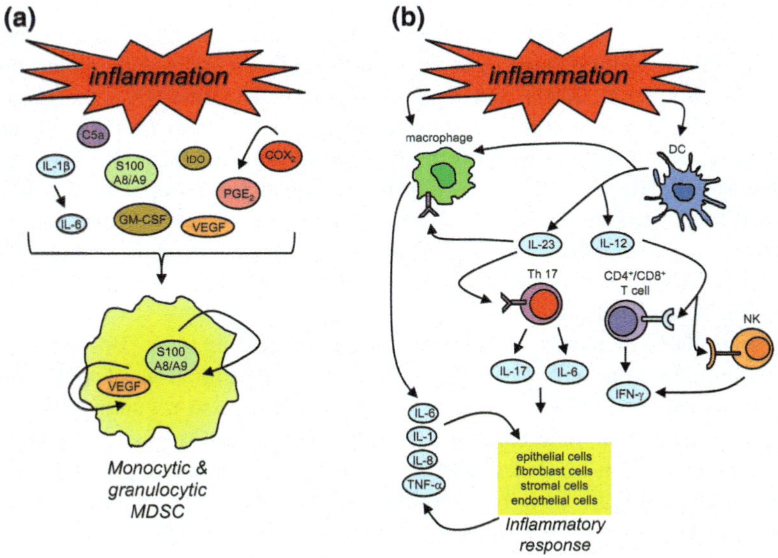

Fig. 7.1 Chronic inflammation drives the accumulation, phenotype and potency of myeloid cells in the tumor microenvironment. **a** Multiple pro-inflammatory mediators regulate MDSC. Pro-inflammatory mediators produced by both tumor cells and tumor-infiltrating host cells increase the levels of MDSC in the blood, spleen, and within the tumor. Some pro-inflammatory mediators, such as VEGF and S100A8/A9 are produced by MDSC themselves and act as autocrine growth factors. **b** Inflammation polarizes macrophages and DC towards a pro-inflammatory phenotype. Cytokines induced by inflammation and produced by macrophages and DC act on host epithelial, fibroblast, stromal, and endothelial cells to amplify inflammation. Inflammation activates DC to produce IL-23 and IL-12. IL-23 activates Th17 cells to produce IL-17 and IL-6 which also act on host epithelial, fibroblast, stromal, and endothelial cells to amplify inflammation. IL-12 polarizes and activates T cells towards a tumor-rejecting type 1 phenotype and activates NK cells, both of which produce IFNγ which further polarizes immunity towards a type 1 response

1 $CD4^+$ T cells and NK cells, which in turn make IFNγ and promote Type 1 immunity. In inflammation settings DC also produce IL-23. IL-23 activates Th17 cells which produce IL-17 and IL-6 (Iwakura and Ishigame 2006). Whether IL-17 promotes tumor progression or regression is controversial. Multiple studies suggest that IL-17 facilitates tumor growth by promoting angiogenesis and inflammation. In contrast, other studies indicate that IL-17 helps activate cytotoxic $CD8^+$ effector cells, although most evidence indicates that IL-17 and IL-23 favor tumor progression (Ji and Zhang 2010). Additional studies are clearly needed to resolve these apparent contradictions.

Cancer patients are frequently deficient for mature DC and their existing DC are frequently ineffective APC. As a result, the ability of cancer patients to generate an immune response is compromised. The deficiency in quantity of DC is principally due to abnormal myelopoiesis. The expansion of MDSC probably contributes because in vitro studies have shown that accumulation of MDSC from myelocytic progenitor cells occurs at the expense of DC (Sinha et al. 2007). At least two factors associated with inflammation contribute to DC dysfunction. Hypoxia within the tumor microenvironment induces expression of the adenosine receptor A2B on DC. Adenosine compromises DC function by reducing APC activity and by inducing expression of VEGF, IL-6, IL-10, COX2, IDO, and TGFβ (Novitskiy et al. 2008). IDO produced by DC in response to the TLR9 ligand CpG increases T reg suppressive activity and prevents the reprogramming of T regs to Th1 cells (Baban et al. 2009). As a result pro-tumor T regs increase while anti-tumor Th1 cells decrease, facilitating tumor progression.

Figure 7.1b shows schematically how inflammation polarizes macrophages and DC towards a pro-inflammatory phenotype. Interestingly, if these cells are in a tumor environment and MDSC are present, then IL-10 production by MDSC converts this type 1 cytokine pattern to a type 2 pro-tumor response (see Sects. 7.7, 7.8 and Figs. 7.2 and 7.3).

7.7 Cross-Talk Between MDSC and Macrophages Produces a Tumor-Promoting Environment

As described in Sect. 7.5, the tumor microenvironment is a complex mixture of tumor cells and different types of host cells. Many of these cells are in close proximity to each other. As a result, there is frequent cell-to-cell contact between the diverse tumor-infiltrating cell populations and between tumor cells and host cells. This topology also gives rise to concentration gradients within solid tumors of cell-secreted factors. If cells have plasticity and the appropriate receptors, such gradients have the potential to significantly modulate target cell activity.

When in close proximity, macrophages and MDSC are reciprocally modulated in response to their secreted factors. MDSC constitutively produce IL-10 and in the presence of macrophages, MDSC production of IL-10 is significantly increased

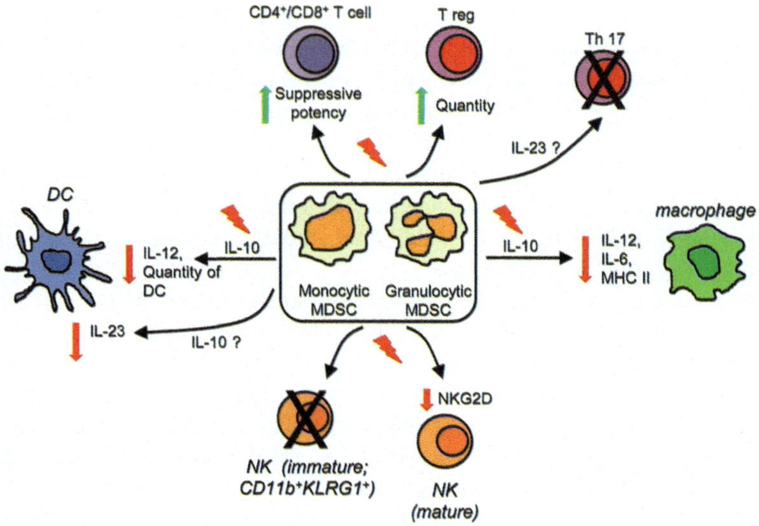

Fig. 7.2 Chronic inflammation increases MDSC suppressive potency. Monocytic and granulocytic MDSC inhibit anti-tumor immunity by (1) inhibiting T cell activation; (2) driving the accumulation of T regulatory cells; (3) polarizing macrophages towards an M2 phenotype and reducing their antigen presentation capability; (4) inhibiting the development and cytotoxic activity of NK cells; and (5) decreasing the quantity of dendritic cells. Chronic inflammation amplifies many of these functions by up-regulating IL-10 production by macrophages

(Sinha et al. 2007). IL-10 is classically considered an anti-inflammatory cytokine; however, it is also a major contributor to tumor progression through multiple mechanisms. IL-10 contributes to the differentiation of T regs, thereby preventing the development of tumoricidal $CD8^+$ T cells (Murai et al. 2009). IL-10 also polarizes $CD4^+$ T cells towards a type 2 phenotype and induces them to secrete IL-4 and IL-13 which prevent the development of cytotoxic $CD8^+$ T cells. MDSC-produced IL-10 directly polarizes macrophages towards an M2 phenotype by inhibiting macrophage production of IL-12 (Sinha et al. 2007). MDSC similarly down-regulate IL-12 production by DC, thereby eliminating another source of IL-12 that could drive the activation of Type 1 $CD4^+$ T cells and tumoricidal NK cells (Hu et al. 2011). When MDSC and macrophages are in close proximity, MDSC production of IL-10 is partially regulated by macrophage production of IL-6 since there is limited cross-talk-induced increase in IL-10 when MDSC are cultured with IL-6-deficient macrophages (Beury and Ostrand-Rosenberg, unpublished). Therefore, IL-10 is a key cytokine that regulates multiple tumor-infiltrating cell populations and thereby drives tumor progression.

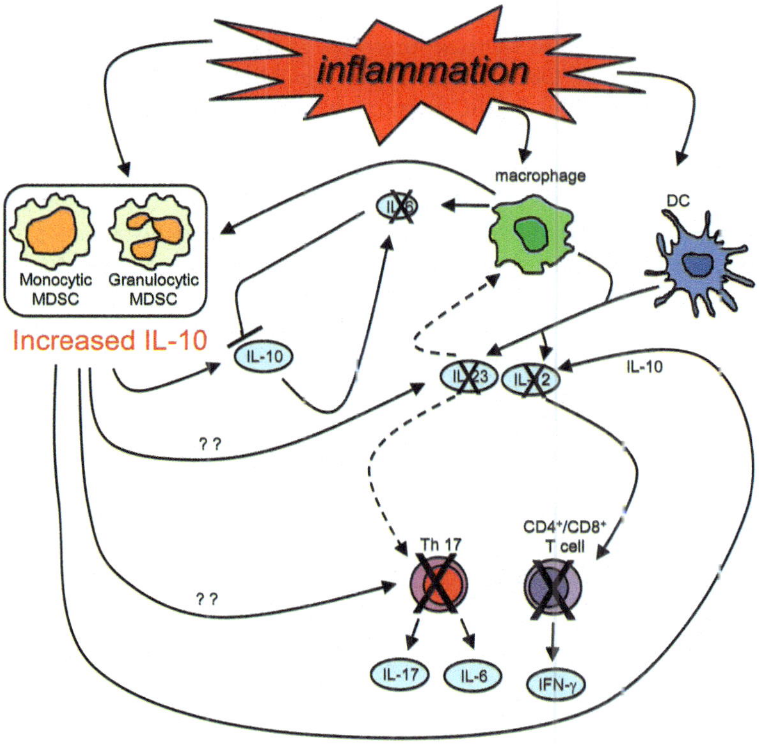

Fig. 7.3 Hypothesis for how inflammation orchestrates the interactions between MDSC, macrophages, and DC to amplify pro-tumor immunity and facilitate tumor progression. As described in Sect. 7.9, chronic inflammation activates MDSC, macrophages, and DC, and increases MDSC production of IL-10. Macrophages increase MDSC production of IL-10 but also produce IL-6 which down-regulates MDSC production of IL-10, and MDSC-produced IL-10 down-regulates macrophage production of IL-6. DC may indirectly regulate MDSC production of IL-10 through their release of IL-23 which stimulates macrophages to make IL-6. As a result, MDSC production of IL-10 is regulated by both macrophages and DC and depends on the levels of IL-23 and IL-6. MDSC may also auto-regulate their production of IL-10 through their interaction with DC since MDSC inhibit DC production of IL-23 which prevents the activation of Th17 cells and the release of IL-6. In the absence of IL-6 there is less down-regulation of IL-10 in MDSC and therefore MDSC suppressive activity is increased leading to tumor progression

7.8 Inflammation Exacerbates Cross-Talk Between MDSC, Macrophages, and DC

In addition to impacting MDSC, macrophages, and DC individually, inflammation also increases and modulates cross-talk between these cell populations (Bunt et al. 2009; Ostrand-Rosenberg et al. 2012). MDSC generated by a tumor environment containing high levels of IL-1β ("inflammatory MDSC") produce elevated levels of IL-10, and production of IL-10 by inflammatory MDSC is further increased by macrophages. The bioactive lipids PGE2 and Butaprost (a PGE2 analog that binds to the EP2 receptor) also increase macrophage-mediated MDSC production of IL-10 (Clements and Ostrand-Rosenberg, unpublished). Not surprisingly, this increase in IL-10 enables MDSC to more effectively down-regulate macrophage production of IL-12.

Inflammation also reduces the ability of macrophages to serve as antigen presenting cells (APC) for the activation of $CD4^+$ and $CD8^+$ T cells. Although macrophages are not as effective APC as DC, their phagocytic ability and co-expression of MHC class I, class II, and costimulatory molecules enable them to endocytose and process and present antigen to T cells. Inflammatory MDSC reduce the APC function of macrophages by down-regulating macrophage expression of MHC II (Clements and Ostrand-Rosenberg, unpublished).

The detrimental effects of MDSC on NK cells are also increased by inflammation (Elkabets et al. 2010; Ostrand-Rosenberg et al. 2012). MDSC that develop in a heightened inflammatory environment decrease the quantity of immature NK cells in the bone marrow and reduce the number of mature $CD11b^+$ $KLRG\text{-}1^+$ NK cells in the spleen. Inflammatory MDSC also compromise the ability of mature NK cells to be activated by reducing expression of the NK activating receptor, NKG2D. Figure 7.2 summarizes how inflammation alters MDSC function and the subsequent effects of MDSC on other immune cells.

7.9 Conclusions

An individual's immune system is capable of recognizing and eliminating malignant cells. However, tumors often generate an inflammatory environment that disrupts the immune system and redirects immune cells to have tumor-promoting functions. Macrophages and DC are particularly susceptible to being co-opted because of their extensive plasticity, while MDSC are readily expanded by inflammation. In addition to the effects of inflammation on individual cell populations, inflammation also alters immune cells by regulating the interactions between cell populations. Some of these interactions have been described in this chapter; others remain unknown. Based on the above information plus information from non-tumor systems, we propose the following hypothesis as a framework for understanding the effects of cross-talk between MDSC, macrophages, and DC.

Monocytic and granulocytic MDSC produce IL-10 which is a key cytokine for promoting tumor progression by inducing immune suppression. Although some recent studies indicate that IL-10 may have beneficial effects (Emmerich et al. 2012; Mumm et al. 2011), most reports, including our own studies with IL-10-deficient mice, demonstrate that the absence of IL-10 promotes anti-tumor immunity (Buery, Parker, and Ostrand-Rosenberg, unpublished results). MDSC production of IL-10 is increased by macrophages and is enhanced by inflammation. However, macrophages also produce IL-6 which down-regulates IL-10. Th17 cells, which are activated by IL-23 produced by DC, also produce IL-6. Since macrophages express the receptor for IL-23, their production of IL-6 is also regulated by IL-23. However, MDSC-produced IL-10 inhibits DC production of IL-23 and macrophage production of IL-6. Therefore, MDSC production of IL-10 is both increased and decreased by macrophages and DC via IL-23 and IL-6. Figure 7.3 schematically illustrates these proposed interactions.

Drug targeting this myeloid network may delay or eliminate tumor growth. However, it is also possible that the complex reciprocal relationship between these three myeloid cell populations is just one of many networks that promote tumor progression through immune and non-immune mechanisms.

Acknowledgments Original studies were supported by NIH RO1CA115880, RO1CA84232 (SOR), and American Cancer Society IRG-97-153-07 (PS).

References

Almand B, Clark JI, Nikitina E, van Beynen J, English NR, Knight SC et al (2001) Increased production of immature myeloid cells in cancer patients: a mechanism of immunosuppression in cancer. J Immunol 166:678–689

Ameille J, Brochard P, Letourneux M, Paris C, Pairon JC (2011) Asbestos-related cancer risk in patients with asbestosis or pleural plaques. Rev Mal Respir 28:e11–e17

Andreu P, Johansson M, Affara NI, Pucci F, Tan T, Junankar S et al (2010) FcRgamma activation regulates inflammation-associated squamous carcinogenesis. Cancer Cell 17:121–134

Baban B, Chandler PR, Sharma MD, Pihkala J, Koni PA, Munn DH et al (2009) IDO activates regulatory T cells and blocks their conversion into Th17-like T cells. J Immunol 183:2475–2483

Balkwill F, Mantovani A (2001) Inflammation and cancer: back to Virchow? Lancet 357:539–545

Balkwill F, Charles KA, Mantovani A (2005) Smoldering and polarized inflammation in the initiation and promotion of malignant disease. Cancer Cell 7:211–217

Biswas SK, Mantovani A (2010) Macrophage plasticity and interaction with lymphocyte subsets: cancer as a paradigm. Nat Immunol 11:889–896

Bronte V, Zanovello P (2005) Regulation of immune responses by L-arginine metabolism. Nat Rev Immunol 5:641–654

Bronte V, Wang M, Overwijk WW, Surman DR, Pericle F, Rosenberg SA et al (1998) Apoptotic death of CD8 + T lymphocytes after immunization: induction of a suppressive population of Mac-1 +/Gr-1 + cells. J Immunol 161:5313–5320

Bronte V, Chappell DB, Apolloni E, Cabrelle A, Wang M, Hwu P et al (1999) Unopposed production of granulocyte-macrophage colony-stimulating factor by tumors inhibits CD8 + T cell responses by dysregulating antigen-presenting cell maturation. J Immunol 162:5728–5737

Bunt SK, Sinha P, Clements VK, Leips J, Ostrand-Rosenberg S (2006) Inflammation induces myeloid-derived suppressor cells that facilitate tumor progression. J Immunol 176:284–290

Bunt SK, Yang L, Sinha P, Clements VK, Leips J, Ostrand-Rosenberg S (2007) Reduced inflammation in the tumor microenvironment delays the accumulation of myeloid-derived suppressor cells and limits tumor progression. Cancer Res 67:10019–10026

Bunt SK, Clements VK, Hanson EM, Sinha P, Ostrand-Rosenberg S (2009) Inflammation enhances myeloid-derived suppressor cell cross-talk by signaling through Toll-like receptor 4. J Leukoc Biol 85:996–1004

Cheng P, Corzo CA, Luetteke N, Yu B, Nagaraj S, Bui MM et al (2008) Inhibition of dendritic cell differentiation and accumulation of myeloid-derived suppressor cells in cancer is regulated by S100A9 protein. J Exp Med 205:2235–2249

Chornoguz O, Grmai L, Sinha P, Artemenko KA, Zubarev RA, Ostrand-Rosenberg S (2011) Proteomic pathway analysis reveals inflammation increases myeloid-derived suppressor cell resistance to apoptosis. Mol cell proteomics 10:M110 002980

Coffelt SB, Lewis CE, Naldini L, Brown JM, Ferrara N, De Palma M (2010) Elusive identities and overlapping phenotypes of proangiogenic myeloid cells in tumors. Am J Pathol 176:1564–1576

Corzo CA, Cotter MJ, Cheng P, Cheng F, Kusmartsev S, Sotomayor E et al (2009) Mechanism regulating reactive oxygen species in tumor-induced myeloid-derived suppressor cells. J Immunol 182:5693–5701

De Santo C, Serafini P, Marigo I, Dolcetti L, Bolla M, Del Soldato P et al (2005) Nitroaspirin corrects immune dysfunction in tumor-bearing hosts and promotes tumor eradication by cancer vaccination. Proc Natl Acad Sci USA 102:4185–4190

de Visser KE, Korets LV, Coussens LM (2005) De novo carcinogenesis promoted by chronic inflammation is B lymphocyte dependent. Cancer Cell 7:411–423

DeNardo DG, Barreto JB, Andreu P, Vasquez L, Tawfik D, Kolhatkar N et al (2009) CD4(+) T cells regulate pulmonary metastasis of mammary carcinomas by enhancing protumor properties of macrophages. Cancer Cell 16:91–102

Diaz-Montero CM, Salem ML, Nishimura MI, Garrett-Mayer E, Cole DJ, Montero AJ (2008) Increased circulating myeloid-derived suppressor cells correlate with clinical cancer stage, metastatic tumor burden, and doxorubicin-cyclophosphamide chemotherapy. Cancer Immunol Immunother

Dinarello CA (2011a) A clinical perspective of IL-1beta as the gatekeeper of inflammation. Eur J Immunol 41:1203–1217

Dinarello CA (2011b) Interleukin-1 in the pathogenesis and treatment of inflammatory diseases. Blood 117:3720–3732

Dumitru CA, Moses K, Trellakis S, Lang S, Brandau S (2012) Neutrophils and granulocytic myeloid-derived suppressor cells: immunophenotyping, cell biology and clinical relevance in human oncology. Cancer Immunol Immunother 61:1155–1167

Elkabets M, Ribeiro VS, Dinarello CA, Ostrand-Rosenberg S, Di Santo JP, Apte RN et al (2010) IL-1beta regulates a novel myeloid-derived suppressor cell subset that impairs NK cell development and function. Eur J Immunol 40:3347–3357

Emmerich J, Mumm JB, Chan IH, Laface D, Truong H, McClanahan T et al (2012) IL-10 directly activates and expands tumor-resident CD8 + T cells without de novo infiltration from secondary lymphoid organs. Cancer Res 72:3570–3581

Ezernitchi AV, Vaknin I, Cohen-Daniel L, Levy O, Manaster E, Halabi A et al (2006) TCR zeta down-regulation under chronic inflammation is mediated by myeloid suppressor cells differentially distributed between various lymphatic organs. J Immunol 177:4763–4772

Fujita M, Kohanbash G, Fellows-Mayle W, Hamilton RL, Komohara Y, Decker SA et al (2011) COX-2 blockade suppresses gliomagenesis by inhibiting myeloid-derived suppressor cells. Cancer Res 71:2664–2674

Gabrilovich DI, Velders MP, Sotomayor EM, Kast WM (2001) Mechanism of immune dysfunction in cancer mediated by immature Gr-1 + myeloid cells. J Immunol 166:5398–5406

Gabrilovich DI, Bronte V, Chen SH, Colombo MP, Ochoa A, Ostrand-Rosenberg S, et al (2007) The terminology issue for myeloid-derived suppressor cells. Cancer Res 67(1):425 (author reply 6)

Gabrilovich DI, Ostrand-Rosenberg S, Bronte V (2012) Coordinated regulation of myeloid cells by tumours. Nat Rev Immunol 12:253–268

Garrity T, Pandit R, Wright MA, Benefield J, Keni S, Young MR (1997) Increased presence of CD34 + cells in the peripheral blood of head and neck cancer patients and their differentiation into dendritic cells. Int J Cancer 73:663–669

Hagemann T, Wilson J, Burke F, Kulbe H, Li NF, Pluddemann A et al (2006) Ovarian cancer cells polarize macrophages toward a tumor-associated phenotype. J Immunol 176:5023–5032

Hanahan D, Weinberg RA (2011) Hallmarks of cancer: the next generation. Cell 144:646–674

Hanson EM, Clements VK, Sinha P, Ilkovitch D, Ostrand-Rosenberg S (2009) Myeloid-derived suppressor cells down-regulate L-selectin expression on CD4 + and CD8 − T cells. J Immunol 183:937–944

Hoechst B, Ormandy LA, Ballmaier M, Lehner F, Kruger C, Manns MP et al (2008) A new population of myeloid-derived suppressor cells in hepatocellular carcinoma patients induces CD4(+)CD25(+)Foxp3(+) T cells. Gastroenterology 135:234–243

Hu CE, Gan J, Zhang RD, Cheng YR, Huang GJ (2011) Up-regulated myeloid-derived suppressor cell contributes to hepatocellular carcinoma development by impairing dendritic cell function. Scand J Gastroenterol 46:156–164

Huang B, Pan PY, Li Q, Sato AI, Levy DE, Bromberg J et al (2006) Gr-1 + CD115 + Immature Myeloid Suppressor Cells Mediate the Development of Tumor-Induced T Regulatory Cells and T-Cell Anergy in Tumor-Bearing Host. Cancer Res 66:1123–1131

Iwakura Y, Ishigame H (2006) The IL-23/IL-17 axis in inflammation. J Clin Invest 116:1218–1222

Ji Y, Zhang W (2010) Th17 cells: positive or negative role in tumor? Cancer Immunol Immunother 59:979–987

Josefowicz SZ, Lu LF, Rudensky AY (2012) Regulatory T cells: mechanisms of differentiation and function. Annu Rev Immunol 30:531–564

Kusmartsev S, Gabrilovich DI (2003) Inhibition of myeloid cell differentiation in cancer: the role of reactive oxygen species. J Leukoc Biol 74:186–196

Kusmartsev S, Nefedova Y, Yoder D, Gabrilovich DI (2004) Antigen-specific inhibition of CD8 + T cell response by immature myeloid cells in cancer is mediated by reactive oxygen species. J Immunol 172:989–999

Kusmartsev S, Su Z, Heiser A, Dannull J, Eruslanov E, Kubler H et al (2008) Reversal of myeloid cell-mediated immunosuppression in patients with metastatic renal cell carcinoma. Clin Cancer Res 14:8270–8278

Li Q, Pan PY, Gu P, Xu D, Chen SH (2004) Role of immature myeloid Gr-1 + cells in the development of antitumor immunity. Cancer Res 64:1130–1139

Lin EY, Li JF, Gnatovskiy L, Deng Y, Zhu L, Grzesik DA et al (2006) Macrophages regulate the angiogenic switch in a mouse model of breast cancer. Cancer Res 66:11238–11246

Lu T, Ramakrishnan R, Altiok S, Youn JI, Cheng P, Celis E et al (2011) Tumor-infiltrating myeloid cells induce tumor cell resistance to cytotoxic T cells in mice. J Clin Invest 121:4015–4029

Mantovani A, Sica A (2010) Macrophages, innate immunity and cancer: balance, tolerance, and diversity. Curr Opin Immunol 22:231–237

Mantovani A, Sozzani S, Locati M, Allavena P, Sica A (2002) Macrophage polarization: tumor-associated macrophages as a paradigm for polarized M2 mononuclear phagocytes. Trends Immunol 23:549–555

Marigo I, Bosio E, Solito S, Mesa C, Fernandez A, Dolcetti L et al (2010) Tumor-induced tolerance and immune suppression depend on the C/EBPbeta transcription factor. Immunity 32:790–802

Matsushita H, Vesely MD, Koboldt DC, Rickert CG, Uppaluri R, Magrini VJ et al (2012) Cancer exome analysis reveals a T-cell-dependent mechanism of cancer immunoediting. Nature 482:400–404

Melani C, Chiodoni C, Forni G, Colombo MP (2003) Myeloid cell expansion elicited by the progression of spontaneous mammary carcinomas in c-erbB-2 transgenic BALB/c mice suppresses immune reactivity. Blood 102:2138–2145

Mirza N, Fishman M, Fricke I, Dunn M, Neuger AM, Frost TJ et al (2006) All-trans-retinoic acid improves differentiation of myeloid cells and immune response in cancer patients. Cancer Res 66:9299–9307

Montero AJ, Diaz-Montero CM, Kyriakopoulos CE, Bronte V, Mandruzzato S (2012) Myeloid-derived suppressor cells in cancer patients: a clinical perspective. J Immunother 35:107–115

Movahedi K, Guilliams M, Van den Bossche J, Van den Bergh R, Gysemans C, Beschin A et al (2008) Identification of discrete tumor-induced myeloid-derived suppressor cell subpopulations with distinct T cell-suppressive activity. Blood 111:4233–4244
Movahedi K, Laoui D, Gysemans C, Baeten M, Stange G, Van den Bossche J et al (2010) Different tumor microenvironments contain functionally distinct subsets of macrophages derived from Ly6C (high) monocytes. Cancer Res 70:5728–5739
Mumm JB, Emmerich J, Zhang X, Chan I, Wu L, Mauze S et al (2011) IL-10 elicits IFNgamma-dependent tumor immune surveillance. Cancer Cell 20:781–796
Murai M, Turovskaya O, Kim G, Madan R, Karp CL, Cheroutre H et al (2009) Interleukin 10 acts on regulatory T cells to maintain expression of the transcription factor Foxp3 and suppressive function in mice with colitis. Nat Immunol 10:1178–1184
Nagaraj S, Gupta K, Pisarev V, Kinarsky L, Sherman S, Kang L et al (2007) Altered recognition of antigen is a mechanism of CD8 + T cell tolerance in cancer. Nat Med 13:828–835
Neurath MF, Finotto S (2011) IL-6 signaling in autoimmunity, chronic inflammation and inflammation-associated cancer. Cytokine Growth Factor Rev 22:83–89
Novitskiy SV, Ryzhov S, Zaynagetdinov R, Goldstein AE, Huang Y, Tikhomirov OY et al (2008) Adenosine receptors in regulation of dendritic cell differentiation and function. Blood 112:1822–1831
Okada S, Strober S (1982a) Spleen cells from adult mice given total lymphoid irradiation (TLI) or from newborn mice have similar regulatory effects in the mixed leukocyte reaction (MLR). II. Generation of antigen-specific suppressor cells in the MLR after the addition of spleen cells from newborn mice. J Immunol 129:1892–1897
Okada S, Strober S (1982b) Spleen cells from adult mice given total lymphoid irradiation or from newborn mice have similar regulatory effects in the mixed leukocyte reaction. I. Generation of antigen-specific suppressor cells in the mixed leukocyte reaction after the addition of spleen cells from adult mice given total lymphoid irradiation. J Exp Med 156:522–538
Ostrand-Rosenberg S, Sinha P (2009) Myeloid-derived suppressor cells: linking inflammation and cancer. J Immunol 182:4499–4506
Ostrand-Rosenberg S, Sinha P, Beury DW, Clements VK (2012) Cross-talk between myeloid-derived suppressor cells (MDSC), macrophages, and dendritic cells enhances tumor-induced immune suppression. Sem Cancer Biol
O'Sullivan T, Saddawi-Konefka R, Vermi W, Koebel C, Arthur C, White JM et al (2012) Cancer immunoediting by the innate immune system in the absence of adaptive immunity. J Exp Med 209(10):1869–1882
Qian BZ, Pollard JW (2010) Macrophage diversity enhances tumor progression and metastasis. Cell 141:39–51
Ralainirina N, Poli A, Michel T, Poos L, Andres E, Hentges F et al (2007) Control of NK cell functions by CD4 + CD25 + regulatory T cells. J Leukoc Biol 81:144–153
Rodriguez PC, Ochoa AC (2008) Arginine regulation by myeloid derived suppressor cells and tolerance in cancer: mechanisms and therapeutic perspectives. Immunol Rev 222:180–191
Rodriguez PC, Hernandez CP, Quiceno D, Dubinett SM, Zabaleta J, Ochoa JB et al (2005) Arginase I in myeloid suppressor cells is induced by COX-2 in lung carcinoma. J Exp Med 202:931–939
Rodriguez PC, Quiceno DG, Ochoa AC (2007) L-arginine availability regulates T-lymphocyte cell-cycle progression. Blood 109:1568–1573
Sinha P, Clements VK, Miller S, Ostrand-Rosenberg S (2005a) Tumor immunity: a balancing act between T cell activation, macrophage activation and tumor-induced immune suppression. Cancer Immunol Immunother 54:1137–1142
Sinha P, Clements VK, Ostrand-Rosenberg S (2005b) Reduction of myeloid-derived suppressor cells and induction of M1 macrophages facilitate the rejection of established metastatic disease. J Immunol 174:636–645
Sinha P, Clements VK, Fulton AM, Ostrand-Rosenberg S (2007a) Prostaglandin E2 promotes tumor progression by inducing myeloid-derived suppressor cells. Cancer Res 67:4507–4513

Sinha P, Clements VK, Bunt SK, Albelda SM, Ostrand-Rosenberg S (2007b) Cross-talk between myeloid-derived suppressor cells and macrophages subverts tumor immunity toward a type 2 response. J Immunol 179:977–983

Sinha P, Okoro C, Foell D, Freeze HH, Ostrand-Rosenberg S, Srikrishna G (2008) Proinflammatory S100 proteins regulate the accumulation of myeloid-derived suppressor cells. J Immunol 181:4666–4675

Sinha P, Chornoguz O, Clements VK, Artemenko KA, Zubarev RA, Ostrand-Rosenberg S (2011) Myeloid-derived suppressor cells express the death receptor Fas and apoptose in response to T cell-expressed FasL. Blood 117:5381–5390

Smith C, Chang MY, Parker KH, Beury DW, DuHadaway JB, Flick HE et al (2012) IDO is a nodal pathogenic driver of lung cancer and metastasis development. Cancer Discov 2:722–735

Solito S, Falisi E, Diaz-Montero CM, Doni A, Pinton L, Rosato A et al (2011) A human promyelocytic-like population is responsible for the immune suppression mediated by myeloid-derived suppressor cells. Blood 118:2254–2265

Sonda N, Chioda M, Zilio S, Simonato F, Bronte V (2011) Transcription factors in myeloid-derived suppressor cell recruitment and function. Curr Opin Immunol 23:279–285

Song X, Krelin Y, Dvorkin T, Bjorkdahl O, Segal S, Dinarello CA et al (2005) CD11b +/Gr-1 + immature myeloid cells mediate suppression of T cells in mice bearing tumors of IL-1beta-secreting cells. J Immunol 175:8200–8208

Srivastava MK, Bosch JJ, Thompson JA, Ksander BR, Edelman MJ, Ostrand-Rosenberg S (2008) Lung cancer patients' CD4(+) T cells are activated in vitro by MHC II cell-based vaccines despite the presence of myeloid-derived suppressor cells. Cancer Immunol Immunother 57:1493–1504

Srivastava MK, Sinha P, Clements VK, Rodriguez P, Ostrand-Rosenberg S (2010) Myeloid-derived suppressor cells inhibit T-cell activation by depleting cystine and cysteine. Cancer Res 70:68–77

Steidl C, Lee T, Shah SP, Farinha P, Han G, Nayar T et al (2010) Tumor-associated macrophages and survival in classic Hodgkin's lymphoma. New Engl J Med 362:875–885

Strober S (1984) Natural suppressor (NS) cells, neonatal tolerance, and total lymphoid irradiation: exploring obscure relationships. Annu Rev Immunol 2:219–237

Talmadge JE, Hood KC, Zobel LC, Shafer LR, Coles M, Toth B (2007) Chemoprevention by cyclooxygenase-2 inhibition reduces immature myeloid suppressor cell expansion. Int Immunopharmacol 7:140–151

Terabe M, Berzofsky JA (2008) The role of NKT cells in tumor immunity. Adv Cancer Res 101:277–348

Terabe M, Swann J, Ambrosino E, Sinha P, Takaku S, Hayakawa Y et al (2005) A nonclassical non-Valpha14Jalpha18 CD1d-restricted (type II) NKT cell is sufficient for down-regulation of tumor immunosurveillance. J Exp Med 202:1627–1633

Tiemessen MM, Jagger AL, Evans HG, van Herwijnen MJ, John S, Taams LS (2007) CD4 + CD25 + Foxp3 + regulatory T cells induce alternative activation of human monocytes/macrophages. Proc Natl Acad Sci U S A 104:19446–19451

Torroella-Kouri M, Silvera R, Rodriguez D, Caso R, Shatry A, Opiela S et al (2009) Identification of a subpopulation of macrophages in mammary tumor-bearing mice that are neither M1 nor M2 and are less differentiated. Cancer Res 69:4800–4809

Trinchieri G (2012) Cancer and inflammation: an old intuition with rapidly evolving new concepts. Annu Rev Immunol 30:677–706

Veltman JD, Lambers ME, van Nimwegen M, Hendriks RW, Hoogsteden HC, Aerts JG et al (2010) COX-2 inhibition improves immunotherapy and is associated with decreased numbers of myeloid-derived suppressor cells in mesothelioma Celecoxib influences MDSC function. BMC Cancer 10(1):464

Wiers KM, Lathers DM, Wright MA, Young MR (2000) Vitamin D3 treatment to diminish the levels of immune suppressive CD34 − cells increases the effectiveness of adoptive immunotherapy. J Immunother 23:115–124

Wyckoff JB, Wang Y, Lin EY, Li JF, Goswami S, Stanley ER et al (2007) Direct visualization of macrophage-assisted tumor cell intravasation in mammary tumors. Cancer Res 67:2649–2656

Youn JI, Collazo M, Shalova IN, Biswas SK, Gabrilovich DI (2012) Characterization of the nature of granulocytic myeloid-derived suppressor cells in tumor-bearing mice. J Leukoc Biol 91:167–181

Young MR, Ihm J, Lozano Y, Wright MA, Prechel MM (1995) Treating tumor-bearing mice with vitamin D3 diminishes tumor-induced myelopoiesis and associated immunosuppression, and reduces tumor metastasis and recurrence. Cancer Immunol Immunother 41:37–45

Young MR, Wright MA, Lozano Y, Matthews JP, Benefield J, Prechel MM (1996) Mechanisms of immune suppression in patients with head and neck cancer: influence on the immune infiltrate of the cancer. Int J Cancer 67:333–338

Young MR, Wright MA, Lozano Y, Prechel MM, Benefield J, Leonetti JP et al (1997) Increased recurrence and metastasis in patients whose primary head and neck squamous cell carcinomas secreted granulocyte-macrophage colony-stimulating factor and contained CD34 + natural suppressor cells. Int J Cancer 74:69–74

Young MR, Kolesiak K, Wright MA, Gabrilovich DI (1999) Chemoattraction of femoral CD34 + progenitor cells by tumor-derived vascular endothelial cell growth factor. Clin Exp Metas 17:881–888

Young MR, Petruzzelli GJ, Kolesiak K, Achille N, Lathers DM, Gabrilovich DI (2001) Human squamous cell carcinomas of the head and neck chemoattract immune suppressive CD34(+) progenitor cells. Human Immunol 62:332–341

Chapter 8
Pleiotropic and Differential Functions of IL-1α and IL-1β Shape the Tumor Microenvironment and Affect the Outcome of Malignancies

Ron N. Apte and Elena Voronov

Abstract Interleukin-1 (IL-1) is a major "alarm" upstream pro-inflammatory cytokine that also affects immunity and hemopoiesis by inducing cytokine cascades. In the tumor arena, IL-1 is produced by malignant or microenvironmental cells. As a pleiotropic cytokine, IL-1 is involved in tumorigenesis and tumor invasiveness and also in the activation of anti-tumor immunity IL-1α and IL-1β are the major agonists of IL-1, while IL-1Ra is a physiological inhibitor of pre-formed IL-1. In their secreted form, IL-1α and IL-1β bind to the same receptors and induce the same biological functions. However, IL-1α and IL-1β differ in their compartmentalization within the producing cell or the microenvironment. IL-1β is only active in its secreted form and mediates inflammation, which promotes carcinogenesis, tumor invasiveness and immunosuppression. On the other hand, IL-1α is mainly cell-associated; in the context of tumors, host- and tumor cell-derived IL-1α stimulates anti-tumor immunity, rather than inflammation. Recent breakthroughs in inflammasome biology and IL-1β processing/secretion have spurred the development of novel anti-IL-1 agents which are being used in clinical trials in patients with diverse diseases with inflammatory manifestations. Better understanding of the integrative role of IL-1α and IL-1β in the malignant process will enable the application of novel IL-1 modulation approaches at the bedside, in cancer patients with minimal residual disease (MRD), as an adjunct to conventional approaches to reduce the tumor burden.

Keywords IL-1 · IL-1Ra · Carcinogenesis · Tumor invasiveness · Tumor-host interactions · Immunogenicity · Anti-tumor immunity · Immunotherapy

---

R. N. Apte (✉) · E. Voronov
The Shraga Segal Department of Microbiology, Immunology and Genetics, Faculty of Health Sciences, Ben-Gurion University of the Negev, 84105 Beer-Sheva, Israel
e-mail: rapte@bgu.ac.il

M. R. Shurin et al. (eds.), *The Tumor Immunoenvironment*,
DOI: 10.1007/978-94-007-6217-6_8,

8.1 The Interleukin-1 Family in the Context of Tumors

The Interleukin-1 (IL-1) family consists of a few members that have differential and sometimes conflicting effects in the tumor microenvironment. IL-1 molecules are involved in carcinogenesis and malignant transformation, tumor growth, invasion and metastasis, but some of them can activate innate and specific immune effector mechanisms that limit the growth of tumors [reviewed in (Apte et al. 2006a, b; Apte and Voronov 2002, 2008)]. In the tumor arena, IL-1 is produced by host cells, such as stromal cells, in response to factors secreted by the malignant cells or by infiltrating leukocytes, which frequently accompany tumor growth, as part of the local inflammatory response. IL-1 can also be produced by malignant cells. At tumor sites, there is a local cytokine network that is determined by the array of expressed cytokines, their relative concentrations and the expression pattern of their receptors. This cytokine network dictates the dominant "net cytokine effect" and it fluctuates at various phases of tumor development. This review will discuss the plethora of effects of IL-1 on malignant processes, including carcinogenesis, tumor invasiveness and angiogenesis, as well as tumor-host interactions.

8.1.1 The biology of IL-1α and IL-1β

IL-1 represents a family of agonists, antagonists and receptors [reviewed in (Apte and Voronov 2008; Auron 1998; Dinarello 1996, 2009; Stylianou and Saklatvala 1998; Arend et al. 2008; Garlanda et al. 2009; O'Neill 2008; Gabay et al. 2010; Martin and Wesche 2002; Sims and Smith 2010)]. Here, we will mainly focus on the two major IL-1 agonistic molecules i.e., IL-1α and IL-1β, and IL-1 receptor antagonist (IL-Ra), which is a physiological inhibitor of IL-1 signaling. IL-1α and IL-1β are synthesized as precursors of 31 kD that are further processed by proteases to their mature secreted 17 kD forms. IL-1 differs from most other cytokines by lack of a signal sequence, thus not passing through the endoplasmic reticulum-Golgi pathway; its mechanisms of secretion are not yet completely understood (Eisenbarth and Flavell 2009; Franchi et al. 2009; Martinon et al. 2009). IL-1Ra has a signal peptide that is secreted in the ER-Golgi exocytic pathway. IL-1 is produced and secreted by various types of cells upon inflammatory or stress conditions, predominantly by myeloid cells, which display the strongest capacity to produce and secrete IL-1. Stimulation of IL-1 production occurs through signaling of germ-line encoded Toll-like receptors (TLRs), which recognize conserved microbial molecules, termed pathogen-associated molecular patterns (PAMP), PAMPs [reviewed in (Iwasaki and Medzhitov 2010; Palm and Medzhitov 2009; Takeuchi and Akira 2010; Franchi et al. 2008; Williams et al. 2010)] as well as endogenous molecules, which are released from stressed or damaged cells and are collectively termed danger-associated molecular patterns (DAMPs) or danger signals (Bianchi 2007;

Srikrishna and Freeze 2009; Carta et al. 2009; Rock and Kono 2008). The activation of TLR signaling via the NF-κB pathway leads to generation of IL-1. Signaling through surface IL-1RI and most of the TLRs is common and converges from MyD88 to NF-κB activation and induction of an inflammatory response, including expression and secretion of IL-1β.

8.1.2 Processing of IL-1β

IL-1β-converting enzyme (ICE), or caspase-1, is a cysteine protease that cleaves the inactive precursors of IL-1β, IL-18 and IL-33 (Dinarello 2009; Eisenbarth and Flavell 2009; Franchi et al. 2009; Hoffman and Wanderer 2010; Martinon 2010; Schroder et al. 2010; Latz 2010; Schroder and Tschopp 2010). Proteolytic activation of pro-caspase-1 occurs in the cytosol on the inflammasome platform, which consists of Nod-like receptors (NLRs), the cytosolic equivalents of TLRs, adaptive proteins and pro-caspase 1 (Keller et al. 2008).

8.1.3 Processing of IL-1α

The processing of IL-1α has not been studied as extensively as that of IL-1β. The precursor of IL-1α (proIL-1α) is processed by the Ca^{2+}—dependent protease calpain into the mature 17 kD form and the 16 kD N-terminal cleavage product-the propiece of IL-1α, also termed IL-1α N-terminal peptide (IL-1NTP). The latent form of calpain is activated in cells under inflammatory conditions. However, many cells contain calpain inhibitors and are thus unable to process and secrete IL-1α and they contain only intracellular proIL-1α.

A biologically active membrane form of IL-1α (23 kD), which is anchored to the membrane via a mannose-like receptor has been demonstrated in activated cells that express and produce the cytokine. Although the biological activity of membrane-associated IL-1α on activated cells can be easily demonstrated using cell fixation techniques, the mechanisms to process pro IL-1α into the membrane form have not yet been elucidated. Using oncogene-transformed fibroblasts that express IL-1α and cells transfected with proIL-1α, co-expression of these forms of IL-1α has been detected (Douvdevani et al. 1991, 1992; Dvorkin et al. 2006; Song et al. 2003, 2005).

8.1.4 IL-1 Receptors

IL-1Rs, which belong to the immunoglobulin (Ig) supergene family, are abundantly expressed on many cell types. IL-1R of type I (IL-1RI) (80 kD) is a signaling receptor, whereas the IL-1R of type II (IL-1RII) (68 kD) serves as a decoy target, acting to reduce excessive amounts of IL-1 [reviewed in (Apte and Voronov 2008; Auron 1998; Dinarello 1996, 2009; Stylianou and Saklatvala 1998; Arend et al. 2008; Garlanda et al. 2009; O'Neill 2008; Gabay et al. 2010; Martin and Wesche 2002; Sims and Smith 2010)]. Following the binding of IL-1 to IL-1RI, a second chain, i.e., the IL-1R acceptor protein (IL-1RAcP) is recruited. This heterodimeric complex triggers IL-1 signal transduction, which is initiated by activation of the IL-1 receptor-associated kinase (IRAK) and ultimately leads to activation of nuclear genes, mainly through NF-κB. On the contrary, IL-1RII and the IL-1Ra fail to form this heterodimeric complex with the IL-1RAcP and to recruit IRAK. Signaling through surface IL-1RI represents an evolutionary conserved mechanism homologous to the Drosophila Toll pathway.

8.1.5 Similarities and Differences of IL-1α and IL-1β

In their mature secreted form, IL-1α and IL-1β bind to the same receptors and exert the same biological activities. However, biological activities of IL-1α and IL-1β differ dramatically [reviewed in (Apte et al. 2006; Apte and Voronov 2008; Auron 1998; Dinarello 1996, 2009; Stylianou and Saklatvala 1998; Sims and Smith 2010; Arend 2002; Mantovani et al. 1998; Martin and Falk 1997)]. IL-1β is not present in homeostatic conditions; it is induced and secreted only upon inflammatory signals and its secretion is tightly controlled at the levels of transcription, mRNA stability, translation and processing. On the other hand, IL-1α is present in the cytosol, nucleus or cell membrane in homeostatic states, as well as in inflammation, when its expression is upregulated. Importantly, IL-1α is only rarely secreted by living cells and in most cases is undetectable in body fluids. Previously, we demonstrated that in vivo, in steady-state homeostasis and in inflammation, IL-1α and IL-1β are differentially expressed in tissues, possibly pointing to their different physiological roles (Hacham et al. 1996, 2002).

IL-1α and IL-1β also differ dramatically in the sub-cellular compartments in which they are active. IL-1β is solely active as a secreted product, while its precursor is inactive and there is no membrane-associated form of IL-1β. On the other hand, IL-1α is mainly active in its cell-associated forms (proIL-1α, IL-1NTP and membrane-associated form), but is only marginally active in its mature from, due to limited processing in cells, with the exception of activated myeloid cells [reviewed in (Apte et al. 2006a, b; Apte and Voronov 2002, 2008; Dinarello 1996, 2009)]. As indicated, proIL-1α binds to IL-1R1 and is biologically active, similar to the secreted mature form of IL-1α, while IL-1NTP does not bind to IL-1R1 and very little is known about its biological significance. Intracellular forms of IL-1α were shown to translocate to the nucleus, due to a nuclear localization sequence (NLS) located within the structure of proIL-1α and IL-1NTP, but not present in the mature form of IL-1α. In cells that express proIL-1α, but do not secrete it, the cytokine possibly acts in an intracrine manner from within the cell, without the need to be secreted, via signaling pathways that are not yet fully characterized.

We have hypothesized that intracellular forms of IL-1α evolved as intracellular effector molecules undertaking important homeostatic regulatory functions beyond the realm of immunity and inflammation. These include effects on gene expression, transcription control, cell growth and differentiation, which were demonstrated in tissue-resident cells, such as endothelial cells, fibroblasts, smooth muscle cells, keratinocytes, epithelial cells and brown fat cells [reviewed in (Apte et al. 2006a, b; Apte and Voronov 2002, 2008; Dinarello 1996, 2009)]. IL-1α belongs to a group of "dual function" cytokines (i.e., HMBG1 and IL-33) that are expressed in the nucleus and cytosol, where they perform homeostatic functions, but upon cell necrosis, they are released into the microenvironment and serve as pro-inflammatory cytokines [reviewed in (Lotze and Tracey 2005; Raucci et al. 2007; Ulloa and Messmer 2006)].

We have hypothesized that the localization of the IL-1 molecules in the context of the producing cell and its microenvironment dictates their biological function in normal homeostasis and also in the malignant process [reviewed in (Apte et al. 2006a, b; Apte and Voronov 2002, 2008)]. Thus, as will be shown below, membrane-associated IL-1α is immunostimulatory, while cytosolic proIL-1α controls intracrine homeostatic functions, such as gene expression and control of proliferation and differentiation (Werman et al. 2004; Cohen et al. 2010). However, when cytosolic proIL-1α is released from damaged cells, it acts as an alarmin to initiate inflammation. The exact role of intracellular forms of IL-1α expressed in malignant cells has not yet been elucidated.

Secreted IL-1 (mainly IL-1β), at low local doses, induces limited inflammatory responses followed by activation of specific immune mechanisms, while at high doses, broad inflammation accompanied by tissue-damage and tumor invasiveness are evident.

In spite of these differences between IL-1α and IL-1β, both molecules have been considered identical and their patterns of expression and biological activities have not been extensively studied in parallel in health and disease.

8.2 Biological Activities of IL-1 Related to Malignant Processes

8.2.1 Effects of IL-1 on the Inflammatory Responses

IL-1α and IL-1β are defined as "alarm cytokines" that are secreted by macrophages and initiate inflammatory responses [reviewed in (Apte et al. 2006b; Apte and Voronov 2008; Auron 1998; Dinarello 1996, 2009; Stylianou and Saklatvala 1998; Sims and Smith 2010; Arend 2002; Mantovani et al. 1998; Martin and Falk 1997)]. IL-1β is easily detected in body fluids during inflammation and has been considered as the major IL-1 pro-inflammatory agonistic molecule. Secreted IL-1 induces a cascade of other pro-inflammatory genes during inflammation. Of major importance are cyclooxygenase type 2 (COX-2), inducible nitric oxide synthase (iNOS), chemokines/cytokines and matrix metalloproteinases (MMPs). The IL-1 molecules stimulate their own and each other's production; this represents an important amplification loop of the inflammatory response. Also, IL-1 increases the expression of high-affinity adhesion molecules (integrins and adhesion molecules of the Ig supergene family) on endothelial cells, stromal cells and leukocytes and by this promotes infiltration of inflammatory cells from the blood into inflammed tissues.

Recently, the alarmin function of intracellular IL-1α, which is released into the microenvironment upon cell death in sterile inflammation, has been described by us and others. In tissue cells, such as epithelial cells, endothelial cells and fibroblasts, IL-1α is located in the cytosol and nucleus. Upon stress induction, most of cytosolic IL-1α translocates into the nucleus, where it is bound to the chromatin in a highly dynamic manner. In cells undergoing necrotic cell death, i.e., in hypoxic conditions, proIL-1α is released and induces inflammation, similar to other nuclear-residing cytokines, such as HMBG1 and IL-33 (Raucci et al. 2007; Hreggvidsdottir et al. 2009; Gadina and Jefferies 2007; Bianchi and Manfredi 2007). However, following apoptotic death, the mobility of IL-1α is greatly reduced and it concentrates in dense nuclear foci and is not released into the environment. This may represent a novel mechanism that explains why inflammatory responses are not generated upon apoptosis. Necrotic cells lacking IL-1α failed to induce this early inflammatory response. We found that the early infiltrate in Matrigel plugs containing lysates of dead cells consisted mainly of neutrophils and myeloid progenitor cells. The recruitment of these cells occurs via IL-1R1 signaling (Cohen et al. 2010). We found that IL-1NTP that contains the NLS but cannot bind to surface IL-1RI cannot induce this effect. We have further shown that proIL-1α released from dying cells is essential to initiate sterile inflammation, characterized by infiltration of neutrophis. Macrophages infiltrate the affected site later and secrete IL-1β, which propagates and possibly terminates the inflammatory response. Thus, in sterile inflammation, IL-1α and IL-1β are produced by different cells and acts in a sequential manner (Rider et al. 2011).

Altogether, IL-1 is a major molecule in the recruitment of leukocytes to the site of tissue-damage, including tumor sites. However, further mechanistic studies will be needed to characterize the differential role of IL-1α and IL-1β in malignant versus stromal or infiltrating tumor cells, in order to understand how to inhibit IL-1-mediated inflammatory responses at tumor sites.

8.2.2 Effects of IL-1 on Immune Responses

As a pleiotropic cytokine, IL-1 has diverse potentiating effects on the proliferation, differentiation and function of various innate (NK cells, macrophages, granulocytes etc.), as well as specific immunocompetent cells (T and B cells) [reviewed in (Apte et al. 2006a, b; Apte and Voronov 2002, 2008; Dinarello 1996, 2009; Sims and Smith 2010)]. IL-1 is a major amplifier of adaptive immunity, affecting mainly

T cell responses; however, its effect on B cell proliferation/differentiation has also been described. IL-1 is instrumental for the activation of the different subsets of CD4+ T cells and possibly also CD8+ T cells by diverse mechanisms operating on the level of T cells, APCs or by modulating the cytokine milieu in the microenvironment. Recent elegant studies by Paul (Ben-Sasson et al. 2009) have demonstrated the role of members of the IL-1 family in T cell activation. It was shown that IL-1β induces a robust and durable expansion of naïve and memory CD4 T cells (Th1, Th2 and Th17) in response to antigen stimulation, which is indispensable by other cytokines. The responding T cells have to bear IL-1RI; WT TcR transgenic T cells could be successfully primed in IL-1RI$^{-/-}$ recipients. The IL-1Ra reduced T cell responses to antigen. Furthermore, the mechanisms by which IL-1 and other members of its family reinforce T cell polarization were defined (Guo et al. 2009). Thus, distinct members of the IL-1 family, together with other cytokines can activate particular STATs. The correct combination of STAT and an IL-1R signal leads to expression of the relevant subset-specific transcription factors and reinforces the polarized phenotype; IL-33 and STAT5 induce Th2, IL-1β and STAT3 induce Th17 and IL-18 and STAT4 induce Th1 cells (Guo et al. 2009).

The effects of IL-1 on T regulatory cells (Treg), which mediate suppression and tolerance, are not clear, although Treg can express IL-1R (Carroll et al. 2008). We recently observed a correlation between Tregs and IL-1β expression in the tumor microenvironment (Carmi et al. 2010). Recent studies have characterized IL-1β as an "endogenous adjuvant" that is generated following immunization with adjuvants, such as CFA and aluminum hydroxide (Alum) (Eisenbarth et al. 2008; Li et al. 2007; Franchi and Nunez 2008). The adjuvanticity of IL-1 in T cell activation possibly stems from its ability to serve as a danger signal, recruiting inflammatory cells to the site of antigen application and inducing maturation and activation of professional APCs. The control of anti-tumor immune responses by IL-1 manipulation is a challenge for immunotherapy.

8.2.3 Effects of IL-1 on Hemopoiesis

Multiple hemopoietic functions have been attributed to IL-1, especially to IL-1β [reviewed in (Apte et al. 2006a, b; Apte and Voronov 2002, 2008; Dinarello 1996, 2009)]. The in vivo importance of IL-1 in stimulating hemopoiesis is best demonstrated by its ability to rescue mice after lethal irradiation or chemotherapy, mainly through induction of recovery of the myeloid compartment (Neta et al. 1986). IL-1 was characterized as hemopoietin-1, a factor that is essential for hemopoiesis by inducing the expression of receptors for colony stimulating factors (CSFs) on primitive precursor cells (Mochizuki et al. 1987).

Of special relevance to the malignant process are immature CD11b+Gr-1+ myeloid cells, also termed myeloid-derived suppressor cells (MDSCs), which consist of cells committed to differentiate into granulocytes, macrophages, myeloid BM-derived dendritic cells as well as other early myeloid precursors [reviewed in (Ostrand-Rosenberg 2010; Gabrilovich and Nagaraj 2009; Ostrand-Rosenberg and Sinha 2009; Bronte and Mocellin 2009)]. Cytokines secreted by the malignant cells, such as VEGF, M-CSF, IL-6, IL-10, TGFβ and IL-1β have been implicated in increased myelopoiesis, as well as immune suppression and increased invasiveness. MDSCs suppress T cell-mediated immunity by interfering with T cell signaling, as evidenced by the down-regulation of CD3/TCRζ, $p56^{lck}$ and $p59^{fyn}$ (Ostrand-Rosenberg 2010; Gabrilovich and Nagaraj 2009; Ostrand-Rosenberg and Sinha 2009; Bronte and Mocellin 2009; Baniyash 2006). MDSCs contribute to tumor progression by secreting factors, such as MMP9 and VEGF, which promote tumor angiogenesis and invasiveness [reviewed in (Ostrand-Rosenberg 2010; Gabrilovich and Nagaraj 2009; Ostrand-Rosenberg and Sinha 2009; Bronte and Mocellin 2009)]. In view of the increased suppressive potential of MDSCs from tumor-bearing mice compared to such cells from WT mice, the existence of MDSC subsets has been suggested (Youn and Gabrilovich 2010). We have recently characterized a new subset of $Ly6C^-$ MDSCs in tumor-bearing mice, which over-express and secrete IL-1β; such MDSCs especially suppress NK cell development and function (Youn and Gabrilovich 2010; Peranzoni et al. 2010; Elkabets et al. 2010).

The effects of IL-1 agonistic molecules on inflammation, immunity and hemopoiesis, described herein, indicate that IL-1 plays a complex role in the malignant process; its elucidation will enable the development of anti-tumor approaches based on modulating molecules of the IL-1 family.

8.3 Overexpression of IL-1α and IL-1β in Malignant Cells Alters Their Invasive Phenotype

In order to compare the effects of IL-1α and IL-1β on tumor invasiveness versus induction of anti-tumor immunity, we used violent fibrosarcoma cells that had been transfected with active forms of IL-1α and IL-1β, i.e., the precursor of IL-1α and the mature form of IL-1β.

8.3.1 Anti-tumor Effects of Cell-Associated IL-1α

We demonstrated the anti-tumor effects of IL-1α expression by malignant cells in different experimental systems (Douvdevani et al. 1992; Dvorkin et al. 2006; Song et al. 2003; Zoller et al. 1992a, b; Voronov et al. 1999; Marhaba et al. 2008; Nazarenko et al. 2008). Some oncogene-transformed fibroblasts constitutively express IL-1α due to alterations in the control of IL-1α expression, which may occur as a result of the transformation process. Fibrosarcoma cells, transfected with cDNA of the proIL-1α, express IL-1α in the cytosol or on the cell membrane, but do not secrete it, and lose their tumorigenicity. They either do not develop tumors in intact mice or tumors start to grow and subsequently regress. Regression of tumors involves their early infiltration by mononuclear cells, mainly CD8+ T cells, NK cells and macrophages, and only a few CD4+ T cells; the tumor's mass is ultimately replaced by fibrotic scar tissue. Regression of IL-1α-positive fibrosarcomas involves the development of a long-term specific immune memory that protects mice against a challenge with violent parental cells. We observed that tumor regression in this model was CD8+ T cell dependent (Song et al. 2003). Tumor cell-associated IL-1α may directly activate CD8+ cytotoxic T lymphocyte precursors (pCTLs) in the absence of CD4+ T cells. Membrane-associated IL-1α, which ligates to IL-1Rs on pCTLs, can possibly serve as a second co-stimulatory signal for direct activation of CD8+ pCTLs. Indeed, it was demonstrated that target cells, which present tumor peptides in the context of MHC class I molecules and concomitantly express co-stimulatory molecules (i.e., B7), can efficiently stimulate CD8+ pCTLs to generate cytokines that will autocrinically stimulate their own proliferation/differentiation (Chambers and Allison 1999). Also, in T cells, ligation of IL-1 to its receptors, similar to B7-CD28 ligation, activates NF-κB, which leads to enhanced IL-2 transcription and secretion (Kalli et al. 1998). In the intact host, cytokines secreted by CD4+ Th cells possibly also contribute to the induction of anti-tumor immunity, as the most effective anti-tumor immunity is induced when both CD4+ and CD8+ Th cells are activated. Tumor cell-associated IL-1α also potentiates antigen presentation by the tumor cells themselves, possibly through IFNγ-induced MHC class II expression, and also via cross-presentation by professional APCs. Non-adaptive effector cells, such as NK cells and activated macrophages, also play a role in the eradication of IL-1α-positive fibrosarcomas.

Tumor cell-associated IL-1α was shown to be effective when used to intervene in the growth of the parental tumor cells and induced anti-tumor immunity and regression of violent fibrosarcomas when applied at a critical "therapeutic window"—5–10 days (single application of Mitomycin-C-treated tumor cells) after inoculation of the malignant cells (Dvorkin et al. 2006).

The "natural" membrane-associated form of IL-1α is important for exerting anti-tumor effects, as it may serve as an adhesion-molecule, allowing efficient cell-to-cell interactions between malignant and immune effector cells that bear IL-1Rs, and is also effective as a focused adjuvant that efficiently acts at low levels of expression, below those which are toxic to the host. Other studies have also emphasized the effectiveness of membrane-associated cytokines expressed on engineered tumor cells (i.e., IFNγ, GM-CSF, M-CSF, TNFα and IL-12) (Thompson et al. 1996; Marr et al. 1997; El-Shami et al. 1999 Hoo et al. 1999).

As there are technical and ethical constraints about using transduction methods for human malignant cells, we decided to express IL-1α in tumor cells in a transient manner. In cells of a malignant T lymphoma, transient IL-1α activity expressed in the cytosol and on the surface membrane was induced by treating the cells with accessory cells/mitogens, similar to activation protocols in primary cells (Voronov et al. 1999). IL-1α expression persists for a few days and was shown to be sufficient to reduce the malignancy of invasive T lymphoma cells, which metastasize to the spine. The efficiency of tumor cell vaccines, transiently expressing IL-1α, was also demonstrated by their ability to intervene in the growth of violent lymphoma cells. Similar findings were observed by us using fibrosarcoma cell lines induced to provide transient expression of IL-1α. Transient expression of cytokines by isolated tumor cells is quite easy to induce and it may be potentially applicable in immunotherapy.

8.3.2 IL-1β Secreted by Malignant Cells Increases Their Invasive Potential and Induces Tumor-Mediated Suppression

To assess effects of tumor cell-associated IL-1β on tumorigenicity patterns, we transfected violent fibrosarcoma cells with constructs bearing the cDNAs of the mature form of IL-1β or the mature form of IL-1β ligated to a signal sequence (ssIL-1β), to induce potent secretion of IL-1β through the endoplasmic reticulum-Golgi pathway (Song et al. 2003, 2005). We found that IL-1β and ssIL-1β transfected fibrosarcoma tumors were more invasive than the violent parental cells or mock-transfected cells. The invasiveness of the malignant cells approximately correlated with the amount of IL-1β that was secreted by them. In addition, only the ssIL-1β transfectants, which secrete relatively large levels of the cytokine, exhibited a metastatic potential, as manifested by the development of experimental metastases in the lungs after *i.v.* inoculation. Increased angiogenesis patterns, as evidenced by high vessel density in tumors and increased secretion of VEGF by the malignant cells, were observed in tumors secreting IL-1β. Similar observations were described in other experimental systems using IL-1β-transfected tumor cells (Nakao et al. 2005; Bunt et al. 2006; Saijo et al. 2002). No anti-tumor effector cells or cytokines that potentiate anti-tumor immunity (i.e., IFNγ and IL-2) could be detected in spleens of mice injected with IL-1β or ssIL-1β transfectants or in spleens from mice injected with the violent parental cells or mock-transfected cells. In contrast, effective anti-tumor cell immune responses were observed in mice injected with fibrosarcoma cells transfected with proIL-1α, as indicated above.

Further studies have shown that in mice bearing tumors of IL-1β secreting cells, general anergy mediated by MDSCs develops (Elkabets et al. 2010; Bunt et al. 2006). In tumor-bearing mice, a systemic inflammatory response induced by tumor cell-derived IL-1β, characterized by leukocytosis, cachexia, liver necrosis and interstitial pneumonia, was observed. Resection of large tumors of IL-1β secreting cells completely restored immune reactivity and reversed hematological alterations within 7–10 days. Treatment of tumor-bearing mice with the IL-1Ra reduced tumor growth and attenuated the hematological alterations.

In spite of tumor-mediated suppression, resection of large tumors of IL-1β secreting cells, followed by a challenge with the violent parental cells, induce resistance to the tumor; protection was not observed in mice bearing tumors of mock-transfected fibrosarcoma cells. Thus, in mice bearing tumors of IL-1β secreting cells, anti-tumor cell specific immunity is activated, due to the adjuvant-like effects of IL-1β; however, protective immunity is not manifested, due to suppression of immune effector mechanisms. It is notable that when tumor cells expressing membrane-associated IL-1α are injected into mice, anti-tumor immune responses occur without concomitant tumor-mediated suppression and thus the malignant cells are rejected.

8.4 Effects of Host-Derived IL-1 on Chemical Carcinogenesis

We have demonstrated the role of host-derived IL-1 molecules on susceptibility to chemical carcinogenesis. We studied the role of the IL-1 molecules in chemical carcinogenesis induced by 3-methylcholantrene (3-MCA), which acts both as an initiator and a tumor promoter, using IL-1 KO mice, i.e., IL-1$\alpha^{-/-}$, IL-1$\beta^{-/-}$, IL-1$\alpha/\beta^{-/-}$(double KO mice) or IL-1Ra$^{-/-}$ mice in comparison to wild-type (WT) mice. We found that deficiency of IL-1β leads to delayed 3-MCA-induced fibrosarcoma development. In WT and IL-1$\alpha^{-/-}$ mice, carcinogenesis patterns were similar and all mice developed tumors. In mice deficient in IL-1β, i.e., in IL-1$\beta^{-/-}$ and IL-1$\alpha/\beta^{-/-}$ (double KO) mice, tumors appeared only after a prolonged lag period (about 110 days) and developed only in part of the treated mice. In IL-1Ra$^{-/-}$ mice, in which unattenuated levels of the IL-1 molecules exist, tumor development was more rapid than in WT mice. Our results indicated for the first time that 3-MCA-induced carcinogenesis is inflammation-dependent, as previously it had been suggested that tumor development is controlled by immune surveillance mechanisms that eliminate the arising malignant cells (Smyth et al. 2006). The tissue reaction at the site of carcinogen injection was further studied at early (10 days) and late time intervals after injection of 3-MCA dissolved in olive oil. Droplets of lipid containing the carcinogen, encapsulated by fibrotic tissue with very little infiltration of leukocytes, were observed in WT and in the various IL-1 KO mice on day 10. On the contrary, in IL-1Ra$^{-/-}$ mice, large multilayer granulomas surround the lipid droplets containing the carcinogen. These granulomas were very heavily infiltrated by neutrophils, with some macrophages that had the morphology of foam cells, and some of the granulomas turned into abscesses. Almost no lymphocytes were observed in these granulomas, indicating their innate nature. Immunohistochemical stainings revealed that the granulomas consist of cells expressing pro-inflammatory molecules, such as IL-1β, COX-2 and to a lesser extent IL-1α and TNFα. The early local inflammatory response in IL-1Ra$^{-/-}$ mice consisted mainly of neutrophils and could be neutralized by treatment with the IL-1Ra, but not with the TNF binding protein, indicating that the early inflammatory response is induced and mediated by IL-1 rather than by IL-1-induced TNFα. On approximately day 70, microscopic tumors were detected in IL-1Ra$^{-/-}$ mice and macrophages predominated in the infiltrate, which correlates with their role in tumor promotion and invasiveness. The inflammatory response around encapsulated lipid droplets containing the carcinogen appeared later in WT BALB/c and IL-1$\alpha^{-/-}$ mice (around day 50), but was not apparent in IL-1β deficient mice (i.e., IL-1$\beta^{-/-}$ and IL-1α/IL-1$\beta^{-/-}$ mice). In mice lacking IL-1β, where the inflammatory response at the site of carcinogen injection was not evident, the incidence of tumor development was lower, but some of the mice did develop tumors. This may suggest that tumor development can also occur in the absence of an evident inflammatory response at the site of tumorigenesis, but if, in addition, the carcinogen evokes a potent local inflammatory response, it enhances tumorigenesis, by shortening the time of tumor development, increasing tumor incidence and promoting the generation of more malignant tumor cell variants.

The role of host-derived IL-1β in promoting tumorigenesis has also been demonstrated in mice transgenic for human IL-1β, encoded by a construct expressing a signal peptide fused to the IL-1β gene (ssIL-1β), which is specifically expressed in the stomach. In these mice, spontaneous gastric inflammation, pre-neoplastic lesions and in some cases also tumors were observed in correlation to recruitment of MDSCs to the stomach and their in situ activation through the IL-1RI/NF-κB pathway (Tu et al. 2008). The IL-1Ra inhibits gastric pre-neoplasia and suppresses MDSC mobilization. Use of this ssIL-1β construct, driven by the elastase promoter, enabled the generation of mice that develop severe chronic pancreatitis. In such mice, the severity of lesions and local inflammation correlated to the extent of human IL-1β expression. Older mice displayed acinar-ductal metaplasia, but did not develop tumors (Marrache et al. 2008). This ssIL-1β construct has also been used in our experiments (Song et al. 2003, 2005; Elkabets et al. 2010). In a model of mesothelioma formation following exposure to asbestos, acute IL-1β production and recruitment of immune cells into the peritoneal cavity were significantly decreased in NLRP3-deficient mice (Chow et al. 2012). The NLRP3 inflammasome is the host sensor for asbestos-induced inflammation. However, NLRP3-deficient mice displayed a similar incidence of malignant mesothelioma and survival times as WT mice. The authors conclude that early inflammatory reactions triggered by asbestos are NLRP3-dependent, while NLRP3 is not critical in the chronic development of asbestos-induced mesothelioma. In a two-stage carcinogenesis (DMBA/PMA)-induced papilloma model, NLRP3-deficient mice showed a resistant phenotype in two different strain backgrounds, suggesting a tumor-promoting role for NLRP3, including IL-1β-induced inflammation, in certain chemically-induced cancer types (Chow et al. 2012).

IL-1β, at sites of tumorigenesis is mainly produced by infiltrating myeloid cells, while upon cell death, carcinogen-affected parenchymal cells can release homeostatically-expressed IL-1α, which may act as an alarmin and trigger local inflammation. Indeed, the relevance of IL-1α released from necrotic hepatocytes has been demonstrated in inflammation-associated liver carcinogenesis. By using mice that lack p38α in hepatocytes and exhibit elevated ROS accumulation after exposure to diethylnitrosamine (DEN), IL-1α was shown to stimulate the compensatory proliferative response that contributes to development of hepatocellular carcinoma (HCC) (Sakurai et al. 2008). Inhibition of IL-1α or ablation of IL-1RI prevents HCC development. In further studies, the stress response to Helicobacter felis infection was assessed in mice with a conditional knockout of IKKβ (lack of NF-κB signaling) in gastric epithelial cells (GECs) and myeloid cells (Shibata et al. 2010). Ablation of IKKβ for extended periods in GECs led to dysplasia followed by increased apoptosis, induction of reactive oxygen species and subsequent cellular necrosis and release of IL-1α, resulting in pronounced inflammation and more rapid progression to gastric pre-neoplasia. Loss of IKKβ in myeloid cells inhibited development of gastric atrophy. These results substantiate the significance of the NF-κB/IKKβ pathway as a key link between inflammation and cancer, inducing pro-inflammatory cytokines in myeloid cells and anti-apoptotic pathways in epithelial cells [reviewed in (Grivennikov and Karin 2010; Hagemann et al. 2009; Pikarsky and Ben-Neriah 2006)]. In a model of two-stage skin carcinogenesis, which involves RAS-mediated tumor formation in association with up-regulation of cytokines/chemokines that mediate an inflammation that supports oncogenesis, IL-1R$^{-/-}$ or MyD88$^{-/-}$ mice were shown to be less sensitive than WT mice to tumor development. Ras expressing keratinocytes from IL-1R$^{-/-}$ or MyD88$^{-/-}$ mice are hyperproliferative and fail to up-regulate proinflammatory genes or down-regulate differentiation markers characteristic of RAS-expressing WT keratinocytes. The differentiation and proinflammatory effects of oncogenic RAS in keratinocytes require the establishment of an autocrine loop through IL-1α, IL-1R, and MyD88, leading to phosphorylation of IκBα and NF-κB activation. Blocking IL-1α-mediated NF-κB activation in RAS-expressing WT keratinocytes reverses the differentiation defect and inhibits proinflammatory gene expression. Collectively, these results demonstrate that MyD88 exerts a cell-intrinsic function in RAS-mediated transformation of keratinocytes (Cataisson et al. 2012).

Altogether, the results of this section indicate that both IL-1β of myeloid cell origin, as well as IL-1α of tissue-resident cells, especially those that are the target cells of the carcinogenesis process, are essential for tumorigenesis. The mechanisms of differential induction of IL-1α and IL-1β in tumor-mediated inflammation and the mode of interaction between the IL-1 molecules has still to be elucidated, in order to devise novel chemoprevention approaches based on IL-1 neutralization.

8.5 The Role of IL-1 in Tumor Invasiveness

8.5.1 Host-Derived IL-1α Affects the Immunogenicity of the Arising Malignant Cells During Carcinogenesis

The "IL-1 milieu" where the carcinogenesis process occurs affects the characteristics of the arising malignant cells. Transplantable fibrosarcoma cell lines obtained from IL-1$\alpha^{-/-}$ mice failed to induce tumors in immune intact mice, whereas tumors developed in sublethally irradiated mice, indicating that the cells had not lost their invasiveness (Elkabets et al. 2009). Despite the fact that tumor incidence was comparable in 3-MCA-treated IL-1$\alpha^{-/-}$ mice and WT mice, the tumor cells were immunogenic in mice deficient in IL-1α, while malignant cells from WT mice induced progressive tumors in intact mice. This is in accordance with reports on immunogenic tumor cells that arise in different immunodeficient 3-MCA-treated mice that lack critical components essential for the development of anti-tumor cell immunity. These include mice lacking immunosurveillance cells, such as Rag2$^{-/-}$ mice, which lack T cells and B cells, nude mice, CD1d$^{-/-}$ mice, which lack CD1d-restricted T cells, and Ja18$^{-/-}$ mice, lacking semi-invariant NKT cells or mice deficient in cytokines critical for anti-tumor immunity, such as IFNγ and IL-12 [reviewed in (Bui and Schreiber 2007; Swann et al. 2007; Zitvogel and Kroemer 2009)]. In WT immune intact mice, these immunogenic variants are eradicated during tumor progression and low- or non-immunogenic tumor cells arise in carcinogen-treated mice, in a process termed "immunoediting".

The process of immunoediting in immunodeficient hosts allows the survival of malignant cell variants, which are "universally immunogenic", because they express surface adhesion or co-stimulatory molecules (i.e., ICAM-1 or 2, LFA-1 or 3, CD1d, VLA-4, B7 etc.). These adhesion molecules mediate high-affinity binding to immune effector cells and also serve as cell-associated co-stimulatory molecules, resulting in the activation of innate or specific immune surveillance mechanisms and efficient eradication of the malignant cells. 3-MCA immunogenic cell lines derived from IL-1$\alpha^{-/-}$ mice were shown to express more MHC class I molecules, co-stimulatory molecules (i.e., B7.1 and B7.2) and more adhesion molecules, such as L-selectin and ICAM-1 on their surface compared to fibrosarcoma cells from WT mice. Immunogenic cells from IL-1$\alpha^{-/-}$ mice are rejected in intact mice by conventional innate and specific anti-tumor immune effector cells, including NK cells, CD4^{+} and CD8^{+} T cells. We have also shown different immune deficiencies in IL-1$\alpha^{-/-}$ mice as compared to WT mice. Thus, in IL-1$\alpha^{-/-}$ mice, we observed fewer mature splenic NK cells, reduced expression of T-bet, a transcription factor that controls NK cell maturation and reduced expression of perforin in purified splenic NK cells. In addition, fewer liver NKT cells were detected in IL-1$\alpha^{-/-}$ mice. Finally, innate and specific anti-tumor effector cells from IL-1$\alpha^{-/-}$ mice, i.e., NK cells, LAK cells and CTLs, displayed reduced killing capacity. IL-1$\alpha^{-/-}$ mice have not been characterized as immunocompromized or of high susceptibility to infection by microbes.

It may be possible that the immune impairments in IL-1$\alpha^{-/-}$ mice are subtle and not manifested when the mice are confronted by a large inoculum of microbes or malignant cells, when multiple immune effector cells are recruited to combat the invader. However, these immune impairments are possibly significant when single immunogenic tumor cells arise and only a few local effector cells confront them, without recruiting a massive immune or inflammatory response. It has to be yet established whether cell-associated forms of host-derived IL-1α or reduced secretion levels of the cytokine contribute to immunosurveillance in WT mice. Supporting the role of IL-1α in immunosurveillance mechanisms, it was shown that in transgenic mice over-expressing IL-1α in the skin, DMBA/TPA treatment resulted in reduced incidence of skin tumors compared to WT mice, due to the rapid eradication of arising malignant cells by innate effector cells activated by local IL-1α in the skin (Murphy et al. 2003). In further studies, described below, we have shown that expression of cell-associated IL-1α in malignant cells increases their immunogenicity and leads to their rejection by immunosurveillance mechanisms.

8.5.2 The Role of Tumor Cell- or Microenvironment-Derived IL-1 in Tumor Invasiveness

In experimental tumor models and in cancer patients, increased local levels of IL-1 usually correlate with tumor invasiveness and a bad prognosis [reviewed in (Apte et al. 2006a, b; Apte and Voronov 2002, 2008; Dinarello 2010a)].

Fibrosarcoma cell lines isolated from 3-MCA-treated WT and IL-1/IL-Ra$^{-/-}$ mice, have been used in transplantation assays, to discriminate the role of malignant cell- or host-derived IL-1 in determining the invasiveness potential of the malignant cells (Marhaba et al. 2008; Nazarenko et al. 2008; Elkabets et al. 2009; Krelin et al. 2007; Voronov et al. 2009).

Tumor cells obtained from mice deficient in IL-1β were shown to be non-tumorigenic in WT mice. They also expressed impaired angiogenesis patterns in Matrigel plugs, possibly due to their inability to recruit a local inflammatory response, which is essential for tumor invasiveness (Marhaba et al. 2008; Nazarenko et al. 2008; Voronov et al. 2009). We could not detect constitutive expression of IL-1β by growing 3-MCA-induced fibrosarcoma cells in vitro and the stimuli which activate IL-1 production by malignant cells are as yet unknown and may include products of damaged or stressed tumor cells or cytokines expressed in the microenvironment. Moreover, expression of IL-1α in 3-MCA tumor cells from IL-1β KO mice activated a strong T helper and CTL response, which also contributed to the reduced in vivo growth of the cells in intact mice. This occurred as MDSCs were not recruited to tumor sites and thus immunosuppression was prevented. Nevertheless, the injection of fibrosarcoma cell lines from IL-1β KO mice together with LPS can induce progressive tumor growth.

These findings indicate the significance of inflammation in tumor development. Cell lines that originated in 3-MCA-treated IL-1Ra$^{-/-}$ mice were very invasive and even metastatic, while fibrosarcoma cells originating in WT mice induced only local tumors. This is due to high-unattenuated levels of IL-1 that are expressed in the malignant cells and facilitate their invasiveness and also promote induction of immunosuppressive mechanisms. Thus, fibrosarcoma cell-derived IL-1α and IL-1β do not act in concert and each IL-1 molecule has unique effects on tumor invasiveness, as well as on anti-tumor immune responses. At tumor sites, immunosuppressive effector cells, mainly induced by tumor cell- or microenvironment-derived IL-1β, usually dominate the function of anti-tumor immunosurveillance mechanisms induced by tumor cell-derived IL-1α.

The unique patterns of invasiveness of 3-MCA-induced tumor cells from IL-1 KO mice possibly stem from the lack of the relevant molecules of IL-1 family molecules in the malignant cells. However, it has yet to be ruled out that genetic changes, imprinted in the cells through effects of the microenvironment during carcinogenesis, may control their invasive features. Our results also clearly indicate that attenuated inflammatory responses at the tumor site, in addition to limiting tumor growth, enable the development of anti-tumor cell immunity, in the absence of tumor-mediated suppression. This has also been shown for low- or non-immunogenic epithelial tumors that develop in IL-1 KO mice. Thus, the inflammatory microenvironment is an important factor that determines the immunogenicity of specific tumors.

Expression patterns of IL-1 by the malignant cells synergized with those of the host. Thus, 3-MCA-induced fibrosarcoma cell lines from BALB/c WT mice manifested low invasiveness in IL-1 deficient mice, intermediate invasiveness in WT mice and high invasiveness in IL-1Ra$^{-/-}$ mice. Furthermore, invasive 3-MCA-induced fibrosarcoma cells from IL-1Ra KO mice were only weakly tumorigenic in IL-1 deficient mice. Thus, both host- and malignant cell-derived IL-1 contribute to tumor invasiveness. We suggest that initially, upon injection of tumor cells into mice, minute amounts of IL-1 are produced by the stressed malignant cells, enabling the initial seeding and growth of the malignant cells. Subsequently, IL-1 of malignant cell origin induces more IL-1 production in the microenvironment, activating a broad pro-inflammatory cytokine network and promoting invasiveness.

In search for a signature of genes involved in the invasiveness of malignant cells, it has been shown that invasive tumor cells constitutively secrete IL-1β. Using an experimental model of stepwise in vitro manipulation of telomerase (hTERT)-immortalized primary human fibroblasts, sequential p53 inactivation and expression of oncogenic Ras resulted in tumor forming cells. Gene profiling of cells capable of forming invasive tumors in nude mice revealed a unique genetic signature, consisting of chemokines/cytokines, ECM modulators and genes involved in cell death. These include IL-1β as well as other cytokines, which are known to be induced by it, such as IL-8, CSF-2, IL-6 and chemokines that play a significant role in tumor invasiveness (Milyavsky et al. 2005). Okamoto et al. (2010) showed constitutively active NLRP3 inflammasome and IL-1β secretion in melanoma cell lines derived from only late stage patients, which can explain the increased invasive phenotype of progressive melanoma tumors. It was also reported that tumor invasiveness is only dependent on IL-1 in certain tissues. Based on patterns of IL-1Ra inhibition of B16 melanoma metastasis following inoculation of malignant cells into mice treated with LPS to induce inflammation, organs were classified as to whether metastasis was dependent on microenvironmental IL-1 (Anasagasti et al. 1997). In the bone marrow, spleen, liver, lung, pancreas, skeletal muscle, adrenal gland and heart, IL-1Ra inhibited metastasis formation, while in the kidney, testis, brain, skin and GI tract; metastases were not affected by IL-1Ra. Thus, inherent features of the tumor cells, which determine their invasive potential and patterns of interaction of the malignant cells with the microenvironment, determine the necessity of IL-1β in invasiveness and dictate whether it will be produced by the malignant cells or the environment.

We have extensively assessed the effects of host-derived IL-1 on tumor-mediated angiogenesis, as a major parameter that affects tumor invasiveness Tumor-mediated angiogenesis is a critical and rate limiting event in the malignant process; the "angiogenic switch" enables malignant cells to progress from in situ tumors to metastatic tumors (Folkman 2003; Kerbel 2008). Tumor vascularization consists of angiogenesis, i.e., the formation of new capillaries from pre-existing vessels by tissue-resident endothelial cells, and is assisted by vasculogenesis, which involves the recruitment of cells of myeloid and endothelial lineages from the BM [reviewed in (Ferrara 2010; Wels et al. 2008; Coffelt et al. 2010; Murdoch et al. 2008)]. Thus, we have demonstrated that microenvironment-derived IL-1β plays a dominant role in in vivo tumor angiogenesis and invasiveness of B16 melanoma cells. In IL-1$\beta^{-/-}$ mice, no local tumor development or lung experimental metastases were observed following intrafootpad or intravenous inoculation, respectively (Voronov et al. 2003) Furthermore, in IL-1$\beta^{-/-}$ mice no recruitment of a blood vessel network into Matrigel plugs containing B16 cells was observed, whereas in WT C57BL/6 mice, potent vascularization of plugs was evident. In IL-1$\alpha^{-/-}$ mice, tumor growth and angiogenesis in Matrigel plugs were observed to a lesser extent than in WT mice but significantly higher than those in IL-1$\beta^{-/-}$ mice (Voronov et al. 2003). Addition of recombinant IL-1 into Matrigel plugs containing B16 cells in IL-1$\beta^{-/-}$ mice partially restores the angiogenic response, while addition of IL-1Ra to B16-containing Matrigel plugs in WT mice inhibited the ingrowth of the blood vessel network into the plugs. Therefore, neutralization of a single molecule, i.e., IL-1, may inhibit the generation of the cascade of downstream (effector) pro-inflammatory molecules with redundant functions in tumor angiogenesis. Indeed, continuous delivery systems of recombinant IL-1Ra or cells over-expressing IL-1Ra, encapsulated within alginate-poly (L-lysine)-alginate (APA) microspheres, reduced the tumor burden and inhibited tumor-mediated angiogenesis when implanted into tumor-bearing mice (Bar et al. 2004; Lavi et al. 2007). Similarly, over-expression of IL-1Ra in human melanoma cell lines inhibits tumor growth and metastasis in human melanoma xenografts in nude mice (Weinreich et al. 2003; Elaraj et al. 2006).

Interestingly, effects of the IL-1Ra were observed on the growth of melanoma cells which produce high levels of IL-1, but not on the growth of cell lines which produce low levels of IL-1. It was shown that IL-1 is frequently expressed in metastases from patients with different types of human cancer. Treatment with IL-1Ra induced regression of xenograft growth of IL-1 producing tumors in nude mice, but did not affect their proliferation rate in culture, indicating that tumor cell-derived IL-1 acts on the microenvironment. For example, high levels of intra-tumoral pro-angiogenic cytokines, IL-8 and VEGF, were detected in tumors producing IL-1. Inhibition of xenograft growth of IL-1 producing tumors also reduced the production of IL-8 and VEGF, which may be an early surrogate of IL-1Ra-mediated anti-tumor activity. Altogether, these results provide pre-clinical support for the use of IL–1Ra as an adjunct therapy in combination with tumor resection and chemotherapy, to attenuate the growth and invasiveness of tumors. This possibility was recently reviewed in (Apte et al. 2006; Apte and Voronov 2008; Dinarello 2010a, b).

8.5.3 Microenvironment IL-1 can also Promote Anti-Tumor Cell Immunity Under Specific Circumstances

As IL-1 is a pleiotropic cytokine, which also participates in induction of adaptive immunity, some initial studies probing the anti-tumor effects of injected IL-1 have been performed, with positive indications, but inconsistent results [reviewed by us (Apte et al. 2006a, b; Apte and Voronov 2002, 2008)]. This is due to the severe side effects induced by IL-1, concomitant to its effects on tumor invasiveness. Injection of recombinant IL-1β into tumor-bearing subjects is risky, since it can induce an unexpected cytokine storm in certain individuals. Recently, an excellent example of the feasibility of microenvironment-derived IL-1β in stimulating anti-tumor immunity has been reported (Ghiringhelli et al. 2009). It was demonstrated that tissue-damage, following cancer treatment with some chemotherapeutical drugs, activates DCs in the tumor microenvironment to present tumor antigens and further stimulate anti-tumor immunity that synergizes with the chemotherapy. In the milieu of anthracycline-treated tumors, the NLRP3 inflammasome is activated and stimulates IL-1β production, which is essential for activating IFNγ producing CD8 T cells. Most interestingly, patients with breast cancer with a loss-of-function allele of P2X7R, which is essential for activation of the NLRP3 inflammasome and IL-1β processing, develop a more metastatic and rapid disease than individuals with the normal allele. This may represent a unique scenario in which a low tumor burden, possibly accompanied by low levels of IL-1β expression in the microenvironment, activates local immunity, without concomitant "destructive" inflammation that promotes tumor invasiveness and induces immunosuppression. These results will hopefully open new avenues for use of IL-1β in cancer immunotherapy, possibly after debulking the primary tumor.

8.6 Future Prospects of IL-1 Manipulation in Anti-Tumor Therapies

The network of cytokines and immune/inflammatory cells in the tumor microenvironment controls patterns of invasiveness [reviewed in (Grivennikov and Karin 2010a, b; Colotta et al. 2009; Demaria et al. 2010; Mantovani et al. 2008; Joyce and Pollard 2009; Solinas et al. 2010; Egeblad et al. 2010; Kenny et al. 2007; Peinado et al. 2008; Witz 2008; Whiteside 2008)]. In the tumor microenvironment, the balance between the "wound healing" type of inflammation, which promotes tumor progression and immune escape, and "favorable" limited inflammatory responses, in which professional APCs are activated and induce anti-tumor adaptive immunity, determines the direction of the malignant process [reviewed in (Bui and Schreiber 2007; Biswas and Mantovani 2010; DeNardo et al. 2010; Ostrand-Rosenberg 2008; Ghiringhelli et al. 2007)]. Immunosurveillance mechanisms also operate in growing tumors and potentially can induce tumor shrinkage and in rare cases, regression. It has recently been shown that survival of cancer patients, even with metastatic disease, correlates with the number of activated CTL and Th1 cells, as well as the presence of lymphoid follicles inside the tumor (Fridman et al. 2010; Galon et al. 2006). This has been shown in patients with diverse types of tumors, such as invasive colon cancer, melanoma, multiple myeloma and pancreatic cancer. Intervention in regulatory circuits in tumors may thus represent a suitable approach to tilt the balance between inflammation and immunity, in order to improve in situ insufficient anti-tumor immune responses.

Due to the plethora of activities of IL-1 in the malignant process, frequently of opposing nature, and its dominant role in local cytokine networks at tumor sites, neutralization of IL-1 as a single target molecule may tilt the balance between destructive inflammation and protective anti-tumor immunity in the tumor microenvironment. Cytokines act in vivo in specific and complex networks; interference in a single critical link of the network can abrogate the entire network. This has been shown for neutralization of either IL-1 or TNFα in alleviating symptoms in rheumatoid arthritis patients [reviewed in (Sims and Smith 2010; Dinarello 2010b)]. The IL-1Ra, also called Anakinra (Kineret; Amgen/Biovitrum) is FDA-approved and has been shown to be safe and efficient in alleviating symptoms of rheumatoid arthritis. Newly developed anti-IL-1 agents include anti-IL-1β antibodies: Canakinumab (Ilaris, Novartis) and XOMA 052 (XOMA), an antibody directed against IL-1RI (AMG 108, Amgen) and the fusion protein (IL-1 trap) of IL-1RI and IL-1RAcP, Rilonacept (Arcalyst, Regeneron).

Major developments in the biology of inflammasomes and IL-1β processing/secretion, have prompted the initiation of multiple clinical trials with initial promising results, to examine the efficiency of these agents in diverse diseases with inflammatory manifestations, such as cryopyrin-associated periodic syndromes, rheumatoid arthritis, systemic onset juvenile arthritis, gout, type 2 diabetes[reviewed in (Sims and Smith 2010; Dinarello 2010b)]. The exact conditions for utilizing these biological agents in cancer patients and how to integrate them with conventional anti-tumor therapies will have to be carefully studied. Optimally, IL-1 neutralization treatments will probably be most efficient in patients with minimal residual disease (MRD), in which the bulk of the primary tumor has been removed by conventional therapy (surgery, chemotherapy or irradiation) to prevent tumor recurrence and metastasis. In such patients, as described above (Zitvogel and Kroemer 2009; Apetoh et al. 2007), tumor-mediated immunosuppression, as well as "destructive" inflammatory responses are hopefully reduced and conditions in the tumor microenvironment will favor development of protective anti-tumor immune responses, with the potential to destroy residual tumor cells that escaped the debulking therapy. Neutralization of IL-1β in patients with local chronic inflammatory diseases, especially those with the genetic background which makes them prone to developing tumors at the site of inflammation, can also be envisioned using chemopreventive approaches. Neutralization of microenvironment IL-1, especially IL-1β, should not be complete, in order to not immunocompromize the patient. However, the homeostatic levels of IL-1 versus cytokines involved in different types of inflammatory responses should first be determined.

In addition, due to the adjuvanticity of cell-associated IL-1α, tumor cell vaccines based on constitutive or transient IL-1α expression have the potential to induce anti-tumor cell immunity in patients with MRD. In such patients, one can also envision systemic neutralization of IL-1β followed by local expression of IL-1α. In conclusion, a better understanding of the role of molecules of the IL-1 family will hopefully result in the implementation of IL-1 modulation in cancer patients. IL-1 modulating agents should be safe, appropriately targeted to tumor deposits, efficient, cost-manageable and easily applied to cancer patients.

Acknowledgments Ron N. Apte was supported by the Israel Ministry of Science (MOS) jointly with the Deutsches Krebsforschungscentrum (DKFZ), Heidelberg, Germany, the Israel Science Foundation funded by the Israel Academy of Sciences and Humanities, the Israel Cancer Association and the Israel Ministry of Health Chief Scientist's Office, FP7: "Cancer and Inflammation" (INFLA-CARE) and The Binational (Israel-USA) Science Foundation. Prof. Ron N. Apte is an incumbent of the Irving Isaac Sklar Chair in Endocrinology and Cancer.

Elena Voronov was supported by the Israel Cancer Association, the Israel Ministry of Health Chief Scientist's Office and the Concern Foundation, FP7: "Cancer and Inflammation" (INFLA-CARE) and ISF (the last two in cooperation with Prof. Apte).

The authors would like to thank their collaborators: Prof. Charles A. Dinarello, the University of Colorado, Denver, USA; Profs. Margot Zoller and Margareta Mueller, the German Cancer Center (DKFZ), Heidelberg, Germany; Prof. Michael Martin, Justus Liebig University, Giessen, Germany; Prof. Yoichiro Iwakura, University of Tokyo, Tokyo, Japan and the late Prof. Shraga Segal, Ben-Gurion University, Beer-Sheva, Israel. We would also like to thank all our wonderful students who contributed to the research on these issues and to Mrs. Rosalyn M. White for her devoted help in preparation of this MS.

References

Anasagasti MJ, Olaso E, Calvo F, Mendoza L, Martin JJ, Bidaurrazaga J et al (1997) Interleukin 1-dependent and -independent mouse melanoma metastases. J Natl Cancer Inst 89:645–651

Apetoh L, Ghiringhelli F, Tesniere A, Obeid M, Ortiz C, Criollo A et al (2007) Toll-like receptor 4-dependent contribution of the immune system to anticancer chemotherapy and radiotherapy. Nat Med 13:1050–1059

Apte RN, Voronov E (2002) Interleukin-1-a major pleiotropic cytokine in tumor-host interactions. Semin Cancer Biol 12:277–290

Apte RN, Voronov E (2008) Is interleukin-1 a good or bad 'guy' in tumor immunobiology and immunotherapy? Immunol Rev 222:222–241

Apte RN, Dotan S, Elkabets M, White MR, Reich E, Carmi Y et al (2006a) The involvement of IL-1 in tumorigenesis, tumor invasiveness, metastasis and tumor-host interactions. Cancer Metastasis Rev 25:387–408

Apte RN, Krelin Y, Song X, Dotan S, Recih E, Elkabets M et al (2006b) Effects of micro-environment- and malignant cell-derived interleukin-1 in carcinogenesis, tumour invasiveness and tumour-host interactions. Eur J Cancer 42:751–759

Arend WP (2002) The balance between IL-1 and IL-1Ra in disease. Cytokine Growth Factor Rev 13:323–340

Arend WP, Palmer G, Gabay C (2008) IL-1, IL-18, and IL-33 families of cytokines. Immunol Rev 223:20–38

Auron PE (1998) The interleukin 1 receptor: ligand interactions and signal transduction. Cytokine Growth Factor Rev 9:221–237

Baniyash M (2006) Chronic inflammation, immunosuppression and cancer: new insights and outlook. Semin Cancer Biol 16:80–88

Bar D, Apte RN, Voronov E, Dinarello CA, Cohen S (2004) A continuous delivery system of IL-1 receptor antagonist reduces angiogenesis and inhibits tumor development. Faseb J 18:161–163

Ben-Sasson SZ, Hu-Li J, Quiel J, Cauchetaux S, Ratner M, Shapira I et al (2009) IL-1 acts directly on CD4 T cells to enhance their antigen-driven expansion and differentiation. Proc Natl Acad Sci U S A 106:7119–7124

Bianchi ME (2007) DAMPs, PAMPs and alarmins: all we need to know about danger. J Leukoc Biol 81:1–5

Bianchi ME, Manfredi AA (2007) High-mobility group box 1 (HMGB1) protein at the crossroads between innate and adaptive immunity. Immunol Rev 220:35–46

Biswas SK, Mantovani A (2010) Macrophage plasticity and interaction with lymphocyte subsets: cancer as a paradigm. Nat Immunol 11:889–896

Bronte V, Mocellin S (2009) Suppressive influences in the immune response to cancer. J Immunother 32:1–11

Bui JD, Schreiber RD (2007) Cancer immunosurveillance, immunoediting and inflammation: independent or interdependent processes? Curr Opin Immunol 19:203–208

Bunt SK, Sinha P, Clements VK, Leips J, Ostrand-Rosenberg S (2006) Inflammation induces myeloid-derived suppressor cells that facilitate tumor progression. J Immunol 176:284–290

Carmi Y, Rinott G, Dotan S, Elkabets M, Rider P, Voronov E et al (2010) Microenvironment-derived IL-1 and IL-17 interact in the control of lung metastasis. J Immunol 186:3462–3471

Carroll RG, Carpenito C, Shan X, Danet-Desnoyers G, Liu R, Jiang S et al (2008) Distinct effects of IL-18 on the engraftment and function of human effector CD8 T cells and regulatory T cells. PLoS ONE 3:e3289

Carta S, Castellani P, Delfino L, Tassi S, Vene R, Rubartelli A (2009) DAMPs and inflammatory processes: the role of redox in the different outcomes. J Leukoc Biol 85:549–555

Cataisson C, Salcedo R, Hakim S, Moffitt BA, Wright L, Yi M et al (2012) IL-1R-MyD88 signaling in keratinocyte transformation and carcinogenesis. J Exp Med 209:1689–1702

Chambers CA, Allison JP (1999) Costimulatory regulation of T cell function. Curr Opin Cell Biol 11:203–210

Chow MT, Tschopp J, Moller A, Smyth MJ (2012) NLRP3 promotes inflammation-induced skin cancer but is dispensable for asbestos-induced mesothelioma. Immunology and cell biology 90(10):983–986

Coffelt SB, Lewis CE, Naldini L, Brown JM, Ferrara N, De Palma M (2010) Elusive identities and overlapping phenotypes of proangiogenic myeloid cells in tumors. Am J Pathol 176:1564–1576

Cohen I, Rider P, Carmi Y, Braiman A, Dotan S, White MR et al (2010) Differential release of chromatin-bound IL-1alpha discriminates between necrotic and apoptotic cell death by the ability to induce sterile inflammation. Proc Natl Acad Sci U S A 107:2574–2579

Colotta F, Allavena P, Sica A, Garlanda C, Mantovani A (2009) Cancer-related inflammation, the seventh hallmark of cancer: links to genetic instability. Carcinogenesis 30:1073–1081

Demaria S, Pikarsky E, Karin M, Coussens LM, Chen YC, El-Omar EM et al (2010) Cancer and inflammation: promise for biologic therapy. J Immunother 33:335–351

DeNardo DG, Andreu P, Coussens LM (2010) Interactions between lymphocytes and myeloid cells regulate pro- versus anti-tumor immunity. Cancer Metastasis Rev 29:309–316

Dinarello CA (1996) Biologic basis for interleukin-1 in disease. Blood 87:2095–2147

Dinarello CA (2009) Immunological and inflammatory functions of the interleukin-1 family. Annu Rev Immunol 27:519–550

Dinarello CA (2010a) Why not treat human cancer with interleukin-1 blockade? Cancer Metastasis Rev 29:317–329

Dinarello CA (2010b) Anti-inflammatory agents: present and future. Cell 140:935–950

Douvdevani A, Huleihel M, Segal S, Apte RN (1991) Aberrations in interleukin-1 expression in oncogene-transformed fibrosarcoma lines: constitutive interleukin-1 alpha transcription and manifestation of biological activity. Eur Cytokine Netw 2:257–264

Douvdevani A, Huleihel M, Zoller M, Segal S, Apte RN (1992) Reduced tumorigenicity of fibrosarcomas which constitutively generate IL-1 alpha either spontaneously or following IL-1 alpha gene transfer. Int J Cancer 51:822–830

Dvorkin T, Song X, Argov S, White RM, Zoller M, Segal S et al (2006) Immune phenomena involved in the in vivo regression of fibrosarcoma cells expressing cell-associated IL-1{alpha}. J Leukoc Biol 80:96–106

Egeblad M, Nakasone ES, Werb Z (2010) Tumors as organs: complex tissues that interface with the entire organism. Dev Cell 18:884–901

Eisenbarth SC, Flavell RA (2009) Innate instruction of adaptive immunity revisited: the inflammasome. EMBO Mol Med. 1:92–98

Eisenbarth SC, Colegio OR, O'Connor W, Sutterwala FS, Flavell RA (2008) Crucial role for the Nalp3 inflammasome in the immunostimulatory properties of aluminium adjuvants. Nature 453:1122–1126

Elaraj DM, Weinreich DM, Varghese S, Puhlmann M, Hewitt SM, Carroll NM et al (2006) The role of interleukin 1 in growth and metastasis of human cancer xenografts. Clin Cancer Res 12:1088–1096

Elkabets M, Krelin Y, Dotan S, Cerwenka A, Porgador A, Lichtenstein RG et al (2009) Host-derived interleukin-1alpha is important in determining the immunogenicity of 3-methylcholantrene tumor cells. J Immunol 182:4874–4881

Elkabets M, Ribeiro VSG, Dinarello CA, Ostrand-Rosenberg S, Di Santo J, Apte RN, Vosshenrich CAJ (2010) IL-1β regulates a novel myeloid-derived suppressor cell subset that impairs NK cell development and function. Eur J immunol 40:3347–3357

El-Shami KM, Tzehoval E, Vadai E, Feldman M, Eisenbach L (1999) Induction of antitumor immunity with modified autologous cells expressing membrane-bound murine cytokines. J Interferon Cytokine Res 19:1391–1401

Ferrara N (2010) Role of myeloid cells in vascular endothelial growth factor-independent tumor angiogenesis. Curr Opin Hematol 17:219–224

Folkman J (2003) Fundamental concepts of the angiogenic process. Curr Mol Med 3:643–651

Franchi L, Nunez G (2008) The Nlrp3 inflammasome is critical for aluminium hydroxide-mediated IL-1beta secretion but dispensable for adjuvant activity. Eur J Immunol 38:2085–2089

Franchi L, Park JH, Shaw MH, Marina-Garcia N, Chen G, Kim YG et al (2008) Intracellular NOD-like receptors in innate immunity, infection and disease. Cell Microbiol 10:1–8

Franchi L, Eigenbrod T, Munoz-Planillo R, Nunez G (2009) The inflammasome: a caspase-1-activation platform that regulates immune responses and disease pathogenesis. Nat Immunol 10:241–247

Fridman WH, Galon J, Dieu-Nosjean MC, Cremer I, Fisson S, Damotte D et al (2010) Immune infiltration in human cancer: prognostic significance and disease control. Curr Top Microbiol Immunol 344:1–24

Gabay C, Lamacchia C, Palmer G (2010) IL-1 pathways in inflammation and human diseases. Nat Rev Rheumatol 6:232–241

Gabrilovich DI, Nagaraj S (2009) Myeloid-derived suppressor cells as regulators of the immune system. Nat Rev Immunol 9:162–174

Gadina M, Jefferies CA (2007) IL-33: a sheep in wolf's clothing? Sci STKE 2007:pe31
Galon J, Costes A, Sanchez-Cabo F, Kirilovsky A, Mlecnik B, Lagorce-Pages C et al (2006) Type, density, and location of immune cells within human colorectal tumors predict clinical outcome. Science 313:1960–1964
Garlanda C, Anders HJ, Mantovani A (2009) TIR8/SIGIRR: an IL-1R/TLR family member with regulatory functions in inflammation and T cell polarization. Trends Immunol 30:439–446
Ghiringhelli F, Apetoh L, Housseau F, Kroemer G, Zitvogel L (2007) Links between innate and cognate tumor immunity. Curr Opin Immunol 19:224–231
Ghiringhelli F, Apetoh L, Tesniere A, Aymeric L, Ma Y, Ortiz C et al (2009) Activation of the NLRP3 inflammasome in dendritic cells induces IL-1beta-dependent adaptive immunity against tumors. Nat Med 15:1170–1178
Grivennikov SI, Karin M (2010a) Dangerous liaisons: STAT3 and NF-kappaB collaboration and crosstalk in cancer. Cytokine Growth Factor Rev 21:11–19
Grivennikov SI, Karin M (2010b) Inflammation and oncogenesis: a vicious connection. Curr Opin Genet Dev 20:65–71
Guo L, Wei G, Zhu J, Liao W, Leonard WJ, Zhao K et al (2009) IL-1 family members and STAT activators induce cytokine production by Th2, Th17, and Th1 cells. Proc Natl Acad Sci U S A. 106:13463–13468
Hacham M, Cristal N, White RM, Segal S, Apte RN (1996) Complementary organ expression of IL-1 vs. IL-6 and CSF-1 activities in normal and LPS-injected mice. Cytokine 8:21–31
Hacham M, Argov S, White RM, Segal S, Apte RN (2002) Different patterns of interleukin-1alpha and interleukin-1beta expression in organs of normal young and old mice. Eur Cytokine Netw 13:55–65
Hagemann T, Biswas SK, Lawrence T, Sica A, Lewis CE (2009) Regulation of macrophage function in tumors: the multifaceted role of NF-kappaB. Blood 113:3139–3146
Hoffman HM, Wanderer AA (2010) Inflammasome and IL-1beta-mediated disorders. Curr Allergy Asthma Rep 10:229–235
Hoo WS, Lundeen KA, Kohrumel JR, Pham NL, Brostoff SW, Bartholomew RM et al (1999) Tumor cell surface expression of granulocyte-macrophage colony-stimulating factor elicits antitumor immunity and protects from tumor challenge in the P815 mouse mastocytoma tumor model. J Immunol 162:7343–7349
Hreggvidsdottir HS, Ostberg T, Wahamaa H, Schierbeck H, Aveberger AC, Klevenvall L et al (2009) The alarmin HMGB1 acts in synergy with endogenous and exogenous danger signals to promote inflammation. J Leukoc Biol 86:655–662
Iwasaki A, Medzhitov R (2010) Regulation of adaptive immunity by the innate immune system. Science 327:291–295
Joyce JA, Pollard JW (2009) Microenvironmental regulation of metastasis. Nat Rev Cancer 9:239–252
Kalli K, Huntoon C, Bell M, McKean DJ (1998) Mechanism responsible for T-cell antigen receptor- and CD28- or interleukin 1 (IL-1) receptor-initiated regulation of IL-2 gene expression by NF-kappaB. Mol Cell Biol 18:3140–3148
Keller M, Ruegg A, Werner S, Beer HD (2008) Active caspase-1 is a regulator of unconventional protein secretion. Cell 132:818–831
Kenny PA, Lee GY, Bissell MJ (2007) Targeting the tumor microenvironment. Front Biosci 12:3468–3474
Kerbel RS (2008) Tumor angiogenesis. N Engl J Med 358:2039–2049
Krelin Y, Voronov E, Dotan S, Elkabets M, Reich E, Fogel M et al (2007) Interleukin-1beta-driven inflammation promotes the development and invasiveness of chemical carcinogen-induced tumors. Cancer Res 67:1062–1071
Latz E (2010) The inflammasomes: mechanisms of activation and function. Curr Opin Immunol 22:28–33
Lavi G, Voronov E, Dinarello CA, Apte RN, Cohen S (2007) Sustained delivery of IL-1Ra from biodegradable microspheres reduces the number of murine B16 melanoma lung metastases. J Control Release 123:123–130

Li H, Nookala S, Re F (2007) Aluminum hydroxide adjuvants activate caspase-1 and induce IL-1beta and IL-18 release. J Immunol 178:5271–5276
Lotze MT, Tracey KJ (2005) High-mobility group box 1 protein (HMGB1): nuclear weapon in the immune arsenal. Nat Rev Immunol 5:331–342
Mantovani A, Muzio M, Ghezzi P, Colotta C, Introna M (1998) Regulation of inhibitory pathways of the interleukin-1 system. Ann N Y Acad Sci 840:338–351
Mantovani A, Allavena P, Sica A, Balkwill F (2008) Cancer-related inflammation. Nature 454:436–444
Marhaba R, Nazarenko I, Knofler D, Reich E, Voronov E, Vitacolonna M et al (2008) Opposing effects of fibrosarcoma cell-derived IL-1alpha and IL-1beta on immune response induction. Int J Cancer 123:134–145
Marr RA, Addison CL, Snider D, Muller WJ, Gauldie J, Graham FL (1997) Tumour immunotherapy using an adenoviral vector expressing a membrane-bound mutant of murine TNF alpha. Gene Ther 4:1181–1188
Marrache F, Pendyala S, Bhagat G, Betz KS, Song Z, Wang TC (2008) Role of bone marrow-derived cells in experimental chronic pancreatitis. Gut 57:1113–1120
Martin MU, Falk W (1997) The interleukin-1 receptor complex and interleukin-1 signal transduction. Eur Cytokine Netw 8:5–17
Martin MU, Wesche H (2002) Summary and comparison of the signaling mechanisms of the Toll/interleukin-1 receptor family. Biochim Biophys Acta 1592:265–280
Martinon F (2010) Signaling by ROS drives inflammasome activation. Eur J Immunol 40:616–619
Martinon F, Mayor A, Tschopp J (2009) The inflammasomes: guardians of the body. Annu Rev Immunol 27:229–265
Milyavsky M, Tabach Y, Shats I, Erez N, Cohen Y, Tang X et al (2005) Transcriptional programs following genetic alterations in p53, INK4A, and H-Ras genes along defined stages of malignant transformation. Cancer Res 65:4530–4543
Mochizuki DY, Eisenman JR, Conlon PJ, Larsen AD, Tushinski RJ (1987) Interleukin 1 regulates hematopoietic activity, a role previously ascribed to hemopoietin 1. Proc Natl Acad Sci U S A 84:5267–5271
Murdoch C, Muthana M, Coffelt SB, Lewis CE (2008) The role of myeloid cells in the promotion of tumour angiogenesis. Nat Rev Cancer 8:618–631
Murphy JE, Morales RE, Scott J, Kupper TS (2003) IL-1 alpha, innate immunity, and skin carcinogenesis: the effect of constitutive expression of IL-1 alpha in epidermis on chemical carcinogenesis. J Immunol 170:5697–5703
Nakao S, Kuwano T, Tsutsumi-Miyahara C, Ueda S, Kimura YN, Hamano S et al (2005) Infiltration of COX-2-expressing macrophages is a prerequisite for IL-1 beta-induced neovascularization and tumor growth. J Clin Invest 115:2979–2991
Nazarenko I, Marhaba R, Reich E, Voronov E, Vitacolonna M, Hildebrand D et al (2008) Tumorigenicity of IL-1alpha- and IL-1beta-deficient fibrosarcoma cells. Neoplasia 10:549–562
Neta R, Douches S, Oppenheim JJ (1986) Interleukin 1 is a radioprotector. J Immunol 136:2483–2485
Okamoto M, Liu W, Luo Y, Tanaka A, Cai X, Norris DA et al (2010) Constitutively active inflammasome in human melanoma cells mediating autoinflammation via caspase-1 processing and secretion of interleukin-1beta. J Biol Chem 285:6477–6488
O'Neill LA (2008) The interleukin-1 receptor/Toll-like receptor superfamily: 10 years of progress. Immunol Rev 226:10–18
Ostrand-Rosenberg S (2008) Immune surveillance: a balance between protumor and antitumor immunity. Curr Opin Genet Dev 18:11–18
Ostrand-Rosenberg S (2010) Myeloid-derived suppressor cells: more mechanisms for inhibiting antitumor immunity. Cancer Immunol Immunother 59:1593–1600
Ostrand-Rosenberg S, Sinha P (2009) Myeloid-derived suppressor cells: linking inflammation and cancer. J Immunol 182:4499–4506

Palm NW, Medzhitov R (2009) Pattern recognition receptors and control of adaptive immunity. Immunol Rev 227:221–233
Peinado H, Rafii S, Lyden D (2008) Inflammation joins the "niche". Cancer Cell 14:347–349
Peranzoni E, Zilio S, Marigo I, Dolcetti L, Zanovello P, Mandruzzato S et al (2010) Myeloid-derived suppressor cell heterogeneity and subset definition. Curr Opin Immunol 22:238–244
Pikarsky E, Ben-Neriah Y (2006) NF-kappaB inhibition: a double-edged sword in cancer? Eur J Cancer 42:779–784
Raucci A, Palumbo R, Bianchi ME (2007) HMGB1: a signal of necrosis. Autoimmunity 40:285–289
Rider P, Carmi Y, Guttman O, Braiman A, Cohen I, Voronov E et al (2011) IL-1{alpha} and IL-1{beta} Recruit Different Myeloid Cells and Promote Different Stages of Sterile Inflammation. J Immunol. 187:4835–4843
Rock KL, Kono H (2008) The inflammatory response to cell death. Annu Rev Pathol 3:99–126
Saijo Y, Tanaka M, Miki M, Usui K, Suzuki T, Maemondo M et al (2002) Proinflammatory cytokine IL-1 beta promotes tumor growth of Lewis lung carcinoma by induction of angiogenic factors: in vivo analysis of tumor-stromal interaction. J Immunol 169:469–475
Sakurai T, He G, Matsuzawa A, Yu GY, Maeda S, Hardiman G et al (2008) Hepatocyte necrosis induced by oxidative stress and IL-1 alpha release mediate carcinogen-induced compensatory proliferation and liver tumorigenesis. Cancer Cell 14:156–165
Schroder K, Tschopp J (2010) The inflammasomes. Cell 140:821–832
Schroder K, Zhou R, Tschopp J (2010) The NLRP3 inflammasome: a sensor for metabolic danger? Science 327:296–300
Shibata W, Takaishi S, Muthupalani S, Pritchard DM, Whary MT, Rogers AB et al (2010) Conditional deletion of IkappaB-kinase-beta accelerates helicobacter-dependent gastric apoptosis, proliferation, and preneoplasia. Gastroenterology 138:1022–1034 e1-10
Sims JE, Smith DE (2010) The IL-1 family: regulators of immunity. Nat Rev Immunol 10:89–102
Smyth MJ, Dunn GP, Schreiber RD (2006) Cancer immunosurveillance and immunoediting: the roles of immunity in suppressing tumor development and shaping tumor immunogenicity. Adv Immunol 90:1–50
Solinas G, Marchesi F, Garlanda C, Mantovani A, Allavena P (2010) Inflammation-mediated promotion of invasion and metastasis. Cancer Metastasis Rev 29:243–248
Song X, Voronov E, Dvorkin T, Fima E, Cagnano E, Benharroch D et al (2003) Differential effects of IL-1 alpha and IL-1 beta on tumorigenicity patterns and invasiveness. J Immunol 171:6448–6456
Song X, Krelin Y, Dvorkin T, Bjorkdahl O, Segal S, Dinarello CA et al (2005) CD11b+/Gr-1+ immature myeloid cells mediate suppression of T cells in mice bearing tumors of IL-1{beta}-secreting cells. J Immunol 175:8200–8208
Srikrishna G, Freeze HH (2009) Endogenous damage-associated molecular pattern molecules at the crossroads of inflammation and cancer. Neoplasia 11:615–628
Stylianou E, Saklatvala J (1998) Interleukin-1. Int J Biochem Cell Biol 30:1075–1079
Swann JB, Coquet JM, Smyth MJ, Godfrey DI (2007) CD1-restricted T cells and tumor immunity. Curr Top Microbiol Immunol 314:293–323
Takeuchi O, Akira S (2010) Pattern recognition receptors and inflammation. Cell 140:805–820
Thompson RC, Pardoll DM, Jaffee EM, Ewend MG, Thomas MC, Tyler BM et al (1996) Systemic and local paracrine cytokine therapies using transduced tumor cells are synergistic in treating intracranial tumors. J Immunother Emphasis Tumor Immunol 19:405–413
Tu S, Bhagat G, Cui G, Takaishi S, Kurt-Jones EA, Rickman B et al (2008) Overexpression of interleukin-1beta induces gastric inflammation and cancer and mobilizes myeloid-derived suppressor cells in mice. Cancer Cell 14:408–419
Ulloa L, Messmer D (2006) High-mobility group box 1 (HMGB1) protein: friend and foe. Cytokine Growth Factor Rev 17:189–201

Voronov E, Weinstein Y, Benharroch D, Cagnano E, Ofir R, Dobkin M et al (1999) Antitumor and immunotherapeutic effects of activated invasive T lymphoma cells that display short-term interleukin 1alpha expression. Cancer Res 59:1029–1035

Voronov E, Shouval DS, Krelin Y, Cagnano E, Benharroch D, Iwakura Y et al (2003) IL-1 is required for tumor invasiveness and angiogenesis. Proc Natl Acad Sci U S A 100:2645–2650

Voronov E, Reich E, Dotan S, Dransh P, Cohen I, Huszar M et al (2009) Effects of IL-1 molecules on growth patterns of 3-MCA-induced cell lines: an interplay between immunogenicity and invasive potential. J Immunotoxicol 7(1):27–38

Weinreich DM, Elaraj DM, Puhlmann M, Hewitt SM, Carroll NM, Feldman ED et al (2003) Effect of interleukin 1 receptor antagonist gene transduction on human melanoma xenografts in nude mice. Cancer Res 63:5957–5961

Wels J, Kaplan RN, Rafii S, Lyden D (2008) Migratory neighbors and distant invaders: tumor-associated niche cells. Genes Dev 22:559–574

Werman A, Werman-Venkert R, White R, Lee JK, Werman B, Krelin Y et al (2004) The precursor form of IL-1alpha is an intracrine proinflammatory activator of transcription. Proc Natl Acad Sci U S A 101:2434–2439

Whiteside TL (2008) The tumor microenvironment and its role in promoting tumor growth. Oncogene 27:5904–5912

Williams A, Flavell RA, Eisenbarth SC (2010) The role of NOD-like receptors in shaping adaptive immunity. Curr Opin Immunol 22:34–40

Witz IP (2008) Tumor-microenvironment interactions: dangerous liaisons. Adv Cancer Res 100:203–229

Youn JI, Gabrilovich DI (2010) The biology of myeloid-derived suppressor cells: the blessing and the curse of morphological and functional heterogeneity. Eur J Immunol 40:2969–2975

Zitvogel L, Kroemer G (2009) Anticancer immunochemotherapy using adjuvants with direct cytotoxic effects. J Clin Invest 119:2127–2130

Zoller M, Douvdevani A, Segal S, Apte RN (1992a) Interleukin-1 production by transformed fibroblasts. II. Influence on antigen presentation and T-cell-mediated anti-tumor response. Int J Cancer 50:450–457

Zoller M, Douvdevani A, Segal S, Apte RN (1992b) Interleukin-1 produced by tumorigenic fibroblasts influences tumor rejection. Int J Cancer 50:443–449

Chapter 9
Impact of Obesity and Aging on the Tumor Immuno-Environment

Annie Mirsoian, Gail D. Sckisel, Anthony E. Zamora and William J. Murphy

Abstract Cancer is a disease of the aged with the average age of diagnosis being over the age of 55 years in the U.S.A. Significant physiologic changes occur with aging such as the redistribution of body mass towards increased lipid deposition and loss of lean body mass. There has recently been an increased appreciation on the ability of adipose tissue to affect immunological responses. Although the correlation between adiposity and inflammation has been best studied within the setting of diabetes, current research indicates that adipose tissue and its effects on inflammation has tremendous implications in cancer biology. Development of esophageal, colon, rectal, kidney, pancreatic, endometrial, and breast cancer has been associated with being overweight. The link between cancer, aging, adiposity, and the immune system is becoming an increasingly important area of study, and the observable effects that these three factors may exert on one another may be seen sooner rather than later within our society as the onset of obesity is on the rise, meanwhile those who are obese are being affected at younger ages. The CDC reports that an estimated 36 % of adults in the U.S. are considered obese, while the percentage of adults who are overweight but not obese is estimated at an additional 33 %. By 2030, these figures will rise to about 75 % of the U.S population being overweight with 42 % of the U.S. population considered obese and 11 % severely obese. Thus, the potential impact of healthy aging, adipose tissue accumulation, and obesity upon affecting the immune response to cancerous cells as well as responses in general to immunotherapy merit a detailed examination and is the focus of this chapter.

A. Mirsoian · G. D. Sckisel · A. E. Zamora · W. J. Murphy (✉)
Department of Dermatology, Davis School of Medicine, University of California,
2921 Stockton Blvd, Suite 1614, Sacramento, CA 95817, USA
e-mail: wmjmurphy@ucdavis.edu

W. J. Murphy
Department of Internal Medicine, Davis School of Medicine, University of California,
2921 Stockton Blvd, Suite 1614, Sacramento, CA 95817, USA

M. R. Shurin et al. (eds.), *The Tumor Immunoenvironment*,
DOI: 10.1007/978-94-007-6217-6_9,

Keywords Obesity · Adiposity · Cancer · Aging · Inflammaging

Abbreviation	*Term*
IFN ($-\gamma$, $-\alpha$)	Interferon (gamma, alpha)
IL (-1β, -2, -6, -12, -15)	Interleukin (-1 beta)
IT	Immunotherapy
FDA	Food and drug administration
HCL	Hairy cell leukemia
AIDS	Acquired immunodeficiency syndrome
CML	Chronic myelogenous leukemia
RCC	Renal cell carcinoma
CTLA-4	Cytotoxic T-lymphocyte antigen 4
TNF ($-\alpha$)	Tumor necrosis factor alpha
PPARγ	Peroxisome proliferator-activated receptor gamma
NF-κB	Nuclear factor-kappa B
FFA	Free fatty acids
ATD	Adipose tissue dysregulation
CRP	C-reactive proteins
MCP-1	Monocyte chemotactic protein-1
VEGF	Vascular endothelial growth factor
STAT	Signal transducer and activator of transcription
EGFR	Epidermal growth factor receptor
CARD11	Caspase recruitment domain-containing protein 11
TIL	Tumor infiltrating lymphocyte
TCR	T cell receptor
Aire	Autoimmune regulator
TECs	Thymic epithelial cells
CR	Caloric restriction
CDC	Center for disease control
NCI	National cancer Institute
FoxP3	Forkhead box P3
M1	Classical macrophages
M2	Wound healing macrophages
TAMs	Tumor associated macrophages
DIO	Diet induced obese
MIF	Migration inhibitory factor
TA-DCs	Tumor associated-dendritic cells
GM-CSF	Granulocyte macrophage-colony stimulating factor
PGE_2	Prostaglandin E2
M-CSF	Monocyte- colony stimulating factor
LN	Lymph node
DC-SIGN	Dendritic cell-specific intracellular adhesion molecule-3-grabbing non-integrin
TAA	Tumor associated antigens
APC	Antigen presenting cell

LIF	Leukemia inhibitory factor
ALT	Alanine aminotransferase

9.1 Cancer and Inflammation

9.1.1 Underlying Principles and Hallmarks of Cancer

The complexity of cancer and the dynamic processes orchestrated and shaped by the influence of inflammatory cells, cancerous cells, stroma, and the extracellular matrix on the tumor environment, and conversely, the adaptations made by the tumor in response to environmental pressures, makes studying cancer an arduous task, particularly with the impact of variables such as age and body mass. Both aging and fat content is associated with inflammation. Inflammation may not only initiate, but may also facilitate and promote cancer progression. Under normal conditions, the inflammatory response is self-limiting with anti-inflammatory cytokines being produced shortly after pro-inflammatory cytokines (Philip et al. 2004; Coussens and Werb 2002). If the response persists, chronic inflammation may ensue, thus potentiating an increased risk of dysplasia and cancer (Grivennikov et al. 2010; Rakoff-Nahoum 2006). Inflammation and genetic instability allow cancer cells to sustain proliferation (RAS, MYC, RAF), evade growth suppressors (RB, p53), resist cell death (Bcl-2, Bcl-X_L, Bax, Bim, Puma), replicate indefinitely (telomerase), induce angiogenesis (VEGF-A, TSP-1), and activate invasion and metastasis (E-cadherin, EMT-Snail, Slug Twitst, Zeb 1/2) (Gerlinger et al. 2012; Hanahan and Weinberg 2011). The physiological changes which occur in response to aging and obesity contribute towards the development of an aberrant and persistent inflammatory response. As such, these respective changes that occur within both obesity and aging along with their consequences has led to increased interest in exploring their potential links with cancer, which will be the focus of this review.

9.1.2 Immunosurveillance and Immunoediting

For over a century, much debate has arisen surrounding the idea proposed by Paul Ehrlich in 1909 that the immune system could eliminate the growth of carcinomas (Dunn et al. 2004). The concept of immunosurveillance maintained that adaptive immunity was responsible for preventing cancer development in immunocompetent hosts (Dunn et al. 2002). Further evidence supporting the idea of immunosurveillance came from studies in the 1990s and into the early 2000s. In 1994, Schreiber and colleagues showed the importance of IFN-γ in promoting immunologically-induced rejection of fibrosarcoma tumor cells in syngeneic mice

(Dighe et al. 1994). In 2001, Shankaran and colleagues more specifically showed that mice lacking adaptive immunity ($RAG2^{-/-}$ mice lacking T, B, and NKT cells) were more susceptible to carcinogen-induced sarcomas and spontaneous epithelial carcinomas (Shankaran et al. 2001). While both of these studies were instrumental in rekindling interest in and validating the concept of cancer immunosurveillance, the latter expanded on this concept, showing the immune system also functions in immunoselection or immunoediting, which effectively reduces immunogenicity and promotes tumor survival. Cancer immunoediting, unlike immunosurveillance, is a dynamic process characterized by three distinct phases: elimination, equilibrium, and escape (Dunn et al. 2002). The elimination phase can be seen as a revised version of cancer immunosurveillance, taking into consideration the impact of both the innate and adaptive immune branches in locating, detecting and eliminating developing tumors (Schreiber et al. 2011). The elimination of detectable tumor cells, those that are strongly immunogenic, results in the initiation of the second phase, equilibrium. During equilibrium, weakly immunogenic cancer cells survive, resulting in their outgrowth. Likewise, this second phase allows the occurrence of the third phase, escape. As cancer cells are able to mutate and transform the natural selection process allows for phenotypic characteristics that allow immune evasion to predominate (Schreiber et al. 2011). While the specifics of each of these three phases are beyond the scope of the current review, they highlight the central role that the immune system plays in the development and progression of tumors, establishing their phenotypic identity, and determining their antigenic expression particularly when immunotherapy is applied in cancer therapies.

9.1.3 Immunotherapy Toxicity Limits Efficacy: Scrutiny of Preclinical Models

Within the past two decades, the field of immunotherapy (IT) has experienced much advancement in the development of immunomodulatory agents but only a few have progressed into clinical trials, with an even lesser number of these agents gaining FDA approval as therapeutic regimens (National Institutes of Health Clinical Trials 2012) (Fig. 9.1). The efficacy and application of immunotherapeutic regimens, clinically, has been limited by the induction of systemic, and at times limiting, toxicities in patients. In particular, this has been most severely observed within immunomodulatory cytokine-based therapies that include IL-2, type I Interferons (IFN-α) and IL-12. Clinical application of high dose IL-2, targeting NK cells and $CD8^{+}$T cells, for treatment of metastatic melanoma and metastatic renal cell carcinoma (RCC) resulted in a 10 % response rate for increased overall median survival (approximately 11 months), and about 2 % of patients experiencing complete regression (Youm et al. 2010; Rosenberg et al. 1994; Schwartzentruber et al. 2011). However, these promising results are limited

Perspectives on Cancer Immunotherapy

Given the dramatic changes that occur within the immune system over a lifespan and during obesity, there is a paucity of data regarding how these alterations may affect the outcome of immunotherapeutic interventions. During the past decade, our understanding of the immune system and cancer has increased exponentially, only a small portion of which has yielded positive translational results. While there are nearly 800 clinical trials (71) involving immunotherapy and cancer, the FDA has only approved four immunotherapeutic interventions for treatment: IL-2, IFN-α, Sipuleucil-T (DC based vaccine), and ipilimumab (anti-CTLA4). Even though these treatments have received approval by the FDA, it is important to note that such therapies have yielded modest results, at best. While several factors likely contribute to this discrepancy, one glaring explanation may be the use of unrepresentative preclinical models, namely with regard to age and body fat content. The majority of preclinical studies use relatively young mice that are 8-12 weeks old, the equivalent of a young adult in humans, while the average age at the onset of cancer is 65. Throughout the chapter we discuss how the aged and/or obesity dysregulates the immune system with permanent alterations that may affect the outcomes immunotherapeutic regimens and their clinical trials.

Type of Treatment	Status of treatment	Pros with age/high body fat	Cons with age/high body fat
Direct Vaccination	No FDA-approved therapies; Various phases of clinical trials over various cancer types; No significant benefit in phase III; Some trials experimental group worse than controls	Increased expression of co-inhibitory molecules allows for potential target with inhibitory checkpoint blockade combination therapy	Decreased T cell diversity and function limits naïve T cell responses to tumor antigens; Increased T regulatory cells limits new T cell responses
DC-based Vaccination	1 FDA-approved therapy: Sipuleucil-T; modest increases in survival	Improved antigen presentation potential of DCs	Defective DC function may limit efficacy; defective T cell diversity and function limits novel T cell responses to antigen
CARs	No FDA-approved therapies; Currently in Phase I/II testing for variety of malignancies; Promising early results	State of T cell likely does not matter as T cell is genetically modified to become activated and recognize antigen; Possible increase in efficacy when derived from memory cells which are more abundant in aged/high body fat with skewed memory/naïve ratios	Increased inhibitory cell types such as T regulatory cells may hinder efficacy
Immunostimulants (ie., cytokines, agonistic antibodies)	2 FDA-approved therapies: IL-2 and IFN-α; Numerous others attempted clinically with limited success; often accompanied with gross systemic toxicities	Increased involvement of antigen-nonspecific memory T cells (which may act "NK-like") due to skewed memory-naïve ratio	Increased toxicities due to chronic, low grade inflammation
Inhibitory Checkpoint Blockade	1 FDA-approved therapy: Ipilimumab; anti-PD-1 undergoing phase I/II clinical trials showing promising results in skin and lung cancer	Increased efficacy as T cells from aged individuals tend to have higher expression of co-inhibitory molecules	Increased toxicities due to chronic, low grade inflammation

Fig. 9.1 Perspectives on Cancer Immunotherapy

by the observation that patients undergoing treatment with IL-2 require intensive management due to severe toxicities including capillary leak syndrome and severe pulmonary edema, comparable to the induction of systemic septic shock. Similarly, IFN-α is approved for the treatment of stage II and III melanoma, AIDS related Kaposi's sarcoma, hairy cell leukemia (HCL), chronic myelogenous leukemia (CML), and co-administration with bevacizumab for advance RCC. IFN-α's efficacy is best seen in HCL where low-dose therapy resulted in an overall response rate of 77 % and a complete recovery response rate in 5 % of patients (Fabrizi et al. 2011). Similarly, IFN-α treatment has led to systemic side effects in the form of hematological toxicities, elevated transaminases associated with liver damage, and neuropsychiatric side effects (mania, depression) requiring the usage of anti-psychotics (Flanigan et al. 1999; Vignaux et al. 1999). Additionally, adverse immune reactions have been documented in anti-CTLA-4 (Ipilumimab) treatment administration in various cancers, which have most notably resulted in autoimmune-associated disorders, such as colitis. Systemic IL-12 administration resulted in dyspnea, stomatitis, leukopenia, hyperbilirubinemia, elevations in transaminases, and thrombocytopenia that occurred in greater than 20 % of patients (Kantoff et al. 2012; Weber 2007). Thus, while significant responses can be observed with IT application, significant toxicities can also result.

The advancement of drugs into phase I trials is dependent upon their safety and efficacy confirmation through usage of preclinical models. Often, studies evaluating the efficacy of these treatments are carried out in young, lean, mouse models that are most commonly housed in specific pathogen-free environments. Likewise, preclinical experiments employ the usage of fast-growing, aggressive, immortalized cell lines that assist to mimic cancer models but may not necessarily be reflective of the long-term progression of human cancers. The importance of evaluating the usage of proper preclinical models was observed in 2010 during the phase I evaluation of CD28 superagonist (TGN1412); an antibody initially directed as a potential immunosuppressant of T cells. Six healthy volunteers recruited into the trial received a single low-dose intravenous injection of CD28 that initiated life-threatening cytokine-release syndrome in all participants. The adverse reaction resulted in their transfer into intensive care within 16 h of the initiation of the trial (Suntharalingam et al. 2006). Investigations aimed at determining the cause of the severe reaction exposed the usage of inappropriate preclinical models (mouse, non-human primate and in vitro effects on human T cells) as root causes. The observed reaction was the result of hyper-stimulation of $CD4^{+}T$ effector memory cells present within the tissues of humans, which are lacking in mouse models due to their clean housing conditions and species differences. Furthermore, follow-up preclinical experiments evaluating TGN1412 employed non-human primates, macaques, whose T-cells, contrary to human T cell physiology, lose CD28 expression during their differentiation into effector memory cells (Hunig 2012). The failure to induce a toxic reaction within these animal models was mistakenly interpreted as evidence that no such reaction would result. The severity of consequences observed in the TGN1412 trial emphasizes the critical need to evaluate common used experimental models. Interestingly, the impact of age as a variable was not assessed.

In line with studies highlighting the role for appropriate models, our laboratory has recently demonstrated marked differences in toxicities following strong immune stimulation using IT regimens that could induce potent anti-tumor effects between young and aged mice (Murphy et al. 2003). Studies using aged mice resulted in rapid lethality within two days of treatment, a marked increase in systemic cytokines, and multi-organ damage. In stark contrast, young mice were able to tolerate treatment without evidence of sustained systemic toxicity or multi-organ damage. Within these experiments, it was noted that aged mice exhibit increased body mass indexes, in comparison to their younger counterparts. Aged mice exhibited increased systemic cytokines –particularly IL-6, IFN-γ, TNF-α– whereas caloric-restricted aged mice had much less (manuscript submitted) suggesting distinct roles for both age and fat in systemic toxicities observed after strong immune stimulation.

Taken together, our findings indicate that both age and adiposity contribute to the observed toxicity and intolerance to IT regimen administration. Importantly, these observations also indicate that IT efficacy can be negated by the induction of severe toxicities in aged and obese models, calling into question the appropriateness of using young, lean, models in determining immune modulating parameters. When viewed in conjunction with the results of the TGN1412 trial, our findings highlight the need for an in-depth analysis of the immunological consequences of aging and obesity as these findings are suggestive that toxicity will also limit IT dosing and thereby may affect anti-tumor efficacy in elderly or obese cancer patients. Throughout this chapter we will discuss how the aged and/or obese immune system results in skewed responses and how, together, these alterations may affect the outcomes of the different cancer immunotherapeutic regimens that are attempted clinically.

9.2 Lipid Deposition, Aging, Inflammation and Cancer: Three Factors Influencing One Disease

9.2.1 Aging and Obesity Affect Immune Cell Distribution and Function Within Tissues

The immune system functions in an orchestrated manner to establish and maintain homeostatic conditions that allow for the destruction of foreign invaders and abnormal cells. With healthy aging, the function and diversity of immunocytes is altered, leading towards a diminished ability to respond to antigen stimulation as well as decreased repertoire diversity. The observed correlation between cancer and age has long been established, but the impact that an aging immune system exerts on the development of cancer, as well as in its response, remains to be

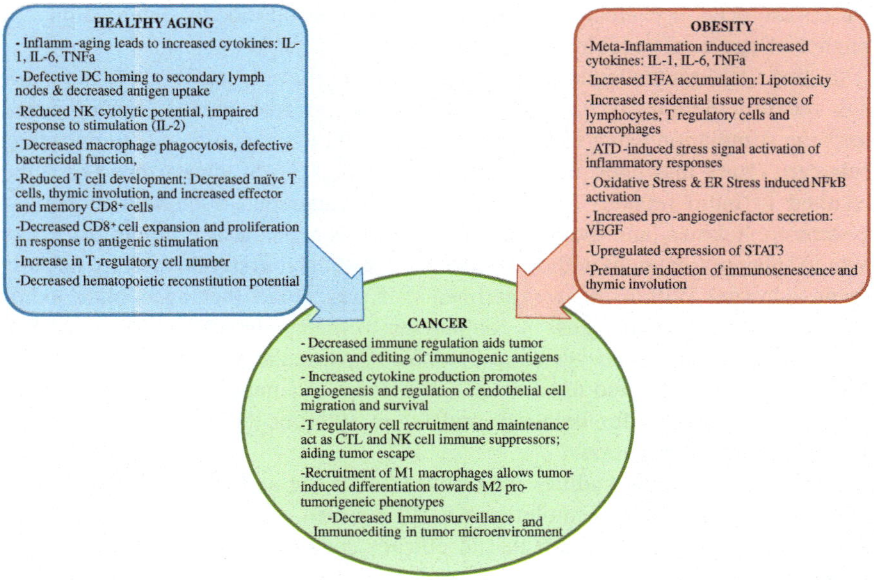

Fig. 9.2 Obesity and Aging Share Pro-Cancerous Characteristics. Although aging and obesity are two separate phenomenons, their resulting altered immune system and characteristics are not mutually exclusive in their pro-tumorigeneic potential. Both share characteristics that aid in tumor growth and cancer disease progression

further elucidated. Aging is a dynamic process characterized by physiological changes and decreased immunological functions over time that can ultimately favor tumorigenesis. In this section we will discuss various processes that occur during aging, their association with the tumor environment, and how they affect patient prognosis (Fig. 9.2).

Like the aged microenvironment, which is characterized by heightened expression of pro-inflammatory cytokines, the obese environment also contains increased levels of pro-inflammatory cytokines IL-1β, IL-6, TNF-α, decreased secretion of anti-inflammatory adiponectin, and decreased expression of anti-inflammatory peroxisome proliferator-activated receptor gamma (PPARγ) (Wu et al. 2007). Conjointly, build-up of lipid deposition is strongly linked to increased oxidative stress and endoplasmic reticulum stress which, in turn, activates the NFkB pathway leading to increased expression of proinflammatory cytokines (Wu et al. 2007). Increases in adipose tissue with aging may thus play an amplifying role in the development and progression of age-related diseases such as cardiovascular disease, type 2 diabetes mellitus, and cancer (Tzanetakou et al. 2012) (Fig. 9.2).

Build-Up of Lipid Deposition Leads to Increased Production of Pro-Inflammatory Factors Favoring a Tumor Promoting Immuno-Environment

Adipose tissue serves as a highly active endocrine and metabolic organ. Its main role is to regulate glucose and lipid metabolism with energy balance (Maiorana et al. 2007). Previously, adipose tissue was thought to simply be an energy storehouse that responds to insulin levels by allowing adipocytes to either store or mobilize triglyercides (Spiegelman and Flier 2001; Hill et al. 2000). Thus, in this respect, obesity was traditionally defined as an imbalance between energy expenditure and energy intake.

The link between adipose deposition and inflammation was first proposed by H.P. Himsworth in the 1930s in which he proposed that there were two different types of diabetics, those that were insulin sensitive and those who were insulin resistant, thus starting a field of study paving the way towards identifying the root causes of diabetes and the differences between these two states. Further studies in the late 1950s into the early 1960s proposed that blood obtained from diabetics contained circulating "antagonists", serum antibodies against insulin, which shed light on the possible link between inflammation and increased adipose deposition (Field and Rigby 1959; Vallance-Owen 1960). In 1993 Hotamisligil et al. established the first strong link with inflammation in a study conducted using obese and non-obese mice. He reported that TNF-α mRNA expression and protein levels were increased in obese mice, locally and systemically. By neutralizing TNF-α levels in the obese models, his study also found that glucose uptake and response to insulin was restored (Hotamisligil et al. 1993; Hotamisligil and Spiegelman 1994). With the subsequent discovery of adipokines and the ability of adipocytes to actively produce pro-inflammatory markers, adipose tissue is now considered an active endocrine organ with immunomodulatory capabilities (Saltiel 2001). As a result, researchers are now documenting the link between inflammation and obesity, and how the increased inflammatory state contributes towards neoplastic diseases.

In obesity, the cross-talk between adipocytes, non-adipose cells, and immune cells is thought to lead towards the development and continuation of a "meta-inflammatory" state, defined as a state of chronic low-grade systemic inflammation, which is responsible for metabolic diseases. The dysregulation of adipokine production, pro-inflammatory cytokine release, decreased production of adiponectin, and loss of ability to properly store free fatty acids (FFA) are hallmarks of adipose tissue dysregulation (ATD) (Skurk et al. 2007; Coppack et al. 1992). It has become well documented that obesity also corresponds with high serum levels of C-reactive protein (CRP), and increased serum levels of pro-inflammatory cytokines TNF-α, IL-1β, IL-6, and increased chemoattractants CCL2 and MCP-1 (Franks 2006; Cindik et al. 2005). These hallmarks together indicate that obesity leads towards an active immune state and induction of inflammatory responses (Fig. 9.3). Obesity has been attributed as a possible causal agent to at least nine cancer types: esophageal, pancreas, colon, rectal, breast, endometrial, kidney,

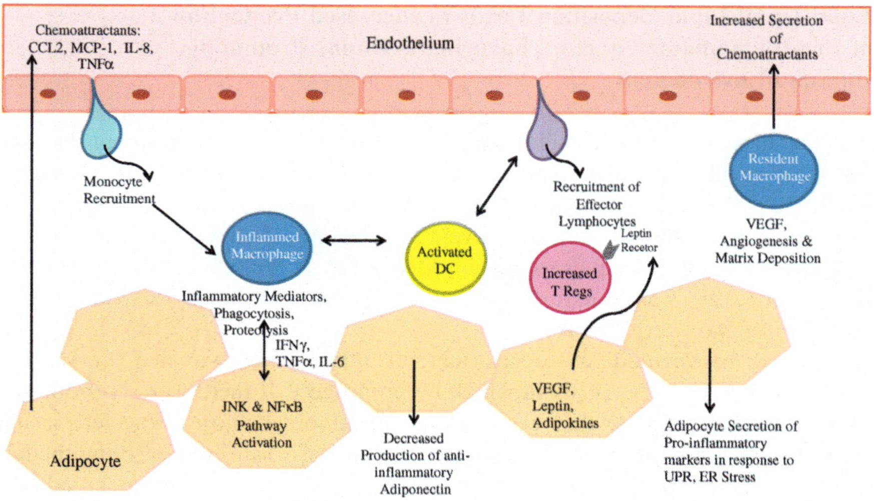

Fig. 9.3 Obesity and Inflammation. In obesity, adipocytes secrete pro-inflammatory factors that result in activation of resident macrophages and the recruitment of immune effector cells into the tissues, namely monocytes, T cells, and DCs. A secondary consequence is the activation of the NF-κB and JNK pathways, which result in the further production and secretion of pro-inflammatory cytokines: IL-1, IL-6, TNF-α. Together, the recruitment of effector cells, activation of resident macrophages, and activation of pro-inflammatory pathways results in the chronic low-grade inflammation that leads to tissue destruction and disease promotion

thyroid, and gall bladder. Although the percentages of cases that were attributed to obesity varied within each tumor type, some were as high as 40 %. In part, the relationship between obesity and cancer is due to the inflammatory rich environment induced by obesity, which allows for tumor promotion.

ATD results in the secretion of dysregulated proliferative signals, such as increased tyrosine kinases, that aid in the promotion of downstream cascades and can alter negative-feedback mechanisms in cell signaling. Moreover, tumor survival and growth is dependent on the ability to induce angiogenesis (Hanahan and Weinberg 2011). In obesity, levels of VEGF, a pro-angiogenic signal protein, increases with the expansion of adipose tissue and is actively secreted by adipocytes for the promotion of vascularization, providing a possible role for adipocytes in tumor promotion. Studies involving melanoma B16F10 cells injected into lean WT, obese melanocortin receptor 4 knockout (MC4R$^{-/-}$), obese leptin-deficient (ob/ob$^{-/-}$), and lean ob ($^{+/-}$) mice resulted in the observation that both obese MC4R$^{-/-}$ and obese ob/ob$^{-/-}$ mice had the highest protein expression of VEGF receptor 1 and 2, supporting the relationship between obesity and angiogenesis (Brandon et al. 2009). Likewise, within this same study, tumors were much larger in the two obese models, when compared to their lean counterparts (Brandon et al. 2009). Not only is obesity associated with an increased risk of cancer, the usage of Metformin® (biguanide), an oral drug prescribed to treat insulin dysregulation in obesity-induced type 2 diabetes, has shown potential to decrease cancer incidences in obese patients. It is hypothesized that biguanide directly inhibits cancer cell growth via inhibition of AMP-activated protein kinase on the mTOR pathway, and does so in a dose-dependent manner with greater length of treatment conferring greater benefits (Bo et al. 2012). Currently, randomized clinical trials are on-going evaluating the usage of biguanide as an adjuvant in breast cancer.

Furthermore, obesity-induced activation of inflammatory cell signaling pathways led to increased NF-κB signal induction and STAT3 activation. Since obesity induces a meta-inflammatory environment, NF-κB signaling is constitutively active within adipose tissues and surrounding immunocytes fostering an autocrine response of continual activation and cytokine production. This prolonged activity has become associated with various necessary steps in cancer progression, including promoting tumor proliferation, preventing apoptosis, and increasing its metastatic potential. Moreover NF-κB activation has been described as necessary for the survival and progression of various solid tumors. NF-κB was first discovered to be pro-tumorigeneic with the discovery of v-Rel, a homolog of the NF-κB subunit c-Rel (Gilmore 2003). Since the discovery of v-Rel, NF-κB has been well documented to be constitutively expressed in many human malignancies, including colorectal cancer (Dolcet et al. 2005; Ben-Neriah and Karin 2011). In 2007, Scartozzi et al. used NF-κB expression on colorectal tumor cells not only as a new molecular marker to diagnose cancerous cells but also as a predictive tool to determine patient response and prognosis to treatment with Cetuximab®, a chimeric monoclonal antibody that functions as an epidermal growth factor receptor (EGFR) inhibitor (Scartozzi et al. 2007).

Additional studies have shown NF-κB constitutive activation to induce greater B cell proliferation and survival, ultimately inducing uncontrolled accumulation of B cells and mutations to occur in CARD11, implicating NF-κB in the development of B cell lymphomas (Lenz et al. 2008). Given this information it is not surprising that NF-κB signaling has, therefore, been linked to leukemia, lymphoma, bone, breast, liver, endometrial and colon cancers, though this list is not all-inclusive (Karin et al. 2002). Similarly, adipocyte production of IL-6 in obesity leads to the upregulation of STAT3. In cancer, STAT3 has been well documented to be overexpressed and phosphorylated in tumor cells and, like NF-κB, affects cellular proliferation, survival, and signaling. Unique to STAT3 is the induction of a stem-cell like phenotype, inducing the expansion of cells susceptible to mutagens (Chen et al. 2008). Support for STAT3's role in tumor stem cell expansion is seen in studies where STAT3 inactivation results in a disruption of angiogenesis and decreased tumor growth (Joyce and Pollard 2009). Conversely, STAT3 has been shown to play a central role in cancer progression by affecting the recruitment of tumor-infiltrating cells and the production of chemoattractants. Kasmi et al. documents that within the tumor microenvironment, expression of STAT3 via IL-6 favors tumor progression by down-regulating immune activation that could have negative effects against the tumor, serving as a negative regulator of Th-1 mediated inflammation. STAT3 alters chemokine and cytokine production towards the maturation of alternatively activated M2 macrophages which promote tumor growth (El Kasmi et al. 2007). Thus, STAT3 promotes continual immune evasion through an immunosuppressive role and allows the progression of an established neoplastic disease by producing a favorable tumor microenvironment. The simultaneous activation of the NF-κB pathway results in the production of pro-inflammatory cytokines, such as IL-6, IL-1β, TNF-α correlating with the hallmark associated infiltration of $CD8^+$ T cells, macrophages, NKT cells, and immature myeloid cells into the adipose tissues.

9.2.2 Immunosenescence

Immunologically, parallel observations are observed on the impact of age with respect to decrease in cellular repertoires, decreased response to stimulation, and overall decreased hematopoietic output as a person ages. These immune system alterations that occur as a result of healthy aging are known as "immunosenescence". Within the tumor environment, one's ability to appropriately induce immunosurveillance and immunoediting first begins with the ability to recognize an abnormal cell, the sensitivity of which may be decreased as one ages. Furthermore, T cells, NK cells, macrophages, and dendritic cells are known to infiltrate solid tumors and their presence directly correlate with patient cancer prognosis. Thus a diminished response brought upon by aging puts one into a disadvantaged immunological state that may favor tumor formation and tumor growth.

Although immunosenescence is associated with aging, emerging data is providing evidence that a chronic pro-inflammatory state within an obese model can prematurely age the immune system. It was previously discussed in this chapter that an obese model is characteristically chronically inflamed, suffers high production of stress signals, and results in an oxidative environment. Together, these factors may contribute towards an immune system that is less able to respond to cues, either internal or external, and can prematurely age the immune system by inducing parallel phenotypic changes that are seen with aging. Thus, where immunosenescence is traditionally defined as a process of aging, new data may be suggestive that immune dysfunction in a young obese model may lead to more severe consequences as an obese person ages since their immune system is at a disadvantaged, less functioning, state earlier in life.

The Lymphocyte Compartment and Cancer

It is widely recognized that tumor infiltrating lymphocytes (TILs) are central in delaying tumor progression and orchestrating a regressive stage in some cancers. The crucial role of TILs is evidenced by their association with a positive prognosis in solid tumors, with the ratios between T cell subsets being widely used as a predictive prognostic tool (Zhang et al. 2003; Leffers et al. 2009). $CD8^+$ cytotoxic T cells (CTLs) play a crucial role in direct anti-tumor responses as they represent the main effector cell capable of targeting and killing cancerous cells. Within the tumor environment, CTLs are crucial in releasing perforin, a cytotoxic factor that allows membrane permeabilization, as well as initiation of Fas mediated cell death. Conversely, CD4$^+$ T-helper cells play an orchestrating role via secretion of effector molecules supporting CTL activity and skewing the Th1/Th2 balance towards a Th1 phenotype for effective anti-tumor responses. This skewing affects the differentiation and maturation of not only CTLs but also mononuclear phagocytes (MPCs) within the tumor microenvironment, maximizing the killing potential of both cell types (Pawelec et al. 2002). Importantly, Th1 skewing results in the production and release of the immunomodulatory cytokine IFN-γ. IFN-γ has been shown to facilitate tumor surveillance and promote immune responses against tumors by arresting tumor growth and inducing the expression of MHC-I for antigen presentation (Taub and Longo 2005). In the following sections we will discuss how aging and obesity contributes towards alterations in the lymphocyte compartment, which may aid in driving tumorigenesis and establishing cancer progression.

A. *Thymic Involution: T cell Compartment Remodeling Alters the Memory to Naïve Ratio and Can Result in Aging-Associated Decreases in Function*

With age, the most evident and well-studied change within the immune system is seen in the T cell repertoire. In 1985 Steinman described a phenomenon where the thymus progressively decreases in size, beginning immediately after birth and continuing throughout one's life. Today, this process has been coined as "thymus involution". This change in structural organization of the thymus led to the crucial observation that, with age, there is a decrease in naïve T cell output, with a concurrent increase in the number of memory T cells (Lynch et al. 2009; Goronzy et al. 2007; Steinmann et al. 1985; Taub and Longo 2005).

Thymic remodeling and involution has also been documented to occur prematurely as a result of obesity, adding an additional level of complexity to the direct cross-talk between adipose tissue and the immune system. In experiments using diet-induced obese mice, obesity resulted in reduced thymocyte counts, acceleration of the age-related reduction in T cell receptor (TCR) repertoire, reduction in naïve T cell output, and increased frequency of effector-memory phenotypes (Youm et al. 2010; Luedke et al. 2012). Furthermore, obesity resulted in reduced expression of IL-7 and autoimmune regulator (Aire) within the thymic epithelial cells (TECs). These findings suggest that obesity induces similar effects upon the thymus, thus mimicking senescence and compromising the T cell compartment independent of age. More importantly, this data suggests that adipose-immune interactions also occur within specific tissue microenvironments and impose greater deficiencies in immune system responses that could be associated with increased disease susceptibility. In contrast, caloric restriction (CR) has been shown to slow down the process of thymic involution, be coupled with maintenance of thymic function, and play an immune enhancing role on NK and CTL activity (Yang et al. 2009). Additionally, CR led to a decreased incidence of tumors, reduced expression of oncogenes, and enhanced expression of tumor suppressor genes (Dirx et al. 2003; Engelman et al. 1995; Cuenca et al. 2001; Fernandes et al. 1995). The link between adiposity and the immune system may lead to premature aging in immune cell compartments and function, thus contributing to cancer incidence and progression.

Furthermore, maintenance of T cell homeostatic proliferation, to maintain the size of the T cell compartment, is also hindered. Researchers have shown that age-related changes include a loss in ability to rebuild a diverse repertoire by the age of 50, and a loss in the ability to maintain homeostasis by the age of 70 that can lead to a decreased diversity up to 100-fold between the ages of 65–75 (Goronzy and Weyand 2005). In general, it should be noted that the CDC and NCI report that the average age of cancer diagnosis occurs over the age of 55. These age-related changes and decreased proliferative potential come at a crucial time, and their correlation with cancer onset and diagnosis merit further study. Taken the importance of T cell activation and response in cancer with the effects of aging on the T cell compartment, it becomes evident that loss of T cell memory responses through aging can be linked to the increased incidence of cancer in elderly populations. Diminished repertoires and responses may affect the ability of T cells to appropriately conduct surveillance and editing, allowing for opportunistic tumor growth. Nonetheless, the importance of T cell coordinated activity in the tumor microenvironment is currently gaining momentum as researchers strive to perfect adoptive T cell transfers as novel cancer therapeutic agents. Importantly, this variable of age can have a dramatic impact when attempting to apply IT in cancer patients.

The overall shift of T cell phenotype from naïve ($CD44^{Lo}$) to memory ($CD44^{Hi}$) as a result of thymic involution and the decreased diversity found in the T cell compartment diminishes the spectrum of antigens that T cells can recognize, thereby limiting the induction of antigen-specific responses (Taub and Longo 2005). Studies have demonstrated that aging is associated with decreased ability to stimulate antigen-specific $CD8^+$ T cell responses due to aging induced $CD44^{Hi}$ $CD8^+$ T cell expression of inhibitory receptors including PD1, LAG3, 2B4, and CD160 (Decman et al. 2012). These aged $CD44^{Hi}$ $CD8^+$ T cells resulted in a phenotype similar to exhausted $CD8^+$ T cells that are found during chronic infections.

Recent studies evaluating the effects of aging on the $CD4^+$ population have indicated that the aged immunoenvironment negatively affects $CD4^+$ T cell functions in their ability to respond to antigen stimulation. They highlight that three steps are affected: hindered recruitment of $CD4^+$ cells into the spleen, reduced priming potential of $CD4^+$s by DCs, and reduced transition to a T follicular helper cell phenotype, which in turn impairs creation of germinal centers (Lefebvre et al. 2012; Thoman 1997). This reduced ability to function is also seen within the $CD8^+$ T cell population as a result of health aging. With aging, $CD8^+$ T cells show resistance to apoptosis, permanent loss of CD28 expression, altered cytokine profiles, and reduced ability to respond to stress (Pappo et al. 2001; Spaulding et al. 1999). CD28 is an important co-stimulatory receptor that is responsible for antigen-mediated T cell activation, proliferation and survival of T cells. Thus, the observed loss of CD28 expression may account for the documented observation that these cells also exhibit reduced antigen receptor diversity and a shorter replicative lifespan while showing enhanced cytotoxicity (Weng et al. 2009). These findings may also, in part, explain the observation that elderly populations with high proportion of $CD8^+$ T cells undergoing replicative senescence correlate with reduced antibody responses to vaccines. Additionally, $CD8^+CD28^-$ T cells also accumulate in patients with certain types of cancer, demonstrating a link between this age-associated phenomenon and development of neoplastic disease. Thus, aging not only shifts the distribution of T cells but also may affect their total numbers, proliferative potential, and activity.

The limited naïve repertoire along with skewing towards memory phenotype during aging may implicate an expanded role for bystander activated T cells in aging as the ability to generate primary T cell responses diminishes with thymic involution. The expansion of $CD8^+$ T cells with a memory phenotype in response to cytokine stimulation in the absence of TCR-MHC engagement has been described extensively in response to bacterial and viral infections (Tough et al. 1996), though their functional role during infection, if any, remains controversial. We and others have more recently characterized bystander activation following cancer immunotherapy. Tietze et al. describes memory CD8 bystander cells becoming activated during immunotherapeutic stimulation as lacking CD25 and PD-1, markers associated with TCR engagement, and significantly upregulating granzyme B and NKG2D (Tietze et al. 2012), further suggesting a functional role for these cells. Furthermore, NKG2D blocking studies confirmed a role for bystander affiliated NKG2D in controlling tumor growth during immunotherapy. Therefore, we propose that bystander cells may act to control tumor growth in an NK-like fashion during periods of strong cytokine stimulation, such as occurs during immunotherapy (IL-15, IL-2, IL-12, CpG treatment) or infection, in addition to antigen specific responses (Tietze et al. 2012). Although the impact of aging and obesity on bystander T cells has not been well characterized, studies have implicated the high cytokine, inflammatory, environment present in aging to induce hyperactivation of bystander T cells and implicate them within the progression of autoimmune arthritis (Kobayashi et al. 2004). Their expansion during immunotherapeutic stimulation in aged or obese hosts may be even more dramatic leading to enhanced antitumor responses.

B. *T Regulatory Cells*

T regulatory cells (T-regs) are defined by their phenotypic expression of being forkhead box P3 $(Foxp3)^+$ $CD25^+$ and $CD4^+$ cells. Their main function is to serve in the maintenance of immunological self-tolerance but also play critical roles in the control of antitumor immune responses. Within the tumor environment, T-regs are recruited to the tumor tissues via cytokine and chemokines, namely CCL22 binding to CCR4, that can be produced by the tumor or other T-regs, where they receive further activating signals, which encourage their expansion. T-regs actively inhibit the activity of $CD4^+$ and $CD8^+$ T cells, dendritic cells, NK cells, NK/T cells, and B cells through secretion of inhibitory cytokines such as IL-10 and TGF-β, and cell-to-cell contact (Murakami et al. 2002; Chen 2006; Nishikawa and

Sakaguchi 2010). Given these properties, the presence of large proportions of T-regs and their activation is advantageous for tumor progression as it directly limits anti-tumor immunity. T-regs, therefore, represent one mechanism of immune evasion used by tumors for their growth and disease progression.

Studies focusing upon aging and its effects on human T regs have found no difference in function between young and elderly subjects. The frequency and capacity to suppress inflammatory cytokine production is seemingly preserved throughout the aging process (Hwang et al. 2009; Pawelec et al. 2002). Yet, observations have been noted of increased proportions of T-regs within the elderly and higher expression of inhibitory co-stimulatory markers such as CTLA-4 (Lages et al. 2008). In obesity, T-regs have been found to express both leptin (Ob) and leptin receptors (ObR). This finding links T-regs ability to respond directly to immune stimulation, with their potential to become activated by various other organ systems. The finding that T-regs express surface receptors for leptin indicates that $FoxP3^{+}$ T-regs cells increase with response to increasing leptin levels, as seen in obesity, suggesting a possible protective role against obesity induced inflammation and obesity associated diseases (Feuerer 2009; Winer 2009). A recently published study similarly showed that T-regs within adipose tissue actively express PPAR-γ to engage in actively suppressing adipose tissue inflammation (Hamaguchi and Sakaguchi 2012). Follow-up studies have resulted in conflicting data where PPAR-γ is preferentially expressed in tumors but its activation leads to growth inhibition, possible activation of apoptosis, and promotion of tumor cell differentiation. Although at first glance this mechanism is seemingly protective in the obese model, it could also be problematic in the co-incidence of cancer and obesity in the elderly. The dual role of T-regs in suppressing inflammation in obesity to protect against disease progression, could aid cancer progression by the very same mechanism of inhibiting immune responses and allowing the tumor to grow with little to no detection by the immune system.

Effects of Aging and Obesity in the Innate Immune Compartment

The Innate compartment of immunity represents the first line of defense against pathogens and disease progression. Several cell types contribute to innate immunity with mononuclear phagocytes playing a pivotal role. Another crucial innate cell involved in the detection and deletion of cancerous cells is the NK cell, which will be the focus of our next section.

A. *NK Cells*

A balance between activating and inhibitory signals helps facilitate NK cell function. Inhibitory receptors interacting directly with MHC class I molecules present on the host "self" cells, provide inhibitory signals, preventing detrimental killing of self-cells. Lack of inhibitory signals from self-cells results in stronger activating signals, which allows NK cells to actively kill pathogenic cells (Moretta et al. 1996).

Previous studies looking at the effects of aging on NK cell function led to contradictory results, but more recent studies suggest that NK cell function, phenotype, and number may be affected by immunosenescence. Analysis of total numbers of NK cells in the elderly have shown an increase in the total number present, with respect to age, as well as a shift in subsets toward a decline in $CD56^{bright}$ NK cells, subtype characterized as weakly cytotoxic but high producer of cytokines, and an increase in $CD56^{dim}CD57^{+}$ NK cells, the cytotoxic subtype (Tarazona et al. 2000; Solerte et al. 1997). Contradictory reports on the total number of NK cells should be taken with caution as this increase may be due to the overall decrease in total number of $CD3^{+}$ cells as a result of the changes in the T cell repertoire that occur with aging, as discussed in previous sections. Furthermore, some studies have suggested that the function of NK cells are impaired leading to decreased responses to IL-2 and IL-12, resulting in a decreased production of IFNγ (Solana and Mariani 2000). There is also an increase in the production of pro-inflammatory cytokines, namely IL-1, IL-6, IL-8, and TNFα in the elderly (Rink et al. 1998). Nonetheless, the importance of NK cell killing of cancerous cells is well defined and their alteration as a result of age could be implicated as a causal factor in the increased susceptibility towards infection and cancer in the aged population. Further analysis of NK cell subsets and markers other than CD56 expression need to be further investigated to determine subset-specific roles in tumor progression in the elderly population.

B. *Immunosenescence: Antigen Presenting Cells and Macrophage Activity*

Activation of an inflammatory response depends on its initiation through the innate branch of immunity. During infection, macrophages and dendritic cells are pivotal for the initiation of an adaptive response and ultimately pathogenic clearance. Macrophages can be polarized into two subtypes: classically activated and wound-healing. Classically activated macrophages (M1) are characterized by their high expression of chemokines and pro-inflammatory cytokines, thus they play a direct role in fighting infections. The second type, wound-healing (M2), is associated with regulating homeostatic conditions and injury repair. These cells are characterized by the ability to secrete various proteases that facilitate remodeling of tissues and repair, thus promoting angiogenesis (Martinez et al. 2008).

Macrophages play a dual role within the tumor microenvironment. M1 phenotype macrophages are associated with anti-tumor responses. Generally, as neoplastic cells proliferate, their expansion results in surrounding tissue damage and remodeling. The resulting pro-inflammatory "danger" signals result in recruitment of M1 macrophages, dendritic cells, and NK cells where copious amounts of IL-12 and IFN-γ are secreted (Tsung et al. 2002). Whereas the general presence of macrophages is necessary for recovery from "danger" signals, their long-term presence within cancer is also associated with an opposite, negative, prognosis for recovery. Within most solid tumors, macrophages account for the majority of infiltrating immunocytes and are consequently termed "tumor-associated macrophages" (TAMs). These macrophages are phenotypically of the M2 class and illustrate a causal relationship between tumor growth progression and inflammation. TAMs are able to secrete various chemokines, cytokines, and growth factors that promote tumor growth and angiogenesis (Lamagna et al. 2006; Kataki et al. 2002). Furthermore, research has long established the ability for tumors to actively recruit monocytes into neoplastic sites, via CCL2 (MCP-1) and CCL5 (RANTES), where they take residence and are promoted to differentiate into an M2 phenotype. These resident TAMs are capable of responding in hypoxic conditions within the tumor microenvironment and serve to secrete growth factors such as VEGF, thereby promoting angiogenic environments that are favorable for tumorigenesis and cancer spreading (Murdoch et al. 2004).

Defective activation and responses have been noted within the innate branch of immunity as a result of aging. During an acute response, the ability of macrophages to engage in antigen presentation is markedly reduced and is associated with decreased expression of MHCII and TLR expression (Gomez et al. 2007; Fortin et al. 2006; Herrero et al. 2002). Much research evidence shows that elevated levels of MIP-2, CXCL2, KC/CXCL1, and IL-1β within the aged microenvironment creates a pro-inflammatory rich niche that supports the observation of aging being characterized as a low-grade inflammatory state (Wu et al. 2007). Furthermore, studies in rat liver tumor models have demonstrated the large infiltration of macrophages into the liver in the presence of neoplastic tumors with a defective capability to perform and engaging in tumor surveillance (Hilmer et al. 2007). Thus, if aging results in a chronic inflammatory state where macrophages are continuously activated into producing pro-inflammatory factors, cells that undergo mutations may be encouraged towards a neoplastic state. Furthermore, this change within the immune environment of the elderly shows an immune system that is chronically under stimulation and begs the question: at what point do these cells become anergic from overstimulation and could this provide an advantageous environment for tumors to grow without threat of attack?

Macrophages also take center-stage within adipose tissues, thus being implicated as the main initiators of obesity-related inflammation. The ability of adipocytes to secrete macrophage-associated chemoattractant proteins MCP-1 and CCL-2 has been studied for the past several decades. Likewise, as both of these chemokines increase proportionally with adiposity, they are implicated in macrophage phenotype switching towards an M1 pro-inflammatory state and glucose homeostasis (Takahashi et al. 2003; Weisberg et al. 2006). DNA microarray analysis of white adipose tissue of normal and high fat diet-induced obese (DIO) mice demonstrated MCP-1 mRNA expression to be increased as much as sevenfold in DIO mice when compared to normal fed mice, and correlated with serum MCP-1 levels (Takahashi et al. 2003). Moreover, in vivo studies in obese mice have suggested MCP-1 to be an insulin-responsive gene, responding to the hyperinsulinemia that is common in obesity through over expression of MCP-1 genes with adipose tissue serving as a major source of MCP-1 production (Sartipy and Loskutoff 2003). Importantly, studies with both obese (db/db) and A-ZIP-Tg (MCP-1 deficient) mice demonstrated that absence of MCP-1 ameliorated insulin resistance and increased expression of the M2 phenotype macrophages (Tamura et al. 2008; Nio et al. 2012) .

The link between macrophages and adipocytes was strengthened with the discovery of macrophage migration inhibitory factor (MIF) expression in adipocytes; a lymphokine involved in macrophage regulation and suppression of anti-inflammatory glucocorticoids (Nishihira 1998). It should be noted that recent data suggests a controversial role of MIF in obesity as trials have produced variable results affected by age and gender. Yet, clinical trials have provided some support for the relationship between MIF and metainflammation as was evidence in Metformin® (an oral anti-hyperglycemic drug) clinical trials, which lowered MIF plasma levels with the resolution of hyperglycemic states (Ghanim et al. 2004). Together, these data suggest that macrophages play a central role in the induction and maintenance of an inflammatory state within the aged environment, adipose tissues, and within cancer.

C. *Dendritic Cells*

Similarly, the roles of dendritic cells as the main antigen-presenting cells of the immune system characterize them as a key player within the tumor environment. Their central role is observed in the fact that dendritic cells are key targets in the development of vaccines and immunotherapeutic anti-cancer agents; most famously Sipuleucel-T (Provenge®). Tumor-associated dendritic cells (TA-DCs) are immunosuppressive in nature and suffer a decreased ability to induce immune responses by impaired abilities to perform adequate antigen presentation. In part, TA-DCs display lower levels of co-stimulatory molecules and are able to produce pro-tumorigenesis factors similar to macrophages (Liu et al. 2009; Mantovani et al. 2004). It has recently become well characterized that cytokines and growth factors secreted by tumors include GM-CSF, IL-6, TGF-β, IL-10, M-CSF, VEGF, and PGE_2; all of which affect both function and differentiation of DCs into low performing phenotypes (Bennaceur et al. 2008; Bennaceur et al. 2009).

Within aging, both DC phenotype and activity are altered. Although total numbers of DCs are largely unaffected in the spleen, the spread of DC subsets endure a shift that results in a decrease in plasmacytoid DCs and $CD11c^+CD8^+$ DCs (Wong et al. 2010). Moreover, a study using melanoma models found that aged DCs experience an impaired ability to migrate to the LNs, were less efficient at inducing anti-tumor immunity through T cell activation, and characteristically displayed decreased DC-SIGN expression (Sunderkotter et al. 1997; Vukmanovic-Stejic et al. 2011). With altered ability to perform proper presentation, the ability of elderly patients to properly detect and present neoplastic antigens could easily become inhibited. As it stands, neoplastic cells generally are characterized as being weakly immunogeneic and poor stimulators of T cell responses, even when antigen is fittingly loaded onto APCs. Thus, the ability to successfully mount an adaptive anti-tumor response is dependent on successful cross-presentation of tumor antigens to $CD8^+$ T cells and the inhibition of this process could greatly hinder the process of immunosurveillance (Petersen et al. 2010). This observation of decrease in functionality may account for one of the central factors in the increased incidence of cancer among the elderly, with a correlation that increases with increases in age. Additionally, this process has become the target of antigen-loaded DC vaccines, namely Provenge®, which functions by pulsing dendritic cells with a respective TAA and infusing them into the tumor microenvironment. This pre-loading allows for DC priming and better activation of presentation (Wesley et al. 2012).

Within the obese model, dendritic cells are elevated in obesity and contribute to the activation and promotion of macrophage infiltration into adipose tissues. Furthermore, studies using obese mice have accounted that these dendritic cells were less able to induce potent T cell proliferation and thereby reduced T cell responses. Most crucially, a recent publication has observed that obesity models with renal cell carcinoma (RCC) displayed impaired dendritic cell function. These researchers indicate that obese tumor bearing mice displayed altered serum cytokine profiles, upregulating IL-17 and LIF, had elevated percentages of splenic DCs but yet these DCs were impaired in their ability to stimulate naïve T cell expansion. Furthermore, obese tumor bearing mice failed to respond to DC-dependent immunotherapy, had decreased local infiltration of IFN producing $CD8^+$ T cells and overall had progressive tumor outgrowth (James et al. 2012). This data suggests that although DC function is not outwardly impaired at first glance, when challenged with a tumor environment, they may produce impaired responses with detrimental consequences in their ability to initiate T cell responses. Furthermore, this study implicates that antigen-specific vaccination based immunotherapeutic regimens may fail within an obese model, further limiting the usage of FDA approved methods such as Provenge®.

"Inflammaging", TNF-α and Tumor Environment

In previous sections it has been discussed that aging leads to a disruption in cellular distribution and function. However, not all immune activities are decreased with aging. The aged environment is also characterized as engaging in a low-grade, chronic, systemic inflammatory state that was termed "inflammaging". It is thought that inflammaging is a direct consequence of immunesenescence and is characterized as a set of five conditions: being low-grade, controlled, asymptomatic, chronic, and systemic (Giunta 2006; Franceschi 2007). This phenomenon leads to increased cytokine production, namely IL-1, IL-6, and TNF-α. Although the exact mechanism for its initiation remains unclear, it is likely a mixture of environmental factors, oxidative stress, and genetic factors. Within the obese microenvironment, inflammaging can be prematurely induced through the production of reactive oxygen species which cause oxidative damage and illicit the production and release of pro-inflammatory cytokines that can then be maintained in an autocrine manner through recruitment and retention of innate immune effectors, namely macrophages (Giunta 2006). It is hypothesized that inflammaging promotes tumor development by promoting immunoediting within the tumor microenvironment. In turn, tumors experience selective pressures that allow for alterations in their immunogenicity and thereby allow escape (Bonafe et al. 2012). Within breast cancer, it is hypothesized that overexpression of IL-6 within the inflammaging environment contributes to tumor initiation and progression by inducing epigenetic reprogramming of cellular chromatin, which exerts a cancer stem cell phenotype resembling TP53 cells (Iliopoulos et al. 2009; D'Anello et al. 2010).

As mentioned previously, changes within the innate compartment of immunity, either through aging or obesity, results in a pro-inflammatory rich environment characterized by high expression of IL-1β, TNF-α, and IL-6. Most importantly, TNF-α is known to be a major mediator of cancer inflammation. In 1975, Carswell et al. first described TNF-α as a tumor suppressive factor that could directly cause tumor necrotizing effects (Carswell et al. 1975). More recently, TNF- α has been shown to play a dual role in cancer, as either a tumor-promoting or tumor-resolving factor within the tumor microenvironment.

Acute high expression of TNF-α has been associated with anti-tumor effects directly and through its activation of effector cells, yet has also been linked to increased toxicity in immunotherapy clinical trials. This chapter has discussed the many ways in which TNF-α is able to activate and influence NK cells, T cells, DCs, and macrophages towards stimulating an inflammatory environment consequentially also full of other pro-inflammatory markers. Data arising from preclinical studies have also implicated tumor vasculature destruction as a side mechanism for the observed TNF-α mediated anti-tumor responses, indicating the variety of ways in which TNF-α can affect the tumor environment in a negative manner (Balkwill 2006).

Conversely, chronic expression of low-dose TNF-α within the tumor environment has been shown to favor tumor cell tissue invasion, migration, and increases metastatic potential (Balkwill 2006), confusing the dual role of this effector molecule. The first lines of evidence that TNF-α could be tumor promoting came from studies showing that treatment of tumor bearing environments with TNF-α antagonists, be it through depletion antibodies or soluble fusion receptor molecules, led to anti-tumor effects in melanoma, colorectal, liver, ovarian, and pancreatic cancer (Scott et al. 2003; Popivanova et al. 2008; Egberts et al. 2008; Madhusudan et al. 2005; Balkwill 2009). Likewise, mouse studies within our laboratory stimulated aged and obese (ob/ob) mice with αCD40/IL-2 combined immunotherapy with or without macrophage depletion and observed that macrophage depletion resolved the observed lethal toxicity. Mice displayed a marked decreased in systemic cytokines, decreased ALT levels and did not develop multi-organ damage from treatment. Furthermore, treatment of mice with TNF-α blockade similarly resulted in rescue from lethality and systemic damage (manuscript submitted). These findings implicate TNF-α as a major contributor to immunotherapy toxicity and indicate the complex nature of this cytokine.

Substantial evidence now shows that TNF-α is involved in the promotion and progression of cancer through activating the NFκB pathways and AP-1 transcription factor complexes (Balkwill 2006). Tumor expression of TNF-α allows for the activation and promotion of the NF-κB pathway that releases chemoattractants targeting macrophages. This process allows tumors to switch macrophage differentiation into M1 phenotypes, and recent studies have noted that TNF-α secreted by breast cancer cells were responsible for inducing fibroblast production of

MMP-9 (Hagemann et al. 2004). Furthermore, as a consequence to these abilities TNF-α production by tumors has become associated with a poor prognosis, loss of hormone responsiveness, and cachexia (Szlosarek et al. 2006; Szlosarek and Balkwill 2003). Therefore, the strong link between cancer-related inflammation and its aiding in proliferation and survival of cancerous cells provides insight to how the presence of TNF-α may contribute towards stimulating angiogenesis and metastasis, thereby aiding tumorigenesis and cancer progression.

9.3 Summary

At present, studies have confirmed that both aging and obesity have been implicated in the creation of a chronic pro-inflammatory state that actively secretes pro-tumorigenesis factors. It is imperative to understand the role of the inflammatory state in the tumor immunoenvironment. Taken together, the research described in this section implies that within both aging and obesity, the immune system may be preoccupied in maintaining its inflammatory condition and could thus provide a perfect environment in which a tumor could easily escape detection, by allowing rapid immunogeneic mutations and continuing the spread of neoplastic disease. Furthermore, the environmental pressures of the tumor microenvironment and their impact on the immune system remain to be elucidated—it is still unknown whether a strong immune response could follow after initial tumor detection, or if the chronic inflammatory state serves to cause cellular anergy.

Lastly, obesity is largely becoming an epidemic within the United States and other developed nations. Alarmingly, obesity is occurring in younger populations with the passage of time and it is estimated that childhood obesity has tripled in the last 30 years. This chapter has discussed the parallels between obesity-associated changes within the immune system and the immune dysfunction that could favor tumorigenesis and cancer progression. In many respects, the obese tumor environment mimics the environment seen in elderly patients. It is well known that these age-associated immunological changes result in increased disease occurrences and poor vaccine responses within the elderly. The clinical relevance of the information provided in this chapter together with increasing incidences of obesity in younger populations is suggestive of the possibility that young and obese populations may already be encountering a dysfunctional physiological state that could promote tumor growth at a younger age.

References

Balkwill F (2006) TNF-alpha in promotion and progression of cancer. Cancer Metastasis Rev 25:409–416

Balkwill F (2009) Tumour necrosis factor and cancer. Nat Rev Cancer 9:361–371

Bennaceur K, Chapman J, Brikci-Nigassa L, Sanhadji K, Touraine JL, Portoukalian J (2008) Dendritic cells dysfunction in tumour environment. Cancer Lett 272:185–196
Bennaceur K, Chapman JA, Touraine JL, Portoukalian J (2009) Immunosuppressive networks in the tumour environment and their effect in dendritic cells. Biochim Biophys Acta 1795:16–24
Ben-Neriah Y, Karin M (2011) Inflammation meets cancer, with NF-kappaB as the matchmaker. Nat Immunol 12:715–723
Bo S, Benso A, Durazzo M, Ghigo E (2012) Does use of metformin protect against cancer in Type 2 diabetes mellitus? J Endocrinol Invest 35:231–235
Bonafe M, Storci G, Franceschi C (2012) Inflamm-aging of the stem cell niche: breast cancer as a paradigmatic example: breakdown of the multi-shell cytokine network fuels cancer in aged people. BioEssays : News Rev Mol Cell Dev Biol 34:40–49
Brandon EL, Gu JW, Cantwell L, He Z, Wallace G, Hall JE (2009) Obesity promotes melanoma tumor growth: role of leptin. Cancer Biol Ther 8:1871–1879
Carswell EA, Old LJ, Kassel RL, Green S, Fiore N, Williamson B (1975) An endotoxin-induced serum factor that causes necrosis of tumors. Proc Natl Acad Sci USA 72:3666–3670
Chen W (2006) Dendritic cells and (CD4 +)CD25 + T regulatory cells: crosstalk between two professionals in immunity versus tolerance. Front Biosci: J Virtual Libr 11:1360–1370
Chen X, Xu H, Yuan P, Fang F, Huss M, Vega VB et al (2008) Integration of external signaling pathways with the core transcriptional network in embryonic stem cells. Cell 133:1106–1117
Cindik N, Baskin E, Agras PI, Kinik ST, Turan M, Saatci U (2005) Effect of obesity on inflammatory markers and renal functions. Acta Paediatr 94:1732–1737
Coppack SW, Evans RD, Fisher RM, Frayn KN, Gibbons GF, Humphreys SM et al (1992) Adipose tissue metabolism in obesity: lipase action in vivo before and after a mixed meal. Metab: Clinical Exp 41:264–272
Coussens LM, Werb Z (2002) Inflammation and cancer. Nature 420:860–867
Cuenca AG, Cress WD, Good RA, Marikar Y, Engelman RW (2001) Calorie restriction influences cell cycle protein expression and DNA synthesis during liver regeneration. Exp Biol Med (Maywood) 226:1061–1067
D'Anello L, Sansone P, Storci G, Mitrugno V, D'Uva G, Chieco P et al (2010) Epigenetic control of the basal-like gene expression profile via Interleukin-6 in breast cancer cells. Mol Cancer 9:300
Decman V, Laidlaw BJ, Doering TA, Leng J, Ertl HC, Goldstein DR et al (2012) Defective CD8 T cell responses in aged mice are due to quantitative and qualitative changes in virus-specific precursors. J Immunol 188:1933–1941
Dighe AS, Richards E, Old LJ, Schreiber RD (1994) Enhanced in vivo growth and resistance to rejection of tumor cells expressing dominant negative IFN gamma receptors. Immunity 1:447–456
Dirx MJ, Zeegers MP, Dagnelie PC, van den Bogaard T, van den Brandt PA (2003) Energy restriction and the risk of spontaneous mammary tumors in mice: a meta-analysis. Int J Cancer 106:766–770
Dolcet X, Llobet D, Pallares J, Matias-Guiu X (2005) NF-κB in development and progression of human cancer. Virchows Archiv : An Int J Pathol 446:475–482
Dunn GP, Bruce AT, Ikeda H, Old LJ, Schreiber RD (2002) Cancer immunoediting: from immunosurveillance to tumor escape. Nat Immunol 3:991–998
Dunn GP, Old LJ, Schreiber RD (2004) The immunobiology of cancer immunosurveillance and immunoediting. Immunity 21:137–148
Egberts JH, Cloosters V, Noack A, Schniewind B, Thon L, Klose S et al (2008) Anti-tumor necrosis factor therapy inhibits pancreatic tumor growth and metastasis. Cancer Res 68:1443–1450
El Kasmi KC, Smith AM, Williams L, Neale G, Panopoulos AD, Watowich SS et al (2007) Cutting edge: A transcriptional repressor and corepressor induced by the STAT3-regulated anti-inflammatory signaling pathway. J Immunol 179:7215–7219

Engelman RW, Owens UE, Bradley WG, Day NK, Good RA (1995) Mammary and submandibular gland epidermal growth factor expression is reduced by calorie restriction. Cancer Res 55:1289–1295

Fabrizi F, Dixit V, Martin P, Messa P (2011) Combined antiviral therapy of hepatitis C virus in dialysis patients: meta-analysis of clinical trials. J Viral Hepat 18:e263–e269

Fernandes G, Chandrasekar B, Troyer DA, Venkatraman JT, Good RA (1995) Dietary lipids and calorie restriction affect mammary tumor incidence and gene expression in mouse mammary tumor virus/v-Ha-ras transgenic mice. Proc Natl Acad Sci USA 92:6494–6498

Feuerer M (2009) Lean, but not obese, fat is enriched for a unique population of regulatory T cells that affect metabolic parameters. Nat Med 15:930–939

Field JB, Rigby B (1959) Circulating insulin antagonists in diabetes mellitus: untreated diabetics, diabetics during infections and acromegalics with diabetes. J Endocrinol 19:174–181

Flanigan TP, Rich JD, Spaulding A (1999) HIV care among incarcerated persons: a missed opportunity. AIDS 13:2475–2476

Fortin CF, Larbi A, Lesur O, Douziech N, Fulop T Jr (2006) Impairment of SHP-1 down-regulation in the lipid rafts of human neutrophils under GM-CSF stimulation contributes to their age-related, altered functions. J Leukoc Biol 79:1061–1072

Franceschi C (2007) Inflammaging as a major characteristic of old people: can it be prevented or cured? Nutr Rev 65:S173–S176

Franks PW (2006) Obesity, inflammatory markers and cardiovascular disease: distinguishing causality from confounding. J Hum Hypertens 20:837–840

Gerlinger M, Rowan AJ, Horswell S, Larkin J, Endesfelder D, Gronroos E et al (2012) Intratumor heterogeneity and branched evolution revealed by multiregion sequencing. New England J Med 366:883–892

Ghanim H, Aljada A, Hofmeyer D, Syed T, Mohanty P, Dandona P (2004) Circulating mononuclear cells in the obese are in a proinflammatory state. Circulation 110:1564–1571

Gilmore TD (2003) The Re1/NF-kappa B/I kappa B signal transduction pathway and cancer. Cancer Treat Res 115:241–265

Giunta S (2006) Is inflammaging an auto[innate]immunity subclinical syndrome? Immun Ageing: I & A 3:12

Gomez CR, Hirano S, Cutro BT, Birjandi S, Baila H, Nomellini V et al (2007) Advanced age exacerbates the pulmonary inflammatory response after lipopolysaccharide exposure. Crit Care Med 35:246–251

Goronzy JJ, Weyand CM (2005) T cell development and receptor diversity during aging. Curr Opin Immunol 17:468–475

Goronzy JJ, Lee WW, Weyand CM (2007) Aging and T-cell diversity. Exp Gerontol 42:400–406

Grivennikov SI, Greten FR, Karin M (2010) Immunity, inflammation, and cancer. Cell 140:883–899

Hagemann T, Robinson SC, Schulz M, Trumper L, Balkwill FR, Binder C (2004) Enhanced invasiveness of breast cancer cell lines upon co-cultivation with macrophages is due to TNF-alpha dependent up-regulation of matrix metalloproteases. Carcinogenesis 25:1543–1549

Hamaguchi M, Sakaguchi S (2012) Regulatory T cells expressing PPAR-gamma control inflammation in obesity. Cell Metab 16:4–6

Hanahan D, Weinberg RA (2011) Hallmarks of cancer: the next generation. Cell 144:646–674

Herrero C, Sebastian C, Marques L, Comalada M, Xaus J, Valledor AF et al (2002) Immunosenescence of macrophages: reduced MHC class II gene expression. Exp Gerontol 37:389–394

Hill JO, Melanson EL, Wyatt HT (2000) Dietary fat intake and regulation of energy balance: implications for obesity. J Nutr 130:284S–288S

Hilmer SN, Cogger VC, Le Couteur DG (2007) Basal activity of kupffer cells increases with old age. J Gerontol Ser A, Biol Sci Med Sci 62:973–978

Hotamisligil GS, Spiegelman BM (1994) Tumor necrosis factor alpha: a key component of the obesity-diabetes link. Diabetes 43:1271–1278

Hotamisligil GS, Shargill NS, Spiegelman BM (1993) Adipose expression of tumor necrosis factor-alpha: direct role in obesity-linked insulin resistance. Science 259:87–91

Hunig T (2012) The storm has cleared: lessons from the CD28 superagonist TGN1412 trial. Nat Rev Immunol 12:317–318

Hwang KA, Kim HR, Kang I (2009) Aging and human CD4(+) regulatory T cells. Mech Ageing Dev 130:509–517

Iliopoulos D, Hirsch HA, Struhl K (2009) An epigenetic switch involving NF-kappaB, Lin28, Let-7 MicroRNA, and IL6 links inflammation to cell transformation. Cell 139:693–706

James BR, Tomanek-Chalkley A, Askeland EJ, Kucaba T, Griffith TS, Norian LA (2012) Diet-induced obesity alters dendritic cell function in the presence and absence of tumor growth. J Immunol 189:1311–1321

Joyce JA, Pollard JW (2009) Microenvironmental regulation of metastasis. Nat Rev Cancer 9:239–252

Kantoff PW, Higano CS, Small EJ, Whitmore JB, Frohlich MW, Schellhammer PF (2012) Re: interdisciplinary critique of sipuleucel-T as immunotherapy in castration-resistant prostate cancer. J Natl Cancer Inst 104:1107–1109

Karin M, Cao Y, Greten FR, Li ZW (2002) NF-kappaB in cancer: from innocent bystander to major culprit. Nat Rev Cancer 2:301–310

Kataki A, Scheid P, Piet M, Marie B, Martinet N, Martinet Y et al (2002) Tumor infiltrating lymphocytes and macrophages have a potential dual role in lung cancer by supporting both host-defense and tumor progression. J Lab Clinical Med 140:320–328

Kobayashi M, Yasui N, Ishimaru N, Arakaki R, Hayashi Y (2004) Development of autoimmune arthritis with aging via bystander T cell activation in the mouse model of Sjogren's syndrome. Arthritis Rheum 50:3974–3984

Lages CS, Suffia I, Velilla PA, Huang B, Warshaw G, Hildeman DA et al (2008) Functional regulatory T cells accumulate in aged hosts and promote chronic infectious disease reactivation. J Immunol 181:1835–1848

Lamagna C, Aurrand-Lions M, Imhof BA (2006) Dual role of macrophages in tumor growth and angiogenesis. J Leukoc Biol 80:705–713

Lefebvre JS, Maue AC, Eaton SM, Lanthier PA, Tighe M, Haynes L (2012) The aged microenvironment contributes to the age-related functional defects of CD4 T cells in mice. Aging Cell 11:732–740

Leffers N, Gooden MJ, de Jong RA, Hoogeboom BN, ten Hoor KA, Hollema H et al (2009) Prognostic significance of tumor-infiltrating T-lymphocytes in primary and metastatic lesions of advanced stage ovarian cancer. Cancer Immunol Immunother: CII 58:449–459

Lenz G, Davis RE, Ngo VN, Lam L, George TC, Wright GW et al (2008) Oncogenic CARD11 mutations in human diffuse large B cell lymphoma. Science 319:1676–1679

Liu Q, Zhang C, Sun A, Zheng Y, Wang L, Cao X (2009) Tumor-educated CD11bhighIalow regulatory dendritic cells suppress T cell response through arginase I. J Immunol 182:6207–6216

Luedke E, Jaime-Ramirez AC, Bhave N, Roda J, Choudhary MM, Kumar B et al (2012) Cetuximab therapy in head and neck cancer: Immune modulation with interleukin-12 and other natural killer cell-activating cytokines. Surgery 152:431–440

Lynch HE, Goldberg GL, Chidgey A, Van den Brink MR, Boyd R, Sempowski GD (2009) Thymic involution and immune reconstitution. Trends Immunol 30:366–373

Madhusudan S, Muthuramalingam SR, Braybrooke JP, Wilner S, Kaur K, Han C et al (2005) Study of etanercept, a tumor necrosis factor-alpha inhibitor, in recurrent ovarian cancer. J Clinical Oncol: Official J Am Soc Clinical Oncol 23:5950–5959

Maiorana A, Del Bianco C, Cianfarani S (2007) Adipose tissue: a metabolic regulator. potential implications for the metabolic outcome of subjects born small for gestational age (SGA). Rev Diabet Stud: RDS 4:134–146

Mantovani A, Sozzani S, Locati M, Schioppa T, Saccani A, Allavena P et al (2004) Infiltration of tumours by macrophages and dendritic cells: tumour-associated macrophages as a paradigm for polarized M2 mononuclear phagocytes. Novartis Found Symp 256:137–145 (discussion 46–48, 259–269)

Martinez FO, Sica A, Mantovani A, Locati M (2008) Macrophage activation and polarization. Front Biosci: J Virtual Libr 13:453–461

Medicine NLo. US National Institutes of Health Clinical Trials (2012) [cited 2012 15 Sept 2012]; Available from: http://clinicaltrials.gov

Moretta A, Bottino C, Vitale M, Pende D, Biassoni R, Mingari MC et al (1996) Receptors for HLA class-I molecules in human natural killer cells. Annu Rev Immunol 14:619–648

Murakami M, Sakamoto A, Bender J, Kappler J, Marrack P (2002) $CD25^{+}$ $CD4^{+}$ T cells contribute to the control of memory $CD8^{+}$ T cells. Proc Natl Acad Sci USA 99:8832–8837

Murdoch C, Giannoudis A, Lewis CE (2004) Mechanisms regulating the recruitment of macrophages into hypoxic areas of tumors and other ischemic tissues. Blood 104:2224–2234

Murphy WJ, Welniak L, Back T, Hixon J, Subleski J, Seki N et al (2003) Synergistic anti-tumor responses after administration of agonistic antibodies to CD40 and IL-2: coordination of dendritic and CD8 + cell responses. J Immunol 170:2727–2733

Nio Y, Yamauchi T, Iwabu M, Okada-Iwabu M, Funata M, Yamaguchi M et al (2012) Monocyte chemoattractant protein-1 (MCP-1) deficiency enhances alternatively activated M2 macrophages and ameliorates insulin resistance and fatty liver in lipoatrophic diabetic A-ZIP transgenic mice. Diabetologia 55(12):3350–3358

Nishihira J (1998) Novel pathophysiological aspects of macrophage migration inhibitory factor (review). Int J Mol Med 2:17–28

Nishikawa H, Sakaguchi S (2010) Regulatory T cells in tumor immunity. Int J Cancer 127:759–767

Pappo I, Lotem M, Klein M, Orda R (2001) Bolus high dose interleukin-2 for the treatment of malignant melanoma. Israel Med Assoc J: IMAJ 3:169–173

Pawelec G, Barnett Y, Forsey R, Frasca D, Globerson A, McLeod J et al (2002) T cells and aging, January 2002 update. Fronti Biosci: J Virtual Libr 7:d1056–d1183

Petersen TR, Dickgreber N, Hermans IF (2010) Tumor antigen presentation by dendritic cells. Crit Rev Immunol 30:345–386

Philip M, Rowley DA, Schreiber H (2004) Inflammation as a tumor promoter in cancer induction. Semin Cancer Biol 14:433–439

Popivanova BK, Kitamura K, Wu Y, Kondo T, Kagaya T, Kaneko S et al (2008) Blocking TNF-alpha in mice reduces colorectal carcinogenesis associated with chronic colitis. J Clin Investig 118:560–570

Rakoff-Nahoum S (2006) Why cancer and inflammation? Yale J Biol Med 79:123–130

Rink L, Cakman I, Kirchner H (1998) Altered cytokine production in the elderly. Mech Ageing Dev 102:199–209

Rosenberg SA, Yang JC, Topalian SL, Schwartzentruber DJ, Weber JS, Parkinson DR et al (1994) Treatment of 283 consecutive patients with metastatic melanoma or renal cell cancer using high-dose bolus interleukin 2. J Am Med Assoc 271:907–913

Saltiel AR (2001) You are what you secrete. Nat Med 7:887–888

Sartipy P, Loskutoff DJ (2003) Monocyte chemoattractant protein 1 in obesity and insulin resistance. Proc Natl Acad Sci USA 100:7265–7270

Scartozzi M, Bearzi I, Pierantoni C, Mandolesi A, Loupakis F, Zaniboni A et al (2007) Nuclear factor-κB tumor expression predicts response and survival in irinotecan-refractory metastatic colorectal cancer treated with cetuximab-irinotecan therapy. J Clinical Oncol: Official J Am Soc Clinical Oncol 25:3930–3935

Schreiber RD, Old LJ, Smyth MJ (2011) Cancer immunoediting: integrating immunity's roles in cancer suppression and promotion. Science 331:1565–1570

Schwartzentruber DJ, Lawson DH, Richards JM, Conry RM, Miller DM, Treisman J et al (2011) gp100 peptide vaccine and interleukin-2 in patients with advanced melanoma. New England J Med 364:2119–2127

Scott KA, Moore RJ, Arnott CH, East N, Thompson RG, Scallon BJ et al (2003) An anti-tumor necrosis factor-alpha antibody inhibits the development of experimental skin tumors. Mol Cancer Ther 2:445–451

Shankaran V, Ikeda H, Bruce AT, White JM, Swanson PE, Old LJ et al (2001) IFNgamma and lymphocytes prevent primary tumour development and shape tumour immunogenicity. Nature 410:1107–1111

Skurk T, Alberti-Huber C, Herder C, Hauner H (2007) Relationship between adipocyte size and adipokine expression and secretion. J Clinical Endocrinol Metab 92:1023–1033

Solana R, Mariani E (2000) NK and NK/T cells in human senescence. Vaccine 18:1613–1620

Solerte SB, Fioravanti M, Severgnini S, Cerutti N, Locatelli M, Ferrari E (1997) Variability of natural killer (NK) cell immune function in normal aging and senile dementia: pathophysiological implications. Aging (Milano) 9:32–33

Spaulding C, Guo W, Effros RB (1999) Resistance to apoptosis in human CD8 + T cells that reach replicative senescence after multiple rounds of antigen-specific proliferation. Exp Gerontol 34:633–644

Spiegelman BM, Flier JS (2001) Obesity and the regulation of energy balance. Cell 104:531–543

Steinmann GG, Klaus B, Muller-Hermelink HK (1985) The involution of the ageing human thymic epithelium is independent of puberty. A morphometric study. Scand J Immunol 22:563–575

Sunderkotter C, Kalden H, Luger TA (1997) Aging and the skin immune system. Arch Dermatol 133:1256–1262

Suntharalingam G, Perry MR, Ward S, Brett SJ, Castello-Cortes A, Brunner MD et al (2006) Cytokine storm in a phase 1 trial of the anti-CD28 monoclonal antibody TGN1412. New England J Med 355:1018–1028

Szlosarek PW, Balkwill FR (2003) Tumour necrosis factor alpha: a potential target for the therapy of solid tumours. Lancet Oncol 4:565–573

Szlosarek PW, Grimshaw MJ, Kulbe H, Wilson JL, Wilbanks GD, Burke F et al (2006) Expression and regulation of tumor necrosis factor alpha in normal and malignant ovarian epithelium. Mol Cancer Ther 5:382–390

Takahashi K, Mizuarai S, Araki H, Mashiko S, Ishihara A, Kanatani A et al (2003) Adiposity elevates plasma MCP-1 levels leading to the increased CD11b-positive monocytes in mice. J Biol Chem 278:46654–46660

Tamura Y, Sugimoto M, Murayama T, Ueda Y, Kanamori H, Ono K et al (2008) Inhibition of CCR2 ameliorates insulin resistance and hepatic steatosis in db/db mice. Arterioscler Thromb Vasc Biol 28:2195–2201

Tarazona R, DelaRosa O, Alonso C, Ostos B, Espejo J, Pena J et al (2000) Increased expression of NK cell markers on T lymphocytes in aging and chronic activation of the immune system reflects the accumulation of effector/senescent T cells. Mech Ageing Dev 121:77–88

Taub DD, Longo DL (2005) Insights into thymic aging and regeneration. Immunol Rev 205:72–93

Thoman ML (1997) Effects of the aged microenvironment on $CD4^+$ T cell maturation. Mech Ageing Dev 96:75–88

Tietze JK, Wilkins DE, Sckisel GD, Bouchlaka MN, Alderson KL, Weiss JM et al (2012) Delineation of antigen-specific and antigen-nonspecific CD8(+) memory T-cell responses after cytokine-based cancer immunotherapy. Blood 119:3073–3083

Tough DF, Borrow P, Sprent J (1996) Induction of bystander T cell proliferation by viruses and type I interferon in vivo. Science 272:1947–1950

Tsung K, Dolan JP, Tsung YL, Norton JA (2002) Macrophages as effector cells in interleukin 12-induced T cell-dependent tumor rejection. Cancer Res 62:5069–5075

Tzanetakou IP, Katsilambros NL, Benetos A, Mikhailidis DP, Perrea DN (2012) "Is obesity linked to aging?": adipose tissue and the role of telomeres. Ageing Res Rev 11:220–229

Vallance-Owen J (1960) Insulin antagonists. Br Med Bull 16:214–218

Vignaux O, Legmann P, de Pinieux G, Chaussade S, Spaulding C, Couturier D et al (1999) Hemorrhagic necrosis due to peliosis hepatis: imaging findings and pathological correlation. Eur Radiol 9:454–456

Vukmanovic-Stejic M, Rustin MH, Nikolich-Zugich J, Akbar AN (2011) Immune responses in the skin in old age. Curr Opin Immunol 23:525–531

Weber J (2007) Review: anti-CTLA-4 antibody ipilimumab: case studies of clinical response and immune-related adverse events. Oncologist 12:864–872

Weisberg SP, Hunter D, Huber R, Lemieux J, Slaymaker S, Vaddi K et al (2006) CCR2 modulates inflammatory and metabolic effects of high-fat feeding. J Clin Investig 116:115–124

Weng NP, Akbar AN, Goronzy J (2009) CD28(-) T cells: their role in the age-associated decline of immune function. Trends Immunol 30:306–312

Wesley J, Whitmore J, Trager J, Sheikh N (2012) An overview of sipuleucel-T: Autologous cellular immunotherapy for prostate cancer. Hum Vaccin Immunother 8:520–527

Winer S (2009) Normalization of obesity-associated insulin resistance through immunotherapy. Nat Med 15:921–929

Wong CP, Magnusson KR, Ho E (2010) Aging is associated with altered dendritic cells subset distribution and impaired proinflammatory cytokine production. Exp Gerontol 45:163–169

Wu D, Ren Z, Pae M, Guo W, Cui X, Merrill AH et al (2007) Aging up-regulates expression of inflammatory mediators in mouse adipose tissue. J Immunol 179:4829–4839

Yang H, Youm YH, Dixit VD (2009) Inhibition of thymic adipogenesis by caloric restriction is coupled with reduction in age-related thymic involution. J Immunol 183:3040–3052

Youm YH, Yang H, Amin R, Smith SR, Leff T, Dixit VD (2010) Thiazolidinedione treatment and constitutive-PPARgamma activation induces ectopic adipogenesis and promotes age-related thymic involution. Aging Cell 9:478–489

Zhang L, Conejo-Garcia JR, Katsaros D, Gimotty PA, Massobrio M, Regnani G et al (2003) Intratumoral T cells, recurrence, and survival in epithelial ovarian cancer. New England J Med 348:203–213

Part III
Tumor Escape from Immune Recognition

Chapter 10
MHC Class I Antigens and the Tumor Microenvironment

Natalia Aptsiauri, Teresa Cabrera, Angel Garcia-Lora, Francisco Ruiz-Cabello and Federico Garrido

Abstract Cancer is characterized by the accumulation of multiple genetic events leading to malignant transformation and escape from the immune surveillance. A complex interaction of malignant cells with the surrounding tissues, including epithelial cells, vascular and lymphatic vessels, extracellular matrix, cytokines, chemokines, and infiltrating immune cells, lead to tumor immunoediting and generation of tumor escape variants with an increased ability to disseminate to distant sites. Tumor- infiltrating cytotoxic T-lymphocytes (TILs) are responsible for tumor cell recognition and elimination, while Major Histocompatibility Complex (MHC) class I and II products play a central role in mounting an effective anti-tumor immune response by restricting T cell recognition of foreign antigens (Ags) processed as small peptides. Therefore, tumor cells that fail to express MHC molecules have an advantage that provides them with an escape route from T cell immunity. The molecular identification of human cancer antigens has allowed the development of antigen-specific immunotherapy. Novel cancer vaccines aim to induce tumor-specific effector T cells that can reduce the tumor mass and to induce tumor-specific memory T cells that can control tumor relapse. However, loss of tumor MHC class I expression may compromise the efficacy of the immunotherapy-induced anti-tumor T cell immunity. Early cancer detection and treatment require more effective cancer biomarkers, or molecular signatures, for diagnosis, prognosis and therapeutic efficacy. Analysis of the tumor expression of HLA class I antigens as biomarkers of cancer development might help to choose an appropriate treatment protocol and monitor clinical response to cancer immunotherapy.

N. Aptsiauri · T. Cabrera · A. Garcia-Lora · F. Ruiz-Cabello · F. Garrido (✉)
Department of Clinical Analysis and Immunology, Virgen de las Nieves University Hospital, Avenida de las Fuerzas Armadas s/n 18014 Granada, Spain
e-mail: federico.garrido.sspa@juntadeandalucia.es

T. Cabrera · F. Ruiz-Cabello · F. Garrido
Department of Biochemistry, Molecular Biology and Immunology III, University of Granada Medical School, Granada, Spain

M. R. Shurin et al. (eds.), *The Tumor Immunoenvironment*,
DOI: 10.1007/978-94-007-6217-6_10,

Keywords HLA · Cancer · Metastatic progression · Immune response · Immune escape · Immunotherapy

10.1 Introduction

Numerous innate and adaptive immune effector cells and molecules participate in the recognition and destruction of cancer cells, a process known as cancer immunosurveillance. In addition, tumor cell interaction with tumor microenvironment (TME) which consists of cells, soluble factors, signaling molecules, and extracellular matrix that can promote neoplastic transformation, support tumor growth and invasion, protect the tumor from host immunity, lead to therapeutic resistance, and provide mechanism for dormant metastases to grow (Schreiber et al. 2011). It is known that the immune system is able to mount a specific T cell mediated immune response against solid tumors. Activated cytotoxic T cells can recognize tumor antigens presented by HLA class I-positive tumor cells, thereby performing effective immune surveillance. However, these T cells are frequently unable to reject the tumor, leading to cancer progression. This is attributable to the immune escape and expansion of cells with low immunogenicity and high metastatic capacity (Marincola et al. 2000; Rosenberg et al. 2005; Boon 2006; Drake et al. 2006). Cancer cells are able to escape from an immunosurveillance through the outgrowth of poorly immunogenic tumor cell variants, for instance with abnormal expression of HLA class I antigens (Garrido et al. 1993, 1997). Cell surface expression of HLA class I molecules in tumor cells is required for the recognition of the class I heavy chain/beta2-microglobulin (β2m)/tumor peptide complex by cytotoxic T cells (CTLs) (Aptsiauri et al. 2007). Cells that are highly immunogenic and express high levels of MHC class I are eliminated by CTLs. Malignant cells with total MHC class I loss are susceptible to NK cell lysis because of inactivation of KIRs. Another immunoselection route is provided by the partial loss of HLA class I antigens that allows tumor cells to escape both CTL and NK attack. As a result, different altered HLA class I tumor phenotypes are produced. Consequently, during tumor progression various tumor cells have certain features that in combination with the factors of tumor environment are beneficial for tumor cell proliferation and give growth advantages to certain tumor cells (Khong and Restifo 2002). Tumor peptide-based immunotherapy is an established approach of cancer treatment, with the aim of boosting anti-tumor T cell reactivity by stimulation with tumor specific peptides. However, the overall clinical outcome of this type of treatment is lower than expected (Rosenberg et al. 2004). In many cases the failure of this therapy and the progression of the cancer are associated with total loss of HLA class I tumor expression. Normal expression of HLA class I molecules on tumor cell surface is crucial for the successful outcome of peptide-based cancer therapy, since cytotoxic T cells can only recognize tumor-derived peptides in a complex with self-MHC class I molecules.

Hence, the elimination of tumor cells by the activated immune system and the success of cancer immunotherapy depend on the proper co-presentation of tumor-specific antigens with HLA class I molecules. Moreover, it was recently demonstrated that tumor HLA class I antigen expression in melanoma is an important component of the "immunological constant of rejection", in which similar molecular pathways lead to CTL-mediated tumor rejection, allograft rejection, graft versus host (GVH) disease, or the development of an autoimmune disease (Carretero et al. 2012). The clinical effect of anti-tumor vaccines and cancer immunotherapy remains below expectations despite efforts to enhance immune reactivity against malignancies. This may be associated with the accumulation of HLA-defective tumor targets during cancer development. In order to select an appropriate immunotherapy protocol, it is important to know whether the nature of the tumor HLA-class I defect is regulatory or structural. Some HLA alterations can be recovered after stimulation with cytokines or immunotherapy (so-called "soft" lesions), while tumor cells with irreversible structural defects ("hard" lesions) may escape immune recognition and become a major threat to innate immunity (Garrido et al. 2010).

10.2 The Role of MHC Antigens in T- and NK-Cell-Mediated Immunity

MHC class I molecules are cell surface glycoproteins composed of two non-covalently associated polypeptide subunits, a polymorphic chain of 45 KD (heavy chain or α chain) and a non-polymorphic protein 12 KD, called β2-microglobulin (β2m). The antigenic peptide is assembled with this structure, resulting in a peptide-MHC class I complex. Extracellular region of the heavy chain is divided into three domains, α1, α2 and α3. The peptide binding groove is formed as an interchain dimer by folding of α1 and α2 domains to create a long cleft, where polymorphic amino acid residues cluster in hypervariable regions. Thus, the groove formed by these domains anchor the peptide recognized by the T cell receptor (TCR). MHC molecules in human are called Human Leucocyte Antigens (HLA). The genes encoding heavy chain are located on the short arm of chromosome 6 in humans and in chromosome 17 in mouse. The β2m gene is independent from MHC (or HLA) complex and is located on chromosome 15 in humans and chromosome 2 in mouse. MHC genes in the mouse are known as H-2 genes. Due to the small size of the H-2 complex genetic recombination is rare, and each pair of alleles is transferred from one generation to another as a set. The alleles transferred together as a unit in each chromosome are called a haplotype. The haplotypes of the most commonly used strains are standard and have assigned certain alleles. For example, BALB/c strain has H-2d haplotype, whereas that C57BL/6 mice have H-2b haplotype. There are three H-2 class-I molecules: K, D and L, and are located in K and D regions of the complex. They correspond to

HLA-A, B and C molecules in the human HLA complex and also consist of a heavy α chain and β2-microglobulin. The molecules K and D are expressed in all haplotypes, but the number of L loci expressed in the different haplotypes range from none (H-2b) to three (H-2k).

Antigen presentation on the plasma membrane is only the last step of a chain of events known as antigen processing occurring within the cell, where the MHC molecules are involved in active form. MHC class I molecules are involved mainly in endogenous antigen processing, such as the products of viral components in virus-infected cells (Cresswell et al. 2005). In any case, the first event to consider is the proteolytic degradation of the antigens into small fragments, capable of forming stable complexes with corresponding MHC molecules. Antigens, which are originally found in the cytosol, are degraded by proteolytic enzymes into short length fragments (8–10 amino-acid residues). In this enzymatic fragmentation the complex known as proteasome plays an important role (Pickart and Cohen 2004). Next, peptides are transported into endoplasmic reticulum (ER) by action of TAP-1 and TAP-2 transporters. During the immune response, interferon-γ producing cells (activated T lymphocytes and NK cells) along with an increase in transcription of MHC and TAP genes induce production of three ß catalytic subunits: LMP2 (ß1i), LMP7 (ß5i) and MECL-1 (ß2i). These subunits replace other ß subunits of the proteolytic proteasome forming the immunoproteosome. Substitution of the constitutive by inducible subunits changes proteasome specificity, generating peptides (C-terminal extreme with hydrophobic and basic residues) which are better transported by TAPs and are better presented by MHC class I molecules. In TAP-deficient cells, class I heterodimers cannot assemble with peptides (De la Salle et al. 1994), being unstable and degraded in the cytosol. Therefore, TAP-deficient cells do not present TAP-dependent peptides on the surface in a complex with class I molecules, causing a low response by cytotoxic T lymphocytes (CTLs). However, alternative routes of peptide processing independent from TAP and proteasome generating distinct CTLs responses have been described (Del Val et al. 2011). In the endoplasmic reticulum (ER) lumen each peptide interacts with a heavy chain class I molecule, conferring stability to the heavy chain-β2m heterodimer. A variety of chaperones are involved in heavy chain-β2m and peptide assembly, including calnexin, calreticulin, ERp57/ER60 and tapasin (Lehner and Trowsdale 1998). Tapasin controls MHC molecule assembly with peptide retaining class I molecules in the ER until they acquire peptides with high affinity (Grandea et al. 2000). In the absence of such peptides calreticulin and ERp57 have a weaker bond with class I heterodimers. In tapasin-deficient mice most class I MHC molecules are loaded with low affinity peptides (Garbi et al. 2000).

Provided with appropriate stability, class I molecule is driven through the Golgi to cell surface to presents tumor-associated peptides to CD8+ T cells. In the effector phase of the immune reaction, CD8+ T cells recognize peptide fragments of the antigen through T cell receptor (TCR) as a peptide-MHC class I complex along with some co-stimulatory molecules expressed on a transformed cell. As a result, the CD8+ cells are stimulated for clonal expansion and production of various cytokines, gaining a cytolytic effector activity and killing the antigen-secreting tumor cell. Cytotoxic T cells regularly patrol to check whether any of the presented peptides are non-self.

In order to escape from CD8+ T cell recognition and destruction, viruses and tumor cells have developed strategies to inhibit the expression and/or function of HLA class I antigens. In contrast, cells with downregulated MHC class I surface expression can be potentially recognized by natural killer (NK) cells. NK cells have inhibitory receptors on their surface that recognize MHC class I molecules on the target cell (Lanier 2005). Transformed cells frequently have low or absent MHC class I expression, becoming susceptible to NK cell anti-tumor activity induced by the lack of inhibitory MHC antigens ("missing-self') (Ljunggren and Karre 1990). The discovery of several stimulatory receptors on NK cells revealed that the outcome of NK activation depends on a balance between inhibitory and stimulatory signals (Moretta et al. 2001). It is evident that any alteration in the expression of any of the MHC class I subunits can affect normal MHC cell surface expression and alter both T and NK cell-mediated immunity. These alterations may affect the tumorigenic phenotype and metastatic capacity of human and experimental tumors (Festenstein and Garrido 1986; Festenstein 1987; Villunger and Strasser 1999; Garrido and Algarra 2001).

10.3 Mechanisms of Cancer Immune Escape

Various mechanisms have been discovered that enable malignant tumors to evade immune surveillance (Festenstein and Garrido 1986; Garrido and Algarra 2001; Dunn et al. 2004; Campoli et al. 2005). Loss of antigens, co-stimulatory signals and/or adhesion molecules, expression of immunosuppressive factors like Fas ligand and deficiencies in the signal transduction pathway of CD8+ cytotoxic T cells (CTL) have all been reported. In addition, identification and characterization of the components of HLA class I antigen processing and presentation machinery have provided new tools to investigate how tumor cells evade recognition by CTLs. The immunogenicity of the tumor develops as a result of various genetic and epigenetic events during the natural development of the tumor and its interaction with the host microenvironment. There are many different factors that could influence the outcome of the tumor-host interaction. Currently there are many ongoing studies providing new experimental evidence that supports the cancer immunosurveillance concept first proposed by Burnet and Thomas (1957). In the past two decades many studies have focused on trying to define the molecular basis of tumor escape mechanisms (Ahmad et al. 2004; Rivoltini et al. 2005; Vesely et al. 2011); for instance, myeloid-derived suppressor cells (MDSC) as a major contributor to tumor escape. MDSC accumulate in tumor bearing hosts and can strongly suppress T cell mediated immune responses by production of nitric oxide, reactive oxygen species and TGF-beta as well as by up-regulation of arginase-1 and cyclooxygenase-2 activities (Condamine and Gabrilovich 2011).

Lack of expression of costimulatory molecules on malignant cells is frequently observed in many types of cancer and can lead to failure of recognition of tumor antigen by T cells and suboptimal activation of NK cells (Schultze and Nadler 2003). Defective apoptosis receptor signaling via the caspase cascade is another mechanism of tumor escape. Inhibition of tumor-specific effector T cell activity by tumor-specific regulatory T cells (Treg) in cancer patients have been supported by various studies (Nummer et al. 2007). Factors associated with chronic inflammation in tumor microenvironment also contribute to cancer progression (Ostrand-Rosenberg et al. 2012). Examples of different immune escape mechanisms are presented in Table 10.1.

It is also known that tumors may directly or indirectly inhibit the development of anti-tumor responses. Immunosurveillance is the protective function of the host performed by the immune system which preferably takes place at early stages of malignant transformation. However, the immune system through its continuous interaction with the tumor, can edit the immunogenicity of the tumor, resulting in the appearance of less immunogenic variants. Thus the immune system has a dual nature; it helps to prevent tumor progression in early stages and promotes the selection of less immunogenic variants. This process is known as "cancer immunoediting" (Schreiber et al. 2011; Dunn et al. 2004).

Altered HLA class I expression, as an important route of cancer immune escape, may have significant implications for the induction of anti-tumor cellular immunity.

10.4 Altered HLA Class I Expression in Malignant Cells

Loss or downregulation of HLA class I antigens in tumor cells has been frequently observed in a variety of human malignancies and it represents an important cancer immune escape mechanism (Marincola et al. 2000; Garrido et al. 1997; Aptsiauri et al. 2007; Campoli et al. 2005; Chang et al. 2005). Viruses use similar mechanism to avoid recognition and elimination by the immune system (Ploegh 1998). The first description of MHC class I loss was done in a mouse model (Gardener lymphoma) in Dr. Festenstein laboratory in 1976 (Festenstein and Garrido 1986). The production and characterization of monoclonal antibodies against HLA molecules made possible to analyze HLA expression in human cell lines and solid tumors. At first, the reported percentage of HLA class I loss was low (10–30 %) (Garrido et al. 1993), since only monomorphic monoclonal antibodies (recognizing an epitope common to all HLA class I molecules) were available at that time. These studies were able to detect only total loss of tumor HLA class I expression and due to low incidence these findings did not attract much attention and were not considered to be significant. Later on, with the appearance of more specific monoclonal antibodies (anti-locus-A, -B and anti-allele-specific antibodies), which recognize polymorphic portions of these molecules, the incidence of HLA altered expression in cancer has been found to be much higher, increasing the relevance of

Table 10.1 Factors of tumor microenvironment and intrinsic characteristics of malignant cells in cancer immune escape

Mechanisms	References
Loss of tumor antigen expression	Jager et al. (1996), Berset et al. (2001), Khong et al. (2004)
Defects in antigen-processing and presentation pathways: • Loss of major histocompatibility complex (MHC) class I molecules and/or β2m • Loss of APM molecules: TAP1, LMP2, LMP7 and tapasin	Marincola et al. (2000), Aptsiauri et al. (2007), Restifo et al. (1996), Meissner et al. (2005)
Resistance of tumor cells to IFN-γ or IFN-α/β through either mutation or epigenetic silencing of genes encoding the IFN-γ receptor signaling components (IFNGR1, IFNGR2, JAK1, JAK2, and STAT1)	Rodriguez et al. (2007), Respa et al. (2011), Dunn et al. (2005)
Tumor-derived immunosuppressive factors that inhibit effector immune cell functions: • Transforming growth factor-β (TGF-β), IL-10 • Vascular endothelial growth factor (VEGF) that has an inhibitory effect on dendritic cells • Metabolic enzymes such as indoleamine 2,3 dioxygenase (IDO) and arginase, that can locally inhibit immune responses by depleting amino acids essential for anabolic metabolism of T cells • Gangliosides, soluble MICA • Factors that recruit regulatory cells which generate an immunosuppressive microenvironment: IL-4, IL-13, GM-CSF, IL-1β, VEGF, or PGE2 • Tumor-derived factors that inhibit DC function • Colony-stimulating factors, IL-1β, VEGF, or PGE2 that increase an accumulation of MDSCs	Khong and Restifo (2002), Aruga et al. (1997), Gabrilovich et al. (1998), Uyttenhove et al. (2003), Terabe and Berzofsky (2004), Wrzesinski et al. (2007), Villablanca et al. (2010), Herber et al. (2010)

(continued)

Table 10.1 (continued)

Mechanisms	References
Mechanisms that provide tumors with the ability to escape immune destruction • Upregulating inhibitors of apoptosis (Bcl-XL, FLIP) • Expressing inhibitory cell surface molecules that directly kill cytotoxic T cells (PD-L1, FasL) • Release of pro-apoptotic factors: TRAIL receptor, DR5, and Fas B7-H1 (PD-L1); that inhibit local antitumor T cell responses	Kataoka et al. (1998), Catlett-Falcone et al. (1999), Hinz et al. (2000), Shin et al. (2001), Dong et al. (2002), Takahashi et al. (2006), Zou et al. (2007)
Mechanisms that prevent tumor cell recognition by NK cells or CTLs: • Loss of ligands for NK cell effector molecules (as NKG2D) • Secretion by tumor of soluble ligands (as NKG2D) for effectors molecules that block T cell and NK-cell function • Inhibition of NK cells and CTLs through tumor cell expression of HLA-E or HLA-G • Lack of expression of costimulatory molecules on malignant cells that can lead to failure of recognition of tumor antigens by T cells and suboptimal activation of NK cells • Signaling defects through TCR: decreased expression of CD3ζ or tyrosin-kinases	Aoe et al. (1995), Schultze and Nadler (2003), Derre et al. (2006), Carosella et al. (2003), Tripathi and Agrawal (2006), Garrido et al. (1993), Stern-Ginossar et al. (2008)
The accumulation of regulatory cells decrease anti-tumor response through: • Release of immunosuppressive cytokines, including IL-10 and TGF-β • Altering the nutrient content of the microenvironment • Inhibit effectors T cells through expression of CTLA-4 and PD-L1 • IL-2 consumption	Vesely et al. (2011), Terabe and Berzofsky (2004)
Recruitment and polarization of MDSCs from myeloid precursors that can block T cell function by expressing TGF-β, ARG1, iNOS and IDO	Vesely et al. (2011)

these defects in the immune response against tumor. Using a broad panel of monoclonal antibodies on cryostat tumor tissue sections these alterations have been found in 60–90 % of tumors depending on the histological type of cancer (Blades et al. 1995; Cabrera et al. 1996; Cabrera et al. 1998, 2000; Koopman et al. 2000; Kageshita et al. 2005). Unfortunately, the number of available allele-specific monoclonal antibodies is still limited. Therefore, the true percentage of HLA class I defects, especially allelic losses, probably is much higher in different types of malignancy.

Thus, early studies using immunohistological analysis of different tumors showed a very low frequency of allelic loss. However, with the arrival of other techniques, such as the study of microsatellites to detect loss of heterozygosity (LOH) on chromosome 6, it has been shown that LOH (haplotype loss) is the most frequent alteration of class I expression (Feenstra et al. 2000; Koopman et al. 2000; Maleno et al. 2002, 2004a, 2006). This alteration is caused by various defects in the HLA genomic region (short arm of chromosome 6, 6p21), including chromosomal dysfunction, mitotic recombination and genetic conversion.

Many years of analysis of HLA expression in human tumors and tumor cell lines permitted us to classify HLA class I alterations in seven phenotypes according to the cell surface expression pattern:

Phenotype I: Total loss of HLA class I molecules
Phenotype II: Loss of an HLA class I haplotype
Phenotype III: Loss of an HLA class I locus
Phenotype IV: HLA class I allelic loss
Phenotype V: Compound phenotype
Phenotype VI: Failure to respond to interferon (IFN)
Phenotype VII: Low expression (down-regulation) of classical HLA molecules (Ia) with aberrant expression of non-classical HLA molecules (Ib)

10.5 Reversible and Irreversible Molecular Defects Underlying Altered Expression of HLA Class I Antigens on Tumor Cells

From experimental work it is clear that the malignant behavior of a cancer cell depends not only on the level of tumor MHC class I expression, but also on the molecular mechanisms which cause alterations in the MHC class I expression. Generation of various tumor MHC phenotypes can occur at any step required for the protein synthesis, assembly, transport or expression on cell surface. These defects can occur at the genetic, epigenetic, transcriptional and posttranscriptional levels and represent either regulatory abnormalities that can be recovered with cytokine treatment or more severe structural defects. Thus, MHC alterations can be classified into two main groups: reversible regulatory defects, and irreversible structural defects (Garrido et al. 2010). Although regulatory defects on transcriptional level are more common among various types of malignancy, the structural MHC defects may have profound implications in the T cell mediated rejection of tumor cells in primary or metastatic lesions and in the outcome of cancer immunotherapy. When the mechanism underlying total HLA class I loss is on transcriptional level, the expression of surface HLA class I antigens can be reversed by cytokine treatment and T cell based therapy can be successfully applied. However, peptide-based immunotherapy aimed at augmenting T cell-specific tumor recognition may not be effective in case of irreversible damage of HLA genes. Therefore, development of an adequate diagnostic approach for precise identification of the HLA class I expression phenotype and underlying molecular mechanisms is central.

The reversible MHC class I deficiencies involve all levels of the MHC class I-restricted antigen presentation machinery on transcriptional level. They can be repaired, at least partially and in vitro, by cytokines (IFN-gamma, TNF-alpha). The IFN-mediated upregulation of APM components normally leads to enhanced MHC class I surface expression and improves anti-tumor CTL responses (Martini et al. 2010; Seliger et al. 2008). Thus, it represents a valuable strategy for the treatment of patients with APM deficiencies. However, in some cases, tumors remain insensitive to IFN treatment despite the lack of structural alterations in APM components, suggesting an impaired IFN signal transduction. Downregulation of TAP1/2 and LMP2/7 gene has been demonstrated in different cell lines and tumor lesions (Meissner et al. 2005; Cabrera et al. 2003). LMP7 downregulation was found in correlation with the level of MHC class I expression in various human cancer cell lines (Yoon et al. 2000). A high frequency of LMP2, LMP7 and TAP1 downregulation or loss was observed in tumor lesions and cell lines obtained from head and neck cancer patients, which could be reversed by IFN-gamma treatment (Rodriguez et al. 2007). Impaired expression of immunoproteasome subunits (Cabrera et al. 2003; Miyagi et al. 2003) and tapasin (Cabrera et al. 2005) is involved in different types of HLA class I molecule loss in human colon cancer.

Epigenetic events associated with tumor development and cancer progression have been found to underlie changes in HLA and APM expression and activity. HLA class I gene hypermethylation leading to HLA loss has been demonstrated in various types of cancer. These alterations can be reversed in vitro with pharmacologic agents that induce DNA hypomethylation or inhibit histone deacetylation (Torres et al. 1996).

Total loss of HLA class I expression is caused by various mutations and chromosomal defects involving genes encoding heavy chain or $\beta 2$-microglobulin. After $\beta 2$m gene mutations, HLA haplotype loss is the second most frequent described phenotype. This alteration is caused by the hemizygous loss of HLA-A, -B and -C alleles or by loss of one copy of chromosome 6 (Torres et al. 1996). This type of HLA class I alteration mechanism has been described in different types of malignancy, e.g. renal cell carcinoma (Maleno et al. 2004a), laryngeal tumor

(Maleno et al. 2002), melanoma (Rodriguez et al. 2005), colorectal tumor (Maleno et al. 2004b), non-Hodgkin's lymphoma (Drénou et al. 2004), and pancreatic cancer (Ryschich et al. 2004). Allelic loss of single HLA alleles defines a third HLA phenotype that is caused by a wide array of genetic defects including point mutations, frame shifts or deletions.

LOH in chromosome 15 (β2m gene region) can be frequently detected in tumors (in 40 % of colon, melanomas and laryngeal carcinomas and in 50 % of bladder carcinomas) (Maleno et al. 2011). This lesion in chromosome 15 may be unnoticed since tumor cells might have "normal" HLA class I pattern and it could represent one of the early events in malignant cells leading to generation of pre-committed tumors to become HLA escape variants. LOH in chromosome 15 in tumors can be found more frequently than mutations in β2m gene.

HLA class I gene mutations include somatic recombination within class I genes (Browning et al. 1996), nonsense mutations (Koopman et al. 2000), missense mutations, deletions, and insertions (Serrano et al. 2000; Jiménez et al. 2001; Lehmann et al. 1995).

Mutations in β2m genes range from large deletions to single nucleotide deletions and mutations are distributed randomly among the genes (Restifo et al. 1996; Benitez et al. 1998; Feenstra et al. 1999; Paschen et al. 2003). A mutation hotspot located in the CT repeat region of exon 1 of the β2m gene has been proposed (Pérez et al. 1999), reflecting an increased genetic instability in this region in malignant cells. A summary of β2m mutations discovered in tumor cell lines and tumor specimens has been recently reviewed (Bernal et al. 2012). In most of the cases, two structural defects are necessary to produce the total loss of HLA class I on malignant cells: β2m mutation in one copy of the β2m gene and loss of the other copy associated with loss of heterozygosity (LOH) in chromosome 15 (Paschen et al. 2006).

Mutations in various APM components appear to be a rare event postulating that dysregulation rather than structural alterations is the major cause for aberrant APM component expression. TAP mutation associated with HLA class I loss was described in lung cancer (Chen et al. 1996) and in melanoma (Seliger et al. 2001).

Resistance to IFN-gamma-mediated upregulation of HLA class I expression can be also a mechanism producing tumor escape variants. It is caused by defects in the Jak-STAT components of interferon (IFN)-mediated signaling pathway (Seliger et al. 2008; Rodriguez et al. 2005).

10.6 Correlation Between HLA Class I Defects and Cancer Progression in Humans

Despite the recent advances in the understanding of the role of HLA class I antigen expression in tumors, information regarding its prognostic value or its association with patient outcome remains controversial. There are a large number of publications describing a relationship between traditional pathologic criteria and/ or patient survival and HLA class I expression, but the results are inconsistent. Downregulation or low expression of MHC class I antigens has been demonstrated to have an important cancer prognostic value in various studies (Marincola et al. 2000; Powell et al. 2012). Morabito et al. (2009) observed that downregulation of HLA class I expression in breast cancer has a significant association with adverse prognostic factors. Kaneko et al. (2011) reported that patients with preserved HLA class I expression have significantly better disease-free interval than those with loss of HLA class I. Down-regulation of HLA class I in rectal cancer has been associated with poor prognosis (Speetjens et al. 2008). On the other hand, loss of class I expression has been associated with good prognosis in breast carcinoma and non-small cell lung cancer (Madjd et al. 2005; Ramnath et al. 2006).

Many studies have failed to show a correlation between HLA-expression and patient prognosis. Normal expression of HLA class I in a non-small-cell lung cancer was associated with a favorable prognosis compared with the heterogeneous expression group, but no significant difference was observed between the normal expression and decreased expression groups (Hanagiri et al. 2012). Kikuchi et al. (2007) revealed down-regulation of HLA class I as an independent factor of poor prognosis in stage I patients, but not in late-stage patient. Two studies have found that total absence of HLA class I resulted in a favourable prognosis as compared to patients with low tumor HLA expression. One study describes that high expression of HLA class I in tumor cells associated with better prognosis as compared to the partial down-regulation of HLA class I (Watson et al. 2006), while another report proved totally opposite findings (Menon et al. 2002). Partial HLA class I loss has also been significantly associated with decreased 5 years overall survival in breast cancer (Kaneko et al. 2011). The correlation between HLA expression and clinical outcome cannot be clearly defined without identification of the exact type of tumor HLA defects (which alleles are missing) in each patient, which would predict the ability of CTLs to recognize tumor-associated peptides. Tumor cells with total HLA loss are not recognized by CTL, but NK cells should be able to target them for elimination. Tumors with partial loss may evade both NK- and T cell-mediated immune surveillance; if the allele responsible for peptide presentation is missing, the remaining allele can inhibit NK cells.

Development of a particular tumor is characterized by outgrowth of cancer cells with genetic and phenotypic features that allow them to escape antitumor immune responses. Figure 10.1 illustrates the combination of somatic evolution and immune selection in cancer development, which represents is a modern view of the clonal expansion of tumor cells. According to this theory, heterogeneous primary tumors give rise to different cell clones. T-lymphocyte responses fail to eliminate tumor cells with altered HLA class I antigens, leading to uncontrolled growth and metastatic progression of these cells. Therefore, immune selection and tumor immunoediting lead to the generation of tumor variants with altered HLA expression with better survival properties capable of evading immune response and developing metastatic colonization. New tumor escape variants may appear sequentially with various types of HLA alterations. Figure 10.1 demonstrates an

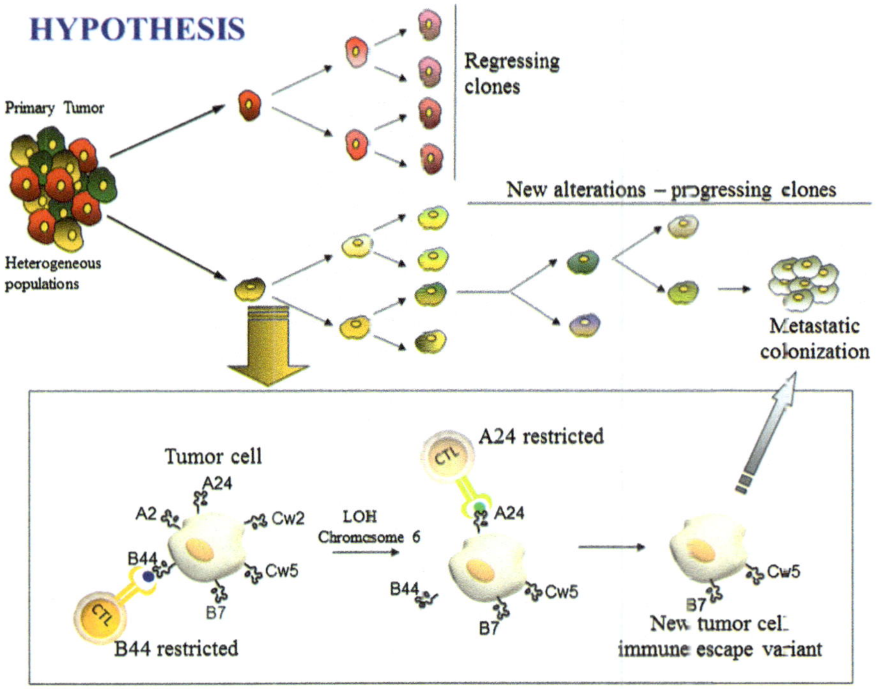

Fig. 10.1 HLA class I-mediated immunoselection of tumor escape variants during cancer progression. Primary tumors consist of heterogeneous populations of cells that give rise to different cell clones undergoing immune selection. The combination of somatic evolution of genetically unstable tumor cells and immune selection during cancer development leads to the generation of tumor variants that have better survival properties. This selective pressure will lead to the expansion of new populations of cells with multiple defects capable of evading different immune responses. In this way, tumor cell with normal HLA class I expression are subjected to T cell cytotoxic response restricted to an HLA class I allele (e.g. B44-restricted CTL reactivity). These cells are destroyed, but new B44-negative tumor cell clones are expanded. LOH in chromosome 6 leading to HLA haplotype loss gives rise to new HLA class I defective tumor cells (e.g. A24 positive cells). CTL response is now A24-restricted. HLA-A24 positive malignant cells are eliminated, and new tumor escape variants emerge

example of cancer development when initially CTLs eliminate tumor cells expressing HLA-B44 allele. However, new genetic alterations lead to emergence of B44-negative cancer cells capable of evading specific immune response. At a later stage of cancer progression, loss of another HLA allele (A24) leads to the emergence of new cancer escape variant.

There is an accumulating body of evidence suggesting that melanoma tumors are heterogeneous, with different molecular mechanisms generating different subsets of cancer cells with distinct metastatic capacity, resulting in distinct clinical courses and variations in the response to therapy. Figure 10.2 depicts a heterogeneous HLA class I expression pattern in cutaneous melanoma metastasis with

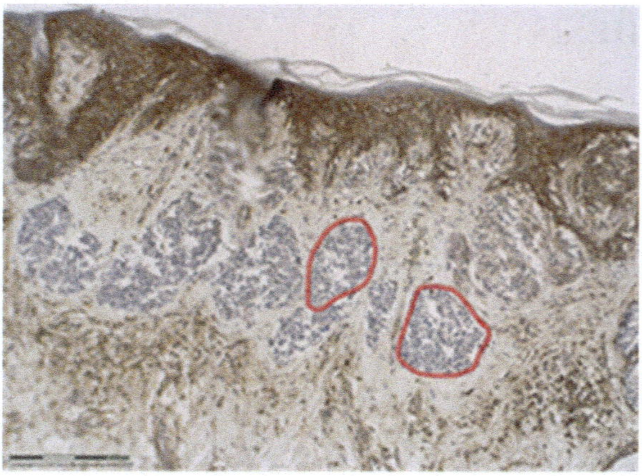

Fig. 10.2 Heterogeneous immunostaining pattern of cutaneous melanoma metastasis (labeled with anti-HLA-A,B,C antibody). HLA class I-negative tumor nests are *circled*

prevalence of class I-negative tumor nodules. Hence, the identification of genetic variants and the characterization of molecular mechanisms underlying the development of aggressive phenotypes could contribute to a better understanding of cancer immune escape. This could be useful to identify novel targets for melanoma treatment and to select the most effective therapy for different melanoma subsets.

In some types of malignancy, such as gastrointestinal carcinomas (GIAC) tumor genetic instability plays an important role in the immune escape and cancer progression. We have previously observed that leukocyte infiltrate in GIAC is strongly associated with tumor microsatellite instability, but not with tumor HLA expression level and the extent of the antitumoral immune response correlates with the type of genetic instability (Bernal et al. 2011). The high density of cytotoxic T lymphocytes in cancer infiltrate has been correlated with a good prognosis (Galon et al. 2006). The type, density, and localization of immune cells in CRC samples were found to be a good predictor of patient survival. Microsatellite instability appears to be the most important factor determining the composition, density, and localization of leukocyte infiltrate, which is independent of other molecular features such expression of HLA class I cells, galectin-3, or programmed death ligand-1 (Bernal et al. 2011). Accordingly, the strong intratumoral CD8 T infiltration of MSI-H tumors may be produced by elevated levels of specific inflammatory chemokines in the tumor microenvironment. In the studied GIAC, microsatellite instability appears to critically influence the composition and density of leukocyte infiltration independently of HLA class I loss. The abundance of CD8 cells detected in the studied MSI-H tumors might favor immunoselection of HLA-negative tumor cells. CD8 T lymphocytes were more frequently detected intratumorally in MSI-H tumors and in the stroma of MSS tumors. In addition, MSI/HLA-class I-negative and MSS/HLA-class I-negative, showed marked differences in the composition and intensity of infiltrating leucocytes, suggesting that their immune escape strategies involve distinct pathways (Bernal et al. 2012b).

We found two possible mechanisms for tumor immune evasion in the MSI tumors studied: a total loss of HLA class I molecules on tumor cell surface, mainly attributable to the accumulation of somatic mutations in the *B2m* gene and cell clonal expansion (Bernal et al. 2011); and inhibition of T cell responses by factors associated with the presence of M2 macrophages (Bernal et al. 2012b).

10.7 Role of HLA Class I Altered Expression in Resistance to Immunotherapy

Malignant transformation is characterized by accumulation of genetic alterations and by epigenetic aberrations in tumor cells leading to expression of atypical proteins called tumor-associated antigens (TAA). Recognition of TAA by HLA class I-restricted CD8+ T cells is fundamental for the detection and destruction of malignant cells (van der Bruggen et al. 1991). The discovery of TAA has changed the field of cancer treatment and introduced a new era of cancer immunotherapy aimed at increasing tumor immunogenicity and T cell-mediated anti-tumor immunity. Unfortunately, while the new protocols of cancer immunotherapy increase the presence of tumor-specific T lymphocytes and/or demonstrated partial responses in patients with certain malignancies, they have not yet delivered significant clinical benefits, such as induction of tumor regression or increased disease-free survival (Rosenberg et al. 2004). The results of early clinical trials were not very promising, but with the introduction of adjuvants and implementation of more innovative monitoring and evaluation criteria (Response Evaluation Criteria in Solid Tumors, RECIST), the outcome of cancer immunotherapy protocols has improved (Klebanoff et al. 2011). In addition, our understanding of the molecular mechanisms of cancer immune escape and the role of complex interaction between tumor and the host has expanded leading to improved novel treatment approaches. In order to counteract immunosuppressive factors of tumor microenvironment novel strategies are being evaluated in both clinical and pre-clinical settings, including combination of immune- and chemotherapy, small-molecule targeted therapies, monoclonal antibodies used to block important immune checkpoint molecules, inhibitors of immune-suppression, etc. (Schlom 2012). Furthermore, initially, many vaccines were tested in patients with advanced metastatic disease treated with other types of cancer therapy. Clinical studies have shown that patients respond better to vaccines when they are treated at early disease stages with only limited previous clinical intervention. Understanding of the possible causes of such poor clinical outcome has become very important for improvement of the existing cancer treatment modalities. In particular, the critical role of HLA class I antigens in the success of T cell based immunotherapy has led to a growing interest in investigating the expression and function of these molecules in metastatic cancer progression and, especially in response to immunotherapy.

The lack of tumor rejection is associated with multiple cancer immune escape mechanisms, including the loss or low expression of tumor HLA class I molecules.

Absence of normal expression of HLA class I molecules on tumor cell surface expression obliterates TAA-peptide presentation to CTLs and leads to tumor progression. Therefore, immunotherapy aimed at increasing anti-tumor immune response may fail and not yield clinical benefit. Various types of T cell-based cancer immunotherapy aimed are currently used in clinical setting. Each of them lead to activation of anti-tumor immune recognition mechanisms, starting with changes in tumor microenvironment, an increase in cytokine production, and induction of DC-mediated tumor-peptide presentation to both CD8 and CD4 T cells. All this leads to HLA-restricted tumor cell recognition by CTLs and consequent elimination. However, if tumor cells lose normal HLA class I expression, they may escape T cell recognition and proliferate. Therefore, the commonly observed MHC-I defects in tumors constitute a potential problem for T cell-based immunotherapy. In addition, the impact of MHC-I defects on the non-responding tumors is largely unknown and corrections of antigen presentation in these tumor types might result in much higher success rates (Lampen and van Hall 2011).

Our group previously reported that the poor clinical response of two melanoma patients to vaccination with HLA-A1-restricted MAGE-derived peptides (BB74-Mel [Me12]) and LB1622-Mel [Me13]) correlated with the loss of HLA Class I surface expression in tumor tissues and cell lines due to the presence of LOH on chromosome 15q21 in combination with b2m gene mutations (Benitez et al. 1998). Likewise, another melanoma patient who did not respond to immunotherapy with IFN-α showed total loss of HLA class I surface expression caused by the concurrence of a β2m gene mutation and LOH on chromosome 15q21 (UKRV-Mel-2b [Me10]). The importance of monitoring tumor HLA class I expression is well illustrated by the report in which a longer overall survival in renal cell cancer was associated with immune responses to multiple tumor-associated peptides (TUMAPs) used for vaccination (Walter et al. 2012). The authors treated HLA-A*02 RCC patients with peptides presented by HLA-A2 without analyzing tumor HLA-A2. Moreover, accumulating evidence suggests that tumor cells that escape immune response during immunomodulating treatment have more dangerous metastatic phenotype due to accumulation of more profound genetic alterations. In this regard, the results that we have obtained recently in our laboratory support this theory.

10.8 HLA, and Resistance to Immunotherapy in Melanoma and Bladder Cancer

We have studied different melanoma metastases from patients with mixed response to immunotherapy and several bladder tumors from patients treated with Bacillus Calmette-Guerin (BCG). We observed a strong correlation between tumor progression/recurrence and response to therapy with defects in tumor HLA class I expression and the nature of underlying mechanisms of these alterations (reversible or irreversible). One melanoma patient developed several metastases after therapy with autologous tumor cell vaccine together and BCG (M-VAX), including three progressing and three regressing lesions. Another melanoma patient was treated first with interferon α2b and later with M-VAX. We studied several progressing and regressing metastases obtained after each of the therapy modalities. All metastases showed HLA class I alterations. However, the progressing metastases developed additional and more profound defects in HLA class I expression. All metastases from the first melanoma patient presented loss of heterozygosity (LOH) in chromosome 6. In addition, progressing metastases showed a weaker expression of HLA class I, loss of HLA-B locus, and LOH in chromosome 15 (Cabrera et al. 2007). None of the metastatic samples from the second melanoma patients showed LOH in chromosomes 6 or 15, although loss of HLA-B we detected in all the samples. Progressing metastases develcped new defects in the HLA system after the therapy. Quantitative expression analysis of HLA-A, B and C genes on microdissected tumor areas demonstrated higher HLA expression in regressing than in progressing metastases (Carretero et al. 2008). A comparative gene expression analysis of these 15 metastases (10 regressing and 5 progressing) obtained from mixed melanoma responders to different types of therapy allowed us to isolate genes differentially expressed in regressing and progressing lesions, with the majority of them being implicated in regulation of the immune response. Upregulation of antigen presentation and immune rejection pathways, including HLA-A, B and C, antigen processing machinery (APM), interferon regulatory factor 1 (IRF-I), signal transducers and activators of transcription 1 (STAT-1), allograft inflammatory factor (AIF-l), granzymes, were found in regressing metastases. In contrast, progressing metastases showed low transcription levels of genes involved in these pathways (Carretero et al. 2012b). These data suggest that regressing tumors are under an acute immune rejection response. The molecular signature of tumor rejection in our case appeared to be similar to those described during allograft rejection, autoimmune disease, graft-versus-host disease and pathogen clearance. Immunostaining with monoclonal antibodies against HLA class II DR indicated that both progressing and regressing tumor cells were negative for HLA-DR expression, while the tumor microenvironment immune infiltrate cells showed strong positive staining with a dramatic difference in the number of infiltrating cells between both type of metastasis; regressing lesions showing a high number of infiltrating cells, mostly of CD3 phenotype, suggesting that the gene expression pattern in regressing metastases is associated with a release of interferon gamma from infiltrating T lymphocytes. The upregulated genes (IRF-1, STAT-1, AIF-1, CCL5, GBP1, GBP2) are mainly involved in Type II interferon response.

We observed that upregulated genes in progressing metastases are mostly involved in cellular metabolism associated with a rapid growth of tumor cells. We did not find markers of inhibitory immune cells, as regulatory T cells (Tregs) or myeloid derived suppressor cells (MDSC), in progressing metastases. The only escape mechanism detected by the 36k genome array analysis was the downregulation of HLA class I molecules.

We believe that immunotherapy promotes a change of tumor microenvironment, leading to a release of immune stimulating factors by immune infiltrating

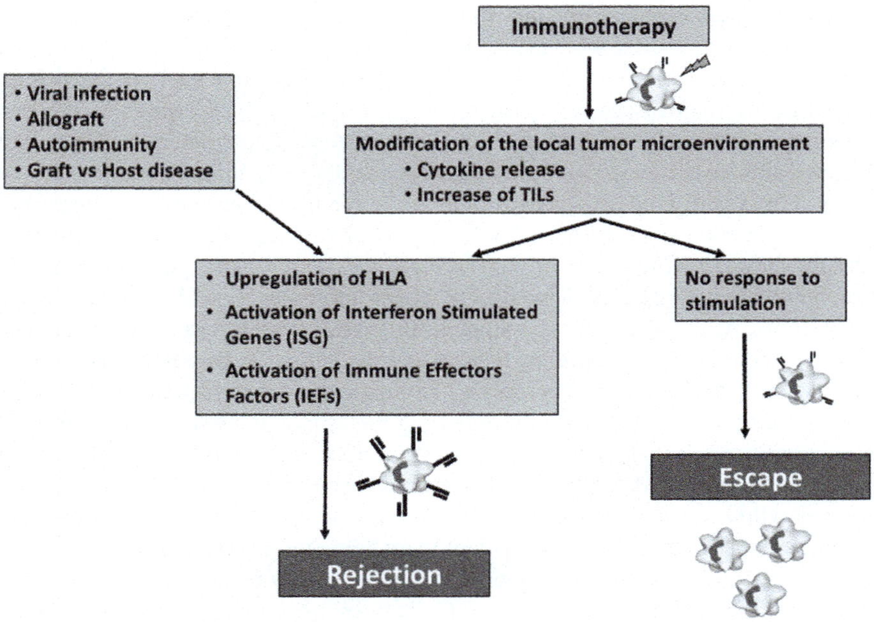

Fig. 10.3 Immunotherapy-mediated changes in tumor microenvironment induce HLA class I upregulation in tumor cells with reversible alterations and immune escape of cancer cells with structural irreversible HLA defects

cells. This immune stimulus will lead to an increase of HLA expression in tumor cells with reversible alterations of HLA Class I expression and, consequently, these tumor cells will be recognized and destroyed by the antigen specific T cells. These T cells, as they recognize tumor cells, produce more proinflammatory factors such as IFN-γ, IL-2, TNF-α and GM-CSF. This, in turn, triggers a positive and self-perpetuating feedback between tumor and immune cells until tumor rejection occurs. In contrast, if cancer cells have irreversible defects in HLA Class I genes, antigen presentation remains defective after immunotherapy impairing the amplification of the local immune response and promoting their escape from immune recognition (Fig. 10.3).

We have also showed that BCG immunotherapy of bladder cancer induces selection of HLA class I-deficient tumor cells (Carretero et al. 2011). We observed a higher incidence of 15q21 chromosomal region loss in high-risk BCG-treated bladder carcinomas that relapsed than in those that did not, suggesting an association between hard b2m lesions and tumor escape. We performed a comparative analysis of HLA class I expression in recurrent bladder tumors in patients treated with mitomycin or BCG. HLA class I expression was studied in 18 bladder cancer patients in total. Among 13 patient treated with BCG, eight were relapse-free, while five patients developed recurrent tumors after the therapy. Five mitomycin-treated patients were used as controls. Both primary and recurrent tumors were studied. More profound alterations in HLA class I expression were found in post-BCG recurrent tumors than in pre-BCG lesions, whereas mitomycin treatment did not change the HLA class I expression pattern. Post-BCG recurrent tumors also showed a higher incidence of structural defects underlying altered HLA class I expression: 80 and 60 % of tumors showed (LOH) in chromosomes 6 and 15, whereas only 25 % of relapse-free patients had LOH in either chromosome. A whole genome transcriptional analysis is also being carried in 13 primary bladder tumors obtained from six relapse-free patients and seven patients with relapse after several years of follow-up. Preliminary results showed that antigen presentation and interferon pathway genes are highly expressed in tumors from relapse-free patients versus patients with recurrence, which showed higher expression level of molecules associated with Th17 lymphocytes as compared to relapse-free patients. The latter group also showed higher expression of Th1-related molecules (unpublished data).

Our results demonstrate that show that the nature of HLA alterations (reversible or irreversible) might determine the success or failure of immunotherapy. For instance, tumor escape variants with low HLA class I expression but soft lesions will recover HLA expression after immunotherapy through effect of cytokines released locally in the tumor microenvironment. In contrast, HLA class I-deficient tumor cells with hard lesions will not recover HLA, regardless of the type of immunotherapy (Fig. 10.3). In fact, additional selective pressure may be exerted during T cell-based immunotherapy, favoring the outgrowth of HLA class I-deficient tumor cells with ''hard lesions''. Therefore, expression of HLA class I alterations in tumor cells is a key factor to be considered during selection of immunotherapy strategy and is a biomarker to be monitored during treatment.

10.9 MHC Class I Expression in Experimental Mouse Models of Cancer: Immunotherapy of Tumors with Different MHC-I Expression Patterns

Evidences obtained in experimental mouse tumor models indicates that the MHC class I phenotype of a metastatic tumor clone can dramatically change depending on the immune status of the host. The correlation between the outcome of immunotherapy as treatment against primary tumor progression and MHC-I cell surface expression on tumor cells has been studied in various murine tumor models. MHC-I cell surface expression level on tumor cells is crucial for the outcome of immunotherapies based on vaccination with peptides derived from tumor associated antigens (TAA). Loss of MHC-I expression or of a specific tumor antigen might lead to the failure of the treatment. Furthermore, it has been reported that IFN-γ limits the effectiveness of melanoma peptide vaccines because tumor cells exposed to IFN-γ evade CTLs by inducing large amounts of noncognate MHC-I molecules, which prevent T cell activation and effector function (Cho et al. 2011). The identification of these noncognate MHC-I molecules and conditions in which they are induced, as well as a correlation with the MHC-I classical expression on tumor cells, seems to be important for the improvement of the success of vaccination with peptides. Recently, it has been reported that tumor-infiltrating myeloid cells (MDSCs) induce tumor cell resistance to CTLs in mice (Lu et al. 2011). MDSCs segregates the free radical peroxynitrite (PNT), which inhibited binding of processed peptides to MHC-I molecules expressed on tumor cells. These last two immunosuppressive mechanisms show as presentation of TAAs by MHC-I molecules may be inhibited in case when MHC-I cell surface expression and TAA expression is not specifically downregulated. Intratumoral electroporation of IL-12 cDNA used in MHC-I negative B16 melanoma cells, produced eradication of established melanomas (Sin et al. 2012). The antitumor effect required the participation of IFN-γ, which upregulated tumor MHC-I expression and increased anti-tumor the activity of specific CD8+ CTLs. Treatment of cervical carcinoma cells with synthetic oligodeoxynucleotidebearing CpG motifs (CpG-ODNs) caused tumor regression which correlated with MHC-I upregulation (Baines and Celis 2003). The antitumor effect was associated with CD8+ T cell activation. In contrast, other studies show that CpG-ODNs immunotherapy significantly reduces the growth of both MHC-I-positive and -deficient tumors (Reinis et al. 2006).

Furthermore, CpG ODN 1585, whose mechanism of action principally involves activation of NK cells, induced regression only of MHC-I-deficient tumors. Combination of CpG with dendritic cell-based vaccines or vaccination with longer peptides resulted in tumor growth inhibition of both MHC-I-positive and -negative tumors (Reinis et al. 2007). Moreover, CpG-ODNs were equally effective in treatment of minimal residual tumor disease with MHC-I-positive and -negative tumors in a murine model after chemotherapy or surgery (Reinis et al. 2007). In these assays, NK1.1+ cells seem to be important for the development of protective immunity against MHC-I-deficient tumors. Furthermore, depletion of T(reg) cells inhibited growth of recurrent tumors after surgery of MHC-I-positive and-deficient tumors transplanted in syngeneic mice (Indrova et al. 2011). Other strategy for treatment of MHC-I deficient tumors by immunotherapycould be recuperation of tumor cell surface MHC-I expression before the treatment. Frequently, MHC-I downregulation or antigen silencing is caused by epigenetic mechanisms. In that case treatment with 5-azacytidine (5AC) or with histone deacetylase inhibitor Trichostatin A may increase tumor MHC-I/antigen expression (Indrova et al. 2006), Manning et al. (2008) Simova et al. have reported an additive therapeutic effect of combination of 5AC with CpG-ODN or with IL-12 producing cellular vaccine, in both MHC + and MHC- tumors (Simova et al. 2011). Chemo-immunotherapy with ifosfamide derivative CBM-4A together with IL-12 also produced asignificant inhibition of growth of both MHC-I- deficient and -positive tumors (Indrova et al. 2006).

Between the 1980s and 1990s, the studies performed by Eisenbach's research group reported an indirect correlation between H-2K tumor cell surface expression and spontaneous metastatic capacity (Eisenbach et al. 1984). Tumor cells clones derived from H-2-K-low or -negative tumors, including 3-Lewis lung carcinoma,

B16 melanoma or BW T lymphoma, have increased spontaneous metastatic capacity. Moreover, recovery of the H-2K expression reverted their metastatic phenotype (Plaksin et al. 1988). Furthermore, injection of the H-2-tranfected cells was successful in eradicating metastases derived from primary tumor originated from parental cells. Therapy with IFN-γ-treated tumor cells or with tumor cells transfected with IFN-γ gene, both promoted induction of MHC-I cell surface expression, and protected against metastatic progression of the parental tumor (Porgador et al. 1993). If the tumor cells are transfected jointly with IFN-γ and allogeneic MHC class I cDNAs a greater antimetastatic effect was achieved.

In our laboratory, we have developed a fibrosarcoma murine model using BALB/c mice treated with 3-methylcholanthrene. The tumor cell line obtained from the tissue culture adapted local tumor was named GR9, and various cancer cell clones have been obtained after serial limiting dilution. All obtained clones have different MHC-I expression patterns and distinct spontaneous metastatic capacity. Using this model, we observed that high tumor cell surface MHC-I expression correlates with higher spontaneous metastatic capacity and slower local tumor growth (Garrido et al. 1986; Garcia-Lora et al. 2001, 2003). GR9 fibrosarcoma cells present intermediate levels of H-2 Kd, Dd and Ld molecules, while the clones derived from it have various MHC expression patterns. Two clones (G2 and A7) present high MHC-I cell surface expression, two other clones (B7 and C5) show intermediate MHC-I levels, and the B11 and B9 fibrosarcoma clones present weak or negative MHC-I cell surface expression. Several cell doses of each fibrosarcoma clones were injected subcutaneously in BALB/c mice and the results depicted an indirect correlation between MHC-I surface expression and local tumor growth rate.

The clone with lower level of MHC-I expression, B11, showed higher local growth rate; followed by B7 and C5 clones and finally A7 clone. In a spontaneous metastasis assay we obtained opposite results, showing a direct correlation between MHC-I phenotype and spontaneous metastatic capacity of the fibrosarcoma clones. The MHC-I positive clone A7 presented highest spontaneous metastatic capacity, whereas that B7 and C5 clones showed a weak spontaneous metastatic capacity, while B11 did not generate spontaneous metastases for any of the injected doses. The original GR9 fibrosarcoma cells showed high metastatic capacity. In GR9 murine tumor model, we have been able to demonstrate to direct relation between successful immunotherapy as antimetastatic treatment and MHC-I cell surface expression on primary tumor cells. The results showed that the success of the immunotherapy against metastatic disease depends on MHC-I cell surface expression on fibrosarcoma cells. Two types of immunotherapy, chemotherapy alone, and chemo-immunotherapy, were applied separately to treat spontaneous metastatic colonization generated by the different fibrosarcoma clones (Garrido et al. 2011). Protein bound polysaccharide K (PSK) and CpG+ irradiated autologous A7 tumor cells were used as immunotherapy treatments, docetaxel as chemotherapy, and PSK+ docetaxel as chemo-immunotherapy.

Two immunotherapy treatments were completely effective eradicating metastatic colonization derived from a high MHC-I positive fibrosarcoma clone, which had a high spontaneous metastatic capacity in non-treated mice. Spontaneous metastatic capacity presented by a highly-positive MHC-I fibrosarcoma clone was completely abrogated by immunotherapies and chemo-immunotherapy. Growth of primary tumor promoted a strong immunosuppression, which could be reverted by the immunotherapy treatments. In contrast, the same immunotherapy treatments only produced partial metastatic inhibition when they were administered to the mice carrying primary tumor originated from a B7 fibrosarcoma clone with intermediate MHC-I expression level. This clone also promoted a strong immunosupression, which only could be reverted partially by immunotherapy treatments. These results indicate that MHC-I cell surface expression in primary tumor may be crucial for the success of immunotherapy against metastatic disease. A MHC-I expression threshold on primary tumor cells is necessary so that immunotherapy can be effective against metastatic disease.

10.10 Potential Therapeutic Approaches for Increasing Tumor Immunogenicity by Upregulation of Tumor MHC Class I Expression

Based on clinical and experimental evidence it has become clear that low or altered expression of HLA class I molecules on tumor cells are likely to have a negative impact on the outcome of cancer immunotherapy, since it provides malignant cells with a mechanism of immune escape from T cell recognition. Various mechanisms, both reversible and irreversible, underlie the MHC class I downregulation. Therapies that lead to MHC class I upregulation on tumour cells might improve outcomes in immune-therapy-based treatments. Attempts are in progress to revert the defects in tumor MHC class I surface expression by introducing the elements of the antigen presentation pathway or by activating transcriptional factors that regulate expression of MHC class I molecules or components of APM machinery. It has been demonstrated that IFNs upregulate the expression of MHC class I molecules on cancer cells in vitro and in vivo (Seliger et al. 2000; Martini et al. 2010), unless tumor is resistant to IFN treatment due to genetic defect in IFN signal transduction pathway. Therefore, INF treatment is a valuable strategy for cancer immunotherapy aimed at increasing tumor cell immunogenicity, but only in cancer cells that do not harbor structural defects in genes coding for MHC molecules causing loss of class I expression. In that case transfer of a wild type MHC gene into tumor cells is necessary to recover normal MHC-I expression.

Epigenetic events associated with tumor development and with cancer progression have been found to underlie changes in HLA antigen and APM components. These alterations can be reversed in vitro with pharmacologic agents that induce DNA hypomethylation or inhibit histone deacetylation (Fonsatti et al. 2007).

In vitro manipulation with β2m gene and with other genes involved in MHC class I complex expression has generated evidence that restoration of normal MHC class I re-expression is important for the tumor cell recognition and elimination by CD8+ T cells. Recovery of MHC class I expression has been employed previously by various investigators to demonstrate the importance of class I molecules in specific tumor lysis by CTL and NK cells. The earliest reports in the 1970s were based on mouse models with known MHC defects. The first description of the loss of an H-2Kk private specificity was reported in Gardener lymphoma derived from a C3H mouse. One particular AKR tumor cell line designated K36.16 had no expression of Kk antigen and was resistant to killing by AKR anti-MuLV cytotoxic lymphocytes in vitro, and always produced tumors in immunocompetent AKR mice (Garrido et al. 1976). In different experimental systems, introduction of MHC class I molecules into MHC class I negative tumor cell lines led to increased immunogenicity of the tumor cells and abrogation of malignancy. The transfection and cell surface expression of an H-2Kk gene in the K36 (H-2Kk negative) lymphoma inhibited the syngeneic growth of this tumor (Hui et al. 1984).

Chen et al. (1996) analyzed breast and lung cancer for β2m downregulation or mutations. They identified 63 tumors without detectable β2m muattions and two neoplasms with β2m mutations; they transfected cells with wild type β2m gene and demonstrated complete restoration of HLA expression. They also observed that mutation in β2m caused cell line H2009 to be resistant to specific lysis by influenza virus-specific CTL from HLA matched donors, and that transfection of the β2m gene restored the cytotoxicity. Tafuro et al. (2001) adopted another approach to reconstitute antigen presentation in HLA class I-negative cancer cell lines. They engineered an HLA-A2 restricted peptide epitope linked to the N terminus of β2 m and delivered this fusion protein to tumor cells using a retroviral vector. The transfected cells were recognized and killed by appropriate CTL clones. Nabel et al. (1996) reported results of a direct transfer of the HLA-B7 gene into HLA-B7-negative patients with advanced melanoma by injection of DNA-liposome complexes (allogeneic vaccination). Plasmid DNA and recombinant HLA-B7 protein were detected in treated tumors. One patient showed a regression of injected nodules after two independent treatments, which was accompanied by regression at distant sites. Bergen et al. (2003) reported preliminary results of the clinical trial of HLA-B7/beta-2-microglobulin plasmid DNA/lipid complex (Allovectin-7((R)) in patients with metastatic melanoma. While the clinical outcome of the gene transfer was not dramatic in this case, Allovectin-7 appears to be a promising agent with a safe toxicity profile. However, the main limitation of this type of vaccine is that it is an allogeneic vaccine, not targeted to restoration of a specific gene defect in a given patient. Experiments by Tsory et al. (2006) suggested that MHC class I glycoproteins may regulate the immune response by modulating the expression and function of other genes essential for proper antigen processing and presentation. They reported that reconstitution of expression of MHC class I glycoproteins in MHC-deficient and highly metastatic B16BL6 melanoma cells augmented the expression of TAP-2 and inducible proteasome subunits, LMP-2 and LMP-7.

Up-regulation of inducible proteasome subunits was also followed by a significant change in the proteolytic activity of the proteasome complex. In APM-deficient mouse lung carcinoma cell line CMT.64, re-expression of TAP1 after infection with TAP1-adenovirus vector led to increased MHC class I surface expression, antigen presentation, and susceptibility to antigen-specific CTLs (Lou et al. 2005). Our group designed an adenoviral vector with wild type β2m gene and was able to restore normal HLA class I expression in human tumor cell lines harboring β2m mutations (del Campo et al. 2009). Reconstitution of β2m expression following transduction with the adenovirus was sufficient to restore total HLA class I expression on different human tumor cells lines recovering the lysis of tumor cells by peptide-stimulated HLA-restricted T-cells and increasing peptide-specific IFN-gamma secretion by these T-cells in HLA-restricted manner (del Campo et al. 2012 and unpublished data).

Epigenetic events associated with tumor development and with cancer progression have been also found to underlie changes in HLA antigen and APM components. DNA methylation was found to be responsable for the MHC class I heavy chain gene inhibition (Serrano et al. 2001; Nie at al. 2001) while both the DNA methylation and histone acetylation changes were associated with inhibition of the antigen presenting machinery (APM) gene expression (Campoli and Ferrone 2008). These alterations can be reversed in vitro with pharmacologic agents that induce DNA hypomethylation or inhibit histone deacetylation (Fonsatti et al. 2007). The therapeutic benefit of such 'epigenetic' agents, including histone deacetylase and DNA methyltransferase inhibitors (DNMTi), has been successfully tested in clinical trials and several compounds, including DNMTi 5-azacytidine (5AC) and 5-aza-20-deoxycytidine (DAC) have been approved for clinical use (Mai and Altucci 2009).

There is an accumulating body of evidence suggesting that a combination of different types of cancer therapy, including chemotherapy, immunotherapy and gene therapy aimed at recovering normal HLA class I expression, gives the best results in both animal models and in clinical trials. Such a combined therapy may have some advantages in combating well-established tumors and metastatic cancer.

10.11 Conclusions

Advances in our understanding of the relationship between the immune system and tumor cells gave rise to new prospects for immunological treatment and the prevention of cancer. Altered MHC Class I expression is a hallmark of malignant transformation and tumor immune escape. Using immunohistochemistry and molecular techniques it has been demonstrated that many types of tumor can lose up to 80 % of normal MHC class I expression. Both reversible and irreversible structural defects of MHC class I have been described in solid tumors, in cancer cell lines and metastatic lesions. As a result, malignant cells develop low immunogenic phenotypes with altered antigen-presentation ability. It leads to the loss of tumor recognition by cytotoxic T lymphocytes, providing an immune escape route for MHC-negative cells, and limits the efficacy of cancer immunotherapy. A growing body of evidence accumulated form both clinical and basic research supports a hypothesis that the irreversible genetic defects underlying abnormal MHC expression are, at least partially, responsible for the emergence of immunotherapy-resistant tumor escape variants. These results indicate that the success of immunotherapy as anti-metastatic treatment may depend strongly of the MHC-I expression level on primary tumor cells. An important task ahead is to find a clinical application for the new knowledge of MHC class I alterations in tumor progression, i.e., to attempt to induce tumor rejection by restoring the altered HLA class I expression. Thus, in the era of "personalized medicine" it is essential to include tumor MHC expression into the list of biomarkers to be closely monitored before, during and after cancer immunotherapy to increase the clinical efficacy of the treatment. Comprehensive information on the tumor and the immune status of an individual could provide a precise picture of the ongoing evolution of the tumor, as well as to yield invaluable information about which strategy will result in optimal therapeutic outcome. We believe that a combination of immunotherapy with chemo- and gene therapy is the most promising approach in fighting cancer.

Acknowledgments We would like to acknowledge contribution of the students and investigators who have been part of the large research group at the Department of Clinical Analysis at the Virgen de las Nieves University Hospital in Granada, Spain. This work has been supported over the years by grants from the Fondo de Investigaciones Sanitarias (FIS), Red Genomica del Cancer (RETIC), Plan Andaluz de Investigación (Group CTS-143), Servicio Andaluz de Salud (SAS), Plan Nacional de Investigación Científica, Desarrollo e Innovación Tecnologica (Proyectos SAF 2007-63262 and SAF 2010-20273) in Spain; and from different European projects.

References

Ahmad M, Rees RC, Ali SA (2004) Escape from immunotherapy: possible mechanisms that influence tumor regression/progression. Cancer Immunol Immunother 53(10):844–854

Aoe T, Okamoto Y, Saito T (1995) Activated macrophages induce structural abnormalities of the T-cell receptor-CD3 complex. J Exp Med 181(5):1881–1886

Aptsiauri N, Cabrera T, Garcia-Lora A, Lopez-Nevot MA, Ruiz-Cabello F, Garrido F (2007) MHC class I antigens and immune surveillance in transformed cells. Int Rev Cytol 256:139–189

Aruga A, Aruga E, Tanigawa K, Bishop DK, Sondak VK, Chang AE (1997) Type 1 versus type 2 cytokine release by V beta T cell subpopulations determines in vivo anti-tumor reactivity -IL-10 mediates a suppressive role. J Immunol 159(2):664–673

Baines J, Celis E (2003) Immune-mediated tumor regression induced by CpG-containing oligodeoxynucleotides. Clin Cancer Res 9(7):2693–2700

Benitez R, Godelaine D, Lopez-Nevot MA, Brasseur F, Jiménez P, Marchand M, Oliva MR, van Baren N, Cabrera T, Andry G, Landry C, Ruiz-Cabello F, Boon T, Garrido F (1998a) Mutations of the beta2-microglobulin gene result in a lack of HLA class I molecules on melanoma cells of two patients immunized with MAGE peptides. Tissue Antigens 52(6):520–529

Benitez R, Godelaine D, Lopez-Nevot MA, Brasseur F, Jimenez P, Marchand M, Oliva MR, van Baren N, Cabrera T, Andry G, Landry C, Ruiz-Cabello F, Boon T, Garrido F (1998b) Mutations of the beta2-microglobulin gene result in a lack of HLA class I molecules on melanoma cells of two patients immunized with MAGE peptides. Tissue Antigens 52:520–529

Bergen M, Chen R, Gonzalez R (2003) Efficacy and safety of HLA-B7/beta-2 microglobulin plasmid DNA/lipid complex (Allovectin-7) in patients with metastatic melanoma. Expert Opin Biol Ther 3(2):377–384

Bernal M, Concha A, Sáenz-López P, Rodríguez AI, Cabrera T, Garrido F, Ruiz-Cabello F (2011) Leukocyte inWltrate in gastrointestinal adenocarcinomas is strongly associated with tumor microsatellite instability but not with tumor immunogenicity. Cancer Immunol Immunother 60:869–882

Bernal M, Ruiz-Cabello F, Concha A, Paschen A, Garrido F (2012a) Implication of the β2-microglobulin gene in the generation of tumor escape phenotypes. Cancer Immunol Immunother 61(9):1359–1371

Bernal M, García-Alcalde F, Concha A, Cano C, Blanco A, Garrido F, Ruiz-Cabello F (2012b) Genome-wide differential genetic profiling characterizes colorectal cancers with genetic instability and specific routes to HLA class I loss and immune escape. Cancer Immunol Immunother 61(6):803–816

Berset M, Cerottini JP, Guggisberg D, Romero P, Burri F, Rimoldi D, Panizzon RG (2001) Expression of Melan-A/MART-1 antigen as a prognostic factor in primary cutaneous melanoma. Int J Cancer 95(1):73–77

Blades RA, Keating PJ, McWilliam LJ, George NJ, Stern PL (1995) Loss of HLA class I expression in prostate cancer: implications for immunotherapy. Urology 46(5):681–686

Boon T, Coulie PG, Van den Eynde BJ, van der Bruggen P (2006) Human T cell responses against melanoma. Annu Rev Immunol 24:175–208

Browning M, Petronzelli F, Bicknell D, Krausa P, Rowan A, Tonks S, Murray N, Bodmer J, Bodmer W (1996) Mechanisms of loss of HLA class I expression on colorectal tumor cells. Tissue Antigens 47(5):364–371

Burnet FM (1957) Cancer—a biological approach. 1. The process of control. Br Med J 1:779–782

Cabrera T, Angustias Fernandez M, Sierra A, Garrido A, Herruzo A, Escobedo A, Fabra A, Garrido F (1996) High frequency of altered HLA class I phenotypes in invasive breast carcinomas. Hum Immunol 50(2):127–134

Cabrera T, Collado A, Fernandez MA, Ferron A, Sancho J, Ruiz-Cabello F, Garrido F (1998) High frequency of altered HLA class I phenotypes in invasive colorectal carcinomas. Tissue Antigens 52(2):114–123

Cabrera T, Salinero J, Fernandez MA, Garrido A, Esquivias J, Garrido F (2000) High frequency of altered HLA class I phenotypes in laryngeal carcinomas. Hum Immunol 61(5):499–506

Cabrera CM, Jiménez P, Cabrera T, Esparza C, Ruiz-Cabello F, Garrido F (2003) Total loss of MHC class I in colorectal tumors can be explained by two molecular pathways: beta2-microglobulin inactivation in MSI-positive tumors and LMP7/TAP2 downregulation in MSI negative tumors. Tissue Antigens 61(3):211–219

Cabrera CM, López-Nevot MA, Jiménez P, Garrido F (2005) Involvement of the chaperone tapasin in HLA-B44 allelic losses in colorectal tumors. Int J Cancer 113(4):611–618

Cabrera T, Lara E, Romero JM, Maleno I, Real LM, Ruiz-Cabello F, Valero P, Camacho FM, Garrido F (2007) HLA class I expression in metastatic melanoma correlates with tumor development during autologous vaccination. Cancer Immunol Immunother 56(5):709–717

Campoli M, Ferrone S (2008) HLA antigen changes in malignant cells: epigenetic mechanisms and biologic significance. Oncogene 27(45):5869–5885

Campoli M, Ferrone S, Zea AH, Rodriguez PC, Ochoa AC (2005) Mechanisms of tumor evasion. Cancer Treat Res 123:61–88

Carosella ED, Moreau P, Le Maoult J, Le Discorde M, Dausset J, Rouas-Freiss N (2003) HLA-G molecules: from maternal-fetal tolerance to tissue acceptance. Adv Immunol 81:199–252

Carretero R, Romero JM, Ruiz-Cabello F, Maleno I, Rodriguez F, Camacho FM, Real LM, Garrido F, Cabrera T (2008) Analysis of HLA class I expression in progressing and regressing metastatic melanoma lesions after immunotherapy. Immunogenetics 60(8):439–447

Carretero R, Cabrera T, Gil H, Saenz-Lopez P, Maleno I, Aptsiauri N, Cozar JM, Garrido F (2011) Bacillus Calmette-Guerin immunotherapy of bladder cancer induces selection of human leukocyte antigen class I-deficient tumor cells. Int J Cancer 129(4):839–846

Carretero R, Wang E, Rodriguez AI, Reinboth J, Ascierto ML, Engle AM, Liu H, Camacho FM, Marincola FM, Garrido F, Cabrera T (2012a) Regression of melanoma metastases after immunotherapy is associated with activation of antigen presentation and interferon-mediated rejection genes. Int J Cancer 31(2):387–395

Carretero R, Wang E, Rodriguez AI, Reinboth J, Ascierto ML, Engle AM, Liu H, Camacho FM, Marincola FM, Garrido F, Cabrera T (2012b) Regression of melanoma metastases after immunotherapy is associated with activation of antigen presentation and interferon-mediated rejection genes. Int J Cancer 131(2):387–395

Catlett-Falcone R, Landowski TH, Oshiro MM, Turkson J, Levitzki A, Savino R, Ciliberto G, Moscinski L, Fernández-Luna JL, Nuñez G, Dalton WS, Jove R (1999) Constitutive activation of Stat3 signaling confers resistance to apoptosis in human U266 myeloma cells. Immunity 10(1):105–115

Chang CC, Campoli M, Restifo NP, Wang X, Ferrone S (2005) Immune selection of hot-spot beta 2-microglobulin gene mutations, HLA-A2 allospecificity loss, and antigen-processing machinery component down-regulation in melanoma cells derived from recurrent metastases following immunotherapy. J Immunol 174(3):1462–1471

Chen HL, Gabrilovich D, Tampé R, Girgis KR, Nadaf S, Carbone DP (1996a) A functionally defective allele of TAP1 results in loss of MHC class I antigen presentation in a human lung cancer. Nat Genet 13(2):210–213

Chen HL, Gabrilovich D, Virmani A, Ratnani I, Girgis KR, Nadaf-Rahrov S, Fernandez-Viña M, Carbone DP (1996b) Structural and functional analysis of beta2 microglobulin abnormalities in human lung and breast cancer. Int J Cancer 67(6):756–763

Cho HI, Lee YR, Celis E (2011) Interferon γ limits the effectiveness of melanoma peptide vaccines. Blood 117(1):135–144

Condamine T, Gabrilovich DI (2011) Molecular mechanisms regulating myeloid-derived suppressor cell differentiation and function. Trends Immunol 32(1):19–25

Cresswell P, Ackerman AL, Giodini A, Reaper D, Wearsch PA (2005) Mechanisms of MHC class I-restricted antigen processing and cross presentation. Immunol Rev 207:145–157

De la Salle H, Hanau D, Fricker D, Urlacher A, Kelly A, Salamero J, Powis SH, Donato L, Bausinger H, Laforet M, Jeras M, Spehner D, Bieber T, Falkenrodt A, Cazenave JP, Trowsdale J, Tongio MM (1994) Homozygous human TAP peptide transporter mutation in HLA class-I deficiency. Science 265:237–241

del Campo AB, Aptsiauri N, Méndez R, Zinchenko S, Vales A, Paschen A, Ward S, Ruiz-Cabello F, González-Aseguinolaza G, Garrido F (2009) Efficient recovery of HLA class I expression in human tumor cells after beta2-microglobulin gene transfer using adenoviral vector: implications for cancer immunotherapy. Scand J Immunol 70(2):125–135

del Campo AB, Carretero J, Aptsiauri N, Garrido F (2012) Targeting HLA class I expression to increase tumor immunogenicity. Tissue Antigens 79(3):147–154

Del Val M, Iborra S, Ramos M, Lazaro S (2011) Generation of MHC class I ligands in the secretory and vesicular pathways. Cell Mol Life Sci 68:1543–1552

Derre L, Corvaisier M, Charreau B, Moreau A, Godefroy E, Moreau-Aubry A, Jotereau F, Gervois N (2006) Expression and release of HLA-E by melanoma cells and melanocytes: potential impact on the response of cytotoxic effector cells. J Immunol 177(5):3100–3107

Dong H, Strome SE, Salomao DR, Tamura H, Hirano F, Flies DB, Roche PC, Lu J, Zhu G, Tamada K, Lennon VA, Celis E, Chen L (2002) Tumor-associated B7-H1 promotes T-cell apoptosis: a potential mechanism of immune evasion. Nat Med 8(8):793–800

Drake CG, Jaffee E, Pardoll DM (2006) Mechanism of immune evasion by tumors. Adv Immunol 90:51–81

Drénou B, Tilanus M, Semana G, Alizadeh M, Birebent B, Grosset JM, Dias P, van Wichen D, Arts Y, De Santis D, Fauchet R, Amiot L (2004) Loss of heterozygosity, a frequent but a non-exclusive mechanism responsible for HLA dysregulation in non-Hodgkin's lymphomas. Br J Haematol 127(1):40–49

Dunn GP, Old LJ, Shreiber RD (2004) Three Es of cancer immunoediting. Annu Rev Immunol 3:991–998

Dunn GP, Sheehan KCF, Old LJ, Schreiber RD (2005) IFN unresponsiveness in LNCaP cells due to the lack of JAK1 gene expression. Cancer Res 65(8):3447–3453

Eisenbach L, Hollander N, Greenfeld L, Yakor H, Segal S, Feldman M (1984) The differential expression of H-2K versus H-2D antigens, distinguishing high-metastatic from lowmetastatic clones, is correlated with the immunogenic properties of the tumor cells. Int J Cancer 34(4):567–573

Feenstra M, Rozemuller E, Duran K, Stuy I, van den Tweel J, Slootweg P, de Weger R, Tilanus M (1999) Mutation in the beta 2m gene is not a frequent event in head and neck squamous cell carcinomas. Hum Immunol 60(8):697–706

Feenstra M, Verdaasdonk M, van der Zwan AW, de Weger R, Slootweg P, Tilanus M (2000) Microsatellite analysis of microdissected tumor cells and 6p high density microsatellite analysis in head and neck squamous cell carcinomas with down-regulated human leukocyte antigen class I expression. Lab Invest 80(3):405–414

Festenstein H (1987) The biological consequences of altered MHC expression on tumours. Br Med Bull 43(1):217–227

Festenstein H, Garrido F (1986) MHC antigens and malignancy. Nature 322 (6079):502–523

Fonsatti E, Nicolay HJ, Sigalotti L, Calabrò L, Pezzani L, Colizzi F, Altomonte M, Guidoboni M, Marincola FM, Maio M (2007) Functional up-regulation of human leukocyte antigen class I antigens expression by 5-aza-2′-deoxycytidine in cutaneous melanoma: immunotherapeutic implications. Clin Cancer Res 13(11):3333–3338

Gabrilovich D, Ishida T, Oyama T, Ran S, Kravtsov V, Nadaf S, Carbone DP (1998) Vascular endothelial growth factor inhibits the development of dendritic cells and dramatically affects the differentiation of multiple hematopoietic lineages in vivo. Blood 92(11):4150–4166

Galon J, Costes A, Sanchez-Cabo F, Kirilovsky A, Mlecnik B,Lagorce-Page's C, Tosolini M, Camus M, Berger A, Wind P,Zinzindohoue′ F, Bruneval P, Cugnenc PH, Trajanoski Z, Fridman WH, Page's F (2006) Type, density, and location of immune cells within human colorectal tumors predict clinical outcome. Science 313:1960–1964

Garbi N, Tan P, Diehl AD, Chambers BJ, Ljunggren HG, Momburg F, Hammerling GJ (2000) Impaired immune responses and altered peptide repertoire in tapasin-deficient mice. Nat Immunol 1:234–238

Garcia-Lora A, Algarra I, Gaforio JJ, Ruiz-Cabello F, Garrido F (2001) Immunoselection by T lymphocytes generates repeated MHC class I-deficient metastatic tumor variants. Int J Cancer 91:109–119

Garcia-Lora A, Martinez M, Algarra I, Gaforio JJ, Garrido F (2003) MHC class I-deficient metastatic tumor variants immunoselected by T lymphocytes originate from the coordinated downregulation of APM components. Int J Cancer 106:521–527

Garrido F, Algarra I (2001) MHC antigens and tumor escape from immune surveillance. Adv Cancer Res 83:117–158

Garrido F, Festenstein H, Schirrmacher V (1976) Further evidence for depression of H-2 and Ia like specificities of foreign haplotypes in mouse tumour cell lines. Nature 261(5562):705–707

Garrido A, Perez M, Delgado C, Garrido ML, Rojano J, Algarra I, Garrido F (1986) Influence of class I H-2 gene expression on local tumor growth. Description of a model obtained from clones derived from a solid BALB/c tumor. Exp Clin Immunogenet 3(2):98–110

Garrido F, Cabrera T, Concha A, Glew S, Ruiz-Cabello F, Stern PL (1993) Natural history of HLA expression during tumour development. Immunol Today 14(10):491–499

Garrido F, Ruiz-Cabello F, Cabrera T, Perez-Villar JJ, LopezBotet M, Duggan-Keen M, Stern PL (1997) Implications for immunosurveillance of altered HLA class I pheno-types in human tumours. Immunol Today 18:89–95

Garrido F, Cabrera T, Aptsiauri N (2010) "Hard" and "soft" lesions underlying the HLA class I alterations in cancer cells: implications for immunotherapy. Int J Cancer 127(2):249–256

Garrido C, Romero I, Berruguilla E, Cancela B, Algarra I, Collado A, Garcia-Lora A, Garrido F (2011) Immunotherapy eradicates metastases with reversible defects in MHC class I expression. Cancer Immunol Immunother 60(9):1257–1268

Grandea AG, Golovina TN, Hamilton SE, Sriram V, Spies T, Brutkiewicz RR, Harty JT, Eisenlohr LC, Van Kaer L (2000) Impaired assembly yet normal trafficking of MHC class I molecules in tapasin mutant mice. Immunity 13:213–222

Hanagiri T, Shigematsu Y, Kuroda K, Baba T, Shiota H, Ichiki Y, Nagata Y, Yasuda M, Uramoto H, So T, Takenoyama M, Tanaka F (2012) Prognostic implications of human leukocyte antigen class I expression in patients who underwent surgical resection for non-small-cell lung cancer. J Surg Res 2012 Jul 31 [Epub ahead of print]

Herber DL, Cao W, Nefedova Y, Novitskiy SV, Nagaraj S, Tyurin VA, Corzo A, Cho HI, Celis E, Lennox B, Knight SC, Padhya T, McCaffrey TV, McCaffrey JC, Antonia S, Fishman M, Ferris RL, Kagan VE, Gabrilovich DI (2010) Lipid accumulation and dendritic cell dysfunction in cancer. Nat Med 16(8):880–886

Hinz S, Trauzold A, Boenicke L, Sandberg C, Beckmann S, Bayer E, Walczak H, Kalthoff H, Ungefroren H (2000) Bcl-X-L protects pancreatic adenocarcinoma cells against CD95-and TRAIL-receptor-mediated apoptosis. Oncogene 19(48):5477–5486

Hui K, Grosveld F, Festenstein H (1984) Rejection of transplantable AKR leukaemia cells following MHC DNA-mediated cell transformation. Nature 311(5988):750–752

Indrova M, Bieblova J, Jandlova T, Vonka V, Pajtasz-Piasecka E, Reinis M (2006) Chemotherapy, IL-12 gene therapy and combined adjuvant therapy of HPV 16-associated MHC class Iproficient and -deficient tumours. Int J Oncol 28(1):253–259

Indrova M, Simova J, Bieblova J, Bubenik J, Reinis M (2011) NK1.1+ cells are important for the development of protective immunity against MHC I-deficient, HPV16-associated tumours. Oncol Rep 25(1):281–288

Jager E, Ringhoffer M, Karbach J, Arand M, Oesch F, Knuth A (1996) Inverse relationship of melanocyte differentiation antigen expression in melanoma tissues and CD8(+) cytotoxic-Tcell responses: evidence for immunoselection of antigen-loss variants in vivo. Int J Cancer 66(4):470–476

Jiménez P, Cabrera T, Méndez R, Esparza C, Cozar JM, Tallada M, López-Nevot MA, Ruiz-Cabello F, Garrido F (2001) A nucleotide insertion in exon 4 is responsible for the absence of expression of an HLA-A*0301 allele in a prostate carcinoma cell line. Immunogenetics 53(7):606–610

Kageshita T, Ishihara T, Campoli M, Ferrone S (2005) Selective monomorphic and polymorphic HLA class I antigenic determinant loss in surgically removed melanoma lesions. Tissue Antigens 65(5):419–428

Kaneko K, Ishigami S, Kijima Y, Funasako Y, Hirata M, Okumura H, Shinchi H, Koriyama C, Ueno S, Yoshinaka H, Natsugoe S (2011) Clinical implication of HLA class I expression in breast cancer. BMC Cancer 11:454

Kataoka T, Schroter M, Hahne M, Schneider P, Irmler M, Thome M, Froelich CJ, Tschopp J (1998) FLIP prevents apoptosis induced by death receptors but not by perforin/granzyme B, chemotherapeutic drugs, and gamma irradiation. J Immunol 161(8):3936–3942

Khong HT, Restifo NP (2002) Natural selection of tumor variants in the generation of "tumor escape" phenotypes. Nat Immunol 3(11):999–1005

Khong HT, Wang QJ, Rosenberg SA (2004) Identification of multiple antigens recognized by tumor-infiltrating lymphocytes from a single patient: tumor escape by antigen loss and loss of MHC expression. J Immunother 27(3):184–190

Kikuchi E, Yamazaki K, Torigoe T, Cho Y, Miyamoto M, Oizumi S, Hommura F, Dosaka-Akita H, Nishimura M (2007) HLA class I antigen expression is associated with a favorable prognosis in early stage non-small cell lung cancer. Cancer Sci 98(9):1424–1430

Klebanoff CA, Acquavella N, Yu Z, Restifo NP (2011) Therapeutic cancer vaccines: are we there yet? Immunol Rev 239(1):27–44

Koopman LA, Corver WE, van der Slik AR, Giphart MJ, Fleuren GJ (2000) Multiple genetic alterations cause frequent and heterogeneous human histocompatibility leukocyte antigen class I loss in cervical cancer. J Exp Med 191(6):961–976

Lampen MH, van Hall T (2011) Strategies to counteract MHC-I defects in tumors. Curr Opin Immunol 23(2):293–298

Lanier LL (2005) NK cell recognition. Annu Rev Immunol 23:225–274

Lehmann F, Marchand M, Hainaut P, Pouillart P, Sastre X, Ikeda H, Boon T, Coulie PG (1995) Differences in the antigens recognized by cytolytic T cells on two successive metastases of a melanoma patient are consistent with immune selection. Eur J Immunol 25(2):340–347

Lehner PJ, Trowsdale J (1998) Antigen presentation: coming out gracefully. Curr Biol 8(17):R605–R608

Ljunggren HG, Karre K (1990) In search of the missing self—MHC molecules and NK cell recognition. Immunol Today 11(7):237–244

Lou Y, Vitalis TZ, Basha G, Cai B, Chen SS, Choi KB, Jeffries AP, Elliott WM, Atkins D, Seliger B, Jefferies WA (2005) Restoration of the expression of transporters associated with antigen processing in lung carcinoma increases tumor-specific immune responses and survival. Cancer Res 65(17):7926–7933

Lu T, Ramakrishnan R, Altiok S, Youn JI, Cheng P, Celis E, Pisarev V, Sherman S, Sporn MB, Gabrilovich D (2011) Tumor-infiltrating myeloid cells induce tumor cell resistance to cytotoxic T cells in mice. J Clin Invest 121(10):4015–4029

Madjd Z, Spendlove I, Pinder SE, Ellis IO, Durrant LG (2005) Total loss of MHC class I is an independent indicator of good prognosis in breast cancer. Int J Cancer 117(2):248–255

Mai A, Altucci L (2009) Epi-drugs to fight cancer: from chemistry to cancer treatment, the road ahead. Int J Biochem Cell Biol 41:199–213

Maleno I, López-Nevot MA, Cabrera T, Salinero J, Garrido F (2002) Multiple mechanisms generate HLA class I altered phenotypes in laryngeal carcinomas: high frequency of HLA haplotype loss associated with loss of heterozygosity in chromosome region 6p21. Cancer Immunol Immunother 51(7):389–396

Maleno I, Lopez Nevot MA, Seliger B, Garrido F (2004a) Low frequency of HLA haplotype loss associated with loss of heterozygocity in chromosome region 6p21 in clear renal cell carcinomas. Int J Cancer 109(4):636–638

Maleno I, Cabrera CM, Cabrera T, Paco L, López-Nevot MA, Collado A, Ferrón A, Garrido F (2004b) Distribution of HLA class I altered phenotypes in colorectal carcinomas: high frequency of HLA haplotype loss associated with loss of heterozygosity in chromosome region 6p21. Immunogenetics 56(4):244–253

Maleno I, Romero JM, Cabrera T, Paco L, Aptsiauri N, Cozar JM, Tallada M, López-Nevot MA, Garrido F (2006) LOH at 6p21.3 region and HLA class I altered phenotypes in bladder carcinomas. Immunogenetics 58(7):503–510

Maleno I, Aptsiauri N, Cabrera T, Gallego A, Paschen A, López-Nevot MA, Garrido F (2011) Frequent loss of heterozygosity in the β2-microglobulin region of chromosome 15 in primary human tumors. Immunogenetics 63(2):65–71

Manning J, Indrova M, Lubyova B, Pribylova H, Bieblova J, Hejnar J, Simova J, Jandlova T, Bubenik J, Reinis M (2008) Induction of MHC class I molecule cell surface expression and epigenetic activation of antigen-processing machinery components in a murine model for human papilloma virus 16-associated tumours. Immunology 123(2):218–227

Marincola FM, Jaffee EM, Hicklin DJ, Ferrone S (2000) Escape of human solid tumors from T-cell recognition: molecular mechanisms and functional significance. Adv Immunol 74:181–273

Martini M, Testi MG, Pasetto M, Picchio MC, Innamorati G, Mazzocco M, Ugel S, Cingarlini S, Bronte V, Zanovello P, Krampera M, Mosna F, Cestari T, Riviera AP, Brutti N, Barbieri O, Matera L, Tridente G, Colombatti M, Sartoris S (2010) IFN-gamma-mediated upmodulation of MHC class I expression activates tumor-specific immune response in a mouse model of prostate cancer. Vaccine 28(20):3548–3557

Meissner M, Reichert TE, Kunkel M, Gooding W, Whiteside TL, Ferrone S, Seliger B (2005) Defects in the human leukocyte antigen class I antigen-processing machinery in head and neck squamous cell carcinoma: association with clinical outcome. Clin Cancer Res 11(7):2552–2560

Menon AG, Morreau H, Tollenaar RA, Alphenaar E, Van Puijenbroek M, Putter H, Janssen-Van Rhijn CM, Van De Velde CJ, Fleuren GJ, Kuppen PJ (2002) Down-regulation of HLA-A expression correlates with a better prognosis in colorectal cancer patients. Lab Invest 82(12):1725–1733

Miyagi T, Tatsumi T, Takehara T, Kanto T, Kuzushita N, Sugimoto Y, Jinushi M, Kasahara A, Sasaki Y, Hori M, Hayashi N (2003) Impaired expression of proteasome subunits and human leukocyte antigens class I in human colon cancer cells. J Gastroenterol Hepatol 18(1):32–40

Morabito A, Dozin B, Salvi S, Pasciucco G, Balbi G, Laurent S, Pastorino S, Carli F, Truini M, Bruzzi P, Del Mastro L, Pistillo MP (2009) Analysis and clinical relevance of human leuko cyte antigen class I, heavy chain, and beta2-microglobulin downregulation in breast cancer. Hum Immunol 70(7):492–495

Moretta A, Bottino C, Vitale M, Pende D, Cantoni C, Mingari MC, Biassoni R, Moretta L (2001) Activating receptors and coreceptors involved in human natural killer cell-mediated cytolysis. Annu Rev Immunol 19:197–223

Nabel GJ, Gordon D, Bishop DK, Nickoloff BJ, Yang ZY, Aruga A, Cameron MJ, Nabel EG, Chang AE (1996) Immune response in human melanoma after transfer of an allogeneic class I major histocompatibility complex gene with DNA-liposome complexes. Proc Natl Acad Sci U S A 93(26):15388–15393

Nie Y, Yang G, Song Y, Zhao X, So C, Liao J, Wang LD, Yang CS (2001) DNA hypermethylation is a mechanism for loss of expression of the HLA class I genes in human esophageal squamous cell carcinomas. Carcinogenesis 22(10):1615–1623

Nummer D, Suri-Payer E, Schmitz-Winnenthal H, Bonertz A, Galindo L, Antolovich D, Koch M, Büchler M, Weitz J, Schirrmacher V, Beckhove P (2007) Role of tumor endothelium in CD4+ CD25+ regulatory T cell infiltration of human pancreatic carcinoma. J Natl Cancer Inst 99(15):1188–1199

Ostrand-Rosenberg S, Sinha P, Chornoguz O, Ecker C (2012) Regulating the suppressors: apoptosis and inflammation govern the survival of tumor-induced myeloid-derived suppressor cells (MDSC). Cancer Immunol Immunother 61(8):1319–1325

Paschen A, Méndez RM, Jimenez P, Sucker A, Ruiz-Cabello F, Song M, Garrido F, Schadendorf D (2003) Complete loss of HLA class I antigen expression on melanoma cells: a result of successive mutational events. Int J Cancer 103(6):759–767

Paschen A, Arens N, Sucker A, Greulich-Bode KM, Fonsatti E, Gloghini A, Striegel S, Schwinn N, Carbone A, Hildenbrand R, Cerwenka A, Maio M, Schadendorf D (2006) The coincidence of chromosome 15 aberrations and beta2-microglobulin gene mutations is causative for the total loss of human leukocyte antigen class I expression in melanoma. Clin Cancer Res 12(11Pt 1):3297–3305

Pérez B, Benitez R, Fernández MA, Oliva MR, Soto JL, Serrano S, López Nevot MA, Garrido F (1999) A new beta 2 microglobulin mutation found in a melanoma tumor cell line. Tissue Antigens 53(6):569–572

Pickart CM, Cohen RE (2004) Proteasomes and their kin: proteases in the machine age. Nat Rev Mol Cell Biol 5:177–187

Plaksin D, Gelber C, Feldman M, Eisenbach L (1988) Reversal of the metastatic phenotype in Lewis lung carcinoma cells after transfection with syngeneic H-2Kb gene. Proc Natl Acad Sci U S A 85(12):4463–4467

Ploegh HL (1998) Viral strategies of immune evasion. Science 10;280(5361):248–253

Porgador A, Bannerji R, Watanabe Y, Feldman M, Gilboa E, Eisenbach L (1993) Anti-metastatic vaccination of tumor-bearing mice with two types of IFN-gamma gene-inserted tumor cells. J Immunol 150(4):1458–1470

Powell AG, Horgan PG, Edwards J (2012) The bodies fight against cancer: is human leucocyte antigen (HLA) class 1 the key? J Cancer Res Clin Oncol 138(5):723–728

Ramnath N, Tan D, Li Q, Hylander BL, Bogner P, Ryes L, Ferrone S (2006) Is downregulation of MHC class I antigen expression in human non-small cell lung cancer associated with prolonged survival? Cancer Immunol Immunother 5(8):891–899

Reinis M, Simova J, Bubenik J (2006) Inhibitory effects of unmethylated CpG oligodeoxynucleotides on MHC class I-deficient and -proficient HPV16-associated tumours. Int J Cancer 118(7):1836–1842

Reinis M, Simova J, Indrova M, Bieblova J, Bubenik J (2007) CpG oligodeoxynucleotides are effective in therapy of minimal residual tumour disease after chemotherapy or surgery in a murine model of MHC class I-deficient, HPV16-associated tumours. Int J Oncol 30(5):1247–1251

Respa A, Bukur J, Ferrone S, Pawelec G, Zhao Y, Wang E, Marincola FM, Seliger B (2011) Association of IFN-gamma signal transduction defects with impaired HLA class I antigen processing in melanoma cell lines. Clin Cancer Res 17(9):2668–2678

Restifo NP, Marincola FM, Kawakami Y, Taubenberger J, Yannelli JR, Rosenberg SA (1996) Loss of functional beta(2)-microglobulin in metastatic melanomas from five patients receiving immunotherapy. J Natl Cancer Inst 88(2):100–108

Rivoltini L, Canese P, Huber V, Iero M, Pilla L, Valenti R, Fais S, Lozupone F, Casati C, Castelli C, Parmiani G (2005) Escape strategies and reasons for failure in the interaction between tumour cells and the immune system: how can we tilt the balance towards immune-mediated cancer control? Expert Opin Biol Ther 5(4):463–476

Rodriguez T, Méndez R, Roberts CH, Ruiz-Cabello F, Dodi IA, López Nevot MA, Paco L, Maleno I, Marsh SG, Pawelec G, Garrido F (2005) High frequency of homozygosity of the HLA region in melanoma cell lines reveals a pattern compatible with extensive loss of heterozygosity. Cancer Immunol Immunother 54(2):141–148

Rodriguez T, Mendez R, Del Campo A, Jimenez P, Aptsiauri N, Garrido F, Ruiz-Cabello F (2007) Distinct mechanisms of loss of IFN-gamma mediated HLA class I inducibility in two melanoma cell lines. BMC Cancer 7:34

Rosenberg SA, Yang JC, Restifo NP (2004a) Cancer immunotherapy: moving beyond current vaccines. Nat Med 10:909–991

Rosenberg SA, Yang JC, Restifo NP (2004b) Cancer immunotherapy: moving beyond current vaccines. Nat Med 10(9):909–915

Rosenberg SA, Sherry RM, Morton KE, Scharfman WJ, Yang JC, Topalian SL, Royal RE, Kammula U, Restifo NP, Hughes MS, Schwartzentruber D, Berman DM et al (2005) Tumor progression can occur despite the induction of very high levels of self/tumor antigen-specific CD8þ T cells in patients with melanoma. J Immunol 175:6169–6176

Ryschich E, Cebotari O, Fabian OV, Autschbach F, Kleeff J, Friess H, Bierhaus A, Büchler MW, Schmidt J (2004) Loss of heterozygosity in the HLA class I region in human pancreatic cancer. Tissue Antigens 64(6):696–702

Schlom J (2012) Recent advances in therapeutic cancer vaccines. Cancer Biother Radiopharm 27(1):2–5

Schreiber RD, Old LJ, Smyth MJ (2011) Cancer immunoediting: integrating immunity's roles in cancer suppression and promotion. Science 331(6024):1565–1570

Schultze JL, Nadler LM (2003) Lack of sufficient B7 expression as a tumor escape mechanism: implications for immunotherapy. In: Ochoa AC (ed) Mechanisms of tumor escape from the immune response. Taylor & Francis, London, pp 66–93

Seliger B, Maeurer MJ, Ferrone S (2000) Antigen-processing machinery breakdown and tumor growth. Immunol Today 21(9):455–464

Seliger B, Ritz U, Abele R, Bock M, Tampé R, Sutter G, Drexler I, Huber C, Ferrone S (2001) Immune escape of melanoma: first evidence of structural alterations in two distinct components of the MHC class I antigen processing pathway. Cancer Res 61(24):8647–8650

Seliger B, Ruiz-Cabello F, Garrido F (2008) IFN inducibility of major histocompatibility antigens in tumors. Adv Cancer Res 101:249–276

Serrano A, Brady CS, Jimenez P, Duggan-Keen MF, Mendez R, Stern P, Garrido F, Ruiz-Cabello F (2000) A mutation determining the loss of HLA-A2 antigen expression in a cervical carcinoma reveals novel splicing of human MHC class I classical transcripts in both tumoral and normal cells. Immunogenetics 51(12):1047–1052

Serrano A, Tanzarella S, Lionello I, Mendez R, Traversari C, Ruiz-Cabello F, Garrido F (2001) Rexpression of HLA class I antigens and restoration of antigen-specific CTL response in melanoma cells following 5-aza-2′-deoxycytidine treatment. Int J Cancer 94(2):243–251

Shin EC, Ahn JM, Kim CH, Choi Y, Ahn YS, Kim H, Kim SJ, Park JH (2001) IFN-gamma induces cell death in human hepatoma cells through a trail/death receptor-mediated apoptotic pathway. Int J Cancer 93(2):262–268

Simova J, Pollakova V, Indrova M, Mikyskova R, Bieblova J, Stepanek I, Bubenik J, Reinis M (2011) Immunotherapy augments the effect of 5-azacytidine on HPV16-associated tumours with different MHC class I-expression status. Br J Cancer 105(10):1533–1541

Sin JI, Park JB, Lee IH, Park D, Choi YS, Choe J, Celis E (2012) Intratumoral electroporation of IL-12 cDNA eradicates established melanomas by Trp2(180–188)-specific CD8+ CTLs in a perforin/granzyme-mediated and IFN-γ-dependent manner: application of Trp2(180–188) peptides. Cancer Immunol Immunother 61(10):1671–1682

Speetjens FM, de Bruin EC, Morreau H, Zeestraten EC, Putter H, van Krieken JH, van Buren MM, van Velzen M, Dekker-Ensink NG, van de Velde CJ, Kuppen PJ (2008) Clinical impact of HLA class I expression in rectal cancer. Cancer Immunol Immunother 57(5):601–609

Stern-Ginossar N, Gur C, Biton M, Horwitz E, Elboim M, Stanietsky N, Mandelboim M, Mandelboim O (2008) Human microRNAs regulate stress-induced immune responses mediated by the receptor NKG2D. Nat Immunol 9:1065–1073

Tafuro S, Meier UC, Dunbar PR, Jones EY, Layton GT, Hunter MG, Bell JI, McMichael AJ (2001) Reconstitution of antigen presentation in HLA class I-negative cancer cells with peptide-beta2m fusion molecules. Eur J Immunol 31(2):440–449

Takahashi H, Feuerhake F, Kutok JL, Monti S, Dal Cin P, Neuberg D, Aster JC, Shipp MA (2006) FAS death domain deletions and cellular FADD-like interleukin 1 beta converting enzyme inhibitory protein (long) overexpression: alternative mechanisms for deregulating the extrinsic apoptotic pathway in diffuse large B-cell lymphoma subtypes. Clin Cancer Res 12(11 Pt 1):3265–3271

Terabe M, Berzofsky JA (2004) Immunoregulatory T cells in tumor immunity. Curr Opin Immunol 16(2):157–162

Torres MJ, Ruiz-Cabello F, Skoudy A, Berrozpe G, Jimenez P, Serrano A, Real FX, Garrido F (1996) Loss of an HLA haplotype in pancreas cancer tissue and its corresponding tumor derived cell line. Tissue Antigens 47(5):372–381

Tripathi P, Agrawal S (2006) Non-classical HLA-G antigen and its role in the cancer progression. Cancer Invest 24(2):178–186

Tsory S, Kellman-Pressman S, Fishman D, Segal S (2006) Reconstitution of expression of H-2K region-encoded murine MHC class I glycoproteins in MHC class I-deficient B16BL6 melanoma cells affects the expression and function of antigen-processing machinery. Immunol Lett 102(2):237–240

Uyttenhove C, Pilotte L, Theate I, Stroobant V, Colau D, Parmentier N, Boon T, Van den Eynde BJ (2003) Evidence for a tumoral immune resistance mechanism based on tryptophan degradation by indoleamine 2,3-dioxygenase. Nat Med 9(10):1269–1274

van der Bruggen P, Traversari C, Chomez P, Lurquin C, De Plaen E, Van den Eynde B, Knuth A, Boon T (1991) A gene encoding an antigen recognized by cytolytic T lymphocytes on a human melanoma. Science 254(5038):1643–1647

Vesely MD, Kershaw MH, Schreiber RD, Smyth MJ (2011) Natural innate and adaptive immunity to cancer. Annu Rev Immunol 29:235–271

Villablanca EJ, Raccosta L, Zhou D, Fontana R, Maggioni D, Negro A, Sanvito F, Ponzoni M, Valentinis B, Bregni M, Prinetti A, Steffensen KR, Sonnino S, Gustafsson JA, Doglioni C, Bordignon C, Traversari C, Russo V (2010) Tumor-mediated liver X receptor-alpha activation inhibits CC chemokine receptor-7 expression on dendritic cells and dampens antitumor responses. Nat Med 16(1):98–105

Villunger A, Strasser A (1999) The great escape: is immune evasion required for tumor progression? Nat Med 5(8):874–875

Walter S, Weinschenk T, Stenzl A, Zdrojowy R, Pluzanska A, Szczylik C, Staehler M, Brugger W, Dietrich PY, Mendrzyk R, Hilf N, Schoor O, Fritsche J, Mahr A, Maurer D, Vass V, Trautwein C, Lewandrowski P, Flohr C, Pohla H, Stanczak JJ, Bronte V, Mandruz-zato S, Biedermann T, Pawelec G, Derhovanessian E, Yamagishi H, Miki T, Hongo F, Takaha N, Hirakawa K, Tanaka H, Stevanovic S, Frisch J, Mayer-Mokler A, Kirner A, Rammensee HG, Reinhardt C, Singh-Jasuja H (2012) Multipeptide immune response to cancer vaccine IMA901 after single-dose cyclophosphamide associates with longer patient survival. Nat Med [Epub ahead of print]

Watson NF, Ramage JM, Madjd Z, Spendlove I, Ellis IO, Scholefield JH, Durrant LG (2006) Immunosurveillance is active in colorectal cancer as downregulation but not complete loss of MHC class I expression correlates with a poor prognosis. Int J Cancer 118(1):6–10

Wrzesinski SH, Wan YY, Flavell RA (2007) Transforming growth factor-beta and the immune response: implications for anticancer therapy. Clin Cancer Res 13(18 Pt 1):5262–5270

Yoon SJ, Kang JO, Park JS, Kim NK, Heo DS (2000) Reduced expression of MHC class I antigen in human cancer cell lines with defective LMP-7. Anticancer Res 20(2A):949–953

Zou W, Chen S, Liu X, Yue P, Sporn MB, Khuri FR, Sun SY (2007) c-FLIP downregulation contributes to apoptosis induction by the novel synthetic triterpenoid methyl-2-cyano-3, 12-dioxooleana-1, 9-dien-28-oate (CDDO-Me) in human lung cancer cells. Cancer Biol Ther 6(10):1614–1620

Chapter 11
Tumor-Produced Immune Regulating Factors

Mads Hald Andersen, Jürgen C. Becker and Per thor Straten

Abstract The tumor exists as an organ comprising a mixture of transformed cells, normal epithelial cells, fibroblast, endothelial cells, cells of the immune system, ect. Obviously, from the viewpoint of cancer as a disease the key cell in the tumor is the mutated cancer cell. However, cancer cells are not autonomous and need stroma to survive and grow, and cancer cells on the other hand express molecules that are key denominators for the function of stroma cells thereby underscoring the bi- or multilateral interactions in the tumor. In this chapter we discuss some important aspects of tumor secreted factors on the environment and in particular on the immune system.

Keywords Stroma · CTLA-4 · MMP · NO · MDSC · Immunosuppression · Arginase · IDO · Hypoxia · Adenosine

11.1 Introduction

Tumors express an array of cell surface molecules as well as soluble factors that influence cells of the immune system. The impact on the environment can be mediated directly by secretion or expression of molecules that modify the actions of cells in the environment; however, it can also be indirectly by attracting cells to

M. H. Andersen · P. thor Straten (✉)
Center for Cancer Immune Therapy (CCIT), Department of Hematology,
University Hospital Herlev, Herlev Ringvej 75, 2730 Herlev, Denmark
e-mail: per.thor.straten@regionh.dk

J. C. Becker
Department of Dermatology, Medical University of Graz, Graz, Austria

M. R. Shurin et al. (eds.), *The Tumor Immunoenvironment*,
DOI: 10.1007/978-94-007-6217-6_11,

the tumor or modifying the function of cells in the environment to influence yet other cells. The tumor exists as an organ comprising a mixture of transformed cells, normal epithelial cells, fibroblast, endothelial cells, cells of the immune system, ect. Although it may seem obvious, it should be stressed that cancer cells are (in most cases) not autonomous—they need stroma cells to survive and grow, and cancer cells on the other hand express molecules that are key denominators for the function of stromal cells thereby underscoring that the cell–cell interactions goes two ways. Herein lies also the key to understand the intrinsic relationship between cancer cells and cells of the stroma, that the co-evolution of the tumor for many entities evolves with inflammation as a tumorigenic signature. This inflammatory signature is present in most cases even prior to key tumorigenic genetic alterations, and represents a substrate for co-evolution of the tumor comprising numerous cell types. To this end, the cancer cells may comprise only 30 or less percent of the cells in the tumor. However, concerning tumor escape by produced immune regulating factors such factors can be secreted by the tumor cells themselves or by other cells at the tumor site—or both.

Another important feature of tumors—beyond the mixed composition and inter-dependency of mutated cancer cells and non-cancerous stoma cells—is the extracellular matrix (ECM), which comprises proteoglycans, hyaluronic acid, and various fibrous proteins; collagen, laminin and fibronectin (Geblad et al. 2010). During tumor development the tumor stroma including ECM undergoes dramatic changes, and again these changes go two ways; the ECM influences the cells and obviously the cells influence the ECM. To the latter point, ECM is produced by cancer cells as well as non-cancerous stromal cells. The dynamics of the ECM during tumor progression is far from fully understood, but it is clear that ECM components play an important role in modifying not only cells of the immune system but also other cells many of which share many properties with the cells of the immune system. To this end, activated fibroblasts (myofibroblasts)—which are the most abundant stroma cells in many tumors—secrete numerous cytokines and chemokines that impact directly on cells in the micro environment or act as attractants of other cells to the tumor site; this includes IL-6, FOX03, transforming growth factor β (TGF-β), COX-2, vascular endothelial growth factor (VEGF), serum derived factor 1 (SDF-1), CXCL1/2, and IL-1β (Pietras and Ostman 2010). As clearly illustrated by the examples listed, these molecules may not be expressed solely by fibroblasts but just as well by cancer cells and/or cell of the immune system. Secreted cytokines and growth factors may in fact be indirectly regulated by ECM by release of sequestered cytokines by proteases (Geblad et al. 2010).

11.2 The “1”s and “2”s of Cells of the Immune System

There is solid evidence that tumors are recognized by cells of the immune system, and also strong indications that the immunological recognition impacts on prognosis (Mlecnik et al. 2011). However, one can ask the question whether tumor produced immune regulating factors play a role in (partial) escape from immune recognition! Expanding on the above the view, tumors should not be viewed as monoclonal cultures but as an organ-like structure that comprise multiple cell types and stromal composition as any organ. As discussed in this chapter, tumors certainly do express immune regulating factors. However, when it comes to the molecular and cellular background it is striking that many of the "suppressor" mechanisms described for tumors are in fact merely induction of what we think is an inappropriate response, i.e., a response that is not anti-tumor but maybe even pro-tumor. The classic example could be induction of a Th2 response instead of a anti-tumor Th1 response that could potentially lead to a CTL response against tumor antigens and tumor clearance.

For almost any cell of the immune system, we now have the "1" and "2" cells; neuthophils as N1 and N2, Macrophages as M1 and M2, Th1 and Th2 T all of them referring to the "good" cells and the "bad" cells (Geblad et al. 2010). Interestingly, as the view on tumors as organs is disseminating a somewhat similar view on the immune system in organ protection has been suggested. Polly Matzinger has suggested what could be called "the tissue in control" hypothesis; that each cell has the capacity to produce immune protective as well as modulatory signals, and thus, the individual cell may have very different actions depending on the tissue or the organ (Matzinger and Kamala 2011). Thus, any cell that enters the tumor microenvironment—be it a T cell, a neutrophil or macrophage, will react to signals on site and respond accordingly.

Importantly, the induced response in any organ has been developed through evolution to most appropriately respond to the microbes that are relevant in that particular tissue. To this end, "tissue" as with tumors are comprised by a variety of cells including cells of the immune system and fibroblasts. To the latter, it has been shown that fibroblast from different tissues are as different in gene expression profile as are the cells of the immune system (Chang et al. 2002), underscoring the important role of the tissue. Cells of the innate immune system play crucial roles in organogenesis, angiogenesis and wound healing and may just as well initiate a "repair response" as a "clearing response" influenced (or educated) by the tissue. Maybe its time to consider whether the "suppressor" role of any given cell type in cancer, was as appropriately studied trying to scrutinize the role of that cell in normal physiology. For example, it has been argued for long that myeloid derived suppressor cells (MDSC) are monocytic or granulocytic cells "captured" at a not fully differentiated stage. However, recent data suggests that these cells play a role—modified by mast cells—in the immune response against parasitic helmint infections (Saleem et al. 2012). More in keeping with the denomination as "suppressors" it was recently shown that MDSC inhibit NK activity during vaccinia infection and that abolishment of the MDSC activity led to increased INF-γ production in response to infection and increased mortality (Fortin et al. 2012). Thus, MDSC in that model seem to protect against a response that is too powerful and leads collateral damage and risk of death. Obviously, this new insight may lead to new knowledge on these cells and new ways to tackle their induction and action in cancer immunotherapy.

11.3 Check Points

Harnessing of the immune system to combat cancer has achieved major breakthroughs over the past few years; FDA approval of the Sipuleucel for the treatment of homone resistant prostate cancer (Madan and Gulley 2011) as well as the FDA approval of cytotoxic T lymphocyte-associated antigen 4 (CTLA-4) antibody ipilimumab for the treatment of metastatic melanoma (Hodi et al. 2010). Concerning the latter strategy—blockade of immune checkpoints—this strategy may or may not be directly related to molecules expressed at the tumor. To this end, CTLA-4 is expressed by cytotoxic T cells upon activation delivers a inhibitory signal to the T cell. CTLA-4 shares ligands with the stimulatory molecule CD28 but has a higher affinity for the ligands CD80 and CD86 or (B7.1 and B7.2, respectively). These molecules—are expressed exclusively by professional APC and thus the mechanism of action of CTLA-4 lies at the initial response to antigen in the sense that CTLA-4 levels on the T cell depends on the signalling strength of the TCR/CD28 signalling. Data from murine models suggest the involvement of CD8 as well as CD4 T cells (Peggs et al. 2009). Concerning responses in humans upon CTLA-4 therapy data suggests that Inducible Costimulator (ICOS) signalling correlates with response and increases the frequency of INF-g CD4 T cells expressing high levels of ICOS and decreases the level of Treg (Liakou et al. 2008). ICOS is a co-stimulatory molecule expressed on activated T cells and the potential impact of ICOS expression suggested by the work from Pam Sharmas lab underscores the complexity but also the potential of targeting immune checkpoints (Fu et al. 2011).

As mentioned the ligands for CTLA-4 are (in most cases) not expressed by cancer cells, and in the present context goes to show that blocking of immune check points may lead to clinically relevant responses. However, several antibodies targeting other immune check-point pathways are in clinical development (Pardoll 2012). In particular, antibodies targeting programmed cell death protein 1 (PD-1) expressed on T cells, or the ligand PDL1—expressed on cancer cells, are in clinical phase I/II trial. PD-1 is expressed on activated T cells and the ligands PDL1 and PDL2 in peripheral tissues at inflammation and thus the main role of PD1 interaction with ligand lies not at the level of induction as is the case with CTLA-4, but at the level of effector site (Pardoll 2012). Moreover, PDL1 and 2 are expressed by tumor cells and blockage may directly unleash a response at the tumor site by blocking the inhibitory interaction. Data were recently published from phase I trials and for both antibodies objective clinical responses were observed (Topalian et al. 2012), however the efficacy as judged by these early data suggest sthat anti-PD1 is the more efficient. To this end, anti-PD1 targets PD1 expressed on the T cell whereas anti-PDL1 only targets one of two ligands on the target cell side. Moreover, given that the PD1 ligands are expressed on tumor cells efficacy of treatment would supposedly depend on ligand expression by tumor cells in the individual patient. Whereas no data are available for administration of PDL1 yet in terms of efficacy in relation to expression of PDL1 on tumor cells, this is indeed strongly suggested concerning PD1. Provided that these data are supported in larger trials it would represent very important selection criteria.

The data demonstrate that targeting of immune checkpoints by antibodies specific for molecules expressed by cancer cells may lead to clinically relevant immune responses, supposedly by unleash of spontaneous T cell responses.

Other molecules of the B7 family of ligands are expressed by tumor cells, e.g., B7-H3 and B7-H4 (Seliger and Quandt 2012). These are expressed by tumor cells as are the PD1 ligands, and thus the regulation by means of these molecules lies at the effector level. Available data suggest that these ligands provide negative signalling; however, the ligands have not been characterized yet and firm evidence for that notion is still lacking (Yi and Chen 2009). Numerous studies have focussed on studying whether the expression of B7-H3 and B7-H4 correlates with prognosis (Seliger and Quandt 2012), however, the picture is unclear and larger prospective studies are needed to be able to answer these questions. However, studies are ongoing based on administration of antibodies targeting B7-H3 and B7-H4. To this end, antibody targeting B7-H3 (MGA-271) is in clinical phase I trial, and B7-H4 is in clinical development (Pardoll 2012).

Summing up, check point blockade is one of several very promising approaches to exploit the immune system in cancer therapy, and strongly underscored by the clinical efficacy, tumor induced check point blockade is a excellent example of escape that can be target therapeutically. Thus, it demonstrates that tumors may hijack such inhibitory mechanisms by expression of molecules for co-inhibitory receptors such as PDL1 and 2, and blockade on either side, i.e., the T cells or tumor cell side may lead to clinical response.

11.4 Stress and DNA Damage Molecules

Tumor cells express a number of molecules directly associated with recognition by cells of the immune system e.g., Hla molecules and accessory molecules linked to T cell recognition of tumor cells. Most tumor cells also express ligands for natural killer group 2, member D (NKG2D) which is an activating receptor on NK cells and activated T cells. In humans the ligands for NKG2D are the UL16 binding protein (ULBP) 1–6, and the stress inducible MHC related protein A (MICA) and B (MICB). Although these ligands seem to be expressed in different tissues throughout the body, NKG2D ligand expression may be induced by viral infection by unknown mechanisms or TLR signalling, cytokines, or DNA damage. Thus, since the DNA damage response is active in cancer cells (Bartkova et al. 2005; Gasser et al. 2005), this may be the primary cause for high expression levels of NKG2D ligands by cancer cells (Textor et al. 2011).

Importantly, this also suggests that the induction of ligand expression is a very early phenomenon since the DNA damage response is believed to be an early transformation block active prior to transformation e.g., in senescent cells. To this end, we have shown that melanocytic lesions are negative for expression of MICA whereas primary melanomas were positive in 31 of 40 primary lesions examined (Vetter et al. 2002). The result of the DNA damage response is cell cycle arrest and subsequent repair or apoptosis. Indeed, most cancer cell lines and tumor cells in biopsies are positive for NKG2D ligands, including both haematological and solid malignancies, and with MICA being the most highly expressed of the ligands (Salih et al. 2003).

Numerous studies have shown that ligand expression leads to recognition and killing of tumor cells by NK cells (Champsaur and Lanier 2010). Moreover, activated T cells express NKG2D and ligand expression seem to be costimulatory to TCR engagement. As a consequence NKG2D ligand expression by cancer cells is prone to lead to improved immune recognition by NK and T cells, and at the same time the expression of ligands seem to be an inherent property of the DNA damage response. However, in addition to membrane bound NKG2D ligands expressed by tumor cells, several of the ligands can be shedded from tumor cells and detected in patient sera. Thus, soluble MICA has been detected in cancer patient sera (Salih et al. 2003) from both haematological and solid cancers, and also MICB and ULBP-2 have been found at elevated levels in cancer patient sera (Champsaur and Lanier 2010).

Importantly, Paschen et al. (2009) recently demonstrated that soluble ULBP-2 is a marker for poor prognosis in melanoma patients pointing to a functional impact of soluble NKG2D ligands in sera from cancer patients. This notion is supported by studies in a murine prostate cancer model in which it was shown that blocking of NKG2D ligand shedding inhibited tumor formation (Wu et al. 2009). Although more data need to substantiate the notion it suggests that the presence of shedded NKG2D ligands may mediate escape from recognition by NK and/or T cells and thereby circumvent the otherwise disadvantage of cancer cells in expressing an immune activating molecule on the cell surface. Importantly, this could lead to therapeutic interventions either by antibodies to NKG2D ligands or by inhibition of ligand shedding. Surprisingly, in the study of Paschen et al., MICA sera levels did not correlate with the clinical course of disease underscoring that very little is known concerning the affinity of individual NKG2D-ligand interactions as well as the impact of chronic versus transient ligand expression (Groh et al. 2002). Another angle added to the tale stems from recent data demonstrating that cancer cells may also express NKG2D (Benitez et al. 2011) and that signalling upon ligand binding stimulates tumor growth.

11.5 The Extracellular Matrix

As already mentioned, ECM have a profound influence on the cells and molecules in the tumor micro environment. To this end, matrix metalloproteinases (MMP) are key enzymes involved with remodelling of the ECM by proteolysis of ECM components (Pytliak et al. 2012), and MMP are crucial in normal physiological processes such as the cyclic remodelling of the endometrium (Caide Chevronnay et al. 2012), angiogenesis, wound healing and cell migration. Cancer cells and non-cancerous cells in the tumor express MMPs, and MMPs are expressed at high level in numerous cancer entities, and are key to tumor growth and metastasis, In fact, MMP expression influence cancer cell growth, diffentiation, metastatic capacity and resistance to apoptosis (Pytliak et al. 2012). For many years the function of MMP was believed to be restricted to ECM remodelling.

However, ECM remodelling obviously also influence cells of immune system and the local immune responses. Moreover, MMPs cleaves not only structural proteins of the ECM. To this end, the bioavailability of TGFβ is regulated by the release from an inactive extracellular complex—a process mediated by MMP2 and -9 (Yu and Stamenkovic 2000). MMPs also cleave growth factor and cytokine receptors (Egeblad and Werb 2002), thus, MMPs may actually cleave the IL-2 receptor (IL-2α) and thereby inhibit proliferation of T cells locally (Sheu et al. 2001). Similarly, MMPs may cleave chemokines as well as chemokine receptors, both of which may influence homing of cells to the tumor site. Obviously, the complex interactions and functions of MMPs in the tumor microenvironment imply that the functions of MMPs could in fact play a role both in the immune response against cancer as well as a role in tumor progression.

The high expression of MMPs in many cancers has prompted therapeutic targeting of MMPs, which has been largely disappointing. However, MMP-2 has been shown to be a tumor antigen as well; a MMP-2 derived peptide is presented in the context of HLA-A2 on the surface of melanoma cells and recognized by cytotoxic CD8 T cells (Godefroy et al. 2005). Recently it was shown that MMP-2 also plays a role in polarizing immune responses. Hence, Godefroy and colleges showed that CD4 T cells recognizing class II restricted peptides derived from MMP-2 were mainly of Th2 type expressing GATA and secreting TNF-α, IL-4 and IL-13 (Godefroy et al. 2011). MMP-2 was shown to be causative for the Th2 bias of CD4 T cells. Interestingly, the mechanism underlying the Th2 polarization was due to both active and inactive MMP-2. Thus, active MMP-2 led to degradation of the type I IFN receptor on DCs, whereas both active and inactive MMP-2 induced upregulation of CD40 ligand on DC by yet unknown mechanism (Godefroy et al. 2011). The above mentioned data are based on studies of melanoma patients but MMP-2 (and MMP-9 which share many properties with MMP-2) is expressed at high levels in other cancers including ovarian, breast, colon, and prostate cancer (Pytliak et al. 2012).

11.6 Immune Suppression by Metabolic Enzymes

The altered tumor metabolism depletes essential nutrients or leads to the accumulation of immune suppressive metabolites in the tumor microenvironment. In the following we highlight the suppressive activity of some of the key metabolic players.

11.6.1 Tryptophan Metabolism by Indoleamine-2, 3-Dioxygenase (IDO)

IDO is an immunoregulatory enzyme that is implicated in suppressing T cell immunity in many settings including cancer. IDO seems to be critical in limiting potentially exaggerated inflammatory reactions in response to danger signals (Romani et al. 2006) and in assisting regulatory T cell effector function (Prendergast et al. 2009). IDO expression can suppress effector T cells directly by degradation of the essential amino acid tryptophan. Effector T cells starved of tryptophan are unable to proliferate and go into G1 cell cycle arrest (Munn et al. 2005; Habibi et al. 2010). Some of the biological effect of IDO is mediated through local depletion of tryptophan, but is in addition mediated via immune modulatory tryptophan metabolites (Habibi et al. 2010; Platten et al. 2005; Bauer et al. 2005). Thus, the metabolites of tryptophan, which have been shown to be directly toxic to $CD8^{+}$ T cells and $CD4^{+}$ Th1 cells (Frumento et al. 2002). Thus, increased IDO activity seems to tilt helper T cell polarization toward a Th2 phenotype (Xu et al. 2008).

Regulation of tryptophan metabolism by IDO in dendritic cells (DC) is a highly flexible modulator of immunity. When IDO^{+} DC are injected in vivo, they create suppression and anergy in antigen-specific T cells in the LN draining the injection site (Munn et al. 2005). Another effect of IDO is mediated through enhancement of local Treg-mediated immune suppression. Constitutive IDO expression in DC imparts T cells with regulatory properties (Munn and Mellor 2007). The B7 receptors on IDO^{+} DC bind to CTLA4 on Tregs causing them to proliferate and induce antigen-specific anergy. Thus, IDO does not only suppress effector T cells directly but also influence Tregs bystander suppressor activity (Prendergast et al. 2009; Sharma et al. 2009). It has been described that exposure to IL-6 may induce reprogramming of mature Tregs to acquire a phenotype resembling of pro-inflammatory Th17 cells (Yang et al. 2008; Zou and Restifo 2010). IDO play a vital role in this conversion by stimulating Treg bystander suppressor activity and simultaneously by blocking the IL-6 production that is required (Sharma et al. 2009). Finally, it was recently shown that IDO has a non-enzymic function that contributes to TGF-β driven tolerance in non-inflammatory contexts (Pallotta et al. 2011).

11.6.2 IDO and Cancer

IDO expression is widely upregulated in cancer patients. Thus, IDO elevation occurs in a subset of plasmacytoid DC in tumor-draining lymph nodes (Munn and Mellor 2007). Thus, IDO-expressing $CD19^{+}$ plasmacytoid DC isolated from tumor-draining LN mediated profound immune suppression and T cell anergy in vivo (Sharma et al. 2009) whereas plasmacytoid DC from normal LNs and spleen did not express IDO. It is believed that constitutive IDO expression in DC in tumor-draining LN is induced by stimulation from Tregs migrating from the tumor to the draining LN. The induction of IDO converts the tumor-draining LN from an immunizing into a tolerizing milieu. In addition, IDO may be expressed within the tumor by tumor cells as well as tumor stromal cells, where it inhibits the effector phase of immune responses (Uyttenhove et al. 2003). In this respect it should be noted that very few cells constitutively express IDO in normal lymphoid tissue except in the gut. Activation of IDO in either tumor cells or nocal regulatory DC each appears to be sufficient to facilitate immune escape of tumors (Munn and Mellor 2007). In accordance with this, it has been described that expression of IDO in tumor cells is associated with decreased serum tryptophan concentration and predicts a bad prognosis (Weinlich et al. 2007). Furthermore, it was described that tumor cells transfected with IDO may become resistant to immune eradication, even in fully protected, immunized mice (Uyttenhove et al. 2003). Recently, it was described that IDO is important for both tumor vascularization and IL-6-dependent MDSC-driven immune espcabe (Smith et al. 2012).

In recent years spontaneous $CD8^+$ (Sorensen et al. 2009) as well as $CD4^+$ (Munir et al. 2012) T cell reactivity against IDO has been detected both in the tumor microenvironment as well as in peripheral blood. IDO reactive $CD8^+$ T cells were cytotoxic effector cells, which were able to recognize and kill IDO-expressing cells, including tumor cells and dendritic cells. Since IDO specific T cells eradicated IDO positive suppressive cells and thereby boosted immunity (Sorensen et al. 2011) and IDO-based immunotherapy may consequently be synergistic with additional immunotherapy. In this respect, IDO has become a very attractive target for the design of new anticancer drugs and several IDO inhibitors are under investigation in preclinical as well as in clinical studies (Lob et al. 2008). In particular, the compound 1-methyl-tryptophan (1MT) has been widely studied as an inhibitor of IDO activity.

11.6.3 IDO2

Interestingly, recent studies have shown that the racemer D-1-MT has superior antitumor activity compared to the racemer L-1-MT (Hou et al. 2007). A novel indoleamine 2, 3-dioxygenase (IDO)-like protein designated IDO2 was recently discovered (Metz et al. 2007). IDO2 seem to function like IDO in tryptophan catabolism, but it has been found that D-1MT but not the L-1MT isomer selectively and potently inhibits IDO2 activity suggesting that IDO2 activity may have a role in the inhibition of immune responses to tumors. In this respect, IDO2 expression has been found in human tumors, including gastric, colon, renal, and in pancreatic tumors IDO2 expression have been found both in tumor cells as well as in immune cells in tumor-draining LN (Witkiewicz et al. 2009). It is not yet known to what extent each isoform of IDO contributes to tumor-related immune suppression and how much clinical benefit (or autoimmune toxicity) targeting one isoform over another confers. Similarly it is unknown whether IDO inhibitors influence other pathways not directly linked to IDO (Lob et al 2008).

11.6.4 Tryptophan 2, 3-Dioxygenase (TDO)

It was recently showed that tumors can use yet another means to degrade tryptophan and resist immune rejection—by the expression of tryptophan 2, 3-dioxygenase (TDO). TDO is a homotetrameric heme-containing cytosolic enzyme encoded by *TDO2*. It is normally almost exclusively expressed in high levels in the lever. It was recently described that tumors of different origin express TDO especially melanoma, bladder cancer, hepatocarinoma as well as human glioblastomas (Pilotte et al. 2012; Opitz et al. 2011). It promotes tumor progression through the production of kynurenine, which is resulting in reduced anti-tumor immune responses. Interestingly, blocking of both TDO and IDO might turn out to be complementary, not redundant. Hence, in a series of 104 human tumor cell lines it was found that 35 % expressed only TDO, 32 % expressed IDO whereas 51 % expressed either one or the other (Pilotte et al. 2012). Finally, it was recently shown that Tryptophan hydroxylase-1 (Tph-1), a synthase that catalyses the conversion of tryptophan to serotonin, was a potent regulator of immunity (Le M 2012) and Tph-1 deficiency induced tumor remission in an animal model.

11.6.5 Arginine Metabolism

IDO inhibits T cell responses by depleting tryptophan and producing kynurenine, which is toxic to lymphocytes. However, tumors may in addition suppress immunity by other enzymatic mechanisms. Hence, another way tumor cells regulate T cells is by manipulating the metabolism of L-arginine, through the enzymes nitric-oxide synthase (NOS) and Arginase (ARG). Thus, many tumors exhibit an increased expression of Arginase and inducible NOS (iNOS) leading to depletion of arginine from the tumor microenvironment (Bronte and Zanovello 2005). Several studies have emphasized the importance of this altered tumor arginine metabolism for the suppression of tumor-specific T cell responses. Hence, Arginase and NOS can induce cell cycle arrest in T cells (Bronte and Zanovello 2005; Bronte et al. 2003). Furthermore, iNOS leads to increased production of nitric oxide (NO), thereby promoting angiogenesis, metastasis, and immune suppression in tumors (Xu et al. 2002). NO leads to nitration of tyrosine and cysteine residues. Blesson et al. showed that this blocks signal transduction in T cells and suppression of IL-2 and granzyme-B production (Blesson et al. 2002). In addition, Rodriguez et al. described that Arginase was induced by cyclooxygenase (COX)-2 in lung carcinoma (Rodriguez et al. 2005). These data may explain why increased concentrations of prostaglandin (PGE)-2 are linked to inhibition of T cell activation and enhanced tumorigenesis. Beside malignant cells, macrophages, granulocytes or myeloid-derived suppressor cells (MDSC) can suppress immunity by the expression of Arginase and NOS. MDSC are often associated with poor prognosis in cancer patients. Inhibition of Arginase/iNOS in both tumor cells and

MDSCs have been described to reconstitute effector functions of T cells, which leads to a decreased tumor growth in animal models (Bronte et al 2003; Rodriguez et al. 2005).

11.7 Tumor Metabolism, Hypoxia, and Adenosine

Over the past years it became increasingly obvious that the altered tumor metabolism depletes essential nutrients or leads to the accumulation of immuno-suppressive metabolites in the tumor microenvironment (Noman et al. 2011). Thus, the tumor microenvironment plays a crucial role in the control of immune protection and contains many overlapping mechanisms to evade antigenic specific immunotherapy. Tumors are not merely masses of neoplastic cells, but instead, are complex tissues composed of both non-cellular (e.g. the extracellular matrix) and cellular components which comprise both neoplastic cells as well as non-transformed host cells.

Hypoxia has an established role in radio-resistance, chemoresistance and tumor stemness; more recently its significance in changing the tumor microenvironment into an immune permissive state had been realized. For example, it has been demonstrated that hypoxia increases tumor cell shedding of MIC molecules through impaired nitric oxide (NO) signaling (Siemens et al. 2008). The primary source of NO is the NOS enzyme, which has three isoforms; two are constitutively expressed, and one is inducible. The inducible form iNOS produces NO for prolonged periods of time in a calcium-independent manner. Levels of NO produced by iNOS in the microenvironment of the cell can range from as low as 10 nM to μM amounts for days. This hypoxia-induced MIC shedding decreases the sensitivity of tumor cells to peripheral blood lymphocyte-mediated killing. Notably, previous studies have shown that hypoxia-induced tumor invasiveness and chemoresistance are also linked to reduced NO signaling (Frederiksen et al. 2007). Moreover, NO also has an impact on other immune competent cells. For example, MDSCs can be activated by NO-mediated increases in cGMP, which in turn, facilitates their binding to CTLs and subsequent inhibition of T cell proliferation (Gallina et al. 2006). Furthermore, when activated by IFN-γ, TNF-α, or IL- $1\alpha/\beta$, mesenchymal stem cells produce both chemokines and iNOS resulting in the suppression of T cells in their vicinity of the MSCs (Li et al. 2012). Notably, NO mimetic treatment attenuates these hypoxia-induced immunosuppressive effects suggesting that reactivation of NO signaling by NO mimetic can be exploited as an immune modulating strategy.

Cancer cells primarily rely on glycolysis for energy production, a phenomenon known as aerobic glycolysis or "Warburg effect" (Dang 2012). This effect is based on the up-regulation of glycolytic enzymes such as pyruvate kinase, hexokinase, and lactate dehydrogenase (LDH). Accordingly, tumor cells are characterized by an increased uptake of glucose. This altered tumor metabolism is both under the control of hypoxia as well as activated oncogenes. For example, expression of the myc oncogene occurs in about 30 % of human cancers; myc signaling leads to up-regulation of glycolytic enzymes. Moreover, myc collaborates with hypoxia inducible factor (HIF), which is stabilized in response to hypoxia and induces the transcription of more than 70 genes including LDH. Genetic alteration or loss of p53 is further associated with decreased oxygen consumption and increased lactate production. Notably, over the past years the immune modulatory properties of high lactate levels have been characterized. Lactic acid inhibits the differentiation of monocytes from dendritic cells, e.g. lactic acid regulates transcription and secretion of IL-23, a tumor-promoting cytokine involved in the generation of Th17 cells (Shime et al. 2008). In addition, lactate exerts negative effects on T cell mediated immune responses, e.g. suppressing the cytotoxic T cell response in vivo (Droge et al. 1987). T cells infiltrating lactic acid-producing multicellular tumor spheroids are characterized by a reduced cytokine production compared with the controls, whereas pre-treatment of tumor spheroids with an inhibitor of lactic acid production could partially abrogate this inhibitory effect. Furthermore, highly proliferating tumor cells but also activated T cells rely on glucose metabolism to provide energy for proliferation and effector functions and may thus compete for glucose in the tumor microenvironment (Cham and Gajewski 2005).

Notably, T cells are exposed to different oxygen tensions, including hypoxic levels, during their development and during migration between blood and tissue; and solid tumors usually are characterized by a particular hypoxic environment. Under hypoxic conditions, T cells increase the expression of genes that are regulated by HIF (e.g., VEGF, glycolytic enzymes) (Chouaib et al. 2012). While hypoxia enhances the transcription of hypoxia-responsive element (HRE)-containing genes, it inhibits the accumulation of non-HRE-containing genes, such as IL-2 and IFN-γ during TCR-driven activation. Thus, T cell activation under hypoxic condition in vivo may lead to different patterns of cytokine secretion (Caldwell et al. 2001).

Lymphocytes express both voltage-dependent potassium and Ca^{2+}-activated potassium channels and their activities are essential for T cell activation. K^+ channels modulate the resting potential of the T cell membrane and indirectly regulate Ca^{2+} signaling which is important for cell proliferation and cytokine production. Blocking voltage-dependent potassium channels inhibits T cell proliferation by inducing membrane depolarization and decreasing calcium influx. Notably, hypoxia selectively inhibits TCR-mediated T cell proliferation by suppressing the expression and activity of these voltage-dependent potassium channels (Conforti et al. 2003).

An additional mechanism for cancer induced immune suppression is the generation of adenosine within the tumor microenvironment. Adenosine is characterized by organ- and cytoprotective functions which also include stimulation of angiogenesis and inhibition of inflammatory reactions and adaptive immune responses (Whiteside et al. 2011). There are several independent sources of adenosine in the tumor microenvironment, e.g., cell death and nucleotide degradation, ischemia and ATP breakdown, ATP/ADP release and subsequent dephosphorylation, AMP release, and *S*-adenosylhomocysteine hydrolysis, which would individually or in combination provide a continuous supply of adenosine.

The balance between adenosine and ATP is crucial in immune homeostasis: While ATP is a danger signal released by damaged and dying cells that acts to prime immune responses adenosine suppresses immune responses (Stagg and Smyth 2010). Consequently, metabolism of ATP into its metabolites ADP, AMP and adenosine is a tightly regulated process. The conversion of ATP into AMP is predominantly catalyzed by nucleoside triphosphate diphosphohydrolase 1 (CD39). Subsequently, adenosine is produced by dephosphorylation of AMP catalyzed by 5'-nucleotidase (5'-NT); several forms of this enzyme have been described, but only two of them, i.e. cytosolic 5'-NT-I and ecto-5'-NT seem to participate in adenosine generation, with ecto-5'-NT (CD73) being most relevant for the tumor microenvironment. Since the conversion of AMP into adenosine by CD73 is reversible only following intracellular transport of adenosine, CD73 is the crucial checkpoint in the conversion of immune activating ATP into immuno-suppressive adenosine. Notably, the activity of CD73 is variable in malignant cells: In contrast to frequently elevated CD73 in malignancies of epithelial origin, very low expression of CD73 was found on hematopoietic malignancies.

Notably, the tumor microenvironment contains factors inducing CD73 expression. The most obvious of these is hypoxia, because a hypoxia-inducible factor (HIF) response element has been identified within the CD73 promoter. In addition, humoral factors such as type I IFNs, TNF-α, IL-1β, PGE2 and TGF-β as well as agonists of the wnt signaling pathway have been demonstrated to up-regulate CD73 expression (Jin et al. 2010).

ATP acts to prime immune responses through the ligation of P_2X and P_2Y purinoreceptors; adenosine suppresses immune responses through the activation of G-protein-coupled receptors (Hasko et al. 2008). Four receptors have been described for adenosine: the pertussis toxin sensitive A_1 and A_3 and the adenylate cyclase activating A_{2A} and A_{2B} receptors; the first two receptors signal via decreased cAMP, whereas the latter two signal via increased cAMP. In addition, ligation of A_1 and A_3 receptors also activates the phosphoinositide 3-kinase (PI3 K) and protein kinase (PK)C pathways. All four receptors stimulate the mitogen-activated protein kinase (MAPK) pathway. Under physiological conditions adenosine acts through the high-affinity A_1 and A_{2A} receptors, but under pathological circumstances such as the tumor microenvironment with high adenosine concentrations the low affinity A_{2B} and A_3 receptors become relevant. Numerous immune competent cells express adenosine receptors including T cells, NK cells, NKT cells, macrophages, DCs, neutrophils, mast cells and B cells (Hasko et al. 2008). Indeed, several lines of evidence indicate the general immunosuppressive and anti-inflammatory properties of adenosine.

Human $CD4^+$ and $CD8^+$ T cells express A_{2A}, A_{2B} and A_3 receptors; their expression is up-regulated upon stimulation with mitogens or through the T cell receptor. Studies with selective agonists/antagonists demonstrated a dominant role for the A_{2A} receptor in the suppression of T cell responses. Stimulation of A_{2A} receptors on T cells inhibits proliferation, cytotoxicity and pro-inflammatory cytokine production, such as IL-2, TNF-α, and macrophage inflammatory protein-1a (Hoskin et al. 2008). The signaling pathway by which A_{2A} receptor activation results in suppression of T cell function is not fully elucidated, however, activation of PKA by A_{2A}/A_{2B} receptor ligation inhibits NF-kB activity and thus reduces T cell activation. Furthermore, adenosine inhibits the adhesion of cytolytic lymphocytes to cancer cells as well as granule exocytosis by natural killer cells (Hausler et al. 2011).

Interestingly, adenosine also modulates T helper (T_{H17}) differentiation: It prevents the development of T_{H17} responses within the tumor microenvironment (Kryczek et al. 2009) Adenosine also inhibits the maturation and pro-inflammatory cytokine production of DCs, and consequently impairs their ability to induce T_{H1} responses; notably, DCs that differentiate in the presence of adenosine have even been reported to suppress anti-tumor immunity (Novitskiy et al. 2008). Furthermore, adenosine promotes expansion and function of MDSCs via ligation of A_{2B} receptors (Novitskiy et al. 2008). MDSCs are present within the tumor microenvironment and tumor draining lymph nodes where they act as potent suppressors of anti-tumor immune responses.

To sum up, there is solid evidence that counter-regulatory responses are important in the immune system as they help limit the intensity and extent of immune responses, which otherwise could cause damage to the host. However, with regard to anti-cancer immunotherapy counter-regulatory responses antagonize the ability to create an intense immune response against the tumor. Counter-regulation differs from tolerance in the sense that counter-regulation is a secondary event, elicited only in response to immune activation. Hence, in cancer immune therapy, the targeting of immune regulation by therapeutic measures could function highly synergistic with additional anti-cancer immune therapy. By definition anti-cancer immune therapies aim at the induction of an immunological activation and inflammation. The therapy aim to induce as much immune activation as possible (within the limits of acceptable toxicity), and, accordingly, immune suppressive counter-regulation is not desired. It goes without saying that the possible introduction of autoimmunity and toxicity are the major worries when targeting molecules involved in immune regulation and suppression.

11.8 Concluding Remarks

The idea of immune surveillance goes back a century initiated by Paul Ehrlich, however, since then the concept and in particular its believers have fluctuated in numbers. Strikingly, it seems that historically there has been little overlap between the views—thought and ideas—of tumor biologists and tumor immunologists until quite recently. To this end, in the 1990ies data were compiling from studies of both mouse and man that the immune system did indeed recognize cancer cells and also that at least in mouse studies a functional immune system is required for protection against spontaneous tumor development (Kaplan et al. 1998; Smyth et al. 2000).

Early data from melanoma using immunhistochemistry pointed to a brisk T cell infiltration as a positive prognostic marker (Clark 1991), and the characterization of the first human tumor antigen by van der Bruggen in (1991) certainly supported the notion of T cells playing a role in immune system/cancer cell interactions. Moreover, the following decade flourished with characterization of tumor antigens recognized by T cells—showing the range of tumor antigens that could be recognized by the immune system, differentiation antigens (Bakker et al. 1994), cancer testis antigens (Boon et al. 1994), regulators of apoptosis proteins (Andersen et al. 2005), and that even a key oncogene like p53 could be recognized by T cells (Ropke et al. 1996).

Nonetheless, one of the most cited reviews of that time—the Hanahan and Weinberg "The Hallmarks of Cancer" had no mentioning of the immune system (Hanahan and Weinberg 2000). Strikingly, at the same time it seemed that immunologists were very little concerned with anything but anti-cancer immune responses and somehow neglected data from inflammation research; even in the late 1990ies data were beginning to accumulate that stroma cells including cells and molecules of the immune system could in fact have a pro-tumor impact (Coussens and Werb 2001) strongly underscored by the preventive actions of nonsteroidal anti-inflammatory drugs (Baron and Sandler 2000). Evidence for an impact of the immune system on tumor growth—making tumor escape an intrinsic necessity of tumor progression has been accumulating over the years, most striking perhaps by the recent data on the prognostic value of T cells infiltrating the tumor (Mlecnik et al. 2011).

As we have discussed in this chapter, the efficacy of check point blockade builds on an unleash of a spontaneous anti-tumor immune response that may or may not stem from blocking of a ligand expressed by cancer cells. Obviously, these breakthroughs are very encouraging and bring hope that a new era has begun. However, maybe the most encouraging phenomena is that there is a much more widespread appreciation between the different research areas that are key to bring us to a higher level of understanding into the biology of tumors. Thus, there is extensive cross fertilizations between scientists in tumor biology, tumor immunology, inflammation, basic immunology, wound healing, cancer prevention, etc. Even when it comes to combining conventional therapies and immune therapy—which even a few years back would have been blasphemy even to consider—this is the focus of tremendous research efforts (Prendergast and Jaffee 2007). To the ones working in the field life is certainly not made any simpler but we have left the somewhat simpleminded tunnel-sight view of what tumors and cancer as a disease is all about.

References

Andersen MH, Becker JC, Thor Straten P (2005) Regulators of apoptosis: Suitable targets for immune therapy of cancer. Nat Rev Drug Discovery 4:399–409

Bakker AB, Schreurs MW, de Boer AJ, Kawakami Y, Rosenberg SA, Adema GJ et al (1994) Melanocyte lineage-specific antigen gp100 is recognized by melanoma-derived tumor-infiltrating lymphocytes. J Exp Med 179:1005–1009

Baron JA, Sandler RS (2000) Nonsteroidal anti-inflammatory drugs and cancer prevention. Annu Rev Med 51:511–23

Bartkova J, Horejsi Z, Koed K, Kramer A, Tort F, Zieger K et al (2005) DNA damage response as a candidate anti-cancer barrier in early human tumorigenesis. Nature 434:864–870

Bauer TM, Jiga LP, Chuang JJ, Randazzo M, Opelz G, Terness P (2005) Studying the immunosuppressive role of indoleamine 2,3-dioxygenase: tryptophan metabolites suppress rat allogeneic T-cell responses in vitro and in vivo. Transpl Int 18:95–100

Benitez AC, Dai Z, Mann HH, Reeves RS, Margineantu DH, Gooley TA et al (2011) Expression, signaling proficiency, and stimulatory function of the NKG2D lymphocyte receptor in human cancer cells. Proc Natl Acad Sci USA 108:4081–4086

Blesson S, Thiery J, Gaudin C, Stancou R, Kolb JP, Moreau JL et al (2002) Analysis of the mechanisms of human cytotoxic T lymphocyte response inhibition by NO. Int Immunol 14:1169–1178

Boon T, Cerottini JC, Van den Eynde BJ, van der Bruggen P, Van Pel A (1994) Tumor antigens recognized by T lymphocytes. Annu Rev Immunol 12:337–365

Bronte V, Zanovello P (2005) Regulation of immune responses by L-arginine metabolism. Nat Rev Immunol 5:641–654

Bronte V, Serafini P, Mazzoni A, Segal DM, Zanovello P (2003) L-arginine metabolism in myeloid cells controls T-lymphocyte functions. Trends Immunol 24:302–306

Caldwell CC, Kojima H, Lukashev D, Armstrong J, Farber M, Apasov SG et al (2001) Differential effects of physiologically relevant hypoxic conditions on T lymphocyte development and effector functions. J Immunol 167:6140–6149

Cham CM, Gajewski TF (2005) Metabolic mechanisms of tumor resistance to T cell effector function. Immunol Res 31:107–118

Champsaur M, Lanier LL (2010) Effect of NKG2D ligand expression on host immune responses. Immunol Rev 235:267–285

Chang HY, Chi JT, Dudoit S, Bondre C, van de Rijn M, Botstein D et al (2002) Diversity, topographic differentiation, and positional memory in human fibroblasts. Proc Natl Acad Sci USA 99:12877–12882

Chouaib S, Messai Y, Couve S, Escudier B, Hasmim M, Noman MZ (2012) Hypoxia promotes tumor growth in linking angiogenesis to immune escape. Front Immunol 3:21(Epub: 2012 Feb 23, p 21)

Clark WH (1991) Tumour progression and the nature of cancer. Br J Cancer 64:631–644

Conforti L, Petrovic M, Mohammad D, Lee S, Ma Q, Barone S et al (2003) Hypoxia regulates expression and activity of Kv1.3 channels in T lymphocytes: a possible role in T cell proliferation. J Immunol 170:695–702

Coussens LM, Werb Z (2001) Inflammatory cells and cancer: think different! J Exp Med 193:F23–F26

Dang CV (2012) Links between metabolism and cancer. Genes Dev 26:877–890

Droge W, Roth S, Altmann A, Mihm S (1987) Regulation of T-cell functions by L-lactate. Cell Immunol 108:405–416

Egeblad M, Werb Z (2002) New functions for the matrix metalloproteinases in cancer progression. Nat Rev Cancer 2:161–174

Fortin C, Huang X, Yang Y (2012) NK Cell Response to Vaccinia Virus Is Regulated by Myeloid-Derived Suppressor Cells. J Immunol 189:1843–1849

Frederiksen LJ, Sullivan R, Maxwell LR, Macdonald-Goodfellow SK, Adams MA, Bennett BM et al (2007) Chemosensitization of cancer in vitro and in vivo by nitric oxide signaling. Clin Cancer Res 13:2199–2206

Frumento G, Rotondo R, Tonetti M, Damonte G, Benatti U, Ferrara GB (2002) Tryptophan-derived catabolites are responsible for inhibition of T and natural killer cell proliferation induced by indoleamine 2,3-dioxygenase. J Exp Med 196:459–68 (19)

Fu T, He Q, Sharma P (2011) The ICOS/ICOSL pathway is required for optimal antitumor responses mediated by anti-CTLA-4 therapy. Cancer Res 71:5445–5454

Gaide Chevronnay HP, Selvais C, Emonard H, Galant C, Marbaix E, Henriet P (2012) Regulation of matrix metalloproteinases activity studied in human endometrium as a paradigm of cyclic tissue breakdown and regeneration. Biochim Biophys Acta 1824:146–156

Gallina G, Dolcetti L, Serafini P, De SC, Marigo I, Colombo MP et al (2006) Tumors induce a subset of inflammatory monocytes with immunosuppressive activity on CD8 + T cells. J Clin Invest 116:2777–2790

Gasser S, Orsulic S, Brown EJ, Raulet DH (2005) The DNA damage pathway regulates innate immune system ligands of the NKG2D receptor. Nature 436:1186–1190

Geblad M, Nakasone ES, Werb Z (2010) Tumors as organs: complex tissues that interface with the entire organism. Dev Cell 18:884–901

Godefroy E, Moreau-Aubry A, Diez E, Dreno B, Jotereau F, Guilloux Y (2005) Alpha v beta3-dependent cross-presentation of matrix metalloproteinase-2 by melanoma cells gives rise to a new tumor antigen. J Exp Med 202:61–72

Godefroy E, Manches O, Dreno B, Hochman T, Rolnitzky L, Labarriere N et al (2011) Matrix metalloproteinase-2 conditions human dendritic cells to prime inflammatory T(H)2 cells via an IL-12- and OX40L-dependent pathway. Cancer Cell 19:333–346

Groh V, Wu J, Yee C, Spies T (2002) Tumour-derived soluble MIC ligands impair expression of NKG2D and T-cell activation. Nature 419:734–738

Habibi D, Jalili RB, Forouzandeh F, Ong CJ, Ghahary A (2010) High expression of IMPACT protein promotes resistance to indoleamine 2,3-dioxygenase-induced cell death. J Cell Physiol 225:196–205

Hanahan D, Weinberg RA (2000) The hallmarks of cancer. Cell 100:57–70

Hasko G, Linden J, Cronstein B, Pacher P (2008) Adenosine receptors: therapeutic aspects for inflammatory and immune diseases. Nat Rev Drug Discov 7:759–770

Hausler SF, Montalban dB I, Strohschein J, Anoop CP, Engel JB, Honig A et al (2011) Ectonucleotidases CD39 and CD73 on OvCA cells are potent adenosine-generating enzymes responsible for adenosine receptor 2A-dependent suppression of T cell function and NK cell cytotoxicity. Cancer Immunol Immunother 60:1405–1418

Hodi FS, O'Day SJ, McDermott DF, Weber RW, Sosman JA, Haanen JB, et al (2010) Improved survival with ipilimumab in patients with metastatic melanoma. N Engl J Med 363:711–23 (19)

Hoskin DW, Mader JS, Furlong SJ, Conrad DM, Blay J (2008) Inhibition of T cell and natural killer cell function by adenosine and its contribution to immune evasion by tumor cells. Int J Oncol 32:527–535 (Review)

Hou DY, Muller AJ, Sharma MD, DuHadaway J, Banerjee T, Johnson M et al (2007) Inhibition of indoleamine 2,3-dioxygenase in dendritic cells by stereoisomers of 1-methyl-tryptophan correlates with antitumor responses. Cancer Res 67:792–801

Jin D, Fan J, Wang L, Thompson LF, Liu A, Daniel BJ et al (2010) CD73 on tumor cells impairs antitumor T-cell responses: a novel mechanism of tumor-induced immune suppression. Cancer Res 70:2245–2255

Kaplan DH, Shankaran V, Dighe AS, Stockert E, Aguet M, Old LJ et al (1998) Demonstration of an interferon gamma-dependent tumor surveillance system in immunocompetent mice. Proc Natl Acad Sci USA 95:7556–7561

Kryczek I, Banerjee M, Cheng P, Vatan L, Szeliga W, Wei S et al (2009) Phenotype, distribution, generation, and functional and clinical relevance of Th17 cells in the human tumor environments. Blood 114:1141–1149

Li W, Ren G, Huang Y, Su J, Han Y, Li J et al (2012) Mesenchymal stem cells: a double-edged sword in regulating immune responses. Cell Death Differ 19:1505–1513

Liakou CI, Kamat A, Tang DN, Chen H, Sun J, Troncoso P et al (2008) CTLA-4 blockade increases IFNgamma-producing CD4 + ICOShi cells to shift the ratio of effector to regulatory T cells in cancer patients. Proc Natl Acad Sci USA 105:14987–14992

Lob S, Konigsrainer A, Schafer R, Rammensee HG, Opelz G, Terness P (2008) Levo- but not dextro-1-methyl tryptophan abrogates the IDO activity of human dendritic cells. Blood 111:2152–2154
Madan RA, Gulley JL (2011) Therapeutic cancer vaccine fulfills the promise of immunotherapy in prostate cancer. Immunotherapy 3:27–31
Matzinger P, Kamala T (2011) Tissue-based class control: the other side of tolerance. Nat Rev Immunol 11:221–230
Metz R, Duhadaway JB, Kamasani U, Laury-Kleintop L, Muller AJ, Prendergast GC (2007) Novel tryptophan catabolic enzyme IDO2 is the preferred biochemical target of the antitumor indoleamine 2,3-dioxygenase inhibitory compound D-1-methyl-tryptophan. Cancer Res 67:7082–7087
Mlecnik B, Bindea G, Pages F, Galon J (2011) Tumor immunosurveillance in human cancers. Cancer Metastasis Rev 30:1–12
Munir S, Larsen SK, Iversen TZ, Donia M, Klausen TW, Svane IM et al (2012) Natural CD4 + T-cell responses against indoleamine 2,3-dioxygenase. PLoS One 7:e34568
Munn DH, Mellor AL (2007) Indoleamine 2,3-dioxygenase and tumor-induced tolerance. J Clin Invest 117:1147–1154
Munn DH, Sharma MD, Baban B, Harding HP, Zhang Y, Ron D et al (2005) GCN2 kinase in T cells mediates proliferative arrest and anergy induction in response to indoleamine 2,3-dioxygenase. Immunity 22:633–642
Noman MZ, Messai Y, Carre T, Akalay I, Meron M, Janji B et al (2011) Microenvironmental hypoxia orchestrating the cell stroma cross talk, tumor progression and antitumor response. Crit Rev Immunol 31:357–377
Novitskiy SV, Ryzhov S, Zaynagetdinov R, Goldstein AE, Huang Y, Tikhomirov OY et al (2008) Adenosine receptors in regulation of dendritic cell differentiation and function. Blood 112:1822–1831
Nowak EC, de Vries VC, Wasiuk A, Ahonen C, Bennett KA, Le M, I, et al (2012) Tryptophan hydroxylase-1 regulates immune tolerance and inflammation. J Exp Med 209:2127–2135
Opitz CA, Litzenburger UM, Sahm F, Ott M, Tritschler I, Trump S et al (2011) An endogenous tumour-promoting ligand of the human aryl hydrocarbon receptor. Nature 478:197–203
Pallotta MT, Orabona C, Volpi C, Vacca C, Belladonna ML, Bianchi R et al (2011) Indoleamine 2,3-dioxygenase is a signaling protein in long-term tolerance by dendritic cells. Nat Immunol 12:870–878
Pardoll DM (2012) The blockade of immune checkpoints in cancer immunotherapy. Nat Rev Cancer 12:252–264
Paschen A, Sucker A, Hill B, Moll I, Zapatka M, Nguyen XD et al (2009) Differential clinical significance of individual NKG2D ligands in melanoma: soluble ULBP2 as an indicator of poor prognosis superior to S100B. Clin Cancer Res 15:5208–5215
Peggs KS, Quezada SA, Chambers CA, Korman AJ, Allison JP (2009) Blockade of CTLA-4 on both effector and regulatory T cell compartments contributes to the antitumor activity of anti-CTLA-4 antibodies. J Exp Med 206:1717–1725
Pietras K, Ostman A (2010) Hallmarks of cancer: interactions with the tumor stroma. Exp Cell Res 316:1324–1331
Pilotte L, Larrieu P, Stroobant V, Colau D, Dolusic E, Frederick R et al (2012) Reversal of tumoral immune resistance by inhibition of tryptophan 2,3-dioxygenase. Proc Natl Acad Sci USA 109:2497–2502
Platten M, Ho PP, Youssef S, Fontoura P, Garren H, Hur EM et al (2005) Treatment of autoimmune neuroinflammation with a synthetic tryptophan metabolite. Science 310:850–855
Prendergast GC, Jaffee EM (2007) Cancer immunologists and cancer biologists: why we didn't talk then but need to now. Cancer Res 67:3500–3504
Prendergast GC, Metz R, Muller AJ (2009) IDO recruits Tregs in melanoma. Cell Cycle 8:1818–1819
Pytliak M, Vargova V, Mechirova V (2012) Matrix metalloproteinases and their role in oncogenesis: a review. Onkologie 35:49–53

Rodriguez PC, Hernandez CP, Quiceno D, Dubinett SM, Zabaleta J, Ochoa JB et al (2005) Arginase I in myeloid suppressor cells is induced by COX-2 in lung carcinoma. J Exp Med 202:931–939

Romani L, Bistoni F, Perruccio K, Montagnoli C, Gaziano R, Bozza S et al (2006) Thymosin alpha1 activates dendritic cell tryptophan catabolism and establishes a regulatory environment for balance of inflammation and tolerance. Blood 108:2265–2274

Ropke M, Hald J, Guldberg P, Zeuthen J, Norgaard L, Fugger L, et al (1996) Spontaneous human squamous cell carcinomas are killed by a human cytotoxic T lymphocyte clone recognizing a wild-type p53-derived peptide. Proc Natl Acad Sci USA 14704–7

Saleem SJ, Martin RK, Morales JK, Sturgill JL, Gibb DR, Graham L et al (2012) Cutting edge: mast cells critically augment myeloid-derived suppressor cell activity. J Immunol 189:511–515

Salih HR, Antropius H, Gieseke F, Lutz SZ, Kanz L, Rammensee HG et al (2003) Functional expression and release of ligands for the activating immunoreceptor NKG2D in leukemia. Blood 102:1389–1396

Seliger B, Quandt D (2012) The expression, function, and clinical relevance of B7 family members in cancer. Cancer Immunol Immunother

Sharma MD, Hou DY, Liu Y, Koni PA, Metz R, Chandler P et al (2009) Indoleamine 2,3-dioxygenase controls conversion of Foxp3 + Tregs to TH17-like cells in tumor-draining lymph nodes. Blood 113:6102–6111

Sheu BC, Hsu SM, Ho HN, Lien HC, Huang SC, Lin RH (2001) A novel role of metalloproteinase in cancer-mediated immunosuppression. Cancer Res 61:237–242

Shime H, Yabu M, Akazawa T, Kodama K, Matsumoto M, Seya T et al (2008) Tumor-secreted lactic acid promotes IL-23/IL-17 proinflammatory pathway. J Immunol 180:7175–7183

Siemens DR, Hu N, Sheikhi AK, Chung E, Frederiksen LJ, Pross H et al (2008) Hypoxia increases tumor cell shedding of MHC class I chain-related molecule: role of nitric oxide. Cancer Res 68:4746–4753

Smith C, Chang MY, Parker KH, Beury DW, Duhadaway JB, Flick HE et al (2012) IDO Is a nodal pathogenic driver of lung cancer and metastasis development. Cancer Discov 2:722–735

Smyth MJ, Thia KY, Street SE, MacGregor D, Godfrey DI, Trapani JA (2000) Perforin-mediated cytotoxicity is critical for surveillance of spontaneous lymphoma. J Exp Med 192:755–760

Sorensen RB, Berge-Hansen L, Junker N, Hansen CA, Hadrup SR, Schumacher TN et al (2009) The immune system strikes back: cellular immune responses against indoleamine 2,3-dioxygenase. PLoS One 4:e6910

Sorensen RB, Hadrup SR, Svane IM, Hjortso MC, Thor Straten P, Andersen MH (2011) Indoleamine 2,3-dioxygenase specific, cytotoxic T cells as immune regulators. Blood 117:2200–2210

Stagg J, Smyth MJ (2010) Extracellular adenosine triphosphate and adenosine in cancer. Oncogene 29:5346–5358

Textor S, Fiegler N, Arnold A, Porgador A, Hofmann TG, Cerwenka A (2011) Human NK cells are alerted to induction of p53 in cancer cells by upregulation of the NKG2D ligands ULBP1 and ULBP2. Cancer Res 71:5998–6009

Topalian SL, Hodi FS, Brahmer JR, Gettinger SN, Smith DC, McDermott DF, et al (2012) Safety, Activity, and immune correlates of anti-PD-1 antibody in cancer. N Engl J Med

Uyttenhove C, Pilotte L, Theate I, Stroobant V, Colau D, Parmentier N et al (2003) Evidence for a tumoral immune resistance mechanism based on tryptophan degradation by indoleamine 2,3-dioxygenase. Nat Med 9:1269–1274

van der Bruggen P, Traversari C, Chomez P, Lurquin C, De Plaen E, Van den Eynde BJ et al (1991) A gene encoding an antigen recognized by cytolytic T lymphocytes on a human melanoma. Science 254:1643–1647

Vetter CS, Groh V, thor Straten P, Spies T, Brocker EB, Becker JC (2002) Expression of stress-induced MHC class I related chain molecules on human melanoma. J Invest Dermatol 118:600–605

Weinlich G, Murr C, Richardsen L, Winkler C, Fuchs D (2007) Decreased serum tryptophan concentration predicts poor prognosis in malignant melanoma patients. Dermatology 214:8–14

Whiteside TL, Mandapathil M, Schuler P (2011) The role of the adenosinergic pathway in immunosuppression mediated by human regulatory T cells (Treg). Curr Med Chem 18:5217–5223

Witkiewicz AK, Costantino CL, Metz R, Muller AJ, Prendergast GC, Yeo CJ et al (2009) Genotyping and expression analysis of IDO2 in human pancreatic cancer: a novel, active target. J Am Coll Surg 208:781–787

Wu JD, Atteridge CL, Wang X, Seya T, Plymate SR (2009) Obstructing shedding of the immunostimulatory MHC class I chain-related gene B prevents tumor formation. Clin Cancer Res 15:632–640

Xu W, Liu LZ, Loizidou M, Ahmed M, Charles IG (2002) The role of nitric oxide in cancer. Cell Res 12:311–320

Xu H, Oriss TB, Fei M, Henry AC, Melgert BN, Chen L et al (2008) Indoleamine 2,3-dioxygenase in lung dendritic cells promotes Th2 responses and allergic inflammation. Proc Natl Acad Sci USA 105:6690–6695

Yang XO, Nurieva R, Martinez GJ, Kang HS, Chung Y, Pappu BP et al (2008) Molecular antagonism and plasticity of regulatory and inflammatory T cell programs. Immunity 29:44–56

Yi KH, Chen L (2009) Fine tuning the immune response through B7–H3 and B7–H4. Immunol Rev 229:145–151

Yu Q, Stamenkovic I (2000) Cell surface-localized matrix metalloproteinase-9 proteolytically activates TGF-beta and promotes tumor invasion and angiogenesis. Genes Dev 14:163–176

Zou W, Restifo NP (2010) T(H)17 cells in tumour immunity and immunotherapy. Nat Rev Immunol 10:248–256

Chapter 12
Roles of Signaling Pathways in Cancer Cells and Immune Cells in Generation of Immunosuppressive Tumor-Associated Microenvironments

Yutaka Kawakami, Tomonori Yaguchi, Hidetoshi Sumimoto, Chie Kudo-Saito, Nobuo Tsukamoto, Tomoko Iwata-Kajihara, Shoko Nakamura, Hiroshi Nishio, Ryosuke Satomi, Asuka Kobayashi, Mayuri Tanaka, Jeong Hoon Park, Hajime Kamijuku, Takahiro Tsujikawa and Naoshi Kawamura

Abstract Cancer cells trigger multiple immunosuppressive cascades and generate immunosuppressive tumor-associated microenvironments including tumor and sentinel lymph nodes. Constitutive activation of various signaling pathways (e.g., MAPK, STAT3, NF-κB, β-catenin) in human cancer cells was found to trigger the multiple immunosuppressive cascades through the production of immunosuppressive cytokines, such as TGF-β, IL-10, IL-6, and VEGF, and induction of immunosuppressive immune cells, such as regulatory T cells, tolerogenic dendritic cells, and myeloid derived suppressor cells. Some of these cancer-derived cytokines impair various immune cells through activation of their signaling molecules such as STAT3 and NF-κB. Inhibitors for these activated signals could inhibit the multiple immunosuppressive cascades by acting on both cancer cells and immune cells. Since common signaling mechanisms are often utilized for some of the hallmarks of cancer (e.g., cell proliferation/survival, invasion/metastasis, and immunosuppression), targeting these common signaling pathways may be an attractive strategy for cancer therapy, including immunotherapy.

Keywords Immunosuppression · BRAF · STAT3 · β-catenin · NF-κB · TGF-β · IL-10 · MAPK · MDSC · Regulatory T cells

Y. Kawakami (✉) · T. Yaguchi · H. Sumimoto · C. Kudo-Saito · N. Tsukamoto · T. Iwata-Kajihara · S. Nakamura · H. Nishio · R. Satomi · A. Kobayashi · M. Tanaka · J. H. Park · H. Kamijuku · T. Tsujikawa · N. Kawamura
Division of Cellular Signaling, Institute for Advanced Medical Research, Keio University School of Medicine, 35 Shinanomachi, Shinjuku, Tokyo, Japan
e-mail: yutakawa@z5.keio.jp

M. R. Shurin et al. (eds.), *The Tumor Immunoenvironment*,
DOI: 10.1007/978-94-007-6217-6_12,

12.1 Introduction

We have previously identified various human tumor antigens recognized by T cells, and developed various antigen specific immunotherapies (Kawakami et al. 2004; Kawakami et al. 1994; Kawakami et al. 1994; Rosenberg et al. 1998). The vaccine using gp100 melanoma antigenic peptide plus IL-2 resulted in 16 % objective response with 9 % CR in the recent multicenter randomized trial (Schwartzentruber et al. 2011), while the recent adoptive immunotherapy using cultured melanoma specific T cells after the myeloablative treatment, which depletes various immunosuppressive cells, resulted in more than 70 % objective response with 21 % durable CR for advanced melanoma patients with multiple metastases (Rosenberg et al. 2011). Immunological analysis of the clinical trials indicated that immunosuppression in cancer patients is one of the major obstacles for development of effective immunotherapy. Thus, understanding of the mechanisms for the immunosuppression in cancer patients and development of strategy to overcome it is important for improvement of cancer therapy.

12.2 Immunopathology of Cancer-Associated Microenvironments

Roles of the immune system in cancer have recently been extensively exploited. The murine studies analyzing the interaction between cancer cells and immune system (and other stromal cells) during cancer development revealed that innate cells such as macrophages and mast cells rather have tumor promoting activity though increase of cancer cell proliferation and invasion ability, as well as induction of angiogenesis. In contrast, T cells and NK cells have the ability to eliminate cancer cells (Immunosurveillance). However, cancer cells having intrinsic genetic instability subsequently evade the immune defense system by losing highly immunogenic tumor antigens and acquiring various immunoresistant and immunosuppressive mechanisms (Immune evasion). This process is also known as Immunoediting (Schreiber et al. 2011). In fact, cancer cells developed from immunocompromised hosts are sensitive to immune cell attack, and cancer cells obtained from patients have a variety of immune-suppressive and resistant features (Zou 2005; Gajewski et al. 2006; Yaguchi et al. 2011). Therefore, the immunological characteristics of cancer cells are defined by both cancer cells' intrinsic nature and immune reactivity of patients.

When investigating the mechanisms of the immunosuppression in cancer patients, it is important to consider tumor-associated microenvironments, including tumor tissues where effector immune cells should eliminate cancer cells, sentinel lymph nodes (SLNs) where tumor specific T cells should be primed, and bone marrow where is a source of various immunosuppressive cells including myeloid-derived suppressor cells (MDSCs) and mesenchymal stem cells (MSCs), as well as a reservoir for tumor specific memory T cells. Analysis of tumor tissues obtained from patients revealed that tumor appears to be under immunosuppressive conditions suggested by the expression of various immunosuppressive molecules in cancer cells (e.g., soluble molecules such as TGF-β, IL-10, IL-6, VEGF, GM-CSF, IL-13, PGE2, sMICA, membrane molecules such as PD-L1, FasL, ILT7L, intracellular molecules such as IDO, COX2), and by accumulation of various immunosuppressive cells (e.g., Treg, MDSCs, M2-like macrophages, tolerogenic dendritic cells (DCs), plasmacytoid DCs, cancer-associated fibroblasts (CAFs), MSCs). Similarly, in the sentinel lymph nodes of cancer patients, accumulation of such immunosuppressive cells was also observed (Kim et al. 2006; Swartz and Lund 2012; Fridman et al. 2011). However, the comprehensive analysis of the molecular and cellular mechanisms for the immunosuppression in the tumor-associated microenvironments remains to be performed.

It has recently been reported that levels of spontaneous $CD8^+$ T cell responses (infiltration of memory $CD8^+$ T cells in the tumor tissue prior to the cancer treatment) are different among patients with various cancers, including colon cancer, ovarian cancer, and melanoma. High infiltration of memory $CD8^+$ T cells in tumor significantly correlated with better prognosis, and its prognostic predictive value appeared to be better than TNM staging (Fridman et al. 2011; Mlecnik et al. 2011). It was also correlated with response to immunotherapy in melanoma and even chemotherapy in colon cancer (Gajewski et al. 2011). Therefore, international collaborative study "Immunoscore validation task force" is currently in progress to confirm the diagnostic value of infiltration of $CD8^+$ T cells in colon cancer (Galon et al. 2012). However, it has not yet been understood what makes the difference of spontaneous $CD8^+$ T cell response among patients. It may be regulated by both cancer cell characteristics and immunological constitution of hosts.

One important point is that immune condition in cancer patients is regulated by complex immune networks and it is first triggered by cancer cells, more specifically genetic or epigenetic alterations in cancer cells. Cancer cells trigger multiple immunosuppressive cascades in which various immunosuppressive molecules such as TGF-β, IL-10, IL-6, VEGF, PD-L1, COX2, and IDO, and immunosuppressive cells, such as tolerogenic DCs, MDSCs and Treg cells, are involved, and finally immunosuppressive conditions are established in the tumor-associate microenvironments.

12.3 Immunosuppressive Cascades Triggered by Gene Alterations in Cancer Cells

To understand the immunosuppressive cascades triggered by cancer cells, we have evaluated the role of TGF-β, which is produced by most human cancer cells, infiltrated immune cells and stromal cells, in the regulation of immunological conditions in tumor and SLN. In our mouse tumor model, increase of TGF-β in the tumor microenvironment by implantation of the TGF-β gene-transduced tumor cells resulted in increased accumulation of CD11b$^+$ Gr-1$^+$ MDSCs and FoxP3$^+$CD4$^+$Treg cells in both tumor and SLN. Numbers of DCs infiltrated the tumor tissue were decreased. Interestingly, numbers of DCs were increased in SLN compared to non-SLN in mice implanted with either TGF-β-transduced tumor cells or control tumor cells, but function of DCs from mice with abundant TGF-β expression in the tumor microenvironment was significantly impaired as assessed by their T cell stimulatory activity. Implantation of TGF-β-producing tumor cells also induced M2-like macrophages, which produced abundant CCL22 in SLN; CCL22 appeared to recruit CCR4$^+$ Tregs into SLN (Tsujikawa et al. 2012). Consequently, induction of tumor antigen specific T cells from SLN was significantly reduced, and, finally, infiltration of CD8$^+$ T cells in tumor appeared to be reduced in the mice with abundant TGF-β expression. It has been reported that inhibition of TGF-β signaling by injection of plasmid DNA encoding TGF-β type II receptor near the tumor sites was reported to enhance tumor antigen specific T cells accompanied by decrease of Treg cells (Fujita et al. 2009). Therefore, these mouse models recapitulate the observations in the analysis of clinical samples and indicate that immunosuppressive molecules, such as TGF-β, may be one of the factors to define the immune status in tumor and SLN, including spontaneous CD8$^+$ T cell response.

We have previously reported that TGF-β-induced-Snail stimulated not only epithelial-to-mesenchymal transition (EMT) of cancer cells, but also production of immunosuppressive cytokines and chemokines, including TGF-β, IL-10, CCL2, and TSP-1, which caused DC impairment and Treg induction. The impaired DCs could also induce Tregs. CCL2 impairs DCs and recruits immunosuppressive MDSCs into tumor. The blockade of Snail in the tumor microenvironment by intratumoral administration of Snail-specific siRNA restored immunocompetence of mice having Snail-transduced tumor and resulted in enhanced induction of tumor antigen specific T cells in vivo (Kudo-Saito et al. 2009). These results illustrate that TGF-β production in the tumor microenvironment by either cancer cells or infiltrated stromal cells, including various immune cells, triggers multiple immunosuppressive cascades involving various immunosuppressive cytokines/chemokines and cells. This reemphasizes that the TGF-β cascade is an attractive target for reversal of cancer-induced immunosuppression (Fig. 12.1).

The molecular mechanisms of the increased production of TGF-β by human cancer cells have not been well understood. In human melanoma, production of TGF-β was not mainly regulated by MAPK and STAT3 pathways as described below. We have recently found that one of the intracellular kinase, which is frequently phoshphorylated in various cancer cells, is involved in TGF-β production by human melanoma cells. This was assessed by screening signaling molecules in melanoma cells involving suppression of DC function by using kinase siRNA library (manuscript in preparation). Therefore, this activated kinase in cancer cells can be an upstream target to inhibit the TGF-β-triggered immunosuppressive cascades. We are currently searching for small molecular drugs which efficiently inhibit this kinase. This is one example of immunosuppressive

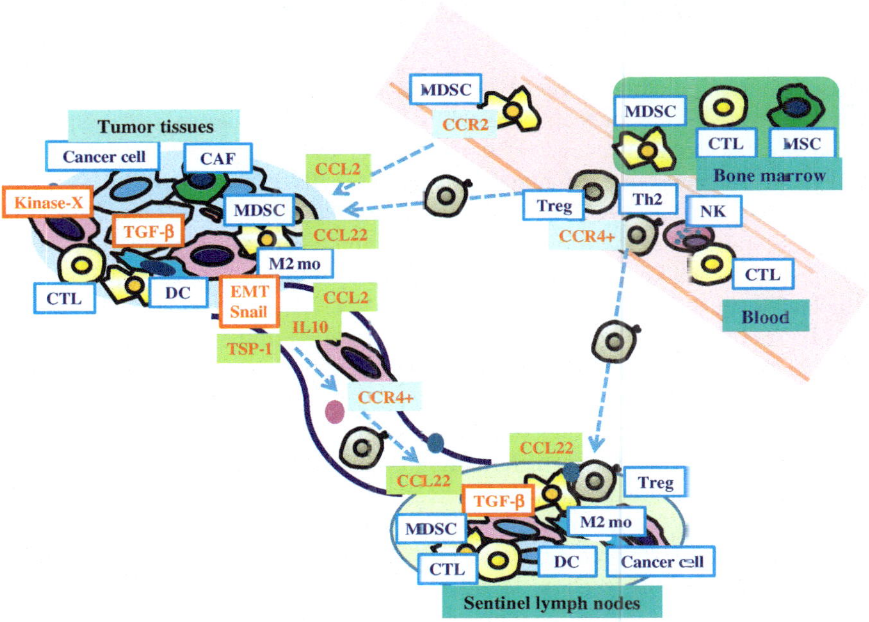

Fig. 12.1 Cancer cell triggered immunosuppressive cascades: TGF-β cascade. TGF-β produced by cancer cells and infiltrated stromal cells including various immune cells triggers immunosuppressive cascade (production of CCL2, IL-10, TGF-β , etc., and subsequent recruitment and induction of various immunosuppressive cells such as MDSC, M2 macrophages, Treg cells via CCL2, CCL22, IL-10, TGF-β , etc.), and generates immunosuppressive condition in the tumor and sentinel lymph nodes. TGF-β-Snail axis induces not only epithelial-to-mesenchymal transition (EMT) which enhances invasion ability of cancer cells, but also immunosuppression which may further enhance metastasis of cancer cells

cascades triggered by gene alterations in cancer cells. Since gene and signal alterations in human cancer cells vary among cancer types, even among patients with the same type of cancer, there are multiple immunosuppressive cascades to be investigated.

12.4 Alterations of Gene and Signal Pathways Involved in the Immunosuppression in Human Cancer

Human cancer cells have various genetic and epigenetic alterations. Thus, understanding of immunosuppressive cascades triggered by each gene/signal alteration is important. Here, we describe some of our recent observations on human cancer cells.

12.4.1 RAS/BRAF/MEK/MAPK Signaling Pathway

When common mutation of BRAF (V600E), a molecule in MAPK signal pathway, was discovered by sequencing signaling molecules in human melanoma cells (Davies et al. 2002), we evaluated the role of the mutant BRAF (V600E) for malignant characteristics of human melanoma cells by using BRAF (V600E)-specific lentiviral shRNA. We found that the BRAF mutation was involved in the cell proliferation and invasion ability of melanoma cells (Sumimoto et al. 2004). We have also found that production of multiple cytokines, IL-6, IL-10, and VEGF, which have the ability to suppress function of DCs, were significantly decreased by BRAF(V600E) shRNA without affecting cell survival of the some melanoma cell lines (Sumimoto et al. 2006). These cytokines suppress DC activity to stimulate T cells mainly through the inhibition of IL-12 and TNF-α production, and augmentation of IL-10 production. Treatment of melanoma cells with BRAF (V600E)-specific shRNA or MEK inhibitors resulted in decrease of immunosuppressive activity of melanoma cells, indicating the MAPK pathway is essential for DC impairment by melanoma cells. MEK inhibitors were also reported to increase susceptibility of melanoma cells to CTL lysis partly due to increased expression of melanoma antigens, such as MART-1/melan-A and gp100 (Kono et al. 2006; Boni et al. 2010). "Avoiding immune destruction" resulting from the loss of highly immunogenic tumor antigens and acquiring immunoresistant and immunosuppressive mechanisms is now generally recognized as one of "the hallmarks of cancer" (Hanahan and Weinberg 2011). These results indicate that the BRAF-MAPK axis is commonly involved in the cancer cell proliferation, invasion, and immunosuppression (Fig. 12.2).

These observations indicate that blockade of the BRAF-MAPK axis may not only suppress proliferation and invasion of cancer cells, but also inhibit immunosuppressive activity and increase susceptibility of melanoma cells to T cells. This suggests that it is a common attractive target for melanoma treatment, particularly in combination with various immunotherapies. Since MAPK signal is also important for T cell proliferation, administration of MAPK pathway inhibitors may also suppress anti-tumor T cell response. However, two BRAF inhibitors, which preferentially inhibit mutant BRAF, have recently been developed, and their administration has already been shown to be effective in patients with melanoma (Chapman et al. 2011; Hauschild et al. 2012). These selective mutant BRAF inhibitors actually cause melanoma cell death in vivo resulting in reduction of tumor sizes in some patients. Therefore, the selective mutant BRAF inhibitors may be useful for combination with immunotherapies through the following mechanisms:

1) Tumor destruction causes release of endogenous tumor antigens which include multiple patient's unique mutated antigens, leading to induction of multiple autologous tumor specific T cells,
2) Reduction of tumor burden via inhibition of cancer cell proliferation and cancer cell death results in reduction of immunosuppressive condition,

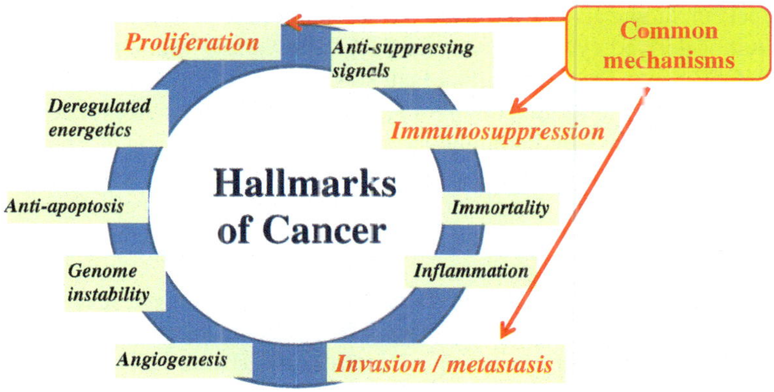

Fig. 12.2 Common mechanisms are sometimes used for hallmarks of cancer such as proliferation, metastasis and immunosuppression. Common signaling pathways such as MAPK signaling are sometimes utilized for some of the hallmarks of cancer including cancer cell proliferation, invasion, and immunosuppression. Therefore, inhibitors against the common pathways may be useful for cancer treatment through simultaneous inhibition of multiple hallmarks of cancer. Combination of molecular targeted drugs and immunotherapy may be an attractive strategy for cancer treatment

3) Decrease of production of multiple immunosuppressive cytokines results in simultaneous inhibition of multiple immunosuppressive cascades,
4) Susceptibility of cancer cells to cytotoxic T cells (CTLs) is increased partly via increased expression of tumor antigens,
5) Mutant BRAF selective inhibitors are less inhibitory for proliferation of anti-tumor T cells, and
6) Invasion and metastatic ability of cancer cells is decreased.

It has recently been reported that administration of mutant BRAF selective inhibitors did not suppress immune response in general (Hong et al. 2012), and actually increased infiltration of T cells, particularly granzyme positive CD8$^+$ T cells, in tumors, which was correlated with tumor reduction and necrosis (Wilmott et al. 2011). In vivo immunological effects of a MEK inhibitor which is also effective for patients with melanoma having mutation of either NRAS or BRAF remain to be investigated (Flaherty et al. 2012). The same strategy may also be applied for other cancers with BRAF mutations, including colon cancer and thyroid cancer.

12.4.2 JAK/STAT3 Signaling Pathway

In human melanoma, in addition to RAS/BRAF mutation, activation of STAT3 is frequently observed. Depletion of STAT3 by lentiviral shRNA in STAT3 active melanoma also resulted in the inhibition of multiple immunosuppressive cytokines, including IL-6, IL-10, and VEGF (Sumimoto et al. 2006). Interestingly, these tumor-derived cytokines activate STAT3 in various immune cells including DCs, MDSCs, and Tregs, and affect their functions. The cytokine-induced STAT3 activation resulted in the generation of low IL-12 and high IL-10 producing human DCs with decreased T cell stimulatory activity.

In the mouse tumor model, STAT3-depleted DCs obtained from myeloid-specific STAT3-conditional knockout mice, were resistant to these tumor-derived immunosuppressive cytokines, and also had strong T cell stimulatory activity along with sustained high IL-12 production. Injection of the STAT3-depleted DCs into tumor, which is under the immunosuppressive condition, showed strong anti-tumor effects accompanied by induction of higher IFN-γ producing tumor antigen specific Th1 cells compared to the injection of control DCs (Iwata-Kajihara et al. 2011). Similarly, STAT3-depleted macrophages were resistant to tumor-derived immunosuppressive cytokines, and induction of immunosuppressive macrophages and MDSCs were partially inhibited by STAT3 depletion. STAT3 was also reported to be involved in expansion of MDSCs (Wu et al. 2011). STAT3 activation was actually observed in $CD14^{+}HLA\text{-}DR^{negative/low}$ MDSCs in blood of cancer patients (Poschke et al. 2010). STAT3 is also important for Treg cells (Pallandre et al. 2007). STAT3 inhibition of anti-tumor $CD8^{+}$T cells was reported to enhance their effects when adoptively transferred into tumor-bearing mice (Kujawski et al. 2010). These results indicate that constitutive activation of STAT3 in cancer cells triggers induction of various immunosuppressive immune cells, including tolerogenic DCs, MDSCs, and Tregs, partly through activation of STAT3 in these immune cells (Kortylewski et al. 2005; Yu et al. 2007). Therefore, STAT3 inhibitors may be useful for reversal of cancer-induced immunosuppression through not only acting on cancer cells, but also acting on various immune cells.

Recently, molecular targeted therapies acting on various signaling molecules in cancer cells have been used for cancer treatment. STAT3 inhibitors are being evaluated in clinical trials. In murine tumor model, various STAT3 inhibitors have been shown to augment anti-tumor immunity (Lee et al. 2011). In addition to STAT3 inhibitors, inhibitors to molecules present at upstream of STAT3, including inhibitors for direct upstream molecule JAK and further upstream molecules EGF-R/VEGF-R (which have already been available for clinical use), may also be useful for reversal of immunosuppression and combination with immunotherapy. JAK inhibitors have been shown to augment anti-tumor immunity and enhance anti-tumor effects in combination with immunotherapies, such as IL-12 administration (Burdelya et al. 2002). We have observed that EGF-R inhibitors suppress production of some of the immunosuppressive cytokines, such as IL-6 and VEGF, from human lung cancer cells with EGF-R mutations. In the murine tumor model, administration of the EGF-R inhibitors along with cancer vaccines showed synergistic anti-tumor effects through indirect (via decrease of immunosuppressive cytokines from cancer cells) and direct enhancement of DC ability to stimulate T cells. Administration of multikinase inhibitor Sunitinib, which also suppresses downstream STAT3 signaling, to RCC patients was reported to result in decrease of MDSCs and Tregs along with increase of IFN-γ producing T cells (Xin et al. 2009; Ozao-Choy et al. 2009; Ko et al. 2009). Another multikinase inhibitor Dasatinib was reported to increase response rate in about half of patients with Ph1^{+}CML and ALL accompanied by LGL lymphocytosis and autoimmune like syndrome, such as pleuritis and colitis; it was reported to inhibit STAT3 signaling in immune cells after administration (Mustjoki et al. 2009; Jalkanen et al. 2010). Therefore, there are various ways of STAT3 signal inhibition for reversal of immunosuppression in cancer patients in clinic. We have recently screened natural compounds contained in the Japanese traditional Kampo medicines, and found that some of the compounds are able to inhibit STAT3 and MAPK pathways, possibly by targeting upstream signaling molecules. Their systemic administration augmented tumor specific T cells accompanied by decrease of Tregs in the tumor in tumor-bearing mice (manuscript in preparation).

12.4.3 NF-κB Signaling Pathway

We have also observed similar phenomenon—involvement of the same signaling pathway in both cancer cells and immune cells for generation of immunosuppressive condition, in human ovarian cancers with constitutively activated NF-κB, which causes high production of IL-6, IL-8, and CCL2. High levels of plasma IL-6 and IL-8 were found to correlate with poor prognosis of cancer patients and poor response to various immunotherapies, including vaccinations with cancer antigen peptides and DCs (manuscript in preparation). NF-κB inhibitor inhibited not only production of these immunosuppressive cytokines and chemokines by cancer cells, but also had direct effects on monocytes: they inhibit their differentiation to immunosuppressive macrophages in the presence of cancer cell-derived factors. Although the cross-talk, such as positive feedback loop between IL-6, STAT3 and NF-κB was previously reported to be involved in chronic inflammation (Yu and Pardoll 2009; Murakami and Hirano 2011), significant role of such cross-talk was not observed in these ovarian cancers. Systemic administration of appropriate dose of a NF-κB inhibitor augmented anti-tumor T cell responses possibly through reversal of immunosuppressive condition in a murine tumor model, although NF-κB signal is also essential for induction of anti-tumor T cells.

NF-κB was found to be involved in the intrinsic expression of ILT7 ligand (ILT7L) in some of human renal cell cancers (RCC), although ILT7L can also be up-regulated by IFN-γ from infiltrated T cells. ILT7L inhibits IFN-α production by plasmacytoid DCs and is possibly involved in immunosuppression in the tumor microenvironments, since type-I IFN was reported to be critical for induction of spontaneous anti-tumor T cell response (Fuertes et al. 2005; Gajewski et al. 2012). NF-κB inhibitor suppressed the intrinsic expression of ILT7L on RCC cells (Tsukamoto et al. 2009). It has recently been reported that expression of PD-L1 on cancer cells was mainly induced by IFN-γ produced by tumor-infiltrating T cells, and the PD-L1 expression on cancer cells and $CD8^{+}$T cell infiltration correlated with significantly better response to anti-PD-1 antibody treatment (Taube et al. 2012). However, some cancer cells intrinsically express PD-L1 partly due to activation of AKT pathway via PTEN deletion (human glioma) (Parsa et al. 2007) or activated MAPK pathway in some other cancers. We have found new inhibitors, which suppresses intrinsic expression of PD-L1. These observations indicate that signal inhibitors may also be useful for inhibition of these immunosuppressive membrane molecules (e.g., ILT7L and PD-L1) intrinsically expressed through altered signaling in human cancer cells.

12.4.4 Wnt/β-Catenin Signal Pathway

Activation of β-catenin pathway (nuclear staining of β-catenin) was observed in about 30 % of human melanoma, and correlated with expression of IL-10 by immunohistochemical analysis. We found that β-catenin directly activated IL-10 transcription in human melanoma (Yaguchi et al. 2012). Supernatant from cultured β-catenin-accumulating melanoma cells induced high IL-10-, low IL-12-producing DCs with low T cell stimulatory activity in vitro, which was IL-10-dependent; these DCs also had the ability to induce FOXP3-positive immunosuppressive Treg cells. Pretreatment of melanoma cells with shRNA for β-catenin reduced their immunosuppressive activities. Interestingly, supernatant from cultured melanoma also inhibited the effector function of melanoma specific cytotoxic T cells in a β-catenin-dependent, but IL-10-independent manner, indicating that other immunosuppressive molecules are also involved in the β-catenin induced immunosuppression.

When β-catenin-activated human melanoma cell lines were implanted in immunodeficient SCID mice, the level of human IL-10 in blood was increased, and mouse DCs in the spleen and tumor were impaired for T cell stimulatory activity. This was likely because human IL10 can also act on mouse cells and suppress mouse DCs. Systemic administration of a β-catenin inhibitor restored mouse splenic DC activity to stimulate T cells along with decrease of human IL-10 in the serum. Interestingly, a β-catenin inhibitor also had the ability to directly enhance T cell stimulatory activity of human DCs partly due to decreased IL-10 production by DCs. β-catenin was also reported to be involved in generation of regulatory DCs (Fu and Jiang 2010; Manicassamy et al. 2010) and survival of Tregs (Ding et al. 2008). These results indicate again that signal inhibitors may be useful for reversal of cancer-induced immunosuppression by acting on both cancer and immune cells.

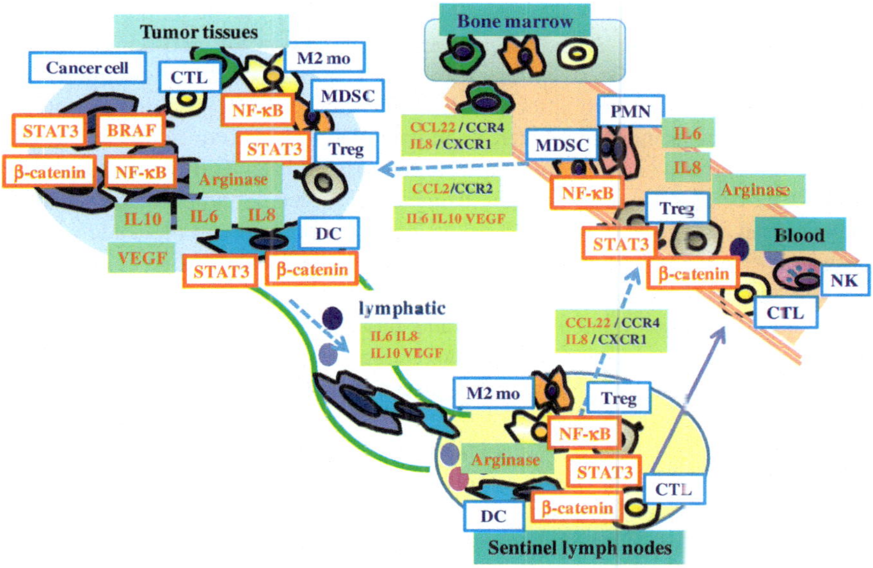

Fig. 12.3 Cancer cell triggered immunosuppressive cascades: MAPK, STAT3, β-catenin, and NF-κB cascades. Alterations of oncogenes and subsequently activated signaling are different among cancer cells even in the same types of cancer. For examples, alterations of MAPK, STAT3, β-catenin, and NF-κB trigger different immunosuppressive cascades via production of immunosuppressive cytokines such as IL-6, IL-8, IL-10, VEGF, etc., and subsequent impairment of DC function, and induction of various immunosuppressive cells such as MDSC and Treg cells

12.5 Clinical Implications of the Immunosuppressive Mechanisms

As described above, multiple immunosuppressive cascades are triggered by gene and signal alternations in human cancer cells and generate immunosuppressive condition particularly in the tumor-associated microenvironments, including the tumor tissues and sentinel lymph nodes (Fig. 12.3). One of the important questions is which molecules and cells, either at upstream or downstream, in the immunosuppressive cascades should be inhibited for the efficient reversal of immunosuppression in patients with cancer. It may depend, at least in part, on cancer types and their genetic alterations.

In general, targeting constitutive active signaling molecules in cancer cells has advantage of direct anti-tumor effects such as inhibition of cancer cell proliferation and direct destruction of malignant cells, which may lead to induction of immune responses to multiple endogenous tumor antigens, including patients' unique antigens (e.g., mutated antigens); this may also lead to simultaneous inhibition of downstream multiple immunosuppressive mechanisms. However, inhibition of upstream molecules may also cause broad adverse effects, including suppression of anti-tumor immune response, although suppression of anti-tumor immune responses may be avoided by the use of appropriate doses of inhibitors. For instance, as was observed after administration of NF-κB inhibitor or mutated molecule selective inhibitors, such as mutant BRAF selective inhibitors. In contrast, targeting downstream molecules and cells, such as TGF-β, IL-10, PD-L1, IDO, Cox2 or MDSCs and Tregs, by using small molecule inhibitors or antibodies may have advantage of high specificity leading to more efficient blockade with less broad adverse effects. However, inhibition of one molecule or one cell type may not be sufficient to overall reversal of immunosuppression in patients with cancer. The combination of signal inhibitors and blockade of major immuosuppressive molecules or cells (e.g., neutralizing or blocking antibodies for TGF-β or PD-1) may also be attractive strategies for strengthening activity to reverse tumor-associated immunosuppression.

Besides inhibition/blocking of tumor-derived immunosuppressive factors, signal inhibition in immune cells may result in the direct activation of immune cells or inhibition of induction of immunosuppressive cells, including Tregs and MDSCs. Altogether, targeting activated signaling molecules involved in triggering of multiple immunosuppressive cascades may be an attractive strategy for reversal of immunosuppressive conditions in the tumor-associated microenvironments for cancer therapies, particularly immunotherapy (Fig. 12.4). Combination treatments utilizing these molecular targeted drugs and various immunotherapies, including cancer vaccines and check point blockade, are particularly appearing and will be evaluated in future clinical trials. One important point is that an appropriate target may be different among patients, since constitutively activated molecules and signaling pathways vary among patients even with the same type of tumor, indicating necessity of personalized strategy (Table 12.1). In the next 10 years, molecular and cellular basis of cancer-induced immunosuppression in the tumor-associated microenvironments will be further understood, and clinical efficacy of combined immunotherapy with molecular targeted drugs will be clinically evaluated.

In addition to the therapeutic implications of altered gene and signaling involved in cancer-induced immunosuppression, they may play a role in diagnostics. As described above, infiltration of memory $CD8^+$ T cells in tumor mass and serum IL-6/IL-8 levels appear to be prognostic markers and response prediction markers for cancer treatment including immunotherapy. Since the immune status may be a reflection of gene/signal alterations as described above, evaluation of the altered gene/signaling status (e.g., pERK, pSTAT, nuclear translocation of NF-κB, or β-catenin) may also serve as diagnostic biomarkers for cancer patients.

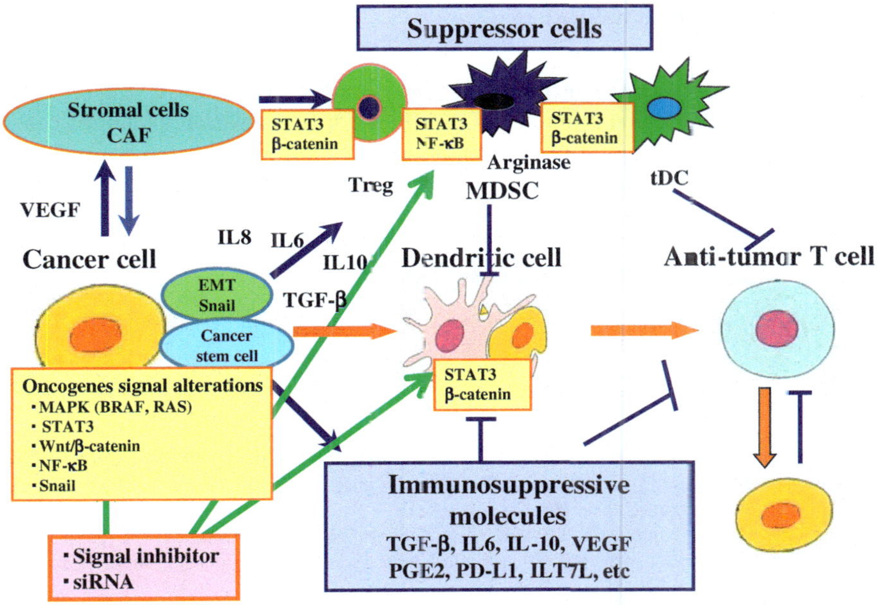

Fig. 12.4 Reversal of immunosuppressive conditions by targeting both cancer cells and immune cells using signal inhibitors. Cancer cell derived factors induce activation of signaling in various immune cells to become immunosuppressive cells. Signal inhibitors (inhibitors for STAT3, *β*-catenin, NF-*κ*B, etc.) may be useful for reversal of cancer induced immunosuppression by acting on both cancer cells and various immune cells such as DC, MDSC, and Treg cells

Table 12.1 Appropriate targets for reversal of cancer-induced immunosuppression may be different among cancer patients

Active signaling molecules	Immunosuppressive molecules	Cancer type
NRAS/BRAF/MAPK	IL-10, VEGF, IL-6	Melanoma
BRAF/MAPK	IL-10	Colon cancer
KRAS/MAPK	IL-8, VEGF	Pancreatic cancer
MAPK	PD-L1	Ovarian cancer
MAPK	VEGF, IL-8	Renal cell cancer
EGFR/MAPK	IL-6, VEGF	Lung cancer
PI3K/AKT	VEGF, IL-8	Renal cell cancer
PTEN/AKT	PD-L1	Glioma
STAT3	IL-10, IL-6, VEGF	Melanoma
β-catenin	IL-10	Melanoma
β-catenin	IL-10	Colon cancer
NF-*κ*B	IL-6, IL-8, CCL2	Ovarian cancer
NF-*κ*B	IL-6, IL-8, ILT7L	Renal cell cancer
Kinase-X	TGF-*β*, IL-10, CCL2	Melanoma

12.6 Concluding Remarks

Understating of molecular mechanisms of the immunopathological features of the tumor-associated microenvironments is critical for further development of cancer diagnostics and therapy; not only immunotherapy but also other types of cancer treatments including chemotherapy. In particular, combination therapy utilizing molecular targeted drugs, which are currently used as single agents, and immunotherapy, such as cancer vaccine and check point blockers, is a promising strategy to be exploited in near future clinical trials.

References

Boni A, Cogdill AP, Dang P, Udayakumar D, Njauw CN, Sloss CM et al (2010) Selective BRAFV600E inhibition enhances T-cell recognition of melanoma without affecting lymphocyte function. Cancer Res 70:5213–5219

Burdelya L, Catlett-Falcone R, Levitzki A, Cheng F, Mora LB, Sotomayor E et al (2002) Combination therapy with AG-490 and interleukin 12 achieves greater antitumor effects than either agent alone. Mol Cancer Ther 1:893–899

Chapman PB, Hauschild A, Robert C, Haanen JB, Ascierto P, Larkin J et al (2011) Improved survival with vemurafenib in melanoma with BRAF V600E mutation. N Engl J Med 364:2507–2516

Davies H, Bignell GR, Cox C, Stephens P, Edkins S, Clegg S et al (2002) Mutations of the BRAF gene in human cancer. Nature 417:949–954

Ding Y, Shen S, Lino AC, De Curotto Lafaille MA, Lafaille JJ (2008) Beta-catenin stabilization extends regulatory T cell survival and induces anergy in nonregulatory T cells. Nat Med 14:162–169

Flaherty KT, Robert C, Hersey P, Nathan P, Garbe C, Milhem M et al (2012) Improved survival with MEK inhibition in BRAF-mutated melanoma. N Engl J Med 367:107–114

Fridman WH, Galon J, Dieu-Nosjean MC, Cremer I, Fisson S, Damotte D et al (2011) Immune infiltration in human cancer: prognostic significance and disease control. Curr Top Microbiol Immunol 344:1–24

Fu C, Jiang A (2010) Generation of tolerogenic dendritic cells via the E-cadherin/β-catenin-signaling pathway. Immunol Res 46:72–78

Fuertes MB, Kacha AK, Kline J, Woo SR, Kranz DM, Murphy KM et al (2005) Host type I IFN signals are required for antitumor CD8+ T cell responses through CD8{alpha}+ dendritic cells. J Exp Med 208:2005–2016

Fujita T, Teramoto K, Ozaki Y, Hanaoka J, Tezuka N, Itoh Y et al (2009) Inhibition of transforming growth factor-beta-mediated immunosuppression in tumor-draining lymph nodes augments antitumor responses by various immunologic cell types. Cancer Res 69:5142–5150

Gajewski TF, Meng Y, Blank C, Brown I, Kacha A, Kline J, et al. (2006) Immune resistance orchestrated by the tumor microenvironment. Immunol Rev, 213:131–145

Gajewski TF, Fuertes M, Spaapen R, Zheng Y, Kline J (2011) Molecular profiling to identify relevant immune resistance mechanisms in the tumor microenvironment. Curr Opin Immunol 23:286–292

Gajewski TF, Fuertes MB, Woo SR (2012) Innate immune sensing of cancer: clues from an identified role for type I IFNs. Cancer Immunol Immunother 61:1343–1347

Galon J, Pagès F, Marincola FM, Angell HK, Thurin M, Lugli A et al (2012) Cancer classification using the Immunoscore: a worldwide task force. J Transl Med 10:205

Hanahan D, Weinberg RA (2011) Hallmarks of cancer: the next generation. Cell 144:646–674

Hauschild A, Grob JJ, Demidov LV, Jouary T, Gutzmer R, Millward M et al (2012) Dabrafenib in BRAF-mutated metastatic melanoma: a multicentre, open-label, phase 3 randomised controlled trial. Lancet 380:358–365

Hong DS, Vence L, Falchook G, Radvanyi LG, Liu C, Goodman V et al (2012) BRAF(V600) inhibitor GSK2118436 targeted inhibition of mutant BRAF in cancer patients does not impair overall immune competency. Clin Cancer Res 18:2326–2335

Iwata-Kajihara T, Sumimoto H, Kawamura N, Ueda R, Takahashi T, Mizuguchi H et al (2011) Enhanced cancer immunotherapy using STAT3-depleted dendritic cells with high Th1-inducing ability and resistance to cancer cell-derived inhibitory factors. J Immunol 187:27–36

Jalkanen SE, Vakkila J, Kreutzman A, Nieminen JK, Porkka K, Mustjoki S (2010) Poor cytokine-induced phosphorylation in chronic myeloid leukemia patients at diagnosis is effectively reversed by tyrosine kinase inhibitor therapy. Exp Hematol 39:102–113

Kawakami Y, Eliyahu S, Delgado CH, Robbins PF, Rivoltini L, Topalian SL et al (1994a) Cloning of the gene coding for a shared human melanoma antigen recognized by autologous T cells infiltrating into tumor. Proc Natl Acad Sci U S A 91:1519–3515

Kawakami Y, Eliyahu S, Delgado CH, Robbins PF, Sakaguchi K, Appella E et al (1994b) Identification of human melanoma antigen recognized by tumor infiltrating lymphocytes associated with in vivo tumor rejection. Proc Natl Acad Sci USA 91:6458–6462

Kawakami Y, Fujita T, Matsuzaki Y, Sakurai T, Tsukamoto M, Toda M et al (2004) Identification of human tumor antigens and its implications for diagnosis and treatment of cancer. Cancer Sci 95:784–791

Kim R, Emi M, Tanabe K, Arihiro K (2006) Immunobiology of the sentinel lymph node and its potential role for antitumour immunity. Lancet Oncol 7:1006–1016

Ko JS, Zea AH, Rini BI, Ireland JL, Elson P, Cohen P et al (2009) Sunitinib mediates reversal of myeloid-derived suppressor cell accumulation in renal cell carcinoma patients. Clin Cancer Res 15:2148–2157

Kono M, Dunn IS, Durda PJ, Butera D, Rose LB, Haggerty TJ et al (2006) Role of the mitogen-activated protein kinase signaling pathway in the regulation of human melanocytic antigen expression. Mol Cancer Res 4:779–792

Kortylewski M, Kujawski M, Wang T, Wei S, Zhang S, Pilon-Thomas S et al (2005) Inhibiting Stat3 signaling in the hematopoietic system elicits multicomponent antitumor immunity. Nat Med 11:1314–1321

Kudo-Saito C, Shirako H, Takeuchi T, Kawakami Y (2009) Cancer metastasis is accelerated through immunosuppression during snail-induced EMT of cancer cells. Cancer Cell 15:195–206

Kujawski M, Zhang C, Herrmann A, Reckamp K, Scuto A, Jensen M et al (2010) Targeting STAT3 in adoptively transferred T cells promotes their in vivo expansion and antitumor effects. Cancer Res 70:9599–9610

Lee H, Pal SK, Reckamp K, Figlin RA, Yu H (2011) STAT3: a target to enhance antitumor immune response. Curr Top Microbiol Immunol 344:41–59

Manicassamy S, Reizis B, Ravindran R, Nakaya H, Salazar-Gonzalez RM, Wang JC et al (2010) Activation of β-catenin in dendritic cells regulates immunity versus tolerance in the intestine. Science 329:849–853

Mlecnik B, Tosolini M, Kirilovsky A, Berger A, Bindea G, Meatchi T et al (2011) Histopathologic-based prognostic factors of colorectal cancers are associated with the state of the local immune reaction. J Clin Oncol 29:610–618

Murakami M, Hirano T (2011) A four-step model for the IL-6 amplifier, a regulator of chronic inflammations in tissue-specific MHC class II-associated autoimmune diseases. Front Immunol 2:22

Mustjoki S, Ekblom M, Arstila TP, Dybedal I, Epling-Burnette PK, Guilhot F et al (2009) Clonal expansion of T/NK-cells during tyrosine kinase inhibitor dasatinib therapy. Leukemia 23:1398–1405

Ozao-Choy J, Ma G, Kao J, Wang GX, Meseck M, Sung M et al (2009) The novel role of tyrosine kinase inhibitor in the reversal of immune suppression and modulation of tumor microenvironment for immune-based cancer therapies. Cancer Res 69:2514–2522

Pallandre JR, Brillard E, Créhange G, Radlovic A, Remy-Martin JP, Saas P et al (2007) Role of STAT3 in CD4+ CD25+ FOXP3+ regulatory lymphocyte generation: implications in graft-versus-host disease and antitumor immunity. JImmunol 179:7593–7604

Parsa AT, Waldron JS, Panner A, Crane CA, Parney IF, Barry JJ et al (2007) Loss of tumor suppressor PTEN function increases B7–H1 expression and immunoresistance in glioma. Nat Med 13:84–88

Poschke I, Mougiakakos D, Hansson J, Masucci GV, Kiessling R (2010) Immature immunosuppressive CD14+ HLA-DR-/low cells in melanoma patients are Stat3hi and overexpress CD80, CD83, and DC-sign. Cancer Res 70:4335–4345

Rosenberg SA, Yang J, Schwartzentruber D, Hwu P, Marincola F, Topalian S et al (1998) Immunologic and therapeutic evaluation of a synthetic peptide vaccine for the treatment of patients with metastatic melanoma. Nat Med 4:321–327

Rosenberg SA, Yang JC, Sherry RM, Kammula US, Hughes MS, Phan GQ et al (2011) Durable complete responses in heavily pretreated patients with metastatic melanoma using T-cell transfer immunotherapy. Clin Cancer Res 17:4550–4557

Schreiber RD, Old LJ, Smyth MJ (2011) Cancer immunoediting: integrating immunity's roles in cancer suppression and promotion. Science 331:1565–1570

Schwartzentruber DJ, Lawson DH, Richards JM, Conry RM, Miller DM, Treisman J et al (2011) gp100 peptide vaccine and interleukin-2 in patients with advanced melanoma. N Eng l J Med 364:2119–2127

Sumimoto H, Miyagishi M, Miyoshi H, Yamagata S, Shimizu A, Taira K et al (2004) Inhibition of growth and invasive ability of melanoma by inactivation of mutated BRAF with lentivirus-mediated RNA interference. Oncogene 23:6031–6039

Sumimoto H, Imabayashi F, Iwata T, Kawakami Y (2006) The BRAF-MAPK signaling pathway is essential for cancer-immune evasion in human melanoma cells. J Exp Med 203:1651–1656

Swartz MA, Lund AW (2012) Lymphatic and interstitial flow in the tumour microenvironment: linking mechanobiology with immunity. Nat Rev Cancer 12:210–219

Taube JM, Anders RA, Young GD, Xu H, Sharma R, McMiller TL et al (2012) Colocalization of inflammatory response with B7–h1 expression in human melanocytic lesions supports an adaptive resistance mechanism of immune escape. Sci Transl Med 4:127

Tsujikawa T, Yaguchi T, Ohmura G, Ohta S, Kobayashi A, Kawamura N, et al. (2012) Autocrine and paracrine loops between cancer cells and macrophages promote lymph node metastasis via CCR4/CCL22 in head and neck squamous cell carcinoma, Int J Cancer, in press

Tsukamoto N, Okada S, Onami Y, Sasaki Y, Umezawa K, Kawakami Y (2009) Impairment of plasmacytoid dendritic cells for IFN production by the ligand for immunoglobulin-like transcript 7 expressed on human cancer cells. Clin Cancer Res 15:5733–5743

Wilmott JS, Long GV, Howle JR, Haydu LE, Sharma RN, Thompson JF et al (2011) Selective BRAF inhibitors induce marked T-cell infiltration into human metastatic melanoma. Clin Cancer Res 18:1386–1394

Wu L, Du H, Li Y, Qu P, Yan C (2011) Signal transducer and activator of transcription 3 (Stat3C) promotes myeloid-derived suppressor cell expansion and immune suppression during lung tumorigenesis. Am J Pathol 179:2131–2141

Xin H, Zhang C, Herrmann A, Du Y, Figlin R, Yu H (2009) Sunitinib inhibition of Stat3 induces renal cell carcinoma tumor cell apoptosis and reduces immunosuppressive cells. Cancer Res 69:2506–2513

Yaguchi T, Sumimoto H, Kudo-Saito C, Tsukamoto N, Ueda R, Iwata-Kajihara T et al (2011) The mechanisms of cancer immunoescape and development of overcoming strategies. Int J Hematol 93:294–300

Yaguchi T, Goto Y, Kido K, Mochimaru H, Sakurai T, Tsukamoto N, Kudo-Saito C, Fujita T, Sumimoto H, Kawakami Y (2012) Immune suppression and resistance mediated by constitutive activation of Wnt/β-catenin signaling in human melanoma cells. J Immunol 189:2110–2117

Yu H, Pardoll D, Jove R (2009) STATs in cancer inflammation and immunity: a leading role for STAT3. Nat Rev Cancer 9:798–809

Yu H, Kortylewski M, Pardoll D (2007) Crosstalk between cancer and immune cells: role of STAT3 in the tumour microenvironment. Nat Rev Immunol 7:41–51

Zou W (2005) Immunosuppressive networks in the tumour environment and their therapeutic relevance. Nat Rev Cancer 5:263–274

Chapter 13
T Cell Mulfunction in the Tumor Environment

Eitan Yefenof

Abstract The cancer microenvironment may be envisaged as a battlefield of a fight between the host's immune system and the tumor. An army of T lymphocytes, which are trained to combat bacterial and viral infections, is mobilized to attack the tumor, but most of the times succumbs to its powers and lose the war. A major reason for this failure is the unique composition of the cancer microenvironment, which is shaped by the tumor in favor of its progression. The immunosuppressive environment of the tumor makes it a hard place for T cells to exert their otherwise powerful effector functions. The study of the cancer microenvironment yields important insights into the nature of the tumor protective shields surrounding the tumor. Recent studies bring new prospects for intervention therapies based on recruitment and activation of T cells at the cancer microenvironment.

Keywords T cells · Cancer · Tumor microenvironment · Immunosuppression · Cytotoxic lymphocytes · Regulatory T cells

13.1 Introduction

The consolidation of the cancer immune surveillance theory in the 1960s (Burnet 1957, 1971; Thomas 1959) coincided with the discovery of T lymphocytes and their pivotal role in the generation of the immune response (Miller 1961). T cytotoxic (Tc) lymphocytes that can effectively reject allografts were identified

E. Yefenof (✉)
The Lautenberg Center for Immunology and Cancer Research, The Hebrew University-Hadassah Medical School, P.O. Box 12272 91120 Jerusalem, Israel
e-mail: eitany@ekmd.huji.ac.il

M. R. Shurin et al. (eds.), *The Tumor Immunoenvironment*,
DOI: 10.1007/978-94-007-6217-6_13,

shortly thereafter (Cerottini et al. 1970; Freedman et al. 1972). These findings evoked much enthusiasm and hope that manipulating anti tumor T cell responses would create revolutionary approaches in the fight against cancer. This view was, in retrospect, naïve and unrealistic. While the management of certain cancers (e.g., melanoma) has been benefited from immunotherapies that use activated T cells, the majority of tumors did not respond to these kind of treatments (Klein and Klein 1977; Smyth et al. 2001). In the present article I shall try to provide an insight into the question why T cells, strong and mighty in combating bacterial and viral infections, fail to exert their powerful effector capabilities when it comes to cancer. Answers to this question are crucial for the future development of effective treatment modalities based on manipulation of anti tumor T cell responses.

13.2 T Cell Receptor Specificity and Avidity

The most important feature ascribed to T cell as potential mediators of tumor rejection is their ability to distinguish between normal (self) and cancerous (non-self) cells. Indeed, numerous investigations using models of animals transplanted with experimental tumors established the concept that specific T cell immunity can be generated against many types of cancers. In many of these cases, the failure of cancer specific T cells to reject the tumor has been attributed to the low avidity of their specific TCRs toward tumor-specific antigens (TSA) or tumor associated antigens (TAA). Indeed, such T cells could be manipulated in vitro or in vivo to express higher avidity TCRs and to bring about tumor rejection (Liddy et al. 2012). These studies have been successfully translated for clinical management of certain human cancers, most notably melanoma and virally induced malignancies, in which the expression TSA or TAA recognized by T cells were detected (Andersen et al. 2012; Bollard et al. 2006). However, for most other tumors, the existence of *bona fide* Tumor specific antigens has not been verified (Klein and Klein 1977).

Lack of antigen specificity does not necessarily indicate that T cell immunity plays no beneficial rule in the host's response to the emerging tumor. Several studies have demonstrated that infiltration of T cells to the site of a tumor correlates with better prognosis and longer survival of cancer patients (Zhang et al. 2003; Fridman et al. 2011). Hence, the tumor environment shapes a population of tumor infiltrating lymphocytes (TIL) that recognize the abnormal tissue and manifest immunological activities in defense of the host (Cho et al. 2003; Fridman et al. 2012). Moreover, TIL could be propagated and manipulated to become effective therapeutic tools in certain cancers (Rosenberg 2012), thus demonstrating their potential as anti cancer effectors even if lacking well defined specificity towards TSA or TAA.

The question ensuing is why T cells infiltrating a tumor do not defeat the cancerous cells as they do effectively when directed against pathogens? Apparently, a tumor or a metastatic lesion creates a microenvironment that, on one hand, attracts immune cells to the tumor site but, on the other hand, compromises their anti tumor activity. In the following sections I shall discuss some of the factors operating within the cancer microenvironment that imperil the activity of T cells against the tumor.

13.3 T Cell Activity

Nearly two decades ago it has been discovered that tumor infiltrating T cells (and NK cells) down-regulate their TCR-zeta chain expression, thus leading to severe defects in their TCR signaling function (Nakagomi et al. 1993). Additional studies indicated that the expression of other signaling proteins, such as LCK, Zap-70 and FYN, is also affected in T cells recovered from the tumor microenvironment (Chiou et al. 2005; Reichert et al. 2002; Lai et al. 1996). Initially it was thought that tumor cells counter-attack T cells by secreting factor(s) that modulate zeta chain expression (Reichert et al. 1998; Taylor et al. 2001). However, later studies demonstrated the same phenomenon at sites of chronic inflammation caused by autoimmune or infectious diseases (Baniyash 2004; Bronstein-Sitton et al. 2003). It seems, thus, that chronic stimulation of immune cells creates a suppressive microenvironment that attenuates the activity of T cells at the tumor vicinity by multiple modes, one of which is zeta chain down-regulation.

Another mode by which micro-environmental T cells fail to exert anti tumor activity is their dependency on co-stimulatory cells and molecules that function sub-optimally at the tumor site (Ochsenbein et al. 1999). CD4+ T cells that fail to secrete sufficient levels of IL2, IFN-γ and other cytokines do not provide stimulatory signals required to sustain cytotoxic activity of CTLs (Valitutti et al. 1996). Likewise, reduced expression of accessory molecules decreases the level of T cell activation and anti tumor activity at the tumor microenvironment. LFA and CD2 are required to establish cellular contact between CTLs and their target cells, while CD28 is a co-stimulatory receptor without which CTLs do not exert their full cytotoxic activities (Valitutti et al. 1996). Both are down-regulated by tumor cells at the microenvironment, thus diminishing their sensitivity to T cell attack (Beal et al. 2008).

Cancer cells exhibit altered metabolic functions that affect the activity of T cells at the tumor microenvironment. Enhanced glycolysis results with depletion of glucose and accumulation of lactic acid, both reducing the activity of CTLs (Fischer et al. 2007). Another hallmark of cancer cells is the constitutive expression of indolamine-2,3-dioxigenase (IDO) (Uyttenhove et al. 2003). IDO catalyzes degradation of the essential amino acid tryptophan into kynurenine, resulting with tryptophan deprived tumor microenvironment. Tryptophan deprivation blocks differentiation of effector T cells and induce their apoptotic death (Fallarino et al. 2006; Lee et al. 2002), thus protecting the tumor against immune attack.

Arginine depletion has been also highlighted as a factor that suppressed anti tumor T cell responses by way of eukaryotic translation initiation factor 2 (EIF2) inhibition (Bronte and Zanovello 2005). Indeed, many tumors express high levels of arginase, thus creating an argenine deficient microenvironment at which T cell activation is compromised (Cederbaum et al. 2004). Another metabolic alteration in cancer cells is the activation of iNOS and production of Nitric Oxide (NO). One deleterious activity of excessive NO at the tumor microenvironment is inhibition of the MAPK pathway, leading to a suppressed production of IL2 and granzyme, both essential for anti-tumor T cell activity (Blesson et al. 2002).

Huang et al. (1997) reported that hypoxia generates high concentrations of adenosine at the tumor site by enhancing ATP metabolism. This, in turn, leads to T cell immuno-suppresion since adenosine binding to its A2 receptors elevates intracellular cAMP which inhibits T cell activation and expansion (Ohta et al. 2006).

Clearly, the unique metabolic activities of cancer cells provide them with intrinsic growth advantage properties that are superior to those of normal counterparts. At the same time, however, they impose an immunosuppressive environment at the tumor site which attenuates potential anti cancer activity of T cells in and around the tumor (Singer et al. 2011).

13.4 Immune Exhaustion

T cell exhaustion results from a chronic exposure of activated T cells to high load of antigen, leading to a gradual loss of their function (Zajac et al. 1998). The main reason for exhaustion of CD8+ effector T cells is the up-regulation of inhibitory receptor expression, such as PD-1, 2B4, CTLA-4 and LAG-3 (Blackburn et al. 2009). While this phenomenon has been initially described in chronic viral infection, T cells at the cancer microenvironment may undergo exhaustion since they are chronically stimulated by the tumor. Indeed, Ahmadzadeh et al. (2009) found that T cells infiltrating a melanoma tumor over-express the inhibitory receptor PD-1 (CD279), and that they become dysfunctional when interacting with PD-1 ligands expressed on melanoma cells. The T cell immunoglobulin and mucin-domain–containing molecule 3 (Tim-3), is another inhibitory receptor up-regulated on melanoma infiltrating T cells and contributing to their dysfunction as anti tumor effectors (Fourcade et al. 2010).

CTLA-4 is a prototype inhibitory receptor of T cells which shares the same ligand, CD28, with the co-stimulatory receptor B7 (Greenwald et al. 2005). Its up-regulation on tumor infiltrating T cells is a major reason for their dysfunction (Leach et al. 1996), since inhibition via its interaction with CD28 dominates co-stimulation via B7 by the same ligand (Carreno and Collins 2002). A proof of principle for this concept was reinforced by studies in which the CTLA4-CD28 interaction has been specifically inhibited by anti CTLA4 antibodies, a treatment that restored their anti tumor activity (Peggs et al. 2006). These observations became a platform for the development of a novel anti cancer therapeutic modality based on blockage of CTLA4 by a humanized anti CTLA4 MoAb (ipilimumab), which holds promise for the treatment of prostate and some other carcinomas (Kwek 2012).

While T cell exhaustion is typically mediated by inhibitory ligands expressed on tumor cells, within the cancer microenvironment it may result from chronic encounter with stromal cells. This is exemplified by the inhibitory interaction between PD1 of T cells with its ligand PD-L1 (CD273) expressed on dendritic cells (DC) populating the tumor site (Currie et al. 2009). Another study have demonstrated that myeloid DC in the tumor environment up-regulate the expression of B7-H1, which is an inhibitory ligand for PD1 expressed on T cells (Curiel et al. 2003). Hence, a tumor can shape microenvironmental T and stromal cells to interact with each other in a manner that will enable immune evasion of the former by way of immune exhaustion.

13.5 Inhibitory T Cell Subsets

$CD4^+CD25^+Foxp3^+$ regulatory T cells (Treg) have emerged as a distinct subset of T lymphocytes that suppress humoral and cellular immune responses (Hori et al. 2003; Walker et al. 2003). Ample evidence exists showing that Treg are suppressing anti tumor immunity, correlated with prognosis and clinical response to therapy (Jones et al. 2003; Lehe et al. 2008; Zhou et al. 2006; Badoual et al. 2009).

Two types of Treg cells have been identified: 1. "Natural" Treg (nTreg) cells that develop in the thymus during the process of T cell selection and maturation (Fontenot et al. 2005), and 2. "Induced" Treg (iTreg) cells that differentiate from peripheral CD4+ T cells in response to antigen and TGFbeta (Lathrop et al. 2008). The tumor microenvironment facilitates differentiation of iTreg cells because it is abundant with myeloid derived dendritic cells that secrete TGF-β (Ghiringhelli et al. 2005). nTreg are attracted to the tumor site by chemokines secreted from tumor or myeloid cells, such as CCL22 and CCL28 (Gobert et al. 2009). The hypoxic microenvironment induces secretion of CCL28 from tumor cells, thus directing nTreg cells to the tumor site.

nTreg and iTreg share several properties that enable them to suppress local and systemic immune responses. Both secrete immunosuppressive cytokines (TGF-β, IL-10, IL-35) and metabolites (adenosine) that act directly on Th and Tc to modulate their effector function (Vignali et al. 2008; Deaglio et al. 2007). They can also lyse effector T cells by secreting granzyme (Boissonnas et al. 2010) or inhibit their activity via direct cellular contact (Wing et al. 2008). A recent study reported that Treg cells recruited to an hypoxic cancer microenvironment enhance tumor growth by secreting VEGF and promoting angiogenesis (Facciabene et al. 2011). Hence, Treg cells seem to dominate the T cell population at the tumor environment both by number and activity. This is a major impediment to T cell based therapy which, indeed, can be greatly improved by depletion or blockage of Treg cells (Shimizu et al. 1999; Chen et al. 2010; Rech and Vonderheide 2009).

Another type of T cells differentiating from peripheral CD4+ T lymphocytes is the recently discovered Th17 cell subset (Harrington et al. 2005; Park et al. 2005). Like iTreg cells, differentiation of Th17 cells requires TGF-β (Mangan et al.

2006), but also IL-6 and IL-23 (Veldhoen et al. 2006; Chizzolini et al. 2008). Since these cytokines are abundant in the tumor microenvironment, Th17 may be differentiated in situ from infiltrating CD4+ T cells, or be attracted from the periphery by chemokines produced at the tumor site (Kryczek et al. 2007). A unique characteristic of Th17 cells is their ability to secrete IL-17A and IL-17F (McAllister et al. 2005). These are pro-inflammatory cytokines playing an important rule in inflammation and autoimmunity (Kolls and Linden 2004; Sutton et al. 2006). It was therefore anticipated that Th17 cells in the cancer microenvironment will promote, rather than inhibit, tumor progression. Indeed, a conspicuous function of IL-17 is to induce angiogenesis by stimulating migration and blood vessel formation of vascular endothelial cells (Takahashi et al. 2005). This activity has been shown to enhance tumor cell growth in vitro and in vivo (Numasaki et al. 2003). Along this line, several studies reported that Th17 cells residing within a tumor or circulating in the periphery, exacerbate the pathology of the malignant disease (Numasaki et al. 2005; Zhang et al. 2008; Wu et al. 2009; Miyahara et al. 2008). This outcome, when studied, has been mostly attributed to the neovascularization induced by the pro-angiogenic IL-17A and IL-17F (Middleton et al. 2012). However, IL-17 can exert its pro-malignant activity by affecting other cytokines or immune cells at the tumor microenvironment. Of a note is the study by (Charles et al. 2009), who showed that IL-17 can synergize with TNF-α in recruiting myeloid suppressor to the tumor site. Ye et al. found that tumor infiltrating Th17 cells can differentiate into FOXP3 positive Treg cells possessing strong immunosuppressive activity. Likewise, IL-2 produced at the tumor microenvironment can induce differentiation of Treg cells into Th17 cells (Leveque et al. 2009).

Despite the strong evidence that Th17 cells promote tumor progression, predominantly by their pro-angiogenic activity, several studies reported anti tumor effects mediated by Th17. Benchetrit et al. (2002) found that IL-17 inhibits tumor cell growth in a T cell dependent manner. Transfection of IL-17 into fibrosarcoma cells induced T cell immunity due to up-regulation of MHC class I and II expression (Hirahara et al. 2001). Benatar et al. (2008; Benatar et al. 2010) discerned a novel family of IL-17, denoted IL-17E, which contrary to IL-17A, possess anti tumor activity in several experimental models. Martin-Orozco et al. (2009) demonstrated that Th17 cells facilitate activation of anti tumor CD8+ T cells by recruiting DC to the tumor microenvironment. Hence, the net effect of Th17 on cancer promotion and progression is ill defined and requires further research before it may be translated to a clinically meaningful approach.

13.6 Suppressor Cells of the Myeloid Lineage

Tumor Associated Macrophages (TAM) play a significant role in the development of cancer. They are involved in several aspects of the tumor microenvironment, including cancer initiation, progression, metastases, angiogenesis, tissue remodeling and immuno-suppression (Nardin and Abastado 2008). Given the nature of the present article, reference will be made to the latter function only. In general, Macrophages can be polarized to M1 and M2, in analogy to Th1 and Th2 polarization (Mantovani et al. 2002). When TAM are identified as M1, they act to inhibit tumor growth by secreting pro-inflammatory cytokines (IFN-γ, IL-1 and IL-6) that activate anti tumor Th1 responses (Ong et al. 2012). However, in most cancers TAM are characterized as anti inflammatory M2 cells which promote tumor progression (Movahedi 2010). The reason for that is the common cytokine profile of the tumor microenvironment and local hypoxia which favor recruitment of monocytes and their differentiation into M2 macrophages Biswas 2010. Such TAM preferentially express CCL17, CCL22 and CCL24, chemokines that help Treg cell recruitment and development (Movahedi et al. 2010). Production of immunosuppressive cytokines such as TGF-β and IL-10 is also a feature of TAM, thus inhibiting both anti tumor CD4 and CD8 T cells at the cancer environment (Mantovani et al. 2004).

Another non-lymphoid population with immunosuppressive activity at the tumor environment is DC with plasmacytoid appearance (Watkins et al. 2011). The role of these cells in antitumor immunity has been well studied in TRAMP mice that develop spontaneous adenocarcinoma of the prostate (Shafer-Weaver et al. 2009). The tumor-associated DC of TRAMP mice has a unique phenotype characteristic of DC with tolerogenic activity on tumor specific CD8+ T cells (Watkins and Hurwitz 2012). They express genes that are classically associated with immuno-suppression such as PD-L1, arginase, indoleamine-2,3-dioxygenase (IDO) and FOXO3 (Hurwitz 2012). These cells could suppress the cytotoxic activity of CTLs both in vitro and in vivo, whereas silencing of FOXO3 by specific siRNA restored the generation of tumor specific T cell immunity (Watkins et al. 2011). Thus, FOXO3 blockage may be useful in future endeavors to develop targeted immunotherapies for cancer.

The most extensively studied immunosuppressive cells at the tumor microenvironment are Myeloid Derived Suppressor Cells (MDSC). They were discovered 25 years ago (Young et al. 1987), but only in recent years their decisive role in regulating immune responses against cancer has been appreciated. MDSC comprise a heterogeneous population of myeloid progenitors and immature myeloid cells that under normal circumstances would differentiate into mature macrophages, granulocytes and DC. However, in certain lympho-proliferative pathologies such as chronic inflammation, autoimmunity and cancer, their differentiation is blocked resulting with MDSC expansion (Kanterman et al. 2012). Moreover, such immature cells acquire immunosuppressive properties, hence their designated acronym.

MDSC are attracted to the cancer microenvironment by pro-inflammatory factors secreted from the tumor and activated immune cells (Sinha et al. 2008). In fact, some of these factors, such as IL1beta and IL6, also activate MDSC to exert their immunosuppressive function (Bunt et al. 2006; Song et al. 2005). MDSC are powerful enhancers of tumor growth because they are engaged in multiple immunosuppressive activities. They suppress both innate (NK) and adaptive (T) immune responses (Ostrand-Rosenberg and Sinha 2009b), affect CD4+ and CD8+ T cells equally well (Ostrand-Rosenberg 2009a), and do so by both MHC restricted and unrestricted manner (Nagaraj et al. 2007; Sinha et al.

2005). A major mechanism of suppression by MDSC is the depletion of L-arginine at the tumor environment due to high rate consumption of that particular amino acid, which is required for optimal T cell activation (Rodriguez and Ochoa 2008). T cell deprived of L-arginine down regulate TCR-zeta chain expression and are growth arrested at G_0/G_1, thus becoming refractory to specific or non specific stimulation signals (Ezernitchi et al. 2006; Rodriguez et al. 2002, 2007). Another suppressive activity of MDSC is elevation of iNOS expression and production of NO which induces apoptosis of T cells (Rivoltini et al. 2002).

MDSC also inhibit indirectly T cell activation and effect by inducing differentiation of Treg cells (Huang et al. 2006; Serafini et al. 2008). Hence, MDSC are recognized today as a major obstacle in generating anti tumor immunity and immunotherapy. More studies are required to advance new strategies for clinical interventions based on attenuation of MDSC immunosuppressive activity.

13.7 Conclusions

The cancer microenvironment is a battlefield at which cells of the immune system and the tumor fight with each other in a zero-sum game (Sullivan 2011). Since cancer cells are, after all, self cells of the host that went astray, they are relatively poor in providing own resources that would directly strike back against a specific and targeted immune attack. This notion is exemplified by the rise and fall of the CD95/FAS-CD95/FASL counter-attack theory, which empowered cancer cells with a potent function that allegedly destroys cytotoxic T cells before the latter are exerting their anti tumor killing activity (Debatin and Krammer 2004; Igney and Krammer 2005). Instead, cancer cells are harnessing immuno-suppressive mechanisms that operate for the benefit of the host in other pathologies (e.g., chronic inflammation and autoimmunity), to tolerize its microenvironment in service of the tumor. T cells are the first to be affected and their malfunction is accountable for the failure of the host to reject the tumor. Consequently, the war usually ends with the surrender of the host to the destructive powers of the tumor. However, as in a conventional war, better knowledge of the battlefield should enable the development of creative intervention modalities that will change the outcome of the host-cancer encounter. When it comes to T cells, there is much hope that their malfunction could be reversed by protecting them from the immunosuppressive factors mobilized by the tumor and dominating its microenvironment.

In the present article I summarized current knowledge of elements accounting for the maintenance of an immunosuppressive cancer environment that prevent T cells from exerting their inherent protective function. Each of these elements is now under the lime light of basic and translational investigations that will verify if and how they could be manipulated to enable better utilization of T cells, weather endogenous or ex vivo manipulated, in the war against cancer. I am optimistic that this war will be eventually won.

Acknowledgments The author would like to acknowledge support from Concern Foundation, Los Angeles, the German-Israel Foundation (GIF) and the Jewish Founders Network

References

Ahmadzadeh M, Johnson LA, Heemskerk B et al (2009) Tumor antigen-specific CD8 T cells infiltrating the tumor express high levels of PD-1 and are functionally impaired. Blood 114:1537–1544

Andersen RS, Thrue CA, Junker N et al (2012) Dissection of T-cell antigen specificity in human melanoma. Cancer Res 72:1642–1650

Badoual C, Hans S, Fridman WH, Brasnu D, Erdman S, Tartour E (2009) Revisiting the prognostic value of regulatory T cells in patients with cancer. J Clin Oncol 27:e5–6; author reply e7

Baniyash M (2004) TCR zeta-chain downregulation: curtailing an excessive inflammatory immune response. Nat Rev Immunol 4:675–687

Beal AM, Anikeeva N, Varma R et al (2008) Protein kinase C theta regulates stability of the peripheral adhesion ring junction and contributes to the sensitivity of target cell lysis by CTL. J Immunol 181:4815–4824

Benatar T, Cao MY, Lee Y et al (2008) Virulizin induces production of IL-17E to enhance antitumor activity by recruitment of eosinophils into tumors. Cancer Immunol Immunother 57:1757–1769

Benatar T, Cao MY, Lee Y et al (2010) IL-17E, a proinflammatory cytokine, has antitumor efficacy against several tumor types in vivo. Cancer Immunol Immunother 59:805–817

Benchetrit F, Ciree A, Vives V et al (2002) Interleukin-17 inhibits tumor cell growth by means of a T-cell-dependent mechanism. Blood 99:2114–2121

Biswas SK, Mantovani A (2010) Macrophage plasticity and interaction with lymphocyte subsets: cancer as a paradigm. Nat Immunol 11:889–896

Blackburn SD, Shin H, Haining WN et al (2009) Coregulation of CD8+ T cell exhaustion by multiple inhibitory receptors during chronic viral infection. Nat Immunol 10:29–37

Blesson S, Thiery J, Gaudin C et al (2002) Analysis of the mechanisms of human cytotoxic T lymphocyte response inhibition by NO. Int Immunol 14:1169–1178

Boissonnas A, Scholer-Dahirel A, Simon-Blancal V (2010) Foxp3+ T cells induce perforin-dependent dendritic cell death in tumor-draining lymph nodes. Immunity 32:266–278

Bollard CM, Huls MH, Buza E et al (2006) Administration of latent membrane protein 2-specific cytotoxic T lymphocytes to patients with relapsed Epstein-Barr virus-positive lymphoma. Clin Lymphoma Myeloma 6:342–347

Bronstein-Sitton N, Cohen-Daniel L, Vaknin I et al (2003) Sustained exposure to bacterial antigen induces interferon-gamma-dependent T cell receptor zeta down-regulation and impaired T cell function. Nat Immunol 4:957–964

Bronte V, Zanovello P (2005) Regulation of immune responses by L-arginine metabolism. Nat Rev Immunol 5:641–654

Bunt SK, Sinha P, Clements VK, Leips J, Ostrand-Rosenberg S (2006) Inflammation induces myeloid-derived suppressor cells that facilitate tumor progression. J Immunol 176:284–290

Burnet M (1957) Cancer: a biological approach. III. Viruses associated with neoplastic conditions. IV. Practical applications. Br Med J 1:841–847

Burnet FM (1971) Immunological surveillance in neoplasia. Transplant Rev 7:3–25

Carreno BM, Collins M (2002) The B7 family of ligands and its receptors: new pathways for costimulation and inhibition of immune responses. Ann Rev Immunol 20:29–53

Cederbaum SD, Yu H, Grody WW, Kern RM, Yoo P, Iyer RK (2004) Arginases I and II: do their functions overlap? Mol Genet Metab 81(Suppl 1):S38–S44

Cerottini JC, Nordin AA, Brunner KT (1970) Specific in vitro cytotoxicity of thymus-derived lymphocytes sensitized to alloantigens. Nature 228:1308–1309

Charles KA, Kulbe H, Soper R et al (2009) The tumor-promoting actions of TNF-alpha involve TNFR1 and IL-17 in ovarian cancer in mice and humans. J Clin Invest 119:3011–3023

Chen CA, Ho CM, Chang MC et al (2010) Metronomic chemotherapy enhances antitumor effects of cancer vaccine by depleting regulatory T lymphocytes and inhibiting tumor angiogenesis. Mol Ther 18:1233–1243

Chiou SH, Sheu BC, Chang WC, Huang SC, Hong-Nerng H (2005) Current concepts of tumor-infiltrating lymphocytes in human malignancies. J Reprod Immunol 67:35–50

Chizzolini C, Chicheportiche R, Alvarez M et al (2008) Prostaglandin E2 synergistically with interleukin-23 favors human Th17 expansion. Blood 112:3696–3703

Cho Y, Miyamoto M, Kato K et al (2003) CD4+ and CD8+ T cells cooperate to improve prognosis of patients with esophageal squamous cell carcinoma. Cancer Res 63:1555–1559

Curiel TJ, Wei S, Dong H et al (2003) Blockade of B7–H1 improves myeloid dendritic cell-mediated antitumor immunity. Nat Med 9:562–567

Currie AJ, Prosser A, McDonnell A et al (2009) Dual control of antitumor CD8 T cells through the programmed death-1/programmed death-ligand 1 pathway and immunosuppressive CD4 T cells: regulation and counterregulation. J Immunol 183:7898–7908

Deaglio S, Dwyer KM, Gao W et al (2007) Adenosine generation catalyzed by CD39 and CD73 expressed on regulatory T cells mediates immune suppression. J Exp Med 204:1257–1265

Debatin KM, Krammer PH (2004) Death receptors in chemotherapy and cancer. Oncogene 23:2950–2966

Ezernitchi AV, Vaknin I, Cohen-Daniel L et al (2006) TCR zeta down-regulation under chronic inflammation is mediated by myeloid suppressor cells differentially distributed between various lymphatic organs. J Immunol 177:4763–4772

Facciabene A, Peng X, Hagemann IS et al (2011) Tumour hypoxia promotes tolerance and angiogenesis via CCL28 and T(reg) cells. Nature 475:226–230

Fallarino F, Grohmann U, You S et al (2006) The combined effects of tryptophan starvation and tryptophan catabolites down-regulate T cell receptor zeta-chain and induce a regulatory phenotype in naive T cells. J Immunol 176:6752–6761

Fischer K, Hoffmann P, Voelkl S et al (2007) Inhibitory effect of tumor cell-derived lactic acid on human T cells. Blood 109:3812–3819

Fontenot JD, Dooley JL, Farr AG, Rudensky AY (2005) Developmental regulation of Foxp3 expression during ontogeny. J Exp Med 202:901–906

Fourcade J, Sun Z, Benallaoua M et al (2010) Upregulation of Tim-3 and PD-1 expression is associated with tumor antigen-specific CD8+ T cell dysfunction in melanoma patients. J Exp Med 207:2175–2186

Freedman LR, Cerottini JC, Brunner KT (1972) In vivo studies of the role of cytotoxic T cells in tumor allograft immunity. J Immunol 109:1371–1378

Fridman WH, Galon J, Pages F, Tartour E, Sautes-Fridman C, Kroemer G (2011) Prognostic and predictive impact of intra- and peritumoral immune infiltrates. Cancer Res 71:5601–5605

Fridman WH, Pages F, Sautes-Fridman C, Galon J (2012) The immune contexture in human tumours: impact on clinical outcome. Nat Rev Cancer 12:298–306

Ghiringhelli F, Puig PE, Roux S et al (2005) Tumor cells convert immature myeloid dendritic cells into TGF-beta-secreting cells inducing CD4+ CD25+ regulatory T cell proliferation. J Exp Med 202:919–929

Gobert M, Treilleux I, Bendriss-Vermare N et al (2009) Regulatory T cells recruited through CCL22/CCR4 are selectively activated in lymphoid infiltrates surrounding primary breast tumors and lead to an adverse clinical outcome. Cancer Res 69:2000–2009

Greenwald RJ, Freeman GJ, Sharpe AH (2005) The B7 family revisited. Ann Rev Immunol 23:515–548

Harrington LE, Hatton RD, Mangan PR et al (2005) Interleukin 17-producing CD4+ effector T cells develop via a lineage distinct from the T helper type 1 and 2 lineages. Nat Immunol 6:1123–1132

Hirahara N, Nio Y, Sasaki S et al (2001) Inoculation of human interleukin-17 gene-transfected Meth-A fibrosarcoma cells induces T cell-dependent tumor-specific immunity in mice. Oncology 61:79–89

Hori S, Nomura T, Sakaguchi S (2003) Control of regulatory T cell development by the transcription factor Foxp3. Science 299:1057–1061

Huang S, Apasov S, Koshiba M, Sitkovsky M (1997) Role of A2a extracellular adenosine receptor-mediated signaling in adenosine-mediated inhibition of T-cell activation and expansion. Blood 90:1600–1610

Huang B, Pan PY, Li Q et al (2006) Gr-1+ CD115+ immature myeloid suppressor cells mediate the development of tumor-induced T regulatory cells and T-cell anergy in tumor-bearing host. Cancer Res 66:1123–1131

Hurwitz AA, Watkins SK (2012) Immune suppression in the tumor microenvironment: a role for dendritic cell-mediated tolerization of T cells. Cancer Immunol Immunother 61:289–293

Igney FH, Krammer PH (2005) Tumor counterattack: fact or fiction? Cancer Immunol Immunother 54:1127–1136

Jones E, Dahm-Vicker M, Golgher D, Gallimore A (2003) CD25+ regulatory T cells and tumor immunity. Immunol Lett 85:141–143

Kanterman J, Sade-Feldman M, Baniyash M (2012) New insights into chronic inflammation-induced immunosuppression. Semin Cancer Biol 22:307–318

Klein G, Klein E (1977) Immune surveillance against virus-induced tumors and nonrejectability of spontaneous tumors: contrasting consequences of host versus tumor evolution. Proc Natl Acad Sci U S A 74:2121–2125

Kolls JK, Linden A (2004) Interleukin-17 family members and inflammation. Immunity 21:467–476

Kryczek I, Wei S, Zou L et al (2007) Cutting edge: Th17 and regulatory T cell dynamics and the regulation by IL-2 in the tumor microenvironment. J Immunol 178:6730–6733

Kwek SS, Cha E, Fong L (2012) Unmasking the immune recognition of prostate cancer with CTLA4 blockade. Nat Rev Cancer 12:289–297

Lai P, Rabinowich H, Crowley-Nowick PA, Bell MC, Mantovani G, Whiteside TL (1996) Alterations in expression and function of signal-transducing proteins in tumor-associated T and natural killer cells in patients with ovarian carcinoma. Clin Cancer Res 2:161–173

Lathrop SK, Santacruz NA, Pham D, Luo J, Hsieh CS (2008) Antigen-specific peripheral shaping of the natural regulatory T cell population. J Exp Med 205:3105–3117

Leach DR, Krummel MF, Allison JP (1996) Enhancement of antitumor immunity by CTLA-4 blockade. Science 271:1734–1736

Lee GK, Park HJ, Macleod M, Chandler P, Munn DH, Mellor AL (2002) Tryptophan deprivation sensitizes activated T cells to apoptosis prior to cell division. Immunology 107:452–460

Lehe C, Ghebeh H, Al-Sulaiman A et al (2008) The Wilms' tumor antigen is a novel target for human CD4+ regulatory T cells: implications for immunotherapy. Cancer Res 68:6350–6359

Leveque L, Deknuydt F, Bioley G et al (2009) Interleukin 2-mediated conversion of ovarian cancer-associated CD4+ regulatory T cells into proinflammatory interleukin 17-producing helper T cells. J Immunother 32:101–108

Liddy N, Bossi G, Adams KJ et al (2012) Monoclonal TCR-redirected tumor cell killing. Nat Med 18:980–7

Mangan PR, Harrington LE, O'Quinn DB et al (2006) Transforming growth factor-beta induces development of the T(H)17 lineage. Nature 441:231–234

Mantovani A, Sozzani S, Locati M, Allavena P, Sica A (2002) Macrophage polarization: tumor-associated macrophages as a paradigm for polarized M2 mononuclear phagocytes. Trends Immunol 23:549–555

Mantovani A, Allavena P, Sica A (2004) Tumour-associated macrophages as a prototypic type II polarised phagocyte population: role in tumour progression. Eur J Cancer 40:1660–1667

Martin-Orozco N, Muranski P, Chung Y et al (2009) T helper 17 cells promote cytotoxic T cell activation in tumor immunity. Immunity 31:787–798

McAllister F, Henry A, Kreindler JL et al (2005) Role of IL-17A, IL-17F, and the IL-17 receptor in regulating growth-related oncogene-alpha and granulocyte colony-stimulating factor in bronchial epithelium: implications for airway inflammation in cystic fibrosis. J Immunol 175:404–412

Middleton GW, Annels NE, Pandha HS (2012) Are we ready to start studies of Th17 cell manipulation as a therapy for cancer? Cancer Immunol Immunother 61:1–7

Miller JF (1961) Immunological function of the thymus. Lancet 2:748–749

Miyahara Y, Odunsi K, Chen W, Peng G, Matsuzaki J, Wang RF (2008) Generation and regulation of human CD4+ IL-17-producing T cells in ovarian cancer. Proc Natl Acad Sci U S A 105:15505–15510

Movahedi K, Laoui D, Gysemans C et al (2010) Different tumor microenvironments contain functionally distinct subsets of macrophages derived from Ly6C(high) monocytes. Cancer Res 70:5728–5739

Nagaraj S, Gupta K, Pisarev V et al (2007) Altered recognition of antigen is a mechanism of CD8+ T cell tolerance in cancer. Nat Med 13:828–835

Nakagomi H, Petersson M, Magnusson I et al (1993) Decreased expression of the signal-transducing zeta chains in tumor-infiltrating T-cells and NK cells of patients with colorectal carcinoma. Cancer Res 53:5610–5612

Nardin A, Abastado JP (2008) Macrophages and cancer. Front Biosci 13:3494–3505

Numasaki M, Fukushi J, Ono M et al (2003) Interleukin-17 promotes angiogenesis and tumor growth. Blood 101:2620–2627

Numasaki M, Watanabe M, Suzuki T et al (2005) IL-17 enhances the net angiogenic activity and in vivo growth of human non-small cell lung cancer in SCID mice through promoting CXCR-2-dependent angiogenesis. J Immunol 175:6177–6189

Ochsenbein AF, Klenerman P, Karrer U et al (1999) Immune surveillance against a solid tumor fails because of immunological ignorance. Proc Natl Acad Sci U S A 96:2233–2238

Ohta A, Gorelik E, Prasad SJ et al (2006) A2A adenosine receptor protects tumors from antitumor T cells. Proc Natl Acad Sci U S A 103:13132–13137

Ong SM, Tan YC, Beretta O et al (2012) Macrophages in human colorectal cancer are pro-inflammatory and prime T cells towards an anti-tumour type-1 inflammatory response. Eur J Immunol 42:89–100

Ostrand-Rosenberg S (2009a) Myeloid-derived suppressor cells: more mechanisms for inhibiting antitumor immunity. Cancer Immunol Immunother 59:1593–1600

Ostrand-Rosenberg S, Sinha P (2009b) Myeloid-derived suppressor cells: linking inflammation and cancer. J Immunol 182:4499–4506

Park H, Li Z, Yang XO et al (2005) A distinct lineage of CD4 T cells regulates tissue inflammation by producing interleukin 17. Nat Immunol 6:1133–1141

Peggs KS, Quezada SA, Korman AJ, Allison JP (2006) Principles and use of anti-CTLA4 antibody in human cancer immunotherapy. Curr Opin Immunol 18:206–213

Rech AJ, Vonderheide RH (2009) Clinical use of anti-CD25 antibody daclizumab to enhance immune responses to tumor antigen vaccination by targeting regulatory T cells. Ann N Y Acad Sci 1174:99–106

Reichert TE, Rabinowich H, Johnson JT, Whiteside TL (1998) Mechanisms responsible for signaling and functional defects. J Immunother 21:295–306

Reichert TE, Strauss L, Wagner EM, Gooding W, Whiteside TL (2002) Signaling abnormalities, apoptosis, and reduced proliferation of circulating and tumor-infiltrating lymphocytes in patients with oral carcinoma. Clin Cancer Res 8:3137–3145

Rivoltini L, Carrabba M, Huber V et al (2002) Immunity to cancer: attack and escape in T lymphocyte-tumor cell interaction. Immunol Rev 188:97–113

Rodriguez PC, Ochoa AC (2008) Arginine regulation by myeloid derived suppressor cells and tolerance in cancer: mechanisms and therapeutic perspectives. Immunol Rev 222:180–191

Rodriguez PC, Zea AH, Culotta KS, Zabaleta J, Ochoa JB, Ochoa AC (2002) Regulation of T cell receptor CD3zeta chain expression by L-arginine. J Biol Chem 277:21123–21129

Rodriguez PC, Quiceno DG, Ochoa AC (2007) L-arginine availability regulates T-lymphocyte cell-cycle progression. Blood 109:1568–1573

Rosenberg SA (2012) Raising the bar: the curative potential of human cancer immunotherapy. Sci Transl Med 4:127ps8

Serafini P, Mgebroff S, Noonan K, Borrello I (2008) Myeloid-derived suppressor cells promote cross-tolerance in B-cell lymphoma by expanding regulatory T cells. Cancer Res 68:5439–5449

Shafer-Weaver KA, Watkins SK, Anderson MJ et al (2009) Immunity to murine prostatic tumors: continuous provision of T-cell help prevents CD8 T-cell tolerance and activates tumor-infiltrating dendritic cells. Cancer Res 69:6256–6264

Shimizu J, Yamazaki S, Sakaguchi S (1999) Induction of tumor immunity by removing CD25+ CD4+ T cells: a common basis between tumor immunity and autoimmunity. J Immunol 163:5211–5218

Singer K, Gottfried E, Kreutz M, Mackensen A (2011) Suppression of T-cell responses by tumor metabolites. Cancer Immunol Immunother 60:425–431

Sinha P, Clements VK, Ostrand-Rosenberg S (2005) Interleukin-13-regulated M2 macrophages in combination with myeloid suppressor cells block immune surveillance against metastasis. Cancer Res 65:11743–11751

Sinha P, Okoro C, Foell D, Freeze HH, Ostrand-Rosenberg S, Srikrishna G (2008) Proinflammatory S100 proteins regulate the accumulation of myeloid-derived suppressor cells. J Immunol 181:4666–4675

Smyth MJ, Godfrey DI, Trapani JA (2001) A fresh look at tumor immunosurveillance and immunotherapy. Nat Immunol 2:293–299

Song X, Krelin Y, Dvorkin T et al (2005) CD11b+/Gr-1+ immature myeloid cells mediate suppression of T cells in mice bearing tumors of IL-1beta-secreting cells. J Immunol 175:8200–8208

Sullivan R, Purushotham AD (2011) Avoiding the zero sum game in global cancer policy: beyond 2011 UN high level summit. Eur J Cancer 47:2375–80

Sutton C, Brereton C, Keogh B, Mills KH, Lavelle EC (2006) A crucial role for interleukin (IL)-1 in the induction of IL-17-producing T cells that mediate autoimmune encephalomyelitis. J Exp Med 203:1685–1691

Takahashi H, Numasaki M, Lotze MT, Sasaki H (2005) Interleukin-17 enhances bFGF-, HGF- and VEGF-induced growth of vascular endothelial cells. Immunol Lett 98:189–193

Taylor DD, Bender DP, Gercel-Taylor C, Stanson J, Whiteside TL (2001) Modulation of TcR/CD3-zeta chain expression by a circulating factor derived from ovarian cancer patients. Br J Cancer 84:1624–1629

Thomas L (1959) In: Lawrence HS (ed) Cellular and humoral aspects of the hypersensitive state. Hober-Harper, New York, p 529

Uyttenhove C, Pilotte L, Theate I et al (2003) Evidence for a tumoral immune resistance mechanism based on tryptophan degradation by indoleamine 2,3-dioxygenase. Nat Med 9:1269–1274

Valitutti S, Muller S, Dessing M, Lanzavecchia A (1996) Different responses are elicited in cytotoxic T lymphocytes by different levels of T cell receptor occupancy. J Exp Med 183:1917–1921

Veldhoen M, Hocking RJ, Atkins CJ, Locksley RM, Stockinger B (2006) TGFbeta in the context of an inflammatory cytokine milieu supports de novo differentiation of IL-17-producing T cells. Immunity 24:179–189

Vignali DA, Collison LW, Workman CJ (2008) How regulatory T cells work. Nat Rev Immunol 8:523–532

Walker MR, Kasprowicz DJ, Gersuk VH et al (2003) Induction of FoxP3 and acquisition of T regulatory activity by stimulated human CD4+ CD25– T cells. J Clin Invest 112:1437–1443

Watkins SK, Hurwitz AA (2012) FOXO3: a master switch for regulating tolerance and immunity in dendritic cells. Oncoimmunology 1:252–254

Watkins SK, Zhu Z, Riboldi E et al (2011) FOXO3 programs tumor-associated DCs to become tolerogenic in human and murine prostate cancer. J Clin Invest 121:1361–1372
Wing K, Onishi Y, Prieto-Martin P et al (2008) CTLA-4 control over Foxp3+ regulatory T cell function. Science 322:271–275
Wu S, Rhee KJ, Albesiano E et al (2009) A human colonic commensal promotes colon tumorigenesis via activation of T helper type 17 T cell responses. Nat Med 15:1016–1022
Ye J, Su X, Hsueh EC et al Human tumor-infiltrating Th17 cells have the capacity to differentiate into IFN-gamma+ and FOXP3+ T cells with potent suppressive function. Eur J Immunol 41:936–51
Young MR, Newby M, Wepsic HT (1987) Hematopoiesis and suppressor bone marrow cells in mice bearing large metastatic Lewis lung carcinoma tumors. Cancer Res 47:100–105
Zajac AJ, Blattman JN, Murali-Krishna K et al (1998) Viral immune evasion due to persistence of activated T cells without effector function. J Exp Med 188:2205–2213
Zhang L, Conejo-Garcia JR, Katsaros D et al (2003) Intratumoral T cells, recurrence, and survival in epithelial ovarian cancer. N Engl J Med 348:203–213
Zhang B, Rong G, Wei H et al (2008) The prevalence of Th17 cells in patients with gastric cancer. Biochem Biophys Res Commun 374:533–537
Zhou G, Drake CG, Levitsky HI (2006) Amplification of tumor-specific regulatory T cells following therapeutic cancer vaccines. Blood 107:628–636

Chapter 14
Signaling of Tumor-Induced Immunosuppression of Dendritic Cells

Yong Lu, Jing Yang and Qing Yi

Abstract Dendritic cells (DCs) are professional antigen-presenting cells that regulate the immune system. In cancers, they uptake tumor-associated antigens, deliver them to T cells, and induce tumor-specific T cell responses. However, tumor cells develop mechanisms to evade the immune system, partly by impairing DC differentiation and function. Functionally deficient DCs may associate with acquisition of tolerogenic/immunosuppressive activities that actively block the development of antitumor immunity, and there is strong evidence supporting the presence of regulatory DCs in different DC subsets. Mechanistic studies reveal that intracellular signaling pathways, such as MAP kinases (MAPKs), JAK/STAT3, PI3 K/Akt, and NF-κB, which are critical to the regulation of DC differentiation, survival, and activity, are found to be hyperactivated both in tumor cells and in DCs in malignancies. The constitutive activation of these pathways in cancer cells leads to tumor cell secretion of cytokines that activate intracellular signaling pathways, particularly p38 MAPKs, in DCs or their progenitor cells and impair DC differentiation and function. In this chapter, we will discuss the dysfunction of DCs and the presence of regulatory DCs in cancer settings. We will focus on the signaling pathways that mediate DC dysfunction, particularly p38 MAPKs, in negatively regulating DC differentiation and function in cancers.

Keywords Dendritic cells · Signaling pathways · Regulatory dendritic cells · Tolerogenic dendritic cells · ERK · MAPK · NF-κB · STAT

Y. Lu · J. Yang
Department of Lymphoma and Myeloma and the Center for Cancer Immunology Research, University of Texas MD Anderson Cancer Center, Houston, TX, USA

Q. Yi (✉)
Department of Lymphoma and Myeloma, MD Anderson Cancer Center, 1515 Holcombe Boulevard Unit 0903, Houston, TX 77030, USA
e-mail: qyi@mdanderson.org

M. R. Shurin et al. (eds.), *The Tumor Immunoenvironment*,
DOI: 10.1007/978-94-007-6217-6_14,

14.1 Introduction

Dendritic cells (DCs) are populations of professional antigen-presenting cells that regulate the immune system (Evel-Kabler and Chen 2006; Santini and Belardelli 2003; Sheng et al. 2005). They originate from $CD34^+$ bone-marrow stem cells, and have high plasticity and common morphological and functional characteristics (Sheng et al. 2005; Gabrilovich et al. 1996). During their development, DCs are classified as immature, semimature and mature cells after migration into the peripheral tissues from their bone marrow precursor cells. In the immature stage, DCs are primarily localized in the peripheral tissues and perform specialized functions of antigen uptake and processing, in which they capture and carry antigens to the lymph nodes. In the lymphoid organs, DCs become mature and subsequently, interact with antigen-specific T cells and initiate immune responses (Di Nicola and Lemoli 2000; Sinkovics and Horvath 2000; Aloysius et al. 2006).

One of the most important findings about DCs is that these cells are endowed with two critical features: subset and functional plasticity (Steinman and Banchereau 2007). This diversity permits the adaptive immune system to mount functionally distinct types of responses. The two major DC subsets are the classic DCs (cDCs) and the plasmacytoid DCs (pDCs). pDCs are the frontline in antiviral immunity because of their capacity to rapidly produce high amounts of type I interferon (IFN) in response to viruses (Liu 2005). In contrast, cDCs are efficient phagocytic cells that reside within lymphoid and nonlymphoid organs. In mice, certain cDC populations such as lymphoid organ $CD11c^{hi}CD11b^{lo}MHCII^{+}CD4^{-}CD8^{+}$ DCs ($CD8^+$ DCs) and tissue-resident $CD11c^{hi}CD11b^{lo}MHCII^{+}CD103^{-}$ DCs ($CD103^+$ DCs) demonstrate efficient antigen cross-presentation ability by using MHC class I to present exogenously derived antigens (den Haan and Bevan 2002; Belz et al. 2004; Beauchamp et al. 2010; Bedoui et al. 2009). Accordingly, $CD8^+$ DCs and $CD103^+$ DCs have important roles in antiviral and antitumor immune responses. Because of their effective antigen presentation properties to deliver tumor-associated MHC class I antigens to $CD8^+$ T cells, human equivalents to murine cDCs are being used for anticancer therapy (Koski et al. 2008).

It is widely accepted that functional properties of DCs are maturation-dependent (Steinbrink et al. 2009). However, recent evidence suggests that both phenotypically immature and mature DCs may be conditioned by the microenvironment to display immune tolerant and/or immunosuppressive functions (Lin et al. 2010; Gregori 2011; Manicassamy and Pulendran 2011). Nonetheless, DCs should be considered to be a specialized group of antigen-presenting cells with high functional plasticity. This plasticity of DCs, including immunostimulating or immunosuppressive potential, or both, depends on the consequence and combination of microenvironmental stimuli affecting DC differentiation, activation and polarization.

14.2 Dendritic Cells in Cancer: Immunosurveillance Versus Tumor Evasion

Cancer immunosurveillance is the inflammatory process whereby the immune system recognizes and eliminates an early developing tumor (Sheng et al. 2011). It is evident that innate leukocytes like DCs, macrophages and natural killer (NK) cells can sense early tissue stress and matrix alteration during cellular transformation (Sheng et al. 2011). Additionally, studies in both animal models and in clinical settings have clearly supported the idea of spontaneous tumor immune surveillance by T cells (Galon et al. 2006; Zhang et al. 2003). The increased susceptibility to spontaneously arising and/or chemically induced murine tumors in IFN-γ, perforin or interleukin (IL)-12 knockout mice is suggestive of the ability of DCs to mature in the tumor microenvironment, effectively uptake tumor-associated antigens for delivery to T cells, and induce tumor-specific T cell responses (Liu et al. 2004; Shankaran et al. 2001; van den Broek et al. 1996).

The cancer "immunoediting" hypothesis implies that all symptomatic tumors represent a failure of the immune system (Schreiber et al. 2011). Tumors can be kept in check for long periods, through a dynamic balance that results in the progressive loss of immunogenicity by tumor cells (Schreiber et al. 2011). In addition, the cancer immunoediting hypothesis has recently evolved to include a role for tumor-induced immunosuppression in accelerated tumor growth, because clinical trials using blocking common immunosuppressive checkpoints (such as CTLA4 or PD-1) demonstrated that preventing tumor-induced T cell paralysis restores protective immunity against established cancers, suggesting that advanced tumors remain somewhat immunogenic (Simeone and Ascierto 2012). However, the role of DCs in the "elimination, equilibrium and escape" stages suggested by immunoediting is still elusive.

Perhaps one of the most successful models to recapitulate the multitude of functional states of DCs in tumor initiation, equilibrium and escape stages is described by Scarlett et al., using a new inducible p53-dependent model of aggressive ovarian carcinoma, which is different from other models that initiate tumors before the development of a mature immune system or use transplantable tumor cell lines (Scarlett et al. 2012). In this model, measurable antitumor immunity from very early stages was driven by infiltrating DCs and prevented steady tumor growth for prolonged periods, indicating a protective role played by DCs in the induction of antitumor T cell-mediated immune responses. However, tumors aggressively progressed to terminal disease in a comparatively short time during which a phenotypic switch in expanding DC infiltrates could be detected. In the escape phase, tumor cells remained immunogenic at advanced stages, whereas antitumor T cells became less responsive and their enduring activity was abrogated by immunosuppressive DCs within the advanced tumor microenvironment. Notably, depleting DCs early in the disease course accelerated tumor expansion, but DC depletion at advanced stages significantly delayed aggressive malignant progression. These results clearly demonstrated that DCs in the tumor microenvironment serve as a double-edged sword: phenotypically divergent DCs drive both immunosurveillance and accelerate malignant growth.

14.3 Dysfunction of Dendritic Cells in Cancer

DCs have been used as biological adjuvants in tumor vaccinations due to their key role in tumor immunity (Steinman and Banchereau 2007; Koido et al. 2010; Palucka et al. 2009; Gabrilovich 2002; Kusmartsev and Gabrilovich 2002). However, antitumor immune responses are often deficient and unsatisfactory, and suppression and re-polarization of DC function in cancer patients are thought to contribute to the failure of antitumor immune responses and consequent disease progression (Palucka et al. 2010a, b; 2011). Subversion of tumor immunity by manipulating the tumor microenvironment and DC subset distribution and/or function is mediated by various tumor-derived and/or stromal factors, many of which remain to be identified (Ma et al. 2012).

Thus far, the abnormalities of DC differentiation and function are considered one of the major factors limiting the success of cancer vaccines in clinical trials. Therefore, studies of the mechanisms of tumor-induced DC dysfunction may be a key point to improve antitumor immune responses in cancer patients. The most common dysfunction of cDCs in the tumor microenvironment is demonstrated as unable to stimulate allogeneic and/or syngeneic T cell proliferation, reduced expression of costimulatory molecules, decreased uptake, processing and presentation of antigens, inefficient motility and migration towards specific chemokines and decreased production of IL-12 (Vicari et al. 2002; Yang and Carbone 2004). This type of functionally deficient DCs is usually not immunosuppressive. However, in specific tumor microenvironment conditions, the loss of function in DCs may, at least in part, be associated with acquisition of tolerogenic/immunosuppressive activities, including actively blockade of antitumor immunity, recruitment and expansion of regulatory Treg (T) cells and support of tumor progression by promoting intratumoral neoangiogenesis and metastases (Lin et al. 2010; Pinzon-Charry et al. 2005; Cools et al. 2007). This type of DCs with tolerogenic properties is termed regulatory DCs (regDCs) or tolerogenic DCs (Shurin et al. 2012). In most cases, the capacity of DCs to coordinate the immune response is not an intrinsic quality of the cell but is rather the result of specific microenvironmental signals for repolarization and/or recruitment, including the local cytokine/chemokine network and the milieu of soluble factors from the neighboring cells. For instance, tumor-derived IL-10, tumor growth factor (TGF)-β, IL-6, vascular endothelial growth factor (VEGF), macrophage colony-stimulating factor (M-CSF), and prostaglandin-E2 (PGE2) can render DCs to acquire regulatory instead of stimulatory capacities (Shurin et al. 2006; Kusmartsev and Gabrilovich 2006a; Lin and Karin 2007).

Strong evidence supports the presence of regDCs in different subsets, including immature and mature myeloid cells, conventional DCs, and pDCs (Gregori 2011; Manicassamy and Pulendran 2011; Shurin et al. 2012). To date, many types of regDCs with different phenotypes have been described. For instance, regDCs have been reported as DCs expressing high levels of CD80 and CD86, producing IL-10 and inducing differentiation of $CD4^+$ Treg cells (Akbari and Umetsu 2005). Other groups suggested that regDCs expressed exceptionally low levels of costimulatory molecules, supported the generation of $CD4^+$ and $CD8^+$ Treg cells and prevented graft-versus-host disease (GVHD) (Isomura et al. 2008). In contrast, it has been also reported that regDCs expressed high levels of costimulatory/inhibitory B7-H1, B7-DC, and B7-H3 molecules and were capable of blocking DTH induction (Zhang et al. 2004). Furthermore, regDCs may be demonstrated by production of IL-10 and nitric oxide (NO), IL-10, TGF-β, cyclooxygenase 2 (COX-2), and indoleamine 2, 3-dioxygenase (IDO) (Kwon et al. 2010). However, many of tumor-associated regDCs were induced in vitro by culturing DCs in the immunosuppressive cytokines or drugs. Importantly, it becomes clear that the normal stromal microenvironment of the spleen, lung, and liver can drive DCs and hemopoietic progenitors to differentiate into regDCs with the phenotype of $CD11c^{low}CD11b^{high}Ia^{low}$ and high secretion of IL-10, NO, and IP-10 but less IL-12 (Zhang et al. 2004; Tang et al. 2006; Xia et al. 2008; Li et al. 2008a). These $CD11^{clow}CD11b^{high}Ia^{low}$ regDCs can be considered as "natural occurring regDCs", favor Th2 type immune responses and also induce Treg cell generation/expansion, and thus suppress type-1 T cell-mediated antitumor immunity and autoimmune diseases (Li et al. 2008b; Liu et al. 2009). We also observed that this type of "natural occurring regDCs" was the majority DC subset in murine lung tumor tissues (Lu et al. 2012). However, under some specific therapy, the recruitment of $CD8\alpha^+$ DCs, which are specialized for cross-presentation of antigen by MHC class I molecules to $CD8^+$ T cells, was associated with significant antitumor CTL responses (Lu et al. 2012).

14.4 Inhibitory Pathways of Immunosuppression Mediated by Cancer-Educated Regulatory Dendritic Cells

As discussed above, dysfunction of DCs within the cancer microenvironment may also associate with tolerogenic and immunosuppressive properties. Such immunosuppressive regDCs can mediate either direct effects on effector T cells or indirect effects on the T cells by induction and/or activation of other immune regulatory cells, such as Treg cells and myeloid-derived suppressor cells (MDSCs). Several soluble regDC-derived factors and membrane-bound or intracellular molecules are also responsible for these regDC-mediated immunosuppressive activities.

Tumor microenvironment-subverted DCs lack effector T cell stimulatory capacity but might be endowed with the ability to promote suppressive Treg cells (Steinman et al. 2003; Hubert et al. 2007; Stoitzner et al. 2008). Several studies provide evidence for the different subsets of regDCs capable of promoting Treg cell expansion and/or function (Hartmann et al. 2003; Wei et al. 2005). In addition to tumor-derived factors which directly induce Treg cell proliferation and/or generation of Treg cells from naive T cells, regDCs that are educated by the tumor microenvironment provide essential signals that contribute to Treg cell expansion and suppressive activity, which include IDO activity, PD-L1, TGF-β, IL-10 and so on (Ni et al. 2012; Ramos et al. 2012; Sharma et al. 2007). Induction of Treg cells by regDCs thus appears to be one of the essential mechanisms employed by tumor cells to generate immunosuppressive Treg cells. Reciprocally, cancer/regDC-induced Treg cells, by restraining DC maturation and by inducing regDC expression and production of immunosuppressive molecules, may further skew DC differentiation towards an inhibitory cell population (Janikashvili et al. 2011). This positive feedback loop by which regDCs induce Treg cells that in turn enhance DC immunosuppressive function may significantly contribute to the persistence of the immune tolerance to cancer, and therefore, targeting the generation and function of these two suppressive cell populations is a desirable goal in immunotherapeutic approaches.

MDSCs are a mixed cell population of myeloid cells including immature granulocytes, macrophages, DCs, and myeloid progenitors (Kusmartsev and Gabrilovich 2006b). In mice, phenotypic $Gr1^{+}CD11b^{+}$ MDSCs were detected in all tested tumor models. Significant accumulation of this cell population has been found in patients with various types of cancer (Almand et al. 2001). MDSCs express high levels of immunosuppressive factors such as IDO, IL-10, arginase, inducible nitric oxide synthase (NOS2), NO, and reactive oxygen species, and use these molecules to suppress T cell-mediated immunity, DC function as well as induce regDCs (Marigo et al. 2008). Tumor/regDC-derived PGE2, in combination with lipopolysaccharide (LPS), IL-1β and IFN-γ induced production of COX-2 by monocytes, and redirected the development of $CD1a^{+}$ DCs to $CD14^{+}CD33^{+}CD34^{+}$ monocytic MDSCs (Obermajer et al. 2011). DCs/regDCs contribute to the induction and persistence of MDSCs, highlighting the potential for its manipulation to enhance immune responses in cancer.

Production of IL-10 and TGF-β by regDCs has been well established, and the role of these two cytokines in polarization of Treg cells has been repeatedly confirmed (Lin et al. 2010; Janikashvili et al. 2011). IL-10 and TGF-β are identified as anti-inflammatory cytokines with immunosuppressive properties and have crucial roles in preventing autoimmunity. They suppress antigen presentation and subsequent T cell proliferation, inhibit Th1 cytokine production and DC maturation. IL-10 and TGF-β expression has been shown to correlate with poor prognoses in many cancers. IL-10 also confers resistance of tumor cells to apoptosis and increases metastatic potential, and promotes angiogenesis by regulating VEGF production by myeloid cells (Zeng et al. 2010; Riboldi et al. 2005). Importantly, IL-10-conditioned tumor cells exhibit decreased expression of MHC class I and are resistant to $CD8^{+}$ CTL-mediated cytotoxicity (Kurte et al. 2004).

The enzyme arginase metabolizes L-arginine to L-ornithine and urea. Besides its fundamental role in the hepatic urea cycle, arginase is also expressed by the immune cells (Munder 2009). L-arginine depletion by arginase profoundly suppresses T cell-mediated immune responses, which has been considered as one of the fundamental mechanisms of inflammation-associated immunosuppression (Munder 2009). However, evidence of arginase expression by tumor-associated regDCs has been obtained recently showing that tumor-infiltrating regDCs can induce $CD8^{+}$ T cell exhaustion via L-arginine metabolism (Norian et al. 2009).

These regDCs are reported to display $CD11c^{+}CD11b^{high}Ia^{low}$ phenotype, and might be educated by tumor-derived factors such as TGF-β and PGE2 (Scarlett et al. 2012; Liu et al. 2009).

IDO catalyzes the degradation of the essential amino acid tryptophan into kynurenine, and can mediate tryptophan deprivation in the T cell microenvironment (Munn et al. 2002). IDO activity has been shown to downregulate the expression of TCR-ζ-chain and lead to the activation of the GCN2 (general control non-repressed 2) kinase pathway that results in T cell G1-phase arrest and apoptosis (Munn et al. 2005). In addition, the byproducts of the tryptophan catabolism such as L-kynurenine, 3-hydroxykynurenie, or 3-hydroxyanthranilic acid may be endowed with inherent suppressive activity (Fallarino et al. 2003). IDO activity can be detected in different subsets of DCs in mouse and humans, and the expression of IDO in DCs was associated with DC-induced immunosuppression (Ghahary et al. 2004; O'Neill et al. 2004). Tryptophan depletion by IDO has been identified as a possible factor involved in regDC-induced Treg cell expansion and activation (Sharma et al. 2007; Fallarino et al. 2006). Treg cell induction and activation by IDO^{+} regDCs require the GCN2 pathway and can be partially prevented by CTLA-4 blockade (Sharma et al. 2007). IDO^{+} regDCs also suppress the conversion of $CD4^{+}Foxp3^{+}$ Tregs cells to Th17-like effector cells in tumor-draining lymph nodes (Sharma et al. 2009). These studies suggested that IDO-expressing regDCs found at the tumor sites and in tumor-draining lymph nodes might help suppress the initiation of immune responses to tumor-associated antigens and create systemic tolerance to tumor cells.

PD-1 and PD-L1 belong to the B7 family of costimulatory molecules and are expressed on activated DCs, monocytes/macrophages, T cells, B cells, as well as tumor cells (Keir et al. 2008; Dong et al. 2002). PD-L1 promotes differentiation and maintains the function of induced Treg cells (Francisco et al. 2009). Blockage of PD-1/PD-L1 interaction increases infiltration of $CD8^{+}$ T cells to tumors, suggesting that PD-L1 induces tumor-specific CTL exhaustion (Zou 2005). PD-L1 and/or PD-1 expression levels on myeloid DCs correlate with poorer cancer prognosis (Zou 2005; Thompson et al. 2007). For instance, it has been reported that ovarian cancer-infiltrating DCs progressively expressed upregulated PD-1 and PD-L1 molecules, and were immunosuppressive to T cell immunity and blocked their infiltration into advanced tumors (Zou 2005).

14.5 Signal Pathways Involved in Dendritic Cell Dysfunction in Cancer

Tumor microenvironment is well known to be immunosuppressive (Kim et al. 2006a; Rabinovich et al. 2007). Tumor cells consistently release many kinds of immunosuppressive and proinflammatory factors such as VEGF, TGF-β, IL-10, PGE2, M-CSF and IL-6, which facilitate tumor immune escape and tumor growth, partially by actively reprogramming DC dysfunction for tumor cell escape of immunological attack (Zou 2005; Bennaceur et al. 2009). Although the list of tumor-derived and stromal factors involving the impaired or repolarized DC function might be getting longer, many of them may utilize similar transcription factors and signaling pathways.

Numerous recent studies have reported that tumor-induced activation of intracellular signaling pathways, such as mitogen-activated protein kinases (MAPKs), JAKs/STATs, and NF-κB, contributes to various defects of the immune system, particularly through compromising DC differentiation and function. Even though these signaling pathways are important for the development of normal hematopoietic cells, activation of these pathways is usually present in both tumor cells and abnormal DCs to support tumor growth and survival (Ade et al. 2007; Nefedova et al. 2004; Philpott et al. 2004). In this section we will discuss signaling pathways mediating DC dysfunction, particularly p38 MAPKs, in negatively regulating DC differentiation and function in cancers.

14.5.1 MAPK Signaling Pathways

MAPKs are proline-directed serine and threonine protein kinases, and are activated by dual-specificity kinases with phosphorylation of threonine and tyrosine in a Thr-Xaa-Tyr motif. Activation of MAPK signaling pathways is through a MAPK-activating phosphorylation cascade, in which upstream kinases phosphorylate their downstream kinases on threonine and tyrosine residues, starting from MAPK kinase kinases (MAPKKKs), to MAPK kinases (MAPKKs), and finally to MAPKs. The activated MAPKs then interact with their cytoplasmic substrates and translocate into the nucleus, where they act as transcription factors and regulate target gene transcription (Nakamura et al. 1996; Ichijo 1999).

MAPK signaling pathways are crucial for diverse cellular functions, including proliferation, differentiation, and apoptosis (Aplin et al. 2002; Budagian et al. 2003; Kawakami et al. 2003; Sigaud et al. 2005). There are three types of MAPKs, extracellular signal-regulated kinases (ERKs), c-jun N-terminal kinases (JNKs), and p38 MAPKs, which are identified by the intervening amino acid. The ERK pathway, activated by polypeptide growth factors through their tyrosine kinase receptors, regulates cellular growth and survival. JNK and p38 signaling pathways are activated by stress stimuli and inflammatory cytokines, and are involved in cellular differentiation, cytokine production, and apoptosis. MAPK signaling pathways have been shown to be frequently activated in cancers, and may contribute to malignant phenotypes and uncontrolled cell growth. In addition, MAPK signaling pathways are involved in the regulation of immune responses, including the initiation phase of innate immunity, activation of adaptive immunity, and cell death after completing immune function (Nakahara et al. 2004; Canesi et al. 2005; Kim et al. 2005; Zou and Hu 2005). Notably, recent studies have indicated that MAPK signaling pathways differentially regulate all aspects of DC phenotypic maturation, cytokine production, and DC functional development (Nakahara et al. 2004; Cruz et al. 1999;

Xie et al. 2005; Wang et al. 2006a). Stimuli such as LPS, TNF-α, haptens, or ultraviolet-B (UVB) induce maturation of DCs via MAPK signaling pathways (Nakahara et al. 2004; Cruz et al. 1999; Tassiulas et al. 2007). On the other hand, tumor-induced abnormalities of DC differentiation and function are also associated with hyperactivation of MAPK signaling pathways (Wang et al. 2006a, b).

14.5.2 p38 MAPKs in Dendritic Cell Differentiation, Maturation, and Activity

There are four p38 MAPKs; α and β, which are 75 % homologous, and γ and δ, which are more distant relatives. All p38 MAPKs can be activated by the same upstream MAP kinase kinases, such as MKK3 or MKK6, upon the stimulation of inflammatory cytokines or stress (Ichijo 1999; Lee et al. 2006). p38 MAPK signaling induces the activation of MAPK-activated protein kinase (MAPKAPK)-2 (Zaru et al. 2007), synthesis of TNF-α (Lee et al. 2006; Park et al. 1999), and phosphorylation of transcription factors such as activating-transcription-factor-2 (ATF-2), Elk-1 and SAP-1.

The p38 MAPK signaling pathway is essential for normal DC maturation and activity (Xie et al. 2005; Ardeshna et al. 2000; Matos et al. 2005a, b; Osawa et al. 2006). LPS-induced maturation and upregulation of surface antigens on DCs such as CD40, CD80, CD83, CD86, and MHC class II molecules require p38 MAPKs (West et al. 2004; Bharadwaj et al. 2005). The p38 MAPK inhibitor SB203580 abrogates the upregulation of surface antigens in the process of DC maturation induced by LPS, $NiCl_2$, $NiSO_4$, and CD40L. Furthermore, LPS-induced DC secretion of cytokines such as TNF-α, IL-6, and IL-12, also depends on the activation of p38 MAPKs, because SB203580 has been shown to inhibit DC secretion of these cytokines (Lee et al. 2006; Randolph et al. 2005; Saito et al. 2006). In addition, LPS-enhanced allostimulatory activity of DCs is abrogated by SB203580 treatment, indicating that p38 MAPKs are required for the endocytotic and allostimulatory functions of DCs (Kang et al. 2004).

However, we have shown that the importance of p38 MAPK signaling pathways in DCs is stage-dependent. While crucial for immature DCs to mature and secrete cytokines, activation of p38 MAPKs is detrimental to the generation and differentiation of DCs from monocytes. During the differentiation of monocytes to immature DCs, p38 MAPK activation induced by LPS impaired DC differentiation and p38 MAPK inhibitor SB203580 restored generation of functional DCs in culture with LPS. Moreover, addition of SB203580 to cultures of normal monocytes accelerated the differentiation of the cells into immature DCs. These results could be explained by the findings that inhibition of p38 MAPKs enhances the phosphorylation of ERK and NF-κB activity and leads to enhanced upregulation of expression of DC-related adhesion and costimulatory molecules and antigen-presentation capacity (Xie et al. 2005; Lee et al. 2006; Ardeshna et al. 2000; Osawa et al. 2006).

14.5.3 p38 MAPKs in Tumor-Induced Dendritic Cell Dysfunction

DCs from cancer patients are functionally defective, however, the underlying molecular mechanisms are poorly understood at the present time. We have used the murine 5TGM1 myeloma model to examine the effects and mechanism of tumor-derived factors on the differentiation and function of DCs. Myeloma cells or tumor culture conditioning medium (TCCM) were shown to inhibit differentiation and function of bone marrow-derived DCs (BMDCs), as evident by the down-regulated expression of DC-related surface molecules, decreased IL-12 secretion, and compromised capacity of the cells to activate allospecific T cells. Moreover, TCCM-treated BMDCs were inferior to normal BMDCs at priming tumor-specific immune responses in vivo. Neutralizing antibodies against IL-6, IL-10, and TGF-β partially abrogated the effects. Our results showed that TCCM treatment activated p38 MAPK and JNK but inhibited ERK. Inhibiting p38 MAPK restored the phenotype, cytokine secretion, and function of TCCM-treated BMDCs. BMDCs from cultures with both TCCM and p38 inhibitor were as efficacious as normal BMDCs at inducing tumor-specific antibody, type-1 T cell, and CTL responses, and prolonging mouse survival. Thus, our results suggest that tumor-induced p38 MAPK activation and ERK inhibition in DCs may be a new mechanism for tumor evasion, and regulating these pathways during DC differentiation provides new strategies for generating potent DC vaccines for immunotherapy in cancer patients (Wang et al. 2006a).

Next, we examined whether the defects can be observed in DCs from patients with myeloma. Previous studies have demonstrated that circulating DCs in myeloma patients are functionally abnormal (Ratta et al. 2002). However, no study had been performed to examine monocyte-derived DCs (MoDCs), which are commonly used for immunotherapy in patients. We found that patient-derived MoDCs are phenotypically and functionally defective. Compared with their normal counterpart, patient-derived mature MoDCs expressed significantly lower levels of CD1a, CD40, CD80, and HLA-DR, and were deficient at activating alloreactive T cells, presenting recall antigen, and activating autologous antigen-specific T cells. These abnormalities may be attributed to elevated production of autocrine cytokines such as IL-6, activated p38 MAPK and STAT3, and inhibited MEK/ERK signaling pathways in the progenitor cells. Treatment with neutralizing IL-6-specific antibody and more importantly, p38 MAPK inhibitor, or both, could correct these abnormalities. Treating patient-derived cells with these agents not only significantly increased cell yield, but also produced MoDCs that were as functional as their normal counterpart (Wang et al. 2006b). Thus, our studies have delineated the mechanistic defects of MoDCs from myeloma patients, and identified ways for restoring the function of the cells to improve the efficacy of DC-based immunotherapy in this disease.

In line with our findings, others showed that constitutive activation of p38 MAPK is responsible for turning off DCs to display a tolerogenic profile during melanoma progression, and suppression of p38 MAPK activity in DCs from tumor-bearing mice could reconstitute their impaired function as shown by normalization of cytokine secretion pattern and T cell stimulation capacity (Zhao et al. 2009). Another recent study also showed that inhibiting p38 MAPK signaling in DCs attenuates Treg cell induction in response to Toll-like receptor (TLR) agonists and enhances their efficacy as vaccine adjuvants and cancer immunotherapeutics (Jarnicki et al. 2008). TLR ligands are commonly used adjuvants that promote type-1 T cell responses against tumor antigens. However, TLR ligands also promote the induction of IL-10-secreting Treg cells through p38 MAPK-induced IL-10 production by DCs. Inhibition of p38 MAPKs by SB203580 suppressed TLR-induced IL-10 and PGE_2 and enhanced IL-12 production in DCs. Inhibition of p38 MAPKs enhanced the antitumor therapeutic efficacy of DCs pulsed with antigen and CpG, which was associated with an enhanced frequency of IFN-γ-secreting T cells and a reduction of Foxp3$^+$ Treg cell infiltration of the tumors. Taken together, these findings indicate that p38 is an important therapeutic target and inhibiting p38 activity in DCs obtained from cancer patients or DCs pulsed with tumor antigens and TLR agonists will enhance the immunogenicity of the cells.

14.5.4 ERK and Dendritic Cell Dysfunction in Cancer

Recent studies have demonstrated that the ERK and p38 MAPK signaling pathways differentially regulate DC maturation and modulate the initial commitment of naïve T-helper (Th) cells toward Th1 or Th2 subsets (Aplin et al. 2002; Lee et al. 2006; Kandilci and Grosveld 2005). The p38 MAPK inhibitor SB203580 suppressed DC maturation, whereas the presence of ERK inhibitors PD98059 or U0126 enhanced LPS-induced phenotypic and functional maturation of DCs, and increased the expression of MHC complex and costimulatory molecules. In a recent study, cDCs derived in vitro from murine ERK1$^{-/-}$ bone marrow progenitors were demonstrated with increased surface expression of activation markers and enhanced T cell stimulation, suggesting that ERK1 negatively regulated functional differentiation of DCs (Bendix et al. 2010). Importantly, ERK signaling in DCs has been shown to suppress the immune response and stimulate the expansion of Treg cells (Escors et al. 2008). Selective ERK activation in both mouse and human DCs generated regDCs with immunosuppressive capacity, leading to Treg cell expansion by secreting bioactive TGF-β1 and IL-10 (Escors et al. 2008; Arce et al. 2011).

However, MAPK pathways, which are frequently activated in cancers, have active roles in immune evasion in cancer. Tumor lysate has been shown to markedly suppress TLR-4-dependent IL-12p40 and p70 production from DCs by hyperactivating ERK signaling in DCs, and these tumor lysate-treated DCs were less able to generate Th1-responses from naïve T cells (Jackson et al. 2008). Blockade of MEK1/2, the upstream kinase for ERK, with U0126 prevented ERK activation, restored IL-12p70 production, and permitted effective generation of Th1-responses (Jackson et al. 2008). In addition, by using ERK inhibitor U0126 and lentiviral $BRAF^{V600E}$ RNA interference, Sumimoto et al. demonstrated that the ERK signaling pathway is essential for production of immunosuppressive factors by human melanoma cells that have constitutively activated ERK due to the $BRAF^{V600E}$ mutation, which can be detected in the majority of patients with melanomas (Sumimoto et al. 2004, 2006; Tanami et al. 2004). These findings indicate that pharmacological intervention in the MEK-ERK axis may be used to render DC resistant to the suppressive effects of tumor microenvironment and may become part of a combination immunotherapy.

14.5.5 Role of JAKs/STATs Signaling in Tumor-Induced Dendritic Cell Dysfunction

Over the past several years, investigators have been working on the JAK/STAT signaling pathways in the context of cancer-mediated evasion of the immune system (Kortylewski et al. 2005a; Nefedova et al. 2005; Kim et al. 2006b). JAK mutations and/or STAT abnormal activation are found in many types of cancers, such as myeloproliferative disorders with acquired JAK2 mutations (Taki and Taniwaki 2006; Jost 2007; Mata et al. 2007), T cell acute lymphoblastic leukemia (Taki and Taniwaki 2006), and leukemia or lymphoma with constitutive phosphorylation of JAK3, STAT1, STAT3, and STAT5 (Aboudola et al. 2007). Among them, constitutive activation of STAT3 is common in a variety of lymphoid or myeloid malignancies and solid tumors, in human tumor cell lines and primary tumor cells from patients, and in virus-transformed cells (Yu et al. 1995; Campbell et al. 1997; Cheng et al. 2004; Park et al. 2005). Recent studies showed that hyperactivation of STAT3 is found in multiple myeloma, breast cancer, and prostate cancer (Wang et al. 2004a).

In addition to STAT3-induced intrinsic oncogenic activities, studies have shed light on STAT3-mediated cancer cell-initiated immune evasion signals in various immune cells (Yu et al. 2007). Soluble factors released from tumor cells, such as IL-10, IL-6, VEGF or M-CSF, induced activation of STAT3 in myeloid cells, leading to systemic accumulation and activation of MDSCs and inhibition of DC differentiation toward immunogenic status (Wang et al. 2004b; Nefedova et al. 2004). Since JAK/STAT3 signaling pathway is a major signaling pathway that can be activated by cytokines binding to their membrane receptors, tumor-derived factors inhibit DC differentiation and function mainly via JAK/STAT3 activation (Li et al. 2007). For instance, treatment of DCs with tumor-conditioned medium reduced expression of IL-12 and MHC II and costimulatory molecules, and promoted transcription of IDO due to activated STAT3-induced inhibition of canonical NF-κB activity (Hoentjen et al. 2005; Kitamura et al. 2005; Nefedova et al. 2005; Sun et al. 2009). In addition, Treg cells hamper DC function by activating STAT3 signaling pathway in DCs (Larmonier et al. 2007). Inactivation of STAT3 signaling in hematopoietic cells by pharmacological inhibitors such as JSI-124 or CPA-7, demonstrated enhanced antitumor immune responses through the activation of various immune cells, especially DCs, and inactivation of immune suppressor cells, such as MDSCs and Treg cells (Nefedova et al. 2005; Kortylewski et al. 2005b). Similarly, STAT3$^{-/-}$ bone marrow progenitor cells were also refractory to tumor-derived inhibitory factor-mediated suppression of DCs differentiation (Wang et al. 2004b). Importantly, DCs derived from STAT3$^{-/-}$ mice displayed higher cytokine production in response to TLR stimulation, and induced effective antitumor effects when used as vaccine through systemic Th1 immune responses (Iwata-Kajihara et al. 2011). Since inhibition of STAT3 abrogated the negative effects of the tumor-derived factors on myeloid cell differentiation, these observations suggest that JAK/STAT signaling pathways may be negative regulators of DC differentiation and function in malignancies.

14.5.6 Other Signaling Pathways in Tumor-Induced Dendritic Cell Dysfunction

It is well known that activation of NF-κB plays an important role in DC maturation and function (Ade et al. 2007; Zou and Hu 2005; Osawa et al. 2006). JAK/STAT, p38 MAPK, and ERK signaling pathways crosstalk with the NF-κB pathway, and factors activating STATs or MAPKs also stimulate NF-κB, which includes members of p50, p52, RelA, RelB, and cRel. The proteins form active hetero- or homodimers, translocate to the nuclei, and initiate the transcription of target genes. NF-κB activity has been shown to be high in DCs, and upregulation of IL-12 expression requires activation of both p38 MAPK and NF-κB (Ade et al. 2007). Our and others previous studies showed that differentiation of immature DCs is accompanied by increased NF-κB activity and that inhibiting p38 MAPK enhances the activity of NF-κB in immature DCs (Ade et al. 2007; Wang et al. 2006a, b). Because high levels of NF-κB activity are frequently found in many types of cancers, NF-κB signaling pathways may also contribute to tumor-induced DC dysfunction in cancer patients.

14.6 Conclusion

DCs play important roles in initiating innate and adaptive immune responses, which are critical for the antitumor immune response. However, hyperactivation of signaling pathways such as JAKs/STATs, MAPKs, and NF-κB in both tumor cells and tumor-infiltrating DCs is critical for tumor-induced immunosuppression of tumor-bearing hosts. The activation of multiple signaling pathways in tumor cells mediates the expression and secretion of tumor-derived factors to the tumor microenvironment. Subsequently, these tumor-derived factors impair DC differentiation and impair their function, resulting in DC-mediated immune tolerance. The concept for revitalizing the capacity of immunogenic DCs to stimulate CTLs is widely accepted to be a critical step to enhancing antitumor immunity.

Blockage of tumor-induced DC dysfunction by targeting signaling molecules or pathways may restore DC function. Inhibitors to signaling molecules are already under investigation in clinical trials as therapeutic agents to treat cancers (Barclay et al. 2007; Chou et al. 2005; Demuth et al. 2007; Do et al. 2004; Jing et al. 2006; Jiang et al. 2007; Kirkwood et al. 2007; McKay et al. 2000; Yoshikawa et al. 2001). These antagonists as anti-cancer drugs are expected to not only improve DC function, but also and more importantly, may boost antitumor immunity in cancer patients. Even though these signaling pathways are pivotally important for normal cell proliferation and survival and blockade of them may possibly lead to toxicity in patients, some encouraging preliminary results have already been obtained from clinical trials that examine the efficacy of the specific inhibitors. In future studies, it will be important to identify novel and specific targets in these signaling pathways for cancer therapy.

References

Aboudola S, Murugesan G, Szpurka H, Ramsingh G, Zhao X, Prescott N et al (2007) Bone marrow phospho-STAT5 expression in non-CML chronic myeloproliferative disorders correlates with JAK2 V617F mutation and provides evidence of in vivo JAK2 activation. Am J Surg Pathol 31:233–239

Ade N, Antonios D, Kerdine-Romer S, Boisleve F, Rousset F, Pallardy M (2007) NF-kappaB plays a major role in the maturation of human dendritic cells induced by NiSO(4) but not by DNCB. Toxicol Sci 99:488–501

Akbari O, Umetsu DT (2005) Role of regulatory dendritic cells in allergy and asthma. Curr Allergy Asthma R 5:56–61

Almand B, Clark JI, Nikitina E, van Beynen J, English NR, Knight SC et al (2001) Increased production of immature myeloid cells in cancer patients: a mechanism of immunosuppression in cancer. J Immunol 166:678–689

Aloysius MM, Takhar A, Robins A, Eremin O (2006) Dendritic cell biology, dysfunction and immunotherapy in gastrointestinal cancers. Surg J Roy Coll Surg Edinb Irel 4:195–210

Aplin AE, Hogan BP, Tomeu J, Juliano RL (2002) Cell adhesion differentially regulates the nucleocytoplasmic distribution of active MAP kinases. J Cell Sci 115:2781–2790

Arce F, Breckpot K, Stephenson H, Karwacz K, Ehrenstein MR, Collins M et al (2011) Selective ERK activation differentiates mouse and human tolerogenic dendritic cells, expands antigen-specific regulatory T cells, and suppresses experimental inflammatory arthritis. Arthritis Rheum 63:84–95

Ardeshna KM, Pizzey AR, Devereux S, Khwaja A (2000) The PI3 kinase, p38 SAP kinase, and NF-kappaB signal transduction pathways are involved in the survival and maturation of lipopolysaccharide-stimulated human monocyte-derived dendritic cells. Blood 96:1039–1046

Barclay JL, Anderson ST, Waters MJ, Curlewis JD (2007) Characterization of the SOCS3 promoter response to prostaglandin E2 in T47D cells. Mol Endocrinol 21:2516–2528

Beauchamp NM, Busick RY, Alexander-Miller MA (2010) Functional divergence among CD103 + dendritic cell subpopulations following pulmonary poxvirus infection. J Virol 84:10191–10199

Bedoui S, Whitney PG, Waithman J, Eidsmo L, Wakim L, Caminschi I et al (2009) Cross-presentation of viral and self antigens by skin-derived CD103 + dendritic cells. Nat Immunol 10:488–495
Belz GT, Smith CM, Kleinert L, Reading P, Brooks A, Shortman K et al (2004) Distinct migrating and nonmigrating dendritic cell populations are involved in MHC class I-restricted antigen presentation after lung infection with virus. Proc Natl Acad Sci USA 101:8670–8675
Bendix I, Pfueller CF, Leuenberger T, Glezeva N, Siffrin V, Muller Y et al (2010) MAPK3 deficiency drives autoimmunity via DC arming. Eur J Immunol 40:1486–1495
Bennaceur K, Chapman JA, Touraine JL, Portoukalian J (2009) Immunosuppressive networks in the tumour environment and their effect in dendritic cells. Biochim Biophys Acta 1795:16–24
Bharadwaj U, Zhang R, Yang H, Li M, Doan LX, Chen C et al (2005) Effects of cyclophilin A on myeloblastic cell line KG-1 derived dendritic like cells (DLC) through p38 MAP kinase activation. J Surg Res 127:29–38
Budagian V, Bulanova E, Brovko L, Orinska Z, Fayad R, Paus R et al (2003) Signaling through P2X7 receptor in human T cells involves p56lck, MAP kinases, and transcription factors AP-1 and NF-kappa B. J Biol Chem 278:1549–1560
Campbell GS, Yu CL, Jove R, Carter-Su C (1997) Constitutive activation of JAK1 in Src-transformed cells. J Biol Chem 272:2591–2594
Canesi L, Lorusso LC, Ciacci C, Betti M, Gallo G (2005) Effects of the brominated flame retardant tetrabromobisphenol-A (TBBPA) on cell signaling and function of Mytilus hemocytes: involvement of MAP kinases and protein kinase C. Aquat Toxicol 75:277–287
Cheng CH, Yu KC, Chen HL, Chen SY, Huang CH, Chan PC et al (2004) Blockade of v-Src-stimulated tumor formation by the Src homology 3 domain of Crk-associated substrate (Cas). FEBS Lett 557:221–227
Chou SD, Khan AN, Magner WJ, Tomasi TB (2005) Histone acetylation regulates the cell type specific CIITA promoters, MHC class II expression and antigen presentation in tumor cells. Int Immunol 17:1483–1494
Cools N, Ponsaerts P, Van Tendeloo VF, Berneman ZN (2007) Balancing between immunity and tolerance: an interplay between dendritic cells, regulatory T cells, and effector T cells. J Leukoc Biol 82:1365–1374
Cruz MT, Duarte CB, Goncalo M, Carvalho AP, Lopes MC (1999) Involvement of JAK2 and MAPK on type II nitric oxide synthase expression in skin-derived dendritic cells. Am J Physiol 277:C1050–C1057
Demuth T, Reavie LB, Rennert JL, Nakada M, Nakada S, Hoelzinger DB et al (2007) MAP-ing glioma invasion: mitogen-activated protein kinase kinase 3 and p38 drive glioma invasion and progression and predict patient survival. Mol Cancer Ther 6:1212–1222
den Haan JM, Bevan MJ (2002) Constitutive versus activation-dependent cross-presentation of immune complexes by CD8(+) and CD8(-) dendritic cells in vivo. J Exp Med 196:817–827
Di Nicola M, Lemoli RM (2000) Dendritic cells: specialized antigen presenting cells. Haematologica 85:202–207
Do Y, Hegde VL, Nagarkatti PS, Nagarkatti M (2004) Bryostatin-1 enhances the maturation and antigen-presenting ability of murine and human dendritic cells. Cancer Res 64:6756–6765
Dong H, Strome SE, Salomao DR, Tamura H, Hirano F, Flies DB et al (2002) Tumor-associated B7–H1 promotes T-cell apoptosis: a potential mechanism of immune evasion. Nat Med 8:793–800
Escors D, Lopes L, Lin R, Hiscott J, Akira S, Davis RJ et al (2008) Targeting dendritic cell signaling to regulate the response to immunization. Blood 111:3050–3061
Evel-Kabler K, Chen SY (2006) Dendritic cell-based tumor vaccines and antigen presentation attenuators. Mol Ther J Am Soc Gene Ther 13:850–858
Fallarino F, Grohmann U, Hwang KW, Orabona C, Vacca C, Bianchi R et al (2003) Modulation of tryptophan catabolism by regulatory T cells. Nat Immunol 4:1206–1212
Fallarino F, Grohmann U, You S, McGrath BC, Cavener DR, Vacca C et al (2006) The combined effects of tryptophan starvation and tryptophan catabolites down-regulate T cell receptor zeta-chain and induce a regulatory phenotype in naive T cells. J Immunol 176:6752–6761

Francisco LM, Salinas VH, Brown KE, Vanguri VK, Freeman GJ, Kuchroo VK et al (2009) PD-L1 regulates the development, maintenance, and function of induced regulatory T cells. J Exp Med 206:3015–3029

Gabrilovich DI (2002) Dendritic cell vaccines for cancer treatment. Curr Opin Mol Ther 4:452–458

Gabrilovich DI, Nadaf S, Corak J, Berzofsky JA, Carbone DP (1996) Dendritic cells in antitumor immune responses. II: dendritic cells grown from bone marrow precursors, but not mature DC from tumor-bearing mice, are effective antigen carriers in the therapy of established tumors. Cell Immunol 170:111–119

Galon J, Costes A, Sanchez-Cabo F, Kirilovsky A, Mlecnik B, Lagorce-Pages C et al (2006) Type, density, and location of immune cells within human colorectal tumors predict clinical outcome. Science 313:1960–1964

Ghahary A, Li Y, Tredget EE, Kilani RT, Iwashina T, Karami A et al (2004) Expression of indoleamine 2,3-dioxygenase in dermal fibroblasts functions as a local immunosuppressive factor. J Invest Dermatol 122:953–964

Gregori S (2011) Dendritic cells in networks of immunological tolerance. Tissue Antigens 77:89–99

Hartmann E, Wollenberg B, Rothenfusser S, Wagner M, Wellisch D, Mack B et al (2003) Identification and functional analysis of tumor-infiltrating plasmacytoid dendritic cells in head and neck cancer. Cancer Res 63:6478–6487

Hoentjen F, Sartor RB, Ozaki M, Jobin C (2005) STAT3 regulates NF-kappaB recruitment to the IL-12p40 promoter in dendritic cells. Blood 105:689–696

Hubert P, Jacobs N, Caberg JH, Boniver J, Delvenne P (2007) The cross-talk between dendritic and regulatory T cells: good or evil? J Leukoc Biol 82:781–794

Ichijo H (1999) From receptors to stress-activated MAP kinases. Oncogene 18:6087–6093

Isomura I, Shintani Y, Yasuda Y, Tsujimura K, Morita A (2008) Induction of regulatory dendritic cells by topical application of NF-kappaB decoy oligodeoxynucleotides. Immunol Lett 119:49–56

Iwata-Kajihara T, Sumimoto H, Kawamura N, Ueda R, Takahashi T, Mizuguchi H et al (2011) Enhanced cancer immunotherapy using STAT3-depleted dendritic cells with high Th1-inducing ability and resistance to cancer cell-derived inhibitory factors. J Immunol 187:27–36

Jackson AM, Mulcahy LA, Zhu XW, O'Donnell D, Patel PM (2008) Tumour-mediated disruption of dendritic cell function: inhibiting the MEK1/2-p44/42 axis restores IL-12 production and Th1-generation. Int J Cancer (J Int Cancer) 123:623–632

Janikashvili N, Bonnotte B, Katsanis E, Larmonier N (2011) The dendritic cell-regulatory T lymphocyte crosstalk contributes to tumor-induced tolerance. Clin Dev Immunol 2011:430394

Jarnicki AG, Conroy H, Brereton C, Donnelly G, Toomey D, Walsh K et al. (2008) Attenuating regulatory T cell induction by TLR agonists through inhibition of p38 MAPK signaling in dendritic cells enhances their efficacy as vaccine adjuvants and cancer immunotherapeutics. J Immunol (Baltimore Md: 1950) 180:3797–3806

Jiang JL, Wang S, Li NS, Zhang XH, Deng HW, Li YJ (2007) The inhibitory effect of simvastatin on the ADMA-induced inflammatory reaction is mediated by MAPK pathways in endothelial cells. Biochem Cell Biol 85:66–77

Jing N, Zhu Q, Yuan P, Li Y, Mao L, Tweardy DJ (2006) Targeting signal transducer and activator of transcription 3 with G-quartet oligonucleotides: a potential novel therapy for head and neck cancer. Mol Cancer Ther 5:279–286

Jost E (2007) do ON, Dahl E, Maintz CE, Jousten P, Habets L, et al. Epigenetic alterations complement mutation of JAK2 tyrosine kinase in patients with BCR/ABL-negative myeloproliferative disorders. Leukemia 21:505–510

Kandilci A, Grosveld GC (2005) SET-induced calcium signaling and MAPK/ERK pathway activation mediate dendritic cell-like differentiation of U937 cells. Leukemia 19:1439–1445

Kang HK, Lee HY, Lee YN, Jo EJ, Kim JI, Aosai F et al (2004) Toxoplasma gondii-derived heat shock protein 70 stimulates the maturation of human monocyte-derived dendritic cells. Biochem Biophys Res Commun 322:899–904

Kawakami Y, Rodriguez-Leon J, Koth CM, Buscher D, Itoh T, Raya A et al (2003) MKP3 mediates the cellular response to FGF8 signalling in the vertebrate limb.[see comment]. Nat Cell Biol 5:513–519

Keir ME, Butte MJ, Freeman GJ, Sharpe AH (2008) PD-1 and its ligands in tolerance and immunity. Annu Rev Immunol 26:677–704

Kim GY, Kim KH, Lee SH, Yoon MS, Lee HJ, Moon DO et al (2005) Curcumin inhibits immunostimulatory function of dendritic cells: MAPKs and translocation of NF-kappa B as potential targets. J Immunol 174:8116–8124

Kim R, Emi M, Tanabe K, Arihiro K (2006a) Tumor-driven evolution of immunosuppressive networks during malignant progression. Cancer Res 66:5527–5536

Kim R, Emi M, Tanabe K (2006b) Cancer immunosuppression and autoimmune disease: beyond immunosuppressive networks for tumour immunity. Immunology 119:254–264

Kirkwood KL, Li F, Rogers JE, Otremba J, Coatney DD, Kreider JM et al (2007) A p38alpha selective mitogen-activated protein kinase inhibitor prevents periodontal bone loss. J Pharmacol Exp Ther 320:56–63

Kitamura H, Kamon H, Sawa S, Park SJ, Katunuma N, Ishihara K et al (2005) IL-6-STAT3 controls intracellular MHC class II alphabeta dimer level through cathepsin S activity in dendritic cells. Immunity 23:491–502

Koido S, Homma S, Hara E, Namiki Y, Takahara A, Komita H et al (2010) Regulation of tumor immunity by tumor/dendritic cell fusions. Clin Dev Immunol 2010:516768

Kortylewski M, Jove R, Yu H (2005a) Targeting STAT3 affects melanoma on multiple fronts. Cancer Metastasis Rev 24:315–327

Kortylewski M, Kujawski M, Wang T, Wei S, Zhang S, Pilon-Thomas S et al (2005b) Inhibiting Stat3 signaling in the hematopoietic system elicits multicomponent antitumor immunity. Nat Med 11:1314–1321

Koski GK, Cohen PA, Roses RE, Xu S, Czerniecki BJ (2008) Reengineering dendritic cell-based anti-cancer vaccines. Immunol Rev 222:256–276

Kurte M, Lopez M, Aguirre A, Escobar A, Aguillon JC, Charo J et al (2004) A synthetic peptide homologous to functional domain of human IL-10 down-regulates expression of MHC class I and Transporter associated with Antigen Processing 1/2 in human melanoma cells. J Immunol 173:1731–1737

Kusmartsev S, Gabrilovich DI (2002) Immature myeloid cells and cancer-associated immune suppression. Cancer Immunol Immunother 51:293–298

Kusmartsev S, Gabrilovich DI (2006a) Effect of tumor-derived cytokines and growth factors on differentiation and immune suppressive features of myeloid cells in cancer. Cancer Metastasis Rev 25:323–331

Kusmartsev S, Gabrilovich DI (2006b) Role of immature myeloid cells in mechanisms of immune evasion in cancer. Cancer Immunol Immunother: CII 55:237–245

Kwon HK, Lee CG, So JS, Chae CS, Hwang JS, Sahoo A et al (2010) Generation of regulatory dendritic cells and CD4 + Foxp3 + T cells by probiotics administration suppresses immune disorders. Proc Natl Acad Sci USA 107:2159–2164

Larmonier N, Marron M, Zeng Y, Cantrell J, Romanoski A, Sepassi M et al (2007) Tumor-derived CD4(+)CD25(+) regulatory T cell suppression of dendritic cell function involves TGF-beta and IL-10. Cancer Immunol Immunother: CII 56:48–59

Lee HJ, Cho JW, Kim SC, Kang KH, Lee SK, Pi SH et al (2006) Roles of p38 and ERK MAP kinases in IL-8 expression in TNF-alpha- and dexamethasone-stimulated human periodontal ligament cells. Cytokine 35:67–76

Li J, Yu B, Song L, Eschrich S, Haura EB (2007) Effects of IFN-gamma and Stat1 on gene expression, growth, and survival in non-small cell lung cancer cells. J Interferon Cytokine Res 27:209–220

Li Q, Guo Z, Xu X, Xia S, Cao X (2008a) Pulmonary stromal cells induce the generation of regulatory DC attenuating T-cell-mediated lung inflammation. Eur J Immunol 38:2751–2761

Li H, Zhang GX, Chen Y, Xu H, Fitzgerald DC, Zhao Z et al (2008b) CD11c + CD11b + dendritic cells play an important role in intravenous tolerance and the suppression of experimental autoimmune encephalomyelitis. J Immunol 181:2483–2493

Lin WW, Karin M (2007) A cytokine-mediated link between innate immunity, inflammation, and cancer. J Clin Investig 117:1175–1183

Lin A, Schildknecht A, Nguyen LT, Ohashi PS (2010) Dendritic cells integrate signals from the tumor microenvironment to modulate immunity and tumor growth. Immunol Lett 127:77–84

Liu YJ (2005) IPC: professional type 1 interferon-producing cells and plasmacytoid dendritic cell precursors. Annu Rev Immunol 23:275–306

Liu J, Xiang Z, Ma X (2004) Role of IFN regulatory factor-1 and IL-12 in immunological resistance to pathogenesis of N-methyl-N-nitrosourea-induced T lymphoma. J Immunol 173:1184–1193

Liu Q, Zhang C, Sun A, Zheng Y, Wang L, Cao X (2009) Tumor-educated CD11bhighIalow regulatory dendritic cells suppress T cell response through arginase I. J Immunol 182:6207–6216

Lu Y, Hong S, Li H, Park J, Hong B, Wang L et al (2012) Th9 cells promote antitumor immune responses in vivo. J Clin Investig 122:4160–4171

Ma Y, Shurin GV, Gutkin DW, Shurin MR (2012) Tumor associated regulatory dendritic cells. Semin Cancer Biol 22:298–306

Manicassamy S, Pulendran B (2011) Dendritic cell control of tolerogenic responses. Immunol Rev 241:206–227

Marigo I, Dolcetti L, Serafini P, Zanovello P, Bronte V (2008) Tumor-induced tolerance and immune suppression by myeloid derived suppressor cells. Immunol Rev 222:162–179

Mata R, Subira D, Garcia-Raso A, Llamas P (2007) JAK2 as a molecular marker in myeloproliferative diseases. Cardiovasc Hematol Agents Med Chemistry. 5:198–203

Matos TJ, Duarte CB, Goncalo M, Lopes MC (2005a) Role of oxidative stress in ERK and p38 MAPK activation induced by the chemical sensitizer DNFB in a fetal skin dendritic cell line. Immunol Cell Biol 83:607–614

Matos TJ, Duarte CB, Goncalo M, Lopes MC (2005b) DNFB activates MAPKs and upregulates CD40 in skin-derived dendritic cells. J Dermatol Sci 39:113–123

McKay DM, Botelho F, Ceponis PJ, Richards CD (2000) Superantigen immune stimulation activates epithelial STAT-1 and PI 3-K: PI 3-K regulation of permeability. Am J Physiol: Gastrointest Liver Physiol 279:G1094–G1103

Munder M (2009) Arginase: an emerging key player in the mammalian immune system. Br J Pharmacol 158:638–651

Munn DH, Sharma MD, Lee JR, Jhaver KG, Johnson TS, Keskin DB et al (2002) Potential regulatory function of human dendritic cells expressing indoleamine 2,3-dioxygenase. Science 297:1867–1870

Munn DH, Sharma MD, Baban B, Harding HP, Zhang Y, Ron D et al (2005) GCN2 kinase in T cells mediates proliferative arrest and anergy induction in response to indoleamine 2, 3-dioxygenase. Immunity 22:633–642

Nakahara T, Uchi H, Urabe K, Chen Q, Furue M, Moroi Y (2004) Role of c-Jun N-terminal kinase on lipopolysaccharide induced maturation of human monocyte-derived dendritic cells. Int Immunol 16:1701–1709

Nakamura K, Zhou CJ, Parente J, Chew CS (1996) Parietal cell MAP kinases: multiple activation pathways. Am J Physiol 271:G640–G649

Nefedova Y, Huang M, Kusmartsev S, Bhattacharya R, Cheng P, Salup R et al (2004) Hyperactivation of STAT3 is involved in abnormal differentiation of dendritic cells in cancer. J Immunol 172:464–474

Nefedova Y, Cheng P, Gilkes D, Blaskovich M, Beg AA, Sebti SM et al (2005) Activation of dendritic cells via inhibition of Jak2/STAT3 signaling. J Immunol 175:4338–4346

Ni XY, Sui HX, Liu Y, Ke SZ, Wang YN, Gao FG (2012) TGF-beta of lung cancer microenvironment upregulates B7H1 and GITRL expression in dendritic cells and is associated with regulatory T cell generation. Oncol Rep 28:615–621

Norian LA, Rodriguez PC, O'Mara LA, Zabaleta J, Ochoa AC, Cella M et al (2009) Tumor-infiltrating regulatory dendritic cells inhibit CD8 + T cell function via L-arginine metabolism. Cancer Res 69:3086–3094

Obermajer N, Muthuswamy R, Lesnock J, Edwards RP, Kalinski P (2011) Positive feedback between PGE2 and COX2 redirects the differentiation of human dendritic cells toward stable myeloid-derived suppressor cells. Blood 118:5498–5505

O'Neill DW, Adams S, Bhardwaj N (2004) Manipulating dendritic cell biology for the active immunotherapy of cancer. Blood 104:2235–2246

Osawa Y, Iho S, Takauji R, Takatsuka H, Yamamoto S, Takahashi T et al (2006) Collaborative action of NF-kappaB and p38 MAPK is involved in CpG DNA-induced IFN-alpha and chemokine production in human plasmacytoid dendritic cells. J Immunol 177:4841–4852

Palucka K, Ueno H, Fay J, Banchereau J (2009) Harnessing dendritic cells to generate cancer vaccines. Ann N Y Acad Sci 1174:88–98

Palucka K, Ueno H, Roberts L, Fay J, Banchereau J (2010a) Dendritic cells: are they clinically relevant? Cancer J 16:318–324

Palucka K, Ueno H, Zurawski G, Fay J, Banchereau J (2010b) Building on dendritic cell subsets to improve cancer vaccines. Curr Opin Immunol 22:258–263

Palucka K, Ueno H, Fay J, Banchereau J (2011) Dendritic cells and immunity against cancer. J Intern Med 269:64–73

Park SM, Kim HS, Choe J, Lee TH (1999) Differential induction of cytokine genes and activation of mitogen-activated protein kinase family by soluble CD40 ligand and TNF in a human follicular dendritic cell line. J Immunol 163:631–638

Park S, Kim D, Kaneko S, Szewczyk KM, Nicosia SV, Yu H et al (2005) Molecular cloning and characterization of the human AKT1 promoter uncovers its up-regulation by the Src/Stat3 pathway. J Biol Chem 280:38932–38941

Philpott NJ, Nociari M, Elkon KB, Falck-Pedersen E (2004) Adenovirus-induced maturation of dendritic cells through a PI3 kinase-mediated TNF-alpha induction pathway. Proc Natl Acad Sci USA 101:6200–6205

Pinzon-Charry A, Maxwell T, Lopez JA (2005) Dendritic cell dysfunction in cancer: a mechanism for immunosuppression. Immunol Cell Biol 83:451–461

Rabinovich GA, Gabrilovich D, Sotomayor EM (2007) Immunosuppressive strategies that are mediated by tumor cells. Annu Rev Immunol 25:267–296

Ramos RN, Chin LS, Dos Santos AP, Bergami-Santos PC, Laginha F Barbuto JA (2012) Monocyte-derived dendritic cells from breast cancer patients are biased to induce CD4 + CD25 + Foxp3 + regulatory T cells. J Leukoc Biol 92:673–682

Randolph GJ, Sanchez-Schmitz G, Angeli V (2005) Factors and signals that govern the migration of dendritic cells via lymphatics: recent advances. Springer Semin Immunopathol 26:273–287

Ratta M, Fagnoni F, Curti A, Vescovini R, Sansoni P, Oliviero B et al (2002) Dendritic cells are functionally defective in multiple myeloma: the role of interleukin-6. Blood 100:230–237

Riboldi E, Musso T, Moroni E, Urbinati C, Bernasconi S, Rusnati M et al (2005) Cutting edge: proangiogenic properties of alternatively activated dendritic cells. J Immunol 175:2788–2792

Saito Y, Yanagawa Y, Kikuchi K, Iijima N, Iwabuchi K, Onoe K (2006) Low-dose lipopolysaccharide modifies the production of IL-12 by dendritic cells in response to various cytokines. J Clin Exp Hematopathology 46:31–36

Santini SM, Belardelli F (2003) Advances in the use of dendritic cells and new adjuvants for the development of therapeutic vaccines. Stem Cells 21:495–505

Scarlett UK, Rutkowski MR, Rauwerdink AM, Fields J, Escovar-Fadul X, Baird J et al (2012) Ovarian cancer progression is controlled by phenotypic changes in dendritic cells. J Exp Med 209:495–506

Schreiber RD, Old LJ, Smyth MJ (2011) Cancer immunoediting: integrating immunity's roles in cancer suppression and promotion. Science 331:1565–1570

Shankaran V, Ikeda H, Bruce AT, White JM, Swanson PE, Old LJ et al (2001) IFNgamma and lymphocytes prevent primary tumour development and shape tumour immunogenicity. Nature 410:1107–1111

Sharma MD, Baban B, Chandler P, Hou DY, Singh N, Yagita H et al (2007) Plasmacytoid dendritic cells from mouse tumor-draining lymph nodes directly activate mature Tregs via indoleamine 2, 3-dioxygenase. J Clin Investig 117:2570–2582

Sharma MD, Hou DY, Liu Y, Koni PA, Metz R, Chandler P et al (2009) Indoleamine 2,3-dioxygenase controls conversion of Foxp3 + Tregs to TH17-like cells in tumor-draining lymph nodes. Blood 113:6102–6111

Sheng KC, Pietersz GA, Wright MD, Apostolopoulos V (2005) Dendritic cells: activation and maturation–applications for cancer immunotherapy. Curr Med Chem 12:1783–1800

Sheng KC, Wright MD, Apostolopoulos V (2011) Inflammatory mediators hold the key to dendritic cell suppression and tumor progression. Curr Med Chem 18:5507–5518

Shurin MR, Shurin GV, Lokshin A, Yurkovetsky ZR, Gutkin DW, Chatta G et al (2006) Intratumoral cytokines/chemokines/growth factors and tumor infiltrating dendritic cells: friends or enemies? Cancer Metastasis Rev 25:333–356

Shurin GV, Ouellette CE, Shurin MR (2012) Regulatory dendritic cells in the tumor immunoenvironment. Cancer Immunol Immunother: CII 61:223–230

Sigaud S, Evelson P, Gonzalez-Flecha B (2005) H2O2-induced proliferation of primary alveolar epithelial cells is mediated by MAP kinases. Antioxid Redox Signal 7:6–13

Simeone E, Ascierto PA (2012) Immunomodulating antibodies in the treatment of metastatic melanoma: the experience with anti-CTLA-4, anti-CD137, and anti-PD1. J Immunotoxicol 9:241–247

Sinkovics JG, Horvath JC (2000) Vaccination against human cancers (review). Int J Oncol 16:81–96

Steinbrink K, Mahnke K, Grabbe S, Enk AH, Jonuleit H (2009) Myeloid dendritic cell: From sentinel of immunity to key player of peripheral tolerance? Hum Immunol 70:289–293

Steinman RM, Banchereau J (2007) Taking dendritic cells into medicine. Nature 449:419–426

Steinman RM, Hawiger D, Nussenzweig MC (2003) Tolerogenic dendritic cells. Annu Rev Immunol 21:685–711

Stoitzner P, Green LK, Jung JY, Price KM, Atarea H, Kivell B et al (2008) Inefficient presentation of tumor-derived antigen by tumor-infiltrating dendritic cells. Cancer Immunol Immunother: CII 57:1665–1673

Sumimoto H, Miyagishi M, Miyoshi H, Yamagata S, Shimizu A, Taira K et al (2004) Inhibition of growth and invasive ability of melanoma by inactivation of mutated BRAF with lentivirus-mediated RNA interference. Oncogene 23:6031–6039

Sumimoto H, Imabayashi F, Iwata T, Kawakami Y (2006) The BRAF-MAPK signaling pathway is essential for cancer-immune evasion in human melanoma cells. J Exp Med 203:1651–1656

Sun Y, Chin YE, Weisiger E, Malter C, Tawara I, Toubai T et al (2009) Cutting edge: Negative regulation of dendritic cells through acetylation of the nonhistone protein STAT-3. J Immunol 182:5899–5903

Taki T, Taniwaki M (2006) Chromosomal translocations in cancer and their relevance for therapy. Curr Opin Oncol 18:62–68

Tanami H, Imoto I, Hirasawa A, Yuki Y, Sonoda I, Inoue J et al (2004) Involvement of overexpressed wild-type BRAF in the growth of malignant melanoma cell lines. Oncogene 23:8796–8804

Tang H, Guo Z, Zhang M, Wang J, Chen G, Cao X (2006) Endothelial stroma programs hematopoietic stem cells to differentiate into regulatory dendritic cells through IL-10. Blood 108:1189–1197

Tassiulas I, Park-Min KH, Hu Y, Kellerman L, Mevorach D, Ivashkiv LB (2007) Apoptotic cells inhibit LPS-induced cytokine and chemokine production and IFN responses in macrophages. Hum Immunol 68:156–164

Thompson RH, Dong H, Lohse CM, Leibovich BC, Blute ML, Cheville JC et al (2007) PD-1 is expressed by tumor-infiltrating immune cells and is associated with poor outcome for patients with renal cell carcinoma. Clin Cancer Res Official J Am Assoc Cancer Res 13:1757–1761

van den Broek ME, Kagi D, Ossendorp F, Toes R, Vamvakas S, Lutz WK et al (1996) Decreased tumor surveillance in perforin-deficient mice. J Exp Med 184:1781–1790

Vicari AP, Caux C, Trinchieri G (2002) Tumour escape from immune surveillance through dendritic cell inactivation. Semin Cancer Biol 12:33–42

Wang T, Niu G, Kortylewski M, Burdelya L, Shain K, Zhang S et al. (2004) Regulation of the innate and adaptive immune responses by Stat-3 signaling in tumor cells. Nat Med 10:48–54 (erratum appears in Nat Med 2004 Feb; 10(2):209)

Wang T, Niu G, Kortylewski M, Burdelya L, Shain K, Zhang S et al (2004b) Regulation of the innate and adaptive immune responses by Stat-3 signaling in tumor cells. Nat Med 10:48–54

Wang S, Yang J, Qian J, Wezeman M, Kwak LW, Yi Q (2006a) Tumor evasion of the immune system: inhibiting p38 MAPK signaling restores the function of dendritic cells in multiple myeloma. Blood 107:2432–2439

Wang S, Hong S, Yang J, Qian J, Zhang X, Shpall E et al (2006b) Optimizing immunotherapy in multiple myeloma: Restoring the function of patients' monocyte-derived dendritic cells by inhibiting p38 or activating MEK/ERK MAPK and neutralizing interleukin-6 in progenitor cells. Blood 108:4071–4077

Wei S, Kryczek I, Zou L, Daniel B, Cheng P, Mottram P et al (2005) Plasmacytoid dendritic cells induce CD8 + regulatory T cells in human ovarian carcinoma. Cancer Res 65:5020–5026

West MA, Wallin RP, Matthews SP, Svensson HG, Zaru R, Ljunggren HG et al (2004) Enhanced dendritic cell antigen capture via toll-like receptor-induced actin remodeling. Science 305:1153–1157

Xia S, Guo Z, Xu X, Yi H, Wang Q, Cao X (2008) Hepatic microenvironment programs hematopoietic progenitor differentiation into regulatory dendritic cells, maintaining liver tolerance. Blood 112:3175–3185

Xie J, Qian J, Yang J, Wang S, Freeman ME 3rd, Yi Q (2005) Critical roles of Raf/MEK/ERK and PI3 K/AKT signaling and inactivation of p38 MAP kinase in the differentiation and survival of monocyte-derived immature dendritic cells. Exp Hematol 33:564–572

Yang L, Carbone DP (2004) Tumor-host immune interactions and dendritic cell dysfunction. Adv Cancer Res 92:13–27

Yoshikawa H, Matsubara K, Qian GS, Jackson P, Groopman JD, Manning JE et al (2001) SOCS-1, a negative regulator of the JAK/STAT pathway, is silenced by methylation in human hepatocellular carcinoma and shows growth-suppression activity.[see comment]. Nat Genet 28:29–35

Yu CL, Meyer DJ, Campbell GS, Larner AC, Carter-Su C, Schwartz J et al (1995) Enhanced DNA-binding activity of a Stat3-related protein in cells transformed by the Src oncoprotein. Science 269:81–83

Yu H, Kortylewski M, Pardoll D (2007) Crosstalk between cancer and immune cells: role of STAT3 in the tumour microenvironment. Nat Rev Immunol 7:41–51

Zaru R, Ronkina N, Gaestel M, Arthur JS, Watts C (2007) The MAPK-activated kinase Rsk controls an acute Toll-like receptor signaling response in dendritic cells and is activated through two distinct pathways. Nat Immunol 8:1227–1235

Zeng L, O'Connor C, Zhang J, Kaplan AM, Cohen DA (2010) IL-10 promotes resistance to apoptosis and metastatic potential in lung tumor cell lines. Cytokine 49:294–302

Zhang L, Conejo-Garcia JR, Katsaros D, Gimotty PA, Massobrio M, Regnani G et al (2003) Intratumoral T cells, recurrence, and survival in epithelial ovarian cancer. N Engl J Med 348:203–213

Zhang M, Tang H, Guo Z, An H, Zhu X, Song W et al (2004) Splenic stroma drives mature dendritic cells to differentiate into regulatory dendritic cells. Nat Immunol 5:1124–1133

Zhao F, Falk C, Osen W, Kato M, Schadendorf D, Umansky V (2009) Activation of p38 mitogen-activated protein kinase drives dendritic cells to become tolerogenic in ret transgenic mice spontaneously developing melanoma. Clin Cancer Res Official J Am Assoc Cancer Res 15:4382–4390

Zou W (2005) Immunosuppressive networks in the tumour environment and their therapeutic relevance. Nat Rev Cancer 5:263–274

Zou GM, Hu WY (2005) LIGHT regulates CD86 expression on dendritic cells through NF-kappaB, but not JNK/AP-1 signal transduction pathway. J Cell Physiol 205:437–443

Chapter 15
Tumor Microenvironment may Shape the Function and Phenotype of NK Cells Through the Induction of Split Anergy and Generation of Regulatory NK Cells

Anahid Jewett and Han-Ching Tseng

Abstract Cytotoxic function of NK cells is suppressed in the tumor microenvironment by a number of distinct effectors. Furthermore, decreased peripheral blood NK cell cytotoxicity has been documented in cancer patients. We have previously demonstrated evidence for the role of NK cells in specific elimination of stem cells and not their differentiated counterparts. In this regard, NK cells were found to mediate significant cytotoxicity against primary oral squamous carcinoma stem cells (OSCSCs) as compared to their more differentiated oral squamous carcinoma cells (OSCCs). In addition, human embryonic stem cells (hESCs), human mesenchymal stem cells (hMSCs), human dental pulp stem cells (hDPSCs) and induced human pluripotent stem cells (hiPSCs) were all significantly more susceptible to NK cell mediated cytotoxicity than their differentiated counterparts or parental cells from which they were derived. There is also a stage wise susceptibility to NK cell mediated cytotoxicity in pancreatic tumors in which case the poorly differentiated tumors are lysed much more than their moderately differentiated tumors. The well differentiated pancreatic tumors were lysed the least when compared to either the moderately differentiated tumors or to poorly differentiated tumors. We have also reported that inhibition of differentiation or reversion of cells to a less-differentiated stage by blocking NF-κB or gene deletion of COX2 significantly augmented NK cell cytotoxicity against both transformed and healthy cells. Therefore, we propose that the two stages of NK cell maturation namely CD16 + CD56^{dim}CD69-NK cells are important for the selection of stem

A. Jewett · H.-C. Tseng
The Jane and Jerry Weintraub Center for Reconstructive Biotechnology, and Division of Oral Biology and Medicine, UCLA School of Dentistry, University of California, Los Angeles, CA, USA

A. Jewett (✉)
Division of Oral Biology and Medicine, UCLA School of Dentistry, 10833 Le Conte Ave, Los Angeles, CA 90095-1668, USA
e-mail: ajewett@ucla.edu

M. R. Shurin et al. (eds.), *The Tumor Immunoenvironment*,
DOI: 10.1007/978-94-007-6217-6_15,

cells whereas the CD16^{dim}/-CD56^{dim}/+ CD69 + NK cells are important for differentiation and eventual regeneration of the tissues and the resolution of inflammation, thus serving as regulatory NK cells (NK_{reg}). The concept of split anergy in NK cells and generation of NK_{reg} and its contribution to cell differentiation, tissue repair and regeneration and in tumor resistance will be discussed in this chapter.

Keywords Apoptosis · NF-κB · NK · Cancer stem cells · Differentiation · Regulation

15.1 Introduction

Advances in our understanding of anti-tumor immune responses and cancer biology have revealed a complex dynamic interaction between the immune effectors and the tumor cells. Effectors of the immune system are thought to shape the survival and maturation of tumor cells and to select for cancers with reduced immunogenicity. However, recent data from our laboratory indicted that the same effector mechanisms are likely responsible for shaping the survival and maturation of healthy stem cells for the ultimate goal of the regeneration of damaged tissues and the resolution of inflammation. Although, immunosuppression and tumor escape from immune recognition are thought to be major factors responsible for the establishment and progression of cancer, neither their underlying physiological significance nor the exact mechanisms by which immunosuppression occurs are completely understood.

NK cells arise from the bone marrow and constitute 5–15 % of total lymphocytes in the peripheral blood. They are known to mediate direct natural cytotoxicity as well as antibody-dependent cellular cytotoxicity (ADCC). By producing key cytokines and chemokines NK cells are known to regulate the functions of other immune cells (Fildes et al. 2008; Farag and Caligiuri 2006). Conventional human NK cells are identified by the expression of CD16 and CD56, and by the lack of surface CD3 expression. NK cells mediate their function through a number of important activating and inhibitory cell receptors listed in Table 15.1 (Pegram et al. 2011). It is thought that the balance between activating and inhibitory signals which NK cells receive from their surface receptors determines their functional fate (Pegram et al. 2011). Many of the receptors listed in Table 15.1 including CD16, killer immunoglobulin like receptors (KIR), NKG2 family of receptors which form a heterodimer with CD94, NKG2D and natural cytotoxicity receptors (NCR) have all been the subject of many studies. Likewise, several key cytokines, chemokines and adhesion molecules are found to have significant roles in maturation, differentiation, and effector function of NK cells. Much less is known regarding the function of Toll Like Receptors (TLRs), NOD-Like Receptors (NLRs) and RIG like Receptors (RLRs) in NK cell effector function.

Table 15.1 List of NK cell Activating and Inhibitory surface receptors and their ligands

Receptors	Ligands
Activating/inhibitory receptors	
FcγRIII (CD16)	Fc of antibodies
CD2	CD58 (LFA-3)
LFA-1	ICAM-1
2B4	CD48
CD69	Unknown
DNAM-1 (CD226)	CD112, CD155
NKp80	AICL
Tactile (CD96)	CD155, CD111
TIGIT	CD112, CD113, CD155
CRTAM	TSLC1
C-type lectin receptors—activating/inhibitory	
CD94/NKG2A/B	HLA-E
NKG2D	MICA, MICB, ULBP-1, ULBP-2, ULBP-3, ULBP-4, ULBP-5, ULBP-6
CD94/NKG2C	HLA-E
CD94/NKG2E/H	HLA-E, Qa-1b
Natural cytotoxicity receptors (NCR)	
NKp46 (NCR1)	Viral Hemagglutinin
NKp44 (NCR2)	Viral Hemagglutinin
NKp30 (NCR3)	B7h6, HCMV-pp65
Killer IG-like (KIR)—activating/Inhibitory	
KIR2DLs, KIR3DLs, KIR2DS	HLA-C, HLA-B, HLA-A, HLA-G
Cytokines, growth factors, chemokines and other adhesion receptors	Cytokines, growth factors, chemokines and other adhesion ligands
Toll-like receptors (TLR), NOD-like receptors (NLR) and RIG-I-like receptors (RLR)	Bacterial DNA, LPS, peptidoglycan, teichoic acids, flagellin, pilin, viral dsRNA and fungi zymosan

The extracellular domains of KIRs are homologous, irrespective of whether they are transmitting activating or inhibiting signals. However, the functional fate of these receptors are determined by their intracellular domains for which they either have an immunoreceptor tyrosine-based activation motif (ITAM) or immunoreceptor tyrosine-based inhibition motif (ITIM) responsible for delivering an activating or inhibitory signals respectively.

The association of distinct effector functions with certain NK cell subsets is thought to be developmentally regulated (Haas et al. 2011; Strowig et al. 2011). In this regard, previous studies have identified two distinct subsets of NK cells namely $CD56^{bright}CD16^{dim}$ and $CD56^{dim}CD16^{bright}$ subpopulations based on their phenotypic and functional analysis (Farag and Caligiuri 2006). The $CD56^{dim}CD16^{bright}$ NK subset is the major subset in the peripheral blood which mediates cytotoxicity whereas the $CD56^{bright}CD16^{dim}$ subset constitutes a minor subpopulation of NK cells in the peripheral blood and its role is in secretion of cytokines (Farag and Caligiuri 2006). $CD56^{bright}CD16^{dim}$ subset does not mediate cytotoxicity. The $CD56^{bright}CD16^{dim}$ NK cells are thought to be precursors to the $CD56^{dim}CD16^{bright}$ NK subset (Romagnani et al. 2007).

Although a lot is known about the inhibitory and activating receptors that modulate the function of NK cells, and many previous studies have indicated that NK cells may recognize and become activated by irradiated or stressed cells (Farag and Caligiuri 2006; Pegram et al. 2011), no previous studies have shown the role of NK cells in recognition, selection and differentiation of stem cells and their potential role in the resolution of inflammation. In this chapter we provide data regarding the factors and mechanisms involved in shaping the function of NK cells in cancer, and after interaction with healthy stem cells, and furthermore we discuss the emerging view from our laboratory which indicates that the NK cells may behave as the effectors of selection, differentiation and resistance of undifferentiated or stem like cells. The concept of split anergy in NK cells and its role in the switch of NK cell function from effector to regulatory cell function and its contribution to cell differentiation, tissue repair and regeneration and in tumor resistance will be discussed in this chapter.

15.2 Cancer Stem-Like Tumors or Poorly Differentiated Tumors as well as Non-transformed Stem Cells are Lysed Significantly More by the NK Cells when Compared to Their Differentiated Counterparts

Increased NK cell cytotoxicity and augmented secretion of IFN-γ were observed when NK cells were co-incubated with OSCSCs which released significantly lower levels of GM-CSF, IL-6 and IL-8 and demonstrated decreased expression of phospho-Stat3, B7H1 and EGFR, and much lower constitutive NF-κB activity when compared to differentiated OSCCs (Fig. 15.1a) and (Tseng et al. 2010). More importantly, OSCSCs expressed CD133 and $CD44^{bright}$ oral stem cell markers (Tseng et al. 2010), whereas differentiated OSCCs express lower CD44 surface receptors. To assess whether the stage of differentiation of other tumor types also correlated with their sensitivity to NK cell mediated lysis we selected five pancreatic lines at different stages of differentiation based on a number of criteria including sphere formation and immunohistochemical analysis. Panc-1 and MP-2, two poorly differentiated, BXPC3 and HPAF, two moderately differentiated and CAPAN-1, a well differentiated pancreatic tumors were co-cultured with the NK cells and NK cell mediated cytotoxicity were determined in a 4 h ^{51}Cr release assay. There was a significant correlation between the stage of differentiation of the tumors and the level of NK cell mediated lysis (Fig. 15.1b). The highest NK cell cytotoxicity was obtained against poorly differentiated tumors Panc-1 and MP-2, intermediate lysis against moderately differentiated BXPC3 and HPAF and the lowest lysis was obtained against well differentiated CAPAN-1 cells (Fig. 15.1b). Both untreated and IL-2 treated NK cells lysed poorly differentiated tumors much more than the differentiated tumors. Anti-CD16mAb treatment of NK cells

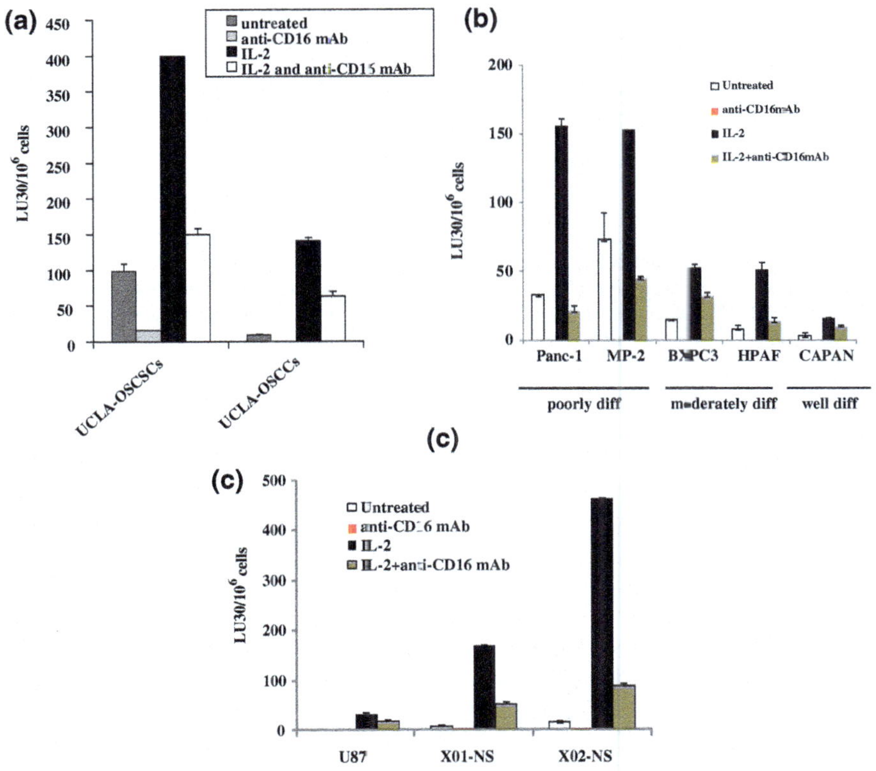

Fig. 15.1 **a** Augmented NK cell cytotoxicity against OSCSCs as compared to differentiated OSCCs; NK cells were left untreated or treated with IL-2 (1,000 u/ml) or anti-CD16 mAb (3 μg/ml) or a combination of IL-2 (1,000 u/ml) and anti-CD16 mAb (3 μg/ml) for 12–24 h before they were added to ^{51}Cr labeled primary oral tumors. NK cell cytotoxicity was determined using a standard ^{51}Cr release assay and the lytic units $30/10^6$ cells were calculated using the inverse number of effectors required to lyse 30 % of the tumor cells $\times 100$. Differences between untreated, anti- CD16 mAb treated or IL-2 and/or anti-CD16 mAb treated NK cell cytotoxicity between OSCCs and OSCSCs were significant at a p value of < 0.05. One of four representative experiments is shown in this figure. **b** Correlation between the stage of differentiation of pancreatic tumors and susceptibility to NK cell mediated cytotoxicity. Treatments of NK cells were carried out as described in Fig. 15.1a before they were used in cytotoxicity assay against poorly differentiated Panc-1 and MP2, moderately differentiated BXPC3 and HPAF and well differentiated CAPAN-1 pancreatic tumors. The lytic units $30/10^6$ were determined as described above. Differences between untreated, IL-2 treated or IL-2 and anti-CD16mAb treated NK cell killing between poorly differentiated as compared to moderately differentiated and well differentiated tumors are significant at a p value of < 0.05. One of five representative experiments is shown in this figure. **c** XO1-NS and XO2-NS GBM stem-like tumors are lysed significantly more than differentiated U87 GBMs. Treatments of NK cells were carried out as described in Fig. 15.1a before they were washed and used in cytotoxicity assay against XO1-NS, XO2-NS and U87 GBMs. The lytic units $30/10^6$ were determined as described above. Differences between untreated, anti-CD16mAb treated or IL-2 and/or anti-CD16mAb treated NK cell killing between XO1-NS or XO2-NS and U87 GBMs were significant at a p value of < 0.05. One of four representative experiments is shown in this figure

Table 15.2 Susceptibility of healthy stem cells but not their differentiated counterparts to NK cell mediated cytotoxicity

Type of cells	Stem cells	Differentiated	Parental cells
hESCs	++++++	+	
hMSCs	++++++	+	
hDPSCs	++++++	+	
hiPSCs	++++++		+
GBMs	++++++	+	

The susceptibility of a number of stem cells and their differentiated counterparts and the parental cells from which hiPSCs were derived to NK cell mediated cytotoxicity was determined using standard 4 h 51Cr release assay (Tseng et al. 2010). The higher the number of + signs the greater the susceptibility of stem cells to NK cell mediated cytotoxicity

abolished NK cell mediated cytotoxicity against all tumors and the combination of IL-2 and anti-CD16mAb significantly reduced IL-2 mediated lysis as expected (Fig. 15.1b). In addition, two Glioblastoma Multiforme (GBM) stem-like tumors XO1-NS and XO2-NS which were previously isolated and characterized (Oka et al. 2007; Inagaki et al. 2007; Soeda et al. 2008) were found to be significantly more susceptible to NK cell mediated cytotoxicity when compared to differentiated U87 GBM tumors (Fig. 15.1c). Since most stem-like tumors or poorly differentiated cells were significantly more susceptible to NK cell mediated cytotoxicity we reasoned that healthy, non-transformed primary stem cells may also be susceptible to NK cell mediated cytotoxicity. We demonstrated previously that NK cells lysed hMSCs, hDPSCs and hESCs significantly (Table 15.2). All different types of stem cells became resistant to NK cell mediated cytotoxicity once they were differentiated (Table 15.2) (Tseng et al. 2010). In addition, higher sensitivity of hiPSCs to NK cell mediated lysis was also observed when compared to parental line from which they were derived (Table 15.2). Differentiation of XO1-NS and XO2-NS also rendered them more resistant to NK cell mediated cytotoxicity (Table 15.2, and manuscript submitted). Increased lysis of cancer stem cells or non-transformed healthy stem cells may be attributed to the use of allogeneic NK cells, however, our previous work using autologous NK cells exhibited similar levels of cytotoxicity against hDPSCs when compared to lysis by allogeneic NK cells (Jewett et al. 2010). Taken together these results indicated that undifferentiated cells are targets of both allogeneic and autologous NK cells. Thus, the stage of differentiation of the cells is predictive of their susceptibility to NK cell mediated cytotoxicity.

15.3 De-Differentiation of Epithelial Cells Activates NK Cell Function

Since the degree of differentiation in the cells is predictive of their sensitivity to NK cell mediated cytotoxicity, we reasoned that blocking NF-κB in the cells may de-differentiate and subsequently revert the cells to more of an undifferentiated phenotype, resulting in their increased susceptibility to NK cell mediated cytotoxicity. Indeed, blocking NF-κB in oral tumors was found to increase CD44 surface receptor expression, which is one of the hallmarks of stem cells (unpublished results). In addition, blocking of NF-κB nuclear function in a primary Oral tumor OSCCs and in a non-tumorigenic oral cells (HOK-16B) as well as in an established tumor line, HEp-2 cells known to be Hela contaminant (Jewett et al. 2003; Murakami et al. 2004, 2005; Abdulkarim et al. 2002), augmented cytotoxicity and the release of key cytokines such as IFN-γ from the NK cells (Jewett et al. 2003, 2006). Similarly, inhibition of NF-κB by Sulindac increased the functional activation of NK cells and enhanced anti-tumor cytotoxic activity (Jewett et al. 2003, 2006).

In agreement with our studies, targeted deletion of IKK-β in epidermis of mice has previously been shown in one study to lead to inflammatory skin manifestations (Pasparakis et al. 2002). Elevated levels of cytokines and chemokines have also been demonstrated in the epidermis of patients and animals with Iκκγ and Iκκβ deletions (Pasparakis et al. 2002; Berlin et al. 2002). Mice with a keratinocyte-specific deletion of Iκκβ demonstrated decreased proliferation of epidermal cells, and developed TNF-α dependent inflammatory skin disease (Pasparakis et al. 2002). In contrast, in other studies in which NF-κB function was blocked in dermal keratinocytes by a mutant IκB-α an increased proliferation and hyperplasia (Seitz et al. 1998) and eventual development of cutaneous squamous cell carcinomas of skin were seen if mice were allowed to survive and reach adulthood (van Hogerlinden et al. 1999, 2004). It is of interest to note that in these studies with diverse functional outcomes in keratinocytes, blocking TNF-α function resulted in the prevention of both the neoplastic transformation and the inflammatory skin disease (Pasparakis et al. 2002, van Hogerlinden et al. 2004). Elevated numbers of immune inflammatory cells recruited to the site of epidermis are likely responsible for the increased secretion of TNF-α. Indeed, we have demonstrated that synergistic induction of TNF-α could be observed when NF-κB knock down oral tumors were cultured with either PBMCs or NK cells (Jewett et al. 2006).

Since tumorigenic and non-tumorigenic human oral keratinocytes acquire sensitivity to NK cell mediated lysis when NF-κB is inhibited, it is likely that this phenomenon is not specific to cancer or oral keratinocytes, and it may occur in other healthy non-transformed cell types. Indeed, when human primary monocytes were differentiated to dendritic cells they too became more resistant to NK cell mediated cytotoxicity (Tseng et al. 2010). Moreover, knock down of COX2 in primary mouse monocytes (Tseng et al. 2010), or in mouse embryonic fibroblasts (unpublished observations), resulted in the reversion or de-differentiation of the monocytes and fibroblasts respectively, and the activation of NK cell cytotoxicity. Indeed, it is likely that any disturbance in cellular differentiation may predispose the cells to NK cell mediated cytotoxicity. Since STAT3 is an important factor increased during differentiation, blocking STAT3 is also critical in the activation of immune effectors (Wang et al. 2004). In support of a critical role of STAT3 in immune evasion of tumor cells in humans, we and others have recently shown that

GBM tumors display constitutive activation of STAT3 (Cacalano and Jewett, unpublished observation) (Rahaman et al. 2002), and poorly induce activating cytokines and tumor-specific cytotoxicity in human peripheral blood mononuclear cells (PBMCs) and NK cells. Ectopic expression of dominant-negative STAT3 in the GBM tumors increased lysis of the tumor cells by the immune effectors and induced production of IFN-γ by the interacting immune effectors (unpublished observations).

Since NF-κB is shown to regulate IL-6 secretion in OSCCs, HOK-16B and HEp2 cells and secreted IL-6 in tumors is known to activate STAT3 expression and function, increase in NF-κB nuclear function could in turn induce STAT3 activation and result in a significant resistance of tumors to NK cell mediated cytotoxicity. Indeed, inhibition of NF-κB in oral tumors resulted in a significant decrease in IL-6 secretion by the tumor cells and the induction of IFN-γ secretion by the NK cells (Tseng et al. 2010; Jewett et al. 2006). Therefore, targeted knock down of STAT3 or signaling pathways upstream of STAT3, such as NF-κB, may de-differentiate the cells and predispose the cells to NK cell mediated cytotoxicity.

15.4 Split Anergy in IL-2 Treated NK Cells is Induced After Their Binding to Sensitive but not Resistant Tumors and After the Triggering of CD16 Receptors

We have previously shown that K562, an NK sensitive tumor, causes loss of NK cell cytotoxicity while it triggers significant induction of TNF-α and IFN-γ from the NK cells (Jewett and Bonavida 1995, 1996). In contrast, NK resistant tumors such as RAJI cells induce very little loss of NK cell cytotoxicity or secretion of cytokines (Jewett and Bonavida 1995, 1996). Moreover, following NK cell cultures with sensitive tumor-target cells but not resistant tumors, the target binding NK cells undergo phenotypic and functional changes. Target cell inactivated NK cells express $CD16^{-}CD56^{dim/-}CD69^{+}$ phenotype (Jewett and Bonavida 1995, 1996). This phenotype has also been observed in several disease manifestations including HIV infection (Hu et al. 1995). Significant down-modulation of CD16 receptor expression and decreased NK cell cytotoxic function were also seen in several cancer patients including those of the oral and ovarian cancer patients (Lai et al. 1996; Kuss et al. 1999).

In addition, down-regulation of CD16 surface receptors on NK cells was also observed when NK cells were treated with CA125 isolated from ovarian tumor cells (Patankar et al. 2005). The decrease in CD16 surface receptors was accompanied by a major decrease in NK cell killing activity against K562 tumor cells (Patankar et al. 2005). These observations suggested that CD16 receptors could likely play an important role in target cell induced loss of NK cell cytotoxicity. Indeed, CD16:Ig fusion proteins were shown to bind to a variety of tumor-target cells indicating the existence of specific ligands for CD16 receptors on tumor cells (Mandelboim et al. 1999). Furthermore, we have previously shown that the triggering of CD16 on untreated or IL-2 treated NK cells was found to result in down-modulation of CD16 receptors and in a great loss of cytotoxicity in NK cells.

In addition, a small subset of NK cells was programmed to undergo apoptosis (Jewett and Bonavida 1995, 1996; Jewett et al. 1996, 1997). Cell death of NK cells was shown to be regulated, in part, by endogenously secreted TNF-α from the NK cells (Jewett et al. 1997). Previous studies by other groups have also shown that a subset of IL-2 activated NK cells undergo cell death following cross-linking of the CD16 receptor (Ortaldo et al. 1995; Azzoni et al. 1995). Addition of antibodies to CD56 or LFA-1 did not cause any decrease in NK cell cytotoxicity demonstrating the specificity of CD16 mAb signaling in mediating inhibition of NK cell cytotoxicity (Jewett and Bonavida 2000).

Thus, we had coined the term "split anergy" for the responses observed by NK cells after their interaction with sensitive target cells or after the triggering of CD16 receptors by the antibody in combination with IL-2 treatment (Jewett and Bonavida 1995, 1996; Jewett et al. 1997, 2006, 2008). Indeed, three subpopulations of NK cells; namely Free, Binder and Killer NK cells with varying degrees of loss of cytotoxicity were identified after the formation of conjugates with K562 targets (Jewett et al. 1996; Bonavida et al. 1993, 1994; Jewett and Bonavida 1994, 1995). Free cells which did not bind or form conjugates with target cells were inactivated less, or exhibited the most cytotoxicity, whereas Binder cells, those that bound but did not kill their bound tumors, and Killer subsets which bound and killed their bound tumors exhibited significant loss of cytotoxicity. In contrast, Binder and Killer subsets but not Free NK subset secreted significant levels of cytokines and exhibited $CD16^{-}CD56^{dim/-}CD69^{+}$ phenotype (Jewett et al. 1996; Bonavida et al. 1993, 1994; Jewett and Bonavida 1994, 1995). Treatment of NK cells with IL-2 and anti-CD16mAb also induced split anergy by significantly decreasing the NK cell cytotoxicity while increasing the cytokine secretion capabilities of NK cells. Furthermore, NK cells exhibited $CD16^{-}CD56^{dim/-}CD69^{+}$ phenotype after treatment with the combination of IL-2 and anti-CD16mAb (Jewett et al. 1997, 2006, 2008; Jewett and Bonavida 2000). Loss of cytotoxicity in NK cells was significantly exacerbated when NK cells were either treated with $F(ab)'_2$ fragment of anti-CD16 mAb with IL-2 or treated with a combination of MHC-Class I and anti-CD16 mAbs in combination with IL-2 while the same treatments resulted in an increased secretion of cytokines (Jewett and Bonavida 2000; 2008). Based on our recent results NKp46 mAb was also able to induce significant NK cell anergy in the presence and absence of IL-2 correlating with their increased expression on untreated and IL-2 treated NK cells (manuscript submitted). Moreover, addition of bacteria or their extracts in the presence of CD16 receptor signaling and IL-2 was able to induce synergistic decrease in NK cell cytotoxicity while increasing the induction of cytokine release substantially (manuscript in prep). Because anergy in NK cells is an active process and it is induced via signaling receptors such as CD16 and NKp46 and not through LFA1 or LFA3 or CD56 it is likely that binding of agonistic antibodies or ligands to signaling receptors induce tolerance in NK cells in a manner similar to that obtained when anti-CD3 antibody is administered to T cells (Chatenoud 2003). In addition, the magnitude of signaling through the receptors on NK cells may determine the extent and levels of anergy induced in NK cells. Therefore, these results suggested that receptor signaling in NK cells via key surface receptors in the presence of IL-2 is likely to result in a rapid loss of NK cell cytotoxicity while continuing to increase secretion of cytokines by the NK cells.

15.5 Split Anergy in NK Cells is Induced by Monocytes

When hMSCs or hDPSCs were cultured with either viable or irradiated monocytes before they were exposed to IL-2-treated NK cells a significant decrease in NK cell mediated cytotoxicity could be observed against hMSCs or hDPSCs. Interestingly, significant lysis of hMSCs and hDPSCs by untreated NK cells was also reproducibly blocked by the addition of monocytes (Jewett et al. 2010). We then determined whether decreased lysis of stem cells by NK cells was due to a competitive lysis of monocytes by the NK cells. We confirmed that monocytes were also lysed by the NK cells significantly. Furthermore, when we co-cultured stem cells with monocytes and sorted to remove the monocytes from the stem cells before assessing the killing function of NK cells, we could still observe significant inhibition of NK cell mediated lysis, arguing against the protection of stem cell lysis by NK cells being solely on the bases of competitive lysis of monocytes (Jewett et al. 2010). Therefore, even though lysis of monocytes by the NK cells may in part contribute to the prevention of NK cell lysis of stem cells, interaction of monocytes with stem cells can also provide resistance of stem cells against NK cell cytotoxicity.

Decrease in NK cell lysis of hMSCs and hDPSCs was paralleled with a significant induction of IFN-γ. Indeed, when hMSCs or hDPSCs were cultured with IL-2 treated NK cells alone we could observe significant induction of IFN-γ secretion. However, the highest increase was seen when IL2-treated NK cells were cultured with hMSCs or hDPSCs in the presence of monocytes. Therefore, although decreased killing of stem cells by the NK cells could be observed in the presence of monocytes, synergistic secretion of IFN-γ by the NK cells in the presence of monocytes and stem cells could be observed, indicating an inverse relationship between cytotoxicity and IFN-γ secretion (split anergy). This was similar to the profiles which we had seen when NK cells were treated with IL-2 and anti-CD16 mAb in which significant decrease in cytotoxicity of NK cells could be observed in parallel with increased secretion of IFN-γ (Jewett et al. 1997).

15.6 Induction of Split Anergy in NK Cells is a Potential Mechanism for the Switch from Effector to Regulatory Function

Induction of split anergy in NK cells could be an important conditioning step responsible for the repair of tissues during pathological processes irrespective of the type of pathology. In tumors since the generation and maintenance of cancer stem cells is higher, the majority, if not all of the NK cells, may be conditioned to support differentiation and repair of the tissues and as such the phenotype of NK cells in tumor microenvironment as well as in the peripheral blood may resemble that of the anergic NK cells, i.e., decreased NK cell cytotoxicity, acquisition of $CD16^{-/dim}CD56^{dim/+}CD69^{+}$ phenotype and augmented ability to secrete inflammatory cytokines. Of course, the degree of the loss of NK cell cytotoxicity may be directly proportional to the load of cancer stem cells. Therefore, our results suggest two very important functions for the NK cells. One function is to limit the numbers of proliferating stem cells and immune inflammatory cells by selecting those with a greater potential for differentiation for the repair of the tissues and second to support differentiation of the stem cells and subsequent regeneration of the tissues.

To achieve these tasks NK cells have to acquire two different phenotypes, and be conditioned to carry out both functions successfully. $CD16^{+}CD56^{dim}CD69^{-}$ subsets of NK cells are cytotoxic and will mediate cytotoxicity depending on which cells they encounter first. In respect to the oral squamous cell carcinomas since the majority of immune effectors can be found at the connective tissue area the chances are that they may first encounter and interact with either the other immune effectors or the effectors of connective tissue such as fibroblasts. However, there is also the possibility that NK cells may first encounter the stem cells at the base of the epithelial layer, in which case by eliminating their bound stem cells, they too can become anergized.

Surprisingly, allogeneic CTLs were also found to target Glioblastoma stem-like cells and not their differentiated counterparts (unpublished observation). By eliminating a subset of stem cells or after their interaction with other immune inflammatory cells or effectors of connective tissue NK cells could then be in a position to support differentiation of selected population of stem cells since they will be conditioned to lose cytotoxicity, induce cytokine and growth factor secretion and gain the $CD16^{-/dim}CD56^{dim/+}CD69^{+}$ phenotype. It is interesting to note that all of the immune effectors isolated from oral gingival tissues of healthy as well as diseased gingiva have CD69+ phenotype, with the exception that the numbers of immune effectors are much less in the healthy oral gingival tissues when compared to diseased tissues (unpublished observation). Therefore, our results suggest two very important functions for the NK cells. One function is to kill and the other function is to support differentiation for the repair and regeneration of the tissues.

In vivo physiological relevance of above-mentioned observations could be seen in a subpopulation of NK cells in peripheral blood, uterine and liver NK cells which express low or no CD16 receptors, have decreased capacity to mediate cytotoxicity and is capable of secreting significant amounts of cytokines (Cooper et al. 2001; Nemeth et al. 2009). In addition, 70 % of NK cells become CD16 dim or negative immediately after allogeneic or autologous bone marrow transplantation (Cooper et al. 2001). Since NK cells lose their cytotoxic function and gain in cytokine secretion phenotype and down modulate CD16 receptors after their interaction with tumor cells or the stem cells (Jewett and Bonavida 1996; Jewett et al. 1997), it is tempting to speculate that in vivo identified CD16- NK cells and in vitro tumor induced CD16- NK cells have similar developmental pathways since they have similar if not identical functional properties.

The proof of concept in support of this model was recently obtained in our laboratory. We observed that anergized NK cells were directly responsible for the increased differentiation and resistance of a number of different stem cells including cancer stem cells and dental pulp stem cells against cytotoxic effectors (Fig. 15.2) (submitted). As presented, when OSCSCs were cultured with supernatants (Fig. 15.2) or paraformaldehyde fixed NK cells (submitted) treated with IL-2 and anti-CD16 mAb, they became resistant to cytotoxicity mediated by freshly isolated untreated or IL-2 treated NK cells. IL-2 treated NK cell supernatants or paraformaldehyde fixed NK cells were also able to impart some resistance to OSCSCs but the highest levels of resistance were achieved when NK cells were treated with IL-2 and anti-CD16mAb (Fig. 15.2). No significant differences in resistance of OSCSCs to NK cell mediated lysis can be achieved by either culturing the OSCSCs with the supernatants or cells from untreated or anti-CD16mAb treated NK cells (Fig. 15.2). The resistance in OSCSCs induced by anergized NK cells correlated with a decrease in CD44 receptor expression and in an increase in B7H1 expression (data not shown), two surface receptors which were inversely expressed in differentiated OSCCs and in OSCSCs (Tseng et al. 2010). In addition, we now have evidence which supports the notion that the induction of anergy in NK cells is an active process which is induced by the triggering of CD16 receptor on the NK cells and is not due to degranulation and exhaustion of cytotoxic granules (unpublished results).

Our work collectively suggests that anergized NK cells are as important as the non-anergized NK cells in their effector functions. NK cells are not only important for the removal and shaping of the size of the stem cells which is mediated by the effector NK cells but also their differentiation, and the eventual regeneration of the new tissues which is mediated by their switch to regulatory NK cells. The task of NK cells in this regard goes above and beyond their most appreciated function of being the effectors of first line defense against viral infection and malignancies. They too can be effectors of differentiation and tissue regeneration.

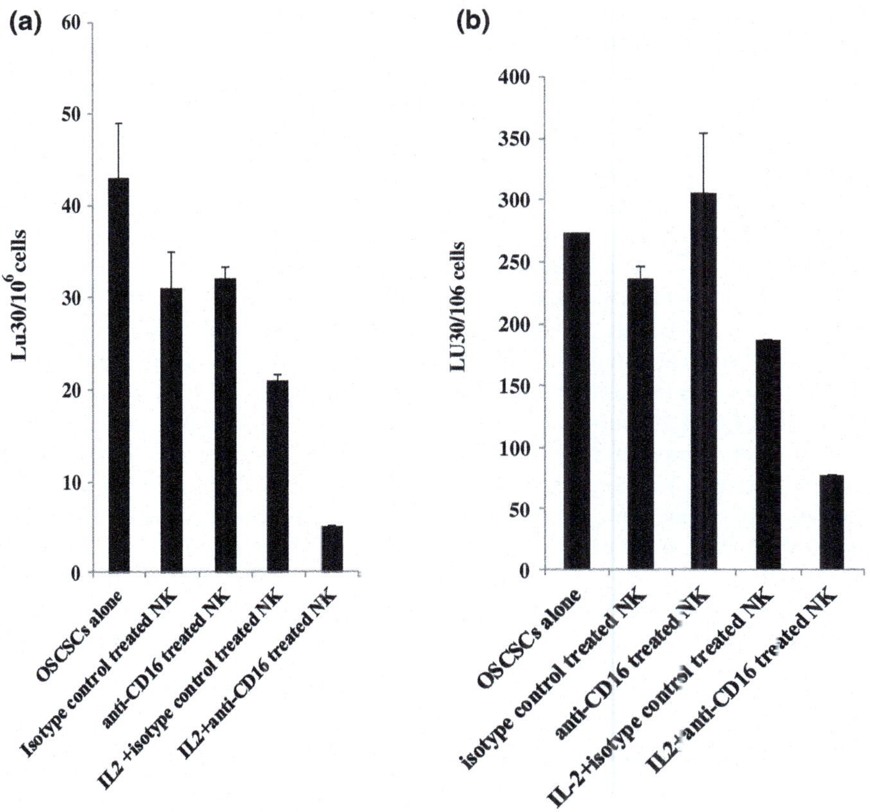

Fig. 15.2 Supernatants from anergized NK cells induce the highest resistance of OSCSCs against NK cell mediated cytotoxicity. Highly purified NK cells at (1 × 10^6 cells/ml) were either left untreated or treated with IL-2 (1,000 u/ml), anti-CD16 mAb (3 μg/ml) or a combination of IL-2 (1,000 u/ml) and anti-CD16 mAb (3 μg/ml) for 24 h before they were harvested and used to induce differentiation of OSCSCs. OSCSCs at 1 × 10^6 cells were added to each plate in 10 ml of media and the cells were allowed to adhere before the NK cell supernatants were added to each plate. A total of 180 microliter of supernatants were added at day 1, 3 and 5 and the levels of NK cell cytotoxicity were determined using freshly isolated untreated (Fig. 15.2a) and IL-2 treated (1,000 u/ml) (Fig. 15.2b) NK cells in a 4 h 51Cr release assay on the 6th day

15.7 Similarities in Immune Cell Effector Function in Inflammatory Tumor Microenvironment and in Non-transformed Inflammatory Microenvironment

The concept of tumor immunosurveillance has previously been expanded to include immunoediting as an important mechanism for the development of cancer (Dunn et al. 2002, 2004). It was suggested that cancer immunoediting comprises of three phases: elimination, equilibrium and escape (Dunn et al. 2004). Elimination represents the classical concept of immunosurveillance. However, during equilibrium and escape the interaction and cross signaling between the immune effectors including NK cells, the tumor cells, and perhaps the effectors of the connective tissue in the tumor microenvironment may result in the generation of tumors which are capable of gradual suppression of the NK cell cytotoxic function. The final stages of cancer development may result in the induction of resistant tumors in the presence of fewer immune effectors capable of lysing the tumors (Dunn et al. 2004).Thus, pressures exerted by the tumor cells and immune effectors may eventually shape the microenvironment for the growth, expansion and invasion of tumors. Similarly, a variation of such interactions may also be observed during the interaction of NK cells with healthy non-transformed human stem cells in non-transformed inflammatory microenvironment in which case the three phases of interaction may include elimination which marks the decrease in the numbers of proliferating stem cells or other immune effectors in the inflammatory microenvironment, potentially resulting in the selection of stem cells by the NK cells, induction of tolerance or anergy which denotes the conditioning of NK cells by the stem cells and/or by the other effectors of microenvironment to become regulatory cells and support maturation and differentiation of remaining stem cells, and finally the resolution phase which denotes the elimination of anergized NK cells and generation of less immunogenic differentiated cells.

15.8 Immunosuppressive Effectors in Tumor and in Non-Transformed Inflammatory Microenvironment

Both the tumor microenvironment as well as non-transformed inflammatory microenvironment consists of a number of heterogeneous cell populations with ability to suppress and limit the function of cytotoxic immune effectors. Patients with cancer often have higher numbers of immature monocytes serving as Myeloid Derived Suppressor Cells (MDSCs) expressing CD14 + HLADR- phenotype (Vuk-Pavlovic 2008; Greten et al. 2010). Tumor associated Macrophages (TAMs) were previously shown to significantly influence and limit immune activation in the tumor microenvironment (Coffelt et al. 2009; Mantovani and Sica 2010). In addition, MDSCs which are comprised of a number of distinct cell populations of myeloid origin and whose roles in immunosuppression have received significant attention in recent years are major cells capable of suppressing the cytotoxic function of T and NK cells (Greten et al. 2010). T cell dysfunction is shown to be induced by MDSCs by the increased secretion of IL-10, TGF-β, induction of reactive oxygen species (ROS), and increased expression of arginase-1 and inducible nitric oxide synthase (iNOS). T regulatory (Treg) and DC regulatory (DCreg) cells were also recently shown to have significant immunosuppressive roles in the tumor microenvironment (Greten et al. 2010). Perhaps one of the most intriguing observations regarding the immunosuppressive effectors is the identification of Cancer Associated Fibroblasts (CAFs) and Mesenchymal Stem Cells (MSCs) as two potential tumor promoters. Fibroblasts from tumor tissues demonstrate an activated phenotype and have the ability to secrete many immunosuppressive factors such as TGF-β and VEGF (Yaguchi et al. 2010, 2011). We have also found that undifferentiated fibroblasts, as well as MSCs and CD14 + HLA-DR- monocytes are significantly more susceptible to NK cell mediated cytotoxicity (Jewett et al. 2010), therefore, these cells may condition NK cells to undergo split anergy and become regulatory NK cells. Indeed, in oral epithelial tumors the majority of recruited immune effectors are usually found in the connective tissue area where through cell–cell interaction with the immunosuppressive cells such as fibroblasts, monocytes-macrophages and to a lesser extent T and B cells (Jewett et al. 2010) can generate regulatory NK cells, resulting in differentiation and resistance of oral epithelial tumors.

15.9 Tumor Microenvironment may Shape the Function and Phenotype of the NK Cells

Based on the work presented in this chapter, it is possible that the resident and recruited immune effectors in tumor microenvironment such as monocytes may serve as shields against NK cell lysis of stem cells (Fig. 15.3). Monocytes can shield stem cells from killing by the NK cells by increasing the total IFN-γ release from the NK cells while decreasing the cytotoxic function of NK cells, resulting in an increased protection and differentiation of stem cells. Indeed, monocytes also increase TNF-α, IL-6 and VEGF secretion in the co-cultures of stem cells with NK cells which could augment NF-κB and increase differentiation of stem cells. The shielding effect of monocytes could be a more generalized function of other effectors since NK cells can also target fibroblasts (Fig. 15.3) (Jewett et al. 2010). Whether other MDSCs such as PMNs can also be targeted by the NK cells awaits future investigation. This may have significant implications for the role of NK cells in not only limiting inflammation, but also the significance of other immune effectors in shielding and limiting the cytotoxic function of NK cells against cancer or healthy stem cells in order to raise maximally the secretion of key cytokines for speedy and optimal differentiation of stem cells during inflammation (Fig. 15.3). This is precisely what is observed in cancer patients in whom global decrease in NK, cytotoxic T cells and monocytes have all been reported (Jewett et al. 2006).

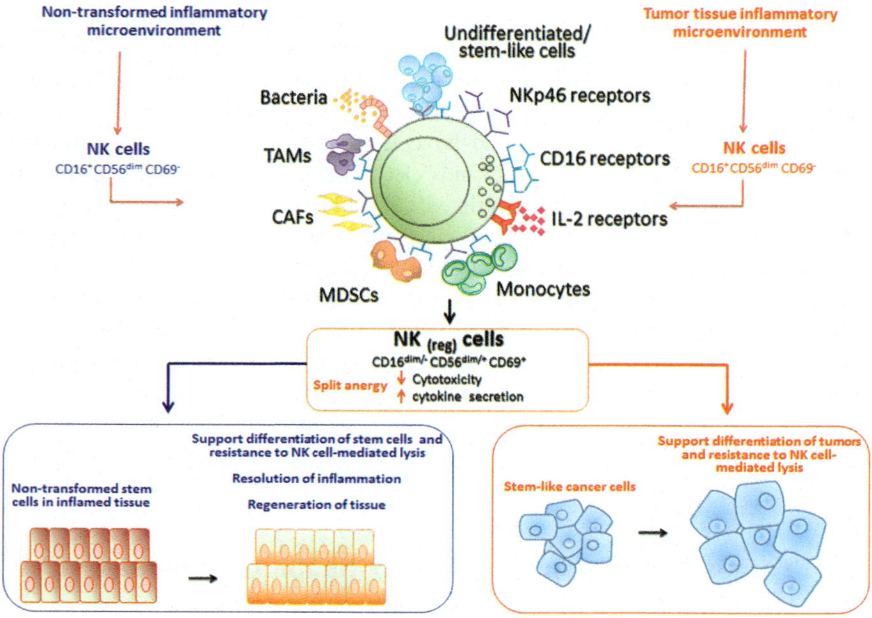

Fig. 15.3 Hypothetical model of induction of regulatory NK cells by immune inflammatory cells and by the effectors of connective tissue to support differentiation of non-transformed stem cells and cancer stem cells. CD16+CD56dimCD69-NK cells are likely to encounter and interact with the other immune effectors such as monocytes, tissue-associated macrophages (TAMs), and/or other myeloid-derived suppressor cells (MDSCs), and/or connective tissue-associated fibroblasts (CAF) in order to be conditioned to become regulatory NK cells (NK_{reg}). NK cells may also directly interact with the stem cells at the base of the epithelial layer, in which case they can alsobecome conditioned to support differentiation of other stem cells. In addition, bacteria through the binding to Toll like receptors can further aid in the generation of NK_{reg}. All of the above mentioned mechanisms may be operational during inflammatory processes in the tumor microenvironment or in healthy non-transformed inflammatory microenvironment. NK cell-differentiated epithelial cells will no longer be killed or induce cytokine secretion by the NK cells, therefore, resulting in the resolution of inflammation

15.10 Potential Functional Similarities Between Induced Regulatory NK Cells and T Regulatory Cells

Because of their ability to drive differentiation, anergized NK cells may have the ability to halt inflammation since differentiated cells are no longer targeted by the NK cells and they do not induce cytokine secretion by the NK cells (Fig. 15.3) (Tseng et al. 2010; Jewett et al. 2010). This function of NK cells is similar to T regulatory cells (Tregs) since Tregs are inhibitory and are capable of decreasing the magnitude of inflammation reported in a number of previous studies. Therefore, although immunosuppression in the tumor microenvironment is not advantageous for the patient, it is indeed, an important function which may not only stimulate differentiation, but it may also halt inflammation.

The majority if not all of the effectors in the mucosal immune system including in the oral cavity are of activated phenotype, i.e., they express CD69 early activation antigen. These cells, including NK cells are likely conditioned in the mucosa to support differentiation and resistance of the epithelial cells. Such environment is anti-inflammatory since the majority of immune cells is tolerant of ingested food particles and self-tissues and is known to contain many regulatory cells including T cells, Dendritic cells and likely regulatory NK cells. However, once the threshold which keeps the inflammation at bay is decreased in the mucosa, immune effectors are activated and may cause tissue damage and establishment of chronic inflammation. Indeed, in this regard our preliminary in vivo observations in humans consuming a combination of proprietary probiotic bacterial strains with potent ability to condition NK cells to support differentiation of OSCSCs and hDPSCs (in preparation) was able to relieve chronic inflammation and pain, and resulted in the resolution of inflammatory mouth ulcers, and oral edema. Furthermore, the number of neutrophils in the blood of a patient who had chronically decreased levels of neutrophils rose to the normal levels and both the numbers and function of NK cells in the blood improved substantially after the consumption of probiotic bacterial strain.

Our results in the conditioning of NK cells to become regulatory NK cells with receptor signaling in the presence of LPS is in line with the numerous anti-inflammatory benefits which are achieved by the consumption of probiotic bacteria in the gut.

15.11 Conclusions

Much work has been done to identify strategies by which tumor cells evade the function of immune system. Altered expression of MHC molecules which block recognition and activation of T and NK cells are examples of mechanisms by which tumor cells evade the function of immune system. In addition, tumor cells by releasing immunosuppressive factors such as Fas, VEGF, IL-6, IL-10, TNF-α, GM-CSF and IL-1β, induce T and NK cell apoptosis, block lymphocyte homing and activation, and dampen macrophage and dendritic cell function. However, the same effector functions are also important in tissue repair.

Based on the accumulated work presented in this chapter, we suggest that NK cells may have two significant functions; one that relates to the removal of excess proliferating stem cells and their selection. In this regard, NK cells could also lyse other effectors in the connective tissue area in order to not only decrease inflammation but also to be conditioned to promote differentiation and resistance of selected stem cells and eventual regeneration of the tissues (Fig. 15.3). The second important task for NK cells is therefore, to support differentiation and promote tissue regeneration after altering their phenotype to cytokine secreting cells (Fig. 15.3). This process will not only remove cells that are perhaps damaged and have flaws in the differentiation process or in general are more than needed, but it will also ensure the regeneration of tissues and the resolution of inflammation. Thus, any disturbance in the NK cell function or in the process of selection and differentiation of stem cells may result in chronic inflammation, causing continual tissue damage and recruitment of immune effectors to aid in tissue regeneration.

The inability of patient NK cells to contain cancer stem cells due to the flooding of NK cells by proliferating cancer stem cells and conversion of NK cells to regulatory NK cells may likely be one mechanism by which cancer may progress and metastasize. Therefore, there should be two distinct strategies by the NK cells to eliminate tumors, one which targets stem cells and the other which targets differentiated cells. Since cancer stem cells were found to be more resistant to certain chemotherapeutic drugs but sensitive to NK cell mediated killing while differentiated oral tumors were more resistant to NK cell mediated killing but relatively more sensitive to chemotherapeutic drugs, combination therapy should be considered for the elimination of both undifferentiated and differentiated tumors. In addition, since a great majority of patient NK cells may have switched to regulatory function to support differentiation of the proliferating cancer stem cells, they may not be effective in eliminating the cancer stem cells. Therefore, these patients may benefit from repeated allogeneic NK cell transplantation for elimination of cancer stem cells. In this regard, depletion of immunosuppressive effectors in the tumor microenvironment, which condition NK cells to become regulatory cells, via radiation or chemotherapeutic drugs should in theory provide a better strategy for successful targeting of tumors by the NK cells.

References

Abdulkarim B, Sabri S, Deutsch E et al (2002) Antiviral agent cidofovir restores p53 function and enhances the radiosensitivity in HPV-associated cancers. Oncogene 21(15):2334–2346

Azzoni L, Anegon I, Calabretta B, Perussia B (1995) Ligand binding to Fc gamma R induces c-myc-dependent apoptosis in IL-2-stimulated NK cells. J Immunol 154(2):491–499

Berlin AL, Paller AS, Chan LS (2002) Incontinentia pigmenti: a review and update on the molecular basis of pathophysiology. J Am Acad Dermatol 47(2):169–187; quiz 88–90

Bonavida B, Lebow LT, Jewett A (1993) Natural killer cell subsets: maturation, differentiation and regulation. Nat Immun 12(4–5):194–208

Bonavida B, Braquet M, Lebow LT, Jewett A (1994) Qualitative and quantitative analysis of subpopulations of cytotoxic effector cells by flow cytometry. J Lipid Mediat Cell Signal 9(1):19–25

Chatenoud L (2003) CD3-specific antibody-induced active tolerance: from bench to bedside. Nat Rev Immunol 3(2):123–132

Coffelt SB, Hughes R, Lewis CE (2009) Tumor-associated macrophages: effectors of angiogenesis and tumor progression. Biochim Biophys Acta 1796(1):11–18

Cooper MA, Fehniger TA, Caligiuri MA (2001) The biology of human natural killer-cell subsets. Trends Immunol 22(11):633–640

Dunn GP, Bruce AT, Ikeda H, Old LJ, Schreiber RD (2002) Cancer immunoediting: from immunosurveillance to tumor escape. Nat Immunol 3(11):991–998

Dunn GP, Old LJ, Schreiber RD (2004) The immunobiology of cancer immunosurveillance and immunoediting. Immunity 21(2):137–148

Farag SS, Caligiuri MA (2006) Human natural killer cell development and biology. Blood Rev 20(3):123–137

Fildes JE, Yonan N, Leonard CT (2008) Natural killer cells and lung transplantation, roles in rejection, infection, and tolerance. Transpl Immunol 19(1):1–11

Greten TF, Manns MP, Korangy F (2010) Myeloid derived suppressor cells in human diseases. Int Immunopharmacol 11(7):802–806

Haas P, Loiseau P, Tamouza R et al (2011) NK-cell education is shaped by donor HLA genotype after unrelated allogeneic hematopoietic stem cell transplantation. Blood 117(3):1021–1029

Hu PF, Hultin LE, Hultin P et al (1995) Natural killer cell immunodeficiency in HIV disease is manifest by profoundly decreased numbers of CD16 + CD56 + cells and expansion of a population of CD16dimCD56- cells with low lytic activity. J Acquir Immune Defic Syndr Hum Retrovirol 10(3):331–340

Inagaki A, Soeda A, Oka N et al (2007) Long-term maintenance of brain tumor stem cell properties under at non-adherent and adherent culture conditions. Biochem Biophys Res Commun 361(3):586–592

Jewett A, Bonavida B (1994) Activation of the human immature natural killer cell subset by IL-12 and its regulation by endogenous TNF-alpha and IFN-gamma secretion. Cell Immunol 154(1):273–286

Jewett A, Bonavida B (1995a) Target-induced anergy of natural killer cytotoxic function is restricted to the NK-target conjugate subset. Cell Immunol 160(1):91–97

Jewett A, Bonavida B (1995b) Interferon-alpha activates cytotoxic function but inhibits interleukin-2-mediated proliferation and tumor necrosis factor-alpha secretion by immature human natural killer cells. J Clin Immunol 15(1):35–44

Jewett A, Bonavida B (1996) Target-induced inactivation and cell death by apoptosis in a subset of human NK cells. J Immunol 156(3):907–915

Jewett A, Bonavida B (2000) MHC-Class I antigens regulate both the function and the survival of human peripheral blood NK cells: role of endogenously secreted TNF-alpha. Clin Immunol 96(1):19–28

Jewett A, Gan XH, Lebow LT, Bonavida B (1996) Differential secretion of TNF-alpha and IFN-gamma by human peripheral blood-derived NK subsets and association with functional maturation. J Clin Immunol 16(1):46–54

Jewett A, Cavalcanti M, Bonavida B (1997) Pivotal role of endogenous TNF-alpha in the induction of functional inactivation and apoptosis in NK cells. J Immunol 159(10):4815–4822

Jewett A, Wang MY, Teruel A, Poupak Z, Bostanian Z, Park NH (2003) Cytokine dependent inverse regulation of CD54 (ICAM1) and major histocompatibility complex class I antigens by nuclear factor kappaB in HEp2 tumor cell line: effect on the function of natural killer cells. Hum Immunol 64(5):505–520

Jewett A, Cacalano NA, Teruel A et al (2006a) Inhibition of nuclear factor kappa B (NFkappaB) activity in oral tumor cells prevents depletion of NK cells and increases their functional activation. Cancer Immunol Immunother 55(9):1052–1063

Jewett A, Cacalano NA, Head C, Teruel A (2006b) Coengagement of CD16 and CD94 receptors mediates secretion of chemokines and induces apoptotic death of naive natural killer cells. Clin Cancer Res 12(7 Pt 1):1994–2003

Jewett A, Head C, Cacalano NA (2006c) Emerging mechanisms of immunosuppression in oral cancers. J Dent Res 85(12):1061–1073

Jewett A, Teruel A, Romero M, Head C, Cacalano N (2008) Rapid and potent induction of cell death and loss of NK cell cytotoxicity against oral tumors by F(ab′)2 fragment of anti-CD16 antibody. Cancer Immunol Immunother 57(7):1053–1066

Jewett A, Arasteh A, Tseng HC et al (2010) Strategies to rescue mesenchymal stem cells (MSCs) and dental pulp stem cells (DPSCs) from NK cell mediated cytotoxicity. PLoS ONE 5(3):e9874

Kuss I, Saito T, Johnson JT, Whiteside TL (1999) Clinical significance of decreased zeta chain expression in peripheral blood lymphocytes of patients with head and neck cancer. Clin Cancer Res 5(2):329–334

Lai P, Rabinowich H, Crowley-Nowick PA, Bell MC, Mantovani G, Whiteside TL (1996) Alterations in expression and function of signal-transducing proteins in tumor-associated T and natural killer cells in patients with ovarian carcinoma. Clin Cancer Res 2(1):161–173

Mandelboim O, Malik P, Davis DM, Jo CH, Boyson JE, Strominger JL (1999) Human CD16 as a lysis receptor mediating direct natural killer cell cytotoxicity. Proc Natl Acad Sci USA 96(10):5640–5644

Mantovani A, Sica A (2010) Macrophages, innate immunity and cancer: balance, tolerance, and diversity. Curr Opin Immunol 22(2):231–237

Murakami J, Asaumi J, Maki Y et al (2004) Influence of CpG island methylation status in O6-methylguanine-DNA methyltransferase expression of oral cancer cell lines. Oncol Rep 12(2):339–345

Murakami J, Asaumi J, Kawai N et al (2005) Effects of histone deacetylase inhibitor FR901228 on the expression level of telomerase reverse transcriptase in oral cancer. Cancer Chemother Pharmacol 56(1):22–28

Nemeth E, Baird AW, O'Farrelly C (2009) Microanatomy of the liver immune system. Semin Immunopathol 31(3):333–343

Oka N, Soeda A, Inagaki A et al (2007) VEGF promotes tumorigenesis and angiogenesis of human glioblastoma stem cells. Biochem Biophys Res Commun 360(3):553–559

Ortaldo JR, Mason AT, O'Shea JJ (1995) Receptor-induced death in human natural killer cells: involvement of CD16. J Exp Med 181(1):339–344

Pasparakis M, Courtois G, Hafner M et al (2002) TNF-mediated inflammatory skin disease in mice with epidermis-specific deletion of IKK2. Nature 417(6891):861–866

Patankar MS, Yu J, Morrison JC et al (2005) Potent suppression of natural killer cell response mediated by the ovarian tumor marker CA125. Gynecol Oncol 1(4):513–523

Pegram HJ, Andrews DM, Smyth MJ, Darcy PK, Kershaw MH (2011) Activating and inhibitory receptors of natural killer cells. Immunol Cell Biol 89(2):216–24

Rahaman SO, Harbor PC, Chernova O, Barnett GH, Vogelbaum MA, Haque SJ (2002) Inhibition of constitutively active Stat3 suppresses proliferation and induces apoptosis in glioblastoma multiforme cells. Oncogene 21(55):8404–8413

Romagnani C, Juelke K, Falco M et al (2007) CD56brightCD16- killer Ig-like receptor- NK cells display longer telomeres and acquire features of CD56dim NK cells upon activation. J Immunol 178(8):4947–4955

Seitz CS, Lin Q, Deng H, Khavari PA (1998) Alterations in NF-kappaB function in transgenic epithelial tissue demonstrate a growth inhibitory role for NF-kappaB. Proc Natl Acad Sci USA 95(5):2307–2312

Soeda A, Inagaki A, Oka N et al (2008) Epidermal growth factor plays a crucial role in mitogenic regulation of human brain tumor stem cells. J Biol Chem 283(16):10958–10966

Strowig T, Chijioke O, Carrega P et al (2011) Human NK cells of mice with reconstituted human immune system components require preactivation to acquire functional competence. Blood 116(20):4158–4167

Tseng HC, Arasteh A, Paranjpe A et al (2010) Increased lysis of stem cells but not their differentiated cells by natural killer cells; de-differentiation or reprogramming activates NK cells. PLoS One 5(7):e11590

Tseng HC, Arasteh A, Paranjpe A et al (2010b) Increased lysis of stem cells but not their differentiated cells by natural killer cells; de-differentiation or reprogramming activates NK cells. PLoS ONE 5(7):e11590 van Hogerlinden M, Rozell BL, Ahrlund-Richter L, Toftgard R (1999) Squamous cell carcinomas and increased apoptosis in skin with inhibited Rel/nuclear factor-kappaB signaling. Cancer Res 59(14):3299–3303

van Hogerlinden M, Rozell BL, Toftgard R, Sundberg JP (2004) Characterization of the progressive skin disease and inflammatory cell infiltrate in mice with inhibited NF-kappaB signaling. J Invest Dermatol 123(1):101–108

Vuk-Pavlovic S (2008) Rebuilding immunity in cancer patients. Blood Cells Mol Dis 40(1):94–100

Wang T, Niu G, Kortylewski M et al (2004) Regulation of the innate and adaptive immune responses by Stat-3 signaling in tumor cells. Nat Med 10(1):48–54

Yaguchi T, Sumimoto H, Kudo-Saito C et al (2010) The mechanisms of cancer immunoescape and development of overcoming strategies. Int J Hematol 93(3):294–300

Part IV
Immune Regulators in the Tumor Immunoenvironment

Chapter 16
The Role of Myeloid Derived Suppressor Cells in Cancer

Jonathan M. Weiss

Abstract Inflammation is established as one of the central hallmarks of cancer. Although the immune system plays an indispensable role in immunosurveillance against cellular transformation, ample evidence demonstrates that certain immune cells can be unwitting conspirators in the promotion of tumors. Immature myeloid cells frequently accumulate in the tumor microenvironment and peripheral organs of cancer patients and in mouse tumor models and correlate with tumor progression and poor survival. Myeloid-derived suppressor cells are a heterogeneous population of immature myeloid cells with profound immunosuppressive abilities that contribute to immune dysfunction and tumor progression. In this chapter, the phenotypic and functional characteristics of MDSC, and the mechanisms underlying their development, accumulation and suppressor functions in murine and human cancers are described. The molecular and immunotherapeutic targeting of those pathways contributing to MDSC expansion and/or function are highlighted for their potential to overcome MDSC-mediated immunosuppression in tumor-bearing hosts and improve cancer treatment strategies.

Keywords Tumor microenvironment · Immunosuppression · T cell dysfunction · Immature myeloid cells · Immunotherapy

J. M. Weiss (✉)
Cancer and Inflammation Program, Frederick National Laboratory for Cancer Research, Building 560, Room 31-18, Frederick, MD 21702, USA
e-mail: weissjo@mail.nih.gov

M. R. Shurin et al. (eds.), *The Tumor Immunoenvironment*,
DOI: 10.1007/978-94-007-6217-6_16,

16.1 Classification of MDSC in Mice and Humans

16.1.1 MDSC in Mouse Models of Cancer

MDSC have been extensively reported to accumulate in mouse models of cancer. Murine MDSC express CD11b and have been divided into two main classifications according to their expression of Ly6G and Ly6C (Fig. 16.1). Granulocytic MDSC are identified as being $CD11b^{+}Ly6G^{+}Ly6C^{Lo}$, whereas monocytic MDSC are $CD11b^{+}Ly6C^{Hi}Ly6G^{Lo}$. A large number of murine studies utilize the Gr1 antigen to identify MDSC, and as such the most readily, albeit somewhat simplified, denotation of MDSC has been a $CD11b^{+}$ myeloid population with low to intermediate Gr1 expression. This classification, however, is rather incomplete, since both CD11b and Gr1 can be expressed by other myeloid cells, as well as NK cells and neither marker is associated with actual suppressor phenotype. Studies characterizing murine MDSC should therefore endeavor to include markers of functional suppressive ability, such as the IL-4 receptor alpha chain (CD124), which mediates IL-4 and IL-13 dependent suppressor pathways by MDSC (Bronte et al. 2003a). Although granulocytic MDSC (G-MDSC) share a common morphology and phenotype as conventional neutrophils, the two cell types can be distinguished since G-MDSC have reduced phagocytic activity, increased expression of arginase, myeloperoxidase and reactive oxygen species (ROS), and, in stark contrast to neutrophils, functionally suppress T cell activation (Youn et al. 2012). On a per cell basis, monocytic MDSC exert an equivalent, indeed in some cases greater, degree of immunosuppression on a per cell basis through higher levels of nitric oxide (NO), rather than ROS (Youn et al. 2008). However, granulocytic MDSC are more prevalent in many mouse tumor models, and might represent the more immunosuppressive MDSC subset in vivo because of their increased frequency relative to monocytic MDSC.

16.1.2 MDSC in Human Cancers

Since the identification of MDSC, their frequency in cancer patients has been investigated. Patients with a variety of solid tumors have an increased frequency of MDSC both in their circulating peripheral blood, as well as in various organs and the tumor microenvironment itself. The challenge with these studies however, has been a lack of consensus as to the actual clinical definition of what markers MDSC express in human cancer patients. Unlike murine studies, in which the reliance upon Ly6C, Ly6G and Gr1 antigens appears sufficient, cancer patients present myeloid populations with a diverse phenotype of surface antigens (Fig. 16.1). This may be due to the fact that human cancers express a greater variety of growth factors, cytokines, and molecules than homogeneous mouse tumors do, and myeloid cell populations, including MDSC, are exquisitely sensitive to this wide

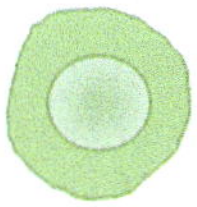

Murine Monocytic MDSC

Marker	Comments
CD11b	Common leukocyte integrin, adhesion
$Ly6C^{HI}$ / $Ly6G^{-o}$	Hematopoietic cell differentiation antigens
IL-4Rα, IL-13Rα	Mediates IL-4 and IL-13 dependent suppression in some systems
CD115, CD116	M-CSF and GM-CSF mediated development
Arginase	Immunosuppression
iNOS	Immunosuppression
Nitrotyrosine	TCR nitration, T cell tolerance

Murine Granulocytic MDSC

Marker	Comments
CD11b	Common leukocyte integrin, adhesion
$Ly6G^{+}$ / $Ly6C^{Lo}$	Hematopoietic cell differentiation antigens
CD15, CD33	Granulocytic adhesion molecules
CD114	G-CSF mediated development, function
Arginase	Immunosuppression
ROS	TCR nitration, T cell tolerance
Nitrotyrosine	TCR nitration, T cell tolerance

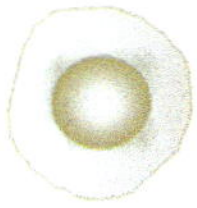

Human MDSC

Marker	Comments
$HLA\text{-}DR^{Neg/Lo}$	Poor antigen presentation; Induction of tolerance
CD14, CD11b	LPS receptor, leukocyte integrin
CD15, CD33	Granulocytic adhesion molecules
CD116	GM-CSF levels often correlate with MDSC frequency
Arginase	Immunosuppression
iNOS	Immunosuppression
Nitrotyrosine	TCR nitration, T cell tolerance

Fig. 16.1 Characteristics of murine and human MDSC populations. The classification of MDSC based upon surface or intracellular markers is complicated. The principal markers and their role in MDSC function are listed, however murine and human MDSC do not always express each of these markers. These markers are listed to indicate the principal developmental and suppressor mechanisms frequently utilized by different MDSC populations. The listed markers may also be expressed by non-MDSC leukocytes, even though those cells do not share suppressor phenotype with MDSC. (Portions reprinted from Chan et al. 2011, with permission from Elsevier)

milieu of immunomodulatory agents. Heterogeneous populations of myeloid cells, including MDSC, are noted in numerous human cancers. Although the identification of functionally suppressive MDSC is controversial, their role in cancer-related immune suppression and tumor progression is becoming more readily appreciated.

One of the earliest accounts of MDSC in human cancer involved head and neck squamous cell carcinoma patients (Pak et al. 1995). These cells were identified as $CD34^+$ hematopoietic progenitor cells that secreted GM-CSF (a cytokine later associated with MDSC accumulation) and were associated with the suppression of intra-tumoral T cells. The depletion of $CD34^+$ cells resulted in a restoration of IL-2 production by T cells. In another study, Almand et al. noted a small percentage (<2 %) of hemopoietic progenitor cells from cancer patients that were negative for CD14 and HLA-DR expression and impaired the ability of DC to stimulate allogeneic T cells (Almand et al. 2001). The addition of all-trans-retinoic acid induced the differentiation of these cells into mature DC and their T cell stimulatory capacity. Numerous subsequent studies of patients with solid tumors identified circulating MDSC, with quite variable phenotypes, which may be suggestive of tumor-specific influences upon surface marker expression by MDSC. A near universal phenotype of MDSC in cancer patients is their low to negligible expression of HLA-DR, indicating their poor degree of antigen presentation capabilities. Frequent MDSC markers include leukocyte adhesion molecules, such as CD33 (Siglec-3), CD34 (Mucosialin), and CD11b (αM integrin), and a general absence of other cell lineage markers, namely CD3, CD19, CD20 and CD56 (Fig. 16.1). The myeloid marker, CD14, is not always a reliable MDSC marker. As can be surmised from these results, a large array of cell surface antigens can be employed in an attempt to identify MDSC, and yet these molecules still do not gauge any actual functional suppressor phenotype by putative MDSC. This can most reliably be ascertained by classic, albeit ex vivo T cell co-culture assays, and by measuring their expression of potentially immunosuppressive molecules such as TGF-β, arginase, and inducible nitric oxide synthase (iNOS) (as described later).

16.2 Mechanisms Whereby MDSC Contribute to Tumor Progression

16.2.1 MDSC-Mediated Immunosuppression

MDSC impair $CD4^+$ and $CD8^+$ T cell responses through the production of an array of immunosuppressive soluble factors as well via direct cell contact-mediated mechanisms (Fig. 16.2). Many of these are shared between monocytic and granulocytic MDSC, although the relative production of immunosuppressive mediators can vary between the two main MDSC subgroups. ROS generation is a hallmark feature of granulocytic MDSC, whereas monocytic MDSC produce increased levels of iNOS (Youn et al. 2008). MDSC also produce large amounts of arginase, indoleamine 2,3-dioxygenase (IDO) as well as immunosuppressive cytokines, such as TGF-β and IL-10. Arginine metabolism via arginase represents a major impediment to T cell responses (Bronte et al. 2003a). Arginase was induced by IL-4 (via CD124 signaling) and elicited increased superoxide production (Bronte et al. 2003b). Although IL-4 and IL-13 are key inducers of arginase and nitric oxide synthase (NOS2), it was shown recently that IL-4Rα (CD124) deficient mice had MDSC with an equivalent suppressive phenotype as wildtype mice (Sinha et al. 2010). Nitric oxide production was also shown to be a major suppressor mechanism. Using a rat glioma model, iNOS inhibition reversed MDSC-induced T cell apoptosis, whereas blockade of arginase, TGF-β or IDO had no effect (Jia et al. 2010). In the C26 murine adenocarcinoma model, iNOS expression was also shown to inhibit IFN responsiveness by T cells as well as NK cells, a process attributed to reduced Stat1 phosphorylation (Mundy-Bosse et al. 2011). Although MDSC express inhibitory molecules, namely PD-L1, PD-L2 and CD124 on their surface, broad studies involving 10 different mouse tumor models showed these did not always correlate with suppressor activity (Youn et al. 2008). These conflicting results are likely related to the relative importance of suppressor pathways utilized by different MDSC subsets and in different tumor models. Contradictory reports also exist as to the ability of MDSC to inhibit NK cells. Membrane-bound expression of TGFβ on MDSC contributed to reduced IFNγ expression, NKG2D and cytotoxicity by NK cells (Li et al. 2009). MDSC depletion, using Gr-1 depleting antibody, restored NK cell activity. However, in a lymphoma tumor model system (RMA-S), MDSC from tumor-bearing mice expressed the NK-cell NKG2D activating receptor, RAE1 and could potently activate NK cells (Nausch et al. 2008).

Antigen-specific T cell tolerance is a major mechanism for tumor escape. MDSC are antigen presenting cells, however, they can potently induce tolerance to the presented peptide (Fig. 16.2). In a series of elegant studies involving double TCR transgenic CD8$^+$ T cells, Nagaraj and colleagues showed this was due to the ability of MDSC to nitrosylate the T cell receptor such that it becomes dissociated from CD3 zeta molecules necessary for downstream signaling (Nagaraj et al. 2007, 2010). No tolerance was induced to the TCR which was unaffected. TCR nitration is induced through the hyperproduction of ROS and peroxynitrite during MDSC-T cell contacts and represents an important mechanism of inducing antigen-specific CD8$^+$ T cell tolerance that circumvent anti-tumor responses (Nagaraj et al. 2007). CD4$^+$ T cell tolerance, on the other hand, could only be induced when MDSC expressed significantly higher levels of MHC class II molecules on their surface. However, direct cell contact between CD4$^+$ T cells and MHC class II molecules expressed on the MDSC elicited a Cox-2- and PGE$_2$-dependent, ability for MDSC to nonspecifically inhibit T cell responses (Nagaraj et al. 2012). Taken together, activated CD4$^+$ T cells can enhance MDSC immunosuppression, further contributing to immune dysfunction in cancer responses.

MDSC also contribute to immunosuppression by impeding T cell activation. One mechanism involves the depletion of cysteine, an essential amino acid for T cell activation (Srivastava et al. 2010). Normally, T cells depend on APCs to export cysteine, however, MDSC compete with immunocompetent APCs for extracellular cysteine and they potently deplete the available pool of cysteine necessary for T cell activation and function. Another mechanism contributing to

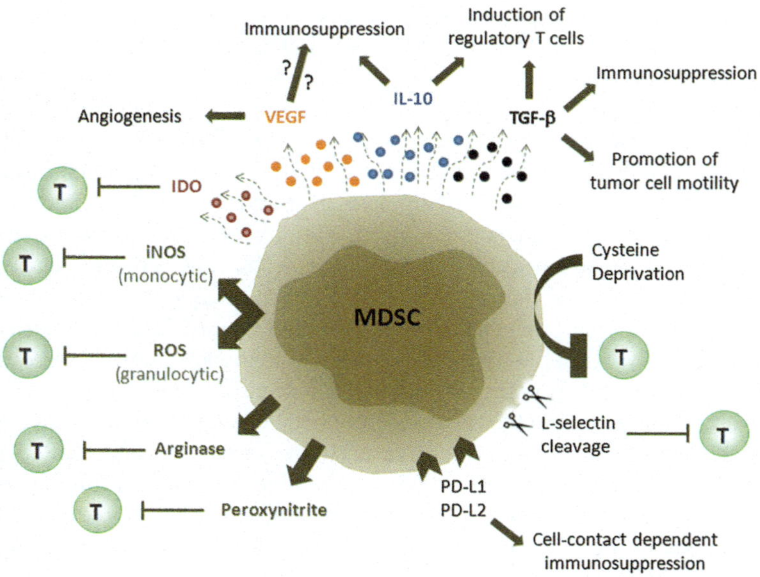

Fig. 16.2 Schematic of the mechanisms whereby MDSC induce immunosuppression and promote tumor progression. Different types of MDSC express surface or soluble mediators to varying levels. The immunosuppressive properties of all these molecules are shown, but it is not intended to suggest that MDSC express these molecules equivalently or that the degree of suppression achieved by each molecule is equivalent. MDSC induce $CD8^+$ T cell tolerance and impair global T cell trafficking and homeostasis via many different mechanisms. (Portions reprinted from Chan et al. 2011, with permission from Elsevier)

reduced T cell activation is the down-regulation of L-selectin expression on $CD4^+$ and $CD8^+$ T cells (Hanson et al. 2009). MDSC express metalloproteinases capable of cleaving L-selectin, a process that could have profound effects on T cell activation and function. L-selectin is a major homing molecule, normally directing T cells to lymph nodes, and the tumor microenvironment, where they become antigen-educated, activated and release important immunomodulatory mediators.

16.2.2 Other Tumor-Promoting Effects of MDSC

Although immunosuppression is the hallmark characteristic of MDSC, these cells are also capable of influencing tumor progression through other means as well (Fig. 16.2). Growth factors such as VEGF and GM-CSF are well associated with MDSC accumulation; however, it is evident that MDSC are capable of promoting tumor angiogenesis via the expression of VEGF and other growth factors as well. MDSC are directly associated with the expression of VEGF-A and G-CSF (Donkor et al. 2009). Tumor exosomes induce the accumulation of MDSC expressing Cox-2, arginase-1 and VEGF (Xiang et al. 2009). In human solid tumors, a population of $CD33^{+}CD11b^{+}CD66b^{+}HLA\text{-}DR^{lo}IL13R\alpha^{int}$ granulocytic-like cells were found to express iNOS, TGFβ, arginase-1, as well as high levels of VEGF (Lechner et al. 2010). These findings establish an amplifying link between increasing tumor size, MDSC frequency, and growth factor production.

Intriguingly, tumor-infiltrating MDSC were shown to promote cancer cell dissemination by inducing epithelial-mesenchymal transition (EMT) (Toh et al. 2011). A pre-requisite for tumor metastasis is their acquisition of a motile phenotype. In a murine model of melanoma, MDSC-derived TGF-β, epidermal growth factor and hepatocyte growth factor were all used to induce EMT in cancer cells, providing them with an early motile phenotype and enhancing their metastatic potential. These findings provide evidence for the link between inflammation and tumor metastasis.

16.3 Factors Contributing to MDSC Accumulation

16.3.1 MDSC Accumulation in the Tumor Microenvironment

MDSC accumulate in most cancer patients and experimental animals with cancer (Youn et al. 2008; Ostrand-Rosenberg 2010), where they can limit the efficacy of host and therapy-mediated anti-tumor responses. The direct correlation between tumor burden and frequency of MDSC strongly supports the conclusion that tumor-derived factors may promote MDSC accumulation and differentiation. The establishment of a hypoxic tumor microenvironment is a major mechanism contributing to the suppressor phenotype of tumor-associated MDSC. MDSC isolated from tumors, but not peripheral lymphoid organs, are capable of suppressing both antigen-specific and nonspecific T cell activity (Corzo et al. 2010). Corzo et al. mechanistically demonstrated that activation of the hypoxia-responsive transcription factor, HIF-1α is responsible for this contrasting phenotype of peripheral versus tumor MDSC (Corzo et al. 2010). Stat3 is another transcription factor frequently dysregulated in many cancers and shown to promote MDSC expansion and suppressor phenotype. In a mouse model of lung carcinogenesis, the transgenic overexpression of constitutively active Stat3 in lung alveolar epithelial cells, resulted in MDSC accumulation in the bronchioalveolar lavage fluid and plasma that could profoundly inhibit T cell proliferation and function (Wu et al. 2011).

Numerous cytokines and growth factors associated with the tumor microenvironment also contribute to MDSC accumulation. One key tumor-associated cytokine is GM-CSF, which supports the generation of $CD11b^{+}Ly6G^{-}Ly6C^{+}$ suppressor subsets capable of inhibiting T cell proliferation and anti-tumor function (Morales et al. 2010). Importantly, since GM-CSF is commonly used for ex vivo expansion of dendritic cells in cell-based immunotherapies, the adverse side-effect of MDSC expansion indicates that GM-CSF based therapies should be carefully re-evaluated. Tumor-derived GM-CSF also appears capable of regulating MDSC suppressor function, in addition to the recruitment of these cells. Dolcetti et al. showed that GM-CSF induced the preferential expansion of $CD11b^{+}Gr1^{int}$ and $CD11b^{+}Gr1^{Lo}$ subsets of MDSC that were potent suppressors of $CD8^{+}$ T cell activation (Dolcetti et al. 2010). Tumor-derived G-CSF was also shown to correlate with the degree of granulocytic MDSC accumulation and function (Waight et al. 2011). Through a series of loss- and gain-of function approaches, that study identified G-CSF as a key contributor to the suppressive properties of granulocytic MDSC. TGF-β is another key immunoregulatory cytokine frequently over-expressed in the tumor microenvironment. Recently, Liu et al. demonstrated that TGF-β1 was the main tumor-derived factor responsible for the upregulation of microRNA-494 in MDSC (Liu et al. 2012). Expression of miR-494 by MDSC not only enhanced chemokine-mediated recruitment of MDSC, it profoundly resulted in the activation of Akt and phosphatase and tensin homolog (PTEN) signaling pathways, resulting in a population of MDSC that were more resistant to apoptosis.

MDSC accumulation within tumors can also be caused by pro-inflammatory cytokines such as IL-1β (Elkabets et al. 2010) and IL-6 (Bunt et al. 2009, 2007; Song et al. 2005) underscoring the complex mechanisms whereby inflammation can promote subversion of the host immune system and tumor progression. IL-6 signals through Stat3, which as previously described promotes MDSC accumulation (Wu et al. 2011). In the case of IL-1β, tumor-derived IL-1β can be secreted into the local environment and contribute to MDSC accumulation. In the absence of IL-1β-induced inflammation, the $Ly6C^{-}$ subgroup of granulocytic MDSC were present at low frequency in tumor-bearing mice, however, these cells predominated under inflammatory conditions (Elkabets et al. 2010). The S100 family of inflammatory mediators also regulates MDSC accumulation. S100A8/A9 proteins bind carboxylated N-glycans expressed on MDSC-associated receptors, signal through the NF-κB pathway and promote MDSC migration (Cheng et al. 2008; Sinha et al. 2008; Ichikawa et al. 2011). In turn, MDSC are capable of producing S100A8/A9 proteins to promote an autocrine feedback loop favoring MDSC accumulation. Recent evidence in a mouse model of colitis-associated colon cancer has shown S100A8/A9 interact with RAGE and carboxylated glycans on colon tumor cells themselves to promote MAPK and NF-κB activation and a wide range of downstream genes involved in leukocyte recruitment, angiogenesis, tumor migration, wound healing and tumor metastasis (Ichikawa et al. 2011). These important findings underscore the importance of therapeutically targeting cancer-associated inflammation that is capable of exerting a wide range of tumor-promoting effects.

Tumors reorient the differentiation of myeloid cells into alternatively activated macrophages that express increased levels of TGF-β, IL-10, VEGF and COX-2 and promote MDSC accumulation. VEGF production was recently identified as a key factor in the iNOS-dependent induction of $CD11b^{+}Gr1^{+}$ MDSC in murine melanoma (Jayaraman et al. 2012). VEGF depletion resulted in Stat3 normalization, reduced reactive oxygen species production by MDSC, and reversed tumor-mediated immunosuppression. Increased COX-2 and PGE_2 expression are also frequently over-expressed in the tumor microenvironment (Eruslanov et al. 2010), where they contribute to reduced antigen-presentation and Th1 cytokine production (Harizi et al. 2002; Sharma et al. 2003). PGE_2 further contributes to immune suppression by upregulating Th2 cytokine production, FoxP3 expression in Tregs (Baratelli et al. 2005) and arginase expression in myeloid cells (Rodriguez et al. 2005). PGE_2 has been implicated in MDSC recruitment by acting directly on cell surface receptors of MDSC (Sinha et al. 2007) and Fas-dependent accumulation of MDSC (Zhang et al. 2009). PGE_2 present within ovarian cancer patient ascites fluid was recently shown to be essential for functional CXCR4 expression in cancer-associated MDSC, production of the CXCR4 ligand, CXCL12, and directly correlated with the frequency of $CD14^{+}CD33^{+}CXCR4^{+}$ MDSC in these patients (Obermajer et al. 2011). These MDSC migrated in a CXCR4-, COX2- and PGE_2-dependent manner toward ovarian cancer ascites. PGE_2 contained within tumor exosomes can also be secreted by the tumor and taken up by bone marrow myeloid cells, where they may also contribute to MDSC accumulation by switching the development of these cells towards the MDSC pathway (Xiang et al. 2009). Thus tumor-associated accumulation of COX2 and PGE_2 are important components of the re-orientation of TAM towards arginase-expressing M2 macrophages and the accumulation of MDSC populations which promote tumor development.

16.3.2 MDSC Accumulation in Other Tissues

Although accumulations of MDSC are found within tumors, the increase is also observed in distant peripheral sites such as the spleen, blood and bone marrow. The liver is a particularly preferred site for the homing and expansion of MDSC (Ilkovitch and Lopez 2009). The accumulation of MDSC in the liver with tumors originating from the abdominal/gastrointestinal region such as early pre-invasive pancreatic neoplasia and advanced colorectal cancers may not be as surprising due to proximity of the tumor to the liver (Connolly 2010). However, hepatic MDSC further accelerated the formation of liver metastasis. In addition to abdominal and gastrointestinal tumors, MDSC accumulation was also observed in subcutaneous tumors of different origins (Ilkovitch and Lopez 2009). Increased MDSC migration, as well as hematopoiesis, appears to be involved with MDSC expansion within the liver. The local expression the granulocytic chemoattractant CXCL1/KC (Connolly 2010), GM-CSF (Dolcetti et al. 2010), or stem cell factor (SCF) (Pan et al. 2008) can each play a role in MDSC accumulation in the liver. The liver is also the principal location for the production of acute-phase proteins such as serum amyloid A (SAA), in response to infection and inflammation. Hepatic signaling via gp130, a common receptor for IL-6 cytokine family members, and STAT3 are required for hepatic control of innate immune responses to SAA. In this regard, hepatic gp130-STAT3 activation resulted in the mobilization and hepatic accumulation of MDSC that were essential for controlling SAA-induced inflammation in the liver and protecting mice from sepsis-associated mortality (Sander et al. 2010). Although the accumulation of MDSC in the liver may have evolved for the necessary control of inflammatory responses during infection, it is also becoming increasingly evident that they may also play an undesirable role in the progression of liver metastases.

16.4 Strategies to Overcome MDSC Suppression in Tumors

16.4.1 Targeting MDSC Development

Hopefully, as the list of factors which promote the development of MDSC expands, this will result in the availability of new therapeutic targets for redirecting the differentiation of these cells into more mature myeloid cells which lack immunosuppressive properties (Fig. 16.3). One promising pathway is the blockade of receptor tyrosine kinases, such as SCF/c-kit ligand. SCF plays an important role in the regulation of hematopoiesis in the bone marrow. SCF is expressed by many human and murine tumors and its blockade inhibited MDSC development, Treg development and tumor-specific T cell anergy (Pan et al. 2008; Kao et al. 2010). Interestingly this blockade also prevented tumor angiogenesis, underscoring the potential role for MDSC in blood vessel formation within the tumor. More recently, the receptor tyrosine kinase inhibitor Sunitinib (Sutent) similarly prevented MDSC accumulation in tumor-bearing mice (Ko et al. 2010; Ozao-Choy et al. 2009) and renal cell carcinoma patients (Ko et al. 2009). SCF blockade (Kao et al. 2010) and Sutent (Ozao-Choy et al. 2009; Finke et al. 2008) also reduced Treg development and their associated production of IL-10 and TGF-β. Cao et al. showed that Sorafenib, another multi-kinase inhibitor, reduced liver cancer growth and significantly decreased MDSC populations, although the mechanism underlying Sorafenib-induced MDSC loss was not determined (Cao et al. 2011). Sutent and other receptor tyrosine kinase inhibitors thus can be used, potentially in combination with additional immunotherapies, for the reversal of immune suppression within the tumor microenvironment and promotion of cell-mediated immune responses. Another approach for promoting the differentiation of MDSC into mature granulocytes is all-*trans*-retinoic acid (ATRA), a derivative of vitamin A which promotes the differentiation of myeloid progenitor cells into mature dendritic cells and macrophages (Kusmartsev et al. 2008). Administration of ATRA into sarcoma-bearing mice induced the differentiation of MDSC into mature myeloid DCs capable of presenting antigen and inducing effector T cell responses (Gabrilovich et al. 2001). The treatment of MDSC isolated from renal cell carcinoma patients with ATRA also promoted the ex vivo differentiation of these cells into fully competent antigen-presenting cells (Kusmartsev et al. 2008).

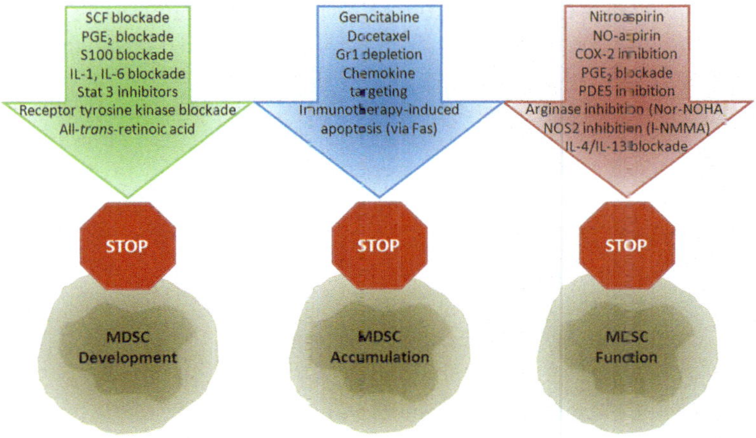

Fig. 16.3 A number of therapies have been utilized to inhibit MDSC development, accumulation within tumors and function. The systemic administration of these agents has been shown to regulate the frequency and/or function of MDSC in murine tumor models and some patient studies. Some of the listed agents may affect more than one of these pathways. The listed agents may have other effects on non-MDSC host cells or tumor cells which are not described. These lists are not intended to be exhaustive, but rather to represent the most commonly described agents used to date and their principal mode of action upon MDSC. (Portions reprinted from Chan et al. 2011, with permission from Elsevier)

These findings demonstrate that MDSC-mediated immune suppression can be reversible.

Other promising approaches for the therapeutic targeting of MDSC development are anti-inflammatory therapies, since pro-inflammatory cytokines such as IL-1β and IL-6 are frequently present in the tumor microenvironment and promote MDSC accumulation (Bunt et al. 2007, Song et al. 2005; Ostrand-Rosenberg and Sinha 2009). The reduction of inflammation through the use of the naturally occurring IL-1 receptor antagonist, IL-1 receptor blockade (Bunt et al. 2007), or PGE_2 blockade (Sinha et al. 2007; Zhang et al. 2009) can reverse MDSC development and accumulation. COX-2 inhibitors prevent the production of PGE_2, and their use in mouse models of glioma delayed tumor development in association with reduced systemic MDSC development and MCP-1 mediated accumulation (Fujita et al. 2011). The involvement of IL-6 and other cytokines in MDSC development has underscored the role for common signaling by downstream transcription factors (e.g., STAT family). Stat3 is constitutively active in MDSC and a key regulator of MDSC development and function, by mediating the upregulation of anti-apoptotic, proliferative, and pro-angiogenic molecules (Niu et al. 2002; Wang et al. 2004). Stat3 inhibition, either through the use of small molecule inhibitors (Nefedova et al. 2005), blocking peptides, peptidomimetics or platinum complexes (Turkson et al. 2005) could be of therapeutic benefit, provided the biologic requirement for Stat3 signaling in a diverse array of normal biologic pathways is not adversely affected. The removal of MDSC following

Sutent therapy (Ozao-Choy et al. 2009; Ko et al. 2009) may also be related to its ability to abrogate Stat3 signaling. The involvement of S100 inflammatory proteins (Cheng et al. 2008; Sinha et al. 2008), not only in the accumulation of MDSC, but also via autocrine production by MDSC and tumor cells, are also attractive candidates for therapy. Blocking antibodies against these proteins and their carboxylated glycan ligands reduce MDSC levels in tumors (Sinha et al. 2008) and have been noted for anti-tumor efficacy in murine tumorigenesis (Turovskaya et al. 2008).

16.4.2 Targeting MDSC Accumulation

An improved understanding of the factors which contribute to MDSC accumulation within tumors will hopefully lead to the development of improved strategies for mitigating this process (Fig. 16.3). Recruitment of MDSC is principally mediated by two chemokine axes: CXCL5/ENA-78 binding to the CXCR2 receptor or CXCL12/SDF-1 binding to the CXCR4 receptor (Yang et al. 2008). These chemokines are produced by M2 macrophages and tumor cells themselves, thereby achieving a high level within the tumor microenvironment serving to recruit MDSC and further amplify this process. The negation of specific chemokine axes is attractive for several reasons. First, it tends to elicit the more selective targeting of MDSC cells while avoiding substantial impact on T effector cells and other leukocytes (Mantovani et al. 2004). Second, the therapeutic modulation of chemokine profiles has potential for the rapid amplification of more desirable M1 macrophage populations. We and others have shown, for example, immunotherapeutic regimens which elicit strong levels of Th1 cytokines such as IL-12 and IFN-γ, dramatically restructure the chemokine and myeloid composition of the tumor microenvironment so that the IFN-γ-dependent chemokines (RANTES, MIG, IP-10, and MIP-1γ) and M1 phenotype of macrophages predominate concomitant with the reversal of MDSC frequency and function (Weiss et al. 2009; Stout et al. 2009). Interestingly, our work showed the combination of IL-2 and agonistic anti-CD40 achieve the dramatic removal of MDSC from the tumor microenvironment, not only via the reorientation of MDSC-promoting chemokines (Weiss et al. 2009), but also through Fas-dependent cell death (our unpublished findings). In other studies, the anti-cancer drug trabectedin was also shown to be capable of inhibiting the expression of tumor-promoting chemokines, macrophage recruitment and tumor-associated vascularization (Germano et al. 2010). Combination immunotherapy in the form of CCL16 chemokine administration plus the injection of CpG and anti-IL10 receptor antibody similarly polarized tumor-infiltrating myeloid populations from M2 into M1 which paralleled innate and adaptive immune cell-mediated anti-tumor responses (Guiducci et al. 2005). An added benefit of these approaches is that reorientation of TAMs towards the M1 phenotype helps remove potential sources of Treg-recruiting chemokines (CCL17 and CCL20), which predominately originate from M2-polarized macrophages (Weiss et al. 2009; Umemura et al. 2008).

Several chemotherapeutic drugs have shown promise for removing MDSC populations. Docetaxel was reported recently to inhibit MDSC accumulation in 4T1-Neu mammary tumor-bearing mice (Kodumudi et al. 2010). Interestingly, docetaxel treatment preferentially targeted M2/mannose receptor positive MDSC while sparing M1 macrophages, further supporting investigation of docetaxel in combination with other immunotherapeutic strategies. The mechanism for paclitaxel-mediated removal of MDSC was recently shown to involve their differentiation into immune competent dendritic cells (Michels et al. 2012). Gemcitabine also removes MDSC (Sinha et al. 2007; Nagaraj et al. 2010) although its mechanism appears to be through the selective induction of apoptosis in these cells (Suzuki et al. 2005). Gemcitabine has been effectively used either as a single agent or in combination with cisplatin, paclitaxel or anti-inflammatory agents in numerous clinical trials (Comella and Phase 2001; Natale 2004) and is considered among the primary treatment options for the treatment of non-small cell lung cancer.

MDSC can be depleted using antibodies which recognize the Gr1 antigen (Bronte et al. 1999). It is apparent, however, that such strategies are not selective for MDSC, since neutrophils, eosinophils and pDC also have variable yet constitutive expression of Gr1 and MDSC eventually rebound. Nevertheless, Gr1 depletion studies have demonstrated the potential for improved anti-tumor responses [(Bronte et al. 1999) and our unpublished observations using orthotopically implanted Renca tumors].

16.4.3 Targeting MDSC Suppressor Functions

Although MDSC induce T cell tolerance and mediate immunosuppression via a multitude of molecular mechanisms, considerable efforts have shown promise for interfering with MDSC suppressor activity (Fig. 16.3). Given the role for IL-4/IL-13 in regulating MDSC suppression (Bronte et al. 2003a; Highfill et al. 2010), a promising strategy is blockade of this signaling pathway. Roth et al. recently reported on the generation of IL-4Rα aptamers, which not only impaired MDSC suppressor function, but unexpectedly promoted their apoptosis and elimination (Roth et al. 2012). This study reveals that IL-4Rα signaling may be critical for MDSC survival, in addition to their function. Other immunotherapeutic strategies have highlighted the promise for blocking MDSC suppression by maturing them into immune competent cells. In one recent study, the prototypic anti-tumor cytokine IL-12 was highlighted for its potential for blocking MDSC suppression. In that study, it was found that IL-12 treatment could mature MDSC, with upregulated antigen presenting capacity and reduced NOS2 production (Steding et al. 2011). Another approach for maturing MDSC into competent APCs involves TLR9 activation via CpG administration. Two different studies showed CpG administration into tumor-bearing mice promoted the maturation and differentiation of MDSC with tumoricidal and Th1 cytokine capabilities (Shirota et al. 2012; Zoglmeier et al. 2011).

The principal targets of reversing the MDSC suppressor phenotype is the removal of arginase or iNOS, which comprise critical components of MDSC immunosuppressive activity (Bronte et al. 2003a, b; Highfill et al. 2010; Rodriguez et al. 2009). Nitroaspirin is a classic aspirin molecule covalently linked to a NO donor group currently under evaluation in phase I/II clinical trials. Orally administered nitroaspirin inhibited the enzymatic activities of MDSC, normalized the immune status of tumor-bearing mice and functioned as an effective adjuvant for cancer vaccination (De Santo et al. 2005). The principal mechanism whereby NO-aspirin achieves these effects is through the feedback inhibition of NOS and arginase expression and activity. NO-aspirin also inhibited protein nitration, within the tumor microenvironment, thus inhibiting antigen binding to the TCR (Nagaraj et al. 2007, 2010). The inhibition of either COX-2 or PGE_2 can also reverse MDSC-mediated suppression, since these enzymes are important for tumor promotion via a number of different mechanisms, arginase expression and MDSC suppressor function (Rodriguez et al. 2005; Sinha et al. 2007; Zhang et al. 2009). Other anti-inflammatory agents, such as IL-1 receptor antagonist or triterpenoid compounds have been shown to reduce MDSC levels and function, in part via the reduction of peroxynitrite and reactive oxygen species generation (Bunt et al. 2007; Nagaraj et al. 2010). Phosphodiesterase-5 (PDE5) inhibitors were elegantly shown to augment antitumor immune responses by interfering with the arginase and NOS-dependent suppressor machinery of MDSC (Serafini et al. 2006). Treatment of tumor-bearing mice with the PDE5 inhibitor sildenafil, in particular, downregulated arginase and NOS2 expression in MDSC isolated from different organs and led to the dramatic restoration of effector $CD4^+$ and $CD8^+$ T cells. The use of other selective arginase or NOS inhibitors, namely Nor-NOHA and l-NMMA respectively, similarly enhanced effector T cell responses. PDE5 inhibitors are currently in clinical use for nonmalignant conditions, such as erectile dysfunction, cardiac hypertrophy and pulmonary hypertension. The demonstration of their anti-tumor potential (Serafini et al. 2006) further supports investigation for their applicability as cancer therapeutics.

16.5 Conclusions, Future Perspectives

The augmentation of MDSC in cancer patients represents a major obstacle to cancer treatment strategies. These cells exert potent immunosuppressive effects that short-circuit nascent host anti-tumor responses and circumvent tumor vaccination and immune-stimulating approaches. Despite the ongoing progress made in the identification of factors contributing to MDSC accumulation and suppressor function, considerable obstacles in addressing their persistence in tumors remain. First, and foremost, the heterogeneity of immature myeloid cells represents a challenge towards efforts to appropriately and functionally identify these cells in patients with diverse tumor types as well as mouse tumor models. MDSC consist of a broad spectrum of myeloid phenotypes that vary with respect to surface marker expression, antigen-presenting and costimulatory abilities, production of immunosuppressive cytokines and effector molecules such as iNOS, arginase, ROS and peroxynitrites. As more recent characterizations of MDSC in cancer patients have been reported, it may be surmised that the tumor microenvironment, indeed the tumor itself, may be influencing the types of MDSC. Cancer treatment strategies should be tailored to target the predominating MDSC (e.g., granulocytic or monocytic) with their accompanying predominant suppressor mechanisms (e.g., ROS versus iNOS). However, another major challenge relates to the multitude of cellular and molecular mechanisms whereby MDSC elicit T cell tolerance, immunosuppression and dysregulated leukocyte homeostasis. Single therapies are unlikely to overcome MDSC suppression, and efforts should be made to continue targeting MDSC development concomitant with suppressor pathways. It is the goal of this review to highlight the expanding list of MDSC-promoting factors, so that new therapeutic targets may emerge. For example, the recent identification of the relative selectiveness for MDSC susceptibility to Fas-mediated cell death represents a promising opportunity for their removal and improvement of cancer treatment strategies. Recent studies have also highlighted the potential to differentiate MDSC into more mature cells with improved capacity for antigen-presentation. These findings signify that we are beginning to uncover more effective treatment strategies aimed at overcoming local immunosuppression and that more effective cancer therapies are on the horizon.

References

Almand B, Clark JI, Nikitina E, van Beynen J, English NR, Knight SC et al (2001) Increased production of immature myeloid cells in cancer patients: a mechanism of immunosuppression in cancer. J Immunol 166:678–689

Baratelli F, Lin Y, Zhu L, Yang SC, Heuze-Vourc'h N, Zeng G et al (2005) Prostaglandin E2 induces FOXP3 gene expression and T regulatory cell function in human $CD4^+$ T cells. J Immunol 175:1483–1490

Bronte V, Chappell DB, Apolloni E, Cabrelle A, Wang M, Hwu P et al (1999) Unopposed production of granulocyte-macrophage colony-stimulating factor by tumors inhibits $CD8^+$ T cell responses by dysregulating antigen-presenting cell maturation. J Immunol 162:5728–5737

Bronte V, Serafini P, Mazzoni A, Segal DM, Zanovello P (2003a) L-arginine metabolism in myeloid cells controls T-lymphocyte functions. Trends Immunol 24:302–306

Bronte V, Serafini P, De Santo C, Marigo I, Tosello V, Mazzoni A et al (2003b) IL-4-induced arginase 1 suppresses alloreactive T cells in tumor-bearing mice. J Immunol 170:270–278

Bunt SK, Yang L, Sinha P, Clements VK, Leips J, Ostrand-Rosenberg S (2007) Reduced inflammation in the tumor microenvironment delays the accumulation of myeloid-derived suppressor cells and limits tumor progression. Cancer Res 67:10019–10026

Bunt SK, Clements VK, Hanson EM, Sinha P, Ostrand-Rosenberg S (2009) Inflammation enhances myeloid-derived suppressor cell cross-talk by signaling through Toll-like receptor 4. J Leukoc Biol 85:996–1004

Cao M, Xu Y, Youn JI, Cabrera R, Zhang X, Gabrilovich D et al (2011) Kinase inhibitor Sorafenib modulates immunosuppressive cell populations in a murine liver cancer model. Lab Invest 91:598–608

Chan T, Wiltrout RH, Weiss JM (2011) Immunotherapeutic modulation of the suppressive liver and tumor microenvironments. Portions reprinted from international immunopharmacology, vol 11(7), pp 879–889 (with permission from Elsevier)

Cheng P, Corzo CA, Luetteke N, Yu B, Nagaraj S, Bui MM et al (2008) Inhibition of dendritic cell differentiation and accumulation of myeloid-derived suppressor cells in cancer is regulated by S100A9 protein. J Exp Med 205:2235–2249

Comella P, Phase III (2001) trial of cisplatin/gemcitabine with or without vinorelbine or paclitaxel in advanced non-small cell lung cancer. Semin Oncol 28:7–10

Connolly MK, Mallen-St Clair J, Bedrosian AS, Malhotra A, Vera V, Ibrahim J et al (2010) Distinct populations of metastases-enabling myeloid cells expand in the liver of mice harboring invasive and preinvasive intra-abdominal tumor. J Leukoc Biol 87:713–725

Corzo CA, Condamine T, Lu L, Cotter MJ, Youn JI, Cheng P et al (2010) HIF-1alpha regulates function and differentiation of myeloid-derived suppressor cells in the tumor microenvironment. J Exp Med 207:2439–2453

De Santo C, Serafini P, Marigo I, Dolcetti L, Bolla M, Del Soldato P et al (2005) Nitroaspirin corrects immune dysfunction in tumor-bearing hosts and promotes tumor eradication by cancer vaccination. Proc Natl Acad Sci USA 102:4185–4190

Dolcetti L, Peranzoni E, Ugel S, Marigo I, Fernandez Gomez A, Mesa C et al (2010) Hierarchy of immunosuppressive strength among myeloid-derived suppressor cell subsets is determined by GM-CSF. Eur J Immunol 40:22–35

Donkor MK, Lahue E, Hoke TA, Shafer LR, Coskun U, Solheim JC et al (2009) Mammary tumor heterogeneity in the expansion of myeloid-derived suppressor cells. Int Immunopharmacol 9:937–948

Elkabets M, Ribeiro VS, Dinarello CA, Ostrand-Rosenberg S, Di Santo JP, Apte RN et al (2010) IL-1beta regulates a novel myeloid-derived suppressor cell subset that impairs NK cell development and function. Eur J Immunol 40:3347–3357

Eruslanov E, Daurkin I, Ortiz J, Vieweg J, Kusmartsev S (2010) Pivotal Advance: tumor-mediated induction of myeloid-derived suppressor cells and M2-polarized macrophages by altering intracellular PGE2 catabolism in myeloid cells. J Leukoc Biol 88:839–848

Finke JH, Rini B, Ireland J, Rayman P, Richmond A, Golshayan A et al (2008) Sunitinib reverses type-1 immune suppression and decreases T-regulatory cells in renal cell carcinoma patients. Clin Cancer Res 14:6674–6682

Fujita M, Kohanbash G, Fellows-Mayle W, Hamilton RL, Komohara Y, Decker SA et al (2011) COX-2 blockade suppresses gliomagenesis by inhibiting myeloid-derived suppressor cells. Cancer Res 71:2664–2674

Gabrilovich DI, Velders MP, Sotomayor EM, Kast WM (2001) Mechanism of immune dysfunction in cancer mediated by immature Gr-1+ myeloid cells. J Immunol 166:5398–5406

Germano G, Frapolli R, Simone M, Tavecchio M, Erba E, Pesce S et al (2010) Antitumor and anti-inflammatory effects of trabectedin on human myxoid liposarcoma cells. Cancer Res 70:2235–2244

Guiducci C, Vicari AP, Sangaletti S, Trinchieri G, Colombo MP (2005) Redirecting in vivo elicited tumor infiltrating macrophages and dendritic cells towards tumor rejection. Cancer Res 65:3437–3446

Hanson EM, Clements VK, Sinha P, Ilkovitch D, Ostrand-Rosenberg S (2009) Myeloid-derived suppressor cells down-regulate L-selectin expression on $CD4^+$ and $CD8^+$ T cells. J Immunol 183:937–944

Harizi H, Juzan M, Pitard V, Moreau JF, Gualde N (2002) Cyclooxygenase-2-issued prostaglandin e(2) enhances the production of endogenous IL-10, which down-regulates dendritic cell functions. J Immunol 168:2255–2263

Highfill SL, Rodriguez PC, Zhou Q, Goetz CA, Koehn BH, Veenstra R et al (2010) Bone marrow myeloid-derived suppressor cells (MDSC) inhibit graft-versus-host (GVHD) disease via an arginase-1 dependent mechanism that is upregulated by IL-13. Blood 116:5738–5747

Ichikawa M, Williams R, Wang L, Vogl T, Srikrishna G (2011) S100A8/A9 activate key genes and pathways in colon tumor progression. Mol Cancer Res 9:133–148

Ilkovitch D, Lopez DM (2009) The liver is a site for tumor-induced myeloid-derived suppressor cell accumulation and immunosuppression. Cancer Res 69:5514–5521

Jayaraman P, Parikh F, Lopez-Rivera E, Hailemichael Y, Clark A, Ma G et al (2012) Tumor-Expressed Inducible Nitric Oxide Synthase Controls Induction of Functional Myeloid-Derived Suppressor Cells through Modulation of Vascular Endothelial Growth Factor Release. J Immunol 188:5365–5376

Jia W, Jackson-Cook C, Graf MR (2010) Tumor-infiltrating, myeloid-derived suppressor cells inhibit T cell activity by nitric oxide production in an intracranial rat glioma + vaccination model. J Neuroimmunol 223:20–30

Kao J, Ko EC, Eisenstein S, Sikora AG, Fu S, Chen SH (2010) Targeting immune suppressing myeloid-derived suppressor cells in oncology. Crit Rev Oncol/Hematol 77:12–19

Ko JS, Zea AH, Rini BI, Ireland JL, Elson P, Cohen P et al (2009) Sunitinib mediates reversal of myeloid-derived suppressor cell accumulation in renal cell carcinoma patients. Clin Cancer Res 15:2148–2157

Ko JS, Rayman P, Ireland J, Swaidani S, Li G, Bunting KD et al (2010) Direct and differential suppression of myeloid-derived suppressor cell subsets by sunitinib is compartmentally constrained. Cancer Res 70:3526–3536

Kodumudi KN, Woan K, Gilvary DL, Sahakian E, Wei S, Djeu JY (2010) A novel chemoimmuno modulating property of docetaxel: suppression of myeloid-derived suppressor cells in tumor bearers. Clin Cancer Res 16:4583–4594

Kusmartsev S, Su Z, Heiser A, Dannull J, Eruslanov E, Kubler H et al (2008) Reversal of myeloid cell-mediated immunosuppression in patients with metastatic renal cell carcinoma. Clin Cancer Res 14:8270–8278

Lechner MG, Liebertz DJ, Epstein AL (2010) Characterization of cytokine-induced myeloid-derived suppressor cells from normal human peripheral blood mononuclear cells. J Immunol 185:2273–2284

Li H, Han Y, Guo Q, Zhang M, Cao X (2009) Cancer-expanded myeloid-derived suppressor cells induce anergy of NK cells through membrane-bound TGF-beta 1. J Immunol 182:240–249

Liu Y, Lai L, Chen Q, Song Y, Xu S, Ma F et al (2012) MicroRNA-494 is required for the accumulation and functions of tumor-expanded myeloid-derived suppressor cells via targeting of PTEN. J Immunol 188:5500–5510

Mantovani A, Allavena P, Sozzani S, Vecchi A, Locati M, Sica A (2004) Chemokines in the recruitment and shaping of the leukocyte infiltrate of tumors. Semin Cancer Biol 14:155–160

Michels T, Shurin GV, Naiditch H, Sevko A, Umansky V, Shurin MR (2010) Paclitaxel promotes differentiation of myeloid-derived suppressor cells into dendritic cells in vitro in a TLR4-independent manner. J Immunotoxicol 9(3):292–300

Morales JK, Kmieciak M, Knutson KL, Bear HD, Manjili MH (2010) GM-CSF is one of the main breast tumor-derived soluble factors involved in the differentiation of CD11b-Gr1- bone marrow progenitor cells into myeloid-derived suppressor cells. Breast Cancer Res Treat 123:39–49

Mundy-Bosse BL, Lesinski GB, Jaime-Ramirez AC, Benninger K, Khan M, Kuppusamy P et al (2011) Myeloid-derived suppressor cell inhibition of the IFN response in tumor-bearing mice. Cancer Res 71:5101–5110

Nagaraj S, Gupta K, Pisarev V, Kinarsky L, Sherman S, Kang L et al (2007) Altered recognition of antigen is a mechanism of CD8⁺ T cell tolerance in cancer. Nat Med 13:828–835

Nagaraj S, Schrum AG, Cho HI, Celis E, Gabrilovich DI (2010a) Mechanism of T cell tolerance induced by myeloid-derived suppressor cells. J Immunol 184:3106–3116

Nagaraj S, Youn JI, Weber H, Iclozan C, Lu L, Cotter MJ et al (2010b) Anti-inflammatory triterpenoid blocks immune suppressive function of MDSCs and improves immune response in cancer. Clin Cancer Res 16:1812–1823

Nagaraj S, Nelson A, Youn JI, Cheng P, Quiceno D, Gabrilovich DI (2012) Antigen-specific CD4(+) T cells regulate function of myeloid-derived suppressor cells in cancer via retrograde MHC class II signaling. Cancer Res 72:928–938

Natale RB. Gemcitabine-containing regimens vs others in first-line treatment of NSCLC. Oncology (Williston Park, N.Y 2004; 18: 27–31
Nausch N, Galani IE, Schlecker E, Cerwenka A (2008) Mononuclear myeloid-derived "suppressor" cells express RAE-1 and activate natural killer cells. Blood 112:4080–4089
Nefedova Y, Nagaraj S, Rosenbauer A, Muro-Cacho C, Sebti SM, Gabrilovich DI (2005) Regulation of dendritic cell differentiation and antitumor immune response in cancer by pharmacologic-selective inhibition of the janus-activated kinase 2/signal transducers and activators of transcription 3 pathway. Cancer Res 65:9525–9535
Niu G, Wright KL, Huang M, Song L, Haura E, Turkson J et al (2002) Constitutive Stat3 activity up-regulates VEGF expression and tumor angiogenesis. Oncogene 21:2000–2008
Obermajer N, Muthuswamy R, Odunsi K, Edwards RP, Kalinski P (2011) PGE(2)-induced CXCL12 production and CXCR4 expression controls the accumulation of human MDSCs in ovarian cancer environment. Cancer Res 71:7463–7470
Ostrand-Rosenberg S (2010) Myeloid-derived suppressor cells: more mechanisms for inhibiting antitumor immunity. Cancer Immunol Immunother 59:1593–1600
Ostrand-Rosenberg S, Sinha P (2009) Myeloid-derived suppressor cells: linking inflammation and cancer. J Immunol 182:4499–4506
Ozao-Choy J, Ma G, Kao J, Wang GX, Meseck M, Sung M et al (2009) The novel role of tyrosine kinase inhibitor in the reversal of immune suppression and modulation of tumor microenvironment for immune-based cancer therapies. Cancer Res 69:2514–2522
Pak AS, Wright MA, Matthews JP, Collins SL, Petruzzelli GJ, Young MR (1995) Mechanisms of immune suppression in patients with head and neck cancer: presence of CD34(+) cells which suppress immune functions within cancers that secrete granulocyte-macrophage colony-stimulating factor. Clin Cancer Res 1:95–103
Pan PY, Wang GX, Yin B, Ozao J, Ku T, Divino CM et al (2008) Reversion of immune tolerance in advanced malignancy: modulation of myeloid-derived suppressor cell development by blockade of stem-cell factor function. Blood 111:219–228
Rodriguez PC, Hernandez CP, Quiceno D, Dubinett SM, Zabaleta J, Ochoa JB et al (2005) Arginase I in myeloid suppressor cells is induced by COX-2 in lung carcinoma. J Exp Med 202:931–939
Rodriguez PC, Ernstoff MS, Hernandez C, Atkins M, Zabaleta J, Sierra R et al (2009) Arginase I-producing myeloid-derived suppressor cells in renal cell carcinoma are a subpopulation of activated granulocytes. Cancer Res 69:1553–1560
Roth F, De La Fuente AC, Vella JL, Zoso A, Inverardi L, Serafini P (2012) Aptamer-mediated blockade of IL4R alpha triggers apoptosis of MDSCs and limits tumor progression. Cancer Res 72:1373–1383
Sander LE, Sackett SD, Dierssen U, Beraza N, Linke RP, Muller M et al (2010) Hepatic acute-phase proteins control innate immune responses during infection by promoting myeloid-derived suppressor cell function. J Exp Med 207:1453–1464
Serafini P, Meckel K, Kelso M, Noonan K, Califano J, Koch W et al (2006) Phosphodiesterase-5 inhibition augments endogenous antitumor immunity by reducing myeloid-derived suppressor cell function. J Exp Med 203:2691–2702
Sharma S, Stolina M, Yang SC, Baratelli F, Lin JF, Atianzar K et al (2003) Tumor cyclooxygenase 2-dependent suppression of dendritic cell function. Clin Cancer Res 9:961–968
Shirota Y, Shirota H, Klinman DM (2012) Intratumoral injection of CpG oligonucleotides induces the differentiation and reduces the immunosuppressive activity of myeloid-derived suppressor cells. J Immunol 188:1592–1599
Sinha P, Parker KH, Horn L, Ostrand-Rosenberg S (2012) Tumor-induced myeloid-derived suppressor cell function is independent of IFN gamma and IL-4R alpha. Eur J Immunol
Sinha P, Clements VK, Fulton AM, Ostrand-Rosenberg S (2007a) Prostaglandin E2 promotes tumor progression by inducing myeloid-derived suppressor cells. Cancer Res 67:4507–4513

Sinha P, Clements VK, Bunt SK, Albelda SM, Ostrand-Rosenberg S (2007b) Cross-talk between myeloid-derived suppressor cells and macrophages subverts tumor immunity toward a type 2 response. J Immunol 179:977–983

Sinha P, Okoro C, Foell D, Freeze HH, Ostrand-Rosenberg S, Srikrishna G (2008) Proinflammatory S100 proteins regulate the accumulation of myeloid-derived suppressor cells. J Immunol 181:4666–4675

Song X, Krelin Y, Dvorkin T, Bjorkdahl O, Segal S, Dinarello CA et al (2005) CD11b+/Gr-1+ immature myeloid cells mediate suppression of T cells in mice bearing tumors of IL-1beta-secreting cells. J Immunol 175:8200–8208

Srivastava MK, Sinha P, Clements VK, Rodriguez P, Ostrand-Rosenberg S (2010) Myeloid-derived suppressor cells inhibit T-cell activation by depleting cystine and cysteine. Cancer Res 70:68–77

Steding CE, Wu ST, Zhang Y, Jeng MH, Elzey BD, Kao C (2011) The role of interleukin-12 on modulating myeloid-derived suppressor cells, increasing overall survival and reducing metastasis. Immunol 133:221–238

Stout RD, Watkins SK, Suttles J (2009) Functional plasticity of macrophages: in situ reprogramming of tumor-associated macrophages. J Leukoc Biol 86:1105–1109

Suzuki E, Kapoor V, Jassar AS, Kaiser LR, Albelda SM (2005) Gemcitabine selectively eliminates splenic Gr-1+/CD11b^{+} myeloid suppressor cells in tumor-bearing animals and enhances antitumor immune activity. Clin Cancer Res 11:6713–6721

Toh B, Wang X, Keeble J, Sim WJ, Khoo K, Wong WC et al (2011) Mesenchymal transition and dissemination of cancer cells is driven by myeloid-derived suppressor cells infiltrating the primary tumor. PLoS Biol 9:e1001162

Turkson J, Zhang S, Mora LB, Burns A, Sebti S, Jove R (2005) A novel platinum compound inhibits constitutive Stat3 signaling and induces cell cycle arrest and apoptosis of malignant cells. J Biol Chem 280:32979–32988

Turovskaya O, Foell D, Sinha P, Vogl T, Newlin R, Nayak J et al (2008) RAGE, carboxylated glycans and S100A8/A9 play essential roles in colitis-associated carcinogenesis. Carcinogenesis 29:2035–2043

Umemura N, Saio M, Suwa T, Kitoh Y, Bai J, Nonaka K et al (2008) Tumor-infiltrating myeloid-derived suppressor cells are pleiotropic-inflamed monocytes/macrophages that bear M1- and M2-type characteristics. J Leukoc Biol 83:1136–1144

Waight JD, Hu Q, Miller A, Liu S, Abrams SI (2011) Tumor-derived G-CSF facilitates neoplastic growth through a granulocytic myeloid-derived suppressor cell-dependent mechanism. PLoS ONE 6:e27690

Wang T, Niu G, Kortylewski M, Burdelya L, Shain K, Zhang S et al (2004) Regulation of the innate and adaptive immune responses by Stat-3 signaling in tumor cells. Nat Med 10:48–54

Weiss JM, Back TC, Scarzello AJ, Subleski JJ, Hall VL, Stauffer JK et al (2009) Successful immunotherapy with IL-2/anti-CD40 induces the chemokine-mediated mitigation of an immunosuppressive tumor microenvironment. Proc Natl Acad Sci USA 106:19455–19460

Wu L, Du H, Li Y, Qu P, Yan C (2011) Signal transducer and activator of transcription 3 (Stat3C) promotes myeloid-derived suppressor cell expansion and immune suppression during lung tumorigenesis. Am J Pathol 179:2131–2141

Xiang X, Poliakov A, Liu C, Liu Y, Deng ZB, Wang J et al (2009) Induction of myeloid-derived suppressor cells by tumor exosomes. Int J Cancer 124:2621–2633

Yang L, Huang J, Ren X, Gorska AE, Chytil A, Aakre M et al (2008) Abrogation of TGF beta signaling in mammary carcinomas recruits Gr-1+CD11b^{+} myeloid cells that promote metastasis. Cancer Cell 13:23–35

Youn JI, Nagaraj S, Collazo M, Gabrilovich DI (2008) Subsets of myeloid-derived suppressor cells in tumor-bearing mice. J Immunol 181:5791–5802

Youn JI, Collazo M, Shalova IN, Biswas SK, Gabrilovich DI (2012) Characterization of the nature of granulocytic myeloid-derived suppressor cells in tumor-bearing mice. J Leukoc Biol 91:167–181

Zhang Y, Liu Q, Zhang M, Yu Y, Liu X, Cao X (2009) Fas signal promotes lung cancer growth by recruiting myeloid-derived suppressor cells via cancer cell-derived PGE2. J Immunol 182:3801–3808

Zoglmeier C, Bauer H, Norenberg D, Wedekind G, Bittner P, Sandholzer N et al (2011) CpG blocks immunosuppression by myeloid-derived suppressor cells in tumor-bearing mice. Clin Cancer Res 17:1765–1775

Chapter 17
Macrophage Differentiation and Activation States in the Tumor Microenvironment

Jo A. Van Ginderachter

Abstract Macrophages are amongst the most plastic cells of the body, contributing to organogenesis and tissue homeostasis and regulating the balance between pro- and anti-inflammatory reactions. To accommodate these different functions, macrophages are notoriously heterogeneous and are able to adopt different activation states in response to a changing microenvironment. Accumulating evidence exists that macrophages contribute to all phases of the cancer process. These cells are central players in inflammation-associated carcinogenesis, participate in tumor immunosurveillance, and are involved in tumor progression and metastasis. Inside tumors, tumor-associated macrophages (TAM) are confronted with different tumor microenvironments, leading to TAM subsets with distinct activation states and specialized functions. A better refinement of the molecular and functional heterogeneity of tumor-associated macrophages might pave the way for novel cancer therapies that directly target these tumor-supporting cells.

Keywords Tumor-associated macrophages · M1 · M2 · Myeloid-derived suppressor cells · Tie2-expressing monocytes · Hypoxia · ER stress · Exosomes

17.1 Introduction

For a long time, the main focus in cancer research was centered on discovering the activating (oncogenes) or deactivating (tumor suppressor genes) mutations that drive the transformation of cells. Currently, it is clear that both cancer cell-intrinsic and –extrinsic mechanisms are equally important in driving cancer progression and that tumors should be considered as organ-like structures in which a complex

J. A. Van Ginderachter (✉)
Laboratory of Cellular and Molecular Immunology, Myeloid Cell Immunology Laboratory, VIB-Vrije Universiteit Brussel, Building E, Level 8, Pleinlaan 2, 1050 Brussels, Belgium
e-mail: jvangind@vub.ac.be

M. R. Shurin et al. (eds.), *The Tumor Immunoenvironment*,
DOI: 10.1007/978-94-007-6217-6_17,

bidirectional interplay exists between transformed and non-transformed cells. These normal tumor-infiltrating cells include fibroblasts, adipocytes, endothelial cells and cells of the innate and adaptive immune system. As a matter of fact, immune cells, in particular cells belonging to the mononuclear phagocyte system (including lineage committed bone marrow precursors, monocytes, and macrophages), are reported to take part in every aspect of a tumor's natural history, from tumor initiation and immunoediting to primary tumor growth and metastasis. In this respect, macrophages are known drivers of chronic inflammatory diseases predisposing to cancer, including *Helicobacter pylori*-driven gastric carcinoma (Kaparakis et al. 2008), colitis-associated colon carcinoma (Grip et al. 2003) and hepatitis-mediated hepatocellular carcinoma (Heymann et al. 2009). Once tumors are established, the normal physiological functions of macrophages are often hijacked by the tumor to promote its progression, resulting in extracellular matrix reorganization, immune suppression, angiogenesis and the stimulation of cancer cell proliferation, motility and invasiveness (Mantovani et al. 1992, 2002; Qian and Pollard 2010). Notably, macrophages are confronted with entirely different microenvironments at sites of chronic inflammation versus the tumor site and so is the macrophage phenotype in these distinct regions, reflecting the adaptability of these cells to a changing environment.

17.2 Macrophage Activation States

When studying macrophages, including tumor-associated macrophages (TAM), one should be aware of the remarkable plasticity of these cells (Van Ginderachter et al. 2006). Heterogeneity is at the heart of the macrophage's polyvalency, making them indispensable for functions as diverse as development and tissue homeostasis (trophic macrophages), pathogen clearance, Th1 or Th2 cytokine-driven inflammation (reflected by the classical versus alternative model of macrophage activation), and wound healing (Gordon and Taylor 2005; Mosser and Edwards 2008).

A popular working scheme for macrophage activation is the M1/M2 dichotomy. Classically activated (or M1) macrophages are elicited in an environment dominated by Th1 cytokines, such as IFN-γ and TNF-α, and/or by recognition of pathogen–associated molecular patterns or endogenous danger signals. This is a prototypical pro-inflammatory type of macrophage, that is implicated in the initiation and propagation of inflammation and pathogen clearance. For a long time, macrophages exposed to anti-inflammatory or Th2-associated mediators were considered as merely deactivated. However, it is clear by now that macrophage behaviour is also significantly altered by the prototypical Th2 cytokines IL-4 and IL-13, often mediated by enhanced intracellular polyamine levels (Van den Bossche et al. 2012), inducing so-called *bona fide* alternatively activated macrophages or M2 (Martinez et al. 2009). Moreover, multiple anti-inflammatory cues, such as IL-10, TGF-β, glucocorticoids, immune complexes and apoptotic cells, modulate the macrophage phenotype. Given the diversity of these signals different authors have proposed different macrophage classification systems (Mosser and Edwards 2008; Martinez et al. 2009; Goerdt and Orfanos 1999; Mantovani et al. 2004). However, a unifying feature of all non-M1 macrophages is their capacity to suppress Th1 cytokine-driven inflammation, to regulate adaptive immune responses and to contribute to wound healing. On the other hand, it should be emphasized that M2 can be considered as pro-inflammatory in Th2-driven pathologies, such as helminth infections and asthma (Martinez et al. 2009). Though the M1/M2 concept provides a useful working scheme, it should be realized that the in vivo situation is far more complex. Accordingly, macrophages are mostly exposed to a mixture of stimuli and will adopt mixed functional profiles. As an illustration of this complexity, the consensus gene signature for in vivo induced alternatively activated macrophages in different pathologies not only contains genes that are strictly IL-4/IL-13-inducible such as E-cadherin (Van den Bossche et al. 2009), but also genes that are not inducible in vitro by any of the known M2 inducing stimuli (Hassanzadeh Ghassabeh et al. 2006).

Remarkably, most studies on human macrophages disconnect the M1/M2 concept from Th1/Th2 cytokines. Instead, the hematopoietic growth factors M-CSF and GM-CSF are used to differentiate more M2-like or M1-like macrophages from peripheral blood monocytes, respectively (Puig-Kröger et al. 2009). M-CSF-generated macrophages probably resemble more the trophic type of macrophage, which do not secrete the pro-inflammatory cytokines IL-12 and IL-23, but instead produce high levels of IL-10 (Pollard 2009). Notably, an M-CSF response signature in the tumor significantly correlates with worse prognosis in breast carcinoma patients (Beck et al. 2009), suggesting that trophic macrophages might also be prominent in the breast tumor microenvironment and mediate tumor progression. In this respect, cancer cell-derived M-CSF was found to upregulate DC-SIGN expression on $CD14^{+}CD68^{+}$ TAM from human breast adenocarcinomas, which upon cross-linking induced the secretion of IL-10 and the overall anti-inflammatory phenotype of TAM (Dominguez-Soto et al. 2011). Conversely, GM-CSF treatment of mouse mammary tumors inhibits tumor growth and metastasis by invoking an antitumoral program in TAM (Eubank et al. 2009). Hence, these data suggest a picture whereby M1-like macrophages are antitumoral, while M2-oriented macrophages exert protumoral activities.

17.3 Macrophages in Oncogenesis

Meta-analysis studies have clearly established that patients suffering from chronic microbial infections or autoimmune diseases have a higher risk of developing tumors at the site of inflammation. This is in line with the chemopreventive effect of nonsteroidal anti-inflammatory drugs and ligands for the anti-inflammatory nuclear receptor PPARγ (Grivennikov et al. 2010; Karin et al. 2006; Balkwill et al. 2005; Coussens and Werb 2002; Van Ginderachter et al. 2008). It is clear by now that macrophages play a non-redundant role in these phenomena. For example, blocking the recruitment of macrophages to the inflamed colon by inhibiting the activity of the CCL2 chemokine resulted in a reduced colitis-associated carcinogenesis (Popivanova et al. 2009). Along the same line, *Helicobacter* infection and prostaglandin E_2 are needed to induce CCL2 expression and recruit macrophages to the stomach, resulting in gastric tumorigenesis (Oshima et al. 2011).

In the case of skin carcinogenesis, fibroblasts were shown to be crucial players, mainly through mediating the CCL2-dependent recruitment of macrophages to the inflammatory site (Zhang et al. 2011). Once arrived in the inflamed tissue, several inflammatory signalling pathways impart the macrophages with oncogenic properties. The inflammatory, M1-inducing transcription factor NF-κB is instrumental, since a myeloid cell-specific defect in this pathway (through deletion of IKKβ in these cells) leads to the absence of inflammatory mediators that act as tumor-promoting paracrine factors and consequently reduced tumor formation (Greten et al. 2004). Conversely, a myeloid cell-specific deficiency of STAT3—an anti-inflammatory M2-inducing transcription factor that antagonizes NF-κB (Yu et al. 2007; Kortylewski et al. 2009)—results in the spontaneous development of colitis triggered by the gut microflora and leads to an enhanced rate of tumor formation in inflamed regions (Deng et al. 2010). Notably, sterile tissue damage can also predispose to cancer, as is exemplified by DEN-induced hepatocarcinogenesis. Also in that case, macrophages and NF-κB activation drive the development of liver tumors through the secretion of the hepatomitogens TNF-α and IL-6, leading to compensatory hepatocyte proliferation and eventually transformation (Maeda et al. 2005). In this scenario, macrophages possibly become activated by endogenous danger signals produced during cell necrosis or extracellular matrix degradation. This observation is in accordance with data demonstrating an important role for MyD88, the adaptor molecule in Toll-like receptor (TLR) and IL-1R signaling and an upstream activator of NF-κB, in inflammation-associated or non-inflammatory carcinogenesis alike (Naugler et al. 2007; Rakoff-Nahoum and Medzhitov 2007; Swann et al. 2008). Besides microbial or endogenous danger signals, macrophages also recognize a variety of opsonising substances, such as antibodies. The relevance of this property for oncogenesis is demonstrated in a transgenic model of skin carcinogenesis, whereby peripheral B cell activation results in the production of autoantibodies and deposition of immune complexes in the premalignant lesion. These complexes bind to activating FcγRs on locally residing macrophages and trigger their protumoral functions (de Visser et al. 2005; Andreu et al. 2010).

From a mechanistic point of view, inflammatory mediators are major culprits for macrophage-mediated tumor stimulation. Inflammatory macrophage products that have been shown to contribute to oncogenesis include COX-2 (Oshima et al. 2011, 1996), MMP9 (Coussens et al. 2000) and reactive nitrogen and oxygen species (Jaiswal et al. 2001; Kim et al. 2003). Most evidence exists however for the role of the inflammatory cytokines IL-6 and TNF. For example, the estrogen-regulated difference in IL-6 production by Kupffer cells from male versus female mice entirely accounts for the gender differences in tumor incidence in the model of DEN-induced hepatocarcinogenesis (Naugler et al. 2007). During hepatitis,

Kupffer cells produce TNF that initiates hepatocyte NF-κB activation and hyperactivation of oval cells, aiding to tumorigenesis (Pikarsky et al. 2004; Knight et al. 2000). Also in gastric carcinogenesis, macrophage-derived TNF promotes tumor formation by stimulating pro-tumorigenic Wnt signaling in epithelial cells (Oshima et al. 2011). Finally, autocrine TNF-R1 signaling in macrophages creates an essential inflammatory loop for colitis-associated colon carcinogenesis (Popivanova et al. 2008).

Overall, these data convincingly demonstrate that M1-oriented pro-inflammatory macrophages play an important role in the earliest phases of carcinogenesis.

17.4 Macrophages in Tumor Immunosurveillance

The immunosurveillance theory states that newly formed cancer cells are continuously under attack by the immune system. In support of this theory, mice with genetic deficiencies in various immune functions more readily develop tumors, be it spontaneously or upon treatment with carcinogens (Vesely et al. 2011). Also in human cancer patients, the "immune contexture" of a tumor, representing the quantity, functional orientation and location of immune cells in the tumor microenvironment, is one of the strongest prognostic factors for recurrence and overall survival (Galon et al. 2006; Bindea et al. 2010).

Classically activated M1 macrophages can be cytotoxic for cancer cells, mostly through the production of reactive oxygen and nitrogen species (such as NO). This capacity has been therapeutically exploited in preclinical tumor models, whereby macrophages act as principal anti-tumoral effector cells following the provision of macrophage stimulators such as IFN-γ, anti-CD40 or TLR ligands (Wu et al. 2009; Vicetti Miguel et al. 2010; Beatty et al. 2011; Guiducci et al. 2005). A combination of anti-CD40 and the chemotherapeutic drug gemcitabine even causes tumor regression in some patients with pancreatic ductal adenocarcinoma (Beatty et al. 2011). However, whether naturally occurring M1 macrophages contribute to early immune surveillance is less well documented. In one mouse model of MHC II-negative multiple myeloma, macrophages are rapidly recruited to the incipient tumor site, present tumor antigens to CD4$^+$ T cells, which subsequently trigger the tumoricidal capacity of local macrophages through IFN-γ (Corthay et al. 2005). Indirect evidence for the role of macrophages in immunosurveillance comes from the frequent downregulation of the tumor suppressor genes p53 and IRF-8. As a matter of fact, experimental re-expression of these genes in cancer cells leads to macrophage-dependent tumor attack, suggesting that macrophages impose a selective pressure on transformed cells (Greeneltch et al. 2007; Xue et al. 2007). Another consequence of this selective pressure appears to be the general expression of the "don't eat me" signal CD47 on human cancer cells, both from solid tumors (Willingham et al. 2012) and haematological malignancies (Kim et al. 2012). CD47 interacts with the ITIM-containing SIRPα receptor on macrophages to inhibit phagocytosis and counterbalance the pro-phagocytic signal delivered by calreticulin (Jaiswal et al. 2009; Chao et al. 2010). Therefore, CD47 is a prime target for blocking antibody-mediated therapy, resulting in cancer cell elimination (Majeti et al. 2009).

17.5 Monocytes and Macrophages in Established Tumors

Tumors function as complex organs, containing multiple interacting cell types and a multitude of specialized microenvironments instructing different characteristics on the cells present (Egeblad et al. 2010). This complexity is reflected by the heterogeneity of tumor-associated mononuclear phagocytes, both at the level of differentiation and activation. One of the difficulties when studying these populations is that they are highly related, often express similar markers and are in some cases able to perform similar functions (Coffelt et al. 2010).

Here, we describe the mechanisms via which monocytes and macrophages are attracted to the tumor site in the first place, which distinct phenotypes they adopt there and which microenvironmental cues influence their behaviour.

17.5.1 Recruitment of Monocytes to the Tumor

Already early after the initiation of the oncogenic programme, monocytes are actively recruited to the tumor site from the bloodstream (Fukuda et al. 2011). Multiple factors have been reported to play a role in monocyte recruitment, several of which are also known tumor-derived angiogenic mediators such as VEGF, PDGF and M-CSF or pro-inflammatory molecules like S100A8/9 (Ostrand-Rosenberg and Sinha 2009). Tie2-expressing monocytes (TEM), a population of constitutively angiogenic monocytes, are attracted under the influence of angiopoietin-2, which orients these cells near blood vessels (Mazzieri et al. 2011) and stimulates their immunoregulatory and angiogenic capacity (Coffelt et al. 2010, 2011).

However, chemokines can be considered as the "professional" cell recruiting proteins. It has been reported that myeloid cell-recruiting chemokines can be expressed de novo in cancer cells under the influence of oncogenes (Borrello et al. 2005). One of these chemokines, CCL2, is a major chemoattractant for monocytes and macrophages in several tumor types and is involved in shaping the functions of these cells (e.g. expression of MMP-9) (Ueno et al. 2000). However, some controversy exists on whether the CCL2/CCR2 axis is implicated in monocyte traffic to the primary tumor or to the metastatic sites. Thus, CCR2 is clearly important for the infiltration of Ly6C^{high} (classical) monocytes to primary tumors in several transplantable models (Movahedi et al. 2012; Hart et al. 2009) and in the K14-HPV/E_2 transgenic model of cervical carcinogenesis (Pahler et al. 2008) and the Kras$^{LSL/G12D/+}$;p53 $^{fl/fl}$ conditional genetic mouse model of lung adenocarcinoma (Cortez-Retamozo et al. 2012). In contrast, Ly6C^{low} (non-classical) monocytes preferentially accumulate in the primary tumors of PyMT mouse breast carcinomas, while CCR2 attracts $Ly6C^{high}$ monocytes to pulmonary metastases (Qian et al. 2011). Nevertheless, in the same PyMT model, injury upon chemotherapy leads to stromal CCL2 expression and acute recruitment of CCR2-expressing monocytic cells to regions of necrotic cell death, that contribute to tumor regrowth after treatment (Nakasone et al. 2012). These data might suggest that the level of danger molecules present within the tumor could be decisive for the type of monocyte that is being recruited. Interestingly, when the CCR2-mediated influx of monocytes is hampered, a compensatory infiltration of angiogenic neutrophils is seen in several models, resulting in sustained or even enhanced tumor growth (Movahedi et al. 2012; Pahler et al. 2008; Sawanobori et al. 2008). Recent findings furthermore suggest that the majority of the tumor-infiltrating monocytes and neutrophils are derived from a splenic reservoir, which seems to release these cells upon acute demand (Cortez-Retamozo et al. 2012). A final twist to the involvement of CCL2–CCR2 in cancer biology, is the regulation of CCL2 activity by nitration/nitrosylation under the influence of intratumorally produced reactive nitrogen species (RNS). As a result, modified CCL2 could no longer attract tumor-specific CTLs, but could still recruit suppressive myeloid cells to the tumor, thereby skewing the response in favor of tumor growth (Molon et al. 2011).

Recently, an interesting new pathway of myeloid cell recruitment was described in a mouse ovarian carcinoma model. High levels of intratumoral TNF-α maintained TNFR1-dependent IL-17 production by $CD4^+$ T cells, which in turn attracts myeloid cells to the tumor microenvironment (Charles et al. 2009).

17.5.2 Diversity of Monocytes and Macrophages at the Tumor Site

1. Myeloid-derived suppressor cells (MDSC)

$CD11b^+Gr\text{-}1^+$ MDSC are not a separate lineage of myeloid cells, but encompass $CD11b^+Ly6C^{hi}Ly6G^{neg}$ monocytic (MO-MDSC) and $CD11b^+Ly6C^{int}Ly6G^{hi}$ granulocytic (PMN-MDSC) cells in an immature differentiation state. These MDSC subsets share the common characteristic of being immunosuppressive (MDSC is a definition describing a function of cells rather than a lineage of cells), although they use different mechanisms (Van Ginderachter et al 2006; Movahedi et al. 2008; Youn et al. 2008). These cells are not only induced by tumors, but actually represent an immunosuppressive feedback mechanism upon a wide range of insults, including bacterial, fungal and parasitic infections, trauma and transplantation (Nagaraj et al. 2009; Van Ginderachter et al. 2010) (Fig. 17.1).

Tumors elicit MDSC accumulation in the bone marrow, blood, spleen and at the tumor site under the control of the STAT3 and c/EBPβ transcription factors and tumor-derived inflammatory mediators and cytokines, such as S100A9, prostaglandins, GM-CSF and IL-6 (Nefedova et al. 2004; Sinha et al. 2007; Cheng et al.

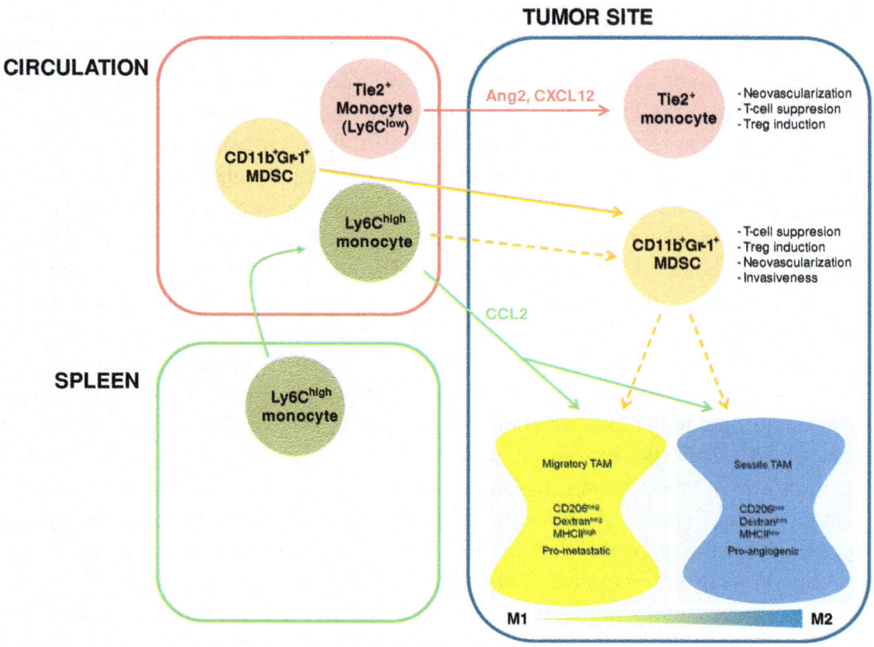

Fig. 17.1 Diversity of monocytes and macrophages at the tumor site. Major monocyte/macrophage populations in the tumor microenvironment include Tie2-expressing monocytes, CD11b$^+$Gr-1$^+$ MDSC and at least two major TAM subsets, which can be discriminated based on migratory behavior, activation state/molecular profile and functions. Precursors of these cells are attracted from the circulation, and at least in the case of monocytes, from the spleen. Each of these myeloid cell population exert functions that are in favor of tumor progression

2008; Dolcetti et al. 2010; Marigo et al. 2010; Bayne et al. 2012; Pylayeva-Gupta et al. 2012). To become suppressive, MDSC need to be activated further via transcription factors such as STAT1 or NF-κB (Movahedi et al. 2008; Greifenberg et al. 2009). As a consequence, these cells deploy a wide range of T cell suppressive mechanisms [reviewed in (Marigo et al. 2008; Ostrand-Rosenberg 2010)].

It is important to realize that not all CD11b$^+$Gr-1$^+$ cells are immunosuppressive (Dolcetti et al. 2010; Greifenberg et al. 2009). For example, in the microenvironment of mouse transplantable tumors, cells that classify as CD11b$^+$Gr-1$^+$ are diverse and are present in different ratios depending on the model: Ly6C^{hi}MHC II^{neg} monocytes, Ly6C^{hi}MHC II^{hi} immature macrophages, Ly6C^{int}MHC II^{neg} immature macrophages, Ly6C^{int}MHC II^{neg}Ly6G^{hi} neutrophils and even Ly6C^{int}MHC II^{neg}CCR3^{hi} eosinophils (Movahedi et al. 2010). As a matter of fact, not many studies have provided direct evidence for the presence of T cell suppressive CD11b$^+$Gr-1$^+$ cells inside tumors. One study on different transplantable mouse tumor models revealed that CD11b$^+$Gr-1$^+$ cells from the tumor site are more suppressive than their splenic counterparts. Mechanistically, this can be explained by the upregulation of the transcription factor HIF-1α under the influence of tumor hypoxia, which results in the induction of iNOS and arginase-1 (Corzo et al. 2010). Similarly, in a transgenic model of pancreatic ductal adenocarcinoma, MDSC from the pancreas are superior suppressors than those from spleen (Bayne et al. 2012). Notably, MDSC from the tumor microenvironment differentiated more rapidly to macrophages (Corzo et al. 2010), which is consistent with their fast turnover rate in tumors and their role as precursors for tumor-associated macrophages, which are strong suppressors in their own right (Movahedi et al. 2010; Kusmartsev 2005). From a therapeutic point of view, MDSC from different locations might differ in their drug sensitivity. For example, the STAT-3 inhibitor Sunitinib efficiently eliminates MDSC from the spleen, but not from the tumor. In the latter location, MDSC survival appears to be mediated by STAT5 signaling triggered by high intratumoral GM-CSF levels (Ko et al. 2010). More recently, 5-fluoroacil was proposed as a compound with the capacity to selectively induce MDSC apoptosis in spleen and tumor, resulting in strongly enhanced anti-tumor immunity (Vincent et al. 2010). Also the synthetic triterpenoid CCDO-Me has shown anti-tumor activity, which depends on a shutdown of the MDSC suppressive activity (but not their survival) and a consequent re-activation of the immune system (Nagaraj et al. 2010).

Finally, it is important to note that $CD11b^{+}Gr\text{-}1^{+}$ cells, present within the tumor microenvironment, mediate pro-tumoral functions which are not clearly linked with immune suppression (are these MDSC or MDSC-like cells?). For example, these cells were shown to infiltrate tumors via the action of CXCL12 and the endothelial adhesion molecule VAP-1 (Liu et al. 2010; Marttila-Ichihara et al. 2009) and to contribute to tumor neovascularisation, thereby mediating the tumor refractoriness to anti-VEGF treatment. Angiogenic molecules released by these cells include MMP9 (Yang et al. 2004) and Bv8 (Shojaei et al. 2007a, 2007b). In addition, $CD11b^{+}Gr\text{-}1^{+}$ cells can increase invasiveness and metastasis in mouse mammary tumors with defective TGF-β signalling, thanks to their attraction to the invasive edge by CXCL5 and CXCL12. Increase of cancer cell invasiveness by $CD11b^{+}Gr\text{-}1^{+}$ cells can be reproduced in vitro and is independent from T cell activity (Yang et al. 2008). Finally, $CD11b^{+}Gr\text{-}1^{+}$ cells are attracted to the premetastatic niche by S100A9, where they clear the route and optimize conditions for cancer cell arrival (Yan et al. 2010; Hiratsuka et al. 2006).

2. Tie2-expressing monocytes/macrophages (TEM)

A relatively low frequency of peripheral blood monocytes expresses Tie2, the receptor for Angiopoietin-2, and are hence termed Tie2-expressing monocytes or TEM (De Palma et al. 2005). In terms of gene expression signature and surface marker expression, these cells share many characteristics with the "patrolling" or "non-classical" type of monocytes, in the sense that they are $Ly6C^{low}$ $CCR2^{neg}$ $CD62L^{neg}$ in mouse and $CD14^{low}CD16^{hi}$ in human (De Palma et al. 2005; Venneri et al. 2007; Pucci et al. 2009). Under the influence of angiopoietin-2 production by tumor endothelium, TEM are recruited to tumors where they differentiate into M2-like macrophages with a non-redundant function in tumor neovascularization (Coffelt et al. 2010; De Palma et al. 2005; Venneri et al. 2007 Pucci et al. 2009;

Murdoch et al. 2007). An alternative pathway of TEM recruitment is via the CXCL12/CXCR4 axis, whereby CXCL12 is induced upon administration of vascular-disrupting agents to the tumor. Infiltration of TEM, as prime proangiogenic cells, is responsible for the failure of this type of therapy (Welford et al. 2011). Notably, angiopoietin-2 is a crucial regulator and amplifier of TEM's angiogenic potential (Coffelt et al. 2010; Murdoch et al. 2007), mainly by stimulating their Tie2 expression and retention near blood vessels (Mazzieri et al. 2011). Moreover, recent data also ascribe a T cell suppressive and Treg-inducing capacity to TEM, again under the influence of angiopoietin-2 (Coffelt et al. 2011).

From a therapeutic point of view, the efficient infiltration of TEM into tumors can be exploited for the targeted delivery of proteins to the tumor site. For example, TEM have been transduced with (IFN-alpha) and semaphorin 3A, both of which lead to reduced tumor growth and metastasis (De Palma et al. 2008; Casazza et al. 2011).

3. Tumor-associated macrophages (TAM)

The majority of studies up to now have revealed a correlation between a high TAM density and poor prognosis for many different cancer types, highly suggesting the importance of macrophages for tumor progression (Bingle et al. 2002; Lewis and Pollard 2006). Moreover, the intratumoral overexpression of macrophage chemoattractants and growth factors, such as CCL2 and M-CSF, is predicitive of poor outcome (Qian and Pollard 2010; Mantovani and Sica 2010) and so is the presence of a M-CSF response gene signature in the primary tumor and corresponding metastases (Sharma et al. 2010; Webster et al. 2010; Espinosa et al. 2009).

In line with these correlative findings, inhibiting M-CSF or M-CSFR function, either genetically (M-CSF-deficient mice $Csf1^{op}/Csf1^{op}$ mice) or pharmacologically, suppresses transplantable tumor growth (Aharinejad et al. 2004; Nowicki et al. 1996; Kubota et al. 2009; Priceman et al. 2010). The same holds true for the transgenic MMTV-PyMT model in a M-CSF-deficient background, whereby primary tumor growth remains unaltered, but progression to invasiveness and metastasis is significantly delayed (Lin et al. 2001). Together, these data unequivocally established TAMs as important players in the tumor microenvironment. The mechanisms by which TAM stimulate tumor growth, including invasion, angiogenesis, metastasis and immunosuppression have been extensively reviewed elsewhere (Qian and Pollard 2010; Mantovani and Sica 2010).

TAM heterogeneity. Intratumoral macrophages occupy different areas of the tumor—including areas of invasion, in the stroma, in perivascular areas, or in avascular and perinecrotic areas—and different functions for these TAM subpopulations have been predicted (Lewis and Pollard 2006). In mouse mammary carcinomas, the majority of macrophages are found at the tumor margins, but those deeper in the tumor are functionally also very important (Wyckoff et al. 2007). Indeed, they are found in close proximity of blood vessels as single cells or in clusters and guide the cancer cells towards intravasation. This depends on a paracrine positive feedback loop, whereby the cancer cells produce M-CSF and the macrophages produce EGF, as such establishing a coordinated migration of both cell types (Wyckoff et al. 2007; Goswami et al. 2005). Notably, other growth factors such as heregulin-$\beta 1$ and CXCL12 can also trigger this paracrine loop (Hernandez et al. 2009). Along the same line, Kedrin et al. (2008) followed the behaviour of photoswitched cancer cells in mammary tumors via intravital imaging, and concluded that cancer cells only migrate in a vascular environment containing perivascular macrophages. Overall, the density of the tripartite interaction between invasive cancer cells, macrophages and endothelial cells is predictive for the occurrence of distant metastases in breast cancer patients (Robinson et al. 2009).

These data clearly illustrate the existence of distinct tumor microenvironments, but to what extent does this relate to the occurrence of distinct macrophage subpopulations? A recent study described two TAM subpopulations in orthotopically growing mouse mammary tumors, which differed at the level of MHC II expression (MHC II^{low} versus MHC II^{high}) and intratumoral localization (Movahedi et al. 2010). MHC II^{low} TAMs reside in more hypoxic regions, are overall more M2-oriented (e.g. higher expression levels of the Macrophage Mannose Receptor or CD206), and display a superior pro-angiogenic activity as compared to the more M1-like MHC II^{hi} TAMs. This finding could be exploited in several ways: (1) imaging of this population, as was recently performed using nanobodies against CD206 (Movahedi et al. 2012), might allow the visualization of hypoxic regions, and (2) elimination of this population could hamper tumor growth. Moreover, these murine findings appear to be clinically relevant, since MHC II^{high} and MHC II^{low} TAM subsets also occur in different regions of human hepatocellular carcinomas, the latter of which being IL-10^{+}, suggestive of a more M2-like phenotype (Kuang et al. 2007). Using spinning disk confocal microscopy on mouse mammary tumors, M-$CSFR^{+}CD68^{+}CD206^{+}$ highly phagocytic cells (dextran uptake) were identified as sessile cells at the tumor border, while $CD68^{+}CD206^{neg}$ myeloid cells that do not ingest intravenously injected dextran (monocytes?) were migratory (Egeblad et al. 2008). In addition, sessile cells that had infiltrated the tumor mass were also mostly $CD68^{+}$ $CD206^{neg}$ $dextran^{neg}$.

This concept of TAM heterogeneity is further supported by extended gene expression analysis on distinct TAM populations. Indeed, when highly phagocytic TAMs were compared to the TAM population that co-migrates with cancer cells in an in vivo migration assay, these two TAM subsets turned out to be quite divergent at the gene expression level (Ojalvo et al. 2010). Interestingly, reminiscent of the dichotomy between MHC II^{high} $CD206^{neg}$ M1-like TAM and the MHC II^{low} $CD206^{+}$ M2-like TAM subpopulations from orthotopically grown mammary tumors (Movahedi et al. 2010), invasive TAM expressed lower levels of all typical M2-associated genes (Hassanzadeh Ghassabeh et al. 2006), including CD206, as compared to the $dextran^{+}$ sessile TAM. In this context, an Ets2-driven transcriptional program in TAM was shown to promote tumor metastasis, and it would be interesting to pinpoint the Ets2 effect to the migratory, pro-metastatic TAM subset (Zabuawala et al. 2010). Overall, the data from different labs seem to be consistent with the existence of at least two major TAM subpopulations: (1) M-$CSFR^{-}$

Gr-1^{neg}Dextran^{neg}CD206^{neg}MHC II^{high} TAM migratory TAM in the neighbourhood of blood vessels. These cells are less M2-oriented an aid cancer cells to intravasate, (2) sessile M-CSFR^{+}Gr-1^{neg}Dextran^{+}CD206^{+}MHC II^{low} TAM found at tumor-stroma borders and/or hypoxic regions that resemble more M2-like or "trophic" macrophages. A possible exception to this apparent dichotomy could be the location of TEM, which also express high levels of CD206 and are rather M2-oriented (Pucci et al. 2009), but stay attached to blood vessels in the presence of high angiopoietin-2 and Tie2 levels (Mazzieri et al. 2011).

Finally, these novel findings warrant against an overinterpretation of data taken from total TAM populations. Indeed, gene expression profiles on total TAM often revealed mixed M1 and M2 characteristics, probably due to the presence of distinct TAM subsets (Biswas et al. 2006; Umemura et al. 2008; Doedens et al. 2010).

Microenvironmental cues potentially regulating TAM heterogeneity

Hypoxia: TAM are known to be attracted to hypoxic tumor regions via several mechanisms (Murdoch et al. 2004). In those regions, the hypoxia-inducible transcription factors HIF-1 and HIF-2 were shown to be implicated in the hypoxic response of primary human and mouse macrophages (Fang et al. 2009). However, even under normoxic conditions, several triggers cause HIF-1α gene transcription and stabilization in macrophages, including NF-κB (Rius et al. 2008) and the secretion of sphingosine-1-phosphate (S1P) and TGF-β by apoptotic cells (Herr et al. 2009). Moreover, Th1 cytokines induce HIF-1α in M1 macrophages, while Th2 cytokines rather upregulate HIF-2α in M2 (Takeda et al. 2010). Importantly, all these stimuli are potentially relevant in the tumor microenvironment. For example, tumors that are unable to synthesize S1P grow slower and are infiltrated by anti-tumoral TAM (Weigert et al. 2009) (Fig. 17.2).

Currently, evidence exists for the involvement of both HIF isoforms in TAM regulation. The conditional deletion of HIF-2α in myeloid cells reduces expression of M-CSFR and CXCR4 by these cells, resulting in a diminished recruitment of macrophages to tumors and a retarded tumor progression in two models of inflammation-induced cancer (Imtiyaz et al. 2010). Conversely, myeloid cell-specific HIF-1α deletion in the MMTV-PyMT background mainly reduces iNOS and arginase-1 expression in TAM resulting in less T cell suppression (Doedens et al. 2010). In that study, no effects on VEGF levels or vascularisation were observed. However, when Werno et al. (2010) applied HIF-1$\alpha^{-/-}$ macrophages to tumor spheroids in vitro, these cells were less angiogenic but expressed higher levels of M2 markers. The impact of hypoxia on TAM has also been exploited therapeutically, whereby macrophages were engineered to carry oncolytic adenovirus genes under the control of HIF-regulatory elements (Muthana et al. 2011).

Cytokines: VEGF, either through the interaction with the VEGFR1 or VEGFR2, is an important chemoattractant for mature macrophages in established tumors and contributor to angiogenesis (Hiratsuka et al. 2011; Muramatsu et al. 2010; Dineen et al. 2008). Hence, modulating VEGF levels could have important consequences for tumor growth. This is illustrated by the fact that M-CSF indirectly contributes to angiogenesis in vivo by inducing VEGF in macrophages through the MAPK/Erk pathway (Curry et al. 2008). Conversely, GM-CSF re-educates macrophages to

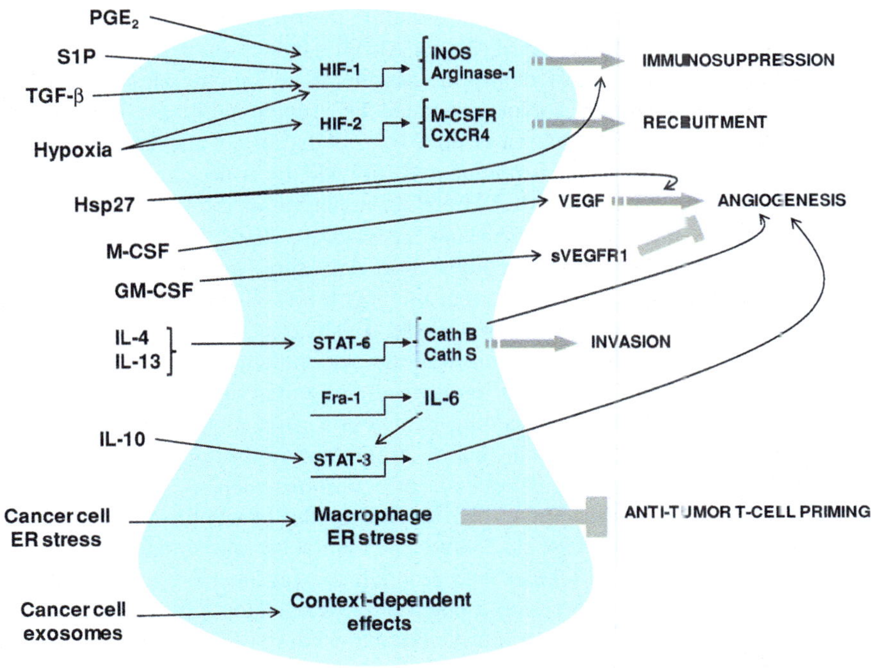

Fig. 17.2 Microenvironmental stimuli influencing the behaviour of tumor-associated macrophages. Distinct mediators, secreted by cancer cells or other cell types in the tumor microenvironment, instruct a tumor-promoting phenotype on tumor-associated macrophages, including immunosuppressive, angiogenic and pro-metastatic functions. Several signalling pathways and transcription factors are implicated in this phenomenon

become anti-angiogenic through the secretion of soluble VEGFR1 (Eubank et al. 2009). Interestingly, the p38 MAPK signalling pathway, which is amenable for drug-mediated intervention, is an integrator of chemoattractant signals, not only by VEGF, but also by CXCR4 (Hiratsuka et al. 2011).

In the tumor microenvironment, macrophages are further influenced by several cytokines. For example, IL-4 produced by $CD4^+$ T cells (DeNardo et al. 2009), or IL-13 produced by NKT cells (Sinha et al. 2005), were shown to induce M2-like TAM that cause invasion or immunosuppression. IL-4 upregulates Cathepsins B and S via STAT6 which appears to mediate these processes in several tumor types (Gocheva et al. 2010). The importance of this pathway is further highlighted by the increased tumor immunosurveillance in mice with a STAT6-deficiency in the hematopoietic compartment (Ostrand-Rosenberg et al. 2002). In addition, the broadly anti-inflammatory cytokines TGF-β and IL-10, derived from various sources, are known to influence the TAM phenotype (Flavell et al. 2010; Sica et al. 2000; Wong et al. 2010).

Prostaglandin E_2: PGE_2 is secreted at relatively high levels by many cancer cell types. This eicosanoid was shown to induce IL-10 by macrophages and to upregulate arginase-1 in TAM, as such initiating their immunosuppressive activity (Huang et al. 1998; Rodriguez et al. 2005). Also the induction of HIF-1α in tumor-infiltrating MDSC, and consequently their higher T cell suppressing capacity, as well as the suppression of tumor-associated DC functionality are mediated by PGE_2 (Corzo et al. 2010; Sharma et al. 2003). PGE_2 also works in an autocrine fashion, since its production is augmented in TAM by a downregulation of the PGE_2-catabolizing enzyme 15-PGDH and an induction of the PGE_2-synthesizing enzymes COX-2 and mPGES1 (Eruslanov et al. 2009, 2010). Further evidence for the importance of this pathway comes from data showing that the absence of $cPLA_2$ in macrophages, an enzyme that initiates the PGE_2 biosynthesis pathway, results in reduced tumor progression (Weiser-Evans et al. 2009).

Tumor ER stress: ER stress comprises the accumulation of unfolded or misfolded proteins in the ER lumen as a result of environmental stress signals, resulting in the activation of signaling pathways known as the unfolded protein response (UPR) (Mahadevan and Zanetti 2011). Within the tumor microenvironment, ER stress can be the consequence of conditions such as hypoxia, glucose deprivation and low extracellular pH, leading to the production of pro-inflammatory cytokines such as IL-6, IL-23 and TNF. Notably, inflammatory mediators, including inflammatory cytokines and reactive oxygen species, can themselves initiate the UPR response creating a positive feedback loop (Zhang et al. 2006). Such cytokines are potentially tumorigenic and a possible functional connection between the UPR and tumor progression has been suggested by the observation that fibrosarcoma cells, in which the UPR-associated chaperone Grp-78 was silenced, have a reduced tumor growth in vivo (Jamora et al. 1996). Remarkably, ER stress in several cancer cell lines is transmissible to macrophages, which involves soluble factors that at least partly mediate their effect via TLR4 (Mahadevan et al. 2011). A possible consequence in tumor-bearers is a diminished priming of anti-tumor T cells, due to a UPR-dependent remodeling of the antigen processing and presentation machinery in antigen-presenting cells (Mahadevan and Zanetti 2011).

Exosomes: In recent years, it has become increasingly clear that exosomes, nano-sized membranous vesicles that develop from exophytic budding of the cellular membrane, can be isolated from tumors and peripheral fluids from cancer patients. These vesicles may contain lipids, peptides, proteins, microRNAs, and mRNAs (Théry et al. 2001) and are therefore involved in autocrine and paracrine signaling (Zhang and Grizzle 2011). In this context, exosomes were suggested to modulate the immune system, through carrying immunoregulatory molecules such as FasL, MMPs and PD-1 as cargo. Consequently, these nano-structures induce signaling in the target cell without the need for direct cell–cell contact. In addition, exosomes may fuse with a recipient cell, leading to the acquisition of novel molecules and the delivery of mRNA and miRNA that may alter the behavior of the target cells (Taylor and Gercel-Taylor 2011). Given these characteristics, it lies within expectation that exosomes also influence myeloid cell behavior. Indeed, blood–derived exosomes from melanoma patients hamper dendritic cell differentiation from peripheral blood monocytes or bone marrow, resulting in an accumulation of immature MDSC-like cells (Valenti et al. 2006; Yu et al. 2007). Similarly, breast carcinoma-derived exosomes in mice promote the generation of MDSCs in an exosome-associated PGE2 and TGF-β-dependent manner (Xiang et al. 2009). Not only MDSC accumulation, but also MDSC suppressive function can be triggered by tumor-derived exosomes and Hsp72/TLR-2 have been suggested to take part in this (Chalmin et al. 2010). Effects of tumor-derived exosomes on macrophages are not so clear-cut and might depend on the tumor type and stage. Thus, melanoma cell-derived exosomes activate macrophages as measured by enhanced NF-κB activity (Marton et al. 2012) and exosome-associated fibronectin induces IL1β production (Takizawa et al. 1995). On the other hand, microvesicles released by late stage mouse melanoma cells were found to exert a dose-dependent suppression of MHC II expression in these cells, which would rather suggest an alternative type of activation (Poutsiaka et al. 1985). Interestingly, exosome-mediated crosstalk works bidirectionally, but again the effect of macrophage-derived exosomes on tumor might be context-dependent. For example, human macrophage-derived exosomes contain ADAM15 which exerts an antitumoral function (Lee et al. 2012), but, conversely, macrophage vesicles were also reported to shuttle invasion-promoting microRNAs into breast cancer cells (Yang et al. 2011). In addition, it has been proposed that miRNAs from tumor exosomes may play a role in mobilizing myeloid cells to pre-metastatic niches, while fibrosarcoma-derived exosomes can also contribute to efficient DC-mediated priming of anti-tumor T cell responses in vivo, again contrasting pro- and antitumoral functions of exosomes (Zeelenberg et al. 2008).

Other stimuli and signaling pathways: The tumor microenvironment is very complex, and it is therefore to be expected that several other TAM-modulating molecules have been described. Interesting examples include serotonin (Nocito et al. 2008), decoy receptor 3 (Chan et al. 2008) and Heat-shock protein (Hsp) 27 (Banerjee et al. 2011). A central signalling pathway in determining the TAM phenotype is NF-κB. Since a strong activation of this pathway leads to overt proinflammatory and antitumoral responses, the activity of this transcription factor is usually kept in check in the tumor microenvironment. For example, TAM display a nuclear accumulation of NF-κB p50 homodimers, which preclude transcription of NF-κB p50/p65-regulated inflammatory and antitumoral genes (Saccani et al. 2006). Moreover, IRAK-M, an inhibitor of TLR signalling, is overexpressed in TAM and its ablation results in more M1-oriented TAM and a reduced tumor growth (Standiford et al. 2011). Nevertheless, some basal NF-κB activity is required to maintain the M2-like phenotype of TAM, by inhibiting STAT1 signaling (Hagemann et al. 2008). Another regulator of the M1/M2 balance in TAM is Notch signalling (Wang et al. 2010), whereby a forced overexpression of Notch leads to M1 activation. Consequently, TAM usually have reduced levels of Notch, biasing these cells towards M2. The importance of M2 polarization for tumor growth and metastasis is recently further highlighted by the effects of the serum protein histidine-rich glycoprotein (HRG). This protein is usually rapidly degraded in the tumor microenvironment because its presence switches the TAM activation state from M2 to M1, leading to enhanced anti-tumor immune responses and vessel normalization (Rolny et al. 2011).

Anti-tumor immune responses are also kept under control by the transcription factor STAT-3, which is often overexpressed in TAM and contributes to their anti-inflammatory properties. As a matter of fact, elimination of STAT3 in the myeloid compartment leads to a multicomponent antitumor immune reaction (Kortylewski et al. 2005). Several cytokines activate STAT3, one of which being IL-6. Therefore, transcription factors that stimulate IL-6 production, such as Fra-1 that is often found to be overexpressed in TAM, contribute indirectly to tumor growth (Luo et al. 2009).

17.6 Concluding Remarks

All available evidence suggests that tumor-associated myeloid cells—including MDSC, TEM and TAM—are valid targets for therapeutic intervention. Several strategies have been proposed, of which attempts to reconvert TAM to antitumoral effector cells (Beatty et al. 2011; Guiducci et al. 2005; Stout et al. 2009) or attempts to eliminate TAM (Luo et al. 2006; Song et al. 2009) have met some success in animal models. Further advancement in the field is expected to come from the identification of additional myeloid cell-dependent mechanisms of tumor promotion and the identification of specialized TAM subsets.

References

Aharinejad S, Paulus P, Sioud M, Hofmann M, Zins K, Schäfer R et al (2004) Colony-stimulating factor-1 blockade by antisense oligonucleotides and small interfering RNAs suppresses growth of human mammary tumor xenografts in mice. Cancer Res 64:5378–5384

Andreu P, Johansson M, Affara NI, Pucci F, Tan T, Junankar S et al (2010) FcRγ activation regulates inflammation-associated squamous carcinogenesis. Cancer Cell 17:121–134

Balkwill F, Charles KA, Mantovani A (2005) Smoldering and polarized inflammation in the initiation and promotion of malignant disease. Cancer Cell 7:211–217

Banerjee S, Lin CF, Skinner KA, Schiffhauer LM, Peacock J, Hicks DG et al (2011) Heat shock protein 27 differentiates tolerogenic macrophages that may support human breast cancer progression. Cancer Res 71:318–327

Bayne LJ, Beatty GL, Jhala N, Clark CE, Rhim AD, Stanger BZ et al (2012) Tumor-derived granulocyte–macrophage colony-stimulating factor regulates myeloid inflammation and T cell immunity in pancreatic cancer. Cancer Cell 21:822–835

Beatty GL, Chiorean EG, Fishman MP, Saboury B, Teitelbaum UR, Sun W et al (2011) CD40 agonists alter tumor stroma and show efficacy against pancreatic carcinoma in mice and humans. Science 331:1612–1616

Beck AH, Espinosa I, Edris B, Li R, Montgomery K, Zhu S (2009) The macrophage colony-stimulating factor 1 response signature in breast carcinoma. Clin Cancer Res 15:778–787

Bindea G, Mlecnik B, Fridman WH, Pagès F, Galon J (2010) Natural immunity to cancer in humans. Curr Op Immunol 22:215–222

Bingle L, Brown NJ, Lewis CE (2002) The role of tumour-associated macrophages in tumour progression: implications for new anticancer therapies. J Pathol 196:254–265

Biswas SK, Gangi L, Paul S, Schioppa T, Saccani A, Sironi M et al (2006) A distinct and unique transcriptional programme expressed by tumor-associated macrophages (defective NF-κB and enhanced IRF-3/STAT1 activation). Blood 107:2112–2122

Borrello MG, Alberti L, Fischer A, Degl'innocenti D, Ferrario C, Gariboldi M et al (2005) Induction of a proinflammatory program in normal human thyrocytes by the RET/PTC1 oncogene. Proc Natl Acad Sci USA 102:14825–14830

Casazza A, Fu X, Johansson I, Capparuccia L, Andersson F, Giustacchini A et al (2011) Systemic and targeted delivery of semaphorin 3A inhibits tumor angiogenesis and progression in mouse tumor models. Arterioscler Thromb Vasc Biol 31:741–749

Chalmin F, Ladoire S, Mignot G, Vincent J, Bruchard M, Remy-Martin JP et al (2010) Membrane-associated Hsp72 from tumor-derived exosomes mediates STAT3-dependent immunosuppressive function of mouse and human myeloid-derived suppressor cells. J Clin Invest 120:457–471

Chan Y-C, Chen T-C, Lee C-T, Yang C-Y, Wang H-W, Wang C-C et al (2008) Epigenetic control of MHC-II expression in tumor-associated macrophages by decoy receptor 3. Blood 111:5054–5063

Chao MP, Jaiswal S, Weissman-Tsukamoto R, Alizadeh AA, Gentles AJ, Volkmer J et al (2010) Calreticulin is the dominant pro-phagocytic signal on multiple human cancers and is counterbalanced by CD47. Sci Transl Med 2:63ra94

Charles KA, Kulbe H, Soper R, Escorcio-Correia M, Lawrence T, Schultheis A et al (2009) The tumor-promoting actions of TNF-α involve TNFR1 and IL-17 in ovarian cancer in mice and humans. J Clin Invest 119:3011–3023

Cheng P, Corzo CA, Luetteke N, Yu B, Nagaraj S, Bui MM et al (2008) Inhibition of dendritic cell differentiation and accumulation of myeloid-derived suppressor cells in cancer is regulated by S100A9 protein. J Exp Med 205:2235–2249

Coffelt SB, Lewis CE, Naldini L, Brown JM, Ferrara N, De Palma M (2010a) Elusive identities and overlapping phenotypes of proangiogenic myeloid cells in tumors. Am J Pathol 176:1564–1576

Coffelt SB, Tal AO, Scholz A, De Palma M, Patel S, Urbich C et al (2010b) Angiopoietin-2 regulates gene expression in TIE2-expressing monocytes and augments their inherent proangiogenic functions. Cancer Res 70:5270–5280

Coffelt SB, Chen YY, Muthana M, Welford AF, Tal AO, Scholz A et al (2011) Angiopoietin 2 stimulates TIE2-expressing monocytes to suppress T cell activation and to promote regulatory T cell expansion. J Immunol 186:4183–4190

Cortez-Retamozo V, Etzrodt M, Newton A, Rauch PJ, Chudnovskiy A, Berger C et al (2012) Origins of tumor-associated macrophages and neutrophils. Proc Natl Acad Sci USA 109:2491–2496

Corthay A, Skovseth DK, Lundin KU, Rosjo E, Omholt H, Hofgaard PO et al (2005) Primary antitumor immune response mediated by $CD4^+$ T cells. Immunity 22:371–383

Corzo CA, Condamine T, Lu L, Cotter MJ, Youn JI, Cheng P et al (2010) HIF-1α regulates function and differentiation of myeloid-derived suppressor cells in the tumor microenvironment. J Exp Med 207:2439–2453

Coussens LM, Werb Z (2002) Inflammation and cancer. Nature 420:860–867

Coussens LM, Tinkle CL, Hanahan D, Werb Z (2000) MMP-9 supplied by bone marrow-derived cells contributes to skin carcinogenesis. Cell 103:481–490

Curry JM, Eubank TD, Roberts RD, Wang Y, Pore N, Maity A et al (2008) M-CSF signals through the MAPK/ERK pathway via Sp1 to induce VEGF production and induces angiogenesis in vivo. PLoS One 3:e3405

De Palma M, Venneri MA, Galli R, Sergi Sergi L, Politi LS, Sampaolesi M et al (2005) Tie2 identifies a hematopoietic lineage of proangiogenic monocytes required for tumor vessel formation and a mesenchymal population of pericyte progenitors. Cancer Cell 8:211–226

De Palma M, Mazzieri R, Politi LS, Pucci F, Zonari E, Sitia G et al (2008) Tumor-targeted interferon-alpha delivery by Tie2-expressing monocytes inhibits tumor growth and metastasis. Cancer Cell 14:299–311

de Visser KE, Korets LV, Coussens LM (2005) De novo carcinogenesis promoted by chronic inflammation is B lymphocyte dependent. Cancer Cell 7:411–423

DeNardo DG, Barreto JB, Andreu P, Vasquez L, Tawfik D, Kolhatkar N et al (2009) CD4+T cells regulate pulmonary metastasis of mammary carcinomas by enhancing protumor properties of macrophages. Cancer Cell 16:91–102

Deng L, Zhou JF, Sellers RS, Li JF, Nguyen AV, Wang Y et al (2010) A novel mouse model of inflammatory bowel disease links mammalian target of rapamycin-dependent hyperproliferation of colonic epithelium to inflammation-associated tumorigenesis. Am J Pathol 176:952–967

Dineen SP, Lynn KD, Holloway SE, Miller AF, Sullivan JP, Shames DS et al (2008) Vascular endothelial growth factor receptor 2 mediates macrophage infiltration into orthotopic pancreatic tumors in mice. Cancer Res 68:4340–4346

Doedens AL, Stockmann C, Rubinstein MP, Liao D, Zhang N, DeNardo DG et al (2010) Macrophage expression of hypoxia-inducible factor-1α suppresses T cell function and promotes tmor progression. Cancer Res 70:7465–7475

Dolcetti L, Peranzoni E, Ugel S, Marigo I, Fernandez Gomez A, Mesa C et al (2010) Hierarchy of immunosuppressive strength among myeloid-derived suppressor cell subsets is determined by GM-CSF. Eur J Immunol 40:22–35

Dominguez-Soto A, Sierra-Filardi E, Puig-Kröger A, Perez-Maceda B, Gomez-Aguado F, Corcuera MT et al (2011) Dendritic cell-specific ICAM-3-grabbing nonintegrin expression on M2-polarized and tumor-associated macrophages is macrophage-CSF dependent and enhanced by tumor-derived IL-6 and IL-10. J Immunol 186:2192–2200

Egeblad M, Ewald AJ, Askautrud HA, Truitt ML, Welm BE, Bainbridge E et al (2008) Visualizing stromal cell dynamics in different tmor microenvironments by spinning disk confocal microscopy. Dis Mod Mech 1:155–167

Egeblad M, Nakasone ES, Werb Z (2010) Tumors as organs: complex tissues that interface with the entire organism. Dev Cell 18:884–901

Eruslanov E, Kaliberov S, Daurkin I, Kaliberova L, Buchsbaum D, Vieweg J et al (2009) Altered expression of 15-hydroxyprostaglandin dehydrogenase in tumor-infiltrated CD11b myeloid cells: a mechanism for immune evasion in cancer. J Immunol 182:7548–7557

Eruslanov E, Daurkin I, Ortiz J, Vieweg J, Kusmartsev S (2010) Tumor-mediated induction of myeloid-derived suppressor cells and M2-polarized macrophages by altering intracellular PGE_2 catabolism in myeloid cells. J Leukoc Biol 88:839–848

Espinosa I, Beck AH, Lee CH, Zhu S, Montgomery KD, Marinelli RJ et al (2009) Coordinate expression of colony-stimulating factor-1 and colony-stimulating-factor-1-related proteins is associated with poor prognosis in gynaecological and nongynecological leiomyosarcoma. Am J Pathol 174:2347–2356

Eubank TD, Roberts RD, Khan M, Curry JM, Nuovo GJ, Kuppusamy P et al (2009) Granulocyte macrophage colony-stimulating factor inhibits breast cancer growth and metastasis by invoking an anti-angiogenic program in tumor-educated macrophages. Cancer Res 69:2133–2140

Fang H-Y, Hughes R, Murdoch C, Coffelt SB, Biswas SK, Harris AL et al (2009) Hypoxia inducible factors 1 and 2 are important transcriptional effectors in primary macrophages experiencing hypoxia. Blood 114:844–859

Flavell RA, Sanjabi S, Wrzesinski SH, Licona-Limon P (2010) The polarization of immune cells in the tumour environment by TGFβ. Nat Rev Immunol 10:554–567

Fukuda A, Wang SC, Morris JP 4th, Folias AE, Liou A, Kim GE et al (2011) Stat3 and MMP7 contribute to pancreatic ductal adenocarcinoma initiation and progression. Cancer Cell 19:441–455

Galon J, Costes A, Sanchez-Cabo F, Kirilovsky A, Mlecnik B, Lagorce-Pagès C et al (2006) Type, density, and location of immune cells within human colorectal tumors predict clinical outcome. Science 313:1960–1964

Gocheva V, Wang H-W, Gadea BB, Shree T, Hunter KE, Garfall AL et al (2010) IL-4 induces cathepsin protease activity in tumor-associated macrophages to promote cancer growth and invasion. Genes Dev 24:241–255

Goerdt S, Orfanos CE (1999) Other functions, other genes: alternative activation of antigen-presenting cells. Immunity 10:137–142

Gordon S, Taylor PR (2005) Monocyte and macrophage heterogeneity. Nat Rev Immunol 5:953–964

Goswami S, Sahai E, Wyckoff JB, Cammer M, Cox D, Pixley FJ et al (2005) Macrophages promote the invasion of breast carcinoma cells via a colony-stimulating factor-1/epidermal growth factor paracrine loop. Cancer Res 65:5278–5283

Greeneltch KM, Schneider M, Steinberg SM, Liewehr DJ, Stewart TJ, Liu K et al (2007) Host immunosurveillance controls tumor growth via IFN regulatory factor-8-dependent mechanisms. Cancer Res 67:10406–10416

Greifenberg V, Ribechini E, Rössner S, Lutz MB (2009) Myeloid-derived suppressor cell activation by combined LPS and IFN-gamma treatment impairs DC development. Eur J Immunol 39:2865–2876

Greten FR, Eckmann L, Greten TF, Park JM, Li ZW, Egan LJ et al (2004) IKKbeta links inflammation and tumorigenesis in a mouse model of colitis-associated cancer. Cell 118:285–296

Grip O, Janciauskiene S, Lindgren S (2003) Macrophages in inflammatory bowel disease. Curr Drug Targets Inflamm Allergy 2:155–160

Grivennikov SI, Greten FR, Karin M (2010) Immunity, inflammation, and cancer. Cell 140:883–899

Guiducci C, Vicari AP, Sangaletti S, Trinchieri G, Colombo MP (2005) Redirecting in vivo elicited tumor infiltrating macrophages a nd dendritic cells towards tumor rejection. Cancer Res 65:3437–3446

Hagemann T, Lawrence T, McNeish I, Charles KA, Kulbe H, Thompson RG et al (2008) "Re-educating" tumor-associated macrophages by targeting NF-kappaB. J Exp Med 205: 1261–1268

Hart KM, Bak SP, Alonso A, Berwin B (2009) Phenotypic and functional delineation of murine CX(3)CR1 monocyte-derived cells in ovarian cancer. Neoplasia 11:564–573

Hassanzadeh Ghassabeh G, De Baetselier P, Brys L, Noël W, Van Ginderachter JA, Meerschaut S et al (2006) Identification of a common gene signature for type II cytokine-associated myeloid cells elicited in vivo during different pathologies. Blood 108:575–583

Hernandez L, Smirnova T, Kedrin D, Wyckoff J, Zhu L, Stanley ER et al (2009) The EGF/CSF-1 paracrine invasion loop can be triggered by heregulin beta 1 and CXCL12. Cancer Res 69:3221–3227

Herr B, Zhou J, Werno C, Menrad H, Namgaladze D, Weigert A et al (2009) The supernatant of apoptotic cells causes transcriptional activation of HIF-1(alpha) in macrophages via sphingosine-1-phosphate and transforming growth factor-(beta). Blood 114:2140–2148

Heymann F, Trautwein C, Tacke F (2009) Monocytes and macrophages as cellular targets in liver fibrosis. Inflamm Allergy Drug Targets 8:307–318

Hiratsuka S, Watanabe A, Aburatani H, Maru Y (2006) Tumour-mediated upregulation of chemoattractants and recruitment of myeloid cells predetermines lung metastasis. Nat Cell Biol 8:1369–1375

Hiratsuka S, Duda DG, Huang Y, Goel S, Sugiyama T, Nagasawa T et al (2011) C-X-C receptor type 4 promotes metastasis by activating p38 mitogen-activated protein kinase in myeloid differentiation antigen (Gr-1)-positive cells. Proc Natl Acad Sci USA 108:302–307

Huang M, Stolina M, Sharma S, Mao JT, Zhu L, Miller PW et al (1998) Non-small cell lung cancer cyclooxygenase-2-dependent regulation of cytokine balance in lymphocytes and macrophages: up-regulation of interleukin 10 and down-regulation of interleukin 12 production. Cancer Res 58:1208–1216

Imtiyaz HZ, Williams EP, Hickey MM, Patel SA, Durham AC, Yuan L-J et al (2010) Hypoxia-inducible factor 2α regulates macrophage function in mouse models of acute and tmor inflammation. J Clin Invest 120:2699–2714

Jaiswal M, LaRusso NF, Gores GJ (2001) Nitric oxide in gastrointestinal epithelial cell carcinogensis: linking inflammation to oncogenesis. Am J Physiol 281:G626–G634

Jaiswal S, Jamieson CH, Pang WW, Park CY, Chao MP, Majeti R et al (2009) CD47 is upregulated on circulating hematopoietic stem cells and leukemia cells to avoid phagocytosis. Cell 138:271–285

Jamora C, Dennert G, Lee AS (1996) Inhibition of tumor progression by suppression of stress protein GRP78/BiP induction in fibrosarcoma B/C10ME. Proc Natl Acad Sci. U S A 93:7690–7694

Kaparakis M, Walduck AK, Price JD, Pedersen JS, van Rooijen N, Pearse MJ et al (2008) Macrophages are mediators of gastritis in acute Helicobacter pylori infection in C57BL/6 mice. Infect Immun 76:2235–2239

Karin M, Lawrence T, Nizet V (2006) Innate immunity gone awry: linking microbial infections to chronic inflammation and cancer. Cell 124:823–835

Kedrin D, Gligorijevic B, Wyckoff J, Verkhusha VV, Condeelis J, Segall JE et al (2008) Intravital imaging of metastatic behavior through a mammary imaging window. Nat Methods 5:1019–1021

Kim HW, Mrakami A, Williams MV, Ohigashi H (2003) Mutagenicity of reactive oxygen and nitrogen species as detected by co-culture of activated inflammatory leukocytes and AS52 cells. Carcinogenesis 24:235–241

Kim D, Wang J, Willingham SB, Martin R, Wernig G, Weissman IL (2012) Anti-CD47 antibodies promote phagocytosis and inhibit the growth of human myeloma cells. Leukemia, 30 May 2012. doi: 10.1038/leu.2012.141. (Epub ahead of print)

Knight B, Yeoh GC, Hsk KL, Ly T, Abraham LJ, Yu C et al (2000) Impaired preneoplastic changes and liver tumor formation in tumor necrosis factor receptor type 1 knockout mice. J Exp Med 192:1809–1818

Ko JS, Rayman P, Ireland J, Swaidani S, Li G, Bunting KD et al (2010) Direct and differential suppression of myeloid-derived suppressor cell subsets by Sunitinib is compartmentally constrained. Cancer Res 70:3526–3536

Kortylewski M, Kujawski M, Wang T, Wei S, Zhang S, Pilon-Thomas S et al (2005) Inhibiting Stat3 signaling in the hematopoietic system elicits multicomponent antitumor immunity. Nat Med 11:1314–1321

Kortylewski M, Xin H, Kujawski M, Lee H, Liu Y, Harris T et al (2009) Regulation of the IL-23 and IL-12 balance by STAT3 signaling in the tumor microenvironment. Cancer Cell 15:114–123

Kuang D-M, Wu Y, Chen N, Cheng J, Zhuang S-M, Zheng L (2007) Tumor-derived hyaluronan induces formation of immunosuppressive macrophages through transient early activation of monocytes. Blood 110:587–595

Kubota Y, Takubo K, Shimizu T, Ohno H, Kishi K, Shibuya M et al (2009) M-CSF inhibition selectively targets pathological angiogenesis and lymphangiogenesis. J Exp Med 206:1089–1102

Kusmartsev S (2005) Gabrilovich D.I. STAT1 signaling regulates tumor-associated macrophage-mediated T cell deletion. J Immunol 174:4880–4891

Lee HD, Koo BH, Kim YH, Jeon OH, Kim DS (2012) Exosome release of ADAM15 and the functional implications of human macrophage-derived ADAM15 exosomes. FASEB J 6:3084–3095

Lewis CE, Pollard JW (2006) Distinct roles of macrophages in different tumor microenvironments. Cancer Res 66:605–612

Lin EY, Nguyen AV, Russell RG, Pollard JW (2001) Colony-stimulating factor 1 promotes progression of mammary tumors to malignancy. J Exp Med 193:727–739

Liu BY, Soloviev I, Chang P, Lee J, Huang X, Zhong C et al (2010) Stromal cell-derived factor-1/CXCL12 contributes to MMTV-Wnt1 tumor growth involving Gr+CD11b+ cells. PLoS One 5:e8611

Luo Y, Zhou H, Krueger J, Kaplan C, Lee S-H, Dolman C et al (2006) Targeting tumor-associated macrophages as a novel strategy against breast cancer. J Clin Invest 116:2132–2141

Luo YP, Zhou H, Krueger J, Kaplan C, Liao D, Markowitz D et al (2009) The role of proto-oncogene Fra-1 in remodeling the tumor microenvironment in support of breast tumor cell invasion and progression. Oncogene 29:662–673

Maeda S, Kamata H, Luo J-L, Leffert H, Karin M (2005) IKKβ couples hepatocyte death to cytokine-driven compensatory proliferation that promotes chemical hepatocarcinogenesis. Cell 121:977–990

Mahadevan NR, Zanetti M (2011) Tumor stress inside out: cell-extrinsic effects of the unfolded protein response in tumor cells modulate the immunological landscape of the tumor microenvironment. J Immunol 187:4403–4409

Mahadevan NR, Rodvold J, Sepulveda H, Rossi S, Drew AF, Zanetti M (2011) Transmission of endoplasmic reticulum stress and pro-inflammation from tumor cells to myeloid cells. Proc Natl Acad Sci U S A 108:6561–6566

Majeti R, Chao MP, Alizadeh AA, Pang WW, Jaiswal S, Gibbs KD Jr et al (2009) CD47 is an adverse prognostic factor and therapeutic antibody target on human acute myeloid leukemia stem cells. Cell 138:286–299

Mantovani A, Sica A (2010) Macrophages, innate immunity and cancer: balance, tolerance, and diversity. Curr Op Immunol 22:231–237

Mantovani A, Bottazzi B, Colotta F, Sozzani S, Ruco L (1992) The origin and function of tumor-associated macrophages. Immunol Today 13:265–270

Mantovani A, Sozzani S, Locati M, Allavena P, Sica A (2002) Macrophage polarization: tumor-associated macrophages as a paradigm for polarized M2 mononuclear phagocytes. Trends Immunol 23:549–555

Mantovani A, Sica A, Sozzani S, Allavena P, Vecchi A, Locati M (2004) The chemokine system in diverse forms of macrophage activation and polarization. Trends Immunol 25:677–686

Marigo I, Dolcetti L, Serafini P, Zanovello P, Bronte V (2008) Tumor-induced tolerance and immune suppression by myeloid derived suppressor cells. Immunol Rev 222:162–179

Marigo I, Bosio E, Solito S, Mesa C, Fernandez A, Dolcetti L et al (2010) Tumor-induced tolerance and immune suppression depend on the C/EBPβ transcription factor. Immunity 32:790–802

Martinez FO, Helming L, Gordon S (2009) Alternative activation of macrophages: an immunologic functional perspective. Annu Rev Immunol 27:451–483

Marton A, Vizler C, Kusz E, Temesfoi V, Szathmary Z, Nagy K et al (2012) Melanoma cell-derived exosomes alter macrophage and dendritic cell functions in vitro. Immunol Lett. 8 Aug (E-pub ahead of print)

Marttila-Ichihara F, Auvinen K, Elima K, Jalkanen S, Salmi M (2009) Vascular adhesion protein-1 enhances tumor growth by supporting recruitment of Gr-1+CD11b+ myeloid cells into tumors. Cancer Res 69:7875–7883

Mazzieri R, Pucci F, Moi D, Zonari E, Ranghetti A, Berti A et al (2011) Targeting the ANG2/TIE2 axis inhibits tumor growth and metastasis by impairing angiogenesis and disabling rebounds of proangiogenic myeloid cells. Cancer Cell 19:512–526

Molon B, Ugel S, Del Pozzo F, Soldani C, Zilio S, Avella D et al (2011) Chemokine nitration prevents intratumoral infiltration of antigen-specific T cells. J Exp Med 208:1949–1962

Mosser DM, Edwards JP (2008) Exploring the full spectrum of macrophage activation. Nat Rev Immunol 8:958–969

Movahedi K, Guilliams M, Van den Bossche J, Van den Bergh R, Gysemans C, Beschin A et al (2008) Identification of discrete tumor-induced myeloid-derived suppressor cell subpopulations with distinct T cell suppressive activity. Blood 111:4233–4244
Movahedi K, Laoui D, Gysemans C, Baeten M, Stangé G, Van den Bossche J et al (2010) Different tumor microenvironments contain functionally distinct subsets of macrophages derived from Ly6C(high) monocytes. Cancer Res 70:5728–5739
Movahedi K, Schoonooghe S, Laoui D, Houbracken I, Waelput W, Breckpot K et al (2012) Nanobody-based targeting of the macrophage mannose receptor for effective in vivo imaging of tumor-associated macrophages. Cancer Res 72:4165–4177
Muramatsu M, Yamamoto S, Osawa T, Shibuya M (2010) Vascular endothelial growth factor receptor-1 signaling promotes mobilization of macrophage lineage cells from bone marrow and stimulates solid tumor growth. Cancer Res 70:8211–8221
Murdoch C, Giannoudis A, Lewis CE (2004) Mechanisms regulating the recruitment of macrophages into hypoxic areas of tmors and other ischemic tissues. Blood 104:2224–2234
Murdoch C, Tazzyman S, Webster S, Lewis CE (2007) Expression of Tie-2 by human monocytes and their responses to angiopoietin-2. J Immunol 178:7405–7411
Muthana M, Giannoudis A, Scott SD, Fang H-Y, Coffelt SB, Morrow FJ et al (2011) Use of macrophages to target therapeutic adenovirus to human prostate tumors. Cancer Res 71:1805–1815
Nagaraj S, Collazo M, Corzo CA, Youn JI, Ortiz M, Quiceno D et al (2009) Regulatory myeloid suppressor cells in health and disease. Cancer Res 69:7503–7506
Nagaraj S, Youn JI, Weber H, Iclozan C, Lu L, Cotter MJ et al (2010) Anti-inflammatory triterpenoid blocks immune suppressive function of MDSCs and improves immune response in cancer. Clin Cancer Res 16:1812–1823
Nakasone ES, Askautrud HA, Kees T, Park JH, Plaks V, Ewald AJ et al (2012) Imaging tumor-stroma interactions during chemotherapy reveals contributions of the microenvironment to resistance. Cancer Cell 21:488–503
Naugler WE, Sakurai T, Kim S, Maeda S, Kim K, Elsharkawy AM et al (2007) Gender disparity in liver cancer due to sex differences in MyD88-dependent IL-6 production. Science 317:121–124
Nefedova Y, Huang M, Kusmartsev S, Bhattacharya R, Cheng P, Salup R et al (2004) Hyperactivation of STAT3 is involved in abnormal differentiation of dendritic cells in cancer. J Immunol 172:464–474
Nocito A, Dahm F, Jochum W, Jang JH, Georgiev P, Bader M et al (2008) Serotonin regulates macrophage-mediated angiogenesis in a mouse model of colon cancer allografts. Cancer Res 68:5152–5158
Nowicki A, Szenajch J, Ostrowska G, Wojtowicz A, Wojtowicz K, Kruszewski AA et al (1996) Impaired tumor growth in colony-stimulating factor 1 (CSF-1)-deficient, macrophage-deficient op/op mouse: evidence for a role of CSF-1-dependent macrophages in formation of tumor stroma. Int J Cancer 65:112–119
Ojalvo LS, Whittaker CA, Condeelis JS, Pollard JW (2010) Gene expression analysis of macrophages that facilitate tumor invasion supports a role for Wnt-signaling in mediating their activity in primary mammary tumors. J Immunol 184:702–712
Oshima M, Dinchuk JE, Kargman SL, Oshima H, Hancock B, Kwong E et al (1996) Suppression of intestinal polyposis in Apc delta716 knockout mice by inhibition of cyclooxygenase 2 (COX-2). Cell 87:803–809
Oshima H, Hioki K, Popivanova BK, Oguma K, van Rooijen N, Ishikawa TO, Oshima M (2011) Prostaglandin E2 signaling and bacterial infection recruit tumor-promoting macrophages to mouse gastric tumors. Gastroenterology 140:596–607
Ostrand-Rosenberg S (2010) Myeloid-derived suppressor cells: more mechanisms for inhibiting antitumor immunity. Cancer Immunol Immunother 59:1593–1600
Ostrand-Rosenberg S, Sinha P (2009) Myeloid-derived suppressor cells: linking inflammation and cancer. J Immunol 182:4499–4506

Ostrand-Rosenberg S, Clements VK, Terabe M, Park JM, Berzofsky JA, Dissanayake SK (2002) Resistance to metastatic disease in STAT6-deficient mice requires hemopoietic and non-hemopoietic cells and is IFN-gamma dependent. J Immunol 169:5796–5804

Pahler JC, Tazzyman S, Erez N, Chen YY, Murdoch C, Nozawa H et al (2008) Plasticity in tumor-promoting inflammation: impairment of macrophage recruitment evokes a compensatory neutrophil response. Neoplasia 10:329–340

Pikarsky E, Porat RM, Stein I, Abramovitch R, Amit S, Kasem S et al (2004) NF-kappaB functions as a tumour promoter in inflammation-associated cancer. Nature 431:461–466

Pollard JW (2009) Trophic macrophages in development and disease. Nat Rev Immunol 9:259–270

Popivanova BK, Kitamura K, Wu Y, Kondo T, Kagaya T, Kaneko S et al (2008) Blocking TNF-alpha in mice reduces colorectal carcinogenesis associated with chronic colitis. J Clin Invest 118:560–570

Popivanova BK, Kostadinova FI, Furuichi K, Shamekh MM, Kondo T, Wada T et al (2009) Blockade of a chemokine, CCL2, reduces chronic colitis-associated carcinogenesis in mice. Cancer Res 69:7884–7892

Poutsiaka DD, Schroder EW, Taylor DD, Levy EM, Black PH (1985) Membrane vesicles shed by murine melanoma cells selectively inhibit the expression of Ia antigen by macrophages. J Immunol 134:138–144

Priceman SJ, Sung JL, Shaposhnik Z, Burton JB, Torres-Collado AX, Moughon DL et al (2010) Targeting distinct tumor-infiltrating myeloid cells by inhibiting CSF-1 receptor: combating tumor evasion of anti-angiogenic therapy. Blood 115:1461–1471

Pucci F, Venneri MA, Biziato D, Nonis A, Moi D, Sica A et al (2009) A distinguishing gene signature shared by tumor-infiltrating Tie2-expressing monocytes, blood "resident" monocytes, and embryonic macrophages suggests common functions and developmental relationships. Blood 114:901–914

Puig-Kröger A, Sierra-Filardi E, Dominguez-Soto A, Samaniego R, Corcuera MT, Gomez-Aguado F et al (2009) Folate receptor β is expressed by tumor-associated macrophages and constitutes a marker for M2 anti-inflammatory/regulatory macrophages. Cancer Res 69:9395–9403

Pylayeva-Gupta Y, Lee KE, Hajdu CH, Miller G, Bar-Sagi D (2012) Oncogenic Kras-induced GM-CSF production promotes the development of pancreatic neoplasia. Cancer Cell 21:836–847

Qian B-Z, Pollard JW (2010) Macrophage diversity enhances tumor progression and metastasis. Cell 141:39–51

Qian BZ, Li J, Zhang H, Kitamura T, Zhang J, Campion LR et al (2011) CCL2 recruits inflammatory monocytes to facilitate breast-tumour metastasis. Nature 475:222–225

Rakoff-Nahoum S, Medzhitov R (2007) Regulation of spontaneous intestinal tumorigenesis through the adaptor protein MyD88. Science 317:124–127

Rius J, Guma M, Schachtrup C, Akassoglou K, Zinkernagel AS, Nizet V et al (2008) NF-kappaB links innate immunity to the hypoxic response through transcritional regulation of HIF-1alpha. Nature 453:807–811

Robinson BD, Sica GL, Liu YF, Rohan TE, Gertier FB, Condeelis JS et al (2009) Tumor microenvironment of metastasis in human breast carcinoma: a potential prognostic marker linked to hematogenous dissemination. Clin Cancer Res 15:2433–2441

Rodriguez PC, Hernandez CP, Quiceno D, Dubinett SM, Zabaleta J, Ochoa JB et al (2005) Arginase I in myeloid suppressor cells is induced by COX-2 in lung carcinoma. J Exp Med 202:931–939

Rolny C, Mazzone M, Tugues S, Laoui D, Johansson I, Coulon C et al (2011) HRG inhibits tumor growth and metastasis by inducing macropage polarization and vessel normalization through downregulation of PlGF. Cancer Cell 19:1–14

Saccani A, Schioppa T, Porta C, Biswas SK, Nebuloni M, Vago L et al (2006) p50 nuclear factor-kappaB overexpression in tumor-associated macrophages inhibits M1 inflammatory responses and antitumor resistance. Cancer Res 66:11432–11440

Sawanobori Y, Ueha S, Kurachi M, Shimaoka T, Talmadge JE, Abe J et al (2008) Chemokine-mediated rapid turnover of myeloid-derived suppressor cells in tumor-bearing mice. Blood 111:5457–5466

Sharma S, Stolina M, Yang SC, Baratelli F, Lin JF, Atianzar K et al (2003) Tumor cyclooxygenase 2-dependent suppression of dendritic cell function. Clin Cancer Res 9:961–968

Sharma M, Beck AH, Webster JA, Espinosa I, Montgomery K, Varma S et al (2010) Analysis of stromal signatures in the tumor microenvironment of ductal carcinoma in situ. Breast Cancer Res Treat 123:397–404

Shojaei F, Wu X, Malik AK, Zhong C, Baldwin ME, Schanz S et al (2007a) Tumor refractoriness to anti-VEGF treatment is mediated by CD11b+Gr-1+ myeloid cells. Nat Biotechnol 25:911–920

Shojaei F, Wu X, Zhong C, Yu L, Liang XH, Yao J et al (2007b) Bv8 regulates myeloid-cell-dependent tumour angiogenesis. Nature 450:825–831

Sica A, Saccani A, Bottazzi B, Polentarutti N, Vecchi A, Van Damme J et al (2000) Autocrine production of IL-10 mediates defective IL-12 production and NF-κB activation in tumor-associated macrophages. J Immunol 164:762–767

Sinha P, Clements VK, Ostrand-Rosenberg S (2005) IL-13-regulated M2 macrophages in combination with myeloid suppressor cells block immune surveillance against metastasis. Cancer Res 65:11743–11751

Sinha P, Clements VK, Fulton AM, Ostrand-Rosenberg S (2007) Prostaglandin E2 promotes tumor progression by inducing myeloid-derived suppressor cells. Cancer Res 67:4507–4513

Song L, Asgharzadeh S, Salo J, Engell K, Wu H-W, Sposto R et al (2009) Vα24-invariant NKT cells mediate antitumor activity via killing of tumor-associated macrophages. J Clin Invest 119:1524–1536

Standiford TJ, Kuick R, Bhan U, Chen J, Newstead M, Keshamouni VG (2011) TGF-β-induced IRAK-M expression in tumor-assoiated macrophages regulates lung tumor growth. Oncogene 30:2475–2484

Stout RD, Watkins SK, Suttles J (2009) Functional plasticity of macrophages: in situ reprogramming of tumor-associated macrophages. J Leukoc Biol 86:1105–1109

Swann JB, Vesely MD, Silva A, Sharkey J, Akira S, Schreiber RD et al (2008) Demonstration of inflammation-induced cancer and cancer immunoediting during primary tumorigenesis. Proc Natl Acad Sci USA 105:652–656

Takeda N, O'Dea EL, Doedens A, Kim J-W, Weidemann A, Stockmann C et al (2010) Differential activation and antagonistic function of HIF-α isoforms in macrophages are essential for NO homeostasis. Genes Dev 24:491–501

Takizawa T, Nishinarita S, Kitamura N, Hayakawa J, Kang H, Tomita Y et al (1995) Interaction of the cell-binding domain of fibronectin with VLA-5 integrin induces monokine production in cultured human monocytes. Clin Exp Immunol 101:376–382

Taylor DD, Gercel-Taylor C (2011) Exosomes/microvesicles: mediators of cancer-associated immunosuppressive microenvironments. Semin Immunopathol 33:441–454

Théry C, Boussac M, Véron P, Ricciardi-Castagnoli P, Raposo G, Garin J et al (2001) Proteomic analysis of dendritic cell-derived exosomes: a secreted subcellular compartment distinct from apoptotic vesicles. J Immunol 166:7309–7318

Ueno T, Toi M, Saji H, Muta M, Bando H, Kuroi K et al (2000) Significance of macrophage chemoattractant protein-1 in macrophage recruitment, angiogenesis, and survival in human breast cancer. Clin Cancer Res 6:3282–3289

Umemura N, Saio M, Suwa T, Kitoh Y, Bai J, Nonaka K et al (2008) Tumor-infiltrating myeloid-derived suppressor cells are pleiotropic-inflamed monocytes/macrophages that bear M1- and M2-type characteristics. J Leukoc Biol 83:1–9

Valenti R, Huber V, Filipazzi P, Pilla L, Sovena G, Villa A et al (2006) Human tumor-released microvesicles promote the differentiation of myeloid cells with transforming growth factor-beta-mediated suppressive activity on T lymphocytes. Cancer Res 66:9290–9298

Van den Bossche J, Bogaert P, van Hengel J, Guérin CJ, Berx G, Movahedi K et al (2009) Alternatively activated macrophages engage in homotypic and heterotypic interactions through IL-4 and polyamine-induced E-cadherin/catenin complexes. Blood 114:4664–4674
Van den Bossche J, Lamers WH, Koehler ES, Geuns JM, Alhonen L, Uimari A et al (2012) Pivotal advance: arginase-1-independent polyamine production stimulates the expression of IL-4-induced alternatively activated macrophage markers while inhibiting LPS-induced expression of inflammatory genes. J Leukoc Biol 91:685–699
Van Ginderachter JA, Movahedi K, Hassanzadeh Ghassabeh G, Meerschaut S, Beschin A, Raes G et al (2006a) Classical and alternative activation of mononuclear phagocytes: picking the best of both worlds for tumor promotion. Immunobiology 211:487–501
Van Ginderachter JA, Meerschaut S, Liu Y, Brys L, De Groeve K, Hassanzadeh Ghassabeh G et al (2006b) Peroxisome proliferator-activated receptor (PPAR) ligands reverse CTL suppression by alternatively activated (M2) macrophages in cancer. Blood 108:525–535
Van Ginderachter JA, Movahedi K, Van den Bossche J, De Baetselier P (2008) Macrophages, PPARs, and cancer. PPAR Res 2008:169414
Van Ginderachter JA, Beschin A, De Baetselier P, Raes G (2010) Myeloid-derived suppressor cells in parasitic infections. Eur J Immunol 40:2976–2985
Venneri MA, De Palma M, Ponzoni M, Pucci F, Scielzo C, Zonari E et al (2007) Identification of proangiogenic TIE2-expressing monocytes (TEMs) in human peripheral blood and cancer. Blood 109:5276–5285
Vesely MD, Kershaw MH, Schreiber RD, Smyth MJ (2011) Natural innate and adaptive immunity to cancer. Annu Rev Immunol 29:235–271
Vicetti Miguel RD, Cherpes TL, Watson LJ, McKenna KC (2010) CTL induction of tumoricidal nitric oxide production by intratumoral macrophages is critical for tumor elimination. J Immunol 185:6706–6718
Vincent J, Mignot G, Chalmin F, Ladoire S, Bruchard M, Chevriaux A et al (2010) 5-fluoroacil selectively kills tumor-associated myeloid-derived suppressor cells resulting in enhanced T cell-dependent antitumor immunity. Cancer Res 70:3052–3061
Wang YC, He F, Feng F, Liu XW, Dong GY, Qin HY et al (2010) Notch signalling determines the M1 versus M2 polarization of macrophages in antitumor immune responses. Cancer Res 70:4840–4849
Webster JA, Beck AH, Sharma M, Espinosa I, Weigelt B, Schreuder M et al (2010) Variations in stromal signatures in breast and colorectal cancer metastases. J Pathol 222:158–165
Weigert A, Schiffmann S, Sekar D, Ley S, Menrad H, Werno C et al (2009) Sphingosine kinase 2 deficient tumor xenografts show impaired growth and fail to polarize macrophages towards an anti-inflammatory phenotype. Int J Cancer 125:2114–2121
Weiser-Evans MCM, Wang X-Q, Amin J, Van Putten V, Choudhary R, Winn RA et al (2009) Depletion of cytosolic phospholipase A_2 in bone marrow-derived macrophages protects against lung cancer progression and metastasis. Cancer Res 69:1733–1738
Welford AF, Biziato D, Coffelt SB, Nucera S, Fisher M, Pucci F et al (2011) TIE2-expressing macrophages limit the therapeutic efficacy of the vascular-disrupting agent combretastatin A4 phosphate in mice. J Clin Invest 121:1969–1973
Werno C, Menrad H, Weigert A, Dehne N, Goerdt S, Schledzewski K et al (2010) Knockout of HIF-1alpha in tumor-associated macrophages enhances M2 polarization and attenuates their pro-angiogenic responses. Carcinogenesis 31:1863–1872
Willingham SB, Volkmer JP, Gentles AJ, Sahoo D, Dalerba P, Mitra SS et al (2012) The CD47-signal regulatory protein alpha (SIRPa) interaction is a therapeutic target for human solid tumors. Proc Natl Acad Sci USA 109:6662–6667
Wong S-C, Puaux A-L, Chittezath M, Shalova I, Kajiji TS, Wang X et al (2010) Macrophage polarization to a unique phenotype driven by B cells. Eur J Immunol 40:2296–2307
Wu Q-L, Buhtoiarov IN, Sondel PM, Rakhmilevich AL, Ranheim EA (2009) Tumoricidal effects of activated macrophages in a mouse model of chronic lymphocytic leukemia. J Immunol 182:6771–6778

Wyckoff JB, Wang Y, Lin EY, Li J-F, Goswami S, Stanley ER et al (2007) Direct visualization of macrophage-assisted tumor cell intravasation in mammary tumors. Cancer Res 67:2649–2656

Xiang X, Poliakov A, Liu C, Liu Y, Deng ZB, Wang J et al (2009) Induction of myeloid-derived suppressor cells by tumor exosomes. Int J Cancer 124:2621–2633

Xue W, Zender L, Miething C, Dickins RA, Hernando E, Krizhanovsky V et al (2007) Senescence and tumour clearance is triggered by p53 restoration in murine liver carcinomas. Nature 445:656–660

Yan HH, Pickup M, Pang Y, Gorska AE, Li Z, Chytil A et al (2010) Gr-1+CD11b+ myeloid cells tip the balance of immune protection to tumor promotion in the premetastatic lung. Cancer Res 70:6139–6149

Yang L, DeBusk LM, Fukuda K, Fingleton B, Green-Jarvis B, Shyr Y et al (2004) Expansion of myeloid immune suppressor Gr+CD11b+ cells in tumor-bearing host directly promotes tumor angiogenesis. Cancer Cell 6:409–421

Yang L, Huang J, Ren X, Gorska AE, Chytil A, Aakre M et al (2008) Abrogation of TGF signaling in mammary carcinomas recruits Gr-1+CD11b+ myeloid cells that promote metastasis. Cancer Cell 13:23–35

Yang M, Chen J, Su F, Yu B, Su F, Lin L et al (2011) Microvesicles secreted by macrophages shuttle invasion-potentiating microRNAs into breast cancer cells. Mol Cancer 10:117

Youn JI, Nagaraj S, Collazo M, Gabrilovich DI (2008) Subsets of myeloid-derived suppressor cells in tumor-bearing mice. J Immunol 181:5791–5802

Yu H, Kortylewski M, Pardoll D (2007a) Crosstalk between cancer and immune cells: role of STAT3 in the tumour microenvironment. Nat Rev Immunol 7:41–51

Yu S, Liu C, Su K, Wang J, Liu Y, Zhang L et al (2007b) Tumor exosomes inhibit differentiation of bone marrow dendritic cells. J Immunol 178:6867–6875

Zabuawala T, Taffany DA, Sharma SM, Merchant A, Adair B, Srinivasan R et al (2010) An Ets2-driven transcriptional program in tumor-associated macrophages promotes tumor metastasis. Cancer Res 70:1323–1333

Zeelenberg IS, Ostrowski M, Krumeich S, Bobrie A, Jancic C, Boissonnas A et al (2008) Targeting tumor antigens to secreted membrane vesicles in vivo induces efficient antitumor immune responses. Cancer Res 68:1228–1235

Zhang HG, Grizzle WE (2011) Exosomes and cancer: a newly described pathway of immune suppression. Clin Cancer Res 17:959–964

Zhang K, Shen X, Wu J, Sakaki K, Saunders T, Rutkowski DT et al (2006) Endoplasmic reticulum stress activates cleavage of CREBH to induce a systemic inflammatory response. Cell 124:587–599

Zhang J, Chen L, Xiao M, Wang C, Xin Z (2011) FSP1(+) fibroblasts promote skin carcinogenesis by maintaining MCP-1-mediated macrophage infiltration and chronic inflammation. Am J Pathol 178:382–390

Chapter 18
Dendritic Cells and Cancer: Development, Dysfunction and Therapeutic Targets

Stephanie K. Watkins and Arthur A. Hurwitz

Abstract Dendritic cells serve a unique and diverse role in cancer pathogenesis and therapy. On the one hand, dendritic cells are recruited to developing tumors and like many myeloid cells that infiltrate tumors, they exert immune-suppressive effects. On the other hand, dendritic cells have been employed as a tool to initiate anti-tumor immune responses. This chapter discusses these seemingly opposing roles in cancer and identifies some potential strategies that may ensure more durable anti-tumor immune responses.

Keywords Dendritic cells · Immune tolerance · Vaccine · Tumor microenvironment · Immune suppression · Antitumor immunity

18.1 Introduction to Dendritic Cells

Dendritic cells (DCs) were discovered in 1973 and were initially observed to be a population of cells in the peripheral lymphoid organs of mice that were distinct from macrophages and monocytes based on their morphology, size, density and rate of turn over (Steinman and Cohn 1973; Steinman et al. 1974). Later, it was

---

S. K. Watkins · A. A. Hurwitz
Tumor Immunity and Tolerance Section, Laboratory of Molecular Immunoregulation, Cancer and Inflammation Program, Frederick National Laboratory for Cancer Research, Frederick, MD 21072, USA
e-mail: hurwitza@mail.nih.gov

S. K. Watkins (✉)
Department of Surgery, Loyola University Medical Center, Cardinal Bernardin Cancer Center, Room 305, Maywood, IL 60153, USA
e-mail: swatkins1@luc.edu

M. R. Shurin et al. (eds.), *The Tumor Immunoenvironment*,
DOI: 10.1007/978-94-007-6217-6_18,

recognized that DCs are the dominant antigen presenting cells (APCs) of the immune system. Immature DCs are strategically dispersed throughout the epidermis and interstitial tissues as a sentinel population that can sample their environment and capture invading antigens. During antigen uptake, DCs are activated by pathogen- or damage-associated molecular patterns (PAMPs or DAMPs) interacting with pattern recognition receptors (PRRs). This activation ignites DC maturation followed by migration to lymphatic tissue where they present antigen to T cells. Antigen is presented via MHC class I to CD8+ T cells or via class II to CD4+ T cells and requires co-stimulatory signals, most efficiently delivered by CD80 and CD86, resulting in the initiation of an adaptive immune response (Romani et al. 1989; Schuler et al. 1985) (Fig. 18.1).

As opposed to their role in activating immunity, DCs can also be responsible for controlling excessive immune activation and autoimmunity by limiting T cell reactivity to self-antigens. Rendering T cells unresponsive to self-antigen is achieved by the induction of immune tolerance (Steinman et al. 2003). While efficient initiation of an adaptive immune response requires three signals: occupancy of the T cell antigen receptor (TcR) by antigen/MHC complexes, ligation of CD28 by co-stimulatory ligands and pro-inflammatory cytokines such as IL-12,

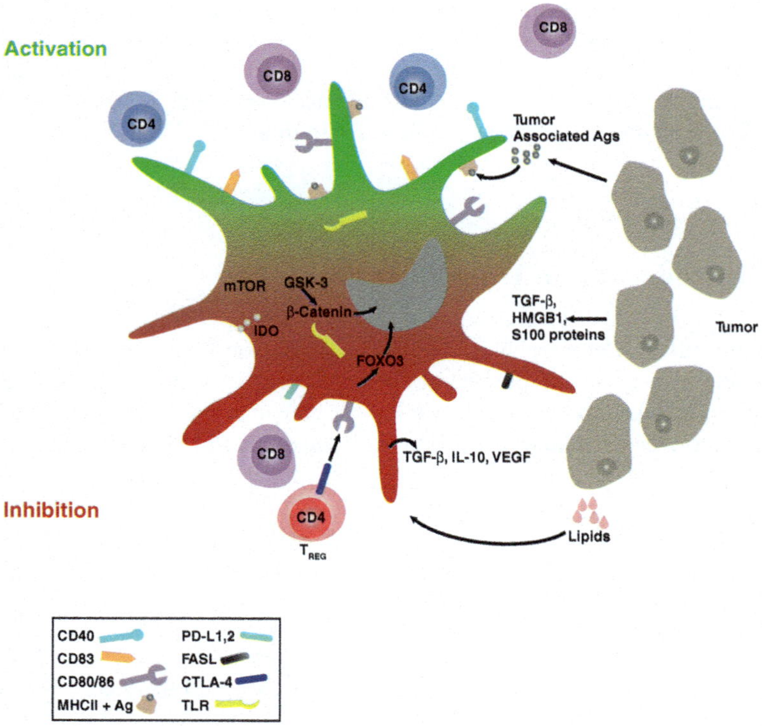

Fig. 18.1 Dendritic cells can activate and inhibit T cell responses. This cartoon identifies the many mechanisms that DCs use to initiate and suppress anti-tumor immune responses

the absence of any one of these signals may result in the induction of T cell tolerance (Curtsinger et al. 2003; Jenkins and Schwartz 1987). The regulation of these critical processes that includes sampling the environment for antigens and initiating immunity or tolerance depict DCs as the commanders of immune surveillance. Therefore, DCs are a promising target for controlling the immune response through immunotherapy for many diseases, including cancer.

18.2 Dendritic Cells and Cancer

It is now well established that tumor growth is associated with both inflammation and immune suppression (Grivennikov et al. 2010; Schreiber et al. 2011). Chronic inflammation has been shown to increase the risk of cancer. For example, hepatitis, inflammation of the lung and inflammatory bowel disease (IBD) predispose to the development of hepatocellular carcinoma, lung cancer and colon cancer respectively (Morson 1985; O'Callaghan et al. 2010; Stauffer et al. 2012). During tumorigenesis, abundant levels of cytokines, chemokines, growth factors, arachidonic acid metabolites, and lipids are produced by both tumor cells and inflammatory cells. All of these factors trigger immune cell activation, migration and eventual infiltration into the tumor (Fujii et al. 2004; Herber et al. 2010). Upon tumor infiltration, these factors continue to impact and manipulate cellular functions to support tumor growth.

Suppressive proteins are more abundant within the tumor microenvironment (TME). They are also produced by other tumor infiltrating immune cell populations such as, macrophages and myeloid derived suppressor cells (MDSC) (Bronte et al. 1997; Condeelis and Pollard 2006). Additionally, tumor cells shed tumor-associated antigens (TAAs) and dendritic cells are capable of capturing, processing and presenting TAAs to T cells. However, there have been conflicting reports as to whether DCs initiate an anti-tumor immune response or whether the TAAs are recognized as "self" and result in the induction of immune tolerance (Chaput et al. 2008; Dhodapkar and Steinman 2002; Hagymasi et al. 2007; Preynat-Seauve et al. 2006; Shurin 1996; Sotomayor et al. 2001).

Due to the paradoxically important role of inflammation in both tumor growth and tumor rejection, the ability to determine prognosis based upon immune cell infiltration has garnered much attention. The results of these studies are variable depending upon a variety of factors including the origin of the tumor. Colorectal adenocarcinoma, gastric-esophageal and nasopharyngeal (Ambe et al. 1989; Imai and Yamakawa 1993; Tsujitani et al. 1990) all reported a more favorable prognosis upon detection of higher frequencies of intratumoral DCs. In contrast, patients with cancers such as breast, hepatocellular, and non-small cell lung carcinoma were associated with poorer overall survival rates when DCs were found within the primary tumor (Perrot et al. 2007; Tang et al. 2006). These data demonstrate the significant impact of the TME on the development and/or effectiveness of DC function and, in turn, the outcome of anti-tumor immunity or the induction of immune tolerance. However, they also imply that the tumor cells themselves may have an important influence on the nature of TME and the resulting clinical outcome. Thus, targeting DC and their role in generating effective anti-tumor immune responses may be an inherent feature of tumor-type.

18.3 Tumor Impact on DC Maturation

In some instances, the TME can activate DC (Fig. 18.1). For example, tumors often contain areas of necrotic cells. Cells that become necrotic lose membrane integrity and release intracellular contents that serve as "danger" signals referred to as "Alarmins", which serve as DAMPs and, as previously mentioned, can activate DCs through PRRs (Oppenheim et al. 2007). DAMPs include endogenous molecules such as DNA, RNA and other nucleic acid-associated proteins. Two DAMPs with known associations to cancer include High Mobility Group Box 1 (HMGB1) and S100 proteins. HMGB1 is a transcriptional regulator that, when released from necrotic cells, can activate DC maturation and migration through its interaction with TLR4 or the Receptor for Advanced Glycation End products (RAGE) (Sims et al. 2010; Andersson and Tracey 2011). In fact, it was found that radiotherapy and chemotherapy treatments for cancer cause release of DAMPs from apoptotic cells that trigger DC activation and promote an anti-tumor immune response (Apetoh et al. 2007). However, HMGB1 was also shown to suppress DC activation (Popovic et al. 2006) and induce apoptosis in macrophages (Kuniyasu et al. 2005). Thus, further investigation of the complex mechanisms by which HMGB1 affects DC function, especially in the context of cancer, is required to understand its role in the regulation of immunity and tolerance.

S100 proteins, another variety of DAMPs, are also found to be up-regulated in many cancers. S100 proteins are expressed by precursor hematopoietic cells and are down-regulated during DC differentiation. However, tumors may promote increased S100 expression that, in effect, prevents DC maturation and promotes the accumulation of myeloid derived suppressor cells (MDSC) that further suppress anti-tumor immune responses (Cheng et al. 2008; Sinha et al. 2008). Another study found that S100+ DCs at the inflammatory ridge of squamous cell carcinomas expressed increased levels of indoleamine 2, 3-dioxygenase (IDO), an enzyme associated with chronic inflammation and T cell tolerance (that will be discussed later) (Kuales et al. 2011) (Fig. 18.1).

Tumors use a variety of mechanisms to inhibit or suppress immune responses. Myeloid cells, including immature DCs, are recruited by tumor-produced chemotactic factors including CCL2, MIP3a, CCL25 and CCL5. The recruitment of different myeloid subsets varies depending upon the chemokine receptor expression profile (Shurin et al. 2006; Wang et al. 1998). Typically, the recruitment of immune cells is accompanied with the subsequent prevention or distortion of the DC maturation process. Maintaining DCs in the immature phenotype results in decreased co-stimulatory molecule (CD80 and CD86) expression that prevents proper initiation of T cell activation (Dhodapkar and Steinman 2002; Mahnke et al. 2002). Several studies show that the majority of DCs residing in tissues peripheral to the tumor are mature but the intratumoral DCs are immature as measured by expression of MHC, CD83 and other co-stimulatory molecules (CD80, CD86, and CD40) (Perrot et al. 2007; Mahnke et al. 2002; Stoitzner et al. 2008; Bell et al. 1999). A study by Bell et.al. demonstrated that mature DCs seem to preferentially localize to peri-tumoral areas whereas immature DC infiltrate the tumor (Bell et al. 1999). These studies all suggest that stromal factors, possibly adhesion or chemotactic factors, may also be partially responsible for DC localization and function within the TME.

DC maturation can be inhibited by the tumor directly as a result of exposure to an array of suppressive proteins including: TGF-β, IL-10, IL-6 CSF, VEGF and PGE2 (Yang et al. 2010, 2009; Greenhough et al. 2009; Rabinovich et al. 2007; Igney and Krammer 2002). While the exact mechanisms for blocking maturation are not clear, the effect of treating DCs with several of these growth factors and cytokines, *in vitro* and *in vivo,* has been reported. VEGF was shown to prevent IL-12 secretion by DCs (Michielsen et al. 2011) and an over-abundance of VEGF in the tumor microenvironment resulted in a decrease in DC frequency. Upon blocking VEGF, increased DC frequency was detected, and was associated with increased co-stimulatory molecule expression and pro-inflammatory functions (Gabrilovich et al. 1999). TGF-β prevented up-regulation of MHC class II and CD86 in response to a powerful pro-inflammatory stimuli that typically induce maturation, such as LPS or TNF-α (Geissmann et al. 1999). In addition, it was reported that TGF-β drives human monocytes toward becoming less stimulatory Langerhans cells rather than mature DC capable of driving T cell activation (Geissmann et al. 1999). Similarly, IL-10 was also shown to halt and redirect monocyte development from DCs to macrophages that resulted in reduced T cell stimulatory capability (Allavena et al. 1998).

Aside from production of suppressive proteins, as mentioned, tumors shed TAAs. Although DCs acquire these antigens and process them for presentation to T cells, the antigens can also interact with DC surface molecules, such as DC-SIGN. The interaction of TAAs with DC-SIGN was found to halt or block antigen processing and impede DC maturation (Igney and Krammer 2002; Nonaka et al. 2008). Thus, an apparently stimulatory activity of TAA uptake can be converted into generation of more suppressive DC populations.

CD83 is the best-known marker for maturation in human DCs. While murine DCs also express CD83, less is known regarding the impact of CD83 expression on maturation and function (Lechmann et al. 2002). The ligand for CD83 is also unknown, but expression of CD83 by DCs is tightly correlated with expression of CD80 and CD86. Functionally, CD83 is required for DC adhesion to monocytes and CD8+ T cells and is therefore important for cell–cell communication (Scholler et al. 2002). One study showed that patients with an increased frequency of tumor infiltrating CD83+ DCs had a better survival rate when compared to patients with higher numbers of CD83− DCs (Iwamoto et al. 2003). Interestingly, another study confirmed those findings and also reported that CD83 expression was found to be inversely proportional to the level of TGF- produced by the tumor (Ananiev et al. 2011). Taken together, these data support the possibility that TGF-β inhibits DC maturation, and that these DCs expressing lower levels of CD83 contribute to a poorer clinical outcome due to reduced immune responses.

Another mechanism by which tumors may inhibit DC maturation is through activation of the glycogen synthase kinase -3 (GSK-3) pathway. Under physiological conditions, GSK-3 is constitutively active in immature DCs and this signal suppresses spontaneous maturation (Alessandrini et al. 2011; Rodionova et al. 2007). In the tumor microenvironment, GSK-3 is often up-regulated and is involved in promoting cell survival (Kubic et al. 2012). It stands to reason that the same mechanisms that promote sustained activation of GSK-3 in tumors cells can similarly activate GSK-3 in DC, thus trapping them in an immature state. However, the precise role of GSK-3 and the pathways it regulates, remain to be elucidated in tumor-associated DCs.

Recently, the role of lipogenesis in the TME was associated with inhibition of DC processing and presentation of antigen (Fig. 18.1). Tumors produce and store lipids as an energy source for rapid and continued growth (Nieman et al. 2011). DCs infiltrating tumors express scavenger receptors and quickly take up lipids. As a result, tumor-infiltrating DCs become saturated with triglycerides (Herber et al. 2010). Interestingly, lipid-containing DCs maintain the ability to present antigens and express co-stimulatory molecules, yet fail to stimulate effector T cells (Herber et al. 2010). Clearly, the role of lipids in inflammation has been well established in diabetes and atherosclerosis, but the role of lipid-driven inflammation in cancer and its effect on anti-tumor immunity require further investigation. This does, however, provide a novel avenue for cancer prevention and therapy (Fig. 18.2).

18.4 Tumor-Induced Tolerogenicity in Mature DC

Although maturation is a critical factor in DC function, tumors can also manipulate mature DCs to support immune tolerance and suppression. Tumor infiltrating DCs are known to become producers of increased levels of suppressive cytokines such as IL-10 and TGF-β (Ghiringhelli et al. 2005; Watkins et al. 2011). IL-10- and TGF-β-producing DCs, in turn, promote "anti-inflammatory" immune responses, the production of Th2 cytokines, and support the expansion of suppressive T regulatory (Treg) cells. An increased frequency of Treg cells has been found to occur in both the TME and in the lymph nodes (Ghiringhelli et al. 2005; Cools et al. 2008; Enk 2006; Yamazaki et al. 2007) of tumor-bearing hosts and is associated with poor clinical outcome (Curiel et al. 2004). Treg cells are found to be associated with all most all types of cancer and are a major TME component known to suppress immune responses.

Another mechanism by which tumors induce DC tolerogenicity is through the up-regulation of transcription factors that program DCs. We recently reported that FOXO3 was up-regulated in DCs that infiltrated prostate, melanoma, and kidney

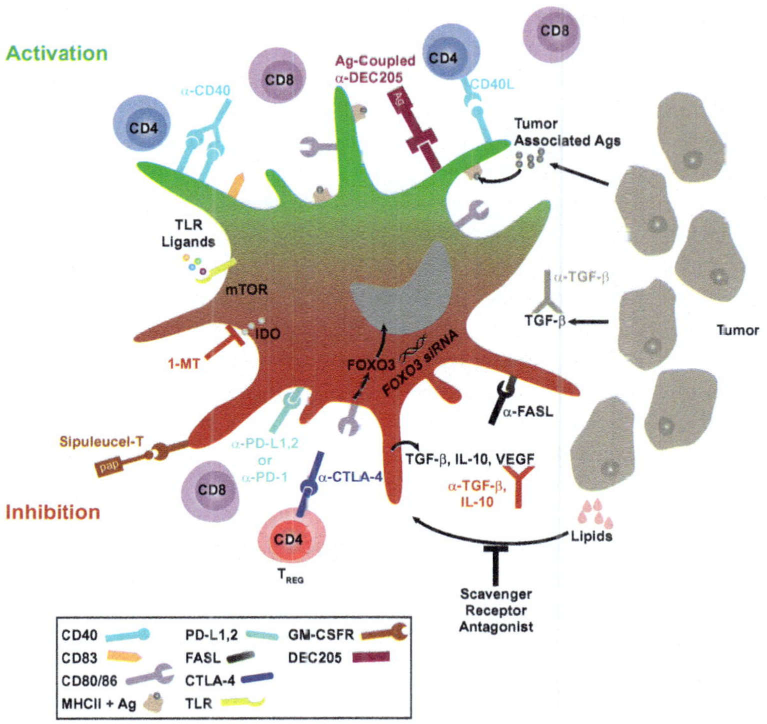

Fig. 18.2 Targeting regulatory pathways to improve anti-tumor immune responses. This cartoon illustrates the many approaches currently employed or proposed that convert tolerogenic and suppressive DCs into immune-stimulatory DCs

tumors (Watkins et al. 2011). FOXO3 is a member of the forkhead box family of transcription factors that was initially found to be a tumor-suppressor gene due to its role in regulating expression of pro-apoptotic genes such as TRAIL, FASL, and Bim (Modur et al. 2002; Paik et al. 2007; Stahl et al. 2002). While expression of FOXO3 is desirable in tumor cells, recent studies provide evidence that FOXO3 expression in DCs is detrimental to promoting an effective anti-tumor immune response and is involved in regulating T cell responses (Watkins et al. 2011; Dejean et al. 2009). FOXO3+ tumor-associated DCs had increased expression of indoleamine 2, 3-dioxygnase (IDO), arginase-I (ARG), TGF-β, and PD-L1, all of which are well known to be associated with the induction of immune tolerance (Fig. 18.1). Upon silencing FOXO3 expression using siRNA, production of tolerogenic mediators was abrogated in conjunction with increased expression of CD80 and IL-12; DCs became immune stimulatory (Watkins et al. 2011) (Fig. 18.2). The mechanisms regulating FOXO3-associated tolerance have not been clearly defined. However, one potential explanation regarding increased FOXO3 expression by DCs in the TME may be extrapolated from a study by Dejean et al. This study showed that CTLA-4-Ig stimulation through B7 molecules

(CD80 and CD86) expressed by DCs led to increased FOXO3 expression (Dejean et al. 2009). CTLA-4 is also expressed by Treg cells, is critical for regulating immune suppression, and has been an effective target in alleviating immune suppression in the TME (Hurwitz et al. 2000; Hurwitz et al. 1998). Taken together, it seems feasible that Treg cells that infiltrate the TME may use CTLA-4 to trigger B7 on the surface of DCs and induce suppressive activity through up-regulation of FOXO3 expression. Thus, targeting CTLA-4 for cancer therapy may also inhibit this suppressive signal to DCs in the TME.

A second transcription factor induced by tumors that has known associations with tolerogenic DCs is β-catenin (Berthon et al. 2012; Fu and Jiang 2010; Holcombe et al. 2002; Manicassamy et al. 2010). Signaling through β-catenin was reported to be critical for the regulation of tolerogenic DCs in the intestine. Activation of β-catenin induced expression of factors such as retinoic acid, IL-10 and TGF-β that assisted in the maintenance of tolerance (Manicassamy et al. 2010). β-catenin activation can also result from disruption of E-cadherin interactions (experimentally described as "cluster disruption") (Fu and Jiang 2010). Although β-catenin+ DCs were reported to be more tolerogenic than pro-inflammatory, β-catenin is involved in and important for the maturation of DCs (Alessandrini et al. 2011; Jiang et al. 2007). While DCs expressing β-catenin displayed increased co-stimulatory ligand expression, they were unable to secrete pro-inflammatory cytokines and induced Treg cells rather than effector T cells (Jiang et al. 2007).

Aside from cluster disruption, tumors may also activate β-catenin through the canonical pathway, which is activated following ligation of Frizzled receptors by members of the Wnt family of ligands. Many tumors produce Wnt ligands that can activate the canonical pathway (Berthon et al. 2012; Holcombe et al. 2002), including Wnt3 (Fig. 18.1). Alternatively, some Wnt ligands, such as Wnt5a, can trigger the non-canonical pathway which was also reported to result in the induction of tolerogenic DCs independent of β-catenin (Widelitz et al. 2005; Valencia et al. 2011). DCs activated by Wnt5a had a reduced ability to stimulate T_H1 cells and produce pro-inflammatory cytokines like TNF-α and IL-12p70, and expressed elevated levels of IL-6 and IL-10 production (Valencia et al. 2011). Collectively, these studies suggest that tumors not only utilize normal developmental and regulatory pathways like Wnt/β-Catenin and FOXO3 to promote cell survival and growth, but they can also trigger the activation of these pathways in DCs to impede immune stimulatory functions and initiate and sustain immune tolerance.

18.5 Development of Tolerogenic DCs *in vitro*

To understand better the developmental pathways of DCs, previous studies have generated different DC subtypes *in vitro*. These subtypes are acquired by stimulating surface receptors utilizing various cytokines and growth factors or by targeting intracellular signaling pathways. Initially GM-CSF and IL-4 were used to generate DCs from monocytes *in vitro* (Inaba et al. 1992). These monocyte-derived cells become functionally active, mature dendritic cells that can process and present Antigen and stimulate naïve T cell proliferation around day 7 of culture (Shurin 1996; Mayordomo et al. 1995). Later, it was reported that Flt3 ligand (Flt3L) played an important role in the differentiation and survival of hematopoietic pre-cursor cells and for the generation of DCs in particular. Further studies confirmed that Flt3L could be used *in vivo* to generate DCs. Following systemic administration of Flt3L to mice, increased frequencies of DCs were found in the blood, spleen and lymph nodes (Shurin et al. 1997). Inflammatory DCs can be generated by culturing blood monocytes in GM-CSF and cytokines such as IFN-γ, TNF-α or IL-15 in addition to IL-4 (Mohty et al. 2003; Chomarat et al. 2003; Palucka and Banchereau 2012). DCs generated in this manner are potent pro-inflammatory cytokine producers and are proficient at activating naïve T cells.

Tolerogenic DCs can also be generated *in vitro* by the addition of IL-10, vitamins A or D3 or through the manipulation of E-cadherin-mediated signaling (Fu and Jiang 2010; Ureta et al. 2007). There are conflicting reports on the effects of Rapamycin on DC development. One recent study demonstrated that Rapamycin treatment promoted a tolerogenic phenotype on *in vitro* cultured dendritic cells (Silk et al. 2012; Sathaliyawala et al. 2010; Turnquist et al. 2007). In contrast, another study suggested that priming of immature DCs in the presence of Rapamycin and TLR ligands generates more stimulatory DC (Amiel et al. 2012). Rapamycin is used to suppress immunity for the acceptance of allografts in transplantation (Eng et al. 1991; Stepkowski et al. 1991). However, it also enhances the generation CD8+ T cell memory (Rao et al. 2010). Interestingly, Rapamycin also has anti-tumor effects. These anti-tumor actions of Rapamycin are mediated through inhibition of the mTOR signaling pathway and dysregulation of the cell cycle (Roberts et al. 2012; Wang et al. 2012). In light of these recent developments, further studies are needed to address the ability to utilize these potent anti-tumor effects without additionally promoting the generation of tolerogenic DCs that prevent successful anti-tumor immunity.

18.6 Tumor-Associated DC Sub-Populations

Several subsets of DCs infiltrate murine and human tumors. These populations include a mix of conventional myeloid derived DCs, Langerhans cells and plasmacytoid DCs (O'Donnell et al. 2007; Perez et al. 2005). Originally, only a small number of DCs was found in most types of cancer due to the limitations of detection. More recently, antigen retrieval immunohistochemistry has enhanced our ability to detect different DC populations within and around tumor tissues (Perez et al. 2005) and there is now a greater appreciation for the diversity of intratumoral DCs.

Plasmacytoid DCs (pDCs) are one of the major sub-populations of dendritic cells and are defined by the surface markers CD11c^{int}/B220/PDCA-1 in mouse or CD123/ILT-7/BDCA-2 in human. pDCs arise from the same progenitor cells as conventional DCs, but display a unique developmental profile that is regulated by the E-box protein E2-2, similar to lymphoid cells (Esashi and Liu 2008) and [reviewed in (Matta et al. 2010)]. pDCs are best known as type I interferon (IFN)-producing cells that can stimulate CD4+ and CD8+ T cells. Type I IFN production by pDCs is critical for boosting immune responses for the clearance of viruses and to promote the survival of activated T cells (Marrack et al. 1999; Delale et al. 2005). Additionally, IFN-α has several anti-tumor properties, including the ability to inhibit proliferation, angiogenesis and metastasis (Liu et al. 2012; von Marschall et al. 2003; McCarty et al. 2002). Human pDCs that express CD303 are also potent producers of IFN-α. The production of IFN-α and expression of MHC class I was shown to produce rapid CD8+ T cell activation in the clearance of viruses (Siegal et al. 1999). However, IFN-α production is often ablated in the tumor microenvironment (Tsukamoto et al. 2009).

In tumor-bearing hosts, pDCs are found in tumors, tumor draining lymph nodes and lymph node metastases (Tsukamoto et al. 2009; Sharma et al. 2007; Gerlini et al. 2007). pDCs associated with cancer often have altered functions. For example, reduced expression of type I IFNs and inflammatory cytokines may be a consequence of the interaction between ILT7 on pDCs and Tetherin (also known as BST2) expressed by tumor cells (Tsukamoto et al. 2009). In addition to reduced cytokine secretion, tumor infiltrating pDCs were reported to promote immune tolerance to tumor antigens through the production of tolerogenic factors and the induction of regulatory T cells (Watkins et al. 2011; Matta et al. 2010; Labidi-Galy et al. 2011). Specific tolerogenic properties of pDCs include the expression of indolemine2, 3-deoxygenase (IDO), TGF-β and IL-10 (Fig. 18.1). pDCs are a potent source of IDO which catabolizes tryptophan, an amino acid critical for maintaining durable T cell response, and produces kynurenine, which is also immune-suppressive (Munn et al. 2004; Munn 2006; Hwu et al. 2000). IDO is expressed by pDCs in the tumor and in the tumor draining lymph nodes, thereby supporting a systemic suppression of T cells to revert anti-tumor immunity. IDO+ pDCs are found in high frequency in tumors that express elevated levels of PGE-2 (von Bergwelt-Baildon et al. 2006; Braun et al. 2005). IDO expression by pDCs may also be responsible for the induction of Treg cells (Sharma et al. 2007; Munn et al. 2004). In addition, CTLA-4 expression by Treg cells may induce IDO expression by pDCs (Baban et al. 2005; Fallarino et al. 2003), revealing a complex, counter-regulatory mechanism strongly favoring suppression of T cells over activation. Some tumor infiltrating pDCs also express increased levels of the enzyme Arginase (Watkins et al. 2011), which depletes L-arginine that, in-turn, leads to reduced expression of CD3− in T cells, inhibiting T cell proliferation and activation (Kuang et al. 2008; Norian et al. 2009). However, in one study, blocking arginase *in vitro* had minimal effect in reducing pDC-induced tolerance, suggesting that arginase is secondary to other dominant suppressive mechanisms imparted by tolerogenic pDCs (Watkins et al. 2011).

Conventional or myeloid-derived DCs is the other major sub-population of DCs. These DCs have been further characterized into subtypes based upon origin (tissue-resident or circulating monocyte-derived) and patterns of surface marker expression. Conventional immature DCs typically express the CCR6 chemokine receptor and infiltrate tumors (Bell et al. 1999). These DCs express the phenotypic markers CD11c, CD11b, and lack CD8 expression. In mice, a second population of tumor-infiltrating DCs is the lymphoid-like CD11c+ CD11b-CD8+ DC (Grohmann et al. 2000). CD8+ DCs have the ability to cross-present MHC class I-restricted antigens (Azuma et al. 2012). CD8+ DCs can be further sub-divided into CD103+ CD207+ and CD103-CD207− (Qiu et al. 2009). It was found that the CD103+ CD207+ CD8+ DCs were responsible for phagocytosis of apoptotic cells and promoting tolerance to cellular antigens; depletion of this DC population resulted in a loss of tolerance to self-antigen (Qiu et al. 2009). There is also evidence that CD8+ DCs can negatively regulate antigen presentation and thereby reduce T cell activation (Grohmann et al. 2000). While it was initially suggested that IL-12 treatment of CD8+ DCs could overcome this suppression (Grohmann et al. 1998), it was subsequently reported that treatment of CD8+ DCs with IFN-γ induced IDO expression that prevented IL-12 rescue of CD8+ DC function (Grohmann et al. 2000), indicating another level of complexity in the interactions between the two populations. In humans, CD141+ DCs are characteristically similar to murine CD8+ DCs in that they have a high capacity to capture exogenous antigen and present to T cells via MHC class I (Bachem et al. 2010; Villadangos and Shortman 2010). CD141+ DCs were shown to produce IL-10 and induce Treg cells to maintain skin homeostasis (Chu et al. 2012). However, their migratory potential, as evidenced by expression of the chemokine receptor XCR1, may allow them to promote anti-tumor immunity in the lymph nodes and TME (Chu et al. 2012; Silk et al. 2011).

Langerhans cells (LCs) are the third major sub-population of DCs. Due to their location throughout all epithelial tissues, LCs are key antigen presenting cells responsible for immune surveillance against tumors (Merad et al. 2008; Ishida et al. 1998). LCs can be further sub-divided based on expression of Langerin (Langerin^{hi} or Langerin^{low}), CD103 and CD8. The origin of these sub-populations, whether dermal or non-dermal, impacts the expression of these phenotypic markers as well as their chemokine receptor profile which, in-turn, affects their migratory abilities (Merad et al. 2008). Dermal LCs (langerin^{hi}, CD8−, CD103+) require TGF-β for development, thus it is not surprising that some solid tumors rich in TGF-β become populated with LCs. Tumors that are heavily infiltrated by LCs typically express S100 proteins (Hammar et al. 1986) (Nakajima et al. 1985). While, as described above, S100 proteins are normally associated with tolerogenic potential, such as expression of IDO and reduced maturation, one report showed that patients with lung tumors containing higher numbers of S100+ LCs had a higher overall survival rate than patients with a lower number of S100+ LCs (Zeid and Muller 1993). Otherwise, LCs have been implicated in supporting immune tolerance. Epidermal LCs can potentiate and activate skin-resident Treg cells (Seneschal et al. 2012). Specifically, one study showed that in response to RANKL, LCs down-regulate co-stimulatory ligand expression and promote suppression by activating Treg cells (Loser et al. 2006). These recent studies describing various populations of LCs and potential roles in promoting immune tolerance suggest that further investigations are required to detail their role in promoting or suppressing anti-tumor immunity.

18.7 Additional Tolerogenic Properties of DC

Irrespective of origin, populations of tumor-infiltrating DCs share some common suppressive mechanisms. Surface ligands such as PD-L1, PD-L2, FASL, and ICOSL are up-regulated upon tumor infiltration or as a result of entrapment in the TME and are known to impact T cell proliferation, effector functions and survival.

The programmed cell death 1 (PD-1) family of molecules has been a major focus in the study of T cell-DC interactions. Ligation of PD-1 on T cells by DCs expressing PD-L1 leads to inhibition of T cell proliferation (Freeman et al. 2000; Selenko-Gebauer et al. 2003) and can also promote conversion of naïve T cells into a Treg cell phenotype, characterized by expression of CD4 and Foxp3 (Wang et al. 2008). DCs can also express PD-L2, the other PD-1 ligand that is more selectively expressed on hematopoietic cells (Selenko-Gebauer et al. 2003). Zhang et al. reported that PD-L2-deficient DCs displayed improved priming of CD4+ and CD8+ T cells. The impact of this observation was further supported by the demonstration that antigen-specific T cell tolerance was lost in PD-L2-deficient mice (Zhang et al. 2006). Both PD-1 ligands are up-regulated in response to inflammatory signals, resulting in limited tissue damage following clearance of an infection (Keir et al. 2008). PD-L1/2 expression is likely up-regulated in the TME as a result of several cytokines including IL-4, IL-10 and Interferon– (Topalian et al. 2012). Blocking PD-1 *in vitro* and *in vivo* can prevent DC-induced T cell tolerance and will be discussed further in the next section (Watkins et al. 2011; Topalian et al. 2012).

DCs can also down regulate or limit T cell activation by the expression of FASL. FASL expressed by DCs triggers FAS on T cells and induce activation-induced cell death (AICD) (Alderson et al. 1995). Some studies reporting FASL expression by DCs refer to them as a subclass of "killer" DCs that can induce apoptosis of FAS-expressing activated CD4+ and CD8+ T cells (Suss and Shortman 1996; Hoves et al. 2004). Several DC subsets were found to upregulate FASL following CD40 activation (Shibaki and Katz 2001). Interestingly, tumor cells also express FAS; therefore, the killer capacity of these DCs may be beneficial in assisting with tumoricidal activities, making them an intriguing target for future investigations (Ioachim et al. 2005).

18.8 DCs and Immunotherapy

Because of their roles in regulating both innate and adaptive immunity and for their ability to migrate in an out of tissue and lymph nodes, DCs have been one of the major targets of immunotherapy. Utilizing DCs to initiate anti-tumor immunity has been attempted in numerous ways with varying degrees of success. One method is the vaccination strategy. A DC vaccine is given to a patient to induce tumor-specific effector T cells that can target the tumor cells in an antigen-specific mechanism. Additionally, the clearance of tumor ideally would be followed up by induction of immunological memory to prevent regrowth of the tumor. This type of therapy was developed by pulsing autologous monocyte-derived DCs *in vitro* with relevant tumor antigen and an adjuvant. The antigen-loaded, mature DCs are then administered to the patient where they prime effector T cells. This strategy has been tested for many years and was found to be safe in terms on immediate health effects on the patient. Recent DC-based therapies include direct and indirect methods of targeting APCs. One example is Sipuleucel-T, an autologous PBMC-derived DC vaccine pulsed with a chimeric molecule consisting of GM-CSF and prostatic acid phosphatase (PAP). Sipuleucel-T was FDA approved for treatment of metastatic prostate cancer (Higano et al. 2009; Kantoff et al. 2010). A more indirect approach of utilizing DC is PROSTVAC, is a poxvirus-based prostate-specific antigen (PSA)-encoding vaccine that is designed to infect APCs and confer expression of tumor antigens that will initiate a targeted T cell-mediated anti-tumor response directed against castration-resistant prostate cancer (Kantoff et al. 2010). The use of these two different vaccines resulted in prolonged median survival of approximately 8 and 4 months, respectively, during the Phase III trials. Owing to these advances, immunotherapy, and peptide vaccines in particular, are now being used in combination with the chemotherapy treatment regime (Chu et al. 2012; Bose et al. 2012; Kaida et al. 1997).

One of the obstacles of vaccine-based therapy is the identification or selection of appropriate tumor antigens for loading the DCs. Two possible avenues include targeting mutated or non-mutated self-antigens to DCs (Somasundaram et al. 2006). Selecting mutated self-antigens can induce a much more potent anti-tumor response. However, mutated self-antigens can be more challenging to identify. Mutations can be difficult to detect and vary widely from patient to patient. Therefore, targeting a mutated self-antigen requires a personalized treatment protocol (Duan et al. 2009) and was reviewed in (Palucka and Banchereau 2012). In contrast, using non-mutated, self-antigens that are more readily available to generate vaccines against a broader range of cancers may be a more feasible approach. Examples include targeting Her-2, PSA, and melanocyte differentiation antigens, all of which have been moderately successful in inducing anti-tumor immunity (Kantoff et al. 2010; Saha and Chatterjee 2010; Ponti et al. 2012). However, targeting these antigens expressed by tumors cells may ultimately fail due to insufficient numbers of high avidity, tumor antigen-specific T cells, as they are depleted through negative selection, or due to the induction of T cell tolerance, as reviewed in (Hurwitz and Watkins 2012; Quezada et al. 2011; Horna and Sotomayor 2007).

An alternative immunotherapeutic method is to target tumor antigen to DCs *in vivo* based on expression of DC-specific surface molecules. Several protocols have been utilized to promote tumor antigen up-take and presentation by DCs. One approach is to use a chimeric molecule consisting of a tumor antigen or dominant epitope that is coupled to a ligand or antibody that targets a DC surface receptor. This approach has been tested in murine models using DEC-205 as the targeted receptor and is being prepared for clinical trials (Bonifaz et al. 2004).

Other approaches have been employed to activate or "license" DCs and improve internalization of tumor antigens. DCs can be licensed through cytokine treatment, interaction with activated helper T cells or stimulation of surface receptors with agonistic antibodies (e.g., anti-CD40) (Shafer-Weaver et al. 2009; Lemoine et al. 2010; Murphy et al. 2003). Pro-inflammatory cytokines can be a good strategy to license DC. IFN-γ was reported to promote DC activation capable of initiating CD8+ T cell responses (Lemoine et al. 2010). However, systemic administration of IFN-γ was reported to have toxic side-effects (Sriskandan et al. 1986) and was reviewed in (Jonasch and Haluska 2001). Therefore, targeting surface receptors on tumor-associated DCs, such as CD40, may be a more feasible approach. Previous studies have demonstrated that ligation of CD40 led to up-regulation of co-stimulatory ligands, increased pro-inflammatory cytokine production, and an increase in the ability to stimulate tumor antigen-specific cytotoxic T cells (Staveley-O'Carroll et al. 2003; Nguyen et al. 2002; Diehl et al. 1999). Triggering CD40 on DCs can be a consequence of delivering CD40L+, antigen-specific CD4+ helper T cells to the tumor or tumor-draining lymph nodes (Shafer-Weaver et al. 2009; Martin-Fontecha et al. 2008; Grohmann et al. 2001) or as a result of administration of a monoclonal antibody that cross-links CD40 (Murphy et al. 2003).

Another method of stimulating DC activation can be achieved by toll like receptor (TLR) ligands that promote DC maturation that results in enhanced CD8+ T cell responses. Because TLRs are expressed on many types of inflammatory cells, similar to directed antigen delivery, TLR ligands can be packaged in nanoparticles and targeted to DCs by labeling the nanoparticles with DC-specific molecules (Tacken et al. 2011). This type of directed delivery is beneficial to avoid off-target excessive pro-inflammatory responses, such as cytokine storm, where excessive levels of pro-inflammatory cytokines induce sepsis-like systemic responses. Imiquimod and Gardiquimod are TLR7 agonists that have been used as vaccine adjuvants to successfully promote an effective antitumor immune response (Ma et al. 2010). Interestingly, Imiquimod was also shown to promote the conversion of tolerogenic pDCs into tumor-killing effector cells (Drobits et al. 2012). Another recent study reported that tolerogenic tumor-associated DCs could also be converted to immune-stimulatory DCs following TLR-9 signaling in response to CpG oligodeoxynucleotides (Zhang et al. 2010). The data from these studies imply that DCs in the TME maintain their plasticity and continue to respond to environmental signals that can enhance their immune stimulatory potential.

In addition to activating DCs, another avenue for improving immunotherapy is to inhibit or block immunosuppressive factors expressed or produced by DCs *in vivo* (Fig. 18.2). One study by Kim et. al. demonstrated that silencing the IL-10R via siRNA enhanced a DC vaccine effect in the HPV E7-expressing TC-1 tumor model (Kim et al. 2011). Similarly, antibody blockade of IL-10-mediated signals can target DC stimulatory capacity and improve anti-tumor responses (Guiducci et al. 2005; Vicari et al. 2002). As mentioned in the previous sections, other studies have sought to alleviate T cell tolerance by blocking PD-1 interactions with its ligands, PD-L1 and/or PD-L2 (Watkins et al. 2011; Topalian et al. 2012; Iwamura K et al. 2011; Rosenblatt et al. 1997). However, recent studies suggest that the efficacy of blocking PD-1 in patients may be more dependent upon tumor cell, rather than DC, expression of PD-L1 (Topalian et al. 2012). Targeting IDO enzymatic activity using inhibitors such as 1-methyl tryptophan can enhance tumor antigen presentation by DCs, delay DC tolerogenicity and reduce the induction of Treg cells (Watkins et al. 2011; Li et al. 2010; Hou et al. 2007). While some attempts have been made to block combinations of DC inhibitory signals to achieve additive or synergistic effects (Watkins et al. 2011; Hou et al. 2007; Vacca et al. 2005), due to the vast array of possible combinations, it may be more effective to target upstream regulators of inhibitory signaling pathways.

As discussed above, we recently reported that FOXO3 regulates tolerogenicity of prostate tumor-associated DCs. Silencing FOXO3 expression using siRNA's abrogated DC tolerogenicity and converted the tumor-associated DC's into stimulatory DCs (Watkins et al. 2011). Another study demonstrated that silencing PTEN and activating the PI3 K/AKT pathway enhanced DC maturation and improved survival of tumor bearing hosts (Kim et al. 2010). Collectively, these studies demonstrate that blocking immunosuppressive factors or pathways can provide a boost in immune-stimulatory capacity of DCs and delay the induction of immune tolerance.

18.9 Conclusions

Dendritic cells are a complex and multi-functional population of APCs. While their principal role seems to be to elicit immune responses, tumors coopt this function and reverse their role, converting them into immune-suppressive cells that promote tumor development and limit tumor immunity. A variety of suppressive pathways are exploited by DCs in the tumor microenvironment. New technologies allowing for the development of small molecule inhibitors may be an effective strategy for targeting these pathways. However, because the TME is a highly tolerogenic environment comprised of multitudes of different suppressive mechanisms, a more complete understanding of the intercellular communication between DCs, T cells, and other regulatory inflammatory cells will undoubtedly contribute to the identification of better targets for the enhancement of immunotherapy against cancer.

Acknowledgments The authors appreciate the patience and skill of Ms. Tammy Schroyer and her assistance in preparing the Figures for this Chapter. Some work described in this Chapter was supported in-part by the Intramural Research Program of the NCI, NIH.

References

Alderson MR, Tough TW, Davis-Smith T, Braddy S, Falk B, Schooley KA et al (1995) Fas ligand mediates activation-induced cell death in human T lymphocytes. J Exp Med 181:71–77

Alessandrini A, De Haseth S, Fray M, Miyajima M, Colvin RB, Williams WW et al (2011) Dendritic cell maturation occurs through the inhibition of GSK-3beta. Cell Immunol 270:114–125

Allavena P, Piemonti L, Longoni D, Bernasconi S, Stoppacciaro A, Ruco L et al (1998) IL-10 prevents the differentiation of monocytes to dendritic cells but promotes their maturation to macrophages. Eur J Immunol 28:359–369

Ambe K, Mori M, Enjoji M (1989) S-100 protein-positive dendritic cells in colorectal adenocarcinomas. Distribution and relation to the clinical prognosis. Cancer 63:496–503

Amiel E, Everts B, Freitas TC, King IL, Curtis JD, Pearce EL et al (2012) Inhibition of mechanistic target of rapamycin promotes dendritic cell activation and enhances therapeutic autologous vaccination in mice. J Immunol 189:2151–2158

Ananiev J, Gulubova MV, Manolova IM (2011) Prognostic significance of CD83 positive tumor-infiltrating dendritic cells and expression of TGF-beta 1 in human gastric cancer. Hepatogastroenterology 58:1834–1840

Andersson U, Tracey KJ (2011) HMGB1 is a therapeutic target for sterile inflammation and infection. Annu Rev Immunol 29:139–162

Apetoh L, Ghiringhelli F, Tesniere A, Obeid M, Ortiz C, Criollo A et al (2007) Toll-like receptor 4-dependent contribution of the immune system to anticancer chemotherapy and radiotherapy. Nat Med 13:1050–1059

Azuma M, Ebihara T, Oshiumi H, Matsumoto M, Seya T (2012) Cross-priming for antitumor CTL induced by soluble Ag+ polyI: C depends on the TICAM-1 pathway in mouse CD11c(+)/CD8alpha(+) dendritic cells. Oncoimmunology 1:581–592

Baban B, Hansen AM, Chandler PR, Manlapat A, Bingaman A, Kahler DJ et al (2005) A minor population of splenic dendritic cells expressing CD19 mediates IDO-dependent T cell suppression via type I IFN signaling following B7 ligation. Int Immunol 17:909–919

Bachem A, Guttler S, Hartung E, Ebstein F, Schaefer M, Tannert A et al (2010) Superior antigen cross-presentation and XCR1 expression define human CD11c+ CD141+ cells as homologues of mouse CD8+ dendritic cells. J Exp Med 207:1273–1281

Bell D, Chomarat P, Broyles D, Netto G, Harb GM, Lebecque S et al (1999) In breast carcinoma tissue, immature dendritic cells reside within the tumor, whereas mature dendritic cells are located in peritumoral areas. J Exp Med 190:1417–1426

Berthon A, Martinez A, Bertherat J, Val P (2012) Wnt/beta-catenin signalling in adrenal physiology and tumour development. Mol Cell Endocrinol 351:87–95

Bonifaz LC, Bonnyay DP, Charalambous A, Darguste DI, Fujii S, Soares H et al (2004) In vivo targeting of antigens to maturing dendritic cells via the DEC-205 receptor improves T cell vaccination. J Exp Med 199:815–824

Bose A, Lowe DB, Rao A, Storkus WJ (2012) Combined vaccine+axitinib therapy yields superior antitumor efficacy in a murine melanoma model. Melanoma Res 22:236–243

Braun D, Longman RS, Albert ML (2005) A two-step induction of indoleamine 2,3 dioxygenase (IDO) activity during dendritic-cell maturation. Blood 106:2375–2381

Bronte V, Serafini P, Apolloni E, Zanovello P (2001) Tumor-induced immune dysfunctions caused by myeloid suppressor cells. J Immunother 24:431–446

Chaput N, Conforti R, Viaud S, Spatz A, Zitvogel L (2008) The Janus face of dendritic cells in cancer. Oncogene 27:5920–5931

Cheng P, Corzo CA, Luetteke N, Yu B, Nagaraj S, Bui MM et al (2008) Inhibition of dendritic cell differentiation and accumulation of myeloid-derived suppressor cells in cancer is regulated by S100A9 protein. J Exp Med 205:2235–2249

Chomarat P, Dantin C, Bennett L, Banchereau J, Palucka AK (2003) TNF skews monocyte differentiation from macrophages to dendritic cells. J Immunol 171:2262–2269

Chu CC, Ali N, Karagiannis P, Di Meglio P, Skowera A, Napolitano L et al (2012a) Resident CD141 (BDCA3)+ dendritic cells in human skin produce IL-10 and induce regulatory T cells that suppress skin inflammation. J Exp Med 209:935–945

Chu CS, Boyer J, Schullery DS, Gimotty PA, Gamerman V, Bender J et al (2012b) Phase I/II randomized trial of dendritic cell vaccination with or without cyclophosphamide for consolidation therapy of advanced ovarian cancer in first or second remission. Cancer Immunol Immunother: CII 61:629–641

Condeelis J, Pollard JW (2006) Macrophages: obligate partners for tumor cell migration, invasion, and metastasis. Cell 124:263–266

Cools N, Van TV, Smits EL, Lenjou M, Nijs G, Van Bockstaele DR et al (2008) Immunosuppression induced by immature dendritic cells is mediated by TGF-beta/IL-10 double-positive CD4+ regulatory T cells. J Cell Mol Med 12:690–700

Curiel TJ, Coukos G, Zou L, Alvarez X, Cheng P, Mottram P et al (2004) Specific recruitment of regulatory T cells in ovarian carcinoma fosters immune privilege and predicts reduced survival. Nat Med 10:942–949

Curtsinger JM, Lins DC, Mescher MF (2003) Signal 3 determines tolerance versus full activation of naive CD8 T cells: dissociating proliferation and development of effector function. J Exp Med 197:1141–1151

Dejean AS, Beisner DR, Ch'en IL, Kerdiles YM, Babour A, Arden KC et al (2009) Transcription factor Foxo3 controls the magnitude of T cell immune responses by modulating the function of dendritic cells. Nat Immunol 10:504–513

Delale T, Paquin A, Asselin-Paturel C, Dalod M, Brizard G, Bates EE et al (2005) MyD88-dependent and -independent murine cytomegalovirus sensing for IFN-alpha release and initiation of immune responses in vivo. J Immunol 175:6723–6732

Dhodapkar MV, Steinman RM (2002) Antigen-bearing immature dendritic cells induce peptide-specific CD8(+) regulatory T cells in vivo in humans. Blood 100:174–177

Diehl L, den Boer AT, Schoenberger SP, van der Voort EI, Schumacher TN, Melief CJ et al (1999) CD40 activation in vivo overcomes peptide-induced peripheral cytotoxic T-lymphocyte tolerance and augments anti-tumor vaccine efficacy. Nat Med 5:774–779

Drobits B, Holcmann M, Amberg N, Swiecki M, Grundtner R, Hammer M et al (2012) Imiquimod clears tumors in mice independent of adaptive immunity by converting pDCs into tumor-killing effector cells. J Clin Invest 122:575–585

Duan F, Lin Y, Liu C, Engelhorn ME, Cohen AD, Curran M et al (2009) Immune rejection of mouse tumors expressing mutated self. Cancer Res 69:3545–3553

Eng CP, Gullo-Brown J, Chang JY, Sehgal SN (1991) Inhibition of skin graft rejection in mice by rapamycin: a novel immunosuppressive macrolide. Transpl Proc 23:868–869

Enk AH (2006) DCs and cytokines cooperate for the induction of tregs. Ernst Schering Research Foundation workshop, pp 97-106

Esashi E, Liu YJ (2008) E-box protein E2–2 is a crucial regulator of plasmacytoid DC development. Eur J Immunol 38:2386–2388

Fallarino F, Grohmann U, Hwang KW, Orabona C, Vacca C, Bianchi R et al (2003) Modulation of tryptophan catabolism by regulatory T cells. Nat Immunol 4:1206–1212

Freeman GJ, Long AJ, Iwai Y, Bourque K, Chernova T, Nishimura H et al (2000) Engagement of the PD-1 immunoinhibitory receptor by a novel B7 family member leads to negative regulation of lymphocyte activation. J Exp Med 192:1027–1034

Fu C, Jiang A (2010) Generation of tolerogenic dendritic cells via the E-cadherin/beta-catenin-signaling pathway. Immunol Res 46:72–78

Fujii S, Liu K, Smith C, Bonito AJ, Steinman RM (2004) The linkage of innate to adaptive immunity via maturing dendritic cells in vivo requires CD40 ligation in addition to antigen presentation and CD80/86 costimulation. J Exp Med 199:1607–1618

Gabrilovich DI, Ishida T, Nadaf S, Ohm JE, Carbone DP (1999) Antibodies to vascular endothelial growth factor enhance the efficacy of cancer immunotherapy by improving endogenous dendritic cell function. Clin Cancer Res: J Am Assoc Cancer Res 5:2963–2970

Geissmann F, Revy P, Regnault A, Lepelletier Y, Dy M, Brousse N et al (1999) TGF-beta 1 prevents the noncognate maturation of human dendritic Langerhans cells. J Immunol 162:4567–4575

Gerlini G, Urso C, Mariotti G, Di Gennaro P, Palli D, Brandani P et al (2007) Plasmacytoid dendritic cells represent a major dendritic cell subset in sentinel lymph nodes of melanoma patients and accumulate in metastatic nodes. Clin Immunol 125:184–193

Ghiringhelli F, Puig PE, Roux S, Parcellier A, Schmitt E, Solary E et al (2005) Tumor cells convert immature myeloid dendritic cells into TGF-beta-secreting cells inducing CD4+ CD25+ regulatory T cell proliferation. J Exp Med 202:919–929

Greenhough A, Smartt HJ, Moore AE, Roberts HR, Williams AC, Paraskeva C et al (2009) The COX-2/PGE2 pathway: key roles in the hallmarks of cancer and adaptation to the tumour microenvironment. Carcinogenesis 30:377–386

Grivennikov SI, Greten FR, Karin M (2010) Immunity, inflammation, and cancer. Cell 140:883–899

Grohmann U, Belladonna ML, Bianchi R, Orabona C, Ayroldi E, Fioretti MC et al (1998) IL-12 acts directly on DC to promote nuclear localization of NF-kappaB and primes DC for IL-12 production. Immunity 9:315–323

Grohmann U, Bianchi R, Belladonna ML, Silla S, Fallarino F, Fioretti MC et al (2000) IFN-gamma inhibits presentation of a tumor/self peptide by CD8 alpha- dendritic cells via potentiation of the CD8 alpha+ subset. J Immunol 165:1357–1363

Grohmann U, Fallarino F, Silla S, Bianchi R, Belladonna ML, Vacca C et al (2001) CD40 ligation ablates the tolerogenic potential of lymphoid dendritic cells. J Immunol 166:277–283

Guiducci C, Vicari AP, Sangaletti S, Trinchieri G, Colombo MP (2005) Redirecting in vivo elicited tumor infiltrating macrophages and dendritic cells towards tumor rejection. Cancer Res 65:3437–3446

Hagymasi AT, Slaiby AM, Mihalyo MA, Qui HZ, Zammit DJ, Lefrancois L et al (2007) Steady state dendritic cells present parenchymal self-antigen and contribute to, but are not essential for, tolerization of naive and Th1 effector CD4 cells. J Immunol 179:1524–1531

Hammar S, Bockus D, Remington F, Bartha M (1986) The widespread distribution of Langerhans cells in pathologic tissues: an ultrastructural and immunohistochemical study. Hum Pathol 17:894–905

Herber DL, Cao W, Nefedova Y, Novitskiy SV, Nagaraj S, Tyurin VA et al (2010) Lipid accumulation and dendritic cell dysfunction in cancer. Nat Med 16:880–886

Higano CS, Schellhammer PF, Small EJ, Burch PA, Nemunaitis J, Yuh L et al (2009) Integrated data from 2 randomized, double-blind, placebo-controlled, phase 3 trials of active cellular immunotherapy with sipuleucel-T in advanced prostate cancer. Cancer 115:3670–3679

Holcombe RF, Marsh JL, Waterman ML, Lin F, Milovanovic T, Truong T (2002) Expression of Wnt ligands and Frizzled receptors in colonic mucosa and in colon carcinoma. Mol Pathol: MP 55:220–226

Horna P, Sotomayor EM (2007) Cellular and molecular mechanisms of tumor-induced T-cell tolerance. Curr Cancer Drug Targets 7:41–53

Hou DY, Muller AJ, Sharma MD, DuHadaway J, Banerjee T, Johnson M et al (2007) Inhibition of indoleamine 2,3-dioxygenase in dendritic cells by stereoisomers of 1-methyl-tryptophan correlates with antitumor responses. Cancer Res 67:792–801

Hoves S, Krause SW, Herfarth H, Halbritter D, Zhang HG, Mountz JD et al (2004) Elimination of activated but not resting primary human CD4+ and CD8+ T cells by Fas ligand (FasL/CD95L)-expressing Killer-dendritic cells. Immunobiology 208:463–475

Hurwitz AA, Watkins SK (2012) Immune suppression in the tumor microenvironment a role for dendritic cell-mediated tolerization of T cells. Cancer Immunol Immunother: CII 61:289–293
Hurwitz AA, Yu TF, Leach DR, Allison JP (1998) CTLA-4 blockade synergizes with tumor-derived granulocyte-macrophage colony-stimulating factor for treatment of an experimental mammary carcinoma. Proc Natl Acad Sci USA 95:10067–10071
Hurwitz AA, Foster BA, Kwon ED, Truong T, Choi EM, Greenberg NM et al (2000) Combination immunotherapy of primary prostate cancer in a transgenic mouse model using CTLA-4 blockade. Cancer Res 60:2444–2448
Hwu P, Du MX, Lapointe R, Do M, Taylor MW, Young HA (2000) Indoleamine 2,3-dioxygenase production by human dendritic cells results in the inhibition of T cell proliferation. J Immunol 164:3596–3599
Igney FH, Krammer PH (2002) Immune escape of tumors: apoptosis resistance and tumor counterattack. J Leukoc Biol 71:907–920
Imai Y, Yamakawa M (1993) Dendritic cells in esophageal cancer and lymph node tissues. In Vivo 7:239–248
Inaba K, Inaba M, Romani N, Aya H, Deguchi M, Ikehara S et al (1992) Generation of large numbers of dendritic cells from mouse bone marrow cultures supplemented with granulocyte/macrophage colony-stimulating factor. J Exp Med 176:1693–1702
Ioachim HL, Decuseara R, Giancotti F, Dorsett BH (2005) FAS and FAS-L expression by tumor cells and lymphocytes in breast carcinomas and their lymph node metastases. Pathol Res Pract 200:743–751
Ishida T, Oyama T, Carbone DP, Gabrilovich DI (1998) Defective function of Langerhans cells in tumor-bearing animals is the result of defective maturation from hemopoietic progenitors. J Immunol 161:4842–4851
Iwamoto M, Shinohara H, Miyamoto A, Okuzawa M, Mabuchi H, Nohara T et al (2003) Prognostic value of tumor-infiltrating dendritic cells expressing CD83 in human breast carcinomas. Int J Cancer 104:92–97
Iwamura K, Kato T, Miyahara Y, Naota H, Mineno J, Ikeda H et al (2011) siRNA-mediated silencing of PD-1 ligands enhances tumor-specific human T-cell effector functions. Gene Ther 9(10):959–966
Jenkins MK, Schwartz RH (1987) Antigen presentation by chemically modified splenocytes induces antigen-specific T cell unresponsiveness in vitro and in vivo. J Exp Med 165:302–319
Jiang A, Bloom O, Ono S, Cui W, Unternaehrer J, Jiang S et al (2007) Disruption of E-cadherin-mediated adhesion induces a functionally distinct pathway of dendritic cell maturation. Immunity 27:610–624
Jonasch E, Haluska FG (2001) Interferon in oncological practice: review of interferon biology, clinical applications, and toxicities. Oncologist 6:34–55
Kaida M, Morita-Hoshi Y, Soeda A, Wakeda T, Yamaki Y, Kojima Y et al (2011) Phase 1 trial of Wilms tumor 1 (WT1) peptide vaccine and gemcitabine combination therapy in patients with advanced pancreatic or biliary tract cancer. J Immunother 34:92–99 (Hagerstown, Md: 1997)
Kantoff PW, Higano CS, Shore ND, Berger ER, Small EJ, Penson DF et al (2010a) Sipuleucel-T immunotherapy for castration-resistant prostate cancer. N Engl J Med 363:411–422
Kantoff PW, Schuetz TJ, Blumenstein BA, Glode LM, Bilhartz DL, Wyand M et al (2010b) Overall survival analysis of a phase II randomized controlled trial of a Poxviral-based PSA-targeted immunotherapy in metastatic castration-resistant prostate cancer. J Clin Oncol: J Am Soc Clin Oncol 28:1099–1105
Keir ME, Butte MJ, Freeman GJ, Sharpe AH (2008) PD-1 and its ligands in tolerance and immunity. Annu Rev Immunol 26:677–704
Kim JH, Kang TH, Noh KH, Kim SH, Lee YH, Kim KW et al (2010) Enhancement of DC vaccine potency by activating the PI3 K/AKT pathway with a small interfering RNA targeting PTEN. Immunol Lett 134:47–54
Kim JH, Kang TH, Noh KH, Bae HC, Ahn YH, Lee YH et al (2011) Blocking the immunosuppressive axis with small interfering RNA targeting interleukin (IL)-10 receptor enhances dendritic cell-based vaccine potency. Clin Exp Immunol 165:180–189

Kuales MA, Wenzel J, Schmid-Wendtner MH, Bieber T, von Bubnoff D (2011) Myeloid CD11c+ S100+ dendritic cells express indoleamine 2,3-dioxygenase at the inflammatory border to invasive lower lip squamous cell carcinoma. Histol Histopathol 26:997–1006

Kuang DM, Zhao Q, Xu J, Yun JP, Wu C, Zheng L (2008) Tumor-educated tolerogenic dendritic cells induce CD3epsilon down-regulation and apoptosis of T cells through oxygen-dependent pathways. J Immunol 181:3089–3098

Kubic JD, Mascarenhas JB, Iizuka T, Wolfgeher D, Lang D (2012) GSK-3 Promotes Cell Survival, Growth, and PAX3 Levels in Human Melanoma Cells. Mol Cancer Res: MCR 10:1065–1076

Kuniyasu H, Yano S, Sasaki T, Sasahira T, Sone S, Ohmori H (2005) Colon cancer cell-derived high mobility group 1/amphoterin induces growth inhibition and apoptosis in macrophages. Am J Pathol 166:751–760

Labidi-Galy SI, Sisirak V, Meeus P, Gobert M, Treilleux I, Bajard A et al (2011) Quantitative and functional alterations of plasmacytoid dendritic cells contribute to immune tolerance in ovarian cancer. Cancer Res 71:5423–5434

Lechmann M, Zinser E, Golka A, Steinkasserer A (2002) Role of CD83 in the immunomodulation of dendritic cells. Int Arch Allergy Immunol 129:113–118

Lemoine R, Velge-Roussel F, Herr F, Felix R, Nivet H, Lebranchu Y et al (2010) Interferon gamma licensing of human dendritic cells in T-helper-independent CD8+ alloimmunity. Blood 116:3089–3098

Li Y, Xu J, Zou H, Wang C (2010) 1-MT enhances potency of tumor cell lysate-pulsed dendritic cells against pancreatic adenocarcinoma by downregulating the percentage of tregs. J Huazhong Univ Sci Technol Med Sci=Hua zhong ke ji da xue xue bao Yi xue Ying De wen ban=Huazhong keji daxue xuebao Yixue Yingdewen ban 30:344–348

Liu X, Lu J, He ML, Li Z, Zhang B, Zhou LH et al (2012) Antitumor effects of interferon-alpha on cell growth and metastasis in human nasopharyngeal carcinoma. Curr Cancer Drug Targets 12:561–570

Loser K, Mehling A, Loeser S, Apelt J, Kuhn A, Grabbe S et al (2006) Epidermal RANKL controls regulatory T-cell numbers via activation of dendritic cells. Nat Med 12:1372–1379

Ma F, Zhang J, Zhang J, Zhang C (2010) The TLR7 agonists imiquimod and gardiquimod improve DC-based immunotherapy for melanoma in mice. Cell Mol Immunol 7:381–388

Mahnke K, Schmitt E, Bonifaz L, Enk AH, Jonuleit H (2002) Immature, but not inactive: the tolerogenic function of immature dendritic cells. ImmunolCell Biol 80:477–483

Manicassamy S, Reizis B, Ravindran R, Nakaya H, Salazar-Gonzalez RM, Wang YC et al (2010) Activation of beta-catenin in dendritic cells regulates immunity versus tolerance in the intestine. Science 329:849–853

Marrack P, Kappler J, Mitchell T (1999) Type I interferons keep activated T cells alive. J Exp Med 189:521–530

Martin-Fontecha A, Baumjohann D, Guarda G, Reboldi A, Hons M, Lanzavecchia A et al (2008) CD40L+ CD4+ memory T cells migrate in a CD62P-dependent fashion into reactive lymph nodes and license dendritic cells for T cell priming. J Exp Med 205:2561–2574

Matta BM, Castellaneta A, Thomson AW (2010) Tolerogenic plasmacytoid DC. Eur J Immunol 40:2667–2676

Mayordomo JI, Zorina T, Storkus WJ, Zitvogel L, Celluzzi C, Falo LD et al (1995) Bone marrow-derived dendritic cells pulsed with synthetic tumour peptides elicit protective and therapeutic antitumour immunity. Nat Med 1:1297–1302

McCarty MF, Bielenberg D, Donawho C, Bucana CD, Fidler IJ (2002) Evidence for the causal role of endogenous interferon-alpha/beta in the regulation of angiogenesis, tumorigenicity, and metastasis of cutaneous neoplasms. Clin Exp Metastasis 19:609–615

Merad M, Ginhoux F, Collin M (2008) Origin, homeostasis and function of Langerhans cells and other langerin-expressing dendritic cells. Nat Rev Immunol 8:935–947

Michielsen AJ, Hogan AE, Marry J, Tosetto M, Cox F, Hyland JM et al (2011) Tumour tissue microenvironment can inhibit dendritic cell maturation in colorectal cancer. PLoS ONE 6:e27944

Modur V, Nagarajan R, Evers BM, Milbrandt J (2002) FOXO proteins regulate tumor necrosis factor-related apoptosis inducing ligand expression. Implications for PTEN mutation in prostate cancer. J Biol Chem 277:47928–47937

Mohty M, Vialle-Castellano A, Nunes JA, Isnardon D, Olive D, Gaugler B (2003) IFN-alpha skews monocyte differentiation into Toll-like receptor 7-expressing dendritic cells with potent functional activities. J Immunol 171:3385–3393

Morson BC (1985) Precancer and cancer in inflammatory bowel disease. Pathology 17:173–180

Munn DH (2006) Indoleamine 2,3-dioxygenase, tumor-induced tolerance and counter-regulation. Curr Opin Immunol 18:220–225

Munn DH, Sharma MD, Hou D, Baban B, Lee JR, Antonia SJ et al (2004) Expression of indoleamine 2,3-dioxygenase by plasmacytoid dendritic cells in tumor-draining lymph nodes. J Clin Invest 114:280–290

Murphy WJ, Welniak L, Back T, Hixon J, Subleski J, Seki N et al (2003) Synergistic anti-tumor responses after administration of agonistic antibodies to CD40 and IL-2: coordination of dendritic and CD8+ cell responses. J Immunol 170:2727–2733

Nakajima T, Kodama T, Tsumuraya M, Shimosato Y, Kameya T (1985) S-100 protein-positive Langerhans cells in various human lung cancers, especially in peripheral adenocarcinomas. Virchows Arch A Pathol Anat Histopathol 407:177–189

Nguyen LT, Elford AR, Murakami K, Garza KM, Schoenberger SP, Odermatt B et al (2002) Tumor growth enhances cross-presentation leading to limited T cell activation without tolerance. J Exp Med 195:423–435

Nieman KM, Kenny HA, Penicka CV, Ladanyi A, Buell-Gutbrod R, Zillhardt MR et al (2011) Adipocytes promote ovarian cancer metastasis and provide energy for rapid tumor growth. Nat Med 17:1498–1503

Nonaka M, Ma BY, Murai R, Nakamura N, Baba M, Kawasaki N et al (2008) Glycosylation-dependent interactions of C-type lectin DC-SIGN with colorectal tumor-associated Lewis glycans impair the function and differentiation of monocyte-derived dendritic cells. J Immunol 180:3347–3356

Norian LA, Rodriguez PC, O'Mara LA, Zabaleta J, Ochoa AC, Cella M et al (2009) Tumor-infiltrating regulatory dendritic cells inhibit CD8+ T cell function via L-arginine metabolism. Cancer Res 69:3086–3094

O'Callaghan DS, O'Donnell D, O'Connell F, O'Byrne KJ (2010) The role of inflammation in the pathogenesis of non-small cell lung cancer. J Thorac Oncol: Int Assoc Study Lung Cancer 5:2024–2036

O'Donnell RK, Mick R, Feldman M, Hino S, Wang Y, Brose MS et al (2007) Distribution of dendritic cell subtypes in primary oral squamous cell carcinoma is inconsistent with a functional response. Cancer Lett 255:145–152

Oppenheim JJ, Tewary P, de la Rosa G, Yang D (2007) Alarmins initiate host defense. Adv Exp Med Biol 601:185–194

Paik JH, Kollipara R, Chu G, Ji H, Xiao Y, Ding Z et al (2007) FoxOs are lineage-restricted redundant tumor suppressors and regulate endothelial cell homeostasis. Cell 128:309–323

Palucka K, Banchereau J (2012) Cancer immunotherapy via dendritic cells. Nat Rev Cancer 12:265–277

Perez L, Shurin MR, Collins B, Kogan D, Tourkova IL, Shurin GV (2005) Comparative analysis of CD1a, S-100, CD83, and CD11c human dendritic cells in normal, premalignant, and malignant tissues. Histol Histopathol 20:1165–1172

Perrot I, Blanchard D, Freymond N, Isaac S, Guibert B, Pacheco Y et al (2007) Dendritic cells infiltrating human non-small cell lung cancer are blocked at immature stage. J Immunol 178:2763–2769

Ponti G, Tomasi A, Pellacani G (2012) Overwhelming response to Dabrafenib in a patient with double BRAF mutation (V600E; V600M) metastatic malignant melanoma. J Hematol Oncol 5:60

Popovic PJ, DeMarco R, Lotze MT, Winikoff SE, Bartlett DL, Krieg AM et al (2006) High mobility group B1 protein suppresses the human plasmacytoid dendritic cell response to TLR9 agonists. J Immunol 177:8701–8707

Preynat-Seauve O, Schuler P, Contassot E, Beermann F, Huard B, French LE (2006) Tumor-infiltrating dendritic cells are potent antigen-presenting cells able to activate T cells and mediate tumor rejection. J Immunol 176:61–67

Qiu CH, Miyake Y, Kaise H, Kitamura H, Ohara O, Tanaka M (2009) Novel subset of CD8{alpha} + dendritic cells localized in the marginal zone is responsible for tolerance to cell-associated antigens. J Immunol 182:4127–4136

Quezada SA, Peggs KS, Simpson TR, Allison JP (2011) Shifting the equilibrium in cancer immunoediting: from tumor tolerance to eradication. Immunol Rev 241:104–118

Rabinovich GA, Gabrilovich D, Sotomayor EM (2007) Immunosuppressive strategies that are mediated by tumor cells. Annu Rev Immunol 25:267–296

Rao RR, Li Q, Odunsi K, Shrikant PA (2010) The mTOR kinase determines effector versus memory CD8+ T cell fate by regulating the expression of transcription factors T-bet and Eomesodermin. Immunity 32:67–78

Roberts PJ, Usary JE, Darr DB, Dillon PM, Pfefferle AD, Whittle MC et al (2012) Combined PI3K/mTOR and MEK inhibition provides broad antitumor activity in faithful Murine cancer models. Clin Cancer Res: J Am Assoc Cancer Res 18(19):5290–5303

Rodionova E, Conzelmann M, Maraskovsky E, Hess M, Kirsch M, Giese T et al (2007) GSK-3 mediates differentiation and activation of proinflammatory dendritic cells. Blood 109:1584–1592

Romani N, Koide S, Crowley M, Witmer-Pack M, Livingstone AM, Fathman CG et al (1989) Presentation of exogenous protein antigens by dendritic cells to T cell clones. Intact protein is presented best by immature, epidermal Langerhans cells. J Exp Med 169:1169–1178

Rosenblatt J, Glotzbecker B, Mills H, Vasir B, Tzachanis D, Levine JD et al (2011) PD-1 blockade by CT-011, anti-PD-1 antibody, enhances ex vivo T-cell responses to autologous dendritic cell/myeloma fusion vaccine. J Immunother 34:409–418

Saha A, Chatterjee SK (2010) Dendritic cells pulsed with an anti-idiotype antibody mimicking Her-2/neu induced protective antitumor immunity in two lines of Her-2/neu transgenic mice. Cell Immunol 263:9–21

Sathaliyawala T, O'Gorman WE, Greter M, Bogunovic M, Konjufca V, Hou ZE et al (2010) Mammalian target of rapamycin controls dendritic cell development downstream of Flt3 ligand signaling. Immunity 33:597–606

Scholler N, Hayden-Ledbetter M, Dahlin A, Hellstrom I, Hellstrom KE, Ledbetter JA (2002) Cutting edge: CD83 regulates the development of cellular immunity. J Immunol 168:2599–2602

Schreiber RD, Old LJ, Smyth MJ (2011) Cancer immunoediting: integrating immunity's roles in cancer suppression and promotion. Science 331:1565–1570

Schuler G, Romani N, Steinman RM (1985) A comparison of murine epidermal Langerhans cells with spleen dendritic cells. J Invest Dermatol 85:99s–106s

Selenko-Gebauer N, Majdic O, Szekeres A, Hofler G, Guthann E, Korthauer U et al (2003) B7–H1 (programmed death-1 ligand) on dendritic cells is involved in the induction and maintenance of T cell anergy. J Immunol 170:3637–3644

Seneschal J, Clark RA, Gehad A, Baecher-Allan CM, Kupper TS (2012) Human epidermal Langerhans cells maintain immune homeostasis in skin by activating skin resident regulatory T cells. Immunity 36:873–884

Shafer-Weaver KA, Watkins SK, Anderson MJ, Draper LJ, Malyguine A, Alvord WG et al (2009) Immunity to murine prostatic tumors: continuous provision of T-cell help prevents CD8 T-cell tolerance and activates tumor-infiltrating dendritic cells. Cancer Res 69:6256–6264

Sharma MD, Baban B, Chandler P, Hou DY, Singh N, Yagita H et al (2007) Plasmacytoid dendritic cells from mouse tumor-draining lymph nodes directly activate mature Tregs via indoleamine 2,3-dioxygenase. J Clin Invest 117:2570–2582

Shibaki A, Katz SI (2001) Activation through CD40 ligation induces functional Fas ligand expression by Langerhans cells. Eur J Immunol 31:3006–3015

Shurin MR (1996) Dendritic cells presenting tumor antigen. Cancer Immunol Immunother 43:158–164

Shurin MR, Pandharipande PP, Zorina TD, Haluszczak C, Subbotin VM, Hunter O et al (1997) FLT3 ligand induces the generation of functionally active dendritic cells in mice. Cell Immunol 179:174–184

Shurin MR, Shurin GV, Lokshin A, Yurkovetsky ZR, Gutkin DW, Chatta G et al (2006) Intratumoral cytokines/chemokines/growth factors and tumor infiltrating dendritic cells: friends or enemies? Cancer Metastasis Rev 25:333–356

Siegal FP, Kadowaki N, Shodell M, Fitzgerald-Bocarsly PA, Shah K, Ho S et al (1999) The nature of the principal type 1 interferon-producing cells in human blood. Science 284:1835–1837

Silk KM, Silk JD, Ichiryu N, Davies TJ, Nolan KF, Leishman AJ et al (2011) Cross-presentation of tumour antigens by human induced pluripotent stem cell-derived CD141(+)XCR1(+) dendritic cells. Gene Ther 10:1035–1040

Silk KM, Leishman AJ, Nishimoto KP, Reddy A, Fairchild PJ (2012) Rapamycin conditioning of dendritic cells differentiated from human ES cells promotes a tolerogenic phenotype. J Biomed Biotechnol 2012:172420

Sims GP, Rowe DC, Rietdijk ST, Herbst R, Coyle AJ (2010) HMGB1 and RAGE in inflammation and cancer. Annu Rev Immunol 28:367–388

Sinha P, Okoro C, Foell D, Freeze HH, Ostrand-Rosenberg S, Srikrishna G (2008) Proinflammatory S100 proteins regulate the accumulation of myeloid-derived suppressor cells. J Immunol 181:4666–4675

Somasundaram R, Swoboda R, Caputo L, Otvos L, Weber B, Volpe P et al (2006) Human leukocyte antigen-A2-restricted CTL responses to mutated BRAF peptides in melanoma patients. Cancer Res 66:3287–3293

Sotomayor EM, Borrello I, Rattis FM, Cuenca AG, Abrams J, Staveley-O'Carroll K et al (2001) Cross-presentation of tumor antigens by bone marrow-derived antigen-presenting cells is the dominant mechanism in the induction of T-cell tolerance during B-cell lymphoma progression. Blood 98:1070–1077

Sriskandan K, Garner P, Watkinson J, Pettingale KW, Brinkley D, Calman FM et al (1986) A toxicity study of recombinant interferon-gamma given by intravenous infusion to patients with advanced cancer. Cancer Chemother Pharmacol 18:63–68

Stahl M, Dijkers PF, Kops GJ, Lens SM, Coffer PJ, Burgering BM et al (2002) The forkhead transcription factor FoxO regulates transcription of p27Kip1 and Bim in response to IL-2. J Immunol 168:5024–5031

Stauffer JK, Scarzello AJ, Jiang Q, Wiltrout RH (2012) Chronic inflammation, immune escape, and oncogenesis in the liver: A unique neighborhood for novel intersections. Hepatology 4:1567–1574 (Baltimore, Md)

Staveley-O'Carroll K, Schell TD, Jimenez M, Mylin LM, Tevethia MJ, Schoenberger SP et al (2003) In vivo ligation of CD40 enhances priming against the endogenous tumor antigen and promotes CD8+ T cell effector function in SV40 T antigen transgenic mice. J Immunol 171:697–707

Steinman RM, Cohn ZA (1973) Identification of a novel cell type in peripheral lymphoid organs of mice. I. Morphology, quantitation, tissue distribution. J Exp Med 137:1142–1162

Steinman RM, Lustig DS, Cohn ZA (1974) Identification of a novel cell type in peripheral lymphoid organs of mice. 3. Functional properties in vivo. J Exp Med 139:1431–1445

Steinman RM, Hawiger D, Liu K, Bonifaz L, Bonnyay D, Mahnke K et al (2003) Dendritic cell function in vivo during the steady state: a role in peripheral tolerance. Ann NY Acad Sci 987:15–25

Stepkowski SM, Chen H, Daloze P, Kahan BD (1991) Rapamycin, a potent immunosuppressive drug for vascularized heart, kidney, and small bowel transplantation in the rat. Transplantation 51:22–26

Stoitzner P, Green LK, Jung JY, Price KM, Atarea H, Kivell B et al (2008) Inefficient presentation of tumor-derived antigen by tumor-infiltrating dendritic cells. Cancer Immunol Immunother 57:1665–1673

Suss G, Shortman K (1996) A subclass of dendritic cells kills CD4 T cells via Fas/Fas-ligand-induced apoptosis. J Exp Med 183:1789–1796

Tacken PJ, Zeelenberg IS, Cruz LJ, van Hout-Kuijer MA, van de Glind G, Fokkink RG et al (2011) Targeted delivery of TLR ligands to human and mouse dendritic cells strongly enhances adjuvanticity. Blood 118:6836–6844

Tang H, Guo Z, Zhang M, Wang J, Chen G, Cao X (2006) Endothelial stroma programs hematopoietic stem cells to differentiate into regulatory dendritic cells through IL-10. Blood 108:1189–1197

Topalian SL, Drake CG, Pardoll DM (2012a) Targeting the PD-1/B7-H1(PD-L1) pathway to activate anti-tumor immunity. Curr Opin Immunol 24:207–212

Topalian SL, Hodi FS, Brahmer JR, Gettinger SN, Smith DC, McDermott DF et al (2012b) Safety, activity, and immune correlates of anti-PD-1 antibody in cancer. N Engl J Med 366:2443–2454

Tsujitani S, Kakeji Y, Watanabe A, Kohnoe S, Maehara Y, Sugimachi K (1990) Infiltration of dendritic cells in relation to tumor invasion and lymph node metastasis in human gastric cancer. Cancer 66:2012–2016

Tsukamoto N, Okada S, Onami Y, Sasaki Y, Umezawa K, Kawakami Y (2009) Impairment of plasmacytoid dendritic cells for IFN production by the ligand for immunoglobulin-like transcript 7 expressed on human cancer cells. Clin Cancer Res: J Am Assoc Cancer Res 15:5733–5743

Turnquist HR, Raimondi G, Zahorchak AF, Fischer RT, Wang Z, Thomson AW (2007) Rapamycin-conditioned dendritic cells are poor stimulators of allogeneic CD4+ T cells, but enrich for antigen-specific Foxp3+ T regulatory cells and promote organ transplant tolerance. J Immunol 178:7018–7031

Ureta G, Osorio F, Morales J, Rosemblatt M, Bono MR, Fierro JA (2007) Generation of dendritic cells with regulatory properties. Transpl Proc 39:633–637

Vacca C, Fallarino F, Perruccio K, Orabona C, Bianchi R, Gizzi S et al (2005) CD40 ligation prevents onset of tolerogenic properties in human dendritic cells treated with CTLA-4-Ig. Microbes Infect 7:1040–1048

Valencia J, Hernandez-Lopez C, Martinez VG, Hidalgo L, Zapata AG, Vicente A et al (2011) Wnt5a skews dendritic cell differentiation to an unconventional phenotype with tolerogenic features. J Immunol 187:4129–4139

Vicari AP, Chiodoni C, Vaure C, Ait-Yahia S, Dercamp C, Matsos F et al (2002) Reversal of tumor-induced dendritic cell paralysis by CpG immunostimulatory oligonucleotide and anti-interleukin 10 receptor antibody. J Exp Med 196:541–549

Villadangos JA, Shortman K (2010) Found in translation: the human equivalent of mouse CD8+ dendritic cells. J Exp Med 207:1131–1134

von Bergwelt-Baildon MS, Popov A, Saric T, Chemnitz J, Classen S, Stoffel MS et al (2006) CD25 and indoleamine 2,3-dioxygenase are up-regulated by prostaglandin E2 and expressed by tumor-associated dendritic cells in vivo: additional mechanisms of T-cell inhibition. Blood 108:228–237

von Marschall Z, Scholz A, Cramer T, Schafer G, Schirner M, Oberg K et al (2003) Effects of interferon alpha on vascular endothelial growth factor gene transcription and tumor angiogenesis. J Natl Cancer Inst 95:437–448

Wang JM, Deng X, Gong W, Su S (1998) Chemokines and their role in tumor growth and metastasis. J Immunol Methods 220:1–17

Wang L, Pino-Lagos K, de Vries VC, Guleria I, Sayegh MH, Noelle RJ (2008) Programmed death 1 ligand signaling regulates the generation of adaptive Foxp3+ CD4+ regulatory T cells. Proc Natl Acad Sci USA 105:9331–9336

Wang C, Gao D, Guo K, Kang X, Jiang K, Sun C et al (2012) Novel synergistic antitumor effects of rapamycin with bortezomib on hepatocellular carcinoma cells and orthotopic tumor model. BMC Cancer 12:166

Watkins SK, Zhu Z, Riboldi E, Shafer-Weaver KA, Stagliano KE, Sklavos MM et al (2011) FOXO3 programs tumor-associated DCs to become tolerogenic in human and murine prostate cancer. J Clin Invest. 121:1361–1372

Widelitz R (2005) Wnt signaling through canonical and non-canonical pathways: recent progress. Growth Factors 23:111–116 (Chur, Switzerland)

Yamazaki S, Bonito AJ, Spisek R, Dhodapkar M, Inaba K, Steinman RM (2007) Dendritic cells are specialized accessory cells along with TGF- for the differentiation of Foxp3+ CD4+ regulatory T cells from peripheral Foxp3 precursors. Blood 110:4293–4302

Yang DH, Park JS, Jin CJ, Kang HK, Nam JH, Rhee JH et al (2009) The dysfunction and abnormal signaling pathway of dendritic cells loaded by tumor antigen can be overcome by neutralizing VEGF in multiple myeloma. Leuk Res 33:665–670

Yang L, Pang Y, Moses HL (2010) TGF-beta and immune cells: an important regulatory axis in the tumor microenvironment and progression. Trends Immunol 31:220–227

Zeid NA, Muller HK (1993) S100 positive dendritic cells in human lung tumors associated with cell differentiation and enhanced survival. Pathology 25:338–343

Zhang Y, Chung Y, Bishop C, Daugherty B, Chute H, Holst P et al (2006) Regulation of T cell activation and tolerance by PDL2. Proc Natl Acad Sci USA 103:11695–11700

Zhang X, Munegowda MA, Yuan J, Wei Y, Xiang J (2010) Optimal TLR9 signal converts tolerogenic CD4-8- DCs into immunogenic ones capable of stimulating antitumor immunity via activating CD4+ Th1/Th17 and NK cell responses. J Leukoc Biol 88:393–403

Chapter 19
The Role of Tumor Associated Neutrophils in Cancer

Zvi G. Fridlender

Abstract Neutrophils play an established role in host defense and in killing invading microorganisms. As so, they are traditionally considered in the context of their anti-bacterial functions. It is becoming increasingly clear that tumor-associated neutrophils (TAN) play a major role in cancer biology, and especially in the tumor immune-microenvironment. Neutrophils, the major immune cell in the blood in humans, make up also a significant portion of the inflammatory cell infiltrate in many types of cancer. Like all other leukocytes, they move into tissues under the influence of specific chemokines, cytokines and cell adhesion molecules, most of them coming from the tumor microenvironment, being responsible for their recruitment into the tumor. We have found that TAN are a distinct population of neutrophils, differing markedly in their transcriptomic profile from both naïve neutrophils and the granulocytic fraction of myeloid-derived suppressor cells (G-MDSC). Furthermore, we recently found that neutrophils in the tumor develop to have a pro-tumorigenic phenotype during tumor progression. Studies have demonstrated specific examples of tumor-mediated signals (such as transforming growth factor-β [TGF-β]) that induce the formation of a pro-tumorigenic (N2) phenotype capable of supporting tumor growth and suppressing the anti-tumor immune response. Other studies show that TAN can also have an anti-tumorigenic (N1) phenotype. We explore here the literature on the different mechanisms of TAN-recruitment to tumors, the unique characteristics of TAN in animals and humans, and what shapes their pro- and/or anti-tumor effects.

Z. G. Fridlender (✉)
Center for COPD and Smoking Damages, Laboratory of Lung Cancer Research,
Institute of Pulmonary Medicine, Hadassah-Hebrew University Medical Center,
P.O. Box 12000, 91120 Jerusalem, Israel
e-mail: fridlender@hadassah.org.il

Z. G. Fridlender
Thoracic Oncology Research Laboratory, University of Pennsylvania,
Philadelphia, PA, USA

M. R. Shurin et al. (eds.), *The Tumor Immunoenvironment*,
DOI: 10.1007/978-94-007-6217-6_19,

Keywords N1 · N2 neutrophils · MDSC · TAN · TGF-β · VEGF · Neoangiogenesis · Monocytic · Granulocytic · Chemokines

19.1 Introduction: Neutrophils in Cancer

Neutrophils are the predominant circulating leukocyte population in humans, accounting for 50–70 % of circulating leukocytes. They are absolutely essential to protect humans and animals from microbial pathogens. When examining tumors by immuno-histo-chemistry (IHC), neutrophils can be seen in close association with tumor cells and within tumor vasculature (Welch et al. 1989). The exact role of neutrophils in the tumor cell microenvironment, however, is the subject of controversy.

The interest in the role of neutrophils in cancer increased during the late 80s and early 90s, as can be seen from an increase in the percentage of publications related to neutrophils out of all cancer-related publications (Fig. 19.1). However, with the rise of interest in T cell biology, and because no clear roles for neutrophils in cancer were defined, a gradual decrease in the interest in these major cells of the immune system was noted in the following decade, even as new immunotherapy modalities were developing. Interest has increased lately, as recent data suggests more important and significant roles for neutrophils in tumor biology than generally reflected in the literature (Piccard et al. 2011; Mantovani et al. 2011). A decade-long dominance of studies on the myeloid cells, dealing exclusively with the contribution of macrophages to tumor progression, has occurred despite the abundant and long-standing data indicating that intra-tumoral neutrophils are correlated with poorer prognosis in cancer patients (Quigley and Deryugina 2012).

Neutrophils play a well-established role in host defense, where they extravasate from the circulation and enter tissues (Heifets 1982). There, they phagocytose and kill invading microorganisms (such as bacteria and fungi) by releasing reactive oxygen species, activating cytokines (e.g., TNF-α, IL-1, interferons, etc.,) and

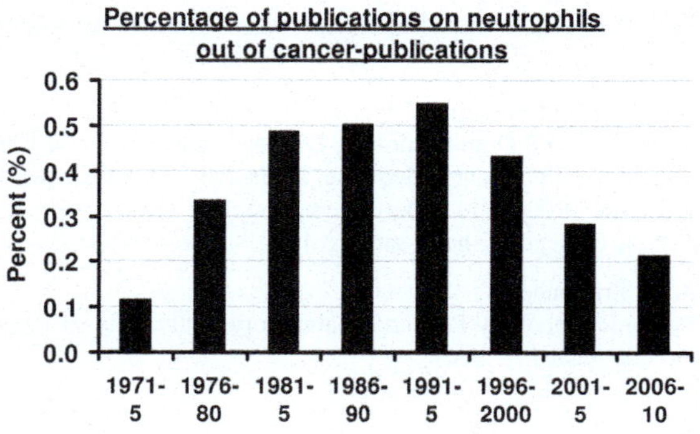

Fig. 19.1 The percentage of publications that include the MESH terms neutrophils or granulocytes out of all publications related to cancer for every five years between the years 1971 and 2010

defensins. Although neutrophils are traditionally considered in the context of their anti-bacterial functions, it is becoming clear that neutrophils in tumors and their myeloid precursors in the spleen, bone marrow and blood, play an important role in cancer biology (Piccard et al. 2011; Mantovani et al. 2011; Shojaei et al. 2008; Fridlender et al. 2009; Gregory and Houghton 2011; Peranzoni et al. 2010). Due to their very short life span, it is challenging to isolate and study tumor neutrophils in general, and particularly human TAN. For that reason, most of the data on the characteristics, phenotype and function of TAN was elucidated using animal models. The current knowledge on human TAN is summarized at the last part of this review.

In contrast to the well-described ability of inflammatory neutrophils to engulf bacteria, activate the immune system, and induce tissue damage in infections (Ashtekar and Bhaskar 2003), it has now become apparent that myeloid cells can also directly kill tumor cells (Fridlender et al. 2009; Granot et al. 2011), but also function as immunosuppressive cells in the context of tumors (Nagaraj et al. 2010). This property has been very well described in recent years for the so called "myeloid-derived suppressor cells" (MDSC) found in large quantities in the spleens of tumor-bearing animals (Nagaraj et al. 2010; Youn et al. 2008; Movahedi et al. 2008; Gabrilovich and Nagaraj 2009) and for tumor-associated macrophages which can develop an "M2" or tumor-supportive phenotype (Mantovani et al. 2002; Yunping et al. 2006). More recently, we have demonstrated that in untreated tumors, neutrophils can also assume a pro-tumorigenic state, which by analogy to macrophages, we called the "N2" phenotype (Fridlender et al. 2009). The full range of mechanisms responsible for this pro-tumorigenic activity have not yet been elucidated, but neutrophils are known to impact angiogenesis, immune surveillance, as well as to secrete chemokines, cytokines and reactive oxygen species (Gregory and Houghton 2011). We, however, also noted that under certain conditions (for example, after TGF-β blockade), TAN can take on an "N1" phenotype, which is pro-inflammatory and anti-tumorigenic. When interpreting the literature, it is important to consider which of these two possible phenotypes predominate.

19.2 Recruitment of Neutrophils into Tumors

As discussed above, neutrophils can make up a significant portion of the inflammatory cell infiltrate in many models of cancer, and it has been shown that tumor cells themselves mediate neutrophil recruitment to the tumor (Gregory and Houghton 2011). When neutrophils traffic into tumors, they are usually referred to as tumor-associated neutrophils, TAN. In mice, TAN can be defined by the specific surface markers CD11b and Ly6G with low expression of macrophage markers such as F4/80 (Wang et al. 2009)

Neutrophils, like all other leukocytes, move into tissues from the blood under the influence of specific chemokines (e.g., KC/CXCL-1 and MIP2α/CXCL2), cytokines (e.g., TNF-α and IFN-γ), and cell adhesion molecules located on their own surface (e.g., CD11b) and on the surface of endothelial cells (e.g., selectins, ICAM-1 and PECAM-1) (Kobayashi 2008). Growth factors may also have a role in the chemoattraction of neutrophils into tumors. G-CSF has been shown to indirectly support the extravasation of neutrophils from the blood, mostly by activating the CXCL2-CXCR2 axis (Wengner et al. 2008; Eash et al. 2010). Interestingly, production of G-CSF by human carcinomas has been reported, but seems to be a rare event (Joshita et al. 2009). GM-CSF is well known to have priming and anti-apoptotic effects on neutrophils, inducing neutrophilia. More recent, studies demonstrated that GM-CSF can also be a strong chemoattractant for neutrophils in vivo (Khajah et al. 2011; Gomez-Cambronero et al. 2003), although this has not yet been shown specifically for their recruitment into tumors.

Ueha et al. (2011) have recently summarized the dynamics of myeloid cells, including neutrophils, from the bone marrow to the circulation and into mouse tumors. CXCR2 and CXCR4 were shown to cooperatively regulate the release of neutrophils from bone marrow (Eash et al. 2010). Ueha further suggested that tumor-infiltration by neutrophils was at-least partly mediated by autocrine CXCL2 production. Indeed, in our recent work using microarrays (Fridlender et al. 2012) (see below), we found that the expression of CXCL2 was up-regulated by 188 fold in TAN compared to bone marrow neutrophils. Similar results were found for two other known neutrophil chemoattractants—CXCL1 (140-fold compared to bone marrow neutrophils), and CCL3 (76-fold in TAN compared to bone marrow neutrophils) (Fridlender et al. 2012). It seems therefore, that the neutrophils initiate a positive feedback loop by secreting neutrophil chemoattractants that recruit more neutrophils into the tumor, as previously described in infections (Appelberg 1992; Kobayashi 2006).

The most potent neutrophil chemoattractant into tumors in humans seems to be interleukin-8 (IL-8/CXCL8), promoting neutrophil recruitment through the receptors CXCR1 and CXCR2 (Brandau and Dumitru 2012; Trellakis et al. 2011). Recently, Zhou et al. (2012) found CXCL-5, another chemokine, to be over-expressed in hepatocellular carcinoma patients with recurrent disease, and shorter overall survival. This over-expression was correlated with high neutrophil infiltration into the tumor, and the data was supported by direct demonstration of CXCL-5 dependent recruitment of neutrophils into murine tumors. Another chemokine recently shown to be important in the recruitment of neutrophils to melanoma tumors is GCP-2 (CXCL6). Specific anti-CXCL6 mAbs not only reduced recruitment of neutrophils to tumor sites in mice, but also caused eventual reduction in tumor growth (Verbeke et al. 2011).

We have shown that TGF-β receptor blockade increases the number of neutrophils in tumors and that this effect occurs through all three parts of the recruitment pathway including increased expression of mRNA for CXC chemokines, CC chemokines and activating cytokines within the tumor, as well as up-regulating ICAM-1 message and protein expression on endothelial cells (Fridlender et al. 2009; Kim et al. 2008). Our preliminary data shows that macrophages, as well as endothelial cells, are an important player in neutrophil recruitment, as previously shown in lung inflammation (Maus et al. 2003). Interestingly, TGF-β

appears to inhibit endothelial adhesiveness for neutrophils and neutrophil transmigration through endothelium in vitro (Smith et al. 1996) and in different inflammatory disease states (Allen et al. 2008).

Although it has been shown that $CD8^+$ T cell depletion decreases the tissue influx of neutrophils in infectious diseases (Appelberg 1992), there are surprisingly few studies examining the effect of $CD8^+$ T cells on the recruitment of neutrophils in cancer. In the only tumor study we were able to identify, a marked decrease in TAN following $CD8^+$ T cell depletion was shown in a model in which CT26 colon carcinoma cells transduced to express G-CSF were placed into mice (Stoppacciaro et al. 1993). The mechanisms by which T cells might attract and/or activate neutrophils are not known for certain, but include the ability of tumor-stimulated activated T cells to produce GM-CSF (Aruga et al. 1997), KC/CXCL-1 and MIP2α/CXCL2 (Sherwood et al. 2004), or cytokines such as TNF-α and IFN-γ. These cytokines may act to recruit neutrophils by stimulating tumor macrophages or endothelial cells to produce appropriate chemokines and cell adhesion molecules (Maus et al. 2003; Iking-Konert et al. 2008). Recently Richards et al. demonstrated that T-regulatory cells also can play a role in the inhibition of neutrophil-recruitment to a site of tumor inoculation. This effect was shown to be mediated by decreased expression of the neutrophil chemoattractants CXCL1 and CXCL2 (Richards et al. 2010). In contrast, Himmel et al. (2011) showed that human T-regulatory cells can actually promote recruitment of neutrophils by secretion of IL-8. A simplified scheme of the factors influencing recruitment of neutrophils into the tumor is shown in Fig. 19.2.

19.3 Unique Characteristics of TAN

"Immunosculpting", i.e., the crosstalk between immune and tumor cells changing the phenotype of tumor biology, is widely recognized (Reiman et al. 2007). However, until recently, the role of neutrophils in this cross-talk has been underestimated. In recent years, several studies have shown specific examples of tumor-mediated signals eliciting pro-tumor responses from neutrophils (Gregory and Houghton 2011). One interesting illustration of these effects was shown by Queen et al. (2005), who demonstrated that cancer cells can stimulate neutrophils to produce oncostatin-M, which in turn increases secretion of VEGF by tumors.

It is worth considering the relationship between TAN and the granulocytic fraction of myeloid-derived suppressor cells (G-MDSC). MDSC, a heterogeneous population of immune suppressive cells that are produced at high levels in cancer, are defined in mice on the basis of expression of the surface markers CD11b and Gr-1 and by their ability to inhibit T-lymphocyte activation. The $CD11b^+/GR1^+$ MDSC population is comprised of at least two subsets—granulocytic (Ly6G+) and monocytic cells (Ly6C+), possibly with different immunosuppressive properties (Peranzoni et al. 2010).

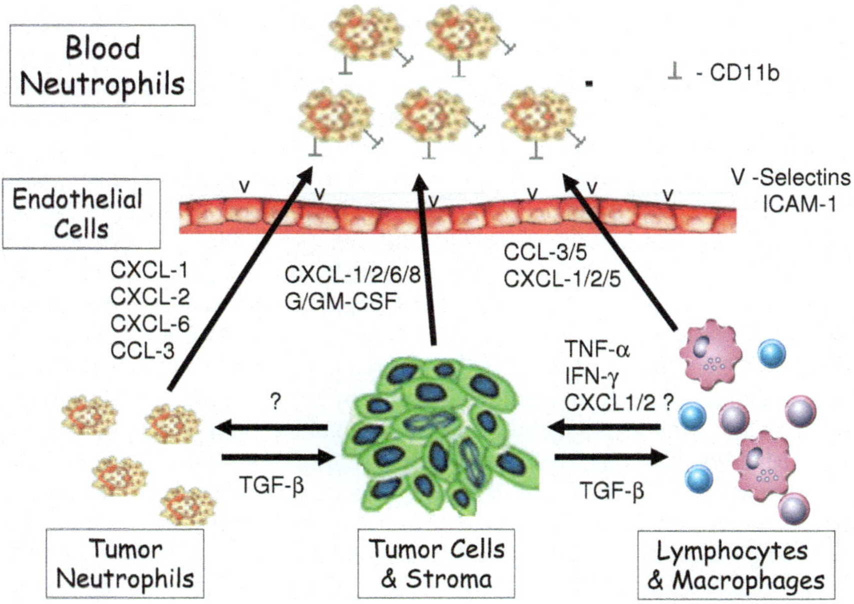

Fig. 19.2 A simplified schematic representation of the cells and factors influencing recruitment of neutrophils into the tumor

There is substantial agreement on the immunosuppressive activity of the monocytic MDSC subset. However, there is still contrasting evidence on the effects of the granulocytic fraction. Whereas some have shown that granulocytic MDSC have immunosuppression properties similar to the monocytic fraction (Youn et al. 2008; Movahedi et al. 2008; Gabrilovich and Nagaraj 2009), others have recently demonstrated that they are less immunosuppressive (Peranzoni et al. 2010; Morales et al. 2010; Dolcetti et al. 2010). It has been previously shown that adoptively transferred MDSC can enter tumors and differentiate to mature macrophages (TAM) or neutrophils (TAN) (Kusmartsev et al. 2005). However, little is known about whether MDSC leave the spleen and circulate. It is thus not clear whether the majority of TAN are actually G-MDSC that were attracted to the tumor or whether they are bone marrow/blood-derived neutrophils that were then converted to N2 TAN by the tumor microenvironment. In order to further evaluate the specific characteristics of TAN in relationship to other populations of neutrophils, we recently used a transcriptomics approach, comparing the phenotype of TAN to naïve neutrophils from the bone marrow (NN) and to the granulocytic fraction of myeloid derived suppressor cells (G-MDSC) (Fridlender et al. 2012). In our microarray study, we clearly show that TAN are not "tissue-based G-MDSC", but are a distinct population of neutrophils, differing markedly in their genetic profile from both NN and G-MDSC, with the NN and G-MDSC being more closely related to each other than to TAN. Some of the unique characteristics of TAN compared to NN and G-MDSC that were described in this work, and by others are:

Structural Genes: Genes related to cytoskeleton organization and biogenesis, as well as in pathways related to actin binding and polymerization were down regulated in TAN, consistent with their loss of ability to leave the tumor microenvironment after infiltrating the tumor.

Cytotoxic and Phagocytic Genes: The two cytotoxic pathways of neutrophils, granule proteins production and the respiratory burst (Eash et al. 2010), are dramatically down-regulated in TAN. Interestingly, Shen et al. (2007) have shown that TGF-β, of which high levels are found in the tumor microenvironment, can inhibit neutrophils degranulation. An alternative explanation for these findings could be that the mature neutrophils have finished producing granule contents, and the relevant mRNAs are not needed (Theilgaard-Monch et al. 2005). In contrast to these findings, we noted no clear changes in the pathways and genes related to phagocytosis, another major function of neutrophils. In a recent comparison between neutrophils from tumor-free mice and G-MDSC, Youn et al. (2012) found the neutrophils to be more mature, activated cells with high phagocytic activity, and high expression of lysosomal proteases.

Apoptosis: Despite data suggesting that TAN may be longer lasting cells than circulating neutrophils (Sawanobori et al. 2008), we found that most genes related to apoptosis were expressed at similar levels. However, several anti-apoptotic members of the NF-κB family were up-regulated in TAN. NF-κB may be, therefore, an important regulator of the apoptotic machinery in TAN and it is possible that this pathway is responsible for the notable longevity of TAN compare to other neutrophils. In another recent work, neutrophils were shown to express more anti-apoptotic Mcl-1 but less pro-apoptotic Bax, shortly after exposure to supernatant from tumor cells, resulting in sustained survival of them (Wu et al. 2011).

Immune system: It has become increasingly clear that the contribution of neutrophils to host defense and natural immunity extends well beyond their traditional role as professional phagocytes (Ashtekar and Bhaskar 2003). Neutrophils and their myeloid precursors can be induced to express a number of genes whose products lie at the core of inflammatory and immune responses, suggesting a potential role for these cells in orchestrating the sequential recruitment and activation of distinct leukocyte types to the inflamed tissue (Scapini et al. 2000). The effects of neutrophils on other components of the immune system in cancer was reviewed recently (Mantovani et al. 2011). Neutrophils in tumors, either spontaneously or following appropriate stimulation, have been shown to express and/or produce numerous cytokines, chemokines and angiogenic factors. These include CC- and CXC-chemokines (e.g., CXCL-1/2/5/10, CCL-2/4/17), pro- and anti-inflammatory cytokines (e.g., IL-1α/β, IL-6, IL-4), immunoregulatory cytokines (e.g., IL-12, IFN-α/γ), colony stimulating factors, angiogenic factors, and members of the TNF superfamily (Mantovani et al. 2011). However, as mentioned above, until recently the role of neutrophils in this cross-talk between immune and tumor cells has been under-estimated. Accumulating data shows that neutrophil can participate in MHC class I and class II restricted antigen presentation, are capable of collecting and cleaving antigens, of forming complexes with MHC-II molecules, and of expressing co-stimulatory molecules (Ashtekar and Bhaskar 2003; Van Gisbergen et al. 2005; Abi Abdallah et al. 2011). In our work, we also found that TAN show increased expression, compared to NN, of gene pathways needed to present antigens, suggesting an enhanced capability of functioning as antigen-presenting cells (APCs) (Fridlender et al. 2012). Indeed, it has been recently shown that mature neutrophils can function as professional antigen-presenting cells capable of priming a Th-1 and Th-17-acquired immune response (Abi Abdallah et al. 2011). In inflamed tissues, neutrophils have been shown to engage in complex bi-directional interactions with macrophages, dendritic cells (DC), natural killer (NK) cells, lymphocytes and mesenchymal stem cells, affecting proliferation, activation, differentiation and survival (Mantovani et al. 2011). This has not been appreciated significantly in the tumor microenvironment.

The most prominent difference that we found between TAN and the other populations of neutrophils was the significant up-regulation of cytokines and chemokines, suggesting an important role of tumor neutrophils in the recruitment of immunocytes and in the balance between activation and suppression of the immune system. Among the broad group of chemokines whose mRNAs were upregulated in TAN were the CCL chemokines 2, 3, 4, 8, 12, and 17 and the CXCL chemokines 1, 2, 9, and 16. The up-regulation of chemokines in TAN suggests that they have a pivotal role in recruiting other cells of the immune system to the tumor. This is similar to the role that "classical neutrophils" would have in wound healing. At least some of the recruited cells are known to support tumor growth, such as macrophages (by CCL2 and CCL7) and T-regulatory cells (by CCL17) (Curiel et al. 2004).

Table 19.1 highlights the major pathways and group genes that we found to be significantly different in one of the 3 neutrophil populations examined (NN, G-MDSC and TAN) compared to the others (Fridlender et al. 2012).

Table 19.1 Major pathways and gene-groups differing Neutrophil populations

Neutrophil function		Naïve neutrophils	G-MDSC	TAN
Granule proteins	Primary	High	Low	Low
	Secondary	High	Mod	Low
	Tertiary	High	Mod	Low
Respiratory burst	Peroxidase	High	High	Low
	NADPH complex	High	Mod	Low
	TLR	Low	High	Mod
Structural genes	Actin binding	Mod	Mod	Low
	Cytoskeleton	Mod	Mod	Low
Apoptosis	Intrinsic (BCL2) pathway	Mod	High (BH-3)	Low
	NF-KB—Anti apoptotic	High	Low	Low
Immune system	Chemokines	Low	Mod	High
	Cytokine activity	Low	Mod	High
	APC genes	Low	High	High

Mod = Moderate

19.4 The Tumor-Supportive Roles of TAN

As previously mentioned, we and others have noted that in untreated tumors, TAN appear to develop a pro-tumorigenic phenotype that we have termed "N2 TAN" in analogy to the M2 macrophage phenotype that appears to contribute to tumor growth (Shojaei et al. 2008; Fridlender et al. 2009; Pekarek et al. 1995; Tazawa et al. 2003) and suppression of the anti-tumor immune response (Schmielau and Finn 2001). Depletion of these "pro-tumorigenic" N2 neutrophils, therefore, inhibits tumor growth (Fridlender et al. 2009; Pekarek et al. 1995; Nozawa et al. 2006) and reduces the level of immunosuppression in the tumor microenvironment, allowing, for example, increased activity of $CD8^+$ cytotoxic T-lymphocytes (CTL) (Fridlender et al. 2009). TAN appear to be involved in tumorigenesis and tumor growth through multiple mechanisms:

Initiation, carcinogenesis, and tumor growth: Neutrophils have a dual role in the initiation process of tumors, mainly by affecting the extra-cellular matrix (ECM), and the neoplastic cells microenvironment (Piccard et al. 2011). A key mediator secreted from neutrophils and involved in carcinogenesis is MMP9. It was demonstrated to be involved in skin tumorigenesis, by regulation of oncogene-induced keratinocyte hyperproliferation and their progression to invasive cancer (Coussens and Werb 2002). In addition to its effects on matrix, it has been shown that MMP-9 secreted from neutrophils (among other bone-marrow cells) prevents apoptosis of tumor cells in the lung (Acuff et al. 2006). Other factors secreted from neutrophils that were demonstrated to indirectly support tumor growth by additional recruitment of leukocytes are KC/CXCL1 (using the CXCR2 axis) (Loukinova et al. 2000), as well as CXCL8 and CCL3 (Hirose et al. 1995). Another potential direct effect of neutrophils on tumor growth is through secretion of neutrophil elastase (NE) which enters tumor cells, where it binds IRS-1 allowing increased activation of AKT (Houghton et al. 2010).

Angiogenesis: Mounting evidence supports the role of neutrophils in the important pro-tumorigenic process of angiogenesis and neovascularization (Tazzyman et al. 2009), at least partially mediated by the secretion of chemokines and MMPs (Jablonska et al. 2010). Neutrophils have been implicated in angiogenesis in studies where passive immunization against IL-8 attenuated tumor growth and angiogenic response in mice with lung tumors (De Larco et al. 2003, 2004). MMP-9 secreted from neutrophils, in addition to its role in carcinogenesis, was shown to have a pivotal role in switching on angiogenesis, by counteracting anti-angiogenic molecules, and possibly also by promoting the release of VEGF (Nozawa et al. 2006; Kuang et al. 2011), although Shojaei et al. (2007) elegantly demonstrated that neutrophils can render tumors refractory to anti-VEGF Abs, by mediating a pathway of angiogenesis that bypasses VEGF. It has even been suggested that MMP-9 secreted from neutrophils is mandatory for vascularization in vivo (Masson et al. 2005). Other factors in neutrophils that could contribute to angiogenesis include CXCL6 (Gijsbers et al. 2005), and as previously mentioned, CXCL1/KC, signaling through the chemokine receptor CXCR2 on endothelial cells

(Loukinova et al. 2000). Jablonska et al. (2010) have recently shown that these pro-angiogenic factors of TAN might be regulated by levels of IFN-β. Angiogenesis appears to be increased in the absence of IFN-β, and is reduced back to normal levels following re-exposure to IFN-β, possibly explaining the therapeutic effect of IFN treatment during the early stages of cancer development. It is possible that this is a major difference when N2 TAN is changed to N1 TAN (Piccard et al. 2011). Interestingly, our data show that the level of MMP9 mRNA, which is high in naïve neutrophils, is actually lost in later TAN, possibly suggesting that the important role of MMP9 is mainly at the early, rather than late stages of tumor development (Fridlender et al. 2012). As suggested by Quigley and Deryugina (2012) neutrophils appear to be functionally positioned upstream of VEGF-induced angiogenesis, and therefore targeting neutrophil influx into the tumor at early stages, could be an anti-cancer approach to inhibit MMP-9 triggered angiogenesis.

Extravasation and metastases: Studies have shown indirectly that neutrophils are associated with more lung metastases in skin squamous cell carcinoma (Loukinova et al. 2000) and in melanoma (Schaider et al. 2003). More than two decades ago, Welch et al. (1989) demonstrated that neutrophils elicited from tumors (nowadays called TAN) secrete high levels of the basement-membrane (BM) degrading enzymes collagenase-IV and heparanase, assisting tumor cell extravasation during the metastatic process. TAN caused a dose-dependent increase in invasion through a reconstituted BM barrier in an in vitro invasion assay, and significantly increased metastases, whereas circulating normal PMNs did not significantly alter invasive or metastatic potential. Subsequently, it has been shown in humans that neutrophils can augment the capability of tumor cells to extravasate through endothelium (EC) (Wu et al. 2001), allowing, for example, adherence of tumor cells in the lungs (Orr et al. 2001). One important protease that has been attributed to this function is the major inducer of tissue damage, neutrophil elastase (NE), with an inverse correlation between its levels and prognosis observed (Sato et al. 2006). High levels of NE were shown to promote invasion and metastasis of several cancers by degrading extracellular matrix and promoting tumor invasion (Sun and Yang 2004). Furthermore, inhibition of NE was shown to reduce hepatic metastases (Doi et al. 2002). Neutrophils were further shown to release growth factors which enhance invasion, such as oncostatin-M, that except for stimulating the secretion of VEGF as detailed above, stimulated tumor cells to detach and invade the surrounding Matrigel (Queen et al. 2005), and HGF, which enhanced the invasive properties of the hepatocellular tumor cells in a feedback manner (Imai et al. 2005).

Another mechanism by which neutrophils promote metastasis has been demonstrated recently in vivo by Huh et al. (2010). Circulating tumor cells were shown to directly anchor to the vascular endothelium, facilitating transendothelial migration of tumor cells, extravasation and formation of new metastases. Injection of neutrophils in this model increased significantly cancer cell retention.

Suppression of the adaptive immune system: The ability of neutrophils to influence $CD8^+$ T cells has been suggested in infections (Tvinnereim et al. 2004) and in cancer (Di Carlo et al. 2001a; Kousis et al. 2007; Colombo et al. 1992a).

However, the activation state of the TAN needs to be considered carefully. Whereas N1 TAN were shown to promote recruitment and activation of immunocytes (see below), our data demonstrated that N2 neutrophils can inhibit T cell effector functions; neutrophil depletion of untreated tumor-bearing animals (i.e., removal of N2 TAN) increased the activation status of $CD8^+$ T cells, supporting the idea that N2 TAN can function in an immunosuppressive fashion (Fridlender et al. 2009) in the same way that has been proposed for M2 TAMs (Movahedi et al. 2008; Rodriguez et al. 2004). A possible suggested mechanism for this suppression of T cell proliferation and responsiveness to stimulation is by the secretion of stored arginase-1 (ARG1) which degrades extracellular arginine, a factor needed for the proper activity of T cells (Rotondo et al. 2009). Effects of neutrophils on the immune system were also recently demonstrated in head and neck tumors. Tumor cells induced secretion of CCL4, CXCL-8 and MMP9 by neutrophils, through activation of p38-MAPK (Dumitru et al. 2012a).

19.5 The Antitumor Effects of N1 TAN

Despite the broad literature on the pro-tumor effects of TAN reviewed above, there are several studies reporting anti-tumor roles for these cells, mostly with engineered tumor cell lines, or following specific therapies (Gregory and Houghton 2011). Interpretation of these studies in the light of the idea of differential neutrophil activation status within tumors is instructive. In contrast to the studies described above, in which depletion of neutrophils inhibited tumor growth, we found that neutrophils can assume a more tumor-cytotoxic N1 phenotype, for example, during TGF-β inhibition (Fridlender et al. 2009) or after immunologic or cytokine activation, where they have the potential to kill tumor cells and inhibit growth (Di Carlo et al. 2001a; Colombo et al. 1992b; Hicks et al. 2006). Depletion of these N1 TAN thus either augments tumor growth and/or blunts the anti-tumor effects of immunologic treatments (Fridlender et al. 2009; Stoppacciaro et al. 1993; Kousis et al. 2007; Suttmann et al. 2006).

In general, most of the anti tumor effects of neutrophils were described in association to anti-cancer therapies, as summarized recently by Brandau and colleagues (Brandau and Dumitru 2012):

1. During antibody-mediated tumor therapy, when TAN are activated via their Fc receptors, releasing anti-tumoricidal mediators. Several receptors have been implicated as mediating these effects, such as CD32 (FCγRII), and the antibody-dependent target cell lysis via FcαRI (CD89). Interestingly these 2 are up-regulated during the maturation of neutrophils, compared to immature or younger neutrophils. Antibody-dependent cellular cytotoxicity (ADCC), as a mechanism of anti-tumoricidal activity of neutrophils was demonstrated in several systems (Piccard et al. 2011; Hubert et al. 2011), for example as part of the mechanism of the epidermal growth factor receptor (EGF-R) antibodies—panitumumab and zalutumumab (Schneider-Merck et al. 2010).

2. In response to different pathogen-derived biologics used in cancer immunotherapy, by their reaction to pattern recognition receptors. Based on this mechanism in neutrophils and other cells of the immune system, synthetic and natural compounds, which mimic a microbial pathogen are being explored for cancer therapy (Mantovani et al. 2011; Kandasamy et al. 2011). An impressing example for this concept is the well known Bacillus Calmette–Guerin (BCG) immunotherapy of bladder cancer, Neutrophils are required for this therapy to be effective (Suttmann et al. 2006).
3. In several gene therapy approaches, probably by an acute inflammatory response.
4. Pharmacologic and immunomodulatory intervention directly affecting the phenotype of neutrophils, such as in the inhibition of TGF-β.

TAN have been shown to have effective anti-tumoricidal properties in several phases of tumor progression:

Tumorigenesis: Neutrophils were shown to be capable of protecting against tumor development, by secreting MMP-8. Loss of MMP-8, mainly arising from neutrophils, increased skin susceptibility to chemical carcinogens in mice. Ironically, this effect seems to be mediated by preventing an influx of neutrophils to the carcinogen injection site (Balbin et al. 2003).

Tumor cytotoxicity and inhibition of tumor growth: The direct killing of tumor cells by neutrophils was demonstrated in vitro (Gerrard et al. 1981) and in vivo (Katano and Torisu 1982) almost three decades ago. Neutrophils from tumor-bearing animals were shown to have an enhanced cytotoxicity profile as measured by superoxide anion generation and phagocytosis, inducing a marked decrease in the size and number of metastatic foci in the lung (Ishihara et al. 1998; Ishihara and Matsunaga 1998). We and others have demonstrated that oxidative damage caused by reactive oxygen species secreted from neutrophils are capable of inducing tumor cell lysis (Fridlender et al. 2009; Lichtenstein et al. 1989; Zivkovic et al. 2007). Interestingly, there seems to be a difference between the cytotoxicity of neutrophils to primary versus metastatic cells, the latter being less affected (Schaider et al. 2003). Furthermore, it is possible, as recently shown by Granot et al. (2011) that tumor-entrained neutrophils can actually inhibit metastatic seeding in the lungs, inducing a neutrophil-mediated inhibitory process at the metastatic site.

A second mechanism by which neutrophils were shown to be capable of directly inhibiting tumor cells is by mediating Fas-Ligand associated apoptosis (Chen et al. 2003). Forced expression of FasL in the tumor microenvironment led to apoptosis of immune cells, and recruitment of neutrophils (Dupont and Warrens 2007). This mechanism is also in line with our observation that an increased percentage of N1 TAN are Fas-positive (Fridlender et al. 2009).

Activation and proper direction of the adaptive immune system and tumor rejection: The ability of the adaptive immune system, and specifically the $CD8^+$ CTLs, to reject tumors is a key process for the success of any immunotherapy. As suggested above, N2 neutrophils can be major inhibitors of T cell effector functions in a similar way previously proposed for M2 TAM (Fridlender et al. 2009;

Movahedi et al. 2008; Kousis et al. 2007; Colombo et al. 1992a; Rodriguez et al. 2004; Di Carlo et al. 2001b). However, we and others have shown that N1 neutrophils can actually be immunostimulatory, supporting tumor rejection. These proinflammatory N1 neutrophils can promote CD8$^+$ recruitment and activation by producing T cell attracting chemokines (e.g., CCL3, CXCL9, and CXCL10) and pro-inflammatory cytokines (e.g., IL-12, TNF-α, and GM-CSF) (Fridlender et al. 2009; Scapini et al. 2000). Furthermore, neutrophils have been shown to cross-present antigens in vitro, and antigen-pulsed neutrophils promoted the activation of CD8$^+$ T cells (Beauvillain et al. 2007). Additional examples of this neutrophil-CD8$^+$ lymphocytes interaction include studies showing that photodynamic therapy-induced CD8$^+$ T cell induction required the presence of neutrophils (Kousis et al. 2007) and, as mentioned above, in the BCG immunotherapy of bladder cancer, where depletion of neutrophils significantly impaired T cell trafficking and reduced the efficacy of this immunotherapy (Suttmann et al. 2006).

There is also evidence that TAN can activate dendritic cells via cell–cell contact and through secretion of TNF-α (Van Gisbergen et al. 2005), activate CD4$^+$ T cells, promote anti-tumoral memory (Cavallo et al. 1992), and forward IL-12 induced tumor regression (Medina-Echeverz et al. 2011). Depletion of these N1 TAN thus either augments tumor growth and/or blunts the anti-tumor effects of immunologic treatments (Fridlender et al. 2009; Stoppacciaro et al. 1993; Kousis et al. 2007; Suttmann et al. 2006). Neutrophils can also be involved in the known "bystander" effect of anti-tumor treatment with oncolytic viruses. Breitbach et al. (2007) demonstrated that in vivo, most of the tumor killing activity of vesicular stomatitis (VSV) and vaccinia viruses is caused by indirect killing of uninfected tumor cells, mediated by an influx of neutrophils to the tumor, and that depletion of neutrophils inhibited their anti-tumor effects.

Figure 19.3 summarizes, in a simplified scheme, the pro-tumor and anti-tumor effects that have been described in neutrophils, ascribing these effects to N1 versus N2 TANs.

19.6 Neutrophils in Human Cancer

As mentioned above, relatively little is known about neutrophils in human cancers. Many patients with advanced cancer show high levels of blood neutrophilia (Schmidt et al. 2005). The mechanisms by which neutrophilia is induced by tumors is uncertain, although GM-CSF production has been implicated in some tumor systems, such as lung, melanoma, pancreas, and breast (McGary et al. 1995). Several additional cytokines secreted from tumors and stroma cells have been suggested to contribute to neutrophilia, and to the induction of suppressive properties of these neutrophils. These include, among others, G-CSF, VEGF, IL1-β and IL-6 (Lechner et al. 2010). In several large clinical trials it was shown that not only the number, but also the activity of peripheral neutrophils is altered in cancer patients (Schmidt et al. 2005; Teramukai et al. 2009; Dumitru et al. 2012b).

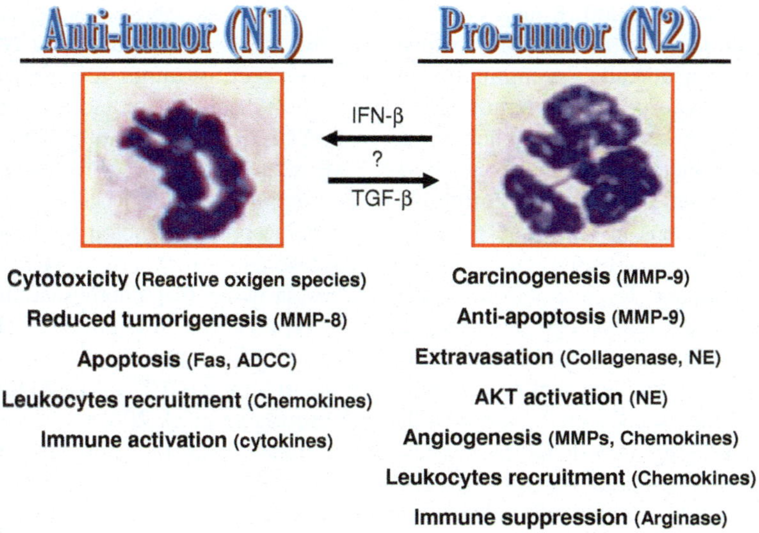

Fig. 19.3 A simplified schematic representation summarizing the pro-tumor and anti-tumor effects that have been described in neutrophils

Peripheral neutrophils in cancer patients, for example, were shown to have lower ROS production and oxidative burst capacity, as well as reduced spontaneous apoptosis (Trellakis et al. 2011; Uehara and Sato 1994). Others, however, have shown accelerated apoptosis rates, and an elevated caspase-8 activity in neutrophils isolated from cancer patients (Jablonska et al. 2009). The neutrophils themselves were also shown to have a changed secretion pattern of several cytokines, including increased secretion of IL-1, VEGF and IL-17, and reduced secretion of IL-18 and sTRAIL (Jablonska et al. 2005, 2008). Interestingly, in a recent evaluation of activation markers in neutrophils, Cheung et. al. found that activation of neutrophils with GM-CSF was associated with improved outcome in patients with neuroblastoma following immunotherapy with anti-GD2 monoclonal antibodies (Cheung et al. 2012).

In general, neutrophilia has been associated with poorer prognosis in many cancers, including bronchoalveolar carcinoma (Bellocq et al. 1998), metastatic melanoma (Schmidt et al. 2005) and renal carcinoma (Atzpodien and Reitz 2008). The neutrophil to lymphocyte ratio (NLR) has been introduced as a prognostic factor in many tumor types, including colorectal cancer (Walsh et al. 2005) and non-small cell lung cancer (Sarraf et al. 2009). In a recent work, circulating neutrophils were also shown to be an important cell in the pro-thrombotic stage accompanying cancer (Demers et al. 2012).

There is surprisingly little data about the presence of neutrophils within human tumors. In humans, IL-8 secreted by the tumor cells probably plays an important role in attracting neutrophils to the tumor microenvironment (De Larco et al. 2003,

2004; Sparmann and Bar-Sagi 2004). Intratumoral neutrophils were shown to be a strong, independent prognostic factor for recurrence-free, as well as cancer-specific and overall survival in metastatic (Donskov and von der Maase 2006) and in localized (Jensen et al. 2009) clear cell renal cell carcinoma, and in head and neck squamous cell carcinoma (HNSCC) (Trellakis et al. 2011). Infiltration of neutrophils was found to correlate with tumor grade in human gliomas (Fossati et al. 1999) and to be related to more aggressive types of pancreatic tumors (Reid et al. 2011). In gastric cancer, a positive correlation was noted between the density of tumor-infiltrating neutrophils and staging (Zhao et al. 2012), although high neutrophil count in this type of cancer has been associated by others with a favorable prognosis (Caruso et al. 2002). In a similar way as the peripheral blood NLR, it has been shown that the ratio of neutrophils to CD8$^+$ T cells inside the tumor, are also correlated with incidence of relapse and overall prognosis in non-small cell lung cancer (Llie et al. 2012).

We have recently examined a large number of human lung cancers for the presence of neutrophils (myeloperoxidase- positive cells) (Fig. 19.4). Striking heterogeneity was noted, with some tumors heavily infiltrated (Fig. 19.4a), some with moderate infiltration (Fig. 19.4b), and some with no neutrophils (Fig. 19.4c). The prognostic implication of neutrophil infiltration in these patients is under study.

Several effects of human tumor cells on neutrophils were demonstrated in vitro. HNSCC tumor-derived factors modulated cellular functions of PMNs and increased their inflammatory activity (e.g., secretion of MMP-9 and CCL-4 (Trellakis et al. 2011). This was shown to be at least partially mediated by Tumor-derived macrophage migration inhibitory factor (MIF) (Dumitru et al. 2011). The neutrophil-tumor interactions in humans at the molecular level have been only scarcely addressed. Several studies showed that tumor cells and neutrophils interact with direct cell–cell contact and/or release of soluble factors proposed as potential mechanisms (Dumitru et al. 2012b). Molecular interactions between neutrophils and tumor cells seems very complex and characterization of these mechanisms is still required.

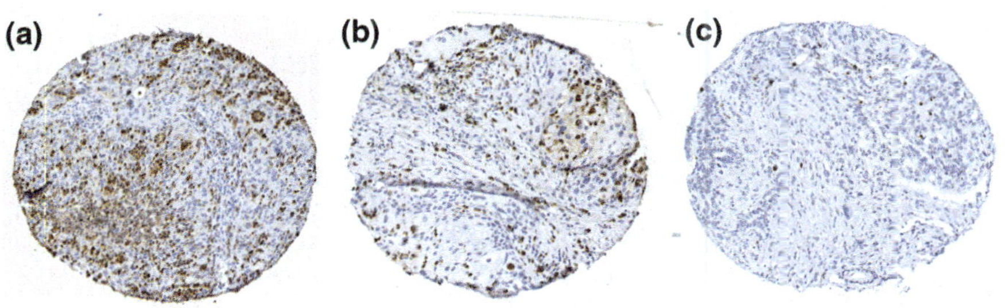

Fig. 19.4 Imunohistochemistry staining of human lung cancer tumors with MPO, demonstrating the large heterogeneity in the amount of TAN present in the tumors

19.7 N1 and N2: Polarization or Hyperactivation?

In their recent review on TAN as targets for cancer therapy Gregory and Houghton (2011) raised the interesting question whether the differences between N1 and N2 TAN were due to two unique transcriptional programs as suggested in our work (Fridlender et al. 2009) or instead represented two states of activation, that is, that N1 TANs produce the same mediators, but at higher levels. In our unpublished data, comparing the mRNA expression of N1 versus N2 TAN, we found that a vast majority of the changes were indeed up-regulation of the same genes and pathways in N1 TAN compared to N2 TAN. However, there were some clear exceptions. For example, we noted the chemokine CCL17 (which attracts T regulatory cells) was much more highly expressed in N2 TAN than in N1 TAN. The important question whether TAN can be manipulated to undergo frank irreversible polarization or possibly reversible activation states remains unresolved, and should be a matter of further research.

19.8 Conclusions

It is becoming increasingly clear that tumor-associated neutrophils (TAN) play a major role in cancer biology. TAN are a distinct population of neutrophils, which in their basic un-manipulated state are induced by the tumor microenvironment (by TGF-β and likely other factors) to elicit pro-tumor responses (N2 TAN). However, recent evidence shows that these cells can be altered to assume anti-tumor roles (N1 TAN). Neutrophils are thus an important under-appreciated cell population in cancer biology, and their functions need to be better characterized. A more complete understanding of the way these cells support or fight cancer will be important to develop strategies to direct the immune system against tumors.

References

Abi Abdallah DS, Egan CE, Butcher BA, Denkers EY (2011) Mouse neutrophils are professional antigen-presenting cells programmed to instruct Th1 and Th17 T-cell differentiation. Int Immunol 23:317–326

Acuff HB, Carter KJ, Fingleton B, Gorden DL, Matrisian LM (2006) Matrix metalloproteinase-9 from bone marrow-derived cells contributes to survival but not growth of tumor cells in the lung microenvironment. Cancer Res 66:259–266

Allen SS, Mackie JT, Russell K, Jeevan A, Skwor TA, McMurray DN (2008) Altered inflammatory responses following transforming growth factor-[beta] neutralization in experimental guinea pig tuberculous pleurisy. Tuberculosis 88:430–436

Appelberg R (1992) Mycobacterial infection primes T cells and macrophages for enhanced recruitment of neutrophils. J Leukoc Biol 51:472–477

Aruga A, Aruga E, Cameron MJ, Chang AE (1997) Different cytokine profiles released by CD4+ and CD8+ tumor-draining lymph node cells involved in mediating tumor regression. J Leukoc Biol 61:507–516

Ashtekar A, Bhaskar S (2003) Poly's plea: membership to the club of APC's. Trends Immunol 24:485–490

Atzpodien J, Reitz M (2008) Peripheral blood neutrophils as independent immunologic predictor of response and long-term survival upon immunotherapy in metastatic renal-cell carcinoma. Cancer Biother Radiopharm 23:129–134

Balbin M, Fueyo A, Tester AM et al (2003) Loss of collagenase-2 confers increased skin tumor susceptibility to male mice. Nat Genet 35:252–257

Beauvillain C, Delneste Y, Scotet M et al (2007) Neutrophils efficiently cross-prime naive T cells in vivo. Blood 110:2965–2973

Bellocq A, Antoine M, Flahault A et al (1998) Neutrophil alveolitis in bronchioloalveolar carcinoma: induction by tumor-derived interleukin-8 and relation to clinical outcome. Am J Pathol 152:83–92

Brandau S, Dumitru CA, Lang S (2012) Protumor and antitumor functions of neutrophil granulocytes. Semin Immunopathol [E-pub ahead of print]

Breitbach CJ, Paterson JM, Lemay CG et al (2007) Targeted inflammation during oncolytic virus therapy severely compromises tumor blood flow. Mol Ther 15:1686–1693

Caruso RA, Bellocco R, Pagano M, Bertoli G, Rigoli L, Inferrera C (2002) Prognostic value of intratumoral neutrophils in advanced gastric carcinoma in a high-risk area in northern Italy. Mod Pathol 15:831–837

Cavallo F, Giovarelli M, Gulino A et al (1992) Role of neutrophils and CD4+ T lymphocytes in the primary and memory response to nonimmunogenic murine mammary adenocarcinoma made immunogenic by IL-2 gene. J Immunol 149:3627–3635

Chen YL, Chen SH, Wang JY, Yang BC (2003) Fas ligand on tumor cells mediates inactivation of neutrophils. J Immunol 171:1183–1191

Cheung IY, Hsu K, Cheung NK (2012) Activation of peripheral-blood granulocytes is strongly correlated with patient outcome after immunotherapy with anti-GD2 monoclonal antibody and granulocyte-macrophage colony-stimulating factor. J Clin Oncol 30:426–432

Colombo MP, Modesti A, Parmiani G, Forni G (1992a) Local cytokine availability elicits tumor rejection and systemic immunity through granulocyte-T-lymphocyte cross-talk. Cancer Res 52:4853–4857

Colombo MP, Lombardi L, Stoppacciaro A et al (1992b) Granulocyte colony-stimulating factor (G-CSF) gene transduction in murine adenocarcinoma drives neutrophil-mediated tumor inhibition in vivo. Neutrophils discriminate between G-CSF-producing and G-CSF-nonproducing tumor cells. J Immunol 149:113–119

Coussens LM, Werb Z (2002) Inflammation and cancer. Nature 420:860–867

Curiel TJ, Coukos G, Zou L et al (2004) Specific recruitment of regulatory T cells in ovarian carcinoma fosters immune privilege and predicts reduced survival. Nat Med 10:942–949

De Larco JE, Wuertz BRK, Yee D, Rickert BL, Furcht LT (2003) Atypical methylation of the interleukin-8 gene correlates strongly with the metastatic potential of breast carcinoma cells. Proc Natl Acad Sci 100:13988–13993

De Larco JE, Wuertz BRK, Furcht LT (2004) The potential role of neutrophils in promoting the metastatic phenotype of tumors releasing interleukin-8. Clin Cancer Res 10:4895–4900

Demers M, Krause DS, Schatzberg D et al (2012) Cancers predispose neutrophils to release extracellular DNA traps that contribute to cancer-associated thrombosis. Proc Natl Acad Sci U S A 109:13076–13081

Di Carlo E, Forni G, Lollini P, Colombo MP, Modesti A, Musiani P (2001a) The intriguing role of polymorphonuclear neutrophils in antitumor reactions. Blood 97:339–345

Di Carlo E, Iezzi M, Pannellini T et al (2001b) Neutrophils in anti-cancer immunological strategies: old players in new games. J Hematother Stem Cell Res 10:739–748

Doi K, Horiuchi T, Uchinami M et al (2002) Neutrophil elastase inhibitor reduces hepatic metastases induced by ischaemia-reperfusion in rats. Eur J Surg 168:507–510

Dolcetti L, Peranzoni E, Ugel S et al (2010) Hierarchy of immunosuppressive strength among myeloid-derived suppressor cell subsets is determined by GM-CSF. Eur J Immunol 40:22–35
Donskov F, von der Maase H (2006) Impact of immune parameters on long-term survival in metastatic renal cell carcinoma. J Clin Oncol 24:1997–2005
Dumitru CA, Gholaman H, Trellakis S et al (2011) Tumor-derived macrophage migration inhibitory factor modulates the biology of head and neck cancer cells via neutrophil activation. Int J Cancer 129:859–869
Dumitru CA, Fechner MK, Hoffmann TK, Lang S, Brandau S (2012a) A novel p38-MAPK signaling axis modulates neutrophil biology in head and neck cancer. J Leukoc Biol 91:591–598
Dumitru CA, Moses K, Trellakis S, Lang S, Brandau S (2012b) Neutrophils and granulocytic myeloid-derived suppressor cells: immunophenotyping, cell biology and clinical relevance in human oncology. Cancer Immunol Immunother 61:1155–1167
Dupont PJ, Warrens AN (2007) Fas ligand exerts its pro-inflammatory effects via neutrophil recruitment but not activation. Immunology 120:133–139
Eash KJ, Greenbaum AM, Gopalan PK, Link DC (2010) CXCR2 and CXCR4 antagonistically regulate neutrophil trafficking from murine bone marrow. J Clin Invest 120:2423–2431
Fossati G, Ricevuti G, Edwards SW, Walker C, Dalton A, Rossi ML (1999) Neutrophil infiltration into human gliomas. Acta Neuropathol 98:349–354
Fridlender ZG, Sun J, Kim S et al (2009) Polarization of tumor-associated neutrophil phenotype by TGF-beta: "N1" versus "N2" TAN. Cancer Cell 16:183–194
Fridlender ZG, Sun J, Mishalian I et al (2012) Transcriptomic analysis comparing tumor-associated neutrophils with granulocytic myeloid-derived suppressor cells and normal neutrophils. PLoS ONE 7:e31524
Gabrilovich DI, Nagaraj S (2009) Myeloid-derived suppressor cells as regulators of the immune system. Nat Rev Immunol 9:162–174
Gerrard TL, Cohen DJ, Kaplan AM (1981) Human neutrophil-mediated cytotoxicity to tumor cells. J Natl Cancer Inst 66:483–488
Gijsbers K, Gouwy M, Struyf S et al (2005) GCP-2/CXCL6 synergizes with other endothelial cell-derived chemokines in neutrophil mobilization and is associated with angiogenesis in gastrointestinal tumors. Exp Cell Res 303:331–342
Gomez-Cambronero J, Horn J, Paul CC, Baumann MA (2003) Granulocyte-macrophage colony-stimulating factor is a chemoattractant cytokine for human neutrophils: involvement of the ribosomal p70 S6 kinase signaling pathway. J Immunol 171:6846–6855
Granot Z, Henke E, Comen EA, King TA, Norton L, Benezra R (2011) Tumor entrained neutrophils inhibit seeding in the premetastatic lung. Cancer Cell 20:300–314
Gregory AD, Houghton AM (2011) Tumor-associated neutrophils: new targets for cancer therapy. Cancer Res 71:2411–2416
Heifets L (1982) Centennial of Metchnikoff's discovery. J Reticuloendothel Soc 31:381–391
Hicks AM, Riedlinger G, Willingham MC et al (2006) Transferable anticancer innate immunity in spontaneous regression/complete resistance mice. Proc Natl Acad Sci 103:7753–7758
Himmel ME, Crome SQ, Ivison S, Piccirillo C, Steiner TS, Levings MK (2011) Human CD4+ FOXP3+ regulatory T cells produce CXCL8 and recruit neutrophils. Eur J Immunol 41:306–312
Hirose K, Hakozaki M, Nyunoya Y et al (1995) Chemokine gene transfection into tumour cells reduced tumorigenicity in nude mice in association with neutrophilic infiltration. Br J Cancer 72:708–714
Houghton AM, Rzymkiewicz DM, Ji H et al (2010) Neutrophil elastase-mediated degradation of IRS-1 accelerates lung tumor growth. Nat Med 16:219–223
Hubert P, Heitzmann A, Viel S et al (2011) Antibody-dependent cell cytotoxicity synapses form in mice during tumor-specific antibody immunotherapy. Cancer Res 71:5134–5143
Huh SJ, Liang S, Sharma A, Dong C, Robertson GP (2010) Transiently entrapped circulating tumor cells interact with neutrophils to facilitate lung metastasis development. Cancer Res 70:6071–6082

Iking-Konert C, Vogl T, Prior B et al (2008) T lymphocytes in patients with primary vasculitis: expansion of CD8+ T cells with the propensity to activate polymorphonuclear neutrophils. Rheumatology 47:609–616
Ilie M, Hofman V, Ortholan C et al (2012) Predictive clinical outcome of the intratumoral CD66b-positive neutrophil-to-CD8-positive T-cell ratio in patients with resectable nonsmall cell lung cancer. Cancer 118:1726–1737
Imai Y, Kubota Y, Yamamoto S et al (2005) Neutrophils enhance invasion activity of human cholangiocellular carcinoma and hepatocellular carcinoma cells: an in vitro study. J Gastroenterol Hepatol 20:287–293
Ishihara YIH, Matsunaga K (1998) Contribution of cytokines on the suppression of lung metastasis. Biotherapy 11:267–275
Ishihara Y, Fujii T, Iijima H, Saito K, Matsunaga K (1998) The role of neutrophils as cytotoxic cells in lung metastasis: suppression of tumor cell metastasis by a biological response modifier (PSK). In Vivo 12:175–182
Jablonska E, Puzewska W, Grabowska Z, Jablonski J, Talarek L (2005) VEGF, IL-18 and NO production by neutrophils and their serum levels in patients with oral cavity cancer. Cytokine 30:93–99
Jablonska E, Jablonski J, Marcinczyk M, Grabowska Z, Piotrowski L (2008) The release of soluble forms of TRAIL and DR5 by neutrophils of oral cavity cancer patients. Folia Histochem Cytobiol 46:177–183
Jablonska E, Garley M, Jablonski J (2009) The expressions of intrinsic and extrinsic apoptotic pathway proteins in neutrophils of oral cavity cancer patients: a preliminary study. Arch Immunol Ther Exp 57:229–234 (Warsz)
Jablonska J, Leschner S, Westphal K, Lienenklaus S, Weiss S (2010) Neutrophils responsive to endogenous IFN-beta regulate tumor angiogenesis and growth in a mouse tumor model. J Clin Invest 120:1151–1164
Jensen HK, Donskov F, Marcussen N, Nordsmark M, Lundbeck F, von der Maase H (2009) Presence of intratumoral neutrophils is an independent prognostic factor in localized renal cell carcinoma. J Clin Oncol 27:4709–4717
Joshita S, Nakazawa K, Sugiyama Y et al (2009) Granulocyte-colony stimulating factor-producing pancreatic adenosquamous carcinoma showing aggressive clinical course. Intern Med 48:687–691
Kandasamy M, Bay BH, Lee YK, Mahendran R (2011) Lactobacilli secreting a tumor antigen and IL15 activates neutrophils and dendritic cells and generates cytotoxic T lymphocytes against cancer cells. Cell Immunol 271:89–96
Katano M, Torisu M (1982) Neutrophil-mediated tumor cell destruction in cancer ascites. Cancer 50:62–68
Khajah M, Millen B, Cara DC, Waterhouse C, McCafferty DM (2011) Granulocyte-macrophage colony-stimulating factor (GM-CSF): a chemoattractive agent for murine leukocytes in vivo. J Leukoc Biol 89:945–953
Kim S, Buchlis G, Fridlender ZG et al (2008) Systemic blockade of transforming growth factor-beta signaling augments the efficacy of immunogene therapy. Cancer Res 68:10247–10256
Kobayashi Y (2006) Neutrophil infiltration and chemokines. Crit Rev Immunol 26:307–315
Kobayashi Y (2008) The role of chemokines in neutrophil biology. Front Biosci 13:2400–2407
Kousis PC, Henderson BW, Maier PG, Gollnick SO (2007) Photodynamic therapy enhancement of antitumor immunity is regulated by neutrophils. Cancer Res 67:10501–10510
Kuang DM, Zhao Q, Wu Y et al (2011) Peritumoral neutrophils link inflammatory response to disease progression by fostering angiogenesis in hepatocellular carcinoma. J Hepatol 54:948–955
Kusmartsev S, Nagaraj S, Gabrilovich DI (2005) Tumor-associated CD8+ T cell tolerance induced by bone marrow-derived immature myeloid cells. J Immunol 175:4583–4592
Lechner MG, Liebertz DJ, Epstein AL (2010) Characterization of cytokine-induced myeloid-derived suppressor cells from normal human peripheral blood mononuclear cells. J Immunol 185:2273–2284

Lichtenstein A, Seelig M, Berek J, Zighelboim J (1989) Human neutrophil-mediated lysis of ovarian cancer cells. Blood 74:805–809

Loukinova E, Dong G, Enamorado-Ayalya I et al (2000) Growth regulated oncogene-alpha expression by murine squamous cell carcinoma promotes tumor growth, metastasis, leukocyte infiltration and angiogenesis by a host CXC receptor-2 dependent mechanism. Oncogene 19:3477–3486

Mantovani A, Sozzani S, Locati M, Allavena P, Sica A (2002) Macrophage polarization: tumor-associated macrophages as a paradigm for polarized M2 mononuclear phagocytes. Trends Immunol 23:549–555

Mantovani A, Cassatella MA, Costantini C, Jaillon S (2011) Neutrophils in the activation and regulation of innate and adaptive immunity. Nat Rev Immunol 11:519–531

Masson V, de la Ballina LR, Munaut C et al (2005) Contribution of host MMP-2 and MMP-9 to promote tumor vascularization and invasion of malignant keratinocytes. FASEB J 19:234–236

Maus UA, Waelsch K, Kuziel WA et al (2003) Monocytes are potent facilitators of alveolar neutrophil emigration during lung inflammation: role of the CCL2-CCR2 axis. J Immunol 170:3273–3278

McGary CT, Mele M, Welch DR (1995) Highly metastatic 13762NF rat mammary adenocarcinoma cell clones stimulate bone marrow by secretion of granulocyte-macrophage colony-stimulating factor/interleukin-3 activity. Am J Pathol 147:1668–1681

Medina-Echeverz J, Fioravanti J, Zabala M, Ardaiz N, Prieto J, Berraondo P (2011) Successful colon cancer eradication after chemoimmunotherapy is associated with profound phenotypic change of intratumoral myeloid cells. J Immunol 186:807–815

Morales J, Kmieciak M, Knutson K, Bear H, Manjili M (2010) GM-CSF is one of the main breast tumor-derived soluble factors involved in the differentiation of CD11b-Gr1- bone marrow progenitor cells into myeloid-derived suppressor cells. Breast Cancer Res Treat 123:39–49

Movahedi K, Guilliams M, Van den Bossche J et al (2008) Identification of discrete tumor-induced myeloid-derived suppressor cell subpopulations with distinct T cell-suppressive activity. Blood 111:4233–4244

Nagaraj S, Schrum AG, Cho HI, Celis E, Gabrilovich DI (2010) Mechanism of T cell tolerance induced by myeloid-derived suppressor cells. J Immunol 184:3106–3116

Nozawa H, Chiu C, Hanahan D (2006) Infiltrating neutrophils mediate the initial angiogenic switch in a mouse model of multistage carcinogenesis. Proc Natl Acad Sci 103:12493–12498

Orr FW, Wang HH (2001) Tumor cell interactions with the microvasculature: a rate-limiting step in metastasis. Surg Oncol Clin N Am 10:357–381, ix–x

Pekarek LA, Starr BA, Toledano AY, Schreiber H (1995) Inhibition of tumor growth by elimination of granulocytes. J Exp Med 181:435–440

Peranzoni E, Zilio S, Marigo I et al (2010) Myeloid-derived suppressor cell heterogeneity and subset definition. Curr Opin Immunol 22:238–244

Piccard H, Muschel RJ, Opdenakker G (2011) On the dual roles and polarized phenotypes of neutrophils in tumor development and progression. Crit Rev Oncol Hematol 82:296–309

Queen MM, Ryan RE, Holzer RG, Keller-Peck CR, Jorcyk CL (2005) Breast cancer cells stimulate neutrophils to produce oncostatin M: potential implications for tumor progression. Cancer Res 65:8896–8904

Quigley JP, Deryugina EI (2012) Combating angiogenesis early: potential of targeting tumor-recruited neutrophils in cancer therapy. Future Oncol 8:5–8

Reid MD, Basturk O, Thirabanjasak D et al (2011) Tumor-infiltrating neutrophils in pancreatic neoplasia. Mod Pathol 24:1612–1619

Reiman JM, Kmieciak M, Manjili MH, Knutson KL (2007) Tumor immunoediting and immunosculpting pathways to cancer progression. Semin Cancer Biol 17:275–287

Richards H, Williams A, Jones E et al (2010) Novel role of regulatory T cells in limiting early neutrophil responses in skin. Immunology 131:583–592

Rodriguez PC, Quiceno DG, Zabaleta J et al (2004) Arginase I production in the tumor microenvironment by mature myeloid cells inhibits T-cell receptor expression and antigen-specific T-cell responses. Cancer Res 64:5839–5849

Rotondo R, Barisione G, Mastracci L et al (2009) IL-8 induces exocytosis of arginase 1 by neutrophil polymorphonuclears in nonsmall cell lung cancer. Int J Cancer 125:887–893
Sarraf KM, Belcher E, Raevsky E, Nicholson AG, Goldstraw P, Lim E (2009) Neutrophil/lymphocyte ratio and its association with survival after complete resection in non-small cell lung cancer. J Thorac Cardiovasc Surg 137:425–428
Sato T, Takahashi S, Mizumoto T et al (2006) Neutrophil elastase and cancer. Surg Oncol 15:217–222
Sawanobori Y, Ueha S, Kurachi M et al (2008) Chemokine-mediated rapid turnover of myeloid-derived suppressor cells in tumor-bearing mice. Blood 111:5457–5466
Scapini P, Lapinet-Vera JA, Gasperini S, Calzetti F, Bazzoni F, Cassatella MA (2000) The neutrophil as a cellular source of chemokines. Immunol Rev 177:195–203
Schaider H, Oka M, Bogenrieder T et al (2003) Differential response of primary and metastatic melanomas to neutrophils attracted by IL-8. Int J Cancer 103:335–343
Schmidt H, Bastholt L, Geertsen P et al (2005) Elevated neutrophil and monocyte counts in peripheral blood are associated with poor survival in patients with metastatic melanoma: a prognostic model. Br J Cancer 93:273–278
Schmielau J, Finn OJ (2001) Activated granulocytes and granulocyte-derived hydrogen peroxide are the underlying mechanism of suppression of T-cell function in advanced cancer patients. Cancer Res 61:4756–4760
Schneider-Merck T, Lammerts van Bueren JJ, Berger S et al (2010) Human IgG2 antibodies against epidermal growth factor receptor effectively trigger antibody-dependent cellular cytotoxicity but, in contrast to IgG1, only by cells of myeloid lineage. J Immunol 184:512–520
Shen L, Smith JM, Shen Z, Eriksson M, Sentman C, Wira CR (2007) Inhibition of human neutrophil degranulation by transforming growth factor-beta. Clin Exp Immunol 149:155–161
Sherwood ER, Enoh VT, Murphey ED, Lin CY (2004) Mice depleted of CD8+ T and NK cells are resistant to injury caused by cecal ligation and puncture. Lab Invest 84:1655–1665
Shojaei F, Wu X, Malik AK et al (2007) Tumor refractoriness to anti-VEGF treatment is mediated by CD11b+ Gr1+ myeloid cells. Nat Biotech 25:911–920
Shojaei F, Singh M, Thompson JD, Ferrara N (2008) Role of Bv8 in neutrophil-dependent angiogenesis in a transgenic model of cancer progression. Proc Natl Acad Sci 105:2640–2645
Smith WB, Noack L, Khew-Goodall Y, Isenmann S, Vadas MA, Gamble JR (1996) Transforming growth factor-beta 1 inhibits the production of IL-8 and the transmigration of neutrophils through activated endothelium. J Immunol 157:360–368
Sparmann A, Bar-Sagi D (2004) Ras-induced interleukin-8 expression plays a critical role in tumor growth and angiogenesis. Cancer Cell 6:447–458
Stoppacciaro A, Melani C, Parenza M et al (1993) Regression of an established tumor genetically modified to release granulocyte colony-stimulating factor requires granulocyte-T cell cooperation and T cell-produced interferon gamma. J Exp Med 178:151–161
Sun Z, Yang P (2004) Role of imbalance between neutrophil elastase and alpha 1-antitrypsin in cancer development and progression. Lancet Oncol 5:182–190
Suttmann H, Riemensberger J, Bentien G et al (2006) Neutrophil granulocytes are required for effective bacillus Calmette-Guerin immunotherapy of bladder cancer and orchestrate local immune responses. Cancer Res 66:8250–8257
Tazawa H, Okada F, Kobayashi T et al (2003) Infiltration of neutrophils is required for acquisition of metastatic phenotype of benign murine fibrosarcoma cells: implication of inflammation-associated carcinogenesis and tumor progression. Am J Pathol 163:2221–2232
Tazzyman S, Lewis CE, Murdoch C (2009) Neutrophils: key mediators of tumour angiogenesis. Int J Exp Pathol 90:222–231
Teramukai S, Kitano T, Kishida Y et al (2009) Pretreatment neutrophil count as an independent prognostic factor in advanced non-small-cell lung cancer: an analysis of Japan Multinational Trial Organisation LC00-03. Eur J Cancer 45:1950–1958
Theilgaard-Monch K, Jacobsen LC, Borup R et al (2005) The transcriptional program of terminal granulocytic differentiation. Blood 105:1785–1796

Trellakis S, Farjah H, Bruderek K et al (2011) Peripheral blood neutrophil granulocytes from patients with head and neck squamous cell carcinoma functionally differ from their counterparts in healthy donors. Int J Immunopathol Pharmacol 24:683–693
Tvinnereim AR, Hamilton SE, Harty JT (2004) Neutrophil involvement in cross-priming CD8+ T cell responses to bacterial antigens. J Immunol 173:1994–2002
Ueha S, Shand FH, Matsushima K (2011) Myeloid cell population dynamics in healthy and tumor-bearing mice. Int Immunopharmacol 11:783–788
Uehara M, Sato N (1994) Impaired ability of neutrophils to produce oxygen-derived free radicals in patients with chronic liver disease and hepatocellular carcinoma. Hepatology 20:326–330
Van Gisbergen KP, Geijtenbeek TB, Van Kooyk Y (2005) Close encounters of neutrophils and DCs. Trends Immunol 26:626–631
Verbeke H, Struyf S, Berghmans N et al (2011) Isotypic neutralizing antibodies against mouse GCP-2/CXCL6 inhibit melanoma growth and metastasis. Cancer Lett 302:54–62
Walsh SR, Cook EJ, Goulder F, Justin TA, Keeling NJ (2005) Neutrophil-lymphocyte ratio as a prognostic factor in colorectal cancer. J Surg Oncol 91:181–184
Wang LC, Thomsen L, Sutherland R et al (2009) Neutrophil influx and chemokine production during the early phases of the antitumor response to the vascular disrupting agent DMXAA (ASA404). Neoplasia 11:793–803
Welch DR, Schissel DJ, Howrey RP, Aeed PA (1989) Tumor-elicited polymorphonuclear cells, in contrast to "normal" circulating polymorphonuclear cells, stimulate invasive and metastatic potentials of rat mammary adenocarcinoma cells. Proc Natl Acad Sci U S A 86:5859–5863
Wengner AM, Pitchford SC, Furze RC, Rankin SM (2008) The coordinated action of G-CSF and ELR + CXC chemokines in neutrophil mobilization during acute inflammation. Blood 111:42–49
Wu QD, Wang JH, Condron C, Bouchier-Hayes D, Redmond HP (2001) Human neutrophils facilitate tumor cell transendothelial migration. Am J Physiol Cell Physiol 280:C814–C822
Wu Y, Zhao Q, Peng C, Sun L, Li XF, Kuang DM (2011) Neutrophils promote motility of cancer cells via a hyaluronan-mediated TLR4/PI3 K activation loop. J Pathol 225:438–447
Youn J-I, Nagaraj S, Collazo M, Gabrilovich DI (2008) Subsets of myeloid-derived suppressor cells in tumor-bearing mice. J Immunol 181:5791–5802
Youn JI, Collazo M, Shalova IN, Biswas SK, Gabrilovich DI (2012) Characterization of the nature of granulocytic myeloid-derived suppressor cells in tumor-bearing mice. J Leukoc Biol 91:167–181
Yunping L, He Z, Krueger J et al (2006) Targeting tumor-associated macrophages as a novel strategy against breast cancer. J Clin Invest 116:2132–2141
Zhao JJ, Pan K, Wang W et al (2012) The prognostic value of tumor-infiltrating neutrophils in gastric adenocarcinoma after resection. PLoS ONE 7:e33655
Zhou SL, Dai Z, Zhou ZJ et al (2012) Overexpression of CXCL5 mediates neutrophil infiltration and indicates poor prognosis for hepatocellular carcinoma. Hepatology 56:2242–2254
Zivkovic M, Poljak-Blazi M, Zarkovic K, Mihaljevic D, Schaur RJ, Zarkovic N (2007) Oxidative burst of neutrophils against melanoma B16–F10. Cancer Lett 246:100–108

Chapter 20
Mast Cell Modulation of the Tumor Microenvironment

Sharon A. Oldford and Jean S. Marshall

Abstract Mast cells are myeloid derived immune effector cells that have been most widely studied in the context of allergic disease. However, over recent years it has become apparent that they also play critical roles in the regulation of tissue remodelling and host defence. In solid tumors, mast cells are abundant at the tumor periphery in close proximity to blood vessels and are frequently considered to function in a tumor promoting capacity. They release potent angiogenic cytokines that augment tumor blood vessel formation, tumor enhancing growth factors and tissue-degrading enzymes that enable tumor metastasis. Mast cells can also release mediators in the tumor microenvironment that enhance aspects of immune suppression, such as interleukin 10. These observations have led to the consideration of inhibiting mast cells as an approach to cancer therapy. In marked contrast to this negative picture of mast cells in tumors, some studies of human disease have suggested that increased mast cell numbers can be associated with an improved prognosis. It has also been demonstrated, in mouse models, that mast cells might be important targets for immune activation during immunotherapy. Since mast cells are resistant to radiation and normally serve a sentinel cell role recruiting effector cells such as natural killer cells and T cells to sites of infection, they might have substantial potential as an additional target for therapy in the context of more

S. A. Oldford · J. S. Marshall
Dalhousie Inflammation Group and Department of Microbiology and Immunology, Dalhousie University, Halifax, NS, Canada

S. A. Oldford (✉) · J. S. Marshall (✉)
Department of Microbiology and Immunology, Faculty of Medicine, Dalhousie University, 5850 College Street Room 7-C, P.O. BOX 15000, Halifax, NS B3H 4R2, Canada
e-mail: sharon.oldford@dal.ca

J. S. Marshall
e-mail: jean.marshall@dal.ca

M. R. Shurin et al. (eds.), *The Tumor Immunoenvironment*,
DOI: 10.1007/978-94-007-6217-6_20,

traditional treatments. Our understanding of mast cells in human tumor settings is limited, but there remain excellent opportunities to attempt to modulate mast cell function as a new approach to therapy.

Keywords Mast cells · Angiogenesis · Tissue remodelling · Innate immunity · Adaptive immunity

Abbreviations

APC	Adenomatous polyposis coli
bFGF	basic fibroblast growth factor
CTMC	Connective tissue mast cell
DAMPs	Danger associated molecular patterns
DC	Dendritic cells
ECM	Extracellular matrix
EPC	Endothelial progenitor cells
H_1	Histamine receptor 1
H_2	Histamine receptor 2
HIF-1α	Hypoxia inducible factor alpha
HMGB1	High-mobility group protein B1
IFN-γ	Interferon gamma
Ig	Immunoglobulin
IL	Interleukin
LT	Leukotriene
MCP	Mast cell protease
MDSC	Myeloid derived suppressor cells
MMC	Mucosal mast cell
MMP	Matrix metalloproteinase
NK	Natural killer cells
NKT	Natural killer T cells
PAF	Platelet activating factor
pDC	Plasmacytoid Dendritic cells
PG	Prostaglandin
ROS	Reactive oxygen species
SCF	Stem cell factor
SMA	Smooth muscle actin
TGF-β	Transforming growth factor beta
TLR	Toll-like receptor
TNF	Tumor necrosis factor
T_H	Helper T cells
T_{reg}	Regulatory T cells
VEGF	Vascular endothelial growth factor

20.1 Introduction

Mast cells are highly granulated, tissue resident cells that were first described in 1878 by Paul Erhlich. They were identified by the reactivity of their metachromatic granules with aniline dyes and found in abundance around the blood vessels in loose connective tissue. Ehrlich was also the first to document large numbers of mast cells in solid tumors and subsequent work by his doctoral student, Westphal, demonstrated that mast cells predominately accumulated in the tumor stroma (reviewed in Crivellato et al. 2003). In more recent years, many histological studies have confirmed an abundance of mast cells in the tumor periphery where they often correlate with increased microvessel density (Esposito et al. 2004; Toth-Jakatics et al. 2000).

Interest in mast cells exploded in the 1950s, following the discovery by Riley and West that tissue mast cells serve as a repository for histamine (Riley and West 1953). Investigations linking histamine release from mast cells with anaphylactic response and studies examining the roles of other mediators released from degranulating mast cells (e.g., leukotrienes) in allergic disease took centre stage. Thus, mast cells are best known for their effector function following IgE/antigen mediated activation in allergic disease and response to parasitic infection. However, we have also come to recognize the multifaceted roles of mast cells in diverse biological processes including maintaining tissue homeostasis, regulating tissue remodelling events, and host defence where mast cells are particularly important as sentinel cells that recruit innate and adaptive immune cells during infection (Galli and Tsai 2008; Theoharides et al. 2012; Dawicki and Marshall 2007).

The aim of this chapter is to review the complex and sometimes conflicting literature on mast cells and the tumor microenvironment, highlighting the tumor enhancing roles of mast cells in promoting angiogenesis, tumor metastasis and immune suppression in the tumor microenvironment. In contrast, we will then discuss the sentinel role of mast cells in eliciting effective innate and adaptive immune responses and the therapeutic potential of targeting mast cells at tumor sites.

20.2 Mast Cell Biology

20.2.1 Mast Cell Origin

Mast cells are bone marrow derived, tissue resident myeloid cells that contain cytoplasmic granules with preformed, stored mediators (histamine, proteoglycans, neutral serine proteases and cytokines). In contrast to other myeloid-derived cells, which differentiate in the bone marrow and circulate as mature cells, mast cells are released from the bone marrow as committed progenitor cells that only fully differentiate within the tissues (reviewed in Gurish and Austen 2012). They are found throughout the body in close proximity to blood vessels and in large numbers at sites vulnerable to infection such as the skin, airways and gastrointestinal tract.

20.2.2 Mast Cell Mediators

Mast cells are unique immune cells that possess an array of mediators that can be selectively released, depending on the activation stimulus. Activated mast cells produce three major classes of mediators: (1) pre-formed granule associated mediators, (2) *de novo* synthesized lipid derived mediators, and (3) cytokines and chemokines (Fig. 20.1). Mast cells tightly control granule stored mediator release which can occur through complete degranulation or a slower and potentially more

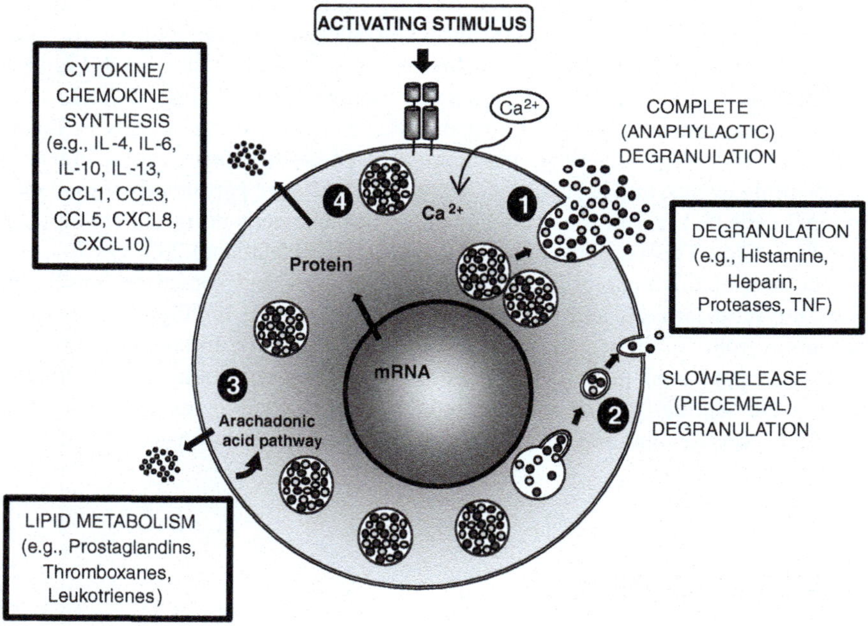

Fig. 20.1 Mast cell mediator release. Following activation, mast cells release mediators via different mechanisms: (1) complete (anaphylactic) calcium dependent, degranulation occurs immediately following activation and results in fusion of granule membranes to each other and to the plasma membrane, resulting in release of granule contents into extracellular spaces; (2) piecemeal degranulation involves vesicular transport of mediators from cytoplasmic granules to the cell membrane to allow for slow, sustained release of selective mediators; (3) lipid mediators are synthesized within minutes of membrane perturbation following the release of arachadonic acid from membrane phospholipids; and (4) within the hours following activation mast cells undergo transcription and translation events to produce and secrete multiple cytokines, chemokines and growth factors

selective process of 'piecemeal' degranulation. A feature of mast cells is their ability to selectively release different mediator profiles in response to different stimuli. For example, human cord blood derived mast cells have been demonstrated to release IL-6 in the absence of degranulation following activation with the pro-inflammatory cytokine IL-1 (Kandere-Grzybowska et al. 2003) and display a highly selective release of cytokines and chemokines following activation through different Toll-like receptors (TLR) (McCurdy et al. 2003). Selective mediator release has also been demonstrated by rodent mast cells. Mast cells secrete a multitude of mediators which may have important functions in the tumor microenvironment (Table 20.1).

Proteases released from mast cell granules have important roles in the promotion of angiogenesis, extracellular matrix (ECM) remodelling and the release of ECM sequestered enzymes and growth factors. They also can degrade local cytokines and chemokines and act through protease activated receptors on multiple cell types (Stevens and Adachi 2007; Younan et al. 2010; Caughey 2011; Conti et al. 2007; Coussens et al. 1999). Proteoglycan–protease complexes and interactions of proteases with the coagulation cascade add to the influence of these potent mediators on the tissue microenvironment. Histamine has a substantial influence on tumor immunity, discussed later in this chapter, as well as on the local vasculature. In addition, certain preformed and stored cytokines such as tumor necrosis factor (TNF), add to the potent inflammatory response following mast cell degranulation.

Activated mast cells can synthesize lipid mediators from arachadonic acid. Prostanoids can add to the pro-angiogenic environment local to tumor sites while others lipid mediators, such as the leukotriene LTB_4 and platelet activating factor (PAF), can have both vascular and immune modulatory effects in addition to providing a chemotactic signal for anti-tumor effector cells (Ott et al. 2003; Boyce 2007).

The vast array of different cytokines and chemokines that can be produced by mast cells, in many cases selectively, without granule-associated mediator release,

Table 20.1 MC mediators which may have important roles in the tumor microenvironment[1]

Mediator class	Examples
Pre-formed granule associated	Proteases (tryptase, chymase, serine proteases, carboxypeptidase A), histamine, proteoglycans (chondroitin sulphate containing and heparin sulphate containing)
Lipid-derived	LTB_4, LTC_4, PGE_2, PGD_2, PAF
Cytokines	IL-1α,IL-1β, IL-2, IL-3, IL-4, IL-5, IL-6, IL-9, IL-10, IL-11, IL-12, IL-13, IL-14, IL-15, IL-16, IL-17, IL-18, IL-22, IL-25, IL-33, IFN-α, IFN-β, IFN-γ, LIF, MIF, SCF, TNF, TGF-β1, M-CSF, GM-CSF, TSLP
Chemokines	CCL1, CCL2, CCL3, CCL3L1, CCL4, CCL5, CCL7, CCL8, CCL11, CCL13, CCL16, CCL17, CCL20, CCL22
	CXCL1, CXCL2, CXCL3, CXCL8, CXCL9, CXCL10, CXCL11, CXCL12, XCL1
Growth factors	VEGF-A, bFGF, EGF, IGF-1, NGF, PDGF
Other mediators	NO, superoxide, substance P, MMP-9, cathelicidins

[1] Adapted from Galli et al. (2005), Sarchio et al. (2012)

provides enormous opportunity for mast cells to influence tumor growth and metastasis as well as the immune response to tumors. We are only at the beginning of exploring the importance of such mast cell mediators to host responses and the tumor microenvironment.

20.2.3 Mast Cell Subsets and Localization

MC progenitor cells differentiate in tissues into subsets with distinct functional characteristics. Classically these were referred to as connective tissue mast cells (CTMC) and mucosal mast cells (MMC). CTMC are found typically adjacent to blood vessels, lymphatic vessels and nerve endings in the connective tissue. MMC predominately associate with the epithelial lining of the mucosal tissue of the respiratory, digestive and urogenital tracts (reviewed in Gurish and Austen 2012).

Human mast cell subsets are often characterized based on the protease content of their cytoplasmic granules. MC^{TC} cells contain tryptase together with chymase, cathepsin-G like protease, and mast cell carboxypeptidase and are found in the skin and submucosa. MC^{T} cells contain tryptase, but lack the other neutral proteases present in MC^{TC} cells and are localized to mucosal tissues (Irani et al. 1986, 1991). Importantly, mast cell protease content may be altered by the cytokine milieu and cell interactions within the tumor microenvironment. In human intestinal mast cell cultures IL-4 selectively increased tryptase positive mast cells (Bischoff et al. 1999), whereas co-culture with endothelial cells preferentially induced tryptase/chymase double positive cells (Mierke et al. 2000). Mast cell subsets display marked heterogeneity with respect to granule contents and differences in lipid mediator production in different tissues (Katz et al. 1985). Functional heterogeneity of mast cell subsets is also observed in mast cells isolated from different microenvironments within the same tissue (Finotto et al. 1994). Classical mast cell designations, such as the examples above, represent two extremes of a broad spectrum of mast cell subsets which display a large degree of phenotypic heterogeneity and functional plasticity in different microenvironments. Immature mast cells and those undergoing chronic activation may also display altered characteristics and further complicate the picture regarding heterogeneity within the tumor microenvironment.

It is important to note that mast cell infiltrates in tumor samples have sometimes been assessed using c-Kit specific antibodies. Mast cells express high levels of c-Kit and signalling through c-Kit is required for mast cell survival. However, c-Kit is also expressed at high levels on hematopoietic stem cells and common myeloid progenitors and has been reported on other cell types including innate lymphoid cells, endothelial cell precursors and some tumor cells. Care therefore has to be taken in interpreting the results of c-Kit based analyses. However, immunohistochemical or other analyses that rely on mast cell granule content can also be misleading. Macrophages may take up granule products and immature mast cells with few granules or degranulated mature mast cells may not stain sufficiently for detection. For many routine histological stains, formalin fixation of tissues can reduce metachromatic staining and alternate fixation approaches need to be employed.

20.3 Mast Cells in the Tumor Microenvironment

20.3.1 Experimental Models to Study the Role of Mast Cells

Commonly used in vitro models to study mast cells include mast cell lines such as the murine MC9, the rat RM3 or the human HMC-1 and LAD-2. Primary mast cell cultures typically derived from bone marrow for murine studies or from progenitor cells in cord blood or peripheral blood for human studies are also widely employed. These cells are often used in co-culture systems to assess mast cell interactions with tumor, stromal and immune cells. However, in vitro systems cannot fully replicate the complex structure of the tumor microenvironment in vivo. Much of what has been learned about the roles of mast cells in the tumor microenvironment has been gleaned from murine studies of mast cell deficient *Kit* mutant mice, ideally with complimentary tissue reconstitution of mast cell subsets. *Kit* mutant mice exhibit profound mast cell deficiency but possess other phenotypic abnormalities. $Kit^{W/W-v}$ mice have a compound deletional and truncated *Kit* mutation and are sterile, anemic, neutropenic, lack intestinal interstitial cells of Cajal and have a marked decrease in $\gamma\delta$ T cells (Grimbaldeston et al. 2005). $Kit^{W-sh/W-sh}$ mice have an inversion mutation in the regulatory locus of the *Kit* gene and are neither sterile nor anemic, but exhibit neutrophilia, thromobocytosis, splenomegaly and cardiomegaly (Nigrovic et al. 2008).

To circumvent the functional defects observed in mast cell deficient models which are based on mutations of the *c-Kit* or stem cell factor (SCF) genes, newer models have been developed that target mast cell specific proteases. Inducible and constitutive CTMC deficiency was obtained by breeding the *Mcpt5-Cre* transgenic mice to the Cre-inducible diphtheria toxin receptor (*iDTR*) mice and to the *R-DTA* line which expresses diptheria toxin A in Cre-expressing cells, respectively. These models confirm the involvement of mast cells in contact mediated hypersensitivity (Dudeck et al. 2011). Mast cells were unexpectedly deleted in mice which express the Cre recombinase gene under the control of regulatory elements of the mast cell protease carboxypeptidase A3 gene as a result of Cre-mediated genotoxicity. Studies performed in these mice showed defects in passive local and systemic IgE-mediated anaphylaxis, but could not confirm previously reported roles for mast cells in passive autoantibody induced arthritis or experimentally induced autoimmune encephalitis (Feyerabend et al. 2011). While these discrepant findings may in part be influenced by differences in experimental design as suggested by others (Brown et al. 2012), they highlight that it will be crucial to utilize multiple models of mast cell deficiency in conjunction with other relevant and well established experimental approaches to both confirm and reveal roles for mast cells and mast cell specific mediators in tumorigenesis.

20.3.2 Mechanisms of Mast Cell Activation at Tumor Sites

The solid tumor microenvironment contains a number of stimuli which could activate mast cells (Fig. 20.2). Substantial evidence from both human in vitro and mouse in vivo studies indicates local mast cell activation with evidence of traditional and/or piecemeal degranulation processes and other indicators of mast cell activity, such as diffuse distribution of mast cell proteases and cytokine expression.

Mast cells are known to respond to locally hypoxic conditions, found in the vicinity of growing solid tumors, directly through the function of hypoxia

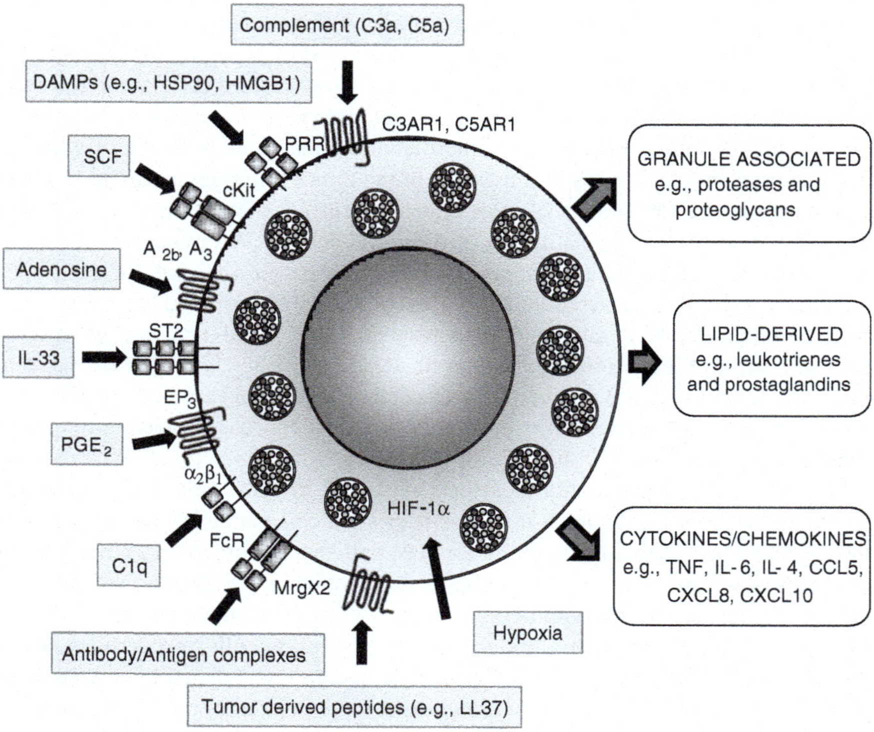

Fig. 20.2 Mast cell activation within the tumor microenvironment. Several mast cell activation events are likely to occur within the tumor microenvironment. The tumor microenvironment is rich in damage associated molecular patterns and other endogenous alarmins such as IL-33 and tumor derived peptides, which can activate mast cells via ST2 and G-protein coupled receptors, respectively. Mast cells express Fc receptors and complement receptors and can be activated by circulating immune complexes and complement components. Mast cells can also be activated in the tumor microenvironment following engagement of receptors for SCF, adenosine and PGE_2 and up-regulation of the transcription factor HIF-1α in the hypoxic tumor microenviroment. Mast cells secrete 3 classes of mediators: preformed granule associated mediators, *de novo* synthesized lipid mediators, and cytokines and chemokines. Different activators stimulate selective mediator release with some mediators primarily enhancing degranulation while others enhance eicosanoid and/or cytokine/chemokine production

inducible factor (HIF)-1α (Sumbayev et al. 2012; Gulliksson et al. 2010). It has recently been demonstrated that HIF-1α mediates the antimicrobial impact of mast cells through regulation of several mechanisms, including antimicrobial peptide release and the development of extracellular traps (Branitzki-Heinemann et al. 2012). HIF-1α also plays a role in regulating responses to TLR-mediated mast cell activation, which could result from antimicrobial peptide release or tissue breakdown products (Sumbayev et al. 2012).

Adenosine and prostaglandin (PG)E_2 are also observed in increased concentrations within the tumor microenvironment. PGE_2 has been shown to be an effective stimulus for mast cell cytokine production including the generation of immunoregulatory cytokines such as IL-6 (Leal-Berumen et al. 1995) and the production of vascular endothelial growth factor (VEGF) (Abdel-Majid and Marshall 2004). Both of these events occur in the absence of mast cell degranulation.

Adenosine is found in increased concentrations within tumor sites as a consequence of local hypoxia and can be produced by some types of mast cells in response to IgE receptor-mediated activation (Lloyd et al. 1998). Mast cells express multiple adenosine receptors and have been described to respond to adenosine stimulation, especially through their A_{2b} and A_3 adenosine receptors. This leads to a pro-inflammatory response from mast cells that has been extensively studied in the context of allergic disease (recently reviewed in Rudich et al. 2012).

Mast cells can be activated via TLRs and other innate receptor systems to produce substantial amounts of cytokines including TNF and IL-6 as well as chemokines such as CCL3 and CXCL8. These types of responses may be initiated in certain tissue microenvironments by tissue breakdown products such as hyaluronan, the release of intracellular cell contents via necrotic cell death, or by the release of antimicrobial peptides by other cells that can signal via TLRs. Additional damage associated molecular patterns (DAMPs) that can be actively released from viable tumor cells and passively released from necrotic tumor cells within the tumor microenvironment include HSP60, 70 and 90 (Basu et al. 2000; Ciocca and Calderwood 2005).

Alarmins are up-regulated in a number of cancers and include such factors as tumor derived peptides (e.g., LL37), S100A8/S100A9 proteins, HMGB1, and cytokines such as IL-33. Some alarmins can act directly on cancer cells to induce tumor cell proliferation, migration and angiogenesis (reviewed in Chan et al. 2012). However, they can also act indirectly via immune effector cells such as mast cells. Adding to the complexity of this interaction, mast cells themselves have been shown to produce cathelicidins such as LL37 (Wang et al. 2012; Di Nardo et al. 2003) and the IL-1 family member IL-33 (Hsu et al. 2010). IL-33 is released by many cell types, especially epithelial cells, in response to tissue damage. Human skin explants and normal human skin derived fibroblasts and keratinocytes up-regulate IL-33 following exposure to physiological doses of UVB radiation. Murine studies have demonstrated that IL-33 is up-regulated in squamous cell carcinoma cells that evade immunological destruction (Byrne et al 2011). Both human and murine mast cells express functional IL-33 receptors

IL-33 released from necrotic stromal cells induces murine mast cells to secrete pro-inflammatory cytokines (TNF and IL-6) and leukotrienes (Enoksson et al. 2011). Interestingly, mast cells associate with IL-33 expressing fibroblasts in UV-exposed murine skin samples (Byrne et al. 2011), suggesting mast cells may release such mediators in the tumor microenvironment of skin cancers and possibly other tumor subtypes. Since IL-33 is generally pro-inflammatory, the impact of such secretion would likely be to enhance some aspects of immune cell recruitment, with impacts on both the immune and angiogenic response to the tumor.

Many of the mechanisms whereby macrophages in the tumor microenvironment can become activated are likely shared by mast cells. These include the actions of several cytokines and chemokines (e.g., SCF, CXCL12) as well as aspects of innate immune signaling. Large amounts of SCF have been reported in tumor microenvironments and in other inflammatory settings. This enhances local mast cell survival and differentiation as well as potentially activating mast cells for enhanced secretion of VEGF and basic fibroblast growth factor (bFGF). CXCL12 has numerous impacts on mast cells including activation and chemotaxis. Enhanced levels of this chemokine may contribute particularly to immature mast cell recruitment as discussed in the context of mast cell impacts on tumor angiogenesis.

During allergic disease the predominant mechanism for mast cell activation is via allergen specific IgE bound to FcεRI. A number of studies have suggested that high serum levels of IgE are associated with a decreased cancer risk. The concept that IgE mediated mast cell activation may play a role in host defence against cancer has begun to emerge during recent years (Jensen-Jarolim et al. 2008). A recent rigorous examination of this concept using samples from a Norwegian cohort demonstrated a clear relationship between elevated allergen specific IgE and decreased risk of developing glioma (Schwartzbaum et al. 2012). These studies build on previous examinations of the potential for decreased cancer risk in allergic individuals. While such studies have had varying conclusions, evidence for a beneficial effect of allergy has been found for colorectal cancer (Negri et al. 1999) and a substantial meta-analysis concluded that there was an overall beneficial effect of allergic disease (Sherman et al. 2008). Notably however, a study of operable breast cancer concluded that lower IgE concentrations were also associated with longer survival, but only in patients whose tumors were E_2R positive (Ownby et al. 1985), so care needs to be taken in developing general conclusions regarding cancer and IgE from current information.

Rodent mast cells normally express Fcγ receptors and human mast cells can also express a full range of IgG receptors in certain microenvironments, especially those rich in IFN-γ (Tkaczyk et al. 2004). IgG containing immune complexes therefore have the ability to activate mast cells leading to degranulation, lipid mediator production and cytokine/chemokine generation. Complement products produced as a result of IgG or IgM containing immune complex deposition may also induce mast cell activation with degranulation. The contribution of these processes to mast cell activation in tumor microenvironments is unknown. The impact of such mechanisms on mast cell activation needs to be considered, especially in the context of the increased use of antibody mediated therapies and fusion proteins containing Fc structures.

Mast cells are most efficiently activated via high affinity IgE receptors. The use of monoclonal IgE antibodies directed against tumor associated antigens has shown some success in experimental tumor models (Jensen-Jarolim et al. 2008). IgE-mediated mast cell degranulation can overcome allograft tolerance and elicit effector $CD4^+$ and $CD8^+$ T cell mediated rejection (de Vries et al. 2009). Therefore, local mast cell degranulation may be advantageous to subvert the immune suppressive microenvironment of solid tumors and elicit effective anti-tumor immunity. However, there are a number of safety concerns, such as the risk of anaphylaxis, that have been raised regarding the use of IgE antibodies in human therapy.

Notably, adhesion interactions, the cytokine and chemokine microenvironment and multiple other factors can affect not only the type and number of mast cells present within a given tumor setting or tissue location, but also the mediator response that will ensue following activation. It can therefore be difficult to predict from in vitro studies the full nature of the mast cell response. The consequences of chronic mast cell activation with specific mediators and adhesion interactions within a tissue setting may differ considerably from those of acute activation models used in the majority of experimental studies.

20.3.3 Mast Cell Recruitment to Solid Tumor Sites

Mast cells express a wide array of chemokine and growth factor receptors and migrate to sites of inflammation in response to chemotactic stimuli (reviewed in Halova et al. 2012). Mast cells are among the first cells recruited to tumor sites, are increased in precancerous lesions, further increase with cancer progression and are positively associated with microvessel density (Benitez-Bribiesca et al. 2001; Mohtasham et al. 2010; Kankkunen et al. 1997). Mediators released within the tumor microenvironment including SCF, transforming growth factor (TGF)-β, CCL5 and CXCL12 are thought to be important in the recruitment of mast cells to tumor sites.

SCF expression is highly elevated in human breast, gastric, colorectal, lung, ovarian, liver and esophageal tumor tissue as compared to adjacent normal tissue (Huang et al. 2008). SCF is important for the differentiation and activation of human mast cells, but has also been implicated in their recruitment. SCF has been demonstrated to be important for the recruitment of mast cells in several experimental tumor models including breast carcinoma and hepatocellular carcinoma (Huang et al. 2008; Zhang et al. 2000). Another factor found in abundance in solid tumors is TGF-β. TGF-β isoforms are chemotactic to murine cultured mast cells, rat peritoneal mast cells, the human mast cell line HMC-1 and cord blood derived cultured human mast cells (Olsson et al. 2000; Gruber et al. 1994). Other mediators expressed at high levels in tumors that are likely to be chemotactic to mast cells include VEGF isoforms, adenosine, complement components and lipid mediators (reviewed in Halova et al. 2012).

Chemokines likely play a crucial role in mediating the recruitment of mast cells to tumor sites. Chemokine receptor expression varies between immature and mature mast cells and among mast cell subsets. CCL5 induces chemotaxis of human cord blood derived mast cells via CCR1 and CCR4 in vitro (Juremalm et al. 2002). CCL5 is expressed in Hodgkin's lymphoma tumor tissue samples and CCL5 released from Hodgkin/Reed-Sternberg cell lines induces human mast cell chemotaxis in vitro (Fischer et al. 2003). Human mast cells also migrate to CCL5 released from human keratinocytes following exposure to UVB radiation (Van Nguyen et al. 2011). CCL5 expression by human uterine smooth muscle tumor cells correlates with mast cell density in the tumor tissue and the majority of infiltrating mast cells express the CCR3 receptor for CCL5 (Zhu et al. 2007). These data suggest CCL5 may be an important mast cell chemotactic mediator in several tumor settings.

Mast cells may also be attracted to the tumor microenvironment through the CXCL12/CXCR4 axis. CXCL12 protein is detected in solid tumors (Kryczek et al. 2005) and can be produced by both tumor cells and stromal cells. In human glioblastoma multiformes large numbers of $CXCR4^+$ mast cells surround CXCL12 producing tumor cells (Polajeva et al. 2011). CXCL12 induces transendothelial migration of human cord blood derived mast cells in vitro and their selective release of the pro-angiogenic chemokine CXCL8 (Lin et al. 2001). In a xenograft model, carcinoma associated fibroblasts mobilize early endothelial progenitor cells (EPC), recruit them to implanted breast tumors and enhance tumor angiogenesis in a CXCL12 dependent manner (Orimo et al. 2005). Physiological concentrations of CXCL12 or VEGF alone were not sufficient to induce significant angiogenesis in an in vivo xenograft Matrigel model using ovarian cancer patient ascites. However, combined, the two mediators synergistically induced significant microvessel formation (Kryczek et al. 2005). Similarly, CXCL8 produced from pancreatic cancer cells acts cooperatively with fibroblast derived CXCL12 to enhance human endothelial cell angiogenesis in vitro (Matsuo et al. 2009). Thus, CXCL12 may coordinate the recruitment of EPC and mast cells to the leading edge of solid tumors whereby mast cell mediators such as VEGF and CXCL8 may synergize with CXCL12 in the tumor microenvironment to promote angiogenesis.

20.3.4 Prognostic Significance of Mast Cells in Solid Tumors

Mast cell density has been reported to associate with a poor outcome in many solid tumor types including Hodgkin's lymphoma, melanoma, endometrial, cervical, esophageal, lung, gastric, colorectal and prostate carcinomas. The poor prognosis associated with mast cell infiltration has been linked to the positive correlations with microvessel density and metastasis (reviewed in Groot Kormelink et al. 2009).

While most studies suggest that mast cell infiltration of solid tumors is a poor prognostic indicator, others have failed to correlate mast cell infiltration with angiogenesis, metastasis and decreased survival (Xia et al. 2011). Moreover, in some studies, mast cell infiltration is correlated with improved survival (Fleischmann et al. 2009; Hedstrom et al. 2007; Rajput et al. 2008). Mast cells have demonstrated anti-tumor activities in vitro and in experimental tumor models. In vitro studies have demonstrated mast cells can mediate direct TNF-mediated tumor cell cytotoxicity (Benyon et al. 1991; Dery et al. 2000) and inhibit human breast tumor cell clonogenic growth in the presence of fibroblasts (Samoszuk et al. 2005). Anti-tumor roles for other mast cell derived mediators have also been described. In the tumor microenvironment, mast cells are a major source of histamine which can protect against tumor development as demonstrated by increased susceptibility to carcinogen-induced colorectal and skin tumors in histamine deficient mice (Yang et al. 2011).

The location of mast cells within the tumor microenvironment of different tumor subtypes may also be a crucial factor in their ability to provide a protective role against the growth of solid tumors. Studies in non-small cell carcinoma and prostate carcinoma have demonstrated that intratumoral, but not peritumoral, mast cells independently predict improved patient survival (Welsh et al. 2005; Johansson et al. 2010). In a large study of 4,444 breast tumors increased numbers of stromal mast cells independently predicted improved survival (Rajput et al. 2008). It is highly plausible that the prognostic significance of mast cells will be largely dependent on their activation status within the tumor microenvironment, as reflected by the diverse array of mediators that can be released from mast cells in a selective manner depending on the activation stimulus (Kandere-Grzybowska et al. 2003; McCurdy et al. 2003) and their functional interactions with blood vessels.

20.3.5 Mast Cells and Early Tumor Development

The contribution of mast cells to early tumor development is unclear. Only a few studies have evaluated the influence of mast cells on cancer incidence, with conflicting results. Mast cell deficient $Kit^{W/W-v}$ mice display decreased susceptibility to chemically induced intestinal tumors when compared with congenic WBB6F1-$Kit^{+/+}$ wild-type mice, which could be normalized via wild type bone marrow reconstitution (Wedemeyer and Galli 2005). Similarly, studies using polyposis prone $APC^{\Delta 468}$ mice, which have a mutation in the adenomatous polyposis coli (*APC*) gene, demonstrated that following lethal irradiation reconstitution with bone marrow from $Kit^{W-sh/W-sh}$ mice resulted in decreased polyp development compared to mice reconstituted with wild type bone marrow (Gounaris et al. 2007). These studies suggest an important role for mast cells in intestinal cancer development. In contrast, when mast cell deficient $Kit^{W-sh/W-sh}$ mice were crossed to the multiple intestinal neoplasia ($APC^{Min/+}$) mice these mice developed significantly more adenomas than littermate controls (Sinnamon et al. 2008), suggesting a protective role for mast cells in this model of early-stage intestinal tumorigenesis. Differences in these studies may be attributable to genetic differences in mouse strains and developing tumors, additional immunological defects in c-Kit mutant mice or differences in gut microflora contributing to mast cell activation status. It will be useful to examine tumorigenesis in the recently developed, non c-Kit mutant, mast cell deficient models and other tumor subtypes to determine the true impact of mast cells on tumor incidence.

20.3.6 Mast Cells and Angiogenesis

Activated mast cells can secrete angiogenesis promoting factors including VEGF, bFGF, TGF-β, TNF and CXCL8 and release extracellular matrix (ECM) bound pro-angiogenic factors via action of granule associated proteases (Norrby 2002). Angiogenesis is required for macroscopic tumor expansion and metastasis (Folkman 1990). Mast cells are increased in most solid carcinomas and their numbers correlate with increased microvessel density, increased invasiveness and poor clinical outcome (reviewed in Groot Kormelink et al. 2009). Within the tumor microenvironment, mast cells produce many pro-angiogenic mediators including VEGF and bFGF and their expression correlates with microvessel density (Esposito et al. 2004; Toth-Jakatics et al. 2000). Several experimental model systems have demonstrated the pro-angiogenic activity of mast cells. Early animal studies that compared melanoma and bladder cancer cell growth in mast cell deficient $Kit^{W/Wv}$ mice and wild-type littermates demonstrated tumor angiogenesis was decreased in mast cell deficient mice (Starkey et al. 1988; Dethlefsen et al. 1994). In a xenograft model of human thyroid cancer, transferred human mast cells were recruited to tumor sites and enhanced tumor growth and angiogenesis (Melillo et al. 2010). In murine tumor models, co-transfer of bone marrow derived mast cells significantly enhanced angiogenesis in a murine plasmacytoma model via production of angiopoietin-1 (Nakayama et al. 2004) and mast cells were demonstrated to be critical for macroscopic tumor expansion in experimental models of human papilloma virus induced sarcoma (Coussens et al. 1999) and *Myc*-induced pancreatic carcinoma (Soucek et al. 2007).

Early studies using the chick chorioallantoic membrane assay demonstrated that degranulating rodent mast cells increased the angiogenic response via production of bFGF and VEGF (Ribatti et al. 2001; Rizzo and DeFouw 1996). Primary human lung mast cells, cord blood derived human mast cells and the HMC-1 and LAD-2 mast cell lines express several isoforms of VEGF (Abdel-Majid and Marshall 2004). Blocking antibody studies using a chick chorio allantoic membrane assay demonstrated human lung mast cells enhance the angiogenic response in a VEGF-A dependent manner (Detoraki et al. 2009) and have also identified pro-angiogenic roles for mast cell tryptase and chymase (Ribatti et al. 2011).

Mast cell degranulation at tumor sites has been detected immunohistochemically using anti-tryptase antibodies and ultrastructurally by electron microscopy

(Samoszuk et al. 2005; Caruso et al. 2004). Mediators released by degranulating mast cells enhance angiogenesis in vitro and in experimental tumor models. Several in vitro experiments have identified roles for mast cell mediators in enhancing the angiogenic response. Mast cell tryptases (mMCP-6 and mMCP-7) induce endothelial spreading and tube formation in an in vitro angiogenesis assay (de Souza et al. 2012). Mast cell derived proteases can also contribute to the angiogenic process in the tumor microenvironment through their enzymatic release of ECM sequestered pro-angiogenic factors. For example, mMCP-4 chymase released from mast cells can enhance the angiogenic response in hyperplastic skin via activation of pro-MMP-9, which can liberate angiogenic factors such as VEGF and bFGF from neoplastic tissues (Coussens et al. 1999). Mast cell modulation of angiogenesis is likely altered throughout the course of tumorigenesis Indeed, levels of the mast cell chymase (mMCP-5), mast cell tryptases (mMCP-6 and mMCP-7) and carboxypeptidase A increased progressively with tumorigenesis and correlated with increased angiogenic response in a murine model of chemically induced skin carcinoma (de Souza et al. 2012).

While there is an abundance of experimental data to support the angiogenesis promoting activities of mast cells in the context of tumorigenesis, experimental models have also identified anti-angiogenic factors that can be released from mast cells in the tumor microenvironment. Anti-angiogenic effects of mast cell derived PGD_2 have been observed in a murine model of lung carcinoma where mast cell derived PGD_2 inhibits production of the pro-angiogenic cytokine TNF and decreases vascular permeability in the developing tumor (Murata et al. 2011). These data provide additional support for the concept that mast cells can play a dual role in the tumor microenvironment (Theoharides and Conti 2004).

20.3.7 Mast Cells, Tissue Remodelling and Metastasis

Mast cells are important regulators of tissue remodelling events in the tumor microenvironment through their interactions with stromal cells, such as fibroblasts and myofibroblasts and their release of tissue degrading enzymes (Fig. 20.3). Several mast cell mediators likely contribute to the increased fibroblast activity in the tumor microenvironment (Dvorak 1986). In vitro co-culture studies have demonstrated that mast cell tryptase can enhance human fibroblast chemotaxis, proliferation and pro-collagen synthesis (Gruber et al. 1997). Murine studies have demonstrated that mMCP-6 tryptase release from mast cells induces fibroblast proliferation in vitro and that tryptase expressing mast cell numbers correlate with increased fibroblast numbers and synthesis of $\alpha1$ type I procollagen in dysplastic skin lesions (Coussens et al. 1999). Additional mediators, released from mast cells, are likely to contribute to fibroplasia in the tumor microenvironment. For example, human mast cells secrete bioactive TGF-$\beta1$ (Kanbe et al. 1999), which is a well-documented enhancer of fibrogenesis and ECM molecule synthesis in the tumor microenvironment (Bissell 2001). In addition to mast cell derived TGF-β, MMP

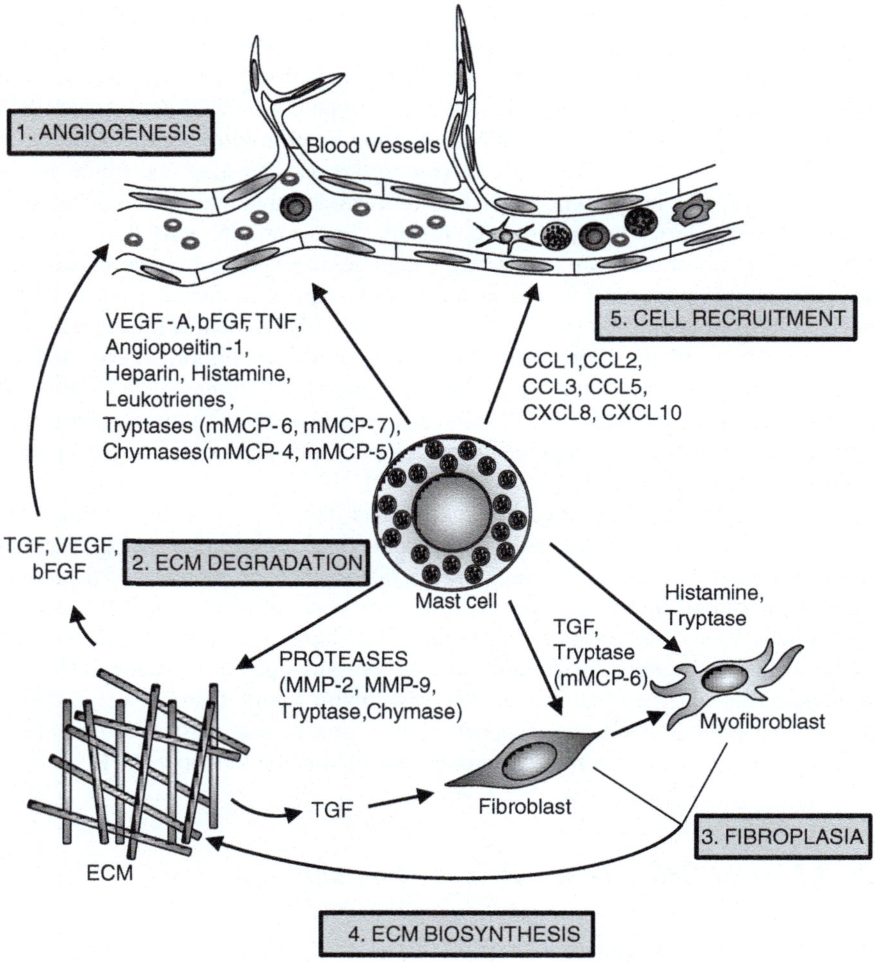

Fig. 20.3 Mast cells contribute to de-regulated tissue homeostasis in the tumor microenvironment. (1) Activated mast cells release many pro-angiogenic factors, which enhance endothelial cell migration, proliferation and blood vessel formation. (2) Mast cells secrete multiple proteases, which degrade the extracellular matrix (ECM) and release growth factors that have been sequestered in the ECM to enhance fibroblast proliferation and the angiogenic response. (3) Mast cells also enhance fibroplasia in the tumor microenvironment through their release of mediators that induce fibroblast proliferation and differentiation into myofibroblasts. (4) Fibroblasts and myofibroblasts activated by mast cell mediators synthesize ECM molecules, contributing to tissue remodelling events. (5) Mast cells indirectly modulate the tumor microenvironment through their release of multiple chemokines, which mediate the recruitment of additional immune effector cells to tumor sites

activation by mast cell proteases is likely to induce the activation of ECM-bound forms of latent TGF-β in the tumor microenvironment potentiating the response.

Myofibroblasts are contractile fibroblasts which express features of smooth muscle differentiation. They are important contributors to tissue remodelling through their release of mediators that interact with epithelial cells, production of ECM components and their ability to provide contractile force to facilitate wound closure. Tissue remodelling events in the tumor microenvironment are likely influenced via the direct actions of mast cell mediators on myofibroblasts. Numbers of α-smooth muscle actin (SMA) expressing myofibroblasts correlate with tryptase positive mast cells in human breast cancer tissue (Mangia et al. 2011). In vitro, human mast cell tryptase and histamine induce α-SMA expression in dermal fibroblasts and tryptase also stimulates their ability to contract a collagen matrix (Gailit et al. 2001). Thus, through direct release of ECM degrading enzymes and interactions with stromal cells mast cells are important contributors to the remodelling events that occur during tumorigenesis.

Tumor metastasis involves tissue remodelling events which enable tumor invasiveness and spread. Migration and invasiveness of tumor cells are considered prerequisites for tumor metastasis and a high mast cell density often correlates with lymph node metastases (Esposito et al. 2004; Cai et al. 2011; Elpek et al. 2001). Early evidence for a role of mast cells in tumor metastasis comes from experimental tumor models where metastasis incidence is decreased in mast cell deficient mice (Starkey et al. 1988; Dethlefsen et al. 1994). Several groups have demonstrated that mast cell mediators can increase tumor cell migration and invasiveness in vitro. For example, chymase degrades fibronectin and enhances cervical cancer cell detachment, while tryptase increases invasion and metastasis of breast cancer cells and activates the ECM degrading gelatinase pro-MMP-2 (Diaconu et al. 2011; Xiang et al. 2010). Many mediators released from mast cells (e.g., histamine, leukotrienes, TNF, VEGF, tryptase) can increase vascular permeability (Kunder et al. 2011) and may thereby enhance tumor cell extravasation and spread.

20.4 Mast Cell Regulation of Tumor Immunity Within the Tumor Microenvironment

20.4.1 Mast Cell Regulation of Anti-Tumor Immune Responses

Mast cells can have profound effects on immune regulation relevant to the tumor microenvironment through mediator production and potentially through direct interactions with effector cells. Mast cells are sentinel cells with the ability to alert and mobilize immune responses through both the recruitment of effector cells and the initiation of immune processes. In the context of a tumor microenvironment, however, mast cells are often not activated by pathways that lead to a full complement of mediator release. As a consequence, depending upon the available microenvironmental or therapy-derived signals mast cells can both enhance and inhibit effective anti-tumor immune responses through distinct pathways.

A number of pathways exist whereby mast cells can have immunosuppressive functions. These include the direct release of immunosuppressive cytokines, the enhancement of activity or recruitment of suppressive cell types such as regulatory T cells (T_{reg}) and myeloid derived suppressor cells (MDSC) as well as the impact of preformed granule products such as proteases and histamine. Human and rodent mast cells can be substantial sources of a number of cytokines that could limit local or systemic anti-tumor immunity. These include both TGF-β and IL-10. While we have little information on microenvironmental factors that induce selective production of such mediators, the potential for them playing a role in limiting local immunity is clear. Vitamin D_3 has been demonstrated to be an important factor in inducing selective mast cell production of IL-10 in the skin (Grimbaldeston et al. 2007) and intensive research is underway to identify other relevant factors with similar activity. The action of mast cell proteases in activating latent TGF-β could add to their immunosuppressive role following mast cell degranulation. An additional immunosuppressive activity of proteases may be in the degradation of other bioactive cytokines and chemokines rendering them inactive. This could limit the long term recruitment of certain cell types, although in murine models, mast cell proteases have been shown to be important for acute selective inflammatory cell recruitment by a separate mechanism (Shin et al. 2009; Huang et al. 1998).

Mast cell interactions with MDSCs may also impact effective tumor immunity. In a murine hepatocellular carcinoma model it was observed that mast cells could mobilize the infiltration of MDSCs and induce their production of IL-17. In turn, MDSC-derived IL-17 acted to mobilize T_{reg} cells, enhancing their suppressor function and inducing IL-9 production. The IL-9 produced locally then had the capacity to promote the survival of mast cells in the tumor microenvironment (Yang et al. 2010). In marked contrast, a more recent study has demonstrated that the ability of monocyte–like MDSCs to exert an immune suppressive effect in a murine model of B16 melanoma metastasis was highly dependent on interactions with mast cells. This was further contrasted by a mast cell dependent immunostimulatory effect of granulocyte related MDSCs in a nematode parasite model (Saleem et al. 2012). Whether these mechanisms occur in a human tumor setting remains unclear.

Histamine has been widely studied as a potential inhibitor of effective anti-tumor immunity as in some situations it may reduce the development of effective acquired immune responses to tumors. In addition, histamine may impact more directly on tumor growth, since several tumor cell types express histamine receptors (Lieberman 2011). The high histamine content of many human tumors, largely as a result of associated tissue mast cells has been cited as having both pro-

and anti-tumorigenic effects. Histamine is one of the major mediators leading to the pathologic events that follow allergic activation. Important immunomodulatory roles for histamine have been observed, controlling both innate and adaptive immune responses, through actions on H_1 and H_2 receptors on immune cells. Histamine acting via H_2 on monocytes and phagocytes can block their cell-contact dependent immune suppression of natural killer (NK) cells to enhance natural and antibody mediated NK cell cytotoxicity (Hellstrand and Hermodsson 1991; Hellstrand et al. 1994) and prevent phagocyte derived reactive oxygen species down-regulation of NKp46 and NKG2D activating receptors (Romero et al. 2006). Histamine is also an important regulator of adaptive immune responses via its actions on dendritic cell (DC) and T cell subsets. Studies in $H1R^{-/-}$ mice have demonstrated that H_1 engagement on DC is necessary for DC activation and subsequent priming of IFN-γ producing $CD8^+$ T cells (Vanbervliet et al. 2011). Histamine, via H_2, can modulate the differentiation, activation and functional activities of monocyte derived DC subsets (Simon et al. 2011), enhance antigen uptake by immature DC, induce the recruitment of $CD11b^+$ DC and plasmacytoid DC (pDC) to draining lymph nodes (Dawicki et al. 2010) and modulate cytokine production in activated pDC (Mazzoni et al. 2003). Histamine is also a potent T cell regulator, which has been demonstrated to enhance T_H1-type responses and inhibit T_{reg} suppression via the H_1 receptor but suppress both T_H1 and T_H2 T cell responses through the H_2 receptor (Noubade et al. 2007; Forward et al. 2009; Jutel et al. 2001).

Levels of circulating histamine are three times higher in some groups of newly diagnosed cancer patients and remain high for two months post-surgical removal (Moriarty et al. 1988). In a long term follow-up study of solid malignancies, a progressive decrease in histamine blood levels preceded clinical relapse or detection of metastasis (Burtin et al. 1983). Clinically, histamine has shown promise as an immune stimulant when combined with IL-2 immunotherapy in clinical trials of acute myeloid leukemia and melanoma patients (Hellstrand et al. 1997). The enhanced efficacy of IL-2 immunotherapy in the presence of histamine has been attributed to enhanced NK cell mediated killing of tumor cells (Hellstrand et al. 1997; Brune et al. 1996) and enhanced type 1 T cell responses (Asemissen et al. 2005). The aforementioned immunomodulatory effects of histamine indicate histamine can enhance NK cytotoxicity indirectly in a H_2 dependent manner and enhance type 1 T cell responses via interactions with H_1 and H_2 on T cells and DC. These findings are in keeping with the results of some animal models studies which demonstrate a role for histamine in regulating colorectal cancer development following treatment with carcinogenic agents (Yang et al. 2011). Within the tumor microenvironment, histamine and histamine receptor interactions have not been fully investigated and there is much work to be done to fully understand the role of histamine within the tumor setting. However, it is clear that there is great potential for mast cells to modulate tumor immunity and microenvironmental changes through histamine dependent mechanisms.

Mast Cells and Dendritic Cells Within the Tumor Microenvironment

A number of studies have shown that both histamine and mast cells have an important role in regulating DC populations and their mobilization. This work has been mainly carried out using models of infection or allergic disease. However, it has important implications for the potential impact on mast cells in the tumor microenvironment. A number of mast cell mediators including histamine and TNF have substantial effects on DC mobilization and polarization. DC migration from the tissue into the draining lymph node is a critical first step for the generation of an effective immune response. Early studies of Langerhan's cells in the skin demonstrated that mast cell activation with IgE/antigen can lead to the mobilization of DC out of the epidermis and into the draining lymph node by a mechanism that was dependent on mast cells, TNF and histamine (Jawdat et al. 2004). Similarly, bacterial peptidoglycan treatment of the skin led to the mobilization of Langerhan's cells by a mast cell dependent mechanism, although in this case the response was TNF independent. From more recent studies it has become apparent that mast cells also have important roles in the mobilization of other DC subsets of particular importance for the development of anti-tumor immunity including $CD8^+$ DC and pDC (Dawicki et al. 2010). Interaction between mast cells and DC has also been shown to enhance T_H1 and T_H17 responses (Dudeck et al. 2011) which, depending upon the stage of tumor development and location, could either aid in effective tumor immunity development or promote inflammatory changes that enhance local angiogenesis. In other settings, it has been demonstrated that mast cell activators can serve as effective vaccine adjuvants with potential for cancer therapy (McLachlan et al. 2008). The recent recognition of important roles for c-Kit in DC biology may suggest that the direct impact of c-Kit on DCs needs to be considered in interpreting findings from mast cell deficient mouse models.

Mast Cell Effects on Local T Cell Populations

The role of mast cells may not always be to enhance tumor immunity through interactions with DCs. For example, prostanoids produced by mast cells in response to IgE-mediated activation, while potentially limiting angiogenesis (Murata et al. 2011) could act on DP1 receptors on dendritic cells to promote T_{reg} development (Hammad et al. 2007). PGD_2 has also been shown to inhibit IL-12 responses by DC leading to increased T_H2 polarization (Theiner et al. 2006), potentially further restricting the development of an effective anti-tumor T cell response.

Mast cells may modulate T cell responses by mechanisms that are independent of effects on dendritic cell function (Nakae et al. 2006). Such interactions can include the direct effects of mast cell mediators such as TNF and histamine as well as cell contact dependent events. Mast cell modulation of local T_{reg} populations within the tumor microenvironment may be of particular importance in dictating the effectiveness of T cell responses to tumors. Mast cells are an excellent source of IL-6 following activation with multiple different stimuli. In particular, mast cell IL-6 can be produced selectively following TLR-mediated activation or following IgE mediated mast cell activation, in the latter case associated with degranulation. IL-6 plays a critical role in regulating the function of T_{reg} cells and can alter the activities of such cells so that they develop pro-inflammatory T_H17 characteristics. Since an abundance of T_{reg} cells within the tumor microenvironment is one mechanism whereby immune responses to tumors are thought to be frequently suppressed, activation of mast cells could provide opportunities to reduce local T_{reg} activities.

Cell Recruitment via Chemokine Production

During early infection and inflammation mast cells bring effector cells to local tissue sites (reviewed in Marshall 2004). This occurs as a consequence of a multi-step recruitment process that involves increased adhesion interactions, altered vascular permeability as well as direct chemoattractant actions. Although mast cells can influence all of these steps, the extent to which these processes occur in a tumor microenvironment, in the absence of immunotherapy, remains unclear. Mast cell influence on many immune responses has frequently been associated with production of TNF. This cytokine is produced following mast cell activation and has also been reported to be preformed and associated with mast cell granules and released immediately upon degranulation. One of the major actions of TNF is the up-regulation of adhesion molecules on the vascular endothelium. IL-1 produced by mast cells would also contribute to such vascular effects.

Multiple chemokines can be produced by mast cells, however, they are known to be a particularly potent tissue source of CXCL8, CCL3, CCL4, CCL5 and CXCL10 and also produce other key chemokines such as CCL2 and CXCL9. Mast cell derived CXCL8 can not only recruit neutrophils, but has more recently been shown to be a potent and selective inducer of human NK cell recruitment in the context of virally activated mast cells (Burke et al. 2008). In contrast, CCR3 and CCR5 ligands have been shown to be more important for the recruitment of $CD56^+$ T cells, including invariant NKT cells as well as cytotoxic T cells (McAlpine et al. 2012). In granulomas, neutrophils recruited as a result of mast cell derived TNF production, have also been implicated as an important source of CCL3 (von Stebut et al. 2003). This illustrates the types of complex cell–cell interactions that can lead to a chemokine rich microenvironment. In the B16 mouse melanoma model CCL2 and CCL3 have been demonstrated to have important roles in host immunity and prevention of metastasis (Nakasone et al. 2012). Notably CCL2 is also active in mast cell recruitment. CCL3 has also been shown to be critical for the development of immune responses following tumor cell apoptosis (Iida et al. 2008). Both mast cells and macrophages probably contribute to such critical chemokine responses.

CXCR3 ligands (CXCL9, CXCL10 and CXCL11) are all found in tissue microenvironments and contribute to angiogenic responses as well as to the recruitment of T cells to tumor sites. Human mast cells can produce CXCL10 in response to IFN-γ or to viral infection. IFN-γ is frequently increased at local tumor sites, as are CXCR3 ligands. Recently, it has been shown that CXCR3 ligands can also cause partial mast cell degranulation and activate several signaling pathways that could result in cytokine and chemokine production from mast cells (Willox et al. 2010). This could be an important mechanism for local mast cell activation within a tumor setting with CXCR3 ligands providing a regulatory link between the processes of angiogenesis and cell recruitment.

20.4.2 Mast Cells and Tumor Immunotherapy

Mast cells represent attractive targets for anti-cancer immunotherapy due to their abundance at the periphery of many solid tumors and ideal location in close proximity to blood vessels. In concordance with the well documented pro-tumorigenic roles of mast cells, several groups have suggested that therapeutic strategies that are designed to inhibit mast cells and their mediators are viable approaches for the treatment of solid tumors (Groot Kormelink et al. 2009; Maltby et al. 2009). Indeed, therapeutic success has been achieved in experimental tumor models using anti-inflammatory approaches that have impact on mast cell numbers. For example infliximab treatment of experimental colitis reduced both mast cell responses and the development of later tumors (Kim et al. 2010). In a model of mast cell enhanced human thyroid tumor growth mast cell stabilization was also shown to effectively reduce tumor growth (Melillo et al. 2010). We have recently demonstrated that mast cells can play a critical role in mediating the anti-tumor effects of TLR2 targeted immunotherapy in a murine model of melanoma and that TLR2 activated mast cells can recruit NK cells and T cells in a CCL3-dependent manner (Oldford et al. 2010). Mast cells have also been demonstrated to be crucial for the recruitment of pDC via CCL2 production and regulation of anti-tumor immunity following TLR7 targeted immunotherapy of murine melanoma (Drobits et al. 2012). These data suggest an alternate approach that harnesses the immune potential of mast cells and specifically targets the sentinel role of mast cells using innate immune activator based immunotherapy strategies. In contrast to other immune cells, mast cells are relatively radioresistant (Soule et al. 2007), which makes them prime candidates for combined treatment modalities. Furthermore, widespread inhibition of mast cell function may not be advantageous for all tumor types. Inhibition of mast cell function by administration of imatinib mesylate (GLEEVEC™), a receptor tyrosine kinase inhibitor which inhibits c-Kit, resulted in increased mammary cancer development and peritumoral blood clotting in a murine model of breast carcinoma (Samoszuk and Corwin 2003). In a murine model of prostate carcinoma imatinib administration decreased the incidence of prototypical prostate carcinoma, but increased the incidence of prostate carcinomas with the more aggressive neuroendocrine phenotype (Pittoni et al. 2011). Thus, it is evident that the influence of mast cells on the tumor microenvironment is complex and is likely dependent on the state of mast cell activation at the tumor site. Selective mast cell activation therefore provides a novel opportunity to modify the tumor microenvironment for successful cancer immunotherapy.

The use of monoclonal IgE antibodies directed against tumor associated antigens has shown some success in experimental tumor models (Jensen-Jarolim et al. 2008). The effectiveness of this treatment may be the result of a variety of mechanisms including the enhanced recruitment of effector cells, the direct impact of mediators on tumor cells and impact on immune regulation. IgE-mediated mast cell degranulation can overcome allograft tolerance and elicit effector $CD4^+$ and $CD8^+$ T cell mediated rejection (de Vries et al. 2009). Therefore, local mast cell degranulation may be advantageous to subvert the immune suppressive microenvironment of solid tumors and elicit effective anti-tumor immunity. However, IgE mediated mast cell activation does not always overcome tolerance and in models of oral tolerance it has been shown to be ineffective in altering T_{reg} responses (Tunis et al. 2012). More work is required to directly assess the impact of such mechanisms on T cell responses to tumors.

20.5 Concluding Remarks

From the discussion above, it is clear that mast cells can have multiple roles within the tumor microenvironment. These extend from the most widely known pro-tumorigenic effects of mast cells in the promotion of angiogenesis through to immune activating impacts of mast cells that could enhance effective host defence and even the use of mast cells as potential triggers for immunotherapy. The challenges we currently face in understanding and exploiting these responses therapeutically are twofold. First, we need to determine which interactions are of importance in a human tumor setting, since much of the available data has been obtained from animal models. Second, to develop strategies to exploit or alter human mast cell responses within the tumor microenvironment appropriately and selectively such that their positive roles can be enhanced and their pro-tumorigenic and pro-metastatic effects can be reduced. Mast cells have enormous potential as targets to selectively alter both tissue remodelling events and immune responses if we are willing to take on these challenges.

References

Abdel-Majid RM, Marshall JS (2004) Prostaglandin E2 induces degranulation-independent production of vascular endothelial growth factor by human mast cells. J Immunol 172:1227–1236

Asemissen AM, Scheibenbogen C, Letsch A, Hellstrand K, Thoren F, Gehlsen K et al (2005) Addition of histamine to interleukin 2 treatment augments type 1 T-cell responses in patients with melanoma in vivo: immunologic results from a randomized clinical trial of interleukin 2 with or without histamine (MP 104). Clin Cancer Res 11:290–297

Basu S, Binder RJ, Suto R, Anderson KM, Srivastava PK (2000) Necrotic but not apoptotic cell death releases heat shock proteins, which deliver a partial maturation signal to dendritic cells and activate the NF-kappa B pathway. Int Immunol 12:1539–1546

Benitez-Bribiesca L, Wong A, Utrera D, Castellanos E (2001) The role of mast cell tryptase in neoangiogenesis of premalignant and malignant lesions of the uterine cervix. J Histochem Cytochem Off J Histochem Soc 49:1061–1062

Benyon RC, Bissonnette EY, Befus AD (1991) Tumor necrosis factor-alpha dependent cytotoxicity of human skin mast cells is enhanced by anti-IgE antibodies. J Immunol 147:2253–2258

Bischoff SC, Sellge G, Lorentz A, Sebald W, Raab R, Manns MP (1999) IL-4 enhances proliferation and mediator release in mature human mast cells. Proc Natl Acad Sci U S A 96:8080–8085

Bissell DM (2001) Chronic liver injury, TGF-beta, and cancer. Exp Mol Med 33:179–190

Boyce JA (2007) Mast cells and eicosanoid mediators: a system of reciprocal paracrine and autocrine regulation. Immunol Rev 217:168–185

Branitzki-Heinemann K, Okumura CY, Vollger L, Kawakami Y, Kawakami T, Naim HY et al (2012) A novel role for the transcription factor HIF-1alpha in the formation of mast cell extracellular traps. Biochem J 446:159–163

Brown MA, Hatfield JK, Walker ME, Sayed BA (2012) A game of kit and mouse: the Kit is still in the bag. Immunity 36:891–892 (author reply 3-4)

Brune M, Hansson M, Mellqvist UH, Hermodsson S, Hellstrand K (1996) NK cell-mediated killing of AML blasts: role of histamine, monocytes and reactive oxygen metabolites. Eur J Haematol 57:312–319

Burke SM, Issekutz TB, Mohan K, Lee PW, Shmulevitz M, Marshall JS (2008) Human mast cell activation with virus-associated stimuli leads to the selective chemotaxis of natural killer cells by a CXCL8-dependent mechanism. Blood 111:5467–5476

Burtin C, Noirot C, Paupe J, Scheinmann P (1983) Decreased blood histamine levels in patients with solid malignant tumours. Br J Cancer 47:367–372

Byrne SN, Beaugie C, O'Sullivan C, Leighton S, Halliday GM (2011) The immune-modulating cytokine and endogenous Alarmin interleukin-33 is upregulated in skin exposed to inflammatory UVB radiation. Am J Pathol 179:211–222

Cai SW, Yang SZ, Gao J, Pan K, Chen JY, Wang YL et al (2011) Prognostic significance of mast cell count following curative resection for pancreatic ductal adenocarcinoma. Surgery 149:576–584

Caruso RA, Ieni A, Fabiano V, Basile G, Inferrera C (2004) Perivascular mast cells in advanced gastric adenocarcinomas: an electron microscopic study. Anticancer Res 24:2257–2263

Caughey GH (2011) Mast cell proteases as protective and inflammatory mediators. Adv Exp Med Biol 716:212–234

Chan JK, Roth J, Oppenheim JJ, Tracey KJ, Vogl T, Feldmann M et al (2012) Alarmins: awaiting a clinical response. J Clin Investig 122:2711–2719

Ciocca DR, Calderwood SK (2005) Heat shock proteins in cancer: diagnostic, prognostic, predictive, and treatment implications. Cell Stress Chaperones 10:86–103

Conti P, Castellani ML, Kempuraj D, Salini V, Vecchiet J, Tete S et al (2007) Role of mast cells in tumor growth. Ann Clin Lab Sci 37:315–322

Coussens LM, Raymond WW, Bergers G, Laig-Webster M, Behrendtsen O, Werb Z et al (1999) Inflammatory mast cells up-regulate angiogenesis during squamous epithelial carcinogenesis. Genes Dev 13:1382–1397

Crivellato E, Beltrami C, Mallardi F, Ribatti D (2003) Paul Ehrlich's doctoral thesis: a milestone in the study of mast cells. Br J Haematol 123:19–21

Dawicki W, Marshall JS (2007) New and emerging roles for mast cells in host defence. Curr Opin Immunol 19:31–38

Dawicki W, Jawdat DW, Xu N, Marshall JS (2010) Mast cells, histamine, and IL-6 regulate the selective influx of dendritic cell subsets into an inflamed lymph node. J Immunol 184:2116–2123 de Souza DA Jr, Toso VD, Campos MR, Lara VS, Oliver C, Jamur MC (2012) Expression of mast cell proteases correlates with mast cell maturation and angiogenesis during tumor progression. PloS ONE 7:e40790

de Vries VC, Wasiuk A, Bennett KA, Benson MJ, Elgueta R, Waldschmidt TJ et al (2009) Mast cell degranulation breaks peripheral tolerance. Am J Transpl Off J Am Soc Transpl Am Soc Transpl Surg 9:2270–2280

Dery RE, Lin TJ, Befus AD, Milne CD, Moqbel R, Menard G et al (2000) Redundancy or cell-type-specific regulation? Tumour necrosis factor in alveolar macrophages and mast cells. Immunology 99:427–434

Dethlefsen SM, Matsuura N, Zetter BR (1994) Mast cell accumulation at sites of murine tumor implantation: implications for angiogenesis and tumor metastasis. Invasion Metastasis 14:395–408

Detoraki A, Staiano RI, Granata F, Giannattasio G, Prevete N, de Paulis A et al (2009) Vascular endothelial growth factors synthesized by human lung mast cells exert angiogenic effects. J Allergy Clin Immunol 123:1142–1149, 9 e1-5

Di Nardo A, Vitiello A, Gallo RL (2003) Cutting edge: mast cell antimicrobial activity is mediated by expression of cathelicidin antimicrobial peptide. J Immunol 170:2274–2278

Diaconu NC, Rummukainen J, Naukkarinen A, Matto M, Harvima RJ, Pelkonen J et al (2011) Mast cell chymase is present in uterine cervical carcinoma and it detaches viable and growing cervical squamous carcinoma cells from substratum in vitro. Arch Dermatol Res 303:499–512

Drobits B, Holcmann M, Amberg N, Swiecki M, Grundtner R, Hammer M et al (2012) Imiquimod clears tumors in mice independent of adaptive immunity by converting pDCs into tumor-killing effector cells. J Clin Investig 122:575–585

Dudeck A, Dudeck J, Scholten J, Petzold A, Surianarayanan S, Kohler A et al (2011) Mast cells are key promoters of contact allergy that mediate the adjuvant effects of haptens. Immunity 34:973–984

Dvorak HF (1986) Tumors: wounds that do not heal. Similarities between tumor stroma generation and wound healing. N Engl J Med 315:1650–1659

Elpek GO, Gelen T, Aksoy NH, Erdogan A, Dertsiz L, Demircan A et al (2001) The prognostic relevance of angiogenesis and mast cells in squamous cell carcinoma of the oesophagus. J Clin Pathol 54:940–944

Enoksson M, Lyberg K, Moller-Westerberg C, Fallon PG, Nilsson G, Lunderius-Andersson C (2011) Mast cells as sensors of cell injury through IL-33 recognition. J Immunol 186:2523–2528

Esposito I, Menicagli M, Funel N, Bergmann F, Boggi U, Mosca F et al (2004) Inflammatory cells contribute to the generation of an angiogenic phenotype in pancreatic ductal adenocarcinoma. J Clin Pathol 57:630–636

Feyerabend TB, Weiser A, Tietz A, Stassen M, Harris N, Kopf M et al (2011) Cre-mediated cell ablation contests mast cell contribution in models of antibody- and T cell-mediated autoimmunity. Immunity 35:832–844

Finotto S, Dolovich J, Denburg JA, Jordana M, Marshall JS (1994) Functional heterogeneity of mast cells isolated from different microenvironments within nasal polyp tissue. Clin Exp Immunol 95:343–350

Fischer M, Juremalm M, Olsson N, Backlin C, Sundstrom C, Nilsson K et al (2003) Expression of CCL5/RANTES by Hodgkin and Reed-Sternberg cells and its possible role in the recruitment of mast cells into lymphomatous tissue. Int J Cancer J Int du Cancer 107:197–201

Fleischmann A, Schlomm T, Kollermann J, Sekulic N, Huland H, Mirlacher M et al (2009) Immunological microenvironment in prostate cancer: high mast cell densities are associated with favorable tumor characteristics and good prognosis. Prostate 69:976–981

Folkman J (1990) What is the evidence that tumors are angiogenesis dependent? J Natl Cancer Inst 82:4–6

Forward NA, Furlong SJ, Yang Y, Lin TJ, Hoskin DW (2009) Mast cells down-regulate CD4+CD25+ T regulatory cell suppressor function via histamine H1 receptor interaction. J Immunol 183:3014–3022

Gailit J, Marchese MJ, Kew RR, Gruber BL (2001) The differentiation and function of myofibroblasts is regulated by mast cell mediators. J Invest Dermatol 117:1113–1119

Galli SJ, Tsai M (2008) Mast cells: versatile regulators of inflammation, tissue remodeling, host defense and homeostasis. J Dermatol Sci 49:7–19

Galli SJ, Nakae S, Tsai M (2005) Mast cells in the development of adaptive immune responses. Nat Immunol 6:135–142

Gounaris E, Erdman SE, Restaino C, Gurish MF, Friend DS, Gounari F et al (2007) Mast cells are an essential hematopoietic component for polyp development. Proc Natl Acad Sci U S A 104:19977–19982

Grimbaldeston MA, Chen CC, Piliponsky AM, Tsai M, Tam SY, Galli SJ (2005) Mast cell-deficient W-sash c-kit mutant Kit W-sh/W-sh mice as a model for investigating mast cell biology in vivo. Am J Pathol 167:835–848

Grimbaldeston MA, Nakae S, Kalesnikoff J, Tsai M, Galli SJ (2007) Mast cell-derived interleukin 10 limits skin pathology in contact dermatitis and chronic irradiation with ultraviolet B. Nat Immunol 8:1095–1104

Groot Kormelink T, Abudukelimu A, Redegeld FA (2009) Mast cells as target in cancer therapy. Current Pharm Design 15:1868–1878

Gruber BL, Marchese MJ, Kew RR (1994) Transforming growth factor-beta 1 mediates mast cell chemotaxis. J Immunol 152:5860–5867

Gruber BL, Kew RR, Jelaska A, Marchese MJ, Garlick J, Ren S et al (1997) Human mast cells activate fibroblasts: tryptase is a fibrogenic factor stimulating collagen messenger ribonucleic acid synthesis and fibroblast chemotaxis. J Immunol 158:2310–2317

Gulliksson M, Carvalho RF, Ulleras E, Nilsson G (2010) Mast cell survival and mediator secretion in response to hypoxia. PLoS ONE 5:e12360

Gurish MF, Austen KF (2012) Developmental origin and functional specialization of mast cell subsets. Immunity 37:25–33

Halova I, Draberova L, Draber P (2012) Mast cell chemotaxis—chemoattractants and signaling pathways. Frontiers Immunol 3:119

Hammad H, Kool M, Soullie T, Narumiya S, Trottein F, Hoogsteden HC et al (2007) Activation of the D prostanoid 1 receptor suppresses asthma by modulation of lung dendritic cell function and induction of regulatory T cells. J Exp Med 204:357–367

Hedstrom G, Berglund M, Molin D, Fischer M, Nilsson G, Thunberg U et al (2007) Mast cell infiltration is a favourable prognostic factor in diffuse large B-cell lymphoma. Br J Haematol 138:68–71

Hellstrand K, Hermodsson S (1991) Cell-to-cell mediated inhibition of natural killer cell proliferation by monocytes and its regulation by histamine H2-receptors. Scand J Immunol 34:741–752

Hellstrand K, Asea A, Hermodsson S (1994) Histaminergic regulation of antibody-dependent cellular cytotoxicity of granulocytes, monocytes, and natural killer cells. J Leukoc Biol 55:392–397

Hellstrand K, Hermodsson S, Brune M, Naredi P, Carneskog J, Mellqvist UH (1997) Histamine in cancer immunotherapy. Scand J Clin Lab Invest 57:193–202

Hsu CL, Neilsen CV, Bryce PJ (2010) IL-33 is produced by mast cells and regulates IgE-dependent inflammation. PLoS ONE 5:e11944

Huang C, Friend DS, Qiu WT, Wong GW, Morales G, Hunt J et al (1998) Induction of a selective and persistent extravasation of neutrophils into the peritoneal cavity by tryptase mouse mast cell protease 6. J Immunol 160:1910–1919

Huang B, Lei Z, Zhang GM, Li D, Song C, Li B et al (2008) SCF-mediated mast cell infiltration and activation exacerbate the inflammation and immunosuppression in tumor microenvironment. Blood 112:1269–1279

Iida N, Nakamoto Y, Baba T, Kakinoki K, Li YY, Wu Y et al (2008) Tumor cell apoptosis induces tumor-specific immunity in a CC chemokine receptor 1- and 5-dependent manner in mice. J Leukoc Biol 84:1001–1010

Irani AA, Schechter NM, Craig SS, DeBlois G, Schwartz LB (1986) Two types of human mast cells that have distinct neutral protease compositions. Proc Natl Acad Sci U S A 83:4464–4468

Irani AM, Goldstein SM, Wintroub BU, Bradford T, Schwartz LB (1991) Human mast cell carboxypeptidase. Selective localization to MCTC cells. J Immunol 147:247–253

Jawdat DM, Albert EJ, Rowden G, Haidl ID, Marshall JS (2004) IgE-mediated mast cell activation induces Langerhans cell migration in vivo. J Immunol 173:5275–5282

Jensen-Jarolim E, Achatz G, Turner MC, Karagiannis S, Legrand F, Capron M et al (2008) AllergoOncology: the role of IgE-mediated allergy in cancer. Allergy 63:1255–1266

Johansson A, Rudolfsson S, Hammarsten P, Halin S, Pietras K, Jones J et al (2010) Mast cells are novel independent prognostic markers in prostate cancer and represent a target for therapy. Am J Pathol 177:1031–1041

Juremalm M, Olsson N, Nilsson G (2002) Selective CCL5/RANTES-induced mast cell migration through interactions with chemokine receptors CCR1 and CCR4. Biochem Biophys Res Commun 297:480–485

Jutel M, Watanabe T, Klunker S, Akdis M, Thomet OA, Malolepszy J et al (2001) Histamine regulates T-cell and antibody responses by differential expression of H1 and H2 receptors. Nature 413:420–425

Kanbe N, Kurosawa M, Nagata H, Saitoh H, Miyachi Y (1999) Cord blood-derived human cultured mast cells produce transforming growth factor beta1. Clin Exp Allergy J British Soc Allergy Clinical Immunol 29:105–113

Kandere-Grzybowska K, Letourneau R, Kempuraj D, Donelan J, Poplawski S, Boucher W et al (2003) IL-1 induces vesicular secretion of IL-6 without degranulation from human mast cells. J Immunol 171:4830–4836

Kankkunen JP, Harvima IT, Naukkarinen A (1997) Quantitative analysis of tryptase and chymase containing mast cells in benign and malignant breast lesions. Int J Cancer J Int du Cancer 72:385–388

Katz HR, Stevens RL, Austen KF (1985) Heterogeneity of mammalian mast cells differentiated in vivo and in vitro. J Allergy Clin Immunol 76:250–259

Kim YJ, Hong KS, Chung JW, Kim JH, Hahm KB (2010) Prevention of colitis-associated carcinogenesis with infliximab. Cancer Prev Res 3:1314–1333

Kryczek I, Lange A, Mottram P, Alvarez X, Cheng P, Hogan M et al (2005) CXCL12 and vascular endothelial growth factor synergistically induce neoangiogenesis in human ovarian cancers. Cancer Res 65:465–472

Kunder CA, St John AL, Abraham SN (2011) Mast cell modulation of the vascular and lymphatic endothelium. Blood 118:5383–5393

Leal-Berumen I, O'Byrne P, Gupta A, Richards CD, Marshall JS (1995) Prostanoid enhancement of interleukin-6 production by rat peritoneal mast cells. J Immunol 154:4759–4767

Lieberman P (2011) The basics of histamine biology. Ann Allergy Asthma Immunol Off Publ Am Coll Allergy Asthma Immunol 106:S2–S5

Lin TJ, Issekutz TB, Marshall JS (2001) SDF-1 induces IL-8 production and transendothelial migration of human cord blood-derived mast cells. Int Arch Allergy Immunol 124:142–145

Lloyd HG, Ross L, Li KM, Ludowyke RI (1998) Evidence that IgE receptor stimulation increases adenosine release from rat basophilic leukaemia (RBL-2H3) cells. Pulm Pharmacol Ther 11:41–46

Maltby S, Khazaie K, McNagny KM (2009) Mast cells in tumor growth: angiogenesis, tissue remodelling and immune-modulation. Biochim Biophys Acta 1796:19–26

Mangia A, Malfettone A, Rossi R, Paradiso A, Ranieri G, Simone G et al (2011) Tissue remodelling in breast cancer: human mast cell tryptase as an initiator of myofibroblast differentiation. Histopathology 58:1096–1106

Marshall JS (2004) Mast-cell responses to pathogens. Nat Rev Immunol 4 787–799

Matsuo Y, Ochi N, Sawai H, Yasuda A, Takahashi H, Funahashi H et al (2009) CXCL8/IL-8 and CXCL12/SDF-1alpha co-operatively promote invasiveness and angiogenesis in pancreatic cancer. Int J Cancer J Int du Cancer 124:853–861

Mazzoni A, Leifer CA, Mullen GE, Kennedy MN, Klinman DM, Segal DM (2003) Cutting edge: histamine inhibits IFN-alpha release from plasmacytoid dendritic cells. J Immunol 170:2269–2273

McAlpine SM, Issekutz TB, Marshall JS (2012) Virus stimulation of human mast cells results in the recruitment of CD56(+) T cells by a mechanism dependent on CCR5 ligands. FASEB J 26:1280–1289

McCurdy JD, Olynych TJ, Maher LH, Marshall JS (2003) Cutting edge: distinct toll-like receptor 2 activators selectively induce different classes of mediator production from human mast cells. J Immunol 170:1625–1629

McLachlan JB, Shelburne CP, Hart JP, Pizzo SV, Goyal R, Brooking-Dixon R et al (2008) Mast cell activators: a new class of highly effective vaccine adjuvants. Nat Med 14:536–541

Melillo RM, Guarino V, Avilla E, Galdiero MR, Liotti F, Prevete N et al (2010) Mast cells have a protumorigenic role in human thyroid cancer. Oncogene 29:6203–6215

Mierke CT, Ballmaier M, Werner U, Manns MP, Welte K, Bischoff SC (2000) Human endothelial cells regulate survival and proliferation of human mast cells. J Exp Med 192:801–811

Mohtasham N, Babakoohi S, Salehinejad J, Montaser-Kouhsari L, Shakeri MT, Shojaee S et al (2010) Mast cell density and angiogenesis in oral dysplastic epithelium and low- and high-grade oral squamous cell carcinoma. Acta Odontol Scand 68:300–304

Moriarty CM, Stucky JL, Hamburger KW, Patil KD, Foley JF, Koefoot RR (1988) Blood histamine and solid malignant tumors. J Cancer Res Clin Oncol 114:588–592

Murata T, Aritake K, Matsumoto S, Kamauchi S, Nakagawa T, Hori M et al (2011) Prostagladin D2 is a mast cell-derived antiangiogenic factor in lung carcinoma. Proc Natl Acad Sci U S A 108:19802–19807

Nakae S, Suto H, Iikura M, Kakurai M, Sedgwick JD, Tsai M et al (2006) Mast cells enhance T cell activation: importance of mast cell costimulatory molecules and secreted TNF. J Immunol 176:2238–2248

Nakasone Y, Fujimoto M, Matsushita T, Hamaguchi Y, Huu DL, Yanaba M et al (2012) Host-derived MCP-1 and MIP-1alpha regulate protective anti-tumor immunity to localized and metastatic B16 melanoma. Am J Pathol 180:365–374

Nakayama T, Yao L, Tosato G (2004) Mast cell-derived angiopoietin-1 plays a critical role in the growth of plasma cell tumors. J Clin Investig 114:1317–1325

Negri E, Bosetti C, La Vecchia C, Levi F, Tomei F, Franceschi S (1999) Allergy and other selected diseases and risk of colorectal cancer. Eur J Cancer 35:1838–1841

Nigrovic PA, Gray DH, Jones T, Hallgren J, Kuo FC, Chaletzky B et al (2008) Genetic inversion in mast cell-deficient (Wsh) mice interrupts corin and manifests as hematopoietic and cardiac aberrancy. Am J Pathol 173:1693–1701

Norrby K (2002) Mast cells and angiogenesis. APMIS: Acta Pathologica, Microbiologica, et Immunologica Scandinavica 110:355–371

Noubade R, Milligan G, Zachary JF, Blankenhorn EP, del Rio R, Rincon M et al (2007) Histamine receptor H1 is required for TCR-mediated p38 MAPK activation and optimal IFN-gamma production in mice. J Clin Investig 117:3507–3518

Oldford SA, Haidl ID, Howatt MA, Leiva CA, Johnston B, Marshall JS (2010) A critical role for mast cells and mast cell-derived IL-6 in TLR2-mediated inhibition of tumor growth. J Immunol 185:7067–7076

Olsson N, Piek E, ten Dijke P, Nilsson G (2000) Human mast cell migration in response to members of the transforming growth factor-beta family. J Leukoc Biol 67:350–356

Orimo A, Gupta PB, Sgroi DC, Arenzana-Seisdedos F, Delaunay T, Naeem R et al (2005) Stromal fibroblasts present in invasive human breast carcinomas promote tumor growth and angiogenesis through elevated SDF-1/CXCL12 secretion. Cell 121:335–348

Ott VL, Cambier JC, Kappler J, Marrack P, Swanson BJ (2003) Mast cell-dependent migration of effector CD8+ T cells through production of leukotriene B4. Nat Immunol 4:974–981

Ownby DR, Ownby HE, Bailey J, Frederick J, Tilley B, Brooks SC et al (1985) Presurgical serum immunoglobulin concentrations and the prognosis of operable breast cancer in women. J Natl Cancer Inst 75:655–663

Pittoni P, Piconese S, Tripodo C, Colombo MP (2011) Tumor-intrinsic and -extrinsic roles of c-Kit: mast cells as the primary off-target of tyrosine kinase inhibitors. Oncogene 30:757–769

Polajeva J, Sjosten AM, Lager N, Kastemar M, Waern I, Alafuzoff I et al (2011) Mast cell accumulation in glioblastoma with a potential role for stem cell factor and chemokine CXCL12. PLoS ONE 6:e25222

Rajput AB, Turbin DA, Cheang MC, Voduc DK, Leung S, Gelmon KA et al (2008) Stromal mast cells in invasive breast cancer are a marker of favourable prognosis: a study of 4,444 cases. Breast Cancer Res Treat 107:249–257

Ribatti D, Crivellato E, Candussio L, Nico B, Vacca A, Roncali L et al (2001) Mast cells and their secretory granules are angiogenic in the chick embryo chorioallantoic membrane. Clin Exp Allergy J British Soc Allergy Clin Immunol 31:602–608

Ribatti D, Ranieri G, Nico B, Benagiano V, Crivellato E (2011) Tryptase and chymase are angiogenic in vivo in the chorioallantoic membrane assay. Int J Dev Biol 55:99–102

Riley JF, West GB (1953) The presence of histamine in tissue mast cells. J Physiol 120:528–537

Rizzo V, DeFouw DO (1996) Mast cell activation accelerates the normal rate of angiogenesis in the chick chorioallantoic membrane. Microvasc Res 52:245–257

Romero AI, Thoren FB, Brune M, Hellstrand K (2006) NKp46 and NKG2D receptor expression in NK cells with CD56dim and CD56bright phenotype: regulation by histamine and reactive oxygen species. Br J Haematol 132:91–98

Rudich N, Ravid K, Sagi-Eisenberg R (2012) Mast cell adenosine receptors function: a focus on the a3 adenosine receptor and inflammation. Frontiers Immunol 3:134

Saleem SJ, Martin RK, Morales JK, Sturgill JL, Gibb DR, Graham L et al (2012) Cutting edge: mast cells critically augment myeloid-derived suppressor cell activity. J Immunol 189:511–515

Samoszuk M, Corwin MA (2003) Acceleration of tumor growth and peri-tumoral blood clotting by imatinib mesylate (Gleevec). Int J Cancer J Int du Cancer 106:647–652

Samoszuk M, Kanakubo E, Chan JK (2005) Degranulating mast cells in fibrotic regions of human tumors and evidence that mast cell heparin interferes with the growth of tumor cells through a mechanism involving fibroblasts. BMC Cancer 5:121

Sarchio SN, Kok LF, O'Sullivan C, Halliday GM, Byrne SN (2012) Dermal mast cells affect the development of sunlight-induced skin tumours. Exp Dermatol 21:241–248

Schwartzbaum J, Ding B, Johannesen TB, Osnes LT, Karavodin L, Ahlbom A et al (2012) Association between prediagnostic IgE levels and risk of glioma. J Natl Cancer Inst 104:1251–1259

Sherman PW, Holland E, Sherman JS (2008) Allergies: their role in cancer prevention. Q Rev Biol 83:339–362

Shin K, Nigrovic PA, Crish J, Boilard E, McNeil HP, Larabee KS et al (2009) Mast cells contribute to autoimmune inflammatory arthritis via their tryptase/heparin complexes. J Immunol 182:647–656

Simon T, Laszlo V, Falus A (2011) Impact of histamine on dendritic cell functions. Cell Biol Int 35:997–1000

Sinnamon MJ, Carter KJ, Sims LP, Lafleur B, Fingleton B, Matrisian LM (2008) A protective role of mast cells in intestinal tumorigenesis. Carcinogenesis 29:880–886

Soucek L, Lawlor ER, Soto D, Shchors K, Swigart LB, Evan GI (2007) Mast cells are required for angiogenesis and macroscopic expansion of Myc-induced pancreatic islet tumors. Nat Med 13:1211–1218

Soule BP, Brown JM, Kushnir-Sukhov NM, Simone NL, Mitchell JB, Metcalfe DD (2007) Effects of gamma radiation on FcepsilonRI and TLR-mediated mast cell activation. J Immunol 179:3276–3286

Starkey JR, Crowle PK, Taubenberger S (1988) Mast-cell-deficient W/Wv mice exhibit a decreased rate of tumor angiogenesis. Int J Cancer J Int du Cancer 42:48–52

Stevens RL, Adachi R (2007) Protease–proteoglycan complexes of mouse and human mast cells and importance of their beta-tryptase-heparin complexes in inflammation and innate immunity. Immunol Rev 217:155–167

Sumbayev VV, Yasinska I, Oniku AE, Streatfield CL, Gibbs BF (2012) Involvement of hypoxia-inducible factor-1 in the inflammatory responses of human LAD2 mast cells and basophils. PLoS ONE 7:e34259

Theiner G, Gessner A, Lutz MB (2006) The mast cell mediator PGD2 suppresses IL-12 release by dendritic cells leading to Th2 polarized immune responses in vivo. Immunobiology 211:463–472

Theoharides TC, Conti P (2004) Mast cells: the Jekyll and Hyde of tumor growth. Trends Immunol 25:235–241

Theoharides TC, Alysandratos KD, Angelidou A, Delivanis DA, Sismanopoulos N, Zhang B et al (2012) Mast cells and inflammation. Biochim Biophys Acta 1822:21–33

Tkaczyk C, Okayama Y, Metcalfe DD, Gilfillan AM (2004) Fcgamma receptors on mast cells: activatory and inhibitory regulation of mediator release. Int Arch Allergy Immunol 133:305–315

Toth-Jakatics R, Jimi S, Takebayashi S, Kawamoto N (2000) Cutaneous malignant melanoma: correlation between neovascularization and peritumor accumulation of mast cells overexpressing vascular endothelial growth factor. Hum Pathol 31:955–960

Tunis MC, Dawicki W, Carson KR, Wang J, Marshall JS (2012) Mast cells and IgE activation do not alter the development of oral tolerance in a murine model. J Allergy Clin Immunol 130:705–15 e1

Van Nguyen H, Di Girolamo N, Jackson N, Hampartzoumian T, Bullpitt P, Tedla N et al (2011) Ultraviolet radiation-induced cytokines promote mast cell accumulation and matrix metalloproteinase production: potential role in cutaneous lupus erythematosus. Scand J Rheumatol 40:197–204

Vanbervliet B, Akdis M, Vocanson M, Rozieres A, Benetiere J, Rouzaire P et al (2011) Histamine receptor H1 signaling on dendritic cells plays a key role in the IFN-gamma/IL-17 balance in T cell-mediated skin inflammation. J Allergy Clin Immunol 127:943–953 e1-10

von Stebut E, Metz M, Milon G, Knop J, Maurer M (2003) Early macrophage influx to sites of cutaneous granuloma formation is dependent on MIP-1alpha/beta released from neutrophils recruited by mast cell-derived TNFalpha. Blood 101:210–215

Wang Z, Macleod DT, Di Nardo A (2012) Commensal bacteria lipoteichoic acid increases skin mast cell antimicrobial activity against vaccinia viruses. J Immunol 189:1551–1558

Wedemeyer J, Galli SJ (2005) Decreased susceptibility of mast cell-deficient Kit(W)/Kit(W−v) mice to the development of 1, 2-dimethylhydrazine-induced intestinal tumors. Lab Investig J Tech Methods Pathol 85:388–396

Welsh TJ, Green RH, Richardson D, Waller DA, O'Byrne KJ, Bradding P (2005) Macrophage and mast-cell invasion of tumor cell islets confers a marked survival advantage in non-small-cell lung cancer. J Clin Oncol 23:8959–8967

Willox I, Mirkina I, Westwick J, Ward SG (2010) Evidence for PI3K-dependent CXCR3 agonist-induced degranulation of human cord blood-derived mast cells. Mol Immunol 47:2367–2377

Xia Q, Wu XJ, Zhou Q, Jing Z, Hou JH, Pan ZZ et al (2011) No relationship between the distribution of mast cells and the survival of stage IIIB colon cancer patients. J Transl Med 9:88

Xiang M, Gu Y, Zhao F, Lu H, Chen S, Yin L (2010) Mast cell tryptase promotes breast cancer migration and invasion. Oncol Rep 23:615–619

Yang Z, Zhang B, Li D, Lv M, Huang C, Shen GX et al (2010) Mast cells mobilize myeloid-derived suppressor cells and Treg cells in tumor microenvironment via IL-17 pathway in murine hepatocarcinoma model. PLoS ONE 5:e8922

Yang XD, Ai W, Asfaha S, Bhagat G, Friedman RA, Jin G et al (2011) Histamine deficiency promotes inflammation-associated carcinogenesis through reduced myeloid maturation and accumulation of CD11b+Ly6G+ immature myeloid cells. Nat Med 17:87–95

Younan G, Suber F, Xing W, Shi T, Kunori Y, Abrink M et al (2010) The inflammatory response after an epidermal burn depends on the activities of mouse mast cell proteases 4 and 5. J Immunol 185:7681–7690
Zhang W, Stoica G, Tasca SI, Kelly KA, Meininger CJ (2000) Modulation of tumor angiogenesis by stem cell factor. Cancer Res 60:6757–6762
Zhu XQ, Lv JQ, Lin Y, Xiang M, Gao BH, Shi YF (2007) Expression of chemokines CCL5 and CCL11 by smooth muscle tumor cells of the uterus and its possible role in the recruitment of mast cells. Gynecol Oncol 105:650–656

Chapter 21
Regulatory T Cells in Patients with Cancer

Theresa L. Whiteside

Abstract Current monitoring for the presence of regulatory T cells (Treg) in cancer patients' body fluids is challenging, largely because of the lack of markers specific for this minor subset of CD4+ T cells. Further, current evidence is conflicted by observations that $CD4^{+}FOXP3^{+}CD25^{high}$ T cells accumulating at tumor sites may have either beneficial or unfavorable effects on prognosis. The presence in tumor-associated inflammatory infiltrates of two Treg subsets with distinct phenotypic and functional profiles might, in part, explain these results. The division of labor exists between natural (n) Treg, normally responsible for maintaining peripheral tolerance and inducible (i) Treg arising by tumor-driven conversion of conventional CD4+ T cells, which are responsible for down-regulating anti-tumor immune responses and promoting tumor progression. This suggests that Treg might be a highly diverse subpopulation of CD4+ T cells. In cancer, the type, frequency and suppression levels of accumulating Treg are regulated by the tumor. Thus, it is iTreg that play a critical role in disease, and attention needs to be focused on monitoring their frequency and activity.

Keywords Natural (n) Treg · Inducible (i) Treg · FOXP3 · CD39 · Tumor microenvironment

---

T. L. Whiteside (✉)
Departments of Pathology, Immunology and Otolaryngology, University of Pittsburgh School of Medicine; University of Pittsburgh Cancer Institute, Pittsburgh, PA, USA
e-mail: whitesidetl@upmc.edu

T. L. Whiteside
University of Pittsburgh Cancer Institute, Pittsburgh, PA, USA

M. R. Shurin et al. (eds.), *The Tumor Immunoenvironment*,
DOI: 10.1007/978-94-007-6217-6_21,

21.1 Introduction

Regulatory T cells (Treg) have been under intense scrutiny in recent years. The realization that Treg play a key role in maintaining the balance of immune responses provides a strong motivation for obtaining insights into their origins, phenotype and importance in immune regulation. This small subset of CD4+ T cells plays a critical role in maintaining the immune balance in health, and disturbances in Treg are associated with various diseases. While in healthy individuals, thymus-derived CD4+FOXP3+ T cells, which largely utilize contact-dependent suppression mechanisms, are responsible for peripheral tolerance, the emergence of inducible or adaptive Treg in response to physiologic or pathologic signals changes the regulatory landscape and adapts it to the microenvironmental needs. For example, accumulations of iTreg in tissues or peripheral blood of patients with cancer (Curiel et al. 2004; Mougiakakos et al. 2010) represent an attempt to regulate immune responses induced by the presence of cancer. That this attempt may be responsible for suppression of anti-tumor immune effector cells in situ and in the periphery is the consequence of disease-induced changes in immunoregulatory mechanisms. This suggests that Treg operating in health may be phenotypically and functionally distinct from their counterparts induced by disease. It is likely that disease-associated changes in the immune system might require more powerful regulation to restore the immune equilibrium to its normal levels.

The question of whether a single subset of Treg can meet the challenge of controlling diverse immune responses in health and disease has been frequently asked. Recently, Lanzavecchia and collaborators have described a complex system of immunoregulation in the peripheral circulation of normal donors involving subsets of Treg, each regulating a functionally-distinct subset of T effector (Teff) cells (Duhen et al. 2012). This study suggests that human Treg are functionally diverse and that in disease, this diversity might be even greater. The more we learn about Treg, the more difficult it becomes to define their precise phenotypic and functional profiles and to sort out the mechanisms these cells use to mediate suppression. Most likely, the difficulties we are faced with are related to the tremendous plasticity of CD4+ T cells in general (Zhou et al. 2009) and Treg in particular (Duarte et al. 2009), and to the critical role environmental factors play in the generation, activation and functional regulation of Treg (Nishikawa and Sakaguchi 2010).

The objective of this brief review is to compare phenotypic and functional properties of Treg recovered from the peripheral circulation of healthy donors with those of Treg in the blood and tumor tissues of patients with cancer. In addition, given the existing controversy regarding the role these cells play in cancer progression and their clinical significance, this review will summarize the current information available for Treg involvement in promoting or inhibiting tumor progression and their participation in cancer-associated inflammation.

21.2 Phenotypic Properties of Human Treg

The availability of monoclonal antibodies (Abs) specific for surface or intracytoplasmic markers expressed on Treg and the use of multiparameter flow cytometry have greatly facilitated Treg phenotypic characterization (Mandapathil et al. 2009; Schuler et al. 2012). Treg represent about 5 % of circulating CD4+ T cells in man (Schuler et al. 2012). At least two distinct subsets of Treg have been recognized: (a) natural (n) Treg, which originate in the thymus, mediate suppression via the contact-dependent mechanisms involving the granzyme B/perforin or Fas/FasL pathways and represent the major Treg subset responsible for maintaining peripheral tolerance (Whiteside 2012) and (b) indicible or adaptive (i)Treg, also referred to as type 1 Treg or Tr1, which are induced in the periphery by environmental signals, such as antigens plus IL-2, TGF-β1 or IL-10 (Mandapathil et al. 2010). Tr1 mediate suppression via contact-independent mechanisms through the production of various immunosuppressive factors such as TGF-β1, IL-10, adenosine (ADO), PGE_2 and others (Whiteside et al. 2011).

In contrast to murine Treg, the phenotype of human Treg is not yet firmly defined. None of the above listed markers are specific for Treg. Consequently, the phenotypic Treg definition depends on a constellation of surface and cytoplasmic markers and may be problematic, because of up-regulation or down-regulation of these markers depending on the state of Treg activation. Mouse Treg and most human nTreg are usually characterized by expression of the intracytoplasmic transcription factor forkhead box p3 (FOXP3), belonging to the forkhead/winghead-helix family (Zheng et al. 2010). While most nTreg are FOXP3+, Tr1 may not be, suggesting that these two Treg subsets represent different lineages of CD4+ T cells. Importantly, neither FOXP3 nor CD25 are markers specific for Treg, as they may be found on activated CD4+ or CD8+ T effector cells, which have no suppressor activity (Roncarolo and Gregori 2008). Only CD4–CD25^{high} T cells mediating suppressor function are considered as Treg (Strauss et al. 2007). This has resulted in a somewhat arbitrary distinction between human Treg as CD4+CD25^{high} T cells and activated T conventional (conv) as CD4+CD25+ cells based on flow cytometry (Whiteside et al. 2012). An urgent need exists for additional and hopefully specific markers which could reliably identify human Treg and allow for their separation from other CD4+ T cells in the blood and tissues. While the various markers listed in Table 21.1 cannot be used to distinguish or isolate Treg from other CD4+ T cells, they endow Treg with special functions. For example, the chemokine receptors are critical for Treg migration, while signaling via GITR down-regulates Treg functions and it may serve as a co-stimulatory factor for activated T cells (Azuma 2010; Kleinewietfeld et al. 2005; Yuan et al. 2007; Menning et al. 2007). A relatively new marker, HELIOS, an Ikaros family transcription factor, is said to be present on nTreg but not on iTreg (Elkord et al. 2011), although some investigators report the opposite results (Gottschalk et al. 2012).

Table 21.1 Phenotypic characteristics of human nTreg and iTreg

Markers[a]	nTreg	iTreg
CD3	+	+
CD4	+	+
CD25	High	±
CD123/CD132	±	++
FOXP3[d]	+	±
CTLA-4	+	+
GITR	+	+
ICOS	±	+
HELIOS	+	-[c]
PD-1	±	+
PD-L1	±	+
CCR4	+	+
CCR6	+	+
CCR7	+	+
CD127	Low	Low
CD49d	Low	Low
CD39	+	++
CD73[d]	±	+
GARP/LAP[a]	−	++
TGF-β[d]	+	+
IL-10[d]	+	±
GrB/perforin	±	+
COX-2	−	±

[a] The list includes most commonly studied markers. The asterisks indicate intracytoplasmic expression. For a more in depth commentary on expression of the listed markers on nTreg and iTreg see Ref. (Whiteside et al. 2012)

[b] GARP (Glycoprotein A Repetitions Predominant) or garpin and LAP (latency-associated peptide) are TGF-β associated membrane-bound molecules

[c] Expression of HELIOS on nTreg and not on iTreg had been reported (Elkord et al. 2011). However, other studies suggest HELIOS is present on iTreg (Gottschalk et al. 2012)

[d] Expression levels variable and dependent on the state of Treg activation

Treg are also characterized by the absence of certain surface markers that are expressed on Tconv. These include IL-7-receptor, CD127 (Liu et al. 2006), and an integrin alpha subunit, CD49d (Kleinewietfeld et al. 2009). Negative selection of Treg based on the absence of these markers, followed by confirmatory expression of FOXP3, has become a method of choice for Treg isolation (Peters et al. 2008). But, in common with CD25, these two surface markers do not provide a distinct cut-off in expression levels between Treg and Tconv in flow cytometry, so that gate setting for these markers is also an arbitrary decision.

The presence of ectonucleotidases, CD39 and CD73, has been reported on murine Treg (Borsellino et al. 2007; Deaglio et al. 2007), and our recent data confirm the presence of these enzymes in human Treg (Mandapathil et al. 2010).

CD39 hydrolyzes exogenous ATP to ADP and 5'AMP, which is further hydrolyzed to adenosine (ADO) by CD73 (Bynoe and Viret 2008). Murine Treg express CD73 on the cell surface, while its presence on the surface of human nTreg is difficult to demonstrate by conventional methods (Whiteside et al. 2012) and may require vigorous cell permeabilization (Schuler and Whiteside, unpublished). In contrast, CD73 is readily demonstrated on the surface of many iTreg by flow cytometry (Whiteside et al. 2012). Because CD39 expression defines a subset of CD4+ T cells which mediate suppression in vitro (Mandapathil et al. 2010) and in vivo, as shown in studies with murine Treg (Ohta et al. 2012), this marker is considered suitable for positive selection of Treg from CD4+ T cells (Schuler et al. 2011). We have recently reported that human Treg selected from the blood by surface expression of CD39 consist of two closely interacting cell subsets, a subset of CD39+CD25^{+}FOXP3^{+} cells which mediate suppression and a subset of CD39+CD25^{neg}FOXP3^{neg} cells which are not able to suppress T-cell proliferation but always accompany FOXP3+ T cells, perhaps serving as precursor cells (Schuler et al. 2011, 2012). Because cells in both subsets are CD39+ and are capable of ATP-hydrolysis to AMP and eventually to ADO, they are operationally considered as suppressor cells.

Adenosine is known to modulate functions of a variety cell types via adenosine receptors and cAMP (Zarek and Powell 2007). Human nTreg as well as iTreg express mRNA for A1R, A2aR and A3R (our unpublished data). Thus, Treg not only produce adenosine but may also be responsive to autocrine adenosine-mediated signals. In addition, iTreg express PGE$_2$ receptors Ep2 and Ep4 and PGE2 (Mandapathil and Whiteside 2011).

21.3 Functional Characteristics of Human Treg Cells

In view of the lack of markers that are specific for human Treg, their ability to mediate suppression of other immune cells remains the best approach to their identification. It is always more important, although more difficult, to measure suppressor function in addition to phenotyping Treg. Under the best of circumstances, the presence of CD4+CD25^{high}FOXP3+ Treg should be supported by evidence of suppression, although the frequency of these cells might not always correlate with suppression levels. Table 21.2 lists some of the most widely used in vitro suppression assays for human Treg.

T cell co-cultures with Treg or Treg cultures require substantial numbers of preferably freshly-harvested and purified cells. For this reason, they are not practical, and alternative suppressor assays are often used. Among them, flow cytometry-based cytokine assays, which can be performed with relatively small numbers of cryopreserved and thawed lymphocytes, lend themselves well to measurements of inhibitory cytokine expression levels in responder cells (Schuler et al. 2012). These intracytoplasmic cytokine assays, which require cell permeabilization steps, could serve as surrogate functional markers for Treg, especially

Table 21.2 Suppression assays for human Treg[a]

Assay type	Responder cell	Result	Reference
CFSE co-cultures	CD4+CD25neg Teff	Proliferation inhibition	(Strauss et al. 2007)
	CD8+CD25neg Teff	Proliferation inhibition	(Canavan et al. 2012)
Multiparameter flow cytometry	Ex vivo activated T cells (6–12h co-culture)	Intracellular expression of 1–5 cytokines	(Schuler et al. 2012)
Co-cultures	CD4+ Tconv	Cellular cAMP levels	(Mandapathil et al. 2010)
Co-cultures	CD4+ Tconv	Inhibition of Ca2+, NFKB, NFAT signaling	(Schmidt et al. 2011)
	CD8+ Tconv		
Flow cytometry	nTreg or iTreg	GARP/LAP expression	(Schuler et al. 2012)
Supernatants	nTreg or iTreg	ADO production	(Schuler et al. 2012)
		PGE2 production	(Mandapathil and Whiteside 2011)
		TGF-β, IL-10 production	(Strauss et al. 2007)
FOXP3 demethylation	nTreg	Detection of Treg-specific demethylation region (TSDR) by MS-QPCR	(Polansky et al. 2010)

[a] Assays requiring co-incubation of responder cells with Treg have to be run at several different T suppressor/T responder (T_S/T_R) cell ratios (1:1; 1:2; 1:5; 1:10). These assays are usually performed with autologous T cell subsets to avoid confounding effects of allostimulation. Purified Treg are used in co-culture or Treg culture assays; in the flow-cytometry-based assays, Treg purification may not be essential, if they can be discriminated by their phenotype and shown to express immunosuppressive or inhibitory factors or cytokines

iTreg, which use TGF-β or IL-10 for suppression and express TGF-β-associated membrane-tethered and functionally-active GARP and LAP (Schuler et al. 2012). Another alternative is to measure increases in cAMP levels as an indicator of suppression in responder cells co-incubated with Treg, using classical biochemical methods. Finally, demethylation of FOXP3 in the Treg-specific demethylation region (TSDR) at intron 1 by a MS-QPCR method has been used to differentiate Treg from activated CD4+ non-Treg (Polansky et al. 2010). Evidence has accumulated indicating that stable expression of FOXP3 is dependent on the demethylation status of the *foxp3* gene. Of note, TSDR is only found in nTreg and not in ex vivo TGF-β-induced Tr1 (Polansky et al. 2010). There is also evidence to suggest that various activation signals delivered to dividing Treg can modulate the *foxp3* gene demethylation and FOXP3 expression in Treg (Schenk et al. 2011). In human tumors, which are enriched in ATP and in accumulating or expanding Treg, conditions might favor the loss of FOXP3. Incidentally, such a loss was shown to promote conversion of Treg to IL-17+ TT helper (TH17) cells in mice (Schenk et al. 2011). Overall, it appears that local environment can regulate FOXP3 expression in Treg recruited to the tumor.

It is also important to remember that iTreg might be involved in multiple cellular functions not involving immune suppression. For example, Treg can hydrolyze ATP disposing of its excess during inflammation, and attenuating its toxic effects (Schenk et al. 2011). ATP could also serve as a recruiting signal for Treg and may be responsible for their accumulations at tumor sites.

21.4 Accumulations of Treg in the Blood and Tumor Tissues of Cancer Patients

Figure 21.1 illustrates events occurring at the tumor site that potentially lead to accumulations of Treg. Human tumors are known to produce a variety of chemokines, including CCL22, a ligand of CCR7 which is expressed on Treg (Nishikawa et al. 2005). A gradient of chemokines favors Treg migration and accumulation in the tumor. In addition, the tumor microenvironment rich in TGF-β and tumor-derived antigens encourages conversion of Tconv to Treg. Indeed, our studies and those of others have documented increased percentages of CD4+CD25^{high}FOXP3+ T cells in the tumor or in the peripheral circulation of patients with cancer (Curiel et al. 2004; Bergmann et al. 2008). Furthermore, suppressor function of Treg isolated from tumor sites was significantly higher than that of Treg from the blood of cancer patients or of normal donors (Bergmann et al 2008). While the accumulation of Treg at tumor sites is generally acknowledged the role these cells play in tumor progression remains unclear, as recently discussed (Whiteside 2012). Their potential to inhibit Th1 responses by interfering with anti-tumor functions of effector T cells could contribute to poor outcome, as seems to be the case in some human solid tumors (Curiel et al. 2004; Ladoire et al. 2011). On the

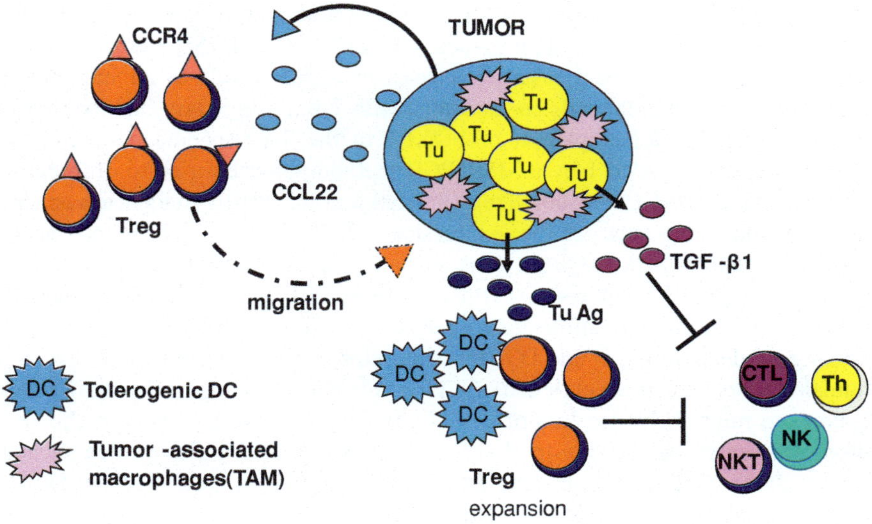

Fig. 21.1 One potential mechanism responsible for Treg accumulations in the tumor involves the recruitment of $CCR4^+$ Treg from the periphery mediated by CCL22. This chemokine, secreted by tumor cells or tumor-associate macrophages (*TAM*), is a ligand for CCR4. Treg, responding to the chemokine gradient, arrive at the tumor site and, in the microenvironment rich in TGF-β as well as tumor-associated antigens being processed by dendritic cells (*DC*), they expand, accumulate and suppress the generation of anti-tumor immune responses thus promoting tumor escape

other hand, the presence of Treg among inflammatory infiltrates in colorectal cancer, for example, has been associated with an improved prognosis (Tosolini et al. 2011), presumably because accumulating Treg down-regulate inflammatory responses which contribute to tumor progression (Ladoire et al. 2011). The inflammatory milieu the tumor creates and the presence of bacterial and perhaps viral infections in tissues undergoing carcinogenesis appear to be the main factors influencing the role of Treg in promoting or suppressing the tumor progression via immune regulation.

21.5 Biologic and Clinical Significance of the Increased Treg Frequency/Activity in Cancer

Many recent studies have suggested that infiltrations of human solid tumors with CD8+ T cells are associated with improved prognosis (Fridman et al. 2011). Therefore, Treg through their suppression of CD4+ T helper and CD8+ effector cell functions could have an impact on disease progression. This suggests that a ratio of CD8+ T cells/Treg in situ or in the periphery could serve as a marker of favorable prognosis in cancer. Based on this premise, the CD8/Treg ratio has been used in some clinical studies to follow changes in the tumor-infiltrating or circulating lymphocytes during therapy. Changes in this ratio monitored over time are then linked to outcome after therapy. As this monitoring strategy depends on the accuracy of phenotyping Treg, it may not be reliable at this time, and thus it does not allow for a definitive conclusion to be made about the prognostic utility of this ratio.

It appears that as CD8+ effector and CD4+ helper T cells accumulate in response to local or systemic signals, so do nTreg, presumably to prevent potential tissue damage by activated T cells. In cancer, the tumor-driven conversion of Tconv to iTreg occurs, thus creating a pool of highly activated suppressor cells which interfere with functions of immune cells, including anti-tumor effector T cells. These tumor-associated iTreg produce adenosine, PGE_2 and TGF-β, all of which are immunosuppressive and down-regulate anti-tumor effects of immune cells (Whiteside et al. 2011). Interestingly, PGE_2 was recently identified as a single inflammatory signal that triggers DNA methylation, thereby shutting off tumor suppressor and DNA repair genes in murine models of colon cancer (Xia et al. 2012). Tumor-associated iTreg appear to be resistant to radio-, chemo- and immunotherapy and significantly increase in the frequency following these therapies potentially contributing to persistence of immune suppression (Kachikwu et al. 2011). The tissue damage and tumor cell death resulting from therapy also contribute to increased ATP and TGF-β tissue levels, conditions which favor iTreg generation. Post-therapy increases and persistence of iTreg at the tumor site and the periphery could contribute to tumor recurrence. Studies are currently in progress to evaluate the clinical significance of Treg persistence after cancer therapies, for example, their role in tumor recurrence.

The role of Treg in cancer therapy, similar to the role they play in prognosis, remains controversial. Although it is widely believed that in vivo depletion of Treg via the use of cyclophosphamide, daclizumab (anti-CD25 Ab), denileukin diftitox (ONTAC) or tyrosine kinase inhibitors such as Sunitinib (Rech and Vonderheide 2009; Hobeika et al. 2011; Finke et al. 2008) in part restores anti-tumor immune responses, experience with these Treg-depleting agents in humans with cancer have yielded mixed and unclear results (de Vries et al. 2011). The problem relates to only transient and incomplete depletion of Treg by these agents and the rapid Treg recovery after immune therapies such as anti-tumor vaccines, preventing a rational or consistent estimate of their impact on tumor progression (de Vries et al. 2011).

Further concern about depleting Treg in patient with cancer arises in view of preliminary reports suggesting that their presence may improve, not reduce, anti-tumor effects of cancer therapies (Ladoire et al. 2011). If this proves to be the case, then Treg depletion would be counter-indicated. Clearly, immunotherapies able to induce robust anti-tumor immune responses also induce expansion of Treg (Strauss et al. 2007; Mandapathil et al. 2009). Their presence is beneficial presumably due to the Treg-mediated control of pro-tumor inflammatory responses. However, it is unclear at this time whether this is a valid assumption and whether Treg depletion is necessary for improving results of cancer immunotherapies. It is entirely possible that Treg expansion, e.g., by the delivery of IL-2, rather than their depletion might be a more effective anti-cancer strategy. Future clinical studies will undoubtedly provide insights into this critical question.

Meanwhile, a better understanding of Treg phenotypic and functional heterogeneity is necessary. In cancer, where iTreg generated in the tumor microenvironment predominantly regulate anti-tumor immunity, emphasis has to be on iTreg and not on nTreg. The former have distinct phenotypic and functional attributes and differ from FOXP3+ nTreg normally in charge of peripheral tolerance (Whiteside et al. 2011). It is, therefore, possible that a selective depletion of iTreg could be therapeutically more effective than elimination of nTreg. In fact, depletion of FOXP3+ nTreg might be detrimental to health and probably should be avoided. iTreg are recruited to and conditioned by the tumor to produce numerous immunosuppressive factors, e.g., adenosine (Mandapathil and Whiteside 2011). We have suggested that targeting this pathway, which leads to up-regulation of 3'–5' cAMP levels and immune suppression in responder effector T cells, with pharmacologic inhibitors, could facilitate oncologic therapies (Whiteside et al. 2011; Whiteside 2010). This strategy of "inhibiting the inhibitors" appears to be currently in clinical use for blocking the critical T cell checkpoints with specific Abs (e.g. Zhou et al. 2009). The success of this immune strategy is encouraging for its future application of selective depletion of iTreg in patients with cancer.

References

Azuma M (2010) Role of the glucocorticoid-induced TNFR-related protein (GITR)-GITR ligand pathway in innate and adaptive immunity. Crit Rev Immunol 30:547–557

Bergmann C, Strauss L, Wang Y, Szczepanski MJ, Lang S, Johnson JT et al (2008) T regulatory type 1 cells (Tr1) in squamous cell carcinoma of the head and neck: mechanisms of suppression and expansion in advanced disease. Clin Cancer Res 14:3706–3715

Borsellino G, Kleinewietfeld M, DiMitri D, Sternjak A, Diamantini A, Giometto R et al (2007) Expression of ectonucleotidase CD39 by Foxp3 + Treg cells: hydrolysis of extracellular ATP and immune suppression. Blood 110:1225–1232

Brahmer JR, Tykodi SS, Chow LQ, Hwu WJ, Topalian SL, Hwu P et al (2012) Safety and activity of anti-PD-L1 antibody in patients with advanced cancer. N Engl J Med 366:2455–2465

Bynoe MS, Viret C (2008) Foxp3+CD4+ T cell-mediated immunosuppression involves extracellular nucleotide catabolism. Trends Immunol 29:99–102

Canavan JB, Afzali B, Scotta C, Fazekasova H, Edozie FC, Macdonald TT et al (2012) A rapid diagnostic test for human regulatory T-cell function to enable regulatory T-cell therapy. Blood 119:e57–e66

Curiel TJ, Coukos G, Zou L, Alvarez X, Cheng P, Mottram P et al (2004) Specific recruitment of regulatory T cells in ovarian carcinoma fosters immune privilege and predicts reduced survival. Nat Med 10:942–949

de Vries IJ, Castelli C, Huygens C, Jacobs JF, Stockis J, Schuler-Thurner B et al (2011) Frequency of circulating Tregs with demethylated FOXP3 intron 1 in melanoma patients receiving tumor vaccines and potentially Treg-depleting agents. Clin Cancer Res 17:841–848

Deaglio S, Dwyer KM, Gao W, Friedman D, Usheva A, Erat A et al (2007) Adenosine generation catalyzed by CD39 and CD73 expressed on regulatory T cells mediates immune suppression. J Exp Med 204:1257–1265

Duarte JH, Zelenay S, Bergman ML, Martins AC, Demengeot J (2009) Natural Treg cells spontaneously differentiate into pathogenic helper cells in lymphogenic conditions. Eur J Immunol 39:948–955

Duhen T, Duhen R, Lanzavecchia A, Sallusto F, Campbell DJ (2012) Functionally distinct subsets of human FOXP3+ Treg cells that phenotypically mirror effector TH cells. Blood 119:4430–4440

Elkord E, Sharma S, Burt DJ, Hawkins RE (2011) Expanded subpopulation of Foxp3+ T regulatory cells in renal cell carcinoma co-express Helios, indicating they could be derived from natural but not induced Tregs. Clin Immunol 140:218–222

Finke JH, Rini B, Ireland J, Rayman P, Richmond A, Golshayan A et al (2008) Sunitinib reverses type-1 immune suppression and decreases T-regulatory cells in renal cell carcinoma patients. Clin Cancer Res 14:6674–6682

Fridman WH, Galon J, Pages F, Tartour E, Sautes-Fridman C, Kroemer G (2011) Prognostic and predictive impact of intra- and peritumoral immune infiltrates. Cancer Res 71:5601–5605

Gottschalk RA, Corse E, Allison JP (2012) Expression of helios in peripherally induced Foxp3+ regulatory T cells. J Immunol 188:976–980

Hobeika AC, Morse MA, Osada T, Peplinski S, Lyerly HK, Clay TM (2011) Depletion of human regulatory T cells. Meth Mol Biol 707:219–231

Kachikwu EL, Iwamoto KS, Liao YP, DeMarco JJ, Agazaryan N, Economou JS et al (2011) Radiation enhances regulatory T cell representation. Int J Radiat Oncol Biol Phys 81:1128–1135

Kleinewietfeld M, Puentes F, Borsellino G, Battistini L, Rotzschke O, Falk K (2005) CCR6 expression defines regulatory effector/memory-like cells within the CD25(+)CD4+ T-cell subset. Blood 105:2877–2886

Kleinewietfeld M, Starke M, Di Mitri D, Borsellino G, Battistini L, Rotzschke O et al (2009) CD49d provides access to "untouched" human Foxp3 + Treg free of contaminating effector cells. Blood 113:827–836

Ladoire S, Martin F, Ghiringhelli F (2011) Prognostic role of FOXP3− regulatory T cells infiltrating human carcinomas: the paradox of colorectal cancer. Cancer Immunol Immunother 60:909–918

Liu W, Putnam AL, Xu-Yu Z, Szot GL, Lee MR, Zhu S et al (2006) CD127 expression inversely correlates with FoxP3 and suppressive function of human CD4+ T reg cells. J Exp Med 203:1701–1711

Mandapathil M, Whiteside TL (2011) Targeting human inducible regulatory T cells (Tr1) in patients with cancer: blocking of adenosine-prostaglandin E2 cooperation. Exp Opin Biol Ther 11:1203–1214

Mandapathil M, Lang S, Gorelik E, Whiteside TL (2009a) Isolation of functional human regulatory T cells (Treg) from the peripheral blood based on the CD39 expression. J Immunol Meth 346:55–63

Mandapathil M, Szczepanski MJ, Szajnik M, Ren J, Lenzner DE, Jackson EK et al (2009b) Increased ectonucleotidase expression and activity in regulatory T cells of patients with head and neck cancer. Clin Cancer Res 15:6348–6357

Mandapathil M, Szczepanski MJ, Szajnik M, Ren J, Jackson EK, Johnson JT et al (2010a) Adenosine and prostaglandin E_2 cooperate in the suppression of immune responses mediated by adaptive regulatory T cells. J Biol Chem 285:27571–27580

Mandapathil M, Hilldorfer B, Szczepanski MJ, Czystowska M, Szajnik M, Ren J et al (2010b) Generation and accumulation of immunosuppressive adenosine by human $CD4^{+}CD25^{high-}$ $FOXP3^{+}$ regulatory T cells (Treg). J Biol Chem 285:7176–7186

Menning A, Hopken UE, Siegmund K, Lipp M, Hamann A, Huehn J (2007) Distinctive role of CCR7 in migration and functional activity of naive- and effector/memory-like Treg subsets. Eur J Immunol 37:1575–1583

Mougiakakos D, Choudhury A, Lladser A, Kiessling R, Johansson CC (2010) Regulatory T cells in cancer. Adv Cancer Res 107:57–117

Nishikawa H, Sakaguchi S (2010) Regulatory T cells in tumor immunity. Int J Cancer 127:759–767

Nishikawa H, Kato T, Tawara I, Saito K, Ikeda H, Kuribayashi K et al (2005) Definition of target antigens for naturally occurring CD4(+) CD25(+) regulatory T cells. J Exp Med 201:681–686

Ohta A, Kini R, Ohta A, Subramanian M, Madasu M, Sitkovsky M (2012) The development and immunosuppressive functions of CD4(+) CD25(+) FoxP3(+) regulatory T cells are under influence of the adenosine-A2A adenosine receptor pathway. Front Immunol 3:190

Peters JH, Preijers FW, Woestenenk R, Hilbrands LB, Koenen HJ, Joosten I (2008) Clinical grade Treg: GMP isolation, improvement of purity by CD127 Depletion, Treg expansion, and Treg cryopreservation. PLoS ONE 3:e3161

Polansky JK, Schreiber L, Thelemann C, Ludwig L, Kruger M, Baumgrass R et al (2010) Methylation matters: binding of Ets-1 to the demethylated Foxp3 gene contributes to the stabilization of Foxp3 expression in regulatory T cells. J Mol Med (Berl) 88:1029–1040

Rech AJ, Vonderheide RH (2009) Clinical use of anti-CD25 antibody daclizumab to enhance immune responses to tumor antigen vaccination by targeting regulatory T cells. Ann N Y Acad Sci 1174:99–106

Roncarolo MG, Gregori S (2008) Is FOXP3 a bona fide marker for human regulatory T cells? Eur J Immunol 38:925–927

Schenk U, Frascoli M, Proietti M, Geffers R, Traggiai E, Buer J, et al (2011) ATP inhibits the generation and function of regulatory T cells through the activation of purinergic P2X receptors, 4(162)

Schmidt A, Oberle N, Weis EM, Vobis D, Frischbutter S, Baumgrass R, et al. Human regulatory T cells rapidly suppress T cell-receptor-induced Ca2+, NF-κB, and NFAT signaling in conventional T cells. Sci Signal 2011; 4:ra90

Schuler PJ, Harasymczuk M, Schilling B, Lang S, Whiteside TL (2011) Separation of human CD4(+)CD39(+) T cells by magnetic beads reveals two phenotypically and functionally different subsets. J Immunol Meth 369:59–68

Schuler PJ, Schilling B, Harasymczuk M, Hoffmann TK, Johnson J, Lang S et al (2012) Phenotypic and functional characteristics of $CD4^+CD39^+FOXP3^+$ and $CD4^+CD39^+FOXP3^{neg}$ T-cell subsets in cancer patients. Eur J Immunol 42:187685

Strauss L, Bergmann C, Whiteside TL (2007a) Functional and phenotypic characteristics of CD4+$CD25^{high}$Foxp3+ Treg clones obtained from peripheral blood of patients with cancer. Int J Cancer 121:2473–2483

Strauss L, Bergmann C, Gooding W, Johnson JT, Whiteside TL (2007b) The frequency and suppressor function of CD4+$CD25^{high}$Foxp3+ T cells in the circulation of patients with squamous cell carcinoma of the head and neck. Clin Cancer Res 13:6301–6311

Strauss L, Bergmann C, Szczepanski M, Gooding W, Johnson JT, Whiteside TL (2007c) A unique subset of CD4+$CD25^{high}$Foxp3+ T cells secreting IL-10 and TGF-β_1 mediates suppression in the tumor microenvironment. Clin Cancer Res 13:4345–4354

Tosolini M, Kirilovsky A, Mlecnik B, Fredriksen T, Mauger S, Bindea G et al (2011) Clinical impact of different classes of infiltrating T cytotoxic and helper cells (Th1, Th2, Treg, Th17) in patients with colorectal cancer. Cancer Res 71:1263–1271

Whiteside TL (2010) Inhibiting the inhibitors: evaluating agents targeting cancer immunosuppression. Expert Opin Biol Ther 10:1019–1035

Whiteside TL (2012) What are regulatory T cells (Treg) regulating in cancer and why? Semin Cancer Biol 22:327–334

Whiteside TL, Mandapathil M, Schuler P (2011) The role of the adenosinergic pathway in immunosuppression mediated by human regulatory T cells (Treg). Curr Med Chem 18:5217–5223

Whiteside TL, Schuler P, Schilling B (2012) Induced and natural regulatory T cells in human cancer. Expert Opin Biol Ther 12:1383–1397

Xia D, Wang D, Kim SH, Katoh H, DuBois RN (2012) Prostaglandin E2 promotes intestinal tumor growth via DNA methylation. Nat Med 18:224–226

Yuan Q, Bromley SK, Means TK, Jones KJ, Hayashi F, Bhan AK et al (2007) CCR4-dependent regulatory T cell function in inflammatory bowel disease. J Exp Med 204:1327–1334
Zarek PE, Powell JP (2007) Adenosine and anergy. Autoimmunity 40:425–432
Zheng Y, Josefowicz S, Chaudhry A, Peng XP, Forbush K, Rudensky AY (2010) Role of conserved non-coding DNA elements in the FOXP3 gene in regulatory T-cell fate. Nature 463:808–812
Zhou L, Chong MM, Littman DR (2009) Plasticity of CD4+ T cell lineage differentiation. Immunity 30:646–655

Chapter 22
Tumor-Evoked Regulatory B Cells as Important Mediators of Cancer Escape

Catalina Lee-Chang, Monica Bodogai and Arya Biragyn

Abstract Our understanding of immune regulation was drastically changed by Sakaguchi's seminal discovery of regulatory T cells (Tregs). To date, the cells that exert regulatory activity are found among almost every type of immune cells. Their primary function is to control immune homeostasis and prevent aberrant autoimmune responses, making them an attractive target of cancer. In fact, cancer actively hijacks Tregs and myeloid-derived suppressive cells (MDSCs) by corrupting them in order to abrogate antitumor effector responses and to promote a cancer-benefiting milieu. However, the role of regulatory B cells (Bregs) in this process is poorly understood, despite the existence of protective IL-10-producing B cells and Bregs in autoimmune diseases. Here in this review, we discuss the pros and cons and the latest evidence of the importance of Bregs in mediation of immunosuppression required in cancer escape and metastasis.

Keywords B cells in autoimmunity and cancer · Bregs · Cancer escape · Tregs · Tumor-evoked regulatory B cells · tBregs · IL-10 · Immunosuppression

22.1 Introduction

A common feature between cancer escape and autoimmune diseases is an inappropriate involvement of the regulatory immune system, albeit often for opposing purposes. While autoimmune disease is a reflection of the failure of

C. Lee-Chang · M. Bodogai · A. Biragyn (✉)
Immunoregulations Section, Laboratory of Molecular Biology and Immunology,
National Institute on Aging, 251 Bayview Blvd, Suite 100, Baltimore, MD, USA
e-mail: biragyna@mail.nih.gov

M. R. Shurin et al. (eds.), *The Tumor Immunoenvironment*,
DOI: 10.1007/978-94-007-6217-6_22,

controls to self, cancer is a result of an exaggerated use of these controls to abrogate antitumor effector responses. The machinery controlling autoimmune responses is well conserved between rodents and humans despite more than 100 million years of separate evolution. It is mediated by specialized subsets of regulatory immune cells, such as myeloid suppressive and myeloid-derived suppressive cells (MSCs and MDSCs) and regulatory T cells (Tregs). As key controllers of immunological homeostasis and potent suppressors of adaptive and innate immune responses (Cheng et al. 2008; Nagaraj and Gabrilovich 2008; Sakaguchi 2000), they are actively hijacked by cancer in order to evade immunosurveillance. As a result, increased presence of MDSCs and Tregs often is a sign of bad disease outcome in mice and humans with cancer (Woo et al. 2002; Beyer et al. 2005; Curiel et al. 2004). However, the role of other immune cells such as B cells in cancer escape remains poorly understood and even debatable. This is despite the fact that B cells exerting immunosuppression have been known for more than 30 years and, recently, unique subsets of regulatory B cells (Bregs, the definition first used by Mizoguchi to describe B cells exerting protection from colitis in mice (Mizoguchi et al. 1997)) were shown to exist in both mice and humans.

To date, a number of Breg subsets have been reported to exist in mice and humans, such as murine IL-10-producing B10 and B1b Bregs (Mizoguchi et al. 1997; Yanaba et al. 2008); human IL-10 producing memory $CD24^{hi}CD27^{+}$ B cells (Iwata et al. 2011), or $CD25^{hi}$ $CD27^{hi}$ $CD86^{hi}$ $CD1d^{hi}$ B cells (Kessel et al. 2012), $CD19^{+}CD24^{high}CD38^{high}$ B cells (Blair et al. 2010), and Tim-1^{+} Bregs (Ding et al. 2011; Xiao et al. 2012). Their impairments or loss, for example after depletion of B cells, can be associated with exacerbated ulcerative colitis in patients with non-Hodgkins lymphoma, colitis, and Graves disease and increases incidences of psoriasis in patients with psoriatic arthropathy (Mielke et al. 2008; Dass et al. 2007; Goetz et al. 2007). However, the involvement of Bregs in cancer escape is mostly unknown and remains a topic of debate. To the best of our knowledge, only two clearly defined examples of cancer escape-promoting Bregs are reported. First, IL-10-producing B10 cells reduce the therapeutic efficacy of anti-CD20 antibody against lymphoma by inhibiting monocyte activity and surface expression of FcγR in IL-10-dependent fashion (Horikawa et al. 2011). Second, we recently found that 4T1 carcinoma cells actively convert normal B cells into TGF-β-producing Bregs, designated tumor-evoked Bregs (tBregs) in order to successfully metastasize by down-regulating protective effector antitumor immune responses (Olkhanud et al. 2011). The pros of their existence is the importance of B cells in carcinogenesis and tumor progression. Mice deficient in B cells often poorly support growth of implanted syngeneic tumor cells unless replenished with $B220^{+}$ B cells (Olkhanud et al. 2011; Qin et al. 1998); and, conversely, the presence of B cells is also associated with carcinogenesis of methylcholanthrene-induced (Brodt and Gordon 1978, 1982) or transplanted tumors (Monach et al. 1993). The cons indicate that mice neonatally depleted of B cells have impaired antitumor $CD4^{+}$ T helper, and $CD8^{+}$ cytotoxic T cells (Schultz et al. 1990). Also, B cells, as immunoglobulin producers and antigen-presenting cells, can amplify adaptive antitumor immune responses by cooperating with other professional antigen-presenting cells (Candolfi et al. 2011). Similarly, symptoms of human autoimmune diseases, such as rheumatoid arthritis (RA), type 1 diabetes mellitus (T1D), multiple sclerosis (MS) and systemic lupus erythematosus (SLE) (Townsend et al. 2010) can be ameliorated by depleting B cells. We believe that these controversies can be presumably explained by a dual functional nature of B cells. Thus, as for many immune cells, the antagonistic B cell functions, such as activating/pathogenic and regulatory activities in cancer, can presumably be attributed to its specialized subsets. In analogy with Tregs, that represent a small proportion of $CD4^+$ T cells (up to 5 % (Baatar et al. 2007)), the overwhelming responses of pathogenic B cells may conceal Bregs, if their numbers are low.

However, to understand the role of Bregs in cancer escape, they probably need to be discriminated from other functions of B cells, such as their tumor survival and angiogenesis-inducing properties or/and suppression and inflammation induced via production of immunoglobulins and various pro-inflammatory factors and cytokines. Although the majority of Bregs involved in autoimmunity expresses and uses IL-10 (Ding et al. 2011; Byrne and Halliday 2005; Sun et al. 2008), an immunomodulatory cytokine that can induce Th2-type skewed cytokine responses and inhibit antitumor $CD8^+$ CTLs and NK cells (Moore et al. 2001), its role in cancer escape remains debatable. Interestingly, although LPS and other TLR ligands upregulate IL-10 production and the immune tolerance-inducing activity of IL-10 -producing Bregs and B cells involved in autoimmune responses (Ding et al. 2011; Byrne and Halliday 2005; Sun et al. 2008), TLR activation can instead reduce functions of cancer-associated Bregs and enhance anti-B16 melanoma immune responses in mice (Bodogai et al. 2012, manuscript in submission). In fact, LPS-treated B cells failed to enhance metastasis of mice with highly metastatic 4T1 breast cancer (Olkhanud et al. 2011), a widely used model to study cancer escape mediated by regulatory immune cells. Thus, tBregs appear to differ from the immune tolerance-inducing IL-10-producing Bregs and B cells both phenotypically and functionally (Ding et al. 2011; Byrne and Halliday 2005; Sun et al. 2008).

The role of Bregs in autoimmune responses has been extensively reviewed in a number of excellent papers (Klinker and Lundy 2012; DiLillo et al. 2010; Lund and Randall 2010), and, as such, we will only provide a simplified overview of B cells and Bregs and the mechanism of their involvement in autoimmune responses. The focus of this article is mostly to discuss the role of Bregs in cancer escape and to provoke new ideas and quests in the search of other tBreg-like cells that promote cancer escape. However, since suppressive B cell activities often overlap with Breg functions, we attempted to segregate cancer-associated Bregs, as a discrete cell subset, from an activation state of B cells by providing a simple readout, such as their ability to induce non-cytolytic suppression of proliferation of TCR-stimulated T cells and to promote conversion of $FoxP3^+$ Tregs from non-Treg $CD4^+$ T cells. This segregation also allows us to explain some controversies in the field, in particular unexpected results after depletion/inactivation of B cells in tumor-bearing mice and cancer patients.

22.2 B Cell Development

Our understanding of B cells as immunoglobulin producers and important players of adaptive immune protection is evolving together with the development of new tools and animal models. Although some B cell subsets and functions may not be complementary in mice and humans, as they diverged more than 100 million years ago (i.e., expression of CD27 and CD5), the majority of findings in mice can be extrapolated to humans. This evolutionary conservatism is probably a reflection of their biological importance: the majority of factors and mechanisms that govern development, survival and migration of B cells are conserved. The development of B cells starts in bone marrow from hematopoietic totipotent stem cells that are retained and nurtured by factors of the stromal environment, such as chemokines (CXCL12) and adhesion molecules. It involves a sequential differentiation of discrete types of cells characterized by expression of unique genes and cell surface markers (Fig. 22.1). First, expression of transcription factors Ikaros and PU.1 leads to upregulation of interleukin (IL) 7-receptor α to enable the use of IL-7 and the thymic stromal lymphopoietin produced from the stromal cells (Kikuchi et al. 2005; DeKoter and Singh 2000). Expression of the early B cell transcription factor (EBF) disables the T or NK cell differentiation pathways in pre-proB cells (Kikuchi et al. 2005; Maeda et al. 2007), while Pax5 and Stat5 induce differentiation into pre-B cells. The pre-B cells express pre-BCR composed of two μH chain and two surrogated light chains $\lambda5$ and V-preB (Vettermann and Jack 2010). Its engagement with self-antigen in large-preB cells induces differentiation into small pre-B cells, which substitute pre-BCR with monomeric IgM (mature BCR) consisting of the surrogate light κ or λ chains associated with the μH chain (Lazorchak et al. 2006). The surface expression of mature BCR marks the end of the central development and the passage to the immature B cells.

B cells migrate to the spleen as transitional 1 (T) B cells through the periarteriolar lymphoid sheaths to home into the follicle as T2 B cells. There, B cells begin their final maturation process. Most mature naïve B cells consist of follicular (FO) B cells expressing chemokine receptor CXCR5 to migrate into the B cell zone of the follicle producing CXCL13 (CXCR5 ligand). Upon encounter with dendritic cells (DCs) and macrophages, B cells up regulate chemokine receptor CCR7 and migrate to CCL19 and CCL21 (CCR7 ligands) produced in the T cell zone (Phan et al. 2009; Suzuki et al. 2009; Reif et al. 2002). At this stage, B cells either differentiate into short-lived extra follicular plasmocytes that can rapidly produce antibodies (William et al. 2002) or engage in the germinal center (GC) reaction to ameliorate their antigen-specific affinity. GC committed B cells migrate first into the dark zone of the follicle to proliferate as centroblasts expressing chemokine receptor CXCR4 and the Bcl6 protein (Allen et al. 2004). They also express activation-induced cytidine deaminase (AID) enzyme to initiate affinity maturation of their BCR, a process also known as the somatic hypermutation. Centroblasts move then into the light zone of the GC to become non-proliferating centrocytes upon encounter with follicular DC. The fate of centrocytes (reactivity

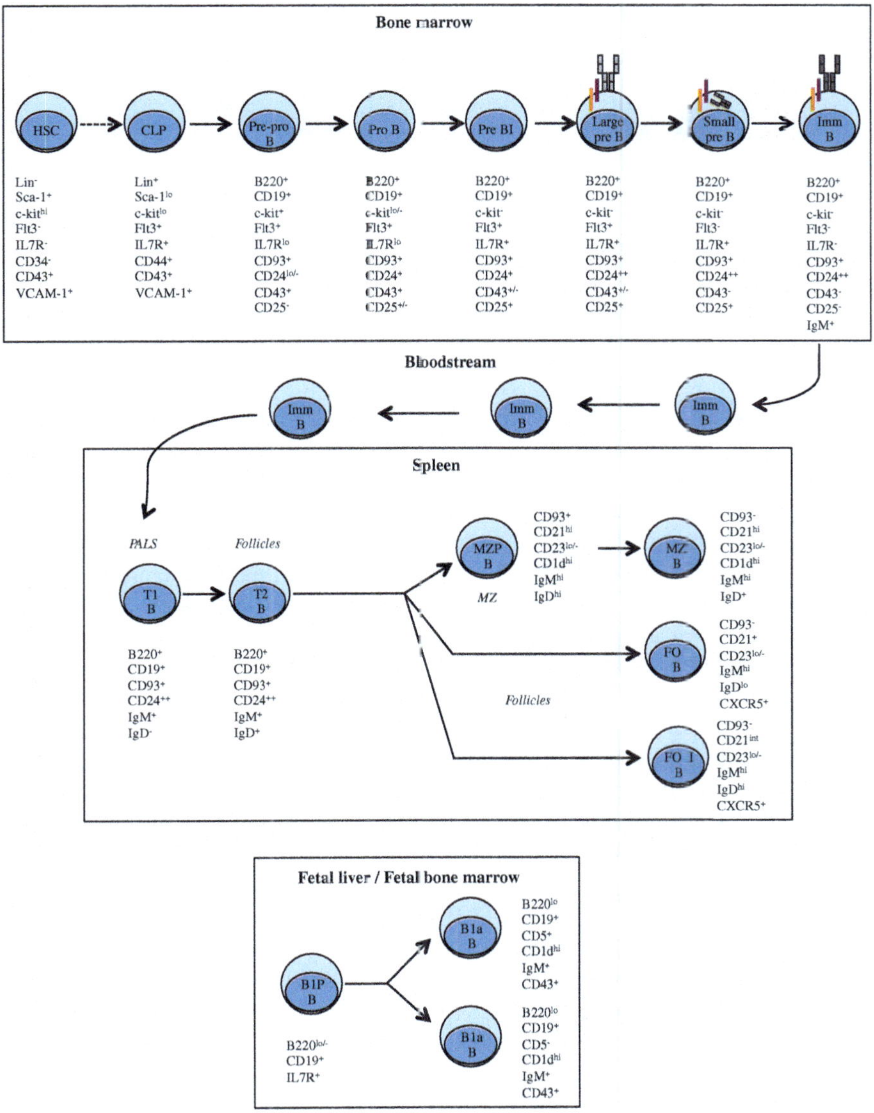

Fig. 22.1 Expression of surface markers and stages of differentiation of B cells. Hematopoietic stem cell: HSC; common lymphoid progenitor: *CLP* immature B cell, *Imm B* transitional 1 and 2 B cell, *T1* and *T2* marginal zone progenitor, *MZP* marginal zone B cell, *MZ B* follicular B cell, FO B

to antigen) is controlled by both negative and positive selections depending on the interaction with follicular T cells (Tfh). The encounter with Tfh induces Fas-mediated apoptosis (negative selection) (Takahashi et al. 2001), while signaling induced by CD40/CD40L axis or IL-4, IL-21 and interferon (IFN) γ induces

Table 22.1 Phenotype of B cell subsets in humans

B cell subset	Phenotype
Transitional	IgM^{Hi} IgD^{+} $CD38^{Hi}$ $CD24^{+}$
Naïve	IgM^{Lo} IgD^{+} $CD27^{-}$ $CD38^{-}$
Marginal zone (MZ)[a]	IgM^{+} IgD^{+} $CD27^{+}$
B1	$CD20^{+}$ $CD27^{+}$ $CD43^{+}$ $CD70^{-}$ Griffin et al. (2011)
Memory[b]	IgM^{Hi} $IgD^{-}CD27^{+}$ (IgM only)
	IgM^{-} IgD^{-} $IgG/A/E^{+}$ $CD27^{+}$ (switched-memory)

[a] In humans, MZ-like B cells are considered memory B cells ($CD27^{+}$) with low levels of mutation. They are also known as unswitched IgM^{+} memory B cells

[b] Even though the expression of CD27 is a hallmark of the memory compartment, a population of $CD27^{-}$ memory B cells has been described in patients with lupus disease (Wei et al. 2007)

cell survival and, together with AID activity, immunoglobulin class switch (positive selection) (Vinuesa et al. 2005; Bryant et al. 2007; Reinhardt et al. 2009). The positively selected B cells leave the follicle as a small group of long-lived memory B cells that express a high affinity BCR. In humans, most of memory B cells express surface marker CD27 and can be classified into switched memory B cells ($CD27^{+}IgM^{-}IgD^{-}IgG/E/A^{+}$) and unswitched IgM memory B cells ($CD27^{+}IgM^{+}IgD^{+/-}$) (Table 22.1). In rodents, unmutated and mutated memory B cells are classified based on expression of activation markers CD80 and CD35, $IgM^{+}CD80^{+}CD35^{high}$ and $IgG^{+}CD80^{+}CD35^{lo}$, respectively (Anderson et al. 2007). In rodents, T2 B cells can give rise to an independent B-cell population that exclusively locates in the marginal zone (MZ) of the spleen. Their homing and retention in the MZ is facilitated by the expression of integrins ($\alpha L\beta 2$ and $\alpha 4\beta 1$) and the sphingosine-1 phosphate (S1P) 1 and 3 (Lu and Cyster 2002; Goetzl and Rosen 2004; Girkontaite et al. 2004). The MZ B cell precursors (also known as T2-MZP) differentiate into mature MZ B cells utilizing Notch2 signaling (Saito et al. 2003) and are positively selected after weak BCR-ligation with self-antigen (Wen et al. 2005). Unlike naïve FO B cells, murine MZ B cells are in a "pre-activated" state expressing high levels of costimulatory molecules, such as CD80 and CD86, to rapidly respond to T cell-independent antigens from blood-borne pathogens. They secrete low affinity and poly-reactive IgM and IgG3 immunoglobulins (Zandvoort and Timens 2002). In contrast, human MZ-like B cells are considered memory B cells despite their low mutational levels, although their ontogeny is not fully understood.

In addition to conventional B cells (also so-called B2 cells), a small population of $IgM^{+}IgD^{+}$ B cells that preferentially reside in the pleural and peritoneal cavities (also at lesser extent in the spleen and lymph nodes) are known as B1 were first described in rodents (Hayakawa et al. 1983; Caligaris-Cappio et al. 1982). The B1 cells consist of two subsets discriminated by expression of CD5, $CD5^{+}$ B1a and $CD5^{-}$ B1b cells (Baumgarth et al. 1999). However most of studies are focused on CD5-expressing B1 B cells, also called B1a B cells. These cells contribute to innate immune responses in the peritoneum via Ag-independent natural antibody production (mostly IgM isotype) (Baumgarth et al. 1999). B1a cells efficiently present antigen to T cells (Zhong et al. 2007) and rapidly migrate to the spleen to proliferate and differentiate into antibody-secreting cells after activation, for example, with lipopolysaccharides (LPS) (Yang et al. 2007). In humans, however, CD5 expression is an activation marker of B cells. The biology of human B1 cells is poorly understood, although a small population of murine B1-cell-like cells was recently reported in human cord and peripheral blood (Table 22.1). These cells ($CD20^{+}CD27^{+}CD43^{+}$) spontaneously secrete natural antibody in an Ag-independent manner and activate T cell responses. Although they express the precursor marker CD43 (as in mice) and the memory marker CD27, they are not considered as activated memory B cells (Griffin et al. 2011). Overall, development of B cells is a tightly regulated process. Its improper function or impairment can lead to various B cell malignancies or even autoimmune diseases. For example, mistakes at the centroblast/centocyte selection step promote spontaneous formation of ectopic GC and production of autoantibodies by pathogenic IL-17R-expressing B cells (Hsu et al. 2008) and induce systemic lupus erythematous (SLE) (Vinuesa et al. 2005) and inflammation of central nervous system (SNC) (Peters et al. 2011).

22.3 Regulatory B Cells

As potent antigen-presenting cells, B cells induce immune responses to xeno and self-antigens (von Bergwelt-Baildon et al. 2002). Their aberrant activation promotes autoimmune diseases, such as rheumatoid arthritis (RA), type 1 diabetes mellitus (T1D), multiple sclerosis (MS) and SLE. As such, the depletion of B cells with anti-CD20 antibody rituximab [which mostly targets mature and memory B cells, but not long-lived plasma cells (Ahuja et al. 2008)] impairs antigen-specific $CD4^{+}$ T cell activation (Bouaziz et al. 2007) and ameliorates RA, MS and T1D (Townsend et al. 2010). Surprisingly, treatment with rituximab also exacerbates the disease in some patients with ulcerative colitis, or even induces other diseases, such as psoriasis with psoriatic arthropathy and colitis in patients with Graves disease and non-Hodgkin lymphoma, respectively (Mielke et al. 2008; Dass et al. 2007; Goetz et al. 2007). Conversely, the increase in numbers of B cells in peripheral blood of transplant patients is positively associated with a rare but long-term drug-free clinical tolerance (Newell et al. 2010; Sagoo et al. 2010; Pallier et al. 2010). Taken together, these clinical observations suggest that B cells also may be needed for protection from autoimmune diseases. It appears that both effector/pathogenic and suppressive B cells and regulatory B cells (Bregs) are induced in autoimmune responses. However, the conundrum is that the existence and importance of Bregs is still not fully understood. This is probably due to difficulties to detect Bregs, which (in analogy with Tregs) can represent a small proportion of cells and, as such, may be concealed by the sheer numbers and overwhelming responses of effector B cells. This issue is further complicated by hitherto function of B cells, as first proposed by Morris and Muller in late 1960s (Morris and Moller 1968): the production of natural antibody (IgM) from B1 cells

(Yang et al. 2007; Herzenberg et al. 2000) and suppressive IgG from B cells that can trigger ITIM-mediated suppressive signaling in target cells upon binding with inhibitory FcγRIIB (Ravetch and Bolland 2001). Also, IgG can indirectly participate in immune suppression by modulating dendritic cells (DCs) via activating FcγR (Siragam et al. 2006). These "non-specific" suppressive properties of immunoglobulin have been also used to control and ameliorate various mouse and human diseases (Siragam et al. 2005, 2006; Mimouni et al. 2000; Bruhns et al. 2003). Injection of non-specific immune serum was reported to reduce incidences of skin diseases in military recruits (Mimouni et al. 2000). Autoantibody (mostly poly-reactive IgM and IgA) can also promote tissue regeneration/remyelination in the SNC improving the disease severity in mice with experimental encephalomyelitis (EAE) (Miller et al. 1997).

The first evidence of suppressive B cells (Bregs?) that functioned independently of their immunoglobulin was shown by Shimamura and colleagues about 30 years ago, experimenting with mice immunized with a high dose of sheep erythrocytes (Shimamura et al. 1982). This also appears to be a first example of the ability of suppressive B cells to induce suppressive T cells (Tregs?) (Shimamura et al. 1982), a key mechanism of immunoregulation utilized by various Bregs (see below). Not long after this report, Janeway and colleagues demonstrated the importance of B cells in the control of autoimmune responses in μMT mice (deficient in mature B cells due to a deletion in the transmembrane domain of the IgM heavy chain). The absence of B cells exacerbated EAE in these mice after immunization with myelin basic protein (Wolf et al. 1996). Similar exacerbated autoimmune responses were induced in other models, such as in mice genetically deficient in CD19 (Yanaba et al. 2008). The absence of B cells in μMT mice also leads to increased numbers of T cells, macrophages, microglial cells and neutrophils associated with larger volumes, higher mortality and more severe functional deficits upon middle cerebral artery occlusion (Ren et al. 2011). The majority of these cases appears to be associated with the loss of IL-10-producing B cells (Yanaba et al. 2008; Mizoguchi et al. 2002), although IL-10 is expressed by a variety of immune cells. As such, the B cell deficiency-associated autoimmune symptoms can be ameliorated by adoptive transfer of IL-10-producing B cells or by inducing Th2-type skewed cellular responses (Fillatreau et al. 2002). However, since recently, these protective activities are mostly linked with the existence of unique and specialized Breg subsets. They include the so-called CD1d^{high} B1b cells (CD5^{-}B220^{low}CD11b^{+}IgM^{+}CD1d^{high}) that protect mice from colitis (Mizoguchi et al. 2002) and IL-10-producing B10 regulatory cells (CD1d^{high}CD5^{+} B cells) (Yanaba et al. 2008) and Tim-1^{+} (T cell Ig domain and mucin domain protein 1) CD1d^{high}CD5^{+} Bregs (Ding et al. 2011). As for FoxP3^{+} Tregs, they appear to exist in small numbers (B10 cells, 1–2 % of spleen B220^{+} cells) but efficiently prevent T cell-mediated autoimmune responses (Yanaba et al. 2008). Overall, a wealth of experimental data unequivocally indicates the importance and existence of unique types of IL-10-producing Bregs regulating autoimmune responses in mice. In contrast, little is known about their role in humans, although their existence was indirectly suggested by increased incidences in autoimmune diseases in some humans treated with rituximab, as discussed above. Confirming this, protection from systemic lupus erythematosus was recently linked with a discrete subset of $CD19^{+}CD24^{high}CD38^{high}$ B cells (Blair et al. 2010). Unlike B cells from healthy humans, which acquired regulatory capacity and suppressed the differentiation of Th1 cells after CD40 activation, the patients' $CD19^{+}CD24^{high}CD38^{high}$ B cells were impaired (Blair et al. 2010). Moreover, a rare subset of IL-10-producing memory $CD24^{high}CD27^{+}$ B cells that functions like murine B10 cells was also shown to exist in humans (Iwata et al. 2011). Humans also have IL10 and TGF-β-producing $CD25^{high}CD27^{high}CD86^{high}CD1d^{high}$ B cells that can suppress proliferation of autologous T cells and induce the generation of $Foxp3^{+}CTLA4^{+}$ Tregs (Kessel et al. 2012).

22.4 Mechanism of Breg-Mediated Suppression

The majority of Bregs that protect against autoimmune responses requires IL-10 production (Yanaba et al. 2008; Blair et al. 2010; Byrne and Halliday 2005; Matsushita et al. 2008) presumably to inhibit Th1 cells (Fillatreau et al. 2002) and control monocyte/macrophage differentiation (Moulin et al. 2000) (Fig. 22.2), as by hitherto immunomodulatory function of IL-10 suggests so. The loss of ability to produce IL-10 in B cells or IL-10-producing $Tim1^{+}$ Bregs (as a result of Tim1 deletion) drastically decreases systemic IL-10 with age and induces development of spontaneous autoimmunity associated with hyperactive T cells (Xiao et al. 2012). Interestingly, treatment of B cells with LPS alone can induce IL-10 production and protect from autoimmune responses in mice by rendering T cells anergic (Parekh et al. 2003; Lampropoulou et al. 2008) and tolerogenic (Fuchs and Matzinger 1992). Moreover, IL-10 production is a hallmark of $CD5^{+}$ B1 cells (O'Garra and Howard 1992) and the MZ B cells that, by producing large amounts of IL-10, can ameliorate collagen-induced arthritis in mice (Brummel and Lenert 2005; Lenert et al. 2005; Evans et al. 2007). Although IL-10 production and protective activity of B10 Bregs from EAE are augmented by IL-21-dependent cognate interactions (Yoshizaki et al. 2012), the question still remains whether IL-10 is a primary mediator of Bregs' suppressive activity.

The role of IL-10 on Bregs themselves, such as promoting their survival and homeostasis, has not been studied. In fact, B1 cells are autoregulated by IL-10 in an autocrine fashion (Sindhava et al. 2010), which acting alone or together with other factors, such as SDF1, synergistically enhances survival, proliferation and chemotaxis of murine B1a and human B1 cells (Balabanian et al. 2002; Gary-Gouy et al. 2002). This may explain a recent finding in mice with IL-10 gene deletion in the B cell lineage, which did not reveal evidence of suppressive IL-10-producing Bregs in protection from lupus (Teichmann et al. 2012). On the other hand, to exert suppression at full power, Bregs appear to require additional activation, for example, induced by chronic inflammation (Miller et al. 1997), ligands of toll-like receptors (TLR) (Fillatreau et al. 2002), apoptotic cells (Gray et al. 2007), or IL-21

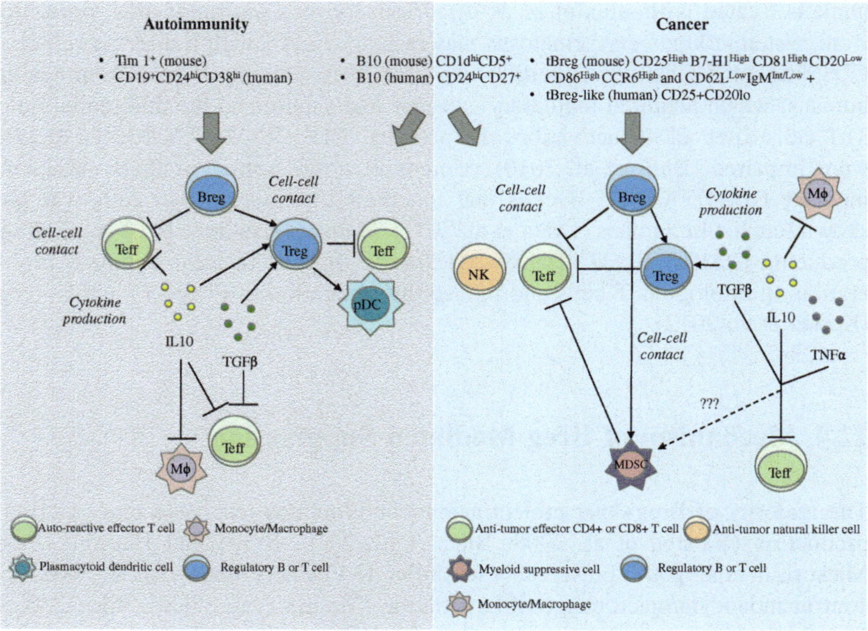

Fig. 22.2 The role of regulatory B cells in autoimmunity and cancer. In autoimmunity, Bregs (Tim1$^+$, B10 or CD19$^+$CD24^{hi}CD38$^+$) can exert their regulatory function by inhibiting effector auto-reactive T cells in the MHCII-TCR and CD40-CD40L-dependent fashions, or activation of Tregs through the CD80/CD86-CTLA4 (cell–cell contact). They can generate Tregs and thereby suppress auto-reactive effector T cells, inducing apoptosis and anergy. Tregs can also induce the expansion of tolerogenic plasmacytoid dendritic cell activating the immunoregulatory enzyme indoleamine 2,3-dioxygenase (IDO). Bregs can also produce immunosuppressive cytokines such as IL-10 and TGF-β. Among their pleiotropic functions, they suppress auto-reactive T cells and inflammatory innate immune cells such as macrophages (MΦ). In cancer, IL-10-producing Bregs (B10 cells) inhibit macrophages and monocytes (MΦ/Mo) activation, a critical step for the anti-CD20-mediated B cell lymphoma clearance. These innate immune cells produce TNF-α and up regulate CD20 expression on the surface of B cell lymphoma, rendering them more susceptible to depletion with CD20-recognizing antibody. On the other hand, tumor-evoked Bregs (tBreg) exert direct and indirect immunosuppressive activity and promote cancer escape. For example, tBregs can directly suppress activity of effector antitumor immune cells (Teff) in cell-contact-dependent fashion. Also, they produce TGF-β to induce the conversion of naïve CD4 T cells into metastasis-promoting Foxp3$^+$ Tregs, which suppress effector antitumor T cells and natural killers (NK) cells. The role of tBregs in cross-talk and regulation of other regulatory immune cells remains to be shown (???), although proposed

and CD40 (Yoshizaki et al. 2012). This raises an interesting possibility that stimuli of innate and adaptive immune responses control/induce Bregs as a feedback response to limit inflammation.

Besides IL-10, activated B cells produce various immunomodulatory factors, including TGF-β and galectin-1, and express high levels of surface antigens, like PD-1 and CTLA-4, that directly induce apoptosis and anergy of effector Th1 cells and $CD8^+$ T cells (Parekh et al. 2003; Tretter et al. 2008; Frommer et al. 2008; Zuniga et al. 2001). Some IL-10-producing $CD1d^+$ Bregs reversed allergic airway inflammation in mice independently of TGFβ (Amu et al. 2010; Scapini et al. 2011). They indirectly suppressed immune responses by inducing $FoxP3^+$ Tregs (Amu et al. 2010; Scapini et al. 2011). However, the ability of naïve B cells and Bregs, activated in response to inflammation or even ultraviolet irradiation-induced damage, to initiate suppressive responses by inducing other suppressive immune cells is not limited to the generation of Tregs (Sun et al. 2008; Scapini et al. 2011; Reichardt et al. 2007; Sayi et al. 2011), as they can inhibit activity of DCs and thereby suppressing antitumor $CD8^+$ T cells in MHC class I-dependent way (Byrne and Halliday 2005; Watt et al. 2007). However, the majority of IL-10-producing Bregs appears to mostly induce iTregs ($FoxP3^+CD25^+CD4^+$) (Sun et al. 2008; Amu et al. 2010), although the possibility exists that Bregs mediate protection from autoimmune responses via the generation of other types of IL-10-dependent Tregs, such as Tr1 cells and effector-memory Tregs (T_{REM}) (Levings and Roncarolo 2000; Kleinewietfeld et al. 2005). For example, the TLR2—activated B cells are shown to utilize IL-10 to suppress Helicobacter–mediated gastritis in mice by generating Tr1 cells (Sayi et al. 2011). On the other hand, the mechanism and use of other regulatory factors is less known, although some activated B cells and Bregs utilize TGF-β (Kessel et al. 2012; Parekh et al. 2003; Lampropoulou et al. 2008), a potent inducer of iTregs conversion from non-Treg $CD4^+$ T cells (Zheng et al. 2007).

22.5 B Cells in Cancer: Anti-tumor vs Regulatory B Cell Responses

As important controllers of cellular immune responses, cancers cleverly target Tregs and MDSCs to mediate escape from immunosurveillance. As a result, increased presence of Tregs and MDSCs is a sign of bad disease outcome in cancer patients (Woo et al. 2002; Beyer et al. 2005; Curiel et al. 2004). Although, as discussed above, B cells/Bregs actively mediate protection from autoimmune diseases, their role in cancer escape is poorly understood. Intratumoral tertiary lymphoid structures contain a significant proportion of proliferating ($Ki67^+$) germinal center-like B cells, besides T cells and mature dendritic cells (Dieu-Nosjean et al. 2008). The presence of $CD20^+$ B cells (as well as $CD8^+$ T cells) in metastatic lymph nodes is a sign of favorable outcome in patients with head and neck cancer (Pretscher et al. 2009). Antitumor $CD4^+$ T helper and $CD8^+$ cytotoxic T cells failed to be elicited in mice depleted neonatally of B cells (Schultz et al. 1990), and injection of CpG-ODN-activated B cells retards experimental lung metastasis of B16 melanoma in mice (Sorrentino et al. 2011). Overall, B cells are mostly known for their antitumor properties and their ability to present tumor antigens and amplify adaptive antitumor immune responses by cooperating with other professional antigen-presenting cells (Candolfi et al. 2011; Lanzavecchia 1985), which is further enhanced by stimulation of B cells with anti-CD40 antibody (Jackaman et al. 2011) or TLR ligands (Sorrentino et al. 2011). This probably confuses the search for Bregs promoting cancer-escape, as Bregs involved in autoimmune responses often become more suppressive after stimulations with TLR ligands [as discussed above and (Lampropoulou et al. 2008)].

B cells are shown to promote carcinogenesis of methylcholanthrene-induced (Brodt and Gordon 1978; Brodt and Gordon 1982) and progression of transplanted tumors (Monach et al. 1993); and syngeneic tumors poorly progress in μMT mice deficient in B cells unless replenished with B220$^+$ B cells (Olkhanud et al. 2011; Qin et al. 1998). The advanced stage of melanoma and other solid tumors is associated with preferential loss of CD27$^+$ and CD19$^+$ B cells (Carpenter et al. 2009; Barbera-Guillem et al. 2000). As B cells and Bregs that control autoimmunity, cancer-promoting B cells exert a multitude of functions, such as production of immunoglobulins and immunomodulatory factors and cytokines, and the antigen presentation to control regulatory T cells (Townsend et al. 2010). As in induction of RA (Firestein 2003) and insulin resistance (Winer et al. 2011), the immunoglobulin deposition induces FcR- and complement-mediated chronic inflammation needed for cancers (Zusman et al. 1996). Immunoglobulin produced by infiltrating B cells induced inflammation in premalignant tissues, enabling subsequent tumor growth in the HPV16-induced spontaneous model of carcinogenesis (de Visser et al. 2005). Activated B cells produce TGF-β (Parekh et al. 2003; Lampropoulou et al. 2008) that alone or as a complex with immunoglobulin is delivered to various immune cells to mediate suppression of cellular immune responses (Stach and Rowley 1993; Rowley and Stach 1998). Besides immunoglobulin, tumor-infiltrating B cells also produce lymphotoxin α/β and promote androgen-independent growth of prostate cancer cells by inducing the nuclear translocation of IKKα and activation of STAT3 (Ammirante et al. 2010). Protumorigenic activity of B cells also requires production of IL-10 (Inoue et al. 2006) and TNF-α (Schioppa et al. 2011) to presumably mediate Th2 polarization and inhibition of CD8$^+$ T and NK cells cytotoxic activity. Overall, the studies of last 30 years clearly underscore the importance of B cells in cancer progression and escape from immune surveillance. However, to date, the existence and the role of cancer escape-promoting Bregs remain poorly understood. Only few studies suggest the existence of unique subsets of Bregs that promote escape. For example, adoptive transfer of IL-10-producing B10 cells enhances B16 melanoma progression (DiLillo et al. 2010). Murine B10 cells modulate anti-B cell lymphoma responses in mice in IL-10-dependent fashion by reducing surface expression of FcγR and abrogating monocyte activity (Horikawa et al. 2011). As a result, the presence of B10 cells inhibited the therapeutic efficacy of anti-CD20 antibody (Ab) against lymphoma. B cells isolated from tumor-bearing mice inhibit CD4$^+$ T cell-mediated help for CTLs (Qin et al. 1998); and B220$^+$ B cells from nasal antigen tolerized mice not only induce T cell anergy, but also render these cells suppressive to other T cells (Wu et al. 2006). B cells from patients with advanced stage solid tumors have reduced ability to activate T cells to express IFN-γ and IL-2 (Carpenter et al. 2009).

To address the existence and role of specialized and unique Breg subsets in cancer escape, we speculated that, if cancer induces both protective/pathogenic B cells and Bregs, the overwhelming response from the former could conceal the detection of Bregs. However, to segregate them, first we sought for B cell changes in cancer bearing immune competent BALB/c mice bearing 4T1 carcinoma cells, a model that mimics breast cancer-induced immunosuppression (Lelekakis et al. 1999). Mice subcutaneously challenged with 4T1 cells in the mammary gland quickly succumb to massive metastasis in the lungs, lymph nodes, liver, bone and other sites (Lelekakis et al. 1999). The tumor-induced suppressive microenvironment and GM-CSF induce massive expansion of CD11b$^+$ MDSCs that are thought to facilitate metastasis of 4T1 cells. Since in this tumor model, lung metastasis (as a state of cancer cells vulnerable to immune attack) requires CCR4$^+$FoxP3$^+$ Tregs to inactivate antitumor NK cells utilizing β-galactoside -binding protein (galectin-1) (Olkhanud et al. 2009), the potential Breg candidates were tested for ability to induce non-cytotoxic suppression of T cell proliferation stimulated with anti-CD3 Ab and to convert FoxP3$^+$ Tregs from non-Treg CD4$^+$ T cells. As a result, we found that the Tregs by themselves were controlled by a previously unknown small subset of B cells, designated tumor-evoked Bregs (tBregs) (Olkhanud et al. 2011). Phenotypically resembling immature B2 cells (IgD^{high}), tBregs represent a unique and small population of B cells that express constitutively active Stat3. Although they are Tim-1$^+$, tBregs do not express CD1d^{high} and CD5. We define tBregs as pStat3$^+$CD19$^+$ B cells that are CD25$^+$B7-H1^{high}CD81^{high}CD20^{low}CD86^{high} CCR6^{high} CD62L^{low}IgM$^{int/low}$. They promote metastasis and cancer escape by suppressing effector antitumor immune responses acting directly or indirectly by converting FoxP3$^+$ Tregs from non-Treg CD4$^+$ T cells utilizing TGF-β (Olkhanud et al. 2011). We believe that other tBreg-like cells may exist, as the progression and escape of B cell lymphoma in mice also utilize Tregs (Elpek et al. 2007); and the absence of B cells induces an anti-tumor immune response by impairing Tregs in mice with the EMT-6 breast cancer (Tadmor et al. 2011). Similarly CD25$^+$ B cells that induce anergy of activated T cells by competing for IL-2 upon treatment with *S. aureus* Cowan 1 antigen were recently reported (Tretter et al. 2008), suggesting that tBregs may also be induced by parasite infection. Unlike them, tBregs efficiently inhibit both resting and activated T cells, including CD4$^+$ and CD8$^+$ T cells without induction of cell death or use of IL-2. In humans, tBreg-like cells also exist, as we readily generated them ex vivo by treating normal human donor B cells with conditioned media of human cancer lines, such as breast, ovarian and colon carcinomas (Olkhanud et al. 2011). However, tBregs differ phenotypically and functionally from other Bregs involved in autoimmune responses (Yanaba et al. 2008; Mizoguchi et al. 2002; Matsushita et al. 2008) and LPS- or BCR-activated B cells (Fuchs and Matzinger 1992; Hussain and Delovitch 2007). tBregs poorly proliferate and do not express CD27 and CD5 or up regulate CD1d, and their regulatory activity does not require suppressive pathways, such as B7-H1-PD1, Fas-FasL, and IL-27/IL-35. The requirement and use of IL-10 by tBregs remains unknown, although their suppressive activity was retained in vitro despite the importance of IL-10 in cancer escape.

The mechanism of tBreg generation remains poorly understood. Both murine and human tBregs are readily generated from normal B cells by treating them with conditioned media of various human cancer lines (Olkhanud et al. 2011). Interestingly, at least in mice, tBregs are mostly induced by the non-metastastic cancer cell subsets (Olkhanud et al. 2011) utilizing metabolites of 5-lipoxigenese pathway (5-LO) (Wejksza et al. manuscript submitted). For example, we demonstrated that, to do this, 5-LO metabolites activate the proliferator-activated receptor alpha (PPARα) in B cells. Non-metastatic cancer cell subsets also express B cell survival factor BAFF to actively maintain tBregs viable, which contrasts with the ability of BAFF to induce murine IL-10-producing CD1d^{high}CD5^{+} Bregs from MZ B cells (Yang et al. 2010). BAFF and APRIL are often found expressed in human solid tumors, including breast carcinomas (Roosnek et al. 2009; Pelekanou et al. 2008), suggesting an interesting possibility that this is to support survival of tBregs or/and induce B10-like Bregs in humans. It should be noted that, in contrast to the widely accepted view, MDSCs appear not to play a primary immunoregulatory role needed for lung metastasis. The depletion of tBregs or Tregs alone successfully abrogates lung metastasis of 4T1 cancer cells in syngeneic immunocompetent mice (Olkhanud et al. 2009, 2011). Despite the presence of MDSCs and unaltered progression of 4T1 cancer cells at the primary site, the mammary gland, they failed to metastasize into the lungs of NOD/SCID mice deficient in T and B cells (Olkhanud et al. 2009, 2011). Similarly, although MDSCs thought to promote conversion of Tregs (Nagaraj and Gabrilovich 2008; Youn et al. 2008; Liu et al. 2007; DuPre et al. 2007), adoptive transfer of CD25^{-}CD4^{+} T cells (non-Treg cells depleted of Tregs) alone in NOD/SCID mice did not affect this. In contrast, these mice succumbed to massive lung metastasis if they were transferred with Tregs alone or tBregs together with non-Treg (to supply a source of tBreg-mediated Treg conversion) (Olkhanud et al. 2009, 2011). Conversely, the loss or inactivation of B cells in WT mice leads to the reduction of MDSCs and Tregs (Olkhanud et al. 2009, 2011; Sorrentino et al. 2011), suggesting that B cells may also mediate suppression of immune responses by regulating other immune cells. tBregs may cross-talk with MDSCs by providing or receiving survival benefits to further enhance immune suppression. Furthermore, tBregs may also act like B1 cells promoting M2 polarization of macrophages and TAMs utilizing IL-10 (Wong et al. 2010).

The clinical implication of tBregs is that, as long as cancer persists, it will induce tBregs and thereby initiate the chain of suppressive events. Thus, strategies that abrogate any step of this process are expected to inhibit cancer escape and metastasis. Indeed, the depletion of Tregs and tBregs with PC61 anti-CD25 antibody (Ab) successfully abrogates 4T1 breast cancer lung metastasis (Olkhanud et al. 2009, 2011). However, tumor burden in the lungs of mice intravenously injected with B16-F10 melanoma was also enhanced after anti-CD20 Ab-mediated B cell depletion (DiLillo et al. 2010), and depletion of B cells with rituximab (anti-CD20 Ab) did not provide clinical benefit in patients with renal cell carcinoma and melanoma (Aklilu et al. 2004). This discrepancy can probably be explained if anti-CD20 Ab treatment only depleted "normal functioning" or "immune stimulatory"

$CD20^+$ B cells that participate in the induction of adaptive antitumor immune responses (Candolfi et al. 2011) promoting favorable disease outcome in some cancer patients (Pretscher et al. 2009). Indeed, we recently demonstrated that anti-CD20 Ab enhances cancer escape in mice with 4T1.2 breast cancer by depleting antitumor B cells and by enriching for $CD20^{low}$ tBregs (Bodogai et al. manuscript submitted). It remains to be seen whether anti-CD20 Ab depletion-induced increase in Tregs and protection from arthritis (Hamel et al. 2011) and reverses autoimmune diseases (Hu et al. 2007) may also be linked with enrichment for $CD20^{low}$ tBreg-like cells.

22.6 Concluding Remarks

Cancer induces activation of both pathogenic/cancer-eradicating B cells and cancer-promoting regulatory B cells (Fig. 22.2). Although the overwhelming response of the former may conceal Bregs, modeling studies in mice and clinical observations unequivocally indicate the existence of cancer-promoting B cells and tBregs. The clinical implication of the ability of cancer to actively convert tBregs from normal B cells is that as long as cancer persists, it will induce their generation and thereby initiate the chain of suppressive events. Thus, strategies that abrogate any step of this process are expected to inhibit cancer escape and metastasis, a primary cause of patients' bad disease outcome. However, the success of a strategy will also depend on the use of tailored approaches, such as ones that only inactivate tBregs, while protecting or promoting "good" B cells needed for optimal cancer eradication. For example, 4T1 breast cancer metastasis is abrogated by antibody that targets IL-2Rα (Olkhanud et al. 2009, 2011). In contrast, confirming a recent failure of anti-CD20 Ab rituximab in patients with renal cell carcinoma (Aklilu et al. 2004), tBregs express low levels of CD20 and thereby cannot be depleted with anti-CD20 antibody treatment (Bodogai et al., manuscript submitted). Treatment with anti-CD20 antibody preferentially depletes "good" and activated B cells, while enriching for tBregs and thereby enhancing cancer escape and metastasis.

Acknowledgments This research was supported by the Intramural Research Program of the National Institute on Aging, NIH. The authors work for the US government and declare no competing financial interests.

References

Ahuja A, Anderson SM, Khalil A, Shlomchik MJ (2008) Maintenance of the plasma cell pool is independent of memory B cells. Proc Natl Acad Sci USA 105:4802–4807

Aklilu M, Stadler WM, Markiewicz M, Vogelzang NJ, Mahowald M, Johnson M et al (2004) Depletion of normal B cells with rituximab as an adjunct to IL-2 therapy for renal cell carcinoma and melanoma. Ann Oncol Official J Eur Soc Med Oncol/ESMO 15:1109–1114

Allen CD, Ansel KM, Low C, Lesley R, Tamamura H, Fujii N et al (2004) Germinal center dark and light zone organization is mediated by CXCR4 and CXCR5. Nat Immunol 5:943–952

Ammirante M, Luo JL, Grivennikov S, Nedospasov S, Karin M (2010) B-cell-derived lymphotoxin promotes castration-resistant prostate cancer. Nature 464:302–305

Amu S, Saunders SP, Kronenberg M, Mangan NE, Atzberger A, Fallon PG (2010) Regulatory B cells prevent and reverse allergic airway inflammation via FoxP3-positive T regulatory cells in a murine model. J Allergy Clin Immunol 125(1114):1124e8

Anderson SM, Tomayko MM, Ahuja A, Haberman AM, Shlomchik MJ (2007) New markers for murine memory B cells that define mutated and unmutated subsets. J Exp Med 204:2103–2114

Baatar D, Olkhanud P, Sumitomo K, Taub D, Gress R, Biragyn A (2007) Human peripheral blood T regulatory cells (Tregs), functionally primed CCR4$^+$ Tregs and unprimed CCR4- Tregs, regulate effector T cells using FasL. J Immunol 178:4891–4900

Balabanian K, Foussat A, Bouchet-Delbos L, Couderc J, Krzysiek R, Amara A et al (2002) Interleukin-10 modulates the sensitivity of peritoneal B lymphocytes to chemokines with opposite effects on stromal cell-derived factor-1 and B-lymphocyte chemoattractant. Blood 99:427–436

Barbera-Guillem E, Nelson MB, Barr B, Nyhus JK, May KF Jr, Feng L et al (2000) B lymphocyte pathology in human colorectal cancer. Experimental and clinical therapeutic effects of partial B cell depletion. Cancer Immunol Immunother CII 48:541–549

Baumgarth N, Herman OC, Jager GC, Brown L, Herzenberg LA (1999) Innate and acquired humoral immunities to influenza virus are mediated by distinct arms of the immune system. Proc Natl Acad Sci USA 96:2250–2255

Beyer M, Kochanek M, Darabi K, Popov A, Jensen M, Endl E et al (2005) Reduced frequencies and suppressive function of CD4$^+$ CD25hi regulatory T cells in patients with chronic lymphocytic leukemia after therapy with fludarabine. Blood 106:2018–2025

Blair PA, Norena LY, Flores-Borja F, Rawlings DJ, Isenberg DA, Ehrenstein MR et al (2010) CD19(+)CD24(hi)CD38(hi) B cells exhibit regulatory capacity in healthy individuals but are functionally impaired in systemic Lupus Erythematosus patients. Immunity 32:129–140

Bouaziz JD, Yanaba K, Venturi GM, Wang Y, Tisch RM, Poe JC et al (2007) Therapeutic B cell depletion impairs adaptive and autoreactive CD4$^+$ T cell activation in mice. Proc Natl Acad Sci USA 104:20878–20883

Brodt P, Gordon J (1978) Anti-tumor immunity in B lymphocyte-deprived mice. I. Immunity to a chemically induced tumor. J Immunol 121:359–362

Brodt P, Gordon J (1982) Natural resistance mechanisms may play a role in protection against chemical carcinogenesis. Cancer Immunol Immunother CII 13:125–127

Bruhns P, Samuelsson A, Pollard JW, Ravetch JV (2003) Colony-stimulating factor-1-dependent macrophages are responsible for IVIG protection in antibody-induced autoimmune disease. Immunity 18:573–581

Brummel R, Lenert P (2005) Activation of marginal zone B cells from lupus mice with type A(D) CpG-oligodeoxynucleotides. J Immunol 174:2429–2434

Bryant VL, Ma CS, Avery DT, Li Y, Good KL, Corcoran LM et al (2007) Cytokine-mediated regulation of human B cell differentiation into Ig-secreting cells: predominant role of IL-21 produced by CXCR5$^+$ T follicular helper cells. J Immunol 179:8180–8190

Byrne SN, Halliday GM (2005) B cells activated in lymph nodes in response to ultraviolet irradiation or by interleukin-10 inhibit dendritic cell induction of immunity. J Invest Dermatol 124:570–578

Caligaris-Cappio F, Gobbi M, Bofill M, Janossy G (1982) Infrequent normal B lymphocytes express features of B-chronic lymphocytic leukemia. J Exp Med 155:623–628

Candolfi M, Curtin JF, Yagiz K, Assi H, Wibowo MK, Alzadeh GE et al (2011) B cells are critical to T-cell-mediated antitumor immunity induced by a combined immune-stimulatory/conditionally cytotoxic therapy for glioblastoma. Neoplasia 13:947–960

Carpenter EL, Mick R, Rech AJ, Beatty GL, Colligon TA, Rosenfeld MR et al (2009) Collapse of the CD27^{+} B-cell compartment associated with systemic plasmacytosis in patients with advanced melanoma and other cancers. Clin Cancer Res 15:4277–4287

Cheng P, Corzo CA, Luetteke N, Yu B, Nagaraj S, Bui MM et al (2008) Inhibition of dendritic cell differentiation and accumulation of myeloid-derived suppressor cells in cancer is regulated by S100A9 protein. J Exp Med 205:2235–2249

Curiel TJ, Coukos G, Zou L, Alvarez X, Cheng P, Mottram P et al (2004) Specific recruitment of regulatory T cells in ovarian carcinoma fosters immune privilege and predicts reduced survival. Nat Med 10:942–949

Dass S, Vital EM, Emery P (2007) Development of psoriasis after B cell depletion with rituximab. Arthritis Rheum 56:2715–2718

de Visser KE, Korets LV, Coussens LM (2005) De novo carcinogenesis promoted by chronic inflammation is B lymphocyte dependent. Cancer Cell 7:411–423

DeKoter RP, Singh H (2000) Regulation of B lymphocyte and macrophage development by graded expression of PU.1. Science 288:1439–1441

Dieu-Nosjean MC, Antoine M, Danel C, Heudes D, Wislez M, Poulot V et al (2008) Long-term survival for patients with non-small-cell lung cancer with intratumoral lymphoid structures. J Clin Oncol 26:4410–4417

DiLillo DJ, Matsushita T, Tedder TF (2010) B10 cells and regulatory B cells balance immune responses during inflammation, autoimmunity, and cancer. Ann N Y Acad Sci 1183:38–57

Ding Q, Yeung M, Camirand G, Zeng Q, Akiba H, Yagita H et al (2011) Regulatory B cells are identified by expression of TIM-1 and can be induced through TIM-1 ligation to promote tolerance in mice. J Clin Investig 121:3645–3656

DuPre SA, Redelman D, Hunter KW Jr (2007) The mouse mammary carcinoma 4T1: characterization of the cellular landscape of primary tumours and metastatic tumour foci. Int J Exp Pathol 88:351–360

Elpek KG, Lacelle C, Singh NP, Yolcu ES, Shirwan H (2007) CD4^{+} CD25^{+} T regulatory cells dominate multiple immune evasion mechanisms in early but not late phases of tumor development in a B cell lymphoma model. J Immunol 178:6840–6848

Evans JG, Chavez-Rueda KA, Eddaoudi A, Meyer-Bahlburg A, Rawlings DJ, Ehrenstein MR et al (2007) Novel suppressive function of transitional 2 B cells in experimental arthritis. J Immunol 178:7868–7878

Fillatreau S, Sweenie CH, McGeachy MJ, Gray D, Anderton SM (2002) B cells regulate autoimmunity by provision of IL-10. Nat Immunol 3:944–950

Firestein GS (2003) Evolving concepts of rheumatoid arthritis. Nature 423:356–361

Frommer F, Heinen TJ, Wunderlich FT, Yogev N, Buch T, Roers A et al (2008) Tolerance without clonal expansion: self-antigen-expressing B cells program self-reactive T cells for future deletion. J Immunol 181:5748–5759

Fuchs EJ, Matzinger P (1992) B cells turn off virgin but not memory T cells. Science 258:1156–1159

Gary-Gouy H, Harriague J, Bismuth G, Platzer C, Schmitt C, Dalloul AH (2002) Human CD5 promotes B-cell survival through stimulation of autocrine IL-10 production. Blood 100:4537–4543

Girkontaite I, Sakk V, Wagner M, Borggrefe T, Tedford K, Chun J et al (2004) The sphingosine-1-phosphate (S1P) lysophospholipid receptor S1P3 regulates MAdCAM-1^{+} endothelial cells in splenic marginal sinus organization. J Exp Med 200:1491–1501

Goetz M, Atreya R, Ghalibafian M, Galle PR, Neurath MF (2007) Exacerbation of ulcerative colitis after rituximab salvage therapy. Inflamm Bowel Dis 13:1365–1368

Goetzl EJ, Rosen H (2004) Regulation of immunity by lysosphingolipids and their G protein-coupled receptors. J Clin Investig 114:1531–1537

Gray M, Miles K, Salter D, Gray D, Savill J (2007) Apoptotic cells protect mice from autoimmune inflammation by the induction of regulatory B cells. Proc Natl Acad Sci USA 104:14080–14085

Griffin DO, Holodick NE, Rothstein TL (2011) Human B1 cells in umbilical cord and adult peripheral blood express the novel phenotype CD20+ CD27+ CD43+ CD70. J Exp Med 208:67–80

Hamel KM, Cao Y, Ashaye S, Wang Y, Dunn R, Kehry MR et al (2011) B cell depletion enhances T regulatory cell activity essential in the suppression of arthritis. J Immunol 187:4900–4906

Hayakawa K, Hardy RR, Parks DR, Herzenberg LA (1983) The "Ly-1 B" cell subpopulation in normal immunodefective, and autoimmune mice. J Exp Med 157:202–218

Herzenberg LA, Baumgarth N, Wilshire JA (2000) B-1 cell origins and VH repertoire determination. Curr Top Microbiol Immunol 252:3–13

Horikawa M, Minard-Colin V, Matsushita T, Tedder TF (2011) Regulatory B cell production of IL-10 inhibits lymphoma depletion during CD20 immunotherapy in mice. J Clin Investig 121:4268–4280

Hsu HC, Yang P, Wang J, Wu Q, Myers R, Chen J et al (2008) Interleukin 17-producing T helper cells and interleukin 17 orchestrate autoreactive germinal center development in autoimmune BXD2 mice. Nat Immunol 9:166–175

Hu CY, Rodriguez-Pinto D, Du W, Ahuja A, Henegariu O, Wong FS et al (2007) Treatment with CD20-specific antibody prevents and reverses autoimmune diabetes in mice. J Clin Investig 117:3857–3867

Hussain S, Delovitch TL (2007) Intravenous transfusion of BCR-activated B cells protects NOD mice from type 1 diabetes in an IL-10-dependent manner. J Immunol 179:7225–7232

Inoue S, Leitner WW, Golding B, Scott D (2006) Inhibitory effects of B cells on antitumor immunity. Cancer Res 66:7741–7747

Iwata Y, Matsushita T, Horikawa M, Dilillo DJ, Yanaba K, Venturi GM et al (2011) Characterization of a rare IL-10-competent B-cell subset in humans that parallels mouse regulatory B10 cells. Blood 117:530–541

Jackaman C, Cornwall S, Graham PT, Nelson DJ (2011) CD40-activated B cells contribute to mesothelioma tumor regression. Immunol Cell Biol 89:255–267

Kessel A, Haj T, Peri R, Snir A, Melamed D, Sabo E et al (2012) Human CD19(+)CD25(high) B regulatory cells suppress proliferation of CD4(+) T cells and enhance Foxp3 and CTLA-4 expression in T-regulatory cells. Autoimmun Rev 11:670–677

Kikuchi K, Lai AY, Hsu CL, Kondo M (2005) IL-7 receptor signaling is necessary for stage transition in adult B cell development through up-regulation of EBF. J Exp Med 201:1197–1203

Kleinewietfeld M, Puentes F, Borsellino G, Battistini L, Rotzschke O, Falk K (2005) CCR6 expression defines regulatory effector/memory-like cells within the CD25(+)CD4+ T-cell subset. Blood 105:2877–2886

Klinker MW, Lundy SK (2012) Multiple mechanisms of immune suppression by B lymphocytes. Mol Med 18:123–137

Lampropoulou V, Hoehlig K, Roch T, Neves P, Calderon Gomez E, Sweenie CH et al (2008) TLR-activated B cells suppress T cell-mediated autoimmunity. J Immunol 180:4763–4773

Lanzavecchia A (1985) Antigen-specific interaction between T and B cells. Nature 314:537–539

Lazorchak AS, Schlissel MS, Zhuang Y (2006) E2A and IRF-4/Pip promote chromatin modification and transcription of the immunoglobulin kappa locus in pre-B cells. Mol Cell Biol 26:810–821

Lelekakis M, Moseley JM, Martin TJ, Hards D, Williams E, Ho P et al (1999) A novel orthotopic model of breast cancer metastasis to bone. Clin Exp Metastasis 17:163–170

Lenert P, Brummel R, Field EH, Ashman RF (2005) TLR-9 activation of marginal zone B cells in lupus mice regulates immunity through increased IL-10 production. J Clin Immunol 25:29–40

Levings MK, Roncarolo MG (2000) T-regulatory 1 cells: a novel subset of CD4 T cells with immunoregulatory properties. J Allergy Clin Immunol 106:S109–S112

Liu VC, Wong LY, Jang T, Shah AH, Park I, Yang X et al (2007) Tumor evasion of the immune system by converting $CD4^{+}$. J Immunol 178:2883–2892

Lu TT, Cyster JG (2002) Integrin-mediated long-term B cell retention in the splenic marginal zone. Science 297:409–412

Lund FE, Randall TD (2010) Effector and regulatory B cells: modulators of CD4(+) T cell immunity. Nat Rev Immunol 10:236–247

Maeda T, Merghoub T, Hobbs RM, Dong L, Maeda M, Zakrzewski J et al (2007) Regulation of B versus T lymphoid lineage fate decision by the proto-oncogene LRF. Science 316:860–866

Matsushita T, Yanaba K, Bouaziz JD, Fujimoto M, Tedder TF (2008) Regulatory B cells inhibit EAE initiation in mice while other B cells promote disease progression. J Clin Invest 118:3420–3430

Mielke F, Schneider-Obermeyer J, Dorner T (2008) Onset of psoriasis with psoriatic arthropathy during rituximab treatment of non-Hodgkin lymphoma. Ann Rheum Dis 67:1056–1057

Miller DJ, Bright JJ, Sriram S, Rodriguez M (1997) Successful treatment of established relapsing experimental autoimmune encephalomyelitis in mice with a monoclonal natural autoantibody. J Neuroimmunol 75:204–209

Mimouni D, Gdalevich M, Mimouni FB, Grotto I, Eldad A, Shpilberg O (2000) Does immune serum globulin confer protection against skin diseases? Int J Dermatol 39:628–631

Mizoguchi A, Mizoguchi E, Smith RN, Preffer FI, Bhan AK (1997) Suppressive role of B cells in chronic colitis of T cell receptor alpha mutant mice. J Exp Med 186:1749–1756

Mizoguchi A, Mizoguchi E, Takedatsu H, Blumberg RS, Bhan AK (2002) Chronic intestinal inflammatory condition generates IL-10-producing regulatory B cell subset characterized by CD1d upregulation. Immunity 16:219–230

Monach PA, Schreiber H, Rowley DA (1993) CD4$^+$ and B lymphocytes in transplantation immunity. II. Augmented rejection of tumor allografts by mice lacking B cells. Transplantation 55:1356–1361

Moore KW, de Waal MR, Coffman RL, O'Garra A (2001) Interleukin-10 and the interleukin-10 receptor. Annu RevImmunol 19:683–765

Morris A, Moller G (1968) Regulation of cellular antibody synthesis effect of adoptively transferred antibody-producing spleen cells on cellular antibody synthesis. Journal of immunology. 101:439–445

Moulin V, Andris F, Thielemans K, Maliszewski C, Urbain J, Moser M (2000) B lymphocytes regulate dendritic cell (DC) function in vivo: increased interleukin 12 production by DCs from B cell-deficient mice results in T helper cell type 1 deviation. J Exp Med 192:475–482

Nagaraj S, Gabrilovich DI (2008) Tumor escape mechanism governed by myeloid-derived suppressor cells. Cancer Res 68:2561–2563

Newell KA, Asare A, Kirk AD, Gisler TD, Bourcier K, Suthanthiran M et al (2010) Identification of a B cell signature associated with renal transplant tolerance in humans. J Clin Investig 120:1836–1847

O'Garra A, Howard M (1992) Cytokines and Ly-1 (B1) B cells. Int Rev Immunol 8:219–234

Olkhanud PB, Baatar D, Bodogai M, Hakim F, Gress R, Anderson RL et al (2009) Breast cancer lung metastasis requires expression of chemokine receptor CCR4 and regulatory T cells. Cancer Res 69:5996–6004

Olkhanud PB, Damdinsuren B, Bodogai M, Gress RE, Sen R, Wejksza K et al (2011a) Tumor-evoked regulatory B cells promote breast cancer metastasis by converting resting CD4 T cells to T-regulatory cells. Cancer Res 71:3505–3515

Olkhanud PB, Damdinsuren B, Bodogai M, Gress RE, Sen R, Wejksza K et al (2011b) Tumor-evoked regulatory B cells promote breast cancer metastasis by converting resting CD4$^+$ T cells to T regulatory cells. Cancer Res 15:3505–3515

Pallier A, Hillion S, Danger R, Giral M, Racape M, Degauque N et al (2010) Patients with drug-free long-term graft function display increased numbers of peripheral B cells with a memory and inhibitory phenotype. Kidney Int 78:503–513

Parekh VV, Prasad DV, Banerjee PP, Joshi BN, Kumar A, Mishra GC (2003) B cells activated by lipopolysaccharide, but not by anti-Ig and anti-CD40 antibody, induce anergy in CD8$^+$ T cells: role of TGF-beta 1. J Immunol 170:5897–5911

Pelekanou V, Kampa M, Kafousi M, Darivianaki K, Sanidas E, Tsiftsis DD et al (2008) Expression of TNF-superfamily members BAFF and APRIL in breast cancer: immunohistochemical study in 52 invasive ductal breast carcinomas. BMC Cancer 8:76

Peters A, Pitcher LA, Sullivan JM, Mitsdoerffer M, Acton SE, Franz B et al (2011) Th17 cells induce ectopic lymphoid follicles in central nervous system tissue inflammation. Immunity 35:986–996

Phan TG, Gray EE, Cyster JG (2009) The microanatomy of B cell activation. Curr Opin Immunol 21:258–265

Pretscher D, Distel LV, Grabenbauer GG, Wittlinger M, Buettner M, Niedobitek G (2009) Distribution of immune cells in head and neck cancer: $CD8^+$ T-cells and $CD20^+$ B-cells in metastatic lymph nodes are associated with favourable outcome in patients with oro- and hypopharyngeal carcinoma. BMC Cancer 9:292

Qin Z, Richter G, Schuler T, Ibe S, Cao X, Blankenstein T (1998) B cells inhibit induction of T cell-dependent tumor immunity. Nat Med 4:627–630

Ravetch JV, Bolland S (2001) IgG Fc receptors. Annu Rev Immunol 19:275–290

Reichardt P, Dornbach B, Rong S, Beissert S, Gueler F, Loser K et al (2007) Naive B cells generate regulatory T cells in the presence of a mature immunologic synapse. Blood 110:1519–1529

Reif K, Ekland EH, Ohl L, Nakano H, Lipp M, Forster R et al (2002) Balanced responsiveness to chemoattractants from adjacent zones determines B-cell position. Nature 416:94–99

Reinhardt RL, Liang HE, Locksley RM (2009) Cytokine-secreting follicular T cells shape the antibody repertoire. Nat Immunol 10:385–393

Ren X, Akiyoshi K, Dziennis S, Vandenbark AA, Herson PS, Hurn PD et al (2011) Regulatory B cells limit CNS inflammation and neurologic deficits in murine experimental stroke. J Neurosci 31:8556–8563

Roosnek E, Burjanadze M, Dietrich PY, Matthes T, Passweg J, Huard B (2009) Tumors that look for their springtime in APRIL. Crit Rev Oncol Hematol 72:91–97

Rowley DA, Stach RM (1998) B lymphocytes secreting IgG linked to latent transforming growth factor-beta prevent primary cytolytic T lymphocyte responses. Int Immunol 10:355–363

Sagoo P, Perucha E, Sawitzki B, Tomiuk S, Stephens DA, Miqueu P et al (2010) Development of a cross-platform biomarker signature to detect renal transplant tolerance in humans. J Clin Investig 120:1848–1861

Saito T, Chiba S, Ichikawa M, Kunisato A, Asai T, Shimizu K et al (2003) Notch2 is preferentially expressed in mature B cells and indispensable for marginal zone B lineage development. Immunity 18:675–685

Sakaguchi S (2000) Regulatory T cells: key controllers of immunologic self-tolerance. Cell 101:455–458

Sayi A, Kohler E, Toller IM, Flavell RA, Muller W, Roers A et al (2011) TLR-2-activated B cells suppress Helicobacter-induced preneoplastic gastric immunopathology by inducing T regulatory-1 cells. J Immunol 186:878–890

Scapini P, Lamagna C, Hu Y, Lee K, Tang Q, DeFranco AL et al (2011) B cell-derived IL-10 suppresses inflammatory disease in Lyn-deficient mice. Proc Natl Acad Sci USA 108:E823–E832

Schioppa T, Moore R, Thompson RG, Rosser EC, Kulbe H, Nedospasov S et al (2011) B regulatory cells and the tumor-promoting actions of TNF-alpha during squamous carcinogenesis. Proc Natl Acad Sci USA 108:10662–10667

Schultz KR, Klarnet JP, Gieni RS, HayGlass KT, Greenberg PD (1990) The role of B cells for in vivo T cell responses to a Friend virus-induced leukemia. Science 249:921–923

Shimamura T, Hashimoto K, Sasaki S (1982) Feedback suppression of the immune response in vivo. I. Immune B cells induce antigen-specific suppressor T cells. Cell Immunol 68:104–113

Sindhava V, Woodman ME, Stevenson B, Bondada S (2010) Interleukin-10 mediated autoregulation of murine B-1 B-cells and its role in Borrelia hermsii infection. PLoS ONE 5:e11445

Siragam V, Brinc D, Crow AR, Song S, Freedman J, Lazarus AH (2005) Can antibodies with specificity for soluble antigens mimic the therapeutic effects of intravenous IgG in the treatment of autoimmune disease? J Clin Investig 115:155–160
Siragam V, Crow AR, Brinc D, Song S, Freedman J, Lazarus AH (2006) Intravenous immunoglobulin ameliorates ITP via activating Fc gamma receptors on dendritic cells. Nat Med 12:688–692
Sorrentino R, Morello S, Forte G, Montinaro A, De Vita G, Luciano A et al (2011) B cells contribute to the antitumor activity of CpG-oligodeoxynucleotide in a mouse model of metastatic lung carcinoma. Am J Respir Crit Care Med 183:1369–1379
Stach RM, Rowley DA (1993) A first or dominant immunization. II. Induced immunoglobulin carries transforming growth factor beta and suppresses cytolytic T cell responses to unrelated alloantigens. J Exp Med 178:841–852
Sun JB, Flach CF, Czerkinsky C, Holmgren J (2008) B lymphocytes promote expansion of regulatory T cells in oral tolerance: powerful induction by antigen coupled to cholera toxin B subunit. J Immunol 181:8278–8287
Suzuki K, Grigorova I, Phan TG, Kelly LM, Cyster JG (2009) Visualizing B cell capture of cognate antigen from follicular dendritic cells. J Exp Med 206:1485–1493
Tadmor T, Zhang Y, Cho HM, Podack ER, Rosenblatt JD (2011) The absence of B lymphocytes reduces the number and function of T-regulatory cells and enhances the anti-tumor response in a murine tumor model. Cancer Immunol Immunother CII 60:609–619
Takahashi Y, Ohta H, Takemori T (2001) Fas is required for clonal selection in germinal centers and the subsequent establishment of the memory B cell repertoire. Immunity 14:181–192
Teichmann LL, Kashgarian M, Weaver CT, Roers A, Muller W, Shlomchik MJ (2012) B cell-derived IL-10 does not regulate spontaneous systemic autoimmunity in MRL.Fas(lpr) mice. J Immunol 188:678–685
Townsend MJ, Monroe JG, Chan AC (2010) B-cell targeted therapies in human autoimmune diseases: an updated perspective. Immunol Rev 237:264–283
Tretter T, Venigalla RK, Eckstein V, Saffrich R, Sertel S, Ho AD et al (2008) Induction of CD4$^+$ T-cell anergy and apoptosis by activated human B cells. Blood 112:4555–4564
Vettermann C, Jack HM (2010) The pre-B cell receptor: turning autoreactivity into self-defense. Trends Immunol 31:176–183
Vinuesa CG, Tangye SG, Moser B, Mackay CR (2005) Follicular B helper T cells in antibody responses and autoimmunity. Nat Rev Immunol 5:853–865
von Bergwelt-Baildon MS, Vonderheide RH, Maecker B, Hirano N, Anderson KS, Butler MO et al (2002) Human primary and memory cytotoxic T lymphocyte responses are efficiently induced by means of CD40-activated B cells as antigen-presenting cells: potential for clinical application. Blood 99:3319–3325
Watt V, Ronchese F, Ritchie D (2007) Resting B cells suppress tumor immunity via an MHC class-II dependent mechanism. J Immunother 30:323–332
Wei C, Anolik J, Cappione A, Zheng B, Pugh-Bernard A, Brooks J et al (2007) A new population of cells lacking expression of CD27 represents a notable component of the B cell memory compartment in systemic lupus erythematosus. J Immunol 178:6624–6633
Wen L, Brill-Dashoff J, Shinton SA, Asano M, Hardy RR, Hayakawa K (2005) Evidence of marginal-zone B cell-positive selection in spleen. Immunity 23:297–308
William J, Euler C, Christensen S, Shlomchik MJ (2002) Evolution of autoantibody responses via somatic hypermutation outside of germinal centers. Science 297:2066–2070
Winer DA, Winer S, Shen L, Wadia PP, Yantha J, Paltser G et al (2011) B cells promote insulin resistance through modulation of T cells and production of pathogenic IgG antibodies. Nat Med 17:610–617
Wolf SD, Dittel BN, Hardardottir F, Janeway CA Jr (1996) Experimental autoimmune encephalomyelitis induction in genetically B cell-deficient mice. J Exp Med 184:2271–2278
Wong SC, Puaux AL, Chittezhath M, Shalova I, Kajiji TS, Wang X et al (2010) Macrophage polarization to a unique phenotype driven by B cells. Eur J Immunol 40:2296–2307

Woo EY, Yeh H, Chu CS, Schlienger K, Carroll RG, Riley JL et al (2002) Cutting edge: regulatory T cells from lung cancer patients directly inhibit autologous T cell proliferation. J Immunol 168:4272–4276

Wu HY, Monsonego A, Weiner HL (2006) The mechanism of nasal tolerance in lupus prone mice is T-cell anergy induced by immature B cells that lack B7 expression. J Autoimmun 26:116–126

Xiao S, Brooks CR, Zhu C, Wu C, Sweere JM, Petecka S et al (2012) Defect in regulatory B-cell function and development of systemic autoimmunity in T-cell Ig mucin 1 (Tim-1) mucin domain-mutant mice. Proc Natl Acad Sci USA 109:12105–12110

Yanaba K, Bouaziz JD, Haas KM, Poe JC, Fujimoto M, Tedder TF (2008) A regulatory B cell subset with a unique CD1dhiCD5^{+} phenotype controls T cell-dependent inflammatory responses. Immunity 28:639–650

Yang Y, Tung JW, Ghosn EE, Herzenberg LA (2007) Division and differentiation of natural antibody-producing cells in mouse spleen. Proc Natl Acad Sci USA 104:4542–4546

Yang M, Sun L, Wang S, Ko KH, Xu H, Zheng BJ et al (2010) Novel function of B cell-activating factor in the induction of IL-10-producing regulatory B cells. J Immunol 184:3321–3325

Yoshizaki A, Miyagaki T, DiLillo DJ, Matsushita T, Horikawa M, Kountikov EI et al (2012) Regulatory B cells control T-cell autoimmunity through IL-21-dependent cognate interactions. Nature 491:264–268

Youn JI, Nagaraj S, Collazo M, Gabrilovich DI (2008) Subsets of myeloid-derived suppressor cells in tumor-bearing mice. J Immunol 181:5791–5802

Zandvoort A, Timens W (2002) The dual function of the splenic marginal zone: essential for initiation of anti-TI-2 responses but also vital in the general first-line defense against blood-borne antigens. Clin Exp Immunol 130:4–11

Zheng SG, Wang J, Wang P, Gray JD, Horwitz DA (2007) IL-2 is essential for TGF-beta to convert naive CD4+ CD25- cells to CD25+ Foxp3+ regulatory T cells and for expansion of these cells. J Immunol 178:2018–2027

Zhong X, Gao W, Degauque N, Bai C, Lu Y, Kenny J et al (2007) Reciprocal generation of Th1/Th17 and T(reg) cells by B1 and B2 B cells. Eur J Immunol 37:2400–2404

Zuniga E, Rabinovich GA, Iglesias MM, Gruppi A (2001) Regulated expression of galectin-1 during B-cell activation and implications for T-cell apoptosis. J Leukoc Biol 70:73–79

Zusman T, Lisansky E, Arons E, Anavi R, Bonnerot C, Sautes C et al (1996) Contribution of the intracellular domain of murine Fc-gamma receptor type IIB1 to its tumor-enhancing potential. Int J Cancer 68:219–227

Part V
Tumor Escape and Cancer Immunotherapy

Chapter 23
Cancer Immunotherapy: Overview in Brief

Philipp Beckhove

Abstract Within the recent years, the indispensable contribution of immune responses and inflammation during control and progression of malignant diseases has become apparent and a focus of scientific investigation, revealing new insights into the highly complex cellular and molecular mechanisms determining the fate of tumor-driven and tumor specific immune responses in individual patients. On this basis, cancer immunotherapy has been provided with a broad variety of novel tool and therapeutic strategies that not only allow inducing stronger and better T cell responses but that also allow interfering with regulatory mechanisms within the tumor microenvironment. Exploiting these approaches, cancer immunotherapy has recently provided proof of clinical efficacy for a number of different therapeutic approaches which pave the way towards broader exploration of more advanced strategies combining systemic and local immunologic interventions on different mechanistic levels. Here, respective major strategies of cancer immunotherapy and their future perspectives are introduced and discussed.

Keywords Cancer immunotherapy · Cancer vaccine · CTL · Immune response · Antitumor immunity · Sipuleucel-T · Tumor antigen · Dendritic cells · Immune modulation · TLR

P. Beckhove (✉)
Translational Immunology Division, German Cancer Research Center and National Center of Tumor Diseases, Heidelberg, INF 460, Heidelberg 69120, Germany
e-mail: p. beckhove@dkfz.de

M. R. Shurin et al. (eds.), *The Tumor Immunoenvironment*,
DOI: 10.1007/978-94-007-6217-6_23,

23.1 Introduction

Cancer Immunotherapy has entered a fascinating era. After decades of multiple unsuccessful attempts to exploit the basic principles of immune protection against pathogens for treatment of cancer patients, clinical efficacy of immunotherapeutic approaches has been finally demonstrated in a number of controlled clinical trials and many promising approaches are currently under advanced clinical investigation. While proof of therapeutic efficacy has meanwhile been provided, the field of cancer immunotherapy faces the challenge of translating their concepts into successful treatment protocols for a plethora of tumor entities and indications. On this way, multiple challenges need to be overcome. Some of them, such as the identification and selection of suitable target antigens or the systemic induction of functionally competent tumor antigen specific CTL have been addressed extensively in the past with some (although not yet satisfactory) success. Others, such as the targeting of tumor specific T cells to the tumor tissue, counter regulatory mechanisms of the immune system or intrinsic resistance of tumor cells to T cell mediated lysis, remain critical hurdles that need to be overcome before cancer immunotherapy can be successfully implemented as a third therapeutic pillar besides conventional and targeted tumor therapies.

A challenge particularly imposed to cancer immunotherapy derives from the tremendous inter individual heterogeneity in regard to the complex interactions between tumor cells, tumor stroma components (including vasculature, connective tissue and cells of the innate immune system) and the adaptive immune response. This heterogeneity results in various different response patterns to immunotherapeutic interventions and may often be responsible for insufficient clinical efficacy but, on the other hand, may explain clinical benefit in some patients. Major forces that determine the outcome of immune response against tumors are driven by or active within the tumor microenvironment which therefore demands our full scientific attention. Our understanding of criteria and prerequisites of successful and unsuccessful tumor immune control will thus finally determine our capacity to develop and clinically implement immunotherapeutic protocols into future standard of care cancer treatments. Three major therapeutic strategies are currently being pursued either alone or in combinations; namely active vaccination, adoptive transfer of tumor reactive effector T cells and the modulation of intrinsic immune response by blockade of immune inhibitory molecules expressed on immune cells or tumor cells.

23.2 Vaccines

Among the different immunotherapeutic strategies that have been introduced so far, cancer vaccination represents the one that has been applied most often in various preclinical and clinical settings. Therefore, ample experience has been accumulated throughout decades and has promoted a gradual improvement of protocols finally resulting not only in successful induction of antigen specific T cell responses but also in improved overall survival of patients in some cases. A first proof of concept was provided in large controlled phase III studies with patients suffering from metastatic prostate cancer who were vaccinated with Sipuleucel-T. Sipuleucel-T comprises preparations of autologous antigen presenting cells activated and pulsed with a fusion protein of GM-CSF with prostate acid phosphatase. The treatment improved median survival by roughly 4 months and has been approved as first antigen specific cancer immunotherapy by the FDA for treatment of metastatic prostate cancer in 2009 (Kantoff et al. 2010). Meanwhile, another randomized (phase IIb) study in patients with metastatic prostate cancer revealed an improved overall survival after vaccination with a prostate specific antigen (PSA) encoding poxviral vector (PROSTVAC) of more than 8 months compared to the control vector group (Kantoff et al. 2010).

Vaccination with a single peptide derived from the melanoma associated antigen gp100, applied together with high does IL-2 revealed a significantly improved progression free survival in a cohort of metastatic melanoma patients in a controlled phase III trial (Schwartzentruber et al. 2011). In a fully individualized and cancer specific approach (BiovaxID) the unique B cell receptors expressed on the clones of malignant B cell lymphoma cells were used in a phase III trial of idiotype vaccination in patients with follicular B cell lymphoma after chemotherapy induced clinical remission. BioVaxID treatment resulted in a strongly prolonged duration of clinical remission (Schuster et al. 2011).

Together, these trials pave the way towards new investments into industrial and academic development of further improved vaccination procedures. Currently, a series of promising clinical phase II and –III trials employing various technical approaches and targeting different tumor indications are being conducted and expected to reveal positive results within the upcoming years.

The basic concept underlying therapeutic cancer vaccination was based on the assumption that in cancer patients a repertoire of functionally competent cytotoxic effector and memory T cells with specificity for tumor cell antigens does not exist at sufficient frequencies to control tumor progression and thus needs to be established. Meanwhile, numerous studies have demonstrated the presence of spontaneously generated tumor antigen reactive CD4 and CD8 T cells with therapeutic potential in many cancer patients and correlated these and the accumulation of effector/memory T cells within tumor tissue to improved patient prognosis (Domschke et al. 2009; Galon et al. 2006; Halama et al. 2011). While the functional competence of this rather heterogeneous populations of preexisting T cells is a matter of debate and intensely investigated, their presence opens the possibility that cancer vaccines not only induce de novo a newly primed T cell repertoire but may also reactivate and potentially re-educate the repertoire of preexisting tumor reactive memory T cells. In this case, the presence or absence of preexisting effector/memory T cell responses might represent an important prognostic biomarker for response to vaccination.

Based on our current knowledge, successful vaccination strategies need to achieve the induction of high quality CD4 and CD8 effector and memory T cells responses against tumor associated antigens and to circumvent the problem of selection of immune escape variants of cancer cells.

Several vaccination strategies are currently being pursued. Short peptide vaccines consist of one or more HLA-I or –II restricted epitopes derived from tumor antigens (Rammensee 2006; Dutoit et al. 2012). Such peptides bind to the HLA-peptide binding groove on the cell surface of antigen presenting cells without the need of being taken up and processed. The use of short peptides bears the advantage to select only the immune dominant epitopes of an antigen which potentially increases the efficiency of T cell priming. Through elution, sequencing and quantification of MHC bound peptides presented by primary tumor cells this approach ensures that the selected epitope is presented at sufficient amount on the surface of the targeted tumor cells. Moreover, short peptides facilitate the precise assessment of subsequent immune response, e.g., by fluorescent HLA-peptide multimers. Disadvantages of the approach derive from the restriction to patients carrying appropriate HLA-alleles, the induction of an only narrow T cell repertoire bearing a high risk of selection of immune escape variants and from lack of CD4 T cell help in case the vaccine is based on HLA-I restricted epitopes only. Recent strategies aim at combining multiple short HLA-I and –II restricted short peptides derived from many different tumor antigen to overcome these restrictions (Dutoit et al. 2012).

Within the past years, several clinical vaccination trials used synthetic long overlapping peptides of 20–30 amino acids, covering entire sequences of a targeted tumor antigen. Long peptides are taken up by antigen presenting cells, processed and presented by both HLA-I and –II molecules. Therefore, they intrinsically provide CD4 T cell help that supports the induction of cytotoxic CD8 T cell responses and is a key requirement for the induction of long lived memory T cells (Bonertz et al. 2009; Khazaie et al. 2009). Moreover, the use of long overlapping peptides allows inclusion of patients irrespective of their individual HLA haplotype. Long peptides exerted a superior capacity to induce T cell responses against defined HLA-I restricted epitopes when compared to the respective short counterparts which were not taken up and processed by the antigen presenting cell, suggesting that the quality of T cell priming by an antigen presenting cell may be influenced by antigen uptake and processing (Melief 2008). With this approach, robust protective T cell responses could be induced in patients with premalignant HPV lesions of the vulva/cervix (Kenter et al. 2009). Conceptually similar to the use of long overlapping peptides is the application of recombinant full length proteins and such approaches, e.g. with MUC1, are currently investigated in controlled clinical trials by large companies (Powell and Chow 2008).

Alternatively, immunogenic tumor antigen expression can be induced through gene therapy by vaccination with DNA or RNA encoding for single tumor antigens. After intradermal or intramuscular injection naked nucleic acids are not entirely degraded but to some extent locally expressed by a variety of somatic cells, among them antigen presenting cells such as dendritic cells. These become activated through stimulation of TLRs, e.g., TLR9, by naked nucleic acids which causes the establishment of a local inflammatory environment on the one hand, and on the other hand of increased antigen processing and presentation of dendritic cells and their migration into regional lymph nodes where T cell responses are efficiently primed. Thus, without need of further adjuvants, protein expression occurs in an immunogenic context resulting in the development of complex immune responses comprising not only cytotoxic and T helper cells but specific antibodies as well. Since major proportions of applied nucleic acids, are rapidly degraded by intra and extracellular nucleases, several modifications have been introduced that have strongly improved protein expression. These involve the introduction of strong promoters driving gene translation, optimizations of the codon sequences or of the cap structure and poly A tails in mRNA and the formulation of RNA complexes that shield the nucleic acids from nucleases and thus strongly increase their half-lifes in vivo.

Easy production and high stability in vitro are major advantages of DNA and RNA when compared to proteins or peptides and the first cancer immunotherapy that has been approved was a DNA based tyrosinase vaccine for treatment of malignant melanoma in dogs (Grosenbaugh et al. 2011). Major disadvantages are their rapid degradation in vivo particularly of RNA and, in case of DNA, the potential integration into the genome which is associated with a certain risk of genomic alterations. Therefore, requirements for preclinical safety assessments are higher for DNA vaccines when compared to vaccines with peptides or proteins.

Whatever vaccine formulation is used, be it peptides, proteins or nucleic acids, it needs to be taken up and presented by activated dendritic cells in order to prime a T cell response. The strength and type of the immune response is critically determined by the amount of antigens taken up and presented and by the activation status of the dendritic cells.

23.3 Dendritic Cell-Based Vaccines

Dendritic cells are the only antigen presenting cells that are capable of inducing de novo T cell responses (Banchereau and Steinman 1998; Yewdell and Haeryfar 2005) and are therefore indispensable for the induction of protective immunity. On the other hand, under steady state conditions non activated immature dendritic cells are critical for maintenance of peripheral tolerance as they confer through various mechanisms an irreversible state of unresponsiveness to self reactive naïve T cells upon their encounter (Steinman et al. 2003).

Thus, induction of protective immunity requires the preceding activation of dendritic cells which occurs through stimulation of danger associated molecular pattern (DAMP) receptors. Many molecular stimuli, most of them associated with infection, cellular stress or tissue damage can induce the activation of dendritic cells and their molecular composition allows the dendritic cell to sense the nature of an infectious agent. Such information results in translation of distinct programs that regulate T cell activation, proliferation and the generation of different subsets of cytotoxic and/or T helper cell subsets with distinct functional properties that together precisely shape the induced immune response towards efficient elimination of a respective pathogen (Finkelman et al. 1996).

In order to better control these circumstances, antigens have been loaded for more than a decade unto ex vivo generated dendritic cells under standardized conditions in vitro and patients were subsequently treated with subcutaneous or intra nodal injections of antigen pulsed, activated dendritic cells (Melief 2008; Palucka et al. 2010).

A broad variety of different dendritic cell subsets populating peripheral and lymphoid tissues are equipped with various functions of which the most important ones regard their differentially developed capacity to cross present exogenous antigens via MHC I molecules to CD8 T cells and their differential capacity to exert proinflammatory or tolerogenic stimulation.

Two dendritic cell subsets are of major therapeutic interest. Myeloid dendritic cells differentiate from monocytes upon exposure to GM-CSF and IL-4 into immature dendritic cells which, upon appropriate exposure to toll like receptor ligands or to certain cytokines mature and acquire a proinflammatory phenotype that can efficiently prime naïve CD8 and CD4 T cell responses. Plasmacytoid dendritic cells represent a differentiated and mature dendritic cell subset present that is present at low frequencies in the peripheral blood. Plasmacytoid dendritic cells secrete high amounts of interferon-alpha in response to viral infections and thereby play a crucial role in the establishment of cytotoxic T cell responses. While they may have developed primarily for defense against viral infections (Siegal et al. 1999), they can also confer protective immunity against tumors.

While plasmacytoid dendritic cells are commonly isolated at relatively low numbers from the peripheral blood, efficient protocols for clinical grade ex vivo generation of high numbers of myeloid dendritic cells from their monocyte precursors have been established in the past. Therefore, myeloid dendritic cells represent the most commonly used dendritic cell type for cellular therapy.

Myeloid derived dendritic cells have proven their ability to prime and expand functional T cell responses against tumor associated antigens in numerous clinical trials of dendritic cell vaccination. In most of these trials, clinical efficacy was associated with successful induction of T cell immunity but within the subgroup of patients who developed tumor antigen specific T cell responses to vaccination, only a minority also showed clinical response to treatment. More recent analyses suggested that clinical efficacy of vaccination may be correlated to an induced phenotype of multi cytokine producing T cells which are characterized by simultaneous production of IFN-γ, IL-2 and TNF-α (Yuan et al. 2011). Therefore, the exploration of therapeutic protocols that support the induction of multi cytokine secreting T cells is a major goal of current vaccine development. A key aspect in this respect is the use of appropriate combinations of cytokines for dendritic cell differentiation and maturation. Immature dendritic cells differentiated in the presence of GM-CSF and IL-4 alone without further maturation promote the induction of immune suppressive T cell populations rather than that of therapeutic ones (Dhodapkar et al. 2001). In contrast, exposure to proinflammatory cytokines such as IFN-α or IL-15 during dendritic cell differentiation confers the capacity of priming functionally potent tumor antigen specific cytotoxic T cells (Lapenta et al. 2003; Dubsky et al. 2007). In addition to appropriate differentiation stimuli, optimal maturation stimuli are required that induce antigen processing, presentation and dendritic cell migration to secondary lymphoid organs. While in the past dendritic cells have been often maturated by a cocktail of prostaglandin E2, IL-1ß and TNF-α, more recent maturation protocols favour the addition of type I interferons, such as IFN-α, or respective TLR agonists such as poly I:C together with IFN-γ and TNF-α (Giermasz et al. 2009).

Dendritic cells express surface receptors such as DEC-205 that induce antigen uptake and processing when cross linked by their respective ligands or specific antibodies. Coupling of antigens to such ligands allows their selective targeting to dendritic cells in vivo without the requirement of adoptive cellular transfer and could improve the strength and accuracy of induced T cell responses in tumor mouse models when combined with appropriate maturation stimuli (Bonifaz et al. 2002; Hawiger et al. 2001). Clinical trials exploiting such strategies are ongoing and may form the basis of a new era of targeted dendritic cell vaccination.

23.4 Adoptive T Cell Transfer

In cancer patients, vaccines are challenged by multiple hurdles of systemic and local immune suppression that limit their capacity to generate sufficient numbers of fully functional tumor antigen specific effector and memory T cells Indeed, vaccination has revealed promising results particularly in patients with limited residual disease, while patients with established large tumors may benefit less from vaccination. In these cases, adoptive transfer of ex vivo generated, fully activated tumor-specific effector T cells represent a promising treatment alternative.

In the context of donor lymphocyte infusions, adoptive T cell therapy has become a standard treatment for some leukemias, particularly CML, after allogeneic stem cell transplantation–possessing the ability to eradicate malignant clones and to cure malignant disease. Clinical efficacy in such setting goes along with establishment of long lasting immunity against tumor cell associated antigens such as BCR-abl or WT1 (Rezvani et al. 2007). Effective anti tumor T cell responses in the allogeneic setting are often based on the recognition of mismatched minor HLA antigens that are expressed on the leukemic cells but are not expressed in the donor—thus, the T cell repertoire of the donor is not subjected to central tolerance mechanisms against these antigens. Moreover, leukemic cells often possess T cell co-stimulatory abilities of professional antigen presenting cells and the efficiency of donor lymphocyte infusion seems to correlate with this ability. Thus, adoptive T cell therapy for treatment of autologous, solid tumors needs to be based on a repertoire of T cells equipped with T cell receptors of sufficient avidity to selected tumor antigens and needs to overcome the general lack of co-stimulation provided by solid tumor cells.

Attempts to exploit T cell populations enriched with tumor antigen specific T cell receptors have been made already decades ago by isolation and ex vivo expansion of tumor infiltrating T cells (Wang and Rosenberg 1996) and revealed long lasting clinical remissions in advanced stage tumor patients (Dudley et al. 2010; Bollard et al. 2007). Although the endogenous repertoire of tumor reactive T cells can contain high avidity T cells with specificity to tumor specific mutated antigens (Lennerz et al. 2005) the majority of tumor reactive T cells and of tumor associated antigens are non mutated self antigens that are subjected to central tolerance or peripheral anergy induction. Therefore, the endogenous repertoire of tumor antigen specific T cells is based on a majority of low to intermediate affinity T cell receptors which may limit their therapeutic efficacy. While it is not yet clear, whether this, or other restrictions, such as inefficient homing to the tumor tissue, short half life of transferred T cells, or their suppression in situ by local immune suppressive mechanisms are the key restrictions for therapeutic efficacy, approaches to genetically equip autologous effector T cells with high avidity antigen receptors have been developed throughout the past years. These involve transduction with transgenic TCR α and β chains or with chimeric antigen receptors (Rosenberg et al. 2008).

Genomic sequencing and subsequent cloning of the α and β chains of potent T cell receptors (TCR) provided the basis for TCR transduction of human T cells. While initially candidate TCRs were derived from spontaneously induced highly potent CTL clones of patients (Lennerz et al. 2005), today more systematic approaches systematically generate high avidity human TCRs either in vitro through cocultures of naïve T cells with allogeneic, peptide pulsed antigen presenting cells (Wilde et al. 2012) or through vaccination of humanized mice expressing human HLA alleles together with the human TCR α and β chains (Li et al. 2010). Candidate TCRs are sequenced, cloned and transferred to fresh T cells using appropriate vector systems, such as lentiviruses (Voss et al. 2010) conferring the same T cell specificity to all transduced recipient cells (Rosenberg et al. 2008). With this approach, high numbers of antigen specific, activated CTL can be generated and therapeutically applied. A series of first clinical trials using this methodology provided highly promising results in advanced stage melanoma patients with complete clinical remission rates of up to 50 % (Restifo et al. 2012).

The transfer of high avidity TCRs specific for self antigens into large numbers of functionally potent CTL imposes two major risks. The first one regards cytotoxic "on target" activity against non malignant cells that also express the respective antigen. In the case of melanoma specific T cells, such toxicity can lead to vitiligo or, more seriously, to retinitis (Restifo et al. 2012). Therefore, gene transfer of high avidity TCRs requires a target antigen expression that is restricted to tumor cells or at least spares broadly abundant or vitally important cell types. A second caveat lays in the common occurrence of hybrid TCRs containing pairs of endogenous and transferred TCR chains. These hybrid TCRs bear a certain risk of high avidity binding to self antigens. In mouse models, spontaneous formation of such hybrid TCRs caused fatal graft versus host disease (Bendle et al. 2010). Strategies to overcome such risk involve improvement of the pairing probabilities of transduced TCR chains by inserting murine sequences, insertion of inducible suicide genes or inhibition of endogenous TCR chain expression (Uckert and Schumacher 2009).

A major disadvantage of TCR gene transfer lies in its restriction to single HLA alleles. Thereby, such approaches can be applied to limited groups of patents, only. This limitation can be overcome by genetic equipment of T cells with chimeric antigen receptors (CARs). CARs consist of high affinity antigen receptors based on the light and heavy chains of single chain antibodies that are coupled to transmembrane and cytoplasmic signaling domains of the TCR complex mediating T cell activation upon antigen encounter. CARs thus recognize antigens as antibodies independent of HLA molecules on the surface of target cells, thereby overcoming several immune evasion strategies of tumor cells, such as loss of HLA expression or antigen processing (Kershaw et al. 2006). While the CAR technology allows a broader application of single CARs, their high affinity nature, imposes similar risks of toxicity as TCR gene transfer. Moreover, since only cell surface molecules can be targeted by CARs, the selection of target molecules that are exclusively expressed on tumor cells represents a major challenge–particularly since the risk of inducing severe side effects such as cytokine storms through generalized activation appear to be increased compared to TCR gene transfer. Although upon first clinical exploration clinical success was modest in subjects with ovarian or renal cancer, most likely due to lack of co-stimulation and subsequent T cell expansion (Lamers et al. 2006), CAR-transduced T cells mediated striking tumor remissions in some patients—suggesting that this approach might represent a promising alternative to TCR gene transfer. In order to improve CAR efficiency, second and third generation CARs have been developed that include domains for single co-stimulatory signals or their combinations, such as CD28, CD40 or 41BB (Zhao et al. 2009).

A similar strategy as used by CARs, namely the exploitation of tumor antigen specific antibodies for T cell activation is also exploited by bispecific T cell engaging antibodies (BITEs) (Thakur and Lum 2010). Bispecific antibodies bind with one arm to a tumor antigen expressed on the tumor cell surface. With the other arm they crosslink TCR-associated signaling molecules, such as CD3. Thereby, effector T cells become engaged and activated at the tumor cell surface and reject tumor cells irrespective of the specificities of their endogenous TCRs. Besides simple cross linking of T cells and target cells, BITEs may also recruit proinflammatory or tumor toxic cells of the innate immune system such as macrophages or NK cells, through specifically designed constant fragment domains. Numerous formats of bispecific antibodies have been introduced and are currently investigated in clinical trials using a plethora of selected tumor antigens and signaling molecules expressed on various immune cell populations, particularly T cells and NK cells. This strategy has in the last years provided convincing proof of concept in a clinical trial in patients with advanced, treatment resistant B cell lymphoma who showed strong clinical responses after treatment with a bispecific antibody cross linking T cells via CD3 to CD19 positive lymphoma cells (Bargou et al. 2008).

Although antigen specificity and receptor avidity of both natural and genetically engineered T cells certainly play an important role for therapeutic efficacy, thorough analyses of patient outcomes after adoptive T cell therapy suggest that in vivo expansion, long term maintenance and efficient T cell infiltration into tumor lesions might be even more important for clinical efficacy. T cells require regular access to lymphoid niches in which they are provided with survival stimuli that enable clonal maintenance and homeostatic expansion over time. Thus, T cells compete for access to such niches and this competition strongly interferes with maintenance and clonal expansion of adoptively transferred T cells. Therefore, non-myeloablative lymphodepletion through cytostatic drugs or gamma irradiation was introduced as a pre-conditioning treatment before the T cell transfer to reduce the numbers of lymphocytes competing with the transferred T cells for the required niches. Implementation of such regimens has since then strongly improved T cell maintenance and clinical efficacy of T cell transfer (Restifo et al. 2012). In addition, the cytokines IL-7 and IL-15 have been identified as being required for homeostatic proliferation of memory CD4 and CD8 T cells, respectively Berger et al. (2009) suggesting their therapeutic application in the context of adoptive T cell transfer (Sportes et al. 2008). Still, in many tumor patients, adoptively transferred T cells are rapidly lost from the circulation.

Throughout the last years, studies in tumor mouse models and humanized xenograft models suggested that distinct T cell subsets strongly differ in their capacities to proliferate in vivo, to survive for long term and to maintain long term protective immunity after transfer.

While antigen inexperienced naïve T cells posses a high proliferative potential and can differentiate into functionally potent effector T cells (Hinrichs et al. 2009), they have high requirements for efficient priming in order to give rise not only to short lived effector but also to long lived memory T cells. Long term persistence in vivo and superior anti tumor efficacy was reported for the subset of CD8+ central memory T cells (Berger et al. 2008). More recently, a Wnt-β catenin expressing stem cell like memory T cell subset has been described in humans and in mice that showed improved proliferative activity, in vivo persistence and therapeutic activity in a tumor mouse model than all other tested T cell subsets (Gattinoni et al. 2009), suggesting that the presence and frequencies of stem cell like memory T cells among the population of adoptively transferred T cells may determine clinical response to T cell therapy.

23.5 Immune Modulation

Induction and effector of T cell responses take place in and are massively influenced by a complex inflammatory context that is driven by the patient‘s tumor and active not only in the tumor microenvironment but also systemically. As part of "self", tumor tissue is protected against self destruction as any other normal tissue by a plethora of immune check point mechanisms. At the same time tumors, unlike infectious pathogens, do not fully activate the defense mechanisms of the innate immune response that are required for the establishment of long lasting, strong and protective immunity. Both aspects need to be addressed by immune modulatory interventions in order to support anti tumor T cell responses in vivo and to achieve satisfying clinical responses of cancer immunotherapy. Throughout the past years, such interventions have been developed and clinically tested that help to either support T cell function or to inhibit immune suppression. These aim at achieving either one of the following critical goals, namely improving the quality of induced CTL and T helper cells (Appay et al. 2008; Antony et al. 2005; Quezada et al. 2010), inhibition of regulatory T cells and reversion of the immune suppressive tumor microenvironment. Besides the pattern of cytokines and cytotoxins produced, the quality of tumor specific T cells is determined by their panel of adhesion molecules and chemokine receptors that influence their ability to immigrate into the tumor tissue (Harlin et al. 2009).

The quantity and quality of induced T cell responses can be increased by application of cytokines, TLR ligands and co-stimulatory molecules together with active or passive T cell therapy. For example, adjuvant application of distinct cytokine combinations during T cell induction can determine the function of resulting T cell responses: Through expansion of antigen presenting cells and concomitant T cell stimulation, GM-CSF and IL-12 synergize in quantitatively increasing CD8 T cell responses (Ahlers et al. 1997). In contrast, combined application of tumor necrosis factor-alpha and IL-12 resulted in predominant IFN-gamma secretion of induced T cells (Ahlers et al. 2001). Stimulation of CD40 on antigen presenting cells synergized with application of GM-CSF in efficient induction of antigen specific effector and memory T cell responses (Ahlers et al. 2002).

Agonistic antibodies, such as anti-CD40 antibodies, which are efficiently exploited for activation of antigen presenting cells, can also be used to provide direct co-stimulatory signals to T cells and thereby to increase their proliferative or functional activity. A promising example is anti-CD137, a ligand for 4-1BB which strongly increases CD8 T cell expansion and activity in vivo (Maus et al. 2002; Watts 2005).

The functional avidity of CTL is strongly influenced by their expression of TCR associated molecules, involving the CD8 co-receptor (Cawthon and Alexander-Miller 2002). Since their expression can be therapeutically influenced, therapeutic improvement of the functional avidity of CD8 T cells might in the future increase the clinical outcome after T cell based cancer immunotherapy. It was recently demonstrated that functional avidity of CTL and their capacity to recognize and reject cancer cells strongly increases with increased provision of co-stimulatory signals during T cell induction, e.g. through construction of vectors encoding not only for the respective tumor antigen but also for immune modulator molecules such as B7-1 or ICAM-1 (Hodge et al. 2005). Besides co-stimulatory molecules, certain cytokines such as IL-15 or combinations of TLR3, −2 and −9 agonists can exert similar positive effects on functional avidity of tumor specific CD8 T cells

(Zhu et al. 2010) and besides reduce the dependence of CD8 T cell responses on CD4 T cell help to form long lasting protective memory. Thus, combinations of immune modulatory cytokines, co-stimulatory agonistic antibodies and TLR agonists may in future clinical trials help to overcome a critical threshold for quality and quantity of tumor specific T cell responses required to link the induced T cell response to durable clinical responses.

In addition to providing positive signals to stimulate appropriate T cell responses against cancer, multiple immune suppressive mechanisms must be overcome to allow for efficient immune rejection by either endogenous or therapeutically induced tumor specific T cells. A highly successful example is blockade of the inhibitory receptor CTLA-4 on activated T cells and regulatory T cells by the anti CTLA-4 monoclonal antibody ipilimumab which in metastasized melanoma patients significantly improved their overall survival without further T cell therapy in controlled phase III clinical trials and has been meanwhile approved for treatment of metastatic melanoma (Peggs et al. 2009). Another highly promising approach to improve T cell mediated tumor rejection is the blockade of an inhibitory ligand PDL1 which is expressed on the cell surface of many tumors, or of its inhibitory receptor, PD-1 which his expressed on activated T cells (Gajewski 2007). Respective blocking antibodies are currently under advanced clinical investigation.

The success of ipilimumab demonstrates the clinical relevance of immune suppressive mechanisms and thus future T cell therapy will require combination with other therapeutic strategies that interfere with tumor stroma driven inflammation and angiogenesis, to offset and reprogram the immune suppressive environment (Dougan and Dranoff 2009).

Within the tumor stroma, soluble and cellular agents contribute to T cell suppression. Soluble targets of therapeutic immune modulation are mainly immune suppressive cytokines that can be blocked by either monoclonal antibodies or their soluble receptors. Among the many candidates described, TGF-ß, IL-10, IL-13 and IL-35 may be the most promising ones and clinical studies to assess the effect of their inhibition in the context of specific T cell therapy have been initiated for some of them (Terabe et al. 2000; Moore et al. 2001).

TGF-ß can potently induce the conversion of immune suppressive regulatory T cells from conventional CD4 T cells as well as the recruitment of immune suppressive CD11b+GR1+ myeloid cells into the tumor microenvironment TGFß1 and also the Treg-derived cytokine IL-35 inhibit directly the activation and function of CD8 and CD4 effector/memory T cells (Yang et al. 2008; Thomas and Massague 2005). IL-13 is secreted by a subset of tumor infiltrating regulatory NKT cells. IL-13, similar to IL-10, activates a variety of immune suppressive functions in myeloid suppressor cells (Fichtner-Feigl et al. 2008) and in tumor associated immune suppressive M2 macrophages (Marigo et al. 2008; Gordon and Martinez 2010) that play a detrimental role in the establishment of a pro angiogenic, tumor promoting and immune suppressive tumor microenvironment.

Besides neutralization of soluble molecules, deletion or inhibition of immune suppressive cell types represent an additional strategy of immune intervention. Based on their constitutive expression of the high affinity IL-2 receptor CD25, regulatory T cells are particularly sensitive to deprivation of IL-2. Based on promising tumor mouse models, this was exploited by treatments with cytostatic drugs that interfere with IL-2 signaling, such as cyclophosphamide and clinical trials using this strategy have provided encouraging results demonstrating efficient induction of tumor specific T cell responses after conditioning of patients with low dose cyclophosphamide (Ge et al. 2012). Alternatively, toxin-conjugated monoclonal antibodies targeting CD25 have been used in clinical trials with some success to deplete regulatory T cells prior to vaccination of melanoma patients (Jacobs et al. 2010). A strategy to inhibit the activity of myeloid suppressor cells is the inhibition of their major effector molecule iNOS, by an iNOS inhibitor. While efficient in preclinical tumor mouse models to enhance T cell based tumor immune rejection (Meyer et al. 2011), clinical exploration of this strategy is ongoing. Taken together, multiple different mechanisms act on the level of the tumor microenvironment to suppress even high frequent anti tumor T cell responses of high functional quality. Depending on their respective activation in individual cancer patients, personalized immunotherapies may need to individually target these mechanisms by combinations of different strategies in addition to the induction or promotion of antigen specific T cell immunity.

23.6 Conclusion

Cancer immunotherapy has entered an exciting era in which growing understanding of the complex interplay between multiple components of the immune system with tumor cells is provided by an increasing number of clinical trials and can be integrated into new concepts of protective immunity based on molecularly defined activation of innate immune cells. Novel tools to modulate these interactions open a broad window of opportunities for development of clinically more efficient cancer immunotherapies.

These will rely on CD4+ and CD8+ T cells with thoroughly defined functional capacities and long term persistence to prevent tumor relapse and will imply personalized combinations with immune modulatory agents targeting dominant immune suppressive mechanisms within the tumor microenvironment to overcome local T cell suppression and to allow for rejection of non resectable tumors.

References

Ahlers JD, Dunlop N, Alling DW, Nara PL, Berzofsky JA (1997) Cytokine-in-adjuvant steering of the immune response phenotype to HIV-1 vaccine constructs: granulocyte-macrophage colony-stimulating factor and TNF-alpha synergize with IL-12 to enhance induction of cytotoxic T lymphocytes. J Immunol 158:3947–3958

Ahlers JD, Belyakov IM, Matsui S, Berzofsky JA (2001) Mechanisms of cytokine synergy essential for vaccine protection against viral challenge. Int Immunol 13:897–908

Ahlers JD, Belyakov IM, Terabe M, Koka R, Donaldson DD, Thomas EK et al (2002) A push-pull approach to maximize vaccine efficacy: abrogating suppression with an IL-13 inhibitor while augmenting help with granulocyte/macrophage colony-stimulating factor and CD40L. Proc Natl Acad Sci USA 99:13020–13025

Antony PA, Piccirillo CA, Akpinarli A, Finkelstein SE, Speiss PJ, Surman DR et al (2005) CD8+ T cell immunity against a tumor/self-antigen is augmented by CD4+ T helper cells and hindered by naturally occurring T regulatory cells. J Immunol 174:2591–2601

Appay V, Douek DC, Price DA (2008) CD8+ T cell efficacy in vaccination and disease. Nat Med 14:623–628

Banchereau J, Steinman RM (1998) Dendritic cells and the control of immunity. Nature 392:245–252

Bargou R, Leo E, Zugmaier G, Klinger M, Goebeler M, Knop S et al (2008) Tumor regression in cancer patients by very low doses of a T cell-engaging antibody. Science 321:974–977

Bendle GM, Linnemann C, Hooijkaas AI, Bies L, de Witte MA, Jorritsma A et al (2010) Lethal graft-versus-host disease in mouse models of T cell receptor gene therapy. Nature Med 16:565–570

Berger C, Jensen MC, Lansdorp PM, Gough M, Elliott C, Riddell SR (2008) Adoptive transfer of effector CD8+ T cells derived from central memory cells establishes persistent T cell memory in primates. J Clin Investig 118:294–305

Berger C, Berger M, Hackman RC, Gough M, Elliott C, Jensen MC et al (2009) Safety and immunologic effects of IL-15 administration in nonhuman primates. Blood 114:2417–2426

Bollard CM, Gottschalk S, Leen AM, Weiss H, Straathof KC, Carrum G et al (2007) Complete responses of relapsed lymphoma following genetic modification of tumor-antigen presenting cells and T-lymphocyte transfer. Blood 110:2838–2845

Bonertz A, Weitz J, Pietsch DH, Rahbari NN, Schlude C, Ge Y et al (2009) Antigen-specific tregs control T cell responses against a limited repertoire of tumor antigens in patients with colorectal carcinoma. J Clin Investig 119:3311–3321

Bonifaz L, Bonnyay D, Mahnke K, Rivera M, Nussenzweig MC, Steinman RM (2002) Efficient targeting of protein antigen to the dendritic cell receptor DEC-205 in the steady state leads to antigen presentation on major histocompatibility complex class I products and peripheral CD8+ T cell tolerance. J Exp Med 196:1627–1638

Cawthon AG, Alexander-Miller MA (2002) Optimal colocalization of TCR and CD8 as a novel mechanism for the control of functional avidity. J Immunol 169:3492–3498

Dhodapkar MV, Steinman RM, Krasovsky J, Munz C, Bhardwaj N (2001) Antigen-specific inhibition of effector T cell function in humans after injection of immature dendritic cells. J Exp Med 193:233–238

Domschke C, Schuetz F, Ge Y, Seibel T, Falk C, Brors B et al (2009) Intratumoral cytokines and tumor cell biology determine spontaneous breast cancer-specific immune responses and their correlation to prognosis. Cancer Res 69:8420–8428

Dougan M, Dranoff G (2009) Immune therapy for cancer. Annu Rev Immunol 27:83–117

Dubsky P, Saito H, Leogier M, Dantin C, Connolly JE, Banchereau J et al (2007) IL-15-induced human DC efficiently prime melanoma-specific naive CD8+ T cells to differentiate into CTL. Eur J Immunol 37:1678–1690

Dudley ME, Gross CA, Langhan MM, Garcia MR, Sherry RM, Yang JC et al (2010) CD8+ enriched "young" tumor infiltrating lymphocytes can mediate regression of metastatic melanoma. Clin Cancer Res: Official J Am Assoc Cancer Res 16:6122–6131

Dutoit V, Herold-Mende C, Hilf N, Schoor O, Beckhove P, Bucher J et al (2012) Exploiting the glioblastoma peptidome to discover novel tumour-associated antigens for immunotherapy. Brain:J Neurol 135:1042–1054

Fichtner-Feigl S, Terabe M, Kitani A, Young CA, Fuss I, Geissler EK et al (2008) Restoration of tumor immunosurveillance via targeting of interleukin-13 receptor-alpha 2. Cancer Res 68:3467–3475

Finkelman FD, Lees A, Birnbaum R, Gause WC, Morris SC (1996) Dendritic cells can present antigen in vivo in a tolerogenic or immunogenic fashion. J Immunol 157:1406–1414

Gajewski TF (2007) Failure at the effector phase: immune barriers at the level of the melanoma tumor microenvironment. Clin Cancer Res: Official J Am Assoc Cancer Res 13:5256–5261

Galon J, Costes A, Sanchez-Cabo F, Kirilovsky A, Mlecnik B, Lagorce-Pages C et al (2006) Type, density, and location of immune cells within human colorectal tumors predict clinical outcome. Science 313:1960–1964

Gattinoni L, Zhong XS, Palmer DC, Ji Y, Hinrichs CS, Yu Z et al (2009) Wnt signaling arrests effector T cell differentiation and generates CD8+ memory stem cells. Nat Med 15:808–813

Ge Y, Domschke C, Stoiber N, Schott S, Heil J, Rom J et al (2012) Metronomic cyclophosphamide treatment in metastasized breast cancer patients: immunological effects and clinical outcome. Cancer Immunol, Immunother: CII 61:353–362

Giermasz AS, Urban JA, Nakamura Y, Watchmaker P, Cumberland RL, Gooding W et al (2009) Type-1 polarized dendritic cells primed for high IL-12 production show enhanced activity as cancer vaccines. Cancer Immunol, Immunother: CII 58:1329–1336

Gordon S, Martinez FO (2010) Alternative activation of macrophages: mechanism and functions. Immunity 32:593–604

Grosenbaugh DA, Leard AT, Bergman PJ, Klein MK, Meleo K, Susaneck S et al (2011) Safety and efficacy of a xenogeneic DNA vaccine encoding for human tyrosinase as adjunctive treatment for oral malignant melanoma in dogs following surgical excision of the primary tumor. Am J Vet Res 72:1631–1638

Halama N, Michel S, Kloor M, Zoernig I, Benner A, Spille A et al (2011) Localization and density of immune cells in the invasive margin of human colorectal cancer liver metastases are prognostic for response to chemotherapy. Cancer Res 71:5670–5677

Harlin H, Meng Y, Peterson AC, Zha Y, Tretiakova M, Slingluff C et al (2009) Chemokine expression in melanoma metastases associated with CD8+ T-cell recruitment. Cancer Res 69:3077–3085

Hawiger D, Inaba K, Dorsett Y, Guo M, Mahnke K, Rivera M et al (2001) Dendritic cells induce peripheral T cell unresponsiveness under steady state conditions in vivo. J Exp Med 194:769–779

Hinrichs CS, Borman ZA, Cassard L, Gattinoni L, Spolski R, Yu Z et al (2009) Adoptively transferred effector cells derived from naive rather than central memory CD8+ T cells mediate superior antitumor immunity. Proc Natl Acad Sci USA 106:17469–17474

Hodge JW, Chakraborty M, Kudo-Saito C, Garnett CT, Schlom J (2005) Multiple costimulatory modalities enhance CTL avidity. J Immunol 174:5994–6004

Jacobs JF, Punt CJ, Lesterhuis WJ, Sutmuller RP, Brouwer HM, Scharenborg NM et al (2010) Dendritic cell vaccination in combination with anti-CD25 monoclonal antibody treatment: a phase I/II study in metastatic melanoma patients. Clin Cancer Res: Official J Am Assoc Cancer Res 16:5067–5078

Kantoff PW, Higano CS, Shore ND, Berger ER, Small EJ, Penson DF et al (2010a) Sipuleucel-T immunotherapy for castration-resistant prostate cancer. New Engl J Med 363:411–422

Kantoff PW, Schuetz TJ, Blumenstein BA, Glode LM, Bilhartz DL, Wyand M et al (2010b) Overall survival analysis of a phase II randomized controlled trial of a Poxviral-based PSA-targeted immunotherapy in metastatic castration-resistant prostate cancer. J Clin Oncol: Official J Am Soc Clin Oncol 28:1099–1105

Kenter GG, Welters MJ, Valentijn AR, Lowik MJ, der Meer Berends-van DM, Vloon AP et al (2009) Vaccination against HPV-16 oncoproteins for vulvar intraepithelial neoplasia. New Engl J Med 361:1838–1847

Kershaw MH, Westwood JA, Parker LL, Wang G, Eshhar Z, Mavroukakis SA et al (2006) A phase I study on adoptive immunotherapy using gene-modified T cells for ovarian cancer. Clin Cancer Res Official J Am Assoc Cancer Res 12:6106–6115

Khazaie K, Bonertz A, Beckhove P (2009) Current developments with peptide-based human tumor vaccines. Curr Opin Oncol 21:524–530

Lamers CH, Sleijfer S, Vulto AG, Kruit WH, Kliffen M, Debets R et al (2006) Treatment of metastatic renal cell carcinoma with autologous T-lymphocytes genetically retargeted against carbonic anhydrase IX: first clinical experience. J Clin Oncol: Official J Am Soc Clin Oncol 24:e20–e22

Lapenta C, Santini SM, Logozzi M, Spada M, Andreotti M, Di Pucchio T et al (2003) Potent immune response against HIV-1 and protection from virus challenge in hu-PBL-SCID mice immunized with inactivated virus-pulsed dendritic cells generated in the presence of IFN-alpha. J Exp Med 198:361–367

Lennerz V, Fatho M, Gentilini C, Frye RA, Lifke A, Ferel D et al (2005) The response of autologous T cells to a human melanoma is dominated by mutated neoantigens. Proc Natl Acad Sci USA 102:16013–16018

Li LP, Lampert JC, Chen X, Leitao C, Popovic J, Muller W et al (2010) Transgenic mice with a diverse human T cell antigen receptor repertoire. Nat Med 16:1029–1034

Marigo I, Dolcetti L, Serafini P, Zanovello P, Bronte V (2008) Tumor-induced tolerance and immune suppression by myeloid derived suppressor cells. Immunol Rev 222:162–179

Maus MV, Thomas AK, Leonard DG, Allman D, Addya K, Schlienger K et al (2002) Ex vivo expansion of polyclonal and antigen-specific cytotoxic T lymphocytes by artificial APCs expressing ligands for the T-cell receptor, CD28 and 4–1BB. Nat Biotechnol 20:143–148

Melief CJ (2008) Cancer immunotherapy by dendritic cells. Immunity 29:372–383

Meyer C, Sevko A, Ramacher M, Bazhin AV, Falk CS, Osen W et al (2011) Chronic inflammation promotes myeloid-derived suppressor cell activation blocking antitumor immunity in transgenic mouse melanoma model. Proc Natl Acad Sci USA 108:17111–17116

Moore KW, Moore KW, de Waal Malefyt R, Coffman RL, O'Garra A (2001) de Interleukin-10 and the interleukin-10 receptor. Annu Rev Immunol 19:683–765

Palucka K, Ueno H, Roberts L, Fay J, Banchereau J (2010) Dendritic cells: are they clinically relevant? Cancer J 16:318–324

Peggs KS, Quezada SA, Chambers CA, Korman AJ, Allison JP (2009) Blockade of CTLA-4 on both effector and regulatory T cell compartments contributes to the antitumor activity of anti-CTLA-4 antibodies. J Exp Med 206:1717–1725

Powell E, Chow LQ (2008) BLP-25 liposomal vaccine: a promising potential therapy in non-small-cell lung cancer. Expert Rev Respir Med 2:37–45

Quezada SA, Simpson TR, Peggs KS, Merghoub T, Vider J, Fan X et al (2010) Tumor-reactive CD4(+) T cells develop cytotoxic activity and eradicate large established melanoma after transfer into lymphopenic hosts. J Exp Med 207:637–650

Rammensee HG (2006) Some considerations on the use of peptides and mRNA for therapeutic vaccination against cancer. Immunol Cell Biol 84:290–294

Restifo NP, Dudley ME, Rosenberg SA (2012) Adoptive immunotherapy for cancer: harnessing the T cell response. Nat Rev Immunol 12:269–281

Rezvani K, Yong AS, Savani BN, Mielke S, Keyvanfar K, Gostick E et al (2007) Graft-versus-leukemia effects associated with detectable Wilms tumor-1 specific T lymphocytes after allogeneic stem-cell transplantation for acute lymphoblastic leukemia. Blood 110:1924–1932

Rosenberg SA, Restifo NP, Yang JC, Morgan RA, Dudley ME (2008) Adoptive cell transfer: a clinical path to effective cancer immunotherapy. Nat Rev Cancer 8:299–308

Schuster SJ, Neelapu SS, Gause BL, Janik JE, Muggia FM, Gockerman JP et al (2011) Vaccination with patient-specific tumor-derived antigen in first remission improves disease-free survival in follicular lymphoma. J Clin Oncol: Official J Am Soc Clin Oncol 29:2787–2794

Schwartzentruber DJ, Lawson DH, Richards JM, Conry RM, Miller DM, Treisman J et al (2011) gp100 peptide vaccine and interleukin-2 in patients with advanced melanoma. New Engl J Med 364:2119–2127

Siegal FP, Kadowaki N, Shodell M, Fitzgerald-Bocarsly PA, Shah K, Ho S et al (1999) The nature of the principal type 1 interferon-producing cells in human blood. Science 284:1835–1837

Sportes C, Hakim FT, Memon SA, Zhang H, Chua KS, Brown MR et al (2008) Administration of rhIL-7 in humans increases in vivo TCR repertoire diversity by preferential expansion of naive T cell subsets. J Exp Med 205:1701–1714

Steinman RM, Hawiger D, Nussenzweig MC (2003) Tolerogenic dendritic cells. Annu Rev Immunol 21:685–711

Terabe M, Matsui S, Noben-Trauth N, Chen H, Watson C, Donaldson DD et al (2000) NKT cell-mediated repression of tumor immunosurveillance by IL-13 and the IL-4R-STAT6 pathway. Nat Immunol 1:515–520

Thakur A, Lum LG (2010) Cancer therapy with bispecific antibodies: Clinical experience. Curr Opin Mol Ther 12:340–349

Thomas DA, Massague J (2005) TGF-beta directly targets cytotoxic T cell functions during tumor evasion of immune surveillance. Cancer Cell 8:369–380

Uckert W, Schumacher TN (2009) TCR transgenes and transgene cassettes for TCR gene therapy: status in 2008. Cancer Immunol, Immunother: CII 58:809–822

Voss RH, Thomas S, Pfirschke C, Hauptrock B, Klobuch S, Kuball J et al (2010) Coexpression of the T-cell receptor constant alpha domain triggers tumor reactivity of single-chain TCR-transduced human T cells. Blood 115:5154–5163

Wang RF, Rosenberg SA (1996) Human tumor antigens recognized by T lymphocytes: implications for cancer therapy. J Leukoc Biol 60:296–309

Watts TH (2005) TNF/TNFR family members in costimulation of T cell responses. Annu Rev Immunol 23:23–68

Wilde S, Geiger C, Milosevic S, Mosetter B, Eichenlaub S, Schendel DJ (2012) Generation of allo-restricted peptide-specific T cells using RNA-pulsed dendritic cells: a three phase experimental procedure. Oncoimmunology 1:129–140

Yang L, Huang J, Ren X, Gorska AE, Chytil A, Aakre M et al (2008) Abrogation of TGF beta signaling in mammary carcinomas recruits Gr-1+CD11b+ myeloid cells that promote metastasis. Cancer Cell 13:23–35

Yewdell JW, Haeryfar SM (2005) Understanding presentation of viral antigens to CD8+ T cells in vivo: the key to rational vaccine design. Annu Rev Immunol 23:651–682

Yuan J, Adamow M, Ginsberg BA, Rasalan TS, Ritter E, Gallardo HF et al (2011) Integrated NY-ESO-1 antibody and CD8+ T-cell responses correlate with clinical benefit in advanced melanoma patients treated with ipilimumab. Proc Natl Acad Sci USA 108:16723–16728

Zhao Y, Wang QJ, Yang S, Kochenderfer JN, Zheng Z, Zhong X et al (2009) A herceptin-based chimeric antigen receptor with modified signaling domains leads to enhanced survival of transduced T lymphocytes and antitumor activity. J Immunol 183:5563–5574

Zhu Q, Egelston C, Gagnon S, Sui Y, Belyakov IM, Klinman DM et al (2010) Using 3 TLR ligands as a combination adjuvant induces qualitative changes in T cell responses needed for antiviral protection in mice. J Clin Investig 120:607–616

Chapter 24
Programming of MDSC: New Opportunities for Targeted Therapy

Peter Svider, Shu-Hsia Chen, Andrew G. Sikora and Wen-Chin Yang

Abstract Myeloid-derived suppressor cells have recently been widely recognized as major players in tumor-mediated immunosuppression. MDSCs act to suppress anti-tumor immunity through increased production of Arginase, Reactive Oxygen Species (ROS), up regulation of inducible Nitric Oxide Synthase (iNOS), and expression of negative costimulatory molecules, leading to inhibition of both antigen-specific and bystander T cell responses. These immature myeloid precursors are recruited peripherally and to tumors by inflammation, which retards their differentiation into mature myeloid lineages. MDSCs can be subdivided into Granulocytic (G-MDSC) and Monocytic (M-MDSC) forms. Both lineages can further differentiate into phenotypic forms analogous to those previously described in macrophages and neutrophils, including a so-called Type 1 and Type 2 phenotypes, which are functionally tumoricidal or promote tumor growth and progression, respectively. While eliminating MDSCs or inducing their differentiation into conventional myeloid lineages are potential strategies to restore anti-tumor immunity, the existence of opposing functional phenotypes presents an opportunity to drive MDSC towards

S.-H. Chen (✉)
Department of Oncological Sciences, Mount Sinai School of Medicine, Icahn Medical Institute, 1425 Madison Avenue, NY 10029, USA
e-mail: shu-hsia.chen@mssm.edu

P. Svider · A. G. Sikora (✉)
Department of Otolaryngology, Dermatology, Immunology Institute, Tisch Cancer Institute, Mount Sinai School of Medicine, Icahn Medical Institute, 1425 Madison Avenue, NY 10029, USA
e-mail: Andrew.Sikora@mssm.edu

W.-C. Yang
Department of Oncological Sciences,and Immunology Institute, Mount Sinai School of Medicine, Icahn Medical Institute, 1425 Madison Avenue, NY 10029, USA

W.-C. Yang
Agricultural Biotechnology Research Center, Academia Sinica, Taipei 115, Taiwan, China

M. R. Shurin et al. (eds.), *The Tumor Immunoenvironment*,
DOI: 10.1007/978-94-007-6217-6_24,

Type 1 polarization to reverse tumor-mediated immunosuppression. Interference with the PIR (paired immunoglobulin like receptor) pathway and TGF-β receptor signaling has been shown to cause such a reorientation of M-MDSCs and G-MDSCs in murine models. Exploitation of this principle in humans, including the use of small molecule inhibitors and biological therapies which alter the balance of Type 1 and Type 2 polarization may lead to novel therapeutic approaches, particularly, in combination with existing immunomodulatory strategies focused on enhancing antitumor T cell responses (such as vaccination and immune checkpoint blockade).

Keywords MDSC · Granulocytic MDSC · Monocytic MDSC · iNOS · Immunotherapy · Biological therapy · Polarization · Immunosuppression

24.1 Basic Biology of Myeloid-Derived Suppressor Cells

24.1.1 Phenotypic Characteristics

Myeloid-derived suppressor cells (MDSC) have only recently been recognized to play an integral role in the immune response to diverse disease processes, ranging from autoimmunity to immunosuppression. Comprised of immature progenitors of myeloid cells, this heterogeneous population is derived from hematopoietic stem cells and, in the absence of inflammation, ultimately differentiates towards conventional granulocytic (G-MDSC) or monocytic (M-MDSC) hematopoietic lineages. Inflammatory mediators induced by pathologic processes such as cancer, including stem cell factor (SCF), Vascular Endothelial Growth Factor (VEGF), IL-6, GM-CSF, and other soluble molecules impede normal differentiation to mature myeloid cells and promote recruitment of MDSCs from the bone marrow into the periphery, where they home to tumor and secondary lymphoid organs.

Over the past decade, a relative consensus has emerged describing the phenotypic characteristics of murine MDSC. CD11b and Gr-1 are the principal cell surface-markers used to identify these cells, and can be used to classify MDSC into granulocytic and myeloid lineages (Montero et al. 2012). $CD11b^{+}Gr1^{high}$ ($CD11b^{+}Ly\text{-}6G^{+}Ly6C^{low}$) cells are considered Granulocytic MDSCs (G-MDSCs), while $CD11b^{+}Gr1^{low}$ ($CD11b^{+}Ly\text{-}6G^{low}Ly6C^{high}$) are considered Monocytic-MDSCs (M- MDSCs) (Poschke and Kiessling 2012). Although G-MDSCs can comprise up to 70–80 % of all MDSCs in tumor-bearing hosts, depending on the tumor type, both forms play a pivotal role in tumor-mediated immunosuppression (Youn et al. 2012).

Although there has been extensive characterization of MDSCs in mice, there is much less consensus on markers identifying human MDSC populations. MDSCs have been described in patients with many different types of cancer, including but not limited to head and neck cancer, lung cancer, breast cancer, renal cell carcinoma, and colon cancer (Poschke and Kiessling 2012; Pak et al. 1995; Corzo et al. 2009; Almand et al. 2001; Kusmartsev et al. 2008; Mundy-Bosse et al. 2011; Corzo et al. 2010). In nearly all reports, varying phenotypic markers have been used to define human MDSCs, though many studies use $CD15^{+}CD14^{-}CD33^{+}$, $CD11b^{+}CD14^{+}CD33^{+}HLA\text{-}DR^{-/low}$ or $Lin^{-}CD33^{+}HLA\text{-}DR^{-}$ to define human MDSCs (Ostrand-Rosenberg and Sinha 2009; Raychaudhuri et al. 2011). The diversity of surface markers identifying functionally suppressive MDSC suggest that MDSC are a phenotypically heterogeneous population of cells sharing common functional attributes of inducibility by tumor-derived mediators, and the ability to suppress anti-tumor immune responses.

24.1.2 MDSC-Mediated T Cell Suppression

Familiarity with the diverse mechanisms of MDSC-mediated T cell suppression is critical for understanding the concept of MDSC programming. MDSCs play an immunosuppressive role within the tumor microenvironment by suppressing cytotoxic T-lymphocytes and helper T cell function through a variety of processes. G-MDSC controls T cell suppression chiefly through reactive oxygen species resulting in T cell toxicity (Ochando and Chen 2012; Cheng et al. 2008; Kusmartsev et al. 2004), while M-MDSC suppresses T cell function through increased expression of arginase and iNOS (Ochando and Chen 2012; Huang et al. 2006; Dolcetti et al. 2010; Gallina et al. 2006; Kusmartsev et al. 2000). Arginine is essential for T cell function, and its depletion by arginase impairs progression of the cell cycle through the G0/G1 stage in T cells by interfering with cyclin D3 and cyclin-dependent kinase (CDK-4) function (Dilek 2012; Rodriguez et al. 2002, 2004, 2007; Highfill et al. 2010).

Another important mechanism of MDSC-mediated immunosuppression is iNOS-mediated NO production by MDSCs (Jayaraman et al. 2012). iNOS has previously been shown to be overexpressed in cancer, with several groups demonstrating that the direct effects of high-output NO production and resulting peroxynitrite formation had suppressive effects on T cell activation and survival (Mazzoni et al. 2002; Nagaraj et al. 2007). Up regulation of iNOS by M-MDSCs leads to production of nitric oxide (NO), which in combination with ROS can form the highly reactive intermediate peroxynitrite. Peroxynitrite-mediated damage can impair the capacity of T cell receptors to recognize specific peptide antigens (Ochando and Chen 2012; Nagaraj et al. 2007).

Increased NO production has also been suggested to interfere with other signal transduction pathways responsible for T cell function, including the JAK3 and STAT5 pathways (Bingisser et al. 1998). iNOS inhibition has been shown to restore T cell-mediated immunity in several systems, including a recent publication by our group in a MT-RET melanoma murine model system, in which iNOS inhibition decreased MDSC accumulation and increased tumor-infiltrating CD8+ T cells with a subsequent decrease in tumor growth (Jayaraman et al. 2012).

In addition to direct effects on T cells, MDSCs also contribute to T cell suppression through indirect mechanisms. Regulatory T cells, discussed in further detail later, play an important immunosuppressive role in the tumor microenvironment. MDSCs in some cases can directly promote Treg expansion, which further antagonizes T cell mediated antitumor immunity. This phenomenon is thought to occur mostly through IFN-gama dependent mechanisms (Ochando and Chen 2012; Dilek 2012; Mazzoni et al. 2002; Garcia et al. 2010).

24.1.3 MDSCs Programming by Inflammatory Signaling

MDSCs and inflammation can in many cases be mutually reinforcing. For example, while VEGF is a soluble mediator known to induce MDSC accumulation, MDSCs have been shown to promote angiogenesis via increasing the availability of VEGF. Yang et al. found that in the tumor microenvironment, MDSCs produce MMP9, thus increasing bioavailability of VEGF (Yang et al. 2004). In addition to this critical for MDSC in VEGF modulation, the same study noted some MDSCs actually differentiated into endothelial cells during angiogenesis. Other in vivo studies have shown that VEGF infusion significantly increases MDSCs (Gabrilovich et al. 1998; Gabrilovich 2004; Melani et al. 2003), and that modulation of VEGF release is an important mechanism through which iNOS controls MDSC induction (Jayaraman et al. 2012). Thus, there is a reciprocally positive relationship between MDSC induction and VEGF.

Prostaglandin E2, another canonical inflammatory mediator, is elevated in cancer due to COX2 overexpression. PGE2 has been reported to promote MDSC-mediated immunosuppression and up regulate arginase levels (Ochando and Chen 2012; Ochoa et al. 2007; Obermajer et al. 2011; Bronte and Zanovello 2005). A recent analysis illustrated that a positive feedback loop between PGE2 and COX2 is necessary to promote differentiation of immature myeloid cells towards MDSCs rather than conventional dendritic cells(Obermajer et al. 2011). PGE2 exposure induced expression of arginase, iNOS, and IL-10, while COX2 inhibition by Celecoxib impeded production of these factors and decreased the suppressive capacities of MDSCs. Another study demonstrated that COX2 inhibitors down regulate MDSC development from bone marrow cultures (Eruslanov et al. 2010). In another study using a murine model of mesothelioma, tumor-bearing mice consuming celecoxib had fewer MDSCs, and the remaining MDSCs produced less NO and ROS, resulting in impaired MDSC function and decreased immunosuppression (Veltman et al. 2010).

While soluble mediators increase MDSC induction and recruitment to the periphery, inflammation also increases MDSC levels by promoting their survival. In vitro experiments have shown that Fas-mediated apoptosis normally regulates MDSC survival and is essential for anti-tumor immunity (Ostrand-Rosenberg et al. 2012). Thus, in murine studies, activated T cells expressing Fas-Ligand induce apoptosis of MDSCs and are required for control of metastatic disease. One way inflammation boosts MDSC survival is by increasing resistance to this mechanism of apoptosis (Zhao et al. 2012).

Soluble mediators are responsible for signaling the presence of aberrant inflammatory processes, such as cancer, that ultimately recruit MDSCs towards sites of inflammation. Tumor-induced MDSC induction compromises immune surveillance and promotes an immunosuppressive environment, enabling tumor growth and progression. While inhibiting tumor-induced inflammation to decrease MDSCs is certainly one potential therapeutic strategy, manipulation of MDSCs towards less immunosuppressive and potentially even immunostimulatory phenotypes by controlling their polarization is an alternative and potentially complementary therapeutic strategy.

24.2 MDSC Polarization

24.2.1 Polarization of Tumor-Associated Macrophages

Macrophage polarization can serve as a model for understanding analogous (but less well-understood) processes occurring in MDSCs. Macrophages demonstrate considerable heterogeneity in vivo and in vitro, including polarization to an M1 ("classical") or M2 ("alternative") activation state (Sica and Mantovani 2012; Medrek et al. 2012). The role macrophages play differs during tumor initiation and further tumor growth, and is reflected in the polarization profile at these various points of tumorigenesis (Ruffell et al. 2012).

M1-Macrophages are robust producers of inflammatory mediators and are present in high proportions at the initiation of inflammation. These processes are dependent on the presence of IFN-γ and LPS, both Th1-mediated cytokines (Ruffell et al. 2012). These M1-macrophages are an integral component of anti-tumor immunity, disrupting tumors through varied mechanisms including production of high levels of reactive oxygen species and expression of iNOS and Tumor Necrosis Factor α (TNF-α) (Sica and Mantovani 2012; Mantovani et al. 2002; Umemura et al. 2008; Ma et al. 2011). Consistent with this paradigm, M1-polarized macrophages promote T cell function in a way which beneficially interferes with tumor growth and progression, participating in a positive feedback system in which Th1 cytokines cause further macrophage activation (Sica and Mantovani 2012; Kang et al. 2011; Schreiber et al. 2011).

M2-macrophages, present in higher numbers during chronic inflammation, play an important role in tissue remodeling and immunoregulation, promoting tumor growth and progression as a result (Martinez et al. 2009). In contrast to M1 macrophages, M2-cells are induced by Th2 cytokines, including IL-4 and IL-13 (Sica and Mantovani 2012).

As tumors progress, the distribution of Tumor-Associated Macrophages (TAMs) have been shown to "switch" from a predominantly M1 to an M2 phenotype, with concomitant increases in angiogenesis and tissue remodeling and high levels of IL-10 expression (Sica and Mantovani 2012; Zaynagetdinov et al. 2011). Viewing macrophage activity in terms of polarization is a framework that is helpful for understanding the role of alternatively polarized MDSCs in antitumor immunity.

24.2.2 The Fate of MDSCs in the Tumor Microenvironment

The observation that TAMs principally exhibit an M2-like phenotype that suppress anti-tumor T cell responses raises questions of whether monocytic MDSCs, M-MDSCs, can also be differentiated into an "M1-like" versus "M2-like" phenotype (Sica and Mantovani 2012; Mantovani et al. 2002, 2009). Both TAMs and M-MDSCs, at least in some models, express the monocytic lineage markers F4/F80 and CD115 and, as noted, have increased expression of iNOS and ARG1 (Ma et al. 2011; Lewis and Pollard 2006). Multiple studies have described the presence of M-MDSCs with M2-like characteristics, promoting immunosuppression in the tumor microenvironment, although as with TAMs, M-MDSCs with both M1 and M2 phenotypes have been observed in murine tumor models (Huang et al. 2006; Dolcetti et al. 2010; Umemura et al. 2008; Marigo et al. 2008; Solito et al. 2011; Pan et al. 2008a, 2008b). As M-MDSCs resemble a population of immature, undifferentiated myeloid cells, they likely have the potential to differentiate into TAMs under certain conditions. Upon exposure to the tumor microenvironment, several authors have suggested MDSCs differentiate into TAMs in the presence of M-CSF and GM-CSF, inducing increased IL-10 and IL-4 production, promoting angiogenesis via VEGF modulation, and contributing to suppression of T cell immunity (Yang et al. 2004; Ma et al. 2011; Wilcox 2010).

As mentioned above, in tumor-bearing mice, MDSC can also promote development of Regulatory T Cells (Treg), another immunosuppressive population with an important role in compromising mechanisms of antitumor immunity (Huang et al. 2006; Ma et al. 2011; Kao et al. 2011). Identified by the presence of the nuclear transcription factor FOXP3, Tregs normally play an important role in immune self-tolerance (Centuoriet al. 2012). Effectors mechanisms specific to M2-phenotype MDSCs are necessary for MDSC-mediated Treg expansion (Ma et al. 2011). In contrast, recent findings suggest that G-MDSCs can interfere with TGF-β1 signaling necessary for Treg activation (Centuori et al. 2012). Interactions between Treg and MDSCs are likely to be complex and highly context-dependent, highlighting the need for further study.

24.2.3 Repolarization of Monocytic MDSCs

The identification of both M1- and M2-like M-MDSCs represents an interesting framework for understanding the role of MDSCs in antitumor immunity and tumor-mediated immunosuppression. The ability to manipulate polarization of M-MDSCs from a "pro-tumor" M2-phenotype to an M1-polarized MDSCs could reverse immunosuppression caused by M2-polarized MDSCs. Paired immunoglobulin-like receptors (PIRs) may be important regulators of MDSC polarization (Ma et al. 2011). Two molecules in the PIR family, PIR-A and PIR-B, may have an important immunoregulatory function in this regard. Both of these molecules are transmembrane receptors that interact with major histocompatibility complex (MHC) class I glycoproteins with low affinity (Takai 2005). PIR-A has intracellular interactions phosphorylating immunoreceptor tyrosine-based activation motifs (ITAMs) that have an activating influence, while PIR-Bs have analogous activity on inhibitory motifs that promote inhibitory signals (Ma et al. 2011; Kubagawa et al. 1999). In chronically inflamed environments, illustrated in tumor-bearing murine models, PIR-B inhibits PIR-A expression (Ma et al. 2011).

Our recent study examining tumor-bearing mice deficient in PIR-B found that the M-MDSCs adopted M1 characteristics (Ma et al. 2011). Most notably, tumor progression and growth of a lung cancer cell line was impeded in PIR-B deficient mice relative to their wild-type counterparts, and MDSCs in this model were found to express increased amounts of iNOS and TNF-α, and decreased amounts of ARG1 and IL-10, suggesting the predominance of an M1-like phenotype. Further confirming this polarization were decreased amounts of cell-surface markers known to be expressed in M2-like myeloid cells, such as CD36, CD206, Tie2, and IL-4R (Sica and Mantovani 2012; Mantovani et al. 2002, 2009; Guruvayoorappan 2008). Furthermore, along with decreased PIR-B activity, the authors demonstrated a concomitant increase in PIR-A activity, suggesting a balance in activity of these two molecules is critical in regulating the immunosuppressive capacity of MDSCs in the tumor environment.

MDSCs in this PIR-B deficient model inhibited growth of tumors and metastases, restoring anti-tumor immunity by polarization and reprogramming of M-MDSCs to an M1-like phenotype. The role of PIR was further characterized through blockade of SHP, a downstream effector molecule of PIR-B, producing a similar re-orientation of MDSCs towards the M1-phenotype with decreased growth in tumors and metastases.

24.2.4 Repolarization of G-MDSCs

Although polarization of G-MDSCs is less well understood than in their monocytic counterparts, they represent over 75 % of MDSCs in many murine tumor models (Youn et al. 2012). Their immunosuppressive effects are mediated primarily through arginase up regulation and ROS production, with consequent suppression of T cell immunity (Movahedi et al. 2008; Youn et al. 2008). Their phenotypic characteristics are morphologically similar to neutrophils, leading some immunologists to believe G-MDSCs represent an "activated" version of Neutrophils, sometimes referred to in the literature as "tumor associated neutrophils" (TAN) (Youn et al. 2012; Fridlender et al. 2009). While, most consider G-MDSC to be a distinct subset, it is clear that there can be both phenotypic and functional overlap between TAN and G-MDSC, making their distinction in the literature somewhat confusing.

TGF-β is a pleiotropic molecule that is overexpressed in many tumors and has been shown to play an important role in tumor-mediated immunosuppression. Inhibition of TGF-β has been shown in multiple settings to have antitumor activity, partly through restoration of T cell activity compromised by immunosuppressive mechanisms found in tumors. However, studies have shown that TGF-β blockade becomes ineffective in the setting of tumor-bearing neutrophil-depleted and/or CD8+ T cell depleted mice, suggesting that TAN, G-MDSCs, or both can be a critical target of TGF-β regulation in tumor immunosuppression (Fridlender et al. 2009; Suzuki et al. 2007). TGF-β receptor inhibition with the kinase inhibitor SM16 also considerably increased infiltration of CD8+ T cells in the tumor microenvironment in murine models of mesothelioma and lung cancer (Fridlender et al. 2009; Suzuki et al. 2007; Kim et al. 2008). TGF-β inhibition decreased arginase levels in G-MDSCs, while increasing TNF-α expression. These findings strongly suggest that TGF-blockade re-orients the immune system from a protumor towards an antitumor environment, with TGF-β blockade preferentially affecting granulocytic myeloid lineages such as G-MDSCs. Another study using murine colon cancer and glioma models also reported similar findings with an anti-TGF-β antibody, which was able to repolarize G-MDSCs from a predominantly G2–G1 phenotype (Umemura et al. 2008).

Fridlander et al. attributed reprogramming of granulocytic cells in the tumor-bearing context to TANs (G-MDSCs) adopting an "N1-phenotype" rather than a protumor "N2-phenotype." (Fridlender et al. 2009). As there are no definitive phenotypic markers differentiating TANs from G-MDSCs, this can also be cautiously interpreted as evidence that TGF-β blockade repolarizes G-MDSCs from a "G2"-like pro-tumor phenotype to a "G1"-like phenotype, similar to repolarization of M-MDSCs by PIR-B manipulation.

24.3 Emerging Targets to Direct MDSC Polarization

The PIR and TGFβ-R pathways have shown excellent promise as critical regulatory points in the polarization of M-MDSC and G-MDSC respectively. Further examination of associated ligands, alternative receptors, and downstream mediators of these pathways may elucidate the mechanisms governing MDSC polarization. SHP-1, a tyrosine phosphatase in the PIR-B pathway, normally mediates inhibition of JAK-STAT pathways in activated M-MDSCs (Ma et al. 2011; Takai 2005). As noted above, SHP inhibition suppresses PIR-B function, promoting a phenotypic switch from M2-like MDSCs to M1-MDSCs (Ma et al. 2011).

Identification of other ligands or effectors in this pathway would enhance our understanding of M-MDSC polarization. Although MDSCs were not specifically examined, a study examining PIR-B deficient mice in a murine Acute Myeloid Leukemia (AML) model found that PIR-B inhibition resulted in a similar anti-tumor effect, promoting differentiation of leukemia cells (Zheng et al. 2012). Another notable finding in this study was that Angiopoietin-like proteins (ANGPTLs), a family of glycoproteins previously found to promote activity and differentiation of hematopoietic stem cells, are ligands for both PIR-B and the analogous receptor in humans, leukocyte immunoglobulin-like receptor B (LILRB). With these findings in mind, ANGPTLs may potentially represent another target for down regulating the PIR-B pathway and inducing a phenotypic switch from M2-like MDSCs to M1-like MDSCs. Certain specific MHC I molecules as well as myelin-derived inhibitors have also been known to act as a PIR-B ligand (Zheng et al. 2012; Shiroishi et al. 2003) and therefore may potentially be therapeutic targets.

Cytokines overexpressed in the tumor and surrounding tissues offer another possible target for manipulation of the M1–M2 MDSC balance. Th2-cell mediated cytokines produced in the tumor microenvironment, such as IL-4 and IL-13 (Verma et al. 2009; Younos et al. 2011) have been shown to increase the induction and suppressive capacity of MDSCs via induction of arginase I with consequent effector T cell suppression (Angulo et al. 2000; Bronte et al. 2003) indicating that the IL-4 and IL-13 receptor pathways may play a key role in maintaining an M2-MDSC phenotype. Blockade of these cytokines is one potential strategy for differentiating MDSCs towards an M1-type M-MDSC phenotype.

Toll like-receptor activation (TLR) and IFN-γ have been shown to promote MDSC-mediated immunosuppression, with TLR promoting accumulation of MDSCs and IFN-γ inducing iNOS expression (Greifenberg et al. 2009; Ozaki and Leonard 2002). Greifenberg et al. examined these cytokines, reporting that while treatment with LPS, a TLR-4 ligand, promoted MDSC induction, treatment with LPS combined with IFN-γ caused an even greater degree of MDSC-mediated immunosuppression (Greifenberg et al. 2009). These results suggest that strategies that interfere with TLR-receptor signaling and/or IFN-γ signal transduction could decrease MDSC numbers and promote an M1-phenotype.

24.4 Therapeutic Applications of MDSC Programming and Polarization

Manipulation of MDSC induction and activation is an appealing strategy for reversal of the tumor-mediated immunosuppression driving tumor growth and progression. In contrast to nearly widespread agreement over general principles of MDSC identification in mouse models, where the CD11b and GR-1 marker

(LY6G/C) are consistently observed, an obstacle limiting clinical applications thus far is the considerable variability and lack of consensus over the phenotypes of MDSCs in humans. MDSCs with different phenotypic definitions have been described in breast cancer, colon cancer, head and neck squamous cell carcinoma, and many other malignancies (Montero et al. 2012). This divergence in MDSC phenotypes suggests that different cancers induce MDSCs by different mechanisms, highlighting the potential context dependence of different mechanisms of MDSC polarization and the importance of a broad selection of potentially therapeutic approaches to drive MDSC repolarization.

Several classes of medications aimed at inhibiting MDSC effector function or differentiation, such as kinase inhibitors, differentiating agents, and chemotherapy drugs have demonstrated successful inhibition of MDSC induction and activation and consequent reversal of immunosuppression in the tumor microenvironment. These mechanisms are reviewed, along with a consideration of a shift in strategy from exclusively targeting (inhibiting) MDSCs, to also repolarizing MDSCs as a therapeutic approach.

24.4.1 Kinase Inhibitors

Biologic agents have become increasingly important cancer therapies. Immunotherapies in particular, however, have had inconsistent clinical success. One of the mechanisms responsible for inconsistent efficacy of immune-based therapy has is tumor-mediated immunosuppression, including MDSCs (Kao et al. 2011). Agents that target or repolarize MDSCs, and reverse other immunosuppressive factors may be a valuable adjunct to active immunotherapies.

Our group study using a murine colon cancer model identified Stem-Cell Growth Factor (SCF), also known as c-kit Ligand, as a signaling molecule secreted by tumor cells that is responsible for MDSC induction and activation by increasing the production of myeloid precursors and also limiting their capacity to differentiate (Pan et al. 2008; Kao et al. 2011). Interference with c-kit ligand function decreases tumor angiogenesis and amount of MDSCs while increasing T cell-mediated antitumor immunity (Pan et al. 2008).

The tyrosine kinase inhibitor Sunitinib has several receptor targets, including VEGFR, and c-kit (Kao et al. 2011). Sunitinib administration has been shown in mice bearing tumors derived from MCA-26 (colon) or LLC1 (lung) cell lines to restore some degree of antitumor immunity through a decrease in MDSC and Treg levels, and—importantly—has been shown to increase survival (Ozao-Choy et al. 2009). One group reporting preliminary results of Sunitinib's effects in patients demonstrated that treatment resulted in decreased circulating MDSCs and Tregs (Kao et al. 2011). Another clinical study in metastatic Renal Cell Carcinoma patients found that Sunitinib therapy also caused considerable decreases in MDSCs (Ko et al. 2009). With these results in mind, Sunitinib has therapeutic potential to minimize MDSC-mediated immunosuppression, and can act as an adjunct to immunotherapy-based treatment. Imatinib is another kinase inhibitor with potential use in MDSC suppression (Montero et al. 2012; Kao et al. 2011; Mumprecht et al. 2006). It is more specific for c-kit than Sunitinib, and has been shown to reduce MDSC levels (Kao et al. 2011; Mumprecht et al. 2006). The prominent role of receptor kinases in driving macrophage polarization, and the promising effects of kinase inhibitors in this system suggest that kinase inhibitors are also a potential strategy to manipulate MDSC polarization (Sica and Mantovani 2012).

24.4.2 Differentiating Agents

CpG oligonucleotides (ODN) can partially reverse tumor-mediated immunosuppression through activation of CTL and NK cells (Murad and Clay 2009; Vollmer and Krieg 2009; Shirota et al. 2012). While studies of CpG ODN administration in humans demonstrate good induction of humoral immune responses, in contrast to their robust efficacy in mouse cancer models they have shown only limited success in slowing tumor growth and progression in human trials (Murad and Clay 2009; Vollmer and Krieg 2009; Shirota and Klinman 2011). A preclinical study specifically examining the effects of direct intratumoral injection of CpG ODNs found increased CTL and NK cells within the tumor microenvironment, diminished M-MDSC immunosuppressive activity via decreased arginase and NO production, a switch from a Th2 to Th1-cytokine production, and increased differentiation into tumoricidal macrophages (Shirota et al. 2012). The findings of decreased arginase and NO production and shift in cytokine milieu in particular suggests that in addition to acting as a differentiating agent, the immunostimulatory effects of ODNs may also contribute to a reorientation of M-MDSCs from an M2 pro-tumor to M1-like tumoricidal phenotype, although this specific possibility was not commented upon. Thus combination of CPG ODN with active immunotherapy approaches may be one strategy for reversing MDSC-mediated immunosuppression and enhancing overall efficacy.

Vitamin D derivatives have also been examined for their potential value as “differentiators,” with the hope that they would promote MDSC differentiation into mature myeloid lineages and restore antitumor T cell immunity. One study using 25-hydroxyvitamin D3 in patients with Head and Neck Cancer found increased CD34+ myeloid cell differentiation, although no clinically significant responses were seen (Lathers et al. 2004). Another Vitamin D derivative, *All-trans*-retinoic acid (ATRA), has had more success in reducing MDSC-mediated immunosuppression. A recent analysis examined ATRA’s effects in ex vivo experiments using MDSCs from patients with metastatic renal cell carcinoma (Kusmartsev et al. 2008). ATRA reversed MDSC-mediated immunosuppression by promoting differentiation of MDSCs into myeloid lineage cells, with consequent restoration of T cell responses, a finding also noted in other clinical studies as well (Montero et al. 2012; Kusmartsev et al. 2008; Mirza et al. 2006).

24.4.3 iNOS Inhibition

Our group has shown in the murine MT-RET-1 melanoma line that, rather than simply acting as an effector mechanism, tumor-expressed iNOS also plays a pivotal role in programming of myeloid precursors to become functional MDSCs (Jayaraman et al. 2012). In this study, iNOS inhibition decreased homing of MDSCs to the tumor, blocked up regulation of reactive oxygen species required for MDSC effector activity, and decreased elevated serum VEGF levels, essentially reversing tumor-mediated immunosuppression. Another study showed that the nonselective NOS inhibitor L-NAME or inhibition of PDE5 (which indirectly decreases iNOS expression) decreased systemic MDSC numbers and suppressed tumor growth in colon cancer and lymphoma models (Capuano et al. 2009). PDE5 inhibition was also shown to suppress MDSC accumulation and tumor growth in a head and neck cancer model (Serafini et al. 2006). Since iNOS plays a pivotal role as both a modulator of MDSC development and effector mechanism of T cell suppression, targeted therapy approaches directed at down regulating iNOS or blocking its activity have significant promise for reversing MDSC-mediated immunosuppression.

24.4.4 Chemotherapy Agents

Several studies have examined the effects of chemotherapeutic agents on MDSC populations. One study showed gemcitabine significantly decreased splenic MDSCs and restored MDSC-mediated immunosuppression in tumor-bearing mice (Suzuki et al. 2005). Another recent study with docetaxel showed decreased tumor burden in a murine breast cancer model, associated with decreased MDSC and increased CTL and CD4+ T cell activity (Kodumudi et al. 2010). In examining the mechanism of Docetaxel's effects on MDSCs, the authors demonstrated that not only did Docetaxel treatment have direct effects on differentiation of of MDSCs, but that the tumor-mediated immunosuppressive environment was modulated both through depletion of M2-like cells and polarization of MDSCs from an M2-like to an M1-like phenotype. In addition to this polarization, M2-like MDSCs appeared to be more sensitive to Docetaxel-induced cytotoxicity, as M2-like MDSCs were preferentially targeted for cell death relative to M1-like MDSCs. By reorienting the MDSC environment to an M1 phenotype, MDSCs appeared to have decreased immunosuppressive capabilities. Reversal of the immunosuppressive response and consequent stimulation of anti-tumor immunity makes Docetaxel, a taxane already used in chemotherapies for a wide-variety of therapeutic regimens, a promising adjunct to increase the efficacy of immunotherapies (Kodumudi et al. 2010).

24.5 Controversies and Future Areas of Research

Derangements in immunity have only recently been widely recognized as a major factor necessary for successful growth and progression of tumors (Hanahan and Weinberg 2011). Myeloid-derived suppressor cells play a major role in coordinating tumor-induced immunosuppression, and a general consensus has emerged that they are partly responsible for antagonizing antitumor T-cell responses. Thus MDSCs naturally present a promising target for restoring endogenous antitumor immunity.

One area that needs further clarification for this concept to mature is development of robust phenotypic markers, which identify MDSC associated with human cancers, and a taxonomy of human MDSC that will allow application of specific reprogramming and repolarization strategies to susceptible MDSC populations. The consistency of MDSC markers in preclinical (mouse) models has led to a wealth of studies focusing on all aspects of the immunoregulatory functions of MDSC, not only in antitumor immunity, but in other states such as autoimmunity, transplant rejection, obesity, and sepsis (Montero et al. 2012; Dilek 2012).

Several strategies have already emerged for targeting MDSCs directly. Future therapeutic directions may build on recent research elucidating mechanisms by which MDSCs mediate immunosuppression. Modulating inflammatory signaling, through inhibition of mediators such as COX2 and iNOS have shown success in preclinical studies and offer one alternative avenue to pursue in clinical trials (Jayaraman et al. 2012; Obermajer et al. 2011). Another exciting possibility is reorienting the MDSC profile from tumorigenic Type 2-like phenotypes to a Type 1-like tumoricidal polarization. By manipulating MDSC polarization, rather than targeting MDSCs for elimination, it may be possible to increase the population Type 1 of tumoricidal MDSCs with therapeutic benefit. As described above, our group has demonstrated a mechanism for repolarization of M-MDSCs in mice to an M1 phenotype, via interference with PIR signaling (Ma et al. 2011). Fridlander et al. also demonstrated that interfering with TGF-β signaling could result in an analogous polarization in murine G-MDSCs. In humans, tyrosine kinase inhibitors and chemotherapeutic drugs have been shown some ability to skew MDSC away from a Type 2 phenotype and maintain a Type 1 phenotype in human MDSCs. These studies show that harnessing already present but aberrantly altered tumor-associated immunocytes through manipulation of MDSC phenotype may be an effective anti-cancer approach.

References

Almand B, Clark JI, Nikitina E, van Beynen J, English NR, Knight SC et al (2001) Increased production of immature myeloid cells in cancer patients: a mechanism of immunosuppression in cancer. J Immunol 166:678–689

Angulo I, Rullas J, Campillo JA, Obregon E, Heath A, Howard M et al (2000) Early myeloid cells are high producers of nitric oxide upon CD40 plus IFN-gamma stimulation through a mechanism dependent on endogenous TNF-alpha and IL-1alpha. Eur J Immunol 30:1263–1271

Bingisser RM, Tilbrook PA, Holt PG, Kees UR (1998) Macrophage-derived nitric oxide regulates T cell activation via reversible disruption of the Jak3/STAT5 signaling pathway. J Immunol 160:5729–5734

Bronte V, Zanovello P (2005) Regulation of immune responses by L-arginine metabolism. Nat Rev Immunol 5:641–654

Bronte V, Serafini P, De Santo C, Marigo I, Tosello V, Mazzoni A et al (2003) IL-4-induced arginase 1 suppresses alloreactive T cells in tumor-bearing mice. J Immunol 170:270–278

Capuano G, Rigamonti N, Grioni M, Freschi M, Bellone M (2009) Modulators of arginine metabolism support cancer immunosurveillance. BMC Immunol 10:1

Centuori SM, Trad M, Lacasse CJ, Alizadeh D, Larmonier CB, Hanke NT et al (2012) Myeloid-derived suppressor cells from tumor-bearing mice impair TGF-beta-induced differentiation of CD4+ CD25+ FoxP3+ Tregs from CD4+ CD25-FoxP3- T cells. J Eukocyte Biol 92:987

Cheng P, Corzo CA, Luetteke N, Yu B, Nagaraj S, Bui MM et al (2008) Inhibition of dendritic cell differentiation and accumulation of myeloid-derived suppressor cells in cancer is regulated by S100A9 protein. J Exp Med 205:2235–2249

Corzo CA, Cotter MJ, Cheng P, Cheng F, Kusmartsev S, Sotomayor E et al (2009) Mechanism regulating reactive oxygen species in tumor-induced myeloid-derived suppressor cells. J Immunol 182:5693–5701

Corzo CA, Condamine T, Lu L, Cotter MJ, Youn JI, Cheng P et al (2010) HIF-1alpha regulates function and differentiation of myeloid-derived suppressor cells in the tumor microenvironment. J Exp Med 207:2439–2453

Dilek N (2012) Vuillefroy de Silly R, Blancho G, Vanhove B. Myeloid-derived suppressor cells: mechanisms of action and recent advances in their role in transplant tolerance. Front Immunol 3:208

Dolcetti L, Peranzoni E, Ugel S, Marigo I (2010) Fernandez Gomez A, Mesa C, et al. Hierarchy of immunosuppressive strength among myeloid-derived suppressor cell subsets is determined by GM-CSF. Eur J Immunol 40:22–35

Eruslanov E, Daurkin I, Ortiz J, Vieweg J, Kusmartsev S (2010) Pivotal Advance: Tumor-mediated induction of myeloid-derived suppressor cells and M2-polarized macrophages by altering intracellular PGE(2) catabolism in myeloid cells. J Leukoc Biol 88:839–848

Fridlender ZG, Sun J, Kim S, Kapoor V, Cheng G, Ling L et al (2009) Polarization of tumor-associated neutrophil phenotype by TGF-beta: "N1" versus "N2" TAN. Cancer Cell 16:183–194

Gabrilovich D (2004) Mechanisms and functional significance of tumour-induced dendritic-cell defects. Nat Rev Immunol 4:941–952

Gabrilovich D, Ishida T, Oyama T, Ran S, Kravtsov V, Nadaf S et al (1998) Vascular endothelial growth factor inhibits the development of dendritic cells and dramatically affects the differentiation of multiple hematopoietic lineages in vivo. Blood 92:4150–4166

Gallina G, Dolcetti L, Serafini P, De Santo C, Marigo I, Colombo MP et al (2006) Tumors induce a subset of inflammatory monocytes with immunosuppressive activity on CD8+ T cells. J Clin Investig 116:2777–2790

Garcia MR, Ledgerwood L, Yang Y, Xu J, Lal G, Burrell B et al (2010) Monocytic suppressive cells mediate cardiovascular transplantation tolerance in mice. J Clin Investig 120:2486–2496

Greifenberg V, Ribechini E, Rossner S, Lutz MB (2009) Myeloid-derived suppressor cell activation by combined LPS and IFN-gamma treatment impairs DC development. Eur J Immunol 39:2865–2876

Guruvayoorappan C (2008) Tumor versus tumor-associated macrophages: how hot is the link? Integr Cancer Ther 7:90–95

Hanahan D, Weinberg RA (2011) Hallmarks of cancer: the next generation. Cell 144:646–674

Highfill SL, Rodriguez PC, Zhou Q, Goetz CA, Koehn BH, Veenstra R et al (2010) Bone marrow myeloid-derived suppressor cells (MDSCs) inhibit graft-versus-host disease (GVHD) via an arginase-1-dependent mechanism that is up-regulated by interleukin-13. Blood 116:5738–5747

Huang B, Pan PY, Li Q, Sato AI, Levy DE, Bromberg J et al (2006) Gr-1+ CD115 + immature myeloid suppressor cells mediate the development of tumor-induced T regulatory cells and T-cell anergy in tumor-bearing host. Cancer Res 66:1123–1131

Jayaraman P, Parikh F, Lopez-Rivera E, Hailemichael Y, Clark A, Ma G et al (2012) Tumor-expressed inducible nitric oxide synthase controls induction of functional myeloid-derived suppressor cells through modulation of vascular endothelial growth factor release. J Immunol 188:5365–5376

Kang TW, Yevsa T, Woller N, Hoenicke L, Wuestefeld T, Dauch D et al (2011) Senescence surveillance of pre-malignant hepatocytes limits liver cancer development. Nature 479:547–551

Kao J, Ko EC, Eisenstein S, Sikora AG, Fu S, Chen SH (2011) Targeting immune suppressing myeloid-derived suppressor cells in oncology. Critical Rev Oncol/Hematol 77:12–19

Kim S, Buchlis G, Fridlender ZG, Sun J, Kapoor V, Cheng G et al (2008) Systemic blockade of transforming growth factor-beta signaling augments the efficacy of immunogene therapy. Cancer Res 68:10247–10256

Ko JS, Zea AH, Rini BI, Ireland JL, Elson P, Cohen P et al (2009) Sunitinib mediates reversal of myeloid-derived suppressor cell accumulation in renal cell carcinoma patients. Clinical Cancer Res: Official J Am Assoc Cancer Res 15:2148–2157

Kodumudi KN, Woan K, Gilvary DL, Sahakian E, Wei S, Djeu JY (2010) A novel chemoimmunomodulating property of docetaxel: suppression of myeloid-derived suppressor cells in tumor bearers. Clinical Cancer Res: Official J Am Assoc Cancer Res 16:4583–4594

Kubagawa H, Chen CC, Ho LH, Shimada TS, Gartland L, Mashburn C et al (1999) Biochemical nature and cellular distribution of the paired immunoglobulin-like receptors, PIR-A and PIR-B. J Exp Med 189:309–318

Kusmartsev SA, Li Y, Chen SH (2000) Gr-1 + myeloid cells derived from tumor-bearing mice inhibit primary T cell activation induced through CD3/CD28 costimulation. J Immunol 165:779–785

Kusmartsev S, Nefedova Y, Yoder D, Gabrilovich DI (2004) Antigen-specific inhibition of CD8 + T cell response by immature myeloid cells in cancer is mediated by reactive oxygen species. J Immunol 172:989–999

Kusmartsev S, Su Z, Heiser A, Dannull J, Eruslanov E, Kubler H et al (2008) Reversal of myeloid cell-mediated immunosuppression in patients with metastatic renal cell carcinoma. Clinical Cancer Res: Official J Am Assoc Cancer Res 14:8270–8278

Lathers DM, Clark JI, Achille NJ, Young MR (2004) Phase 1B study to improve immune responses in head and neck cancer patients using escalating doses of 25-hydroxyvitamin D3. Cancer Immunol Immunother: CII 53:422–430

Lewis CE, Pollard JW (2006) Distinct role of macrophages in different tumor microenvironments. Cancer Res 66:605–612

Ma G, Pan PY, Eisenstein S, Divino CM, Lowell CA, Takai T et al (2011) Paired immunoglobin-like receptor-B regulates the suppressive function and fate of myeloid-derived suppressor cells. Immunity 34:385–395

Mantovani A, Sozzani S, Locati M, Allavena P, Sica A (2002) Macrophage polarization: tumor-associated macrophages as a paradigm for polarized M2 mononuclear phagocytes. Trends Immunol 23:549–555

Mantovani A, Sica A, Allavena P, Garlanda C, Locati M (2009) Tumor-associated macrophages and the related myeloid-derived suppressor cells as a paradigm of the diversity of macrophage activation. Hum Immunol 70:325–330

Marigo I, Dolcetti L, Serafini P, Zanovello P, Bronte V (2008) Tumor-induced tolerance and immune suppression by myeloid derived suppressor cells. Immunol Rev 222:162–179

Martinez FO, Helming L, Gordon S (2009) Alternative activation of macrophages: an immunologic functional perspective. Annu Rev Immunol 27:451–483

Mazzoni A, Bronte V, Visintin A, Spitzer JH, Apolloni E, Serafini P et al (2002) Myeloid suppressor lines inhibit T cell responses by an NO-dependent mechanism. J Immunol 168:689–695

Medrek C, Ponten F, Jirstrom K, Leandersson K (2012) The presence of tumor associated macrophages in tumor stroma as a prognostic marker for breast cancer patients. BMC Cancer 12:306

Melani C, Chiodoni C, Forni G, Colombo MP (2003) Myeloid cell expansion elicited by the progression of spontaneous mammary carcinomas in c-erbB-2 transgenic BALB/c mice suppresses immune reactivity. Blood 102:2138–2145

Mirza N, Fishman M, Fricke I, Dunn M, Neuger AM, Frost TJ et al (2006) All-trans-retinoic acid improves differentiation of myeloid cells and immune response in cancer patients. Cancer Res 66:9299–9307

Montero AJ, Diaz-Montero CM, Kyriakopoulos CE, Bronte V, Mandruzzato S (2012) Myeloid-derived suppressor cells in cancer patients: a clinical perspective. J Immunother 35:107–115

Movahedi K, Guilliams M, Van den Bossche J, Van den Bergh R, Gysemans C, Beschin A et al (2008) Identification of discrete tumor-induced myeloid-derived suppressor cell subpopulations with distinct T cell-suppressive activity. Blood 111:4233–4244

Mumprecht S, Matter M, Pavelic V, Ochsenbein AF (2006) Imatinib mesylate selectively impairs expansion of memory cytotoxic T cells without affecting the control of primary viral infections. Blood 108:3406–3413

Mundy-Bosse BL, Thornton LM, Yang HC, Andersen BL, Carson WE (2011) Psychological stress is associated with altered levels of myeloid-derived suppressor cells in breast cancer patients. Cell Immunol 270:80–87

Murad YM, Clay TM (2009) CpG oligodeoxynucleotides as TLR9 agonists: therapeutic applications in cancer. BioDrugs: Clinical Immunother Biopharmaceuticals Gene Ther 23:361–375

Nagaraj S, Gupta K, Pisarev V, Kinarsky L, Sherman S, Kang L et al (2007) Altered recognition of antigen is a mechanism of CD8+ T cell tolerance in cancer. Nat Med 13:828–835

Obermajer N, Muthuswamy R, Lesnock J, Edwards RP, Kalinski P (2011) Positive feedback between PGE2 and COX2 redirects the differentiation of human dendritic cells toward stable myeloid-derived suppressor cells. Blood 118:5498–5505

Ochando JC, Chen SH (2012) Myeloid-derived suppressor cells in transplantation and cancer. Immunologic Res 54:275

Ochoa AC, Zea AH, Hernandez C, Rodriguez PC (2007) Arginase, prostaglandins, and myeloid-derived suppressor cells in renal cell carcinoma. Clinical Cancer Res: Official J Am Assoc Cancer Res 13:721s–726s

Ostrand-Rosenberg S, Sinha P (2009) Myeloid-derived suppressor cells: linking inflammation and cancer. J Immunol 182:4499–4506

Ostrand-Rosenberg S, Sinha P, Chornoguz O, Ecker C (2012) Regulating the suppressors: apoptosis and inflammation govern the survival of tumor-induced myeloid-derived suppressor cells (MDSC). Cancer Immunol Immunother: CII 61:1319

Ozaki K, Leonard WJ (2002) Cytokine and cytokine receptor pleiotropy and redundancy. J Biol Chem 277:29355–29358

Ozao-Choy J, Ma G, Kao J, Wang GX, Meseck M, Sung M et al (2009) The novel role of tyrosine kinase inhibitor in the reversal of immune suppression and modulation of tumor microenvironment for immune-based cancer therapies. Cancer Res 69:2514–2522

Pak AS, Wright MA, Matthews JP, Collins SL, Petruzzelli GJ, Young MR (1995) Mechanisms of immune suppression in patients with head and neck cancer: presence of CD34(+) cells which suppress immune functions within cancers that secrete granulocyte-macrophage colony-stimulating factor. Clinical Cancer Res: Official J Am Assoc Cancer Res 1:95–103

Pan PY, Ozao J, Zhou Z, Chen SH (2008a) Advancements in immune tolerance. Adv Drug Deliv Rev 60:91–105

Pan PY, Wang GX, Yin B, Ozao J, Ku T, Divino CM et al (2008b) Reversion of immune tolerance in advanced malignancy: modulation of myeloid-derived suppressor cell development by blockade of stem-cell factor function. Blood 111:219–228

Poschke I, Kiessling R (2012) On the armament and appearances of human myeloid-derived suppressor cells. Clin Immunol 144:250–268

Raychaudhuri B, Rayman P, Ireland J, Ko J, Rini B, Borden EC et al (2011) Myeloid-derived suppressor cell accumulation and function in patients with newly diagnosed glioblastoma. Neuro-oncology 13:591–599

Rodriguez PC, Zea AH, Culotta KS, Zabaleta J, Ochoa JB, Ochoa AC (2002) Regulation of T cell receptor CD3zeta chain expression by L-arginine. J Biol Chem 277:21123–21129

Rodriguez PC, Quiceno DG, Zabaleta J, Ortiz B, Zea AH, Piazuelo MB et al (2004) Arginase I production in the tumor microenvironment by mature myeloid cells inhibits T-cell receptor expression and antigen-specific T-cell responses. Cancer Res 64:5839–5849

Rodriguez PC, Quiceno DG, Ochoa AC (2007) L-arginine availability regulates T-lymphocyte cell-cycle progression. Blood 109:1568–1573

Ruffell B, Affara NI, Coussens LM (2012) Differential macrophage programming in the tumor microenvironment. Trends Immunol 33:119–126

Schreiber RD, Old LJ, Smyth MJ (2011) Cancer immunoediting: integrating immunity's roles in cancer suppression and promotion. Science 331:1565–1570

Serafini P, Meckel K, Kelso M, Noonan K, Califano J, Koch W et al (2006) Phosphodiesterase-5 inhibition augments endogenous antitumor immunity by reducing myeloid-derived suppressor cell function. J Exp Med 203:2691–2702

Shiroishi M, Tsumoto K, Amano K, Shirakihara Y, Colonna M, Braud VM et al (2003) Human inhibitory receptors Ig-like transcript 2 (ILT2) and ILT4 compete with CD8 for MHC class I binding and bind preferentially to HLA-G. Proc Natl Acad Sci USA 100:8856–8861

Shirota H, Klinman DM (2011) CpG-conjugated apoptotic tumor cells elicit potent tumor-specific immunity. Cancer Immunol Immunother: CII 60:659–669

Shirota Y, Shirota H, Klinman DM (2012) Intratumoral injection of CpG oligonucleotides induces the differentiation and reduces the immunosuppressive activity of myeloid-derived suppressor cells. J Immunol 188:1592–1599

Sica A, Mantovani A (2012) Macrophage plasticity and polarization: in vivo veritas. J Clin Investig 122:787–795

Solito S, Bronte V, Mandruzzato S (2011) Antigen specificity of immune suppression by myeloid-derived suppressor cells. J Leukoc Biol 90:31–36

Suzuki E, Kapoor V, Jassar AS, Kaiser LR, Albelda SM (2005) Gemcitabine selectively eliminates splenic Gr-1 +/CD11b + myeloid suppressor cells in tumor-bearing animals and enhances antitumor immune activity. Clinical Cancer Res: Official J Am Assoc Cancer Res 11:6713–6721

Suzuki E, Kim S, Cheung HK, Corbley MJ, Zhang X, Sun L et al (2007) A novel small-molecule inhibitor of transforming growth factor beta type I receptor kinase (SM16) inhibits murine mesothelioma tumor growth in vivo and prevents tumor recurrence after surgical resection. Cancer Res 67:2351–2359

Takai T (2005) Paired immunoglobulin-like receptors and their MHC class I recognition. Immunology 115:433–440

Umemura N, Saio M, Suwa T, Kitoh Y, Bai J, Nonaka K et al (2008) Tumor-infiltrating myeloid-derived suppressor cells are pleiotropic-inflamed monocytes/macrophages that bear M1- and M2-type characteristics. J Leukoc Biol 83 1136–1144

Veltman JD, Lambers ME, van Nimwegen M, Hendriks RW, Hoogsteden HC, Aerts JG et al (2010) COX-2 inhibition improves immunotherapy and is associated with decreased numbers of myeloid-derived suppressor cells in mesothelioma celecoxib influences MDSC function. BMC cancer 10:464

Verma VK, Singh V, Singh MP, Singh SM (2009) Effect of physical exercise on tumor growth regulating factors of tumor microenvironment: implications in exercise-dependent tumor growth retardation. Immunopharmacol Immunotoxicol 31:274–282

Vollmer J, Krieg AM (2009) Immunotherapeutic applications of CpG oligodeoxynucleotide TLR9 agonists. Adv Drug Deliv Rev 61:195–204

Wilcox RA (2010) Cancer-associated myeloproliferation: old association, new therapeutic target. Mayo Clinic Proc 85:656–663

Yang L, DeBusk LM, Fukuda K, Fingleton B, Green-Jarvis B, Shyr Y et al (2004) Expansion of myeloid immune suppressor Gr + CD11b + cells in tumor-bearing host directly promotes tumor angiogenesis. Cancer Cell 6:409–421

Youn JI, Nagaraj S, Collazo M, Gabrilovich DI (2008) Subsets of myeloid-derived suppressor cells in tumor-bearing mice. J Immunol 181:5791–5802

Youn JI, Collazo M, Shalova IN, Biswas SK, Gabrilovich DI (2012) Characterization of the nature of granulocytic myeloid-derived suppressor cells in tumor-bearing mice. J Leukoc Biol 91:167–181

Younos I, Donkor M, Hoke T, Dafferner A, Samson H, Westphal S et al (2011) Tumor- and organ-dependent infiltration by myeloid-derived suppressor cells. Int Immunopharmacol 11:816–826

Zaynagetdinov R, Sherrill TP, Polosukhin VV, Han W, Ausborn JA, McLoed AG et al (2011) A critical role for macrophages in promotion of urethane-induced lung carcinogenesis. J Immunol 187:5703–5711

Zhao X, Rong L, Zhao X, Li X, Liu X, Deng J et al (2012) TNF signaling drives myeloid-derived suppressor cell accumulation. J Clin Investig 122:4094–4104

Zheng J, Umikawa M, Cui C, Li J, Chen X, Zhang C et al (2012) Inhibitory receptors bind ANGPTLs and support blood stem cells and leukaemia development. Nature 485:656–660

Chapter 25
Therapeutic Targeting Regulatory T Cells in Tumor

Wei Wang and Weiping Zou

Abstract Cancer cells express a range of tumor associated antigens (TAA) including self-antigens, which should engender TAA-specific immune response to dampen tumor progression and result in tumor eradiation. However, this is rarely happened in patients with cancer. It is thought that active immunosuppressive network in the cancer microenvironment hinders anti-tumor immunity. Regulatory T cells (Tregs) are an important component in the immune suppressive network, suppress TAA-specific T cell anti-tumor immunity and are one of the main obstacles tempering successful immunotherapy and active vaccination. In this chapter, we focus on Tregs, review their characteristics and suppressive mechanisms in the tumor microenvironment, and summarize recent development in the Treg-targeted immunotherapeutic regimens.

Keywords Tregs · TAA · Immunotherapy · Deletion · Function suppression

25.1 Introduction

Treg cells, originally termed suppressive T cells, were initially described by Gershon and Kondo (1970, 1971) in the early 1970s as thymus-derived lymphocytes that tolerized bone marrow-derived lymphocytes to antigenic challenge. After many years suspicion, Drs. Sehon (Fujimoto et al. 1975), North (Berendt and North 1980; Bursuker and North 1984) and colleagues confirmed the existence of

W. Wang · W. Zou (✉)
Department of Surgery, University of Michigan School of Medicine, C560B MSRB II, 1150 W. Medical Center Dr., Ann Arbor, MI 48109-0669, USA
e-mail: wzou@med.umich.edu

M. R. Shurin et al. (eds.), *The Tumor Immunoenvironment*,
DOI: 10.1007/978-94-007-6217-6_25,

this T cell subgroup in mouse model. Later on, Dr. Sakaguchi et al. ascertained that the interleukin-2 (IL-2) receptor α-chain (CD25) could be used to identify Tregs in both human and mouse (Sakaguchi et al. 1995). Consequent studies in the same laboratory (Hori et al. 2003), as well as studies from Dr. Rudensky et al. (Fontenot et al. 2003), defined the transcription factor forkhead box P3 (FoxP3) as not only a key intracellular marker of Tregs but also a fundamental factor for controlling their development and functions. Indeed, in human, the loss-of-function mutation of the *foxp3* gene, impairs Treg development and causes a breach in self-tolerance, and leads to a severe auto-immune syndrome (Bennett et al. 2001). In this chapter we discuss the complex roles of Tregs in the context of cancer and cancer therapy.

25.2 Characteristics of Tregs in the Tumor Microenvironment

Prevalence of Tregs in tumor. As compared to healthy peripheral blood, apparently elevated Treg levels have been observed in all types of tumors, including pancreatic cancer, breast cancer (Gobert et al. 2009; Liyanage et al. 2002), colorectal cancer (Wilke et al. 2010; Somasundaram et al. 2002; Wolf et al. 2003), gastric cancer, esophageal cancer (Ichihara et al. 2003; Woo et al. 2001a; Amedei et al. 2012), leukemia, lymphoma (Rico et al. 2012; Marshall et al. 2004; Wang and Ke 2011), melanoma (Jacobs et al. 2012; Gray et al. 2003), non-small-cell lung cancer (Erfani et al. 2012) and ovarian cancer (Kryczek et al. 2009a; Peng et al. 2012). High levels of Tregs may be associated with advanced tumor stages and poor therapeutic efficacy (Wang et al. 2012). In addition to tumor islets, Tregs are also found in tumor stroma (Wei et al. 2006), and are co-localized with their targets–effector T cells as well as antigen presenting cells (APCs) (Curiel et al. 2004). Their co-localization suggests that Tregs are positioned to target immune cells and to limit their anti-tumor response in the tumor microenvironment (Zou 2006; Berezhnaya 2010).

Tregs may be a heterogeneous population of cells in the tumor microenvironment, which may be arising through distinct pathways of development and mediate disparate functions through a variety of different mechanisms (Zou 2006). Both natural Treg (nTreg) and induced Treg (iTreg) can contribute to tumor immune suppression in tumor bearing mouse model (Zhou and Levitsky 2007). However, there are no studies demonstrating similar characters of nTregs and iTregs in patients with cancer. It is likely that human nTregs and iTregs are phenotypically and functionally indistinguishable, and mediate immune suppression through distinct modes of action in the human tumor microenvironment.

Phenotype of Tregs in tumor. CD4+Foxp3+ Tregs are relatively well defined in human tumors. When naïve $CD4^+$ Tregs enter into the tumor and become activated, a series of surface markers will be changed due to the tumor environment.

Activation makers including CD44, GITA and OX40 are up-regulated. Some surface markers including ICOS (Gobert et al. 2009; Strauss et al. 2008), Helio(Zabransky et al. 2012), CD39 and CD73 (Gobert et al. 2009; Whiteside et al. 2011; Mandapathil et al. 2010) which are associated with Treg function, are also increased. These changes of surface marker consequently lead to functional alteration of Tregs including enhanced secretion of IL-10 and TGF-β as well as stronger suppression. It is generally accepted that suppressive functionality of $CD4^+$ Tregs in tumor patients is similar to that in healthy donors. A few studies pointed out that peripheral blood Tregs in patients with prostate cancer (Yokokawa et al. 2008) and non-small cell lung cancer (Ju et al. 2009) were more suppressive than those in healthy individuals. Treg in some human cancers could selectively induce $CD8^+$, but not $CD4^+$ effector T cell apoptosis through FasL/Fas signaling pathway (Strauss et al. 2009). Notably, regulatory T cells are not restricted to CD4+ T cells. Several studies have reported the existence of $CD8^+$ Tregs in various cancer patients including melanoma, renal and breast cancer (Andersen et al. 2009; Wang and Alexander 2009). These CD8 Tregs isolated from peripheral blood of cancer patients are HO-1 and Qa-1 specific(Lu and Cantor 2008) and inhibited T cell cytokine release, proliferation, and cytotoxicity (Andersen et al. 2009; Mahic et al. 2008).

Treg tumor trafficking. Treg tumor trafficking is one of the important mechanisms whereby Treg accumulate in the tumor microenvironment. In ovarian cancer patients, there are fewer Tregs in the tumor-draining lymph nodes than in lymphoid organs unrelated to the tumor (Peng et al. 2012; Wei et al. 2006; Curiel et al. 2004). This suggests that human Tregs preferentially migrate to and accumulate in the tumors. Treg trafficking relies on the interactions between chemokines/chemokine receptors (Wei et al. 2006). Depending on their activation status and tissue localization, Tregs can express abundant chemokine receptors including CCR2, CCR4, CCR5, CCR7, CCR8, CXCR4 and CXCR5. Therefore, Tregs are actively responsive to a variety of ligands. CCL22 is mainly secreted by tumor cells and myeloid cells, especially local macrophages, and mediates Treg tumor trafficking in patients with ovarian cancer as well as breast carcinoma (Curiel et al. 2004; Menetrier-Caux et al. 2012). CCR4, receptor for CCL22, expressed by Tregs, is accordingly increased during tumor development (Jordan et al. 2008; Mailloux et al. 2010; Schott et al. 2010) and metastasis (Olkhanud et al. 2009), accompanied with reduced lymphoid homing molecules, CD62L (L-selectin) and CCR7 (Huehn and Hamann 2005), as compared to circulating Treg counterparts. CCR4/CCL22 axis attracts and retains Treg in the tumor microenvironment. In addition to CCR4/CCL22, other chemokine and chemokine receptors may be also important for Treg trafficking to specific tumor. Treg pancreatic cancer migration is partly driven by CCR5 (Tan et al. 2009). High levels of CXCR4 are found on activated Tregs. CXCR4 is the receptor for CXCL12. CXCL12 expression is particularly high in ovarian cancer patients (Kryczek et al. 2005) and patients with prostate cancer bone metastasis (Zou et al. 2004). CXCR4/CXCL12 promotes Treg bone marrow homing, which may potentially explain why several human cancers are prone to bone marrow metastasis (Zou et al. 2004; Zhao et al. 2012). Treg cells exhibit active cell cycling in the bone marrow, and bone marrow dendritic cells express high levels of receptor activator of NF-κB (RANK). RANK and its ligand (RANKL) signals promote Treg cell expansion. Apart from chemokines and chemokine receptors, integrin CD103($\alpha_E\beta_7$) is also important for Treg tumor infiltration (Anz et al. 2011). Manipulation of Treg tumor trafficking may be an interesting approach to block Treg suppressive function in the tumor.

25.3 Suppression Mechanism of Tregs in Tumor

Suppressive mechanisms of Tregs have been elaborated using different models both in human and in mouse. Tregs can directly target and suppress T cells and APCs via cell-to-cell contact(Shevach 2009) and soluble factors (Sakaguchi 2005; Josefowicz et al. 2012). Although these proposed suppressive mechanisms are not exclusively investigated in tumor settings, it is likely that these modes of action of Tregs are relevant in the context of tumor immunity.

Soluble inhibitory cytokines. TGF-β and IL-10 are two central soluble inhibitory cytokines. It is thought that TGF-β and IL-10 are essential for Treg development and maintenance (Liu et al. 2008), and suppression in murine tumor models (Ghiringhelli et al. 2005; Chen et al. 2005). High levels of TGF-β and IL-10 can be detected in tumor tissue. However, they can be produced by multiple cell types in the tumor environment. Tregs may not be their major source. Nonetheless, Tregs may express these cytokines and disable the maturation and function of APCs (Zou et al. 1997), and inhibit T cell function (Huber and Schramm 2006). Additionally, Tregs suppress NK cell function in a TGF-β dependent manner in tumor bearing mice (Maloy et al. 2003).

Recently, a cytokine IL-35 was suggested by Vignali and his colleagues (Collison et al. 2007; Collison et al. 2010). IL-35, an Ebi3 (Epstein-Barr-virus-induced gene 3)-IL-12α heterodimer, is highly expressed and constitutively secreted by mouse Foxp3 Tregs, but not by resting or activated effector CD4+ T cells. Ebi3 is a downstream target of Foxp3, therefore it is required for Treg development and function. Both Ebi3$^{-/-}$ and IL-12$\alpha^{-/-}$ Tregs have significantly reduced regulatory activity in vitro and fail to control homeostatic proliferation and to cure inflammatory bowel disease in vivo. Recombinant IL-35 suppresses T cell proliferation in an IL-10 or TGF-β independent manner. Thus, IL-35 may be involved in Treg-mediated suppression.

Competitive consumption of IL-2. Tregs express all three components of the high-affinity IL-2R—CD25, CD122, and CD132. Therefore, they have a relatively high-affinity for IL-2. IL-2 is a cytokine to maintain conventional T cell survival. Competitive consumption of IL-2 by Tregs results in T cell apoptosis (Takahashi et al. 1998) which may depend on the pro-apoptotic factor Bim (Zou 2006; Shevach 2009; Pandiyan et al. 2007).

Perforin and granzyme pathway. Activated human CD4$^+$ Tregs express granzyme A and can kill activated CD4$^+$ and CD8$^+$ T cells and other cell types in a perforin-dependent, Fas-FasL-independent manner (Grossman et al. 2004). Up to 30 % murine CD4$^+$ Tregs in the tumor microenvironment express granzyme B. These Tregs can kill NK cells and CTLs by a perforin-independent, granzyme B-dependent pathway (Cao et al. 2007). Surprisingly, no studies have documented if Tregs mediated cytolysis of myeloid suppressor cells or tumor cells in a similar manner.

CTLA4 induction of indoleamine 2,3-dioxygenase (IDO)$^+$ APCs. Tregs constitutively express CTLA4. IDO is a key immunomodulatory enzyme in the tumor tissue or tumor draining lymph nodes (Mellor and Munn 2004). The close ligation of Treg CTLA-4 and DC CD80/CD86 increases the functional activity of IDO (Chung et al. 2009; Fallarino et al. 2003), consequently mediates the suppressive function of Tregs, and causes immune tolerance in the tumor microenvironment (Mellor and Munn 2004).

Extracellular nucleotide metabolism via *CD39 and CD73*. Extracellular ATP functions as an indicator of tissue destruction and may exert its effects both on DCs and activated T cells. The degradation of ATP by CD39 together with CD73 represents a mechanism by which Tregs induce the production of peri-cellular adenosine that inhibits T cell responses through adenosine receptor (A_2A). Tregs from CD39 knockout mice lost their suppressive properties in vitro and failed to block allograft rejection in vivo (Deaglio et al. 2007). CD39 receptor A_2A-dificient Tregs show 50 % reduction of suppressive capacity as compared to wild type Tregs (Deaglio et al. 2007; Borsellino et al. 2007). Catalytic inactivation of extracellular ATP by CD39 together with CD73 through adenosine receptor (A_2A) may represent a mechanism by which Tregs inhibit T cell responses. For example, human tumor associated Tregs highly express CD39, and inhibited Th17 cell development in the tumor microenvironment via CD39 (Kryczek et al. 2009b; Ye et al. 2011).

B7 family members. PD-L1/B7-H1, a member of the B7 family co-inhibitory molecules, has been detected in the majority of human cancer cells (Ascierto et al. 2010) and myeloid APCs (Curiel et al. 2003). Its engagement with its receptor, programmed death 1(PD-1), leads to the induction of Tregs (Wang et al. 2008) as well as anergy and apoptosis of activated T cells then favoring tumor cells to overcome host response (Curiel et al. 2003).

B7-H4 (B7x, B7S1), another B7 co-stimulatory family member, is strongly expressed by tumor associated macrophages, but not by normal macrophages (Kryczek et al. 2006; Kryczek et al. 2007). B7-H4$^+$ tumor macrophages suppress tumor associated antigen-specific T cell immunity. Interestingly, Tregs convey suppressive activity to APCs by stimulating B7-H4 expression on APCs (Zou 2006). High numbers of Tregs and B7-H4$^+$ macrophages are highly associated and co-localized in the tumor environments, and predict poor patient survival. It suggests that Tregs mediate suppression via B7-H4 induction in tumor patients.

25.4 Tregs and Cancer Therapy

Given that Tregs suppress anti-tumor immunity, targeting Tregs may be an immunotherapeutic modality to subvert immunosuppression in patients with cancer. Systematic Treg depletion may cause autoimmune responses. Immune targeting Tregs may be cautiously realized in patients with cancer (Mougiakakos et al. 2010). Numerous studies are under development or in clinical evaluation to deplete or functional inactivate Tregs.

25.4.1 Treg Depletion

Chemotherapeutic agents. Chemotherapeutic agents were utilized to deplete suppressor T cells. It was reported that the antitumor effect of cyclophosphamide in murine experimental cancer models was due to the depletion of "suppressor T cells" (North 1982; Awwad and North 1988). As an effective tumor chemotherapeutic drug, high doses of cyclophosphamide manifests its immunosuppression (Rollinghoff et al. 1977). Strikingly, low doses of cyclophosphamide act as an immune-stimulus. This has been confirmed in a variety of animal models (Lutsiak et al. 2005) and in patients with metastatic melanoma (Berd and Mastrangelo 1988). Low-dose cyclophosphamide could relatively reduce $CD4^+$ $CD25^+$ Tregs, but not the total T cell population (Le and Jaffee 2012). Depletion of $CD4^+$ Tregs by cyclophosphamide was further confirmed by recent studies in normal mice(Lutsiak et al. 2005; Ercolini et al. 2005), in mice bearing B16 melanoma (Turk et al. 2004) or neu-expressing tumor (Ercolini et al. 2005) and rats bearing a chemically-induced colon cancer (Ghiringhelli et al. 2004). Potential mechanisms of Treg deletion include induction of Treg apoptosis and reduction of Treg proliferation (Taieb et al. 2006). Furthermore, metronomic cyclophosphamide could decreases the level of immunosuppressive cytokines TGF-β, IL-10 (Sharabi and Ghera 2010; Matar et al. 2001), and restores the function of NKT cells and T cells (Ghiringhelli et al. 2007).

Similarly, low-dose of cyclosporine A (CsA) and tacrolimus (FK506) reduce Treg numbers and function (Shibutani et al. 2005; Kawai et al. 2005), probably by suppressing IL-2. Further pre-clinical studies are needed to assess whether low doses of these agents are an option to promote tumor immunity in patients.

*Anti-CD*25. Tregs express high levels of CD25. CD25-specific antibody, PC61, abrogated suppressive function of Tregs in mouse cancer models (Onizuka et al. 1999). Basiliximab (simulect) and daclizumab (zenapex) are two humanized CD25 mAbs. Basiliximab effectively targets Tregs, augments T cell IFN-γ and IL-2 production (Okawaki et al. 2008). In metastatic breast carcinoma patients, daclizumab treatment durably reduced circulating Tregs and improved the emergence of cancer-specific CTL response after vaccination with cancer antigen peptides (hTERT/survivin) (Rech and Vonderheide 2009) (Fig. 25.1).

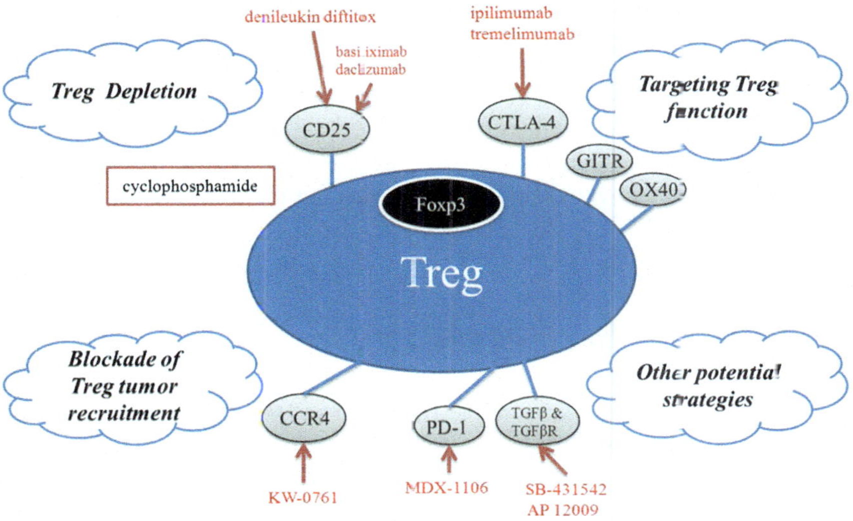

Fig. 25.1 Therapeutic targeting Tregs in tumor—Treg depletion, targeting Treg function, blockade of Treg tumor recruitment and other potential strategies. Several agents (in *red*) discussed in the chapter are in clinical trials. Denileukin difitox is a ligand fusion toxin consisting of full-length IL-2R fused to the enzymatically active and translocating domains of diphtheria toxin, Basiliximab and daclizumab are two humanized CD25 mAbs. Ipilimumab and tremelimumab are humanized anti-CTLA-4 mAb and approved by FDA to treat patients with advanced melanoma. KW-0761 is defucosylated humanized anti-CCR4 antibody. MDX-1106 is an anti-PD-1 mAb. SB-431542 is a TGF-βRI antagonist and AP120009 is a specific TGF-βRII inhibitor

Denileukin diftitox DAB389IL-2 (ONTAK) is a ligand fusion toxin consisting of full-length IL-2R fused to the enzymatically active and translocating domains of diphtheria toxin (Foss 2000). Denileukin diftitox is used to treat CD25^{+} cutaneous T cell and B cell leukemia/lymphoma (Olsen et al. 2001; Dang et al. 2004; Foss et al. 2005; Frankel et al. 2006). It has a short half-life of 60 min and is designed to target cells expressing the high-affinity IL-2R. After internalization via endocytosis, ADP-ribosyltransferase activity of diphtheria toxin is cleaved in the endosome and is translocated into the cytosol where it inhibits protein synthesis, finally leading to apoptosis (Figgitt et al. 2000). The administration of denileukin diftitox in human prior to DC vaccination strategy increased the vaccine-mediated anti-tumor T cell response in conjunction with efficient depletion of peripheral Treg and reduced Treg-associated immunosuppression (Dannull et al. 2005). Denileukin diftitox is also used to treat patients with lung, ovarian or breast cancer, where CD4^{+} Treg numbers are greatly elevated (Curiel et al. 2004; Woo et al. 2001b). A single dose of denileukin diftitox is sufficient to reduce the prevalence and absolute numbers of peripheral CD4^{+} CD25^{+} Foxp3^{+} T cells and increase effector T cell activation in all the patients, lasting for about 1 month. A single denileukin diftitox infusion can sufficiently normalize the CA-125 of patient's blood, which is a marker measuring the growth of ovarian cancer. Multiple weekly denileukin diftitox treatment largely relieved the visceral and lymphatic metastases. The underline mechanism of denileukin diftitox remains controversy. Nonetheless, denileukin diftitox may possibly reduce CD25+ effector T cells. Treg fast turnover may also compromise the effects of denileukin diftitox on Treg reduction (Dannull et al. 2005; Attia et al. 2005; Mahnke et al. 2007). Based on current literature, single therapy may not efficiently lead to objective clinical responses in patients with cancer.

25.4.2 Targeting Treg Function

*Anti-CTLA*4. Cytotoxic T-lymphocyte-associated antigen 4 (CTLA4) is a negative regulator of T cell function. CTLA4-specific antibody treatment resulted in murine tumor rejection (Kwon et al. 1997; Kwon et al. 1999; Leach et al. 1996) and achieved synergistic anti-tumor effects in combination with other therapeutic reagents (Hurwitz et al. 1998; Hurwitz et al. 2000; van Elsas et al. 1999). Humanized anti-CTLA-4 antagonist mAb ipilimumab (MDX-100) and tremelimumab (CP675206) have been approved by the United States Food and Drug administration to treat patients with advanced melanoma (Camacho et al. 2009), renal cell carcinoma and prostate cancers. A phase I trial, demonstrated that a single dose of tremelimumab is sufficient to break peripheral tolerance (Menard et al. 2008). Several phase II trials confirmed that the response rate of ipilimumab used as monotherapy or combined with other therapies was consistently in a range of 10–15 %. Those treated patients showed enhanced anti-tumor response and impressive 1 and 2 years survival (Weber 2007).

Importantly, CTLA4-specific antibody treatment resulted in significant autoimmune side effects, which is not observed in CD25 mAb therapy (Ansell et al. 2009; Maker et al. 2005; Phan et al. 2003), suggesting that the targets of CD25-specific antibody and CTLA4-specific antibody may not be identical. Therefore, the mechanistic link between the effects of treatment with CTLA4-specific antibody and $CD4^+$ Treg function remains to be determined in humans. In fact, CTLA4-specific antibody could both act on Treg and $CD25^-$ effector T cells (Peggs et al. 2009; Jain et al. 2010). Melief and colleagues (Sutmuller et al. 2001) proposed that $CD4^+CD25^+$ Tregs and CTLA4 signaling represent two independent and synergistic regulatory mechanisms for suppression of T cell activation in vivo (Sutmuller et al. 2001). Instead of inducing depletion of peripheral $CD4^+CD25^+$ T cells (Maker et al. 2005), it largely inhibited their suppressive function (Phan et al. 2003). Meanwhile, it targeted on effector T cells directly by favoring the binding of C80/CD86–CD28. Direct evidence came from a study—in the absence of $CD25^+$ cells, CTLA4 blockade could still benefit the induction of TAA-specific effector T cells (Sutmuller et al. 2001). Similarly, $CD4^+CD25^+$ T cells exhibited equivalent suppressive activity no matter in $CTLA4^{-/-}$ mice or $CTLA4^{+/+}$ mice (Tang et al. 2004). To maximize its efficacy and minimize side effect, CTLA-4 mAb treatment needs to be deliberately designed including administration schedule, dose and in combination with other therapies.

Anti-GITR. In addition to CTLA-4-specific antibody, an agonistic antibody against the glucocorticoid-induced TNF receptor (GITR) has also been considered a target for Treg depletion and functional inhibition (Nocentini and Riccardi 2005). GITR is constitutively expressed on nTregs but also at lower levels on activated conventional T cells and APCs (Shimizu et al. 2002; Shevach and Stephens 2006). GITR expression can be rapidly up-regulated during the activation of $CD4^{+}CD25^{-}$ T cells (Menetrier-Caux et al. 2012; Shevach and Stephens 2006). In vitro experiment, engagement of GITR on effectors T cells by its ligand expressed on DC render them resistant to Treg suppression (Shimizu et al. 2002). Strikingly, the agonistic anti-GITR antibody DTA-1 may attenuate suppressive activity of $CD4^{+}$ Tregs as reported in vitro (Shimizu et al. 2002), administration of DTA-1 invoked both potent antitumor immunity and decreased intra-tumoral Treg recruitment, resulting in the regression of small established B16 melanoma tumors in vivo (Cohen et al. 2010). The exact mechanism by which this approach achieves its effects is controversial. GITR-specific antibody treatment has not yet been tested in humans. Future investigations are warranty to evaluate the feasibility and potential synergistic effects of GITR-specific antibody with current immunotherapeutic strategies such as anti-CTRA-4 mAb (Mitsui et al. 2010).

Anti-OX40. OX40(CD134) is a co-stimulatory molecule of the TNF receptor family. It is constitutively expressed on Tregs and transiently expressed on activated T cells. Stimulation of OX40 hinders the suppressive function of Tregs in vitro by down-regulating the expression of FOXP3 (Kitamura et al. 2009). In line with the in vitro studies, anti-OX40 mAb blocks Treg-mediated suppression and results in tumor rejection in murine models (Piconese et al. 2008). However, it is unknown if anti-OX40 has similar effects in patients with cancer.

25.4.3 Blockade of Treg Tumor Recruitment

CCR4/CCL22 axis is critical for Treg tumor migration (Curiel et al. 2004). Altering Treg tumor trafficking would be an attractive strategy to block Treg-mediated suppression in the tumor microenvironment. Therapeutic anti-CCR4 mAb efficacy was actually examined in patients with Hodgkin Lymphoma (Ishida et al. 2006) and adult T Cell Leukemia/Lymphoma (ATLL) in vitro (Ishida et al. 2004). Defucosylated humanized anti-CCR4 antibody (KW-0761) has been in Phase I clinical trial to treat patients with with ATLL and peripheral T cell lymphoma. Five out of fifteen patients achieved objective responses. Subsequent phase II studies are thus warranted (Yamamoto et al. 2010). However, CCR4/CCL22 blockade may potentially impair effector T cell migration (Wei et al. 2006). Therapeutic utility must be chosen more cautiously and dual effects of chemokine/receptor antagonism must be taken into account in strategies to block chemokine and their receptors. Notably, IL-2 therapy enhances CXCR4 expression on Tregs, and may promote immune suppression, and in turn affect tumor development and metastases(Mougiakakos et al. 2010). Altogether, these findings support a potential combinatory therapy by targeting Treg-associated chemokines and chemokine receptors (Byrne et al. 2011).

25.4.4 Other Potential Strategies

Targeting TGF-β signaling pathway. Anti-TGF-β neutralizing antibody 2G7 (Kobie et al. 2003), competitive soluble receptor TβRII (Won et al. 1999) and TGFβRI/II kinase inhibitor LY2109761 can successfully block TGF-β signaling pathway, decrease tumor growth and metastases in mouse models. TGF-β receptor type I antagonist SB-431542, can rescue TGF-β-mediated immune suppression in human malignant glioma cell lines (Hjelmeland et al. 2004). A specific TGF-βRII inhibitor, AP 12009, was assessed in phase I/II open-label dose-escalation studies in patients with recurrent or refractory high-grade glioma. Current animal and clinical studies suggest that targeting TGF-β2 may be a potential therapeutic approach for malignant tumor therapy (Schlingensiepen et al. 2008). However, TGF-β is broadly expressed in different organs by different types of cells. TGF-β signaling pathway is involved in multiple layers of physiological development and biological activities. For example, TGF-β modulates NK cell function, effector T cell development including Th17 cells (Zou 2006). Given the protective roles of NK cells, CD8+ T cells and effector Th17 cells in anti-tumor immunity, targeting TGF-β signaling pathway needs to be carefully examined in preclinical models.

Inhibition of inhibitory B7 family signaling pathway. Preclinical cancer models suggest that interruption of PD-1L/PD-1 ligation augments antigen-specific CTL infiltration and improved antitumor T cell responses (Curiel et al. 2003) and decreased Treg suppressive function (Wang et al. 2009). An anti-PD-1 mAb, MDX-1106 (Ono-4538) is currently being evaluated in a phase I/II clinical trial for various types of cancer including non-small cell lung cancer, renal cell carcinoma, colon cancer, melanoma and prostate cancer. Intermittent dosing of MDX-1106 at 10 mg/kg demonstrated clinical activity against renal cell carcinoma and melanoma without serious toxicity (Brahmer et al. 2010). Recently two important clinical studies on evaluation of anti-PD-1 and anti-PD-L1 were published (Topalian et al. 2012; Brahmer et al. 2012). Important clinical objective responses have been observed. However, it remains to be defined if the treatment affects Treg compartment in patients.

B7-H4, another inhibitory B7 family member, has been studied in patients with cancer. B7-H4$^+$ ovarian cancer-associated macrophages inhibit T cell immunity. The intensity of B7-H4 expression in macrophages was correlated with Treg numbers in the tumor microenvironment (Kryczek et al. 2006), predict poor survival in ovarian cancer (Kryczek et al. 2007) and renal cell carcinoma

(Krambeck et al. 2006). Given that Tregs convey suppressive activity to APCs through B7-H4 induction (Zou 2006), it is possible that targeting B7-H4 may affect Treg function and generate clinical benefits.

25.5 Concluding Remarks

Tumors actively develop various mechanisms to refrain tumor immunity. Immunosuppressive cells, especially Tregs, play a critical role in tumor immune escape and impede the anti-tumor immune responses in cancer patients. Current preclinical studies and clinical trials suggest that Treg manipulation is an attractive strategy for cancer immunotherapy. However, it appears technically challenging to selectively and reliably target Tregs in patients with cancer. Furthermore, compelling evidence indicates that single therapy may not be clinical effective to treat patients with cancer. Novel tumor immunotherapy should target immune suppressive networks including Tregs, in combination with other therapeutic modalities so that effective, reliable and consistent clinical efficacy may be reached.

References

Amedei A, Della Bella C, Silvestri E, Prisco D, D'Elios MM (2012) T cells in gastric cancer: friends or foes. Clin Dev Immunol 2012:690571

Andersen MH, Sorensen RB, Brimnes MK, Svane IM, Becker JC, thor Straten P (2009) Identification of heme oxygenase-1-specific regulatory CD8+ T cells in cancer patients. J Clin Invest. 119:2245–2256

Ansell SM, Hurvitz SA, Koenig PA, LaPlant BR, Kabat BF, Fernando D et al (2009) Phase I study of ipilimumab, an anti-CTLA-4 monoclonal antibody, in patients with relapsed and refractory B-cell non-Hodgkin lymphoma. Clin Cancer Res 15:6446–6453

Anz D, Mueller W, Golic M, Kunz WG, Rapp M, Koelzer VH et al (2011) CD103 is a hallmark of tumor-infiltrating regulatory T cells. Int J Cancer 129:2417–2426

Ascierto PA, Simeone E, Sznol M, Fu YX, Melero I (2010) Clinical experiences with anti-CD137 and anti-PD1 therapeutic antibodies. Semin Oncol 37:508–516

Attia P, Maker AV, Haworth LR, Rogers-Freezer L, Rosenberg SA (2005) Inability of a fusion protein of IL-2 and diphtheria toxin (Denileukin diftitox, DAB389IL-2, ONTAK) to eliminate regulatory T lymphocytes in patients with melanoma. J Immunother 28:582–592

Awwad M, North RJ (1988) Cyclophosphamide (Cy)-facilitated adoptive immunotherapy of a Cy-resistant tumour. Evidence that Cy permits the expression of adoptive T-cell mediated immunity by removing suppressor T cells rather than by reducing tumour burden. Immunology 65:87–92

Bennett CL, Christie J, Ramsdell F, Brunkow ME, Ferguson PJ, Whitesell L et al (2001) The immune dysregulation, polyendocrinopathy, enteropathy, X-linked syndrome (IPEX) is caused by mutations of FOXP3. Nat Genet 27:20–21

Berd D, Mastrangelo MJ (1988) Effect of low dose cyclophosphamide on the immune system of cancer patients: depletion of CD4+, 2H4+ suppressor-inducer T-cells. Cancer Res 48:1671–1675

Berendt MJ, North RJ (1980) T-cell-mediated suppression of anti-tumor immunity. An explanation for progressive growth of an immunogenic tumor. J Exp Med 151:69–80
Berezhnaya NM (2010) Interaction between tumor and immune system: the role of tumor cell biology. Exp Oncol 32:159–166
Borsellino G, Kleinewietfeld M, Di Mitri D, Sternjak A, Diamantini A, Giometto R et al (2007) Expression of ectonucleotidase CD39 by Foxp3+ Treg cells: hydrolysis of extracellular ATP and immune suppression. Blood 110:1225–1232
Brahmer JR, Drake CG, Wollner I, Powderly JD, Picus J, Sharfman WH et al (2010) Phase I study of single-agent anti-programmed death-1 (MDX-1106) in refractory solid tumors: safety, clinical activity, pharmacodynamics, and immunologic correlates. J Clin Oncol 28:3167–3175
Brahmer JR, Tykodi SS, Chow LQ, Hwu WJ, Topalian SL, Hwu P et al (2012) Safety and activity of anti-PD-L1 antibody in patients with advanced cancer. N Engl J Med 366:2455–2465
Bursuker I, North RJ (1984) Generation and decay of the immune response to a progressive fibrosarcoma. II. Failure to demonstrate postexcision immunity after the onset of T cell-mediated suppression of immunity. J Exp Med 159:1312–1321
Byrne WL, Mills KH, Lederer JA, O'Sullivan GC (2011) Targeting regulatory T cells in cancer. Cancer Res 71:6915–6920
Camacho LH, Antonia S, Sosman J, Kirkwood JM, Gajewski TF, Redman B et al (2009) Phase I/II trial of tremelimumab in patients with metastatic melanoma. J Clin Oncol 27:1075–1081
Cao X, Cai SF, Fehniger TA, Song J, Collins LI, Piwnica-Worms DR et al (2007) Granzyme B and perforin are important for regulatory T cell-mediated suppression of tumor clearance. Immunity 27:635–646
Chen ML, Pittet MJ, Gorelik L, Flavell RA, Weissleder R, von Boehmer H et al (2005) Regulatory T cells suppress tumor-specific CD8 T cell cytotoxicity through TGF-beta signals in vivo. Proc Natl Acad Sci U S A 102:419–424
Chung DJ, Rossi M, Romano E, Ghith J, Yuan J, Munn DH et al (2009) Indoleamine 2,3-dioxygenase-expressing mature human monocyte-derived dendritic cells expand potent autologous regulatory T cells. Blood 114:555–563
Cohen AD, Schaer DA, Liu C, Li Y, Hirschhorn-Cymmerman D, Kim SC et al (2010) Agonist anti-GITR monoclonal antibody induces melanoma tumor immunity in mice by altering regulatory T cell stability and intra-tumor accumulation. PLoS ONE 5:e10436
Collison LW, Workman CJ, Kuo TT, Boyd K, Wang Y, Vignali KM et al (2007) The inhibitory cytokine IL-35 contributes to regulatory T-cell function. Nature 450:566–569
Collison LW, Chaturvedi V, Henderson AL, Giacomin PR, Guy C, Bankoti J et al (2010) IL-35-mediated induction of a potent regulatory T cell population. Nat Immunol 11:1093–1101
Curiel TJ, Wei S, Dong H, Alvarez X, Cheng P, Mottram P et al (2003) Blockade of B7-H1 improves myeloid dendritic cell-mediated antitumor immunity. Nat Med 9:562–567
Curiel TJ, Coukos G, Zou L, Alvarez X, Cheng P, Mottram P et al (2004) Specific recruitment of regulatory T cells in ovarian carcinoma fosters immune privilege and predicts reduced survival. Nat Med 10:942–949
Dang NH, Hagemeister FB, Pro B, McLaughlin P, Romaguera JE, Jones D et al (2004) Phase II study of denileukin diftitox for relapsed/refractory B-Cell non-Hodgkin's lymphoma. J Clin Oncol 22:4095–4102
Dannull J, Su Z, Rizzieri D, Yang BK, Coleman D, Yancey D et al (2005) Enhancement of vaccine-mediated antitumor immunity in cancer patients after depletion of regulatory T cells. J Clin Invest 115:3623–3633
Deaglio S, Dwyer KM, Gao W, Friedman D, Usheva A, Erat A et al (2007) Adenosine generation catalyzed by CD39 and CD73 expressed on regulatory T cells mediates immune suppression. J Exp Med 204:1257–1265
Ercolini AM, Ladle BH, Manning EA, Pfannenstiel LW, Armstrong TD, Machiels JP et al (2005) Recruitment of latent pools of high-avidity CD8(+) T cells to the antitumor immune response. J Exp Med 201:1591–1602

Erfani N, Mehrabadi SM, Ghayumi MA, Haghshenas MR, Mojtahedi Z, Ghaderi A et al (2012) Increase of regulatory T cells in metastatic stage and CTLA-4 over expression in lymphocytes of patients with non-small cell lung cancer (NSCLC). Lung Cancer 77:306–311

Fallarino F, Grohmann U, Hwang KW, Orabona C, Vacca C, Bianchi R et al (2003) Modulation of tryptophan catabolism by regulatory T cells. Nat Immunol 4:1206–1212

Figgitt DP, Lamb HM, Goa KL (2000) Denileukin diftitox. Am J Clin Dermatol 1:67–72 (discussion 3)

Fontenot JD, Gavin MA, Rudensky AY (2003) Foxp3 programs the development and function of CD4+ CD25+ regulatory T cells. Nat Immunol 4:330–336

Foss FM (2000) DAB(389)IL-2 (ONTAK): a novel fusion toxin therapy for lymphoma. Clin Lymphoma 1:110–117

Foss F, Demierre MF, DiVenuti G (2005) A phase-1 trial of bexarotene and denileukin diftitox in patients with relapsed or refractory cutaneous T-cell lymphoma. Blood 106:454–457

Frankel AE, Surendranathan A, Black JH, White A, Ganjoo K, Cripe LD (2006) Phase II clinical studies of denileukin diftitox diphtheria toxin fusion protein in patients with previously treated chronic lymphocytic leukemia. Cancer 106:2158–2164

Fujimoto S, Greene M, Sehon AH (1975) Immunosuppressor T cells in tumor bearing host. Immunol Commun 4:201–217

Gershon RK, Kondo K (1970) Cell interactions in the induction of tolerance: the role of thymic lymphocytes. Immunology 18:723–737

Gershon RK, Kondo K (1971) Infectious immunological tolerance. Immunology 21:903–914

Ghiringhelli F, Larmonier N, Schmitt E, Parcellier A, Cathelin D, Garrido C et al (2004) CD4+ CD25+ regulatory T cells suppress tumor immunity but are sensitive to cyclophosphamide which allows immunotherapy of established tumors to be curative. Eur J Immunol 34:336–344

Ghiringhelli F, Menard C, Terme M, Flament C, Taieb J, Chaput N et al (2005) CD4+ CD25+ regulatory T cells inhibit natural killer cell functions in a transforming growth factor-beta-dependent manner. J Exp Med 202:1075–1085

Ghiringhelli F, Menard C, Puig PE, Ladoire S, Roux S, Martin F et al (2007) Metronomic cyclophosphamide regimen selectively depletes CD4+ CD25+ regulatory T cells and restores T and NK effector functions in end stage cancer patients. Cancer Immunol Immunother 56:641–648

Gobert M, Treilleux I, Bendriss-Vermare N, Bachelot T, Goddard-Leon S, Arfi V et al (2009) Regulatory T cells recruited through CCL22/CCR4 are selectively activated in lymphoid infiltrates surrounding primary breast tumors and lead to an adverse clinical outcome. Cancer Res 69:2000–2009

Gray CP, Arosio P, Hersey P (2003) Association of increased levels of heavy-chain ferritin with increased CD4+ CD25+ regulatory T-cell levels in patients with melanoma. Clin Cancer Res 9:2551–2559

Grossman WJ, Verbsky JW, Barchet W, Colonna M, Atkinson JP, Ley TJ (2004) Human T regulatory cells can use the perforin pathway to cause autologous target cell death. Immunity 21:589–601

Hjelmeland MD, Hjelmeland AB, Sathornsumetee S, Reese ED, Herbstreith MH, Laping NJ et al (2004) SB-431542, a small molecule transforming growth factor-beta-receptor antagonist, inhibits human glioma cell line proliferation and motility. Mol Cancer Ther 3:737–745

Hori S, Nomura T, Sakaguchi S (2003) Control of regulatory T cell development by the transcription factor Foxp3. Science 299:1057–1061

Huber S, Schramm C (2006) TGF-beta and CD4+ CD25+ regulatory T cells. Front Biosci 11:1014–1023

Huehn J, Hamann A (2005) Homing to suppress: address codes for Treg migration. Trends Immunol 26:632–636

Hurwitz AA, Yu TF, Leach DR, Allison JP (1998) CTLA-4 blockade synergizes with tumor-derived granulocyte-macrophage colony-stimulating factor for treatment of an experimental mammary carcinoma. Proc Natl Acad Sci U S A 95:10067–10071

Hurwitz AA, Foster BA, Kwon ED, Truong T, Choi EM, Greenberg NM et al (2000) Combination immunotherapy of primary prostate cancer in a transgenic mouse model using CTLA-4 blockade. Cancer Res 60:2444–2448

Ichihara F, Kono K, Takahashi A, Kawaida H, Sugai H, Fujii H (2003) Increased populations of regulatory T cells in peripheral blood and tumor-infiltrating lymphocytes in patients with gastric and esophageal cancers. Clin Cancer Res 9:4404–4408

Ishida T, Iida S, Akatsuka Y, Ishii T, Miyazaki M, Komatsu H et al (2004) The CC chemokine receptor 4 as a novel specific molecular target for immunotherapy in adult T-Cell leukemia/lymphoma. Clin Cancer Res 10:7529–7539

Ishida T, Ishii T, Inagaki A, Yano H, Komatsu H, Iida S et al (2006) Specific recruitment of CC chemokine receptor 4-positive regulatory T cells in Hodgkin lymphoma fosters immune privilege. Cancer Res 66:5716–5722

Jacobs JF, Nierkens S, Figdor CG, de Vries IJ, Adema GJ (2012) Regulatory T cells in melanoma: the final hurdle towards effective immunotherapy? Lancet Oncol 13:e32–e42

Jain N, Nguyen H, Chambers C, Kang J (2010) Dual function of CTLA-4 in regulatory T cells and conventional T cells to prevent multiorgan autoimmunity. Proc Natl Acad Sci U S A 107:1524–1528

Jordan JT, Sun W, Hussain SF, DeAngulo G, Prabhu SS, Heimberger AB (2008) Preferential migration of regulatory T cells mediated by glioma-secreted chemokines can be blocked with chemotherapy. Cancer Immunol Immunother 57:123–131

Josefowicz SZ, Lu LF, Rudensky AY (2012) Regulatory T cells: mechanisms of differentiation and function. Annu Rev Immunol 30:531–564

Ju S, Qiu H, Zhou X, Zhu B, Lv X, Huang X et al (2009) CD13+ CD4+ CD25hi regulatory T cells exhibit higher suppressive function and increase with tumor stage in non-small cell lung cancer patients. Cell Cycle 8:2578–2585

Kawai M, Kitade H, Mathieu C, Waer M, Pirenne J (2005) Inhibitory and stimulatory effects of cyclosporine A on the development of regulatory T cells in vivo. Transplantation 79:1073–1077

Kitamura N, Murata S, Ueki T, Mekata E, Reilly RT, Jaffee EM et al (2009) OX40 costimulation can abrogate Foxp3 + regulatory T cell-mediated suppression of antitumor immunity. Int J Cancer 125:630–638

Kobie JJ, Wu RS, Kurt RA, Lou S, Adelman MK, Whitesell LJ et al (2003) Transforming growth factor beta inhibits the antigen-presenting functions and antitumor activity of dendritic cell vaccines. Cancer Res 63:1860–1864

Krambeck AE, Thompson RH, Dong H, Lohse CM, Park ES, Kuntz SM et al (2006) B7–H4 expression in renal cell carcinoma and tumor vasculature: associations with cancer progression and survival. Proc Natl Acad Sci U S A 103:10391–10396

Kryczek I, Lange A, Mottram P, Alvarez X, Cheng P, Hogan M et al (2005) CXCL12 and vascular endothelial growth factor synergistically induce neoangiogenesis in human ovarian cancers. Cancer Res 65:465–472

Kryczek I, Zou L, Rodriguez P, Zhu G, Wei S, Mottram P et al (2006) B7–H4 expression identifies a novel suppressive macrophage population in human ovarian carcinoma. J Exp Med 203:871–881

Kryczek I, Wei S, Zhu G, Myers L, Mottram P, Cheng P et al (2007) Relationship between B7–H4, regulatory T cells, and patient outcome in human ovarian carcinoma. Cancer Res 67:8900–8905

Kryczek I, Liu R, Wang G, Wu K, Shu X, Szeliga W et al (2009a) FOXP3 defines regulatory T cells in human tumor and autoimmune disease. Cancer Res 69:3995–4000

Kryczek I, Banerjee M, Cheng P, Vatan L, Szeliga W, Wei S et al (2009b) Phenotype, distribution, generation, and functional and clinical relevance of Th17 cells in the human tumor environments. Blood 114:1141–1149

Kwon ED, Hurwitz AA, Foster BA, Madias C, Feldhaus AL, Greenberg NM et al (1997) Manipulation of T cell costimulatory and inhibitory signals for immunotherapy of prostate cancer. Proc Natl Acad Sci U S A 94:8099–8103

Kwon ED, Foster BA, Hurwitz AA, Madias C, Allison JP, Greenberg NM et al (1999) Elimination of residual metastatic prostate cancer after surgery and adjunctive cytotoxic T lymphocyte-associated antigen 4 (CTLA-4) blockade immunotherapy. Proc Natl Acad Sci U S A 96:15074–15079

Le DT, Jaffee EM (2012) Regulatory T-cell modulation using cyclophosphamide in vaccine approaches: a current perspective. Cancer Res 72:3439–3444

Leach DR, Krummel MF, Allison JP (1996) Enhancement of antitumor immunity by CTLA-4 blockade. Science 271:1734–1736

Liu Y, Zhang P, Li J, Kulkarni AB, Perruche S, Chen W (2008) A critical function for TGF-beta signaling in the development of natural CD4+ CD25+ Foxp3+ regulatory T cells. Nat Immunol 9:632–640

Liyanage UK, Moore TT, Joo HG, Tanaka Y, Herrmann V, Doherty G et al (2002) Prevalence of regulatory T cells is increased in peripheral blood and tumor microenvironment of patients with pancreas or breast adenocarcinoma. J Immunol 169:2756–2761

Lu L, Cantor H (2008) Generation and regulation of CD8(+) regulatory T cells. Cell Mol Immunol 5:401–406

Lutsiak ME, Semnani RT, De Pascalis R, Kashmiri SV, Schlom J, Sabzevari H (2005) Inhibition of CD4(+)25+ T regulatory cell function implicated in enhanced immune response by low-dose cyclophosphamide. Blood 105:2862–2868

Mahic M, Henjum K, Yaqub S, Bjornbeth BA, Torgersen KM, Tasken K et al (2008) Generation of highly suppressive adaptive CD8(+)CD25(+)FOXP3(+) regulatory T cells by continuous antigen stimulation. Eur J Immunol 38:640–646

Mahnke K, Schonfeld K, Fondel S, Ring S, Karakhanova S, Wiedemeyer K et al (2007) Depletion of CD4+ CD25+ human regulatory T cells in vivo: kinetics of Treg depletion and alterations in immune functions in vivo and in vitro. Int J Cancer 120:2723–2733

Mailloux AW, Clark AM, Young MR (2010) NK depletion results in increased CCL22 secretion and Treg levels in Lewis lung carcinoma via the accumulation of CCL22-secreting CD11b+ CD11c+ cells. Int J Cancer 127:2598–2611

Maker AV, Attia P, Rosenberg SA (2005) Analysis of the cellular mechanism of antitumor responses and autoimmunity in patients treated with CTLA-4 blockade. J Immunol 175:7746–7754

Maloy KJ, Salaun L, Cahill R, Dougan G, Saunders NJ, Powrie F (2003) CD4+ CD25+ T(R) cells suppress innate immune pathology through cytokine-dependent mechanisms. J Exp Med 197:111–119

Mandapathil M, Hilldorfer B, Szczepanski MJ, Czystowska M, Szajnik M, Ren J et al (2010) Generation and accumulation of immunosuppressive adenosine by human CD4+ CD25high-FOXP3+ regulatory T cells. J Biol Chem 285:7176–7186

Marshall NA, Christie LE, Munro LR, Culligan DJ, Johnston PW, Barker RN et al (2004) Immunosuppressive regulatory T cells are abundant in the reactive lymphocytes of Hodgkin lymphoma. Blood 103:1755–1762

Matar P, Rozados VR, Gervasoni SI, Scharovsky OG (2001) Down regulation of T-cell-derived IL-10 production by low-dose cyclophosphamide treatment in tumor-bearing rats restores in vitro normal lymphoproliferative response. Int Immunopharmacol 1:307–319

Mellor AL, Munn DH (2004) IDO expression by dendritic cells: tolerance and tryptophan catabolism. Nat Rev Immunol 4:762–774

Menard C, Ghiringhelli F, Roux S, Chaput N, Mateus C, Grohmann U et al (2008) Ctla-4 blockade confers lymphocyte resistance to regulatory T-cells in advanced melanoma: surrogate marker of efficacy of tremelimumab? Clin Cancer Res 14:5242–5249

Menetrier-Caux C, Curiel T, Faget J, Manuel M, Caux C, Zou W (2012) Targeting regulatory T cells. Target Oncol 7:15–28

Mitsui J, Nishikawa H, Muraoka D, Wang L, Noguchi T, Sato E et al (2010) Two distinct mechanisms of augmented antitumor activity by modulation of immunostimulatory/inhibitory signals. Clin Cancer Res 16:2781–2791

Mougiakakos D, Choudhury A, Lladser A, Kiessling R, Johansson CC (2010) Regulatory T cells in cancer. Adv Cancer Res 107:57–117

Nocentini G, Riccardi C (2005) GITR: a multifaceted regulator of immunity belonging to the tumor necrosis factor receptor superfamily. Eur J Immunol 35:1016–1022

North RJ (1982) Cyclophosphamide-facilitated adoptive immunotherapy of an established tumor depends on elimination of tumor-induced suppressor T cells. J Exp Med 155:1063–1074

Okawaki M, Yamaguchi Y, Okita R, Ohara M, Okada M (2008) Dose-finding study of anti-CD25 antibody for targeting regulatory T cells in locoregional immunotherapy of malignant effusion. Hiroshima J Med Sci 57:37–46

Olkhanud PB, Baatar D, Bodogai M, Hakim F, Gress R, Anderson RL et al (2009) Breast cancer lung metastasis requires expression of chemokine receptor CCR4 and regulatory T cells. Cancer Res 69:5996–6004

Olsen E, Duvic M, Frankel A, Kim Y, Martin A, Vonderheid E et al (2001) Pivotal phase III trial of two dose levels of denileukin diftitox for the treatment of cutaneous T-cell lymphoma. J Clin Oncol 19:376–388

Onizuka S, Tawara I, Shimizu J, Sakaguchi S, Fujita T, Nakayama E (1999) Tumor rejection by in vivo administration of anti-CD25 (interleukin-2 receptor alpha) monoclonal antibody. Cancer Res 59:3128–3133

Pandiyan P, Zheng L, Ishihara S, Reed J, Lenardo MJ (2007) CD4+ CD25+ Foxp3+ regulatory T cells induce cytokine deprivation-mediated apoptosis of effector CD4+ T cells. Nat Immunol 8:1353–1362

Peggs KS, Quezada SA, Chambers CA, Korman AJ, Allison JP (2009) Blockade of CTLA-4 on both effector and regulatory T cell compartments contributes to the antitumor activity of anti-CTLA-4 antibodies. J Exp Med 206:1717–1725

Peng DJ, Liu R, Zou W (2012) Regulatory T cells in human ovarian cancer. J Oncol 2012:345164

Phan GQ, Yang JC, Sherry RM, Hwu P, Topalian SL, Schwartzentruber DJ et al (2003) Cancer regression and autoimmunity induced by cytotoxic T lymphocyte-associated antigen 4 blockade in patients with metastatic melanoma. Proc Natl Acad Sci U S A 100:8372–8377

Piconese S, Valzasina B, Colombo MP (2008) OX40 triggering blocks suppression by regulatory T cells and facilitates tumor rejection. J Exp Med 205:825–839

Rech AJ, Vonderheide RH (2009) Clinical use of anti-CD25 antibody daclizumab to enhance immune responses to tumor antigen vaccination by targeting regulatory T cells. Ann N Y Acad Sci 1174:99–106

Rico MJ, Rozados VR, Mainetti LE, Zacarias Fluck MF, Matar P, Scharovsky OG (2012) Regulatory T cells but not NKT I cells are modulated by a single low-dose cyclophosphamide in a B cell lymphoma tumor-model. Exp Oncol 34:38–42

Rollinghoff M, Starzinski-Powitz A, Pfizenmaier K, Wagner H (1977) Cyclophosphamide-sensitive T lymphocytes suppress the in vivo generation of antigen-specific cytotoxic T lymphocytes. J Exp Med 145:455–459

Sakaguchi S (2005) Naturally arising Foxp3-expressing CD25+ CD4+ regulatory T cells in immunological tolerance to self and non-self. Nat Immunol 6:345–352

Sakaguchi S, Sakaguchi N, Asano M, Itoh M, Toda M (1995) Immunologic self-tolerance maintained by activated T cells expressing IL-2 receptor alpha-chains (CD25). Breakdown of a single mechanism of self-tolerance causes various autoimmune diseases. J Immunol 155:1151–1164

Schlingensiepen KH, Fischer-Blass B, Schmaus S, Ludwig S (2008) Antisense therapeutics for tumor treatment: the TGF-beta2 inhibitor AP 12009 in clinical development against malignant tumors. Recent Results Cancer Res 177:137–150

Schott AK, Pries R, Wollenberg B (2010) Permanent up-regulation of regulatory T-lymphocytes in patients with head and neck cancer. Int J Mol Med 26:67–75

Sharabi A, Ghera NH (2010) Breaking tolerance in a mouse model of multiple myeloma by chemoimmunotherapy. Adv Cancer Res 107:1–37

Shevach EM (2009) Mechanisms of foxp3+ T regulatory cell-mediated suppression. Immunity 30:636–645

Shevach EM, Stephens GL (2006) The GITR-GITRL interaction: co-stimulation or contrasuppression of regulatory activity? Nat Rev Immunol 6:613–618
Shibutani S, Inoue F, Aramaki O, Akiyama Y, Matsumoto K, Shimazu M et al (2005) Effects of immunosuppressants on induction of regulatory cells after intratracheal delivery of alloantigen. Transplantation 79:904–913
Shimizu J, Yamazaki S, Takahashi T, Ishida Y, Sakaguchi S (2002) Stimulation of CD25(+)CD4(+) regulatory T cells through GITR breaks immunological self-tolerance. Nat Immunol 3:135–142
Somasundaram R, Jacob L, Swoboda R, Caputo L, Song H, Basak S et al (2002) Inhibition of cytolytic T lymphocyte proliferation by autologous CD4+/CD25+ regulatory T cells in a colorectal carcinoma patient is mediated by transforming growth factor-beta. Cancer Res 62:5267–5272
Strauss L, Bergmann C, Szczepanski MJ, Lang S, Kirkwood JM, Whiteside TL (2008) Expression of ICOS on human melanoma-infiltrating CD4+ CD25highFoxp3+ T regulatory cells: implications and impact on tumor-mediated immune suppression. J Immunol 180:2967–2980
Strauss L, Bergmann C, Whiteside TL (2009) Human circulating CD4+ CD25highFoxp3+ regulatory T cells kill autologous CD8+ but not CD4+ responder cells by Fas-mediated apoptosis. J Immunol 182:1469–1480
Sutmuller RP, van Duivenvoorde LM, van Elsas A, Schumacher TN, Wildenberg ME, Allison JP et al (2001) Synergism of cytotoxic T lymphocyte-associated antigen 4 blockade and depletion of CD25(+) regulatory T cells in antitumor therapy reveals alternative pathways for suppression of autoreactive cytotoxic T lymphocyte responses. J Exp Med 194:823–832
Taieb J, Chaput N, Schartz N, Roux S, Novault S, Menard C et al (2006) Chemoimmunotherapy of tumors: cyclophosphamide synergizes with exosome based vaccines. J Immunol 176:2722–2729
Takahashi T, Kuniyasu Y, Toda M, Sakaguchi N, Itoh M, Iwata M et al (1998) Immunologic self-tolerance maintained by CD25+ CD4+ naturally anergic and suppressive T cells: induction of autoimmune disease by breaking their anergic/suppressive state. Int Immunol 10:1969–1980
Tan MC, Goedegebuure PS, Belt BA, Flaherty B, Sankpal N, Gillanders WE et al (2009) Disruption of CCR5-dependent homing of regulatory T cells inhibits tumor growth in a murine model of pancreatic cancer. J Immunol 182:1746–1755
Tang Q, Boden EK, Henriksen KJ, Bour-Jordan H, Bi M, Bluestone JA (2004) Distinct roles of CTLA-4 and TGF-beta in CD4+ CD25+ regulatory T cell function. Eur J Immunol 34:2996–3005
Topalian SL, Hodi FS, Brahmer JR, Gettinger SN, Smith DC, McDermott DF et al (2012) Safety, activity, and immune correlates of anti-PD-1 antibody in cancer. N Engl J Med 366:2443–2454
Turk MJ, Guevara-Patino JA, Rizzuto GA, Engelhorn ME, Sakaguchi S, Houghton AN (2004) Concomitant tumor immunity to a poorly immunogenic melanoma is prevented by regulatory T cells. J Exp Med 200:771–782
van Elsas A, Hurwitz AA, Allison JP (1999) Combination immunotherapy of B16 melanoma using anti-cytotoxic T lymphocyte-associated antigen 4 (CTLA-4) and granulocyte/macrophage colony-stimulating factor (GM-CSF)-producing vaccines induces rejection of subcutaneous and metastatic tumors accompanied by autoimmune depigmentation. J Exp Med 190:355–366
Wang YM, Alexander SI (2009) CD8 regulatory T cells: what's old is now new. Immunol Cell Biol 87:192–193
Wang J, Ke XY (2011) The four types of Tregs in malignant lymphomas. J Hematol Oncol 4:50
Wang L, Pino-Lagos K, de Vries VC, Guleria I, Sayegh MH, Noelle RJ (2008) Programmed death 1 ligand signaling regulates the generation of adaptive Foxp3+ CD4+ regulatory T cells. Proc Natl Acad Sci U S A 105:9331–9336

Wang W, Lau R, Yu D, Zhu W, Korman A, Weber J (2009) PD1 blockade reverses the suppression of melanoma antigen-specific CTL by CD4+ CD25(Hi) regulatory T cells. Int Immunol 21:1065–1077

Wang W, Hodkinson P, McLaren F, Mackinnon A, Wallace W, Howie S et al (2012) Small cell lung cancer tumour cells induce regulatory T lymphocytes, and patient survival correlates negatively with FOXP3(+) cells in tumour infiltrate. Int J Cancer 131:E928–E937

Weber J (2007) Review: anti-CTLA-4 antibody ipilimumab: case studies of clinical response and immune-related adverse events. Oncologist 12:864–872

Wei S, Kryczek I, Zou W (2006) Regulatory T-cell compartmentalization and trafficking. Blood 108:426–431

Whiteside TL, Mandapathil M, Schuler P (2011) The role of the adenosinergic pathway in immunosuppression mediated by human regulatory T cells (Treg). Curr Med Chem 18:5217–5223

Wilke CM, Wu K, Zhao E, Wang G, Zou W (2010) Prognostic significance of regulatory T cells in tumor. Int J Cancer 127:748–758

Wolf AM, Wolf D, Steurer M, Gastl G, Gunsilius E, Grubeck-Loebenstein B (2003) Increase of regulatory T cells in the peripheral blood of cancer patients. Clin Cancer Res 9:606–612

Won J, Kim H, Park EJ, Hong Y, Kim SJ, Yun Y (1999) Tumorigenicity of mouse thymoma is suppressed by soluble type II transforming growth factor beta receptor therapy. Cancer Res 59:1273–1277

Woo EY, Chu CS, Goletz TJ, Schlienger K, Yeh H, Coukos G et al (2001a) Regulatory CD4(+)CD25(+) T cells in tumors from patients with early-stage non-small cell lung cancer and late-stage ovarian cancer. Cancer Res 61:4766–4772

Woo EY, Chu CS, Goletz TJ, Schlienger K, Yeh H, Coukos G et al (2001b) Regulatory CD4(+)CD25(+) T cells in tumors from patients with early- stage non-small cell lung cancer and late-stage ovarian cancer. Cancer Res 61:4766–4772

Yamamoto K, Utsunomiya A, Tobinai K, Tsukasaki K, Uike N, Uozumi K et al (2010) Phase I study of KW-0761, a defucosylated humanized anti-CCR4 antibody, in relapsed patients with adult T-cell leukemia-lymphoma and peripheral T-cell lymphoma. J Clin Oncol 28:1591–1598

Ye ZJ, Zhou Q, Zhang JC, Li X, Wu C, Qin SM et al (2011) CD39+ regulatory T cells suppress generation and differentiation of Th17 cells in human malignant pleural effusion via a LAP-dependent mechanism. Respir Res 12:77

Yokokawa J, Cereda V, Remondo C, Gulley JL, Arlen PM, Schlom J et al (2008) Enhanced functionality of CD4+ CD25(high)FoxP3+ regulatory T cells in the peripheral blood of patients with prostate cancer. Clin Cancer Res 14:1032–1040

Zabransky DJ, Nirschl CJ, Durham NM, Park BV, Ceccato CM, Bruno TC et al (2012) Phenotypic and functional properties of Helios + regulatory T cells. PLoS ONE 7:e34547

Zhao E, Wang L, Dai J, Kryczek I, Wei S, Vatan L et al (2012) Regulatory T cells in the bone marrow microenvironment in patients with prostate cancer. Oncoimmunology 1:152–161

Zhou G, Levitsky HI (2007) Natural regulatory T cells and de novo-induced regulatory T cells contribute independently to tumor-specific tolerance. J Immunol 178:2155–2162

Zou W (2006) Regulatory T cells, tumour immunity and immunotherapy. Nat Rev Immunol 6:295–307

Zou W, Lackner AA, Simon M, Durand-Gasselin I, Galanaud P, Desrosiers RC et al (1997) Early cytokine and chemokine gene expression in lymph nodes of macaques infected with simian immunodeficiency virus is predictive of disease outcome and vaccine efficacy. J Virol 71:1227–1236

Zou L, Barnett B, Safah H, Larussa VF, Evdemon-Hogan M, Mottram P et al (2004) Bone marrow is a reservoir for CD4+ CD25+ regulatory T cells that traffic through CXCL12/CXCR4 signals. Cancer Res 64:8451–8455

Chapter 26
ChemoImmunoModulation: Focus on Myeloid Regulatory Cells

Michael R. Shurin and Viktor Umansky

Abstract Cancer is the second leading cause of death in the United States and Europe: One in two men and women will be diagnosed during their lifetime, and a large proportion of them will require chemotherapy. Chemotherapy-induced immunosuppression can result in reactivation or appearance of new viral, bacterial or fungal infection, causing severe morbidity and mortality. On the other side, a key feature of the clinical course of many types of cancer is an induction of immunosuppression, leading to increased susceptibility to infections and failure of anti-tumor immune responses. Although cytotoxic chemotherapy still forms the mainstream of most current treatment regimens, it is not curative, and its lack of specificity means that it also targets normal immune cells, exacerbating this immunosuppression. This can result in the limitation of the treatment effeciency by infectious complications, particularly in the elderly who comprise the majority of patients with cancer. The adjuvant/neoadjuvant use of chemotherapeutic agents in low-dose and ultra low-dose regimen potentially offers a way out of this dilemma, due to its low toxicity and ability to enhance immune responses to the tumor antigens. There has been a dramatic increase in the range of available low-dose chemotherapeutic options over the past decade, and many are now in the process

M. R. Shurin
Departments of Pathology and Immunology, University of Pittsburgh Medical Center, University of Pittsburgh Cancer Institute, Pittsburgh, PA, USA

V. Umansky
Skin Cancer Unit, German Cancer Research Center, Heidelberg, Germany

V. Umansky
Department of Dermatology, Venereology and Allergology, University Medical Center Mannheim, Ruprecht-Karl University of Heidelberg, Mannheim, Heidelberg, Germany

V. Umansky (✉)
Clinical Cooperation Unit Dermato-Oncology (G300), German Cancer Research Center, Im Neuenheimer Feld 280, 69120 Heidelberg, Germany
e-mail: V.Umansky@dkfz-heidelberg.de

M. R. Shurin et al. (eds.), *The Tumor Immunoenvironment*,
DOI: 10.1007/978-94-007-6217-6_26,

of making the transition to the clinic. However, the immunostimulatory and antitumor properties of ultra low noncytotoxic doses of certain chemotherapeutic drugs remain to be confirmed and accepted by clinicians before this new neoadjuvant approach termed chemoimmunomodulation can find its way to current clinical practice.

Keywords Chemotherapy · Chemomodulation · Chemoimmunomodulation · Immunosuppression · Chronic inflammatory factory · Dendritic cells · Myeloid-derived suppressor cells

26.1 Introduction: Cancer Chemotherapy

Despite the fact that the modern era of cancer chemotherapy began following the World War II, the origins of cancer treatment are recorded in ancient documents.The medicinal utilization of different chemical substances can be dated back to the ancient Indian system of traditional medicine called Ayurveda (the 'science of life') that recommended various herbs and metals for patients with different diseases. The Ebers papyrus, the Edwin Smith papyrus and the Ramayana describe malignant diseases and their treatments. Dioscorides, in the first century A.D., compiled a listing of medicinal herbs and botanicals, including topical applications for treatment of tumors and carcinomata. Later, in the 9–10th century, Persian physician and alchemist, Muhammad ibn Zakariya al-Razi, known as Rhazes or Rasis, introduced sulfate, mercuric and arsenic salts, gold, copper, chalk, clay, coral, pearl and other natural agents for medical purposes. In the 11th century an Arabic physician, Ibn Sina, known in the West as Avicenna, used the arsenical therapy systemically, although it was found to be dangerous and received a little attention. Arsenical preparations known as *Unguentum Aegypticum* were used topically until the 16th century. Many consider the use of potassium arsenite to treat chronic myelogenous leukemia in 1865 by Lissauer as the first instance of effective chemotherapy for malignant disease (Timkin 1942; Papac 2001; DeVita and Chu 2008; Shurin et al. 2012).

The origins of effective chemotherapy for cancer date to World War I when mustard gas (sulfur mustard) was used. The blood and bone marrow findings in cases of mustard gas poisoning were described by Krumbhaar, a captain in the US Medical Corps, noted the development of profound leukopenia in those individuals who survived for several days (Krumbhaar 1919). This probably was the very first suggestion that mustard gas might have the effect on the immune cells and therapeutic possibilities for treating lymphomas (Hirsch 2006). The advent of World War II stimulated further research on chemical warfare. A series of analogues of sulfur mustards were produced as potential offensive agents. It was recognized that β-chlorethyl amines could exert cytotoxic actions on a variety of tissues, particularly related to the degree of their proliferative activity. Since the data were classified during the wartime, these findings were reported following World War II by Louis S. Goodman and Alfred Gilman from the Pharmacology Section, Medical Division of the Chemical Warfare Service of the US Army (Goodman et al. 1946).

Interestingly, major textbooks of cancer medicine have attributed the initial use of nitrogen mustard to findings of exposure to mustard gas following an explosion that occurred in the harbor of Bari, Italy, where a very few survivors developed a profound lymphoid and myeloid suppression (Hirsch 2006). The Bari attack occurred in December 1943, but according to several references, the initial clinical trial occurred at Yale in May 1942 (Papac 2001; Goodman et al. 1946; Gilman 1963). 66 patients with Hodgkin's disease, lymphosarcoma, leukemia or other types of cancer were treated with either a nitrogen mustard called tris(β-chloroethyl)amine hydrochloride (HN3) or another nitrogen mustard, methyl-bis(β-chloroethyl)amine hydrochloride (HN2) in the early 1940s. The most important observation was that patients with Hodgkin's disease often showed a prominent response to the nitrogen mustards (Joensuu 2008). Although this effect lasted only a few weeks, this was the first step to realization that cancer could be treated by pharmacological agents (Goodman et al. 1946; Goodman et al. 1984).

Following the introduction of nitrogen mustard into clinical practice, other types of alkylating agents were developed many of which are in the clinical practice today, most notably chlorambucil, melphalan, busufan and cyclophosphamide (Papac 2001). Interestingly, the clinical spectrum of their effectiveness is minimally changed from the initial report, although the toxicities differ. The use of an antifolate compound was reported by Sidney Farber et al. (1948), initiating the development of antimetabolite therapy. The authors used folate analogues aminopterin and amethopterin (now methotrexate) that are antagonists of folic acid, and reported in 1948 that when administered to children with acute lymphoblastic leukemia (ALL), these agents induced remission of the disease. This was followed by the use of purine and pyrimidine analogs. Subsequently, anti-tumor antibiotics, platinum compounds, imidizole compounds, vinca rosea alkaloids, taxols, camptothecin analogs and biologic agents have become a part of the therapeutic schedule for neoplastic diseases (Papac 2001; DeVita and Chu 2008).

26.2 Cancer Chemotherapy and Immunosuppression

Immunosuppression in cancer patients is a common clinical condition that is induced by the cancer itself and high dose chemotherapeutic agents. Many commonly used chemotherapeutic regimens decrease white blood cell counts in addition to their intended effects (Wijayahadi et al. 2007; Rasmussen and Arvin 1982). Many chemotherapeutic agents are myelosuppressive by inhibiting the bone marrow production of blood cells and resulting in a diminished absolute blood count. The result of myelosuppression is evident during the chemotherapy-induced neutropenia (Kim and Demetri 1996). Immune toxicity, in particular neutropenia, can lead to a considerable morbidity and mortality by predisposing patients with cancer to life-threatening bacterial infections and sepsis. Bacterial infection is one of the most common consequences of neutropenia in cancer patients and is directly related to the degree and duration of neutropenia: the more intensive the chemotherapy regimen, the greater the neutropenia and the higher the risk of infection with accompanying fever and potential sepsis (Mackall 2000; Steele 2002).

In addition to myelosuppression, chemotherapy-induced immunosuppression (Goldman 2000) is commonly detected in treated patients and reflects a reduced ability to mount an immune response due to either decreased lymphocyte numbers or the absence of antigen-specific lymphocyte populations (Steele 2002). For instance, treatment of cancer patients with 2-chlorodeoxyadenosine resulted in severe depletion of both $CD4^+$ and $CD8^+$ T cells (Dmoszynska et al. 1999). While $CD8^+$ T cells recover reasonably fast, $CD4^+$ T cells may recover to only 35 % of pre-therapy levels at three months post-chemotherapy (Mackall et al. 1997). However, rapid up-regulation of the number of $CD8^+$ T cell following chemotherapy does not reflect recovery of their functionality, since low expression of CD28 molecules may be observed during the first year after the cessation of chemotherapy (Mackall 2000). B lymphocytes are also sensitive to dose-intensive chemotherapy, which results in the suppression of immunoglobulin levels: almost 90 % of treated patients were reported to have low serum antibody concentrations following chemotherapy for at least one year post-chemotherapy (Mackall 2000; Mustafa et al. 1998). NK cell numbers are also decreased following chemotherapy, but rebound in numbers and cytotoxic activity relatively quickly as compared to other lymphocyte populations (Alanko et al. 1995). Based on the established role for NK cells in anti-infectious and anti-tumor responses and in regulating an induction of adaptive immunity, the chemotherapy-induced depletion of NK cells is an important component of the general immunosuppression in cancer therapy. Similarly, depletion of NKT and $\gamma\delta$T cells can also be involved in diminished immunity induced by high dose chemotherapy (Zocchi and Poggi 2004).

Chemotherapeutic agents affect immune cells by different mechanisms. For instance, paclitaxel, docetaxel and doxorubicin have been shown to achieve their immunosuppressive effects through inhibition of the function or proliferation of immune cells (Ferrari et al. 2003; Nakashima et al. 2005; Ferraro et al. 2000). Human NK cells showed a dose-dependent decrease in cytotoxicity against tumor cell lines upon an incubation with clinically relevant concentrations of paclitaxel, as did peripheral blood mononuclear cells (PBMC), MHC-unrestricted T cells and cytotoxic T cells (CTL) (Chuang et al. 1993; Chuang et al. 1994). In contrast to microtubule stabilizing drugs, microtubule destabilizing drugs, such as colchicines and vinblastine, in addition to inducing apoptosis in dividing cells, functionally repress CTL by inhibiting microtubule-organizing centre translocation (Kupfer and Dennert 1984) with the subsequent killing of target cells (Wolberg et al. 1984). Colchicine also reduces the stability of the immunological synapse (Bunnell et al. 2001) and the continuous recycling of T cell receptors at the immunological synapse (Das et al. 2004). Anthracyclines doxorubicin and daunorubicin could induce apoptosis in both mitogen-activated and non-activated human peripheral blood lymphocytes and induce a marked depletion of B and T cells after the injection in mice (Ferraro et al. 2000). Thus, conventional high-dose chemotherapy is often associated with a profound inhibition of immune cell generation and activity, leading to an acute or prolonged immunosuppression.

26.3 Cancer Chemotherapy and Immunostimulation

Several recent studies demonstrated that certain chemotherapeutic regimens were able to increase the efficacy of anti-cancer immunotherapy (Nowak et al. 2006; Lake and Robinson 2005), allowing to speculate that the effect of chemotherapy on the immune response depends on the dose of chemotherapeutic agents, their mechanisms of action on immune cells, the type of malignancy and the level of tumor-induced immunosuppression or tolerance (Shurin et al. 2012). For instance, cyclophosphamide could enhance vaccine-induced tumor-specific antibody responses in patients with breast cancer, while in higher doses it suppressed immunity (Emens et al. 2009). Analysis of immune alterations in breast cancer patients undergoing chemotherapy indicated that patients responded to the treatment with paclitaxel or docetaxel dysplayed an increase in serum levels of IFN-γ, IL-2, IL-6 and GM-CSF as well as an enhancement of PBMC, NK and LAK cell activity (Tsavaris et al. 2002). Experimental data revealed that the doxorubicin treatment could enhance tumor antigen-specific proliferation of CD8 T cells in tumor-draining lymph nodes and promote tumor infiltration by activated IFN-γ-producing CD8$^+$ T cells (Mattarollo et al. 2011). Furthermore, macrophages isolated from doxorubicin or cyclophosphamide treated mice displayed increased cytotoxicity for tumor cells (Stoychkov et al. 1979), confirming data on enhanced activity and tumoricidal potential of macrophage and NK cells treated with doxorubicin and cyclophosphamide in vitro (Mihich and Ehrke 2000; Ewens et al. 2006; Ujhazy et al. 2003).

Tumor cell death induced by antineoplastic chemotherapy might be accompanied by the release of "danger" signals assisting engulfment of tumor antigens (Nowak et al. 2003). For instance, chemotherapy with paclitaxel up-regulated CD8$^+$ T cell function, which was attributed to drug-induced apoptosis of malignant cells and accessibility of tumor antigens (Emens and Jaffee 2005; Coleman et al. 2005). Increased immunogenicity of tumor cells due to their apoptosis was also reported for doxorubicin, 5-fluorouracil and gemcitabine (Ehrke et al. 2000; Casares et al. 2005). Anthracyclins, for instance, have been shown to induce the rapid translocation of calreticulin to the surface of dying cells which promoted phagocytosis of cancer cells by DC (Obeid et al. 2007). HMGB 1 alarmin protein secreted by dying tumor cells interacts with TLR4 on DC to cause immunogenic cross-presentation (Apetoh et al. 2007). Also, many chemotherapeutic agents induce the production of uric acid caused by tumor cell death. Uric acid, at least supposedly in crystal form, may activate DC (Melief 2008). Thus, the augmented efficacy of successful combinations of conventional and moderate-dose cytotoxic chemotherapy and DC vaccines was in part due to the accessibility of antigens, expression of calreticulin and release of alarmin and uric acid from dying tumor cells; i.e., this success relies on the cytotoxic effect of chemotherapy on tumor cells (Melief 2008; Liu et al. 2007; Yu et al. 2003).

Interestingly, certain chemotherapeutic agents such as cisplatin, doxorubicin, mitomycin C, fluorouracil and camptothecin, which exert direct cytotoxic effects, also exhibit an indirect cytotoxic activity by "preparing" tumor cells to their distraction by NK cells or cytotoxic T lymphocytes using Fas or TRAIL-dependent pathways (Lacour et al. 2001; Micheau et al. 1997). For example, human colon carcinoma cell lines treated with 5-fluorouracil acquired CD95 and ICAM1 expressions and became more sensitive to lysis by CTL (Tanaka et al. 2002).

Chemotherapy-induced lymphodepletion can also be beneficial because it may eliminate immunosuppressive regulatory T cells (Treg) and poorly functional anti-tumor T cells (Nowak et al. 2006). In the late 60s, a few reports demonstrated that cyclophosphamide might improve the anti-tumor activity of allogeneic tumor-specific splenocytes and prolong the animal survival in leukemia models (Glynn et al. 1969; Fefer 1969). The authors speculated that this effect was due to the immunosuppression induced by cyclophosphamide in the allogeneic system. At the same time, Mihich et al. (1969) demonstrated that other chemotherapeutic agents might up-regulate the formation of anti-tumor lymphocytes and synergized in the anti-tumor activity with serum or spleen cells obtained from mice pretreated with irradiated tumor cells. These findings were confirmed when it was reported that cyclophosphamide in combination with tumor-specific lymphocytes prolonged animal survival in different models of murine leukemia (Vadlamudi et al. 1971; Fass and Fefer 1972; Berenson et al. 1975). Data showing that tumor-bearing animals have T lymphocytes that can inhibit the immune response to tumor antigens (Fujimoto et al. 1976; Takei et al. 1977) allowed hypothesizing that cyclophosphamide might eliminate tumor-induced T suppressor cells and thus potentiate the effect of immunotherapy (North 1982; Mastrangelo et al. 1986), which has been eventually proven in experimental setting (Liu et al. 2007; Lutsiak et al. 2005; Ghiringhelli et al. 2004) and clinical trials (Berd et al. 1986; Berd and Mastrangelo 1988). New data revealed that low-dose cyclophosphamide rendered Treg cells susceptible to apoptosis by down-regulating Bcl-xL and CTLA-4 in these cells and decreased production of IL-2 by effector CD4 T cells (Sharabi et al. 2010).

In addition to elimination and suppression of Treg cells in the tumor environment, chemotherapeutic agents may also affect other subpopulations of immune regulators. Gemcitabine, an antimetabolite agent, has been reported to deplete myeloid-derived suppressor cells (MDSC) (Suzuki et al. 2005; Ko et al. 2007). Docetaxel can polarize MDSC toward the macrophage M1-like phenotype, resulting in the expression of CCR7 (M1 marker) in 40 % of MDSC in vivo and in vitro (Kodumudi et al. 2010). Therefore, alleviation of Treg- and MDSC-mediated immunosuppression and elicitation of immunogenic tumor cell death are involved in known immunopotentiating activity of several chemotherapeutic regimens (Chan and Yang 2000; Javeed et al. 2009). In this context, the administration of selected chemotherapy agents before immunotherapy can have multiple effects, eliminating both immune suppressors and regulators, as well as facilitating tumor antigen cross-presentation (Liao et al. 2007; Emens 2006; Sinkovics and Horvath 2006). Additional mechanisms through which conventional chemotherapeutics may affect the immune system have been recently reviewed (Galluzzi et al. 2012).

26.4 Chemoimmunomodulation: The Effect of Extra low Noncytotoxic Noncytostatic Doses of Chemotherapeutic Agents on the Immune System

Recently a new approach to immune cell modulation by chemotherapeutic agents has been introduced and termed 'chemomodulation' or 'chemoimmunomodulation', indicating that this approach aims primarily at the alteration of the tumor immunoenvironment rather than at direct elimination of malignant cells. It was based on new findings demonstrated that different chemotherapeutic agents displayed an unpredicted ability to regulate signal transduction pathways in immune cells without inhibiting cell cycle or inducing cell death when used in ultra low concentrations. For instance, it has been shown that several chemotherapeutic drugs from different groups in low nanomolar concentrations were able to regulate activity of classical and non-classical members of the small Rho GTPase family of proteins, including Rac, RhoA, Cdc42 and RhoE in DC (Shurin et al. 2008). These results suggested for the first time that different classes of chemotherapeutic drugs in ultra low concentrations may directly activate signaling pathways in immune cells and alter their function or differentiation without inducing their apoptosis. This possibility has been tested both in vitro and in vivo by several groups.

For example, it has been reported that several chemotherapeutic drugs in low nontoxic concentrations affect DC maturation in vitro. The authors treated murine bone marrow-derived DC with different agents and revealed that paclitaxel, methotrexate and doxorubicin in ultra low noncytotoxic noncytostatic concentrations significantly up-regulated expression of phenotypic markers of DC maturation (such as CD80, CD86, CD40 or MHC class II), while vinblastin, vincristin, 5-azacytidine and bleomycin did not significantly alter the expression of these molecules on DC (Shurin et al. 2009a). Furthermore, the authors found that doxorubicin significantly increased both spontaneous and induced migration of DC toward CCL3 (MIP-1α) and CCL19 (MIP-3β), and that paclitaxel, methotrexate and doxorubicin increased antigen presentation by DC. Interestingly, all these chemotherapeutics appeared to be the strongest inducers of IL-12 expression and marked inhibitors of IL-10 production by DC (Shurin et al. 2009a, b).

Analysis of chemomodulation phenomenon in human DC prepared from CD14$^+$ monocytes obtained from PBMC of healthy volunteers revealed that several agents (including paclitaxel, doxorubicin and methotrexate) in noncytotoxic concentrations up-regulated phenotypic maturation of DC and stimulate APC function of human DC (Kaneno et al. 2009). It has been also shown that paclitaxel, doxorubicin and methotrexate protected APC function of human DC from inhibition induced by primary HNSCC cells in an allogeneic MLR assay. Furthermore, the authors demonstrated that human DC loaded with tumor lysates prepared from paclitaxel-pretreated colon carcinoma cells, induced CTL with significantly higher cytotoxic activity against tumors than DC loaded with lysates from control (nontreated) tumor cells (Kaneno et al. 2011). Interestingly, CTL induced by DC loaded with nontreated tumor cells displayed significantly higher cytotoxicity against paclitaxel-pretreated tumor cells than against nontreated tumor cells. Altogether, these results suggest that the presence of chemotherapeutic agents in nontoxic concentrations in the tumor microenvironment may augment the development of DC-mediated antitumor immune responses in vivo (Shurin et al. 2012).

Our new data revealed another interesting phenomenon associated with chemoimmunomodulation in the tumor microenvironment. We and others have reported that the development of lung cancer in mice is associated with a rapid accumulation of regulatory DC (regDC) in the lung and lymphoid tissues (Liu et al. 2009; Shurin et al. 2011). Using this model, and several in vitro and in vivo approaches, we demonstrated that paclitaxel in ultra-low noncytotoxic doses abrogated the polarization of cDC into immunosuppressive protumorigenic regDC through the small Rho GTPase signaling and increased the anti-tumor potential of DC vaccine in lung cancer-bearing animals (Zhong et al., submitted). These findings offer novel insights into the therapeutic efficacy and immunomodulatory capacity of chemotherapeutic agents used in noncytotoxic doses.

Besides DC, noncytotoxic paclitaxel application was also studied with regards to its effects on the major myeloid cell subset with immunosuppressive functions, myeloid-derived suppressor cells. This extremely heterogeneous population of immature myeloid cells representing precursors of granulocytes, macrophages, and DC (Gabrilovich et al. 2012; Condamine and Gabrilovich 2011; Peranzoni et al. 2010) has been described to inhibit anti-tumor T cell reactions via different mechanisms including an increased production of nitric oxide (NO) and reactive oxygen species, as well as an enhanced arginase (ARG)-1 activity (Gabrilovich et al. 2012; Condamine and Gabrilovich 2011; Ostrand-Rosenberg 2010; Marigo et al. 2008; Raber et al. 2012). The administration of paclitaxel in noncytotoxic doses into C57BL/6 mice resulted in the decrease in the frequencies of $Gr1^{+}CD11b^{+}$ immature myeloid cells (IMC) in the spleen and lymph nodes (Sevko et al. 2012). This subset is considered as an MDSC counterpart in healthy mice (Gabrilovich et al. 2012; Gabrilovich and Nagaraj 2009). The observed changes might be due to an augmented differentiation of IMC in the bone marrow into mature granulocytes, macrophages and DC (Gabrilovich and Nagaraj 2009; Naiditch et al. 2011), resulting in reduced migration of IMC to the secondary lymphoid tissue. Interestingly, the levels of NO production and ARG-1 expression in $Gr1^{+}CD11b^{+}$ IMC in treated mice were not significantly different from those in untreated animals indicating that, in the absence of a growing tumor, IMC are not able to exert the immunosuppressive activity (Youn and Gabrilovich 2010).

To investigate the effect of noncytotoxic paclitaxel in tumor-bearing mice, we applied *ret* transgenic mouse model that, in contrast to transplantation models, closely resemble human skin malignant melanoma with regard to etiology, genetic alterations, histopathology and clinical development (Kato et al. 1998; Umansky et al. 2008; Lengagne et al. 2011). In this model, mouse melanocytes express the human *ret* transgene controlled by the mouse metallothionein I promoter-enhancer (Kato et al. 1998). A constant activation of Ret kinase, a member of the receptor tyrosine kinase family (Eng 1999) resulted in the stimulation of other downstream kinases leading to the spontaneous development of skin melanoma with metastases in lymph nodes, lungs, liver, brain, and the bone marrow (Kato et al. 1998; Umansky et al. 2008; Lengagne et al. 2011). Such metastatic profile shows a high similarity to metastasis observed in malignant melanoma patients (Houghton and Polsky 2002). It has been demonstrated by us and others that the melanoma progression in these transgenic mice was characterized by a strong accumulation of highly immunosuppressive MDSC and immature, non-functional DC in skin tumors, metastatic lymph nodes, spleen and bone marrow (Lengagne et al. 2011; Meyer et al. 2011; Zhao et al. 2009).

Upon the paclitaxel administration in tumor-bearing *ret* transgenic mice, we detected a significant reduction of MDSC frequencies in melanoma lesions (Sevko et al., 2013, in press). Furthermore, we revealed a reduction in frequencies of MDSC producing NO. Importantly, tumor-infiltrating MDSC also displayed a significantly lower ability to suppress the proliferation of stimulated normal T cells indicating their decreased immunosuppressive function after paclitaxel application.

Analysis of signaling pathways in MDSC affected by the paclitaxel-based chemoimmunomodulation revealed a strong down-regulation of the phosphorylation of p38 MAPK in these cells located in skin tumors, the bone marrow and spleen as compared to the untreated group (Sevko et al. 2013, in press). Signaling through p38 MAPK in myeloid cells was previously reported to induce a reduction of DC immunostimulatory capacities in *ret* transgenic melanoma-bearing mice (Zhao et al. 2009) and to stimulate the production of chronic inflammatory mediators TNF-α, IL-1β, IL-10, TGF-β and IL-6 (Perdiguero et al. 2011; Kim et al. 2008; Ho et al. 2004; Baldassare et al. 1999; Ajibade et al. 2012). Indeed, we have found that a down-regulation of the p38 MAPK signaling in MDSC was strongly associated with reduction of TNF-α production by these cells and decreased levels of the abovementioned mediators of chronic inflammation (Sevko et al. 2013, in press). An autocrine regulation of p38 MAPK activation could also be at work since TNF-α and IL-1β were found to activate p38 MAPK signaling (Zhou et al. 2006; Gao and Bing 2011), stimulating thereby the IL-6 production, an activation of inducible NO synthase (Chae et al. 2001; Chen et al. 1999) and the MDSC accumulation (Lokuta and Huttenlocher 2005). In agreement with these publications, our findings demonstrated that paclitaxel-mediated abrogation of p38 MAPK activation in MDSC resulted in the decrease in their numbers in tumors. Since p38 MAPK is also activated by secreted S100A8/A9 in autocrine manner (Hermani et al. 2006; Gebhardt et al. 2006; Sunahori et al. 2006), we evaluated the intracellular expression of S100A9 in MDSC. A significant reduction in the number of tumor-infiltrating MDSCs expressing S100A9 was found after administration of paclitaxel (Sevko et al. 2013, in press) additionally highlighting the importance of the stimulation of p38 MAPK-S100A8/A9 pathways in MDSC-mediated immunosuppression.

Reduced amounts of tumor-infiltrating MDSC could be also due to the inhibition of MDSC generation, induction of MDSC apoptosis or enhanced MDSC differentiation. We found that the generation of MDSC from the bone marrow precursors in vitro was not altered paclitaxel treatment (Michels et al. 2012). Under these conditions, the rate of apoptosis in MDSC cultures increased. However, our results revealed that paclitaxel in ultra-low concentrations can promote the MDSC differentiation into functional conventional DC (Michels et al. 2012). In addition, we and others provided evidence that the stimulation of MDSC differentiation into DC by paclitaxel in vitro is not mediated by TLR4 signaling (Kodumudi et al. 2010; Michels et al. 2012).

Taking into account a critical role of MDSC and chronic inflammation in tumor progression, we also assessed the anti-tumor potential of non-toxic application of paclitaxel and revealed a significant delay in tumor development indicated by a prolonged survival of treated animals (Sevko et al. 2013, in press). Investigating the mechanism of this effect, we found an increased numbers of tumor infiltrating $CD8^+$ T lymphocytes. Importantly, the expression of TCR ζ-chain in $CD8^+$ cells, which was usually strikingly reduced in tumor-bearing hosts (Meyer et al. 2011; Baniyash 2006), was found to be increased upon the treatment. Importantly, in the metastatic lymph nodes, frequencies of $CD8^+$ T cells specific for melanoma associated antigen tyrosinase related protein (TRP)-2 were also significantly higher, although the total numbers of $CD8^+$ T cells were not changed. Along this line, we have also reported that in healthy C57BL/6 mice, administration of ultra-low dose paclitaxel strongly increased the frequencies of TRP-2 specific spleen T cells upon the vaccination with the respective peptide (Sevko et al. 2012). These results are in agreement with data reported by us and others that the reduction of MDSC numbers and/or immunosuppressive functions resulted in the restoration of effector functions of tumor-specific $CD8^+$ T cells (Meyer et al. 2011; Apetoh et al. 2011). Furthermore, a selective depletion of $CD8^+$ T cells led to the complete abrogation of the beneficial anti-tumor activity of paclitaxel, underlying a critical role of $CD8^+$ T cells in the mechanism of its therapeutic efficiency.

Analyzing effects of paclitaxel chemoimmunomodulation on various lymphoid cell subsets in healthy C57BL/6 mice, we found a significant reduction in numbers of Treg in treated animals, which displayed also an enhanced TRP-2 specific T cell response after peptide vaccination (Sevko et al. 2012). Elimination of Treg has been reported to be essential for the development of immune responses upon vaccination with TRP-2 antigen in mice (Grauer et al. 2008). Moreover, immunopotentiating effects of ultra-low-dose paclitaxel were demonstrated to be associated with a substantial elevation in frequencies of $CD8^+$ and $CD4^+$ T cells, NK, and NKT cells as well as enhanced IFN-γ production in NK cells. Interestingly, it has been reported that paclitaxel in low concentrations stimulates the cytotoxicity of human NK cells against breast carcinoma cells in vitro by increasing the concentration of perforin both at mRNA and protein levels (Kubo et al. 2005).

26.5 Conclusions

It is well-documented that chemotherapeutic agents can significantly modulate numbers and functions of various subsets of immune cells. The direction of this modulation has been found to critically depend on the doses of one and the same agent. Conventional chemotherapy, which is based on high, maximum tolerated doses (MTD), is widely used as an anti-cancer therapy to eliminate quickly proliferating tumor cells. However, such approach is strongly associated with severe side effects including besides general toxicity also myelotoxicity and inhibition of immune effector cells linked with immunosuppression in tumor bearing hosts. It has been recently reported that the application of chemotherapeutics in moderately low doses (20-33 % MTD) can induce immunogenic death of cancer cells involving the cell surface alterations and the release of soluble immunogenic factors from dying tumor cells. Although this approach involves cytotoxic effect on tumor cells, it induces rather stimulation of immune responses than immunosuppression.

Completely different strategy of therapy with chemotherapeutic agents in ultra-low, non-cytotoxic and non-cytostatic doses (1/20–1/40 of MTD) has been recently proposed and termed chemoimmunomodulation. In contrast to "classical" treatments, this novel approach induces no inhibition of proliferation of tumor cells, hematopoietic cells or immune cells in vitro, but supports the development of anti-tumor responses by altering activity of immune cells. In particular, chemomodulation causes a significant reduction of tumor-associated immunosuppression due to 1) the abrogation of conventional DC polarization into immunosuppressive protumorigenic regDC; 2) the decrease in MDSC numbers and functions in the tumor lesions; 3) the stimulation of anti-tumor activities of conventional DC; and 4) the modulation of intratumoral cytokine network. This suggests that administration of chemotherapeutics in ultra low, non-cytotoxic doses can be considered as a novel neoadjuvant approach for decreasing the protumorigenic, immunosuppressive potential of the tumor microenvironment and increasing the efficacy of various anti-cancer treatments including immunotherapy.

References

Ajibade AA, Wang Q, Cui J, Zou J, Xia X, Wang M et al (2012) TAK1 negatively regulates NF-kappaB and p38 MAP kinase activation in Gr-1 + CD11b + neutrophils. Immunity 36(1):43–54

Alanko S, Salmi TT, Pelliniemi TT (1995) Recovery of natural killer cells after chemotherapy for childhood acute lymphoblastic leukemia and solid tumors. Med Pediatr Oncol 24(6):373–378

Apetoh L, Ghiringhelli F, Tesniere A, Obeid M, Ortiz C, Criollo A et al (2007) Toll-like receptor 4-dependent contribution of the immune system to anticancer chemotherapy and radiotherapy. Nat Med 13(9):1050–1059

Apetoh L, Vegran F, Ladoire S, Ghiringhelli F (2011) Restoration of antitumor immunity through selective inhibition of myeloid derived suppressor cells by anticancer therapies. Curr Mol Med 11(5):365–372

Baldassare JJ, Bi Y, Bellone CJ (1999) The role of p38 mitogen-activated protein kinase in IL-1 beta transcription. J Immunol 162(9):5367–5373

Baniyash M (2006) Chronic inflammation, immunosuppression and cancer: new insights and outlook. Semin Cancer Biol 16(1):80–88

Berd D, Mastrangelo MJ (1988) Effect of low dose cyclophosphamide on the immune system of cancer patients: depletion of CD4 +, 2H4 + suppressor-inducer T-cells. Cancer Res 48(6):1671–1675

Berd D, Maguire HC Jr, Mastrangelo MJ (1986) Induction of cell-mediated immunity to autologous melanoma cells and regression of metastases after treatment with a melanoma cell vaccine preceded by cyclophosphamide. Cancer Res 46(5):2572–2577

Berenson JR, Einstein AB Jr, Fefer A (1975) Syngeneic adoptive immunotherapy and chemoimmunotherapy of a friend leukemia: requirement for T cells. J Immunol 115(1):234–238

Bunnell SC, Kapoor V, Trible RP, Zhang W, Samelson LE (2001) Dynamic actin polymerization drives T cell receptor-induced spreading: a role for the signal transduction adaptor LAT. Immunity 14(3):315–329

Casares N, Pequignot MO, Tesniere A, Ghiringhelli F, Roux S, Chaput N et al (2005) Caspase-dependent immunogenicity of doxorubicin-induced tumor cell death. J Exp Med 202(12):1691–1701

Chae HJ, Kim SC, Chae SW, An NH, Kim HH, Lee ZH et al (2001) Blockade of the p38 mitogen-activated protein kinase pathway inhibits inducible nitric oxide synthase and interleukin-6 expression in MC3T3E—1 osteoblasts. Pharmacol Res 43(3):275–283

Chan OT, Yang LX (2000) The immunological effects of taxanes. Cancer Immunol Immunother 49(4–5):181–518

Chen C, Chen YH, Lin WW (1999) Involvement of p38 mitogen-activated protein kinase in lipopolysaccharide-induced iNOS and COX-2 expression in J774 macrophages. Immunology 97(1):124–129

Chuang LT, Lotzova E, Cook KR, Cristoforoni P, Morris M, Wharton JT (1993) Effect of new investigational drug taxol on oncolytic activity and stimulation of human lymphocytes. Gynecol Oncol 49(3):291–298 (Epub 1993/06/01)

Chuang LT, Lotzova E, Heath J, Cook KR, Munkarah A, Morris M et al (1994) Alteration of lymphocyte microtubule assembly, cytotoxicity, and activation by the anticancer drug taxol. Cancer Res 54(5):1286–1291

Coleman S, Clayton A, Mason MD, Jasani B, Adams M, Tabi Z (2005) Recovery of CD8 + T-cell function during systemic chemotherapy in advanced ovarian cancer. Cancer Res 65(15):7000–7006

Condamine T, Gabrilovich DI (2011) Molecular mechanisms regulating myeloid-derived suppressor cell differentiation and function. Trends Immunol 32(1):19–25

Das V, Nal B, Dujeancourt A, Thoulouze MI, Galli T, Roux P et al (2004) Activation-induced polarized recycling targets T cell antigen receptors to the immunological synapse; involvement of SNARE complexes. Immunity 20(5):577–588

DeVita VT Jr, Chu E (2008) A history of cancer chemotherapy. Cancer Res 68(21):8643–8653

Dmoszynska A, Legiec W, Wach M (1999) Attempted reconstruction of the immune system using low doses of interleukin 2 in chronic lymphocytic leukemia patients treated with 2-chlorodeoxyadenosine: results of a pilot study. Leuk Lymphoma 34(3–4):335–340

Ehrke MJ, Verstovsek S, Maccubbin DL, Ujhazy P, Zaleskis G, Berleth E et al (2000) Protective specific immunity induced by doxorubicin plus TNF-alpha combination treatment of EL4 lymphoma-bearing C57BL/6 mice. Int J Cancer 87(1):101–109
Emens LA (2006) Cancer vaccines: toward the next revolution in cancer therapy. Int Rev Immunol 25(5–6):259–268
Emens LA, Jaffee EM (2005) Leveraging the activity of tumor vaccines with cytotoxic chemotherapy. Cancer Res 65(18):8059–8064
Emens LA, Asquith JM, Leatherman JM, Kobrin BJ, Petrik S, Laiko M et al (2009) Timed sequential treatment with cyclophosphamide, doxorubicin, and an allogeneic granulocyte-macrophage colony-stimulating factor-secreting breast tumor vaccine: a chemotherapy dose-ranging factorial study of safety and immune activation. J Clini Oncol 27(35):5911–5918
Eng C (1999) RET proto-oncogene in the development of human cancer. J Clin Oncol 17(1):380–393
Ewens A, Luo L, Berleth E, Alderfer J, Wollman R, Hafeez BB et al (2006) Doxorubicin plus interleukin-2 chemoimmunotherapy against breast cancer in mice. Cancer Res 66(10):5419–5426
Farber S, Diamond LK, Mercer RD, Sylvester RF, Wolff JA (1948) Temporary remissions in acute leukemia in children produced by folic acid antagonist, 4-Aminopteroyl-glutamic acid (Aminopterin). N Engl J Med 238(787):787–793
Fass L, Fefer A (1972) Studies of adoptive chemoimmunotherapy of a friend virus-induced lymphoma. Cancer Res 32(5):997–1001
Fefer A (1969) Immunotherapy and chemotherapy of moloney sarcoma virus-induced tumors in mice. Cancer Res 29(12):2177–2183
Ferrari S, Rovati B, Porta C, Alessandrino PE, Bertolini A, Collova E et al (2003) Lack of dendritic cell mobilization into the peripheral blood of cancer patients following standard- or high-dose chemotherapy plus granulocyte-colony stimulating factor. Cancer Immunol Immunother 52(6):359–66
Ferraro C, Quemeneur L, Prigent AF, Taverne C, Revillard JP, Bonnefoy-Berard N (2000) Anthracyclines trigger apoptosis of both G0-G1 and cycling peripheral blood lymphocytes and induce massive deletion of mature T and B cells. Cancer Res 60(7):1901–1907
Fujimoto S, Greene MI, Sehon AH (1976) Regualtion of the immune response to tumor antigens. I. Immunosuppressor cells in tumor-bearing hosts. J Immunol 116(3):791–799
Gabrilovich DI, Nagaraj S (2009) Myeloid-derived suppressor cells as regulators of the immune system. Nat Rev 9(3):162–174
Gabrilovich DI, Ostrand-Rosenberg S, Bronte V (2012) Coordinated regulation of myeloid cells by tumours. Nat Rev 12(4):253–268
Galluzzi L, Senovilla L, Zitvogel L, Kroemer G (2012) The secret ally: immunostimulation by anticancer drugs. Nat Rev Drug Discov 11(3):215–233
Gao D, Bing C (2011) Macrophage-induced expression and release of matrix metalloproteinase 1 and 3 by human preadipocytes is mediated by IL-1beta via activation of MAPK signaling. J Cell Physiol 226(11):2869–2880
Gebhardt C, Nemeth J, Angel P, Hess J (2006) S100A8 and S100A9 in inflammation and cancer. Biochem Pharmacol 72(11):1622–1631
Ghiringhelli F, Larmonier N, Schmitt E, Parcellier A, Cathelin D, Garrido C et al (2004) CD4 + CD25 + regulatory T cells suppress tumor immunity but are sensitive to cyclophosphamide which allows immunotherapy of established tumors to be curative. Eur J Immunol 34(2):336–344
Gilman A (1963) The initial clinical trial of nitrogen mustard. Am J Surg 105:574–578
Glynn JP, Halpern BL, Fefer A (1969) An immunochemotherapeutic system for the treatment of a transplanted Moloney virus-induced lymphoma in mice. Cancer Res 29(3):515–520
Goldman D (2000) Chronic lymphocytic leukemia and its impact on the immune system. Clin J Oncol Nurs 4(5):233–234,236

Goodman L, Wintrobe M, Dameshek W, Goodman M, Gilman A, McLennan M (1946) Nitrogen mustard therapy: use of methyl-bis(beta-chloroethyl)amine hydrochloride and tris(beta-chloroethyl)amine hydrochloride for Hodgkin's disease, lymphosarcoma, leukemia, and certain allied and miscellaneous disorders. JAMA 132:126–132

Goodman LS, Wintrobe MM, Dameshek W, Goodman MJ, Gilman A, McLennan MT (1984) Landmark article Sept. 21, 1946: Nitrogen mustard therapy. Use of methyl-bis(beta-chloroethyl)amine hydrochloride and tris(beta-chloroethyl)amine hydrochloride for Hodgkin's disease, lymphosarcoma, leukemia and certain allied and miscellaneous disorders. By Louis S. Goodman, Maxwell M. Wintrobe, William Dameshek, Morton J. Goodman, Alfred Gilman and Margaret T. McLennan. JAMA 251(17):2255–2261

Grauer OM, Sutmuller RP, van Maren W, Jacobs JF, Bennink E, Toonen LW et al (2008) Elimination of regulatory T cells is essential for an effective vaccination with tumor lysate-pulsed dendritic cells in a murine glioma model. Int J Cancer 122(8):1794–1802

Hermani A, De Servi B, Medunjanin S, Tessier PA, Mayer D (2006) S100A8 and S100A9 activate MAP kinase and NF-kappaB signaling pathways and trigger translocation of RAGE in human prostate cancer cells. Exp Cell Res 312(2):184–197

Hirsch J (2006) An anniversary for cancer chemotherapy. JAMA 296(12):1518–1520

Ho FM, Lai CC, Huang LJ, Kuo TC, Chao CM, Lin WW (2004) The anti-inflammatory carbazole, LCY-2-CHO, inhibits lipopolysaccharide-induced inflammatory mediator expression through inhibition of the p38 mitogen-activated protein kinase signaling pathway in macrophages. Br J Pharmacol 141(6):1037–1047

Houghton AN, Polsky D (2002) Focus on melanoma. Cancer Cell 2(4):275–278

Javeed A, Ashraf M, Riaz A, Ghafoor A, Afzal S, Mukhtar MM (2009) Paclitaxel and immune system. Eur J Pharm Sci 38(4):283–290

Joensuu H (2008) Systemic chemotherapy for cancer: from weapon to treatment. Lancet Oncol 9(3):304

Kaneno R, Shurin GV, Tourkova IL, Shurin MR (2009) Chemomodulation of human dendritic cell function by antineoplastic agents in low noncytotoxic concentrations. J Transl Med 7:58

Kaneno R, Shurin GV, Kaneno FM, Naiditch H, Luo J, Shurin MR (2011) Chemotherapeutic agents in low noncytotoxic concentrations increase immunogenicity of human colon cancer cells. Cellular oncology 34(2):97–106 (Dordrecht)

Kato M, Takahashi M, Akhand AA, Liu W, Dai Y, Shimizu S et al (1998) Transgenic mouse model for skin malignant melanoma. Oncogene 17(14):1885–1888

Kim SK, Demetri GD (1996) Chemotherapy and neutropenia. Hematol Oncol Clin North Am 10(2):377–395

Kim C, Sano Y, Todorova K, Carlson BA, Arpa L, Celada A et al (2008) The kinase p38 alpha serves cell type-specific inflammatory functions in skin injury and coordinates pro- and anti-inflammatory gene expression. Nat Immunol 9(9):1019–1027

Ko HJ, Kim YJ, Kim YS, Chang WS, Ko SY, Chang SY et al (2007) A combination of chemoimmunotherapies can efficiently break self-tolerance and induce antitumor immunity in a tolerogenic murine tumor model. Cancer Res 67(15):7477–7486

Kodumudi KN, Woan K, Gilvary DL, Sahakian E, Wei S, Djeu JY (2010) A novel chemoimmunomodulating property of docetaxel: suppression of myeloid-derived suppressor cells in tumor bearers. Clin Cancer Res 16(18):4583–4594

Krumbhaar EB (1919) Role of the blood and the bone marrow in certain forms of gas poisoning. JAMA 72:39–41

Kubo M, Morisaki T, Matsumoto K, Tasaki A, Yamanaka N, Nakashima H et al (2005) Paclitaxel probably enhances cytotoxicity of natural killer cells against breast carcinoma cells by increasing perforin production. Cancer Immunol Immunother 54(5):468–476

Kupfer A, Dennert G (1984) Reorientation of the microtubule-organizing center and the Golgi apparatus in cloned cytotoxic lymphocytes triggered by binding to lysable target cells. J Immunol. 133(5):2762–2766

Lacour S, Hammann A, Wotawa A, Corcos L, Solary E, Dimanche-Boitrel MT (2001) Anticancer agents sensitize tumor cells to tumor necrosis factor-related apoptosis-inducing ligand-mediated caspase-8 activation and apoptosis. Cancer Res 61(4):1645–1651

Lake RA, Robinson BW (2005) Immunotherapy and chemotherapy—a practical partnership. Nat Rev Cancer 5(5):397–405

Lengagne R, Pommier A, Caron J, Douguet L, Garcette M, Kato M et al (2011) T cells contribute to tumor progression by favoring pro-tumoral properties of intra-tumoral myeloid cells in a mouse model for spontaneous melanoma. PLoS One 6(5):e20235

Liao YP, Schaue D, McBride WH (2007) Modification of the tumor microenvironment to enhance immunity. Front Biosci 12:3576–3600

Liu JY, Wu Y, Zhang XS, Yang JL, Li HL, Mao YQ et al (2007) Single administration of low dose cyclophosphamide augments the antitumor effect of dendritic cell vaccine. Cancer Immunol Immunother 56(10):1597–1604

Liu Q, Zhang C, Sun A, Zheng Y, Wang L, Cao X (2009) Tumor-educated CD11bhighIalow regulatory dendritic cells suppress T cell response through arginase I. J Immunol 182(10):6207–6216

Lokuta MA, Huttenlocher A (2005) TNF-alpha promotes a stop signal that inhibits neutrophil polarization and migration via a p38 MAPK pathway. J Leukoc Biol 78(1):210–219

Lutsiak ME, Semnani RT, De Pascalis R, Kashmiri SV, Schlom J, Sabzevari H (2005) Inhibition of CD4(+)25 + T regulatory cell function implicated in enhanced immune response by low-dose cyclophosphamide. Blood 105(7):2862–2868

Mackall CL (2000) T-cell immunodeficiency following cytotoxic antineoplastic therapy. Rev Stem cells 18(1):10–18 (Dayton, Ohio)

Mackall CL, Fleisher TA, Brown MR, Andrich MP, Chen CC, Feuerstein IM et al (1997) Distinctions between CD8 +and CD4 + T-cell regenerative pathways result in prolonged T-cell subset imbalance after intensive chemotherapy. Blood 89(10):3700–3707

Marigo I, Dolcetti L, Serafini P, Zanovello P, Bronte V (2008) Tumor-induced tolerance and immune suppression by myeloid derived suppressor cells. Immunol Rev 222:162–179

Mastrangelo MJ, Berd D, Maguire H Jr (1986) The immunoaugmenting effects of cancer chemotherapeutic agents. Semin Oncol 13(2):186–194

Mattarollo SR, Loi S, Duret H, Ma Y, Zitvogel L, Smyth MJ (2011) Pivotal role of innate and adaptive immunity in anthracycline chemotherapy of established tumors. Cancer Res 71(14):4809–4820

Melief CJ (2008) Cancer immunotherapy by dendritic cells. Immunity 29(3):372–383

Meyer C, Sevko A, Ramacher M, Bazhin AV, Falk CS, Osen W et al (2011) Chronic inflammation promotes myeloid-derived suppressor cell activation blocking antitumor immunity in transgenic mouse melanoma model. Proc Natl Acad Sci USA 108(41):17111–17116

Micheau O, Solary E, Hammann A, Martin F, Dimanche-Boitrel MT (1997) Sensitization of cancer cells treated with cytotoxic drugs to fas-mediated cytotoxicity. J Natl Cancer Inst 89(11):783–789

Michels T, Shurin GV, Naiditch H, Sevko A, Umansky V, Shurin MR (2012) Paclitaxel promotes differentiation of myeloid-derived suppressor cells into dendritic cells in vitro in a TLR4-independent manner. J Immunotoxicol 9(3):292–300

Mihich E (1969) Combined effects of chemotherapy and immunity against leukemia L1210 in DBA-2 mice. Cancer Res 29(4):848–854

Mihich E, Ehrke MJ (2000) Anticancer drugs plus cytokines: immunodulation based therapies of mouse tumors. Int J Immunopharmacol 22(12):1077–1081

Mustafa MM, Buchanan GR, Winick NJ, McCracken GH, Tkaczewski I, Lipscomb M et al (1998) Immune recovery in children with malignancy after cessation of chemotherapy. J Pediatr Hematol Oncol 20(5):451–457

Naiditch H, Shurin MR, Shurin GV (2011) Targeting myeloid regulatory cells in cancer by chemotherapeutic agents. Immunol Res 50(2–3):276–285

Nakashima H, Tasaki A, Kubo M, Kuroki H, Matsumoto K, Tanaka M et al (2005) Effects of docetaxel on antigen presentation-related functions of human monocyte-derived dendritic cells. Cancer Chemother Pharmacol 55(5):479–87

North RJ (1982) Cyclophosphamide-facilitated adoptive immunotherapy of an established tumor depends on elimination of tumor-induced suppressor T cells. J Exp Med 155(4):1063–1074

Nowak AK, Robinson BW, Lake RA (2003) Synergy between chemotherapy and immunotherapy in the treatment of established murine solid tumors. Cancer Res 63(15):4490–4496

Nowak AK, Lake RA, Robinson BW (2006) Combined chemoimmunotherapy of solid tumours: improving vaccines? Adv Drug Deliv Rev 58(8):975–990

Obeid M, Tesniere A, Ghiringhelli F, Fimia GM, Apetoh L, Perfettini JL et al (2007) Calreticulin exposure dictates the immunogenicity of cancer cell death. Nat Med 13(1):54–61

Ostrand-Rosenberg S (2010) Myeloid-derived suppressor cells: more mechanisms for inhibiting antitumor immunity. Cancer Immunol Immunother 59(10):1593–1600

Papac RJ (2001) Origins of cancer therapy. Yale J Biol Med 74(6):391–398

Peranzoni E, Zilio S, Marigo I, Dolcetti L, Zanovello P, Mandruzzato S et al (2010) Myeloid-derived suppressor cell heterogeneity and subset definition. Curr Opin Immunol 22(2):238–244

Perdiguero E, Sousa-Victor P, Ruiz-Bonilla V, Jardi M, Caelles C, Serrano AL et al (2011) p38/MKP-1-regulated AKT coordinates macrophage transitions and resolution of inflammation during tissue repair. J Cell Biol 195(2):307–322

Raber P, Ochoa AC, Rodriguez PC (2012) Metabolism of L-arginine by myeloid-derived suppressor cells in cancer: mechanisms of T cell suppression and therapeutic perspectives. Immunol Invest 41(6–7):614–634

Rasmussen L, Arvin A (1982) Chemotherapy-induced immunosuppression. Environ Health Perspect 43:21–25

Sevko A, Kremer V, Falk C, Umansky L, Shurin MR, Shurin GV et al (2012) Application of paclitaxel in low non-cytotoxic doses supports vaccination with melanoma antigens in normal mice. J Immunotoxicol 9(3):275–281

Sevko A, Michels T, Vrohlings M, Umansky L, Beckhove P, Kato M et al (2013) Anti-tumoreffect of paclitaxel is mediated by inhibition of MDSCs and chronic inflammation in thespontaneous melanoma model. J Immunol (in press).

Sharabi A, Laronne-Bar-On A, Meshorer A, Haran-Ghera N (2010) Chemoimmunotherapy reduces the progression of multiple myeloma in a mouse model. Cancer Prev Res (Phila) 3(10):1265–1276

Shurin GV, Tourkova IL, Shurin MR (2008) Low-dose chemotherapeutic agents regulate small Rho GTPase activity in dendritic cells. J Immunother 31(5):491–499

Shurin GV, Tourkova IL, Kaneno R, Shurin MR (2009a) Chemotherapeutic agents in noncytotoxic concentrations increase antigen presentation by dendritic cells via an IL-12-dependent mechanism. J Immunol 183(1):137–144

Shurin GV, Amina N, Shurin MR (2009b) Cancer therapy and dendritic cell immunomodulation. In: R D Salter MRS (eds) Dendritic cells in cancer. Springer, New York pp 201–216

Shurin MR, Naiditch H, Zhong H, Shurin GV (2011) Regulatory dendritic cells: new targets for cancer immunotherapy. Cancer Biol Ther 11(11):988–992

Shurin MR, Naiditch H, Gutkin DW, Umansky V, Shurin GV (2012) ChemoImmunoModulation: immune regulation by the antineoplastic chemotherapeutic agents. Curr Med Chem 19(12):1792–1803

Sinkovics JG, Horvath JC (2006) Evidence accumulating in support of cancer vaccines combined with chemotherapy: a pragmatic review of past and present efforts. Int J Oncol 29(4):765–777

Steele TA (2002) Chemotherapy-induced immunosuppression and reconstitution of immune function. Leuk Res 26(4):411–414

Stoychkov JN, Schultz RM, Chirigos MA, Pavlidis NA, Goldin A (1979) Effects of adriamycin and cyclophosphamide treatment on induction of macrophage cytotoxic function in mice. Cancer Res 39(8):3014–3017

Sunahori K, Yamamura M, Yamana J, Takasugi K, Kawashima M, Yamamoto H et al (2006) The S100A8/A9 heterodimer amplifies proinflammatory cytokine production by macrophages via activation of nuclear factor kappa B and p38 mitogen-activated protein kinase in rheumatoid arthritis. Arthritis Res Ther 8(3):R69

Suzuki E, Kapoor V, Jassar AS, Kaiser LR, Albelda SM (2005) Gemcitabine selectively eliminates splenic Gr-1 +/CD11b + myeloid suppressor cells in tumor-bearing animals and enhances antitumor immune activity. Clin Cancer Res 11(18):6713–6721

Takei F, Levy JG, Kilburn DG (1977) Characterization of suppressor cells in mice bearing syngeneic mastocytoma. J Immunol 118(2):412–417

Tanaka F, Yamaguchi H, Ohta M, Mashino K, Sonoda H, Sadanaga N et al (2002) Intratumoral injection of dendritic cells after treatment of anticancer drugs induces tumor-specific antitumor effect in vivo. Int J Cancer 101(3):265–269

Timkin O (1942) A medieval translation of Rhazes' clinical observations. Bull Hist Med 12:102–117

Tsavaris N, Kosmas C, Vadiaka M, Kanelopoulos P, Boulamatsis D (2002) Immune changes in patients with advanced breast cancer undergoing chemotherapy with taxanes. Br J Cancer 87(1):21–27

Ujhazy P, Zaleskis G, Mihich E, Ehrke MJ, Berleth ES (2003) Doxorubicin induces specific immune functions and cytokine expression in peritoneal cells. Cancer Immunol Immunother 52(7):463–472

Umansky V, Abschuetz O, Osen W, Ramacher M, Zhao F, Kato M et al (2008) Melanoma-specific memory T cells are functionally active in ret transgenic mice without macroscopic tumors. Cancer Res 68(22):9451–9458

Vadlamudi S, Padarathsingh M, Bonmassar E, Goldin A (1971) Effect of combination treatment with cyclophosphamide and isogeneic or allogeneic spleen and bone marrow cells in leukemic (L1210) mice. Int J Cancer 7(1):160–166

Wijayahadi N, Haron MR, Stanslas J, Yusuf Z (2007) Changes in cellular immunity during chemotherapy for primary breast cancer with anthracycline regimens. J Chemother 19(6):716–723

Wolberg G, Stopford CR, Zimmerman TP (1984) Antagonism by taxol of effects of microtubule-disrupting agents on lymphocyte cAMP metabolism and cell function. Proc Natl Acad Sci USA 81(11):3496–3500

Youn JI, Gabrilovich DI (2010) The biology of myeloid-derived suppressor cells: the blessing and the curse of morphological and functional heterogeneity. Eur J Immunol 40(11):2969–2975

Yu B, Kusmartsev S, Cheng F, Paolini M, Nefedova Y, Sotomayor E et al (2003) Effective combination of chemotherapy and dendritic cell administration for the treatment of advanced-stage experimental breast cancer. Clin Cancer Res 9(1):285–294

Zhao F, Falk C, Osen W, Kato M, Schadendorf D, Umansky V (2009) Activation of p38 mitogen-activated protein kinase drives dendritic cells to become tolerogenic in ret transgenic mice spontaneously developing melanoma. Clin Cancer Res 15(13):4382–4390

Zhou FH, Foster BK, Zhou XF, Cowin AJ, Xian CJ (2006) TNF-alpha mediates p38 MAP kinase activation and negatively regulates bone formation at the injured growth plate in rats. J Bone Miner Res 21(7):1075–1088

Zocchi MR, Poggi A (2004) Role of gammadelta T lymphocytes in tumor defense. Front Biosci. 9:2588–2604

Chapter 27
Combining Vaccines with Therapies that Render Tumor Cells more Susceptible to Immune Mediated Killing

Nishith Singh, James Hodge, Ravi Madan and James L. Gulley

Abstract Therapeutic cancer vaccines are an efficient and minimally toxic way to stimulate the host's immune system in a dynamic way such that immune responses persists beyond the period of vaccine administration. The persisting immunity can be utilized by standard of care therapies given concurrently or subsequently to vaccines in the fight against cancer. Emerging pre-clinical and clinical evidence shows that the standard of care anti-cancer therapies can "modulate". Both the Tumor and the immune system of the host such as to potentiate the immune response induced by therapeutic cancer vaccines. This book chapter focuses on these combinatorial approaches to cancer that target the residual cancer cells in the host that have not been contained by definitive procedures such as surgery. Specifically, we discuss the biological basis of anti-tumor immunity, immuno-modulation with standard of care therapies such as radiation, cytotoxic chemo-therapies, hormone abrogation and small molecule inhibitors, pre-clinical and clinical evidence of synergism in combinatorial strategies, clinical trial design and published and ongoing clinical studies in that regard.

Keyword Cancer vaccine · Combination immunotherapy · Immunogenic modulation · Immunogenic cell death · Anti-cancer immunity · Immunomodulation

N. Singh · R. Madan · J. L. Gulley (✉)
Medical Oncology Branch, Center for Cancer Research,
National Cancer Institute, National Institutes of Health,
10 Center Dr., 13N208, MSC-1750, Bethesda, MD 20892, USA
e-mail: gulleyj@mail.nih.gov

J. Hodge · R. Madan · J. L. Gulley
Laboratory of Tumor Immunology and Biology, Center for Cancer Research,
National Cancer Institute, National Institutes of Health, Bethesda, Maryland, USA

M. R. Shurin et al. (eds.), *The Tumor Immunoenvironment*,
DOI: 10.1007/978-94-007-6217-6_27,

27.1 Introduction

Therapeutic cancer vaccines constitute a major emerging strategy for the treatment of a wide range of human malignancies. The goal of this approach is to generate a targeted immune response that ultimately curtails cancer growth. An intense area of research is the combination of specific vaccines with conventional modalities like chemotherapy, radiotherapy, hormone therapy and the small molecular targeted inhibitors. Traditionally considered immune-neutral or immune-inhibitory, some conventional anti-cancer modalities have now been shown to have considerable impact on the phenotype of cancer cells and have been found to synergize with therapeutic cancer vaccines (Rotow et al. 2010). In this chapter we summarize the biological basis and rationale for combining therapeutic vaccines with conventional cancer therapies. We also discuss the ongoing and published clinical studies looking at the combination of conventional anticancer modalities and cancer vaccines.

27.2 Cancer and the Immune System

The relationship between cancer and the immune system is complex and dynamic. Several components of the innate and adaptive immune response are responsible for tumor elimination. They include $\alpha\beta$ and $\gamma\delta$ T-cells, and NK cells (Girardi et al. 2003; Smyth et al. 2000). The process begins by antigen recognition via dendritic cells (DCs) that is followed by a subsequent maturation signal to the DCs. The latter may involve exogenous factors such as toll-like receptors (TLR), endogenous factors such as high mobility group (HMG) proteins or adenosine tri-phosphate (ATP) from the dying cells (Mellman et al. 2011). The next step involves the generation of T-cell responses in the lymphoid organs where the peptide antigen bound to the major histocompatibility complex (MHC)-derived molecule is presented to the T-cell receptor (TCR); this process represents the central event that leads to effector cellular immunity. The activation of naive T-cells and their subsequent differentiation into effector cells requires the primary signal via the TCR and the CD4 or CD8 co-receptors as well as the co-stimulatory signals. The most characterized co-stimulatory signal involves the interaction of B7 with CD28 and cytotoxic T lymphocyte (CTL)-associated antigen 4 (CTLA-4) (Greenwald et al. 2005). Once CTLs are generated, tumor cell killing can take place via three pathways (Andersen and Schrama 2006):

1. Fas ligand, which is expressed on the surface of CTLs, binds to the Fas receptor (Fas, CD95) on the targeT-cell which triggers apoptosis through the caspases.
2. CTL releases perforins and granzymes, which are cytotoxic molecules, into the intercellular space during the cell–cell contact.
3. Indirect killing of targeT-cells by release of cytokines such as tumor necrosis factor-α (TNF-α) and interferon-γ (IFN-γ).

This framework provides for understanding how different aspects of an immune response can be modulated to enhance immune-mediated killing by standard regimens such as radiation, cytotoxic chemotherapy and small molecule inhibitors (Fig. 27.1) (Hodge et al. 2012). They include, but are not limited to, stimulation of DCs, up-regulation of tumor antigens, MHC, Fas and cytokines, differential modulation of immune cell subsets such as effector and regulatory T-cell ratios, which will be discussed in detail in the upcoming sections.

On the other hand, several components of the immune system are known to support tumor growth and metastases and their down-regulation makes for a potential approach in immunotherapeutics. These include transcription factor fork head box P3 positive (FOXP3) regulatory T (Treg) cells, tumor associated macrophages (TAM), myeloid-derived suppressor cells (MDSCs) and the type 2 helper CD4 + (Th2) T-cells (Curiel et al. 2004; Mantovani et al. 2002; Young et al. 1992; DeNardo et al. 2009). They can accumulate at the tumor bed and promote an immunosuppressive environment, thereby having a negative impact on prognosis. In order to evade immune surveillance, the tumor bed itself can display specific morphological alterations. Tumor evasion of cytotoxic T-cells can involve down-regulation of tumor antigen expression, components of the antigen-processing and presentation machinery, and MHC molecules (Marincola et al. 2000).

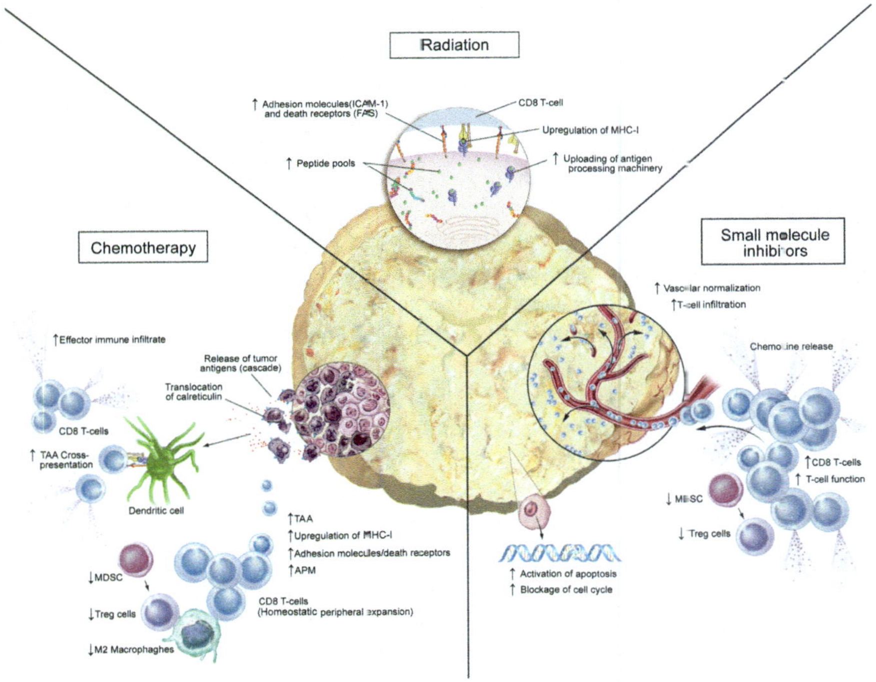

Fig. 27.1 Multiple mechanisms of synergy between radiation therapy, chemotherapy, or small molecule inhibitors and immunotherapy. (Obtained with permission, Hodge et al. Semin Oncol 2012.)

Interestingly, the tumor microenvironment can actively down-regulate the T-cell immune response by means of decreasing the expression of co-stimulatory molecules and increasing the surface expression of molecules that negatively regulate activation of T-cells, such as programmed death-ligand (PD-L1, ligand for PD-1), by secreting soluble factors like transforming growth factor-β (TGF-β) (Driessens et al. 2009; Thomas and Massague 2005), and arginase, PGE2 and indoleamine 2,3-dioxygenase (IDO) (Mellman et al. 2011).

Since the goal of any anti-cancer therapy is to decrease or stabilize tumor burden, in the given context, an immune-based approach should have the capability to alter the morphology of the tumor bed to make it more sensitive to the immune system and/or augment specific immune response with the idea to mitigate tolerance.

27.3 The Biological Basis of Cytotoxic Chemotherapy and Immune Activation

Several cancer chemotherapeutics such as cyclophosphamide and methotrexate are also used as immunosuppressants for the treatment of systemic autoimmune conditions. The primary mechanism involves the induction of lymphopenia via myelosuppression. Accumulating evidence has brought up the possibility that some cytotoxic agents can enhance anti-cancer immune response and potentiate cancer vaccine induced T-cell killing of cancer cells (Schlom 2012; Zitvogel et al. 2008). We begin this section with a brief overview of the chronology in which the effects of the immune system were discovered. In the end we focus on the vital work done on the phenotypic modulation of cancer cells that can be brought about by sub-lethal stimuli. We consider immunogenic modulation as a novel and crucial framework to understand the clinical impact of combinatorial regimens, because ultimately it is the cancer cells not killed by standard therapy that drive host compromise.

The potential of immune therapy to inhibit tumors more efficiently in combination with cyclophosphamide in the pre-clinical setting was first published in 1982 in a landmark observation (North 1982). The work on adoptive immunotherapy showed that cyclophosphamide lowers the numbers of a special T-cell subset (an immune inhibitory subset) that were later characterized as CD4 + CD25 + regulatory T-cells (Tregs). Subsequent work showed that the mechanism not only involves a quantitative decrease in Tregs (apoptosis) but also a functional decrease (down-regulation of GITR and FOXP3 expression) (Lutsiak et al. 2005). A large body of pre-clinical work has identified other contributory factors associated with cyclophosphamide-induced immune stimulation: IFN secretion, Th17 potentiation, expansion of memory T-cells, and potentiation of myeloid DCs (Greenberg et al. 1980; Schiavoni et al. 2000, 2011; Salem et al. 2010;

Moschella et al. 2011; Ghiringhelli et al. 2004). Given in a defined sequence (a few days before whole cell vaccine in a pre-clinical model) and in immuno-modulatory doses, cyclophosphamide augments anti-tumor immunity at least in part by an increase in number and function of antigen-specific T-cells, especially Th1 cells (Machiels et al. 2001).

Immunogenic tumor cell death (ICD) is another model in the context of cytotoxic therapy and immune activation. ICD reflects the capacity of cancer cells to die from lethal stimuli and stimulate an anti-tumor immune response and it proposes apoptotic cell death as a non-tolerogenic event. The molecular determinants of ICD include calreticulin exposure, secretion of HMGB1 and ATP, enhanced cross-priming of the tumor-associated antigens (TAAs) by DCs and subsequent activation of anti-tumor T-cells. ICD has been touted to explain the immunogenic effects of oxaliplatin, mitoxantrone, cyclophosphamide and doxorubicin as well as certain types of radiation on the susceptible host T-cells (Kepp et al. 2011; Tesniere et al. 2010; Zitvogel et al. 2010).

Mounting evidence from pre-clinical studies indicates that certain chemotherapeutic agents can also increase the susceptibility of tumor cells to vaccine-mediated T-cell cytotoxicity (Garnett et al. 2008, 2010; Gameiro et al. 2011, 2012; Gelbard et al. 2006). Tumor cells within a given tumor mass receive sub-lethal doses of therapy due to perfusion limitations or because of the need to limit damage to normal tissues. A number of pre-clinical studies have shown that non-cytolytic levels of chemotherapy are capable of inducing phenotypic changes within tumor cells that ultimately facilitate immune-cell recognition and immune-mediated tumor killing, now termed "immunogenic modulation". This is especially revelatory since the knowledge opens up several venues to translate and validate the combinatorial approaches in which a low-dose standard chemotherapeutic agent is combined with a cancer vaccine in order to achieve synergism. Immunogenic modulation of the tumor cells that have escaped lethal stimuli appears to be the dominant underlying mechanism in this model. For instance, sub-lethal docetaxel can up-regulate one or more of the surface molecules (Fas, ICAM-1, MUC-1, CEA, and MHC class I) in both sensitive and resistant human carcinoma cell lines (Garnett et al. 2010). The up-regulation of the molecules is then associated with enhanced killing by specific HLA-A2-restricted CD8 + CTLs. Similar phenotypic changes, as well as cytokine secretion and increased MHC-restricted T-cell killing, have been shown with cisplatin and vinorelbine (Gameiro et al. 2012). In contrast to ICD, the chemotherapy dose is non-cytolytic in this model and does not seem to involve antigen presenting cells. The molecular bases of the observed biological activities, mostly from pre-clinical studies, including the mechanisms have been listed in Table 27.1.

Table 27.1 Immunomodulation and cytotoxic drugs

Biological activity	Mechanism	Reference
Increased tumor immunogenicity and APC activation	Upregulation of Fas, ICAM-1, MUC-1, CEA, and/or MHC class I	(Garnett et al. 2008; Gameiro et al. 2011; Gelbard et al. 2006; Garnett et al. 2010; Gameiro et al. 2012)
	Up-regulation of cell-surface calreticulin	(Obeid et al. 2007)
	Up-regulation of pattern recognition receptors (MBL1 and C1Q)	(Moschella et al. 2011)
	Increased pro-inflammatory soluble factors (growth factors, cytokines, chemokine receptors, chemokines)	(Moschella et al. 2011)
	Tumor "danger" signal release (HMGB1)	(Apetoh et al. 2007)
	DC activation (through TLR-4, MyD88, HMGB1, defensins)	(Moschella et al. 2011; Apetoh et al. 2007)
	Suppression of Tregs	(North 1982; Lutsiak et al. 2005)
Potentiation of T-cell anti-tumor response	Generation of memory T-cells	(Schiavoni et al. 2000; Ge et al. 2002)
	Induction of T-helper 17 (Th17)-related gene signature (decreased T-cell tolerance)	(Machiels et al. 2001)
	Inhibition of marrow derived suppressor cells	(Greenberg et al. 1980; Schiavoni et al. 2011)

27.4 The Biological Basis of Radiotherapy and Immune Activation

Radiotherapy, in the form of external beam radiotherapy, brachytherapy, radiolabeled monoclonal anti-bodies, and bone-seeking radionuclides, is a standard modality in many cancer types. Combining standard radiation therapy with cancer vaccines in the pre-clinical studies and the early clinical trials has allowed one to make interesting observations in support of a combination approach. While radiation in high doses can cause lymphopenia, in sub-lethal doses it induces phenotypic changes in the tumor cells, including up-regulation of many cell-surface proteins involved in T-cell target recognition, adhesion, and lysis on tumor as well as endothelium (Table 27.2). Some of the proteins include calreticulin, ICAM-1, MHC class I and II, Fas, and multiple TAAs (carcinoembryonic antigen, CEA; mucin 1, MUC-1, cancer antigen 125, CA125; Human Epidermal Growth Factor Receptor 2, Her2-neu; p53, prostate specific antigen, PSA; prostate specific membrane antigen, PSMA; and prostatic acid phosphatase, PAP) (Gaugler et al. 1997; Chakraborty et al. 2004; Garnett et al. 2004). These changes boost T-cell cytolytic activity and T-cell trafficking towards the irradiated sites (Garnett et al. 2004;

Table 27.2 Immunomodulation and radiation

Biological activity	Mechanism	Reference
Increased tumor immunogenicity and APC activation	Up-regulation of cell-surface calreticulin (promotes antigen uptake)	(Obeid et al. 2007)
	Up-regulation of MHC class 1 and 2, TAAs, mTCR activation, Fas, ICAM-1	(Gaugler et al. 1997; Chakraborty et al. 2004; Garnett et al. 2004; Lugade et al. 2005; Reits et al. 2006; Chamoto et al. 2009; Ye et al. 2007)
	Tumor "danger" signal release (HMGB1)	(Reits et al. 2006)
	DC activation (through TLR-4 and MyD88)	(Reits et al. 2006)
Potentiation of T-cell anti-tumor response	Increased cytolytic T-cell traffic to tumor	(Garnett et al. 2004; Lugade et al. 2005)

Lugade et al. 2005). Radiation can also induce novel peptides through destruction of pre-existing proteins, through increased translation and through increased synthesis (Reits et al. 2006). Interestingly, the cross-presentation of novel antigens can contribute to the "abscopal effect" (Reits et al. 2006). In this phenomenon, local irradiation of single tumor is associated with reduction in the size of non-irradiated metastases. The synergism between radiation and several different vaccine strategies has been shown in pre-clinical studies (Chamoto et al. 2009; Ye et al. 2007).

27.5 The Biological Basis of Hormone Abrogation and Immune Activation

Androgens can modulate the immune system through the expression of androgen receptor (AR) that some, but not all, have shown to be present on the lymphocytes (Hsueh et al. 2003). In addition, the androgen-related immune-reactive signals from epithelial cells and macrophages can modulate the effector cells (Hsueh et al. 2003). While the exact molecular mechanism remains to be elucidated, a number of pre-clinical and clinical observations show definite biological activity associated with both hormone replacement and hormone withdrawal. The latter is particularly important in prostate cancer where hormone antagonism is employed as a primary therapeutic modality in multiple settings. At the cellular level, the mechanism of immunomodulation with androgen ablation may involve alteration of CD4 + and CD8 + cell sub-populations, inhibition of regulatory T-cells, increased peripheral and intra-tumoral traffic of effector cells and decreased tolerance (Mercader et al. 2001; Gannon et al. 2009; Drake et al. 2005; Sutherland et al. 2005; Roden et al. 2004; Page et al. 2006). In a murine study, androgen

Table 27.3 Immunomodulation and androgen ablation

Biological activity	Mechanism	Reference
Potentiation of T-cell and NK-cell anti-tumor response	Increased tumor CD4 + and CD8 + T-cell traffic and density	(Mercader et al. 2001; Gannon et al. 2009)
	Mitigation of CD4 + T-cell tolerance to tumor- and tumor cancer-restricted antigen	(Drake et al. 2005)
	Increased thymic output of T-cells and increased peripheral population	(Sutherland et al. 2005; Roden et al. 2004)
	Facilitation of host tissue specific T-cell responses	(Roden et al. 2004)
	Decreased percentage of peripheral CD4 + CD25 + T-cells	(Page et al. 2006)
	Enhanced vaccine-induced ADCC	(Hsueh et al. 2003)
	Enhanced vaccine-induced splenocyte secretion of IFN-γ	(Hsueh et al. 2003)

blockade was shown to enhance splenocyte proliferation, splenocyte stimulated secretion of IFN-γ, and antibody dependenT-cell-mediated toxicity (ADCC) to subsequent administration of an irradiated whole-cell melanoma vaccine (Hsueh et al. 2003). A detailed review of hormone abrogation and its impact on the immune system and the implications for combination therapy has been published. Table 27.3 summarizes the biological effects and mechanisms of androgen deprivation on the immune system (Aragon-Ching et al. 2007).

27.6 The Biological Basis of Small Molecule Inhibitors and Immune Activation

Small molecule targeted inhibitors such as imatinib are current standard of care in various cancers. Logically, it is held that the small molecule inhibitors possess therapeutic efficacy by means of their ability to inhibit a target oncogenic enzyme. Inhibition of the oncogenic target then reverses the tumorigenic state through cell-autonomous activity. However, in contrast to the long-held view, a number of pre-clinical and clinical studies have borne evidence of host immunomodulation with the use of small molecule inhibitors. Interestingly, analysis of peripheral immune cell subsets and cytokines show both immune-potentiation as well as immune-suppression in a clinically heterogeneous population exposed to small molecule inhibitors (Hayashi et al. 2012). Sensitization of the tumor cells by means of increased danger signals and antigenicity to facilitate the immune effector cells has been noted with 5-aza-2'-deocycitidine and bortezomib (Spisek et al. 2007; Serrano et al. 2001). On the other hand, direct increase in number and function of effector cells has been shown with the tyrosine kinase inhibitors (Hayashi et al. 2012; Rohon et al. 2010). In a mouse model we have shown that multiple tyrosine kinase inhibitors such as sunitinib create a permissive environment to vaccinate into by their in vivo effect on reducing Tregs and MDSCs (Farsaci et al. 2012). Pre-clinical work testing the feasibility and efficacy of a combination of pan-BCL-2 (B cell lymphoma 2) inhibitor with vaccine has revealed that the drug inhibits in vivo Tregs (Farsaci et al. 2010). The potential mechanisms of immunomodulation that may have a positive anti-tumor effect either alone or in combination with a cancer vaccine are listed and referenced in Table 27.4.

27.7 Combination Immunotherapies in Clinical Trial: Clinical Evidence

27.7.1 Chemotherapy and Cancer Vaccines

In this section, we will look into the various clinical studies that attempt to test the role of combining vaccine with a given standard anti-cancer modality (cytotoxic drugs, radiotherapy, hormone therapy and small molecule inhibitors). Before we proceed, it is important to sketch out certain vital differences between the former and the latter approaches. The differences arise fundamentally because the target for a cancer vaccine is not the tumor bed but the immune system that eventually targets the tumor. The end result is a dynamic interplay between the growth kinetics of tumor cells, the death of tumor cells from immune-mediated killing, the alterations in tumor microenvironment, and the development of specific immune memory. In a conventional sense, the biological characteristics of tumors undergoing vaccine treatment may seem stagnant, without significant impact on progression-free survival (PFS). With time, however, from the effects of long-term memory T-cells, improvements in overall survival have been seen in prostate cancer vaccine trials as well as in the check-point inhibitor data from melanoma (Hodi et al. 2010; Stein et al. 2011).

Numerous clinical trials have combined conventional anti-cancer agents with cancer vaccines (phase II and phase III studies are referenced in Table 27.4, ongoing trials are identified in Tables 27.5). While a confirmation is still pending in a randomized, controlled clinical trial, some of the published evidence suggests that the conventional chemotherapy can create a synergy with the cancer vaccines either in terms of immune-based endpoints or standard clinical outcomes. In a phase II study, 29 patients with extensive stage non-small cell lung cancer were treated with an autologous peripheral blood mononuclear cell (PBMC) vaccine protocol that involved infecting the autologous cells with adenoviral vector encoding p53 (Antonia et al. 2006). In those patients with tumor progression, salvage chemotherapy produced a greater tumor response rate (51.5 vs. < 8 % in historical controls) and an unexpectedly high 38 % survival rate at 1 year.

Table 27.4 Immunomodulation and small molecule targeted inhibitors

	Mechanism	Reference
Increased tumor immunogenicity and APC activation	Bortezomib induces heat shock protein 90 exposure	(Spisek et al. 2007)
	5-aza-2'-deocycitidine increases MHC class 1 and TAA expression	(Serrano et al. 2001)
	BRAF inhibitor PLX4720 increases TAA expression	(Boni et al. 2010)
Potentiation of T-cell and NK cell antitumor response	Imatinib stimulates DC-NK-cell cross-talk	(Taieb et al. 2006; Borg et al. 2004)
	Imatinib alleviates indoleamine 2,3-dioxygenase (IDO) induced immunosuppression	(Balachandran et al. 2011)
	Imatinib interferes with immunosuppressive functions of Treg cells	(Larmonier et al. 2008)
	Imatinib stimulates the development of anti-leukaemic TNF-secreting CD4 + T-cells	(Chen et al. 2008)
	Dasatinib enhances NK-cell reactivity	(Hayashi et al. 2012)
	Dasatinib increases peripheral CD56 + CD57 + & CD3 + CD57 + cells	(Hayashi et al. 2012)
	Dasatinib increases peripheral NK-cell and NKT-cells	(Serrano et al. 2001)
	Dasatinib increases peripheral CD8 + cells	(Serrano et al. 2001)
	Dasatinib decreases peripheral Tregs	(Taieb et al. 2006)
	Sunitinib inhibits CD33 + HLA-DR − and CD15 + CD14 − MDSCs and Tregs	(Farsaci et al. 2012; Finke et al. 2008)
	Saracitninb (Src family kinase inhibitor) increases CD62L(high)/CD44(high) central memory CD8(+) T-cells	(Takai et al. 2012)
	Pan-BCL-2 inhibitor GX15-070 inhibits Tregs	(Farsaci et al. 2010)
	Erlotinib increases MHC class I/II expression	(Pollack et al. 2011)

Only 30 % of patients who did not develop a p53-specific response to vaccination responded clinically to second-line chemotherapy, whereas 75 % of p53 cellular immune responders had objective clinical responses to chemotherapy after vaccination ($P = 0.08$). The data suggest synergy between cancer vaccine-induced immune response (based on a peripheral surrogate assay) and chemotherapy Table (27.6). Similar clinical synergism with post-progression chemotherapy (docetaxel) has been seen with dendritic cell-based vaccine and pox-viral vector vaccine in metastatic castration-resistant prostate cancer (Petrylak 2006; Arlen et al. 2006; Schlom et al. 2008).

Table 27.5 Examples of immunotherapy combined with conventional therapy in clinical trials with published data

Vaccine	Conventional therapy	Indication	Phase	Reference
Viral vector: rV-PSA/rVB7.1 prime/rF-PSA boost	Radiotherapy	Localized prostate cancer	II	(Gulley et al. 2005)
Dendritic cell: autologous PBMC activated with a PAP-GM-CSF fusion protein	Docetaxel	Metastatic CRPC	III (post hoc data)	(Petrylak 2006)
Viral vector: rV-PSA/rVB7.1 prime/rF-PSA boost	Docetaxel	Metastatic CRPC	III	(Arlen et al. 2006)
Viral vector: rV-PSA/rVB7.1 prime/rF-PSA boost	Nilutamide	Non-metastatic CRPC	II	(Madan et al. 2008)
L-BLP25 MUC-1 tumor associated antigen liposomal vaccine	Cyclophosphamide	Stable or responding stage IIIB or IV NSCLC after first-line chemotherapy	IIB	(Butts et al. 2005)
TG4010 (Mva-Muc1-Il2)	Cisplatin + Vinorelbine	Stage IIIB/IV NSCLC	II	(Ramlau et al. 2008)
TG4010 (Mva-Muc1-Il2)	Cisplatin + Gemcitabine	MUC-1 expressing advanced NSCLC	IIB	(Quoix et al. 2011)
CEA (605–613) peptide pulsed autologous PBMCs and CIK cells	Platinum agent + Vinorelbine	Stage IIIB/IV NSCLC	II	(Zhong et al. 2011)
Murine anti-idiotype monoclonal (MAb) called 1E10	Cyclophosphamide + Methotrexate	Metastatic breast cancer	I/II	(Soriano et al. 2011)
Peptide vaccine	Gemcitabine	Unresectable pancreatic cancer	II	(Yanagimoto et al. 2010)
Dendritic cell: autologous PBMC transduced with wild-type p53 gene, via adenoviral vector	Platinum, cetuximab, docetaxel, paclitaxel, epirubicin, irinotecan	Extensive stage post first line NSCLC	II	(Antonia et al. 2006)

(continued)

Table 27.5 (continued)

Vaccine	Conventional therapy	Indication	Phase	Reference
Autologous/allogeneic vaccine containing IFN alfa/ gamma-treated tumor + GM-CSF	Cyclophosphamide	Advanced solid cancer (multiple cancer types)	I/II	(Wiseman et al. 1989)
rV-PSA/TRICOM (Prostate)/rf-PSA/TRICOM	153Sm-EDTMP radiation	Androgen-insensitive prostate cancer	II	(Heery et al. 2012)
PSA-TRICOM	Docetaxel	Metastatic CRPC	II	(Madan et al. 2009)

rV = recombinant vaccinia, rF = rcombinant fowlfox, PAP = prostatic acid phosphatase, CRPC = castration resistant prostate cancer, NSCLC = non-small cell lung cancer, GM-CSF = granulocyte macrophage colony stimulating factor, CEA = carcinoembryonic antigen, CIK = cytokine induced killer, IFN = interferon, TRICOM = TRIad of COstimulatory Molecules, Sm-EDTMP = Samarium-153-ethylene diamine tetramethylene phosphonate

Table 27.6 Clinical trials with unpublished results (vaccines combined with conventional therapy)

Vaccine	Conventional Therapy	Indication	Trial Phase	Clinical Trials.gov Identifier
Genitourinary Malignancies				
Attenuated modified vaccinia virus Ankara (MVA), 5T4 antigen (TroVax®)	Docetaxel	Hormone refractory prostate cancer (HRPC)	II	NCT01194960
ProstAtak™(AdV-tk + valacyclovir)	Radiation therapy + androgen deprivation therapy	Intermediate-high risk localized prostate cancer	III	NCT01436968
BLP25 liposome vaccine	Cyclophosphamide	High risk prostate cancer	II	NCT01496131
Peptide-based cancer vaccine with 13 tumor-associated peptides (TUMAPs) - IMA910	Cyclophosphamide/sunitinib	Advanced/metastatic renal cell carcinoma	III	NCT01265901
Breast Cancer				
Recombinant vaccinia-CEA-TRICOM, and recombinant fowlpox-CEA-TRICOM (B7.1/ICAM-1/LFA-3)	Doxorubicin/cyclophosphamide	Stage II/III breast cancer	II	NCT00052351
Adenovirus p53 infected DC vaccine	Doxorubicin, cyclophosphamide, paclitaxel, radiotherapy	p53-overexpressing stage III breast cancer	I/II	NCT00082641
Multiepitope DC vaccine	Trastuzumab, vinorelbine	Locally recurrent or metastatic breast cancer	II	NCT00266110
Allogeneic GM-CSF-secreting breast tumor vaccine	Trastuzumab, cyclophosphamide	High risk/metastatic HER-2/Neu- overexpressing breast cancer	II	NCT00847171
PANVAC ™-V (Vaccinia) and PANVAC ™-F (Fowlpox)	Docetaxel	Metastatic breast cancer	II	NCT00179309
Thoracic Malignancies				
BLP25 liposome vaccine	Cyclophosphamide	Stage IIIB or stage IV NSCLC	II	NCT00157209
BLP25 liposome vaccine	Cyclophosphamide	Stage III, unresectable, NSCLC	III	NCT01015443
TGF β-2 antisense gene-modified allogeneic tumor cell vaccine (Lucanix™)	Platinum-combination chemotherapy	Stages III/IV non-small cell lung cancer	III	NCT00676507

(continued)

Table 27.6 (continued)

Vaccine	Conventional Therapy	Indication	Trial Phase	Clinical Trials.gov Identifier
Adenovirus p53 infected DC vaccine	Vaccine with or without all trans-retinoic acid (paclitaxel on progression)	Extensive stage small cell lung cancer	II	NCT00617409
GM-CSF gene-modified autologous tumor vaccine (CG8123)	Cyclophosphamide	Advanced stage NSCLC	II	NCT00089726
TroVax® vaccine	Pemetrexed/cisplatin	Malignant pleural mesothelioma	II	NCT01569919
GastroIntestinal Cancer				
Autologous whole cell vaccine, FANG™, with the rhGMCSF transgene and the bifunctional shRNA furin	Oxaliplatin, fluorouracil	Colorectal carcinoma with liver metastases	II	NCT01505166
G17T immunogen	Cisplatin (CDDP), 5-fluorouracil (5-FU)	Metastatic or locally recurrent gastric/gastroesophageal cancer	III	NCT00020787
Intralymphatic allogeneic pancreatic cancer cells treated with interferon alfa +GM-CSF	Cyclophosphamide	Advanced pancreatic cancer	II	NCT00002773
Allogeneic pancreatic tumor cells transfected with the GM-CSF gene	5-fluorouracil, radiation	Resected stage I/stage II pancreatic adenocarcinoma	II	NCT00084383
Yeast-*ras* (GI-4000)	FOLFOX or FOLFIRI	Stage IV colon cancer	II	NCT01322815
Peptide-based cancer vaccine with 13 tumor-associated peptides (TUMAPs)—IMA910	Cyclophosphamide	Advanced colorectal carcinoma	I/II	NCT00785122
Allogeneic pancreatic tumor cells transfected with GM-CSF gene	Cyclophosphamide, cetuximab	Metastatic or locally advanced pancreatic cancer	II	NCT00305760
BLP25 liposome vaccine	Concomitant with cyclophosphamide, 5-fluorouracil, radiation	Rectal cancer	II	NCT01507103
Yeast-*ras* (GI-4000)	Gemcitabine	Non-metastatic, post-resection pancreatic cancer	II	NCT00300950

(continued)

Table 27.6 (continued)

Vaccine	Conventional Therapy	Indication	Trial Phase	Clinical Trials.gov Identifier
KLH-pulsed autologous DC vaccine	Radiation	Unresectable pancreatic cancer	II	NCT00868114
Central Nervous System Malignancies				
Autologous glioma lysate-derived DC vaccine	Vaccine following adjuvant temozolomide + radiotherapy	Glioblastoma multiforme	II	NCT00323115
Trivax, 5x10e6 autologous interleukin-12 secreting DC with autologous tumor lysate	Temozolomide + radiation	Glioblastoma multiforme	II	NCT01213407
PEP-3-KLH conjugate vaccine	Temozolomide (with versus without daclizumab/basiliximab)	Glioblastoma multiforme	I/II	NCT00626015
Allogeneic tumor cell vaccination	Oral metronomic cytoxan	Neuroblastoma	I/II	NCT01192555
Melanoma				
IDO/survivin peptide vaccine	Imiquimod and temozolomide	Metastatic melanoma	II	NCT01543464
Melanoma peptide-loaded DC vaccine	Cyclophosphamide	Stage IV melanoma	II	NCT00722098
Hematological malignancies				
Hodgkin's antigens-GM-CSF-expressing cell vaccine	Rituximab/cyclophosphamide	Relapsed Hodgkin's lymphoma	I/II	NCT00134082
DC/AML vaccine	AML induction chemotherapy (with PD-1 blocker)	AML	II	NCT01096602
Autologous immunoglobulin idiotype-KLH conjugate vaccine (FavId®)	Etoposide, doxorubicin, vincristine, cyclophosphamide	Mantle cell lymphoma	II	NCT00005780

27.7.2 Radiation and Cancer Vaccines

The clinical proof of concept of synergy between radiation and cancer vaccine comes from a phase II clinical trial in localized prostate cancer that utilized a vaccine with a priming dose of rV-PSA admixed with B7-1, followed by monthly boosts of rF-PSA for a total of eight vaccinations. Around 70 % of the 19 patients randomized to receive vaccine with radiation therapy had a $\geq$ 3-fold increase in PSA-specific T-cells, compared to no change in T-cells in patients treated with radiation alone ($P = 0.0005$) (Gulley et al. 2005).

Synergy in terms of antigen cascade was also seen in another phase II single-arm clinical trial that included 18 patients with localized prostate cancer. The patients were treated with recombinant virus vaccine encoding PSA and co-stimulatory molecule B7-1 combined with radiation therapy. When the PBMCs from 3 patients with HLA-A2 haplotype were evaluated for systemic immune response, a definite evidence of antigen cascade was noted. Two patients developed immunoreactivity to XAGE-1 and a third developed immunoreactivity to PAGE-4, both members of the PAGE/GAGE gene family but not the target of the original vaccine (Lechleider et al. 2008).

The interim results from a phase II, placebo-controlled, multi-center trial of the viral vector vaccine PSA-TRICOM with bone-seeking radionuclide (samarium-153) showed that the combination was well tolerated. Of 34 evaluable patients, six patients had $\geq$ 30 % PSA decline (2 with $\geq$ 50 %) in the combination group compared to none in the samarium only group. PFS was 117 days in the combination group as compared to 60 days in the samarium only group (Heery et al. 2012).

27.7.3 Hormone Ablation and Cancer Vaccines

When used alone, androgen ablation is associated with the induction of dense immunological infiltrates into the prostate gland (Mercader et al. 2001; Gannon et al. 2009). Key neo-adjuvant studies support combining hormone ablation and cancer vaccine. The synergy is most evident in prostate cancer, which is a good model to study hormone ablation and vaccine combination in neo-adjuvant, PSA-recurrent and metastatic settings due to its relatively slow growth curve and lower tumor burden. An attractive approach is to combine androgen ablation with cancer vaccine in men with PSA-recurrent prostate cancer especially after prostatectomy when the tumor burden is low and tumor induced tolerance is expected to be minimal. In a phase II clinical trial in non-metastatic castration resistant prostate cancer (CRPC) treated with recombinant viral vector in a prime and boost strategy, evidence of synergy was noted between vaccine and the androgen receptor blocker nilutamide (Madan et al. 2008). Median overall survival for patients on nilutamide after progression on vaccine was 6.2 years compared to 3.7 years for patients on vaccine following progression on nilutamide ($P = 0.045$).

Biological activity of autologous cellular immunotherapy (sipuleucel-T) has been demonstrated in patients with androgen-dependent prostate cancer (ADPC) on androgen deprivation. In a double-blind, randomized, controlled, multi-center study, 176 patients with biochemical recurrence after prostatectomy were randomized into sipuleucel-T and placebo arms after a 3–4 month run-in period of androgen suppression therapy (Beer et al. 2011). Sipuleucel-T patients had a 48 % increase in PSA doubling time following testosterone recovery (155 vs. 105 days, $P = 0.038$). The impact of sequencing androgen deprivation therapy (before, concurrent or after sipuleucel-T) on systemic immune responses will be investigated in at least two phase 2, randomized clinical trials (ClinicalTrials.gov Identifiers NCT01431391 and NCT01487863).

27.8 Novel Immunotherapies

One of the most exciting of the combination immunotherapy approaches involves combining vaccine therapies with immune checkpoint blockers such as anti-CTLA4 antibodies (e.g., ipilimumab). Ipilimumab has been shown in several preclinical models to enhance the avidity of T-cells and to enhance antitumor effects in combination with vaccines such as CEA-TRICOM recombinant poxviruses (Hodi et al. 2003). In a recent phase 1 dose escalation trial, ipilimumab was shown to be safe and tolerable when used in combination with PSA-TRICOM vaccine in patients with metastatic CRPC (Madan et al. 2012). The programmed death 1 (PD-1) receptor is an inhibitory T-cell receptor that is engaged by its two known ligands, PD-L1 and PD-L2, and leads to immunosuppression. The receptor and its ligand are a different set of immune-checkpoint molecules studied in extensive detail. Evidence of safety and early clinical efficacy of anti–PD-1 antibody and anti–PD-L1 antibody have been published (Topalian et al. 2012; Brahmer et al. 2012). Preclinical data show that PD-1 blockade synergizes with Treg-cell suppression by a single low dose of cyclosphosphamide, leading to an enhanced therapeutic outcome of cancer vaccine (Mkrtichyan et al. 2011).

27.9 Future Directions

As new cytotoxic, hormonal and small molecule targeted therapies become available, an understanding of the optimal combination of various modalities will be paramount in moving the field of combinatorial immunotherapy forward. Because of the established rationale of the approach and demonstrable synergism in the pre-clinical and clinical studies, we envision a multimodal immunotherapeutic approach to be an important way to address the host's latent cancer burden not addressed by primary modalities. Optimization of bio-markers, appropriate patient population, pre-treatment prognostic immune analysis, clinical end-points and the dose, right window and sequence of the combining agent need to be defined and validated in controlled settings. Apart from focusing on the advanced stage cancers around which most studies are primarily designed, the combinatorial approaches will need to be tested in the early stage cancers as well as on a wider scale. Larger clinical trials to test the feasibility of a combinatorial strategy involving immune checkpoint blockade and conventional regimens are also needed. We think that with the early clinical data and with the multiple trials that are ongoing in the combinatorial setting, the outlook for the field is positive.

References

Andersen MH, Schrama D, Thor Straten P, Becker JC (2006) Cytotoxic T-cells. J Invest Dermatol 126:32–41
Antonia SJ, Mirza N, Fricke I et al (2006) Combination of p53 cancer vaccine with chemotherapy in patients with extensive stage small cell lung cancer. Clin Cancer Res 12:878–887
Apetoh L, Ghiringhelli F, Tesniere A et al (2007) Toll-like receptor 4-dependent contribution of the immune system to anticancer chemotherapy and radiotherapy. Nat Med 13:1050–1059
Aragon-Ching JB, Williams KM, Gulley JL (2007) Impact of androgen-deprivation therapy on the immune system: implications for combination therapy of prostate cancer. Front Biosci 12:4957–4971
Arlen PM, Gulley JL, Parker C et al (2006) A randomized phase II study of concurrent docetaxel plus vaccine versus vaccine alone in metastatic androgen-independent prostate cancer. Clin Cancer Res 12:1260–1269
Balachandran VP, Cavnar MJ, Zeng S et al (2011) Imatinib potentiates antitumor T-cell responses in gastrointestinal stromal tumor through the inhibition of Ido. Nat Med 17:1094–1100
Beer TM, Bernstein GT, Corman JM et al (2011) Randomized trial of autologous cellular immunotherapy with sipuleucel-T in androgen-dependent prostate cancer. Clin Cancer Res 17:4558–4567
Boni A, Cogdill AP, Dang P et al (2010) Selective BRAFV600E inhibition enhances T-cell recognition of melanoma without affecting lymphocyte function. Cancer Res 70:5213–5219
Borg C, Terme M, Taieb J et al (2004) Novel mode of action of c-kit tyrosine kinase inhibitors leading to NK cell-dependent antitumor effects. J Clin Invest 114:379–388
Brahmer JR, Tykodi SS, Chow LQ et al (2012) Safety and activity of anti-PD-L1 antibody in patients with advanced cancer. N Engl J Med 366:2455–2465
Butts C, Murray N, Maksymiuk A et al (2005) Randomized phase IIB trial of BLP25 liposome vaccine in stage IIIB and IV non-small-cell lung cancer. J Clin Oncol 23:6674–6681
Chakraborty M, Abrams SI, Coleman CN, Camphausen K, Schlom J, Hodge JW (2004) External beam radiation of tumors alters phenotype of tumor cells to render them susceptible to vaccine-mediated T-cell killing. Cancer Res 64:4328–4337
Chamoto K, Takeshima T, Wakita D et al (2009) Combination immunotherapy with radiation and CpG-based tumor vaccination for the eradication of radio- and immuno-resistant lung carcinoma cells. Cancer Sci 100:934–939
Chen CI, Maecker HT, Lee PP (2008) Development and dynamics of robust T-cell responses to CML under imatinib treatment. Blood 111:5342–5349
Curiel TJ, Coukos G, Zou L et al (2004) Specific recruitment of regulatory T-cells in ovarian carcinoma fosters immune privilege and predicts reduced survival. Nat Med 10:942–949

DeNardo DG, Barreto JB, Andreu P et al (2009) CD4(+) T-cells regulate pulmonary metastasis of mammary carcinomas by enhancing protumor properties of macrophages. Cancer Cell 16:91–102

Drake CG, Doody AD, Mihalyo MA et al (2005) Androgen ablation mitigates tolerance to a prostate/prostate cancer-restricted antigen. Cancer Cell 7:239–249

Driessens G, Kline J, Gajewski TF (2009) Costimulatory and coinhibitory receptors in anti-tumor immunity. Immunol Rev 229:126–144

Farsaci B, Sabzevari H, Higgins JP et al (2010) Effect of a small molecule BCL-2 inhibitor on immune function and use with a recombinant vaccine. Int J Cancer 127:1603–1613

Farsaci B, Higgins JP, Hodge JW (2012) Consequence of dose scheduling of sunitinib on host immune response elements and vaccine combination therapy. Int J Cancer 130:1948–1959

Finke JH, Rini B, Ireland J et al (2008) Sunitinib reverses type-1 immune suppression and decreases T-regulatory cells in renal cell carcinoma patients. Clin Cancer Res 14:6674–6682

Gameiro SR, Caballero JA, Higgins JP, Apelian D, Hodge JW (2011) Exploitation of differential homeostatic proliferation of T-cell subsets following chemotherapy to enhance the efficacy of vaccine-mediated antitumor responses. Cancer Immunol Immunother 60:1227–1242

Gameiro SR, Caballero JA, Hodge JW (2012) Defining the molecular signature of chemotherapy-mediated lung tumor phenotype modulation and increased susceptibility to T-cell killing. Cancer Biother Radiopharm 27:23–35

Gannon PO, Poisson AO, Delvoye N, Lapointe R, Mes-Masson AM, Saad F (2009) Characterization of the intra-prostatic immune cell infiltration in androgen-deprived prostate cancer patients. J Immunol Methods 348:9–17

Garnett CT, Palena C, Chakraborty M, Tsang KY, Schlom J, Hodge JW (2004) Sublethal irradiation of human tumor cells modulates phenotype resulting in enhanced killing by cytotoxic T lymphocytes. Cancer Res 64:7985–7994

Garnett CT, Schlom J, Hodge JW (2008) Combination of docetaxel and recombinant vaccine enhances T-cell responses and antitumor activity: effects of docetaxel on immune enhancement. Clin Cancer Res 14:3536–3544

Garnett CT, Farsaci B, Schlom J, Hodge JW (2010) Docetaxel modulates phenotype of human carcinoma cells, including drug-resistant tumor cells, resulting in enhanced killing by CTLs [poster presentation, AACR Annual Meeting]. Cancer Res, 70:abstr 4790

Gaugler MH, Squiban C, van der Meeren A, Bertho JM, Vandamme M, Mouthon MA (1997) Late and persistent up-regulation of intercellular adhesion molecule-1 (ICAM-1) expression by ionizing radiation in human endothelial cells in vitro. Int J Radiat Biol 72:201–209

Ge Q, Hu H, Eisen HN, Chen J (2002) Different contributions of thymopoiesis and homeostasis-driven proliferation to the reconstitution of naive and memory T-cell compartments. Proc Natl Acad Sci U S A 99:2989–2994

Gelbard A, Garnett CT, Abrams SI et al (2006) Combination chemotherapy and radiation of human squamous cell carcinoma of the head and neck augments CTL-mediated lysis. Clin Cancer Res 12:1897–1905

Ghiringhelli F, Larmonier N, Schmitt E et al (2004) CD4 + CD25 + regulatory T-cells suppress tumor immunity but are sensitive to cyclophosphamide which allows immunotherapy of established tumors to be curative. Eur J Immunol 34:336–344

Girardi M, Glusac E, Filler RB et al (2003) The distinct contributions of murine T-cell receptor (TCR)gammadelta + and TCRalphabeta + T-cells to different stages of chemically induced skin cancer. J Exp Med 198:747–755

Greenberg PD, Cheever MA, Fefer A (1980) Detection of early and delayed antitumor effects following curative adoptive chemo immunotherapy of established leukemia. Cancer Res 40:4428–4432

Greenwald RJ, Freeman GJ, Sharpe AH (2005) The B7 family revisited. Annu Rev Immunol 23:515–548

Gulley JL, Arlen PM, Bastian A et al (2005) Combining a recombinant cancer vaccine with standard definitive radiotherapy in patients with localized prostate cancer. Clin Cancer Res 11:3353–3362

Hayashi Y, Nakamae H, Katayama T et al (2012) Different immuno profiles in patients with chronic myeloid leukemia treated with imatinib, nilotinib or dasatinib. Leuk Lymphoma 53:1084–1089

Heery C, Madan RA, Bilusic M, et al (2012) Interim analysis of a phase II randomized clinical trial of samrium-153 (Sm-153) with or without PSA-TRICOM vaccine in metastatic castration-resistant prostate cancer after docetaxel. J Clin Oncol, 30:abstr 2526

Hodge JW, Ardiani A, Farsaci B, Kwilas AR, Gameiro SR (2012) The tipping point for combination therapy: cancer vaccines with radiation, chemotherapy, or targeted small molecule inhibitors. Semin Oncol 39:323–339

Hodi FS, Mihm MC, Soiffer RJ et al (2003) Biologic activity of cytotoxic T lymphocyte-associated antigen 4 antibody blockade in previously vaccinated metastatic melanoma and ovarian carcinoma patients. Proc Natl Acad Sci U S A 100:4712–4717

Hodi FS, O'Day SJ, McDermott DF et al (2010) Improved survival with ipilimumab in patients with metastatic melanoma. N Engl J Med 363:711–723

Hsueh EC, Gupta RK, Lefor A, Reyzin G, Ye W, Morton DL (2003) Androgen blockade enhances response to melanoma vaccine. J Surg Res 110:393–398

Kepp O, Galluzzi L, Martins I et al (2011) Molecular determinants of immunogenic cell death elicited by anticancer chemotherapy. Cancer Metastasis Rev 30:61–69

Larmonier N, Janikashvili N, LaCasse CJ et al (2008) Imatinib mesylate inhibits CD4 + CD25 + regulatory T-cell activity and enhances active immunotherapy against BCR-ABL- tumors. J Immunol 181:6955–6963

Lechleider RJ, Arlen PM, Tsang KY et al (2008) Safety and immunologic response of a viral vaccine to prostate-specific antigen in combination with radiation therapy when metronomic-dose interleukin 2 is used as an adjuvant. Clin Cancer Res 14:5284–5291

Lugade AA, Moran JP, Gerber SA, Rose RC, Frelinger JG, Lord EM (2005) Local radiation therapy of B16 melanoma tumors increases the generation of tumor antigen-specific effector cells that traffic to the tumor. J Immunol 174:7516–7523

Lutsiak ME, Semnani RT, De Pascalis R, Kashmiri SV, Schlom J, Sabzevari H (2005) Inhibition of CD4 (+)25 + T regulatory cell function implicated in enhanced immune response by low-dose cyclophosphamide. Blood 105:2862–2868

Machiels JP, Reilly RT, Emens LA et al (2001) Cyclophosphamide, doxorubicin, and paclitaxel enhance the antitumor immune response of granulocyte/macrophage-colony stimulating factor-secreting whole-cell vaccines in HER-2/neu tolerized mice. Cancer Res 61:3689–3697

Madan RA, Gulley JL, Schlom J et al (2008) Analysis of overall survival in patients with non-metastatic castration-resistant prostate cancer treated with vaccine, nilutamide, and combination therapy. Clin Cancer Res 14:4526–4531

Madan RA, Dahut W, Figg WD et al (2009) Comparing the overall survival of metastatic castration resistant prostate cancer patients treated with docetaxel, a vaccine admixed with one costimulatory molecule, and a vaccine with three costimulatory molecules [poster]. ASCO Genitourinary Cancers Symposium, abstr 210

Madan RA, Mohebtash M, Arlen PM et al (2012) Ipilimumab and a poxviral vaccine targeting prostate-specific antigen in metastatic castration-resistant prostate cancer: a phase 1 dose-escalation trial. Lancet Oncol 13:501–508

Mantovani A, Sozzani S, Locati M, Allavena P, Sica A (2002) Macrophage polarization: tumor-associated macrophages as a paradigm for polarized M2 mononuclear phagocytes. Trends Immunol 23:549–555

Marincola FM, Jaffee EM, Hicklin DJ, Ferrone S (2000) Escape of human solid tumors from T-cell recognition: molecular mechanisms and functional significance. Adv Immunol 74:181–273

Mellman I, Coukos G, Dranoff G (2011) Cancer immunotherapy comes of age. Nat 480:480–489

Mercader M, Bodner BK, Moser MT et al (2001) T-cell infiltration of the prostate induced by androgen withdrawal in patients with prostate cancer. Proc Natl Acad Sci U S A 98:14565–14570

Mkrtichyan M, Najjar YG, Raulfs EC et al (2011) Anti-PD-1 synergizes with cyclophosphamide to induce potent anti-tumor vaccine effects through novel mechanisms. Eur J Immunol 41:2977–2986

Moschella F, Valentini M, Arico E et al (2011) Unraveling cancer chemo immunotherapy mechanisms by gene and protein expression profiling of responses to cyclophosphamide. Cancer Res 71:3528–3539

North RJ (1982) Cyclophosphamide-facilitated adoptive immunotherapy of an established tumor depends on elimination of tumor-induced suppressor T-cells. J Exp Med 155:1063–1074

Obeid M, Tesniere A, Ghiringhelli F et al (2007) Calreticulin exposure dictates the immunogenicity of cancer cell death. Nat Med 13:54–61

Page ST, Plymate SR, Bremner WJ et al (2006) Effect of medical castration on CD4 + CD25 + T-cells, CD8 + T-cell IFN-gamma expression, and NK cells: a physiological role for testosterone and/or its metabolites. Am J Physiol Endocrinol Metab 290:E856–E863

Petrylak D (2006) Defining the optimal role of immunotherapy and chemotherapy: advanced prostate cancer patients who receive sipuleucel-T (PROVENGE) followed by docetaxel derive greatest survival benefit. Chemotherapy Foundation Symposium 9–13

Pollack BP, Sapkota B, Cartee TV (2011) Epidermal growth factor receptor inhibition augments the expression of MHC class I and II genes. Clin Cancer Res 17:4400–4413

Quoix E, Ramlau R, Westeel V et al (2011) Therapeutic vaccination with TG4010 and first-line chemotherapy in advanced non-small-cell lung cancer: a controlled phase 2B trial. Lancet Oncol 12:1125–1133

Ramlau R, Quoix E, Rolski J et al (2008) A phase II study of Tg4010 (Mva-Muc1-Il2) in association with chemotherapy in patients with stage III/IV Non-small cell lung cancer. J Thorac Oncol 3:735–744

Reits EA, Hodge JW, Herberts CA et al (2006) Radiation modulates the peptide repertoire, enhances MHC class I expression, and induces successful antitumor immunotherapy. J Exp Med 203:1259–1271

Roden AC, Moser MT, Tri SD et al (2004) Augmentation of T-cell levels and responses induced by androgen deprivation. J Immunol 173:6098–6108

Rohon P, Porkka K, Mustjoki S (2010) Immuno profiling of patients with chronic myeloid leukemia at diagnosis and during tyrosine kinase inhibitor therapy. Eur J Haematol 85:387–398

Rotow J, Gameiro SR, Madan RA, Gulley JL, Schlom J, Hodge JW (2010) Vaccines as monotherapy and in combination therapy for prostate cancer. Clin Transl Sci 3:116–122

Salem ML, Al-Khami AA, El-Naggar SA, Diaz-Montero CM, Chen Y, Cole DJ (2010) Cyclophosphamide induces dynamic alterations in the host microenvironments resulting in a Flt3 ligand-dependent expansion of dendritic cells. J Immunol 184:1737–1747

Schiavoni G, Mattei F, Di Pucchio T et al (2000) Cyclophosphamide induces type I interferon and augments the number of CD44(hi) T lymphocytes in mice: implications for strategies of chemo immunotherapy of cancer. Blood 95:2024–2030

Schiavoni G, Sistigu A, Valentini M et al (2011) Cyclophosphamide synergizes with type I interferons through systemic dendritic cell reactivation and induction of immunogenic tumor apoptosis. Cancer Res 71:768–778

Schlom J (2012) Therapeutic cancer vaccines: current status and moving forward. J Natl Cancer Inst 104:599–613

Schlom J, Gulley JL, Arlen PM (2008) Paradigm shifts in cancer vaccine therapy. Exp Biol Med (Maywood) 233:522–534

Serrano A, Tanzarella S, Lionello I et al (2001) Rexpression of HLA class I antigens and restoration of antigen-specific CTL response in melanoma cells following 5-aza-2'-deoxycytidine treatment. Int J Cancer 94:243–251

Smyth MJ, Thia KY, Street SE et al (2000) Differential tumor surveillance by natural killer (NK) and NKT-cells. J Exp Med 191:661–668

Soriano JL, Batista N, Santiesteban E et al (2011) Metronomic cyclophosphamide and methotrexate chemotherapy combined with 1E10 anti-Idiotype vaccine in metastatic breast cancer. Int J Breast Cancer 2011:710292

Spisek R, Charalambous A, Mazumder A, Vesole DH, Jagannath S, Dhodapkar MV (2007) Bortezomib enhances dendritic cell (DC)-mediated induction of immunity to human myeloma via exposure of cell surface heat shock protein 90 on dying tumor cells: therapeutic implications. Blood 109:4839–4845

Stein WD, Gulley JL, Schlom J et al (2011) Tumor regression and growth rates determined in five intramural NCI prostate cancer trials: the growth rate constant as an indicator of therapeutic efficacy. Clin Cancer Res 17:907–917

Sutherland JS, Goldberg GL, Hammett MV et al (2005) Activation of thymic regeneration in mice and humans following androgen blockade. J Immunol 175:2741–2753

Taieb J, Chaput N, Menard C et al (2006) A novel dendritic cell subset involved in tumor immunosurveillance. Nat Med 12:214–219

Takai S, Sabzevari H, Farsaci B, Schlom J, Greiner JW (2012) Distinct effects of saracatinib on memory CD8 + T-cell differentiation. J Immunol 188:4323–4333

Tesniere A, Schlemmer F, Boige V et al (2010) Immunogenic death of colon cancer cells treated with oxaliplatin. Oncogene 29:482–491

Thomas DA, Massague J (2005) TGF-beta directly targets cytotoxic T-cell functions during tumor evasion of immune surveillance. Cancer Cell 8:369–380

Topalian SL, Hodi FS, Brahmer JR et al (2012) Safety, activity, and immune correlates of anti-PD-1 antibody in cancer. N Engl J Med 366:2443–2454

Wiseman CL, Rao VS, Kennedy PS, Presant CA, Smith JD, McKenna RJ (1989) Clinical responses with active specific intralymphatic immunotherapy for cancer–a phase I-II trial. West J Med 151:283–288

Yanagimoto H, Shiomi H, Satoi S et al (2010) A phase II study of personalized peptide vaccination combined with gemcitabine for non-resectable pancreatic cancer patients. Oncol Rep 24:795–801

Ye GW, Park JB, Park YJ, Choi YS, Sin JI (2007) Increased sensitivity of radiated murine cervical cancer tumors to E7 subunit vaccine-driven CTL-mediated killing induces synergistic anti-tumor activity. Mol Ther 15:1564–1570

Young MR, Wright MA, Coogan M, Young ME, Bagash J (1992) Tumor-derived cytokines induce bone marrow suppressor cells that mediate immunosuppression through transforming growth factor beta. Cancer Immunol Immunother 35:14–18

Zhong R, Teng J, Han B, Zhong H (2011) Dendritic cells combining with cytokine-induced killer cells synergize chemotherapy in patients with late-stage non-small cell lung cancer. Cancer Immunol Immunother 60:1497–1502

Zitvogel L, Apetoh L, Ghiringhelli F, Kroemer G (2008) Immunological aspects of cancer chemotherapy. Nat Rev Immunol 8:59–73

Zitvogel L, Kepp O, Senovilla L, Menger L, Chaput N, Kroemer G (2010) Immunogenic tumor cell death for optimal anticancer therapy: the calreticulin exposure pathway. Clin Cancer Res 16:3100–3104

Chapter 28
Prophylactic Cancer Vaccines

Pamela L. Beatty and Olivera J. Finn

Abstract Immunotherapy of cancer has had little impact on overall patient survival. This is largely due to the numerous mechanisms that growing tumors use to inhibit, subvert, suppress, or tolerize an effective antitumor immune responses. The importance of the immune system and vaccines as tools to strengthen it with the aim of cancer prevention has been grossly overlooked in spite of accumulating evidence that many difficulties that challenge vaccines for cancer therapy would not exist in the setting of prevention. The target of immunoprevention in early stage of disease, and preferably in premalignancy, would be cells that did not yet undergo immune selection and their microenvironment has not been shaped by a growing and evolving tumor. Several premalignant conditions have been identified and antigens that characterize premalignant disease have been characterized. This offers to date unexplored opportunity to utilize vaccines to prevent cancer.

Keywords Immunotherapy · Cancer vaccines · Immunization · Antitumor immunity · Immunosurveillance · Tumor antigens

28.1 Rationale for Cancer Prevention

With a majority of the world considered polio-free and the complete eradication of smallpox, it is clear that vaccination as a means to protect against infectious disease has been an extremely effective approach. More recently, vaccines for prevention of certain cancer have been granted FDA approval, including vaccines

P. L. Beatty · O. J. Finn (✉)
Department of Immunology, University of Pittsburgh School of Medicine, E1040 Biomedical Science Tower, University of Pittsburgh, Pittsburgh, PA 15262, USA
e-mail: ojfinn@pitt.edu

M. R. Shurin et al. (eds.), *The Tumor Immunoenvironment*,
DOI: 10.1007/978-94-007-6217-6_28,

against hepatitis B virus (HBV) and human papilloma viruses (HPV), the causative agents of liver and cervical cancers respectively. Unfortunately, cancers that are not caused by viruses, or at least the viral connection has not yet been discovered, represent >80 % of all human cancers and in this context vaccination has been mostly attempted as a means of cancer therapy in patients who already have tumors. This has led to disappointing results and set-backs for the development and testing of prophylactic cancer vaccines (Finn 2003a).

All cancer cells at some point during their growth express tumor antigens derived from genetic alterations and epigenetic dysregulation of normal molecules that can then be recognized as abnormal by specific antibodies and T cells. Analysis of immune responses in patients with cancer has shown that many individuals spontaneously generate tumor antigen-specific immune responses. In the majority of cancer patients these immune responses are too weak and fail to arrest tumor progression. Strong tumor specific immune responses are very important as illustrated by numerous findings that the presence of high numbers of infiltrating T cells into the tumor mass is correlated with a more favorable prognosis in many cancer types (Clemente et al. 1996; Lee et al. 1999; Pages et al. 2005; Schumacher et al. 2001; Zhang et al. 2003). In colorectal cancer, for example, density and location of adaptive immune cells within tumors has been shown to be a better predictor of patient survival than the current histopathological staging (Galon et al. 2006). Many studies over the past several years in cancer patients and in preclinical models have revealed numerous mechanisms that can inhibit, subvert, suppress, or tolerize an effective antitumor T cell immune responses (Rabinovich et al. 2007; Whiteside 2010). The importance of the immune system and the opportunity to elicit strong anti-tumor immunity by vaccination for cancer prevention has been grossly overlooked even though emerging evidence shows that the difficulties that plague vaccines for cancer therapy would not exist in the setting of prevention. Naturally occurring Foxp3$^+$ cell accumulate in the microenvironment of growing tumors and suppress tumor-specific T cell responses and tumor expression of indoleamine 2, 3 dioxygenase (IDO), a tryptophan the catabolizing enzyme can increase the migration of Foxp3$^+$ regulatory cells into the tumor infiltrate. IDO can also down regulate MHC Class I molecules and decreases the expression of the T cell receptor zeta chain resulting in T cell anergy. Defects in antigen-processing machinery confer resistance of tumors to cytotoxic T lymphocyte recognition. This has provided evidence that an immune response to eliminate cancer must be directed at the earliest possible time (i.e. prevention).

28.2 Immune Control of Cancer and Metastasis

In the 1960s Burnet and Thomas proposed the "immunosurveillance" hypothesis (Burnet 1970). It stated that the immune system not only detects the presence of cancer but also destroys it prior to its development into clinically apparent tumor. At the time, the concept was difficult to test experimentally and the immune surveillance hypothesis remained contentious for years. It was not until the past decade with the use of gene-targeted mice that cancer immunosurveillance was confirmed. Rag knockout mice that lack recombinase-activating genes (RAG-1 or RAG-2), are unable to rearrange antigen-specific receptors and lack T and B lymphocytes and natural killer T (NKT) cells, develop chemically induced tumors more rapidly and with greater frequency than immunocompetent controls (Smyth et al. 2001, 2006). In addition they also developed a larger number of spontaneous tumors compared to immunocompetent controls. Other key findings from using Inf-$\gamma^{-/-}$ and perforin$^{-/-}$mice were that both IFN-γ and perforin are crucial components of an effective tumor immunosurveillance (Street et al. 2001).

The immunosurveillance hypothesis underwent a major revision with the discovery that tumors formed in mice that lacked an intact immune system were, as a group, more immunogenic than similar tumors derived from immunocompetent mice. This suggested that not only can the immune system protect the host from tumors but that the immune system exerts continued selective pressure on tumor cells that favors development and survival of cells that can escape the control by the immune system. The immunosurveillance hypothesis was replaced with the cancer immunoediting hypothesis that describes three distinct phases in tumor progression: *elimination, equilibrium and escape* (Dunn et al. 2004). *Elimination phase* encompasses the old cancer immunosurveillance hypothesis, in which the innate and the adaptive immune systems work together to detect the presence of a developing tumor and eliminate it before it becomes clinically apparent. If elimination is not complete, the tumor can enter a long *equilibrium phase* during which the presence of the tumor is not obvious but the adaptive immune system continues to control it by sculpting its immunogenicity. Under this immune pressure rare tumor cell variants can arise leading to the *escape phase* characterized by progressively growing tumors. Progression from *equilibrium* to *escape* can occur either because tumor cells change in response to the immune system's selective pressures and/or because the host immune system changes, as a result of age, for example. As this whole process is understood better, it begins to suggest that it might be possible to manipulate the immune system to favor elimination or life-long equilibrium with cancer and prevent escape. Studies are underway or have already been published showing that vaccination that elicits immunity against one or more tumor antigens expected to be on future tumors, can prevent clinical cancer.

Cancer metastasis represents another area of investigation that highlights an opportunity for prophylactic cancer vaccines. The majority of cancer-related deaths are due to tumor recurrences at sites often different than the primary tumor site. Dissemination of tumor cells from their primary site to distant organs is known as *tumor metastasis*, an orderly series of events starting with tumor cells detachment from the primary tumor, migration through stroma, intravasation into blood or lymph vessels, extravasation into a secondary site, and finally formation of metastatic nodules (Nguyen and Massague 2007). Metastasis has been viewed as a late-stage event in tumor development due to acquisition of additional mutations but recent studies have challenged this concept (Husemann et al. 2008; Podsypanina et al. 2008). Premalignant cells from mammary tissue in a mouse model of breast cancer were shown to disseminate to lungs and bone marrow prior to the appearance of the primary tumors (Husemann et al. 2008). If the earliest transformed cells are able to disseminate and grow distant metastasis then it becomes increasingly important to instruct the immune system to recognize not mature tumors but premalignant changes as well and eliminate them. This suggests that characterization of antigens on such lesions for many different cancer types would yield candidate vaccines for prevention of many cancers.

28.3 Evidence of Immunosurveillance in Human Cancer

It was once assumed that even if a cancer cell expressed tumor antigens, the tumor tissue could not support immune activation because it would be perceived as 'self' by the immune system. We now know that tumor growth is associated with disruption of normal tissue architecture and an important consequence of this disruption is the release of danger signals to alert the immune system (Matzinger 2002). It is now appreciated that the lack of an effective anti-tumor immunity is not a result of immune tolerance to 'self'. Rather, nonresponsiveness is due to multiple mechanisms of active tolerance and suppression of anti-tumor effector cells in the peculiar tumor microenvironment (Matzinger 2007). Early support for this came from paraneoplastic syndromes, a group of rare disorders that are associated with effective antitumor immunity (Albert and Darnell 2004). Patients with paraneoplastic neurological degenerations (PNDs) present initially with neurologic problems that include memory loss, sensory loss, cerebellar dysfunction, motor dysfunction or blindness, but after a more thorough examination they are found to harbor small adenocarcinomas. These tumors, which are most frequently breast, ovarian or small-cell lung cancers, express defined neuronal antigens, and the patients produce high titer antibodies and a high frequency of cytotoxic T cells against these antigens. A retrospective analysis revealed that this immunity, which is destructive to neuronal tissue, appears to have been generated in response to the tumor and to be suppressing the growth of the tumor (Albert et al. 1998; Darnell and Posner 2003). Moreover, evidence suggests that antigen-specific CTL are generated via cross-priming DCs that capture the PND antigen from apoptotic tumor cells for presentation to T cells (Darnell and Posner 2003).

Paraneoplastic myopathies are heterogeneous connective tissue diseases that mainly affect skeletal muscle (Aggarwal and Oddis). Patients present with symptoms of muscle weakness and/or rheumatic-like arthritis symptoms and other signs include rapid-onset of symptoms, atypical age at onset, poor response to corticosteroids or immunosuppressive therapy, atypical distribution of involved joints, and abnormal laboratory tests. Data collected on these patients support the hypothesis that the immune responses generated against muscle antigens is related to anti-tumor responses (Aggarwal and Oddis 2011; Racanelli et al. 2008). The clinical course usually parallels that of the primary tumor and in most cases surgical removal or pharmacological treatment of the tumor resolves the paraneoplastic symptoms. Paraneoplastic inflammatory myopathies have been observed as the early clinical manifestations of such cancers as ovarian, renal, lung and melanoma (Racanelli et al. 2008; Buchbinder et al. 2001; Schiller et al. 2006). The more we learn about the spontaneous relationship between the immune system and tumors as they develop, the more effective prevention methods based on the immune system could be developed.

28.4 Clinical Evidence of Human Immune Response to Premalignancy

The idea of directing the immune system against a tumor at an early stage of progression from normal to neoplastic has distinct advantages. The experience so far with cancer vaccines indicates that greatest barrier to success has been the administration of vaccine to patients in late stage disease where extensive and often long-term interplay between the immune system and the tumor has already occurred. The target of immunoprevention in early stage of disease, and preferably in premalignancy, would be cells that did not yet interact with the immune system and are not in a microenvironment that has been manipulated by a growing and evolving tumor. Cancer takes several years and sometime decades to become a clinically detectable abnormal tissue mass, but it is now known for some cancers that even at the early stages of malignant transformation, preneoplastic cells express antigenic targets for recognition by the immune system (Dhodapkar 2005; Finn 2003b).

Monoclonal gammopathy of undetermined significance (MGUS) is one of the most common premalignant plasma cell disorders and it is a precursor for multiple myeloma. Most patients with MGUS will generally remain non-symptomatic throughout their life, but about 1 % per year will progress to multiple myeloma (Kyle et al. 2002). Freshly isolated T cells from the bone marrow of patients with preneoplastic gammopathy were found to be enriched for T cells secreting IFN-γ in response to autologous tumor-loaded DCs (Dhodapkar et al. 2003). In addition, the T cells were capable of in vitro expansion in response to preneoplastic cells and this response was specific for the pattern of antigens expressed by autologous preneoplastic cells (Dhodapkar et al. 2003). Many of the cytogenetic and genomic changes in tumor cells initially identified in myeloma can now also be detected in the preneoplastic cells in MGUS (Fonseca et al. 2002; Zhan et al. 2002). In particular, aneuploidy as well as chromosome translocations that involve the immunoglobulin heavy chain (IgH) locus are commonly observed in multiple myeloma as well as its precursor MGUS.

28.5 Premalignant Disease and Target Antigens

The National Cancer Institute (NCI) initiated a pilot project to identify tumor antigens that should be prioritized for translation into the clinic (Cheever et al. 2009). The tumor antigens on the list have been well characterized for their tumor-specific expression, immunogenicity, and therapeutic potential. Although antigenic profiles in premalignant disease have not yet been fully studied and characterized, several antigens on the NCI prioritization list have shown promise as targets in premalignant disease (Lollini et al. 2006). In addition, molecular and genetic advancements have contributed to the cancer field with identification of premalignant lesions for many cancers. Premalignant conditions for lung, pancreatic, esophageal, colon, and oral cancers, and the potential target antigens for prevention of these cancers are discussed below.

28.5.1 MUC1 Antigen: Expression in Tumors and Premalignant Lesions

Epithelial mucin MUC1 is a glycoprotein that is expressed at low levels by normal epithelial cells. In the majority of adenocarcinomas MUC1 becomes profoundly overexpressed and hypoglycosylated taking on a very different appearance and function than on normal cells. Tumor MUC1 was the first human tumor antigen to be reported as a target for human cytotoxic T cells (Barnd et al. 1989). The reduced glycosylation exposes the peptide backbone resulting in peptide epitopes as well as truncated glycopeptide epitopes that are processed and presented to the immune system (Ryan et al. 2009; Vlad et al. 2002, 2004). These epitopes are exquisitely tumor specific and the immune system can destroy tumor cells expressing MUC1 while not causing damage to normal tissue expressing MUC1. Tumor forms of MUC1 have been found in many premalignant lesions. MUC1 was also found to have a cancer promoting role in cigarette smoke-induced lung carcinogenesis by potentiating bronchial epithelial transformation through the epidermal growth factor receptor pathway (Xu et al. 2012). More recently, development of genetically engineered mice has provided an opportunity to study in vivo cancer progression and correlation with MUC1expression. The KrasMUC1 mouse model was derived by crossing human MUC1transgenic (MUC1 Tg) mice that carry the human MUC1 transgene expressed under the endogenous MUC1 promoter and have the correct temporal and spatial pattern of MUC1 expression in human cancer (Rowse et al. 1998), with the KrasG12D where the Kras G to D mutation in codon 12 is kept silent by the lox-stop-lox sequence that can be removed by Cre recombinase to initiate tumorigenesis. The KrasMUC1 mouse develops twice as many premalignant and malignant lesions compared with the KrasG12D mice (Finn et al. 2011). Immunostaining of lung tissues in these mice shows increased expression of the tumor form of MUC1 on disorganized epithelia. In addition, the presence of MUC1

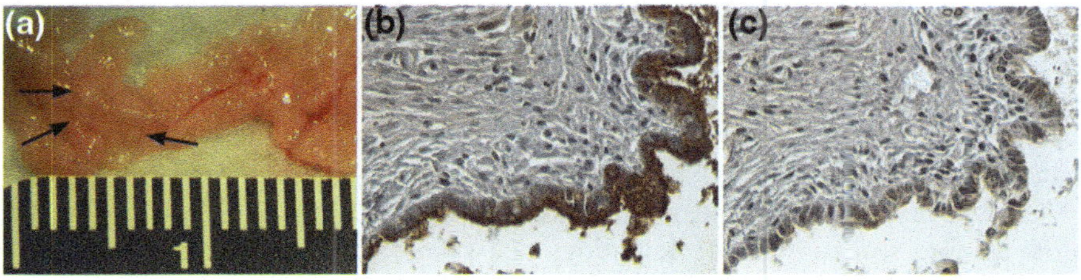

Fig. 28.1 Premalignant pancreatic cysts express the tumor form of MUC1. **a** Gross morphology of mouse pancreas showing several cystic nodules (*arrows*). Scale is in centimeters; **b** Immunostaining of pancreatic cyst with anti-MUC1 antibody HMPV that recognizes both normal and tumor forms of MUC1; **c** Immunostaining of pancreatic cyst with anti-MUC1 antibody 4H5 that preferentially recognizes the tumor form of MUC1

correlates with a large number of infiltrating cells reflecting the well known chemotactic role of tumor MUC1 (Carlos et al. 2005).

MUC1 has been shown to be expressed in pancreatic intraepithelial neoplasia (PanIN), precursors of invasive cancer. PanINs are microscopic lesions found in small (less than 5 mm) pancreatic ducts that accumulate histologic and genetic abnormalities in their progression towards invasive cancer. A multivariate analysis of pancreatic cysts found by MRI in patients with and without pancreatic cancer revealed that cyst presence was a significant risk factor for pancreatic cancer, especially cysts >10 mm in diameter (Matsubara et al. 2012). MUC1 expression has been shown to be increased on the pancreatic cysts from patients as well as in mouse models (Fig. 28.1). Figure 28.1 shows a mouse pancreas with several cystic nodules. The intense brown staining shows high expression of MUC1 and much of it is the hypoglycosylated tumor form (Fig. 28.1c). Molecular assays comparing patient tumor samples associated with short and long-term survival after resection for pancreatic ductal carcinoma revealed that MUC1 is a highly significant predictor of early cancer-specific mortality, and is superior to conventional pathologic features as prognostic markers (Winter et al. 2012). The KrasG12D mouse model has been adapted for the study of pancreatic cancer development as well (Finn et al. 2011; Mukherjee et al. 2000). The Krasp48 mouse activates the Kras mutation in the pancreas by co-expression of Cre recombinase under the pancreas-specific promoter p48. When crossed with MUC1 Tg mice the KrasMUC1p48 mice develop spontaneous pancreatic tumors that express human MUC1, which promotes accelerated tumor development and more numerous lesions compared to Krasp48 mice (Finn et al. 2011). Another spontaneous pancreatic tumor animal model is the MET mouse that is derived from crossing MUC1 Tg mice with ET mice that express the first 127 amino acids of the SV40 large T antigen under the control of the rat elastase promoter. MET mice follow a progression of dysplasia to microadenomas and eventual MUC1-positive pancreatic cancer (Gendler and Mukherjee 2001). Low level anti MUC1 antibody response and low frequency MUC1-specifc CD8 T cells can be detected early on in these mice, but wanes as the tumor progresses (Mukherjee et al. 2001). When MUC1-specific T cells are isolated from these mice and adoptively transferred they are able to prevent the growth of transplanted MUC1-expressing tumor cells (Mukherjee et al. 2001). Importantly this suggests that eliciting MUC1-specific T cells prior to tumor occurrence, as could be done by vaccination, would be effective in preventing MUC1-positive tumors.

A majority of colon cancers are thought to develop from adenomas that over time acquire somatic genetic mutations that stimulate transformation to invasive cancer. This is predicated on the well characterized adenoma-carcinoma sequence model (Fearon and Vogelstein 1990). MUC1 has been found to be significantly overexpressed in polyps, the extent of expression correlating with polyp size, degree of dysplasia and villous histology (Ajioka et al. 1997; Ho et al. 1996). Zotter et al. found that the underglycosylated epitope of MUC1 that is expressed in approximately 90 % of colorectal carcinomas is first expressed in severely dysplastic polyps (Zotter et al. 1987). In addition, patients with the history of adenomatous polyps have high levels of MUC1-specific IgG suggesting that helper T cells and B cells were in response to these premalignant changes.

It is now well established that chronic inflammation can serve as a promoter of cancer development at the affected site and can be considered a premalignant condition. Chronically inflamed sites contain a diverse population of leukocytes that express and secrete an assortment of mediators that foster cell proliferation and genomic instability and progression to malignancy (Balkwill et al. 2005; Coussens and Werb 2002). This association is exemplified in individuals with inflammatory bowel disease (IBD) where patients with chronic uncontrolled colitis have a significantly increased risk of developing colitis-associated colorectal cancer (CACC). The MUC1/IL-$10^{-/-}$ mouse model of spontaneous IBD validated MUC1 as a major player in IBD as well as progression to CACC. MUC1/IL-$10^{-/-}$ mice develop human MUC1-positive IBD as a result of an exaggerated immune response to intestinal microflora due to the complete lack of IL-10 that prevents immunoregulation. By breeding IL-$10^{-/-}$ mice with human MUC1Tg mice, the role of MUC1 can be studied in spontaneous chronic intestinal inflammation. MUC1/IL-$10^{-/-}$ mice developed more severe inflammation and accelerated occurrence of IBD and higher incidence of progression to colon cancer (Beatty et al. 2007). These results were recapitulated in another model, MUC1-DSS mouse, where chronic intestinal inflammation is induced in MUC1 Tg mice by administration of dextran sodium sulfate (DSS) in drinking water (Beatty et al. 2012).

Another chronic inflammatory condition is Barrett's Esophagus (BE), where healing from esophageal mucosal injury is metaplastic, with replacement of damaged squamous cells by columnar epithelium (Wiseman and Ang 2011). Mucosal injury can be triggered among other things by gastro-esophageal reflux disease (GERD) (Wiseman and Ang 2011; Barrett 1950). Esophageal healing typically involves regeneration of squamous cells, and since only a minority of patients with GERD develop BE, it is unclear why their response is metaplastic. Many of the genetic insults driving this metaplasia-dysplasia-carcinoma sequence have recently been characterized, providing targets for candidate biomarkers with potential clinical utility to stratify risk in individual patients. Mariette and

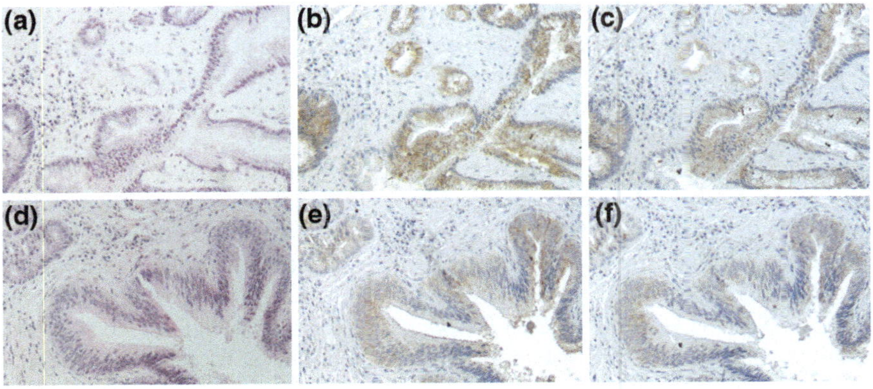

Fig. 28.2 MUC1 on Barrett's esophagus. **a** Histology of nondysplastic lesion; **b** Immunostaining of nondysplastic lesion with anti-MUC1 antibody HMPV that recognizes both normal and tumor forms of MUC1; **c** Immunostaining of nondysplastic lesion with anti-MUC1 antibody 4H5 that preferentially recognizes the tumor form of MUC1; **d** Histology of dysplastic lesion; **e** Immunostaing of high grade dysplasia with anti-MUC1 antibody HMPV; **f** Immunostaing of high grade dysplasia with anti-MUC1 antibody 4H5

colleagues found that bile acids are strong inducers of MUC1 expression in human esophageal tissues and that the regulation occurs at the transcriptional level (Mariette et al. 2008). Moreover, bile acids in patients with Barrett's esophagus may contribute to the metaplastic change in some patients. Several groups have looked at MUC1 immunostaining in the esophagus to see if it may serve as a biomarker of BE patients who are at increased risk for development of esophageal adenocarcinoma. Increased MUC1 expression was found in some but not all human nondysplastic Barrett's tissue samples and expression increased as lesions became more dysplastic (Fig. 28.2). Further studies are needed to determine if MUC1 may serve as a predictor of BE patients that will progress to esophageal adenocarcinoma. Clearly those patients would in the future be good candidates for an anti-MUC1 vaccine and vaccines against other antigens that might be identified in those studies, that would prevent progression from BE to esophageal cancer.

28.5.2 Cyclin B1: Expression in Tumors and Premalignant Lesions

Cyclin B1is expressed transiently in the nucleus at the transition from the G2 to M phase of the cell cycle during normal cell replication. Compared to normal cells, many tumors, including breast, lung, colorectal, head and neck, and lymphoma constitutively overexpress cyclin B1 in the cytoplasm. Cyclin B1 has been identified as a tumor antigen that is recognized by both human antibodies and T cells (Kao et al. 2001). Overexpression of cyclin B1 has been found to be the result of inactivation of p53 function, an early event in the carcinogenesis of many tumors and are found with high frequency in premalignant lesions (Yu et al. 2002). Further support for cyclin B1 as a cancer prevention antigen has come from studies in patients with lung cancer, early premalignant lesions, and patients with a history of smoking. Anti cyclin B1 antibodies where found to be significantly higher in heavy smokers compared to light smokers and non smokers. Importantly, this was a helper T cell-dependent IgG antibody response (Suzuki et al. 2005). Immunostaining of metaplastic and dysplastic lung lesions from heavy smokers showed that cyclin B1 protein was already overexpressed in these lesions at levels equal or higher than in fully transformed cancer cells. Another study found that cyclin B1 expression in resected early stage lung tumors was an adverse prognostic factor of survival (Soria et al. 2000). Cyclin B1 expression has also been found to be high in metaplasia, dysplasia, and carcinoma in the evolution of Barrett's esophagus, a precursor to esophageal cancer. There was a marked increase in the percentage of cyclin B1positive cells in high grade dysplasia and cancer (Geddert et al. 2002). Cyclin B1 overexpression in tissues at risk of developing cancer clearly supports cyclin B1 as an important target for cancer prevention.

28.5.3 Other Tumor Antigens Found on Premalignant Lesions

MAGE-A tumor antigens are a subgroup of cancer/testis antigens with normal expression restricted to male germ cells in the testis but not in adult somatic tissues. MAGE-A was identified from human melanoma tumors and was found to be recognized in vitro by cytotoxic T cells derived from tumor bearing patients (van der Bruggen et al. 1991). The tumor expression of MAGE-A is not restricted to melanoma and has subsequently been found in various types of cancer (Scanlan et al. 2002).

Oral squamous cell carcinoma (OSCC) develops from continuous damage of the oral mucosa and is preceded by precancerous lesions called oral leukoplakia (OLP). Studies have demonstrated that the development of cancer in the oral premalignant lesions was independent of the presence or absence of epithelial dysplasia in a pre-surgical biopsy (Holmstrup et al. 2006, 2007). Several factors are thought to contribute to this conclusion; (1) reading of epithelial dysplasia is very subjective, (2) the biopsy may not be representative of the whole lesion, and (3) epithelial dysplasia itself may not be a significant prognostic factor for future malignant development. MAGE-A antigens have been found on oral squamous cell carcinomas and retrospective analysis of biopsies detected MAGE-A in precancerous oral lesions (Krauss et al. 2011). In another study analysis of OLP lesions that transformed to OSCC and OLP that did not transform to OSCC during a 5 years follow up period, found significant correlation between malignant transformation and MAGE-A expression (Ries et al. 2012).

Carcinoembryonic antigen (CEA) is an oncofetal antigen that is overexpressed in many tumor types of epithelial origin including colon, lung, and breast. CEA is thought to play a role in the process of tumorigenesis, and thus it may be found throughout cancer progression. Melanocytic nevi are precursor lesions of cutaneous melanoma and comprise a spectrum of benign nevi (BE), dysplastic nevi (DN), and superficial spreading melanoma (SSM). CEA protein expression has been found to increase in a stepwise fashion in melanocytic tumors with significantly increased CEA expression in DN and SSM compared with BN (Gambichler et al. 2009).

elan-A, also known as Mart-1 is a melanocyte differentiation antigen. Melan-A vaccines have been developed and tested in clinical trials to induce anti tumor immune responses in melanoma patients but have met with very limited success. Regression of benign and malignant pigment lesions is a well documented phenomenon in about 1 % of the population and recently Melan-A has been found to play a role in this phenomenon. Melan-A-specific T cells have been found in regressing nevus and in the peripheral blood of a patient with multiple regressing nevi (Speeckaert et al. 2011). This is an example of an effective early anti tumor immune response and supports Melan-A as an important target for early cancer prevention.

28.6 Efficacy of Tested Prophylactic Cancer Vaccines

Intranasal administration of MUC1 vaccine has been tested and shown to delay IBD and prevent progression to CACC in the MUC1/IL-10$^{-/-}$ mouse model of CACC (Beatty et al.). The MUC1 peptide vaccine corresponds to five tandem repeats of a 20–amino acid sequence HGVTSAPDTRPAPGSTAPPA from the extracellular tandem repeat region of MUC1. The vaccine induced MUC1-specific adaptive immunity, both humoral (anti-MUC1 IgG) and cellular (MUC1-specific T cells) immunity. The mechanism of action of the preventative vaccine was shown to be elimination of abnormal MUC1-positive cells in IBD colons, which supports the goal of vaccines to strengthen the most important phase of immunosurveillance/immunoediting process, *elimination*. In addition, early administration of the vaccine converted the tumor-promoting environment to a tumor-inhibiting environment by decreasing myeloid-derived suppressor cells in the spleen and infiltration of neutrophils in the colon, both of which can compromise adaptive immunity and facilitate tumor growth. This same effects of the MUC1 vaccine were shown in the MUC1-DSS mouse model of CACC (Beatty et al. 2012).

MUC1 vaccine was tested and shown to prevent lung lesions in the KrasMUC1 spontaneous model of lung cancer (Finn et al. 2011). Immunohistological analysis for tumor formation and MUC1 expression revealed that vaccinated KrasMUC1mice had normal lung morphology and low or no aberrant MUC1 expression. These results suggest that the vaccine induced MUC1-specific adaptive immune response capable of eliminating abnormal MUC1-positive cells in the lung, similar to the above CACC models.

Recently, the first experience with a preventative cancer vaccine in humans, based not on a viral antigen but on a tumor-associated antigen (MUC1), was tested in the premalignant setting in individuals with a history of an advanced adenoma of the colon (Kimura et al., Cancer Prev. Res. In press). Because these patients do not have invasive cancer nor have they undergone immunosuppressive chemotherapy, the response to a vaccine could be assessed in the absence of these and other confounding factors that are present in patients with cancer. The vaccine was shown to elicit both humoral and cellular MUC1-specific immunity. In addition, the vaccine was also shown to promote a memory T cell response and importantly, the vaccine was not associated with autoimmunity in these individuals.

28.7 Target Populations for Prophylactic Cancer Vaccines

Advancements in the genetics of cancer susceptibility have facilitated the identification of several populations that would greatly benefit from prophylactic cancer vaccines. One example is women with *BRCA1* or *BRCA2* mutations that are at greatly increased risk of breast and ovarian cancer with estimates of cancer risk can be as high as 80 and 45 %, respectively. Given the increased risk of cancer, particularly at young ages, screening and prevention are crucial components of medical management for female mutation carriers. Prophylactic mastectomy and/or oopherectomy are main prevention options offered to this group of high-risk individuals. Both retrospective and prospective studies have examined the breast and ovarian cancer risk reduction associated with prophylactic mastectomy and/or oopherectomy and found a significant reduction of breast and ovarian cancer risk with these prophylactic surgeries (Nathanson and Domchek 2011). However, neither of these prophylactic surgeries is void of complications. Physical and psychological ramifications including anxiety, depression, and impaired body image are just a few of the complications associated with these aggressive procedures.

Mutations associated with hereditary familial adenomatous polyposis syndrome (FAP) and non-polyposis colorectal cancer syndrome have been identified and these individuals have greatly increased risk of colorectal cancer. Repeated colonoscopy and polypectomy, both invasive procedures, represent current prevention strategies for this group of high risk individuals.

As highlighted early in the chapter there are now many well defined premalignant lesions and patients presenting with these conditions are another key population that could greatly benefit from prophylactic cancer vaccines.

28.8 Summary: A Revised Paradigm for Prophylactic Cancer Vaccines

Anti-tumor immune responses have traditionally been thought of as immune responses directed against self antigens. When tumor antigens are considered self antigens then it is assumed that they are subject to self tolerance mechanisms. In addition, considering tumor antigens as self molecules continues to raise fear of autoimmunity that currently has prevented more appropriate clinical application based on these antigens in healthy individuals for cancer prevention. Experiments in animal models have shown, however, that when these molecules are made by tumor cells they are not perceived by the immune system as self molecules but rather as abnormal self. More recent evidence has emerged that supports a more general characterization of abnormal self antigens as disease associated antigens. This represents a paradigm shift, a change in the way we think about tumor immunity suggesting that it overlaps in its targets with immunity to external pathogens as well and is part of general immunosurveillance to maintain wellness.

Several experimental approaches provide support for this new paradigm. Ludewig et al. infected mice with lymphocyte choriomenigitis virus and vaccinia virus and found that 79 and 83 % of the antibodies generated respectively were against orthologues of human tumor antigens (Ludewig et al. 2004). Infection of human fibroblasts with varicella zoster virus or human cytomegalovirus were found to induce abnormal expression of cyclin B1 similar to what is seen in cancer cells (Leisenfelder and Moffat 2006; Sanchez et al. 2003).

Additional observations have been made in humans. Healthy individuals with no history of cancer were found to have both cyclin B1-specific antibodies and memory T cells (Vella et al. 2009). The T cells isolated from the PBMC of these healthy individuals produced IFN-γ in response to dendritic cells presenting cyclin B1 and using a cyclin B1 peptide library three peptides spanning amino acids 215–223 were found to stimulate T cells from multiple healthy donors. A study looking at T cell responses in melanoma patients surprisingly found Melan-A-specific cells in a large proportion (60 %) of healthy individuals (Pittet et al. 1999). A group of healthy donors were found to harbor CEA-specific T cells that had been activated in vivo and were able to mount a CEA memory responses when stimulated in vitro (Pickford et al. 2007). Antibodies and T cells for abnormal MUC1 were found in women with a history of mastitis (Jerome et al. 1997) and abnormal MUC1 expression and MUC1-specific T cells have been found in individuals with inflammatory bowel disease and pancreatitis (Beatty et al. 2010; Beatty et al. 2012; Kadayakkara et al. 2010). In a large retrospective case control study of events that reduce life-time risk for ovarian cancer, those that led to generation of anti-MUC1 immunity had been experienced primarily in the control group that remained free of cancer. Events associated with generation of anti-MUC1 antibodies, such as oral contraceptive use, breast mastitis, bone fracture or osteoporosis, pelvic surgeries and nonuse of talc in genital hygiene, all of which could be responsible for generating acute inflammation of the respective epithelia and transient increase of abnormal expression of MUC1. Women that experienced two or more of these events were three times as likely to have antibodies against abnormal MUC1 compared to women that did not have these events and their lifetime risk of ovarian cancer was proportionately and significantly reduced (Cramer et al. 2005).

These observations are consistent with the hypothesis that the general mechanism of immunosurveillance against cancer and other diseases is through generation of immune memory early in life against abnormal self that is a result of a pathogenic event in a particular organ or tissue. During the process of malignant transformation starting with early premalignant lesions, similar changes are recapitulated leading to an antitumor immune response that is actually a memory anti-abnormal self response. Consistent with this hypothesis, studies show that febrile childhood diseases lower the risk for cancer in adulthood. A case control study of stomach, colorectal, breast, and ovarian cancer found lower risk for these cancers associated with childhood chicken pox and pertussis infections as well as with more frequent episodes of cold and influenza in adulthood (Abel et al. 1991). In another study, analysis of serum stored from individuals going through an active mumps infection found a higher level of antibodies against tumor antigen MUC1 that is abnormally expressed on the salivary gland epithelium during mumps infection. A meta-analysis found that mumps parotitis creates effective immune surveillance of ovarian cancer cells that express this form of MUC1 (Cramer et al. 2010).

Genomic research has revealed a common theme that there is a high level conservation of genes and gene product function across organisms and among diseases. Many epidemiological studies have begun to link cancer and neurodegenerative diseases such as a possible association between melanoma and amyotrophic lateral sclerosis (ALS), and case–control and cohort studies have reported a reduced risk of almost all cancers among individuals with Parkinson's disease (Inzelberg and Jankovic 2007; Plun-Favreau et al. 2010). It is likely that more epidemiologic and genetic studies and meta-analysis will reveal additional associations between cancer and many diseases and the antigens connecting them. Prophylactic vaccines against these disease associated antigens could be expected to generate immune memory early in life for a broad protection from all diseases, including cancer, throughout life.

References

Abel U, Becker N, Angerer R et al (1991) Common infections in the history of cancer patients and controls. J Cancer Res Clin Oncol 117:339–344

Aggarwal R, Oddis CV (2011) Paraneoplastic myalgias and myositis. Rheum Dis Clin North Am 37:607–621

Ajioka Y, Watanabe H, Jass JR (1997) MUC1 and MUC2 mucins in flat and polypoid colorectal adenomas. J Clin Pathol 50:417–421

Albert ML, Darnell RB (2004) Paraneoplastic neurological degenerations: keys to tumour immunity. Nat Rev Cancer 4:36–44

Albert ML, Darnell JC, Bender A, Francisco LM, Bhardwaj N, Darnell RB (1998) Tumor-specific killer cells in paraneoplastic cerebellar degeneration. Nat Med 4:1321–1324

Balkwill F, Charles KA, Mantovani A (2005) Smoldering and polarized inflammation in the initiation and promotion of malignant disease. Cancer Cell 7:211–217

Barnd DL, Lan MS, Metzgar RS, Finn OJ (1989) Specific, major histocompatibility complex-unrestricted recognition of tumor-associated mucins by human cytotoxic T cells. Proc Natl Acad Sci USA 86:7159–7163

Barrett NR (1950) Chronic peptic ulcer of the oesophagus and 'oesophagitis'. Br J Surg 38:175–182

Beatty PL, Plevy SE, Sepulveda AR, Finn OJ (2007) Cutting edge: transgenic expression of human MUC1 in IL-10-/- mice accelerates inflammatory bowel disease and progression to colon cancer. J Immunol 179:735–739
Beatty PL, Narayanan S, Gariepy J, Ranganathan S, Finn OJ (2010) Vaccine against MUC1 antigen expressed in inflammatory bowel disease and cancer lessens colonic inflammation and prevents progression to colitis-associated colon cancer. Cancer Prev Res (Phila) 3:438–446
Beatty P, Ranganathan S, Finn OJ (2012) Prevention of colitis-associated colon cancer using a vaccine to target abnormal expression of the MUC1 tumor antigen. Oncoimmunology 1:263–270
Buchbinder R, Forbes A, Hall S, Dennett X, Giles G (2001) Incidence of malignant disease in biopsy-proven inflammatory myopathy. A population-based cohort study. Ann Intern Med 134:1087–1095
Burnet FM (1970) The concept of immunological surveillance. Prog Exp Tumor Res 13:1–27
Carlos CA, Dong HF, Howard OM, Oppenheim JJ, Hanisch FG, Finn OJ (2005) Human tumor antigen MUC1 is chemotactic for immature dendritic cells and elicits maturation but does not promote Th1 type immunity. J Immunol 175:1628–1635
Cheever MA, Allison JP, Ferris AS et al (2009) The prioritization of cancer antigens: a national cancer institute pilot project for the acceleration of translational research. Clin Cancer Res 15:5323–5337
Clemente CG, Mihm MC Jr, Bufalino R, Zurrida S, Collini P, Cascinelli N (1996) Prognostic value of tumor infiltrating lymphocytes in the vertical growth phase of primary cutaneous melanoma. Cancer 77:1303–1310
Coussens LM, Werb Z (2002) Inflammation and cancer. Nature 420:860–867
Cramer DW, Vitonis AF, Pinheiro SP et al (2010) Mumps and ovarian cancer: modern interpretation of an historic association. Cancer Causes Control 21:1193–1201
Cramer DW, Titus-Ernstoff L, McKolanis JR et al (2005) Conditions associated with antibodies against the tumor-associated antigen MUC1 and their relationship to risk for ovarian cancer. Cancer Epidemiol Biomarkers Prev 14:1125–1131
Darnell RB, Posner JB (2003) Paraneoplastic syndromes involving the nervous system. N Engl J Med 349:1543–1554
Dhodapkar MV (2005) Harnessing host immune responses to preneoplasia: promise and challenges. Cancer Immunol Immunother 54:409–413
Dhodapkar MV, Krasovsky J, Osman K, Geller MD (2003) Vigorous premalignancy-specific effector T cell response in the bone marrow of patients with monoclonal gammopathy. J Exp Med 198:1753–1757
Dunn GP, Old LJ, Schreiber RD (2004) The three Es of cancer immunoediting. Annu Rev Immunol 22:329–360
Fearon ER, Vogelstein B (1990) A genetic model for colorectal tumorigenesis. Cell 61:759–767
Finn OJ (2003a) Cancer vaccines: between the idea and the reality. Nat Rev Immunol 3:630–641
Finn OJ (2003b) Premalignant lesions as targets for cancer vaccines. J Exp Med 198:1623–1626
Finn OJ, Gantt KR, Lepisto AJ, Pejawar-Gaddy S, Xue J, Beatty PL (2011) Importance of MUC1 and spontaneous mouse tumor models for understanding the immunobiology of human adenocarcinomas. Immunol Res 50:261–268
Fonseca R, Bailey RJ, Ahmann GJ et al (2002) Genomic abnormalities in monoclonal gammopathy of undetermined significance. Blood 100:1417–1424
Galon J, Costes A, Sanchez-Cabo F et al (2006) Type, density, and location of immune cells within human colorectal tumors predict clinical outcome. Science 313:1960–1964
Gambichler T, Grothe S, Rotterdam S, Altmeyer P, Kreuter A (2009) Protein expression of carcinoembryonic antigen cell adhesion molecules in benign and malignant melanocytic skin lesions. Am J Clin Pathol 131:782–787
Geddert H, Heep HJ, Gabbert HE, Sarbia M (2002) Expression of cyclin B1 in the metaplasia-dysplasia-carcinoma sequence of Barrett esophagus. Cancer 94:212–218
Gendler SJ, Mukherjee P (2001) Spontaneous adenocarcinoma mouse models for immunotherapy. Trends Mol Med 7:471–475

Ho SB, Ewing SL, Montgomery CK, Kim YS (1996) Altered mucin core peptide immunoreactivity in the colon polyp-carcinoma sequence. Oncol Res 8:53–61
Holmstrup P, Vedtofte P, Reibel J, Stoltze K (2006) Long-term treatment outcome of oral premalignant lesions. Oral Oncol 42:461–474
Holmstrup P, Vedtofte P, Reibel J, Stoltze K (2007) Oral premalignant lesions: is a biopsy reliable? J Oral Pathol Med 36:262–266
Husemann Y, Geigl JB, Schubert F et al (2008) Systemic spread is an early step in breast cancer. Cancer Cell 13:58–68
Inzelberg R, Jankovic J (2007) Are Parkinson disease patients protected from some but not all cancers? Neurology 69:1542–1550
Jerome KR, Kirk AD, Pecher G, Ferguson WW, Finn OJ (1997) A survivor of breast cancer with immunity to MUC-1 mucin, and lactational mastitis. Cancer Immunol Immunother 43:355–360
Kadayakkara DK, Beatty PL, Turner MS, Janjic JM, Ahrens ET, Finn OJ (2010) Inflammation driven by overexpression of the hypoglycosylated abnormal mucin 1 (MUC1) links inflammatory bowel disease and pancreatitis. Pancreas 39:510–515
Kao H, Marto JA, Hoffmann TK et al (2001) Identification of cyclin B1 as a shared human epithelial tumor-associated antigen recognized by T cells. J Exp Med 194:1313–1323
Krauss E, Rauthe S, Gattenlohner S et al (2011) MAGE-A antigens in lesions of the oral mucosa. Clin Oral Investig 15:315–320
Kyle RA, Therneau TM, Rajkumar SV et al (2002) A long-term study of prognosis in monoclonal gammopathy of undetermined significance. N Engl J Med 346:564–569
Lee PP, Yee C, Savage PA et al (1999) Characterization of circulating T cells specific for tumor-associated antigens in melanoma patients. Nat Med 5:677–685
Leisenfelder SA, Moffat JF (2006) Varicella-zoster virus infection of human foreskin fibroblast cells results in atypical cyclin expression and cyclin-dependent kinase activity. J Virol 80:5577–5587
Lollini PL, Cavallo F, Nanni P, Forni G (2006) Vaccines for tumour prevention. Nat Rev Cancer 6:204–216
Ludewig B, Krebs P, Metters H, Tatzel J, Tureci O, Sahin U (2004) Molecular characterization of virus-induced autoantibody responses. J Exp Med 200:637–646
Mariette C, Piessen G, Leteurtre E, Hemon B, Triboulet JP, Van Seuningen I (2008) Activation of MUC1 mucin expression by bile acids in human esophageal adenocarcinomatous cells and tissues is mediated by the phosphatidylinositol 3-kinase. Surgery 143:58–71
Matsubara S, Tada M, Akahane M et al (2012) Incidental Pancreatic Cysts found by magnetic resonance imaging and their relationship with pancreatic cancer. Pancreas
Matzinger P (2002) The danger model: a renewed sense of self. Science 296:301–305
Matzinger P (2007) Friendly and dangerous signals: is the tissue in control? Nat Immunol 8:11–13
Mukherjee P, Ginardi AR, Madsen CS et al (2000) Mice with spontaneous pancreatic cancer naturally develop MUC-1-specific CTLs that eradicate tumors when adoptively transferred. J Immunol 165:3451–3460
Mukherjee P, Ginardi AR, Madsen CS et al (2001a) MUC1-specific CTLs are non-functional within a pancreatic tumor microenvironment. Glycoconj J 18:931–942
Mukherjee P, Ginardi AR, Tinder TL, Sterner CJ, Gendler SJ (2001b) MUC1-specific cytotoxic T lymphocytes eradicate tumors when adoptively transferred in vivo. Clin Cancer Res 7:848s–855s
Nathanson KL, Domchek SM (2011) Therapeutic approaches for women predisposed to breast cancer. Annu Rev Med 62:295–306
Nguyen DX, Massague J (2007) Genetic determinants of cancer metastasis. Nat Rev Genet 8:341–352
Pages F, Berger A, Camus M et al (2005) Effector memory T cells, early metastasis, and survival in colorectal cancer. N Engl J Med 353:2654–2666

Pickford WJ, Watson AJ, Barker RN (2007) Different forms of helper tolerance to carcinoembryonic antigen: ignorance and regulation. Clin Cancer Res 13:4528–4537

Pittet MJ, Valmori D, Dunbar PR et al (1999) High frequencies of naive Melan-A/MART-1-specific CD8(+) T cells in a large proportion of human histocompatibility leukocyte antigen (HLA)-A2 individuals. J Exp Med 190:705–715

Plun-Favreau H, Lewis PA, Hardy J, Martins LM, Wood NW (2010) Cancer and neurodegeneration: between the devil and the deep blue sea. PLoS Genet 6:e1001257

Podsypanina K, Du YC, Jechlinger M, Beverly LJ, Hambardzumyan D, Varmus H (2008) Seeding and propagation of untransformed mouse mammary cells in the lung. Science 321:1841–1844

Rabinovich GA, Gabrilovich D, Sotomayor EM (2007) Immunosuppressive strategies that are mediated by tumor cells. Annu Rev Immunol 25:267–296

Racanelli V, Prete M, Minoia C, Favoino E, Perosa F (2008) Rheumatic disorders as paraneoplastic syndromes. Autoimmun Rev 7:352–358

Ries J, Agaimy A, Vairaktaris E, et al (2012) Detection of MAGE-A Expression predicts malignant transformation of oral Leukoplakia. Cancer Invest 30:495–502

Rowse GJ, Tempero RM, VanLith ML, Hollingsworth MA, Gendler SJ (1998) Tolerance and immunity to MUC1 in a human MUC1 transgenic murine model. Cancer Res 58:315–321

Ryan SO, Vlad AM, Islam K, Gariepy J, Finn OJ (2009) Tumor-associated MUC1 glycopeptide epitopes are not subject to self-tolerance and improve responses to MUC1 peptide epitopes in MUC1 transgenic mice. Biol Chem 390:611–618

Sanchez V, McElroy AK, Spector DH (2003) Mechanisms governing maintenance of Cdk1/cyclin B1 kinase activity in cells infected with human cytomegalovirus. J Virol 77:13214–13224

Scanlan MJ, Gure AO, Jungbluth AA, Old LJ, Chen YT (2002) Cancer/testis antigens: an expanding family of targets for cancer immunotherapy. Immunol Rev 188:22–32

Schiller M, Bohm M, Hensen P, Riemann H, Luger TA, Nashan D (2006) Dermatomyositis associated with malignant melanoma–a marker of poor prognosis? J Am Acad Dermatol 54:221–226

Schumacher K, Haensch W, Roefzaad C, Schlag PM (2001) Prognostic significance of activated CD8(+) T cell infiltrations within esophageal carcinomas. Cancer Res 61:3932–3936

Smyth MJ, Crowe NY, Godfrey DI (2001) NK cells and NKT cells collaborate in host protection from methylcholanthrene-induced fibrosarcoma. Int Immunol 13:459–463

Smyth MJ, Dunn GP, Schreiber RD (2006) Cancer immunosurveillance and immunoediting: the roles of immunity in suppressing tumor development and shaping tumor immunogenicity. Adv Immunol 90:1–50

Soria JC, Jang SJ, Khuri FR et al (2000) Overexpression of cyclin B1 in early-stage non-small cell lung cancer and its clinical implication. Cancer Res 60:4000–4004

Speeckaert R, van Geel N, Luiten RM, et al (2011) Melanocyte-specific immune response in a patient with multiple regressing nevi and a history of melanoma. Anticancer Res 31:3697–3703

Street SE, Cretney E, Smyth MJ (2001) Perforin and interferon-gamma activities independently control tumor initiation, growth, and metastasis. Blood 97:192–197

Suzuki H, Graziano DF, McKolanis J, Finn OJ (2005) T cell-dependent antibody responses against aberrantly expressed cyclin B1 protein in patients with cancer and premalignant disease. Clin Cancer Res 11:1521–1526

van der Bruggen P, Traversari C, Chomez P et al (1991) A gene encoding an antigen recognized by cytolytic T lymphocytes on a human melanoma. Science 254:1643–1647

Vella LA, Yu M, Fuhrmann SR, El-Amine M, Epperson DE, Finn OJ (2009) Healthy individuals have T-cell and antibody responses to the tumor antigen cyclin B1 that when elicited in mice protect from cancer. Proc Natl Acad Sci U S A 106:14010–14015

Vlad AM, Muller S, Cudic M et al (2002) Complex carbohydrates are not removed during processing of glycoproteins by dendritic cells: processing of tumor antigen MUC1 glycopeptides for presentation to major histocompatibility complex class II-restricted T cells. J Exp Med 196:1435–1446

Vlad AM, Kettel JC, Alajez NM, Carlos CA, Finn OJ (2004) MUC1 immunobiology: from discovery to clinical applications. Adv Immunol 82:249–293

Whiteside TL (2010) Inhibiting the inhibitors: evaluating agents targeting cancer immunosuppression. Expert Opin Biol Ther 10:1019–1035

Winter JM, Tang LH, Klimstra DS et al (2012) A Novel Survival-Based Tissue Microarray of Pancreatic Cancer Validates MUC1 and Mesothelin as Biomarkers. PLoS ONE 7:e40157

Wiseman EF, Ang YS (2011) Risk factors for neoplastic progression in Barrett's esophagus. World J Gastroenterol 17:3672–3683

Xu X, Bai L, Chen W et al (2012) MUC1 contributes to BPDE-induced human bronchial epithelial cell transformation through facilitating EGFR activation. PLoS ONE 7:e33846

Yu M, Zhan Q, Finn OJ (2002) Immune recognition of cyclin B1 as a tumor antigen is a result of its overexpression in human tumors that is caused by non-functional p53. Mol Immunol 38:981–987

Zhan F, Hardin J, Kordsmeier B et al (2002) Global gene expression profiling of multiple myeloma, monoclonal gammopathy of undetermined significance, and normal bone marrow plasma cells. Blood 99:1745–1757

Zhang L, Conejo-Garcia JR, Katsaros D et al (2003) Intratumoral T cells, recurrence, and survival in epithelial ovarian cancer. N Engl J Med 348:203–213

Zotter S, Lossnitzer A, Hageman PC, Delemarre JF, Hilkens J, Hilgers J (1987) Immunohistochemical localization of the epithelial marker MAM-6 in invasive malignancies and highly dysplastic adenomas of the large intestine. Lab Invest 57:193–199

Part VI
Analyzing Immune Responses in Cancer

Chapter 29
Approaches to Immunologic Monitoring of Clinical Trials

Lisa H. Butterfield, Lazar Vujanovic and Angela D. Pardee

Abstract Performance of thoughtful immunologic monitoring is critical to the conduct of an immune-based clinical trial. In the absence of high frequencies of significant clinical responses, immunologic responses may be the primary basis for understanding the outcomes of the intervention tested. In addition, important immunologic changes may occur from non-immunologic interventions like surgery, chemotherapy and radiotherapy. These standard-of-care treatments may promote or inhibit anti-tumor immunity. Measurement of these changes can yield important insights into the mechanism of clinical responses in patients.

Keywords Immune monitoring · ELISPOT · T cells · Biomarkers · Tumor immunity

---

L. H. Butterfield (✉) · L. Vujanovic · A. D. Pardee
Hillman Cancer Center, University of Pittsburgh Cancer Institute, University of Pittsburgh School of Medicine, 5117 Centre Avenue, Suite 1.27, Pittsburgh, PA 15213, USA
e-mail: butterfieldl@upmc.edu

L. H. Butterfield · L. Vujanovic · A. D. Pardee
Departments of Medicine, University of Pittsburgh, Pittsburgh, PA, USA

L. H. Butterfield
Departments of Immunology, University of Pittsburgh, Pittsburgh, PA, USA

L. H. Butterfield
Departments of Surgery, University of Pittsburgh, Pittsburgh, PA, USA

M. R. Shurin et al. (eds.), *The Tumor Immunoenvironment*,
DOI: 10.1007/978-94-007-6217-6_29,

29.1 Introduction to Immunologic Monitoring

Clinical trials testing an intervention with any immunologic impact greatly benefit from testing patient specimens for the hypothesized immunologic effects. Such data can point towards mechanisms of clinical response or immune-mediated toxicity. However, immunologic monitoring assay data do not often correlate with clinical outcomes. Provocative results may come from individual trials, but in the field of immunotherapy, the results have been uneven. However, in the last 10 years, an increasing number of trials have included well-designed and carefully performed immunologic monitoring which have resulted in identification of important correlates of clinical outcomes, including multiple functional assessments of $CD8^+$ and $CD4^+$ T cells (the cells generally hypothesized to be the most critical effectors), NK cells and antibody responses (Fig. 29.1).

Some examples of these include the vaccination of vulvar intraepithelial neoplasia patients with long peptides from HPV16 (Welters et al. 2010). Investigators tested lymphocyte proliferation and cytokine production to immunizing antigens as well as circulating regulatory T cells (Treg). Their study showed that the patients with the strongest proliferation, positive IFN-γ and IL-5 production, and lowest Treg were those with complete responses (CR) to therapy. A trial testing WT-1 antigen in AML patients indentified significant changes in WT-1 specific $CD8^+$ T cell frequencies, and more dramatic activation of circulating NK cells in patients with CRs (Van Tendeloo et al. 2010). Another example, in a setting where

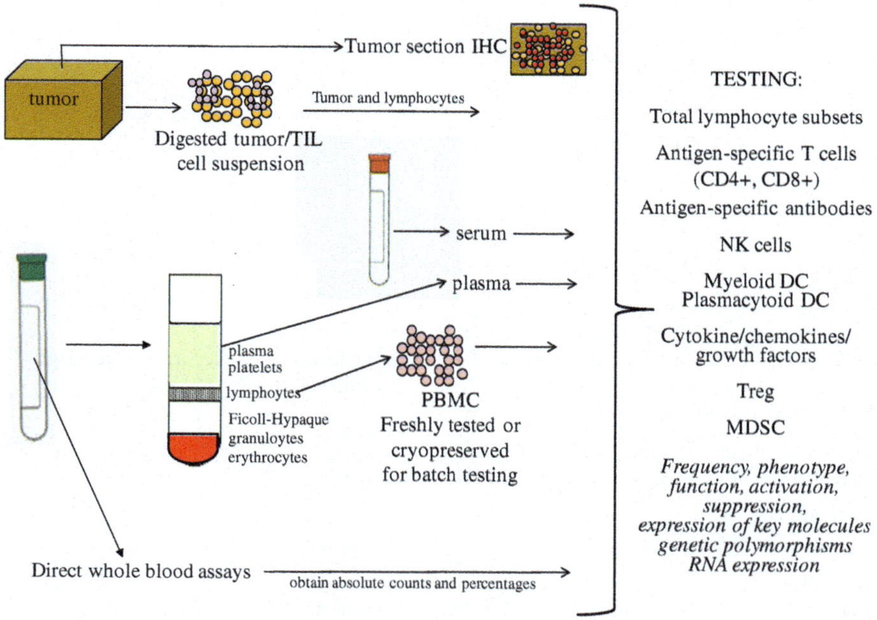

Fig. 29.1 Patient-derived materials used in immunologic monitoring

a central laboratory was utilized for shipping, processing and banking of samples from a large, multi-center clinical vaccine trial, immunity again correlated with clinical outcome. The laboratory tested both MHC tetramer flow cytometry to measure the frequency of vaccine antigen-specific T cells phenotypically, and functional ELISPOT assays of patient PBMC, and it was the functional IFN-γ ELISPOT assay response to the peptide vaccine antigens that correlated with melanoma patient overall survival (OS) (Kirkwood et al. 2009). The phenotypic MHC tetramer assay detected vaccine-specific T cell frequency increases, but those data did not correlate with clinical outcome (Schaefer et al. 2012). A recent study testing a multi-peptide vaccine in renal cancer patients yielded several important insights from immunologic monitoring. The authors demonstrated that inhibition of Treg strengthened antitumor immunity, frequencies of specific subsets of MDSC at baseline were critical, and that improved clinical outcome correlated with a larger number of peptides responded to (Walter et al. 2012). This last observation has been made in previous multi-peptide vaccine trials (Banchereau et al. 2001).

In an effort to identify a correlation between clinical response and immunity in patients treated with ipilimumab, a number of groups have examined humoral and cellular responses from peripheral blood. Recently, integrated humoral and CD8^{+} T cell response to the shared, immunogenic cancer/testes antigen NY-ESO-1 was shown to be a significant immunologic measure correlating with clinical outcomes (Yuan et al. 2011). In a related publication, circulating T cells specific to NY-ESO-1 and MART-1 antigens (but not MAGE-A3 or survivin antigens) were correlated with melanoma patient survival (Weide et al. 2012). Specific immunity to NY-ESO-1 may be of particular importance, as it has been examined in many settings with positive results (for reasons yet to be identified), and this antigen may serve as an indicator of overall positive antitumor immunity.

A final example of an immune measure that several independent groups have correlated to clinical outcome is "epitope spreading" or "determinant spreading". This acquisition of T cell and antibody reactivity to shared antigens or epitopes other than those used in a vaccine, has been observed in melanoma (Butterfield et al. 2003; Ribas et al. 2004; Butterfield et al. 2008), renal cancer (Wierecky et al. 2006) and breast cancer (Disis et al. 2002). While the in vivo mechanism has yet to be identified, tumor cell lysis, in an immunogenic milieu which promotes uptake and subsequent cross-presentation of the additional tumor antigens, is hypothesized to be critical. Detection of increased immunity to non-vaccine shared antigens is an approach which permits identification of this phenomenon. However, shared antigens may not be drivers of the phenomenon, and mutated, private, patient-specific antigens may also be of central importance. The same in vivo mechanism of cross-priming may also result in autoimmunity (Krauze et al. 2011), which has been identified as a biomarker of clinical response to interferon in melanoma patients (Gogas et al. 2006). Such systemic autoimmunity can be reliably measured in hospital clinical lab assays for anti-nuclear antibodies, rheumatoid factors and thyroid-related antibodies.

Immunologic monitoring is also very pertinent to standard of care interventions like surgery, chemotherapy and radiotherapy. There are increasing examples of immune mechanisms playing critical roles in clinical response to these interventions. Most recently, radiotherapy and the abscopal effect were shown to involve changes in NY-ESO-1-specific antibodies, $CD4^+$ T cells, and myeloid-derived suppressor cells (MDSC) (Postow et al. 2012). Dendritic cells (DC) and $CD8^+$ T cells may also be involved (Gupta et al. 2012). Different classes of chemotherapeutic drugs have been shown to be potent activators of antitumor immunity, while other chemotherapy drugs can have an immune-suppressive effect. Such positive anti-tumor immune effects are mediated by changes in immunologic danger signals in tumor cells, including release of danger-associated molecular patterns (DAMPs) like ATP, HMGB1 and modulation of calreticulin (Zitvogel et al. 2011). Both chemo-embolization and radiofrequency ablation have been demonstrated to have at least transient immune activating effects, including increased circulating tumor antigen-specific T cells. This has been observed with circulating AFP-specific T cells in hepatocellular cancer, a tumor which occurs in a particularly immune suppressive organ site (Ayaru et al. 2007; Zerbini et al. 2006). Taken together, multiple measures of immunity are serving as significant prognostic biomarkers of clinical outcome in cancer trials testing a very broad array of interventions. These immunologic monitoring measures can yield important insights into the mechanism of resulting clinical responses, and point towards important areas for further interventions.

29.2 Importance of Standardization, Sample Handling

There have been several important recent movements towards greater harmonization and standardization in the field of immunologic monitoring which will strengthen the data obtained from clinical trials and improve the field.

29.2.1 SITC Taskforce Recommendations

To facilitate further development of effective immunotherapy approaches, there remains a need to standardize and validate immunologic monitoring approaches to identify patients who can benefit from immunotherapy, both before treatment and at early time points after therapy initiation. Despite substantial effort, and the exciting recent advances described above, we do not yet know which parameters of anti-tumor immunity to measure in a specific trial setting and which assays and controls are most reliable for those measurements. Between 2008 and 2010 the Society for Immunotherapy of Cancer (SITC) joined with the FDA and NCI to create a task force to address these issues. Many international immunotherapy societies participated in this effort and these topics were discussed at two meetings (Butterfield et al. 2008; Tahara et al. 2009; Butterfield et al. 2010; Bedognetti et al. 2011; Fox et al. 2011).

Table 29.1 Approaches to addressing inherent variability in immunologic monitoring of clinical trials (Butterfield et al. 2011)

Source of variability	Recommendation
Patient	Save DNA/RNA/cells/tumor to understand host variation; include healthy donor control
Blood draw	Standardized tubes and procedures
Processing/cryopreservation/thaw	Standardized procedures and reagents
Cellular product	Phenotypic and functional assays to characterize the individual product, development of potency assays
Assay choice	Standardized functional tests
Assay conduct	Standardized procedures (SOP)
Assay analysis	Appropriate biostatistical methods
Data reporting	Full details, controls, quality control/assurance (QA/QC); MIATA guidelines
Newest, non-standardized technology	Sufficient blood/tissue to interrogate the samples *now*, as well as *later*, to generate new hypotheses

While specific immune parameters and assays are not yet validated as biomarkers, the Taskforce recommended following standardized (accurate, precise and reproducible) procedures and use of functional assays (not only phenotypic) for the primary immunologic readouts of a trial (Butterfield et al. 2011). There have been many publications regarding blood collection, processing, cryopreservation and thawing, resulting in published protocols, particularly those from the Immunologic Monitoring Consortium. Many are summarized in the Taskforce recommendations (Butterfield et al. 2011). The SITC Taskforce also recommended consideration of central laboratories for immune monitoring of large, multi-institutional trials to reduce the inevitable variables inherent in such trials. Testing of multiple phenotypic, and functional potency assays should be performed for any cellular product, particularly autologous products. To promote a broad analysis of immunity at multiple anatomic sites and at many biologic levels [genetic, transcriptional, protein (cell-associated and secreted)], the Taskforce recommended that in addition to blood cells and serum, that tumor, RNA and DNA samples be banked (under appropriate standardized conditions) for later testing. Perhaps most importantly, sufficient blood should be drawn to allow for not only the planned testing of the primary hypothesis being addressed in the trial (many described below), but also that additional baseline and post-treatment blood and tissue be banked for testing novel hypotheses (or generating new hypotheses) that arise in the field (Table 29.1).

29.2.2 *Minimal Information About T Cell Assays Project*

The effective use of data obtained from immune assays has been hampered not only by variability in those data, but also by inconsistent and incomplete data reporting in publications. To address the latter challenge, between 2009 and 2012, the immune monitoring community led by the Association for Cancer Immunotherapy's CIMT Immunoguiding Program and the Cancer Research Institute's Cancer Immunotherapy Consortium (CIC) groups, has defined a checklist of essential information necessary to more clearly report the results from immunologic monitoring studies focused on T cell assays. Recommendations were derived from an iterative consensus process involving more than 120 scientific experts from the fields of oncology, autoimmunity and infectious diseases which occurred in several forums over several years (Janetzki et al. 2009, 2010; Britten et al. 2011). The resulting "MIATA reporting guidelines" do not create standards on performing T cell assays nor do they promote specific reagents or protocols. Minimal information about T cell assays (MIATA) offers minimum criteria for consistent reporting of T cell experiments to increase data interpretability and experimental reproducibility across the scientific community (Britten et al. 2012). These guidelines will enable better understanding of the strengths and weaknesses of clinical trial immunologic monitoring studies published, and thus, support future immune biomarker identification.

29.2.3 Assay Proficiency Panels

The CIMT and CIC organizations and EQAPOL have also led a series of multi-laboratory, multi-institutional and international proficiency panel programs in which laboratories involved in immunologic monitoring can voluntarily participate. These have tested a variety of assays (ELISPOT, intracellular cytokine staining, Luminex, multimer staining) and assay parameters using standardized, pre-tested PBMC from which expected results are known (Janetzki et al. 2010; Moodie et al. 2010; Mander et al. 2010; Britten et al. 2008). The results have identified important sources of assay variability at multiple steps (cell resting, serum in culture medium, flow cytometry gating strategies), and allowed participants to compare their performance anonymously with many others in the field. These proficiency panels continue to evolve and address new aspects of immunologic monitoring.

29.3 Answering the Trial Hypothesis in the Peripheral Blood

There are a wide variety of immunologic interventions which have been translated from the laboratory to the clinic. These include cancer vaccines designed to promote antigen specific T cell responses, antibody therapies targeting cell surface proteins, adoptive cell transfer to deliver ex vivo activated effector cells, inhibition of immune suppressors like Treg, cytokines and growth factors, (delivered systemically or locally), and small molecule signal transduction inhibitors (Kirkwood et al. 2012).

Each of these focuses on a different aspect of immunity, and tests a different hypothesis. To determine whether these individual approaches have achieved their immunologic goals, different monitoring approaches are needed

Within the area of cancer vaccines, there is a wide array of strategies. Cellular vaccines have been studied for many years, and include autologous and allogeneic tumor cells, often engineered to express immune stimulating molecules (to overcome the natural immune inhibition of tumors). While many of these strategies have fallen out of favor, some have shown clinical responses and have been carried forward. These include GM-CSF-transfected autologous tumors and cell lines (Dranoff et al. 1993; Nemunaitis et al. 2004; Higano et al. 2008). Because cell lines can be tested for shared tumor antigen expression, such vaccines can be tested for the ability to promote antigen-specific immunity across patients.

Optimally immunogenic antigen presenting cells, like dendritic cells (DC) have been used in many trials in many forms (Banchereau and Steinman 1998; Banchereau and Palucka 2005). They have been pulsed with individual peptides (Butterfield et al. 2003; Butterfield et al. 2003), matured with specific cytokine cocktails for specific T cell interaction properties (Mailliard et al. 2004; Kalinski et al. 2001; Palucka et al. 2008) and resulted in positive clinical responses in several trials (O'Rourke et al. 2007; Okada et al. 2011). In most trials, these cells are loaded with some identified antigens which again allow for assessment of specific T cell activation.

Peptide-based vaccines, generally formulated with an adjuvant and/or a cytokine or growth factor, are usually specific MHC-restricted epitopes which most easily lend themselves to monitoring for the expansion and activation of the CD8+ T cell clones targeted. These short (8–10aa) peptide vaccines have been demonstrated to be highly immunogenic (Butterfield et al. 2003; Slingluff et al. 2004), and can be combined into mixtures (Slingluff et al. 2007) to activate not only multiple $CD8^+$ T cells, but also combined with longer peptides activating $CD4^+$ T cells (Slingluff et al. 2008; Kenter et al. 2009). The peptides activating helper T cells can be defined with known MHC class II restriction, or longer to potentially encompass multiple known and unknown epitopes.

Antibody therapeutics have been used for many years. These include antibodies targeting HER-2, CD19, CD20, and CTLA-4. There are number of mechanisms of action which are not yet completely defined for each antibody, which include antibody-dependent cytotoxicity (ADCC), surface target blockade, and signal transduction modulation. Each mechanism can be tested; by NK cell ADCC, examining surface expression of the molecule, and testing downstream signaling molecules or target gene expression. Assessment of human-anti-murine antibody responses (HAMA) is important when the antibodies are not entirely humanized, and can be targets of immunity themselves.

Adoptive transfer of effector cells is an important strategy for ex vivo activation and expansion of TIL, CTL and/or mixtures of $CD8^+$ T cells (Dudley et al. 2005, 2008). The culture conditions used to preferentially expand these cells to have specific attributes have evolved steadily over the last several years. In addition to cells isolated from patients with their endogenous specificities, PBMC are also genetically engineered to express specific T cell receptors or chimeric antigen receptors. Transferred T cells can be tested for activity before transfer, labeled for in vivo trafficking, and tracked based on knowledge of their surface receptors (Roszkowski et al. 2005).

29.3.1 Antigen-Specific Cytotoxic T Lymphocytes

Antigen-specific cytotoxic T lymphocytes (CTL) are the key effector cells of the adaptive immune response. They mediate antigen-specific cellular immune responses against infected cells and, in the case of tumors, mutated or overexpressed self-antigens. Previous studies have implicated tumor-associated antigen (TAA)-CTL as major antitumor effectors in humans and potent mediators of antitumor immunity (Boon and van der Bruggen 1996; Clemente et al. 1996; Old and Chen 1998; Penn 1996). CTL recognize TAA-derived epitopes presented in the context of MHC class I molecules on the surface of tumor cells via T cell receptor (TCR). Upon recognition, CTL release cytolytic granules containing membrane-perturbing proteins (perforin and granulysin), and a family of serine proteases known as granzymes (A, B, C, D, E, F, G, H, K, and M) into the formed immunological synapse to induce target cell apoptosis (Lieberman 2003; Raja et al. 2003). In addition to their cytotoxic properties, these cells also have a helper function, mediated by IFN-γ and TNFα, that support the overall induction of type-1 immunity (Nakamura et al. 2007; Mailliard et al. 2002). The importance of CTL in anti-tumor immunity is substantiated by multiple reports directly correlating high levels of CTL infiltration into various types of neoplasms with positive clinical outcome (Boon et al. 2006; Hamanishi et al. 2007; Hiraoka et al. 2006; Mahmoud et al. 2011).

29.3.2 Antigen-Specific CD4⁺ T Cells

Mature $CD4^+$ are typically known as T helper (Th) cells. $CD4^+$ lymphocytes are believed to polarize the adaptive immune response by secreting a dominant panel of cytokines in response to specific antigen recognition. Based on these cytokine profiles, Th cells can be generally segregated into three major subsets: Th1 (IFN-γ, IL-2), Th2 (IL-4, IL-5, IL-13), and Th3/T-regulatory (IL-10, TGFβ) subsets (Weiner 2001). Studies have shown that the $CD8^+$ T cell response may ultimately depend on a potent helper $CD4^+$ T cell response, which plays a crucial role in the regulation of the quality, magnitude, and durability of $CD8^+$ CTL immunity in vivo (Janssen et al. 2005; Fallarino et al. 2000; Hung 1998; Surman et al. 2000; Friedman et al. 2012; Wiesel and Oxenius 2012). In particular, Th1 responses were shown to mediate effective anti-tumor immunity by inducing responses that can facilitate the cross-presentation of TAA by host antigen presenting cells (APC) and consequent determinant spreading in the antitumor T cell repertoire (Crotzer and Blum 2010; Disis 2011). These studies have suggested that in order to induce objective clinical response in cancer patients, an immunotherapy must effectively induce both $CD8^+$ and $CD4^+$ T cell responses against tumor antigens (Bos and Sherman 2010). Similarly to CTL, infiltration of high numbers of $CD4^+$ T cells (excluding Treg) into various types of neoplasms is associated with better patient prognosis (Badoual et al. 2006; Wakabayashi et al. 2003; Yoshida et al. 2006). In addition to their well-described helper ability, multiple reports have described the existence of cytotoxic $CD4^+$ T cells detected directly from peripheral blood in various human disease settings, like viral infections (HIV, CMV and EBV), rheumatoid arthritis, ankylosing spondylitis and B cell chronic lymphocytic leukaemia (Appay 2004). These cells were reported to have cytolytic granules containing cytotoxic factors (i.e. granzymes and perforin), and that their lytic activity is MHC class II restricted.

29.3.3 Natural Killer Cells

Natural killer (NK) cells are the main effector cells of innate immunity that rapidly recognize and directly eliminate virally-infected and transformed cells without the need for TAA recognition (Andoniou et al. 2008). Recent studies have shown that they are a subset of cytotoxic innate lymphoid cells (ILCs) that express the transcription factor E4BP4 (E4 promoter-binding protein 4; also known as NFIL3) (Vivier et al. 2012). NK cells express a number of inhibitory and activating receptors which help them to efficiently detect and eliminate aberrant and infected cells. In vitro, NK cells can kill a broad range of tumor cells of hematopoietic and non-hematopoietic origin mediated by release of soluble factors (i.e. perforin or granzymes), by TNF family ligands and by cytotoxicity of antibody-coated targets triggered by Fc receptor clustering (ADCC) (Vivier et al. 2012; Vujanovic 2001). Additionally, NK cells secrete a number of proinflammatory and immunoregulatory cytokines (including IFN-γ) with which they mediate inflammation as well as polarization and regulation of both innate and adaptive immune responses (Andoniou et al. 2008; Fernandez et al. 1999; Gerosa et al. 2002).

NK cells can cross-talk with other immune cells, and this interaction impacts the activation state of NK cells. One of the key interacting partners for NK cells are DC. Several lines of evidence suggest that DC/NK cell crosstalk is mediated by cell-to-cell contact (via plasma membrane-bound molecules, such as transmembrane tumor necrosis factor (tmTNF) and trans-IL-15) and by soluble factors produced by DC (e.g. IL-12, IL-15, IL-18, and IFN-α) (Andrews et al. 2003; Borg et al. 2004; Ferlazzo and Munz 2004; Jinushi et al. 2003; Vujanovic et al. 2010). Ultimately, DC/NK cell crosstalk mediated by cell-to-cell contact leads to reciprocal regulation and stimulation of type-1 polarization and cytokine secretion (Fernandez et al. 1999; Vujanovic et al. 2010; Ferlazzo et al. 2002; Piccioli et al. 2002; Xu et al. 2007). NK cells induce DC maturation and secretion of IL-12p70, while DC reciprocally induce NK-cell activation and enhanced IFN-γ secretion, tumoricidal activity and proliferation. This early cellular crosstalk is an essential immunoregulatory mechanism bridging innate and adaptive immunity, and defining the quality and magnitude of immune mechanisms that control viral infections and tumor growth in vivo. In addition to reciprocal stimulation, a subset of activated NK cells are able to kill immature DC (iDC) by triggering NK activating receptor NKp30 and DC TNF-related apoptosis-inducing ligand death receptors (Ferlazzo et al. 2002; Piccioli et al. 2002; Hayakawa et al. 2004; Spaggiari et al. 2001). The elimination of DC by NK cells is thought to be an important control switch that selectively removes DC unsuitable for efficient antigen presentation and initiation of adaptive immune responses. These data support the importance of measuring NK cell activity in clinical trials.

In a clinical setting, NK cell deficiencies are rarely observed (Orange and Ballas 2006), however, an epidemiological study has correlated low peripheral blood NK cell activity with increased cancer risk (Imai et al. 2000). Additionally, NK cell infiltration into tumor tissue is associated with better disease prognosis in a number of malignancies (Platonova et al. 2011; Eckl et al. 2012; Halama et al. 2011). The study of patients with advanced gastrointestinal stromal tumors that received imatinib mesylate (a tyrosine kinase inhibitor) therapy showed that the NK cell IFN-γ production after 2 months of treatment could be considered an independent predictor of long term survival, giving support to the hypothesis that NK cells exert antitumor effects not only through direct cytolytic activity, but also through their helper ability mediated by the production of cytokines such as IFN-γ (Ménard et al. 2009).

29.3.4 Immunologic Monitoring of T and NK Cells in Cancer Patients

A number of phenotypic and functional immune monitoring assessments are used to evaluate T and NK cells in a clinical setting in order to determine the quality and magnitude of the anti-tumor response. Multi-color flow cytometry-based analysis is commonly applied to evaluate the phenotype of both NK and T cells in isolated peripheral blood mononuclear cells (PBMC) as well as tumor-infiltrating lymphocytes (TIL). Using this method, one can evaluate the overall frequencies of antigen-specific T lymphocytes (evaluated by multimer staining), memory status (CD45RA/RO and CCR7 expression), relative activation status (e.g. CD69, HLA-DR and CD25 staining), and expression of immuno-inhibitory checkpoint molecules (e.g. PD-1 and CTLA-4 expression). In regard to NK cells, this method can be used to evaluate the overall frequencies of different NK cell subsets ($CD56^{lo}CD16^{+}$, $CD56^{lo}CD16^{-}$, $CD56^{hi}CD16^{-}$), their migratory capabilities (measuring various chemokine receptors) and their activation levels (CD69, CD25, NKp30, NKp44, NKp46).

Functional assessments of epitope-specific T cells are preferable for monitoring of anti-tumor vaccines. One common method is to evaluate the cytokine production bias (IFN-γ/IL-2, IL-4/IL-5, and TGF-β/IL-10 representing type-1, 2 and 3 immunological responses, respectively) and cytotoxic ability (perforin or granzyme B release) of T lymphocytes in response to antigen recognition by intracellular staining (ICS) (Whiteside et al. 2003). A similar technique can be used to evaluate the functionality of NK cells. These functions can also be tested with the highly sensitive enzyme-linked immunosorbent spot (ELISPOT) assay (Whiteside et al. 2003; Asai et al. 2000; Bennouna et al. 2002; Shafer-Weaver et al. 2003). This is, essentially, a modified version of the enzyme-linked immunosorbent assay (ELISA) which is used to enumerate cells secreting various immune modulators (e.g. IFN-γ, IL-5, granzyme B) in response to a target cell and at the single cell level. When compared to tetramer and ICS assays, the ELISPOT assay may be less sensitive, however, physiologically it is perhaps the most relevant as it measures target-specific T and NK cell effector molecule release and is not plagued by specificity or gating issues found in the other two techniques (Whiteside et al. 2003; Asai et al. 2000).

Cytotoxicity assays are also critical for evaluating the killing ability of T and NK cells. A number of assays have been developed for monitoring cell-mediated killing of tumor targets. The gold standard method is the ^{51}Cr release assay. While it has advantages (direct measure of lymphocyte killing of the target), there are a number of issues which make this assay cumbersome and hazardous (use of a radioactive isotope, poor labeling of primary cells, high effector-to-target ratios which are non-physiologic). As such, alternative methods have been developed to measure cell-mediated cytotoxicity. One such method is the granzyme B ELISPOT which was shown to strongly correlate with the ^{51}Cr release assay, but with greater sensitivity and lack of radioactive waste (Shafer-Weaver et al. 2003). Another way to measure cytotoxicity is by flow cytometry-based killing assays which were also shown to correlate well with the ^{51}Cr release assay (Kim et al 2007), including CD107a upregulation on the effector cell surface which can detect active degranulation (Rubio et al. 2003; Malyguine et al. 2009).

29.3.5 Non-Antigen-Specific T Cell Responses

While antigen-specific measures test the frequencies and functions of T cells reactive with disease targets, there may also be important insights gained from testing absolute counts, frequencies and phenotypes of total circulating CD8$^+$ and CD4$^+$ T cells. Absolute lymphocyte count as determined in a simple clinical lab CBC, has correlated with clinical outcomes from CTLA-4 blockade (Hamid et al. 2003; Wolchok et al. 2010), and is being pursued to further substantiate this potential biomarker in several current immunotherapy clinical trials.

29.3.6 Regulatory T cells

While it has become clear over the past decade that regulatory T cells (Treg) play an important and non-redundant role in tumor progression, the precise phenotypic and functional profiles of these cells have yet to be established. Human Treg are most commonly identified as $CD4^{+}CD25^{hi}Foxp3^{+}$ (Sakaguchi et al. 2010). However, because both CD25 and Foxp3 expression can be induced in activated $CD4^{+}$ effector T cells, this definition does not represent a pure population of Treg (Gavin et al. 2006). Several recent publications support the idea that two populations of circulating Treg with distinct phenotypic and functional properties exist. Advanced-stage cancer patients were found to have elevated levels of IL-10-dependent T regulatory type 1 (Tr1) cells in the peripheral blood compared to patients with early-stage disease or healthy donors (Bergmann et al. 2008; Bergmann et al. 2007). In contrast to $CD4^{+}CD25^{hi}Foxp3^{+}$ natural Treg (nTreg), which are thought to function via contact-dependent suppression of target cells, these Tr1 cells exhibit a $CD4^{+}CD25^{-}Foxp3^{low/neg}$ phenotype and suppress target cells through IL-10 and TGF-β secretion. It can therefore be argued that monitoring of Tr1 rather than nTreg may be more relevant in the setting of cancer clinical trials (Whiteside and Schuler 2012).

Several clinical trials investigating various treatment modalities, both immune- and non immune-based, and involving patients with diverse cancers have revealed the correlative role of peripheral blood Treg during therapy. For example, a decrease in circulating $CD4^{+}CD25^{hi}$ Treg correlated with clinical response in breast cancer patients who received trastuzumab (HER-2/*neu* antibody) therapy (Perez et al. 2007). Overall survival of renal cell carcinoma patients receiving sunitinib, a small molecule tyrosine kinase inhibitor, was also shown to correlate with a reduction (of at least 20 % from baseline) in peripheral blood $CD4^{+}CD25^{hi}Foxp3^{+}$ Treg (Adotevi et al. 2010). The T effector cell-to-Treg ratio is another index that is commonly used to assess clinical response (Welters et al. 2010; Vergati et al. 2011), both in the circulation, as well as in tumor tissue infiltrates. Monitoring Treg function over the course of a clinical trial can provide further insight into a particular therapy's mode of action. Vergati and colleagues report that the ability of $CD4^{+}CD25^{hi}$ Treg to suppress effector T cell proliferation was significantly reduced only in prostate cancer patients who responded to vaccination (Vergati et al. 2011). Surprisingly, the efficacy of some therapeutic modalities may not depend on Treg reduction or inactivation. In a recent Phase III trial, it was shown that clinical response to high-dose IL-2 alone or in combination with a peptide vaccine correlated with higher frequencies of circulating $CD4^{+}Foxp3^{+}$ Treg (Schwartzentruber et al. 2011). It remains to be seen, however, whether these expanded Treg retain their suppressor function. Functional testing of suppressor function, by separation of $CD4^{+}CD25^{-}$ T cells (stimulated with nitrogen to be 'responders') from $CD4^{+}CD25^{-}$ Treg (without intracellular FoxP3 staining) or $CD4^{+/}CD25^{-}$ $CD127^{neg}$ Treg, can require a very large volume of blood. The Treg function is tested over several ratios (between 1:1 and 1:10 or 1:20 Treg: Responders), testing reduction of proliferation, often by

3–5 days CFSE dye dilution flow cytometric assay. The high Treg numbers required make such testing very cumbersome, and often only prioritized highly if the intervention is based on Treg reduction (with cyclophosphamide or denileukin-diftitox).

29.3.7 Myeloid-Derived Suppressor Cells

Myeloid-derived suppressor cells (MDSC), another regulatory cell subset that was identified more recently, have also been shown to play a critical role in tumor development and progression. Similar to Treg, MDSC do not represent a uniform population with a defined mechanism of action, and there is little consensus over how to best monitor these cells in the clinic (Montero et al. 2012). Two distinct MDSC populations, both of which are lineage negative ($Lin1^-$) and $CD11b^+$, have been identified in the peripheral blood of cancer patients: (1) $CD15^{Hi}CD33^{low}$ granulocytic MDSC, and (2) $CD15^{low}CD33^{hi}$ monocytic MDSC (Eruslanov et al. 2012). Additional MDSC subsets related to the $CD11b^+/CD33^+$ phenotype include $lineage^-/HLA\text{-}DR^-/CD33^+/CD11b^+$ cells (where the lineage cocktail allows gating out of $CD3^+/CD4^+/CD8^+/CD14^+/CD19^+/CD56^+$ cells) (Poschke et al. 2010; Gabitass et al 2011; Zea et al. 2005; Diaz-Montero et al. 2009). Another MDSC subset has been more simply defined as $CD14^+/HLA\text{-}DR^{low}$ cells (Filipazzi et al. 2007; Mandruzzato et al. 2009; Filipazzi et al. 2011; Peranzoni et al. 2010; Youn and Gabrilovich 2010). Functionally, human MDSCs can inactivate effector T cells through a variety of mechanisms, including arginase activity (Zea et al. 2005), reactive oxygen species (ROS) production, and TGF-β secretion (Nagaraj and Gabrilovich 2010). Accordingly, investigators have used diverse MDSC definitions and cellular assays when monitoring these cells over the course of a clinical trial, as described below. To complicate the monitoring process even more, it was recently shown that cryopreservation of human MDSC alters both cell frequency and suppressor function, indicating that clinical studies of MDSC should optimally be performed in fresh blood samples (Kotsakis et al. 2012).

The prognostic significance of MDSCs in cancer was demonstrated by Gabitass and colleagues, who report that patients with levels of MDSC >2 % in the peripheral blood had inferior overall survival compared to patients with MDSC levels in the normal range (Gabitass et al. 2011). Correlative alterations in MDSC following various therapies have also been observed. Enhanced levels of circulating $Lin1^-/HLA\text{-}DR^-/CD33^+/CD11b^+$ MDSC were found to correlate with doxorubicin-cyclophosphamide chemotherapy in early stage breast cancer patients (Diaz-Montero et al. 2009). MDSC function was recently assessed in melanoma patients receiving a GM-CSF-based vaccine (Filipazzi et al. 2007). In this case, MDSC isolated from the peripheral blood of post vaccinated (vs. pre-vaccinated) patients produced elevated levels of TGF-β ex vivo. Finally, low pretreatment levels of peripheral blood MDSC were able to discriminate the responder subset of patients receiving high-dose IL-2 from non-responders, indicating that MDSC can serve as a predictive biomarker of IL-2 therapeutic efficacy (Finkelstein et al. 2010).

29.3.8 Antibody Responses

Humoral immune responses against over 100 different cancer antigens have been identified to date, although there is little consensus regarding the role of endogenous antibodies in tumor immunity. Antibodies against p53 are detected at higher prevalence in advanced tumor stages, and p53 humoral immunity is associated with poor prognosis (Lai et al. 1998; Gumus et al. 2004). Some studies suggest that p53 antibodies may accelerate tumor growth by forming immune complexes that inhibit p53 antigen uptake and subsequent presentation by DC (Reuschenbach 2009). Conversely, titers of MUC1-specific antibodies peak at early stages of tumor development, and seropositive patients have a more favorable prognosis than seronegative patients (von Mensdorff-Pouilly et al. 2000). In this case, it has been proposed that MUC1 antibodies may exert a protective effect against tumor development by binding tumor cell surface MUC1 molecules and inhibiting their invasive functions (Reuschenbach 2009; Pinheiro et al. 2010). Currently, it appears that tumor antigen-specific antibodies can have either tumor promoting or inhibiting activity. Alternatively, humoral antitumor immunity may have no functional relevance, and serve rather as a marker of a broader adaptive immune response. Further experimental evidence is needed to address these concepts.

In the setting of cancer clinical trials, monitoring patient serology over the course of treatment has provided insight into therapeutic mode of action. In a recent time course analysis of NY-ESO-1 antibody responses in a patient receiving combination ipilimumab (anti-CTLA-4) and radiotherapy, it was shown that in parallel with therapy, NY-ESO-1 antibody titers increased substantially and reactivity towards additional NY-ESO-1 epitopes was induced (i.e. seroconversion) (Postow et al. 2012). For some therapeutic modalities, induction of humoral immunity can serve as a surrogate endpoint of clinical efficacy. For example, generation of an endogenous HER-2/*neu*-specific antibody response during trastuzumab therapy was found to correlate with therapeutic efficacy in HER-2$^+$ breast cancer (Taylor et al. 2007). Moreover, in a recent study examining the pretreatment immunophenotype of melanoma patients receiving ipilimumab therapy, it was reported that patients who are seropositive for NY-ESO-1 at baseline are more likely to experience clinical response to ipilimumab therapy than seronegative patients (Yuan et al. 2011), indicating that humoral immune responses can also serve as predictive biomarkers of response to certain immunotherapies.

29.3.9 Circulating Dendritic Cells (mDC, pDC)

Through their role as professional APC, DC orchestrate the generation of adaptive anti-tumor immunity. However, in the setting of advanced stage disease, it has been shown that circulating DC have lowered frequencies, altered phenotype and function, and decreased T cell stimulatory activity (Gabrilovich et al. 2012).

Advanced stage breast cancer patients exhibit markedly reduced levels of circulating Lin1^{-}/HLA-DR^{hi} DC compared to early stage patients or healthy controls (Pinzon-Charry et al. 2007). DC functional deficiencies, including diminished IL-12 production or decreased immunostimulatory activity in a mixed-lymphocyte reactions (MLR), have also been reported in the peripheral blood of patients (Pinzon-Charry et al. 2007; Della Bella et al. 2003). Two major subsets of circulating blood DCs have been characterized: myeloid DC (mDC), which are Lin1^{-}/HLA-DR^{hi}/CD11c^{+}/CD123^{-}, and plasmacytoid DC (pDC), which are Lin1^{-}/HLA-DR^{hi}/CD11c^{-}/CD123^{+}. pDC represent a minor population of blood DC and are the main producer of IFN-α in the body. Importantly, while mDC frequencies in the peripheral blood of breast cancer patients were significantly lower compared to healthy donors, pDC frequencies were unaltered (Della Bella et al. 2003), suggesting that monitoring of the mDC subset rather than the pDC subset may more pertinent for cancer clinical trials.

Several therapeutic regimens are known to alter blood DC frequency, phenotype, and/or function as well. Administration of systemic GM-CSF, a DC growth factor and common immunologic adjuvant, and IL-4 expanded circulating DC populations, and modulated the ratio of DC subtypes (Kiertscher et al. 2003). Elevated expression of the maturation and costimulatory markers CD86, CD40, CD83, and CCR7 was observed in circulating DC following adjuvant GM-CSF treatment in patients with resected melanoma (Daud et al. 2008). Moreover, in this trial, patients who experienced an immunologic response (increase in circulating DC) to GM-CSF therapy had a more favorable prognosis than immunologic nonresponders. DC can serve as a predictive biomarker of therapeutic efficacy as well. Above median pretreatment levels of CD1c/BDCA-1^{+} myeloid DC were able to discriminate the responder subset of patients receiving sunitinib from non-responders (van Cruijsen et al. 2008). In another study, high baseline frequencies of peripheral blood DC correlated with a clinical response to high-dose IL-2 (Finkelstein et al. 2010). Together, these data emphasize the importance of DC in endogenous and therapy-induced antitumor immunity.

29.3.10 Circulating Cytokines, Chemokines and Growth Factors

Identification of circulating proteins which could serve to identify patients who are at an early stage of disease, suitable for therapy, or early clinical responders is an elusive goal. While detection of very rare circulating tumor cells would be a direct measure of cancer presence, the characterization of cytokines chemokines and growth factors can potentially be a marker of disease state, toxicity and/or therapeutic response. The literature has many examples of specific data sets supporting circulating molecules associated with disease and response. The field was previously limited by older ELISA technologies in which duplicate or triplicate wells of

100–200 μl of serum or plasma each was required for each analyte tested. Presently, there are several multiplex platforms like Luminex (several manufacturers), Searchlight and Meso Scale, in which $\geq$30 analytes can be measured reliably from a single 50 μl sample. Such assays can be run with plasma (heparin or EDTA), serum or culture supernantents. The reproducibility of these platforms has been tested, as have other technical aspects of these technologies. While the intra-assay variabilities are often quite small, and the range of detection much broader than most ELISA assays, there are some areas for caution which might be negatively impacting the field and limiting the reproducibility of the candidate biomarker data obtained (Chaturvedi et al. 2011; Khan et al. 2004). Absolute values are generally not comparable between platforms, hence a single platform should be used in a given study (Toedter et al. 2008), the Searchlight platform may have more limited recovery and variability (Bastarache et al. 2011), and there are differences between serum or plasma sources (Chaturvedi et al. 2011). Not all analytes are stable over time, with long term storage or after single or multiple freeze–thaw cycles (de Jager et al. 2009; Naranbhai et al. 2011; Butterfield et al. 2011). Biostatistical analysis methods are also critical (Westwell-Roper et al. 2011). Adherence to a single platform, standard procedures, blood processing methods, storage conditions and time, kit lots and analysis methods may help this powerful multiplex technology correctly identify novel circulating biomarkers. This approach can be a useful addition to immune monitoring to detect skewing of immunity, to track the presence of systemically administered cytokines, or to identify toxicities like cytokine storms.

29.4 Testing Immunity at the Tumor

29.4.1 Prognostic Value of the Tumor Microenvironment

While the peripheral blood is easily and inexpensively accessible, knowing the nature of the immune response at the tumor is of primary importance. For the most accessible tumors, like melanoma, the clinical significance of a robust immune infiltrate is well known. More recently, tumor infiltrate character has been more carefully dissected, and the clinical impact of $CD3^+/CD8^+$ T cells, particularly those of memory ($CD45RO^+$) phenotype has been demonstrated (Galon et al. 2006; Fridman et al. 2012). Not only the presence or absence, but the location relative to tumor, stroma and vasculature is also critical in evaluation of intratumoral immunity. Testing for CD3, CD8 and CD45RO by immunohistochemistry or immunofluorescence, and quantifying the number and location of the immune cells relative to tumor, may describe an important "immune score." The markers above were characterized in colorectal cancer but the same infiltrates have been shown to be a positive biomarker in many other solid tumors. A recent study of lymphocytic infiltrates in melanoma metastases revealed three histologic patterns of immune cell infiltration: no immune cell infiltrate, infiltration of immune cells limited only to regions proximal to intratumoral blood vessels, and, the most favorable, a diffuse immune cell infiltrate throughout a metastatic tumor (Erdag et al. 2012). These "immunotypes" were identified in 29, 53, and 8 % of metastases, and the median survival periods for such patients was 15, 23, and 130 months, respectively. The lymphocytic subtypes were identified, and T cells and B cells were most frequent. Higher densities of CD8+ T cells correlated best with survival, a higher density of CD45+ leukocytes, T cells, and B cells also correlated with increased survival. These studies strongly support the importance of assessing the immune-tumor cross-talk at the tumor site.

Among $CD4^+$ T helper cell subsets, a majority of studies indicate that Th1-polarized responses are associated with favorable prognosis, while Th2 responses correlate with negative outcomes (Tatsumi et al. 2002; Wiegering et al. 2011). With regard to Th17 and $Foxp3^+$ Treg subsets, the data are more contradictory (Fridman et al. 2012). While high frequencies of Th17 cells correlate with poor survival in hepatocellular carcinoma, elevated levels of these cells in ovarian cancer positively predict patient outcome (Zhang et al. 2009; Kryczek et al. 2009). Similarly, the presence of Treg in the tumor microenvironment in head and neck or ovarian cancers has been associated with favorable or unfavorable prognosis, respectively (Badoual et al. 2006; Curiel et al. 2004; Martin et al. 2010). Whether Th17 and Treg cells exert tumor promoting or antagonizing activity likely depends on the particular cellular and molecular makeup of each unique lesion, necessitating a functional assessment of isolated cells.

Transcriptional analyses have also been performed, and different classes of tumors can again be identified. A pro-inflammatory chemokine gene expression pattern has correlated with lymphocyte infiltrates and a more favorable clinical outcome (Harlin et al. 2009). While a tumor infiltrate is not a pre-malignancy biomarker, this consistent pattern of a positive impact of immune infiltrate on patient outcome is now well substantiated. There are many technical hurdles, however, to widespread testing of tumors and their infiltrates. In many settings, only baseline timepoint tumor is obtained, primarily for pathologic examination and proper diagnosis. The only tissue available may be formalin-fixed, paraffin-embedded, which may not be suitable for all types of analysis, many specific antibodies or transcriptional testing. Post-treatment biopsies may not be available as standard of care, and institutional review boards may not approve biopsies obtained solely for research. Less invasive biopsies, like fine-needle aspirates and needle core biopsies, may only yield sufficient tissue for molecular assays, and not for cellular assays.

The examples above are descriptive assessments of the identity and location of immune cells, but not a functional confirmation of their capabilities. Obtaining fresh tissue, which is dissociated to single cells, allows for not only identification of the nature of the infiltrate, but also functional testing of effector activity (cytotoxicity, cytokine secretion) and suppression activity (inhibition of T cell proliferation). Despite technical challenges, single cell testing can be informative (Albers et al. 2005). Flow cytometric analysis of the T cell infiltrate in colorectal cancer revealed that lesions with a less advanced pathological stage were enriched in early memory T cells ($CD45RO^+CCR7^-CD28^+CD27^+$) and effector memory T cells ($CD45RO^+CCR7^-CD28^-CD27^-$) (Pages et al. 2005). Recruitment of memory T cell subsets into the colorectal tumor microenvironment appears to be orchestrated by a panel of chemokines (CX3CL1, CXCL9, CXCL10); high levels of these chemokines in situ was associated with a favorable prognosis (Mlecnik et al. 2010).

Analysis of the tumor stroma can provide additional prognostic information. The vasculature of solid tumors is characterized by disorganized, tortuous vessels that restrict the infiltration of nutrients, oxygen, therapeutic drugs, and tumor-primed T cells. Correspondingly, limited angiogenesis and high levels of IP-10, an effector T cell-recruiting chemokine, in cervical cancer lesions was associated with prolonged survival of these patients (Sato et al. 2007).

29.4.2 Therapy-Related Alterations to the Tumor Microenvironment

Several factors within the tumor microenvironment—both host- and tumor-derived—are altered following therapeutic intervention and can distinguish responders from non-responders. Among these factors, the T cell infiltrate has been the most widely studied. Examination of colorectal cancer liver metastases revealed that clinical response to chemotherapy correlated with a robust T cell infiltrate (Halama et al. 2011). A strong association was also reported between high CD8/Foxp3 ratios and response to neoadjuvant chemotherapy, suggesting that as in the peripheral blood, integrated $CD8^+$ T cell and $Foxp3^+$ Treg subset analysis may have more prognostic value than either alone (Ladoire et al. 2011). Not surprisingly, an enrichment in intratumoral T effector cells and diminished levels of $Foxp3^+$ Treg cells is also observed following immune-based therapy (Liakou et al. 2008). Finally, high pretreatment TIL levels were able to discriminate the responder subset of breast cancer patients receiving chemotherapy (docetaxel, doxorubicin, cyclophosphamide) from non-responders, indicating that TIL density can serve as a predictive biomarker of chemotherapeutic efficacy (Denkert et al. 2010).

Testing tumor cells for expression of therapeutic target molecules can yield important information about the mechanism of action or the response of treated patients. For example, testing tumors for MHC class I expression can potentially identify tumors which are more or less susceptible to $CD8^+$ CTL killing, or, conversely NK cell killing (Seliger et al. 2003). In head and neck cancer patients, dysregulation of the antigen processing machinery in tumor cells correlated with poor overall survival (Meissner et al. 2005). Additionally, clinical response to cyclophosphamide chemotherapy correlated with higher HLA Class I expression by breast cancer tumor cells (de Kruijf et al. 2010), and recent data also suggests that clinical responsiveness to PD-L1 or PD-1 antibody therapy may be restricted to patients with PD-L1-positive tumors (Brahmer et al. 2012; Topalian et al. 2012). An inverse correlation between tumor cell expression of PD-L1 patient survival was reported in the setting of ovarian cancer (Hamanishi et al. 2007).

29.5 Patient-Specific Monitoring Considerations

29.5.1 Patient Genetic Background

While much effort has gone into the identification of somatic mutations in tumor cells and the implications of these genetic alterations in therapeutic response, it is becoming clear that patients' immunologic genotype can also dictate response to immune-based therapies (Wang et al. 2012). This is most apparent in the setting of monoclonal antibody (mAb) therapeutics. Tumor antigen-targeted mAbs approved for clinical use include cetuximab (anti-EGFR), rituximab (anti-CD20), and trastuzumab (anti-HER-2/*neu*). However, despite expression of the relevant target antigen by patients' tumors, tumor antigen-targeted mAb therapy is clinically effective in only a minority of patients (Ferris et al. 2010). One of the primary mechanisms through which therapeutic mAbs exert their antitumor activity is via ADCC, whereby antibody-target cell conjugates are recognized by FcγR-expressing NK cells, monocytes, and granulocytes. Accordingly, a number of published reports indicate that polymorphisms in FcγR genes can discriminate therapeutic mAb responders from non-responders in the setting of colorectal and breast cancers (Bibeau et al. 2009; Zhang et al. 2007; Musolino et al. 2008).

Investigators have recently identified several additional immune response genotypes that can predict response to immunotherapy. For example, in melanoma patients receiving adjuvant IFN-α, the presence of certain HLA polymorphisms (HLA-Cw6, HLA-B44) was associated with better overall survival (Wang et al. 2012). Adoptive therapy of melanoma patients with TIL can achieve high response rates, as exemplified by trials at the NCI Surgery Branch (Dudley et al. 2002). Upon genotyping of these TIL, it was recently determined that polymorphisms within the interferon regulatory factor (IRF)-5 gene, which regulates type I interferon signaling, can predict clinical response to adoptive therapy (Uccellini et al. 2012). Cytokine and chemokine gene variants can also predispose patients to immune responsiveness. The IFN-γ (+874A $\rightarrow$ tT) allele correlates with a favorable response to combinational chemo-immunotherapy in melanoma patients, while a 32 base-pair deletion in the CCR5 gene (CCR5Δ32), which produces a non-functional protein, is associated with poor response to immunotherapy in patients who are heterozygous for the CCR5Δ32 gene (Liu et al. 2005; Ugurel et al. 2008). Identification of additional immune-responsive and immune-resistant genotypes is anticipated to greatly aid patient stratification to the most suitable treatment(s).

29.5.2 *Additional Variability from Autologous, Patient-Derived (Cells, Tumor) Treatments*

Cellular therapies for cancer have recently been acknowledged as a legitimate weapon in the oncologist's armamentarium (Kantoff et al. 2010). Autologous cellular products can exhibit considerable patient-to-patient variability, and as described below, recent studies suggest that this variability may be linked to clinical responsiveness. Before cellular products can be administered to patients, several predefined criteria should be met before batch release, which involves sterility, viability, purity, consistency, and potency testing (Butterfield et al. 2011; Nicolette et al. 2007) (Table 29.1). While sterility, viability, purity, and consistency assessments are relatively straightforward, potency assays vary greatly between laboratories and products. In the case of autologous DC vaccines, the two most common potency assays are IL-12p70 production and CD54 expression (Butterfield et al. 2008; Sheikh and Jones 2008). We have recently developed a DC potency assay which measures both the ability of DC to secrete IL-12p70 and respond to CD40L stimulation (Butterfield et al. 2008). Using this assay, it was demonstrated that clinical responses to an autologous DC vaccine in glioma patients correlated with high levels of IL-12 production by the vaccine product (Okada et al. 2011). The first FDA-approved autologous cellular vaccine, sipuleucel-T, is enriched in activated APC that express the adhesion molecule CD54 (ICAM-1). Upregulation of CD54 by APC between cell isolation and reinfusion correlates with enhanced antigen presentation and costimulatory ability (Sheikh and Jones 2008). In three phase three clinical trials, the magnitude of CD54 upregulation and the total number of CD54-expressing cells in the vaccine product was associated with overall survival in prostate cancer patients (Sheikh et al. 2012). Together, these data argue that the efficacy of autologous APC-based vaccines is strongly influenced by batch-to-batch variability, and detection of these variables with phenotypic and functional tests of cell identity, purity and potency assays can further substantiate cellular product efficacy.

29.6 Conclusions

Progress in the field of immunotherapy has been slow, but recent clinical successes have given strong support to the potential of this approach as a treatment strategy in cancer. In order to define biomarkers to identify patients who will have clinical benefit, and to ultimately identify appropriate patient groups to enroll, clinical trials testing immunotherapies should perform more thorough and standardized immunologic assays to fully analyze clinical responders and non-responders. A more thoughtful and detailed focus on immunologic monitoring in these patients will allow us to learn more about the mechanisms of action of the therapeutic approaches as well as the global positive and negative immune responses in treated patients.

References

Adotevi O, Pere H, Ravel P, Haicheur N, Badoual C, Merillon N, Medioni J, Peyrard S, Roncelin S, Verkarre V, Mejean A, Fridman WH et al (2010) A decrease of regulatory T cells correlates with overall survival after sunitinib-based antiangiogenic therapy in metastatic renal cancer patients. J Immunother 33:991–998

Albers AE, Ferris RL, Kim GG, Chikamatsu K, de Leo AB, Whiteside TL (2005) Immune responses to p53 in patients with cancer: enrichment in tetramer +p53 peptide-specific T cells and regulatory T cells at tumor sites. Cancer Immunol Immunother 54:1072–1081

Andoniou CE, Coudert JD, Degli-Esposti MA (2008) Killers and beyond: NK-cell-mediated control of immune responses. Eur J Immunol 38:2938–2942

Andrews DM, Scalzo AA, Yokoyama WM, Smyth MJ, Degli-Esposti MA (2003) Functional interactions between dendritic cells and NK cells during viral infection. Nat Immunol 4:175–181

Appay V (2004) The physiological role of cytotoxic CD4+ T-cells: the holy grail? Clin Exp Immunol 138:10–13

Asai T, Storkus WJ, Whiteside TL (2000) Evaluation of the modified ELISPOT assay for gamma interferon production in cancer patients receiving antitumor vaccines. Clin Diagn Lab Immunol 7:145–154

Ayaru L, Pereira SP, Alisa A, Pathan AA, Williams R, Davidson B, Burroughs AK, Meyer T, Behboudi S (2007) Unmasking of alpha-fetoprotein-specific CD4(+) T cell responses in hepatocellular carcinoma patients undergoing embolization. J Immunol 178:1914–1922

Badoual C, Hans S, Rodriguez J, Peyrard S, Klein C, Agueznay NEH, Mosseri V, Laccourreye O, Bruneval P, Fridman WH, Brasnu DF, Tartour E (2006) Prognostic value of tumor-infiltrating CD4+ T-cell subpopulations in head and neck cancers. Clin Cancer Res 12:465–472

Banchereau J, Palucka AK (2005) Dendritic cells as therapeutic vaccines against cancer. Nat Rev Immunol 5:296–306

Banchereau J, Steinman RM (1998) Dendritic cells and the control of immunity. Nature 392:245–252

Banchereau J, Palucka AK, Dhodapkar M, Burkeholder S, Taquet N, Rolland A, Taquet S, Coquery S, Wittkowski KM, Bhardwaj N, Pineiro L, Steinman R et al (2001) Immune and clinical responses in patients with metastatic melanoma to CD34(+) progenitor-derived dendritic cell vaccine. Cancer Res 61:6451–6458

Bastarache JA, Koyama T, Wickersham NE, Mitchell DB, Mernaugh RL, Ware LB (2011) Accuracy and reproducibility of a multiplex immunoassay platform: a validation study. J Immunol Methods 367:33–39

Bedognetti D, Balwit JM, Wang E, Disis ML, Britten CM, Delogu LG, Tomei S, Fox BA, Gajewski TF, Marincola FM, Butterfield LH (2011) SITC/iSBTc cancer immunotherapy biomarkers resource document: online resources and useful tools—a compass in the land of biomarker discovery. J Transl Med 9:155

Bennouna J, Hildesheim A, Chikamatsu K, Gooding W, Storkus WJ, Whiteside TL (2002) Application of IL-5 ELISPOT assays to quantification of antigen-specific T helper responses. J Immunol Methods 261:145–156

Bergmann C, Strauss L, Zeidler R, Lang S, Whiteside TL (2007) Expansion and characteristics of human T regulatory type 1 cells in co-cultures simulating tumor microenvironment. Cancer Immunol Immunother 56:1429–1442

Bergmann C, Strauss L, Wang Y, Szczepanski MJ, Lang S, Johnson JT, Whiteside TL (2008) T regulatory type 1 cells in squamous cell carcinoma of the head and neck: mechanisms of suppression and expansion in advanced disease. Clin Cancer Res 14:3706–3715

Bibeau F, Lopez-Crapez E, di Fiore F, Thezenas S, Ychou M, Blanchard F, Lamy A, Penault-Llorca F, Frebourg T, Michel P, Sabourin JC, Boissiere-Michot F (2009) Impact of Fc{gamma}RIIa-Fc{gamma}RIIIa polymorphisms and KRAS mutations on the clinical outcome of patients with metastatic colorectal cancer treated with cetuximab plus irinotecan. J Clin Oncol 27:1122–1129

Boon T, van der Bruggen P (1996) Human tumor antigens recognized by T lymphocytes. J Exp Med 183:725–729

Boon T, Coulie PG, van den Eynde BJ, van der Bruggen P (2006) Human T cell responses against melanoma. Annu Rev Immunol 24:175–208

Borg C, Jalil A, Laderach D, Maruyama K, Wakasugi H, Charrier S, Ryffel B, Cambi A, Figdor C, Vainchenker W, Galy A, Caignard A et al (2004) NK cell activation by dendritic cells (DCs) requires the formation of a synapse leading to IL-12 polarization in DCs. Blood 104:3267–3275

Bos R, Sherman LA (2010) CD4+ T-cell help in the tumor milieu is required for recruitment and cytolytic function of CD8+ T lymphocytes. Cancer Res 70:8368–8377

Brahmer JR, Tykodi SS, Chow LQ, Hwu WJ, Topalian SL, Hwu P, Drake CG, Camacho LH, Kauh J, Odunsi K, Pitot HC, Hamid O et al (2012) Safety and activity of anti-PD-L1 antibody in patients with advanced cancer. N Engl J Med 366:2455–2465

Britten CM, Gouttefangeas C, Welters MJ, Pawelec G, Koch S, Ottensmeier C, Mander A, Walter S, Paschen A, Muller-Berghaus J, Haas I, Mackensen A et al (2008) The CIMT-monitoring panel: a two-step approach to harmonize the enumeration of antigen-specific CD8+ T lymphocytes by structural and functional assays. Cancer Immunol Immunother 57:289–302

Britten CM, Janetzki S, van der Burg SH, Huber C, Kalos M, Levitsky HI, Maecker HT, Melief CJ, O'Donnell-Tormey J, Odunsi K, Old LJ, Pawelec G et al (2011) Minimal information about T cell assays: the process of reaching the community of T cell immunologists in cancer and beyond. Cancer Immunol Immunother 60:15–22

Britten CM, Janetzki S, Butterfield LH, Ferrari G, Gouttefangeas C, Huber C, Kalos M, Levitsky HI, Maecker HT, Melief CJ, O'Donnell-Tormey J, Odunsi K et al (2012) T cell assays and MIATA: the essential minimum for maximum impact. Immunity 37:1–2

Butterfield LH, Ribas A, Dissette VB, Amarnani SN, Vu HT, Oseguera D, Wang HJ, Elashoff RM, McBride WH, Mukherji B, Cochran AJ, Glaspy JA et al (2003a) Determinant spreading associated with clinical response in dendritic cell-based immunotherapy for malignant melanoma. Clin Cancer Res 9:998–1008

Butterfield LH, Ribas A, Meng WS, Dissette VB, Amarnani S, Vu HT, Seja E, Todd K, Glaspy JA, McBride WH, Economou JS (2003b) T-cell responses to HLA-A*0201 immunodominant peptides derived from alpha-fetoprotein in patients with hepatocellular cancer. Clin Cancer Res 9:5902–5908

Butterfield LH, Comin-Anduix B, Vujanovic L, Lee Y, Dissette VB, Yang JQ, Vu HT, Seja E, Oseguera DK, Potter DM, Glaspy JA, Economou JS et al (2008a) Adenovirus MART-1-engineered autologous dendritic cell vaccine for metastatic melanoma. J Immunother 31:294–309

Butterfield LH, Disis ML, Fox BA, Lee PP, Khleif SN, Thurin M, Trinchieri G, Wang E, Wigginton J, Chaussabel D, Coukos G, Dhodapkar M et al (2008b) A systematic approach to biomarker discovery; preamble to "the iSBTc-FDA taskforce on immunotherapy biomarkers". J Transl Med 6:81

Butterfield LH, Gooding W, Whiteside TL (2008c) Development of a potency assay for human dendritic cells: IL-12p70 production. J Immunother 31:89–100

Butterfield LH, Disis ML, Khleif SN, Balwit JM, Marincola FM (2010) Immuno-oncology biomarkers 2010 and beyond: perspectives from the iSBTc/SITC biomarker task force. J Transl Med 8:130

Butterfield LH, Palucka AK, Britten CM, Dhodapkar MV, Hakansson L, Janetzki S, Kawakami Y, Kleen TO, Lee PP, Maccalli C, Maecker HT, Maino VC et al (2011a) Recommendations from the iSBTc-SITC/FDA/NCI workshop on immunotherapy biomarkers. Clin Cancer Res 17:3064–3076

Butterfield LH, Potter DM, Kirkwood JM (2011b) Multiplex serum biomarker assessments: technical and biostatistical issues. J Transl Med 9:173

Chaturvedi AK, Kemp TJ, Pfeiffer RM, Biancotto A, Williams M, Munuo S, Purdue MP, Hsing AW, Pinto L, McCoy JP, Hildesheim A (2011) Evaluation of multiplexed cytokine and inflammation marker measurements: a methodologic study. Cancer Epidemiol Biomarkers Prev 20:1902–1911

Clemente CG, Mihm MCJ, Bufalino R, Zurrida S, Collini P, Cascinelli N (1996) Prognostic value of tumor infiltrating lymphocytes in the vertical growth phase of primary cutaneous melanoma. Cancer 77:1303–1310

Crotzer VL, Blum JS (2010) Autophagy and adaptive immunity. Immunology 131:9–17

Curiel TJ, Coukos G, Zou L, Alvarez X, Cheng P, Mottram P, Evdemon-Hogan M, Conejo-Garcia JR, Zhang L, Burow M, Zhu Y, Wei S et al (2004) Specific recruitment of regulatory T cells in ovarian carcinoma fosters immune privilege and predicts reduced survival. Nat Med 10:942–949

Daud AI, Mirza N, Lenox B, Andrews S, Urbas P, Gao GX, Lee JH, Sondak VK, Riker AI, Deconti RC, Gabrilovich D (2008) Phenotypic and functional analysis of dendritic cells and clinical outcome in patients with high-risk melanoma treated with adjuvant granulocyte macrophage colony-stimulating factor. J Clin Oncol 26:3235–3241

de Jager W, Bourcier K, Rijkers GT, Prakken BJ, Seyfert-Margolis V (2009) Prerequisites for cytokine measurements in clinical trials with multiplex immunoassays. BMC Immunol 10:52

de Kruijf EM, van Nes JG, Sajet A, Tummers QR, Putter H, Osanto S, Speetjens FM, Smit VT, Liefers GJ, van de Velde CJ, Kuppen PJ (2010) The predictive value of HLA class I tumor cell expression and presence of intratumoral Tregs for chemotherapy in patients with early breast cancer. Clin Cancer Res 16:1272–1280

Della Bella S, Gennaro M, Vaccari M, Ferraris C, Nicola S, Riva A, Clerici M, Greco M, Villa ML (2003) Altered maturation of peripheral blood dendritic cells in patients with breast cancer. Br J Cancer 89:1463–1472

Denkert C, Loibl S, Noske A, Roller M, Muller BM, Komor M, Budczies J, Darb-Esfahani S, Kronenwett R, Hanusch C, von Torne C, Weichert W et al (2010) Tumor-associated lymphocytes as an independent predictor of response to neoadjuvant chemotherapy in breast cancer. J Clin Oncol 28:105–113

Diaz-Montero CM, Salem ML, Nishimura MI, Garrett-Mayer E, Cole DJ, Montero AJ (2009) Increased circulating myeloid-derived suppressor cells correlate with clinical cancer stage, metastatic tumor burden, and doxorubicin-cyclophosphamide chemotherapy. Cancer Immunol Immunother 58:49–59

Disis ML (2011) Immunologic biomarkers as correlates of clinical response to cancer immunotherapy. Cancer Immunol Immunother 60:433–442

Disis ML, Gooley TA, Rinn K, Davis D, Piepkorn M, Cheever MA, Knutson KL, Schiffman K (2002) Generation of T-cell immunity to the HER-2/neu protein after active immunization with HER-2/neu peptide-based vaccines. J Clin Oncol 20:2624–2632

Dranoff G, Jaffee E, Lazenby A, Golumbek P, Levitsky H, Brose K, Jackson V, Hamada H, Pardoll D, Mulligan RC (1993) Vaccination with irradiated tumor cells engineered to secrete murine granulocyte-macrophage colony-stimulating factor stimulates potent, specific, and long-lasting anti-tumor immunity. Proc Natl Acad Sci U S A 90:3539–3543

Dudley ME, Wunderlich JR, Robbins PF, Yang JC, Hwu P, Schwartzentruber DJ, Topalian SL, Sherry R, Restifo NP, Hubicki AM, Robinson MR, Raffeld M et al (2002) Cancer regression and autoimmunity in patients after clonal repopulation with antitumor lymphocytes. Science 298:850–854

Dudley ME, Wunderlich JR, Yang JC, Sherry RM, Topalian SL, Restifo NP, Royal RE, Kammula U, White DE, Mavroukakis SA, Rogers LJ, Gracia GJ et al (2005) Adoptive cell transfer therapy following non-myeloablative but lymphodepleting chemotherapy for the treatment of patients with refractory metastatic melanoma. J Clin Oncol 23:2346–2357

Dudley ME, Yang JC, Sherry R, Hughes MS, Royal R, Kammula U, Robbins PF, Huang J, Citrin DE, Leitman SF, Wunderlich J, Restifo NP et al (2008) Adoptive cell therapy for patients with metastatic melanoma: evaluation of intensive myeloablative chemoradiation preparative regimens. J Clin Oncol 26:5233–5239

Eckl J, Buchner A, Prinz PU, Riesenberg R, Siegert SI, Kammerer R, Nelson PJ, Noessner E (2012) Transcript signature predicts tissue NK cell content and defines renal cell carcinoma subgroups independent of TNM staging. J Mol Med (Berl) 90:55–66

Erdag G, Schaefer JT, Smolkin ME, Deacon DH, Shea SM, Dengel LT, Patterson JW, Slingluff CL Jr (2012) Immunotype and immunohistologic characteristics of tumor-infiltrating immune cells are associated with clinical outcome in metastatic melanoma. Cancer Res 72:1070–1080

Eruslanov E, Neuberger M, Daurkin I, Perrin GQ, Algood C, Dahm P, Rosser C, Vieweg J, Gilbert SM, Kusmartsev S (2012) Circulating and tumor-infiltrating myeloid cell subsets in patients with bladder cancer. Int J Cancer 130:1109–1119

Fallarino F, Grohmann U, Bianchi R, Vacca C, Fioretti MC, Puccetti P (2000) Th1 and Th2 cell clones to a poorly immunogenic tumor antigen initiate CD8+ T cell-dependent tumor eradication *in vivo*. J Immunol 165:5495–5501

Ferlazzo G, Munz C (2004) NK cell compartments and their activation by dendritic cells. J Immunol 172:1333–1339

Ferlazzo G, Tsang ML, Morretta L, Melioli G, Steinman RM, Munz C (2002) Human dendritic cells activate resting natural killer (NK) cells and are recognized via NKp30 receptor by activated NK cells. J Exp Med 195:343–351

Fernandez NC, Lozier A, Flament C, Ricciardi-Castagnoli P, Bellet D, Suter M (1999) Dendritic cells directly trigger NK cell functions: cross-talk relevant in innate anti-tumor immune responses *in vivo*. Nature Med 5:405–411

Ferris RL, Jaffee EM, Ferrone S (2010) Tumor antigen-targeted, monoclonal antibody-based immunotherapy: clinical response, cellular immunity, and immunoescape. J Clin Oncol 28:4390–4399

Filipazzi P, Valenti R, Huber V, Pilla L, Canese P, Iero M, Castelli C, Mariani L, Parmiani G, Rivoltini L (2007) Identification of a new subset of myeloid suppressor cells in peripheral blood of melanoma patients with modulation by a granulocyte-macrophage colony-stimulation factor-based antitumor vaccine. J Clin Oncol 25:2546–2553

Filipazzi P, Huber V, Rivoltini L (2011) Phenotype, function and clinical implications of myeloid-derived suppressor cells in cancer patients. Cancer Immunol Immunother 61:255–263

Finkelstein SE, Carey T, Fricke I, Yu D, Goetz D, Gratz M, Dunn M, Urbas P, Daud A, DeConti R, Antonia S, Gabrilovich D et al (2010) Changes in dendritic cell phenotype after a new high-dose weekly schedule of interleukin-2 therapy for kidney cancer and melanoma. J Immunother 33:817–827

Fox BA, Schendel DJ, Butterfield LH, Aamdal S, Allison JP, Ascierto PA, Atkins MB, Bartunkova J, Bergmann L, Berinstein N, Bonorino CC, Borden E et al (2011) Defining the critical hurdles in cancer immunotherapy. J Transl Med 9:214

Fridman WH, Pages F, Sautes-Fridman C, Galon J (2012) The immune contexture in human tumours: impact on clinical outcome. Nat Rev Cancer 12:298–306

Friedman KM, Prieto PA, Devillier LE, Gross CA, Yang JC, Wunderlich JR, Rosenberg SA, Dudley ME (2012) Tumor-specific CD4+ melanoma tumor-infiltrating lymphocytes. J Immunother 35:400–408 10.1097/CJI.0b013e31825898c5

Gabitass RF, Annels NE, Stocken DD, Pandha HA, Middleton GW (2011) Elevated myeloid-derived suppressor cells in pancreatic, esophageal and gastric cancer are an independent prognostic factor and are associated with significant elevation of the Th2 cytokine interleukin-13. Cancer Immunol Immunother 60:1419–1430

Gabrilovich DI, Ostrand-Rosenberg S, Bronte V (2012) Coordinated regulation of myeloid cells by tumours. Nat Rev Immunol 12:253–268

Galon J, Costes A, Sanchez-Cabo F, Kirilovsky A, Mlecnik B, Lagorce-Pages C, Tosolini M, Camus M, Berger A, Wind P, Zinzindohoue F, Bruneval P et al (2006) Type, density, and location of immune cells within human colorectal tumors predict clinical outcome. Science 313:1960–1964

Gavin MA, Torgerson TR, Houston E, DeRoos P, Ho WY, Stray-Pedersen A, Ocheltree EL, Greenberg PD, Ochs HD, Rudensky AY (2006) Single-cell analysis of normal and FOXP3-mutant human T cells: FOXP3 expression without regulatory T cell development. Proc Natl Acad Sci U S A 103:6659–6664

Gerosa F, Baldani-Guerra B, Nisii C, Marchesini V, Carra G, Trinchieri G (2002) Reciprocal activating interaction between natural killer cells and dendritic cells. J Exp Med 195:327–333

Gogas H, Ioannovich J, Dafni U, Stavropoulou-Giokas C, Frangia K, Tsoutsos D, Panagiotou P, Polyzos A, Papadopoulos O, Stratigos A, Markopoulos C, Bafaloukos D et al (2006) Prognostic significance of autoimmunity during treatment of melanoma with interferon. N Engl J Med 354:709–718

Gumus E, Erdamar S, Demirel G, Horasanli K, Kendirci M, Miroglu C (2004) Association of positive serum anti-p53 antibodies with poor prognosis in bladder cancer patients. Int J Urol 11:1070–1077

Gupta A, Probst HC, Vuong V, Landshammer A, Muth S, Yagita H, Schwendener R, Pruschy M, Knuth A, van den Broek M (2012) Radiotherapy promotes tumor-specific effector CD8+ T cells via dendritic cell activation. J Immunol 189:558–566

Halama N, Braun M, Kahlert C, Spille A, Quack C, Rahbari N, Koch M, Weitz J, Kloor M, Zoernig I, Schirmacher P, Brand K et al (2011a) Natural killer cells are scarce in colorectal carcinoma tissue despite high levels of chemokines and cytokines. Clin Cancer Res 17:678–689

Halama N, Michel S, Kloor M, Zoernig I, Benner A, Spille A, Pommerencke T, von Knebel DM, Folprecht G, Luber B, Feyen N, Martens UM et al (2011b) Localization and density of immune cells in the invasive margin of human colorectal cancer liver metastases are prognostic for response to chemotherapy. Cancer Res 71:5670–5677

Hamanishi J, Mandai M, Iwasaki M, Okazaki T, Tanaka Y, Yamaguchi K, Higuchi T, Yagi H, Takakura K, Minato N, Honjo T, Fujii S (2007) Programmed cell death 1 ligand 1 and tumor-infiltrating CD8+ T lymphocytes are prognostic factors of human ovarian cancer. Proc Natl Acad Sci 104:3360–3365

Hamid O, Varterasian ML, Wadler S, Hecht JR, Benson A 3rd, Galanis E, Uprichard M, Omer C, Bycott P, Hackman RC, Shields AF (2003) Phase II trial of intravenous CI-1042 in patients with metastatic colorectal cancer. J Clin Oncol 21:1498–1504

Harlin H, Meng Y, Peterson AC, Zha Y, Tretiakova M, Slingluff C, McKee M, Gajewski TF (2009) Chemokine expression in melanoma metastases associated with CD8+ T-cell recruitment. Cancer Res 69:3077–3085

Hayakawa Y, Screpanti V, Yagita H, Grandien A, Ljunggren HG, Smith MJ, Chambers BJ (2004) NK cell TRAIL eliminates immature dendritic cells *in vivo* and limits dendritic cell vaccination efficacy. J Immunol 172:123–129

Higano CS, Corman JM, Smith DC, Centeno AS, Steidle CP, Gittleman M, Simons JW, Sacks N, Aimi J, Small EJ (2008) Phase 1/2 dose-escalation study of a GM-CSF-secreting, allogeneic, cellular immunotherapy for metastatic hormone-refractory prostate cancer. Cancer 113:975–984

Hiraoka K, Miyamoto M, Cho Y, Suzuoki M, Oshikiri T, Nakakubo Y, Itoh T, Ohbuchi T, Kondo S, Katoh H (2006) Concurrent infiltration by CD8+ T cells and CD4+ T cells is a favourable prognostic factor in non-small-cell lung carcinoma. Br J Cancer 94:275–280

Hung KRH, Lafond-Walker A, Lowenstein C, Pardoll D, Levitsky H (1998) The central role of CD4+ T cells in the antitumor immune response. J Exp Med 188:2357–2368

Imai K, Matsuyama S, Miyake S, Suga K, Nakachi K (2000) Natural cytotoxic activity of peripheral-blood lymphocytes and cancer incidence: an 11 year follow-up study of a general population. Lancet 356:1795–1799

Janetzki S, Britten CM, Kalos M, Levitsky HI, Maecker HT, Melief CJ, Old LJ, Romero P, Hoos A, Davis MM (2009) "MIATA"-minimal information about T cell assays. Immunity 31:527–528

Janetzki S, Price L, Britten CM, van der Burg SH, Caterini J, Currier JR, Ferrari G, Gouttefangeas C, Hayes P, Kaempgen E, Lennerz V, Nihlmark K et al (2010) Performance of serum-supplemented and serum-free media in IFNgamma Elispot assays for human T cells. Cancer Immunol Immunother 59:609–618

Janssen EM, Droin NM, Lemmens EE, Pinkoski MJ, Bensinger SJ, Ehst BD, Griffith TS, Green DR, Schoenberger SP (2005) CD4+ T-cell help controls CD8+ T-cell memory via TRAIL-mediated activation-induced cell death. Nature 434:88–93

Jinushi M, Takehara T, Kanto T, Tatsumi T, Groh V, Spies T, Miyagi T, Suzuki T, Sasaki Y, Hayashi N (2003) Critical role of MHC class I-related chain A and B expression on IFN-alpha stimulated dendritic cells in NK cell activation: Impairment in chronic hepatitis C virus infection. J Immunol 170:1249–1256

Kalinski P, Vieira PL, Schuitemaker JH, de Jong EC, Kapsenberg ML (2001) Prostaglandin E(2) is a selective inducer of interleukin-12 p40 (IL-12p40) production and an inhibitor of bioactive IL-12p70 heterodimer. Blood 97:3466–3469

Kantoff PW, Higano CS, Shore ND, Berger ER, Small EJ, Penson DF, Redfern CH, Ferrari AC, Dreicer R, Sims RB, Xu Y, Frohlich MW et al (2010) Sipuleucel-T immunotherapy for castration-resistant prostate cancer. N Engl J Med 363:411–422

Kenter GG, Welters MJ, Valentijn AR, Lowik MJ, van der Berends-Meer DM, Vloon AP, Essahsah F, Fathers LM, Offringa R, Drijfhout JW, Wafelman AR, Oostendorp J (2009) Vaccination against HPV-16 oncoproteins for vulvar intraepithelial neoplasia. N Engl J Med 361:1838–1847

Khan SS, Smith MS, Reda D, Suffredini AF, McCoy JP Jr (2004) Multiplex bead array assays for detection of soluble cytokines: comparisons of sensitivity and quantitative values among kits from multiple manufacturers. Cytometry B Clin Cytom 61:35–39

Kiertscher SM, Gitlitz BJ, Figlin RA, Roth MD (2003) Granulocyte/macrophage-colony stimulating factor and interleukin-4 expand and activate type-1 dendritic cells (DC1) when administered *in vivo* to cancer patients. Int J Cancer 107:256–261

Kim GG, Donnenberg VS, Donnenberg AD, Gooding W, Whiteside TL (2007) A novel multiparametric flow cytometry-based cytotoxicity assay simultaneously immunophenotypes effector cells: Comparisons to a 4 h 51Cr-release assay. J Immunol Methods 325:51–66

Kirkwood JM, Butterfield LH, Tarhini AA, Zarour H, Kalinski P, Ferrone S (2012) Immunotherapy of cancer in 2012. CA Cancer J Clin 62(5):309–335

Kirkwood JM, Lee S, Moschos SJ, Albertini MR, Michalak JC, Sander C, Whiteside T, Butterfield LH, Weiner L (2009) Immunogenicity and antitumor effects of vaccination with peptide vaccine $\pm$ granulocyte-monocyte colony-stimulating factor and/or IFN-alpha2b in advanced metastatic melanoma: Eastern cooperative oncology group phase II trial E1696. Clin Cancer Res 15:1443–1451

Kotsakis A, Harasymczuk M, Schilling B, Georgoulias V, Argiris A, Whiteside TL (2012) Myeloid-derived suppressor cell measurements in fresh and cryopreserved blood samples. J Immunol Methods 381:14–22

Krauze MT, Tarhini A, Gogas H, Kirkwood JM (2011) Prognostic significance of autoimmunity during treatment of melanoma with interferon. Semin Immunopathol 33:385–391

Kryczek I, Banerjee M, Cheng P, Vatan L, Szeliga W, Wei S, Huang E, Finlayson E, Simeone D, Welling TH, Chang A, Coukos G et al (2009) Phenotype, distribution, generation, and functional and clinical relevance of Th17 cells in the human tumor environments. Blood 114:1141–1149

Ladoire S, Mignot G, Dabakuyo S, Arnould L, Apetoh L, Rebe C, Coudert B, Martin F, Bizollon MH, Vanoli A, Coutant C, Fumoleau P et al (2011) In situ immune response after neoadjuvant chemotherapy for breast cancer predicts survival. J Pathol 224:389–400

Lai CL, Tsai CM, Tsai TT, Kuo BI, Chang KT, Fu HT, Perng RP, Chen JY (1998) Presence of serum anti-p53 antibodies is associated with pleural effusion and poor prognosis in lung cancer patients. Clin Cancer Res 4:3025–3030

Liakou CI, Kamat A, Tang DN, Chen H, Sun J, Troncoso P, Logothetis C, Sharma P (2008) CTLA-4 blockade increases IFNgamma-producing CD4+ ICOShi cells to shift the ratio of effector to regulatory T cells in cancer patients. Proc Natl Acad Sci U S A 105:14987–14992

Lieberman J (2003) The ABCs of granule-mediated cytotoxicity: new weapons in the arsenal. Nat Rev Immunol 3:361–370

Liu D, O'Day SJ, Yang D, Boasberg P, Milford R, Kristedja T, Groshen S, Weber J (2005) Impact of gene polymorphisms on clinical outcome for stage IV melanoma patients treated with biochemotherapy: an exploratory study. Clin Cancer Res 11:1237–1246

Mahmoud SMA, Paish EC, Powe DG, Macmillan RD, Grainge MJ, Lee AHS, Ellis IO, Green AR (2011) Tumor-infiltrating CD8+ lymphocytes predict clinical outcome in breast cancer. J Clin Onc 29:1949–1955

Mailliard RB, Egawa S, Cai Q, Kalinska A, Bykovskaya SN, Lotze MT, Kapsenberg ML, Storkus WJ, Kalinski P (2002) Complementary dendritic cell-activating function of CD8+ and CD4+ T cells: helper role of CD8+ T cells in the development of T helper type 1 responses. J Exp Med 195:473–483

Mailliard RB, Wankowicz-Kalinska A, Cai Q, Wesa A, Hilkens CM, Kapsenberg ML, Kirkwood JM, Storkus WJ, Kalinski P (2004) Alpha-type-1 polarized dendritic cells: a novel immunization tool with optimized CTL-inducing activity. Cancer Res 64:5934–5937

Malyguine A, Umansky V, Shurin MR (2009) Cancer immunotherapy and immunomonitoring (CITIM): redefining cancer therapy. J Immunotoxicol 6:205–208

Mander A, Gouttefangeas C, Ottensmeier C, Welters MJ, Low L, van der Burg SH, Britten CM (2010) Serum is not required for *ex vivo* IFN-gamma ELISPOT: a collaborative study of different protocols from the European CIMT immunoguiding program. Cancer Immunol Immunother 59:619–627

Mandruzzato S, Solito S, Falisi E, Francescato S, Chiarion-Sileni V, Mocellin S, Zanon A, Rossi CR, Nitti D, Bronte V, Zanovello P (2009) IL4Ralpha+ myeloid-derived suppressor cell expansion in cancer patients. J Immunol 182:6562–6568

Martin F, Ladoire S, Mignot G, Apetoh L, Ghiringhelli F (2010) Human FOXP3 and cancer. Oncogene 29:4121–4129

Meissner M, Reichert TE, Kunkel M, Gooding W, Whiteside TL, Ferrone S, Seliger B (2005) Defects in the human leukocyte antigen class I antigen processing machinery in head and neck squamous cell carcinoma: association with clinical outcome. Clin Cancer Res 11:2552–2560

Ménard C, Blay J-Y, Borg C, Michiels S, Ghiringhelli F, Robert C, Nonn C, Chaput N, Taïeb J, Delahaye NF, Flament C, Emile J-F et al (2009) Natural killer cell IFN-γ levels predict long-term survival with imatinib mesylate therapy in gastrointestinal stromal tumor-bearing patients. Cancer Res 69:3563–3569

Mlecnik B, Tosolini M, Charoentong P, Kirilovsky A, Bindea G, Berger A, Camus M, Gillard M, Bruneval P, Fridman WH, Pages F, Trajanoski Z et al (2010) Biomolecular network reconstruction identifies T-cell homing factors associated with survival in colorectal cancer. Gastroenterology 138:1429–1440

Montero AJ, Diaz-Montero CM, Kyriakopoulos CE, Bronte V, Mandruzzato S (2012) Myeloid-derived suppressor cells in cancer patients: a clinical perspective. J Immunother 35:107–115

Moodie Z, Price L, Gouttefangeas C, Mander A, Janetzki S, Lower M, Welters MJ, Ottensmeier C, van der Burg SH, Britten CM (2010) Response definition criteria for ELISPOT assays revisited. Cancer Immunol Immunother 59:1489–1501

Musolino A, Naldi N, Bortesi B, Pezzuolo D, Capelletti M, Missale G, Laccabue D, Zerbini A, Camisa R, Bisagni G, Neri TM, Ardizzoni A (2008) Immunoglobulin G fragment C receptor polymorphisms and clinical efficacy of trastuzumab-based therapy in patients with HER-2/neu-positive metastatic breast cancer. J Clin Oncol 26:1789–1796

Nagaraj S, Gabrilovich DI (2010) Myeloid-derived suppressor cells in human cancer. Cancer J 16:348–353

Nakamura Y, Watchmaker P, Urban J, Sheridan B, Giermasz A, Nishimura F, Sasaki K, Cumberland R, Muthuswamy R, Mailliard RB, Larregina AT, Falo LD et al (2007) Helper function of memory CD8+ T cells: heterologous CD8+ T cells support the induction of therapeutic cancer immunity. Cancer Res 67:10012–10028

Naranbhai V, Bartman P, Ndlovu D, Ramkalawon P, Ndung'u T, Wilson D, Altfeld M, Carr WH (2011) Impact of blood processing variations on natural killer cell frequency, activation, chemokine receptor expression and function. J Immunol Methods 366:28–35

Nemunaitis J, Sterman D, Jablons D, Smith JW 2nd, Fox B, Maples P, Hamilton S, Borellini F, Lin A, Morali S, Hege K (2004) Granulocyte-macrophage colony-stimulating factor gene-modified autologous tumor vaccines in non-small-cell lung cancer. J Natl Cancer Inst 96:326–331

Nicolette CA, Healey D, Tcherepanova I, Whelton P, Monesmith T, Coombs L, Finke LH, Whiteside T, Miesowicz F (2007) Dendritic cells for active immunotherapy: optimizing design and manufacture in order to develop commercially and clinically viable products. Vaccine 25(Suppl 2):B47–B60

Okada H, Kalinski P, Ueda R, Hoji A, Kohanbash G, Donegan TE, Mintz AH, Engh JA, Bartlett DL, Brown CK, Zeh H, Holtzman MP et al (2011) Induction of CD8+ T-cell responses against novel glioma-associated antigen peptides and clinical activity by vaccinations with {alpha}-type 1 polarized dendritic cells and polyinosinic-polycytidylic acid stabilized by lysine and carboxymethylcellulose in patients with recurrent malignant glioma. J Clin Oncol 29:330–336

Old LJ, Chen YT (1998) New paths in human cancer serology. J Exp Med 187:1163–1167

Orange JS, Ballas ZK (2006) Natural killer cells in human health and disease. Clin Immunol 118:1–10

O'Rourke MG, Johnson MK, Lanagan CM, See JL, O'Connor LE, Slater GJ, Thomas D, Lopez JA, Martinez NR, Ellem KA, Schmidt CW (2007) Dendritic cell immunotherapy for stage IV melanoma. Melanoma Res 17:316–322

Pages F, Berger A, Camus M, Sanchez-Cabo F, Costes A, Molidor R, Mlecnik B, Kirilovsky A, Nilsson M, Damotte D, Meatchi T, Bruneval P et al (2005) Effector memory T cells, early metastasis, and survival in colorectal cancer. N Engl J Med 353:2654–2666

Palucka AK, Ueno H, Fay J, Banchereau J (2008) Dendritic cells: a critical player in cancer therapy? J Immunother 31:793–805

Penn I (1996) Malignant melanoma in organ allograft recipients. Transplantation 61:274–278

Peranzoni E, Zilio S, Marigo I, Dolcetti L, Zanovello P, Mandruzzato S, Bronte V (2010) Myeloid-derived suppressor cell heterogeneity and subset definition. Curr Opin Immunol 22:238–244

Perez SA, Karamouzis MV, Skarlos DV, Ardavanis A, Sotiriadou NN, Iliopoulou EG, Salagianni ML, Orphanos G, Baxevanis CN, Rigatos G, Papamichail M (2007) CD4+ CD25+ regulatory T-cell frequency in HER-2/neu (HER)-positive and HER-negative advanced-stage breast cancer patients. Clin Cancer Res 13:2714–2721

Piccioli D, Sbrana S, Melandri E, Valiante NM (2002) Contact-dependent stimulation and inhibition of dendritic cells by natural killer cells. J Exp Med 195:335–341

Pinheiro SP, Hankinson SE, Tworoger SS, Rosner BA, McKolanis JR, Finn OJ, Cramer DW (2010) Anti-MUC1 antibodies and ovarian cancer risk: prospective data from the Nurses' Health Studies. Cancer Epidemiol Biomarkers Prev 19:1595–1601

Pinzon-Charry A, Ho CS, Maxwell T, McGuckin MA, Schmidt C, Furnival C, Pyke CM, Lopez JA (2007) Numerical and functional defects of blood dendritic cells in early- and late-stage breast cancer. Br J Cancer 97:1251–1259

Platonova S, Cherfils-Vicini J, Damotte D, Crozet L, Vieillard V, Validire P, André P, Dieu-Nosjean M-C, Alifano M, Régnard J-F, Fridman W-H, Sautès-Fridman C et al (2011) Profound coordinated alterations of intratumoral NK cell phenotype and function in lung carcinoma. Cancer Res 71:5412–5422

Poschke I, Mcugiakakas D, Hansson J, Masucci GV, Klessling R (2010) Immature immunosuppressive CD14+HLA-DR-/low cells in melanoma patients are Stat3hi and overexpress CD80, CD83, and DC-sign. Cancer Res 70:4335–4345

Postow MA, Callahan MK, Barker CA, Yamada Y, Yuan J, Kitano S, Mu Z, Rasalan T, Adamow M, Ritter E, Sedrak C, Jungbluth AA et al (2012) Immunologic correlates of the abscopal effect in a patient with melanoma. N Engl J Med 366:925–931

Raja SM, Metkar SS, Froelich CJ (2003) Cytotoxic granule-mediated apoptosis: unraveling the complex mechanism. Curr Opin Immunol 15:528–532

Reuschenbach M, von Knebel Doeberit M, Wentzensen N (2009) A systematic review of humoral immune responses against tumor antigens. Cancer Immunol Immunother 58:1535–1544

Ribas A, Glaspy JA, Lee Y, Dissette VB, Seja E, Vu HT, Tchekmedyian NS, Oseguera D, Comin-Anduix B, Wargo JA, Amarnani SN, McBride WH et al (2004) Role of dendritic cell phenotype, determinant spreading, and negative costimulatory blockade in dendritic cell-based melanoma immunotherapy. J Immunother 27:354–367

Roszkowski JJ, Lyons GE, Kast WM, Yee C, Van Besien K, Nishimura MI (2005) Simultaneous generation of CD8+ and CD4+ melanoma-reactive T cells by retroviral-mediated transfer of a single T-cell receptor. Cancer Res 65:1570–1576

Rubio V, Stuge TB, Singh N, Betts MR, Weber JS, Roederer M, Lee PP (2003) *Ex vivo* identification, isolation and analysis of tumor-cytolytic T cells. Nat Med 9:1377–1382

Sakaguchi S, Miyara M, Costantino CM, Hafler DA (2010) FOXP3+ regulatory T cells in the human immune system. Nat Rev Immunol 10:490–500

Sato E, Fujimoto J, Toyoki H, Sakaguchi H, Alam SM, Jahan I, Tamaya T (2007) Expression of IP-10 related to angiogenesis in uterine cervical cancers. Br J Cancer 96:1735–1739

Schaefer C, Butterfield LH, Lee S, Kim GG, Visus C, Albers A, Kirkwood JM, Whiteside TL (2012) Function but not phenotype of melanoma peptide-specific CD8(+) T cellsT cells correlate with survival in a multiepitope peptide vaccine trial (ECOG 1696). Int J Cancer 131:874–884

Schwartzentruber DJ, Lawson DH, Richards JM, Conry RM, Miller DM, Treisman J, Gailani F, Riley L, Conlon K, Pockaj B, Kendra KL, White RL et al (2011) gp100 peptide vaccine and interleukin-2 in patients with advanced melanoma. N Engl J Med 364:2119–2127

Seliger B, Abken H, Ferrone S (2003) HLA-G and MIC expression in tumors and their role in anti-tumor immunity. Trends Immunol 24:82–87

Shafer-Weaver K, Sayers T, Strobl S, Derby E, Ulderich T, Baseler M, Malyguine A (2003) The Granzyme B ELISPOT assay: an alternative to the 51Cr-release assay for monitoring cell-mediated cytotoxicity. J Transl Med 1:14

Sheikh NA, Jones LA (2008) CD54 is a surrogate marker of antigen presenting cell activation. Cancer Immunol Immunother 57:1381–1390

Sheikh NA, Petrylak D, Kantoff PW, Dela Rosa C, Stewart FP, Kuan LY, Whitmore JB, Trager JB, Poehlein CH, Frohlich MW, Urdal DL (2012) Sipuleucel-T immune parameters correlate with survival: an analysis of the randomized phase 3 clinical trials in men with castration-resistant prostate cancer. Cancer Immunol Immunother

Slingluff CL Jr, Petroni GR, Yamshchikov GV, Hibbitts S, Grosh WW, Chianese-Bullock KA, Bissonette EA, Barnd DL, Deacon DH, Patterson JW, Parekh J, Neese PY et al (2004) Immunologic and clinical outcomes of vaccination with a multiepitope melanoma peptide vaccine plus low-dose interleukin-2 administered either concurrently or on a delayed schedule. J Clin Oncol 22:4474–4485

Slingluff CL Jr, Petroni GR, Chianese-Bullock KA, Smolkin ME, Hibbitts S, Murphy C, Johansen N, Grosh WW, Yamshchikov GV, Neese PY, Patterson JW, Fink R et al (2007) Immunologic and clinical outcomes of a randomized phase II trial of two multipeptide vaccines for melanoma in the adjuvant setting. Clin Cancer Res 13:6386–6395

Slingluff CL Jr, Petroni GR, Olson W, Czarkowski A, Grosh WW, Smolkin M, Chianese-Bullock KA, Neese PY, Deacon DH, Nail C, Merrill P, Fink R et al (2008) Helper T-cell responses and clinical activity of a melanoma vaccine with multiple peptides from MAGE and melanocytic differentiation antigens. J Clin Oncol 26:4973–4980

Spaggiari GM, Carosio R, Pende D, Marcenaro S, Rivera P, Zocchi MR, Moretta L, Poggi A (2001) NK cell mediated lysis of autologous antigen presenting cells is triggered by the engagement of the phosphatidylinositol 3-kinase upon ligation of the natural cytotoxicity receptors NKp30a nad NKp46. Eur J Immunol 3:1656–1665

Surman D, Dudley M, Overwijk W, Restifo N (2000) Cutting edge: CD4+ T cell control of CD8+ T cell reactivity to a model tumor antigen. J Immunol 164:562–565

Tahara H, Sato M, Thurin M, Wang E, Butterfield LH, Disis ML, Fox BA Lee PP, Khleif SN, Wigginton JM, Ambs S, Akutsu Y et al (2009) Emerging concepts in biomarker discovery; the US-Japan workshop on immunological molecular markers in oncology. J Transl Med 7:45

Tatsumi T, Kierstead LS, Ranieri E, Gesualdo L, Schena FP, Finke JH, Bukowski RM, Mueller-Berghaus J, Kirkwood JM, Kwok WW, Storkus WJ (2002) Disease-associated bias in T helper type 1 (Th1)/Th2 CD4(+) T cell responses against MAGE-6 in HLA-DRB10401(+) patients with renal cell carcinoma or melanoma. J Exp Med 196:619–628

Taylor C, Hershman D, Shah N, Suciu-Foca N, Petrylak DP, Taub R, Vahdat L, Cheng B, Pegram M, Knutson KL, Clynes R (2007) Augmented HER-2 specific immunity during treatment with trastuzumab and chemotherapy. Clin Cancer Res 13:5133–5143

Toedter G, Hayden K, Wagner C, Brodmerkel C (2008) Simultaneous detection of eight analytes in human serum by two commercially available platforms for multiplex cytokine analysis. Clin Vaccine Immunol 15:42–48

Topalian SL, Hodi FS, Brahmer JR, Gettinger SN, Smith DC, McDermott DF, Powderly JD, Carvajal RD, Sosman JA, Atkins MB, Leming PD, Spigel DR et al (2012) Safety, activity, and immune correlates of anti-PD-1 antibody in cancer. N Engl J Med 366:2443–2454

Uccellini L, De Giorgi V, Zhao Y, Tumaini B, Erdenebileg N, Dudley ME, Tomei S, Bedognetti D, Ascierto ML, Liu Q, Simon R, Kottyan L et al (2012) IRF5 gene polymorphisms in melanoma. J Transl Med 10:170

Ugurel S, Schrama D, Keller G, Schadendorf D, Brocker EB, Houben R, Zapatka M, Fink W, Kaufman HL, Becker JC (2008) Impact of the CCR5 gene polymorphism on the survival of metastatic melanoma patients receiving immunotherapy. Cancer Immunol Immunother 57:685–691

van Cruijsen H, van der Veldt AA, Vroling L, Oosterhoff D, Broxterman HJ, Scheper RJ, Giaccone G, Haanen JB, van den Eertwegh AJ, Boven E, Hoekman K, de Gruijl TD (2008) Sunitinib-induced myeloid lineage redistribution in renal cell cancer patients: CD1c+ dendritic cell frequency predicts progression-free survival. Clin Cancer Res 14:5884–5892

van Tendeloo VF, van de Velde A, van Driessche A, Cools N, Anguille S, Ladell K, Gostick E, Vermeulen K, Pieters K, Nijs G, Stein B, Smits EL et al (2010) Induction of complete and molecular remissions in acute myeloid leukemia by Wilms' tumor 1 antigen-targeted dendritic cell vaccination. Proc Natl Acad Sci U S A 107:13824–13829

Vergati M, Cereda V, Madan RA, Gulley JL, Huen NY, Rogers CJ, Hance KW, Arlen PM, Schlom J, Tsang KY (2011) Analysis of circulating regulatory T cells in patients with metastatic prostate cancer pre- versus post-vaccination. Cancer Immunol Immunother 60:197–206

Vivier E, Ugolini S, Blaise D, Chabannon C, Brossay L (2012) Targeting natural killer cells and natural killer T cells in cancer. Nat Rev Immunol 12:239–252

von Mensdorff-Pouilly S, Verstraeten AA, Kenemans P, Snijdewint FG, Kok A, van Kamp GJ, Paul MA, Van Diest PJ, Meijer S, Hilgers J (2000) Survival in early breast cancer patients is favorably influenced by a natural humoral immune response to polymorphic epithelial mucin. J Clin Oncol 18:574–583

Vujanovic NL (2001) Role of TNF family ligands in antitumor activity of natural killer cells. Internat Rev Immunol 20:415–435

Vujanovic L, Szymkowski DE, Vujanovic NL, Butterfield LH (2010) Virally-infected and cytokine-matured human dendritic cells activate natural killer cells via cooperative activity of plasma membrane-bound TNF and IL-15. Blood 116:575–583

Wakabayashi O, Yamazaki K, Oizumi S, Hommura F, Kinoshita I, Ogura S, Dosaka-Akita H, Nishimura M (2003) CD4+ T cells in cancer stroma, not CD8+ T cells in cancer cell nests, are associated with favorable prognosis in human non-small cell lung cancers. Cancer Sci 94:1003–1009

Walter S, Weinschenk T, Stenzl A, Zdrojowy R, Pluzanska A, Szczylik C, Staehler M, Brugger W, Dietrich PY, Mendrzyk R, Hilf N, Schoor O et al (2012) Multipeptide immune response to cancer vaccine IMA901 after single-dose cyclophosphamide associates with longer patient survival. Nat Med

Wang E, Uccellini L, Marincola FM (2012a) A genetic inference on cancer immune responsiveness. Oncoimmunology 1:520–525

Wang E, Zhao Y, Monaco A, Uccellini L, Kirkwood JM, Spyropoulou-Vlachou M, Fanelli MC, Marincola FM, Gogas H (2012b) A multi-factorial genetic model for prognostic assessment of high risk melanoma patients receiving adjuvant interferon. PLoS ONE 7:e40805

Weide B, Zelba H, Derhovanessian E, Pflugfelder A, Eigentler TK, di Giacomo AM, Maio M, Aarntzen EH, de Vries IJ, Sucker A, Schadendorf D, Buttner P et al (2012) Functional T cells targeting NY-ESO-1 or Melan-A are predictive for survival of patients with distant melanoma metastasis. J Clin Oncol 30:1835–1841

Weiner HL (2001) Induction and mechanism of action of transforming growth factor-beta-secreting Th3 regulatory cells. Immunol Rev 182:207–214

Welters MJ, Kenter GG, de Vos van Steenwijk PJ, Lowik MJ, Berends-van der Meer DM, Essahsah F, Stynenbosch LF, Vloon AP, Ramwadhdoebe TH, Piersma SJ, van der Hulst JM, Valentijn AR et al. (2010) Success or failure of vaccination for HPV16-positive vulvar lesions correlates with kinetics and phenotype of induced T-cell responses. Proc Natl Acad Sci U S A 107:11895–11899

Westwell-Roper C, Dai DL, Soukhatcheva G, Potter KJ, van Rooijen N, Ehses JA, Verchere CB (2011) IL-1 blockade attenuates islet amyloid polypeptide-induced proinflammatory cytokine release and pancreatic islet graft dysfunction. J Immunol 187:2755–2765

Whiteside TL, Schuler P (2012) Schilling B. Induced and natural regulatory T cells in human cancer, Expert Opin Biol Ther

Whiteside TL, Zhao Y, Tsukishiro T, Elder EM, Gooding W, Baar J (2003) Enzyme-linked immunospot, cytokine flow cytometry, and tetramers in the detection of T-cell responses to a dendritic cell-based multipeptide vaccine in patients with melanoma. Clin Cancer Res 9:641–649

Wiegering V, Eyrich M, Rutkowski S, Wolfl M, Schlegel PG, Winkler B (2011) TH1 predominance is associated with improved survival in pediatric medulloblastoma patients. Cancer Immunol Immunother 60:693–703

Wierecky J, Muller MR, Wirths S, Halder-Oehler E, Dorfel D, Schmidt SM, Hantschel M, Brugger W, Schroder S, Horger MS, Kanz L, Brossart P (2006) Immunologic and clinical responses after vaccinations with peptide-pulsed dendritic cells in metastatic renal cancer patients. Cancer Res 66:5910–5918

Wiesel M, Oxenius A (2012) From crucial to negligible: functional $CD8^+$ T-cell responses and their dependence on $CD4^+$ T-cell help. Eur J Immunol 42:1080–1088

Wolchok JD, Neyns B, Linette G, Negrier S, Lutzky J, Thomas L, Waterfield W, Schadendorf D, Smylie M, Guthrie T Jr, Grob JJ, Chesney J et al (2010) Ipilimumab monotherapy in patients with pretreated advanced melanoma: a randomised, double-blind, multicentre, phase 2, dose-ranging study. Lancet Oncol 11:155–164

Xu J, Chakrabarti AK, Tan JA, Ge L, Gambotto A, Vujanovic NL (2007) Essential role of the TNF-TNFR2 cognate interaction in mouse dendritic cell-natural killer cell cross-talk. Blood 109:3333–3341

Yoshida N, Abe H, Ohkuri T, Wakita D, Sato M, Noguchi D, Miyamoto M, Morikawa T, Kondo S, Ikeda H, Nishimura T (2006) Expression of the MAGE-A4 and NY-ESO-1 cancer-testis antigens and T cell infiltration in non-small cell lung carcinoma and their prognostic significance. Int J Oncol 28:1089–1098

Youn JI, Gabrilovich DI (2010) The biology of myeloid-derived suppressor cells: the blessing and the curse of morphological and functional heterogeneity. Eur J Immunol 40:2969–2975

Yuan J, Adamow M, Ginsberg BA, Rasalan TS, Ritter E, Gallardo HF, Xu Y, Pogoriler E, Terzulli SL, Kuk D, Panageas KS, Ritter G et al (2011) Integrated NY-ESO-1 antibody and CD8+ T-cell responses correlate with clinical benefit in advanced melanoma patients treated with ipilimumab. Proc Natl Acad Sci U S A 108:16723–16728

Zea AH, Rodriguez PC, Atkins MB, Hernandez C, Signoretti S, Zabaleta J, McDermott D, Quiceno D, Youmans A, O'Neill A, Mier J, Ochoa AC (2005) Arginase-producing myeloid suppressor cells in renal cell carcinoma patients: a mechanism of tumor evasion. Cancer Res 65:3044–3048

Zerbini A, Pilli M, Penna A, Pelosi G, Schianchi C, Molinari A, Schivazappa S, Zibera C, Fagnoni FF, Ferrari C, Missale G (2006) Radiofrequency thermal ablation of hepatocellular carcinoma liver nodules can activate and enhance tumor-specific T-cell responses. Cancer Res 66:1139–1146

Zhang W, Gordon M, Schultheis AM, Yang DY, Nagashima F, Azuma M, Chang HM, Borucka E, Lurje G, Sherrod AE, Iqbal S, Groshen S et al (2007) FCGR2A and FCGR3A polymorphisms associated with clinical outcome of epidermal growth factor receptor expressing metastatic colorectal cancer patients treated with single-agent cetuximab. J Clin Oncol 25:3712–3718

Zhang JP, Yan J, Xu J, Pang XH, Chen MS, Li L, Wu C, Li SP, Zheng L (2009) Increased intratumoral IL-17-producing cells correlate with poor survival in hepatocellular carcinoma patients. J Hepatol 50:980–989

Zitvogel L, Kepp O, Kroemer G (2011) Immune parameters affecting the efficacy of chemotherapeutic regimens. Nat Rev Clin Oncol 8:151–160

Chapter 30
Evaluation of the Tumor Immunoenvironment in Clinical Trials

Anatoli Malyguine, Kimberly Dunham, Thomas J. Sayers and Michael R. Shurin

Abstract Monitoring of immunotherapeutic clinical trials has undergone a considerable change in the last decade resulting in a general agreement that immune monitoring should guide the development of cancer vaccines. The emphasis on immune cell functions and the quantitation of antigen-specific T cells has played a major role in the attempts to establish meaningful correlations between therapy-induced alterations in immune responses and clinical endpoints. However, one significant unresolved issue in modern immunotherapy is that when a tumor-specific cellular immune response is observed following a course of immunotherapy, it does not always lead to clinically proven cancer regression. This disappointing lack of a correlation between the tumor-specific cytotoxic immune responses and the therapeutic benefit may be explained in part by the notion that the analysis of any single immunological parameter is not sufficient to provide detailed information about the complex interactions between different cell subsets in the peripheral blood or immune, tumor, and stromal cells in the tumor milieu itself. Therefore, following administration of vaccines or immunomodulators, a systemic approach is required to improve the quality of a serial monitoring to ensure that it adequately and reliably measures crucial beneficial immunological changes induced in patients. Comprehensive evaluation of the balance between the immunostimulatory and immunosuppressive compartments of the immune

A. Malyguine (✉) · K. Dunham
Laboratory of Cell Mediated Immunity, SAIC-Frederick, Inc., Frederick National Laboratory, Frederick, MD 21702, USA
e-mail: malyguinea@mail.nih.gov

T. J. Sayers
Basic Sciences Program, SAIC-Frederick, Inc., Cancer and Inflammation Program, Frederick National Laboratory, Frederick, MD, USA

M. R. Shurin
Department of Pathology, Division of Clinical Immunopathology, University of Pittsburgh Medical Center, Pittsburgh, PA, USA

M. R. Shurin et al. (eds.), *The Tumor Immunoenvironment*,
DOI: 10.1007/978-94-007-6217-6_30,

system could be critical for obtaining a better understanding of why a given immunotherapy does or does not work in a particular clinical trial. New approaches to characterize tumor-infiltrating leukocytes, including their phenotypic, biochemical and genetic characteristics within the tumor microenvironment need to be developed, validated and standardized for reliability and consistency in order to establish the rigorous performance standards that should additionally complement current monitoring techniques.

Keywords Immunomonitoring · Vaccine clinical trials · Tumor immunoenvironment · Leukocytes

30.1 Introduction

The last decades have been characterized by substantial progress in our understanding of the role of the immune system in tumor progression. There are now a large number of examples in both animal models and cancer patients themselves demonstrating how the immune system is able to recognize tumor antigens and eliminate or control tumor cell growth, progression and metastases. Thus attempts have been made to appropriately manipulate the immune system to generate measurable tumor-specific immune responses. Recently, the Sipuleucel-T vaccine (Provenge^{TM}, Dendreon Corporation, Seattle, WA) was approved by the FDA as the first therapeutic cancer vaccine in humans for the treatment of patients with castration-resistant metastatic prostate cancer. In addition, the monoclonal antibody Ipilimumab (Yervoy, Bristol-Myers Squibb) was approved for the treatment of advanced melanoma as a second-line therapy. These developments raise high expectations among scientists and the general public that immunotherapy may provide further beneficial breakthroughs in cancer treatment.

Immune response profiling and monitoring are the key elements in the development of new biotherapies, and a variety of assays including MHC-tetramers, cytokine release/catch assays, ELISPOT assays and flow cytometric assays have been introduced for assessing different parameters of the immune status in cancer patients. However, clinical studies have demonstrated that the therapy-induced tumor-specific immune responses often did not correlate with clinical responses when high levels of tumor-specific cytotoxic lymphocytes that could recognize and efficiently kill tumor cells ex vivo were detectable (Malyguine et al. 2012). A comprehensive monitoring of the balance between the immunostimulatory and immunosuppressive components of the immune response in the circulation and in the local tumor immunoenvironment could bring new insights to our understanding of the prognostic significance of immune responses associated with cancer therapy.

Recently we have analyzed more than 20 clinical trials in which patients with various cancers were treated with genetically engineered dendritic cells (DC)

(Shurin et al. 2010). Increases in specific CD8 + T cell frequency/activity were detected in 201 of 272 tested patients (74 %) and increased NK cell activity was revealed in 19 of 24 evaluated patients (79 %). However, clinical results were disappointing. From 333 analyzed patients, disease stabilization was only observed in 10.8 % of patients and survival benefits were seen in just 7.5 % of patients after vaccination. Overall, only 8 from all 333 patients (2.4 %) demonstrated a therapeutic response as assessed by standard tumor measurement criteria. Additionally, another group has shown that the mere presence of dramatically expanded numbers of vaccine-induced antigen-specific T cells cannot by themselves be used as a "surrogate marker" for vaccine efficacy (Rosenberg et al. 2005). In a recent clinical trial 74 % of vaccinated glioma patients demonstrated specific responses against targeted glioma-associated antigens, but only two from 19 patients (9 %) had an objective clinical response, with both being non-responders in an ELISPOT assay (Okada et al. 2011). Ribas et al. (2010) reported that from 33 melanoma patients vaccinated with DC vaccine only 9 % had an objective clinical response, while 90 % of 29 tested patients had detectable tumor antigen-specific T cells in the peripheral blood. In a multi-center study by Schwartzentruber et al. (2011), 185 melanoma patients were treated with IL-2 or IL-2 and gp100 peptide vaccine and no correlation between the anti-peptide reactivity and the objective clinical responses was found. These and multiple other data have demonstrated that there is very often a lack of correlation between the tumor-specific cytotoxic immune responses and objective clinical responses to immunotherapy.

Several reasons may be responsible for the lack of correlation between the immune and clinical response to cancer immunotherapy and they have been discussed elsewhere (Whiteside 2010; Butterfield et al. 2011). Based on a growing body of new evidence, two issues should attract a special attention. First of all, the cytotoxic immune response to vaccination represents an extremely complex array of interactions between immune, tumor and stromal cells and therefore cannot be realistically evaluated based on a single immunological parameter or cell type. Second, only the peripheral blood components are typically accessible for serial analysis, yet there is no convincing evidence that immune responses in the peripheral blood are representative of those occurring at the site of the tumor.

30.2 Rationale for the Local Immunomonitoring in Clinical Trials

Much emerging data suggests that the main molecular events that determine tumor fate in face of immune attack occur at the tumor site itself (Shurin et al. 2006). A significant number of identified mechanisms within the tumor microenvironment lead to immune unresponsiveness. These include (1) down-regulation of antigen processing and MHC class I/peptide complex expression on the surface of tumor cells, (2) inhibition of generation, function, activity and survival of many immune effector cells (i.e., T cells, NK cells, neutrophils, conventional DC) induced by cancerous cells, stromal elements and immune regulatory cells and (3) tumor-induced polarization of immune cells into immune regulators, such as Treg and Breg cells, MDSC, M2 TAM, regDC and N2 tumor-associated neutrophils (TAN). Consequently, malignant cells, the surrounding stromal cells and intratumoral immune/inflammatory cells are able to support tumor neoangiogenesis, tumor cell spreading and inhibit development of efficient anti-tumor immunity. Many cells at the tumor site produce a variety of soluble, cytoplasmic and membrane-bound molecules, including transforming growth factor β (TGF-β), interleukin (IL)-10, IL-8, IL-13, indoleamine-2,3-dioxygenase (IDO), inducible nitric oxide synthase (iNOS), prostaglandin E2 (PGE2), arginase-1 (ARG1), neuropeptides, gangliosides, B7-H1, B7-DC, CTLA-4 and other molecules that can limit anti-tumor immune responses and support tumor-specific immune tolerance (Talmadge 2011; Quezada et al. 2011). However, despite robust evidence demonstrating that host immune cells with immunoregulatory function within the tumor milieu are responsible for the establishment of tumor antigen-specific tolerance, and represent a significant hurdle to successful therapy for cancer, analysis of tumor-associated immune cells and factors has not been widely considered as an essential part of immunomonitoring in cancer clinical trials (Pages et al. 2009). In addition to technical difficulties, the limited accessibility of clinical materials and the uncertain significance of these assays, the main problem that still limits incorporation of intratumoral immunomonitoring into clinical practice seems to be limited data available today to prove that number, phenotype or function of specific subsets of intratumoral leukocytes correlate with the disease progression or a patient's response to anti-cancer therapy.

30.3 Local Immunomonitoring of Tumor-Associated Lymphocytes

The infiltration of primary or metastatic tumors by T lymphocytes can contribute in both a positive and negative manner to tumor growth, invasion and patient outcomes. Numerous clinical studies have concluded that in most cases tumor-infiltrating $CD8^+$ T cells had antitumor activity as judged by their favorable effect on patients' survival in melanoma, renal, ovarian, urethral, colorectal, esophageal, head and neck, breast, pancreatic and lung cancer (Sharma et al. 2007; Leffers et al. 2009; Uppaluri et al. 2008; Ohtani 2007; Mahmoud et al. 2011). In most studies the infiltration of tumor mass with $CD8^+$ T cells and the higher $CD8^+$ to $CD4^+$ T cell ratio correlated with improved outcome (Talmadge 2011; Jochems and Schlom 2011). For example, a significant correlation was observed between the extent of $CD8^+$ T cell infiltration, $CD8^+$ T to $CD4^+$ T cell ratio and a high $CD8^+$ T to Treg cell ratio in patients who had tumors that failed to metastasize to the draining lymph nodes (Piersma et al. 2007). In a recent study of patients with stage IV non-small-cell lung cancer, it was reported that patients with a higher frequency of tumor-infiltrating $CD8^+$ T cells in the tumor epithelium, as compared to the tumor stroma, had a significantly better survival (Kawai et al. 2008). Pages et al. (2009) found that five-year survival in colorectal cancer patients with high densities of both $CD8^+$ T and $CD45RO^+$ T cells was 86.2 % with only 4.8 % of patients having tumor recurrence, whereas in the group with low densities of these cells, 75 % of patients had tumor recurrence and only 27.5 % survived. In another study of metastatic melanoma, Murphy et al. (1993) demonstrated that whereas pre-vaccination tumor biopsies failed to reveal significant infiltration by lymphocytes, biopsies obtained after vaccination were markedly infiltrated by T cells with $CD8^+$ phenotype.

In contrast to tumor infiltration by $CD8^+$ T cells, tumor infiltration by $CD4^+$ Treg cells has been reported to have either a negative or positive impact on subsequent clinical outcomes (Talmadge 2011). Regulatory T cells are usually characterized by concurrent expression of CD4, CD25 and FoxP3, essential markers of these cells in humans. Other Treg markers may also include cytotoxic T lymphocyte antigen 4 (CTLA-4), CD127, HLADR, CD45RA, glucocorticoid-induced TNFR-related protein (GITR) and lymphocyte activation gene 3 (LAG-3). Among $CD4^+$ T cells present in the tumor, a subset of $CD4^+CD25^{high}FoxP3^+$ Treg cells may constitute from 5 to 15 % of CD4 T cells in the infiltrate. While $FoxP3^+CD4^+$ T cells have been shown to be increased in the peripheral blood of melanoma patients as compared to healthy donors, they could be even more enriched in the tumor-infiltrated lymph nodes and at the tumor sites (Jandus et al. 2008). In patients with ovarian cancer, recruitment of Treg cells supported tumor growth and was associated with reduced survival. At the later stages of disease, $CD4^+CD25^+FoxP3^+$ cells significantly accumulated at the tumor site with ~75 % of Treg being in proximity to infiltrating $CD8^+$ T cytotoxic cells (Curiel et al. 2004). Sinicrope et al. (2009) have recently reported that a low intraepithelial effector $CD3^+$/regulatory $FoxP3^+$ T cell ratio can predict an adverse outcome in colon carcinoma patients, indicating the importance of an effector to Treg cell ratio in colon cancer prognosis. In contrast, analysis of T cell infiltrates in 967 colorectal tumors showed that patients with higher $FoxP3^+$ Treg density in tumor tissue had improved survival (Salama et al. 2009). However, on understanding the high heterogeneity of tumor-associated T cells, it is important to note that a recent study revealed that $CD8^+FoxP3^+$ T cells mark the presence of tumor-rejecting antigen-specific T cells, and their accumulation serves as a marker for an effective T cell response (Le et al. 2011). The role of FoxP3 + cells infiltration in tumor environment has been recently analyzed in detail by Whiteside (2012), Whiteside et al. (2012), Yoon et al. (2012).

An interesting analysis of the prognostic value of a combined evaluation of intratumoral Treg and $CD8^+$ CTL has been recently reported. Chen et al. (2011) determined tumor infiltration by Tregs and CTL in 141 hepatocellular carcinoma (HCC) patients after tumor resection. The density of intratumoral Tregs and peritumoral CTL were an independent predictive factor for overall survival (OS), but not for disease-free survival (DFS). The combined analysis of Tregs and CTL demonstrated a better prognostic parameter than either measurement alone. Tumor infiltration by tumor-associated $FoxP3^+$ Tregs and $CD8^+$ cytotoxic T lymphocytes was also evaluated in 1270 cases of invasive breast carcinoma (Liu et al. 2011). Within the tumor bed, increased infiltration of Tregs and CTL was significantly more common in patients with unfavorable histological features. High density Treg infiltration was associated with Her2 overexpression and decreased overall as well as progression-free survival. In contrast, in the tumor surrounding tissue, high CTL/Treg ratio was associated with improved overall survival and progression-free survival.

Cancer therapy has been shown to result in marked infiltration of CD3, CD4, CD8, B lymphocytes, DC and NK cells in treated breast lesions, with significantly increased numbers of FasL + , granzyme + and perforin + TIL (Lu et al. 2009; Hornychova et al. 2008). Finally, recent findings extend the impact of monitoring the local immune response on the clinical course of metastatic colorectal cancer by revealing a strong association between increases in TIL number and function with the efficacy of the chemotherapy and the subsequent prognosis (Halama et al. 2011). Therefore, evaluation of intra- and peritumoral T cells should be considered as an important immunomonitoring parameter, which can help selecting patients for a particular immunotherapeutic and chemo-immunotherapeutic treatment or determine patients with differential responses to therapy. Identification of additional molecular markers on tumor-associated lymphocytes should further improve the usefulness of this local immunomonitoring.

30.4 Local Immunomonitoring of Tumor-Associated Macrophages

Up to 50 % of a malignant tumor mass can be composed of TAM. While normal macrophages (M1) take up antigens and play an important role in the control of infections, TAM can be reprogrammed in the tumor microenvironment to M2 cells as a result of tumor-driven 'alternative' activation. M2 are able to inhibit functions of immune cells and promote tumor survival, progression, angiogenesis and metastasis by releasing IL-10, PGE2, NO, high amounts of TGF-β or reactive oxygen species (ROS) (Whiteside 2010; Talmadge 2011). Clinical studies of macrophage infiltration have suggested that high tumor infiltration by TAM often correlates with a poor prognosis. Leek et al. (1996) found a positive correlation between macrophage infiltration of breast carcinoma and reduced overall survival. Analysis of available publications by Bingle et al. (2002) revealed that over 80 % of studies showed a positive correlation between high macrophage density and a patient's poor prognosis. However, in a few reports, macrophage infiltration was associated with good prognosis. For example, Shimura et al. (2000) reported high TAM number to be an independent predictor of longer disease-free survival for prostate cancer patients. The contradiction between studies may reflect differences in the number, grade, stage and size of tumors included in each study. Also, it should be noted that utilization of a wide variety of methods to assess TAM infiltration and difficulties in the identification of M1 and M2 cells in tested specimens could account for the inconsistency of these results. Unfortunately, there is a lack of information about alterations of TAM densities induced by cancer therapy therefore more studies are required before the evaluation of TAM can be included in the list of feasible immunomonitoring procedures (Malyguine et al. 2012).

30.5 Local Immunomonitoring of Myeloid-Derived Suppressor Cells

Myeloid-derived suppressor cells are a heterogeneous cell population composed mainly of myeloid progenitor cells that do not completely differentiate into mature macrophages, DC or granulocytes. Immature bone-marrow-derived myeloid cells (IMC) represent less than 1 % of PBMC in healthy individuals. During cancer development and progression, this subset of cells can be increased by up to 10-fold (possibly due to partially blocked differentiation) and these cells have high levels of immunosuppressive activity (Nagaraj and Gabrilovich 2010). The tumor microenvironment effects the composition of cancer-induced MDSC through the release of various tumor-derived factors, including prostaglandins, granulocyte–macrophage colony stimulating factor (GM-CSF), macrophage CSF (M-CSF), IL-6, IL-10, vascular endothelial growth factor (VEGF), stem-cell factor, IL-3, FMS-related tyrosine kinase 3 (FLT3) and cell-expressed molecules (such as Notch) (Condamine and Gabrilovich 2011). MDSC are characterized by combinations of different phenotypic markers, such as CD11b, CD34, CD33, CD15, CD13, CD14, IL-4Rα and HLA-DR (Whiteside 2010; Talmadge 2011; Jochems and Schlom 2011) and can be divided into 2 major subsets: granulocytic PMN- and monocytic MO-MDSC. Although many reports have revealed an increase of MDSC in the peripheral blood of patients with cancer, analysis of tumor-associated MDSC subsets with few exceptions (Filipazzi et al. 2012) has been limited to experimental animal studies. At present, in spite of numerous data demonstrating that elimination of MDSC may be beneficial for cancer patients (Naiditch et al. 2011), one can only speculate whether an evaluation of MDSC in tumor specimens will be a useful monitoring tool to predict patients' responses to therapy, or confirm the efficacy of therapeutic approaches to cancer.

30.6 Future Directions

We would speculate that assays characterizing and enumerating immune effector and regulatory cells within accessible tumor specimens could be very beneficial when added to the battery of commonly available immunomonitoring assays.

First, therapy-induced alterations of the local tumor immunoenvironment might better reflect the clinical efficacy of the therapy, by showing critical changes in the immunosuppressive potential of the tumor, and any increased function of immune effectors. Second, the analysis of immune responses at the tumor site might help discriminate patients likely to benefit from systemic treatment from those unlikely to respond due to treatment-related side effects. Third, the development of a feasible monitoring of the local tumor immunoenvironment could aid in the development of a new staging system for assessing advanced cancer progression.

Gene expression profiling experiments have suggested that the presence of T cells in metastatic melanomas prior to the vaccination of the patients with tumor antigens could be a biomarker for clinical benefit of the vaccines (Cipponi et al. 2011). Using microarrays, gene-expression patterns of pretreatment biopsy specimens from patients with esophageal squamous cell carcinoma have been analyzed to identify genes correlated with increased survival times (Ashida et al. 2006). The genes involved in the immune response were characteristically up-regulated in the long-term survivors, and an immunohistochemical staining confirmed an increased $CD8^+$ T cell number in the long-term survivors over that in the short-term survivors after chemoradiotherapy.

A promising strategy has been developed for the study of tumor–host interactions in melanoma based on fine-needle aspirates (FNA) and serial gene expression analysis. Though it does not provide enough tissue for cellular functional assays it still allows repeated analysis of a tumor samples over time and in combination with high-throughput biochemical and genetic methods of analysis might help better understand the molecular basis of tumor regression in response to immune modulation. Retrospective correlation of gene expression profiles with the clinical outcome might also help to develop more efficient immunotherapies (Wang and Marincola 2000; Bedognetti et al. 2010). It is possible that analysis of tumor material from individual patients will assist in the identification of predictive biomarkers to select patients who most likely will respond to appropriate immune-based therapies. Identification of patient tumor-specific antigenic epitopes generated through somatic point mutations results in a more specific targeting of tumor antigens in individual patients, and in combination with immunohistochemistry may also reveal any patients phenotype that might be predictive of clinical outcome (Gajewski 2012; Gajewski et al. 2010). This additionally suggests the importance of developing a scoring system that can be used as a predictive tool for response to cancer therapy (Galon et al. 2012).

Although initial attempts to assess both effector and regulatory arms of the antitumor immune response have been made, the data obtained are still controversial and limited to the monitoring of systemic immunity in the peripheral blood. However, for tumor eradication, circulating effector cells have to infiltrate established and often hard to access tumor deposits. At present, our ability to follow this process in the clinical setting is very limited, and the evaluation of 'in situ' immune responses represents a significant technical challenge. New approaches to characterize the tumor-infiltrating leukocytes and intratumoral cytokine network are needed. Such technologies as in situ PCR, FISH analysis, scanning flow cytometry and microarray methods should provide opportunities to study the properties of tumor infiltrating immune cell populations in more detail. There is an urgent need to identify surrogate biomarkers of anti-tumor efficacy. The development and validation of in situ monitoring techniques will allow researchers to combine the analysis of both systemic and local immune responses in order to accomplish a complete monitoring of patients in immunotherapeutic protocols. It is conceivable that a combination of in situ and systemic immune parameters may provide a better understanding of the molecular and cellular basis whereby efficient immunotherapy translates into objective responses. Therefore in the future more sophisticated and detailed immunomonitoring approaches may assist in the development of an improved and more personalized immunotherapy.

Acknowledgments This project was supported by NIH RO1 CA154369 grant (to M.R.S.). This project has been funded in whole or in part with federal funds from the National Cancer Institute, National Institutes of Health, under Contract No. HHSN261200800001E. The content of this publication does not necessarily reflect the views or policies of the Department of Health and Human Services, nor does mention of trade names, commercial products, or organizations imply endorsement by the U.S. Government.

References

Ashida A, Boku N, Aoyagi K, Sato H, Tsubosa Y, Minashi K, Muto M, Ohtsu A, Ochiai A, Yoshida T, Yoshida S, Sasaki H (2006) Expression profiling of esophageal squamous cell carcinoma patients treated with definitive chemoradiotherapy: clinical implications. Int J Oncol 28(6):1345–1352

Bedognetti D, Wang E, Sertoli MR, Marincola FM (2010) Gene-expression profiling in vaccine therapy and immunotherapy for cancer. Expert Rev Vaccines 9(6):555–565

Bingle L, Brown NJ, Lewis CE (2002) The role of tumour-associated macrophages in tumour progression: implications for new anticancer therapies. J Pathol 196(3):254–265

Butterfield LH, Palucka AK, Britten CM, Dhodapkar MV, Hakansson L, Janetzki S, Kawakami Y, Kleen TO, Lee PP, Maccalli C, Maecker HT, Maino VC, Maio M, Malyguine A, Masucci G, Pawelec G, Potter DM, Rivoltini L, Salazar LG, Schendel DJ, Slingluff CL Jr, Song W, Stroncek DF, Tahara H, Thurin M, Trinchieri G, van Der Burg SH, Whiteside TL, Wigginton JM, Marincola F, Khleif S, Fox BA, Disis ML (2011) Recommendations from the iSBTc-SITC/FDA/NCI workshop on immunotherapy biomarkers. Clin Cancer Res 17(10): 3064–3076

Chen KJ, Zhou L, Xie HY, Ahmed TE, Feng XW, Zheng SS (2011) Intratumoral regulatory T cells alone or in combination with cytotoxic T cells predict prognosis of hepatocellular carcinoma after resection. Medical oncology, Northwood, London, England, 2011 (in press)

Cipponi A, Wieers G, van Baren N, Coulie PG (2011) Tumor-infiltrating lymphocytes: apparently good for melanoma patients. But why? Cancer Immunol Immunother 60(8):1153–1160

Condamine T, Gabrilovich DI (2011) Molecular mechanisms regulating myeloid-derived suppressor cell differentiation and function. Trends Immunol 32(1):19–25

Curiel TJ, Cheng P, Mottram P, Alvarez X, Moons L, Evdemon-Hogan M, Wei S, Zou L, Kryczek I, Hoyle G, Lackner A, Carmeliet P, Zou W (2004) Dendritic cell subsets differentially regulate angiogenesis in human ovarian cancer. Cancer Res 64(16):5535–5538

Filipazzi P, Huber V, Rivoltini L (2012) Phenotype, function and clinical implications of myeloid-derived suppressor cells in cancer patients. Cancer Immunol Immunother 61(2):255–263

Gajewski T (2012) Harnessing the immune response. Clin Adv Hematol Oncol 10(1):46–48

Gajewski TF, Fuertes M, Spaapen R, Zheng Y, Kline J (2010) Molecular profiling to identify relevant immune resistance mechanisms in the tumor microenvironment. Curr Opin Immunol 23(2):286–292

Galon J, Pages F, Marincola FM, Thurin M, Trinchieri G, Fox BA, Gajewski TF, Ascierto PA (2012) The immune score as a new possible approach for the classification of cancer. J Transl Med 10:1

Halama N, Michel S, Kloor M, Zoernig I, Benner A, Spille A, Pommerencke T, von Knebel Doeberitz M, Folprecht G, Luber B, Feyen N, Martens UM, Beckhove P, Gnjatic S, Schirmacher P, Herpel E, Weitz J, Grabe N, Jaeger D (2011) Localization and density of immune cells in the invasive margin of human colorectal cancer liver metastases are prognostic for response to chemotherapy. Cancer research

Hornychova H, Melichar B, Tomsova M, Mergancova J, Urminska H, Ryska A (2008) Tumor-infiltrating lymphocytes predict response to neoadjuvant chemotherapy in patients with breast carcinoma. Cancer Invest 26(10):1024–1031

Jandus C, Bioley G, Speiser DE, Romero P (2008) Selective accumulation of differentiated FOXP3(+) CD4(+) T cells in metastatic tumor lesions from melanoma patients compared to peripheral blood. Cancer Immunol Immunother 57(12):1795–1805

Jochems C, Schlom J (2011) Tumor-infiltrating immune cells and prognosis: the potential link between conventional cancer therapy and immunity. Exp Biol Med 236(5):567–579

Kawai O, Ishii G, Kubota K, Murata Y, Naito Y, Mizuno T, Aokage K, Saijo N, Nishiwaki Y, Gemma A, Kudoh S, Ochiai A (2008) Predominant infiltration of macrophages and CD8(+) T Cells in cancer nests is a significant predictor of survival in stage IV nonsmall cell lung cancer. Cancer 113(6):1387–1395

Le DT, Ladle BH, Lee T, Weiss V, Yao X, Leubner A, Armstrong TD, Jaffee EM (2011) CD8 Foxp3 tumor infiltrating lymphocytes accumulate in the context of an effective anti-tumor response. Int J Cancer 129(3):636–647

Leek RD, Lewis CE, Whitehouse R, Greenall M, Clarke J, Harris AL (1996) Association of macrophage infiltration with angiogenesis and prognosis in invasive breast carcinoma. Cancer Res 56(20):4625–4629

Leffers N, Gooden MJ, de Jong RA, Hoogeboom BN, ten Hoor KA, Hollema H, Boezen HM, van der Zee AG, Daemen T, Nijman HW (2009) Prognostic significance of tumor-infiltrating T-lymphocytes in primary and metastatic lesions of advanced stage ovarian cancer. Cancer Immunol Immunother 58(3):449–459

Liu F, Lang R, Zhao J, Zhang X, Pringle GA, Fan Y, Yin D, Gu F, Yao Z, Fu L (2011) CD8(+) cytotoxic T cell and FOXP3(+) regulatory T cell infiltration in relation to breast cancer survival and molecular subtypes. Breast Cancer Res Treat 130(2):645–655

Lu P, Zhu XQ, Xu ZL, Zhou Q, Zhang J, Wu F (2009) Increased infiltration of activated tumor-infiltrating lymphocytes after high intensity focused ultrasound ablation of human breast cancer. Surgery 145(3):286–293

Mahmoud SM, Paish EC, Powe DG, Macmillan RD, Grainge MJ, Lee AH, Ellis IO, Green AR (2011) Tumor-infiltrating CD8 + lymphocytes predict clinical outcome in breast cancer. J Clin Oncol 29(15):1949–1955

Malyguine AM, Strobl SL, Shurin MR (2012) Immunological monitoring of the tumor immunoenvironment for clinical trials. Cancer Immunol Immunother 61(2):239–247

Murphy GF, Radu A, Kaminer M, Berd D (1993) Autologous melanoma vaccine induces inflammatory responses in melanoma metastases: relevance to immunologic regression and immunotherapy. J Invest Dermatol 100(3):335S–341S

Nagaraj S, Gabrilovich DI (2010) Myeloid-derived suppressor cells in human cancer. Cancer J 16(4):348–353

Naiditch H, Shurin MR, Shurin GV (2011) Targeting myeloid regulatory cells in cancer by chemotherapeutic agents. Immunol Res 50(2–3):276–285

Ohtani H (2007) Focus on TILs: prognostic significance of tumor infiltrating lymphocytes in human colorectal cancer. Cancer Immun 7:4

Okada H, Kalinski P, Ueda R, Hoji A, Kohanbash G, Donegan TE, Mintz AH, Engh JA, Bartlett DL, Brown CK, Zeh H, Holtzman MP, Reinhart TA, Whiteside TL, Butterfield LH, Hamilton RL, Potter DM, Pollack IF, Salazar AM, Lieberman FS (2011) Induction of CD8 + T-cell responses against novel glioma-associated antigen peptides and clinical activity by vaccinations with {alpha}-type 1 polarized dendritic cells and polyinosinic-polycytidylic acid stabilized by lysine and carboxymethyl cellulose in patients with recurrent malignant glioma. J Clin Oncol 29(3):330–336

Pages F, Galon J, Dieu-Nosjean MC, Tartour E, Sautes-Fridman C, Fridman WH (2009a) Immune infiltration in human tumors: a prognostic factor that should not be ignored. Oncogene 29(8):1093–1102

Pages F, Kirilovsky A, Mlecnik B, Asslaber M, Tosolini M, Bindea G, Lagorce C, Wind P, Marliot F, Bruneval P, Zatloukal K, Trajanoski Z, Berger A, Fridman WH, Galon J (2009b) In situ cytotoxic and memory T cells predict outcome in patients with early-stage colorectal cancer. J Clin Oncol 27(35):5944–5951

Piersma SJ, Jordanova ES, van Poelgeest MI, Kwappenberg KM, van der Hulst JM, Drijfhout JW, Melief CJ, Kenter GG, Fleuren GJ, Offringa R, van der Burg SH (2007) High number of intraepithelial CD8 + tumor-infiltrating lymphocytes is associated with the absence of lymph node metastases in patients with large early-stage cervical cancer. Cancer Res 67(1):354–361

Quezada SA, Peggs KS, Simpson TR, Allison JP (2011) Shifting the equilibrium in cancer immunoediting: from tumor tolerance to eradication. Immunol Rev 241(1):104–118

Ribas A, Camacho LH, Lee SM, Hersh EM, Brown CK, Richards JM, Rodriguez MJ, Prieto VG, Glaspy JA, Oseguera DK, Hernandez J, Villanueva A, Chmielowski B, Mitsky P, Bercovici N, Wasserman E, Landais D, Ross MI (2010) Multicenter phase II study of matured dendritic cells pulsed with melanoma cell line lysates in patients with advanced melanoma. J Transl Med 8:89

Rosenberg SA, Sherry RM, Morton KE, Scharfman WJ, Yang JC, Topalian SL, Royal RE, Kammula U, Restifo NP, Hughes MS, Schwartzentruber D, Berman DM, Schwarz SL, Ngo LT, Mavroukakis SA, White DE, Steinberg SM (2005) Tumor progression can occur despite the induction of very high levels of self/tumor antigen-specific CD8 + T cells in patients with melanoma. J Immunol 175(9):6169–6176

Salama P, Phillips M, Grieu F, Morris M, Zeps N, Joseph D, Platell C, Iacopetta B (2009) Tumor-infiltrating FOXP3 + T regulatory cells show strong prognostic significance in colorectal cancer. J Clin Oncol 27(2):186–192

Schwartzentruber DJ, Lawson DH, Richards JM, Conry RM, Miller DM, Treisman J, Gailani F, Riley L, Conlon K, Pockaj B, Kendra KL, White RL, Gonzalez R, Kuzel TM, Curti B, Leming PD, Whitman ED, Balkissoon J, Reintgen DS, Kaufman H, Marincola FM, Merino MJ, Rosenberg SA, Choyke P, Vena D, Hwu P (2011) gp100 peptide vaccine and interleukin-2 in patients with advanced melanoma. New Engl J Med 364(22):2119–2127

Sharma P, Shen Y, Wen S, Yamada S, Jungbluth AA, Gnjatic S, Bajorin DF, Reuter VE, Herr H, Old LJ, Sato E (2007) CD8 tumor-infiltrating lymphocytes are predictive of survival in muscle-invasive urothelial carcinoma. Proc Natl Acad Sci U S A 104(10):3967–3972

Shimura S, Yang G, Ebara S, Wheeler TM, Frolov A, Thompson TC (2000) Reduced infiltration of tumor-associated macrophages in human prostate cancer: association with cancer progression. Cancer Res 60(20):5857–5861

Shurin MR, Shurin GV, Lokshin A, Yurkovetsky ZR, Gutkin DW, Chatta G, Zhong H, Han B, Ferris RL (2006) Intratumoral cytokines/chemokines/growth factors and tumor infiltrating dendritic cells: friends or enemies? Cancer Metastasis Rev 25(3):333–356

Shurin MR, Gregory M, Morris JC, Malyguine AM (2010) Genetically modified dendritic cells in cancer immunotherapy: a better tomorrow? Expert Opin Biol Ther 10(11):1539–1553

Sinicrope FA, Rego RL, Ansell SM, Knutson KL, Foster NR, Sargent DJ (2009) Intraepithelial effector (CD3 +)/regulatory (FoxP3 +) T-cell ratio predicts a clinical outcome of human colon carcinoma. Gastroenterology 137(4):1270–1279

Talmadge JE (2011) Immune cell infiltration of primary and metastatic lesions: mechanisms and clinical impact. Semin Cancer Biol 21(2):131–138

Uppaluri R, Dunn GP, Lewis JS Jr (2008) Focus on TILs: prognostic significance of tumor infiltrating lymphocytes in head and neck cancers. Cancer Immun 8:16

Wang E, Marincola FM (2000) A natural history of melanoma: serial gene expression analysis. Immunol Today 21(12):619–623

Whiteside TL (2010) Immune responses to malignancies. J Allergy Clin Immunol 125(2 Suppl 2):S272–S283

Whiteside TL (2012) What are regulatory T cells (Treg) regulating in cancer and why? Semin Cancer Biol 22(4):327–334

Whiteside TL, Schuler P, Schilling B (2012) Induced and natural regulatory T cells in human cancer. Expert Opin Biol Ther 12(10):1383–1397

Yoon HH, Orrock JM, Foster NR, Sargent DJ, Smyrk TC, Sinicrope FA (2012) Prognostic impact of FoxP3 + regulatory T cells in relation to CD8 + T lymphocyte density in human colon carcinomas. PLoS ONE 7(8):e42274

Chapter 31
Analysis of Myeloid-Derived Suppressor Cells in Patients with Cancer

Peiyuan Zhu, Yevgeniya V. Segal, Galina V. Shurin and Michael R. Shurin

Abstract Myeloid-derived suppressor cells (MDSCs) represent a phenotypically heterogeneous population of immature myeloid cells at different stages of differentiation found in tumor-bearing mice and in cancer patients. MDSCs play an important role in maintaining a state of immunological anergy and tolerance in cancer, suppressing antitumor immunity and promoting tumor growth, proliferation and metastases. Human MDSCs have been shown to inhibit immune responses through L-arginine metabolism, induction of oxidative stress, activation and expansion of regulatory T cells, and generation of soluble inhibitory factors. The phenotypes of human MDSCs are not well characterized due to the lack of specific markers. In cancer patients, MDSCs are defined as cells that express the common myeloid marker, CD33, but lack the expression of mature myeloid marker, HLA-DR. MDSCs with the phenotype $Lin^-HLA\text{-}DR^-CD33^+$, $Lin^-HLA\text{-}DR^-CD33^+CD11b^+$ or $CD14^-CD11b^+CD33^+$ have been isolated from the blood of patients with cancer. Human MDSCs can be divided into two main subsets: $CD15^+$ granulocytic and $CD14^+$ monocytic subpopulations, both are $HLA\text{-}DR^{-/low}$ and $CD33^+$. The prognostic significance of circulating MDSCs in human cancers is also described.

Keywords Myeloid-derived suppressor cells · Cancer patients · Immunosuppression · Tumor escape · Immunomonitoring · Myeloid regulatory cells

---

P. Zhu
Department of Blood Transfusion, Jinling Hospital, School of Medicine, Nanjing University, Nanjing, Jiangsu, China

Y. V. Segal · G. V. Shurin · M. R. Shurin (✉)
Department of Pathology and Immunology, University of Pittsburgh Medical Center, Scaife Hall S735, 3550 Terrace Street, Pittsburgh, PA 15261, USA
e-mail: shurinmr@upmc.edu

M. R. Shurin et al. (eds.), *The Tumor Immunoenvironment*,
DOI: 10.1007/978-94-007-6217-6_31,

31.1 Introduction

In the past decade, a large number of experimental cancer immunotherapies have been tested in clinical trials, however, most of these trials have shown minimal impact on the clinical status of the majority of patients (Lesterhuis et al. 2011; Ostrand-Rosenberg et al. 2012). Immune evasion is an emerging hallmark of cancer (Hanahan and Weinberg 2011). Tumors may employ various mechanisms to avoid immune recognition, suppress antitumor immunity, induce tumor-specific tolerance or polarize immune cells into protumorigenic regulatory subsets (Shurin et al. 2006; Rabinovich et al. 2007). Myeloid cells, including polymorphonuclear neutropils (PMNs), monocytes, macrophages, and dendritic cells (DCs), are the most abundant hematopoietic cells in the human body, which are essential for the normal functions of the innate and adaptive immunity. Evidence from numerous studies strongly indicates that the tumor microenvironment can drive myeloid cells into potent immunosuppressive cells (Hurwitz and Watkins 2012; Lechner et al. 2011; Rodrigues et al. 2010). Abnormal accumulation of myeloid regulatory cells (MRCs) and regulatory T lymphocytes (Tregs) in the tumor tissue, lymph nodes, or circulation has been considered as one of the major mechanisms of how tumors escape from an immune attack (Gabrilovich et al. 2012). In order to improve the efficacy of immunotherapy, the immune suppression of these regulatory cells will have to be eliminated or controlled.

The MRC subpopulations, including myeloid-derived suppressor cells (MDSCs), regulatory dendritic cells (regDCs), tumor-associated macrophages (TAMs), TIE2-expressing monocytes (TEMs), and tumor-associated neutrophils (TANs), all constitute a network of immune suppressive cells that are present in most cancer patients and which profoundly inhibit the generation of antitumor immunity (Ostrand-Rosenberg et al. 2012; Naiditch et al. 2011; Shurin et al. 2012). Regulatory DCs have been demonstrated to be induced by tumor-derived factors and be able to inhibit T cell response, which may be involved in how the tumor escapes from an immunological attack (Shurin et al. 2011, 2012; Dumitriu et al. 2009; Ma et al. 2012). Tumor microenvironments have shown to polarize macrophages into an alternatively activated (M2) macrophages, known as TAMs, contributing to tumor progression by inhibiting the proliferation of T cells and supporting the induction of Tregs (Schmieder et al. 2012; Ruffell et al. 2012). Studies have also demonstrated that tumor-mediated signals, such as transforming growth factor-β (TGF-β), can induce the formation of a pro-tumorigenic (N2), TANs, capable of supporting tumor growth and suppressing the antitumor immune response (Fridlender and Albelda 2012).

MDSCs represent a phenotypically heterogeneous population of immature myeloid cells at different stages of differentiation found in tumor-bearing mice and in cancer patients (Filipazzi et al. 2012; Romano et al. 2011). In mice, MDSCs represent approximately 30–40 % of normal bone marrow cells and only 2–4 % of all nucleated normal splenocytes (Nagaraj and Gabrilovich 2010). However, in tumor-bearing mice, as many as 20–40 % of nucleated splenocytes are MDSCs, and MDSCs are also found in tumor tissues and in the lymph nodes (Gabrilovich and Nagaraj 2009). Furthermore, murine MDSCs have been shown to play a pivotal role in suppressing T cell responses using multiple mechanisms, resulting in tumor progression (Gabrilovich and Nagaraj 2009). By contrast, human MDSCs are less understood because of the lack of specific markers. In this review, we will summarize our current understanding of the MDSCs in cancer patients.

31.2 Phenotypes of MDSCs

In mice, MDSCs are most clearly defined as cells that co-express CD11b and Gr-1 (Youn et al. 2012). CD11b is the α-chain of the integrin $\alpha M\beta 2$ and a common myeloid cell marker expressed at the surface of granulocytes, monocytes and macrophages, while Gr-1 a myeloid differentiation antigen, which consists of two epitopes recognized by anti-Ly6G or anti-Ly6C antibodies (Martin et al. 2012). Murine MDSCs can be further subdivided into two major subsets: $CD11b^{+}Gr\text{-}1^{high}$ granulocytic MDSCs (which can also be identified as $CD11b^{+}Ly6G^{+}Ly6C^{low}$ MDSC) and $CD11b^{+}Gr\text{-}1^{low}$ monocytic MDSCs (which can also be identified as $CD11b^{+}Ly6G^{-}Ly6C^{high}$ MDSC) (Wilcox 2012; Lindenberg et al. 2011; Ostrand-Rosenberg 2010). In tumor-bearing mice, the granulocytic MDSC is the prevalent population of circulating MDSCs. However, they are less immunosuppressive than monocytic MDSCs (Gabrilovich et al. 2012; Youn et al. 2008).

In contrast to murine models, the phenotypes of human MDSCs are not very well characterized due to the lack of specific markers. Because lack of a Gr-1 analogue in humans, different combinations of surface molecules (e.g., CD14, CD15, CD34, CD11b, CD33, Lin, and HLA-DR) have been used to define human MDSCs, leading to conflicting results in different groups of cancer patients (Nagaraj and Gabrilovich 2010) (Table 31.1). In patients, MDSCs are seen as cells that express the common myeloid marker, CD33, but lack the expression of mature myeloid marker, HLA-DR (Almand et al. 2001). MDSCs with the phenotype $Lin^{-}HLA\text{-}DR^{-}CD33^{+}$, $Lin^{-}HLA\text{-}DR^{-}CD33^{+}CD11b^{+}$ or $CD14^{-}CD11b^{+}CD33^{+}$ have been isolated from the blood of patients with renal cell carcinoma (RCC) (Kusmartsev et al. 2008; Mirza et al. 2006), melanoma (Daud et al. 2008), breast cancer (Solito et al. 2011; Diaz-Montero et al. 2009), colorectal cancer (Solito et al. 2011), pancreatic cancer (Gabitass et al. 2011) and head and neck squamous cell carcinoma (HNSCC) (Corzo et al. 2009).

As in mice, attempts have been made to divide human MDSCs into two main subsets: granulocytic and monocytic subpopulations. It has been suggested that monocytic MDSCs are $CD14^{+}$ and granulocytic MDSCs express CD15, while both groups of MDSC are $HLA\text{-}DR^{-/low}$ and $CD33^{+}$ (Filipazzi et al. 2012; Greten et al. 2011). For instance, Zea et al. (2005) showed for the first time the existence of suppressor myeloid cells producing arginase in human cancer patients. In this study, peripheral blood mononuclear cells from 123 patients with metastatic RCC prior to treatment were found to have significantly increased arginase activity.

Table 31.1 Phenotypes and activity of MDSCs in patients with cCancer

Cancer type	MDSC subset	Phenotype	Action mechanisms	References
RCC	ND	$Lin^-HLA\text{-}DR^-CD33^+$	Unknown	Mirza et al. 2006
			Secrete ROS and NO	Kusmartsev et al. 2008
		$HLA\text{-}DR^-CD33^+$	Activation and expansion of Tregs	Ko et al. 2009
	Granulocytic	$CD11b^+CD14^-CD15^+HLA\text{-}DR^-$	Produce arginase	Zea et al. 2005
		$CD15^+CD14^-$	Activation and expansion of Tregs	Ko et al. 2009
		$CD11b^+CD14^-CD15^+CD16^{low}CD62L^{low}CD66b^+VEGFR1^+$	Produce arginase	Rodriguez et al. 2009
	Monocytic	$CD14^+HLA\text{-}DR^{-/low}$	Unknown	van Cruijsen et al. 2008
Melanoma	ND	$Lin^-HLA\text{-}DR^-CD33^+$	Unknown	Daud et al. 2008
	Granulocytic	$CD15^+IL\text{-}4\alpha^+$	Unknown	Mandruzzato et al. 2009
	Monocytic	$CD14^+HLA\text{-}DR^{-/low}$	Release TGF-β	Filipazzi et al. 2007
			Produce arginase and generate oxidative stress	Poschke et al. 2010
		$CD14^+IL\text{-}4\alpha^+$	Unknown	Mandruzzato et al. 2009
		$CD14^+CD11b^+HLA\text{-}DR^{-/low}$	Unknown	Kaufman et al. 2010
HNSCC	ND	$Lin^-HLA\text{-}DR^-$	Unknown	Almand et al. 2001
		$CD11b^+CD14^-CD33^+$	Increase ROS production; up-regulation of NADPH oxidase	Corzo et al. 2009
	Granulocytic	$SSC^{high}CD66b^+CD125^-CD33^+HLA\text{-}DR^-$	Unknown	Brandau et al. 2011
		$CD15^+CD16^{low}$	Unknown	Choi et al. 2012
	Monocytic	$CD14^+$	Produce arginase and iNOS	Serafini et al. 2006
		$CD14^+HLA\text{-}DR^-$	Release TGF-β	Chikamatsu et al. 2012
Colon cancer	ND	$Lin^{-/low}HLA\text{-}DR^-CD33^+CD11b^+$	Unknown	Solito et al. 2011
	Granulocytic	$CD15^+ IL\text{-}4\alpha^+$	Unknown	Mandruzzato et al. 2009
	Monocytic	$CD14^+IL\text{-}4\alpha^+$	Unknown	Mandruzzato et al. 2009
		$CD14^+HLA\text{-}DR^{-/low}S100A9^{high}$	Express iNOS	Zhao et al. 2012a
NSCLC	ND	$Lin^-HLA\text{-}DR^-$	Unknown	Almand et al. 2001
		$CD33^+CD11b^+$	Unknown	Srivastava et al. 2008
	Granulocytic	$CD11b^+CD14^-CD15^+CD33^+$	Unknown	Liu et al. 2010
		$SSC^{high}CD66b^+CD125^-CD33^+HLA\text{-}DR^-$	Unknown	Brandau et al. 2011

(continued)

Table 31.1 (continued)

Cancer type	MDSC subset	Phenotype	Action mechanisms	References
Breast cancer	ND	$Lin^{-}HLA\text{-}DR^{-}$	Unknown	Almand et al. 2001
		$Lin^{-/low}HLA\text{-}DR^{-}CD33^{+}CD11b^{+}$	Unknown	Solito et al. 2011, Diaz-Montero et al. 2009
	Granulocytic	$CD15^{+}CD16^{low}$	Unknown	Choi et al. 2012
Gastrointestinal malignancy	Granulocytic	$CD33^{+}HLA\text{-}DR^{-}CD11b^{+}CD15^{+}$	Unknown	Mundy-Bosse et al. 2011
		$CD15^{+}CD16^{low}$	Unknown	Choi et al. 2012
	Monocytic	$CD14^{+}HLA\text{-}DR^{-/low}CD33^{+}$	Unknown	Mundy-Bosse et al. 2011
MM	Monocytic	$CD14^{+}$	Produce arginase and iNOS	Serafini et al. 2006
		$CD14^{+}HLA\text{-}DR^{-/low}$	Unknown	Brimnes et al. 2010
NHL	Monocytic	$CD14^{+}HLA\text{-}DR^{-/low}$	Produce arginase	Lin et al. 2011
		$CD14^{+}HLA\text{-}DR^{-}B7\text{-}H1^{+}$	Unknown	Wilcox et al. 2009
Pancreatic cancer	ND	$Lin^{-/low}HLA\text{-}DR^{-}CD33^{+}CD11b^{+}$	Produce arginase; induce $CD4^{+}CD25^{+}CD127^{low/-}FoxP3^{+}$ Tregs	Gabitass et al. 2011
Esophago-gastric cancer	ND	$Lin^{-/low}HLA\text{-}DR^{-}CD33^{+}CD11b^{+}$	Produce arginase; induce $CD4^{+}CD25^{+}CD127^{low/-}FoxP3^{+}$ Tregs	Gabitass et al. 2011
Bladder cancer	Granulocytic	$CD11b^{+}CD15^{high}CD33^{low}$	Expand $CD4^{+}Foxp3^{+}$ Tregs	Eruslanov et al. 2012
HCC	Monocytic	$CD14^{+}HLA\text{-}DR^{-/low}$	Produce arginase; induce $CD4^{+}CD25^{+}$ $Foxp3^{+}$ Tregs	Hoechst et al. 2008
Prostate cancer	Monocytic	$CD14^{+}HLA\text{-}DR^{-/low}$	Unknown	Vuk-Pavlovic et al. 2010
SCC	Granulocytic	$iNOS^{+}CD11b^{+}CD33^{+}CD11c^{-}HLA\text{-}DR^{-}$	Produce NO, TGF-β,and arginase; express iNOS	Gehad et al. 2012
Glioblastoma	ND	$CD33^{+}HLA\text{-}DR^{-}$	Produce arginase	Raychaudhuri et al. 2011

HCC hepatocellular carcinoma, *HNSCC* head and neck squamous cell carcinoma, *MM* multiple myeloma, *ND* not determined, *NHL* non-Hodgkin lymphoma, *NO* nitric oxide, *NSCLC* non-small-cell lung cancer, *RCC* renal cell carcinoma, *ROS* reactive oxygen species, *SCC* squamous cell carcinoma

Cell separation studies showed that the increased arginase activity was limited to a specific subset of $CD11b^+CD14^-CD15^+$ MDSCs with a polymorphonuclear granulocyte morphology and markers. Further phenotyping demonstrated that this subset was negative for the expression of CD11a, CD80, CD83, CD86, and HLA-DR (Zea et al. 2005). In another study of patients with RCC, increased frequency of granulocytic MDSCs with a similar phenotype, $CD11b^+CD14^-CD15^+$, were also detected (Rodriguez et al. 2009). These cells expressed markers of mature activated granulocytes, including high levels of CD66b and low levels of CD62L and CD16 and vascular endothelial growth factor receptor 1 (VEGFR1). MDSCs with a similar phenotype were also found in the peripheral blood of patients with treatment-naïve, advanced-stage non-small-cell lung cancer (NSCLC) (Liu et al. 2010). Recently, Brandau et al. (2011) identified a subset of $SSC^{high}CD66b^+$ MDSCs using CD66b (CEA-related cell adhesion molecule 8) as a marker for human granulocytic MDSCs in the peripheral blood of 103 patients with head and neck, lung, bladder, and ureteral cancers. This $SSC^{high}CD66b^+$ subpopulation of granulocytic MDSCs were suppressive on T cells and had reduced migratory properties as well as reduced effector cell functions. Further immunophenotyping has determined that this cell population expressed CD33 and was negative for CD125 and HLA-DR (Brandau et al. 2011).

In addition, a subset of monocytic MDSCs with the phenotype $CD14^+HLA\text{-}DR^{-/low}$ were identified in PBMCs of metastatic melanoma patients ($n = 16$) who were treated with a GM-CSF-based antitumor vaccine (Filipazzi et al. 2007). Suppressive activity of these cells on T cells was mediated by TGF-β, whereas no involvement of the arginase and inducible nitric oxide (NO) synthase (iNOS) pathways could be detected (Filipazzi et al. 2007). In another study of patients with hepatocellular carcinoma ($n = 111$), increased levels of circulating $CD14^+HLA\text{-}DR^{-/low}$ MDSCs were also described (Hoechst et al. 2008). However, these MDSCs had high arginase activity, and induced an expansion of an IL-10 secreting $CD4^+CD25^+Foxp3^+$ Treg cells. The prevalence of a $CD14^+/HLA\text{-}DR^{-/low}$ subset of MDSCs has also been found in patients with RCC (van Cruijsen et al. 2008), multiple myeloma (Brimnes et al. 2010), prostate cancer (Vuk-Pavlovic et al. 2010), non-Hodgkin lymphoma (NHL) (Lin et al. 2011), and HNSCC (Chikamatsu et al. 2012). In recent years, additional and more specific molecules have been used to characterize human MDSCs in cancers more accurately. In a study of patients with colon cancer and melanoma, the accumulation of both monocytic ($CD14^+IL\text{-}4R\alpha^+$) and granulocytic MDSCs ($CD15^+IL\text{-}4R\alpha^+$) with immunosuppressive properties in PBMCs were identified (Mandruzzato et al. 2009). Of interest, IL-4Rα was up-regulated in both cell populations but, its presence correlated with an immunosuppressive phenotype only for the monocytic MDSCs (Mandruzzato et al. 2009). IL-4Rα (CD124) is the common receptor for IL-4 and IL-3 but its role as an immunosuppressive molecule is still controversial (Tadmor et al. 2011). In mice, expression of IL-4Rα on a subset of immunosuppressive, $CD11b^+$ monocytic cells was shown to be involved in suppression of CD8 T cells (Gallina et al. 2006). However, IL-4Rα was not shown as a key regulator of murine MDSCs in a recent study using IL-4Rα-deficient mice

(Sinha et al. 2012). Another study also examined circulating $CD14^{+}HLA\text{-}DR^{-/low}$ MDSCs in 34 patients with advanced malignant melanoma (Poschke et al. 2010). Phenotyping results demonstrated that this cell population overexpressed markers associated with the mature phenotype in human, including CD80, CD83, and DC-Sign (CD209). It was also shown that only the subset of $CD14^{+}HLA\text{-}DR^{-/low}$ MDSCs that expressed IL-4Rα was able to suppress T cells. Moreover, no differential expression of S100A9 in $CD14^{+}HLA\text{-}DR^{-/low}$ MDSCs was detected (Poschke et al. 2010). Nevertheless, S100A9 was proposed as an additional useful marker for human monocytic MDSCs in a recent study in patients with colon cancer (Zhao et al. 2012a). In this study, a differential display analysis on $CD14^{+}HLA\text{-}DR^{+}$ monocytes and $CD14^{+}HLA\text{-}DR^{-/low}$ MDSCs was performed to search for specific markers present on human MDSCs. An increase in the population of $CD14^{+}HLA\text{-}DR^{-/low}S100A9^{high}$ MDSCs was observed in the peripheral blood from colon cancer patients in comparison with healthy controls. These cells induced iNOS expression upon lipopolysaccharide (LPS)/IFN-γ stimulation. However, S100A9 is an intracellular protein, which makes it impossible to use this marker to isolate MDSCs and use them in functional studies (Zhao et al. 2012a).

Emerging evidence suggests that multiple MDSCs subsets exist simultaneously in human cancer patients. In 2009, Ko et al. (2009) demonstrated that metastatic RCC patients had elevated levels of $CD33^{+}HLA\text{-}DR^{-}$ and $CD15^{+}CD14^{-}$ MDSCs (mean 5.49 and 5.42 %, respectively) when compared with healthy age-matched normal donors (mean 0.23 and 0.76 %; $p < 0.001$ and $p = 0.002$, respectively). These cells were partially overlapping populations and were positive for myeloid markers CD11b, CD11c, and CD13. In addition, treatment with sunitinib resulted in significant reduction in MDSCs, which was correlated with reversal of type 1 T cell suppression and $CD3^{+}CD4^{+}CD25^{high}Foxp3^{+}$ Tregs elevation (Ko et al. 2009). In another study, PBMCs from 28 patients with newly diagnosed glioblastoma were analyzed for MDSCs by 4-color flow cytometry and were compared with 11 healthy, age-matched donors (Raychaudhuri et al. 2011). The analysis revealed that an increased accumulation of $CD33^{+}HLA\text{-}DR^{-}$ MDSCs in the blood of glioblastoma patients were composed of granulocytic ($CD15^{+}$; >60 %), lineage-negative ($CD15^{-}CD14^{-}$; 31 %), and monocytic ($CD14^{+}$; 6 %) subsets. Depletion of MDSCs from the PBMCs with anti-CD33 and anti-CD15 antibody-coated beads significantly restored the ability of T cells for IFN-γ production. However, the precise suppressive activity of the granulocytic MDSCs was not assessed and additional studies are needed to define the relative contribution that each of these MDSC subsets makes to T cell suppression (Raychaudhuri et al. 2011). Another study also demonstrated that both granulocytic MDSCs ($CD33^{+}HLA\text{-}DR^{-}CD11b^{+}CD15^{+}$) and monocytic ($CD33^{+}HLA\text{-}DR^{-/low}CD14^{+}$) MDSCs were significantly elevated in patients with gastrointestinal malignancies in comparison with normal healthy donors ($p < 0.0001$) (Mundy-Bosse et al. 2011). In this study, the immature myeloid cells with a $CD33^{+}HLA\text{-}DR^{-}$ phenotype was further subclassified by the presence of CD15: $CD33^{+}HLA\text{-}DR^{-}CD11b^{+}CD15^{+}$, $CD33^{+}HLA\text{-}DR^{-}CD11b^{+}CD15^{-}$, or $CD33^{+}HLA\text{-}DR^{-/low}CD14^{+}$. The $CD15^{+}$ and $CD14^{+}$ MDSCs were more prevalent in the PBMCs of patients, and there were significant co-linear relationships between the each of the MDSC subsets. Plasma IL-6 and IL-10 were also significantly higher in patients versus normal donors. Plasma IL-6 level was correlated with $CD33^{+}HLA\text{-}DR^{-}CD15^{+}$ MDSCs ($p = 0.008$) and IL-10 level with $CD33^{+}HLA\text{-}DR^{-}CD15^{-}$ MDSCs ($p = 0.002$). The percentage of $CD15^{+}$ and $CD15^{-}$, but not $CD14^{+}$ MDSC subsets, were inversely correlated with IFN-α-induced STAT1 phosphorylation in $CD4^{+}$ T cells (Mundy-Bosse et al. 2011). Recently, Eruslanov et al. (2012) identified two distinct populations of MDSCs in PBMCs from bladder cancer patients: granulocytic $CD11b^{+}CD15^{high}CD33^{low}$ cells with coexpression of neutrophil markers CD114 (G-CSF receptor) and CD177; and monocytic $CD11b^{+}CD15^{low}CD33^{low}$ cells with the co-expression of monocyte-macrophage markers CD14, CD115 (M-CSF receptor), CD116 (GM-CSF receptor) and CCR2. However, only the number of circulating $CD11b^{+}CD15^{high}CD33^{low}$ cells in cancer patients was markedly increased when compared to healthy individuals. Both MDSC subsets were found to produce substantial amounts of pro-inflammatory chemokines/cytokines including CCL2, CCL3, CCL4, GCSF, IL-8, and IL-6, but only $CD11b^{+}CD15^{high}CD33^{low}$ cells were able to inhibit in vitro T cell proliferation through induction of $CD4^{+}Foxp3^{+}$ Tregs. Further analysis of bladder cancer tissues revealed that tumors were infiltrated with granulocytic $CD11b^{+}CD15^{+}HLA\text{-}DR^{-}$ myeloid cells (Eruslanov et al. 2012).

31.3 Immunoregulatory Mechanisms of MDSCs

In mice, it is well established that MDSCs inhibit immune responses through multiple mechanisms, including production of arginine-metabolizing enzyme arginase I, inducible nitric oxide synthase (iNOS), reactive oxygen species (ROS) and soluble inhibitory factors, and activation and expansion of $CD4^{+}CD25^{+}Foxp3^{+}$ Tregs (Gabrilovich et al. 2012; Gabrilovich and Nagaraj 2009; Lindenberg et al. 2011; Ostrand-Rosenberg 2010; Srivastava et al. 2012). Some of these mechanisms have already been confirmed in human cancers while others are still ongoing (Tadmor et al. 2011; Lees et al. 2011) (Table 31.1).

L-arginine metabolism is a major mechanism used by human MDSCs to impair antitumor immunity. Arginine is an essential amino acid for T cell proliferation because the depletion of arginine leads to a decrease in ζ-chain expression in the T cell receptor (TCR) complex and proliferative arrest of antigen-activated T cells (Gabrilovich et al. 2012). L-arginine serves as a substrate for iNOS (NOS2) and arginase: iNOS generates NO and arginase converts L-arginine into urea and L-ornithine (Nagaraj et al. 2010). Granulocytic MDSCs with a $CD11b^{+}CD14^{-}CD15^{+}$ phenotype isolated from RCC patients have been shown to express arginase, which was accompanied by a decreased CD3ζ chain expression on T cells (Zea et al. 2005). The increased arginase activity resulted in significantly decreased serum levels of arginine and increased ornithine in patients (Zea et al. 2005). Depletion of these MDSCs was able to resort IFN-γ production and T cell proliferation (Ochoa et al. 2007). In contrast to murine MDSCs, human granulocytic MDSCs expressing high levels of CD66b, CD11b, and VEGFR1 in RCC patients do not deplete L-arginine by increasing its uptake, but, instead release arginase I into the circulation (Rodriguez et al. 2009). Monocytic MDSCs in multiple myeloma (MM) and HNSCC patients have also been suggested to employ this mechanism to blunt antitumor immunity (Serafini et al. 2006). In another study of 40 patients with NHL, elevated levels of arginase in their plasma was found, which correlated with increase in $CD14^{+}HLA\text{-}DR^{-/low}$ MDSCs (Lin et al. 2011). In addition, exogenous arginine supplementation partially overcame the suppression of T cell proliferation generated by MDSCs in vitro, which suggested that this suppression was mediated in part through arginine metabolism (Lin et al. 2011). Furthermore, increased plasma arginase activity was also observed in the plasma obtained from patients with glioblastoma that have increased MDSC counts ($CD33^{+}HLA\text{-}DR^{-}$) in their blood (Raychaudhuri et al. 2011). Recently, Zhao et al. demonstrated that $CD14^{+}HLA\text{-}DR^{-/low}S100A9^{high}$ cells in the peripheral blood from patients with cancer expressed high levels of NOS2, which is one of the proposed mediators of the inhibitory properties of MDSCs (Zhao et al. 2012a).

Induction of oxidative stress is also involved in suppression of T cell responses by human MDSCs. MDSCs isolated from RCC patients were shown to suppress antigen-specific T cell responses in vitro through the secretion of ROS and NO upon interaction with CTL (Kusmartsev et al. 2008). MDSC-mediated immune suppression and IFN-γ down-regulation was reversible in vitro by exposing cells to the ROS inhibitors (Kusmartsev et al. 2008). In another study, up-regulation of ROS by MDSCs was observed in patients with stage III HNSCC (Corzo et al. 2009). The further investigations showed that increased ROS levels in MDSCs and the ROS-mediated suppressive function of these cells was caused by the up-regulation of several subunits of NADPH oxidase (NOX2), this up-regulation was controlled by the signal transducer and activator of transcription 3 (STAT3) transcription factor (Corzo et al. 2009). Interestingly, STAT3-dependent oxidative stress in MDSC-mediated T cell suppression was also noted in melanoma MDSCs (Poschke et al. 2010). However, quantitative PCR analyses did not detect the up-regulation of the most important NADPH-oxidase subunits, p47phox and gp91phox, in melanoma $CD14^{+}HLA\text{-}DR^{-/low}$ cells, suggesting an alternative source of oxidative stress (Poschke et al. 2010).

Another mechanism by which human MDSCs are able to suppress T cell activity is through the activation and expansion of Tregs. MDSCs have been suggested to promote the clonal expansion of antigen-specific natural Tregs and also induce the conversion of naive $CD4^{+}$ T cells into induced Tregs (Gabrilovich et al. 2012). The underlying mechanisms are not completely understood, but may employ both TGF-β dependent and independent pathways (Ostrand-Rosenberg 2010; Tadmor et al. 2011; Serafini et al. 2008). In a study of metastatic RCC patients, sunitinib-mediated reduction in MDSCs was correlated with a reversal of $CD3^{+}CD4^{+}CD25^{high}Foxp3^{+}$ Treg cell elevation (Ko et al. 2009). Furthermore, human $CD14^{+}HLA\text{-}DR^{-/low}$ MDSCs were demonstrated to convert Th17 cells into $FoxP3^{+}$ Tregs through a retinoic acid-dependent mechanism (Hoechst et al. 2011).

Interestingly, granulocytic MDSCs with the phenotype $CD11b^{+}CD15^{high}CD33^{low}$ from bladder cancer patients were also shown to inhibit T cell proliferation in vitro through the induction of $CD4^{+}Foxp3^{+}$ Tregs (Eruslanov et al. 2012).

Finally, the suppressor mechanism is related to soluble inhibitory factors produced by MDSCs. For example, TGF-β released by MDSCs has been implicated in MDSC-mediated T cell suppression in melanoma and HNSCC patients (Filipazzi et al. 2007; Chikamatsu et al. 2012). TGF-β is an important immune suppressive cytokine present in many cancer patients, and clinical analysis of tumor samples have demonstrated that TGF-β signalling is strongly implicated in tumor progression (Ikushima and Miyazono 2010).

It should be noted that two or more mechanisms may coexist in a single population of MDSCs in patients with cancer. In a study in HCC patients, monocytic MDSCs with a phenotype of $CD14^{+}HLA\text{-}DR^{-/low}$, capable of suppressing autologous T cell proliferation, were found have high arginase activity (Hoechst et al. 2008). Addition of L-arginine or depletion of MDSCs in the suppression assay resulted in enhanced IFN-γ secretion. Of interest, these MDSCs induced $CD4^{+}CD25^{+}Foxp3^{+}$ Tregs when cocultured with autologous T cells (Hoechst et al. 2008). In another study in melanoma patients, $CD14^{+}HLA\text{-}DR^{-/low}$ MDSCs were demonstrated to employ both arginase and oxidative stress to induce T cell suppression (Poschke et al. 2010). Recently, Gehad et al. 2012 reported that untreated human squamous cell carcinomas (SCCs) were infiltrated by a population of NO-producing $iNOS^{+}CD11b^{+}CD33^{+}CD11c^{-}HLA\text{-}DR^{-}$ MDSCs. Further characterization by flow cytometry and real-time PCR analysis demonstrated that majority of these MDSCs isolated from SCCs expressed TGF-β and arginase I. Analysis of nitrate and nitrite levels from tumor supernatants using the Griess method demonstrated that MDSCs were critical producers of NO in SCCs. By culturing human umbilical vein endothelial cells in the presence of $CD11b^{+}$ cell, NO produced by MDSCs was suggested to impair vascular E-selectin expression, likely impairing T cell migration into tumors (Gehad et al. 2012).

31.4 Prognostic Significance of MDSCs

With the understanding of MDSCs biology, increasing attention is being paid to the clinical implications of circulating MDSCs in human cancers in recent years (Montero et al. 2012). The overall levels of MDSCs with a phenotype $Lin^{-/low}HLA\text{-}DR^{-}CD33^{+}CD11b^{+}$ in the peripheral blood have been shown to correlate with clinical cancer stage in 3 studies (Solito et al. 2011; Diaz-Montero et al. 2009; Gabitass et al. 2011). Diaz-Montero et al. (2009) determined percentages of circulating MDSCs ($Lin^{-/low}HLA\text{-}DR^{-}CD33^{+}CD11b^{+}$) in whole blood collected from 106 newly diagnosed solid tumor patients (stages I–IV). The results showed that there was a significant correlation between both percentage and absolute number of circulating MDSCs and clinical cancer stage. Moreover, in patients with stage IV cancers both percentages and absolute numbers of MDSCs directly

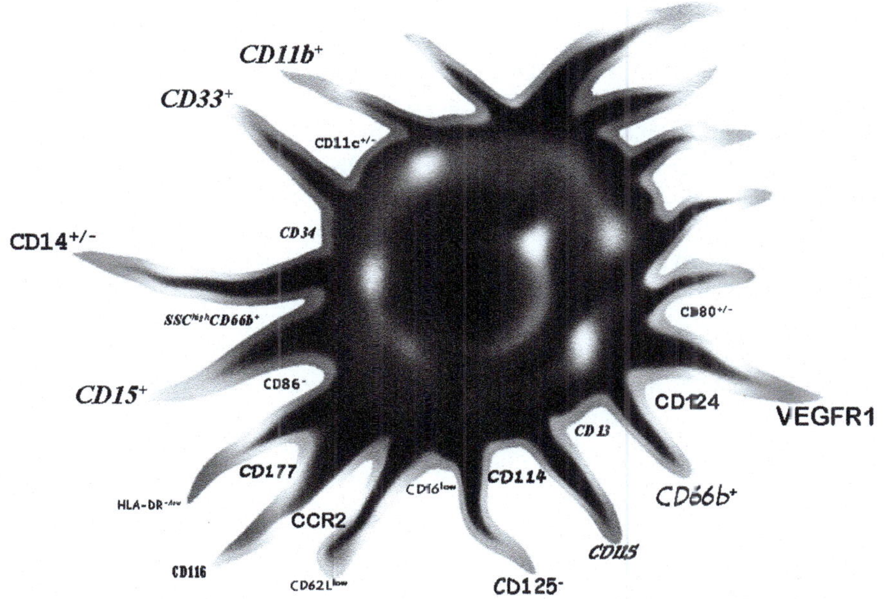

Fig. 31.1 Known markers of human MDSCs. CD11b$^+$ is a positive (*italics*), relatively common (*large font*), cell marker. CD11c$^{+/-}$ can be positive or negative (*Courier New font*), not very common (*small font*), cell marker. CD33$^+$ is a positive (*italics*), relatively common (*large font*), cell marker. CD34 is not a very common (*small font*) cell marker, it only appears on certain cells. CD14$^{+/-}$ can be positive or negative (*Courier New font*), relatively common (*large font*), cell marker. SSC^{high}CD66b$^+$ is a positive (*italics*), relatively new/not very common (*small font*), cell marker. CD15$^+$ is a positive (*italics*), relatively common (*large font*), cell marker. CD86$^-$ is a negative (*Comic Sans font*), not very common (*small font*), cell marker that only appears on certain cells. HLA-DR$^{-/low}$ is a negative (*Comic Sans font*), low expression (*smallest size font*) cell marker. CD177 is a positive (*italics*) cell marker that is a co-expressed neutrophil (*Lucida Handwriting font*).CD116 is a not very common (*small font*) cell marker that is a co-expressed monocyte-macrophage GM-CSF receptor (*Rockwell Condensed font*). CCR2 is a relatively common (*large font*) receptor (*Arial Rounded font*). CD62L^{low} is a low (*smallest font*) cell marker that is expressed by mature activated granulocytes (*Tempus Sans font*). CD16^{low} is a low (*smallest font*) cell marker that is expressed by mature activated granulocytes (*Tempus Sans font*). CD125$^-$ is a negative (*Comic Sans font*) cell marker. CD114 is a positive (*italics*), not relatively common (*small font*), cell marker that is a co-expressed neutrophil. CD115 is a positive (*italics*) cell marker that is a co-expressed monocyte-macrophage M-CSF receptor (*Informal Roman font*). CD13 is a positive (*italics*), not relatively common (*small font*), cell marker. CD66b$^+$ is a positive (*italics*) cell marker that is expressed by mature activated granulocytes (*Tempus Sans font*). CD124 or IL-4Rα is a receptor (*Arial Rounded font*) that is considered a potential cell marker. VEGFR1 is a relatively common (*large font*) receptor (*Arial Rounded font*). CD80$^{+/-}$ can be positive or negative (*Courier New font*), not very common (*small font*) cell marker

correlated with metastatic tumor burden (Diaz-Montero et al. 2009). In a large study of 131 cancer patients, MDSC levels were found to be an independent prognostic factor for survival in patients with pancreatic, esophageal and gastric cancers in a multivariate analysis (Gabitass et al. 2011). A unit increase in MDSC percentage was associated with a 22 % increased risk of death (hazard ratio 1.22, 95 % confidence interval 1.06–1.41) (Gabitass et al. 2011). In another cohort of patients with stage IV breast cancer (n = 25), patients with circulating MDSCs >3.17 % (median) at baseline had a poorer overall survival (OS) than patients with circulating MDSCs ≤3.17 %, with median OS times of 5.5 months and 19.32 months, respectively (p = 0.048) (Solito et al. 2011) (Fig. 31.1).

Similarly, the overall levels of monocytic MDSCs with a phenotype $CD14^{+}HLA\text{-}DR^{-/low}$ were also demonstrated to correlate with clinical cancer stage and aggressiveness of disease in NHL patients (Lin et al. 2011). The OS of patients was analyzed according to the numbers of granulocytic MDSCs in another study of patients with terminal cancer (Choi et al. 2012). Patients with low levels of peripheral blood $CD15^{+}CD16^{low}$ cells (<293/ml) had significantly longer survival times than those with high levels (p = 0.0011, median survival time was 2.6 versus 0.8 months). Moreover, patients with high levels of $CD15^{+}CD16^{low}$ cells also tended to have poor performance status (p = 0.05) (Choi et al. 2012).

Mundy-Bosse et al. (2011) evaluated potential relationships between MDSC subsets and OS in patients with gastrointestinal malignancies using multivariable Cox proportional hazards models. In this study, the level of individual $CD33^{+}HLA\text{-}DR^{-}CD11b^{+}CD15^{+}$ or $CD33^{+}HLA\text{-}DR^{-/low}CD14^{+}$ MDSC subsets was significantly elevated in patients as compared to normal healthy donors and widely distributed across patients with various GI malignancies. The results revealed that an increased percentage of the $CD33^{+}HLA\text{-}DR^{-}CD15^{-}$ MDSC subset was associated with reduced OS (p = 0.049), while an increased percentage of the $CD33^{+}HLA\text{-}DR^{-/low}CD14^{+}$ subset was associated with greater overall survival (p = 0.033).

31.5 Conclusions

MDSCs are bone marrow-derived cells of myeloid origin that are present in blood, lymph nodes, spleen, tumors, and immune-activated tissues (Goedegebuure et al. 2011). Under pathological conditions (e.g., tumor), the normal differentiation process of myeloid cells in bone marrow is blocked and immature myeloid cells are expanded and activated, and become MDSCs, acquiring immunosuppressive activity (Zhao et al. 2012b). MDSCs play an important role in maintaining a state of immunological anergy and tolerance in cancer, suppressing antitumor immunity and promoting tumor growth, proliferation, and metastases. As stated above, it is the heterogeneity and complexity of the characteristics of MDSCs, which results in the limited progress that has been made in understanding the biology of MDSCs in cancer patients. Lack of specific markers is one of the major obstacles in the clinical study of MDSCs in cancer patients. Although several molecules such as IL-4α, VEGFR1, CD66b, and S100A9 have been suggested as more specific markers for human MDSCs, these molecules can only be found on some MDSC subsets.

It should be noted that different phenotypes in a certain type of human cancer were described by different groups. For instance, in RCC patients, granulocytic MDSCs with a phenotype $CD11b^{+}CD14^{-}CD15^{+}HLA\text{-}DR^{-}$ or $CD11b^{+}CD14^{-}CD15^{+}CD16^{low}CD62L^{low}CD66b^{+}VEGFR1^{+}$ were isolated from PBMCs in two studies (Zea et al. 2005; Rodriguez et al. 2009), while only monocytic MDSCs with a phenotype $CD14^{+}HLA\text{-}DR^{-/low}$ were found in another study (van Cruijsen et al. 2008). In the other hand, although several immunosuppressive mechanisms by human MDSCs have been proven, a certain MDSC subset with same phenotype in different cancers may have controversial action mechanisms. For example, $CD14^{+}HLA\text{-}DR^{-/low}$ MDSCs in HCC patients showed not only high arginase activity, but also the induction of the $CD4^{+}CD25^{+}Foxp3^{+}$ Tregs; however, they did not secrete TGF-β (Hoechst et al. 2008). In contrast, in a study of melanoma patients, suppressive activity of $CD14^{-}HLA\text{-}DR^{-/low}$ MDSCs was mediated by TGF-β, whereas no involvement of arginase could be detected (Filipazzi et al. 2007). However, in another study of melanoma patients, the $CD14^{+}HLA\text{-}DR^{-/low}$ MDSCs were found to use arginase to suppress $CD4^{+}$ and $CD8^{+}$ T cells (Poschke et al. 2010).

The criteria for the selection of patients and treatment history may influence the phenotyping results, because MDSCs are driven by tumor-derived factors and different tumors secrete different combinations of molecules; therefore, the characteristics of MDSCs might be different in each malignancy (Chikamatsu et al. 2012). In summary, a better understanding of the phenotypes and immunosuppressive mechanisms will improve the therapeutic approaches targeting human MDSCs. Further studies are needed to identify more specific markers present on human MDSCs, thereby characterizing MDSCs more accurately. Furthermore, multi-central, randomized, controlled clinical trials contribute to accurate evaluation of MDSC phenotypes in a consistent number of patients with comparable disease conditions (Filipazzi et al. 2012).

References

Almand B, Clark JI, Nikitina E, van Beynen J, English NR, Knight SC et al (2001) Increased production of immature myeloid cells in cancer patients: a mechanism of immunosuppression in cancer. J Immunol 166:678–689

Brandau S, Trellakis S, Bruderek K, Schmaltz D, Steller G, Elian M et al (2011) Myeloid-derived suppressor cells in the peripheral blood of cancer patients contain a subset of immature neutrophils with impaired migratory properties. J Leukoc Biol 89:311–317

Brimnes MK, Vangsted AJ, Knudsen LM, Gimsing P, Gang AO, Johnsen HE et al (2010) Increased level of both CD4+FOXP3+ regulatory T cells and CD14+HLA-DR/low myeloid-derived suppressor cells and decreased level of dendritic cells in patients with multiple myeloma. Scand J Immunol 72:540–547

Chikamatsu K, Sakakura K, Toyoda M, Takahashi K, Yamamoto T, Masuyama K (2012) Immunosuppressive activity of CD14+HLA-DR—cells in squamous cell carcinoma of the head and neck. Cancer Sci 103(6):976–983

Choi J, Suh B, Ahn YO, Kim TM, Lee JO, Lee SH et al (2012) CD15+/CD16low human granulocytes from terminal cancer patients: granulocytic myeloid-derived suppressor cells that have suppressive function. Tumour Biol 33:121–129

Corzo CA, Cotter MJ, Cheng P, Cheng F, Kusmartsev S, Sotomayor E et al (2009) Mechanism regulating reactive oxygen species in tumor-induced myeloid-derived suppressor cells. J Immunol 182:5693–5701

Daud AI, Mirza N, Lenox B, Andrews S, Urbas P, Gao GX et al (2008) Phenotypic and functional analysis of dendritic cells and clinical outcome in patients with high-risk melanoma treated with adjuvant granulocyte macrophage colony-stimulating factor. J Clin Oncol 26:3235–3241

Diaz-Montero CM, Salem ML, Nishimura MI, Garrett-Mayer E, Cole DJ, Montero AJ (2009) Increased circulating myeloid-derived suppressor cells correlate with clinical cancer stage, metastatic tumor burden, and doxorubicin-cyclophosphamide chemotherapy. Cancer Immunol Immunother 58:49–59

Dumitriu IE, Dunbar DR, Howie SE, Sethi T, Gregory CD (2009) Human dendritic cells produce TGF-beta 1 under the influence of lung carcinoma cells and prime the differentiation of CD4+CD25+Foxp3+ regulatory T cells. J Immunol 182:2795–2807

Eruslanov E, Neuberger M, Daurkin I, Perrin GQ, Algood C, Dahm P et al (2012) Circulating and tumor-infiltrating myeloid cell subsets in patients with bladder cancer. Int J Cancer 130:1109–1119

Filipazzi P, Valenti R, Huber V, Pilla L, Canese P, Iero M et al (2007) Identification of a new subset of myeloid suppressor cells in peripheral blood of melanoma patients with modulation by a granulocyte-macrophage colony-stimulation factor-based antitumor vaccine. J Clin Oncol 25:2546–2553

Filipazzi P, Huber V, Rivoltini L (2012) Phenotype, function and clinical implications of myeloid-derived suppressor cells in cancer patients. Cancer Immunol Immunother 61:255–263

Fridlender ZG, Albelda SM (2012) Tumor associated neutrophils: friend or foe? Carcinogenesis 33(5):949–955

Gabitass RF, Annels NE, Stocken DD, Pandha HA, Middleton GW (2011) Elevated myeloid-derived suppressor cells in pancreatic, esophageal and gastric cancer are an independent prognostic factor and are associated with significant elevation of the Th2 cytokine interleukin-13. Cancer Immunol Immunother 60:1419–1430

Gabrilovich DI, Nagaraj S (2009) Myeloid-derived suppressor cells as regulators of the immune system. Nat Rev Immunol 9:162–174

Gabrilovich DI, Ostrand-Rosenberg S, Bronte V (2012) Coordinated regulation of myeloid cells by tumours. Nat Rev Immunol 12:253–268

Gallina G, Dolcetti L, Serafini P, De Santo C, Marigo I, Colombo MP et al (2006) Tumors induce a subset of inflammatory monocytes with immunosuppressive activity on CD8+ T cells. J Clin Invest 116:2777–2790

Gehad AE, Lichtman MK, Schmults CD, Teague JE, Calarese AW, Jiang Y et al (2012) Nitric oxide-producing myeloid-derived suppressor cells inhibit vascular E-selectin expression in human squamous cell carcinomas. J Invest Dermatol 132(11):2642–2651

Goedegebuure P, Mitchem JB, Porembka MR, Tan MC, Belt BA, Wang-Gillam A et al (2011) Myeloid-derived suppressor cells: general characteristics and relevance to clinical management of pancreatic cancer. Curr Cancer Drug Targets 11:734–751

Greten TF, Manns MP, Korangy F (2011) Myeloid derived suppressor cells in human diseases. Int Immunopharmacol 11:802–807

Hanahan D, Weinberg RA (2011) Hallmarks of cancer: the next generation. Cell 144:646–674

Hoechst B, Ormandy LA, Ballmaier M, Lehner F, Kruger C, Manns MP et al (2008) A new population of myeloid-derived suppressor cells in hepatocellular carcinoma patients induces CD4(+)CD25(+)Foxp3(+) T cells. Gastroenterology 135:234–243

Hoechst B, Gamrekelashvili J, Manns MP, Greten TF, Korangy F (2011) Plasticity of human Th17 cells and iTregs is orchestrated by different subsets of myeloid cells. Blood 117:6532–6541

Hurwitz AA, Watkins SK (2012) Immune suppression in the tumor microenvironment: a role for dendritic cell-mediated tolerization of T cells. Cancer Immunol Immunother 61:289–293

Ikushima H, Miyazono K (2010) TGFbeta signalling: a complex web in cancer progression. Nat Rev Cancer 10:415–424

Kaufman HL, Kim DW, DeRaffele G, Mitcham J, Coffin RS, Kim-Schulze S (2010) Local and distant immunity induced by intralesional vaccination with an oncolytic herpes virus encoding GM-CSF in patients with stage IIIc and IV melanoma. Ann Surg Oncol 17:718–730

Ko JS, Zea AH, Rini BI, Ireland JL, Elson P, Cohen P et al (2009) Sunitinib mediates reversal of myeloid-derived suppressor cell accumulation in renal cell carcinoma patients. Clin Cancer Res 15:2148–2157

Kusmartsev S, Su Z, Heiser A, Dannull J, Eruslanov E, Kubler H et al (2008) Reversal of myeloid cell-mediated immunosuppression in patients with metastatic renal cell carcinoma. Clin Cancer Res 14:8270–8278

Lechner MG, Megiel C, Russell SM, Bingham B, Arger N, Woo T et al (2011) Functional characterization of human Cd33+ and Cd11b+ myeloid-derived suppressor cell subsets induced from peripheral blood mononuclear cells co-cultured with a diverse set of human tumor cell lines. J Transl Med 9:90

Lees JR, Azimzadeh AM, Bromberg JS (2011) Myeloid derived suppressor cells in transplantation. Curr Opin Immunol 23:692–697

Lesterhuis WJ, Haanen JB, Punt CJ (2011) Cancer immunotherapy–revisited. Nat Rev Drug Discov 10:591–600

Lin Y, Gustafson MP, Bulur PA, Gastineau DA, Witzig TE, Dietz AB (2011) Immunosuppressive CD14+HLA-DR(low)/– monocytes in B-cell non-Hodgkin lymphoma. Blood 117:872–881

Lindenberg JJ, Fehres CM, van Cruijsen H, Oosterhoff D, de Gruijl TD (2011) Cross-talk between tumor and myeloid cells: how to tip the balance in favor of antitumor immunity. Immunotherapy 3:77–96

Liu CY, Wang YM, Wang CL, Feng PH, Ko HW, Liu YH et al (2010) Population alterations of L-arginase- and inducible nitric oxide synthase-expressed CD11b+/CD14/CD15+/CD33+ myeloid-derived suppressor cells and CD8+ T lymphocytes in patients with advanced-stage non-small cell lung cancer. J Cancer Res Clin Oncol 136:35–45

Ma Y, Shurin GV, Gutkin DW, Shurin MR (2012) Tumor associated regulatory dendritic cells, Semin Cancer Biol 22:298–306

Mandruzzato S, Solito S, Falisi E, Francescato S, Chiarion-Sileni V, Mocellin S et al (2009) IL4Ralpha+ myeloid-derived suppressor cell expansion in cancer patients. J Immunol 182:6562–6568

Martin F, Apetoh L, Ghiringhelli F (2012) Role of myeloid-derived suppressor cells in tumor immunotherapy. Immunotherapy 4:43–57

Mirza N, Fishman M, Fricke I, Dunn M, Neuger AM, Frost TJ et al (2006) All-trans-retinoic acid improves differentiation of myeloid cells and immune response in cancer patients. Cancer Res 66:9299–9307

Montero AJ, Diaz-Montero CM, Kyriakopoulos CE, Bronte V, Mandruzzato S (2012) Myeloid-derived suppressor cells in cancer patients: a clinical perspective. J Immunother 35:107–115

Mundy-Bosse BL, Young GS, Bauer T, Binkley E, Bloomston M, Bill MA et al (2011) Distinct myeloid suppressor cell subsets correlate with plasma IL-6 and IL-10 and reduced interferon-alpha signaling in CD4 T cells from patients with GI malignancy. Cancer Immunol Immunother 60:1269–1279

Nagaraj S, Gabrilovich DI (2010) Myeloid-derived suppressor cells in human cancer. Cancer J 16:348–353

Nagaraj S, Youn JI, Weber H, Iclozan C, Lu L, Cotter MJ et al (2010) Anti-inflammatory triterpenoid blocks immune suppressive function of MDSCs and improves immune response in cancer. Clin Cancer Res 16:1812–1823

Naiditch H, Shurin MR, Shurin GV (2011) Targeting myeloid regulatory cells in cancer by chemotherapeutic agents. Immunol Res 50:276–285

Ochoa AC, Zea AH, Hernandez C, Rodriguez PC (2007) Arginase, prostaglandins, and myeloid-derived suppressor cells in renal cell carcinoma. Clin Cancer Res 13:721s–726s

Ostrand-Rosenberg S (2010) Myeloid-derived suppressor cells: more mechanisms for inhibiting antitumor immunity. Cancer Immunol Immunother 59:1593–1600

Ostrand-Rosenberg S, Sinha P, Beury DW, Clements VK (2012) Cross-talk between myeloid-derived suppressor cells (MDSC), macrophages, and dendritic cells enhances tumor-induced immune suppression. Semin Cancer Biol 22(4):275–281

Poschke I, Mougiakakos D, Hansson J, Masucci GV, Kiessling R (2010) Immature immuno-suppressive CD14+HLA-DR−/low cells in melanoma patients are Stat3hi and overexpress CD80, CD83, and DC-sign. Cancer Res 70:4335–4345

Rabinovich GA, Gabrilovich D, Sotomayor EM (2007) Immunosuppressive strategies that are mediated by tumor cells. Annu Rev Immunol 25:267–296

Raychaudhuri B, Rayman P, Ireland J, Ko J, Rini B, Borden EC et al (2011) Myeloid-derived suppressor cell accumulation and function in patients with newly diagnosed glioblastoma. Neuro Oncol 13:591–599

Rodrigues JC, Gonzalez GC, Zhang L, Ibrahim G, Kelly JJ, Gustafson MP et al (2010) Normal human monocytes exposed to glioma cells acquire myeloid-derived suppressor cell-like properties. Neuro Oncol 12:351–365

Rodriguez PC, Ernstoff MS, Hernandez C, Atkins M, Zabaleta J, Sierra R et al (2009) Arginase I-producing myeloid-derived suppressor cells in renal cell carcinoma are a subpopulation of activated granulocytes. Cancer Res 69:1553–1560

Romano A, Vetro C, Adriani M (2011) Advances in understanding regulatory myeloid cells. Cancer Biol Ther 11:923–926

Ruffell B, Affara NI, Coussens LM (2012) Differential macrophage programming in the tumor microenvironment. Trends Immunol 33:119–126

Schmieder A, Michel J, Schonhaar K, Goerdt S, Schledzewski K (2012) Differentiation and gene expression profile of tumor-associated macrophages. Semin Cancer Biol 22:289–297

Serafini P, Meckel K, Kelso M, Noonan K, Califano J, Koch W et al (2006) Phosphodiesterase-5 inhibition augments endogenous antitumor immunity by reducing myeloid-derived suppressor cell function. J Exp Med 203:2691–2702

Serafini P, Mgebroff S, Noonan K, Borrello I (2008) Myeloid-derived suppressor cells promote cross-tolerance in B-cell lymphoma by expanding regulatory T cells. Cancer Res 68:5439–5449

Shurin MR, Shurin GV, Lokshin A, Yurkovetsky ZR, Gutkin DW, Chatta G et al (2006) Intratumoral cytokines/chemokines/growth factors and tumor infiltrating dendritic cells: friends or enemies? Cancer Metastasis Rev 25:333–356

Shurin MR, Naiditch H, Zhong H, Shurin GV (2011) Regulatory dendritic cells: new targets for cancer immunotherapy. Cancer Biol Ther 11:988–992

Shurin GV, Ouellette CE, Shurin MR (2012) Regulatory dendritic cells in the tumor immunoenvironment. Cancer Immunol Immunother 61:223–230

Sinha P, Parker KH, Horn L, Ostrand-Rosenberg S (2012) Tumor-induced myeloid-derived suppressor cell function is independent of IFNgamma and IL-4Ralpha. Eur J Immunol 42:2052–2059

Solito S, Falisi E, Diaz-Montero CM, Doni A, Pinton L, Rosato A et al (2011) A human promyelocytic-like population is responsible for the immune suppression mediated by myeloid-derived suppressor cells. Blood 118:2254–2265

Srivastava MK, Bosch JJ, Thompson JA, Ksander BR, Edelman MJ, Ostrand-Rosenberg S (2008) Lung cancer patients' CD4+ T cells are activated in vitro by MHC II cell-based vaccines despite the presence of myeloid-derived suppressor cells. Cancer Immunol Immunother 57:1493–1504

Srivastava MK, Andersson A, Zhu L, Harris-White M, Lee JM, Dubinett S et al (2012) Myeloid suppressor cells and immune modulation in lung cancer. Immunotherapy 4:291–304

Tadmor T, Attias D, Polliack A (2011) Myeloid-derived suppressor cells–their role in haemato-oncological malignancies and other cancers and possible implications for therapy. Br J Haematol 153:557–567

van Cruijsen H, van der Veldt AA, Vroling L, Oosterhoff D, Broxterman HJ, Scheper RJ et al (2008) Sunitinib-induced myeloid lineage redistribution in renal cell cancer patients: CD1c + dendritic cell frequency predicts progression-free survival. Clin Cancer Res 14:5884–5892

Vuk-Pavlovic S, Bulur PA, Lin Y, Qin R, Szumlanski CL, Zhao X et al (2010) Immunosuppressive CD14+HLA-DRlow/– monocytes in prostate cancer. Prostate 70:443–455

Wilcox RA (2012) Myeloid-derived suppressor cells: therapeutic modulation in cancer. Front Biosci (Elite Ed) 4:838–855

Wilcox RA, Feldman AL, Wada DA, Yang ZZ, Comfere NI, Dong H et al (2009) B7–H1 (PD-L1, CD274) suppresses host immunity in T-cell lymphoproliferative disorders. Blood 114:2149–2158

Youn JI, Nagaraj S, Collazo M, Gabrilovich DI (2008) Subsets of myeloid-derived suppressor cells in tumor-bearing mice. J Immunol 181:5791–5802

Youn JI, Collazo M, Shalova IN, Biswas SK, Gabrilovich DI (2012) Characterization of the nature of granulocytic myeloid-derived suppressor cells in tumor-bearing mice. J Leukoc Biol 91:167–181

Zea AH, Rodriguez PC, Atkins MB, Hernandez C, Signoretti S, Zabaleta J et al (2005) Arginase-producing myeloid suppressor cells in renal cell carcinoma patients: a mechanism of tumor evasion. Cancer Res 65:3044–3048

Zhao F, Hoechst B, Duffy A, Gamrekelashvili J, Fioravanti S, Manns MP et al (2012a) S100A9 a new marker for monocytic human myeloid-derived suppressor cells. Immunology 136:176–183

Zhao E, Xu H, Wang L, Kryczek I, Wu K, Hu Y et al (2012b) Bone marrow and the control of immunity. Cell Mol Immunol 9:11–19

Chapter 32
When Results of T cell Immune Monitoring Match/Do Not Match Clinical Outcomes of Tumor Vaccine Trials: What More Could and Should We Measure?

Paul V. Lehmann and Srividya Sundararaman

Abstract Activation of cancer-specific T cells by tumor vaccines can lead to the rejection of the tumor. However, monitoring the tumor-specific T cell response continues to be a challenge. The magnitude and cytokine effector lineage of the vaccine-induced T cells, their ability to kill, and the avidity of the T cells for the tumor are among the primary parameters that define the T cell's potential impact on the tumor. Due to determinant spreading, immunization with a tumor antigen can result in recruitment of T cells with specificity for unrelated antigens of the tumor. The efficacy of this second wave T cell repertoire will depend on clonal sizes, effector lineages, and their avidities for the tumor. Comprehensive immune monitoring should be able to measure all the above parameters of the primary and secondary T cell responses to the tumor. In this chapter, we propose an ELISPOT-based protocol which can help accomplish the above goal with reasonable cost and effort, requiring only 10 ml of patient blood.

Keywords Immunomonitoring · ELISPOT · T cell activity · CTL · Antigen map · Immunotherapy · Clinical trials

32.1 Introduction

For decades it has been debated whether tumors underlie immune surveillance. Today there is no doubt that tumors can spontaneously induce immune responses directed against themselves (Disis 2011). More importantly, from a therapeutic

P. V. Lehmann · S. Sundararaman (✉)
Cellular Technology Limited, Shaker Heights, Ohio 44122, USA
e-mail: srividya.sundararaman@immunospot.com

M. R. Shurin et al. (eds.), *The Tumor Immunoenvironment*,
DOI: 10.1007/978-94-007-6217-6_32,

perspective, protective anti-tumor responses can be induced by various immunization/vaccination protocols. Several important questions regarding understanding the effect of these immune responses on the tumor cells, and means to measure and interpret them are still left unanswered. One major question pertains to the protective class(es) of immune response versus responses that are neutral to, or promote tumor growth. Traditionally, certain effector classes of T cells, rather than B cells and antibodies, have been thought to mediate protection against cancer. Therefore, in this chapter we will focus on monitoring T cell immunity. However, recent successes in targeting tumor cells, achieved by administering tumor-specific antibodies, imply that active immunizations triggering endogenous antibody responses might hold therapeutic potential as well.

When it comes to T cell-mediated immunity against tumors, the question arises as to which T cell effector class(es) protect the host against the tumor, which ones are inefficient/neutral, and which ones protect the tumor rather than the host (Bates et al. 2006). The detection of IFN-γ producing type1 T cells using an ELISPOT assay frequently correlates with the success of cancer vaccines (Barth et al. 2010; Disis et al. 2009; Gulley et al. 2010; Kenter et al. 2009; Kirkwood et al. 2009; Slingluff et al. 2007). In other cases, cutaneous delayed type hypersensitivity (DTH) reactions were indicative of protective anti-tumor T cell immunity (Baars et al. 2000; de Vries et al. 2005; Disis et al. 2000; Jaffee et al. 2001; Lopez et al. 2009). Using murine studies, we have shown that DTH is mediated by IL-17 producing Th17 cells (IL-17 and can be measured by ELISPOT) while IFN-γ producing T cells do not mediate DTH (Tigno-Aranjuez et al. 2009). In addition, the role of T cell-mediated cytotoxicity along with cytokine and chemokine production by CD8 and CD4 needs to be considered. The scope of immune monitoring, therefore, is not only to establish that an immune response has been induced e.g., after administering a tumor vaccine, and how strong that response is, but also to carefully assess the quality of these T and B cell responses. All too frequently only IFN-γ producing T cells are measured, neglecting other effector classes, and cytotoxicity. In this chapter, we will address options for multiplexing T cell measurements so as to account for the different types of effector classes and T cell responses.

32.2 What Should We and What Can We Measure?

Following the administration of vaccines to target tumor cells, tumor rejection is often assumed to be mediated by T cells specific to the tumor antigen in the vaccine. Therefore, to monitor T cell immunity these tumor antigens are utilized to test frequencies of the induced T cell responses, as well as effector functions. However, following the first wave of effector T cells specific for the administered antigen, a second wave can become engaged that targets other antigens on the tumor (Disis 2011; Hardwick and Chain 2011; Inderberg-Suso et al. 2012). This phenomenon, known as determinant/epitope spreading has been first discovered by our group in the context of autoimmunity in the early 1990s (Lehmann et al. 1992, 1993), and is now gaining importance in tumor immunology. In this chapter, we will provide practical hints to account for determinant spreading in immune monitoring trials.

For practical reasons, immune monitoring typically relies on measuring the T cell response to a single dose of the tumor antigen. Such measurements neglect the functional avidity of the antigen-specific T cells (Targoni and Lehmann 1998). This avidity of the T cell response can define whether the tumor antigen-specific cells can indeed recognize the tumor itself. High numbers of MHC-tumor peptide ligands can be artificially generated on antigen presenting cells (APC) by immunizing with the tumor antigen, or by pulsing dendritic cells *in vitro*. In such cases the number of MHC-tumor peptide complexes on the APC can exceed by far the copy number of these T Cell Receptor (TCR) ligands on the tumor itself. T cells that have low avidity for the tumor antigen will be primed by such immunizations, but these T cells may not be stimulated by the lower copy number of the antigen on the tumor cell. Such T cells will be detected in recall assays using high concentrations of the tumor peptide, revealing that a tumor antigen-specific T cell response was induced. However, these T cells would be irrelevant for combating the tumor.

When tumor peptides are used for vaccinations, they can assume conformations on the APC that differ from the endogenously loaded antigen. This observation was first made by Emil Unanue's group in the context of autoimmunity (Mohan and Unanue 2012). They showed that many T cell clones that were generated by immunization with autoantigen peptides did not mediate autoimmune disease when injected into mice. These clones did recognize the synthetic peptide, but were not stimulated by the peptide/MHC complexes generated through natural processing of the autoantigen. Apparently, the peptide assumed a different conformation in the MHC molecule's binding groove when it was bound directly from the extracellular space which is at a neutral pH, compared to when the peptide was loaded in the acidic environment intracellularly. Thus, as far as T cell recognition goes, the same peptide sequence can be presented as "self" (Type A determinant, assumed after natural processing), or as "foreign" (Type B determinant, assumed after adding soluble peptide). The former is relevant for autoimmunity or tumor immunology, whereas the latter is a third party antigen and thus irrelevant to autoimmunity or tumor recognition. Detecting T cell responses to a tumor peptide after immunizing with it, therefore, does not warrant a protective anti-tumor T cell response.

The issues associated with monitoring T cell immunity against cancer discussed here is only a subset of what needs to be considered. Immune monitoring is typically done using peripheral blood mononuclear cells (PBMC) isolated from blood. Thus, for reasons of feasibility, immunity is assessed at the highway of T cell recirculation, and not at the relevant site which is the tumor itself. Assuming that tumor-specific T cells behave like autoreactive T cells do, and that lymphocyte trafficking is similar between mice and men, then the frequencies and signatures of the T cells studied in the blood is likely to reliably reveal the total clonal size ("effector cell mass"), avidities, and the effector lineages of the tumor antigen-specific T cells in the body (Hofstetter et al. 2005). However, it will not provide information on whether those T cells gain access to the tumor and if they do so, whether they can exert their effector functions in the tumor microenvironment. Several chapters in this book review the mechanisms by which tumors can evade immune recognition. In spite of these drawbacks, measuring T cell immunity in the blood provides a fair assessment of the immune potential, without which immune effector functions cannot be exerted at the site.

For all the above reasons and several more, reliable monitoring of anti-tumor T cell immunity continues to be a technical and intellectual challenge. The community is frequently confronted with the statement: "We measured T cell immunity, and it does not match the clinical outcome". In many of these cases anti-tumor immunity was measured by assessing the frequency of tetramer positive T cells in PBMC. The commonly used tetramer approach does not provide information on effector functions of the T cells, does not account for determinant spreading, and does not measure the avidity of the T cells. In other cases, anti-tumor immunity was determined by measuring the frequency of antigen-specific, IFN-γ producing T cells by intra cytoplasmic cytokine staining or ELISPOT. This method demonstrates IFN-γ secretion, which is just one of several possible effector functions of T cells. Frequently it is implied that IFN-γ producing CD8 cells are also cytolytic, but this inference might not be warranted (Tigno-Aranjuez et al. 2009; Kuerten et al. 2008). Among IFN-γ producing T cells those that co-express IL-2 and TNF-α show increased ability to combat infections (Ding and Zhou 2012; Thakur et al. 2012), and such polyfunctional T cells might also have enhanced anti-tumor properties although there are limited studies to corroborate these claims. Measuring IFN-γ production by tumor antigen-specific T cells using a single tumor antigen and a single antigen dose also does not account for determinant spreading, and does not measure the avidity of these T cells. Therefore, the statement: "We measured T cell immunity, and it does not match the clinical outcome" does not necessarily imply that measuring T cell immunity does not provide a good clinical correlation of protection. It merely states that efforts of immune monitoring are still in its infancy. In other words, to cite Rolf Zinkernagel from one of his talks: "Scientists do not do the experiments they should, but they do what they can".

32.3 Comprehensive Immune Monitoring with Only 10 ml of Blood

Along the same lines, for comprehensive immune monitoring, it is important to measure T cell responses to several tumor antigens to account for determinant spreading. For each of these responses one should measure the cytokine signature for all key effector T cell lineages, as well as the cytolytic potential of these T cells.

In addition, for each response, one should test the antigens in serial dilution to measure avidity. At first sight, one might object that in praxis, the amount of cells needed for such comprehensive testing is far more than what can be feasibly obtained. Even if one could get access to all the cells, e.g. through leucapheresis, one might object that neither work load nor cost would permit such comprehensive testing.

In the remainder of this chapter, we will argue that in theory (that mostly already has been reduced to practice) such comprehensive testing is feasible with only 10 ml of patient blood, and that the cost and effort associated with it are fairly reasonable. We will argue so, hoping that pioneers of immune monitoring for cancer will help bring the last steps to practice, and doing so, they will help the community progress towards comprehensive immune monitoring, and gain better insights into correlates of protection for cancer vaccines.

As the next step towards comprehensive T cell monitoring, we set as our goal to use no more than 10 ml blood (corresponding to about 10 million PBMC) to obtain all the above information on T cell immunity to cancer. With this 10 ml of blood, we propose that dozens of tumor peptides can be tested to determine the T cell response also accounting for determinant spreading. In addition, with this 10 ml of blood, we set the goal to verify and extend the results by repeated testing of a different aliquot of this blood draw. Moreover, we also propose that with this 10 ml of blood, T cell effector functions can be tested comprehensively measuring Granzyme B–producing CD8 effector memory cells, IFN-γ and IL-2 producing Th1, ThX and polyfunctional T cells, IL-4 and IL-5 producing Th2 cells, and IL-17 producing Th17 cells as well as resting but potentially cytolytic CD8 memory cells. Finally, we propose that with this 10 ml of blood it is possible to establish the functional avidity of the each of the positive peptide responses detected. We also propose that this comprehensive testing can be done with relatively minor cost and experimental effort. This may sound provocative but we firmly believe that it can be done routinely since most of the building blocks have already been reduced to practice.

32.4 The Serial ELISPOT Approach

To generate these building blocks, we employed an ELISPOT based approach that permits serial testing of PBMC. Unlike intracytoplasmic staining (ICS) that requires the cells to be fixed and permeabilized for testing, cells survive ELISPOT assays unharmed, permitting their transfer from one ELISPOT assay into another, serially (Herzog et al. 2004). Since the production and secretion of different analytes by T cells peaks at different time points, such serial testing permits us to optimize the time points for measuring different analytes. In contrast, when multiple analytes are measured by ICS, the information gained is representative of one particular time point, that when the cells were killed. This time point may be too early or too late for the detection of asynchronously produced analytes.

The feasibility of serial ELISPOT testing has been demonstrated by us in the past (Herzog et al. 2004). ELISPOT is also particularly suited to meet our goal because of its economy of cell utilization. Unlike flow cytometry based measurements where much of the cells are lost in the tubing of the instrument, in ELISPOT assays every single cell that is plated is also interrogated. In addition, as mentioned above, this economy of cell usage can be multiplied through serial testing. In the above publication (Herzog et al. 2004), we used 5.3 million PBMC per donor to test them for reactivity to 70 individual peptides, serially. In some of our previous work for measuring T cell avidity, we used only 1 million PBMC to test 13 concentrations of a peptide (Hofstetter et al. 2005). In typical 96-well plate-based ELISPOT assays 100,000–1 million PBMC can be interrogated per well, per test condition. In these assays we determined a linear relationship between the number of PBMCs plated per well and the spot counts obtained in that well (Kuerten et al. 2012; Zhang et al. 2009). In the newly available 384 well format, 25,000–250,000 PBMC can be plated in the linear range (Hanson et al. in preparation). Thus, when pushing the limits, with 10 ml of blood (with 10 million PBMC) 400 individual peptides can be tested, repeatedly, at 25,000 PBMC per well.

ELISPOT lends itself to meet our goal for comprehensive immune monitoring because of its unprecedented sensitivity and signal to noise ratio (Lehmann and Zhang 2012). When tests are performed with serum free media, the background for most donors is zero, or close to zero (Kuerten et al. 2012; Ramachandran et al. 2012), permitting us to reliably detect rare antigen-specific T cells, far below the 1 in 10,000 (0.01 %) threshold that is generally considered the detection limit of flow cytometry. When ELISPOT assays are performed with 25,000 PBMC per well in 384-well plates, the detection limit is 1:25,000 and when 1 million cells are tested in 96-well plates, the detection limit is 1 in a million (Hesse et al. 2001). Since tumor antigen-specific T cells typically occur in low frequencies in the blood, high sensitivity is critical for the success of immune monitoring.

ELISPOT assays can be performed with fresh and with cryopreserved cells providing similar results (Kuerten et al. 2012; Kreher et al. 2003). The key to freezing cells without losing function is the use of specialized serum free cryo-media which is warmed to room temperature and added to cells that also are kept in warm medium before starting the cooling process. This practice is contrary to the original protocols that suggest the use of ice cold cryo-media to be added to cells that are chilled prior to freezing them. To prevent cell loss during the thawing process, it is also critical to bring the cells rapidly to 37 °C (Ramachandran et al. 2012). According to a large scale systematic study that we performed, resting PBMC after thawing does not necessarily improve the results of ELISPOT assays (Kuerten et al. 2012). We propagate cryopreserving PBMC aliquots from a blood draw because it permits batch testing of many samples, and to repeat, and extend testing of individual samples.

The studies that we propose here for comprehensive immune monitoring will generate dozens of data points per test subject. Therefore, using an automated test platform would be ideal for streamlined work, and also to introduce objectivity and to eliminate user bias from data analysis. Unlike flow cytometry based analysis,

ELISPOT is such a platform (Hawkins et al. 2006; Zhang and Lehmann 2012). A single well trained and organized technician can set up ELISPOT assays for dozens of PBMC samples. These samples can be tested for a multitude of peptides, provided the peptides are pre-aliquoted for a trial and can be introduced into the assay by multi-channel pipetting. Once the plates are developed, the completely-automated scanning, counting, data documentation, evaluation, management, and archiving occurs at a rate of 1.5 s per well using the latest ImmunoSpot Analyzer models (Zhang and Lehmann 2012). In the above example, in which 400 individual peptides are tested on 10 million PBMC, it would take about 10 min of walk away analyzer time to have all data exported for evaluation into a data base.

Based on the above feasibility studies we think that already today it is feasible to perform comprehensive immune monitoring with cancer patients in which 10 ml of blood suffices (1) to test up to a multitude of tumor antigens to account for determinant spreading, (2) to measure the signature of all key effector cell lineages for each antigen including (3) killing, and (4) to measure the avidity of each positive T cell response. How would we design such an experiment?

32.5 Experimental Design for the Detection of Multiple T cell Lineages

In our example we assume that 10 ml heparinized blood will be obtained from each blood draw from a cancer patient. In order to maintain the functionality of the immune cells, the blood must never be refrigerated, but should be kept at room temperature or at 37 °C for better results. When the blood needs to be transported or shipped in cold climate, it should be protected from chilling. For ideal conditions, it is best to transport blood on warm packs. If kept warm, little to no T cell reactivity will be lost during overnight shipping or storage However, it is important to process the blood within 24 h. For the proposed testing PBMC obtained from the 10 ml of blood are frozen into 2 aliquots. Each aliquot contains 5 million cells in 5 ml of cryo-media. After thawing more than 4 million viable and fully functional PBMC are obtained from each aliquot. Since overnight resting does not improve ELISPOT results and inevitably leads to loss of approximately half of the thawed cells (Kuerten et al. 2012), it is not recommended to apply overnight resting to these cells, unless it has been verified that the particular patient population's results would benefit from resting. For reasons of work economy and inter-sample comparability, PBMC collected from 10 to 30 donors are tested simultaneously in a single assay run. In each test run reference PBMC samples are included as control samples. Reference PBMC samples have established T cell reactivity for the analytes of interest, e.g. PBMC that produce Perforin and Granzyme B, IL-2, IFN-γ, IL-4, IL-5 and IL-17 after challenge with antigen. Such cryopreserved reference PBMC is readily available and can be selected at http://epbmc.immunospot.com.

The well format (96 or 384) and the numbers of cells plated per well depends on the expected frequency of the tumor antigen-reactive T cells. As stated above, up to 1 million and as few as 100,000 PBMC can be plated per well in 96-well plates to accommodate linear frequency measurements in the 1 in a million to 1 in 100,000 resolution range. Unfortunately, larger well formats (e.g. 24 or 48 well PVDF plates) are currently not available. When frequencies of lower than 1 in 1 million measurements are intended, cells are plated at 1 million PBMC per well in replicate wells, e.g. 10 replicate wells tested at 1 million PBMC/well will afford a 1 in 10 million frequency measurement where the spot counts of the replicate wells will be summated. Such cumulative spot counts are permissive, since signatures of individual cells are measured within the test cell population (Dittrich and Lehmann 2012). The use of 384-well plates permit testing of as few as 25,000 PBMC per well (at a resolution of 1:25,000). A 96-well plate format, while testing 400,000 cells per well, would be ideal for the proposed example. Thus, with the 4 million PBMC obtained after thawing one of the aliquots, a 10 well assay can be performed.

32.5.1 Antigen Map: Testing for Determinant Spreading

As to the plate layout, it is recommended to assign three wells for the negative control (media only), whereas, one well would be sufficient for the positive control (in this case, Phytohemagglutinin, PHA). This layout would leave 6 wells for testing reactivity to tumor antigens. Replicate wells are not necessary for antigen-stimulated cultures, since the data will be repeated and extended in the second experiment. Thus, for the proposed experiments one could either test six individual antigens, or test fewer antigens in replicates. For the antigens tested, one tumor antigen would be the one that was used for the tumor vaccination. The other antigen(s) would test whether or not a determinant spreading reaction has occurred.

The simplest method to test all the tumor antigens that could have been recruited into the determinant spreading reaction is by using the entire tumor cell to stimulate the T cells. We have successfully used this approach in mice for all the tumor models we examined (Bartholomae et al. 2004; Lonsdorf et al. 2003; Schlingmann et al. 2009; Stern et al. 2002). In limited, yet successful, experiments we used tumor cells isolated from glioblastoma resection material to stimulate PBMC of glioblastoma patients. Resection of solid tumors yields abundant excess material that can be used to create single cell suspensions for cryopreservation and future immune monitoring of the patient.

An alternative approach for testing the spreading reaction is to include peptides of unrelated antigens on the same tumor. For many types of tumors these antigens/peptides have been defined. The drawback of this approach is that, due to the MHC polymorphism and polygenism of the patients, an accurate prediction of the peptides that are recognized by the tumor cells in a particular donor is difficult.

For example, while most donors who are HLA-A2 positive will generate a CD8 T cell response after immunization with an A-2-restricted peptide, there is only a random chance that a potentially A-2 restricted peptide will become immune dominant within the endogenously primed repertoire (Zhang et al. 2012). Therefore, the safest approach might be working with peptide pools instead of individual peptides. We recently showed that up to 32 peptides can be pooled such that the individual peptide responses can be precisely additive (Zhang et al. 2012). For example, for a donor who responds to 5 of the 32 peptides, the number of spots elicited by the five individual peptides in an ELISPOT assay matches closely to the response triggered by the pool of 32 peptides. Therefore, the use of such larger peptide pools is a promising strategy for revealing T cell responses to potential determinants.

For the initial ELISPOT assays, we recommend the use of a high concentration of the peptides, for example 10 μg/ml. Peptide-dose response curves in ELISPOT assays typically follow sigmoidal curves, reaching a plateau at maximum stimulation, and it is rare that excess peptide would cause a dip in the plateau. When these experiments will be repeated to measure T cell avidity, the dose response at lower peptide concentration will be tested.

32.5.2 Serial Test 1: Detecting *In Vivo* Primed Cytolytic Effector Cells

A key building block of our proposal is serial testing of the PBMC while accommodating the peak production of the different analytes that the stimulated T cells produce. Our proposal for the first assay in the ELISPOT assay series to test the PBMCs would be a dual color Granzyme B/Perforin assay. This is performed at 400,000 PBMC/well in 150 μl of media/well while testing 10 wells per donor with the antigens arranged as specified above. The duration of the cell culture would be 12 h (Rininsland et al. 2000), after which the PBMC from their respective wells are transferred into the next ELISPOT assay that measures IFN-γ and IL-2 in a dual color format. CD8 effector cells have Granzyme B and Perforin present in their cytolytic vesicles. When these cells come in contact with antigen-bearing target cells, they immediately release Granzyme B and Perforin from their cytolytic vesicles. Only CD8 cells that had been stimulated by antigen *in vivo* in the recent past produce Granzyme B and Perforin under these conditions (Nowacki et al. 2007). According to our studies of Vaccinia-virus immunized human volunteers, the virus-specific CD8 cells produce Granzyme B and Perforin in such *ex vivo* ELISPOT assays within the first 30 days after vaccination. However, after one month, they do not secrete Granzyme B and Perforin any more while maintaining their ability to produce IFN-γ. Under conditions of ongoing CD8 cell activity *in vivo*, such as HIV infection, or a re-acerbation of CMV infection, the antigen-specific CD8 cells produce Granzyme B and Perforin in such short term *ex vivo* ELISPOT assays. In contrast, EBV, CMV and Flu-specific memory cells typically do not produce these analytes *ex vivo*, e.g. when stimulated with the CEF peptide pool or the individual peptides. However, after a 3 day restimulation culture, during which the resting memory cells transform into CD8 effector cells, these analytes are produced and secreted. As a result, this first dual color Granzyme B and Perforin ELISPOT assay should serve to identify *in vivo* activated CD8 effector cells. For example, melan-A-reactive CD8 cells produce Granzyme B and Perforin shortly after melan-A vaccination, or if another melanoma vaccine has been administered and T cell reactivity spreads to include melan-A, but these CD8 cells should produce IFN-γ only if they had been primed by endogenous melan-A, e.g. during the last sunburn several months ago. We suggest the measurement of both Granzyme B and Perforin in a dual color ELISPOT format (enzymatic or fluorescent), because the secretion of these two cytolytic molecules is not necessarily linked. CD8 cells can secrete one, or the other, or both (Kuerten et al. 2008), the significance of which still needs to be established.

32.5.3 Serial Test 2: Detecting THx, TH1, and Polyfunctional T Cells

The second ELISPOT assay to which we propose to transfer the PBMC after 12 h is a dual color IL-2/IFN-γ assay. Transfer after 12 h was selected because IL-2 is a critical early signal of T cell activation. When performing this IL-2/IFN-γ dual color assay directly, the absorption of the T cell derived IL-2 by the capture antibody will reduce approximately 50 % of IFN-γ production, and can abrogate the secretion of IL-4 and IL-17 by T cells completely (Quast et al. 2005). This inhibition is not observed if the initial 12 h T cell activation occurs in the absence of an IL-2 capture antibody (Quast et al. 2005), e.g. in a Granzyme B and Perforin assay. The transfer at 12 h accommodates the IL-2 and IFN-γ secretion by T cells that peaks between 18 and 24 h after antigen stimulation, whereas Granzyme B and Perforin production peak before 12 h (Rininsland et al. 2000). The transfer technique can be performed as follows. In the previous assay, the PBMC were plated in 150 μl. Using a 100 μl pipette and wide orifice pipette tips, the cells are gently re-suspended 3 times and 100 μl is transferred to the next assay (which leaves 1/3rd of the non-adherent cells behind in 50 μl). Now, 100 μl of warm medium is added to the previous assay, the cells are resuspended again, and 100 μl is transferred to the new assay, which leaves 1/9th of the originally plated non adherent cells behind. We do not recommend the aspiration of all the 150 μl because the membrane is likely to be damaged. With this simple approach, about 89 % of the non-adherent cells, which include T cells, B cells and most dendritic cells, can be transferred along with the antigen and antigen presenting cells. No further washing or addition of antigen is required at this point. If highly quantitative frequency measurements are desired for such transfer assays, it is advisable to establish the number of cells transferred at each step. This can be done simply and quickly using the following approach. The PBMC are labeled with CFSE (Carboxy-Fluorescein Succinimidyl Ester) before the primary assay, and the ELISPOT assays are done in low auto- fluorescence PVDF (Poly-Vinylidene Fluoride) plates. Just before each transfer, by which time the cells have sedimented, the density of the cells on the bottom of the well is measured by an ImmunoSpot Analyzer with microscopic fluorescence imaging capabilities. Even after three transfers, more than 200,000 of the originally plated 400,000 cells will still be contained in the third assay. Due to the linear relationship between cell numbers plated and spot counts measured, one can normalize the spot counts to accommodate the loss.

The dual color IL-2/IFN-γ ELISPOT assay will provide information on the number of antigen-specific IFN-γ producing T cells, the number of antigen-specific T cells that produce only IL-2 in the absence of IFN-γ. These T cells are thought to be uncommitted precursors to Th1, Th2 or Th17 cells that are engaged in the absence of strong TLR stimulation (Mosmann et al. 2009). Furthermore, the numbers of polyfunctional T cells that co-express IL-2 and IFN-γ can be established. Similar to ICS, Dual color ELISPOT is equally suited to detect cytokine co-expressing cells (Tian et al. 1997). Finally, the size of IFN-γ spots can provide insights into the *in vivo* activation state of T cells, as we observed that T cells that are stimulated *in vivo* in the recent past produce more IFN-γ on a per cell basis than the T cells that have encountered the antigen in the distant past (Schlingmann et al. 2009).

32.5.4 Serial Test 3: Detecting TH2 T Cells

After the IL-2/IFN-γ assay, at 24 h the cells are transferred into an IL-4/IL-5 double color assay. This second transfer follows the same protocol as described for the first transfer, with one difference: 50 µl of the 200 µl culture supernatant is discarded by careful aspiration before the cells are resuspended. Production of IL-4 and IL-5 by antigen-stimulated T cells peaks at 48 h, at which time the cells are transferred into the next assay. It is important to include reference PBMC that do contain antigen-specific IL-4 and IL-5 producing T cells. Unlike responses generated with IFN-γ, many donors (ones who are Th1 biased) do not produce these cytokines even after PHA stimulation, while other donors (ones who are Th2 biased) can have IL-4 and IL-5 producing T cells in high numbers. For immune monitoring in cancer, it is important to consider that determinant spreading can progress along the Th1 or the Th2 pathways, as we observed in murine models in which spreading along the Th1 pathway is associated with accelerated progression of autoimmune disease, while spreading along the Th2 pathway is associated with profound protection from autoimmune disease (Tian et al. 1997). Therefore, it is also likely that when anti-tumor spreading occurs along the Th2 pathway, tumor protection may occur, rather than tumor rejection. We propose to measure IL-4 and IL-5 in a dual color format because we observed that canonical Th2 cells that produce IL-4 and IL-5 simultaneously are rare, and that IL-4 and IL-5 are frequently produced by different T cell subpopulations (Karulin et al. 2000).

32.5.5 Serial Test 4: Detecting TH17 T Cells and Resting CTL Memory Cells

Following the IL-4/IL-5 assay, at 48 h we propose to transfer the cells into a dual color IL-17/Granzyme B assay as described for the previous transfer. IL-17 production by antigen-stimulated Th17 cells peaks on days 3–4. By this time, CD8 cells that were resting memory cells at the time of their isolation (IFN-γ positive, Granzyme B negative), would have transformed into CD8 effector cells that are positive for Granzyme B (Nowacki et al. 2007). Therefore, the third and last transfer of our immune diagnostic transfer series, measures Th17 immunity and the presence of resting CD8 memory cells that have cytolytic potential upon reactivation.

After 92 h (4 days), when the third assay is complete, approximately 50 % of the originally plated cells are still live and functional. At this time, one might decide to have gained sufficient information from the PBMC used so far, or these cells can be further studied: to do B cell ELISPOT assays, to spectra type the TCR repertoire, or to refreeze the cells.

So far we have presented an experimental protocol to perform an eight parameter analysis on 10 experimental conditions of the patient sample, using only 5 million fresh (4 million thawed) PBMC to generate the data. The results provide insights into the antigen-specific T cell response to four tumor antigens at a resolution between 1 in 400,000 and 1 in 200,000. For each of the antigens, we learned about the presence of (1) *in vivo* active CD8 effector cells, (2) Th1 cells, (3) ThX cells, (4) polyfunctional T cells, (5) Th2, and (6) Th17 cells, and (7) in vivo resting CD8 memory cells that have the potential for cytolytic activity upon re-activation. In addition, the experiments were set up to learn whether determinant spreading has occurred, and if so whether the second generation tumor-specific T cells are Th1, Th2 or Th17 and cytolytic. The shortcoming of the data so far is that all measurements were done with a single antigen concentration, which may be too high to have immune relevance for cancer. Therefore, the second set of experiments not only aims at reproducing positive data, but also to establish the functional avidity of these T cell responses.

Performing this series of 4 successive assays for 30 patient samples would require about one week of wet lab work, and two trained technicians dedicated to the task. The analysis and documentation of the data would be automated. For this example, it would take approximately 15 s per patient and assay, which takes about half an hour of automated analyzer time for all four experiments and the 30 patients. Therefore, after one week, the data would be available for identifying positive responses and planning the follow up experiment accordingly.

32.5.6 Reproducing Data and Expanding to Include Avidity Measurements

The aim of the follow up experiment would be to reproduce the data of the first one, and extend it to include avidity measurements. This second set of experiments will be done in a 384 well format. The surface area of the membrane for these 384-well plates is one fourth of the 96-well plate permitting us to scale down the parameters of the assay by one in four. With one fourth of PBMC plated per well into 384-well plates, exactly one fourth of the spot counts will be obtained, compared to the 96-well plates, i.e. the counts are identical when normalized for the cell numbers plated. The range of PBMC for which the cell numbers plated and spot counts obtained per well are linear also translates one in four, thus resulting in the same cell density on the membrane. For the 96-well plate this range is 100,000–1 million PBMC per well, for the 384-well plate it is 25,000–250,000 cells per well. Therefore, when repeating the first experiment in a 384-well plate, 100,000 PBMC are plated per well instead of the 400,000 per well for the 96-well plate. If the spot counts in the first experiment were high, and/or maximizing cell numbers is an issue, fewer cells can be plated into the 384-well plate. For example, a signal of 400 spots over 0 background in the 96-well plate with 400,000 cells per well would result in a signal of 100 spots over 0 background with 100,000 PBMC per well for a corresponding 384-well plate. In this case, the 384-well assay could be done with half the PBMC per well, because the expected average result of 50 spots per well over 0 background still represents a strong and clear signal.

For the follow up experiment in the 384-well plate, the second aliquot of the cryopreserved PBMC is thawed, yielding 4 million viable and fully functional cells. Since serial transfer of cells from 384-well plates is difficult with the present technology, the PBMC are plated directly into the 384-well ELISPOT plate. This time only those antigen/analyte combinations that scored positive in the previous assay are tested. For each of these positive combinations a wide range of antigen concentrations need to be tested. To reproduce the results of the first assay, the highest antigen concentration would again be 10 μg/ml. The second and third concentration would be 1 and 0.1 μg/ml, respectively, thereby covering a range of three logs. Employing such a range of antigen concentration is advisable because we have observed that the avidity of antigen-specific T cells can occur over the above specified range and even beyond (Targoni and Lehmann 1998; Zhang et al. 2012). One well is tested for each of the above mentioned three concentrations with the fourth being a control well with no antigen (medium control). In this way, the 4 million PBMC per donor would permit us to follow up on, reproduce, and extend the results of, up to 10 responses that were positive in the first experiment. High avidity T cell responses that are recalled by low concentration of the antigen are indicative of the T cells being able to recognize low concentration of endogenous antigen on the tumor cells. However, the direct proof is testing the T cells using the tumor cells themselves as APC. This also accounts for Type A/B determinant recognition (Mohan and Unanue 2012).

32.6 The Flight of the Phoenix

Most of the feasibility studies leading to the above proposal have been done with PBMC from healthy donors. Moreover, these studies were facilitated by unlimited access to "ideal", characterized cell material in the form of ePBMC. The CD8 cell responses were determined using well-characterized viral antigens, such as individual peptides of Cytomegalovirus, Epstein Barr virus, or Flu virus. The CD4 cell responses were determined using common recall antigens, such as proteins of mumps, PPD, candida, dust mite, etc. The studies done on tumor immunology were on murine models. We are aware that measuring T cell responses in cancer patients, in a clinical setting, under non-ideal conditions is much more challenging.

This reminds us of the 1965 movie "The Flight of the Phoenix" with James Stewart. In this movie, an airplane crashes in the Sahara, and one of the survivors identifies himself as an airplane engineer who can get the wreckage to fly again. Later, the others find out that he is an "engineer" for airplane models only. Like that engineer, we believe that the challenges may be different, but the basics are similar. We hope our contribution in this chapter will stimulate those of you fighting the real battle to join forces with us to close the gap between what should be measured, and what can be measured for immune monitoring of cancer.

References

Baars A, Claessen AM, van den Eertwegh AJ, Gall HE, Stam AG, Meijer S et al (2000) Skin tests predict survival after autologous tumor cell vaccination in metastatic melanoma: experience in 81 patients. Ann Oncol 11(8):965–970

Barth RJ Jr, Fisher DA, Wallace PK, Channon JY, Noelle RJ, Gui J et al (2010) A randomized trial of ex vivo CD40L activation of a dendritic cell vaccine in colorectal cancer patients: tumor-specific immune responses are associated with improved survival. Clin Cancer Res 16(22):5548–5556

Bartholomae WC, Rininsland FH, Eisenberg JC, Boehm BO, Lehmann PV, Tary-Lehmann M (2004) T cell immunity induced by live, necrotic, and apoptotic tumor cells. J Immunol 173(2):1012–1022

Bates GJ, Fox SB, Han C, Leek RD, Garcia JF, Harris AL et al (2006) Quantification of regulatory T cells enables the identification of high-risk breast cancer patients and those at risk of late relapse. J Clin Oncol 24(34):5373–5380

de Vries IJ, Bernsen MR, Lesterhuis WJ, Scharenborg NM, Strijk SP, Gerritsen MJ et al (2005) Immuno monitoring tumor-specific T cells in delayed-type hypersensitivity skin biopsies after dendritic cell vaccination correlates with clinical outcome. J Clin Oncol 23(24):5779–5787

Ding ZC, Zhou G (2012) Cytotoxic chemotherapy and CD4+ effector T cells: an emerging alliance for durable antitumor effects. Clin Dev Immunol 2012:890178

Disis ML (2011) Immunologic biomarkers as correlates of clinical response to cancer immunotherapy. Cancer Immunol Immunother 60(3):433–442

Disis ML, Schiffman K, Gooley TA, McNeel DG, Rinn K, Knutson KL (2000) Delayed-type hypersensitivity response is a predictor of peripheral blood T-cell immunity after HER-2/neu peptide immunization. Clin Cancer Res 6(4):1347–1350

Disis ML, Wallace DR, Gooley TA, Dang Y, Slota M, Lu H et al (2009) Concurrent trastuzumab and HER2/neu-specific vaccination in patients with metastatic breast cancer. J Clin Oncol 27(28):4685–4692

Dittrich M, Lehmann PV (2012) Statistical analysis of ELISPOT assays. Methods Mol Biol 792:173–183

Gulley JL, Arlen PM, Madan RA, Tsang KY, Pazdur MP, Skarupa L et al (2010) Immunologic and prognostic factors associated with overall survival employing a poxviral-based PSA vaccine in metastatic castrate-resistant prostate cancer. Cancer Immunol Immunother 59(5):663–674

Hardwick N, Chain B (2011) Epitope spreading contributes to effective immunotherapy in metastatic melanoma patients. Immunotherapy 3(6):731–733

Hawkins N, Self S, Wakefield J (2006) The automated counting of spots for the ELISpot assay. J Immunol Methods 316(1–2):52–58

Herzog BA, Ott PA, Dittrich MT, Quast S, Karulin AY, Kalbacher H et al (2004) Increased in vivo frequency of IA-2 peptide-reactive IFNgamma+/IL-4- T cells in type 1 diabetic subjects. J Autoimmun 23(1):45–54

Hesse MD, Karulin AY, Boehm BO, Lehmann PV, Tary-Lehmann M (2001) A T cell clone's avidity is a function of its activation state. J Immunol 167(3):1353–1361

Hofstetter HH, Targoni OS, Karulin AY, Forsthuber TG, Tary-Lehmann M, Lehmann PV (2005) Does the frequency and avidity spectrum of the neuroantigen-specific T cells in the blood mirror the autoimmune process in the central nervous system of mice undergoing experimental allergic encephalomyelitis? J Immunol 174(8):4598–4605

Inderberg-Suso EM, Trachsel S, Lislerud K, Rasmussen AM, Gaudernack G (2012) Widespread CD4+ T-cell reactivity to novel hTERT epitopes following vaccination of cancer patients with a single hTERT peptide GV1001. Oncoimmunology 1(5):670–686

Jaffee EM, Hruban RH, Biedrzycki B, Laheru D, Schepers K, Sauter PR et al (2001) Novel allogeneic granulocyte-macrophage colony-stimulating factor-secreting tumor vaccine for pancreatic cancer: a phase I trial of safety and immune activation. J Clin Oncol 19(1):145–156

Karulin AY, Hesse MD, Tary-Lehmann M, Lehmann PV (2000) Single-cytokine-producing CD4 memory cells predominate in type 1 and type 2 immunity. J Immunol 164(4):1862–1872

Kenter GG, Welters MJ, Valentijn AR, Lowik MJ, Berends-van der Meer DM, Vloon AP, et al (2009) Vaccination against HPV-16 oncoproteins for vulvar intraepithelial neoplasia. N Engl J Med 361(19):1838–1847

Kirkwood JM, Lee S, Moschos SJ, Albertini MR, Michalak JC, Sander C et al (2009) Immunogenicity and antitumor effects of vaccination with peptide vaccine ± granulocyte-monocyte colony-stimulating factor and/or IFN-alpha2b in advanced metastatic melanoma: Eastern Cooperative Oncology Group Phase II Trial E1696. Clin Cancer Res 15(4):1443–1451

Kreher CR, Dittrich MT, Guerkov R, Boehm BO, Tary-Lehmann M (2003) CD4+ and CD8+ cells in cryopreserved human PBMC maintain full functionality in cytokine ELISPOT assays. J Immunol Methods 278(1–2):79–93

Kuerten S, Nowacki TM, Kleen TO, Asaad RJ, Lehmann PV, Tary-Lehmann M (2008) Dissociated production of perforin, granzyme B, and IFN-gamma by HIV-specific CD8(+) cells in HIV infection. AIDS Res Hum Retroviruses 24(1):62–71

Kuerten S, Batoulis H, Recks MS, Kraracsony E, Zhang W, Subbramanian RA et al (2012) Resting of cryopreserved PBMC does not in general benefit the performance of antigen-specific T cell ELISPOT assays. Cells 1(3):409–427

Lehmann PV, Zhang W (2012) Unique strengths of ELISPOT for T cell diagnostics. Methods Mol Biol 792:3–23

Lehmann PV, Forsthuber T, Miller A, Sercarz EE (1992) Spreading of T-cell autoimmunity to cryptic determinants of an autoantigen. Nature 358(6382):155–157

Lehmann PV, Sercarz EE, Forsthuber T, Dayan CM, Gammon G (1993) Determinant spreading and the dynamics of the autoimmune T-cell repertoire. Immunol Today 14(5):203–208

Lonsdorf AS, Kuekrek H, Stern BV, Boehm BO, Lehmann PV, Tary-Lehmann M (2003) Intratumor CpG-oligodeoxynucleotide injection induces protective antitumor T cell immunity. J Immunol 171(8):3941–3946

Lopez MN, Pereda C, Segal G, Munoz L, Aguilera R, Gonzalez FE et al (2009) Prolonged survival of dendritic cell-vaccinated melanoma patients correlates with tumor-specific delayed type IV hypersensitivity response and reduction of tumor growth factor beta-expressing T cells. J Clin Oncol 27(6):945–952

Mohan JF, Unanue ER (2012) Unconventional recognition of peptides by T cells and the implications for autoimmunity. Nat Rev Immunol 12(10):721–728

Mosmann TR, Kobie JJ, Lee FE, Quataert SA (2009) T helper cytokine patterns: defined subsets, random expression, and external modulation. Immunol Res 45(2–3):173–184

Nowacki TM, Kuerten S, Zhang W, Shive CL, Kreher CR, Boehm BO et al (2007) Granzyme B production distinguishes recently activated CD8(+) memory cells from resting memory cells. Cell Immunol 247(1):36–48

Quast S, Zhang W, Shive C, Kovalovski D, Ott PA, Herzog BA et al (2005) IL-2 absorption affects IFN-gamma and IL-5, but not IL-4 producing memory T cells in double color cytokine ELISPOT assays. Cell Immunol 237(1):28–36

Ramachandran H, Laux J, Moldovan I, Caspell P, Lehmann PV (2012) Optimal thawing of cryopreserved peripheral blood mononuclear cells for use in high-throughput human immune monitoring studies. Cells 1(3):313–324

Rininsland FH, Helms T, Asaad RJ, Boehm BO, Tary-Lehmann M (2000) Granzyme B ELISPOT assay for ex vivo measurements of T cell immunity. J Immunol Methods 240(1–2):143–155

Schlingmann TR, Rininsland FH, Bartholomae WC, Kuekrek H, Lehmann PV, Tary-Lehmann M (2009a) Vaccination with tumor cells pulsed with foreign peptide induces immunity to the tumor itself. Clin Immunol 133(1):45–51

Schlingmann TR, Shive CL, Targoni OS, Tary-Lehmann M, Lehmann PV (2009b) Increased per cell IFN-gamma productivity indicates recent in vivo activation of T cells. Cell Immunol 258(2):131–137

Slingluff CL Jr, Petroni GR, Chianese-Bullock KA, Smolkin ME, Hibbitts S, Murphy C et al (2007) Immunologic and clinical outcomes of a randomized phase II trial of two multipeptide vaccines for melanoma in the adjuvant setting. Clin Cancer Res 13(21):6386–6395

Stern BV, Boehm BO, Tary-Lehmann M (2002) Vaccination with tumor peptide in CpG adjuvant protects via IFN-gamma-dependent CD4 cell immunity. J Immunol 168(12):6099–6105

Targoni OS, Lehmann PV (1998) Endogenous myelin basic protein inactivates the high avidity T cell repertoire. J Exp Med 187(12):2055–2063

Thakur A, Pedersen LE, Jungersen G (2012) Immune markers and correlates of protection for vaccine induced immune responses. Vaccine 30(33):4907–4920

Tian J, Lehmann PV, Kaufman DL (1997) Determinant spreading of T helper cell 2 (Th2) responses to pancreatic islet autoantigens. J Exp Med 186(12):2039–2043

Tigno-Aranjuez JT, Jaini R, Tuohy VK, Lehmann PV, Tary-Lehmann M (2009a) Encephalitogenicity of complete Freund's adjuvant relative to CpG is linked to induction of Th17 cells. J Immunol 183(9):5654–5661

Tigno-Aranjuez JT, Lehmann PV, Tary-Lehmann M (2009b) Dissociated induction of cytotoxicity and DTH by CFA and CpG. J Immunother 32(4):389–398

Zhang W, Lehmann PV (2012) Objective, user-independent ELISPOT data analysis based on scientifically validated principles. Methods Mol Biol 792:155–171

Zhang W, Caspell R, Karulin AY, Ahmad M, Haicheur N, Abdelsalam A et al (2009) ELISPOT assays provide reproducible results among different laboratories for T-cell immune monitoring–even in hands of ELISPOT-inexperienced investigators. J Immunotoxicol 6(4):227–234

Zhang W, Moldovan I, Targoni OS, Subbramanian RA, Lehmann PV (2012) How much of virus-specific CD8 T cell reactivity is detected with a peptide pool when compared to individual peptides? Viruses 4(11):2636–2649

Index

M. R. Shurin et al. (eds.), *The Tumor Immunoenvironment*,
DOI: 10.1007/978-94-007-6217-6,

The manufacturer's authorised representative in the EU is Springer Nature Customer Service Centre GmbH, Europaplatz 3, 69115 Heidelberg, Germany. If you have any concerns regarding our products, please contact ProductSafety@springernature.com

Printed and bound by CPI Group (UK) Ltd, Croydon, CR0 4YY
30/09/2026
02225501-0020